*Edited by
Richard Dronskowski,
Shinichi Kikkawa, and
Andreas Stein*

**Handbook of
Solid State Chemistry**

*Edited by
Richard Dronskowski,
Shinichi Kikkawa, and
Andreas Stein*

Handbook of Solid State Chemistry

Volume 5: Theoretical Description

WILEY-VCH Verlag GmbH & Co. KGaA

Editors

Richard Dronskowski
RWTH Aachen
Institute of Inorganic Chemistry
Landoltweg 1
52056 Aachen
Germany

Shinichi Kikkawa
Hokkaido University
Faculty of Engineering
N13 W8, Kita-ku
060-8628 Sapporo
Japan

Andreas Stein
University of Minnesota
Department of Chemistry
207 Pleasant St. SE
Minneapolis, MN 55455
USA

Cover Credit: Sven Lidin, Arndt Simon and Franck Tessier

All books published by **Wiley-VCH** are carefully produced. Nevertheless, authors, editors, and publisher do not warrant the information contained in these books, including this book, to be free of errors. Readers are advised to keep in mind that statements, data, illustrations, procedural details or other items may inadvertently be inaccurate.

Library of Congress Card No.: applied for

British Library Cataloguing-in-Publication Data
A catalogue record for this book is available from the British Library.

Bibliographic information published by the Deutsche Nationalbibliothek
The Deutsche Nationalbibliothek lists this publication in the Deutsche Nationalbibliografie; detailed bibliographic data are available on the Internet at http://dnb.d-nb.de.

© 2017 Wiley-VCH Verlag GmbH & Co. KGaA, Boschstr. 12, 69469 Weinheim, Germany

All rights reserved (including those of translation into other languages). No part of this book may be reproduced in any form – by photoprinting, microfilm, or any other means – nor transmitted or translated into a machine language without written permission from the publishers. Registered names, trademarks, etc. used in this book, even when not specifically marked as such, are not to be considered unprotected by law.

Print ISBN: 978-3-527-32587-0
oBook ISBN: 978-3-527-69103-6

Cover Design   Formgeber
Typesetting   Thomson Digital, Noida, India
Printing and Binding   Markono Print Media Pte Ltd, Singapore

Printed on acid-free paper

Preface

When you do great science, you do not have to make a lot of fuss. This oft-forgotten saying from the twentieth century has served these editors pretty well, so the foreword to this definitive six-volume *Handbook of Solid-State Chemistry* in the early twenty-first century will be brief. After all, is there any real need to highlight the paramount successes of solid-state chemistry in the last half century? – Successes that have led to novel magnets, solid-state lighting, dielectrics, phase-change materials, batteries, superconducting compounds, and a lot more? Probably not, but we should stress that many of these exciting matters were derived from curiosity-driven research — work that many practitioners of our beloved branch of chemistry truly appreciate, and this is exactly why they do it. Our objects of study may be immensely important for various applications but, first of all, they are interesting to us; that is, how chemistry defines and challenges itself. Let us also not forget that solid-state chemistry is a neighbor to physics, crystallography, materials science, and other fields, so there is plenty of room at the border, to paraphrase another important quote from a courageous physicist.

Given the incredibly rich heritage of solid-state chemistry, it is probably hard for a newcomer (a young doctoral student, for example) to see the forest for all the trees. In other words, there is a real need to cover solid-state chemistry in its entirety, but only if it is conveniently grouped into digestible categories. Because such an endeavor is not possible in introductory textbooks, this is what we have tried to put together here. The compendium starts with an overview of materials and of the structure of solids. Not too surprisingly, the next volume deals with synthetic techniques, followed by another volume on various ways of (structural) characterization. Being a timely handbook, the fourth volume touches upon nano and hybrid materials, while volume V introduces the reader to the theoretical description of the solid state. Finally, the sixth volume reaches into the real world by focusing on functional materials. Should we have considered more volumes? Yes, probably, but life is short, dear friends.

This handbook would have been impossible to compile for three authors, let alone a single one. Instead, the editors take enormous pride in saying that they managed to motivate more than a hundred first-class scientists living across the globe, each of them specializing in (and sometimes even shaping) a subfield of solid-state chemistry or a related discipline, and all of these wonderful colleagues did their very best to make our dream come true. Thanks to all of you; we sincerely appreciate your contributions. Thank you, once again, on behalf of solid-state chemistry. The editors also would like to thank Wiley-VCH, in particular Dr. Waltraud Wüst and also Dr. Frank Otmar Weinreich, for spiritually (and practically) accompanying us over a few years, and for reminding us here and there that there must be a final deadline. That being said, it is up to the reader to judge whether the tremendous effort was justified. We sincerely hope that this is the case.

A toast to our wonderful science! Long live solid-state chemistry!

Richard Dronskowski
RWTH Aachen, Aachen, Germany

Shinichi Kikkawa
Hokkaido University, Sapporo, Japan

Andreas Stein
University of Minnesota, Minneapolis, USA

Contents

Volume 1: Materials and Structure of Solids

1 Intermetallic Compounds and Alloy Bonding Theory Derived from Quantum Mechanical One-Electron Models *1*
Stephen Lee and Daniel C. Fredrickson

2 Quasicrystal Approximants *73*
Sven Lidin

3 Medium-Range Order in Oxide Glasses *93*
Hellmut Eckert

4 Suboxides and Other Low-Valent Species *139*
Arndt Simon

5 Introduction to the Crystal Chemistry of Transition Metal Oxides *161*
J.E. Greedan

6 Perovskite Structure Compounds *221*
Yuichi Shimakawa

7 Nitrides of Non-Main Group Elements *251*
P. Höhn and R. Niewa

8 Fluorite-Type Transition Metal Oxynitrides *361*
Franck Tessier

9 Mechanochemical Synthesis, Vacancy-Ordered Structures and Low-Dimensional Properties of Transition Metal Chalcogenides *383*
Yutaka Ueda and Tsukio Ohtani

10	**Metal Borides: Versatile Structures and Properties** *435*	
	Barbara Albert and Kathrin Hofmann	
11	**Metal Pnictides: Structures and Thermoelectric Properties** *455*	
	Abdeljalil Assoud and Holger Kleinke	
12	**Metal Hydrides** *477*	
	Yaoqing Zhang, Maarten C. Verbraeken, Cédric Tassel, and Hiroshi Kageyama	
13	**Local Atomic Order in Intermetallics and Alloys** *521*	
	Frank Haarmann	
14	**Layered Double Hydroxides: Structure–Property Relationships** *541*	
	Shan He, Jingbin Han, Mingfei Shao, Ruizheng Liang, Min Wei, David G. Evans, and Xue Duan	
15	**Structural Diversity in Complex Layered Oxides** *571*	
	S. Uma	
16	**Magnetoresistance Materials** *595*	
	Ichiro Terasaki	
17	**Magnetic Frustration in Spinels, Spin Ice Compounds, $A_3B_5O_{12}$ Garnet, and Multiferroic Materials** *617*	
	Hongyang Zhao, Hideo Kimura, Zhenxiang Cheng, and Tingting Jia	
18	**Structures and Properties of Dielectrics and Ferroelectrics** *643*	
	Mitsuru Itoh	
19	**Defect Chemistry and Its Relevance for Ionic Conduction and Reactivity** *665*	
	Joachim Maier	
20	**Molecular Magnets** *703*	
	J.V. Yakhmi	
21	**Ge–Sb–Te Phase-Change Materials** *735*	
	Volker L. Deringer and Matthias Wuttig	
	Index *751*	

Volume 2: Synthesis

1 **High-Temperature Methods** *1*
Rainer Pöttgen and Oliver Janka

2 **High-Pressure Methods in Solid-State Chemistry** *23*
Hubert Huppertz, Gunter Heymann, Ulrich Schwarz, and Marcus R. Schwarz

3 **High-Pressure Perovskite: Synthesis, Structure, and Phase Relation** *49*
Yoshiyuki Inaguma

4 **Solvothermal Methods** *107*
Nobuhiro Kumada

5 **High-Throughput Synthesis Under Hydrothermal Conditions** *123*
Nobuaki Aoki, Gimyeong Seong, Tsutomu Aida, Daisuke Hojo, Seiichi Takami, and Tadafumi Adschiri

6 **Particle-Mediated Crystal Growth** *155*
R. Lee Penn

7 **Sol–Gel Synthesis of Solid-State Materials** *179*
Guido Kickelbick and Patrick Wenderoth

8 **Templated Synthesis for Nanostructured Materials** *201*
Yoshiyuki Kuroda and Kazuyuki Kuroda

9 **Bio-Inspired Synthesis and Application of Functional Inorganic Materials by Polymer-Controlled Crystallization** *233*
Lei Liu and Shu-Hong Yu

10 **Reactive Fluxes** *275*

11 **Glass Formation and Crystallization** *287*
T. Komatsu

12 **Glass-Forming Ability, Recent Trends, and Synthesis Methods of Metallic Glasses** *319*
Hidemi Kato, Takeshi Wada, Rui Yamada, and Junji Saida

13 **Crystal Growth Via the Gas Phase by Chemical Vapor Transport Reactions** *351*
Michael Binnewies, Robert Glaum, Marcus Schmidt, and Peer Schmidt

14	Thermodynamic and Kinetic Aspects of Crystal Growth *375*
	Detlef Klimm

15	Chemical Vapor Deposition *399*
	Takashi Goto and Hirokazu Katsui

16	Growth of Wide Bandgap Semiconductors by Halide Vapor Phase Epitaxy *429*
	Yuichi Oshima, Encarnación G. Víllora, and Kiyoshi Shimamura

17	Growth of Silicon Nanowires *467*
	Fengji Li and Sam Zhang

18	Chemical Patterning on Surfaces and in Bulk Gels *539*
	Olaf Karthaus

19	Microcontact Printing *563*
	Kiyoshi Yase

20	Nanolithography Based on Surface Plasmon *573*
	Kosei Ueno and Hiroaki Misawa

Index *589*

Volume 3: Characterization

1	Single-Crystal X-Ray Diffraction *1*
	Ulli Englert

2	Laboratory and Synchrotron Powder Diffraction *29*
	R. E. Dinnebier, M. Etter, and T. Runcevski

3	Neutron Diffraction *77*
	Martin Meven and Georg Roth

4	Modulated Crystal Structures *109*
	Sander van Smaalen

5	Characterization of Quasicrystals *131*
	Walter Steurer

6	Transmission Electron Microscopy *155*
	Krumeich Frank

7	Scanning Probe Microscopy *183*
	Marek Nowicki and Klaus Wandelt

8	**Solid-State NMR Spectroscopy: Introduction for Solid-State Chemists** *245* Christoph S. Zehe, Renée Siegel, and Jürgen Senker	
9	**Modern Electron Paramagnetic Resonance Techniques and Their Applications to Magnetic Systems** *279* Andrej Zorko, Matej Pregelj, and Denis Arčon	
10	**Photoelectron Spectroscopy** *311* Stephan Breuer and Klaus Wandelt	
11	**Recent Developments in Soft X-Ray Absorption Spectroscopy** *361* Alexander Moewes	
12	**Vibrational Spectroscopy** *393* Götz Eckold and Helmut Schober	
13	**Mößbauer Spectroscopy** *443* Hermann Raphael	
14	**Macroscopic Magnetic Behavior: Spontaneous Magnetic Ordering** *485* Heiko Lueken and Manfred Speldrich	
15	**Dielectric Properties** *523* Rainer Waser and Susanne Hoffmann-Eifert	
16	**Mechanical Properties** *561* Volker Schnabel, Moritz to Baben, Denis Music, William J. Clegg, and Jochen M. Schneider	
17	**Calorimetry** *589* Hitoshi Kawaji	
	Index *615*	

Volume 4: Nano and Hybrid Materials

1	**Self-Assembly of Molecular Metal Oxide Nanoclusters** *1* Laia Vilà-Nadal and Leroy Cronin
2	**Inorganic Nanotubes and Fullerene–Like Nanoparticles from Layered (2D) Compounds** *21* L. Yadgarov, R. Popovitz-Biro, and R. Tenne

| 3 | Layered Materials: Oxides and Hydroxides 53
Ida Shintaro |
|---|---|
| 4 | Organoclays and Polymer-Clay Nanocomposites 79
M.A. Vicente and A. Gil |
| 5 | Zeolite and Zeolite-Like Materials 97
Watcharop Chaikittisilp and Tatsuya Okubo |
| 6 | Ordered Mesoporous Materials 121
Michal Kruk |
| 7 | Porous Coordination Polymers/Metal–Organic Frameworks 141
Ohtani Ryo and Kitagawa Susumu |
| 8 | Metal–Organic Frameworks: An Emerging Class of Solid-State Materials 165
Joseph E. Mondloch, Rachel C. Klet, Ashlee J. Howarth, Joseph T. Hupp, and Omar K. Farha |
| 9 | Sol–Gel Processing of Porous Materials 195
Kazuki Nakanishi, Kazuyoshi Kanamori, Yasuaki Tokudome, George Hasegawa, and Yang Zhu |
| 10 | Macroporous Materials Synthesized by Colloidal Crystal Templating 243
Jinbo Hu and Andreas Stein |
| 11 | Optical Properties of Hybrid Organic–Inorganic Materials and their Applications – Part I: Luminescence and Photochromism 275
Stephane Parola, Beatriz Julián-López, Luís D. Carlos, and Clément Sanchez |
| 12 | Optical Properties of Hybrid Organic–inorganic Materials and their Applications – Part II: Nonlinear Optics and Plasmonics 317
Stephane Parola, Beatriz Julián-López, Luís D. Carlos, and Clément Sanchez |
| 13 | Bioactive Glasses 357
Hirotaka Maeda and Toshihiro Kasuga |
| 14 | Materials for Tissue Engineering 383
María Vallet-Regí and Antonio J. Salinas |

Index 411

Volume 5: Theoretical Description

1 Density Functional Theory *1*
Michael Springborg and Yi Dong
1.1 Introduction *1*
1.2 Foundations of Density-Functional Theory *2*
1.2.1 Basic Principles *2*
1.2.2 Functionals *5*
1.2.3 Hubbard ($+U$) Corrections *7*
1.2.4 Extended Systems *8*
1.2.5 Descriptors *12*
1.2.6 Excitations *15*
1.3 Examples of Applications *21*
1.4 Open Issues *21*
1.4.1 Functionals *21*
1.4.2 Weak Interactions *22*
1.4.3 Charge (De)localization *23*
1.4.4 Response Properties *24*
1.5 Conclusions *25*
References *26*

2 Eliminating Core Electrons in Electronic Structure Calculations: Pseudopotentials and PAW Potentials *29*
Stefan Goedecker and Santanu Saha
2.1 Generalities *29*
2.2 The Alternative: All-Electron Calculations *31*
2.3 General Advantages of Using Pseudopotentials *34*
2.4 The Basic Concepts of Pseudopotentials *34*
2.5 Mathematical Form of Norm-Conserving Pseudopotentials *42*
2.6 Ultrasoft and Projector Augmented Wave Methods *44*
2.7 Construction Methods for Pseudopotentials *45*
2.8 Numerical Performance of Pseudopotentials *47*
2.9 Accuracy of Pseudopotentials *49*
2.10 Conclusions *54*
References *54*

3 Periodic Local Møller–Plesset Perturbation Theory of Second Order for Solids *59*
Denis Usvyat, Lorenzo Maschio, and Martin Schütz
3.1 Introduction *59*
3.2 Theory *61*
3.2.1 Nomenclature *61*
3.2.2 Periodic LMP2 Wave Function *62*
3.2.3 Periodic LMP2 Energy *63*

3.2.4	Choice of Direct Space Orbitals	65
3.2.5	Local Approximations	68
3.2.5.1	Domain Approximation	68
3.2.5.2	Truncated OSV Approximation	70
3.2.5.3	Pair Approximation	71
3.2.6	Periodic Density Fitting	73
3.2.7	Direct Space Local Density Fitting	75
3.2.8	Multipolar Approximation	76
3.3	A Demonstrative Application: Ibuprofen on Hydroxylated Silica Surface	78
3.4	Summary and Conclusions	81
	References	82
4	**Resonating Valence Bonds in Chemistry and Solid State**	**87**
	Evgeny A. Plekhanov and Andrei L. Tchougréeff	
4.1	Introduction	87
4.2	Historical Overview	88
4.3	Formal Constructs	97
4.3.1	Hückel, Néel, and RVB States of Polyene	98
4.3.2	Trial Variational RVB Wave Function	100
4.3.3	Auxiliary-Boson Theory	105
4.4	Mean Field RVB as Applied to the Heisenberg Model	108
4.5	Discussion and Conclusions	113
	Acknowledgment	114
	References	115
5	**Many Body Perturbation Theory, Dynamical Mean Field Theory and All That**	**119**
	Silke Biermann and Alexander Lichtenstein	
5.1	Introduction: Electronic Correlations from First Principles	119
5.2	Effective Low-Energy Hamiltonians: "Downfolding" the Interaction	122
5.3	Functional Approach: From DFT to DMFT	125
5.4	DMFT and GW + DMFT from an Impurity Reference System	129
5.5	Derivation of GW + DMFT from a Free Energy Functional	133
5.6	LDA + DMFT Scheme for Real Materials	137
5.7	The GW + DMFT Approach in Practice	141
5.8	GW + DMFT: the Example of $SrVO_3$	143
5.9	Dynamical Screening in the Iron Pnictide $BaFe_2As_2$	146
5.10	Screened Exchange Dynamical Mean Field Theory	148
5.11	Perspectives	150
5.12	Conclusions	152
	Acknowledgments	154
	References	155

6	**Semiempirical Molecular Orbital Methods** *159*	
	Thomas Bredow and Karl Jug	
6.1	Early Approaches *159*	
6.2	All-Valence Electron Methods *161*	
6.2.1	Extended Hückel Theory *161*	
6.2.2	NDO Methods *161*	
6.2.2.1	CNDO *162*	
6.2.2.2	INDO *162*	
6.2.2.3	MINDO *163*	
6.2.2.4	SINDO *163*	
6.2.2.5	SINDO1 *163*	
6.2.2.6	MSINDO *164*	
6.2.2.7	ZINDO *165*	
6.2.3	NDDO Methods *166*	
6.2.3.1	MNDO *166*	
6.2.3.2	AM1 *167*	
6.2.3.3	PM3 *167*	
6.2.3.4	MNDOC *167*	
6.2.3.5	OMx *168*	
6.2.3.6	Other Semiempirical Methods *168*	
6.2.4	Recent Developments *169*	
6.2.4.1	Analytic Second Derivatives *169*	
6.2.4.2	Optical Spectra *170*	
6.2.4.3	Solvent Effects *171*	
6.2.4.4	Dispersion Correction *171*	
6.2.4.5	Computational Efficiency *172*	
6.2.4.6	Parametrization Strategies *173*	
6.2.4.7	Implementation in Solid-State Software Packages *173*	
6.3	Applications in Biochemistry and Materials Science *174*	
6.3.1	Biochemistry and Pharmacology *174*	
6.3.1.1	Parametrization *174*	
6.3.1.2	Conformational Analysis *175*	
6.3.1.3	Spectroscopy *175*	
6.3.1.4	Intermolecular Interactions and Reactions *176*	
6.3.2	Solid-State Properties *177*	
6.3.2.1	Quasimolecular LUC Model *178*	
6.3.2.2	Cyclic Cluster Model *179*	
6.3.3	Adsorption and Surface Reactions *180*	
6.3.4	Clusters and Nanoparticles *182*	
6.3.5	Simulation of Liquids *184*	
	References *185*	
7	**Tight-Binding Density Functional Theory: DFTB** *203*	
	Gotthard Seifert	
7.1	Introduction *203*	
7.2	Basic Methodology of DFTB *204*	

7.2.1	Approximate Density Functional Theory	*204*
7.2.2	Weak Interactions – QM/MM Approach	*213*
7.3	Extended Systems	*215*
7.3.1	Periodic Boundary Conditions – PBC	*215*
7.3.2	Open Boundary Conditions	*215*
7.3.3	Order N Methods	*217*
7.4	Time-Dependent DFTB (TD-DFTB) – Excited States	*217*
7.5	Computational Implementations and Extensions	*219*
7.6	Conclusions	*221*
	References	*222*
8	**DFT Calculations for Real Solids**	*227*
	Karlheinz Schwarz and Peter Blaha	
8.1	Introduction	*227*
8.2	Atomic Structure	*227*
8.3	Quantum Mechanics for Treating the Electronic Structure	*230*
8.3.1	Hartree–Fock and Beyond	*230*
8.3.2	Density Functional Theory	*231*
8.4	Solving DFT	*233*
8.4.1	General Considerations for DFT Calculations	*234*
8.4.1.1	All-Electron Treatment	*234*
8.4.1.2	Relativistic Effects	*236*
8.4.1.3	Full Potential and Basis Sets	*237*
8.4.2	The Linearized Augmented Plane Wave (LAPW) Method and WIEN2k	*238*
8.4.3	Various Computer Codes	*241*
8.5	How to Run DFT Calculations	*243*
8.5.1	Idealize the Atomic Structure	*243*
8.5.2	Choice of Quantum Mechanical Treatment (DFT)	*244*
8.5.3	Basis Set and Brillouin Zone Sampling	*245*
8.5.4	Hardware Considerations and Parallelization	*247*
8.6	Examples of DFT Results for Bulk Materials and Surfaces	*247*
8.6.1	Equilibrium Volume Using Different DFT Functionals	*248*
8.6.2	Band Structure and Chemical Bonding	*249*
8.6.3	Verwey Transition in $YBaFe_2O_5$	*251*
8.6.4	Structure Relaxation of $Y_2Nb_2O_7$	*252*
8.6.5	Surfaces and Heterogeneous Catalysis	*254*
8.7	Conclusions	*256*
	References	*257*
9	**Spin Polarization**	*261*
	Dong-Kyun Seo	
9.1	Introduction	*261*
9.2	Spin Polarization Perturbational Orbital Theory (SPPOT)	*263*

9.2.1	Approximate Expressions of Exchange-Correlation Functional and Spin Polarization Energy *264*	
9.2.2	Approximate Expressions of Spin Orbital shape and Energy Changes *265*	
9.3	Applications to Simple Systems *269*	
9.3.1	Spin Polarization in Isolated Atoms and Ions *269*	
9.3.2	magnetic coupling, in transition-metal dimer complexes *271*	
9.3.3	Stoner Condition for Spontaneous Spin Polarization *275*	
9.3.4	Ruderman–Kittel–Kasuya–Yosida (RKKY) Exchange Interactions *278*	
9.4	Prospect *282*	
	References *282*	

10 Magnetic Properties from the Perspectives of Electronic Hamiltonian: Spin Exchange Parameters, Spin Orientation, and Spin-Half Misconception *285*
Myung-Hwan Whangbo and Hongjun Xiang

10.1	Introduction *285*
10.2	Atomic Orbitals and Magnetic Orbitals *289*
10.2.1	Angular Properties of Atomic Orbitals *289*
10.2.2	Crystal Field Split d-States *291*
10.2.2.1	One-Electron States without Electron Correlation *292*
10.2.2.2	One-Electron States with Electron Correlation *294*
10.3	Energy-Mapping Analysis *296*
10.3.1	Use of Eigenstates for an Isolated Spin Dimer *297*
10.3.2	Use of Broken-Symmetry States for an Isolated Spin Dimer *301*
10.3.3	Use of Broken-Symmetry States for General Magnetic Solids *301*
10.3.4	Energy-Mapping Based on Four Ordered Spin States *303*
10.3.5	General Features of Spin Exchanges Numerically Extracted *304*
10.4	Orbital Interactions Controlling Spin Exchanges *304*
10.5	Incorporating the Effect of SOC Indirectly into Spin Hamiltonian *308*
10.5.1	SOC Effect on a Single-Spin Site and Spin-Half Misconception *309*
10.5.2	SOC Effect on Spin Exchange: Mapping Analysis for Anisotropic Spin Exchange *312*
10.5.3	SOC Effect on Two Adjacent Spin Sites *313*
10.5.4	Mapping Analysis for the DM Vector of an Isolated Spin Dimer *315*
10.5.5	Mapping Analysis for the DM Vectors Using the Four-State Method for a General Magnetic Solid *316*
10.6	Uniaxial Magnetism *318*
10.7	Describing SOC Effects with Both Orbital and Spin Degrees of Freedom: Magnetic Anisotropy *320*
10.7.1	Selection Rules for Preferred Spin-Orientation *321*
10.7.2	Degenerate Perturbation and Uniaxial Magnetism *322*
10.7.3	Nondegenerate Perturbation and Weak Magnetic Anisotropy *325*
10.7.4	Magnetic Anisotropy of $S = 1/2$ Systems and Spin-Half Misconception *326*

10.7.5	Ligand-Controlled Spin Orientation	*331*
10.7.6	High-Spin d^5 Systems	*332*
10.8	Magnetic Properties of 5d Ion Oxides	*332*
10.8.1	Spin-Orbit Mott Insulating State and Madelung Potential	*333*
10.8.2	Influence of Intersite Interactions on Crystal Field Split d-States	*333*
10.8.3	Perturbation Theory Analysis of Preferred Spin Orientation	*337*
10.8.4	LS versus jj Coupling Scheme of SOC	*337*
10.9	Concluding Remarks	*338*
	Acknowledgments	*340*
	References	*340*

11 Basic Properties of Well-Known Intermetallics and Some New Complex Magnetic Intermetallics *345*
Peter Entel

11.1	Introduction and Hume-Rothery and Frank–Kasper Phases	*345*
11.2	Laves Phase Intermetalllics and Ni– and Ti-Based Superalloys	*347*
11.3	Ni–Al-Based Intermetallics and Shape Memory Materials	*349*
11.4	Ti–Ni-Based Intermetallics and Shape Memory Materials	*356*
11.5	Martensitically Transforming Ti–Ni Intermetallics, Giant Elastocaloric Effects, and Strain Glass Formation in Ti–Ni–X Alloys	*360*
11.6	Bonding and Chemical Frustration in Ti–Ni–X Intermetallics and Chemical Pressure	*363*
11.7	Magnetic and Functional Properties of Ternary Magnetic Heusler Intermetallics	*369*
11.8	Concluding Remarks and Critical Assessment of Present Work	*390*
	Acknowledgments	*392*
	References	*392*

12 Chemical Bonding in Solids *405*
Gordon J. Miller, Yuemei Zhang, and Frank R. Wagner

12.1	Fundamental Characteristics of the Chemical Bond	*406*
12.2	Types of Solids and Bonding Motifs	*414*
12.2.1	Ionic Solids	*416*
12.2.2	Covalent Solids	*420*
12.2.3	Metallic Solids	*424*
12.3	Quantum Mechanical Calculations and Analyses	*427*
12.3.1	Free-Electron versus Tight-Binding Approximations	*432*
12.3.2	Examples of Electronic Structures	*439*
12.3.3	Bonding Analyses in Position Space: QTAIM and ELF/ELI-D	*445*
12.4	From Calculations to Electron Counting Rules	*456*
12.4.1	Octet or 8–N Rule	*456*
12.4.2	18-Electron Rule	*460*

12.4.3	Wade–Mingos Rules: Polyhedral Skeletal Electron Pair Theory	*464*
12.4.4	Transition Metal Octahedral Clusters	*466*
12.4.5	Shell Model	*467*
12.4.6	Hume–Rothery Electron Compounds	*467*
12.5	Selected Bonding Influences on Structure and Properties	*469*
12.5.1	Structural Influences	*469*
12.5.2	Metals versus Nonmetals	*474*
12.5.3	Magnetism	*477*
12.5.4	High-Pressure Effects	*478*
12.6	Summary	*481*
	Acknowledgments	*481*
	References	*481*
13	**Lattice Dynamics and Thermochemistry of Solid-State Materials from First-Principles Quantum-Chemical Calculations** *491*	
	Ralf Peter Stoffel and Richard Dronskowski	
13.1	Classical Thermodynamics	*491*
13.2	Electronic Structure and Electronic Energy	*494*
13.3	Lattice Dynamics	*495*
13.4	*Ab Initio* Thermochemistry	*501*
13.5	Examples	*504*
13.5.1	$Ag[Ag(C_3H_3)_2]$: Infrared Spectroscopy	*504*
13.5.2	Sb_2Te_3: Partial Densities of Phonon States versus Nuclear Inelastic Scattering	*506*
13.5.3	Barium Cerate ($BaCeO_3$)	*509*
13.5.4	GeSe	*512*
13.5.5	Ab Initio ORTEP	*513*
13.5.6	Tin Pest	*515*
13.5.7	Formation Energies	*518*
13.5.8	Heat Capacities	*520*
13.5.9	Temperature-Dependence of Volume and Bulk Modulus	*522*
13.6	Summary	*523*
	References	*524*
14	**Predicting the Structure and Chemistry of Low-Dimensional Materials** *527*	
	Xiaohu Yu, Artem R. Oganov, Zhenhai Wang, Gabriele Saleh, Vinit Sharma, Qiang Zhu, Qinggao Wang, Xiang-Feng Zhou, Ivan A. Popov, Alexander I. Boldyrev, Vladimir S. Baturin, and Sergey V. Lepeshkin	
14.1	Introduction	*527*
14.2	Methods	*528*
14.2.1	Evolutionary Structure Prediction	*528*
14.2.2	New Developments	*530*
14.2.2.1	Smart Variation Operations and Clustering	*530*
14.2.3	Cluster Structure Prediction	*532*

14.2.4 Polymer Structure Prediction *533*
14.2.5 Prediction of Surface Reconstructions *535*
14.3 Results for Low-Dimensional Systems *537*
14.3.1 0D Systems: Nanoclusters *537*
14.3.1.1 Ce_mO_n *538*
14.3.1.2 Ti_mO_n *540*
14.3.1.3 $Si_{10}H_{2m}$ *543*
14.3.1.4 $Si_{10}O_{4n}$ *548*
14.3.2 1D Systems *548*
14.3.3 2D Systems *551*
14.3.3.1 2D Crystals *551*
14.3.3.2 Surface Reconstructions *559*
14.4 Future Prospects *564*
Acknowledgments *565*
References *566*

15 The Pressing Role of Theory in Studies of Compressed Matter *571*
Eva Zurek
15.1 Introduction *571*
15.2 Techniques for Crystal Structure Prediction *573*
15.2.1 Potential Energy Surfaces *574*
15.2.2 Following Imaginary Phonons *575*
15.2.3 Random (Sensible) Structure Searches *576*
15.2.4 Simulated Annealing *577*
15.2.5 Basin and Minima Hopping *578*
15.2.6 Metadynamics *579*
15.2.7 Particle Swarm Optimization *580*
15.2.8 Evolutionary Algorithms *580*
15.3 Chemistry under Pressure *581*
15.3.1 Unique Electronic Structures *582*
15.3.2 Structure and Bonding *584*
15.3.3 Novel Combinations *587*
15.3.4 Emergent Properties *590*
15.3.5 Melting Behavior *592*
15.4 Conclusions *594*
Acknowledgments *595*
References *595*

16 First-Principles Computation of NMR Parameters in Solid-State Chemistry *607*
Jérôme Cuny, Régis Gautier, and Jean-François Halet
16.1 Introduction *607*
16.2 Theory and Methods *611*
16.2.1 Physical Basis of NMR *611*
16.2.2 Computation of NMR Parameters *615*

16.2.2.1 Density Functional Theory for Extended Systems *615*
16.2.2.2 Electric Field Gradient Tensor *619*
16.2.2.3 Nuclear Shielding Tensor *620*
16.2.2.4 Other Interactions: *J* Coupling and Metallic Shift *622*
16.2.3 Considerations about Computational Details *624*
16.3 Illustrative Examples *626*
16.3.1 Combining Theory and Experiments for Structural Study of Crystalline Compounds *627*
16.3.1.1 The Coloring Problem in Rare-Earth Metal Boron Carbides *627*
16.3.1.2 Characterization of ^{93}Nb Solid-State NMR Parameters *630*
16.3.2 Structural Information in Amorphous Compounds *632*
16.4 Conclusions *634*
References *635*

17 Quantum Mechanical/Molecular Mechanical (QM/MM) Approaches *647*
C. Richard A. Catlow, John Buckeridge, Matthew R. Farrow, Andrew J. Logsdail, and Alexey A. Sokol
17.1 Introduction *647*
17.2 The Hybrid Quantum Mechanical/Molecular Mechanical Approach *649*
17.2.1 Background *650*
17.2.1.1 One-Electron Problem *650*
17.2.1.2 Many-Electron Problem, Electron Groups, Localization, and Structural Elements *651*
17.2.1.3 Electronic Structure of the Defect Containing System and the Choice of the Method *652*
17.2.2 Chemshell Quantum Mechanical/Molecular Mechanical Coupling and Master Equations *653*
17.2.3 Practicalities of Setting up a Quantum Mechanical/Molecular Mechanical Calculation *656*
17.3 Recent Applications of Quantum Mechanical/Molecular Mechanical Approaches *657*
17.3.1 Ionization Potentials *657*
17.3.2 Modeling Point Defects with Hybrid Quantum Mechanical/Molecular Mechanical Methods *661*
17.3.2.1 Defect Chemistry *661*
17.3.2.2 Diffuse States *664*
17.3.3 Active Sites in Zeolites *665*
17.3.3.1 Deprotonation Energies as a Measure of Site Reactivity *665*
17.3.3.2 REDOX Potentials *667*
17.3.3.3 Coordination of Metal Centers *668*
17.4 Final Remarks *670*
Acknowledgments *670*
References *671*

18		**Modeling Crystal Nucleation and Growth and Polymorphic Transitions** *681*
		Dirk Zahn
18.1		Introduction *681*
18.2		Thermodynamics and Kinetics of Classical and Noncrystal Nucleation *682*
18.2.1		Classical and Nonclassical Nucleation *682*
18.2.2		Multistep Nucleation and Crystal Polymorphism *685*
18.2.3		Prenucleation Clusters and Nucleation from Building Blocks *685*
18.2.4		Nucleation Kinetics versus Diffusion-Controlled Crystal Formation *689*
18.2.5		Crystal Growth at Nonconstant Solute Concentration *690*
18.3		Molecular Dynamics Simulation Approaches to Understanding Crystal Nucleation *692*
18.3.1		Molecular Mechanics: Quantum, Classical, and Hybrid Models *692*
18.3.2		Molecular Dynamics Simulations I: Equations of Motion and Numerical Solutions *692*
18.3.3		Molecular Dynamics Simulations II: Bridging Timescale and Length Scale *693*
18.4		Case Studies on ZnO: Understanding Ion Aggregation, Ripening Reactions, Nucleation and Growth, Polymorphic Transitions and Nanoparticle Stabilization from Molecular Simulation *695*
18.5		Conclusion *698*
		References *699*

Index *701*

Volume 6: Functional Materials

1   **Electrical Energy Storage: Batteries** *1*
Eric McCalla

2   **Electrical Energy Storage: Supercapacitors** *25*
Enbo Zhao, Wentian Gu, and Gleb Yushin

3   **Dye-Sensitized Solar Cells** *61*
Anna Nikolskaia and Oleg Shevaleevskiy

4   **Electronics and Bioelectronic Interfaces** *75*
Seong-Min Kim, Sungjun Park, Won-June Lee, and Myung-Han Yoon

5   **Designing Thermoelectric Materials Using 2D Layers** *93*
Sage R. Bauers and David C. Johnson

6	**Magnetically Responsive Photonic Nanostructures for Display Applications** *123* Mingsheng Wang and Yadong Yin	
7	**Functional Materials: For Sensing/Diagnostics** *151* Rujuta D. Munje, Shalini Prasad, and Edward Graef	
8	**Superhard Materials** *175* Ralf Riedel, Leonore Wiehl, Andreas Zerr, Pavel Zinin, and Peter Kroll	
9	**Self-healing Materials** *201* Martin D. Hager	
10	**Functional Surfaces for Biomaterials** *227* Akiko Nagai, Naohiro Horiuchi, Miho Nakamura, Norio Wada, and Kimihiro Yamashita	
11	**Functional Materials for Gas Storage. Part I: Carbon Dioxide and Toxic Compounds** *249* L. Reguera and E. Reguera	
12	**Functional Materials for Gas Storage. Part II: Hydrogen and Methane** *281* L. Reguera and E. Reguera	
13	**Supported Catalysts** *313* Isao Ogino, Pedro Serna, and Bruce C. Gates	
14	**Hydrogenation by Metals** *339* Xin Jin and Raghunath V. Chaudhari	
15	**Catalysis/Selective Oxidation by Metal Oxides** *393* Wataru Ueda	
16	**Activity of Zeolitic Catalysts** *417* Xiangju Meng, Liang Wang, and Feng-Shou Xiao	
17	**Nanocatalysis: Catalysis with Nanoscale Materials** *443* Tewodros Asefa and Xiaoxi Huang	
18	**Heterogeneous Asymmetric Catalysis** *479* Ágnes Mastalir and Mihály Bartók	

| 19 | **Catalysis by Metal Carbides and Nitrides** *511*
Connor Nash, Matt Yung, Yuan Chen, Sarah Carl, Levi Thompson, and Josh Schaidle

| 20 | **Combinatorial Approaches for Bulk Solid-State Synthesis of Oxides** *553*
Paul J. McGinn

Index *573*

1
Density Functional Theory

Michael Springborg[1,2] and Yi Dong[1]

[1]University of Saarland, Physical and Theoretical Chemistry, Campus B2.2, 66123 Saarbrücken, Germany
[2]Tianjin University, School of Materials Science and Engineering, Tianjin 300072, P. R. China

1.1
Introduction

Understanding the connection between composition and structure of materials on the one side and their properties on the other side is of fundamental importance to the development of society with its need for materials with special and optimized properties. For this, electronic-structure calculations play an important role. Such studies have the advantage that well-defined systems are treated and that both existing and nonexisting systems can be studied, whereby any aspect of the relations between composition + structure and property can be studied in detail. On the other hand, computational resources put limitations to the systems that can be treated and approximations in the computational approaches may lead to inaccuracies in the obtained results. Therefore, computational studies of materials properties are most useful when being combined with complementary experimental studies, and then the electronic-structure calculations most often deliver useful information when interpreting, explaining, and extending the experimental studies.

Currently applied electronic-structure methods have been developed first of all for calculating the total energy as a function of structure, thereby giving information on structure and relative energies of, for example, different crystal structures. However, for the development of the above-mentioned relations between composition/structure and property, it is useful to extend these theoretical methods so that other properties can also be calculated, including concepts that hardly are accessible with experiment but useful for interpretative purposes (like, for instance, aromaticity), but also properties that are experimentally accessible. In both cases, the goal is to obtain information that can be used in rationalizing the observations both from the systems that are being studied but, in the ideal case, also for a larger class of systems.

Handbook of Solid State Chemistry, First Edition. Edited by Richard Dronskowski, Shinichi Kikkawa, and Andreas Stein.
© 2017 Wiley-VCH Verlag GmbH & Co. KGaA. Published 2017 by Wiley-VCH Verlag GmbH & Co. KGaA.

In the present contribution, we shall describe the foundations of one of the most important classes of electronic-structure methods for the computational study of properties of solids, that is, density-functional methods. Historically, one may distinguish currently applied electronic-structure methods into two classes, that is, the wavefunction-based and the density-based ones whereby, as mentioned, the latter is at the focus here. Nevertheless, since certain methods combine the two approaches we shall briefly present both methods in Section 1.2. In that section, we shall also describe different strategies for obtaining (approximate) descriptions of the exchange-correlation effects, which is one of the nontrivial issues of density-functional theory. Finally, in that section we shall also present approaches for extracting information on chemical bonding, atomic charges, and so on, which all are concepts that are very useful in rationalizing computational and experimental findings in solid-state chemistry. Section 1.2 is closed by a brief description of how one can calculate excitation energies.

Since this contribution is devoted to the foundations of the theoretical methods, we shall hardly present any results of applications. Instead, we shall, in Section 1.4, discuss various issues that are not resolved completely at the moment of writing. These include the development of approximate but yet sufficiently accurate functionals for exchange-correlation effects and inaccuracies related with weak interactions. Most of the presently applied approximations for those suffer from the problem of charge delocalization that we shall discuss briefly here, too. This can have severe consequences for the theoretical determination of response properties as we shall see. Finally, Section 1.5 contains the conclusions.

1.2
Foundations of Density-Functional Theory

1.2.1
Basic Principles [1]

For a general system with M nuclei and N electrons, the positions of the nuclei are denoted $\vec{R}_1, \vec{R}_2, \cdots, \vec{R}_M$ and those of the electrons $\vec{r}_1, \vec{r}_2, \cdots, \vec{r}_N$. We use Hartree atomic units and set accordingly $m_e = |e| = 4\pi\varepsilon_0 = \hbar \equiv 1$. The mass and charge of the kth nucleus are M_k and Z_k, respectively. The combined coordinate $\vec{x}_i$ denotes the position and spin coordinate of the ith electron. In the absence of external interactions and relativistic effects, the Hamilton operator for this system can be written as a sum of five terms,

$$\hat{H} = \hat{H}_{kn} + \hat{H}_{ke} + \hat{H}_{nn} + \hat{H}_{ee} + \hat{H}_{en}, \tag{1.1}$$

that is, as the kinetic-energy operator for the nuclei, that for the electrons, and the three potential-energy operators for the nucleus–nucleus, the electron–electron, and the electron–nucleus interactions.

1.2 Foundations of Density-Functional Theory

Then, the time-independent Schrödinger equation is

$$\hat{H}\Psi(\vec{R}_1,\vec{R}_2,\cdots,\vec{R}_M,\vec{x}_1,\vec{x}_2,\cdots,\vec{x}_N) = E \cdot \Psi(\vec{R}_1,\vec{R}_2,\cdots,\vec{R}_M,\vec{x}_1,\vec{x}_2,\cdots,\vec{x}_N). \tag{1.2}$$

Except for special cases that are not relevant here, one makes use of the Born–Oppenheimer approximation. Then, $\hat{H}_{kn}$ is ignored and, furthermore, the wavefunction Ψ is written as a product of two functions,

$$\Psi(\vec{R}_1,\vec{R}_2,\cdots,\vec{R}_M,\vec{x}_1,\vec{x}_2,\cdots,\vec{x}_N) = \Psi_n(\vec{R}_1,\vec{R}_2,\cdots,\vec{R}_M) \\ \cdot \Psi_e(\vec{R}_1,\vec{R}_2,\cdots,\vec{R}_M;\vec{x}_1,\vec{x}_2,\cdots,\vec{x}_N). \tag{1.3}$$

Ψ_n is the wavefunction for the nuclei and Ψ_e is the electronic wavefunction. Ψ_e depends functionally on the electronic coordinates but also parametrically on the nuclear coordinates. In practice, the functional dependence of Ψ_e on all electronic coordinates is formulated in terms of functions (orbitals) that depend solely on the coordinates of individual electrons. Ψ_e satisfies the electronic Schrödinger equation

$$\hat{H}_e \Psi_e = E_e \Psi_e \tag{1.4}$$

with

$$\hat{H}_e = \hat{H}_{ke} + \hat{H}_{ee} + \hat{H}_{en}. \tag{1.5}$$

Finally, the total energy becomes

$$E = E_e + \hat{H}_{nn}. \tag{1.6}$$

Within the wavefunction-based methods, Eq. (1.4) can be solved approximately by writing Ψ_e as a single Slater determinant (the Hartree–Fock approximation). Per definition, correlation effects are then ignored, although they may be added subsequently. The N single-particle functions $\psi_1, \psi_2, \cdots, \psi_N$ of the Slater determinant are calculated by solving the Hartree–Fock equations

$$\hat{F}\psi_k(\vec{x}) = \varepsilon_k \psi_k(\vec{x}), \tag{1.7}$$

with

$$\hat{F} = \hat{h}_1 + \sum_{i=1}^{N} (\hat{J}_i - \hat{K}_i). \tag{1.8}$$

Here,

$$\hat{h}_1 = -\frac{1}{2}\nabla^2 - \sum_{k=1}^{M} \frac{Z_k}{|\vec{r}-\vec{R}_k|}, \tag{1.9}$$

represents the kinetic energy and the potential from the nuclei, whereas

$$\sum_{i=1}^{N} \hat{J}_i \psi_k(\vec{x}) = \sum_{i=1}^{N} \int \frac{|\psi_i(\vec{x}_2)|^2}{|\vec{r}_2 - \vec{r}|} \, d\vec{x}_2 \psi_k(\vec{x})$$

$$\sum_{i=1}^{N} \hat{K}_i \psi_k(\vec{x}) = \sum_{i=1}^{N} \int \frac{\psi_i^*(\vec{x}_2) \psi_k(\vec{x}_2)}{|\vec{r}_2 - \vec{r}|} \, d\vec{x}_2 \psi_i(\vec{x}), \quad (1.10)$$

are the Coulomb and exchange terms, respectively. Then,

$$E_e = \sum_{i=1}^{N} \langle \psi_i | \hat{h}_1 | \psi_i \rangle + \sum_{i,j=1}^{N} \left[\left\langle \psi_i \psi_j \left| \frac{1}{|\vec{r}_1 - \vec{r}_2|} \right| \psi_i \psi_j \right\rangle - \left\langle \psi_i \psi_j \left| \frac{1}{|\vec{r}_1 - \vec{r}_2|} \right| \psi_j \psi_i \right\rangle \right]. \quad (1.11)$$

It is immediately observed that the calculation of E_e, and, hence, the total energy, requires the calculation of a very complex object, Ψ_e, that depends on the coordinates of all N electrons. On the other hand, experiments probe properties of the system of interest that depend most often on only the distribution of the electrons in three- or six-dimensional space. Thus, Ψ_e contains very much redundant information.

Hohenberg and Kohn [2] (see Refs [3–5] for a recent account of the development of applied density-functional theory) showed that there is a one-to-one correspondence between the external potential and the electron density in three-dimensional space, $\rho(\vec{r})$. This suggests that, in principle, any ground-state property, including E_e, can be calculated from $\rho(\vec{r})$, that is, that E_e is a functional of $\rho(\vec{r})$,

$$E_e = E_e[\rho(\vec{r})], \quad (1.12)$$

but, unfortunately, the precise form of this functional is unknown. An important step toward a practical scheme was provided by Kohn and Sham [6] who showed that the problem of calculating $E_e[\rho(\vec{r})]$ of Eq. (1.12) can be cast into that of solving a set of single-particle equations,

$$\hat{h}_{\text{eff}} \psi_i(\vec{r}) = \varepsilon_i \psi_i(\vec{r}) \quad (1.13)$$

with

$$\hat{h}_{\text{eff}} = -\frac{1}{2} \nabla^2 + V_{\text{eff}}(\vec{r}). \quad (1.14)$$

In Eq. (1.10), $\sum_i \hat{J}_i$ is a multiplicative operator, but $\sum_i \hat{K}_i$ is a "true" operator. Moreover, correlation effects are completely neglected within the Hartree–Fock approach. In contrast, V_{eff} of Eq. (1.14) is a purely multiplicative operator, although it contains – as in the Hartree–Fock approach – both the external Coulomb potential from the nuclei, the Coulomb potential from the electrons, and the exchange effects, but also correlation effects.

Above, we have used the same notation for the orbitals from the Hartree–Fock and from the Kohn–Sham equations. They will, however, not be identical (although similar in most cases). Nevertheless, both in the Hartree–Fock and in the Kohn–Sham approach the total electron density is given as a sum over the N energetically lowest orbitals,

$$\rho(\vec{r}) = \sum_{i=1}^{N} |\psi_i(\vec{r})|^2. \tag{1.15}$$

Schematically, one may formulate the theorem of Hohenberg and Kohn as

$$\rho \Rightarrow E_e, \tag{1.16}$$

that is, from the electron density one may calculate the electronic energy. The approach of Kohn and Sham amounts to introduce an intermediate step (the calculation of V_{eff}), that is,

$$\rho \Rightarrow V_{\text{eff}} \Rightarrow E_e. \tag{1.17}$$

However, just as the precise form of $E_e[\rho(\vec{r})]$ of Eq. (1.12) is unknown, so is that of the first relation of Eq. (1.17), and one has to restore to approximations. In the Section 1.2.2 we shall present some of the most common approximations, whereas in later sections we shall discuss some recent developments aimed at solving some of the problems related to various approximations for V_{eff}.

1.2.2
Functionals

V_{eff} of Eq. (1.14) contains the external potential (i.e., often only the Coulomb potential from the nuclei), the Coulomb potential from the electrons, and a remainder,

$$V_{\text{eff}}(\vec{r}) = V_{\text{ext}}(\vec{r}) + V_{\text{C}}(\vec{r}) + V_{\text{xc}}(\vec{r}). \tag{1.18}$$

When the external potential is solely the Coulomb potential of the nuclei, V_{ext} becomes

$$V_{\text{ext}}(\vec{r}) = -\sum_{k=1}^{M} \frac{Z_k}{|\vec{r} - \vec{R}_k|}. \tag{1.19}$$

Moreover,

$$V_{\text{C}}(\vec{r}) = \int \frac{\rho(\vec{r}')}{|\vec{r} - \vec{r}'|} d\vec{r}' = \sum_{i=1}^{N} \hat{J}_i. \tag{1.20}$$

According to the theorems of Hohenberg and Kohn, the remaining exchange-correlation potential $V_{\text{xc}}(\vec{r})$ depends on the electron density, but its precise functional form is unknown. Therefore, it has to be approximated and it may not surprise that many approximations have been proposed during the years (see Refs [3,7–9] and references therein).

V_{xc} can be written as the functional derivative of the exchange-correlation energy,

$$V_{xc}(\vec{r}) = \frac{\delta E_{xc}}{\delta \rho(\vec{r})}, \tag{1.21}$$

and E_{xc} is often written as

$$E_{xc} = \int \varepsilon_{xc}(\vec{r}) \rho(\vec{r}) d\vec{r}, \tag{1.22}$$

whereby $\varepsilon_{xc}(\vec{r})$ is approximated. As a first approximation, $\varepsilon_{xc}(\vec{r})$ is assumed to depend on the electron density in only that same point, that is, $\varepsilon_{xc}(\vec{r})$ becomes a function of the electron density $\rho(\vec{r})$,

$$\varepsilon_{xc}(\vec{r}) = \varepsilon_{xc}(\rho(\vec{r})). \tag{1.23}$$

This amounts to the so-called local-density approximation (LDA). For spin-polarized system, all functionals may be generalized to depend not only on the electron density

$$\rho(\vec{r}) = \rho_\uparrow(\vec{r}) + \rho_\downarrow(\vec{r}) \tag{1.24}$$

but also on the spin density

$$m(\vec{r}) = \rho_\uparrow(\vec{r}) - \rho_\downarrow(\vec{r}). \tag{1.25}$$

Here, $\rho_\uparrow(\vec{r})$ and $\rho_\downarrow(\vec{r})$ are the electron density for the spin-up and spin-down electrons, respectively. Actually, these generalizations are beyond the content of the Hohenberg–Kohn theorem, but for the purpose of developing approximate functionals, they provide further flexibility that can improve the accuracy of the calculations.

In order to improve the accuracy of the approximations, one may introduce some nonlocality by letting the functionals depend also on $|\vec{\nabla} \rho(\vec{r})|$ and $\nabla^2 \rho(\vec{r})$, or equivalently on the scaled quantities

$$s = \frac{|\vec{\nabla} \rho(\vec{r})|}{2(3\pi)^{1/3} \rho^{4/3}(\vec{r})}$$
$$q = \frac{\nabla^2 \rho(\vec{r})}{4(3\pi^2)^{2/3} \rho^{5/3}(\vec{r})}. \tag{1.26}$$

Including dependences on s leads to the so-called generalized gradient approximations (GGAs), whereas the inclusion of q, too, results in so-called metageneralized gradient approximations (MGGAs or meta-GGAs) of the form

$$\varepsilon_{xc}(\vec{r}) = \varepsilon_{xc}\left(\rho(\vec{r}), |\vec{\nabla} \rho(\vec{r})|, \nabla^2 \rho(\vec{r}), \cdots\right). \tag{1.27}$$

The term meta-GGAs is, however, often also used for functionals that include dependences on local kinetic energy densities. It may be observed that the solutions ψ_i to the single-particle, so-called Kohn–Sham Eq. (1.13) also can be considered as being functionals of the electron density (in the sense that they can, in principle, be determined once the external potential is known) and, therefore, E_e may include dependences on those, too. The quantity

$$\tau(\vec{r}) = \sum_{i=1}^{N} \frac{1}{2} |\nabla \psi_i(\vec{r})|^2 \qquad (1.28)$$

gives, when integrated over the whole space, the total kinetic energy and is, moreover, everywhere nonnegative. It can therefore be considered a local kinetic energy density and functionals (meta-GGAs) that depend explicitly on τ have been derived.

E_{xc} contains all exchange and correlation effects. As mentioned above, the Kohn–Sham orbitals are functionals of the electron density, so that one may also calculate exchange and/or correlation effects directly with the help of these orbitals. The hybrid methods (see Refs [3,7–9] and references therein) for which a part of the exchange effects are calculated using the Hartree–Fock expression (see Eqs. (1.10) and (1.11)) and the remaining part is calculated using, for example, a GGA expression represent one class of such approaches. Another class is formed by the so-called exact-exchange methods. In all cases, one may write them in the form

$$\varepsilon_{xc}(\vec{r}) = \varepsilon_{xc}\left(\{\psi_i(\vec{r})\}, \{|\vec{\nabla} \psi_i(\vec{r})|\}, \{\nabla^2 \psi_i(\vec{r})\}, \cdots \right). \qquad (1.29)$$

It should be obvious that as the approximate expressions get more complex, also the determination of the exchange-correlation potential (Eqs. (1.21) and (1.22)) becomes increasingly difficult as does the determination of the solutions to the Kohn–Sham equations.

1.2.3
Hubbard (+U) Corrections

The LDA is based on the assumption that the exchange and correlation effects in a given point in space for the system of interest are well described through those of a homogeneous electron gas. Also the GGAs and meta-GGAs take this system as a reference system but include effects due to variations in the density in the closest vicinity of the reference point. This may be a good approximation for delocalized electrons in, for example, metallic systems, but for localized electrons in, for instance, 3d or 4f valence orbitals, the homogeneous electron gas provides a poor reference system. Thus, for the localized electrons one may instead use a Hubbard like model [10–15]. The Hubbard model introduces a parameter U, that quantifies the repulsion between more electrons occupying, for example, 3d or 4f orbitals on the same atom. Therefore, such approaches are often called $+U$ approaches (for instance, LDA$+U$ or GGA$+U$). They introduce an additional energy term so that the total energy takes the form

$$E_{DFT+U} = E_{DFT} + E_U + E_{dc}, \qquad (1.30)$$

where E_{DFT} is the usual density-functional expression evaluated using some approximate functional, E_U is the Hubbard correction, and E_{dc} is a term that shall remove double counting.

A frequently used expression for the last two terms is [16–19]

$$E_U + E_{dc} = \sum_X \frac{U_X}{2} \left\{ \left[\sum_{(m_1,\sigma_1) \neq (m_2,\sigma_2)} n_{X,m_1,\sigma_1} n_{X,m_2,\sigma_2} \right] - n_X(n_X - 1) \right\}. \quad (1.31)$$

Here, X describes the atomic orbitals for which the correction shall be applied (for instance, 3d orbitals on Fe or 4f orbitals on Ce), whereas (m_1, σ_1) and (m_2, σ_2) describe the different orbitals and spin components (e.g., the $3d_{xz}$ and $3d_{xy}$ orbitals as well as spin-up and spin-down). Moreover, $n_{X,m,\sigma}$ is the number of electrons in that orbital, and

$$n_X = \sum_{m,\sigma} n_{X,m,\sigma}. \quad (1.32)$$

With this modification, the effective potential of Eq. (1.18) becomes spin-dependent and equal to

$$V_{\text{eff},U,\sigma} = V_{\text{eff}} + \sum_{X,m} U_X \left(\frac{1}{2} - n_{X,m,\sigma} \right) |\psi_{X,m}\rangle\langle\psi_{X,m}|. \quad (1.33)$$

Here, $\psi_{X,m}$ is the spatial wavefunction of one of the orbitals for which the Hubbard correction shall be included, so that $|\psi_{X,m}\rangle\langle\psi_{X,m}|$ projects any orbital onto the localized one.

Equation (1.33) demonstrates the main problems related to the DFT + U approach. At first, the precise definition of $\psi_{X,m}(\vec{r})$ is not unique. One may use the wavefunctions for the isolated atoms but experience has shown that when passing to some compound also the fairly localized electronic orbitals may change shape slightly. Second, the product of two such functions $\psi_{X_1,m_1}(\vec{r})\psi_{X_2,m_2}(\vec{r})$ for $X_1 \neq X_2$ should vanish for all $\vec{r}$ implying that different localized orbitals are well separated, an assumption that in reality may not be completely satisfied. Third, there is some arbitrariness in the values of the parameters U_X. One may attempt to determine them using constrained density-functional calculations [20] (see also Refs [21,22]) but experience has shown that the best values (in the sense that experimental values are reproduced) depend not only on the system and atom of interest but also on the property that is considered (see Ref. [23]). Thus, even if there are sound theoretical reasons for introducing the +U corrections, their current application does contain some degree of empiricism.

1.2.4 Extended Systems

In this section, we shall discuss approaches that have been developed specifically for extended systems. This includes crystalline materials that are treated as being infinite and consisting of periodically repeated, identical units, but also large, finite systems.

When studying macroscopic, crystalline systems, it is very often a good approximation to consider the system of interest as being infinite and periodic. This amounts to considering large, regular systems consisting of a large number of regularly positioned, identical units where deviations from this regularity are found only near the boundaries of the system. In the thermodynamic limit, these boundaries constitute an infinitessimally small part of the total system and their contribution to most properties can be ignored (see, however, Section 1.2.6).

For the infinite, periodic system with $\vec{a}$, $\vec{b}$, and $\vec{c}$ being the lattice vectors, the translational properties of the single-particle orbitals and their energies can be quantified through a wavevector $\vec{k}$, that is,

$$\psi_i(\vec{r}) \to \psi_i(\vec{k},\vec{r})$$
$$\varepsilon_i \to \varepsilon_i(\vec{k}) \tag{1.34}$$

with

$$\psi_i(\vec{k},\vec{r}+\vec{R}_{n_a,n_b,n_c}) = e^{i\vec{k}\cdot\vec{R}_{n_a,n_b,n_c}} \psi_i(\vec{k},\vec{r})$$
$$\vec{R}_{n_a,n_b,n_c} = n_a\vec{a} + n_b\vec{b} + n_c\vec{c} \tag{1.35}$$

and n_a, n_b, and n_c being integers.

Also the single-particle equations, Eq. (1.13), become $\vec{k}$ dependent,

$$\hat{h}_{\text{eff}}(\vec{k})\psi_i(\vec{k},\vec{r}) = \varepsilon_i(\vec{k})\psi_i(\vec{k},\vec{r}). \tag{1.36}$$

Similar to the case for the finite system, these equations are solved approximately by expanding the eigenfunctions in some set of basis functions. For the infinite, periodic system, one may use atom-centered basis functions,

$$\psi_i(\vec{k},\vec{r}) = \sum_{X,p} c_{i,X,p}(\vec{k}) \chi_{X,p}(\vec{k},\vec{r})$$
$$\chi_{X,p}(\vec{k},\vec{r}) = \lim_{N\to\infty} (2N+1)^{-3/2} \sum_{n_a,n_b,n_c=-N}^{N} e^{i\vec{k}\cdot\vec{R}_{n_a,n_b,n_c}} \chi_{X,n_a,n_b,n_c,p}(\vec{r}), \tag{1.37}$$

where $\chi_{X,n_a,n_b,n_c,p}(\vec{r})$ is the pth basis functions centered at atom X in the unit cell at $\vec{R}_{n_a,n_b,n_c}$. This is equivalent to the traditional approach for finite, molecular systems.

Alternatively, one may use delocalized, plane waves as basis functions,

$$\psi_i(\vec{k},\vec{r}) = \sum_{\vec{G}} c_i(\vec{k}+\vec{G}) e^{i(\vec{k}+\vec{G})\cdot\vec{r}}, \tag{1.38}$$

where $\vec{G}$ is a vector in reciprocal space. It is clear that these basis functions cannot be ascribed individual atoms.

However, there are cases for which it is desirable to study directly the (maybe large) finite system or for which the size of the repeated unit becomes very large. In order to avoid that the computational requirements then become prohibitively large, special methods have to be applied. Here, we shall briefly outline the basic ideas behind the order-N methods as well as some related methods, that is, the divide-and-conquer method and the elongation method. For more details about the former, the reader is referred to Refs [24,25] and references therein.

The basic idea behind the order-N methods is that electronic interactions are short-ranged so that the major part of those can be truncated after some finite cut-off distance. This assumption is closely related to the near-sightedness concept that has been discussed by Kohn [26] and that states that for a system with fixed chemical potential and with a large number of particles (electrons), a localized perturbation in the external potential will not affect the electrons sufficiently far apart from the perturbation.

Through this assumption, the computational costs for the calculation of the matrix elements for these interactions will scale essentially linearly with the size of the system. By also seeking computational methods for solving the secular equation that also scale roughly linearly with system size, one ends up with a so-called order-N method.

As discussed above, density-functional formalism is based on an expression for the total electronic energy as a functional of the electron density of Eq. (1.15). The electron density is the diagonal elements of the first-order reduced density matrix

$$\rho^{(1)}(\vec{r}_1, \vec{r}_2) = \sum_{i=1}^{N} \psi_i^*(\vec{r}_1)\psi_i(\vec{r}_2). \tag{1.39}$$

With the order-N methods, this density matrix is the central quantity. It is known that

$$\left|\rho^{(1)}(\vec{r}_1, \vec{r}_2)\right| \to 0 \quad \text{for} \quad |\vec{r}_1 - \vec{r}_2| \to \infty \tag{1.40}$$

and often a good approximation corresponds to assume that this limit is reached already for small $|\vec{r}_1 - \vec{r}_2|$. Then, the density matrix can be expressed in terms of a basis set of localized basis functions,

$$\rho^{(1)}(\vec{r}_1, \vec{r}_2) = \sum_{i_1, i_2} \chi_{i_1}^*(\vec{r}_1) K_{i_1 i_2} \chi_{i_2}(\vec{r}_2), \tag{1.41}$$

where $K_{i_1 i_2}$ are the sought expansion coefficients that vanish if the basis functions χ_{i_1} and χ_{i_2} are sufficiently well separated.

Subsequently, the total electronic energy is expressed in terms of the density matrix. For the Coulomb and exchange-correlation parts as well as the contributions from the external potential, this is easily done since they depend on only the diagonal parts of the density matrix, that is, on the electron density of Eq. (1.15). Finally, for the kinetic energy we have

$$E_{ke} = -\frac{1}{2}\int \left[\nabla_{\vec{r}_2}^2 \rho^{(1)}(\vec{r}_1, \vec{r}_2)|_{\vec{r}_1 = \vec{r}_2}\right] d\vec{r}_2. \tag{1.42}$$

The expansion coefficients $K_{i_1 i_2}$ are then determined by minimizing the total electronic energy together with the constraint that the electron density gives the correct number of electrons,

$$\int \rho^{(1)}(\vec{r},\vec{r})d\vec{r} = \int \rho(\vec{r})d\vec{r} = N. \tag{1.43}$$

However, for single-particle methods like that of Kohn and Sham, density matrices are known to be idempotent, that is,

$$\int \rho^{(1)}(\vec{r}_1,\vec{r}_3)\rho^{(1)}(\vec{r}_3,\vec{r}_2)d\vec{r}_3 = \rho^{(1)}(\vec{r}_1,\vec{r}_2), \tag{1.44}$$

whereas it is not automatically given that a minimization procedure will lead to a density matrix satisfying this constraint. It has therefore been proposed [27] to apply the so-called purification

$$\rho^{(1)}(\vec{r}_1,\vec{r}_2) \to \tilde{\rho}^{(1)}(\vec{r}_1,\vec{r}_2) = 3\int \rho^{(1)}(\vec{r}_1,\vec{r}_3)\rho^{(1)}(\vec{r}_3,\vec{r}_2)d\vec{r}_3 \\ - 2\int\int \rho^{(1)}(\vec{r}_1,\vec{r}_3)\rho^{(1)}(\vec{r}_3,\vec{r}_4)\rho^{(1)}(\vec{r}_4,\vec{r}_2)d\vec{r}_3 d\vec{r}_4 \tag{1.45}$$

on any given density matrix and then replace $\rho^{(1)}$ by $\tilde{\rho}^{(1)}$ in the expression for the total energy. The purification leads to a density matrix that more closely obeys Eq. (1.44) but also to a more complicated expression for the total energy.

Finally, the minimization of the total energy with respect to the density-matrix expansion coefficients $K_{i_1 i_2}$ can be done through a conjugate-gradient technique, which scales essentially linearly with N instead of solving a matrix eigenvalue problem (which would scale as N^3).

There are other methods based on the observation that most electronic interactions are of relatively short range and, therefore, that (parts of) the computational efforts may be made to scale essentially linearly with the size of the system above a certain smallest size. Among those, we shall briefly mention two.

The divide-and-conquer approach of Yang [28,29] is one such method. With this, the system of interest is separated into smaller fragments that partly can be treated individually. Mathematically, a set of partitioning functions is introduced,

$$1 = \sum_\alpha p_\alpha(\vec{r}), \tag{1.46}$$

where α represents the different parts.

For each subsystem (characterized by α) one introduces a set of localized basis functions, $\{\chi_i^\alpha\}$, with $i = 1, \cdots, M_\alpha$. The Kohn–Sham operator of Eq. (1.14) can now be projected onto the space of these localized basis functions, giving

$$\hat{h}_{\text{eff}}^\alpha = \sum_{j,k,l,m=1}^{M_\alpha} |\chi_j^\alpha\rangle (S^\alpha)^{-1}_{jk} (H^\alpha)_{kl} (S^\alpha)^{-1}_{lm} \langle\chi_m^\alpha| \tag{1.47}$$

with

$$(S^\alpha)_{ij} = \langle \chi_i^\alpha | \chi_j^\alpha \rangle$$
$$(H^\alpha)_{ij} = \langle \chi_i^\alpha | \hat{h}_{\text{eff}} | \chi_j^\alpha \rangle \tag{1.48}$$

This leads to a set of single-particle Kohn–Sham equations for each subsystem separately, that in matrix form becomes

$$\underline{\underline{H}}^\alpha \cdot \underline{C}_i^\alpha = \varepsilon_i^\alpha \cdot \underline{\underline{S}}^\alpha \cdot \underline{C}_i^\alpha. \tag{1.49}$$

Since the Kohn–Sham operator $\hat{h}_{\text{eff}}$ contains information on the complete system, there will be "cross-talking" between the different subsystems. However, the time-consuming part related to the diagonalization of a large matrix has been reduced to that of diagonalizing a number of smaller matrices, each having the size of the number of basis functions of the single subsystem. For large systems this can lead to a significant computer-time saving.

Also the elongation method of Imamura et al. [30–33] is based on the short range of the electronic interactions. It is a method that most easily can be applied to polymeric chains but it has also been extended to three-dimensional crystalline systems. When treating chains with the elongation method, one gradually increases the chain length by adding units one after the other. At any length, the electronic orbitals are transformed into a set that consists of orbitals localized either far apart from the region where the units are added (the so-called A-region) or close to that region (i.e., the so-called B-region). When adding an additional unit, only those orbitals that are localized to the B-region will be perturbed by the additional unit and, accordingly, only those need to be recalculated. From a certain size the length of the B-region will be constant and, thus, a larger part of the calculation will scale roughly linearly with the system size. An additional advantage of this method, and also of the divide-and-conquer method, is that the added units need not be identical, that is, also aperiodic systems can be treated, as can systems with impurities [32,34].

1.2.5
Descriptors

Per construction, density-functional calculations yield the total energy and the electron density as functions of the structure of the system of interest. Depicting those can provide useful information on the system, as also can be the case for the orbitals and their energies. However, for the orbitals it shall be remembered that the ones that are obtained through "standard" calculations most often are fairly delocalized over the complete system instead of being localized to the spatial region close to two or just some few atoms which, however, often is assumed when discussing the concepts of chemical bonds in the system of interest. Moreover, they are not unique and any unitary transformation among the occupied orbitals will leave any observable unchanged. Therefore, in order to relate the results of such calculations to the conventional chemical concepts like bonds and atomic charges and volumes, some additional treatment of the obtained results is necessary. Here, we shall discuss some few proposals.

The electron density itself may be used in attempting to identify atomic contributions and chemical bonds. A useful approach in this direction is the atoms-in-molecules approach by Bader [35]. Starting from the total electron density $\rho(\vec{r})$, one determines the set of (so-called critical) points satisfying

$$\vec{\nabla} \rho(\vec{r}) = \vec{0} . \tag{1.50}$$

Subsequently, starting at any given point, $\vec{r}$, and following the gradient $\vec{\nabla} \rho(\vec{r})$, that is, changing

$$\vec{r} \rightarrow \vec{r} + \tau \cdot \vec{\nabla} \rho(\vec{r}) \tag{1.51}$$

with τ being a positive constant, one will ultimately arrive at one of the so-called attractors, that is, in a point where $\rho(\vec{r})$ has a local maximum. These set of the attractors consists of the positions of the nuclei as well as of points lying somewhere between two nuclei. In the last case one talks about a bond critical point. A curve of the type of Eq. (1.51) consisting of the positions of a pair of nuclei and a bond critical point is defined as a bond path. Finally, all those points $\vec{r}$ for which the gradient path of Eq. (1.51) end at the same nucleus are points that belong to the same atom. Thereby, space has been divided uniquely into atomic, nonoverlapping, space-filling parts. The bond critical points lie on the surfaces separating different atomic parts. A significant advantage of this approach that it can be applied to any electron density irrespectively of its origin. Thus, it does not depend on using, for example, atom-centered basis functions in a calculation, and can even by applied to experimental densities.

With the atoms-in-molecules approach one separates space into atomic contributions. By integration the charge density over those, one obtains estimates of atomic charges. Another approach is based on applying position-dependent weight factors, that is,

$$1 = \sum_X \sum_n w_{X,n}(\vec{r}) \tag{1.52}$$

with n being a short notation for (n_a, n_b, n_c). This resembles the idea behind the divide-and-conquer method (cf. Eq. (1.46)) so that

$$\rho(\vec{r}) = \sum_X \sum_n w_{X,n}(\vec{r}) \rho(\vec{r}) \equiv \sum_X \sum_n \rho_{X,n}(\vec{r}) \tag{1.53}$$

whereby the total electron density is separated into atomic contributions. The total integral of $\rho_{X,n}(\vec{r})$ (modified by the nuclear charge) provides then another measure for an atomic charge. As weight factors one may use a (normalized) superposition of the electron densities of the isolated atoms. This corresponds to the Hirshfeld approach [36].

Another way of obtaining information on the chemical interactions in the system of interest is provided by an analysis related to the one behind Mulliken populations [37,38]. These are closely related to applying atom-centered basis functions in the calculations (as in Eq. (1.37)) so that when instead using plane waves or the like (as in Eq. (1.38)), it is necessary to transform the resulting orbital wavefunctions into linear combinations of atom-centered functions. For atom-centered basis functions, the wavefunctions are written as

$$\psi_i(\vec{k},\vec{r}) = \sum_{X,n,p} c_{i,X,n,p}(\vec{k}) \chi_{X,n,p}(\vec{r}). \tag{1.54}$$

For any single-particle operator $\hat{a}$ we write then

$$\langle \psi_i(\vec{k}) | \hat{a} | \psi_i(\vec{k}) \rangle = \sum_{X,n} \left\{ \sum_{p_1,p_2} c^*_{i,X,n,p_1}(\vec{k}) c_{i,X,n,p_2}(\vec{k}) A_{(X,n,p_1),(X,n,p_2)} \right\}$$

$$+ \frac{1}{2} \sum_{(X_1,n_1) \neq (X_2,n_2)} \left\{ \sum_{p_1,p_2} [c^*_{i,X_1,n_1,p_1}(\vec{k}) c_{i,X_2,n_2,p_2}(\vec{k}) A_{(X_1,n_1,p_1),(X_2,n_2,p_2)} \right.$$

$$+ c^*_{i,X_2,n_2,p_2}(\vec{k}) c_{i,X_1,n_1,p_1}(\vec{k}) A_{(X_2,n_2,p_2),(X_1,n_1,p_1)}] \bigg\}$$

$$\equiv \sum_{X,n} a_{X,n,i}(\vec{k}) + \frac{1}{2} \sum_{(X_1,n_1) \neq (X_2,n_2)} a_{X_1,n_1,X_2,n_2,i}(\vec{k})$$

$$= \sum_{X,n} \left\{ a_{X,n,i}(\vec{k}) + \frac{1}{2} \sum_{(X',n') \neq (X,n)} a_{X',n',X,n,i}(\vec{k}) \right\}$$

$$\equiv \sum_{X,n} A_{X,n,i}(\vec{k}) \tag{1.55}$$

with

$$A_{(X_1,n_1,p_1),(X_2,n_2,p_2)} = \langle \chi_{X_1,n_1,p_1} | \hat{a} | \chi_{X_2,n_2,p_2} \rangle. \tag{1.56}$$

In Eq. (1.55), the property of interest, $\hat{a}$, has been split into net, overlap, and gross contributions from the individual orbitals at the individual $\vec{k}$. By adding the contributions from all occupied orbitals, one obtain the total populations. Alternatively, by plotting the total contributions as a function of orbital energy, one arrives at the crystal orbital overlap populations (for $\hat{a} = 1$) [39] or the crystal orbital Hamilton populations (for $\hat{a} = \hat{h}_{\text{eff}}$) [40].

Another property that often is studied is the electron localization function (ELF) [41]

$$L(\vec{r}) = \left[1 + \left(\frac{D(\vec{r})}{D_h(\vec{r})} \right)^2 \right]^{-1} \tag{1.57}$$

with

$$D(\vec{r}) = \frac{1}{2} \sum_{i=1}^{N} |\vec{\nabla} \psi_i(\vec{r})|^2 - \frac{1}{8} \frac{|\vec{\nabla} \rho(\vec{r})|^2}{\rho(\vec{r})}$$

$$D_h(\vec{r}) = \frac{3}{10} [3\pi^2 \rho(\vec{r})]^{5/3}. \tag{1.58}$$

This is a function in three-dimensional space, but plotting iso-ELF surfaces for a value of $L \simeq 0.8$ turns out to lead to three-dimensional objects that resemble orbitals as we usually employ when rationalizing chemical observations.

1.2.6
Excitations

For the Hartree–Fock approach and for a system with N electrons, Koopmans [42] demonstrated that under certain conditions to be specified below, the energy related to removing an electron from the (occupied) nth orbital or adding an electron to the (unoccupied) mth orbital is

$$E_{HF}(N-1) - E_{HF}(N) = -\varepsilon_n$$
$$E_{HF}(N+1) - E_{HF}(N) = \varepsilon_m \tag{1.59}$$

with $E_{HF}(K)$ being the total energy for the system with K electrons and calculated using the Hartree–Fock approach. It is assumed that relaxation effects (i.e., the redistribution of the other electrons when one electron is removed or added) can be ignored. Thus, Koopmans' theorem offers a rationale for considering the orbital energies, ε_i, or the band structures, $\varepsilon_i(\vec{k})$, as being related to photoelectron spectra and other optical properties as long as relaxation effects are negligible.

For density-functional approaches, Janak's theorem [43] provides a similar framework, except that

$$\lim_{\delta N \to 0} \frac{1}{\delta N}[E_{DFT}(N - \delta N) - E_{DFT}(N)] \simeq -\varepsilon_n$$
$$\lim_{\delta N \to 0} \frac{1}{\delta N}[E_{DFT}(N + \delta N) - E_{DFT}(N)] \simeq \varepsilon_m. \tag{1.60}$$

Again, relaxation effects are neglected and $E_{DFT}(K)$ is the total energy for the system with K electrons as calculated using the density-functional approach. In this case, it is assumed that K can be fractional.

These two theorems make it possible to relate calculated orbital energies or band structures to excitation energies. Although the approximation that relaxation effects can be neglected may have severe impacts on the accuracy of the results, the simplicity of the approach makes it very appealing to make use of it and, therefore, very many studies do so. However, there are cases where more accurate results are desirable. Then, one has to apply more advanced approaches of which two shall be discussed here.

The time-dependent density-functional theory (TD-DFT), first formulated by Runge and Gross [44] (see also [45,46]), provides a mathematically exact scheme with which excitation energies can be calculated within density-functional theory. Although the theory was formulated more than three decades ago, it is only within the last 1–2 decades that it has become of practical use for a larger class of systems, whereby almost all applications are restricted to finite systems like molecules.

The theorems of Hohenberg and Kohn combined with the approach of Kohn and Sham apply to the stationary case in the absence of time-dependent effects. Excitations involve, on the other hand, the response of the system of interest to some external time-dependent perturbation (most often, an electromagnetic field) and, therefore, it becomes necessary to extend the density-functional theory of Hohenberg, Kohn, and Sham to the time-dependent case. Such a theory was presented by Runge and Gross [44].

These authors demonstrated that under certain conditions the total electron density for a general system subject to some external time-dependent interaction can be written as

$$\rho(\vec{r},t) = \sum_{i=1}^{N} \psi_i^*(\vec{r},t)\psi_i(\vec{r},t), \tag{1.61}$$

that is, an expression similar to the stationary case.

The single-particle orbitals are calculated from time-dependent Hartree–Fock or Kohn–Sham-like equations,

$$\left[-\frac{1}{2}\nabla^2 + V_{\text{eff}}(\vec{r},t)\right]\psi_i(\vec{r},t) = i\frac{\partial}{\partial t}\psi_i(\vec{r},t), \tag{1.62}$$

where the effective potential is the sum of the external potential (e.g., the electrostatic potential of the nuclei plus an electromagnetic potential), the Coulomb potential of the electrons, and the remaining time-dependent exchange-correlation potential,

$$V_{\text{eff}}(\vec{r},t) = V_{\text{ext}}(\vec{r},t) + V_{\text{C}}(\vec{r},t) + \frac{\delta A_{\text{xc}}}{\delta \rho(\vec{r},t)}. \tag{1.63}$$

Except for an extra time-dependence, this expression resembles the standard Kohn–Sham expression with, however, the important difference that the last term contains not an energy but an action. For a general quantum–mechanical system with the wavefunction $\Psi(\vec{r},t)$ the action is defined as

$$A(\vec{r}) = \int_{t_0}^{t_1} \left\langle \Psi(\vec{r},t) \left| i\frac{\partial}{\partial t} - \hat{H}(\vec{r},t) \right| \Psi(\vec{r},t) \right\rangle dt. \tag{1.64}$$

$A_{\text{xc}}(\vec{r})$ is the exchange-correlation part of the action. In the general case, $A(\vec{r})$ as well as $A_{\text{xc}}(\vec{r})$ depend on the complete history of the system of interest but a considerable simplification is obtained when considering only the linear response of the system to a time-dependent perturbation. This corresponds to writing

$$\hat{H}(\vec{r},t) = \hat{H}_0(\vec{r}) + \Delta\hat{H}(\vec{r},t), \tag{1.65}$$

where only $\Delta\hat{H}(\vec{r},t)$ is time-dependent. Moreover, $\Delta\hat{H}(\vec{r},t)$ is assumed to be much smaller than $\hat{H}_0$. By letting $\rho_0(\vec{r})$ be the electron density without the perturbation $\Delta\hat{H}(\vec{r},t)$, the exchange-correlation potential can in this case be written as

$$\frac{\delta A_{\text{xc}}}{\delta \rho(\vec{r},t)} = \int \left[f_{\text{xc}}[\rho_0](\vec{r},t,\vec{r}\,',t')\delta\rho(\vec{r}\,',t')d\vec{r}\,'\, dt', \tag{1.66}$$

1.2 Foundations of Density-Functional Theory

where $\delta\rho$ is the change in the electron density due to the perturbation and f_{xc} is the so-called exchange-correlation kernel evaluated for the density ρ_0,

$$f_{xc}[\rho_0](\vec{r},t,\vec{r}',t') = \left.\frac{\delta V_{xc}[\rho](\vec{r},t)}{\delta\rho(\vec{r}',t')}\right|_{\rho=\rho_0}. \tag{1.67}$$

Accordingly, the exchange-correlation kernel is the functional derivative of the exchange-correlation potential which in turn is written with the help of the functional derivative of the exchange-correlation energy density ε_{xc}. All quantities are time-dependent and their precise form is unknown which is identical to the time-independent case. Therefore, also in this case one has to resort to approximations and here much less work in developing accurate approximations has been done compared with the time-independent case.

The simplest and most frequently used approximation is the adiabatic local-density approximation (ALDA). In the static case, the local-density approximation amounts to write $\varepsilon_{xc}(\vec{r})$ as a function of the electron density in the point of interest, $\rho(\vec{r})$. This is also the case in the ALDA and the dependence of the kernel at time t on any other time is ignored (i.e., also in time domain, the response is local). Accordingly,

$$f_{xc}^{ALDA}[\rho](\vec{r},t,\vec{r}',t') = \delta(\vec{r}-\vec{r}')\delta(t-t')\left.\frac{dV_{xc}^{LDA}[\rho']}{d\rho'}\right|_{\rho'=\rho}. \tag{1.68}$$

Through a Fourier transform, the time dependence can be written as a frequency dependence and then one may write the effective potential as

$$V_{eff}(\vec{r},\omega) = V_{ext}(\vec{r},\omega) + \int\frac{\delta\rho(\vec{r}',\omega)}{|\vec{r}-\vec{r}'|}d\vec{r}' + \int f_{xc}(\vec{r},\vec{r}',\omega)\delta\rho(\vec{r}',\omega)d\vec{r}' \tag{1.69}$$

still restricting ourselves to the linear-response case and neglecting the case $\omega = 0$. One may also express $\delta\rho$ in terms of the response function χ,

$$\delta\rho(\vec{r},\omega) = \int\chi(\vec{r},\vec{r}',\omega)V_{ext}(\vec{r}',\omega)d\vec{r}', \tag{1.70}$$

where the response function is written in terms of the Kohn-Sham orbitals in the static case,

$$\chi_s(\vec{r},\vec{r}',\omega) = \sum_{j,k}(f_k-f_j)\frac{\psi_j(\vec{r})\psi_k^*(\vec{r})\psi_j^*(\vec{r}')\psi_k(\vec{r}')}{\omega-(\varepsilon_j-\varepsilon_k)+i\eta}. \tag{1.71}$$

Here, f_j and f_k are the occupation numbers of the orbitals and ε_j and ε_k their energies. η is an infinitesimal, positive energy.

By solving Eqs. (1.70) and (1.71) self-consistently, one obtains the linear response of the electron density to the external perturbation, which could, for example, be an electromagnetic field. From this, various quantities like polarizability and so on can be extracted.

The response function χ_s of Eq. (1.71) has poles at the orbital energy differences $\pm(\varepsilon_j-\varepsilon_k)$ (whereby one of the orbitals ψ_j and ψ_k is occupied and the other empty). Therefore, these poles represent first approximations to the excitation energies. Alternatively, one may apply the TD-DFT formalism so that the linear response to time-dependent, electric fields is calculated directly. The resulting equations for the molecular case were presented by Jamorski et al. [47,48] who showed that the excitation energies ω_I then can be determined from the eigenvalue equations

$$\sum_{ij}\left[\omega_{ij}^2\delta_{ik}\delta_{jl} + 2\sqrt{\omega_{ij}f_{ij}}K_{ij,kl}\sqrt{\omega_{kl}f_{kl}}\right]F_{ij}^I = \omega_I^2 F_{kl}^I \tag{1.72}$$

with

$$\omega_{ij} = \varepsilon_j - \varepsilon_i$$
$$f_{ij} = f_j - f_i \tag{1.73}$$

and with the coefficients F_{ij}^I related to the oscillator strengths. Moreover, the coupling matrix elements equal

$$K_{ij,kl} = \int\int \psi_i^*(\vec{r}_1)\psi_j(\vec{r}_1)\frac{1}{|\vec{r}_1 - \vec{r}_2|}\psi_k(\vec{r}_2)\psi_l^*(\vec{r}_2)\mathrm{d}\vec{r}_1 \mathrm{d}\vec{r}_2$$
$$+ \int\int \psi_i^*(\vec{r}_1)\psi_j(\vec{r}_1)\frac{\delta^2 E_{\mathrm{xc}}}{\delta\rho(\vec{r}_1)\delta\rho(\vec{r}_2)}\psi_k(\vec{r}_2)\psi_l^*(\vec{r}_2)\mathrm{d}\vec{r}_1 \mathrm{d}\vec{r}_2. \tag{1.74}$$

Jamorski et al. [47] showed moreover that their approach also allowed for the calculation of polarizability.

An approach for calculating excitation energies that has been developed for infinite, periodic crystalline systems is the so-called GW approach. Then, one calculates directly the response of the system to the excitation of one electron from a valence orbital to a conduction orbital. The energies of these processes correspond to the combined process of the excitation of an electron and the relaxation of all the other electrons. Thus, an electron is not considered as being an isolated particle, but is affected by all the other electrons. The electron with all these relaxation effects is called a quasiparticle.

It has been shown [49] that the equation that defines the energies of the quasiparticles is

$$\left[-\frac{1}{2}\nabla^2 + V_{\mathrm{n}}(\vec{r}) + V_{\mathrm{C}}(\vec{r})\right]\psi_i\left(\vec{k},\vec{r}\right) + \int \Sigma(\vec{r},\vec{r}',\varepsilon_i)\psi_i\left(\vec{k},\vec{r}'\right)\mathrm{d}\vec{r}' = \varepsilon_i\left(\vec{k}\right)\psi_i\left(\vec{k},\vec{r}\right), \tag{1.75}$$

where $\Sigma(\vec{r},\vec{r}',\varepsilon_i)$ is the so-called electron self-energy operator that contains all the effects of exchange and correlation among the electrons. As mentioned above, the physical idea behind Eq. (1.75) is that of studying what happens when an electron is being excited. However, due to the relaxation effects, the wavefunctions of all electrons change then one is being excited. Therefore, the potential in the point $\vec{r}$ felt by any particular electron (in orbital i) is specific for that orbital and, furthermore, depends not only on $\vec{r}$ but also on the electrons in all other points $\vec{r}'$ of space. All these effects are hidden in the self energy $\Sigma(\vec{r},\vec{r}',\varepsilon_i)$, except for one part, that is, the classical electrostatic potential of the electronic distribution, $V_C(\vec{r})$.

Within the GW approximation, the self-energy is written as [50]

$$\Sigma(\vec{r},\vec{r}',\varepsilon_i) = \frac{i}{2\pi}\int G(\vec{r},\vec{r}',\varepsilon_i - \varepsilon')W(\vec{r},\vec{r}',\varepsilon')e^{-i\delta\varepsilon'}\,d\varepsilon' \quad (1.76)$$

with

$$\delta \to 0^+. \quad (1.77)$$

In Eq. (1.76), G is the Green's function that depends on all orbitals of the system,

$$G(\vec{r},\vec{r}',\varepsilon) = \sum_n \frac{\psi_n(\vec{r})\psi_n^*(\vec{r}')}{\varepsilon - \varepsilon_n - \delta_n}, \quad (1.78)$$

(with n being a compound index consisting of $\vec{k}$ and band index, ε_n being the energy of the nth orbital, and δ_n being a small positive (negative) number if the orbital is occupied (unoccupied)), whereas W is the Fourier transform of the screened interaction between the electrons, that in turn is given by

$$W(\vec{r},\vec{r}') = \int \varepsilon^{-1}(\vec{r},\vec{r}'')\frac{1}{|\vec{r}'' - \vec{r}'|}\,d\vec{r}'', \quad (1.79)$$

where $\varepsilon(\vec{r}_1,\vec{r}_2)$ is the dielectric matrix.

Compared to the "conventional" density-functional calculations, the calculated orbital energies $\varepsilon_i(\vec{k})$ are those for the excitation of an electron combined with the accompanying relaxation effects. Actually, since the orbitals $\psi_i(\vec{k},\vec{r})$ will change, also the Green's function will change. Ignoring the latter leads to the so-called GW0 approximation, where the quasiparticle energies are calculated for the orbitals $\psi_i(\vec{k},\vec{r})$ being those of the "conventional" density-functional calculations. Alternatively, one may solve Eq. (1.76) self-consistently, which corresponds to the GW approach. Depending on the system of interest, the differences between the GW0 and the GW results may be more or less significant.

When the system of interest is exposed to electric fields, the electrons and nuclei change positions due to these fields. This may, for example, be quantified through the dipole moment, that is,

$$\mu_i = \mu_i^{(0)} + \sum_{j=x,y,z}\alpha_{ij}E_j + \frac{1}{2!}\sum_{j,k=x,y,z}\beta_{ijk}E_jE_k + \frac{1}{3!}\sum_{j,k,l=x,y,z}\gamma_{ijkl}E_jE_kE_l + \cdots, \quad (1.80)$$

where $\mu_i^{(0)}$ is the ith component of the permanent dipole moment in the absence of the external field. α is the polarizability, and β, γ, and so on, are hyperpolarizabilities, whereas $\vec{E} = (E_x, E_y, E_z)$ is the electric field vector. As an alternative to Eq. (1.80) one may also determine the polarizabilities and hyperpolarizabilities through the changes in the total energy upon exposure to an electric field,

$$E_{tot} = E_{tot}^{(0)} - \sum_{i=x,y,z}\mu_i E_i. \quad (1.81)$$

In the static case, the frequencies of all field components as well as of the dipole moment are zero, but in the dynamic case they may be nonzero, although they have to satisfy

$$\alpha_{ij} = \alpha_{ij}(-\omega_i; \omega_j)$$
$$\beta_{ijk} = \beta_{ijk}(-\omega_i; \omega_j, \omega_k) \quad (1.82)$$
$$\gamma_{ijkl} = \gamma_{ijkl}(-\omega_i; \omega_j, \omega_k, \omega_l)$$

with

$$\omega_i + \omega_j + \omega_k + \cdots = 0 \quad (1.83)$$

in each expression.

A special case is that of the response of a crystal to an electric field. For simplicity, we shall here focus on an electrostatic field. As the property that quantifies the response we will use the dipole moment or, for a macroscopic system, the polarization being the dipole moment per volume. As demonstrated elsewhere [51], the response as well as the polarization for a large finite system contains a finite contribution from the surfaces of the system. This contribution does not vanish in the limit of the almost infinitely large system. When instead treating the system as being infinite and periodic, the surfaces are ignored. Nevertheless, within the so-called modern theory of polarization [52,53] (see also the equivalent vector-potential approach [54,55]) the effects of the surfaces are included indirectly. Within this theory, the polarization becomes

$$\vec{P} = \vec{P}_e + \vec{P}_n \quad (1.84)$$

with the electronic contribution being

$$\vec{P}_e = -\frac{i}{KV} \sum_k \sum_i \langle \psi_i(\vec{k}, \vec{r}) | e^{i\vec{k}\cdot\vec{r}} \vec{\nabla}_k e^{-i\vec{k}\cdot\vec{r}} | \psi_i(\vec{k}, \vec{r}) \rangle \quad (1.85)$$

and the nuclear contribution being

$$\vec{P}_n = \sum_X Z_X \vec{R}_{X,0}. \quad (1.86)$$

In the last equation, the summation is over the nuclei in the reference (0th) unit cell with Z_X and $\vec{R}_{X,0}$ being the charges and positions, respectively, of those. In Eq. (1.85), the k summation is over all $\vec{k}$ points used in sampling $\vec{k}$ space, and K is their number. V is the volume of one unit cell, and $\vec{\nabla}_k$ is the gradient operator acting on the $\vec{k}$ dependence. As shown elsewhere [56], the dependence of the polarization on the surfaces is "hidden" in phase factors of the orbitals. That is, modifying

$$\psi_i(\vec{k}, \vec{r}) \rightarrow e^{i\phi_i(\vec{k})} \psi_i(\vec{k}, \vec{r}) \quad (1.87)$$

corresponds to modeling a system with different surfaces.

1.3
Examples of Applications

After having introduced the foundations for density-functional theory, it would have been natural to illustrate its potential through various applications. However, the number of systems and/or properties that has been addressed is overwhelming and it is not possible to give just a fair presentation of all those studies. Instead, the interested reader is referred to, for example, the contribution by Schwarz and Blaha (Chapter 22) for some examples of applications. Also in other contributions to these volumes, many examples can be found.

1.4
Open Issues

Roughly 50 years after the density-functional theory was introduced by Hohenberg, Kohn, and Sham [2,6], the theory has become a very important tool in explaining, extending, and understanding the properties of molecular and solid system, although in the second paper the authors wrote: "We do not expect an accurate description of chemical binding" [6]. However, this statement was far too pessimistic as at best is demonstrated through the Nobel Prize in Chemistry that in 1998 was awarded 50% to Walter Kohn [57] "for his development of density-functional theory"[1]. Application of density-functional theory to specific systems will continue to provide important information for those systems.

Nevertheless, despite the overwhelmingly large amount of successful studies based on density-functional theory, there are still some issues that deserve special attention. In this section we shall briefly discuss some few of those for which further, fundamental developments are needed.

1.4.1
Functionals

In the most general form, $\varepsilon_{xc}(\vec{r})$ of Eq. (1.22) is approximated through an expression that resembles that of the wave-function based approaches,

$$\varepsilon_{xc}(\vec{r}) = \int f_{xc}(\vec{r},\vec{r}')d\vec{r}', \qquad (1.88)$$

whereby $f_{xc}(\vec{r},\vec{r}')$ may depend in some way on all the orbitals for the Kohn–Sham system. This expression includes also those for which parts of the exchange interactions are treated using density-functional expressions and other parts using Hartree–Fock expressions. Also so-called range-separated approaches fall into this category and they may be needed in order to circumvent some of the problems that we will discuss below.

---

1) Press release from the Nobel Foundation: http://www.nobelprize.org/nobel_prizes/chemistry/laureates/1998/kohn-facts.html.

For the functionals of the types Eqs. (1.88) various analytical forms for ε_{xc} have been proposed based on chemical or physical considerations. One approach is based on calculating ε_{xc} as exactly as possible for some simple reference systems and subsequently seek analytical functions that can reproduce these results. The expressions for ε_{xc} may contain parameters whose values cannot be fixed from general considerations, which actually is the case for all presently used functionals. One may then either attempt to reduce the number of those parameters whose values are undefined by imposing a maximum number of physically and/or chemically motivated constraints or, alternatively, determine the values of these parameters so that an optimal description of a set of reference information on selected systems and properties is obtained. Independent of the approach, values of parameters have to be determined from fitting to accurate information from other (computational or experimental) studies on some training set.

Accordingly, a serious problem is that the parameter values may depend sensitively on the systems and properties of the training set. Over the years very many functionals (a "zoo" of functionals) have been proposed and, unfortunately, when applying different functionals on the same system or property one may obtain results that differ, sometimes even significantly. Moreover, no currently applied functional is robust in the sense that it "always" provides accurate information on any system or property of interest. Therefore, the choice of the appropriate functional for the scientific question at hand is becoming a scientific discipline in itself. This is unfortunate as it removes a part of the "*ab initio*" character from applied density-functional theory.

1.4.2
Weak Interactions

Covalent bonds of crystals are described well by different density-functionals. Even the simple local-density approximation may be suitable for systems when the valence-electron density varies only slowly in space. However, for systems with noncovalent interactions the situation may be different. The description of these noncovalent interactions, including van der Waals or dispersion interactions, is a challenge for modern computational techniques. On the other hand, electrostatic interactions as in hydrogen bonds are usually described reasonably well by density-functional theory [58].

The dispersion interactions are more difficult. London dispersion decays as R^{-6} with R being the distance between the interacting systems. As local density functionals decay exponentially, they cannot describe these interactions properly [59]. Therefore, empirical dispersion corrections have been introduced. This includes the suggestion by Wu and Yang [60], who introduced a damping function $f_d(R)$ in the description of the dispersion energy,

$$E_{\text{disp}} = -f_d(R) \frac{C_6}{R^6}. \tag{1.89}$$

C_6 is a system-specific constant and the damping function, f_d, equals 1 for large distances and approaces 0 at small values of R so that the asymptotic behavior for large R is correct and its influence vanishes at short distances. Later, Grimme [61,62] has used a similar approach to add empirical dispersion corrections.

An alternative is provided by the so-called van der Waals density functionals [63–66]. Interactions between density fluctuations, $\delta\rho(\vec{r})$, in different parts of space are the physical origin of van der Waals forces. Van der Waals density functionals use the linear response kernel K_{xc} to describe the exchange-correlation energy,

$$E_{xc} = \iint K_{xc}(\vec{r}_1, \vec{r}_2)\delta\rho(\vec{r}_1)\delta\rho(\vec{r}_2)\mathrm{d}\vec{r}_1\mathrm{d}\vec{r}_2. \quad (1.90)$$

A drawback of these van der Waals density functionals is that they have been developed to explicitly describe van der Waals interactions so that it is not obvious whether and how they can be applied to the study of other types of chemical bonding.

In contrast to the empirical dispersion corrections by Wu and Yang [60] and Grimme [61,62], a nonempirical approach was proposed by Becke and Johnson [67,68]. They considered the dipole created by an electron and its exchange hole. Dipole–dipole interactions between two such dipole moments (XDM) located at different atoms result in a dispersion energy term [69,70]. The total (damped) XDM dispersion correction becomes

$$E_{\mathrm{disp}} = -\sum_{i>j} \left(\frac{C_{6,ij}}{R_{\mathrm{vdW},ij}^6 + R_{ij}^6} + \frac{C_{8,ij}}{R_{\mathrm{vdW},ij}^8 + R_{ij}^8} + \frac{C_{10,ij}}{R_{\mathrm{vdW},ij}^{10} + R_{ij}^{10}} \right), \quad (1.91)$$

where $R_{\mathrm{vdW},ij}$ is the sum of effective van der Waals radii of atoms i and j, and the dispersion coefficients $C_{n,ij}$ are calculated from the molecular density and its gradients. This work has been the basis for several later studies [71–74].

There are other approaches for including dispersion corrections in density-functional theory. For more detailed discussions and overviews of the current status of those we refer to Becke's perspective on density-functional theory [75] and to the paper by Klimeš and Michaelides [76].

1.4.3
Charge (De)localization

For a thermodynamically open system, the number of particles (for instance, electrons) can fluctuate, whereas the chemical potential is constant. Accordingly, the time-averaged number of particles may be nonintegral. It has been suggested [77] that also this situation can be treated through a zero-temperature density-functional theory. Under this assumption, it has been demonstrated [77] that the total energy as a function of the number of electrons is piecewise linear and changes slope only at integral number of electrons, that is,

$$E_e(N + \Delta N) = E_e(N) + \Delta N \cdot [E_e(N+1) - E_e(N)] \quad (1.92)$$

with $E_e(Z)$ being the electronic energy for the system with Z electrons, N being an integer, and $0 \leq \Delta N \leq 1$. When approximate functionals are being used, the exchange-correlation energy $E_{xc}(Z)$ becomes a smooth, often differentiable function of the (continuous) number of electrons Z even when passing integral values of Z. For an orbital-based approach, $E_e(Z)$ remains being a smooth function but may be nondifferentiable at integral values of Z. However, in both cases, the linear dependence of Eq. (1.92) is not reproduced.

The incorrect behavior of the density-functional approximations leads to the so-called delocalization error [78]. Thus, with those approximate density functionals, electrons tend to be distributed more or less evenly among weakly interacting subsystems and to become delocalized in large systems. Thereby, errors in charge-transfer processes show up including severe inaccuracies in calculated energy gaps between occupied and unoccupied orbitals. At the moment, no general-purpose approach has been presented for removing this problem. For a further discussion, see, the recent overview by Teale *et al.* [79]

A further consequence of Eq. (1.92) is that the exact exchange-correlation potential V_{xc} should be discontinuous when the total number of electrons changes across integers [80,81]. This has consequences for the calculated HOMO-LUMO gap that turns out to be too small compared with the experimental excitation energies [82]. Even if these problems have been known for many years, no routinely applied, approximate density functional has solved them yet.

1.4.4
Response Properties

As mentioned in Section 1.2, the original theorems of Hohenberg and Kohn [2] state that any ground-state property can be evaluated from some functional of the electron density. The lack of the precise forms for those functionals makes it important to introduce approximations and, thereby, the focus has been on developing accurate descriptions of the electronic energy as a function of structure. Accordingly, density-functional calculations are supposed to provide accurate information on binding energies, relative energies, structures, and so on.

Much less attention has been given to the development of accurate, although still approximate, functionals for the calculation of other properties. In this case, most often the properties of interest are calculated using the orbitals of the Kohn–Sham model system but treating them as were they the electronic orbitals (according to Hartree–Fock theory) of the true system. However, there is no formal justification for this approach and corrections may have to be included. An example is that of the electron density in momentum space. Here, in addition to the sum of the squares of the Fourier-transformed Kohn–Sham orbitals the so-called Lam–Platzman correction [83] shall be included.

For most other experimental observables the precise form of such corrections has not been developed and, accordingly, the calculated values may deviate even significantly from the experimental, "correct" values. Even for a simple case where there is no such correction, one may observe dramatic inaccuracies. The example is that of the responses of larger systems to electric fields. Here, we shall consider the case of a static electric field. The response of the system of interest to a such field can be quantified through the dipole moment $\vec{\mu}$. For a uniform electrostatic field $\vec{E}$, the total field-dependent dipole moment may be expanded in a power series as in Eq. (1.80).

Gonze et al. [84] demonstrated that density-functional calculations may fail in treating extended systems exposed to electric fields. This was confirmed shortly later through calculations on finite oligomers of polyacetylene [85,86] as well as on other quasilinear systems [87] of increasing length. These calculations found a significant overshoot of the response properties to electrostatic fields parallel to the major axis of the system. It was found that the response property per unit converged to the thermodynamic limit for too large systems and at (occasionally much) too large values.

The precise cause for this deficiency, which has been called the DFT catastrophe, is not completely identified. It may be related to the underestimate of the calculated energy gap between occupied and unoccupied orbitals, to inaccuracies related to obtaining an accurate description of the structural properties of those systems, as well as to the charge delocalization discussed in the previous section.

Independent of the cause, most of the currently used functionals do obviously not provide an accurate description of the response properties. Instead, nonlocal responses are needed, whereby the functional $f_{xc}(\vec{r}, \vec{r}')$ includes dependences on the orbitals of the model system at points $\vec{r}$ and any other point $\vec{r}'$. This is, for example, the case for such range-separated functionals where the long-range exchange interactions are treated using a Hartree–Fock like expression, and the short-range exchange interactions are treated using a semi-local approximation. This suggestion has been supported through results of simple model calculations [88].

When an electron is removed from a finite system it will experience a long-range exchange-correlation potential that in the exact case has the form [89,90]

$$\lim_{r \to \infty} V_{xc}(\vec{r}) = -\frac{1}{r} + k \qquad (1.93)$$

with k being a system-specific constant. However, with currently used approximate functionals the long-range potential is exponentially decaying. Therefore, ionization processes are not described correctly with current, commonly used, approximate density functionals.

1.5 Conclusions

Practical use of density-functional calculations has become an indispensable part of chemistry, including solid-state chemistry. Such calculations can provide information that extends, supports, and completements that of the experimental studies. Nevertheless, the ideal case that such calculations can be carried through using the computer programs as black-box programs has not been reached. Instead, in many situations a detailed understanding of the underlying principles of the computational methods and their effects on the calculated properties is very useful if not necessary.

The purpose of this short presentation was to provide the reader with an introduction to the field of density-functional theory including emphasizing the strengths and weaknesses of its current applications. This includes the maybe most fundamental questions of such studies: "When do I choose which exchange-correlation functional?" and: "When do I have to include additional corrections?," although we in the present, fairly short presentation could only indicate answers to some of those questions.

Fortunately, for structural and energetic properties of the ground state of covalently bonded systems, most of the currently used modern density functionals lead to accurate results, although finer details may depend on the calculational approach. As we have discussed here, too, the results can also be used in obtaining insight into chemical bonding and atomic volumes and charges, although, as we saw, it is not obvious how to extract this information from the results.

On the other hand, there are also cases where caution is required. Thus, the calculation of excitation properties and of other responses to external electric fields may be accompanied by severe inaccuracies if not special methods are applied. Also for the description of weakly interacting systems, special methods have to be applied. For the latter, a currently very active field is the study of dispersion/van der Waals interactions for which some special-purpose approaches have been presented. Surprisingly, these corrections are presently also applied to systems for which dispersion forces are not expected to be relevant. The present authors consider this development unfortunate since it does leave the reader with the impression that a computational approach is chosen not according to its foundations but since it turns out to give those results that are closest to some kind of expectations.

We will not close this presentation with the above, somewhat skeptical comment. Instead, we will, once again, emphasize that very useful information can be obtained with the help of density-functional calculations, but that it is important to understand the foundations behind those.

References

1 Springborg, M. (2000) *Methods of Electronic-Structure Calculations*, John Wiley & Sons, Ltd, Chichester, UK.
2 Hohenberg, P. and Kohn, W. (1964) *Phys. Rev.*, **136**, B864.
3 Springborg, M. (2000) *Chemical Modeling: Applications and Theory*, vol. **1** (ed. A. Hinchliffe), The Royal Society of Chemistry, Cambridge, UK, p. 306.
4 Jones, R.O. (2015) *Rev. Mod. Phys.*, **87**, 897.
5 Parr, R.G. and Yang, W. (1989) *Density-Functional Theory of Atoms and Molecules*, Oxford University Press, New York, USA.

6. Kohn, W. and Sham, L.J. (1965) *Phys. Rev.*, **140**, A1133.
7. Perdew, J.P. (1999) *Density Functional Theory: A Bridge Between Chemistry and Physics* (eds P. Geerlings, F. De Proft, and W. Langenaeker), VUB University Press, Brussels, Belgium, p. 87.
8. Perdew, J.P. and Schmidt, K. (2001) *Density Functional Theory and its Applications to Materials* (eds V. E. Van Doren, C. Van Alseroy, and P. Geerlings), AIP Press.
9. Springborg, M. (1999) *Density Functional Theory: A Bridge Between Chemistry and Physics* (eds P. Geerlings, F. De Proft, and W. Langenaeker), VUB University Press, Brussels, Belgium, p. 15.
10. Hubbard, J. (1963) *Proc. R. Soc. Lond. A*, **276**, 238.
11. Hubbard, J. (1964) *Proc. R. Soc. Lond. A*, **277**, 237.
12. Hubbard, J. (1964) *Proc. R. Soc. Lond. A*, **281**, 401.
13. Hubbard, J. (1965) *Proc. R. Soc. Lond. A*, **285**, 542.
14. Hubbard, J. (1967) *Proc. R. Soc. Lond. A*, **296**, 82.
15. Hubbard, J. (1967) *Proc. R. Soc. Lond. A*, **296**, 100.
16. Anisimov, V.I., Zaanen, J., and Andersen, O.K. (1991) *Phys. Rev. B*, **44**, 943.
17. Anisimov, V.I., Solovyev, I.V., Korotin, M.A., Czyzyk, M.T., and Sawatsky, G.A. (1993) *Phys. Rev. B*, **48**, 16929.
18. Anisimov, V.I., Aryasetiawan, F., and Lichtenstein, A.I. (1997) *J. Phys. Condens. Matter.*, **9**, 767.
19. Himmetoglu, B., Floris, A., de Gironcoli, S., and Cococcioni, M. (2013) *Cond. Matt.*, doi: arXiv 1309.3355.
20. Dederichs, P.H., Blügel, S., Zeller, R., and Akai, H. (1984) *Phys. Rev. Lett.*, **53**, 2512.
21. Meider, H. and Springborg, M. (1998) *J. Phys. Condens. Matter.*, **10**, 6953.
22. Meider, H. and Springborg, M. (1999) *Chem. Phys. Lett.*, **300**, 339.
23. Castleton, C.W.M., Kullgren, J., and Hermansson, K. (2007) *J. Chem. Phys.*, **127**, 244704.
24. Goedecker, S. (1999) *Rev. Mod. Phys.*, **71**, 1085.
25. Ordejón, P. (1998) *Comp. Mat. Sci.*, **12**, 157.
26. Kohn, W. (1996) *Phys. Rev. Lett.*, **76**, 3168.
27. McWeeny, R. (1960) *Rev. Mod. Phys.*, **32**, 335.
28. Yang, W. (1991) *Phys. Rev. Lett.*, **66**, 1438.
29. Yang, W. (1991) *Phys. Rev. A*, **44**, 7823.
30. Imamura, A., Aoki, Y., and Maekawa, K. (1991) *J. Chem. Phys.*, **95**, 5419.
31. Maekawa, K. and Imamura, A. (1993) *J. Chem. Phys.*, **98**, 534.
32. Mitani, M., Aoki, Y., and Imamura, A. (1994) *J. Chem. Phys.*, **100**, 2346.
33. Aoki, Y., Suhai, S., and Imamura, A. (1994) *Int. J. Quant. Chem.*, **52**, 267.
34. Mitani, M. and Imamura, A. (1994) *J. Chem. Phys.*, **101**, 7712.
35. Bader, R. Richard (1991) *Atoms in Molecules: A Quantum Theory*, Oxford University Press, New York.
36. Hirshfeld, F.L. (1977) *Theor. Chim. Acta*, **44**, 129.
37. Mulliken, R.S. (1953) *J. Chem. Phys.*, **23**, 1833.
38. Mulliken, R.S. (1953) *J. Chem. Phys.*, **23**, 1841.
39. Hoffmann, R. (1988) *Rev. Mod. Phys.*, **60**, 601.
40. Dronskowski, R. and Blöchl, P.E. (1993) *J. Phys. Chem.*, **97**, 8617.
41. Becke, A.D. and Edgecombe, K.E. (1990) *J. Chem. Phys.*, **92**, 5397.
42. Koopmans, T. (1933) *Physica (Amsterdam)*, **1**, 104.
43. Janak, J.F. (1978) *Phys. Rev. B*, **18**, 7165.
44. Runge, E. and Gross, E.K.U. (1984) *Phys. Rev. Lett.*, **52**, 997.
45. Gross, E.K.U., Dobson, J.F., and Petersilka, M. (1996) *Density Functional Theory* (ed. R.F. Nalewajski), Springer Verlag, Berlin, Germany, p. 81.
46. Petersilka, M., Gross, E.K.U., and Burke, K. (2000) *Int. J. Quant. Chem.*, **80**, 534.
47. Jamorski, C., Casida, M.E., and Salahub, D.R. (1996) *J. Chem. Phys.*, **104**, 5134.
48. Casida, M.E., Jamorski, C., Casida, K.C., and Salahub, D.R. (1998) *J. Chem. Phys.*, **108**, 4439.
49. Hedin, L. and Lundqvist, S. (1969) *Solid State Phys.*, **23**, 1.
50. Hybertsen, M.S. and Louie, S.G. (1986) *Phys. Rev. B*, **34**, 5390.
51. Springborg, M., Tevekeliyska, V., and Kirtman, B. (2010) *Phys. Rev. B*, **82**, 165442.

52 Vanderbilt, D. and King-Smith, R.D. (1993) *Phys. Rev. B*, **48**, 4442.
53 Resta, R. (1994) *Rev. Mod. Phys.*, **66**, 899.
54 Kirtman, B., Gu, F.L., and Bishop, D.M. (2000) *J. Chem. Phys.*, **113**, 1294.
55 Springborg, M. and Kirtman, B. (2008) *Phys. Rev. B*, **77**, 045102.
56 Springborg, M., Kirtman, B., and Mohammad, M. (in press).
57 Kohn, W. (1999) *Rev. Mod. Phys.*, **71**, 1253.
58 Sim, F., St. Amant, A., Papai, I., and Salahub, D.R. (1992) *J. Am. Chem. Soc.*, **114**, 4391.
59 Kristyan, S. and Pulay, P. (1994) *Chem. Phys. Lett.*, **229**, 175.
60 Wu, Q. and Yang, W. (2002) *J. Chem. Phys.*, **116**, 515.
61 Grimme, S. (2004) *J. Comput. Chem.*, **25**, 1463.
62 Grimme, S. (2006) *J. Comput. Chem.*, **27**, 1787.
63 Andersson, Y., Langreth, D.C., and Lundqvist, B.I. (1996) *Phys. Rev. Lett.*, **76**, 102.
64 Dobson, J.F. and Dinte, B.P. (1996) *Chem. Phys. Lett.*, **76**, 1780.
65 Dion, M., Rydberg, H., Schröder, E., Langreth, D.C., and Lundqvist, B.I. (2004) *Phys. Rev. Lett.*, **92**, 246401.
66 Thonhauser, T., Cooper, V.R., Li, S., Puzder, A., Hyldgaard, P., and Langreth, D.C. (2007) *Phys. Rev. B*, **76**, 125112.
67 Becke, A.D. and Johnson, E.R. (2005) *J. Chem. Phys.*, **122**, 154104.
68 Johnson, E.R. and Becke, A.D. (2005) *J. Chem. Phys.*, **123**, 024101.
69 Becke, A.D. and Johnson, E.R. (2006) *J. Chem. Phys.*, **124**, 014104.
70 Becke, A.D. and Johnson, E.R. (2007) *J. Chem. Phys.*, **127**, 154108.
71 Tkatchenko, A. and Scheffler, M. (2009) *Phys. Rev. Lett.*, **102**, 073005.
72 Sato, T. and Nakai, H. (2009) *J. Chem. Phys.*, **131**, 224104.
73 Grimme, S., Antony, J., Ehrlich, S., and Krieg, H. (2010) *J. Chem. Phys.*, **132**, 154104.
74 Grimme, S., Ehrlich, S., and Goerigk, L. (2011) *J. Comput. Chem.*, **32**, 1456.
75 Becke, A.D. (2014) *J. Chem. Phys.*, **140**, 18A301.
76 Klimeš, J. and Michaelides, A. (2012) *J. Chem. Phys.*, **137**, 120901.
77 Perdew, J.P., Parr, R.G., Levy, M., and Balduz, J.L. Jr. (1982) *Phys. Rev. Lett.*, **49**, 1691.
78 Mori-Sánchez, P., Cohen, A.J., and Yang, W. (2008) *Phys. Rev. Lett.*, **100**, 146401.
79 Teale, A.M., De Proft, F., Geerlings, P., and Tozer, D.J. (2014) *Phys. Chem. Chem. Phys.*, **16**, 14420.
80 Perdew, J.P. and Levy, M. (1983) *Phys. Rev. Lett.*, **51**, 1884.
81 Sham, L.J. and Schlüter, M. (1983) *Phys. Rev. Lett.*, **51**, 1888.
82 Lannoo, M., Schlüter, M., and Sham, L.J. (1985) *Phys. Rev. B*, **32**, 3890.
83 Lam, L. and Platzman, P.M. (1974) *Phys. Rev. B*, **9**, 5122.
84 Gonze, X., Ghosez, Ph., and Godby, R.W. (1995) *Phys. Rev. Lett.*, **74**, 4035.
85 Champagne, B., Perpète, E.A., van Gisbergen, S.J.A., Baerends, E.J., Snijders, J.G., Soubra-Ghaoui, C., Robins, K.A., and Kirtman, B. (1998) *J. Chem. Phys.*, **109**, 10489. Erratum 110, 11664 (1999).
86 Champagne, B., Perpète, E.A., Jacquemin, D., van Gisbergen, S.J.A., Baerends, E.J., Soubra-Ghaoui, C., Robins, K.A., and Kirtman, B. (2000) *J. Phys. Chem. A*, **104**, 4755.
87 van Gisbergen, S.J.A., Schipper, P.R.T., Gritsenko, O.V., Baerends, E.J., Snijders, J.G., Champagne, B., and Kirtman, B. (1999) *Phys. Rev. Lett.*, **83**, 694.
88 Vargas, J., Springborg, M., and Kirtman, B. (2014) *J. Chem. Phys.*, **140**, 054117.
89 Perdew, J.P. and Burke, K. (1996) *Int. J. Quant. Chem.*, **57**, 309.
90 Tozer, D.J. and Handy, N.C. (1998) *J. Chem. Phys.*, **108**, 2545.

2
Eliminating Core Electrons in Electronic Structure Calculations: Pseudopotentials and PAW Potentials

Stefan Goedecker and Santanu Saha

University of Basel, Department of Physics, Klingelbergstr. 82, 4056 Basel, Switzerland

2.1
Generalities

Electronic structure calculation is an essential tool in chemistry, solid-state physics, and materials sciences to calculate virtually any property of condensed matter systems. The majority of these calculations are done within density functional theory. Even though density functional theory [1] is in principle an exact theory, approximate functionals have to be used in practice since the explicit form of the exact functional is unknown. These approximate functionals offer nevertheless a good compromise between accuracy and speed and are therefore very popular. In Kohn–Sham density functional theory, N three-dimensional single particle orbitals have to be calculated for a system containing N electrons. The other, numerically much more costly approach, is to solve directly the many-electron Schrödinger equation by methods such as coupled cluster or quantum Monte Carlo methods. The result is in this case a single $3N$-dimensional many-electron wave function. Pseudopotentials can be used in both approaches, but it is conceptually easier to use them within independent particle schemes such as density functional theory, because in this case core orbitals can directly be associated with core electrons and be eliminated. This chapter will therefore mainly concentrate on pseudopotentials within independent particle schemes.

Physical atoms have in general both valence and core electrons. Since only the valence electrons are actively involved in chemical bonding, it is to be expected that the influence of the core electrons is weak on processes such as bond formation and bond breaking. This is the guiding principle exploited by a pseudopotential. The concept of a pseudopotential is fairly old and presumably goes back to a 1935 publication of Hans Hellmann [2]. The notion of a pseudopotential is generally used in the physics literature, whereas in the chemistry literature the notion of an effective core potential is used instead. The pseudopotential is

Handbook of Solid State Chemistry, First Edition. Edited by Richard Dronskowski, Shinichi Kikkawa, and Andreas Stein.
© 2017 Wiley-VCH Verlag GmbH & Co. KGaA. Published 2017 by Wiley-VCH Verlag GmbH & Co. KGaA.

associated with some kind of pseudoatom that contains only valence states. These valence states should have exactly the same behavior as the true valence state of the all-electron atom. It is impossible to quantify the error incurred by neglecting the core electrons in an exact mathematical way. It turns out empirically that for many elements of the periodic table, such as transition metals, it is necessary to include so-called semicore states into the states described explicitly by the pseudopotential to get satisfactory accuracy. These so-called semicore states are the first core states of the atom, that is, the core states that are highest in energy. Core states have to be included in the form of semicore states not only when they are not deep enough in energy but also if there exists a significant spatial overlap between the core state and some valence state. In such a case, the variation of the valence wave function induced by a varying chemical environment will lead to complicated interactions between the valence and semicore states that are hard to describe without an explicit inclusion of these semicore states. For some heavy elements and for very high accuracy, it can even be necessary to use so-called small core pseudopotentials [3] where several shells of core electrons are treated explicitly, that is, as if they were valence orbitals.

Starting from where in the periodic table pseudopotentials are used depends also on the flexibility of the basis set. Gaussian-Type Orbitals (GTOs) can directly handle localized core orbitals by adding a few narrow GTOs. In a plane wave basis set, the resolution is uniform and adding explicit core orbitals requires increasing the resolution everywhere in the computational box, even though high resolution would be needed only in small regions around the nuclei. The resulting dramatic increase in the number of basis functions makes therefore an explicit description of core orbitals impossible. Core orbitals can however implicitly be included by PAW-type schemes in plane wave codes. In practice deep core orbitals are however considered as being frozen and so eliminated as variational degrees of freedom. Thus, plane wave codes [4–6] and real-space codes [7–9] use pseudopotentials or PAW schemes for the entire periodic table and neglect core electrons as variational degrees of freedom starting from lithium or boron. Gaussian codes [10–12] typically use pseudopotentials only for elements heavier than argon.

The general objective of electronic structure calculations is to achieve the so-called chemical accuracy of 1 kcal/mol. If one uses density functional schemes, the errors arising from the use of any state of the art exchange-correlation functional is in general larger than this target accuracy. However, some of the most sophisticated functionals come fairly close to the desired value [13] and further progress is to be expected. So it would be highly desirable that all other sources of error are significantly smaller than this error. Only in this case the "true" results of a calculation with a certain functional can be obtained.

This chapter will try to give an introduction and overview of the field that is easily accessible to practitioners. In particular, it will discuss under a combined perspective the work both in the physics and chemistry community. Concerning pseudopotentials in chemistry, a very comprehensive review article has recently been published by Dolg and Cao [14]. Relativistic effects are described in particularly great detail and the reader interested in this aspect should consult this review article. A more compact review on the chemistry pseudopotentials was also published recently [15]. The review articles on pseudopotential work in the physics community are somewhat outdated [16].

2.2
The Alternative: All-Electron Calculations

Being an approximation, it would seem clear that using a pseudopotential is always less accurate than doing the all-electron calculation for the real atom. Even though this is in principle true, it is not necessarily true in practice, since it turns out that doing an all-electron calculations for molecular systems is a mathematically very difficult thing. In order to describe both the rapid variation of the orbitals near the nucleus and the slow but long-range decay of the valence wave function at larger distances from the nucleus, highly flexible basis sets are needed.

Such a basis set has become available only very recently. It is a multiwavelet basis set [17–19]. Being a systematic basis set, in principle it allows obtaining any desired accuracy and accuracies of 1.e-3 kcal/mol have been achieved in practice [20]. Such multiwavelet calculations are at present still quite expensive and are therefore not an option for realistic large-scale electronic structure calculations, but now for the first time they really allow quantifying the error of standard electronic structure calculations by comparing results for small molecules.

The standard basis set used for all-electron calculations, namely, the Gaussian-type orbitals (GTOs) have severe shortcomings. At the origin they cannot describe the electron–nucleus cusp, that is, the discontinuity of the first derivative. This leads to quite substantial errors in the total energy of atoms. Fortunately, this error cancels out to a large extent in the physically relevant energy differences. In addition, GTOs basis sets have also the wrong long-range behavior. They decay too fast with respect to the radial distance r from the nucleus. Instead of the physical $\exp(-\tilde{a}r)$ decay, they exhibit an $\exp(-ar^2)$ decay. This makes a correct description of the tail very difficult, in particular for weakly bound electrons that have a small decay constant $\tilde{a}$. But presumably the worst limitation of GTOs is that they do not form a systematic basis set. This means that there is no way to enlarge the basis set in such a way that any set threshold accuracy can be reached. Comparisons with multiwavelet benchmark calculations [20] show that only the largest standard GTOs basis sets can on average give the desired chemical precision. However, even for these basis sets, outliers can be found with errors significantly larger than this value. The cost of electronic structure calculations using a GTOs-type basis set increases very steeply with the size of the basis set. For this reason, the majority of calculations are done with much smaller basis sets and the basis set error can be comparable to or even dominate the error arising from the chosen exchange-correlation functional [21].

Another atom-centered basis set are numerical atomic orbitals (NAOs) [22,23] (NAOs). NAOs are the exact wave functions of the free spherical atom, except in the very far tail region where a confining potential leads to a fast decay to zero. Since this far tail region can be shifted to arbitrarily large radial values, the influence of the confining potential can be made arbitrarily small. NAOs have by construction the correct behavior near the nucleus. It is well known that the asymptotic decay of the most extended orbital of a molecular system is given by $\exp(-r\sqrt{-2\varepsilon_{homo}})$, where ε_{homo} is the eigenvalue of the highest-occupied molecular orbital (HOMO) [24]. NAOs have the correct asymptotic behavior in the tail region in the sense that the wave function decays like $\exp(-\alpha r)$. The decay constant α is obtained from the HOMO eigenvalue of the atom and is therefore by construction correct for the atom. It is however in general not correct for a molecule or solid. Comparison with quasi-exact multiwavelet results [20] show that for manageable NAO basis set sizes, the error is smaller than the chemical accuracy and only a few outliers can be found.

In the physics community, LAPW-type methods are widely believed to be most accurate for all-electron calculations. Space is subdivided into the so-called muffin-tin region and the interstitial region (Figure 2.1). The muffin-tin region consists of all the muffin-tin spheres that are centered on all the atoms in the system, the interstitial region consists of the rest.

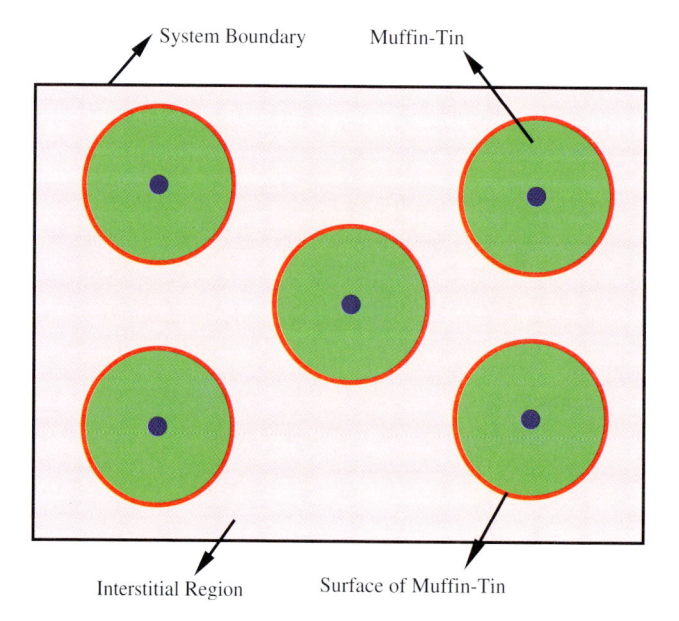

Figure 2.1 Illustration of the subdivision of the computational volume, in which the wave functions are calculated, into the muffin-tin spheres around all atoms (green areas) and the remaining interstitial region (pink area). The system boundary plus the surfaces of the muffin-tin spheres specify the boundary conditions for the wave functions in the interstitial region.

2.2 The Alternative: All-Electron Calculations

In the LAPW method, the wave functions are represented by plane waves in the interstitial region and by so-called partial waves in the muffin-tin region. The partial waves are products of radial functions times spherical harmonics and are solutions to the radial Schrödinger equation of the atom for certain energies, that is, eigenvalues. LAPW methods contain a substantial number of parameters that influence the result in a more or less unpredictable way [20]. LAPW methods constitute nevertheless at present the gold standard for accurate calculations in periodic crystalline materials. For molecular systems, LAPW-type methods are not an option to be considered to obtain chemical accuracy, since, in addition to the above-listed problems, the convergence with respect to the size of the periodic box is quite slow in approaching the limit of free boundary conditions. This is in particular true if the system has a monopole or dipole, leading to long-range interactions of the system with its periodic images.

The LAPW approach is conceptually actually closely related to the pseudopotential approaches to be discussed in this chapter. For this reason, we will also use some nomenclature that comes originally from the LAPW community in the context of our discussion of pseudopotentials. We will, for instance, use the notion of the muffin-tin radius to denote a quantity that is frequently called cutoff radius in the pseudopotential literature. This nonstandard nomenclature is used to stress the close relationship of the underlying principles of LAPW and pseudopotential methods.

The LAPW method can be formulated in such as way that the resulting Hamiltonian and overlap matrices have exactly the same structure as in an ultrasoft pseudopotential approach [25] in a plane wave basis set [26], that is, all the matrices contain only diagonal, convolution, and separable terms. The separable terms arise from a series of projectors that make for different angular momenta l, m the connection between the wave function in the interstitial region and in the muffin-tin sphere. For the standard LAPW approach, these projectors have to give the values and outward radial derivative of the interstitial plane wave expansion on the surface of the muffin-tin sphere [26] for the different angular momenta. As will be discussed later, this information allows determining the wave function inside the muffin-tin sphere. Freezing the potential in the muffin-tin region will fix the coefficients in the separable part of the Hamiltonian matrix and result in a rudimentary pseudopotential [27].

The above-discussed projectors that give information about the plane wave expansion of the interstitial region on the surface of the muffin-tin spheres can easily be calculated in Fourier space but should not be transformed into real space since they have very poor localization properties in real space. Liu *et al.* [28] have introduced another transformation that can be applied in real space. The core of the PAW method [29] is also such a transformation that is based on real-space projectors, allowing determining all-electron wave function inside the muffin-tin sphere from a slowly varying outside plane wave expansion.

Assessing the accuracy of a pseudopotential requires to be able to compare with exact quasi-all-electron values. The previous discussion showed that for an assessment of molecular systems, multiwavelet results should ideally be used.

For crystalline systems, LAPW-type methods are presumably the best ones available at present. If such results are not available, NAO results will under most circumstances also do the job [20].

2.3
General Advantages of Using Pseudopotentials

From a numerical point of view pseudopotentials are extremely useful for the following reasons:

- The number of electrons in the calculation is reduced. Hence, the number of occupied orbitals that have to be calculated is also reduced and the calculations will be faster.
- Both the valence and core orbitals of all-electron calculations have rapid oscillations near the nucleus. Pseudo-orbitals, that is, the solution orbitals of an atom described by a pseudopotential, are much smoother. Hence, it is not necessary to include basis functions into the basis set capable of describing strongly localized oscillations. The smaller basis set size will make the calculations faster.
- The numerical noise is reduced. In all-electron calculations, the contribution from the core electrons to the total energy is much larger than the contribution from the valence electrons. Small relative errors in the core orbitals can therefore spoil the result and introduce considerable errors into the chemically important energy differences that are much smaller in magnitude than typical total energies.
- As shown by Kleinmann [30], relativistic effects are only important close to the nucleus where one finds low energy and therefore high kinetic core electrons that move at speeds comparable to the speed of light. This core region is anyway replaced by the pseudopotential and the pseudopotentials can therefore mimic relativistic effects. As a consequence, one can therefore obtain with a pseudopotential that was fitted to a relativistic atomic reference calculation, relativistic results without performing a relativistic calculation for the target system.

2.4
The Basic Concepts of Pseudopotentials

The conceptual roots of LAPW and pseudopotential methods [31] are identical and their relation is well described in Ref. [32]. Let us assume for the moment that we chose the muffin-tin spheres relatively small such that they just contain the core orbitals. The interstitial region is thus the relatively large remaining part of the computational volume. For the moment we want to solve Schrödinger's equation only in the interstitial region. For simplicity, we will use atomic units in all equations. Let us assume that we solve the single-particle Schrödinger equation:

$$-\frac{1}{2}\nabla^2 \Psi_i(\mathbf{r}) + V(\mathbf{r})\Psi_i(\mathbf{r}) = \varepsilon_i \Psi_i(\mathbf{r})$$

for a fixed potential $V(\mathbf{r})$, in which case the total energy is just the sum over all the occupied eigenvalues ε_i.

The solution depends on the boundary conditions. One part of the boundary is the surface of the computational box. There one has typically either periodic boundary conditions or free boundary conditions, requiring the wave function to tend to zero. The other part of the boundary are the surfaces of the muffin-tin spheres. There one has to know the scattering boundary conditions as given by the logarithmic derivative p_l (Figure 2.2):

$$p_l(\varepsilon) = \frac{R'_l(r_m)}{R_l(r_m)}.$$

p_l is a function of the eigenvalue ε since the radial wave function $R_l(r)$ and its radial derivative $R'_l(r_m)$ depend on the eigenvalue ε. r_m is the radius of the muffin-tin sphere around a certain atom. Radii for different muffin-tin spheres might

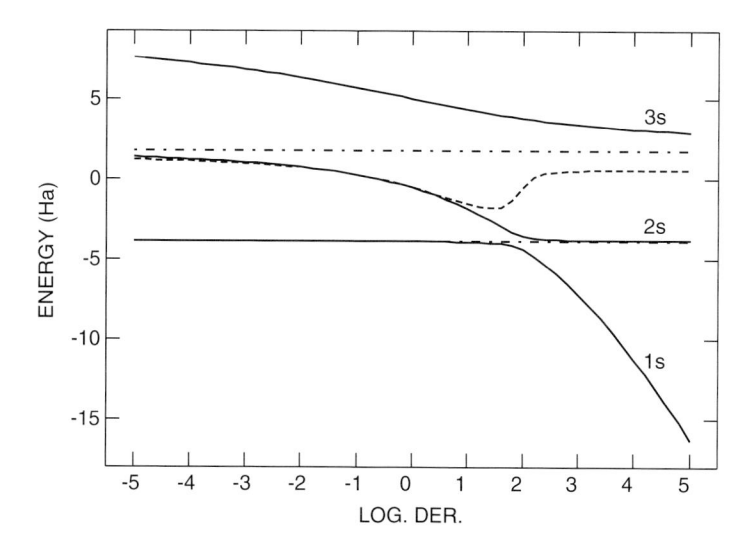

Figure 2.2 The energy versus logarithmic derivative curve for the lowest s states of Be at a radius $r_m = 1.7$. The solid lines represent the exact curve, while the dashed line is the approximation obtained with a pseudopotential. The asymptotics separating the different branches are shown by the dashed–dotted line. The distance between the asymptotics increases for smaller values of r_m. It is therefore not always possible to assign one branch to one atomic eigenvalue. If the radius r_c is small, several atomic eigenvalues may lie on one branch. It is important, to have a good agreement for all the logarithmic derivatives in the region between the atomic reference value and zero, which is the value of the logarithmic derivative taken on by a binding orbital in the middle of the bond.

be different but to keep the notation simple, we will not introduce an additional index that specifies to which atom the muffin-tin sphere and its radius belong.

It is easy to understand why the logarithmic derivative is a valid boundary condition. The radial Schrödinger equation is given by

$$-\left\{\frac{1}{2}R_l''(r) + \frac{R_l'(r)}{r}\right\} + \frac{l(l+1)}{2}R(r) + V(r)R_l(r) = \varepsilon R_l(r). \tag{2.1}$$

If we know $R_l(r)$ and $R_l'(r)$ as well as the eigenvalue ε, we can obtain from the radial Schrödinger equation the second derivative $R_l''(r)$. We can then set up a Taylor expansion that gives us the values of $R_l(r-h)$ and $R_l'(r-h)$ at the new grid point $r-h$. In this way, we can construct the entire radial solution $R_l(r)$ by moving from r_m inward toward the origin. This kind of approach is intended only to demonstrate that the wave function inside the muffin-tin sphere can be constructed from the behavior of the muffin-tin wave function on the surface of the muffin-tin sphere. A numerical implementation of this approach can actually lead to numerical instabilities. Since Schrödinger's equation is linear in $R_l(r)$, a solution remains a solution when multiplied by a constant and therefore $R_l(r)$ and $R_l'(r)$ have not to be specified separately, but only their ratio has to be given. Thus, indeed, only the logarithmic derivative has to be known to calculate the wave functions and to obtain all the valence eigenvalues. If we assume the core orbitals to be frozen, in the sense that they do not respond to changes in the valence orbitals, their eigenvalues will be independent of the chemical environment and including them in the sum over the eigenvalues will only add a constant to the independent particle energy. Hence, the physically relevant energy differences can be obtained only from the eigenvalue sum over the valence orbitals.

Since only the logarithmic derivative has to be known on all the muffin-tin surfaces in order to construct the wave function in the interstitial region, we can ignore the physical processes that lead to a specific logarithmic derivative. It can arise either from the scattering properties of an all-electron atom or of a pseudoatom. So, a good pseudopotential in this context is one that reproduces with high accuracy the logarithmic derivative functions of the valence electrons.

Now, in reality, we of course have to solve Schrödinger's equation in a self-consistent context, where the last two terms of the potential V depend on the charge density

$$V(\mathbf{r}) = V_{\text{ion}}(\mathbf{r}) + V_{\text{Hartree}}(\mathbf{r}) + V_{\text{XC}}(\mathbf{r}).$$

$V_{\text{ion}}(\mathbf{r})$ is the potential arising from the ionic charges of the nuclei. The Hartree potential, $V_{\text{Hartree}}(\mathbf{r}) = \int \frac{\rho(\mathbf{r}')}{|\mathbf{r}'-\mathbf{r}|} d\mathbf{r}'$, as well as the exchange-correlation potential $V_{\text{XC}}(\mathbf{r})$ depend on the charge density $\rho(\mathbf{r})$. If the charge densities in the muffin-tin spheres change, the charge-dependent potential terms will change and the scattering properties of the muffin-tin surfaces will therefore change. The logarithmic derivative, which describes these scattering properties, would become a much more complicated function, which depends not only on the eigenvalue but also on all the parameters that determine the charge density inside the muffin-

2.4 The Basic Concepts of Pseudopotentials | 37

tin sphere. Fortunately, it turns out that the charge density is inert inside a certain region around the core [24]. This is illustrated in Figure 2.3. It shows the difference of the charge distributions that one would obtain by ionizing an atom with and without allowing for a charge relaxation. Ionizing without charge relaxation is of course unphysical and the corresponding charge density is obtained by simply using the orbitals of the neutral nonspin-polarized atom and increasing or decreasing the occupation numbers n_i of these orbitals by one half.

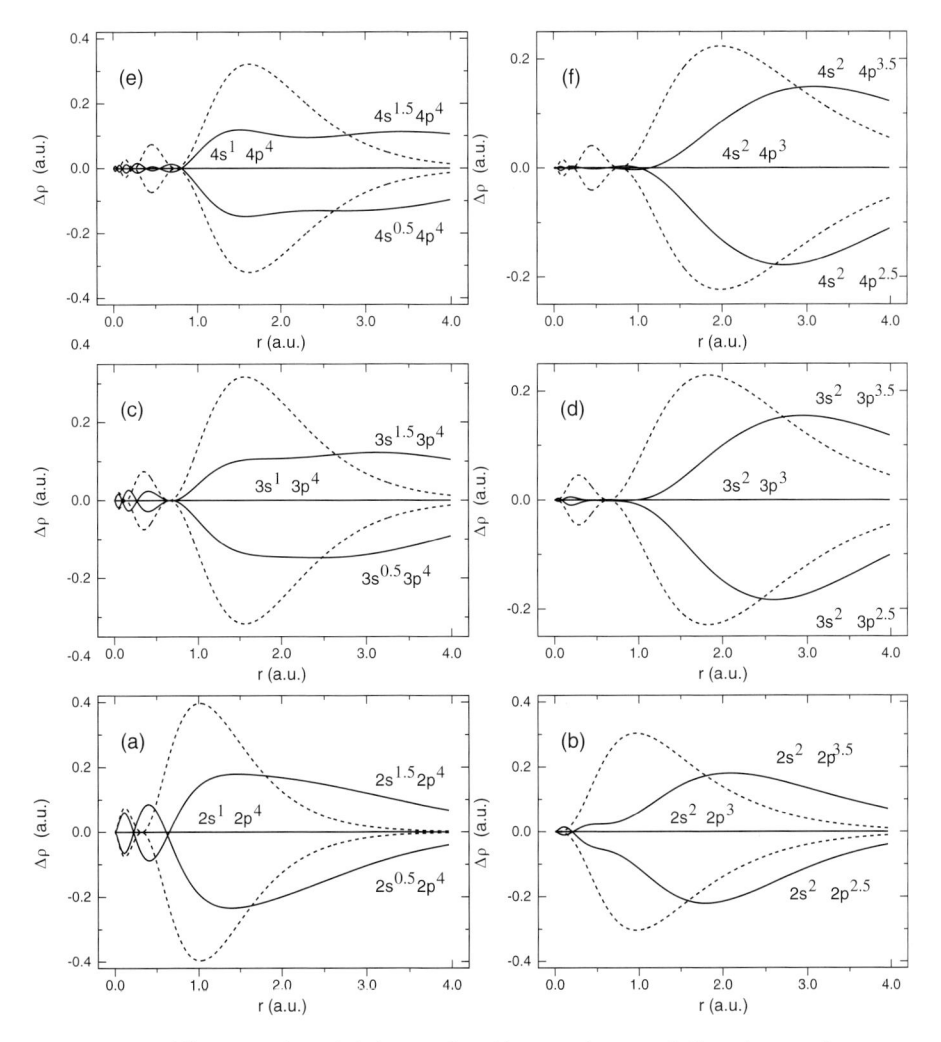

Figure 2.3 The difference in the radial charge $4\pi r^2 \rho$ for different ionic configurations of N (a and b), P (c and d), and As (e and f) specified in the figure. The solid lines are differences of the self-consistent all-electron charge densities. The dotted line is what one would obtain by adding or subtracting half an electron of the atomic reference configuration wave functions without allowing the wave functions to relax to self-consistency. (Reproduced with permission of AIP from Ref. [33].)

The relaxed charge density, also shown in Figure 2.3, is the self-consistent charge density that one obtains by allowing the orbitals to relax in response to the ionization by a fractional charge of 1/2 in a non-spin-polarized calculation. It is clearly seen that this relaxation to a large extent brings back the charge density to the one of the neutral atom. This gives then rise to a small sphere around the nucleus where the charge density is nearly independent of the chemical environment. The extent of this inert charge region is shown in Figure 2.4 for a large part of the periodic table.

Assuming that the charge is completely frozen in the inert region and thus independent of the chemical environment, the only requirement for an accurate pseudopotential would be that the logarithmic derivative of the pseudopotential has to track as closely as possible the logarithmic derivative of the all-electron

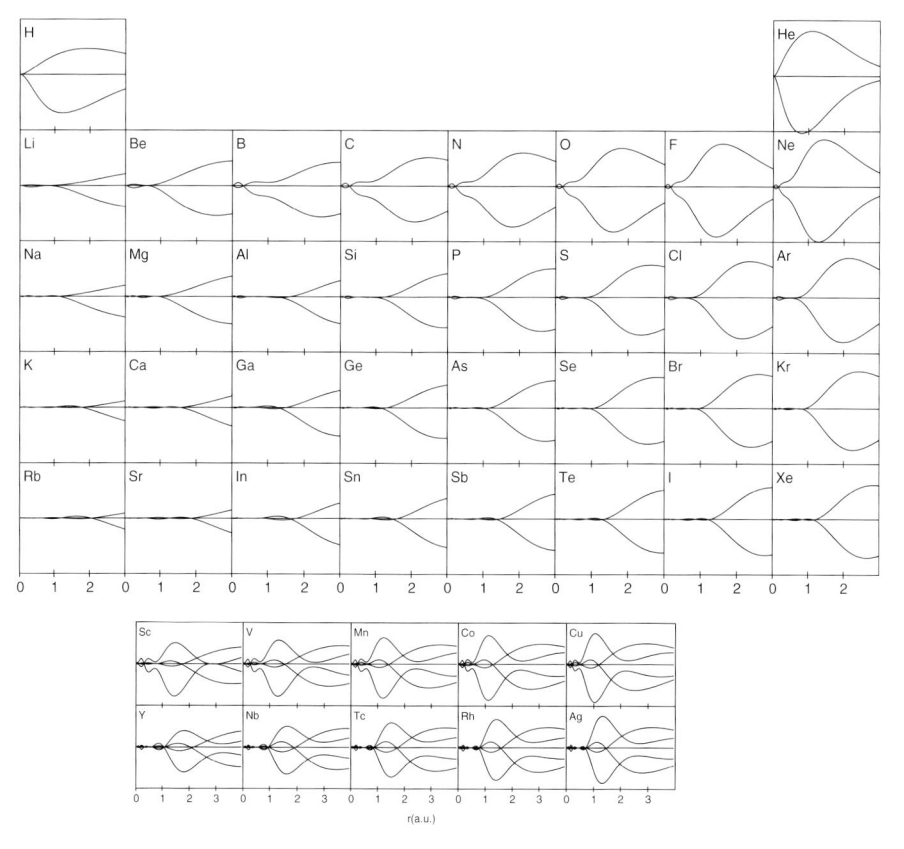

Figure 2.4 The difference in the self-consistent radial all-electron charge density across the periodic table if one changes the occupancy of the most weakly bound orbital by half an electron. For the case of some selected transition metals, changes of the occupancy of both d and s electrons are shown, since both are weakly bound. The former modify the charge density close to the nucleus, while the latter modify it further away from the nucleus. (Reproduced with permission of AIP from Ref. [33].)

atom for a range of eigenvalues typically encountered in electronic structure calculations. In general, this interval is centered around the atomic eigenvalues. It was shown [34] that the derivative of the logarithmic derivative with respect to the eigenvalue energy of a certain orbital is related to the charge of this orbital that is contained in the muffin-tin sphere. This gave rise to the concept of norm-conserving pseudopotentials. If the eigenvalues of the all-electron atom and the pseudoatom match and if in addition the total charge of the all-electron and pseudo-wave function in the muffin-tin sphere are identical, then the logarithmic derivative is correct up to linear order in a small range around the reference value.

If the muffin-tin spheres are small enough, the requirement of a quasi-frozen charge will be satisfied in practice and the scattering properties on the muffin-tin surfaces are virtually independent of the chemical environment, giving rise to a pseudopotential that reproduces the correct eigenvalues. Correct eigenvalues are the most important contribution, but other contributions have to agree as well. This can be seen by writing the density functional total energy E in the usual way in terms of the eigenvalue sum and charge-dependent correction terms

$$E = \sum_{i=\text{occupied}} \varepsilon_i - \frac{1}{2}\int \frac{\rho(\mathbf{r})\rho(\mathbf{r}')}{|\mathbf{r}-\mathbf{r}'|}\,d\mathbf{r}d\mathbf{r}' + E_{\text{xc}} - \int \mu_{\text{xc}}(\rho(\mathbf{r}))\rho(\mathbf{r})\,d\mathbf{r}, \qquad (2.2)$$

where $E_{\text{xc}}(\rho)$ is the exchange-correlation energy and $\mu_{\text{xc}}(\rho)$ the exchange-correlation potential. By separating the total charge distribution ρ,

$$\rho(\mathbf{r}) = \sum_{i=\text{occupied}} n_i |\psi_i(\mathbf{r})|^2,$$

into a muffin-tin, ρ_{MT}, and an interstitial part, ρ_{INT}

$$\rho(\mathbf{r}) = \rho_{\text{MT}}(\mathbf{r}) + \rho_{\text{INT}}(\mathbf{r}),$$

the electrostatic correction term of Eq. (2.2) can obviously be split up into three parts:

$$\frac{1}{2}\int \frac{\rho(\mathbf{r})\rho(\mathbf{r}')}{|\mathbf{r}-\mathbf{r}'|}\,d\mathbf{r}d\mathbf{r}' = \frac{1}{2}\int \frac{\rho_{\text{MT}}(\mathbf{r})\rho_{\text{MT}}(\mathbf{r}')}{|\mathbf{r}-\mathbf{r}'|}\,d\mathbf{r}d\mathbf{r}' + \int \frac{\rho_{\text{INT}}(\mathbf{r})\rho_{\text{MT}}(\mathbf{r}')}{|\mathbf{r}-\mathbf{r}'|}\,d\mathbf{r}d\mathbf{r}' \\ + \frac{1}{2}\int \frac{\rho_{\text{INT}}(\mathbf{r})\rho_{\text{INT}}(\mathbf{r}')}{|\mathbf{r}-\mathbf{r}'|}\,d\mathbf{r}d\mathbf{r}'.$$

Since we assume that the charge density in the muffin-tin sphere is independent of the chemical environment, the interaction of the charge density $\rho_{\text{MT}}(\mathbf{r})$ with itself is a constant and the first term on the right-hand side of the above equation can therefore be omitted. In the interstitial region, the wave functions of the pseudopotential calculation are by construction virtually identical to the all-electron wave functions. Consequently, the charge is the correct one and the last term in the above equation can be evaluated exactly. The most problematic term is the second one, representing the interaction of the charge density in the muffin-tin sphere with the electrostatic potential arising from the charge in the interstitial region. The reference charge density in the muffin-tin region from which the scattering properties of the pseudopotential was calculated was spherically symmetric. If we assume that the charge density generated by the pseudo-wave functions remains spherically symmetric in the muffin-tin region in different chemical environments, then only the resulting monopole will interact with the potential arising from the charge in the muffin-tin region. Since ρ_{INT} is by definition zero in the muffin-tin region, the potential will be a constant or linear function. Then, since the monopole is by construction the correct one for a norm-conserving pseudopotential, this interaction is also correctly described. The used assumptions have however never been studied in detail. The fact that pseudopotentials can have rather large errors [21] when comparing spherically symmetrized atoms with nonspherical atoms suggests that these assumptions are not always well satisfied.

As shown in Figure 2.5, the separate spin-up and spin-down densities change much more in the inert region in response to a spin polarization arising from an ionization than the total charge density itself. The exchange–correlation interaction of the varying spin density with the core charge of the atom has to be accounted for. This can be done by including a core density within a so-called nonlinear core correction. Initial work in this direction [35] tried to represent more or less faithfully the charge distribution of the core. Since the pseudo-valence charge is not identical to the all-electron valence charge, it is questionable whether this is advantageous. In some later pseudopotentials, the amplitude and the shape of the core charge were therefore taken to be fitting parameters [21].

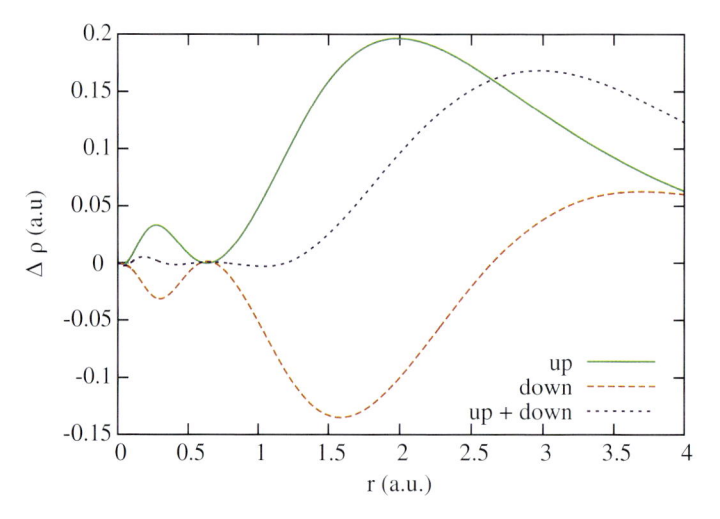

Figure 2.5 Difference in the radial spin densities and total charge density when adding half an electron to a phosphorus atom ([Ne] $3s^2 3p^{3.5}$). The inert core region of the total charge is not observed for the individual spin densities. (Reproduced with permission of AIP from Ref. [21].)

Many elements have a stronger spin polarization in the atomic ground state than in a molecular environment. The presence of a nonlinear core correction charge gives therefore typically the largest improvement for the total energy of atoms.

From all these considerations, it is clear that relying on a good logarithmic derivative is not necessarily sufficient to have good transferability. Transferability means that the pseudopotential can well mimic the behavior of an all-electron atom in widely different chemical environments such as a neutral molecule, an ionized molecule, or a solid. First of all the interval in which the logarithmic derivative has to be highly accurate could well be larger than the region that is well described up to linear order in a norm-conserving pseudopotential. By selecting several reference energies scattered over a larger interval, this region can actually be made larger [25]. A quasi-exact logarithmic derivative however does not guarantee that all the charge-dependent terms that enter into the total energy expression are correctly reproduced. Many traditional norm-conserving pseudopotentials that were widely used for several decades in solid-state calculations were fitted to a single atomic reference configuration. After construction, their transferability was typically tested for a few ionized and excited atomic configurations. If the pseudopotential did too poorly in this test, a new fit was typically attempted. Frequently, the radius of the muffin-tins sphere was then reduced to obtain better transferability. As expected, the typical radii of pseudopotentials correlate with the size of the inert core regions of Figure 2.4. In the context of plane wave calculations, small radii lead however to "harder" pseudopotentials. This means that a larger plane wave basis set containing additional short-wavelength plane waves is needed. The larger basis set makes the calculation of course slower and therefore people constructing pseudopotentials targeted at the use within plane wave codes try to make their pseudopotentials as soft as possible. Since until very recently reliable all-electron reference calculations for molecular systems were not available, it was actually difficult to assess the accuracy of a pseudopotential. Whether a pseudopotential is hard or soft is however immediately visible from convergence tests with respect to the plane wave cutoff, which specifies how many plane waves are used. For this reason, soft but not so accurate pseudopotentials were probably frequently favored in practice. Recently, the accuracy of numerous pseudopotential schemes was systematically tested for the simplest possible case of elementary solids and the most accurate results are displayed on the project web page https:molmod.ugent.bedeltacodesdft.

The above said does not mean that a hard pseudopotential is necessarily accurate. The fact that the pseudo-wave functions inside the muffin-tin region are not unique allows us to make them as smooth as possible inside this region. One can, for instance, minimize the kinetic energy of the wave function in the muffin-tin region based on a decomposition of the radial wave functions into spherical Bessel functions [36]. So, a not too large muffin-tin region together with a smooth wave function in this region is the best compromise between accuracy and efficiency for pseudopotentials used in plane wave and real-space codes.

2.5
Mathematical Form of Norm-Conserving Pseudopotentials

All pseudopotential have a local part V_{loc} that represents the long-range Coulomb potential of the ionic core. If the atom has a total nucleonic charge Z and if Z_{core} electrons are considered as core electrons in the pseudopotential, then the ionic charge Z_{ion} is given by $Z_{ion} = Z - Z_{core}$. In the so-called Stuttgart pseudopotentials, the local part is simply given by

$$V_{loc} = \frac{-Z_{ion}}{r},$$

where r is the distance from the point nucleus and where we have assumed that the nucleus is located at the origin. The singularity of the Coulomb potential at the origin leads to the so-called electron–nucleus cusp, that is, to a discontinuity of the first derivative of the wave function. Neither GTOs nor plane wave basis sets can represent such a cusp. Even if the exact wave function cannot be represented exactly, integrals with the GTOs basis set can nevertheless be calculated and the resulting errors can be absorbed in the parameterization of the pseudopotentials. For plane waves, a cusp is, however, a serious problem. The exponential convergence rate, which can be obtained for a smooth function where all the derivatives are continuous, is replaced by a slow algebraic convergence rate. In addition, the potential is usually calculated on a real-space grid and a potential with a singularity cannot be represented on such a grid.

These problems can be overcome by considering the ionic charge ρ_{ion} to be a Gaussian of finite size σ and monopole Z_{ion}

$$\rho_{ion}(r) = Z_{ion} \frac{1}{(\sigma\sqrt{\pi})^3} \exp\left(-\frac{1}{2}\frac{r^2}{\sigma^2}\right),$$

resulting in a potential

$$V_{loc} = -Z_{ion} \frac{\text{erf}(r/\sigma)}{r}.$$

This potential has the correct asymptotic $-Z_{ion}/r$ behavior, but is finite and differentiable at the origin. In addition, the local potential can still contain other terms that have to be rather localized around the origin to preserve the correct asymptotic behavior.

Local potentials, obtained by inverting the radial Schrödinger equation [37], a method presented in Section 2.7, are typically given in purely numerical form.

Experience has shown that it is not possible to construct accurate pseudopotentials by using only a local potential. Therefore, virtually all pseudopotentials contain an additional nonlocal part V_{nl}. Two major varieties are used for this nonlocal part, the semilocal form

$$V_{nl}(\mathbf{r}) = \sum_{l,m} \int Y_{l,m}(\hat{\mathbf{r}}) \delta(\mathbf{r} - \mathbf{r}') U_l(r') Y_{l,m}(\hat{\mathbf{r}}') d\mathbf{r}'$$

and the separable form

$$V_{\text{nl}}(\mathbf{r}) = \sum_{l,m,i,j} \int Y_{l,m}(\hat{\mathbf{r}}) p_{l,i}(r) h_{i,j} p_{l,j}(r') Y_{l,m}(\hat{\mathbf{r}}') \, d\mathbf{r}', \tag{2.3}$$

where the $Y_{l,m}$ are spherical harmonics, the $p_{l,i}(r)$'s are radial projectors, and the combined $Y_{l,m}(\hat{r}) p_{l,i}(r)$ are three-dimensional projectors. Applying the semilocal form onto a wave function requires a decomposition of the wave function into its different spherical harmonics for all radii within the range where $U_l(r)$ does not vanish. Then each component is multiplied by its potential component $U_l(r)$. Finally, the quantities obtained in this way for all l, m components are multiplied with the spherical harmonics and added up to get the final result. This sequence of operations has to be done for all the atoms in the system. It is a relatively complicated and numerically expensive operation in standard basis sets [14], since it requires many angular integrations. The separable form can be evaluated much more easily because it requires for each three-dimensional projector only a single three-dimensional integration that can be performed on many different types of grids. If the radial projectors themselves are Gaussians [38], the total three-dimensional projector is a Gaussian-type function, that is, a product of a Gaussian and a polynomial. In this case one has to calculate only overlap integrals between Gaussian-type functions, which can be done analytically [39]. For a plane wave basis set, the separable form allows applying the Hamiltonian in operator form, that is, it is not necessary to calculate and store the full Hamiltonian matrix [40]. This is most useful for plane wave codes, since there the diagonalization [6] or minimization [41] is typically done by iterative methods that just require repeated applications of the Hamiltonian operator onto wave functions. Also, in real-space methods and wavelet basis sets, the separable form is simpler and more efficient. The above-discussed terms suffice to specify a norm-conserving pseudopotential. Since in a norm-conserving pseudopotential the charge density is obtained exactly in the same way as in an all-electron method, namely, by squaring and summing all the occupied orbitals, the treatment of the electrostatic interactions is straightforward.

In relativistic calculations the spin–orbit coupling splits up all orbitals with $l > 0$ into a spin-up and spin-down orbital with overall angular momentum $j = l \pm 1/2$. So for each angular momentum $l > 0$, one spin-up and spin-down orbital with different wave functions and two pseudopotentials exist. Most pseudopotentials follow the approach of Bachelet et al. [37] and give a weighted average and a difference potential for these two potentials. The average pseudopotential is defined as

$$V_l(\mathbf{r}, \mathbf{r}') = \frac{1}{2l+1} \left(l V_{l-1/2}(\mathbf{r}, \mathbf{r}') + (l+1) V_{l+1/2}(\mathbf{r}, \mathbf{r}') \right) \tag{2.4}$$

weighted by the different j degeneracies of the $l \pm 1/2$ orbitals. The difference potential describes the spin–orbit coupling and is defined as

$$\Delta V_l^{\text{SO}}(\mathbf{r}, \mathbf{r}') = \frac{2}{2l+1} \left(V_{l+1/2}(\mathbf{r}, \mathbf{r}') - V_{l-1/2}(\mathbf{r}, \mathbf{r}') \right). \tag{2.5}$$

The total pseudopotential is then given by

$$V(\mathbf{r}, \mathbf{r}') = V_{\text{loc}}(r)\delta(\mathbf{r} - \mathbf{r}') + \sum_l V_l(\mathbf{r}, \mathbf{r}') + \Delta V_l^{\text{SO}}(\mathbf{r}, \mathbf{r}')\mathbf{L} \cdot \mathbf{S}, \quad (2.6)$$

where $V_l(\mathbf{r}, \mathbf{r}')$ and $\Delta V_l^{\text{SO}}(\mathbf{r}, \mathbf{r}')$ are of the same form as in the nonrelativistic case, that is, they can be of semilocal or separable form. Neglecting the contributions from $\Delta V_l^{\text{SO}}(\mathbf{r}, \mathbf{r}')$ in Eq. (2.6) gives the average potential that contains all scalar parts of the relativistic pseudopotential, whereas the total potential contains relativistic effects up to order α^{-2}, where α is the fine structure constant.

2.6
Ultrasoft and Projector Augmented Wave Methods

Projector augmented wave (PAW) methods are frequently touted as a type of all-electron method. As discussed in Blöchl's original publication [29], the method becomes identical to an all-electron method within the frozen core approximation within the limit of an infinite number of partial waves in the muffin-tin spheres. In practice, the number of partial waves is finite and rather small (on the order of two for each l, m channel) and PAW methods have a behavior that is very similar to the other pseudopotential methods. In particular, the quality of the wave function in the muffin-tin sphere depends on the size of the muffin-tin sphere (usually called augmentation region in the PAW literature) for a finite number of partial waves. As a consequence, "hard" PAW schemes where the radius of the muffin-tin sphere is small typically give more accurate results than soft schemes, which can give results that differ quite a bit from the exact all-electron results [21]. Ultrasoft pseudopotentials can be considered as a limiting case of a PAW scheme [42]. Both in ultrasoft schemes and in the PAW method, one uses implicitly nonorthogonal basis sets that consist of plane waves in the interstitial region and a combination of plane waves and partial waves in the muffin-tin sphere. As a consequence, an overlap matrix S appears and one has to solve in a self-consistent diagonalization approach a generalized eigenvalue problem:

$$H\Psi_i = \varepsilon_i S\Psi_i$$

instead of the ordinary eigenvalue problem of an orthogonal basis set:

$$H\Psi_i = \varepsilon_i \Psi_i$$

encountered in a norm-conserving pseudopotential. Both the Hamiltonian matrix H and the overlap matrix S have a separable form. The pure plane wave part of the pseudo-wave function in the muffin-tin region forms a physically meaningless pseudo-wave function. In the muffin-tin region, this pseudo-wave function varies typically less than the pseudo-wave functions of a norm-conserving pseudopotential since the constraint of reproducing the physical charge density is absent. As a consequence, the number of plane waves needed to represent the pseudo-wave function in a PAW or ultrasoft [25] scheme is typically significantly reduced compared to the norm-conserving case where the muffin-tin region typically requires the highest resolution. However, the calculation of the charge requires a larger effort. The charge obtained from the unphysical pseudo-wave functions has to be modified to take into account the charge arising from the partial wave expansion. This is done by adding so-called augmentation charges. Since the partial waves represent all-electron orbitals, they vary actually very rapidly and treating the resulting charge density with a plane wave scheme would require a very high plane wave cutoff. Various atoms centered compensating charge distributions that split up the problem into a part that can be treated with a low plane wave cutoff and a remainder that can be treated in real space or analytically have to be used to make the scheme numerically efficient. As a result of the increased complexity of PAW and ultrasoft schemes, there are in most plane wave codes certain advanced features, such as linear response calculations, that are only available for norm-conserving pseudopotentials but not for PAW or ultrasoft schemes. Similar to the LAPW case, the increased complexity also implies numerous tuning parameters that are hard to understand and to control for standard users of electronic structure codes.

2.7
Construction Methods for Pseudopotentials

The semilocal potentials $U_l(r)$ of the seminal pseudopotentials by Bachelet *et al.* were constructed by inverting the radial Schrödinger equation (Eq. (2.1)) for a pseudo-wave function $\tilde{R}_l(r)$ that was identical to the all-electron wave function outside the muffin-tin region, but much smoother inside and that satisfies the norm-conservation condition:

$$\tilde{U}_l(r) = \left[-\frac{1}{2}\tilde{R}_l''(r) - \frac{\tilde{R}_l'(r)}{r} + \frac{l(l+1)}{2}\tilde{R}(r) - \varepsilon\tilde{R}_l(r) \right] / \tilde{R}_l(r). \tag{2.7}$$

The potential $\tilde{U}_l(r)$ obtained from the above equation still contains the electrostatic Hartree and exchange-correlation potentials, which have still to be subtracted to obtain the bare semilocal $U_l(r)$. The norm-conserving pseudopotential then gives by construction the correct reference eigenvalues and their first-order variation. The reference configuration that gave the reference eigenvalues was a single atomic configuration. So for the reference configuration, the scattering properties were reasonably accurate. But as already discussed, this does not necessarily imply good transferability. Also, most of the other pseudopotentials that were later on popular in the solid-state community, such as the Troulier Martins and FHI pseudopotentials, were heavily focused on the scattering properties.

Another limitation of the straightforward Schrödinger equation inversion approach is that it cannot be applied if semicore states are needed for a certain l

channel. In this case, one would obtain two different $V_l(r)$. In addition, the valence pseudo-wave function $\tilde{R}_l(r)$ will have a node and one would divide by zero in Eq. (2.7) at a certain r. These problems can however be overcome by a more general scheme [43].

The inversion of the Schrödinger equation gives a semilocal pseudopotential, whereas a separable form is desired in the context of plane wave calculations. With the method of a Kleinman and Bylander [40], a semilocal Hamiltonian can approximately be converted into a separable form. During this transformation, so-called ghost states can however appear [44]. These ghost states have nodes and should in principle be higher energy scattering states, but they were pulled down in energy by a too small set of projectors.

In the chemistry community, on the other hand, obtaining good transferability is typically already the main objective during the construction stage. The widely used so-called Stuttgart pseudopotentials are constructed in such a way that they reproduce a large number of energies of excited and ionized atomic configurations. For this reason, they are called energy consistent. It turns out that such energy-consistent pseudopotentials satisfy, at least to a certain degree, the norm conservation condition [14]. This means that at a sufficiently large distance from the nucleus where all the other pseudopotential terms have decayed and the potential is a pure coulombic pseudopotential, the valence all-electron wave functions and the pseudo-wave functions coincide. How well the scattering properties are really reproduced by these energy-consistent potentials has however apparently never been examined in detail. Good scattering properties are an important ingredient to obtain high-accuracy band structures in crystalline solids and so the energy-consistent pseudopotentials might be less accurate in this context than the pseudopotentials coming from the physics community. Anyway, as already mentioned, the fact that the Stuttgart pseudopotentials have singularities and are not separable makes them unsuitable for the plane wave programs that are typically used for band structure calculations.

The energy-consistent approach also has the advantage that it can easily be carried over to the many-body context. If the energy of the different configurations that were used for the fitting were obtained by correlated methods, the pseudopotentials should allow us to do correlated calculations with the valence electrons only. Therefore, this approach can not only be used to construct pseudopotentials for independent one-particle schemes, such as density functional theory, but also for correlated many-electron schemes such as coupled cluster methods [45].

Norm-conserving pseudopotentials with a more complicated Hamiltonian than the Stuttgart pseudopotentials that give rise to wavy semilocal potentials and wavy pseudo-wave functions (Figures 2 and 3 of Ref. [46]) were constructed for heavy elements, including all lanthanides and actinides.

Pseudopotentials for correlated methods can however also be constructed based on the scattering picture [47,48]. The basic principle in the independent particle context, namely, that the wave function inside the muffin-tin sphere can be determined if the scattering properties on the surface of the muffin-tins sphere are known, can be carried over to the many-particle framework. In this case, the scattering properties are determined by the values of the many-body density matrix [49] on the surface of the muffin-tin sphere.

The dual-space Gaussian pseudopotentials [38,50] are not constructed by an inversion of the Schrödinger equation but, like the energy-consistent pseudopotentials, by a fitting procedure. The analytical form entering into the fitting process is already a separable form. However, since both occupied and unoccupied atomic states are fitted, no ghost states exist. In the original version, both eigenvalues and the norm conservation were fitted for a single atomic reference configuration. In the latest version of this pseudopotential [21], several atomic reference configurations are used, representing both ionized and excited configurations and both eigenvalues and the norm conservation are fitted for all these configurations. In this way, both accurate scattering properties and high transferability can be achieved.

Since atomic eigenvalues and logarithmic derivatives differ for different exchange-correlation functionals, density functional pseudopotentials should be constructed for a given target exchange-correlation functional, even though attempts have also been made to construct "universal" pseudopotentials [51]. Since there are so many exchange-correlation functionals on the market, the corresponding pseudopotentials might however not always be available and other pseudopotentials are frequently used. It is, for instance, a common practice to use PBE [52] pseudopotentials for calculations with the PBE0 [53] and HSE [54] hybrid functional calculations. As Table 2.1 shows, the error arising from this inconsistency can however be quite large.

The magnitude of the error resulting from inconsistent pseudopotentials depends, of course, on the size of the muffin-tin region in which the pseudopotentials differ from the all-electron potential and on how many electrons are included in the valence. Since the muffin-tin region is typically quite small in the Stuttgart pseudopotentials and since small core versions are available, these inconsistency errors seem to be tolerable in the chemistry community and Hartee–Fock pseudopotentials are in general used for density functional calculations [55].

2.8
Numerical Performance of Pseudopotentials

In plane wave and real-space codes, the form of the pseudopotential or PAW scheme can have a strong influence on the numerical performance. The aim is to find an optimal balance between softness and accuracy. A soft pseudopotential requires only relatively low-resolution plane waves or grids. In addition, the number of floating point operations per basis function should be small. The local potential is the first factor that matters in this context. The potential should be as smooth as possible and vary slowly. In particular, it must not have a singularity. The projectors should also be as smooth as possible. In addition, their analytical form has a strong influence on performance.

Table 2.1 Atomization energy of molecules in kcal/mol for consistent (PBE/PBE and PBE0/PBE0) and inconsistent (PBE/PBE0) use of the exchange-correlation functionals in the molecular calculation and for the generation of the pseudopotential [21].

XC used in calculation	PBE	PBE0	PBE0
XC of pseudopotential used	PBE	PBE	PBE0
$C_2H_2^s$	416.35	399.38	405.79
CH_2^s	178.33	174.87	176.05
CH_3Cl^s	400.13	393.13	395.53
CH_3OH^s	518.58	505.43	508.56
CH_4^s	420.50	415.18	418.13
Cl_2^s	65.97	60.46	61.34
ClF^s	71.75	60.89	61.55
CO_2^s	413.48	384.89	389.52
F_2^s	51.11	33.18	32.82
H_2CO^s	385.02	368.52	371.66
H_2O^s	232.98	225.17	226.07
HCl^s	106.78	105.55	105.89
HF^s	141.39	136.05	136.36
$HOCl^s$	174.07	161.60	162.71
Mg_2^s	2.80	2.41	2.33
MgH^d	32.46	32.93	32.72
O_2^t	140.87	121.28	121.93
O_3^s	180.35	133.54	133.93
OH^d	109.20	104.74	105.14
SiH_3^d	221.86	224.05	224.07

Dual-space Gaussian pseudopotentials were used in the BigDFT [7] code with free boundary conditions.

In plane wave programs, the integral over the product of the wave function and the projector is frequently evaluated in Fourier space in the form of a sum: $\sum_k p_{l,j}(k) Y_{l,m}(\hat{k}) \tilde{\Psi}(k)$, where $p_{l,j}(k)$ is obtained from the Fourier transform of projector in real space and where $\tilde{\Psi}(k)$ are the Fourier coefficients of the wave function. In this case, the number of plane waves is growing linearly with respect to the volume of the cell and all their k vectors have to be included in the summation. Hence, the numerical cost is proportional to the size of the cell. Since these integrals have to be calculated for all wave functions and all atomic projectors, the total numerical cost scales cubically with respect to the cell size. This cubic scaling term typically dominates over the other cubic term that is found in LDA/GGA density functional electronic structure calculations, namely, the orthogonalization term [56] and limits therefore the overall speed of large calculations.

The projector term $\int p_{l,j}(r')Y_{l,m}\left(\hat{\mathbf{r}}'\right)\tilde{\Psi}(\mathbf{r})d\mathbf{r}'$. can however also be evaluated in real space as a sum over grid points. In this case, the integration scales only quadratically with respect to the size of the cell since only grid points from a certain region around the atom to which the projector belongs have to be included in the sum. How big this region has to be depends on the decay properties of the projector. Consequently, for small cells, where the number of plane waves is less than the number of real-space grid points that would be required for the numerical integration, the Fourier space treatment is faster, but then for larger cells the real-space treatment will become significantly faster.

An important property of the dual-space Gaussian pseudopotentials is that the projectors are simple Gaussians times spherical harmonics. The Gaussian form of the projector is optimal, since a Gaussian is the function that can be integrated up to a given high precision with the smallest possible number of equally spaced grid points. If the wave function is slowly varying over the width of the Gaussian projector, the product can also be integrated with a nearly optimal number of grid points. As a consequence, the prefactor of a real-space evaluation is optimal.

2.9
Accuracy of Pseudopotentials

The need for highly accurate pseudopotentials has been widely recognized. Since they are used in so many electronic structure calculations, low-quality pseudopotentials might ruin the overall reputation of electronic structure calculations! Systematic validation tests have recently been started in the physics community. In the delta test [57], leading groups in the development of electronic structure methods have teamed up to investigate the accuracy of different electronic structure codes for the calculation of the energy of elemental solids as a function of the volume of the unit cell. As it should be, all codes using a systematic basis set such as plane waves or wavelets converged essentially to the same result. Major differences were always due to different pseudopotentials. So the delta test became finally to a large extent a test for the accuracy of pseudopotentials. The web page of the delta test (https:molmod.ugent.bedeltacodesdft) lists the accuracy that was achieved by various all-electron- and pseudopotential-based codes. A large number of entries have delta values of around or less than 1, which corresponds roughly to the experimental accuracy. The majority of high-accuracy entries were obtained with PAW or ultrasoft pseudopotentials, but also several norm-conserving pseudopotentials met the accuracy requirements.

The delta test is however a relatively easy test. The chemical environment does not change significantly upon modest compression and expansion of the crystalline cell. Other groups have therefore introduced more demanding solid-state test sets where, for instance, several elemental crystalline structures are examined [58,59]. It is even more difficult to get accurate atomization/formation energies for molecules or crystals with different chemical elements. Especially the ubiquitous compounds with oxygen are difficult since the high electronegativity of oxygen induces a large charge transfer.

This effect is well illustrated by the SG15 [60] pseudopotentials. They were constructed in such a way that they give good results for elemental solids. So it is not surprising that they perform well in the delta test, with Δ values ranging from 1.3 to 1.8 with different codes. As Figure 2.6 shows, they are however not accurate for atomization energies. As a matter of fact, the average error of about 5 kcal/mol, arising from the SG15 pseudopotential, is larger than typical errors coming from the PBE functional [13]. The ONCV-DOJO [61] pseudopotential belongs to the same family as the SG15 pseudopotential and has a slightly lower, but still too large average error of 3.4 kcal/mol.

Figure 2.6 also shows results for the old and widely used FHI Troullier–Martins pseudopotentials [63,66]. They give quite large errors ($\Delta = 13.4$) in the delta test and also large errors in the atomization energies, as summarized in Table 2.2. The GBRV pseudopotentials [59] are a recent representative of the ultrasoft family and were shown to be more accurate for crystalline calculations than some PAW schemes [59]. In the delta test, they reached an high accuracy value of $\Delta = 0.9$. However, Figure 2.6 and Table 2.2 show that the errors in the atomization energies are quite significant. Chemical accuracy for atomization energies can be obtained with the hard PAW sets [67] of VASP (version 5.2.12) [6]. The default PAW set of VASP as well as the PAW set of GPAW (version 0.10.0) [68] also give on average very accurate results for atomization energies and very good delta values of around 1. With the soft PAW VASP version, the error in the atomization energies can however rise to 4.8 kcal/mol [21].

The problem with the SG15 pseudopotentials is that they were optimized for softness by choosing projectors that extend far beyond the inert core region. In this way, the pseudopotential is different from the $-(Z_{ion}/r)$ potential in a region outside the inert region and the scattering properties cannot fully account for potential modifications induced by changes in the charge density. The softness optimization leads to other strange effects. The target plane wave cutoff for this pseudopotential is 60 Ry. Actually beyond this target cutoff at about 110 Ry, the energy can still go up (Figure 2 of Ref. [60]) with further increasing the plane wave cutoff in violation of the variational principle that holds whenever the potential is smooth enough such that all the Hamiltonian matrix elements can be calculated nearly exactly. Real convergence in the mathematical sense could actually only be obtained with an extremely high plane wave cutoff for such a nonsmooth local potential shown in Figure 2.7, which has strong variations near the origin.

Chemical accuracy for atomization energies as well as excellent delta values of about 1 in the delta test have been obtained with the latest version of the dual space Gaussian norm-conserving pseudopotentials for the PBE functional [21]. Chemical accuracy in this context means that the pseudopotential and the exact all-electron values agree on average by less 1 kcal/mol. Since the errors with respect to experiment for atomization energies in the PBE functional are on the order of 4 kcal/mol [13] for main elements, pseudopotential errors of 2 kcal/mol

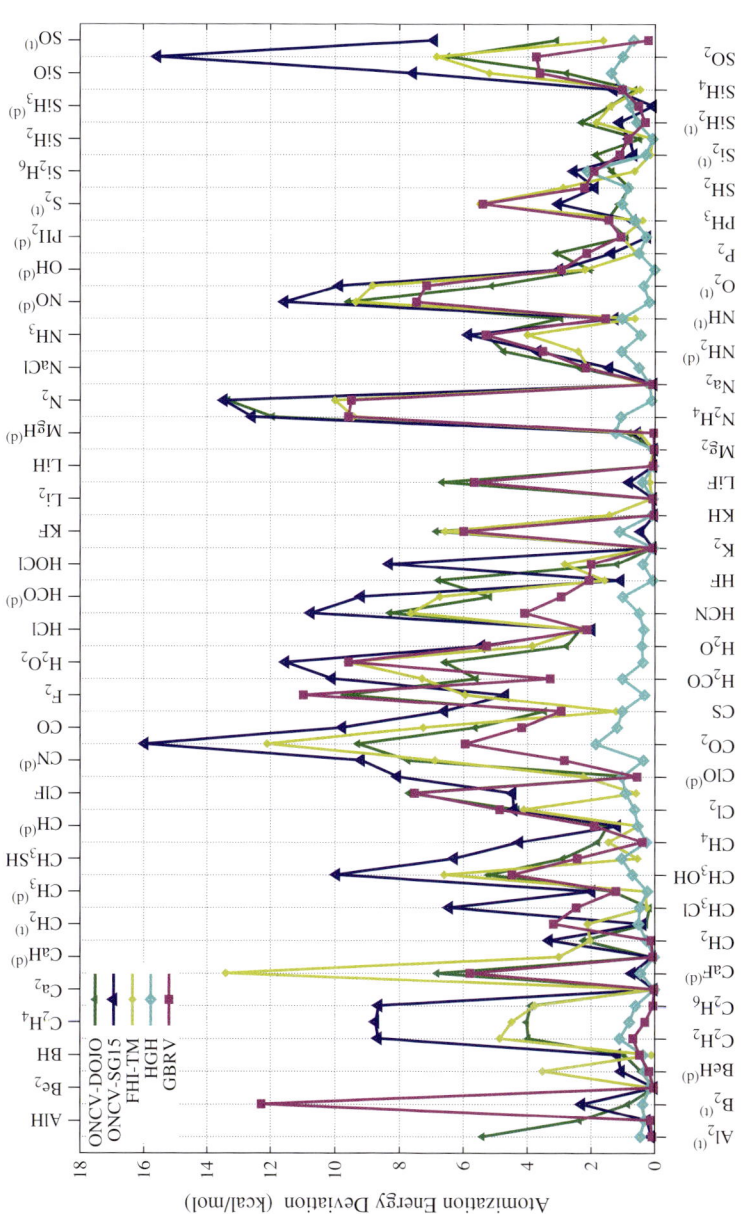

Figure 2.6 The errors in atomization energies of the Gaussian2 [62] test set augmented by some other molecules for the pseudopotentials of references [21,59,60,63,64] calculated with Quantum Espresso using the PBE functional and a dipole correction [65] with a box size of 25 Å and a plane wave cutoff of 200 Ry. For the ultrasoft calculation, the charge density cutoff was eight times the plane wave cutoff. Geometries were taken from the NIST webpage (http:cccbdb.nist.gov). Accurate all-electron reference values were obtained with the FHI-aims [23] all-electron code using a tier4 basis set and free boundary conditions. Even though relativistic effects are negligible for these light elements, the relativistic ZORA scalar approximation was used to be formally compatible with the pseudopotentials that were based on relativistic reference calculations. The atomic energies were obtained for the real, usually nonspherical, ground state with the correct integer spin polarization.

Table 2.2 Analysis of the mean absolute deviation (MAD), root mean square deviation RMSD, and maximum absolute deviation (maxAD) of the atomization energy errors shown in Figure 2.6 in kcal/mol.

	MAD	RMSD	maxAD
ONCV-DOJO	3.40	4.63	13.32
ONCV-SG15	4.34	6.17	15.95
FHI-TM	3.20	4.62	13.40
HGH	0.58	0.73	2.12
GBRV	2.85	4.13	12.27

might still be acceptable and allow for somewhat softer pseudopotentials. For some elements, such softer but nevertheless quite accurate dual-space Gaussian norm-conserving pseudopotentials are available on the BigDFT Web site (http:bigdft.orgWiki)

Chemical accuracy is certainly required when pseudopotentials are used within correlated calculations such as quantum Monte Carlo calculations [69–72] or coupled cluster calculations. Chemical accuracy has recently been demonstrated in this context for the atomization energies of a small test set of molecules [48].

Chemical accuracy should certainly also be aimed at whenever one constructs pseudopotentials for some of the latest exchange-correlation functionals that strive to achieve chemical accuracy.

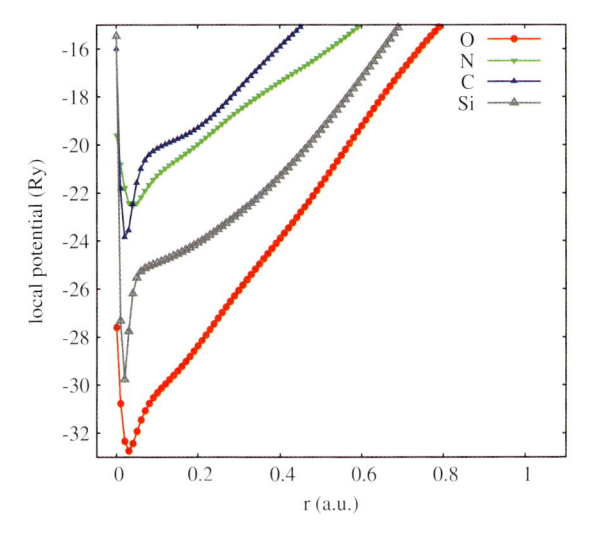

Figure 2.7 Nonsmooth local potential of the SG15 PBE pseudopotentials for different elements as obtained from the webpage http:www.quantum-simulation.orgpotentialssg15_oncv.

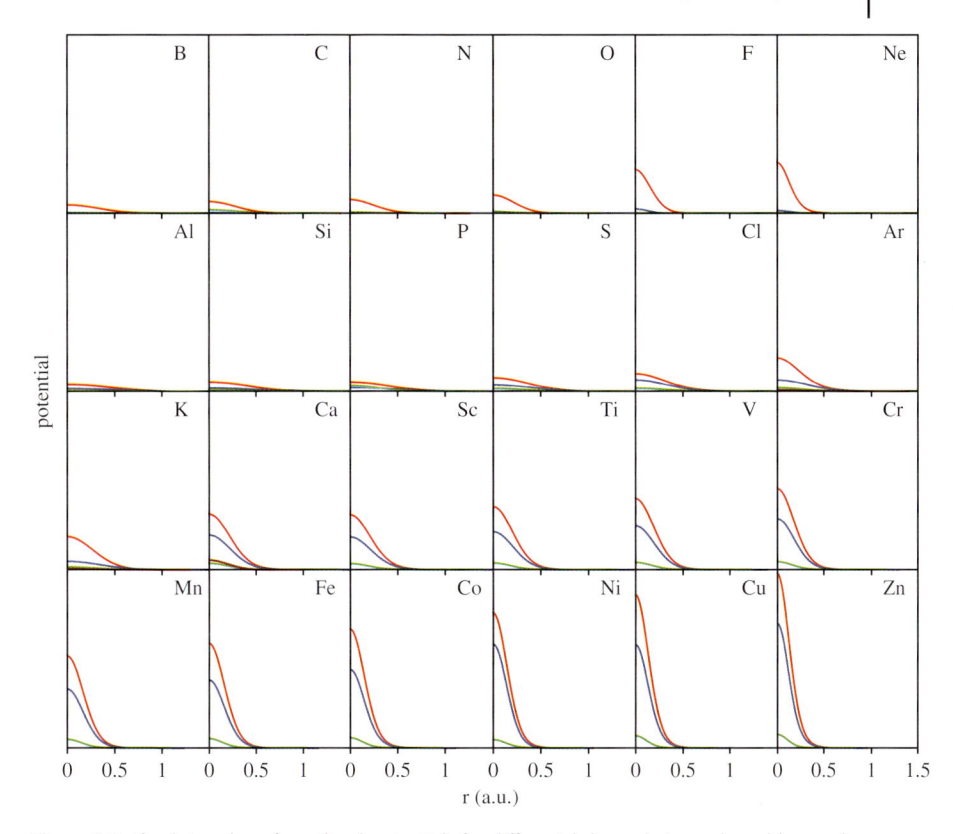

Figure 2.8 Absolute value of semilocal potentials for different l channels (s = red, p = blue, and d = green) without the Z_{ion}/r part for different elements. The pseudopotentials are from http: www.tc.uni-koeln.dePPclickpse.en.html and are described in Refs [76–81].

No systematic tests of the accuracy of atomization energies have been performed for the pseudopotentials used in the chemistry community. However, various comparisons of different quantities with all-electron results such as in Refs [73–75] suggest that they are in general of very high quality. Since these pseudopotentials are generally used with Gaussian basis sets, the softness requirement is absent and they may well be harder in return for higher accuracy. Figure 2.8 shows the magnitude of the semilocal potentials for different l channels excluding the Z_{ion}/r part. One can see that this part of the local pseudopotential is nonvanishing only in a fairly small region around the nucleus, which is contained in the inert core regions depicted in Figure 2.4. This also suggests high accuracy. The absence of any oscillations indicates that no overfitting occurred.

Due to the absence of large and accurate experimental data for molecular systems containing very heavy elements it is very difficult to assess the accuracy of pseudopotentials for very heavy elements for such molecular systems.

2.10
Conclusions

Even though pseudopotentials are widely used for many years, further progress is needed. Many of the pseudopotentials used in the physics and materials science community have errors that are so large that they dominate the error arising from the exchange-correlation functional. Systematic validation efforts have just started with projects such as the delta test [57]. However, a much larger range of testing properties is required to find really high-quality pseudopotentials. The dual-space Gaussian norm-conserving pseudopotentials, available, for instance, in BigDFT, as well as the PAW sets available in VASP and GPAW excel in accuracy tests by giving both high-accuracy atomization energies and a good description of elemental bulk solids as reflected by small delta values. The pseudopotentials used in the chemistry community seem to be quite accurate and the main limitation to high accuracy in real calculations is the Gaussian basis set. The largest basis sets that would give chemical accuracy under most circumstances are numerically too expensive for large-scale calculations. Because of their semilocal form, these pseudopotentials cannot be used within plane wave and wavelet codes. In principle, it should be possible to cast them also in a separable form with a Gaussian nucleus, which would make them suitable for these codes. This functional form would actually also make their use easier within Gaussian basis sets.

References

1 Parr, R.G. and Yang, W. (1989) *Density-Functional Theory of Atoms and Molecules*, vol. **16**, Oxford University Press.

2 Hellmann, H. (1935) A new approximation method in the problem of many electrons. *J. Chem. Phys.*, **3** (1), 61.

3 Küchle, W., Dolg, M., Stoll, H., and Preuss, H. (1994) Energy-adjusted pseudopotentials for the actinides: parameter sets and test calculations for thorium and thorium monoxide. *J. Chem. Phys.*, **100** (10), 7535–7542.

4 Gonze, X., Amadon, B., Anglade, P.-M., Beuken, J.-M., Bottin, F., Boulanger, P., Bruneval, F., Caliste, D., Caracas, R., Cote, M. *et al.* (2009) ABINIT: first-principles approach to material and nanosystem properties. *Comput. Phys. Commun.*, **180** (12), 2582–2615.

5 Giannozzi, P., Baroni, S., Bonini, N., Calandra, M., Car, R., Cavazzoni, C., Ceresoli, D., Chiarotti, GuidoL., Cococcioni, M., Dabo, I. *et al.* (2009) QUANTUM ESPRESSO: a modular and open-source software project for quantum simulations of materials. *J. Phys. Condens. Matter*, **21** (39), 395502.

6 Kresse, G. and Furthmüller, J. (1996) Efficient iterative schemes for *ab initio* total-energy calculations using a plane-wave basis set. *Phys. Rev. B*, **54** (16), 11169.

7 Genovese, L., Neelov, A., Goedecker, S., Deutsch, T., Ghasemi, S.A., Willand, A., Caliste, D., Zilberberg, O., Rayson, M., Bergman, A. *et al.* (2008) Daubechies wavelets as a basis set for density functional pseudopotential calculations. *J. Chem. Phys.*, **129** (1), 014109.

8 Pask, J.E. and Sterne, P.A. (2005) Finite element methods in *ab initio* electronic structure calculations. *Model. Simul. Mater. Sci. Eng.*, **13** (3), R71.

9 Lin, L., Lu, J., Ying, L., and Weinan, E. (2012) Adaptive local basis set for Kohn–Sham density functional theory in a discontinuous Galerkin framework I: total energy calculation. *J. Comput. Phys.*, **231** (4), 2140–2154.

10 Frisch, M.J., Trucks, G.W., Schlegel, H.B., Scuseria, G.E., Robb, M.A., Cheeseman, J.R., Scalmani, G., Barone, V., Mennucci, B., Petersson, G.A., Nakatsuji, H., Caricato, M., Li, X., Hratchian, H.P., Izmaylov, A.F., Bloino, J., Zheng, G., Sonnenberg, J.L., Hada, M., Ehara, M., Toyota, K., Fukuda, R., Hasegawa, J., Ishida, M., Nakajima, T., Honda, Y., Kitao, O., Nakai, H., Vreven, T., Montgomery, J.A., Jr., Peralta, J.E., Ogliaro, F., Bearpark, M., Heyd, J.J., Brothers, E., Kudin, K.N., Staroverov, V.N., Kobayashi, R., Normand, J., Raghavachari, K., Rendell, A., Burant, J.C., Iyengar, S.S., Tomasi, J., Cossi, M., Rega, N., Millam, J.M., Klene, M., Knox, J.E., Cross, J.B., Bakken, V., Adamo, C., Jaramillo, J., Gomperts, R., Stratmann, R.E., Yazyev, O., Austin, A.J., Cammi, R., Pomelli, C., Ochterski, J.W., Martin, R.L., Morokuma, K., Zakrzewski, V.G., Voth, G.A., Salvador, P., Dannenberg, J.J., Dapprich, S., Daniels, A.D., Farkas, O., Foresman, J.B., Ortiz, J.V., Cioslowski, J., and Fox, D.J. (2009) *Gaussian09 Revision E. 01*, Gaussian Inc., Wallingford, CT.

11 Valiev, M., Bylaska, E.J., Govind, N., Kowalski, K., Straatsma, T.P., Van Dam, H.J.J., Wang, D., Nieplocha, J., Apra, E., Windus, T.L. *et al.* (2010) NWChem: a comprehensive and scalable open-source solution for large scale molecular simulations. *Comput. Phys. Commun.*, **181** (9), 1477–1489.

12 Guest, M.F., Bush, I.J., Van Dam, H.J.J., Sherwood, P., Thomas, J.M.H., Van Lenthe, J.H., Havenith, R.W.A., and Kendrick, J. (2005) The GAMESS-UK electronic structure package: algorithms, developments and applications. *Mol. Phys.*, **103** (6–8), 719–747.

13 Peverati, R. and Truhlar, D.G. (2011) Quest for a universal density functional: the accuracy of density functionals across a broad spectrum of databases in chemistry and physics. *Philos. Trans. R. Soc. Lond. A*, **372**, 20120476. 2014.

14 Dolg, M. and Cao, X. (2011) Relativistic pseudopotentials: their development and scope of applications. *Chem. Rev.*, **112** (1), 403–480.

15 Schwerdtfeger, P. (2011) The pseudopotential approximation in electronic structure theory. *ChemPhysChem*, **12** (17), 3143–3155.

16 Pickett, W.E. (1989) Pseudopotential methods in condensed matter applications. *Comput. Phys. Rep.*, **9** (3), 115–197.

17 Harrison, R.J., Fann, G.I., Yanai, T., Gan, Z., and Beylkin, G. (2004) Multiresolution quantum chemistry: basic theory and initial applications. *J. Chem. Phys.*, **121** (23), 11587–11598.

18 Frediani, L., Fossgaard, E., Flå, T., and Ruud, K. (2013) Fully adaptive algorithms for multivariate integral equations using the non-standard form and multiwavelets with applications to the Poisson and bound-state Helmholtz kernels in three dimensions. *Mol. Phys.*, **111** (9–11), 1143–1160.

19 Durdek, A., Jensen, S.R., Juselius, J., Wind, P., Flå, T., and Frediani, L. (2015) Adaptive order polynomial algorithm in a multiwavelet representation scheme. *Appl. Numer. Math.*, **92**, 40–53.

20 Jensen, S., Saha, S., Livas, J.F., Huhn, W., Blum, V., Goedecker, S., and Frediani, L. (2017) The elephant in the room of density functional theory calculations, *J. Chem. Phys. Lett.*, **8**, 1449–1457.

21 Willand, A., Kvashnin, Y.O., Genovese, L., Vázquez-Mayagoitia, Á., Deb, A.K., Sadeghi, A., Deutsch, T., and Goedecker, S. (2013) Norm-conserving pseudopotentials with chemical accuracy compared to all-electron calculations. *J. Chem. Phys.*, **138** (10), 104109.

22 Delley, B. (1990) An all-electron numerical method for solving the local density functional for polyatomic molecules. *J. Chem. Phys.*, **92** (1), 508–517.

23 Blum, V., Gehrke, R., Hanke, F., Havu, P., Havu, V., Ren, X., Reuter, K., and Scheffler, M. (2009) *Ab initio* molecular simulations with numeric atom-centered orbitals. *Comput. Phys. Commun.*, **180** (11), 2175–2196.

24 Morrell, M.M., Parr, R.G., and Levy, M. (1975) Calculation of ionization potentials

from density matrices and natural functions, and the long-range behavior of natural orbitals and electron density. *J. Chem. Phys.*, **62** (2), 549–554.
25 Vanderbilt, D. (1990) Soft self-consistent pseudopotentials in a generalized eigenvalue formalism. *Phys. Rev. B*, **41** (11), 7892.
26 Goedecker, S. and Maschke, K. (1992) Operator approach in the linearized augmented-plane-wave method: efficient electronic-structure calculations including forces. *Phys. Rev. B*, **45** (4), 1597.
27 Goedecker, S. and Maschke, K. (1990) Alternative approach to separable first-principles pseudopotentials. *Phys. Rev. B*, **42** (14), 8858.
28 Liu, A.Y., Singh, D.J., and Krakauer, H. (1994) Linearized-augmented-plane-wave method with Car–Parrinello scaling. *Phys. Rev. B*, **49** (24), 17424.
29 Blöchl, P.E. (1994) Projector augmented-wave method. *Phys. Rev. B*, **50** (24), 17953.
30 Kleinman, L. (1980) Relativistic norm-conserving pseudopotential. *Phys. Rev. B*, **21** (6), 2630.
31 Heine, V. (1970) The pseudopotential concept. *Solid State Physics*, **24**, 1–36.
32 Singh, D.J. and Nordstrom, L. (2006) *Planewaves, Pseudopotentials, and the LAPW Method*, Springer.
33 Goedecker, S. and Maschke, K. (1992) Transferability of pseudopotentials. *Phys. Rev. A*, **45** (1), 88.
34 Hamann, D.R., Schlüter, M., and Chiang, C. (1979) Norm-conserving pseudopotentials. *Phys. Rev. Lett.*, **43** (20), 1494.
35 Louie, S.G., Froyen, S., and Cohen, M.L. (1982) Nonlinear ionic pseudopotentials in spin-density-functional calculations. *Phys. Rev. B*, **26** (4), 1738.
36 Kresse, G., Hafner, J., and Needs, R.J. (1992) Optimized norm-conserving pseudopotentials. *J. Phys. Condens. Matter*, **4** (36), 7451.
37 Bachelet, G.B., Hamann, D.R., and Schlüter, M. (1982) Pseudopotentials that work: from H to Pu. *Phys. Rev. B*, **26** (8), 4199.
38 Goedecker, S., Teter, M., and Hutter, J. (1996) Separable dual-space Gaussian pseudopotentials. *Phys. Rev. B*, **54** (3), 1703.
39 VandeVondele, J., Krack, M., Mohamed, F., Parrinello, M., Chassaing, T., and Hutter, J. (2005) Quickstep: fast and accurate density functional calculations using a mixed Gaussian and plane waves approach. *Comput. Phys. Commun.*, **167** (2), 103–128.
40 Kleinman, L. and Bylander, D.M. (1982) Efficacious form for model pseudopotentials. *Phys. Rev. Lett.*, **48** (20), 1425.
41 Payne, M.C., Teter, M.P., Allan, D.C., Arias, T.A., and Joannopoulos, J.D. (1992) Iterative minimization techniques for *ab initio* total-energy calculations: molecular dynamics and conjugate gradients. *Rev. Mod. Phys.*, **64** (4), 1045.
42 Kresse, G. and Joubert, D. (1999) From ultrasoft pseudopotentials to the projector augmented-wave method. *Phys. Rev. B*, **59** (3), 1758.
43 Titov, A.V., Mitrushenkov, A.O., and Tupitsyn, I.I. (1991) Effective core potential for pseudo-orbitals with nodes. *Chem. Phys. Lett.*, **185** (3), 330–334.
44 Gonze, X., Stumpf, R., and Scheffler, M. (1991) Analysis of separable potentials. *Phys. Rev. B*, **44** (16), 8503.
45 Burkatzki, M., Filippi, C., and Dolg, M. (2007) Energy-consistent pseudopotentials for quantum Monte Carlo calculations. *J. Chem. Phys.*, **126** (23), 234105.
46 Mosyagin, N.S., Zaitsevskii, A.V., Skripnikov, L.V., and Titov, A.V. (2016) Generalized relativistic effective core potentials for actinides. *Int. J. Quantum Chem.*, **116** (4), 301–315.
47 Trail, J.R. and Needs, R.J. (2013) Pseudopotentials for correlated electron systems. *J. Chem. Phys.*, **139** (1), 014101.
48 Trail, J.R. and Needs, R.J. (2015) Correlated electron pseudopotentials for 3D-transition metals. *J. Chem. Phys.*, **142** (6), 064110.
49 Acioli, P.H. and Ceperley, D.M. (1994) Generation of pseudopotentials from correlated wave functions. *J. Chem. Phys.*, **100** (11), 8169–8177.
50 Hartwigsen, C., Gœdecker, S., and Hutter, J. (1998) Relativistic separable dual-space Gaussian pseudopotentials from H to Rn. *Phys. Rev. B*, **58** (7), 3641.

51 Delley, B. (2002) Hardness conserving semilocal pseudopotentials. *Phys. Rev. B*, **66** (15), 155125.

52 Perdew, J.P., Burke, K., and Ernzerhof, M. (1996) Generalized gradient approximation made simple. *Phys. Rev. Lett.*, **77** (18), 3865.

53 Adamo, C. and Barone, V. (1999) Toward reliable density functional methods without adjustable parameters: the PBE0 model. *J. Chem. Phys.*, **110** (13), 6158–6170.

54 Heyd, J., Scuseria, G.E., and Ernzerhof, M. (2003) Hybrid functionals based on a screened Coulomb potential. *J. Chem. Phys.*, **118** (18), 8207–8215.

55 Russo, T.V., Martin, R.L., and Hay, P.J. (1995) Effective core potentials for DFT calculations. *J. Phys. Chem.*, **99** (47), 17085–17087.

56 Goedecker, S. (1999) Linear scaling electronic structure methods. *Rev. Mod. Phys.*, **71** (4), 1085.

57 Lejaeghere, K., Bihlmayer, G., Björkman, T., Blaha, P., Blügel, S., Blum, V., Caliste, D., Castelli, I.E., Clark, S.J., Dal Corso, A. et al. (2016) Reproducibility in density functional theory calculations of solids. *Supramol. Sci.*, **351** (6280), aad3000.

58 Kucukbenli, E., Monni, M., Adetunji, B.I., Ge, X., Adebayo, G.A., Marzari, N., de Gironcoli, S., and Dal Corso, A. (2014) Projector augmented-wave and all-electron calculations across the periodic table: a comparison of structural and energetic properties. doi: arXiv preprint arXiv:1404.3015.

59 Garrity, K.F., Bennett, J.W., Rabe, K.M., and Vanderbilt, D. (2014) Pseudopotentials for high-throughput DFT calculations. *Comp. Mater. Sci.*, **81**, 446–452.

60 Schlipf, M. and Gygi, F. (2015) Optimization algorithm for the generation of ONCV pseudopotentials. *Comput. Phys. Commun.*, **196**, 36–44.

61 Hamann, D.R. (2013) Optimized norm-conserving Vanderbilt pseudopotentials. *Phys. Rev. B*, **88** (8), 085117.

62 Curtiss, L.A., Raghavachari, K., Redfern, P.C., and Pople, J.A. (1997) Assessment of Gaussian-2 and density functional theories for the computation of enthalpies of formation. *J. Chem. Phys.*, **106** (3), 1063–1079.

63 Fuchs, M. and Scheffler, M. (1999) *Ab initio* pseudopotentials for electronic structure calculations of poly-atomic systems using density-functional theory. *Comput. Phys. Commun.*, **119** (1), 67–98.

64 Krack, M. (2005) Pseudopotentials for H to Kr optimized for gradient-corrected exchange-correlation functionals. *Theor. Chem. Acc.*, **114** (1–3), 145–152.

65 Makov, G. and Payne, M.C. (1995) Periodic boundary conditions in *ab initio* calculations. *Phys. Rev. B*, **51**, 4014–4022.

66 Troullier, N. and Martins, J.L. (1991) Efficient pseudopotentials for plane-wave calculations. *Phys. Rev. B*, **43** (3), 1993.

67 Paier, J., Hirschl, R., Marsman, M., and Kresse, G. (2005) The Perdew–Burke–Ernzerhof exchange-correlation functional applied to the g2-1 test set using a plane-wave basis set. *J. Chem. Phys.*, **122** (23), 234102.

68 Enkovaara, J., Rostgaard, C., Jørgen Mortensen, J., Chen, J., Dułak, M., Ferrighi, L., Gavnholt, J., Glinsvad, C., Haikola, V., Hansen, H.A. et al. (2010) Electronic structure calculations with GPAW: a real-space implementation of the projector augmented-wave method. *J. Phys. Condens. Matter*, **22** (25), 253202.

69 Umrigar, C.J. (2015) Observations on variational and projector Monte Carlo methods. *J. Chem. Phys.*, **143** (16), 164105.

70 Toulouse, J., Assaraf, R., and Umrigar, C.J. (2016) Introduction to the variational and diffusion Monte Carlo methods. *Adv. Quantum Chem.*, **73**, 285–314.

71 Umrigar, C.J. (1999) Variational Monte Carlo basics and applications to atoms and molecules in *Quantum Monte Carlo Methods in Physics and Chemistry*, vol. **525**, Springer, pp. 129–160.

72 Foulkes, W.M.C., Mitas, L., Needs, R.J., and Rajagopal, G. (2001) Quantum Monte Carlo simulations of solids. *Rev. Mod. Phys.*, **73** (1), 33.

73 Peterson, K.A. and Yousaf, K.E. (2010) Molecular core-valence correlation effects involving the post-d elements Ga-Rn: benchmarks and new pseudopotential-based correlation consistent basis sets. *J. Chem. Phys.*, **133** (17), 174116.

74 Schwerdtfeger, P., Fischer, T., Dolg, M., IgelMann, G., Nicklass, A., Stoll, H., and Haaland, A. (1995) The accuracy of the pseudopotential approximation: I. An analysis of the spectroscopic constants for the electronic ground states of InCl and InCl3 using various three valence electron pseudopotentials for indium. *J. Chem. Phys.*, **102** (5), 2050–2062.

75 Leininger, T., Nicklass, A., Stoll, H., Dolg, M., and Schwerdtfeger, P. (1996) The accuracy of the pseudopotential approximation: II. A comparison of various core sizes for indium pseudopotentials in calculations for spectroscopic constants of InH, InF, and InCl. *J. Chem. Phys.*, **105** (3), 1052–1059.

76 Bergner, A., Dolg, M., Küchle, W., Stoll, H., and Preuß, H. (1993) *Ab initio* energy-adjusted pseudopotentials for elements of groups 13–17. *Mol. Phys.*, **80** (6), 1431–1441.

77 Bergner, A., Dolg, M., Küchle, W., Stoll, H., and Preuß, H. (1993) *Ab initio* energy-adjusted pseudopotentials for elements of groups 13–17. *Mol. Phys.*, **80** (6), 1431–1441.

78 Nicklass, A., Dolg, M., Stoll, H., and Preuss, H. (1995) *Ab initio* energy-adjusted pseudopotentials for the noble gases Ne through Xe: calculation of atomic dipole and quadrupole polarizabilities. *J. Chem. Phys.*, **102** (22), 8942–8952.

79 Kaupp, M., Schleyer, P.V.R., Stoll, H., and Preuss, H. (1991) Pseudopotential approaches to Ca, Sr, and Ba hydrides: why are some alkaline earth MX_2 compounds bent? *J. Chem. Phys.*, **94** (2), 1360–1366.

80 Leininger, T., Nicklass, A., Küchle, W., Stoll, H., Dolg, M., and Bergner, A. (1996) The accuracy of the pseudopotential approximation: non-frozen-core effects for spectroscopic constants of alkali fluorides XF (X=K, Rb, Cs). *Chem. Phys. Lett.*, **255** (4), 274–280.

81 Andrae, D., Haeussermann, U., Dolg, M., Stoll, H., and Preuss, H. (1990) Energy-adjusted *ab initio* pseudopotentials for the second and third row transition elements. *Theor. Chim. Acta*, **77** (2), 123–141.

3
Periodic Local Møller–Plesset Perturbation Theory of Second Order for Solids

Denis Usvyat,[1] Lorenzo Maschio,[2] and Martin Schütz[1]

[1]*Humboldt-Universität zu Berlin, Department of Chemistry, Brook-Taylor-Str. 2, 12489 Berlin, Germany*
[2]*Università di Torino, Dipartimento di Chimica and Centre of Excellence NIS (Nanostructured Interfaces and Surfaces), via Giuria 5, 10125 Torino, Italy*

3.1
Introduction

Since the early days of quantum mechanics, the tools of theoretical chemistry have become an invaluable complement to experimental studies, providing an unbiased *ab initio* description of electronic structure of molecular systems and related properties. For many years, accurate quantum chemical methods, due to their high computational cost, were not affordable beyond a few atom systems. In the course of the past two decades, development of efficient hardware noticeably expanded the application range for these methods. Standard (canonical) quantum chemical approaches, however, are hampered by a steep scaling of the computational cost with respect to the size of the system, making an accurate treatment of extended molecules still a formidable task. A real breakthrough in this respect was achieved by the development of low- or even linear-scaling quantum chemical models [1–39]. Most of these methods exploit the spatial locality of the correlation phenomena by means of a local orbital representation of the involved quantities. Presently, local correlation models are capable of treating molecular systems with several hundreds of atoms [34], or even up to a few thousands [26] within the scaled-opposite-spin approximation [40].

For crystals, which are bulky and essentially endless, even the simplest post-Hartree–Fock methods are still very expensive. Therefore, for decades, the main computational tool for solid-state first-principle electronic structure studies has been density functional theory (DFT). The Kohn–Sham variant of DFT is, on the one hand, computationally relatively inexpensive (with standard nonhybrid functionals it is even cheaper than HF), and, on the other hand, provides reasonable accuracy, which is acceptable for many applications. Unfortunately, standard

---

Handbook of Solid State Chemistry, First Edition. Edited by Richard Dronskowski, Shinichi Kikkawa, and Andreas Stein.
© 2017 Wiley-VCH Verlag GmbH & Co. KGaA. Published 2017 by Wiley-VCH Verlag GmbH & Co. KGaA.

DFT has several fundamental shortcomings, one of the most important being the notorious difficulty to systematically improve the accuracy of the results.

Another drawback of standard DFT, which is of quite some relevance for bulky systems, is its inability to correctly capture van der Waals dispersion. We note, however, that in recent years several recipes appeared in the literature that to a certain extent fix this problem. An inexpensive and presently very popular technique, often referred to as DFT + D, explicitly adds atomic-pair dispersion energies calculated from fitted or precomputed atomic C_6 coefficients [41–43] (optionally C_8 and even C_9 for the three-body dispersion could also be included [44]). In order to reduce double counting of the energy in such an approach, the contributions from the atom pairs with a small interatomic distance are suppressed by means of a specially chosen damping function. This has proven to be a computationally very efficient and not inaccurate way of including dispersion within DFT. However, the empiricism of the model can lead to artifacts and imposes certain limitations on the predictive power of such an approach. A more rigorous way to capture dispersion in the DFT framework is provided by the random phase approximation [45–48], which is, however, substantially more expensive than standard DFT. For some other approaches to treat dispersion in the framework of DFT we refer to Refs [49,50].

Despite a partial success of DFT-based methods in treating dispersion, the need for an *ab initio* quantum chemical treatment, which captures dispersion and other components of interaction in a balanced and systematically improvable way, is high in the community. Some two decades ago a finite cluster-based technique, known now as the method of increments or the incremental scheme, has been proposed [51–53]. In this method, the HF part of the interaction is supposed to be evaluated within the proper periodic format, and only the correlation contribution is subject to a finite cluster many-body expansion. The correlation part of the interaction is usually of shorter range than that of HF, but at the same time computationally and methodologically much more difficult. Therefore, a finite cluster treatment of the correlation contribution, for which standard molecular implementations of correlated methods can be utilized, is a versatile approach to the problem. The method of increments has been employed in quite a number of applications to solids [54–58], even including metals [59,60], as well as molecular adsorption on surfaces [61–63].

Even though high-level correlation methods and extensive basis sets can be employed in the context of the incremental scheme, its application to solids is not without problems. First, the integration of the R^{-6} decay law of the long-range pair correlation energy (dispersion) over the space beyond a certain sphere yields rather only a slow R_{cut}^{-3} decay of the captured dispersion contribution with respect to the cluster size cut off radius R_{cut}. Therefore, for accurate results the finite cluster-based treatment has to utilize rather extended clusters. Furthermore, a proper representation of a solid with finite clusters might require additional embedding, mimicking this solid. This then introduces an element of ambiguity in the modeling, on the one hand, and further increases the computational cost of the individual cluster calculations, on the other. Therefore, an accurate quantum chemical treatment, employing a decent basis set, can become prohibitively expensive. In the past few years, several alternative fragment-based models for accurate description of solids have appeared in literature [64–74].

In comparison to finite cluster models, a fully periodic quantum chemical treatment has the advantage of (i) an unambiguous and physically correct description of the solid, (ii) avoiding redundant evaluations of intermediate quantities and directly utilizing translation symmetry throughout the calculation, and (iii) a black-box character, implying that there is no need for the additional effort of constructing clusters, embedding, etc. At the same time, the initial formulation and efficient implementation of quantum chemical models for periodic systems is rather complex. Chronologically, first successful attempts to employ quantum chemistry in the periodic format were 1D periodic implementations for polymers [75–84]. A Laplace-transformed Møller–Plesset perturbation theory of second order (MP2) method [85], formally applicable to 3D periodic systems [86,87], was initially reported relatively early in 2001. This method, however, has been applied only to polymers and the 2D BN layer so far.

It was quite natural that one of the first attempts to tackle 3D periodic systems was based on the local correlation methodology that has proven to be efficient for extended molecules. Since the first paper on the local periodic MP2 method in 2005 [88], its efficiency has been substantially improved and the application range of the method extended. In the past 5 years, the local MP2 method has been used by several groups in quite a number of applications to crystals [58,89–107] and molecules adsorbed on surfaces [62,108–112]. Very recently, two other correlated 3D periodic approaches were reported. The first one is based on the plane wave formalism and includes a pilot implementation of coupled cluster singles and doubles with perturbative triples, CCSD(T) [113–117]. The second is a massively parallel Γ-point canonical MP2 and random phase approximation implementation employing atomic orbital basis sets [118,119].

3.2 Theory

3.2.1 Nomenclature

Before presenting the formalism of the LMP2 method, we want to specify the index nomenclature that is used throughout the text: The indices $i, j, \ldots$ and $a, b, \ldots$ denote, respectively, occupied and virtual orbitals. Auxiliary functions used for density fitting are denoted by the indices $P, Q, \ldots$. The Greek symbols $\mu, \nu, \ldots$ represent atomic orbital basis functions.

If not stated otherwise, the specified indices correspond to direct space local orbitals. Each index can be accompanied by a calligraphic font counterpart that defines the cell, the corresponding function is centered in. If no calligraphic font index is given for a direct space orbital, it is assumed to be centered in the reference cell. The symbolical operations $\oplus$ and $\ominus$ applied to the calligraphic cell indices indicate the actual operations over the corresponding translation vectors. Two-electron integrals are written in the chemical (Mulliken) notation.

The above listed indices can also refer to reciprocal space Fourier images of the local orbitals. In this case, they are accompanied by a **k**-vector from the Brillouin zone (BZ). Canonical-occupied and virtual orbitals are decorated with a bar. The same decoration is used for the virtual indices in the pseudocanonical basis that corresponds to a diagonalization of the Fock matrix in some local virtual subspace (e.g., within the pair domains). Since the pseudocanonical space, unlike the actual canonical space, is not translationally invariant, the pseudocanonical orbitals are not eigenfunctions of the translation operator, and thus do not have any **k**-index.

3.2.2
Periodic LMP2 Wave Function

The HF reference determinant is invariant with respect to rotations of the orbitals within the occupied space. Excited determinants and the associated excitation amplitudes, on the other hand, *do* depend on the choice of the orbitals. The use of a spatially localized orbital representation provides in this respect an important advantage that will be utilized throughout this work. In nonconducting periodic systems, localized orbitals possess exponential spatial decay [120–122]. Therefore, as will be discussed in detail in Section 3.2.5, the contribution of the doubles amplitudes to the correlation energy decays as the inverse sixth power of the distance between the centers of the two occupied orbitals in the pair, and exponentially with the distances between occupied and virtual orbital centers. Local approximations (vide infra) exploit this decay property and a priori neglect those amplitudes, whose contribution to the energy is expected to be inessential, as dictated by the spatial proximity of the respective orbitals.

We write the closed-shell periodic LMP2 first-order wave function as

$$\Psi^{[1]}_{\text{LMP2}} = \frac{1}{2} \sum_{i\mathcal{I}j\mathcal{J}} \sum_{a\mathcal{A}b\mathcal{B}} T^{i\mathcal{I}j\mathcal{J}}_{a\mathcal{A}b\mathcal{B}} \Psi^{a\mathcal{A}b\mathcal{B}}_{i\mathcal{I}j\mathcal{J}}$$
$$= \frac{1}{2} \sum_{ij\mathcal{J}} \sum_{a\mathcal{A}b\mathcal{B}} T^{ij\mathcal{J}}_{a\mathcal{A}b\mathcal{B}} \sum_{\mathcal{I}} \Psi^{a(\mathcal{A}\oplus\mathcal{I})b(\mathcal{B}\oplus\mathcal{I})}_{i\mathcal{I}j(\mathcal{J}\oplus\mathcal{I})}, \qquad (3.1)$$

where the $\Psi^{a\mathcal{A}b\mathcal{B}}_{i\mathcal{I}j\mathcal{J}}$ are spin-free configurations [9,123,124], i.e.,

$$\Psi^{a\mathcal{A}b\mathcal{B}}_{i\mathcal{I}j\mathcal{J}} = \sum_{\sigma} \hat{a}^{\dagger}_{a\mathcal{A},\sigma} \hat{a}_{i\mathcal{I},\sigma} \sum_{\sigma'} \hat{a}^{\dagger}_{b\mathcal{B},\sigma'} \hat{a}_{j\mathcal{J},\sigma'} \Psi_0. \qquad (3.2)$$

Ψ_0 is the periodic HF reference determinant and $\hat{a}^{\dagger}$ and $\hat{a}$ are the elementary creation and annihilation operators, respectively, with σ denoting the spin. The orbital indices here correspond to direct space orbitals, which will be specified in Section 3.2.4.

Since the ground state is totally symmetric, the excitation amplitudes T are invariant with respect to translations: $T^{i\mathcal{I}j\mathcal{J}}_{a\mathcal{A}b\mathcal{B}} = T^{ij(\mathcal{J}\ominus\mathcal{I})}_{a(\mathcal{A}\ominus\mathcal{I})b(\mathcal{B}\ominus\mathcal{I})}$. This means that for the translational symmetry unique amplitudes, one of the indices (e.g., i) can be restricted to the zeroth unit cell, as in Eq. (3.1).

The summations in Eq. (3.1) run formally to infinity; no local approximations have yet been introduced to truncate this expansion. Hence, Eq. (3.1) represents the complete MP2 first-order wave function. In order to truncate the summations and to arrive at a practical expansion for the first-order wave function, local approximations are invoked. This will be discussed in detail in Section 3.2.5.

3.2.3
Periodic LMP2 Energy

The LMP2 energy within the local direct space can be defined by utilizing the molecular formalism developed in Ref. [3]. However, its generalization to the periodic case is not without complications due to the formal infiniteness of the system. In order to properly treat infinite quantities (e.g., energy in an infinite system), we consider an infinite model of a crystal as a cyclic cluster, that is, as a cluster with periodic boundary conditions, in the limit of an infinite number of unit cells.

We start with the local MP2 Hylleraas functional E_2 that contains the amplitudes of the wave function of Eq. (3.1) as the variational parameters [124],

$$E_2 = \langle \Psi^{[1]} | \hat{f} - E_0 | \Psi^{[1]} \rangle + 2 \langle \Psi^{[1]} | \hat{H} | \Psi_0 \rangle, \qquad (3.3)$$

where $\hat{f}$ is the Fock operator. At this point, the zero-order energy E_0 as well as E_2, correspond to the full crystal (represented by a cyclic cluster),

$$E_0 = \sum_{i\mathcal{I}} f_{i\mathcal{I},i\mathcal{I}} = \sum_i f_{ii} \sum_{\mathcal{I}} 1, \qquad (3.4)$$

where $\mathbf{f}$ is the Fock matrix, that is, the representation of $\hat{f}$ in the orbital basis.

Following Refs [123,124], we introduce here the contravariant configurations $\tilde{\Psi}^{aAbB}_{i\mathcal{I}j\mathcal{J}}$ that are specified by the conditions

$$\langle \tilde{\Psi}^{aAbB}_{i\mathcal{I}j\mathcal{J}} | \Psi^{a'A'b'B'}_{i'\mathcal{I}'j'\mathcal{J}'} \rangle = \delta_{ii'}\delta_{jj'}\delta_{\mathcal{I}\mathcal{I}'}\delta_{\mathcal{J}\mathcal{J}'} S^{\text{PAO}}_{aAa'A'} S^{\text{PAO}}_{bBb'B'} + \\ \delta_{ij'}\delta_{ji'}\delta_{\mathcal{I}\mathcal{J}'}\delta_{\mathcal{J}\mathcal{I}'} S^{\text{PAO}}_{aAb'B'} S^{\text{PAO}}_{bBa'A'}, \qquad (3.5)$$

with S^{PAO} being the overlap between non-orthogonal virtual functions, e.g. projected atomic orbitals (PAOs), see Section 3.2.4. The contravariant configurations $\tilde{\Psi}^{aAbB}_{i\mathcal{I}j\mathcal{J}}$ can be expressed in terms of the original covariant configurations as

$$\tilde{\Psi}^{aAbB}_{i\mathcal{I}j\mathcal{J}} = \frac{1}{6}\left(2\Psi^{aAbB}_{i\mathcal{I}j\mathcal{J}} - \Psi^{bBaA}_{i\mathcal{I}j\mathcal{J}}\right). \qquad (3.6)$$

Furthermore, the first-order wave function can also be expanded in the space of the contravariant configurations as

$$\Psi^{[1]}_{\text{LMP2}} = \sum_{ij\mathcal{J}} \sum_{aAbB} \tilde{T}^{ij\mathcal{J}}_{aAbB} \sum_{\mathcal{I}} \tilde{\Psi}^{a(A\oplus\mathcal{I})b(B\oplus\mathcal{I})}_{i\mathcal{I}j(\mathcal{J}\oplus\mathcal{I})}, \qquad (3.7)$$

yielding

$$\tilde{T}^{ijJ}_{aAbB} = 2T^{ijJ}_{aAbB} - T^{ijJ}_{bBaA}. \tag{3.8}$$

Now, representing the bra first-order wave function in the Hylleraas functional (3.3) via the contravariant configurations, one obtains

$$\begin{aligned} E_2 &= \sum_{i\mathcal{I}j\mathcal{J}} \sum_{a\mathcal{A}b\mathcal{B}} \tilde{T}^{i\mathcal{I}j\mathcal{J}}_{a\mathcal{A}b\mathcal{B}} \left(K^{a\mathcal{A}b\mathcal{B}}_{i\mathcal{I}j\mathcal{J}} + R^{a\mathcal{A}b\mathcal{B}}_{i\mathcal{I}j\mathcal{J}} \right) \\ &= \left\{ \sum_{ij\mathcal{J}} \sum_{a\mathcal{A}b\mathcal{B}} \tilde{T}^{ij\mathcal{J}}_{a\mathcal{A}b\mathcal{B}} \left(K^{a\mathcal{A}b\mathcal{B}}_{ij\mathcal{J}} + R^{a\mathcal{A}b\mathcal{B}}_{ij\mathcal{J}} \right) \right\} \sum_{\mathcal{I}} 1, \end{aligned} \tag{3.9}$$

where the $K^{a\mathcal{A}b\mathcal{B}}_{i\mathcal{I}j\mathcal{J}} = (i\mathcal{I}a\mathcal{A}|j\mathcal{J}b\mathcal{B})$ are two-electron Coulomb integrals,

$$K^{a\mathcal{A}b\mathcal{B}}_{i\mathcal{I}j\mathcal{J}} = \int d\mathbf{r}_1 \int d\mathbf{r}_2 \, \phi^*_{i\mathcal{I}}(\mathbf{r}_1) \phi_{a\mathcal{A}}(\mathbf{r}_1) \frac{1}{|\mathbf{r}_1 - \mathbf{r}_2|} \phi^*_{j\mathcal{J}}(\mathbf{r}_2) \phi_{b\mathcal{B}}(\mathbf{r}_2), \tag{3.10}$$

and $R^{a\mathcal{A}b\mathcal{B}}_{i\mathcal{I}j\mathcal{J}}$ represents the MP2 residual,

$$\begin{aligned} R^{a\mathcal{A}b\mathcal{B}}_{i\mathcal{I}j\mathcal{J}} = &(i\mathcal{I}a\mathcal{A}|j\mathcal{J}b\mathcal{B}) + \\ &\sum_{a'\mathcal{A}'b'\mathcal{B}'} \left(S^{\text{PAO}}_{a\mathcal{A}a'\mathcal{A}'} T^{i\mathcal{I}j\mathcal{J}}_{a'\mathcal{A}'b'\mathcal{B}'} f^{\text{PAO}}_{b'\mathcal{B}'b\mathcal{B}} \right. \\ &\left. + f^{\text{PAO}}_{a\mathcal{A}a'\mathcal{A}'} T^{i\mathcal{I}j\mathcal{J}}_{a'\mathcal{A}'b'\mathcal{B}'} S^{\text{PAO}}_{b'\mathcal{B}'b\mathcal{B}} \right) \\ &- \sum_{a'\mathcal{A}'b'\mathcal{B}'} S^{\text{PAO}}_{a\mathcal{A}a'\mathcal{A}'} S^{\text{PAO}}_{b'\mathcal{B}'b\mathcal{B}} \sum_{k\mathcal{K}} \left(f_{i\mathcal{I}k\mathcal{K}} T^{k\mathcal{K}j\mathcal{J}}_{a'\mathcal{A}'b'\mathcal{B}'} \right. \\ &\left. + T^{i\mathcal{I}k\mathcal{K}}_{a'\mathcal{A}'b'\mathcal{B}'} f_{k\mathcal{K}j\mathcal{J}} \right). \end{aligned} \tag{3.11}$$

Due to time reversal symmetry, the orbitals in the direct space can be chosen as real, which leads to the additional permutation symmetries in the four-index integrals:

$$(i\mathcal{I}a\mathcal{A}|j\mathcal{J}b\mathcal{B}) = (a\mathcal{A}i\mathcal{I}|j\mathcal{J}b\mathcal{B}) = (a\mathcal{A}i\mathcal{I}|b\mathcal{B}j\mathcal{J}) = (i\mathcal{I}a\mathcal{A}|b\mathcal{B}j\mathcal{J}). \tag{3.12}$$

In order to arrive at Eq. (3.9), we have written the operators in Eq. (3.3) in second quantization and utilized the overlap between the configurations according to Eq. (3.5). For the last equality in Eq. (3.9), the translational invariance of the quantities involved was exploited.

As is seen from Eq. (3.9), the value of the Hylleraas functional in a periodic system is not a well-defined quantity. For a cyclic cluster model of a crystal, it is proportional to the number of unit cells in the supercell, and in the limit of an infinite crystal it becomes infinite. At the same time, the expression within curly brackets is well defined and represents the value of the Hylleraas functional *per unit cell*. Such a definition of size-extensive quantities (e.g., total energies) is commonly employed in periodic studies. In the following, we imply this definition for the energies.

Differentiation of the Hylleraas functional (quadratic in the amplitudes) with respect to a contravariant amplitude $\tilde{T}^{i\mathcal{I}j\mathcal{J}}_{a\mathcal{A}b\mathcal{B}}$ yields the residual $R^{a\mathcal{A}b\mathcal{B}}_{i\mathcal{I}j\mathcal{J}}$ (times a factor of 2) [3]. Therefore, the minimum of the Hylleraas functional, which is equal to the MP2 energy $E^{[2]}$, is obtained with the amplitudes satisfying the system of linear equations

$$R_{iIjJ}^{aAbB} = 0, \tag{3.13}$$

that is,

$$E^{[2]} = \sum_{ij\mathcal{J}} \sum_{a\mathcal{A}b\mathcal{B}} K_{ij\mathcal{J}}^{a\mathcal{A}b\mathcal{B}} \tilde{T}_{a\mathcal{A}b\mathcal{B}}^{ij\mathcal{J}} \bigg|_{R_{ij\mathcal{J}}^{a\mathcal{A}b\mathcal{B}}=0}. \tag{3.14}$$

The Eqs. (3.13) and (3.14) are orbital invariant, that is, any orbital representation in direct space leads to the same energy, as long as the summations are not truncated. We note at this point, that canonical crystalline orbitals live in the reciprocal space that implies Fourier transformations of the equations. In the representation of canonical crystalline orbitals, the Fock matrix in the transformed Eq. (3.13) is diagonal, allowing for its trivial inversion. In a representation of direct space orbitals, on the other hand, Eq. (3.13) has to be solved explicitly with the amplitudes representing the unknowns. Despite this complication, the direct space orbital representation in terms of localized orbitals pays off in extended systems, where the spatial sparsity of the involved objects can be taken into account by effective prescreening or/and introducing a priori local approximations. The introduction of local approximations in the periodic context is discussed in Section 3.2.5.

3.2.4
Choice of Direct Space Orbitals

Now we are in a position to specify the direct space orbital basis for the occupied and virtual spaces. In the periodic context, localized occupied orbitals are referred to as Wannier functions (WFs). They were introduced as early as 1937 [125], but for decades remained a formal construct rather than practical tool, due to the unavailability of robust methods for their generation. The standard molecular techniques for constructing localized orbitals [126–128] are not straightforwardly generalizable to the periodic case. In the last two decades, however, several universal methods for generating Wannier functions have been developed [129–133]. In this chapter, we utilize the method of Zicovich-Wilson et al. as implemented in the CRYSTAL code [133].

General localization procedures and, in particular, the one of Ref. [133] do not guarantee that the resulting WFs are symmetric in the sense that they transform according to irreducible representations of the site groups of their centering points [131]. However, such a property is convenient in calculations, since it allows for the exploitation of point-group symmetry that yields substantial computational savings. To benefit from the point-group symmetry in the LMP2 calculations (see Refs 88,134,135), we employ a scheme of Ref. [136], where the WFs are symmetrized *a posteriori*, that is, after the localization procedure.

Orthogonal local virtual orbitals can, in principle, also be obtained by means of a WF generation procedure. However, due to the much larger number of virtual functions as needed for correlated calculations, such a procedure would be computationally rather expensive, and the "virtual WFs" themselves possess quite extended oscillatory orthogonality tails. In this chapter, we employ a nonorthogonal basis of PAOs to span the virtual space instead. Such functions were originally introduced by Pulay and Saebo for the molecular local correlation scheme, and extensively used thereafter in that context. In the periodic case, they are obtained by projecting the occupied manifold out of the direct space atomic orbitals. The reciprocal space offers here the important advantage of a block-diagonal form of the projector due to the $\mathbf{k}$-symmetry. Therefore, instead of a direct construction of the PAOs $\phi_{a\mathcal{A}}(\mathbf{r})$, their Fourier images $\phi_{a,\mathbf{k}}(\mathbf{r})$ are first obtained,

$$|\phi_{a,\mathbf{k}}\rangle = \left(1 - \hat{P}(\mathbf{k})\right)|\phi_{\mu,\mathbf{k}}\rangle\big|_{a=\mu}. \tag{3.15}$$

Here, $\hat{P}(\mathbf{k})$ is the projection operator onto the part of the occupied space, corresponding to $\mathbf{k}$,

$$\hat{P}(\mathbf{k}) = \sum_{\tilde{i}} \left|\phi_{\tilde{i},\mathbf{k}}\right\rangle \left\langle\phi_{\tilde{i},\mathbf{k}}\right|, \tag{3.16}$$

and $\phi_{\mu,k}(\mathbf{r})$ are the Bloch sums of the corresponding atomic orbitals (AOs),

$$|\phi_{\mu,\mathbf{k}}\rangle = \sum_{\mathcal{M}} \exp(i\mathbf{k}\mathbf{R}_{\mathcal{M}})|\phi_{\mu\mathcal{M}}\rangle. \tag{3.17}$$

The actual direct space PAOs are obtained by the back Fourier transform as

$$|\phi_{a\mathcal{A}}\rangle = \frac{1}{N_{\mathbf{k}}} \sum_{\mathbf{k}} \exp(-i\mathbf{k}\mathbf{R}_{\mathcal{A}})|\phi_{a,\mathbf{k}}\rangle, \tag{3.18}$$

where the sum goes over the $N_{\mathbf{k}}$ k-points sampling the BZ. All the PAOs can be represented by the translationally invariant set of the coefficients $C^{\text{PAO}}_{\mu\mathcal{M},a}$, corresponding to their AO expansions,

$$|\phi_{a\mathcal{A}}\rangle = \sum_{\mu\mathcal{M}} C^{\text{PAO}}_{\mu\mathcal{M},a\mathcal{A}}|\phi_{\mu\mathcal{M}}\rangle = \sum_{\mu\mathcal{M}} C^{\text{PAO}}_{\mu\mathcal{M}\ominus\mathcal{A},a}|\phi_{\mu\mathcal{M}}\rangle. \tag{3.19}$$

The PAO coefficients are obtained as the back Fourier transform of their reciprocal space images $C^{\text{PAO}}_{\mu,a}(\mathbf{k})$,

$$C^{\text{PAO}}_{\mu\mathcal{M},a} = \frac{1}{N_{\mathbf{k}}} \sum_{\mathbf{k}} \exp(i\mathbf{k}\mathbf{R}_{\mathcal{M}}) C^{\text{PAO}}_{\mu,a}(\mathbf{k}). \tag{3.20}$$

The latter are obtained by the projection (3.15), that is,

$$C^{\text{PAO}}_{\nu,a}(\mathbf{k}) = \left(\delta_{\nu,\mu} - \sum_{\tilde{i}} C^{\text{CAN}}_{\nu,\tilde{i}}(\mathbf{k}) \sum_{\nu'} C^{\text{CAN}*}_{\nu',\tilde{i}}(\mathbf{k}) S^{\text{AO}}_{\nu'\mu}(\mathbf{k})\right)\bigg|_{a=\mu}, \tag{3.21}$$

where the $\mathbf{C}^{\text{CAN}}(\mathbf{k})$ are the coefficients of the canonical orbital expansion in the AO Bloch sums, and $S^{\text{AO}}(\mathbf{k})$ are the reciprocal space images of the AO overlap matrix,

$$S^{AO}_{\mu\nu}(\mathbf{k}) = \sum_{\mathcal{N}} exp(i\mathbf{k}\mathbf{R}_{\mathcal{N}}) S^{AO}_{\mu\nu\mathcal{N}}. \qquad (3.22)$$

PAOs have several convenient properties that are exploited in the corresponding formulation of the periodic local LMP2 method. These are (i) strong orthogonality (to the occupied space), (ii) exponential localization, (iii) unambiguous assignment of each PAO to a certain atom, and (iv) each PAO possessing the same symmetry as the parent AO (due to the fact that the projection operator is totally symmetric). On the other hand, the mutual nonorthogonality of the PAOs (which form actually a redundant basis for the virtual space) causes certain (yet not fundamental) complications. These are addressed further in more detail.

An alternative choice of direct space virtual orbitals are the so-called orbital specific virtuals (OSVs). Originally, OSVs have been introduced in the framework of a tensor decomposition of the doubles amplitudes [137], yet they can equivalently be regarded as the pair natural orbitals (PNOs) of diagonal pairs ii of, in the present context, Wannier functions ϕ_i belonging to the reference unit cell, that is, as the orbitals that diagonalize the external part of the MP2 pair density matrices $\mathbf{D}^{ii}$,

$$D^{ii}_{\bar{a}\bar{b}} = \sum_{\bar{c}} T^{ii}_{\bar{a}\bar{c}} T^{ii}_{\bar{b}\bar{c}}, \qquad (3.23)$$

related to diagonal pairs ii. In above equation, the $T^{ii}_{\bar{a}\bar{c}}$ are the MP2 doubles amplitudes in the basis of the pseudocanonical direct space orbitals $\phi_{\bar{a}}$, $\phi_{\bar{b}}$, and $\phi_{\bar{c}}$ (cf. Section 3.2.1) which diagonalize the Fock matrix projected onto the ii-specific local subspace $[ii]$ (domain, cf. Section 3.2.5), that is, the pair-specific Fock matrix $\mathbf{f}^{[ii]}$. In order to construct decent OSVs, this local subspace as spanned by pseudocanonical orbitals should be quite large, which fortunately does not constitute a severe computational burden [138].

Since the amplitude matrix for a diagonal pair is symmetric, the eigenvectors of the density matrix $\mathbf{D}^{ii}$ are identical to those of the amplitudes $T^{ii}_{\bar{a}\bar{b}}$ themselves, hence the latter can be used for the OSV specification as well.

In order to generate a sensible set of OSVs for a subsequent local MP2 calculation, just the required *diagonal pair amplitudes* $\bar{\mathbf{T}}^{ii}$ are calculated at the *uncoupled* LMP2 level [18,33,139,140] (which corresponds to the zeroth LMP2 iteration) according to

$$\bar{T}^{ii}_{\bar{a}\bar{b}} = \frac{(i\bar{a}|i\bar{b})}{\varepsilon_{\bar{a}} + \varepsilon_{\bar{b}} - 2f_{ii}}. \qquad (3.24)$$

In Eq. (3.24), the $\varepsilon_{\bar{a}}$ and $\varepsilon_{\bar{b}}$ are the eigenvalues of the pair-specific Fock matrix $\mathbf{f}^{[ii]}$ in the pair-specific direct space virtual basis, or, in other words, the orbital energies of the respective pseudocanonical orbitals, and f_{ii} is a diagonal element of the internal Fock matrix in the basis of the Wannier functions.

The number of OSVs generated by diagonalizing matrix $\bar{\mathbf{T}}^{ii}$ is identical to the number of pseudocanonical orbitals and spans the same virtual space $[ii]$. In order to arrive at a compact description of the virtual space, the number of OSVs has to be drastically truncated such that an approximate subspace of $[ii]$ is spanned by the OSVs, which is sufficient to obtain the related pair energy with high precision. How exactly the OSV space is truncated is described in more detail in Section 3.2.5.2.

The AO coefficients $\mathbf{C}^{\text{OSV}}$ for OSV $\phi_{\tilde{a}^{[i]}}$ belonging to WF ϕ_i are given by the expression

$$C^{\text{OSV}}_{\mu\mathcal{M},\tilde{a}^{[i]}} = \sum_{a\mathcal{A} \in [ii]_{\text{PAO}}} C^{\text{PAO}}_{\mu\mathcal{M},a\mathcal{A}} \sum_{\bar{a}} W^{[ii]}_{a\mathcal{A},\bar{a}} Q^{ii}_{\bar{a},\tilde{a}^{[i]}}. \tag{3.25}$$

The PAO summation in Eq. (3.25) is restricted to the (large, see above) PAO domain $[ii]$ of the diagonal ii pair. The indices $\bar{a}$ refer to the pseudocanonical orbitals $\phi_{\bar{a}}$ corresponding to this domain, and $\mathbf{W}^{[ii]}$ is the transformation matrix transforming PAOs to pseudocanonical virtuals that is obtained in the course of the construction of the $\phi_{\bar{a}}$ ($\mathbf{W}^{[ii]}$ diagonalizes both the Fock and overlap matrices in the PAO basis projected onto the domain $[ii]$ [9,134,138]). Finally, $Q^{ii}_{\bar{a},\tilde{a}^{[i]}}$ is the eigenvector of the amplitude matrix $\overline{\mathbf{T}}^{ii}$ in the pseudocanonical basis, corresponding to the eigenvalue $\tilde{t}^{ii}_{\tilde{a}^{[i]}}$, that is, $\mathbf{Q}^{ii}$ is the transformation matrix transforming from pseudocanonical virtuals to OSVs.

3.2.5
Local Approximations

Translational symmetry greatly reduces the number of unknowns and thus the equations to be solved: For the symmetry unique amplitudes or residuals, one of the indices, that is, in our notation the index i, is restricted to the reference cell. However, the other three indices are still not restricted and run to infinity, and Eq. (3.13) corresponds to an infinite system of mutually coupled equations. Therefore, for solving these equations in the direct space, additional truncations become inevitable. The exponential localization of the orbitals offers a foundation for this truncation. Importantly, the off-diagonal elements of the PAO overlap matrix and of both the WF–WF and PAO–PAO blocks of the Fock matrix decay exponentially with the distance between the respective orbitals. Therefore, the amplitudes that are solutions of Eq. (3.13) have the same asymptotic decay behavior as the corresponding two-electron integrals on the right-hand side of Eq. (3.11).

3.2.5.1 Domain Approximation

The two-electron integrals $(i\mathcal{I}a\mathcal{A}|j\mathcal{J}b\mathcal{B})$ (and with these the amplitudes $T^{i\mathcal{I}j\mathcal{J}}_{a\mathcal{A}b\mathcal{B}}$) decay exponentially with the distance between the orbitals $i\mathcal{I}$ and $a\mathcal{A}$ or $j\mathcal{J}$ and $b\mathcal{B}$. The contravariant amplitudes $\tilde{T}^{i\mathcal{I}j\mathcal{J}}_{a\mathcal{A}b\mathcal{B}}$, which actually enter the energy expression, also include the amplitudes with interchanged virtual indices (3.7). With that, the decay of $\tilde{T}^{i\mathcal{I}j\mathcal{J}}_{a\mathcal{A}b\mathcal{B}}$ is also exponential, but now with respect to the smallest of the two distances between the orbital $i\mathcal{I}$ and either orbital $a\mathcal{A}$ or $b\mathcal{B}$ (and the same for the orbital $j\mathcal{J}$). Consequently, the amplitudes $\tilde{T}^{i\mathcal{I}j\mathcal{J}}_{a\mathcal{A}b\mathcal{B}}$ for a given pair $ij\mathcal{J}$ are large only if both the PAOs $a\mathcal{A}$ and $b\mathcal{B}$ are spatially close to at least one of the two Wannier functions in the pair. On the basis of that it appears that a contravariant amplitude $\tilde{T}^{iIjJ}_{aAbB}$ may still be sizeable, even though the corresponding amplitude T^{iIjJ}_{aAbB} itself is already small.

The local approximation for the virtual space, which has initially been proposed and used for molecular correlation methods [2], is based on the above analysis: The virtual orbitals for each pair of WFs are a priori restricted to its so-called pair domain; for a given $ij\mathcal{J}$ pair the $[ij\mathcal{J}]$ pair domain comprises only those PAOs that are centered in the spatial vicinity of either one of the two WFs of the pair. All amplitudes with at least one of the two virtuals being beyond the respective domain, are considered to be zero,

$$T^{ij\mathcal{J}}_{aAbB}\Big|_{(a\mathcal{A} \notin [ij\mathcal{J}]) \vee (b\mathcal{B} \notin [ij\mathcal{J}])} \equiv 0, \qquad (3.26)$$

and thus contribute neither to the energy (3.14) nor to the residual (3.11). This pair-specific truncation of the virtual space is the so-called *domain approximation*. A pair domain $[ij\mathcal{J}]$ is constructed as the *union* of the respective two orbital domains $[i]$ and $[j\mathcal{J}]$. The latter can be constructed either by the Boughton–Pulay procedure [6], which ranks the PAO centers on the basis of their Mulliken or Löwdin AO population of a given WF, or by a simple distance criterion, which includes all neighboring PAO atoms closer to the center of a given WF than a certain cutoff distance.

Due to the exponential decay of amplitudes and integrals with the WF–PAO distance the energy contribution in Eq. (3.14), neglected by the domain approximation, decays as the product of the respective exponentials, i.e. very quickly. Therefore, few atom domains are usually sufficient to capture 98–99% of the full MP2 correlation energy at a given basis set level [8]. Thus, this approximation not only conforms a formal infinite virtual space to a finite one, but also reduces it such that only its essential part remains. Compactness of the pair-specific virtual spaces is essential for high computational efficiency: Together with the *pair approximation* (cf. Section 3.2.5.3), the domain approximation leads to linear scaling of the computational cost with unit cell size with a small prefactor [111,141].

Regarding its effect on the correlation energy, the domain approximation is closely related to the incomplete basis set approximation and shares with the latter some important properties. In particular, since the MP2 energy is equal to the minimum of the Hylleraas functional, the LMP2 energy within a given virtual space truncation, whether coming from the basis set restriction or domain approximation, will be an upper bound to the complete basis set MP2 result. This means that the domain error as well as the basis set incompleteness error are systematic and cancel, at least to some extent, in energy differences. Furthermore, an improvement of the basis set reduces the effect of the domain approximation on the result [142]. Likewise, basis set acceleration techniques such as the explicit correlation F12 methods also effectively correct the domain error [143,144].

The domain approximation has another convenient feature: Since it spatially concentrates the virtual functions around respective occupied orbitals, the effect of beyond-domain changes in the geometry on the virtual space is minimal. Hence, even with small basis sets the basis set superposition error is substantially reduced by the local approximation [145]. At the same time, the choice of PAOs to sample the virtual space also has – in the context of the local approximation – a drawback, although in most cases a rather technical one: Because *all* the PAOs belonging to a given atom form an irreducible subspace to be either entirely included in or excluded from a domain, the pair specific virtual space by modification of domains does not change smoothly. Therefore, in energy difference calculations, a balanced description is achieved if both conformations possess the same domain structure. Similarly, to avoid kinks on potential surfaces the domains should be kept frozen throughout potential surface scans.

3.2.5.2 Truncated OSV Approximation

Note that individual PAOs remain identical even if they occur in different pair domains; the pair specificity of individual virtual spaces is just given by *different subsets* of overall the same PAOs. This is an advantage and a disadvantage at the same time. The advantage is that the integral transformations are easy and efficient since the number of different PAOs is equivalent to the number of basis functions and does not depend on individual orbitals or pairs. The disadvantage is that PAOs are actually not the optimal choice, as truly orbital or pair-specific functions like above-mentioned OSVs or PNOs might form an even more compact set to span a certain pair-specific virtual space than PAOs. In order to achieve high accuracy, especially in 3D densely packed systems, PAO pair domains may still comprise a relatively large number of functions (up to a few hundred orbitals per pair) [146] that in turn leads to excessively large doubles amplitude sets. Another disadvantage originates from above-mentioned non-smooth changes of domains when, for example, the geometry is altered. This may lead to discontinuities in potential energy surfaces and to nonphysical artifacts in energy differences (if domains are allowed to respond upon changes in the geometry). OSVs, on the other hand, allow for much smoother changes in pair domains and are not seen to cause any discontinuities in potential energy surfaces [138,147]. Moreover, substantially less OSVs than PAOs per pair are required to reach the same accuracy.

The importance of an individual OSV $\phi_{\tilde{a}^{[i]}}$ (related to Wannier function ϕ_i) is characterized by its contribution

$$e^{ii}_{\tilde{a}^{[i]}} = \tilde{t}^{ii}_{\tilde{a}^{[i]}} (i\tilde{a}^{[i]}|i\tilde{a}^{[i]}) \tag{3.27}$$

to the total pair energy

$$e^{ii} = \sum_{\tilde{a}^{[i]}} e^{ii}_{\tilde{a}^{[i]}}. \tag{3.28}$$

In Eq. (3.27) $\tilde{t}^{ii}_{\tilde{a}^{[i]}}$ is the eigenvalue of the amplitude matrix $\overline{\mathbf{T}}^{ii}$ corresponding to OSV $\phi_{\tilde{a}^{[i]}}$ (cf. Section 3.2.4). In order to span an appropriate approximate subspace of $[ii]$ by a truncated compact set of OSVs, they are ordered according to decreasing $e^{ii}_{\tilde{a}^{[i]}}$. Once the difference between the exact pair energy e^{ii} and the sum over the $e^{ii}_{\tilde{a}^{[i]}}$ falls below a certain threshold $\mathscr{E}_{\text{OSV}}$ all further OSVs are discarded.

As in the PAO case, the amplitudes related to a certain pair of WFs $ij\mathcal{J}$ are truncated according to its pair-domain $[ij\mathcal{J}]$ (cf. Eq. (3.26)). Here, $[ij\mathcal{J}]$ is simply the union of the two sets of OSVs related to WFs ϕ_i and $\phi_{j\mathcal{J}}$; for the AO coefficients of the latter we have

$$C^{\text{OSV}}_{\mu\mathcal{M},\tilde{a}^{[i\mathcal{I}]}} = C^{\text{OSV}}_{\mu\mathcal{M}\ominus\mathcal{I},\tilde{a}^{[i]}}. \tag{3.29}$$

The OSVs naturally allow for a very efficient truncation of the virtual space for diagonal pairs (they are the related natural orbitals). However, this does not imply that they also form an optimal basis for the off-diagonal pairs. Actually, for an accurate description of van der Waals dispersion (which is a form of long-range electron correlation), diffuse high-angular momentum orbitals are required for a proper description of the virtual space [95,148]. On the other hand, such diffuse orbitals are not as essential for the description of short-range correlation close to the cusp of the Coulomb hole. Consequently, those OSVs needed to describe van der Waals correlation contribute very little to the diagonal pair energies and are therefore, usually dropped in the OSV truncation scheme as outlined above, unless an extremely tight $\mathscr{E}_{\text{OSV}}$ is employed. The conclusion of this is that an OSV basis as constructed above has to be augmented by additional diffuse functions. To this end, we add to each individual OSV set related to WF ϕ_i (constructed with a relatively loose $\mathscr{E}_{\text{OSV}}$) the few most diffuse PAOs centered on the atoms related to ϕ_i. It turns out that a single PAO shell per angular momentum for each atom within the minimal domain of ϕ_i (as determined by the BP procedure with a criterion of only 0.9) is sufficient to properly describe van der Waals correlation [138]. The number of PAOs in these minimal domains is rather modest; they contain either just a single atom (for lone-pair WFs) or two atoms (for bonding WFs) and are not altered by geometry changes unless the bonding pattern itself becomes completely different. Hence, augmenting the OSVs by these PAOs does not lead to discontinuities in the potential energy surface [138].

3.2.5.3 Pair Approximation

Despite a substantial reduction of the virtual space for each pair by virtue of the domain approximation (see above), the number of WF pairs in a periodic system remains infinite. Furthermore, the contributions of the long-range pair amplitudes to the LMP2 energy decay much slower with the distance between the two WFs than with the distance between a WF and the corresponding virtual orbital: Due to mutual orthogonality between occupied and virtual orbitals, the integrals $(ia\mathcal{A}|j\mathcal{J}b\mathcal{B})$ asymptotically fade off as R^{-3} with the distance between the WFs i and $j\mathcal{J}$, as dictated by the dipole–dipole interaction. The converged amplitudes follow the same decay pattern, which, according to Eq. (3.14), leads to the standard R^{-6} decay law for the pair energies. Scaling this with the R^2 growth of the number of pairs due to 3D packing slows down the decay of the correlation energy contributions to R^{-4} for a star of pairs with interorbital distance R. Integrating the latter from some (cutoff) distance R to infinity yields an even slower R^{-3} decay for the unaccounted correlation energy from the pairs beyond the cutoff. Due to this slow decay, the pair list in solids has to be extended quite far, as a substantial fraction of the correlation energy would otherwise be omitted. On the other hand, the couplings between pairs in the LMP2 equations via the off-diagonal occupied–occupied Fock matrix elements (fourth and fifth terms in Eq. (3.11)) decay exponentially, implying that the truncation of the pair list has a much weaker effect on individual pair residuals.

For WF pairs $ij\mathcal{J}$ with mutually nonpenetrating WFs i and $j\mathcal{J}$ (denoted as *distant pairs*), the Coulomb integrals $(ia\mathcal{A}|j\mathcal{J}b\mathcal{B})$ can be approximated by the multipole expansion [149] involving the multipole moments of the densities $\rho_{ia\mathcal{A}}(\mathbf{r}) = \phi_i^*(\mathbf{r})\phi_{a\mathcal{A}}(\mathbf{r})$ and $\rho_{j\mathcal{J}b\mathcal{B}}(\mathbf{r})$ (cf. Section 3.2.8). The leading term of this expansion dictates the decay law for the pair energies,

$$E^{[2]}_{ij\mathcal{J}} \approx C^{ij}_6 d^{-6}_{ij\mathcal{J}} \bigg|_{d_{ij\mathcal{J}} > d_{\text{LR}}}, \tag{3.30}$$

for the interorbital distance $d_{ij\mathcal{J}}$ exceeding a certain "long-range" cutoff distance d_{LR}. The coefficients C^{ij}_6 depend on the pair type and express the strength of the dispersion interaction between the electrons in orbitals i and $j\mathcal{J}$, averaged with respect to the mutual orientations of these orbitals. This can be exploited to truncate the pair list more aggressively than dictated by the saturation of the actual long-range contributions, since the energies of the pairs beyond d_{LR} (which we denote as *very distant pairs*) can be added *a posteriori* using Eq. (3.30). The requirement for d_{LR} is such, that the pairs of the intermediate region with $d_{\text{IR}} < d_{ij\mathcal{J}} < d_{\text{LR}}$ (or distant pairs), which are included in the actual LMP2 calculation, should be sufficient to provide an accurate fit for the corresponding C^{ij}_6 coefficients.

After this analysis we are now in a position to specify our pair approximation that includes the two predefined distance tolerances d_{IR} and d_{LR} mentioned above:

1) The long-range cutoff distance d_{LR} defines the truncation of the pair list. Amplitudes beyond the pair list, that is, for very distant pairs, are omitted,

$$T^{ij\mathcal{J}}_{a\mathcal{A}b\mathcal{B}}\bigg|_{d_{ij\mathcal{J}} > d_{\text{LR}}} \equiv 0. \tag{3.31}$$

2) The intermediate-range tolerance d_{IR} is introduced according to the requirement that the mutual penetration of the two WF–PAO product densities $\rho_{ia\mathcal{A}}$ and $\rho_{b\mathcal{B}j\mathcal{J}}$ beyond this distance is small enough to warrant a low-order multipole approximation for the respective two-electron integrals to be accurate [149]. This allows one to neglect in such pairs the integrals $(ia\mathcal{A}|j\mathcal{J}b\mathcal{B})$, in which the PAO $a\mathcal{A}$ belongs to the domain of WF $j\mathcal{J}$ or $b\mathcal{B}$ to the domain of i, and to apply the multipolar approximation

for the remaining integrals, that is, for $(iaA|jJbB)|_{(aA \in [i]) \wedge (bB \in [jJ])}$. This is described in detail in Section 3.2.8.

3) The two-electron integrals for short-range pairs with $d_{ijJ} < d_{IR}$ are calculated for all virtual orbitals in the corresponding pair domain by using the density fitting approximation (see Section 3.2.6). The pairs of this range are usually denoted as "strong/weak" pairs. The difference between the strong and weak pairs becomes relevant only at higher order local treatments like LCCSD [9,12], but we keep this notation here for future reference.

4) The LMP2 energy is defined as

$$E^{[2]} = \sum_{ijJ} E^{[2]}_{ijJ} = \sum_{ijJ}^{\{d_{ijJ} < d_{LR}\}} \sum_{aAbB \in [ijJ]} K^{aAbB}_{ijJ} \tilde{T}^{ijJ}_{aAbB}\bigg|_{R^{aAbB}_{ijJ}=0} + \sum_{ijJ}^{\{d_{ijJ} > d_{LR}\}} C^{ij}_6 d^{-6}_{ijJ}. \quad (3.32)$$

We note that in molecular local correlation calculations the contribution from the very distant pairs (the last term in Eq. (3.32)) is commonly neglected [8].

5) The C^{ij}_6 coefficients are calculated by fitting the actual pair energies $E^{[2]}_{ijJ}$ = of the intermediate range, $d_{IR} < d_{ijJ} < d_{LR}$ to Eq. (3.30).

To summarize, the pair approximation together with the domain approximation reduces a formally infinite periodic problem to a computationally tractable finite one. Moreover, these approximations allow for an efficient computational treatment without compromising the accuracy of the calculated energy differences. The effect of the approximations on the results can be systematically investigated and minimized. In nonconducting systems, the overlap between the WF–PAO product densities drops off exponentially with the distance between the centers of the WFs. Hence, the first tolerance d_{IR}, beyond which the inexpensive multipole approximation for the integral can be activated, does not need to be very large. The second tolerance d_{LR} has to be chosen such that at this interorbital separation the pair energies exhibit a stable $C^{ij}_6 d^{-6}_{ijJ}$ pattern.

3.2.6
Periodic Density Fitting

Efficient evaluation of the Coulomb integrals $(iaA|jJbB)$ for the strong/weak pairs is crucial for the practicability of the periodic LMP2 method. The initial implementation, based on the calculation of such integrals through transformation of four-index AO integrals [88], turned out to be very inefficient despite the efforts of batching the matrix multiplications and prescreening insignificant contributions. The breakthrough in the efficiency of periodic LMP2 calculations was achieved by employing the density fitting approximation for these integrals. Density fitting is very widely used in molecular electronic structure methods [150–158]. However, the adaptation of the density fitting approach to the periodic case [159–165] is not straightforward. In this section, we present the local density fitting approach for periodic systems that is used in the periodic LMP2 [166], LMP2-F12 [135], and CIS [167] methods.

Consider, for example, the four-index two-electron integral:

$$(ia\mathcal{A}|\hat{I}_{12}|j\mathcal{J}b\mathcal{B}) = \int d\mathbf{r}_1 \int d\mathbf{r}_2 \phi_i(\mathbf{r}_1)\phi_{a\mathcal{A}}(\mathbf{r}_1)\hat{I}(\mathbf{r}_1,\mathbf{r}_2)\phi_{j\mathcal{J}}(\mathbf{r}_2)\phi_{b\mathcal{B}}(\mathbf{r}_2), \quad (3.33)$$

with the multiplicative Coulomb kernel $\hat{I}_{12} = \hat{I}(\mathbf{r}_1,\mathbf{r}_2) = |\mathbf{r}_1 - \mathbf{r}_2|^{-1}$. Time reversal symmetry ensures that the direct space orbitals in Eq. (3.33) can be chosen as real-valued. In order to fit this integral, we introduce a direct space one-electron auxiliary basis, spanned by the orbitals $\phi_{\mathcal{PP}}$ with metric

$$I^{\text{met}}_{\mathcal{PP},\mathcal{QQ}} = \left(\mathcal{PP}\middle|\hat{I}^{\text{met}}_{12}\middle|\mathcal{QQ}\right) = \int d\mathbf{r}_1 \int d\mathbf{r}_2 \phi_{\mathcal{PP}}(\mathbf{r}_1)\hat{I}^{\text{met}}(\mathbf{r}_1,\mathbf{r}_2)\phi_{\mathcal{QQ}}(\mathbf{r}_2), \quad (3.34)$$

where $\hat{I}^{\text{met}}_{12}$ may be in general a different kernel than $\hat{I}_{12}$.

The density fitting approximation is usually considered as a fit of the densities involved in the two-electron integrals in the space of the fitting functions $\phi_{\mathcal{PP}}$ that is optimal with respect to a specially chosen fitting functional [15]. Yet in this work we employ a formally different approach that is based on the projector operator technique. Nevertheless, the very same working equations are obtained. We define the operator

$$\hat{P}_{\text{DF}} = \sum_{\mathcal{PP},\mathcal{QQ}} |\mathcal{PP}) \left[I^{\text{met}\,-1}\right]_{\mathcal{PP},\mathcal{QQ}} (\mathcal{QQ}|\hat{I}^{\text{met}}_{12}| \quad (3.35)$$

that projects a one-electron function onto the space spanned by the fitting functions $\phi_{\mathcal{PP}}$ with the metric $\hat{I}^{\text{met}}_{12}$. To approximate the integral in Eq. (3.33,) the projector (3.35) can be applied to the density $\rho_{ia\mathcal{A}}$ or $\rho_{j\mathcal{J}b\mathcal{B}}$, or both. As was proposed originally by Dunlap [168,169], the so-called "robust" density fitting, where the error in the fitted integral is second order with respect to the individual errors in the projected densities, is achieved by the following approximation of the integral,

$$(ia\mathcal{A}|\hat{I}_{12}|j\mathcal{J}b\mathcal{B}) \approx \left(ia\mathcal{A}\hat{P}^{\dagger}_{\text{DF}}\middle|\hat{I}_{12}\middle|j\mathcal{J}b\mathcal{B}\right) + \left(ia\mathcal{A}\middle|\hat{I}_{12}\middle|\hat{P}_{\text{DF}}j\mathcal{J}b\mathcal{B}\right)$$
$$- \left(ia\mathcal{A}\hat{P}^{\dagger}_{\text{DF}}\middle|\hat{I}_{12}\middle|\hat{P}_{\text{DF}}j\mathcal{J}b\mathcal{B}\right) = \sum_{\mathcal{PP}} d^{ia\mathcal{A}}_{\mathcal{PP}} (\mathcal{PP}|\hat{I}_{12}|j\mathcal{J}b\mathcal{B})$$
$$+ \sum_{\mathcal{PP}} (ia\mathcal{A}|\hat{I}_{12}|\mathcal{PP}) d^{j\mathcal{J}b\mathcal{B}}_{\mathcal{PP}} - \sum_{\mathcal{PP},\mathcal{QQ}} d^{ia\mathcal{A}}_{\mathcal{PP}} (\mathcal{PP}|\hat{I}_{12}|\mathcal{QQ}) d^{j\mathcal{J}b\mathcal{B}}_{\mathcal{QQ}}, \quad (3.36)$$

where the (three-index) supermatrix denoted by **d** collects the so-called density fitting coefficients,

$$d^{ia\mathcal{A}}_{\mathcal{PP}} = \sum_{\mathcal{QQ}} \left(ia\mathcal{A}\middle|\hat{I}^{\text{met}}_{12}\middle|\mathcal{QQ}\right) \left[I^{\text{met}-1}\right]_{\mathcal{QQ},\mathcal{PP}}. \quad (3.37)$$

If the metric is chosen such that it coincides with the integral kernel, that is, $\hat{I}^{\text{met}}_{12} = \hat{I}_{12}$, and if the same auxiliary basis is used in both projectors $\hat{P}^{\dagger}_{\text{DF}}$ and $\hat{P}_{\text{DF}}$

of Eq. (3.36), then the expression for the robust density fitting simplifies to

$$(iaA|\hat{I}_{12}|jJbB) \approx \sum_{PP} (iaA|\hat{I}_{12}|PP) d_{PP}^{jJbB}, \tag{3.38}$$

that is, all three terms in Eq. (3.36) are identical. The full direct space density fitting of Eqs. (3.36) or (3.38) is not possible in periodic systems, since the metric matrix defined in Eq. (3.34) is infinite and cannot be inverted in the direct space (Eq. (3.37)). Below we overcome this difficulty by restricting the auxiliary basis in the projectors to *local fit-domains*.

3.2.7
Direct Space Local Density Fitting

In our approach, the auxiliary basis set $\{\phi_{PP}\}$ primarily consists of so-called *Poisson-type orbitals* (PTOs), which possess some specific features that are very beneficial for the fitting of electron repulsion integrals [170–172], in particular in the context of periodic systems [161]. PTOs are like *Gaussian-type orbitals* (GTOs), centered on individual atoms and directly constructed from GTOs. As an important property, PTOs do not carry any moments [172]. In order to accurately fit electron repulsion integrals (which, on the other hand, involve charge densities with dipole and higher moments), the PTO auxiliary basis is commonly augmented by a small set of GTOs (see below).

Consider a set of GTOs $\{g_{QQ}(\mathbf{r})\}$. The respective PTOs of the set $\{p_{PP}(\mathbf{r})\}$ are related to the GTOs as

$$p_{QQ}(\mathbf{r}) = -\frac{1}{4\pi} \nabla^2 g_{QQ}(\mathbf{r}). \tag{3.39}$$

Due to the identity (a form of the Poisson equation) [171]

$$\int d\mathbf{r}' \frac{\nabla^2 g(\mathbf{r}')}{|\mathbf{r} - \mathbf{r}'|} = -4\pi g(\mathbf{r}) \tag{3.40}$$

the bra-vectors of type $p_{QQ}(\mathbf{r})$ reduce to

$$(QQ|\hat{I}_{12}^{\text{met}}| = \int d\mathbf{r}' \frac{p_{QQ}}{|\mathbf{r} - \mathbf{r}'|} = g_{QQ}(\mathbf{r}), \tag{3.41}$$

where we have used specifically the Coulomb kernel for $\hat{I}_{12}^{\text{met}}$. This implies that a three-index Coulomb integral $(PP|iaA) = (iaA|PP)$ with a PTO as the fitting function reduces to an effective one-electron overlap integral involving the corresponding GTO fitting function instead. This simplification remarkably speeds up the evaluation of the three-index integrals needed for the density fitting procedure. Indeed, the effect of the PTO auxiliary basis is twofold, (i) the individual three-index one-electron integrals are much more efficiently computed, and (ii) a three-index overlap integral decays exponentially with distance between the product density and the fitting function rather than polynomically as a Coulomb integral. Consequently, the number of nonnegligible integrals is reduced dramatically.

This fast decay of the electron repulsion integrals involving PTOs as fitting functions originates from above-mentioned fact that PTOs are momentless. Indeed, the Coulomb interaction between two nonoverlapping electron distributions, where one of them is momentless, is naturally zero. However, projecting a density distribution, which does possess multipole moments, onto a momentless basis, would obviously lead to large errors in the respective electron repulsion integrals. Therefore, an auxiliary basis consisting just of PTOs is not appropriate for accurate density fitting. However, augmenting the PTO basis by a few GTOs (one GTO per angular momentum and per center) already fixes this problem.

As already mentioned, the introduction of local fit-domains is a prerequisite for the calculation of the fitting coefficients in the direct space. Otherwise, without local truncation the number of fit coefficients would be infinite. In 3D solids, where a single universal fit-domain would be prohibitively large, it is convenient to employ density-specific fit-domains. This introduces an explicit dependence of the projectors on the density they act upon,

$$\hat{P}_{DF}^{[ia\mathcal{A}]} = \sum_{\mathcal{PP},\mathcal{QQ} \in [ia\mathcal{A}]} |\mathcal{PP}) \left[I_{[ia\mathcal{A}]}^{\text{met}-1} \right]_{\mathcal{PP},\mathcal{QQ}} (\mathcal{QQ}|\frac{1}{r_{12}}|, \qquad (3.42)$$

where $[ia\mathcal{A}]$ denotes the fit-domain related to the density $\rho_{ia\mathcal{A}}$, and $I_{[ia\mathcal{A}]}^{\text{met}}$ is a square block of the metric matrix (cf. Eq. (3.34)) corresponding to the fit-domain-restricted part of the auxiliary basis that is inverted individually. With that, the locally density-fitted electron repulsion integral (cf. Eq. (3.36)) takes the form

$$(ia\mathcal{A}|jb\mathcal{B}) = (ia\mathcal{A}\hat{P}_{DF}^{[ia\mathcal{A}]\dagger}|jb\mathcal{B}) + (ia\mathcal{A}|\hat{P}_{DF}^{[jb\mathcal{B}]}jb\mathcal{B}) - (ia\mathcal{A}\hat{P}_{DF}^{[ia\mathcal{A}]\dagger}|\hat{P}_{DF}^{[jb\mathcal{B}]}jb\mathcal{B})$$

$$= \sum_{\mathcal{PP} \in [ia\mathcal{A}]} \bar{d}_{\mathcal{PP}}^{ia\mathcal{A}} (\mathcal{PP}|jb\mathcal{B}) + \sum_{\mathcal{PP} \in [jb\mathcal{B}]} (ia\mathcal{A}|\mathcal{PP}) \bar{d}_{\mathcal{PP}}^{jb\mathcal{B}}$$

$$- \sum_{\substack{\mathcal{PP} \in [ia\mathcal{A}] \\ \mathcal{QQ} \in [jb\mathcal{B}]}} \bar{d}_{\mathcal{PP}}^{ia\mathcal{A}} (\mathcal{PP}|\mathcal{QQ}) \bar{d}_{\mathcal{QQ}}^{jb\mathcal{B}}, \qquad (3.43)$$

with

$$\bar{d}_{\mathcal{PP}}^{ia\mathcal{A}} = \sum_{\mathcal{QQ} \in [ia\mathcal{A}]} (ia\mathcal{A}|\mathcal{QQ}) \left[I_{[ia\mathcal{A}]}^{\text{met}-1} \right]_{\mathcal{PP},\mathcal{QQ}}. \qquad (3.44)$$

Since the fit-domains $[ia\mathcal{A}]$ and $[jb\mathcal{B}]$ are generally not identical, a simplification of Eq. (3.43) to a one-term DF as in Eq. (3.38) is not possible, and all three terms of Eq. (3.43) have to be calculated.

3.2.8
Multipolar Approximation

Consider the case where the two product densities $\rho_{ia\mathcal{A}}$ and $\rho_{jb\mathcal{B}}$ in the integral $K_{ij\mathcal{J}}^{a\mathcal{A}b\mathcal{B}} = (ia\mathcal{A}|jb\mathcal{B})$ have negligible mutual penetration, that is, where the vector associated to the $\mathcal{J}$ cell index is large enough such that ρ_{iaA} and ρ_{jJbB} can be considered as enclosed in separate spheres. In this case, their interaction can be efficiently described as the interaction between two sets of point multipoles,

$$K_{ij\mathcal{J}}^{aAbB} \approx \sum_{l,m,l',m'} Q_{iaA}^{l,m} V_{\mathbf{R}_{\mathcal{J}}}^{l,m,l',m'} Q_{j\mathcal{J}bB}^{l',m'}, \qquad (3.45)$$

where $Q_{iaA}^{l,m}$ and $Q_{j\mathcal{J}bB}^{l,m}$ are the multipolar moments of order l, m of ρ_{iaA} and ρ_{jJbB}, respectively. The latter are calculated from corresponding multipolar moments of the AO product densities $\rho_{\mu\mathcal{M}\nu\mathcal{N}}$ by transformation with the related AO coefficient matrices $\mathbf{C}$ (cf. Section 3.2.4) according to

$$Q_{iaA}^{l,m} = \sum_{\nu,\mathcal{N}} C_{\nu\mathcal{N}aA} \sum_{\mu,\mathcal{M}} C_{\mu\mathcal{M}i} Q_{\mu\mathcal{M}\nu\mathcal{N}}^{l,m}. \qquad (3.46)$$

This technique is very well known and widely used since decades in quantum chemistry [173]. Nevertheless, some aspects make its adoption in the context of periodic local MP2 especially attractive. In particular, translational invariance can be exploited at different levels that makes the computational effort practically negligible even for large systems. Our computational procedure for calculating the $Q_{iaA}^{l,m}$ used in Eq. (3.45) is as follows:

i) A batch of AO multipoles $Q_{\mu\mu\nu\mathcal{N}}$ – with the first AO index μ always in the reference cell – is computed relative to a conventional (individual) reference point, like the center of (each individual) ϕ_μ.

ii) Before the first half transformation step in Eq. 3.46, that is, with the WF coefficients, all the $Q_{\mu\mathcal{M}\nu\mathcal{N}}$ are obtained simply by shifting the origin of the above set of multipoles in a way that they all share the same origin, for example, the origin of the coordinate system (0,0,0). A common origin is necessary for all the $Q_{\mu\mathcal{M}\nu\mathcal{N}}$ participating in the transformation (3.46) since they all contribute to the multipole moment of the same density ρ_{iaA}. Consequently, the $Q_{\mu\mathcal{M}\nu\mathcal{N}}$ are no longer translationally invariant. Provided that the same origin is shared by all ρ_{iaA} at this stage, origin shifts have to be performed only once for each of the $Q_{\mu\mathcal{M}\nu\mathcal{N}}$.

iii) After the second half transformation (with PAO/OSV coefficients), the individual $Q_{iaA}^{l,m}$ are again translated to a convenient individual center suitable for the approximation of Eq. 3.45 to be reliable, like the center of the WF ϕ_i.

By virtue of this strategy the two computationally costly steps, that is, the calculation of multipoles and the first half transformation, are optimally performed.

Note that the assembly of the integrals $K_{ij\mathcal{J}}^{aAbB}$ according to Eq. (3.45) is extremely fast and can be performed on the fly. Hence, there is no need to store those $K_{i\mathcal{I}j\mathcal{J}}^{aAbB}$ that are obtained via the multipolar approximation (which can be a large fraction of the overall integral distribution). This leads to substantial savings in terms of memory and disk.

3.3
A Demonstrative Application: Ibuprofen on Hydroxylated Silica Surface

The proper description of adsorption of molecules on surfaces is a challenge for quantum chemical methods, for a number of reasons. First, the nature of the interaction is delicate and it is crucial to use a method that is capable of treating hydrogen bonds and dispersive interactions with the same accuracy. Second, due to the cumulative weight of very many weak long-range contributions, a fully periodic treatment is clearly preferable over finite cluster calculations.

Here we consider the interesting case of adsorption of organic or pharmaceutical molecules on biocompatible substrates [174]. A thick slab of hydroxylated SiO_2 is used to represent the surface, and the adsorbate consists of ibuprofen molecules – a well-known drug. The geometry of the system and the basis sets used are the same as in Ref. [174] (cf. Figure 3.1). In that work, the starting geometries were created by maximizing the number of H-bonds between the drug molecule and the surface silanols; these were then subjected to optimizations by relaxing the full molecule and the two most superficial layers of the slab at the DFT+D level.

The basis sets used for the slab are Nada's [175] 88–31G* and 8–411G* sets for Si and O, respectively, and Ahlrichs' 3–11G* VTZd [176] for H. The VTZd basis was also used for the atoms of the ibuprofen molecule. With these specifications our system comprises 144 atoms and 2142 atomic orbitals per unit cell.

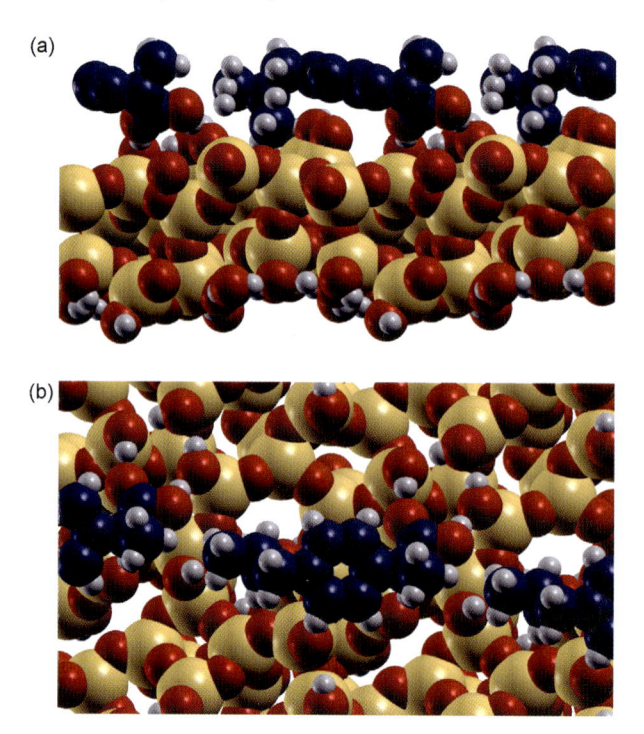

Figure 3.1 Ibuprofen adsorbed on a silica slab: side view (a) and top view (b).

3.3 A Demonstrative Application: Ibuprofen on Hydroxylated Silica Surface

The LMP2 interaction energies E_{int} were corrected for the basis set superposition error (BSSE) according to the counterpoise correction, that is, they were computed as

$$E_{int} = E_{slab+mol} - E_{slab+ghost} - E_{mol+ghost}, \qquad (3.47)$$

where $E_{slab+mol}$ is the energy of the complete system, $E_{slab+ghost}$ the energy of the slab with ghost functions on the monomer, and so on. As the unit call contains only one ibuprofen, Eq. (3.47) automatically provides the interaction energy per ibuprofen molecule. The geometry relaxation correction to the adsorption energy, that is, the contribution to E_{int} due to the optimization of the isolated surface and adsorbate, was not evaluated at the LMP2 level (and is therefore not included in Eq. (3.47)). The geometry change of molecules due to physisorption and, vice versa, of surfaces in the presence of an adsorbate, is commonly governed by electrostatic and exchange interactions between adsorbate and surface. Therefore, already the DFT level is expected to provide a decently accurate account of this contribution. According to Ref. [174], the DFT relaxation energy for our system amounts to 40–50 kJ/mol per ibuprofen molecule, virtually independent of the choice of the functional or inclusion of the dispersion correction.

As shown in earlier works on adsorption [141], the presence of the adsorbate has a rather small and short-ranged effect on the *intra*slab correlation, and vice versa, the slab has a small effect on the *intra*-adsorbate correlation. Hence, *intra*slab and *intra*-adsorbate correlation cancel to large extent in the energy difference (3.47). This can be exploited to further reduce the intraslab and intra-adsorbate WF pair lists $ij\mathcal{J}$: Only pairs up to a distance of 4 Å between the closest atoms of the two minimal domains $[i]$ and $[j\mathcal{J}]$ are included in these lists. These truncations lead to a substantial reduction of the computational cost for solving the LMP2 equations. The list related to molecule slab *inter*pairs, on the other hand, remains untruncated (cf. Eq. (3.32)).

The calculations presented here employ OSVs; the original subspace spanned by the pseudocanonical orbitals stems from large 25-atom domains, and the OSV threshold $\mathscr{E}_{OSV}$ is set to 10^{-5} (cf. Section 3.2.5.2). These OSVs are, as described above, augmented by the most diffuse PAOs of the atoms of the minimal domain. As shown in Table 3.1, the choice of 25-atom domains is rather

Table 3.1 Dependence of the OSV–DF–LMP2 interaction energy E_{int} on the values of the OSV threshold $\mathscr{E}_{OSV}$ and number of atoms in the initial domains N_{AT}^{DOM} defining the size of the PAO virtual space used to construct OSVs.

$\mathscr{E}_{OSV}$	N_{AT}^{DOM}	E_{int}(kJ/mol)	Time (Ibu + silica)
1×10^{-5}	25	−109.7	46 h 12 min
1×10^{-5}	20	−109.7	42 h 18 min
1×10^{-5}	15	−108.2	42 h 07 min
5×10^{-5}	25	−101.2	46 h 01 min
1×10^{-4}	25	−97.3	43 h 21 min

conservative, since a 15-atom domain size yields approximately the same interaction energy – even though we have observed variations in the total energy up to 40 kJ/mol. The reduction in computation time relative to the default setting amounts to less than 10% and is entirely due to savings in the calculation of DF integrals for the OSV construction.

The effect of employing looser values for $\mathscr{E}_{OSV}$ is also shown in Table 3.1, evidently a significant part of the electron correlation contribution is lost, with null or very modest savings in overall time. Therefore, it is not recommended to use $\mathscr{E}_{OSV}$ thresholds greater than 10^{-5}. At the same time, a further tightening of $\mathscr{E}_{OSV}$ to the 10^{-6} range can cause problems with the fitting of electron repulsion integrals (ERIs) that involve OSVs corresponding to small $e^{ii}_{\tilde{a}[i]}$ (Eq. (3.27)) [138]. Larger fitting basis set would be required for such tight $\mathscr{E}_{OSV}$ thresholds, making the calculations substantially more expensive. According to our experience, $\mathscr{E}_{OSV} = 10^{-5}$ is a reasonable compromise between accuracy and efficiency.

We note that a pronounced influence of the choice of $\mathscr{E}_{OSV}$ on the interaction energies is not surprising. The $\mathscr{E}_{OSV}$ tolerance in fact decides on the size of the pair-specific virtual space, and in this respect its effect on the correlation energy is comparable to that of the one-electron basis set.

Parallelization of the OSV code is currently in progress, but the performance of the parallel algorithms is not yet as efficient as for the PAO-based implementation [177]. For this reason, we present here detailed timings for the serial implementation only.

In Table 3.2 we report the timings for the most important steps of the slab + molecule calculation, that is, the most demanding ones. It is seen that the overall time is less than 2 days. The most expensive step is the evaluation of the four-index electron repulsion integral distribution through the density fitting approach. This includes the calculation of two- and three-index quantities as required by Eq. (3.43), solution of DF equations for coefficients, and assembly of the four-index integrals. The three-index integral step takes about 80% of the reported time.

By virtue of above-mentioned truncation of the pair list, the overall number of WF pairs is reduced to about 30 000. The cost for the solution of the

Table 3.2 Computational cost, on a single processor, of the main stages of an OSV–DF–LMP2 calculation of ibuprofen on silica.

Stage	Time
PAO construction	1 h 10 min
DF integrals (for OSV construction)	12 h 02 min
OSV construction	2 h 58 min
Multipoles	0 h 30 min
DF integrals (actual OSV–LMP2 calculation)	25 h 14 min
LMP2 Equations	3 h 41 min
Total	46 h 12 min

LMP2 equations is thus very small compared to the overall cost of the calculation. Furthermore, as discussed in Section 3.2.8, the computation time spent for the evaluation of the multipoles and the related ERIs (distant pair ERIs) is insignificant.

The interaction energy E_{int} as obtained according to Eq. (3.47) amounts to -109.7 kJ/mol at the LMP2 level; the corresponding HF interaction energy is only -41.3 kJ/mol. For comparison, the previously reported DFT values [174] were -161.1 kJ/mol for PBE-D and -137.6 kJ/mol for B3LYP-D* (here, -D stands for Grimme's -D2 empirical correction [42], and -D* for the reparametrization by Civalleri et al. [178]). Although our local MP2 calculations predict a smaller E_{int}, the ratio of 2.7 between the E_{int} obtained at the MP2 and HF levels supports for this system the claim made in Ref. [174] that dispersion effects dominate over H-bonds in such drug–surface interactions – these were estimated to contribute with -73.5 and -75.9 kJ/mol by the -D and -D* corrections, respectively.

We will further and more thoroughly investigate this system later, in particular the effects of basis set size and higher order correlation (via finite cluster corrections). In the present calculations, for example, very diffuse functions, which require a dual basis approach [98], were omitted. With such diffuse functions, usually a larger fraction of dispersion is captured [95,138,147,148].

Nevertheless, these preliminary results presented here show that (i) the algorithms and the implementation as outlined above are robust and capable of handling large systems (note that the present calculations were performed on a single core, and efficient parallelization as achieved in the PAO case will unlock further possibilities regarding systems size). (ii) The local approximations and preset computational thresholds are reliable and transferable from small to large unit cell crystals, thus making the application of the code almost black box for the user.

3.4
Summary and Conclusions

In this chapter we have presented theory and implementation of periodic local Møller–Plesset perturbation theory of second order that can be applied to large unit cells. The method is formulated in direct space; it is based on Wannier functions and either PAOs or OSVs to span, respectively, the occupied and unoccupied orbital space of the Hartree–Fock reference. Robust direct space local density fitting is employed to factorize the electron repulsion integrals of short and intermediate range, while a multipolar approximation is invoked to evaluate integrals between nonoverlapping orbital product densities. As a demonstrative application, we present OSV–LMP2 calculations of ibuprofen adsorbed on a silica slab. Such calculations can be performed within 2 days on a single core. An efficient parallel version of the OSV-LMP2 program is presently under development and will be available soon.

Even though the MP2 level of theory by itself is not highly accurate (it treats dispersion at the uncoupled Hartree–Fock level), a fully periodic MP2 program is a very valuable tool, since it still provides the major fraction of correlation. Hence, the (usually much smaller) higher order correlation corrections can be included on top of the periodic LMP2 result via energy increments on the basis of finite cluster calculations in fully saturated basis sets and with high-level correlation methods. Such a strategy was employed in previous work and provided highly accurate results for very challenging systems [106,111,147,148,179,180].

References

1 Pulay, P. (1983) *Chem. Phys. Lett.*, **100**, 151.
2 Saebø, S. and Pulay, P. (1985) *Chem. Phys. Lett.*, **113**, 13.
3 Pulay, P. and Saebø, S. (1986) *Theor. Chim. Acta*, **69**, 357.
4 Saebø, S. and Pulay, P. (1993) *Annu. Rev. Phys. Chem.*, **44**, 213–236.
5 Saebø, S., Tong, W., and Pulay, P. (1993) *J. Chem. Phys.*, **98**, 2170.
6 Boughton, J.W. and Pulay, P. (1993) *J. Comput. Chem.*, **14**, 736.
7 Saebø, S. and Pulay, P. (2001) *J. Chem. Phys.*, **115**, 3975.
8 Schütz, M., Hetzer, G., and Werner, H.-J. (1999) *J. Chem. Phys.*, **111**, 5691.
9 Hampel, C. and Werner, H.-J. (1996) *J. Chem. Phys.*, **104**, 6286.
10 Schütz, M. (2000) *J. Chem. Phys.*, **113**, 9986.
11 Schütz, M. and Werner, H.-J. (2000) *Chem. Phys. Lett.*, **318**, 370.
12 Schütz, M. and Werner, H.-J. (2001) *J. Chem. Phys.*, **114**, 661.
13 Schütz, M. (2002) *Phys. Chem. Chem. Phys.*, **4**, 3941.
14 Schütz, M. (2002) *J. Chem. Phys.*, **116**, 8772–8785.
15 Werner, H.-J., Manby, F.R., and Knowles, P.J. (2003) *J. Chem. Phys.*, **118**, 8149.
16 Werner, H.-J. and Schütz, M. (2011) *J. Chem. Phys.*, **135**, 144116.
17 Schütz, M., Yang, J., Chan, G.K.L., Manby, F.R., and Werner, H.-J. (2013) *J. Chem. Phys.*, **138**, 054109.
18 Yang, J., Chan, G.K.L., Manby, F.R., Schütz, M., and Werner, H.-J. (2012) *J. Chem. Phys.*, **136**, 144105.
19 Yang, J., Kurashige, Y., Manby, F.R., and Chan, G.K.L. (2011) *J. Chem. Phys.*, **134**, 044123.
20 Flocke, N. and Bartlett, R.J. (2004) *J. Chem. Phys.*, **121**, 10935.
21 Ayala, P.Y. and Scuseria, G.E. (1999) *J. Chem. Phys.*, **110**, 3660–3671.
22 Maslen, P.E., Dutoi, A.D., Lee, M.S., Shao, Y.H., and Head-Gordon, M. (2005) *Mol. Phys.*, **103**, 425.
23 Ochsenfeld, C., Kussmann, J., and Lambrecht, D.S. (2007) Linear-scaling methods in quantum chemistry, in *Reviews in Computational Chemistry*, vol. **23**, Wiley-VCH Verlag GmbH, Weinheim, pp. 1–82.
24 Doser, B., Lambrecht, D., Kussmann, J., and Ochsenfeld, C. (2009) *J. Chem. Phys.*, **130**, 064107.
25 Doser, B., Zienau, J., Clin, L., Lambrecht, D., and Ochsenfeld, C. (2010) *Z. Phys. Chem.*, **224**, 397.
26 Maurer, S.A., Lambrecht, D.S., Kussmann, J., and Ochsenfeld, C. (2013) *J. Chem. Phys.*, **138**, 014101.
27 Li, W., Piecuch, P., Gour, J.R., and Li, S. (2009) *J. Chem. Phys.*, **131**, 114109.
28 Rollik, Z. and Kállay, M. (2011) *J. Chem. Phys.*, **135**, 104111.
29 Rolik, Z., Szegedy, L., Ladjanszki, I., Ladoczki, B., and Kallay, M. (2013) *J. Chem. Phys.*, **139**, 094105.
30 Neese, F., Hansen, A., and Liakos, D.G. (2009) *J. Chem. Phys.*, **131**, 064103.
31 Neese, F., Wennmohs, F., and Hansen, A. (2009) *J. Chem. Phys.*, **130**, 114108.
32 Hansen, A., Liakos, D.G., and Neese, F. (2011) *J. Chem. Phys.*, **135**, 214102.

33 Riplinger, C. and Neese, F. (2013) *J. Chem. Phys.*, **138**, 034106.
34 Riplinger, C., Sandhoefer, B., Hansen, A., and Neese, F. (2013) *J. Chem. Phys.*, **139**, 134101.
35 Tew, D.P., Helmich, B., and Hättig, C. (2011) *J. Chem. Phys.*, **135**, 074107.
36 Kristensen, K., Høyvik, I.-M., Jansìk, B., Jørgensen, P., Kjaergaard, T., Reineb, S., and Jakowskic, J. (2012) *Phys. Chem. Chem. Phys.*, **14**, 15706.
37 Friedrich, J. (2012) *J. Chem. Theory Comput.*, **8**, 1597.
38 Kats, D. and Manby, F.R. (2013) *J. Chem. Phys.*, **138** (14), 144101.
39 Masur, O., Usvyat, D., and Schütz, M. (2013) *J. Chem. Phys.*, **139**, 164116.
40 Jung, Y., Lochan, R.C., Dutoi, A.D., and Head-Gordon, M. (2004) *J. Chem. Phys.*, **121**, 9793.
41 Grimme, S. (2004) *J. Comput. Chem.*, **25**, 1463.
42 Grimme, S. (2006) *J. Comput. Chem.*, **27**, 1787.
43 Tkatchenko, A. and Scheffler, M. (2009) *Phys. Rev. Lett.*, **102**, 73005.
44 Grimme, S., Antony, J., Ehrlich, S., and Krieg, H. (2010) *J. Chem. Phys.*, **132**, 154104.
45 Furche, F. (2008) *J. Chem. Phys.*, **129**, 114105.
46 Harl, J. and Kresse, G. (2009) *Phys. Rev. Lett.*, **103**, 056401.
47 Paier, J., Janesko, B.G., Henderson, T.M., Scuseria, G.E., Grüneis, A., and Kresse, G. (2010) *J. Chem. Phys.*, **132**, 094103.
48 Heßelmann, A. and Görling, A. (2011) *Mol. Phys.*, **109**, 2473.
49 Rydberg, H., Dion, M., Jacobson, N., and Schroeder, E. (2003) *Mater. Sci. Eng. C*, **23**, 721.
50 Schwabe, T. and Grimme, S. (2007) *Phys. Chem. Chem. Phys.*, **9**, 3397.
51 Stoll, H. (1992) *Phys. Rev. B*, **46**, 6700.
52 Stoll, H. (1992) *Chem. Phys. Lett.*, **191**, 548.
53 Stoll, H. (1992) *J. Chem. Phys.*, **97**, 8449.
54 Rościszewski, K., Paulus, B., Fulde, P., and Stoll, H. (1999) *Phys. Rev. B*, **60**, 7905.
55 Rosciszewski, K., Doll, K., Paulus, B., Fulde, P., and Stoll, H. (1998) *Phys. Rev. B*, **57**, 14667.
56 Paulus, B. (2006) *Phys. Rep.*, **428**, 1.
57 Hermann, A. and Schwerdtfeger, P. (2008) *Phys. Rev. Lett.*, **101**, 183005.
58 Müller, C. and Paulus, B. (2012) *Phys. Chem. Chem. Phys.*, **14**, 7605.
59 Voloshina, E., Gaston, N., and Paulus, B. (2007) *J. Chem. Phys.*, **126**, 134115.
60 Voloshina, E. (2012) *Phys. Rev. B*, **83**, 045444.
61 Müller, C., Herschend, B., Hermansson, K., and Paulus, B. (2008) *J. Chem. Phys.*, **128**, 21.
62 Voloshina, E., Usvyat, D., Schütz, M., Dedkov, Y., and Paulus, B. (2011) *Phys. Chem. Chem. Phys.*, **13**, 12041.
63 Hammerschmidt, L., Müller, C., and Paulus, B. (2012) *J. Chem. Phys.*, **136**, 124117.
64 Bygrave, P.J., Allan, N.L., and Manby, F.R. (2012) *J. Chem. Phys.*, **137**, 164102.
65 Hirata, S., Sode, O., Keçeli, M., and Shimazaki, T. (2010) Electron correlation in solids: delocalized and localized orbital approaches, in *Accurate Condensed-Phase Quantum Chemistry*, Computation in Chemistry, vol. **27**, CRC Press.
66 Sode, O., Keçili, M., and Hirata, S. (2013) *J. Chem. Phys.*, **138**, 074501.
67 Rubes, M. and Bludsky, O. (2008) *Phys. Chem. Chem. Phys.*, **10**, 2611.
68 Nolan, S.J., Gillan, M.J., Alfè, D., Allan, N.L., and Manby, F.R. (2009) *Phys. Rev. B*, **80**, 165109.
69 Nolan, S.J., Bygrave, P.J., Allan, N.L., and Manby, F.R. (2010) *J. Phys. Condens. Matter.*, **22**, 074201.
70 Taylor, C.R., Bygrave, P.J., Hart, J.N., Allan, N.L., and Manby, F.R. (2012) *Phys. Chem. Chem. Phys.*, **14**, 7739.
71 Podeszwa, R., Rice, B.M., and Szalewicz, K. (2008) *Phys. Rev. Lett.*, **101**, 115503.
72 Beran, G.J.O. and Nanda, K. (2010) *J. Phys. Chem. Lett.*, **1**, 3480.
73 Wen, S. and Beran, G.J.O. (2011) *J. Chem. Theory Comput.*, **7**, 3733.
74 Wen, S., Nanda, K., Huanga, Y., and Beran, G.J.O. (2012) *Phys. Chem. Chem. Phys.*, **14**, 7578.
75 Suhai, S. and Ladik, J. (1982) *J. Phys. Chem.*, **15**, 4327.
76 Sun, J.-Q. and Bartlett, R.J. (1996) *J. Chem. Phys.*, **104**, 8553.

77 Hirata, S. and Iwata, S. (1998) *J. Chem. Phys.*, **109**, 4147.
78 Hirata, S., Grabowski, I., Tobita, M., and Bartlett, R.J. (2001) *Chem. Phys. Lett.*, **345**, 475.
79 Liegener, C.-M. (1985) *J. Phys. C: Solid State Phys.*, **18**, 6011.
80 Suhai, S. (1994) *Phys. Rev. B*, **50**, 14791.
81 Hirata, S., Podeszwa, R., Tobita, M., and Bartlett, R.J. (2004) *J. Chem. Phys.*, **120**, 2581.
82 Ohnishi, Y. and Hirata, S. (2010) *J. Chem. Phys.*, **133**, 034106.
83 Ohnishi, Y. and Hirata, S. (2011) *J. Chem. Phys.*, **135**, 094108.
84 Shiozaki, T. and Hirata, S. (2010) *J. Chem. Phys.*, **132**, 151101.
85 Almlöf, J. (1991) *Chem. Phys. Lett.*, **181**, 319.
86 Ayala, P.Y., Kudin, K.N., and Scuseria, G.E. (2001) *J. Chem. Phys.*, **115**, 9698.
87 Izmaylov, A.F. and Scuseria, G.E. (2008) *Phys. Chem. Chem. Phys.*, **10**, 3421.
88 Pisani, C., Busso, M., Capecchi, G., Casassa, S., Dovesi, R., Maschio, L., Zicovich-Wilson, C., and Schütz, M. (2005) *J. Chem. Phys.*, **122**, 094113.
89 Casassa, S., Halo, M., and Maschio, L. (2008) *J. of Phys. Conf. Series*, **117**, 012007.
90 Halo, M., Casassa, S., Maschio, L., and Pisani, C. (2009) *Phys. Chem. Chem. Phys.*, **11**, 586.
91 Erba, A., Casassa, S., Maschio, L., and Pisani, C. (2009) *J. Phys. Chem. B*, **113**, 2347.
92 Erba, A., Casassa, S., Dovesi, R., Maschio, L., and Pisani, C. (2009) *J. Chem. Phys.*, **130**, 074505.
93 Pisani, C., Maschio, L., Casassa, S., Halo, M., and Erba, A. (2009) *Theor. Chem. Acc.*, **123**, 327.
94 Halo, M., Pisani, C., Maschio, L., Casassa, S., Schütz, M., and Usvyat, D. (2011) *Phys. Rev. B*, **83**, 035117.
95 Maschio, L., Usvyat, D., Schütz, M., and Civalleri, B. (2010) *J. Chem. Phys.*, **132**, 134706.
96 Erba, A., Pisani, C., Casassa, S., Maschio, L., Schütz, M., and Usvyat, D. (2010) *Phys. Rev. B*, **81**, 165108.
97 Maschio, L., Usvyat, D., and Civalleri, B. (2010) *CrystEngComm*, **12**, 2429.
98 Usvyat, D., Maschio, L., Pisani, C., and Schütz, M. (2010) *Z. Phys. Chem.*, **224**, 441.
99 Pisani, C., Itou, M., Sakurai, Y., Yamaki, R., Ito, M., Erba, A., and Maschio, L. (2011) *Phys. Chem. Chem. Phys.*, **13**, 933.
100 Erba, A., Itou, M., Sakurai, Y., Yamaki, R., Ito, M., Casassa, S., Maschio, L., Terentjevs, A., and Pisani, C. (2011) *Phys. Rev. B*, **83**, 125208.
101 Erba, A., Maschio, L., Salustro, S., and Casassa, S. (2011) *J. Chem. Phys.*, **134**, 074502.
102 Erba, A., Maschio, L., Pisani, C., and Casassa, S. (2011) *Phys. Rev. B*, **84**, 012101.
103 Maschio, L., Civalleri, B., Ugliengo, P., and Gavezzotti, A. (2011) *J. Phys. Chem. A*, **115** (41), 11179–11186.
104 Presti, D., Pedone, A., Menziani, M.C., Civalleri, B., and Maschio, L. (2014) *CrystEngComm*, **16**, 102.
105 Müller, C., Usvyat, D., and Stoll, H. (2011) *Phys. Rev. B*, **83**, 245136.
106 Müller, C. and Usvyat, D. (2013) *J. Chem. Theory Comput.*, **9**, 5590.
107 Constantinescu, G., Kuc, A., and Heine, T. (2013) *Phys. Rev. Lett.*, **111**, 036104.
108 Halo, M., Casassa, S., Maschio, L., Pisani, C., Dovesi, R., Ehinon, D., Baraille, I., Rérat, M., and Usvyat, D. (2011) *Phys. Chem. Chem. Phys.*, **13**, 4434.
109 Martinez-Casado, R., Mallia, G., Usvyat, D., Maschio, L., Casassa, S., Schütz, M., and Harrison, N.M. (2011) *J. Chem. Phys.*, **134**, 014706.
110 Martinez-Casado, R., Mallia, G., Usvyat, D., Maschio, L., Casassa, S., Schutz, M., and Harrison, N.M. (2011) *Phys. Chem. Chem. Phys.*, **13**, 14750.
111 Usvyat, D., Sadeghian, K., Maschio, L., and Schütz, M. (2012) *Phys. Rev. B*, **86**, 045412.
112 Tanskanen, J.T., Maschio, L., Karttunen, A.J., Linnolahti, M., and Pakkanen, T.A. (2012) *Chemphyschem*, **13**, 2361.
113 Marsman, M., Gruneis, A., Paier, J., and Kresse, G. (2009) *J. Chem. Phys.*, **130**, 184103.
114 Grüneis, A., Marsman, M., and Kresse, G. (2010) *J. Chem. Phys.*, **133**, 074107.
115 Grüneis, A., Booth, G.H., Marsman, M., Spencer, J., Alavi, A., and Kresse, G. (2011) *J. Chem. Theory Comput.*, **7**, 2780.

116 Booth, G., Grüneis, A., Kresse, G., and Alavi, A. (2012) *Nature*, **493**, 365.
117 Grüneis, A., Shepherd, J.J., Alavi, A., Tew, D.P., and Booth, G.H. (2013) *J. Chem. Phys.*, **139**, 084112.
118 Ben, M.D., Hutter, J., and VandeVondele, J. (2012) *J. Chem. Theory Comput.*, **8**, 4177.
119 Del Ben, M., Hutter, J., and VandeVondele, J. (2013) *J. Chem. Theory Comput.*, **9**, 2654.
120 Kohn, W. (1959) *Phys. Rev.*, **115**, 809.
121 Des Cloizeaux, J. (1964) *Phys. Rev.*, **135**, A685.
122 Brouder, C., Panati, G., Calandra, M., Mourougane, C., and Marzari, N. (2007) *Phys. Rev. Lett.*, **98**, 046402.
123 Pulay, P., Saebø, S., and Meyer, W. (1984) *J. Chem. Phys.*, **81**, 1901.
124 Knowles, P.J., Schütz, M., and Werner, H.-J. (2000) Ab initio methods for electron correlation in molecules, in *Modern Methods and Algorithms of Quantum Chemistry*, vol. **3**, NIC Series (ed. J. Grotendorst), John von Neumann Institute for Computing (NIC), pp. 97–179.
125 Wannier, G.H. (1937) *Phys. Rev.*, **52**, 191.
126 Foster, J.M. and Boys, S.F. (1960) *Rev. Mod. Phys.*, **32**, 300.
127 Edmiston, C. and Ruedenberg, K. (1963) *Rev. Mod. Phys.*, **35**, 457.
128 Pipek, J. and Mezey, P.G. (1989) *J. Chem. Phys.*, **90**, 4916.
129 Marzari, N. and Vanderbilt, D. (1997) *Phys. Rev. B*, **56**, 12847.
130 Souza, I., Marzari, N., and Vanderbilt, D. (2001) *Phys. Rev. B*, **65**, 035109.
131 Smirnov, V.P. and Usvyat, D.E. (2001) *Phys. Rev. B*, **64**, 245108.
132 Birkenheuer, U. and Izotov, D. (2005) *Phys. Rev. B*, **71**, 125116.
133 Zicovich-Wilson, C.M., Dovesi, R., and Saunders, V.R. (2001) *J. Chem. Phys.*, **115**, 9708.
134 Pisani, C., Maschio, L., Casassa, S., Halo, M., Schütz, M., and Usvyat, D. (2008) *J. Comput. Chem.*, **29**, 2113.
135 Usvyat, D. (2013) *J. Chem. Phys.*, **139**, 194101.
136 Casassa, S., Zicovich-Wilson, C.M., and Pisani, C. (2006) *Theor. Chem. Acc.*, **116**, 726.
137 Yang, J., Kurashige, Y., Manby, F. R., and Chan, G.K.L. (2011) *J. Chem. Phys.*, **134**, 044123.
138 Usvyat, D., Maschio, L., and Schütz, M. (2015) *J. Chem. Phys.*, **143**, 102805.
139 Hättig, C., Tew, D.P., and Helmich, B. (2012) *J. Chem. Phys.*, **136**, 204105.
140 Werner, H.-J., Knizia, G., Schwilk, M., and Dornbach, M. (2015) *J. Chem. Theory Comput.* **11**, 484.
141 Pisani, C., Schütz, M., Casassa, S., Usvyat, D., Maschio, L., Lorenz, M., and Erba, A. (2012) *Phys. Chem. Chem. Phys.*, **14**, 7615.
142 Werner, H.-J. and Pflüger, K. (2006) *Annu. Rep. Comput. Chem.*, **2**, 53.
143 Werner, H.-J. (2008) *J. Chem. Phys.*, **129**, 101103.
144 Adler, T.B., Werner, H.-J., and Manby, F.R. (2009) *J. Chem. Phys.*, **130**, 054106.
145 Schütz, M., Rauhut, G., and Werner, H.-J. (1998) *J. Phys. Chem. A*, **102**, 5997–6003.
146 Usvyat, D., Civalleri, B., Maschio, L., Dovesi, R., Pisani, C., and Schütz, M. (2011) *J. Chem. Phys.*, **134**, 214105.
147 Sansone, G., Maschio, L., Usvyat, D., Schütz, M., and Karttunen, A. (2016) *J. Phys. Chem. Lett.*, 7 (1), 131–136.
148 Usvyat, D. (2015) *J. Chem. Phys.*, **143**, 104704.
149 Hetzer, G., Pulay, P., and Werner, H.-J. (1998) *Chem. Phys. Lett.*, **290**, 143–149.
150 Schrader, D.M. and Prager, S. (1962) *J. Chem. Phys.*, **37**, 3927.
151 Boys, S.F. and Shavitt, I. (1959) Report WIS-AF-13, University of Wisconsin.
152 Baerends, E.J., Ellis, D.E., and Ros, P. (1973) *Chem. Phys.*, **2**, 41.
153 Dunlap, B.I., Connolly, J.W.D., and Sabin, J.R. (1979) *J. Chem. Phys.*, **71**, 3396.
154 Mintmire, J.W. and Dunlap, B.I. (1982) *Phys. Rev. A*, **25**, 88.
155 Vahtras, O., Almlöf, J., and Feyereisen, M.W. (1993) *Chem. Phys. Lett.*, **213**, 514.
156 Weigend, F., Häser, M., Patzelt, H., and Ahlrichs, R. (1998) *Chem. Phys. Lett.*, **294**, 143.
157 Schütz, M. and Manby, F.R. (2003) *Phys. Chem. Chem. Phys.*, **5**, 3349–3358.
158 Polly, R., Werner, H.-J., Manby, F.R., and Knowles, Peter J. (2004) *Mol. Phys.*, **102**, 2311–2321.

159 te Velde, G. and Baerends, E.J. (1991) *Phys. Rev. B*, **44**, 7888.

160 Trickey, S.B., Alford, J.A., and Boettger, J.C. (2004) *Computational Materials Science*, Theoretical and Computational Chemistry, vol. **15** (ed. J. Leszczynski), Elsevier, p. 171.

161 Maschio, L., Usvyat, D., Manby, F.R., Casassa, S., Pisani, C., and Schütz, M. (2007) *Phys. Rev. B*, **76**, 075101.

162 Varga, S., Milko, M., and Noga, J. (2006) *J. Chem. Phys.*, **124**, 034106.

163 Varga, S. (2008) *Int. J. Quantum Chem.*, **108**, 1518.

164 Izmaylov, A.F. and Scuseria, G.E. (2008) *Phys. Chem. Chem. Phys.*, **10**, 3421.

165 Burow, A.M., Sierka, M., and Mohamed, F. (2009) *J. Chem. Phys.*, **131**, 214101.

166 Schütz, M., Usvyat, D., Lorenz, M., Pisani, C., Maschio, L., Casassa, S., and Halo, M. (2010) Density fitting for correlated calculations in periodic systems, in *Accurate Condensed-Phase Quantum Chemistry*, Computation in Chemistry, vol. **27**, CRC Press.

167 Lorenz, M., Maschio, L., Schütz, M., and Usvyat, D. (2012) *J. Chem. Phys.*, **137**, 204119.

168 Dunlap, B.I. (2000) *Phys. Chem. Chem. Phys.*, **2**, 2113–2116.

169 Dunlap, B.I. (2000) *J. Mol. Struct.: Theochem*, **529**, 37–40.

170 Mintmire, J.W. and Dunlap, B.I. (1982) *Phys. Rev. A*, **25**, 88.

171 Manby, F.R. and Knowles, P.J. (2001) *Phys. Rev. Lett.*, **87**, 163001.

172 Manby, F.R., Knowles, P.J., and Lloyd, A.W. (2001) *J. Chem. Phys.*, **115**, 9144.

173 Helgaker, T., Jørgensen, P., and Olsen, J. (2000) *Molecular Electronic Structure Theory*, John Wiley and Sons, Ltd, Chichester, UK.

174 Delle Piane, M., Corno, M., and Ugliengo, P. (2013) *J. Chem. Theory Comput.*, **9** (5), 2404–2415.

175 Nada, R., Nicholas, J.B., McCarthy, M.I., and Hess, A.C. (1996) *Int. J. Quantum Chem.*, **60**, 809–820.

176 Schäfer, A., Horn, H., and Ahlrichs, R. (1992) *J. Chem. Phys.*, **97**, 2571–2577.

177 Maschio, L. (2011) *J. Chem. Theory Comput.*, **7**, 2818.

178 Civalleri, B., Zicovich-Wilson, C.M., Valenzano, L., and Ugliengo, P. (2008) *CrystEngComm*, **10**, 405.

179 Martinez-Casado, R., Usvyat, D., Maschio, L., Mallia, G., Casassa, S., Ellis, J., Schütz, M., and Harrison, N.M. (2014) *Phys. Rev. B*, **89**, 205138.

180 Martinez-Casado, R., Usvyat, D., Maschio, L., Mallia, G., Casassa, S., Ellis, J., Schütz, M., and Harrison, N.M. (2014) *Phys. Chem. Chem. Phys.*, **16**, 21106.

4
Resonating Valence Bonds in Chemistry and Solid State

Evgeny A. Plekhanov[1,3] *and Andrei L. Tchougréeff*[2,3,4]

[1]*King's College London, Theory and Simulation of Condensed Matter (TSCM), The Strand, London WC2R 2LS, UK*
[2]*RWTH Aachen University, Institut für anorganische Chemie, Landoltweg 1, 52056, Aachen, Germany*
[3]*Moscow Center for Continuous Mathematical Education, Bol. Vlasevskiy per., 119002 Moscow, Russia*
[4]*A.N. Frumkin, Institute of Physical Chemistry & Electrochemistry of RAS, Leninsky ave., 31-4, 119071, Moscow, Russia*

4.1
Introduction

This chapter is conceived as a combination of two complementary parts. In the first part, we discuss (largely qualitatively) the relations between the resonance valence bond (RVB) and tight-binding/linear combination of atomic orbitals (LCAO) pictures of the electronic structure using the archetypal example: the π-system of benzene well known to chemists – the principal readers of this volume. The substance of the differences between the two is rarely clearly recognized, although the divergence between them in "molecular" quantum chemistry is well appreciated even as potentially reaching the degree of a "never-ending rivalry" [1]. The second more formal, more physical, and more solid-state part is devoted to extensions of the RVB ideas to solids. This requires a modification of the single determinant form of the electronic wave function based on the tight-binding/LCAO scheme characteristic for solids to enable it to model RVB states. We exemplify these issues by several simple models ranging from the one-dimensional polyene chain: an infinite extension of benzene (hexa*triene*) molecule – through graphene and finish by probably the most spectacular example of the enigmatic quantum RVB state of copper(II) carbodiimide (CuNCN) – a new material, where an assumption of the RVB character of its electronic phases was crucial for explaining its room- and low-temperature physics. The possibility of the RVB phase/state of high-T_c cuprate superconductors is also briefly discussed.

Handbook of Solid State Chemistry, First Edition. Edited by Richard Dronskowski, Shinichi Kikkawa, and Andreas Stein.
© 2017 Wiley-VCH Verlag GmbH & Co. KGaA. Published 2017 by Wiley-VCH Verlag GmbH & Co. KGaA.

4.2
Historical Overview

The concept of resonating bonds is very dear to the hearts of chemists and this is well deserved. Yet at the dawn of theoretical *organic* chemistry, almost right after the concept of an integer valence characteristic for each atom had been introduced [2], the chemists faced the problem of benzene molecule: one composed of six carbon and six hydrogen atoms (C_6H_6) [3–6]. Counting the number of isomers of disubstituted benzenes (C_6H_4XY) led to the idea of the hexagonal-cyclic structure of this molecule and of the sixfold symmetry of the original benzene molecule itself. Such an assumption is incompatible with the formal valence theory ascribing precisely four valences to the carbon atoms: One cannot uniquely distribute three bonds (the fourth is always assigned to one with a hydrogen atom) in a hexagon preserving its sixfold symmetry as shown, for example, in either of bracketed structures of Figure 4.1.

It was August Kekulé who first suggested the real molecule to "oscillate" between two possible distributions of "double" bonds, each possessing a three-fold symmetry so that the sixfold symmetry of the molecule itself is preserved. The nature of this oscillation was pretty much enigmatic and was an occasion for rather caustic jokes like one reproduced in Figure 4.2.[1]

The situation reached its tentative resolution after works by Lewis, Pauling, and Rumer who could convince chemists that it goes not about jumps of monkeys rather about resonance of electronic pairs that, nevertheless, could not be properly (not breaking the sixfold symmetry) distributed among six carbon atoms. The solution based on the early theory of spin valence [8–11] was to represent the delocalized bonds as a superposition of the individual states (the term "resonance" of structures aimed to replace Kekulé's "oscillation" had been coined by Pauling [12,13]), each being a product of electronic pair functions distributed in the molecule, as shown in Figure 4.3.

Switching to a more formal representation necessary for our further discussion, two Kekulé structures denoted as K_1 and K_2 in Figure 4.3 can be written

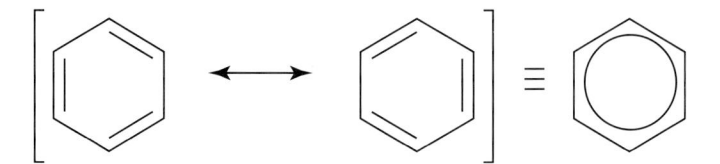

Figure 4.1 Archetypal example of resonating valence bonds: benzene. Six hydrogen atoms each consuming one valency each carbon atom are not shown for simplicity.

---

1) Remarkably, Ladenburg [7] proposed a three-dimensional trigonal prismatic arrangement of six CH units conformant with the number of doubly XY-substituted isomers of benzene, but not with that of XX ones. Such a molecule indeed exists and is known as "prismane."

Figure 4.2 The picture is most probably due to German chemist H. Kolbe. (Berichte der durstigen Chemischen Gesellschaft Unerhörter Jahrgang, Nr. 20 zum 20.9.1886 Sitzung vom 20. September 1886. Vorsitzender: Hr. August Kuleké, Präsident.) The message is pretty clear: first you invent something which you call "bond" and "valence" and then you say that they "oscillate" since the ends do not come together any way. Why not to assume, that there are some invisible monkeys?

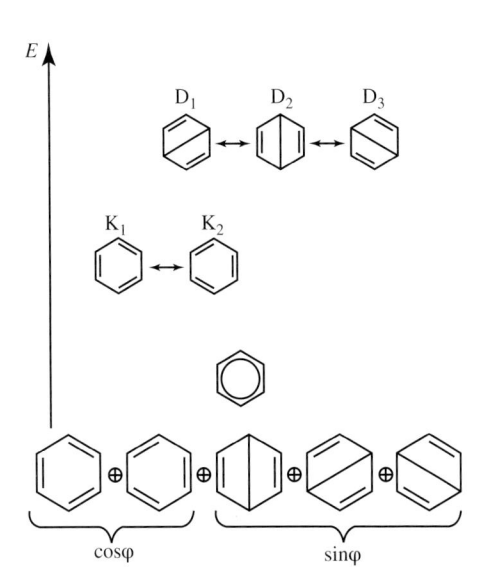

Figure 4.3 Oscillation, mesomery, resonance: three names for the same thing now called quantum mechanical superposition of independent valence states, not necessarily observed as individual compounds. Hexagons themselves are only eyeguides, the additional dashes represent individual electronic pairs that or other way distributed in the molecule [14].

as many-electronic wave functions:

$$|K_1\rangle = \frac{1}{\sqrt{8}}\left[(c_{1\uparrow}^\dagger c_{2\downarrow}^\dagger - c_{1\downarrow}^\dagger c_{2\uparrow}^\dagger)(c_{3\uparrow}^\dagger c_{4\downarrow}^\dagger - c_{3\downarrow}^\dagger c_{4\uparrow}^\dagger)(c_{5\uparrow}^\dagger c_{6\downarrow}^\dagger - c_{5\downarrow}^\dagger c_{6\uparrow}^\dagger)\right]|0\rangle,$$

$$|K_2\rangle = \frac{1}{\sqrt{8}}\left[(c_{1\uparrow}^\dagger c_{6\downarrow}^\dagger - c_{1\downarrow}^\dagger c_{6\uparrow}^\dagger)(c_{2\uparrow}^\dagger c_{3\downarrow}^\dagger - c_{2\downarrow}^\dagger c_{3\uparrow}^\dagger)(c_{4\uparrow}^\dagger c_{5\downarrow}^\dagger - c_{4\downarrow}^\dagger c_{5\uparrow}^\dagger)\right]|0\rangle,$$

(4.1)

and analogous expressions for three "Dewar benzenes" $|D_1\rangle \div |D_3\rangle$ built following the same pattern – as products of the singlet electronic pairs residing in the respective pairs of the site orbitals (here we use the second quantization notation described in all details, for example, in Refs [15,16]). In the expressions above, $|0\rangle$ stands for the state containing no electrons at all (*vacuum* state) and the symbols $c_{n\sigma}^\dagger$ (creation operators) refer to adding to the nth atomic state of the system an electron with the spin projection $\sigma = \uparrow, \downarrow$. The symbols $c_{n\sigma}^\dagger$ are subject to the *anticommutation* relations:

$$\{c_{n\sigma}^\dagger, c_{m\tau}^\dagger\} = c_{n\sigma}^\dagger c_{m\tau}^\dagger + c_{m\tau}^\dagger c_{n\sigma}^\dagger = 0 \qquad (4.2)$$

of which the most transparent consequence is the *nilpotence* of the creation operators, that is, $c_{n\sigma}^\dagger c_{n\sigma}^\dagger = (c_{n\sigma}^\dagger)^2 = 0$, which formalizes the known Pauli principle preventing two electrons with equal spin projection to occur in the same state. The rows of the form $c_{s\sigma}^\dagger c_{tt}^\dagger \ldots c_{rp}^\dagger c_{m\mu}^\dagger |0\rangle$ thus represent Slater determinants uniquely (up to an overall sign) defined by the lists of the occupied one-electron states. Anticommutation of $c^\dagger$'s takes care of (or formalizes) the known property of a determinant to change its sign when two rows (columns) of the latter are exchanged. (The nilpotence is of course the same as the evanescence of the determinant with two coinciding rows or columns).

The above construct was baptized as the resonance valence bond (RVB) state of benzene. It represented a great advantage in describing the electronic structure of benzene and other π-conjugated hydrocarbons and is actually if not the simplest, but for sure best-known example of an RVB state in chemistry. Nowadays one can frequently read that "since in the π-system of benzene the bonds are nonlocalizable and a resonance takes place, it is advisable/convenient/reasonable to introduce delocalized description of electrons in this system." It is thus tacitly assumed that it is not important what is delocalized: bonds or electrons. The delocalization of electrons is formalized by the electronic wave function taken as a single Slater determinant of the occupied MOs, as proposed by Hückel [17–26]. This construction is very well known (see Figure 4.4).

For its more formal representation, first, the linear combinations of the above creation operators $c_{n\sigma}^\dagger$ are formed:

$$\Psi_{6\sigma}^\dagger = \frac{1}{\sqrt{6}}\sum_{n}^{6} e^{i\pi n} c_{n\sigma}^\dagger$$

$$\Psi_{4\sigma}^\dagger = \frac{1}{\sqrt{3}}\sum_{n}^{6} \cos\left(\frac{4\pi}{6}n\right) c_{n\sigma}^\dagger \; ; \; \Psi_{5\sigma}^\dagger = \frac{1}{\sqrt{3}}\sum_{n}^{6} \sin\left(\frac{4\pi}{6}n\right) c_{n\sigma}^\dagger$$

$$\Psi_{2\sigma}^\dagger = \frac{1}{\sqrt{3}}\sum_{n}^{6} \cos\left(\frac{2\pi}{6}n\right) c_{n\sigma}^\dagger \; ; \; \Psi_{3\sigma}^\dagger = \frac{1}{\sqrt{3}}\sum_{n}^{6} \sin\left(\frac{2\pi}{6}n\right) c_{n\sigma}^\dagger$$

$$\Psi_{1\sigma}^\dagger = \frac{1}{\sqrt{6}}\sum_{n}^{6} c_{n\sigma}^\dagger$$

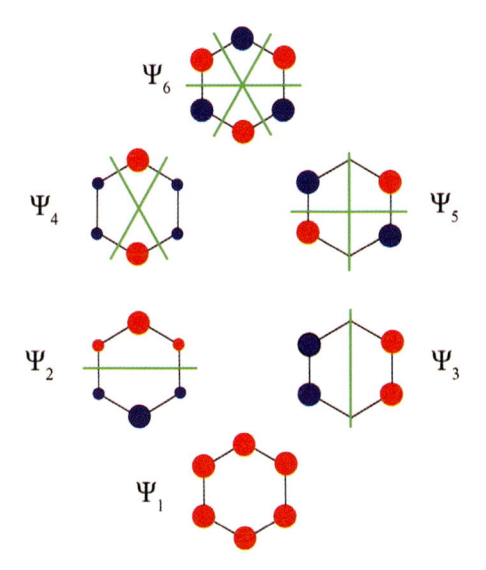

Figure 4.4 The pictorial representation of MO LCAOs of benzene.

Similar to the $c_{n\sigma}^{\dagger}$ ones, these expressions are operators creating electrons with spin projections σ in the *molecular orbitals*, which are the respective linear combinations of the atomic (site) orbitals (MO LCAO).

This wave function of six electrons is the single Slater determinant formed of the orbitals with the smallest available number of nodes (see Figure 4.4), that is, $\Psi_{1\sigma}^{\dagger} \div \Psi_{3\sigma}^{\dagger}$ for both spin projections:

$$\Psi_{1\downarrow}^{\dagger}\Psi_{1\uparrow}^{\dagger}\Psi_{2\downarrow}^{\dagger}\Psi_{2\uparrow}^{\dagger}\Psi_{3\downarrow}^{\dagger}\Psi_{3\uparrow}^{\dagger}|0| \qquad (4.3)$$

It is thus of interest to know what is the relation between the functions Eq. (4.1) (Pauling) and Eq. (4.3) (Hückel). Since, in general, the wave functions themselves are not observable, the only way of comparing the functions is to compare the averages of some observable physical quantities. The simplest one is the electron density – the average over the wave function of the operators $\hat{n}_{i\sigma}$ – those of the number of electrons with the spin projection σ in the ith site (orbital). Operators $\hat{n}_{i\sigma}$ are projection operators, since acting on a Slater determinant with the state $i\sigma$ occupied it produces a factor of unity, whereas for the similar determinant with the empty $i\sigma$ state it yields the factor zero. Anything else is not possible due to the Pauli principle. Opening brackets in (4.1) immediately shows that each of the two Kekulé states contains only singly occupied site (atomic) orbitals. As one can easily check, the number of particles operator $\hat{n}_{1\sigma}$ stumbles only upon one of the terms in only one of the two-electron products (singlets) entering the basis configurations $|K_1\rangle$, $|K_2\rangle$, and $|D_1\rangle \div |D_3\rangle$ used for representing the RVB state (Eq. 4.5). The calculation yields the sought average to be $1/2$. The same is true for all other i's. "Similarly" one sees that in the Hückel solution (Eq. 4.3), the one-center averages of all $\hat{n}_{i\sigma}$ equal to the same $1/2$. It does not seem to be counterintuitive and it is indeed true that the number of electrons per atom (site) – sum of $\langle \hat{n}_{i\sigma} \rangle$ over σ – in both approximations

(LCAO and RVB) equals to unity, Thus, this quantity is absolutely insufficient to distinguish two states of benzene.

In the style characteristic for the solid-state applications, let us consider another average, one related to correlation functions – the observables characterizing different many-electronic states or, respectively, electronic or magnetic phases of solids. The simplest and best of all suitable for our purposes is the average $\langle \hat{n}_{1\downarrow} \hat{n}_{1\uparrow} \rangle$ – a two-electron correlation function – which can also be interpreted as the density of an electronic pair on the single site. Even more general operator products of number of particle operators, $\hat{n}_{i\sigma} \hat{n}_{j\tau}$, are as well projection operators counting electron pairs. The shown operator produces a factor of unity in case when in a Slater determinant both spin orbitals $i\sigma$ and $j\tau$ are occupied and zero otherwise. Operator $\hat{n}_{1\downarrow} \hat{n}_{1\uparrow}$ is a specific case of the previously mentioned projection operator. It produces a factor of unity acting on a Slater determinant with a doubly occupied ith orbital and zero otherwise. One can easily conclude that for the RVB function (Eq. (4.5)), this (and all other similar one-center pair densities) is vanishing. This is very easy to understand since either of the basis configurations $|K_1\rangle$, $|K_2\rangle$, and $|D_1\rangle \div |D_3\rangle$ used for representing the RVB state (Eq. (4.5)) contains in its expansion no more than one creation operator $c^{\dagger}_{1\sigma}$. Thus, the term on which, say, $\hat{n}_{1\downarrow}$ "stumbles" and produces a factor of unity for sure does not contain $c^{\dagger}_{1\uparrow}$ and thus will be killed by the $\hat{n}_{1\uparrow}$ multiplier. Thus, we have $\langle \hat{n}_{1\downarrow} \hat{n}_{1\uparrow} \rangle_{\text{RVB}} = 0$ and the same holds for all i's. The calculation of the average of the same operator over the Hückel wave function cannot be done that easily, but the result is known and reads $\langle \hat{n}_1 \hat{n}_{1\uparrow} \rangle_{\text{Hückel}} = 1/4$ and we see that the RVB and LCAO (respectively, Pauling and Hückel) functions are by far nonequivalent. One may be even surprised by the extent of this nonequivalence. If one performs what in our days is called "wave function VB reading" [27], that is, calculates the projection of the Hückel wave function to the subspace spanned by the homeopolar configurations $|K_1\rangle$, $|K_2\rangle$, and $|D_1\rangle \div |D_3\rangle$ and other (singly, doubly, and triply ionic) configurations (Figure 4.5), one finds that all the homeopolar states together contribute to the Hückel function only 12.5% (1/8 – for more details, see below and Ref. [28]). The rest is supplied by the singly, doubly, and even triply ionic configurations depicted in Figure 4.5 with the weights given next to respective configuration classes.

The result looks pretty counterintuitively at a first glance: In the system where the atomic *charges* are all vanishing, the by far predominant contribution to the wave function is produced by the ionic configurations, which must be (and actually are) well balanced. Based on this analysis, one is forced to conclude that Pauling and Hückel had suggested two pretty much different wave functions for the benzene π-system so that the "common sense explanation" quoted above makes in fact no sense and "a rivalry between the MO and VB pictures" that is believed to be finally conciliated by the prominent authors [1] is in fact more than real. Turning to the solid-state context that is our main occupation in this chapter, we notice that there is a *theorem* [30] expressing the weight (sum of the squared overlaps) of the k-ionic configurations in the Hückel solution of a cyclic polyene (linear chain of carbon atoms supplied by one π-orbital each or a similar

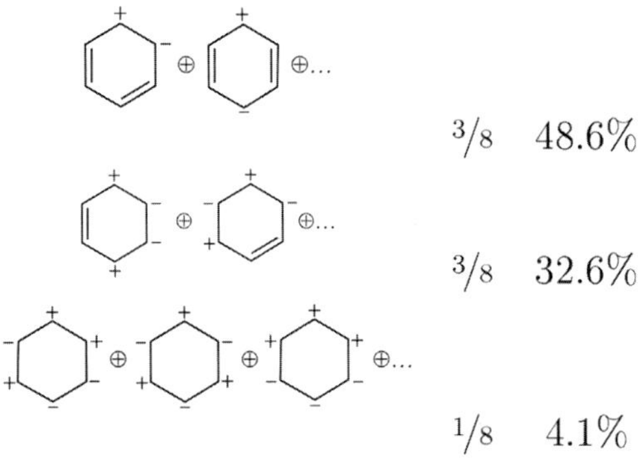

Figure 4.5 Exemplary singly, doubly, and triply ionic configurations of the π-system of benzene molecule with their weights in the expansion of the Hückel wave function and a correlated wave function [29] with a moderate interaction strength U. The weight of the homeopolar (RVB) configurations in the correlated wave function amounts the remaining 14.7%.

chain of the hydrogen atoms with 1s-AO each with periodic boundary conditions) (Figure 4.4) with $2n$ nodes (n-ene; benzene is a *triene*) as follows:

$$w_k \to \binom{n}{k} 2^{-n}; \quad w_k = w_{n-k} \tag{4.4}$$

The estimate for the weight of the RVB state from above is apparently w_0, which flows to zero when it comes to infinite number of nodes as characteristic for the solid state. This creates somewhat inconvenient situation: Although, one may expect that at least under certain conditions, the states of the type proposed by Pauling for benzene might exist in solids, there is no chance to reproduce them by any of available numerical methods of solid-state quantum chemistry since these latter by brute force impose the Hückel view of the electronic structure of solids (tight-binding approximation!) by using the Bloch theorem/sums. Thus, some special moves going beyond the standard tight-binding LCAO scheme are necessary to access these states.[2]

---

2) Currently, the plane waves are widely used for calculating electronic structure of crystals. At a first glance, they represent completely different approach to constructing basis states for such problems and the mentioned concerns seem not to apply to them. Nevertheless, in practice, pure plane waves are used rarely, and the so-called augmentation – that is adding atom-like states inside the so-called Wigner–Seitz spheres around each atom to cheaper mimic their nodal structure close to nuclei, is made. This produces a strange hybrid in which the original *plane waves* of which there was that much buzz in the beginning in fact appear only in the empty space between the spheres where they honestly serve ensuring required flexibility of the electronic states in the interatomic space – the feature very costly in the methods with fixed atomic basis states. Nevertheless, the major part of electronic density resides in the atom-like augmentation parts. It is particularly true when it comes to highly correlated d- or f-shells. The augmentation parts enter in the play effectively exactly as the Bloch sums so that the RVB states have no chance to emerge in the (augmented) plane wave setting either.

One more aspect needs to be mentioned. This is the nature of interactions producing the delocalized structures. The Hückel and Pauling models rely upon completely different types of the coupling. The Hückel model employs hopping of single electrons between the neighbor sites. Such an interaction when applied to either K- or D-benzene structures immediately produces ionic configurations not entering in the original basis of the homeopolar configurations shown in Figure 4.3. By contrast, the states $\Psi_{n\sigma}^{\dagger}|0\rangle$ are the eigenstates of the hopping operator of single electrons that are not changed under the action of the hopping operator. On the other hand, the exchange interaction couples the effective electronic spins residing in each of the sites in the six-membered cycle in each individual Kekulé and Dewar structures. However, as one can check (see, for example, Ref. [31]), the matrix element of the exchange between the Kekulé states $|K_1\rangle$ and $|K_2\rangle$ is vanishing. This means that in fact no resonance (in our days it is usually called a quantum mechanical superposition) of these states is possible – the double arrow connecting these two structures in Figure 4.3 is a fake. One needs additional states, namely, the Dewar structures (states), to ensure the coupling between them to arrive to the RVB state (Eq. (4.5)) to mediate the exchange interaction and thus to lower the energy due to quantum mechanical superposition of all five basis configurations:

$$\frac{\cos\varphi}{\sqrt{2}}(|K_1\rangle + |K_2\rangle) + \frac{\sin\varphi}{\sqrt{3}}(|D_1\rangle + |D_2\rangle + |D_3\rangle) \tag{4.5}$$

and by this generate the observed bond delocalization. It is thus necessary to realize that the Pauling and Hückel models of benzene not only employ very different wave functions but also recognize that this difference is stipulated by the difference of the underlying model of electron dynamics formalized by the Hamiltonian used in either setting: ultimately the antiferromagnetic Heisenberg one for the Pauling model and the one-electronic hopping in the case of Hückel. For that reason, either the very concept of resonance or the value of the resonance (stabilization) energy has absolutely different meaning and value within these two models. Populating six electrons into lowest three states of the one-electron hopping operator (Eq. (4.9)) – see below – we have

$$E = -8t, \tag{4.6}$$

which must be compared with the energy of a hypothetic configuration where the hopping is allowed only within three fixed bonds. In the spirit of the hopping operator (Eq. (4.6)), such a setting is equivalent to three copies of a two-site system with an extremely simple Hamiltonian:

$$H_2 = -t\sum_{\sigma}\left(c_{1\sigma}^{\dagger}c_{2\sigma} + c_{2\sigma}^{\dagger}c_{1\sigma}\right),$$

(actually ethylene or H_2 molecule) having the ground-state energy of $-2t$ so that three copies of it have the energy of $-6t$, which is by $2t$ higher than the Hückel ground state energy (Eq. (4.6)). This gives a specific value of the "resonance" characteristic for the Hückel setting. In the Pauling picture, benzene is described by the Heisenberg Hamiltonian. Its ground state is the RVB type with the energy of $-(\sqrt{13}-1)J$ to be compared with $-9J/4$ for three isolated (nonresonating) singlets in either of the two Kekulé structures. We see that the relative stabilizing effect of delocalization (resonance) of the bonds in the Heisenberg picture is much weaker than in the Hückel one.

If one considers either the Hückel Hamiltonian and Hückel wave function or the Heisenberg exchange Hamiltonian and the Pauling wave function, respectively, together, either is the *exact* solution of the corresponding ground-state problem. From the first glance, only one of the models can be correct, but we cannot decide which one is better. A problem is that the exchange interaction between electrons in all involved site orbitals of the usual Heisenberg form is an effective one: There is no fundamentally antiferromagnetic coupling between electrons/spins occupying otherwise orthogonal site AOs, which would enable them to form *singlet* rather than triplet pairs. A step toward resolving this complication is to imbed both settings into a more general model for the energy. Both Hamiltonians mentioned so far miss a very important ingredient – the electron–electron repulsion. The simplest medication against this flaw is provided by the Hubbard–Anderson model, which is the hopping (Hückel) one supplied by additional terms increasing the energy of ionic configurations. Formally, it is expressed by adding the terms of the form:

$$U\hat{n}_{i\downarrow}\hat{n}_{i\uparrow}$$

for each site. Here U is the energy ascribed to a pair of electrons occupying the same site with opposite spin projections (note that the operator multiplier in the above additive is nothing, but the projection operator counting the number of such pairs). Apparently, with such a term added to the diagonal energy, matrix elements of the k-ionic configurations acquire the contribution of kU. It had been shown yet quite long ago [29] that the weights of the ionic configurations in a correlated wave function are consistently suppressed and the symmetry relation for the weights of the dual ionicity (Eq. (4.4)) does not hold anymore (see rightmost column in Figure 4.5). This observation indicates the trend: The larger the electron–electron repulsion, the higher the weight of the homeopolar configurations eventually yielding the (RVB) state. On the other hand, it had been shown [32] that in the limit of very large U, the Hubbard–Anderson Hamiltonian reduces to the Heisenberg one with $J = 4t^2/U$. This latter can be considered as acting in the subspace of the homeopolar electronic functions: The strong Coulomb repulsion restricts the possibilities of electrons to move around the benzene π-system so that in this limit instead of a "free" hopping of electrons – one restricted solely by the antisymmetry of the many-electronic wave function – only a correlated motion is possible: that is, simultaneous (ex)change of the positions of the two electrons with oppositely oriented spins.

These considerations show that in fact one cannot and should not try to separate the issues of the Hamiltonian used and the method of its treatment since they are pretty much interpenetrating. The problem thus is that for the standard solid-state quantum chemical Hamiltonian containing the one-electron hopping of different magnitudes between different atomic orbitals within and between the unit cells together with the electron–electron repulsion terms, the true electronic wave function may have very little to do with the Hückel (tight-binding) one as characterized by the high contribution of the ionic configurations. Indeed, the parameters of the Hubbard–Anderson Hamiltonian scale at the first approximation as

$$U \propto \frac{Z}{n}; \quad t \propto \exp\left(-\frac{Zr}{n}\right), \tag{4.7}$$

where Z is the atomic number, n is the principal quantum number of an AO, and r is the interatomic separation. In the solid-state quantum chemical setting, each unit cell may contain several atoms, each containing several orbitals. The transition metal or rare-earth atoms contain those with different n values. Thus, different pairs of orbitals may have very different ratios U/t such that some of them fall in the parameters' range where the electronic structure is better described by the Hückel model, whereas for other pairs of orbitals the ratio U/t may get so that the contribution of the homeopolar configurations become dominating.

The problem is thus to implement in a tractable way the reduction/suppression of the ionic configurations or, in other words, to steer the solution into physically substantiated subspace of the wave functions in case it is necessary. The problematic itself is known under the disguise of "taking into account electronic correlation," to which several other chapters of this volume are devoted [33,34]. Technically, this problem can be differently distributed between manipulation with the trial wave function or with the Hamiltonian, which in this case turns into effective one. Transition from the bare Hamiltonian containing the hopping (kinetic energy) and Coulomb repulsion terms to the Heisenberg is one of the examples of reloading the information about the subspace accommodating the ground state upon an effective Hamiltonian, although it seems to be too direct and rigid.

The dominance of the homeopolar configurations in the description of the electronic structure of a solid does not automatically mean that the RVB state represents its best solution. By contrast, in a predominant number of cases, the Heisenberg Hamiltonian results in magnetically ordered structures (e.g., antiferromagnetic or more sophisticated ones). Such behavior is widespread and has well-known experimental manifestations: in magnetic neutron scattering, characteristic magnon contributions to the heat capacity, and, finally, a transition to magnetically unordered high-temperature paramagnetic phases. The magnetically ordered states (phases) form easily only for a restricted class of lattices – so-called bipartite ones when the nodes can be colored in white and black in an alternate manner. In this case, an antiferromagnetic interaction of the spins accordingly arrange them on the nearest-neighbor sites (those occupying the sites of different colors) and produce the so-called Néel state of an antiferromagnet.

For archetypic example of benzene, the Néel state although being unlikely for the real physical conditions can be easily built by setting spin-up electrons on the even sites and spin-down ones on the odd ones:

$$|N_1\rangle = c_{1\downarrow}^\dagger c_{2\uparrow}^\dagger c_{3\downarrow}^\dagger c_{4\uparrow}^\dagger c_{5\downarrow}^\dagger c_{6\uparrow}^\dagger |0\rangle$$

or vice versa ($|N_2\rangle$). Like either of the Kekulé structures, either of the Néel ones breaks the sixfold symmetry, which like in the Kekulé case at least formally reinstalls by taking the wave function of "antiferromagnetic" benzene as a linear combination of the two:

$$|N_1\rangle + |N_2\rangle.$$

Like in case of the Kekulé structures, two equivalent Néel structures are not mixed directly by the Hamiltonian. This means that the energy of the above-symmetrized state can be obtained by performing the calculation of the mean value of the Hamiltonian with either of the components with broken symmetry.

For lattices other than bipartite ones and among them first of all for the lattices featuring odd-membered rings of spins, for example, triangular ones – the spins cannot arrange antiferromagnetically, which leads to frustration, so the competition between the magnetically ordered (with the superstructure wave vector proportional to $4\pi/3$ in the specific case of the triangular lattice) and RVB (spin-liquid) states may lead to the preference of the latter. The Néel state, however, is not the ground state of the Heisenberg Hamiltonian, although it gives a plausible approximation to its energy [35]. It is subject to quantum fluctuations, more important for smaller values of effective spins. These fluctuations may steer the ground state to the RVB segment. This type of state had been assumed on the experimental grounds for a series of halocuprate compounds [36,37]. Corresponding *phases* are assumed to exist at extremely low temperatures. The most spectacular early example as well as fairly established experimentally had been provided by the $NiGa_2S_4$ compound featuring a two-dimensional triangular arrangement of $S = 1$ effective electronic spins residing in the d-shells of the Ni ions [38].

4.3
Formal Constructs

The very successful idea to describe benzene in terms of RVB [13] had been later [12] applied to the description of metals. The result was a very counterintuitive model of a metal that was in apparent contradiction with common sense, experiments, and the band theory. This idea was shortly forgotten and only decades later, Anderson proposed to apply it in a study of the Heisenberg model on triangular lattice [39], where the dominance of singlets was justified by the frustration. Soon after the discovery of high-temperature superconductivity (high-T_c), Anderson has reformulated the RVB ideas proposing a wave function *Ansatz* [40,41]. This wave function became an object of intense studies both numerical and analytical, using a number of approximations. Below we describe three approaches to tackle this wave function. These are (i) a numerically exact variational Monte Carlo study of the RVB wave function; (ii) the formalism allowing for calculation of approximate energy of an RVB (spin-liquid) state, namely, the auxiliary-boson theory; and (iii) its limiting case at the half-filling important for insulating systems with frustrating exchange interactions.

4.3.1
Hückel, Néel, and RVB States of Polyene

The simplest model illustrating the idea of the (resonating) valence bonds in infinite systems (aka "solids") is the model of an N-site long planar hydrocarbon molecule – polyene (Figure 4.6). The (Hückel) Hamiltonian for it in general case reads:

$$H = -t \sum_{n,\sigma} \left(c^{\dagger}_{n\sigma} c_{n+1\sigma} + c^{\dagger}_{n+1\sigma} c_{n\sigma} \right). \tag{4.8}$$

It is habitually expressed in terms of the creation operators $c^{\dagger}_{n\sigma}$ introduced above and their Hermitian conjugates $c_{n\sigma}$. As the creation operators increase the number of electrons in the system by 1, their Hermitean conjugates respectively reduce the number of electrons in the system by 1, thus being destruction (annihilation) operators. The annihilation operator applied to the vacuum state $|0\rangle$ with no electrons yields zero. This is the basis of all calculations that can be performed in the so far informally described second quantization technique. The electronic structure problem for polyene with the Hamiltonian (Eq. (4.8)) can be easily solved by transforming the fermionic operators from coordinate representation to the (quasi)momentum (or wavenumber) one:

$$a_{k\sigma} = \frac{1}{\sqrt{N}} \sum_{n} \exp(-ikn) c_{n\sigma},$$

$$k = \frac{2\pi}{N} j, \quad j \in \left[-\frac{N}{2} + 1, \frac{N}{2} \right],$$

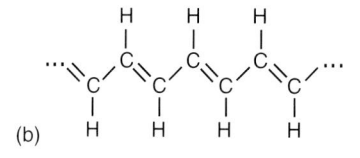

Figure 4.6 Two resonance structures of infinite polyene. (Reprinted with permission from Ref. [42].)

for each spin component σ. Such a transformation leads to the quasi-particle spectrum as follows:

$$\varepsilon_{k\sigma} = -2t \cos k, \tag{4.9}$$

so that the hopping Hamiltonian acquires the diagonal form:

$$H_p = \sum_{k,\sigma} \varepsilon_{k\sigma} a_{k\sigma}^\dagger a_{k\sigma} = \sum_{k,\sigma} \varepsilon_{k\sigma} \hat{n}_{k\sigma}, \tag{4.10}$$

($\hat{n}_{k\sigma}$ is the number of particles operator referring to those occupying the state with quasi-momentum – wavenumber – k and spin projection σ.)

The polyene contains N sites that is $2N$ spin-orbital states of which only N with the lowest energies have to be filled by N electrons in the ground state. These are the orbitals with $|k| < k_F = \pi/2$. Thus, the ground state of the N-site polyene is a Slater determinant (the product sign is an abbreviation of the above-mentioned string of the creation operators):

$$\Psi = \prod_{|k|<k_F,\sigma} a_{k\sigma}^\dagger |0\rangle. \tag{4.11}$$

The wave numbers k are distributed in the interval $[-\pi, \pi]$ of length 2π that play the role of the Brillouin zone in this case. In the case of an infinite (or at least very long) chain, the summation in (4.10) becomes an integral and the ground state energy is $-(4/\pi)\beta$ (per electron) to be compared with $-\beta$ – the similar quantity for a system of isolated bonds like those depicted in Figure 4.6a. We see that the energy of the resonance stabilization counted per electron is somewhat smaller in polyene: $-((4/\pi) - 1)t$, than in benzene: $-(t/3)$.

Turning to the description of the polyene with use of the Heisenberg Hamiltonian, we notice that the Néel state acquires the energy of $-J/4$ per spin, whereas the ground state of two coupled 1/2 spins is $-3J/4$, which produces an amount of $-3J/8$ per spin[3]: a considerable energy gain with respect to the Néel state even without any true resonance (the singlets in either of the Kekulé-like structures in Figure 4.6 remain isolated; the second structure restoring the symmetry does not in fact resonate with the first one, like K_1 and K_2). This reasoning shows that even the components of the RVB state have lower energy than the magnetically ordered state in one dimension. However, this conclusion reverts when applied to a simple square lattice with antiferromagnetic interactions between nearest neighbors. In it the Néel state is preferable ($-J/2$ per spin, whereas the isolated singlets produce as previously $-3J/8$ per spin). This allows one to think that in the lattices of intermediate effective coordination (e.g., featuring frustration), the magnetically ordered and RVB states can compete. Indeed, in graphene represented in Figure 4.7, each atom bears as average 3/2 bonds that produce the energy of $-3J/8$ per spin in its hypothetical Néel state – equal to that of the RVB component states produced by either of the bond distributions (Figure 4.7). Thus, it is not surprising that an RVB state of graphene had

---

3) The same amounts are valid for the benzene molecule.

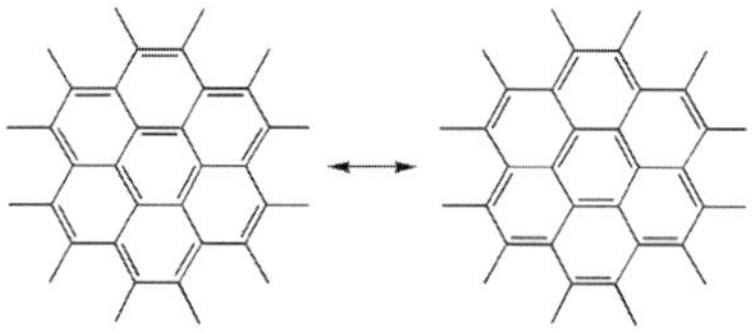

Figure 4.7 Two equivalent (resonating) configurations of valence bonds in the π-system of graphite monolayer (graphene).

been detected numerically [43] (see below). The example of triangular lattice is even more spectacular. In the latter, each site has six nearest neighbors that produce three bonds per site. However, the frustration prevents the usual antiferromagnetic ordering of the 1/2-spins when the mean value of their scalar product equals to $-1/4$. Instead of that a spiral ordered state is formed where the spins are placed in the plane and the neighbor ones form angles of 120° and 240° producing the mean value of the scalar product of the spin operators of $-1/8$. This as well results in an estimate of the energy of such magnetically ordered state of $-3J/8$ per spin degenerate with one of the valence bond ones.

4.3.2
Trial Variational RVB Wave Function

As proposed originally by Anderson, the RVB wave function is essentially a many-body state. It is in general difficult to evaluate analytically the expectation value of energy in such a state. Without employing any approximation, it could only be evaluated numerically. The most widespread technique applied for such studies is variational Monte Carlo (VMC) minimizing the variational energy with stochastic bias.

Before discussing the RVB wave function, it is worth to discuss once again the wave function – Hamiltonian dualism typical for the variational methods. A wave function with a fixed structure and depending on a number of parameters is assumed to approximate the true ground state of a quantum system. The expectation value of the Hamiltonian over this wave function, called (fixed form) variational or trial wave function, is minimized as a function of its parameters, which are called variational parameters (VPs). Subsequently, at minimum, the properties of the least energy variational wave function can be analyzed, thanks to its known structure. Two variational wave functions can be easily compared based on their expectation values. It is clear that a variational wave function can only exhibit features prescribed by its structure. On the other hand, the impossibility to gain energy by introducing a new VP in the trial wave function (e.g., superconductivity) can be used as an indication to rule out the possibility of the corresponding *order parameter* (in variance with VPs, OPs are observable physical quantities and do not necessarily coincide with the VPs) in a given system. In this sense, the RVB wave function was proposed to explain several properties of high-T_c superconductors, while the very properties of the wave function are a subject of active studies nowadays. The Hamiltonian under consideration can vary from Hubbard, t–J, t–U–J, and so on, which fixes the optimal values of VPs. In this setting, the wave function becomes more important than the exact form of the Hamiltonian. The variational approach was shown to correctly reproduce both the weak [44,45] and the strong [46] coupling limits of the Hubbard model.

A typical VMC problem eventually leading to an RVB state consists in searching by, for example, the stochastic minimization methods for the best wave function of the form:

$$|\Psi\rangle = A\hat{P}_N\hat{P}_J\hat{P}_G|\Psi_{BCS}\rangle, \quad (4.12)$$

where A is a normalization factor, $|\Psi_{BCS}\rangle$ is a Bardeen–Cooper–Schrieffer (BCS) wave function written either as

$$|\Psi_{BCS}\rangle = \prod_k \left(u_k + v_k a^\dagger_{k\uparrow} a^\dagger_{-k\downarrow}\right)|0\rangle \quad (4.13)$$

or equivalently:

$$|\Psi_{BCS}\rangle = \prod_{k,\sigma} \left(u_k a_{k\sigma} - 2\sigma v_k a^\dagger_{-k,-\sigma}\right)|0\rangle \quad (4.14)$$

using $\hat{P}_G$:

$$\hat{P}_G = \prod_i \left(1 - \alpha_0 \hat{n}_{i\uparrow}\hat{n}_{i\downarrow}\right), \quad (4.15)$$

the "partial" Gutzwiller projector since it removes the double occupancy not completely, but only to the extent stipulated by the variation parameter α_0, and the Jastrow factor $\hat{P}_J$:

$$\hat{P}_J = e^{-\alpha_1 \sum_{\langle ij \rangle} \hat{n}_i \hat{n}_j - \alpha_2 \sum_{\langle\langle ij \rangle\rangle} \hat{n}_i \hat{n}_j - \cdots}, \quad (4.16)$$

where "..." stands for the summation over next, next-next, and so on nearest neighbor sites (i.e., all those possible on a finite-size sample) applied to suppress the charge density fluctuations at longer distances. In the second form of the BCS wave function, one easily recognizes a single Slater determinant built of (quasi)particles representing superposition of electrons and holes, so that the resulting BCS function does not have definite number of electrons. This is restored by the projection operator $\hat{P}_N$ onto a fixed number of particles. The Slater determinant form of the BCS function considerably simplifies the calculations. On the other hand, it can be easily shown that applying the $\hat{P}_N$ projection operator to the BCS function breaking the particle conservation symmetry produces precisely an RVB-like state. We show here an exemplary derivation to give the reader a flavor of calculations used in our days [47,48]. We start from the BCS function and first factorize the constant terms for each k:

$$\prod_k \left(u_k + v_k a^\dagger_{k\uparrow} a^\dagger_{-k\downarrow}\right)|0\rangle = \left(\prod_k u_k\right) \prod_k \left(1 + \frac{v_k}{u_k} a^\dagger_{k\uparrow} a^\dagger_{-k\downarrow}\right)|0\rangle, \quad (4.17)$$

where the numerical coefficients u_k, v_k are subject to the normalization condition and can be determined in the single-determinant approximation. Nilpotence of the fermion operators allows transforming this expression beyond recognition. Indeed, the functions of operators are defined as Taylor series of analytic functions where the "numeric" argument x is replaced by an operator. Due to nilpotence of the fermion creation operators (its square and in the row all higher powers are vanishing), the following is true:

$$\left(1 + \frac{v_k}{u_k} a^\dagger_{k\uparrow} a^\dagger_{-k\downarrow}\right) = \exp\left(\frac{v_k}{u_k} a^\dagger_{k\uparrow} a^\dagger_{-k\downarrow}\right) \tag{4.18}$$

Moreover, the operator products $a^\dagger_{k\uparrow} a^\dagger_{-k\downarrow}$ for different values of k commute with each other so that we can use the fact that the product of exponents is the exponent of the sum of their powers:

$$\prod_k \left(1 + \frac{v_k}{u_k} a^\dagger_{k\uparrow} a^\dagger_{-k\downarrow}\right) = \prod_k \exp\left(\frac{v_k}{u_k} a^\dagger_{k\uparrow} a^\dagger_{-k\downarrow}\right) = \exp\left(\sum_k \left[\frac{v_k}{u_k} a^\dagger_{k\uparrow} a^\dagger_{-k\downarrow}\right]\right) \tag{4.19}$$

To ensure the singlet character of the original BCS state, the coefficients u_k, v_k must be even functions of k. Then returning from the wavenumber (k) to the site representation, we get

$$\sum_k \left[\frac{v_k}{u_k} a^\dagger_{k\uparrow} a^\dagger_{-k\downarrow}\right] = \frac{1}{2} \sum_{n,m} w_{nm} \left(c^\dagger_{n\uparrow} c^\dagger_{m\downarrow} - c^\dagger_{n\downarrow} c^\dagger_{m\uparrow}\right), \tag{4.20}$$

where

$$w_{nm} = \frac{1}{N} \sum_k \frac{v_k}{u_k} \exp(ik(n-m)).$$

We see precisely the singlet pair states like those entering as multipliers in the Rumer basis states for the benzene RVB wave function Eq. (4.1). Inserting the sum of the singlet pair states back in the exponent, we get

$$|\Psi_{\text{BCS}}\rangle = \left(\prod_k u_k\right) \exp\left[\frac{1}{2} \sum_{n,m} w_{nm} \left(c^\dagger_{n\uparrow} c^\dagger_{m\downarrow} - c^\dagger_{n\downarrow} c^\dagger_{m\uparrow}\right)\right]. \tag{4.21}$$

Taylor expansion of the exponent contains all powers, that is, it is the superposition of the states with all numbers of particles. The projection of such a state on a specified number of particles is obtained by simply taking the term of the corresponding power from that expansion:

$$\hat{P}_N |\Psi_{\text{BCS}}\rangle = \frac{1}{(N/2)!} \left(\prod_k u_k\right) \left(\frac{1}{2} \sum_{n,m} w_{nm} \left(c^\dagger_{n\uparrow} c^\dagger_{m\downarrow} - c^\dagger_{n\downarrow} c^\dagger_{m\uparrow}\right)\right)^{N/2} |0\rangle$$

The Gutzwiller projection operator then takes care about removing the double site occupancies.

This way of doing things is a specific example of a very general trick: first building a Slater determinant with orbitals of broken symmetry (in this case, the

BCS formed of the states being not the particle states, but partially particle, partially hole) and then projecting such broken-symmetry many-electron state on the correct symmetry. This trick is particularly useful since it frequently happens that different components having themselves broken symmetry are not coupled with each other and thus only one of them, namely, a broken symmetry Slater determinant itself, suffices for the energy determination.

Now we exemplify the application of the complete numeric VMC method by considering an extended t–U–J model in two dimensions:

$$\hat{H} = -t \sum_{\langle ij \rangle} \sum_\sigma \left(c_{i\sigma}^\dagger c_{j\sigma}^\dagger + H.c. \right) + U \sum_i \hat{n}_{i\uparrow} \hat{n}_{i\downarrow}$$
$$+ J \sum_{\langle ij \rangle} \left(S_i \cdot S_j - \frac{1}{4} \hat{n}_i \hat{n}_j \right) + V \sum_{\langle ij \rangle} \hat{n}_i \hat{n}_j, \quad (4.22)$$

where, in addition to the on-site repulsion U, we add a nearest-neighbor spin-exchange and a charge-repulsion term, with strengths J and V, respectively. The latter nearest-neighbor interactions compete, J favoring a singlet pairing away from half-filling while V opposes it. For $V = 0$ and U strictly equal to ∞, (4.22) reduces to the standard t–J model, which also corresponds to the large U limit of the pure Hubbard model, in which case $J \to 4t^2/U$. However, contrary to the latter, model (4.22) for $J > V$ is undoubtedly a superconductor at weak coupling (U, V, and J all much smaller than t) also in the Hartree–Fock approximation. For this reason, model (4.22) is more suitable to address electron–electron correlation effects on d-wave superconductivity. Moreover, since V scales charge-density while J spin-density operators, the presence of both gives the opportunity to test the effect of U on charge and spin.

Numerical optimization of the mean value of Eq. (4.22) over the wave function $|\Psi\rangle$ of the form Eq. (4.12) reveals the enhancement of the superconducting correlations (originating from $|\Psi_{BCS}\rangle$ through the attraction in the Cooper channel coming from the Heisenberg term) at large on-site repulsion U due to suppression of the local charge fluctuations (thanks to Gutzwiller projection) and long-range charge ones (thanks to Jastrow factors). These superconducting correlations, however, do not necessarily lead to superconductivity itself, since the mobility of charge carriers is heavily dumped by U. Moreover, at half-filling, and sufficiently large U, the system undergoes a transition to the Mott insulating state, although retaining sizable superconducting fluctuations. The combination of features occurring in this state is the ideal implementation of the Anderson's RVB ideas.

At small doping δ, the system becomes a narrow (proportional to δ) band metal featuring superconducting fluctuations. A further increase of δ enhances the superconductivity with an optimal $\delta_{opt} \approx 0.16$ comparable to the optimal doping in cuprates. It can be seen that spin and charge interactions (J and V, respectively) differently influence the superconducting correlations at large U (see Figure 4.8). In the weak coupling limit, it can be shown that both V and J contribute at the same order of magnitude to the attraction in the superconducting channel (formation of Cooper pairs). That is why, if $J > 0$ and $V < 0$ that is

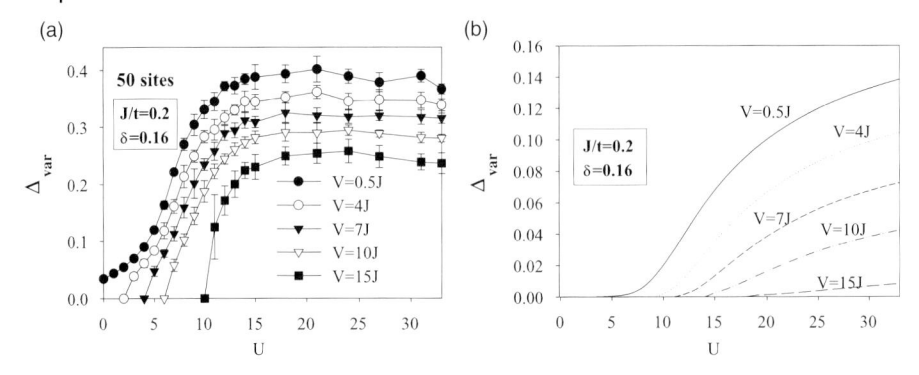

Figure 4.8 Variational gap as function of U for different values of V within variational Monte Carlo (a) and Gutzwiller approximation (b). (Reproduced with permission from Ref. [49]. Copyright 2003, American Physical Society.)

compete, then at $V \sim J$ the superconductivity vanishes. This is not the case in the strong coupling limit (large U and V) when even an order of magnitude larger V is unable to destroy the superconducting correlations (Δ_{var} in Figure 4.8). It happens since at large U the interactions J, V and the kinetic energy acquire different renormalizations: J remains almost unchanged, V scales down by a factor of δ^2, while $t \to t\delta$. Since the interactions enter the BCS gap equations multiplied by the density of states at Fermi level $\rho^* \sim 1/(t\delta)$, we indeed obtain that $\rho^* V \to \delta V/t$, while $\rho^* J \to J/(t\delta)$. It follows that large on-site repulsion U enhances spin-exchange-generated superconductivity by a factor of $1/\delta$, while suppressing density-attraction-generated one by a factor of δ.

Figure 4.8 also suggests that a qualitative analysis of a doped RVB state can be made by using a so-called Gutzwiller approximation. In this approach, the effect of large U is taken into account by ascribing to each term of (4.22) a renormalization factor depending on δ from very beginning (for more details, see Ref. [47]). Such a renormalization factor is determined on the basis of simple combinatorial considerations [49,50].

As shown in Ref. [49], the VP Δ_{var} is not to be confused with the true superconducting order parameter in RVB states (Δ_{LRO}) defined through the long-range four-fermion correlation function f:

$$\Delta_{\text{LRO}} = \frac{1}{2}\sqrt{f - f_{\text{norm}}},$$

and

$$f = \lim_{\Delta \to \infty} \sum_{\sigma,\sigma'} \left(c^\dagger_{r,\sigma} c^\dagger_{r\pm 1,-\sigma} c_{r+\Delta\pm 1,-\sigma'} c_{r+\Delta,\sigma'} \right).$$

The OP Δ_{LRO} goes to zero as $\delta \to 0$ and $U \to \infty$, due to evanescence of electron mobility in that case (f_{norm} stands for the same correlation function as f, but calculated in the normal state at $\Delta_{\text{var}} = 0$). It follows that an insulating RVB state is not identical to a superconductor, although has sizable short-range superconducting correlations without the phase coherency, ensuring its unique properties. It is worth mentioning that a similar approach, involving an RVB-derived antisymmetrized geminal power with the Jastrow factor variational wave function, was also applied at the *ab initio* level to a number of carbon-based molecules (C_2 and C_6H_6) with the conclusions extending to graphene [51], which confirm the possibility of the RVB states in these systems.

4.3.3
Auxiliary-Boson Theory

Numerical results obtained for models with finite numbers of sites and electrons are useful, but do not allow to derive general statements. These latter require analytical treatment that can be performed only under additional assumptions. When speaking about RVB states, one cannot avoid mentioning (and using) the auxiliary-boson theory (ABT), also called slave-boson theory. Here, we briefly sketch its formulation for the case of the single-band *t–J* model. We refer the readers, interested in learning more details of ABT and its generalizations, to the multiorbital case to the dedicated reviews and book chapters [48,52,53].

Within the ABT, one usually needs to reduce the contribution of the ionic (doubly occupied) configurations in the ground state steering the solution into homeopolar subspace. This is done by moving the restrictions on the functional (sub)space from the wave function to the (effective) Hamiltonian. In the most general (VMC) setting, the reduction of the ionic configurations is in a direct way done by the above $\hat{P}_G$, projection operator factor. Comparable effect is achieved by the replacement

$$c_{i\sigma}^\dagger \to \tilde{c}_{i\sigma}^\dagger = c_{i\sigma}^\dagger(1 - \hat{n}_{i-\sigma})$$
$$c_{i\sigma} \to \tilde{c}_{i\sigma} = c_{i\sigma}(1 - \hat{n}_{i-\sigma})$$

performed on the level of the electronic variables. This replacement ensures that the double occupancy is not accessible through the action of the Hamiltonian itself. The problem is that the "tilded" operators do not satisfy the Fermi anti-commutation relations. Nevertheless, the above replacement represents a correct idea: that of renormalizing the original Fermi operators through a multiplier. The ABT method develops these observations by introducing the spin and charge degrees of freedom of each electron explicitly as separate variables. The electron destruction operator $c_{i\sigma}$ is expressed as

$$c_{i\sigma} = b_i^\dagger f_{i\sigma}, \tag{4.23}$$

where $f_{i\sigma}$ is assigned Fermi and $b_i^\dagger$ Bose statistics, respectively, are dubbed a spinon $f_{i\sigma}$ and a holon b_i operators. So redefined operators $c_{i\sigma}$ do not automatically obey the fermionic commutation relation: $c_{i\sigma}^\dagger c_{i\sigma} + c_{i\sigma} c_{i\sigma}^\dagger = 1$, which is recovered by imposing the local constraints:

$$\sum_\sigma f_{i\sigma}^\dagger f_{i\sigma} + b_i^\dagger b_i = 1, \tag{4.24}$$

at each site i. It can be easily shown that once (4.24) is fulfilled, the mapping from $c_{i\sigma}$ to $f_{i\sigma}$ and b_i is exact [54–58]. In these new variables, the Hamiltonian of the t–J model rewrites:

$$H_{ab} = -\sum_{\langle ij \rangle, \sigma} \left(t_{ij} f_{i\sigma}^\dagger f_{j\sigma} b_j^\dagger b_i + \text{h.c.} \right) + H_{ex} - \mu \sum_{i,\sigma} f_{i\sigma}^\dagger f_{i\sigma}$$
$$- \sum_i \lambda_i \left(\sum_\sigma f_{i\sigma}^\dagger f_{i\sigma} + b_i^\dagger b_i - 1 \right),$$

where we included local constraints (4.24) with Lagrange multipliers λ_i and the chemical potential term, while H_{ex} is given by

$$H_{ex} = J \sum_{\langle ij \rangle} \left[\frac{1}{2} \left(f_{i\uparrow}^\dagger f_{i\downarrow} f_{j\downarrow}^\dagger f_{j\uparrow} + f_{i\downarrow}^\dagger f_{i\uparrow} f_{j\uparrow}^\dagger f_{j\downarrow} \right) \right.$$
$$+ \frac{1}{4} \left(f_{i\uparrow}^\dagger f_{i\uparrow} - f_{i\downarrow}^\dagger f_{i\downarrow} \right)\left(f_{j\uparrow}^\dagger f_{j\uparrow} - f_{j\downarrow}^\dagger f_{j\downarrow} \right)$$
$$\left. + \frac{\alpha}{4} \left(f_{i\uparrow}^\dagger f_{i\uparrow} + f_{i\downarrow}^\dagger f_{i\downarrow} \right)\left(f_{j\uparrow}^\dagger f_{j\uparrow} + f_{j\downarrow}^\dagger f_{j\downarrow} \right) \right].$$

In the above formula, the contributions from bosonic variables of the type $(1+b_i^\dagger b_i) \times (1+b_j^\dagger b_j)$ were neglected since $(1+b_i^\dagger b_i)(1+b_j^\dagger b_j) \approx 1$. If $\alpha = -1$, H_{ex} can be rewritten in a more compact way stressing the singlet correlations:

$$H_{ex} = -\frac{J}{2} \sum_{\langle ij \rangle} \left(f_{j\uparrow}^\dagger f_{i\uparrow} + f_{j\downarrow}^\dagger f_{i\downarrow} \right)\left(f_{i\uparrow}^\dagger f_{j\uparrow} + f_{i\downarrow}^\dagger f_{j\downarrow} \right).$$

The mean-field ABA derives from the above by introducing the following averages serving as VPs of the method:

$$\chi_\tau = \langle f_{i\sigma}^\dagger f_{i+\tau\sigma} \rangle,$$
$$\eta_\tau = \langle b_i^\dagger b_{i+\tau} \rangle, \qquad (4.25)$$
$$\Delta_\tau^* = \langle f_{i\uparrow}^\dagger f_{i+\tau\downarrow}^\dagger \rangle.$$

Here, the first two VPs are usually called fermion and boson bond order parameters, respectively, while the third one – the RVB order parameter. A usual simplification made in the ABA is to replace the local constraints (Eq. (4.24)) by the global ones by setting equal all Lagrange multipliers: $\lambda_i = \lambda$, so that λ starts playing the role of chemical potential for (auxiliary) bosons. After that the auxiliary boson approximate (ABA) mean-field Hamiltonian becomes

$$H_{ABMF} = \sum_{k,\sigma} \varepsilon_k f_{k\sigma}^\dagger f_{k\sigma} + \sum_k \omega_k b_k^\dagger b_k - \sum_k \left(F_k f_{k\uparrow}^\dagger f_{-k\downarrow}^\dagger + \text{h.c.} \right) + H_0, \qquad (4.26)$$

where H_0 is a c-number:

$$H_0 = \frac{J}{2}(3+\alpha)\left(\chi_x^2 + \chi_y^2\right) + \frac{J}{2}(3-\alpha)\left(|\Delta_x|^2 + |\Delta_y|^2\right)$$
$$+ 4\sum_\tau t_\tau \eta_\tau \chi_\tau - (\mu_f + \lambda)\delta - \frac{J\alpha}{2}(1-\delta)^2.$$

and the following notations are introduced:

$$\varepsilon_k = -2\sum_\tau t_\tau \eta_\tau \cos k_\tau - \frac{J}{2}(3+\alpha)\left(\chi_x \cos k_x + \chi_y \cos k_y\right) - \mu_f,$$
$$\mu_f = \mu + \lambda - \alpha(1-\delta)J,$$
$$\omega_k = -4\sum_\tau t_\tau \chi_\tau \cos k_\tau - \lambda,$$
$$F_k = \frac{J}{2}(3-\alpha)\left(\Delta_x \cos k_x + \Delta_y \cos k_y\right).$$

The self-consistency conditions within ABA read as follows:

$$\Delta_\tau = \frac{1}{N}\sum_k \frac{F_k}{2E_k} \cos k_\tau \tanh \frac{E_k}{2T},$$
$$\chi_\tau = -\frac{1}{N}\sum_k \frac{\varepsilon_k}{2E_k} \cos k_\tau \tanh \frac{E_k}{2T},$$
$$\eta_\tau = \frac{1}{N}\sum_k \frac{\cos k_\tau}{e^{\omega_k/T} - 1}, \qquad (4.27)$$
$$1-\delta = \frac{1}{N}\sum_k \left[1 - \frac{\varepsilon_k}{E_k} \tanh \frac{E_k}{2T}\right],$$
$$\delta = \frac{1}{N}\sum_k \frac{1}{e^{\omega_k/T} - 1}.$$

Once the self-consistency is reached, the free energy F per site can be represented:

$$\frac{F}{N} = -2\frac{T}{N}\sum_k \ln\left(2\cosh\frac{E_k}{2T}\right) + \frac{T}{N}\sum_k \ln\left(1 - \exp\left(-\frac{\omega_k}{T}\right)\right) + H_0$$

as a sum of entropy contributions of spinons (fermions) and holons (auxiliary bosons) and of the internal energy. Apart from the chemical potentials μ (μ_f), there are three types of VPs: η_τ, χ_τ, and Δ_τ equal for each pair of sites separated by τ. If only $\eta_\tau, \chi_\tau \neq 0$, while $\Delta_\tau = 0$, the system is said to be in the uniform RVB phase (uRVB). If, in addition to η_τ and χ_τ, also $\Delta_\tau \neq 0$, the phase is called a singlet RVB (sRVB) one. Several symmetry choices of the possible solutions should be commented. Since the range of the interaction in the employed Hamiltonian is limited to the next-nearest-neighbor distance, the range of τ in (4.25) does so as well. On a 2D square lattice, τ can be classified according to the lattice symmetries as shown in Table 4.1.

Among other applications, ABT was used in Ref. [43] to describe the graphite individual sheets and graphitic compounds in order to explore the possibility of superconductivity upon doping. The doping was assumed to come from several sources: impurities, intercalation, interlayer coupling, and so on. It was shown

Table 4.1 Classification of nearest-neighbor superconducting order parameter ($\Delta(k_x, k_y)$) and the corresponding phases on a 2D square lattice.

Type	Δ	χ	ϑ
$d_{x^2-y^2}$	$\Delta_y = -\Delta_x$	$\chi_y = \chi_x$	$\vartheta = \pi$
s	$\Delta_y = \Delta_x$	$\chi_y = \chi_x$	$\vartheta = 0$
$s + id_{x^2-y^2}$	$\Delta_y = i\Delta_x$	$\chi_y = \chi_x = 0$	$\vartheta = \frac{\pi}{2}$ a)
1D	$\Delta_y = 0, \quad \Delta_x \neq 0$	$\chi_y = 0, \quad \chi_x \neq 0$	—
Flux	$\Delta_x = \Delta_y = 0$	$\chi_\tau = \chi_0 e^{i\vartheta_\tau}$	b)

Here the angle ϑ is defined as $\Delta_y = e^{i\vartheta}\Delta_x$.
a) In case of $s + id_{x^2-y^2}$ phase, an alternative (and more general) definition of the order parameter reads as $\Delta(k_x, k_y) = s(\cos k_x + \cos k_y) + d(\cos k_x - \cos k_y)$ with two arbitrary complex numbers s and d (see Ref. [59]).
b) More details on flux phase are reported in Refs [48,60].

that the d-wave order parameter is favored over the s-wave one. Moreover, it was shown that the d-wave superconducting solution may also break the time-reversal symmetry, thus opening the possibility to the coexistence with ferro-electricity (charge density waves [61]). However, it is not clear to which extent the results of a mean-field-like theory can be trustfully applied to a quasi-2D material with nonnegligible interaction.

4.4
Mean Field RVB as Applied to the Heisenberg Model

Exactly at half-filling and large U, the Hubbard model reduces to the Heisenberg one. Considering the equations of ABT (4.27), one can easily see significant simplifications in this limit. Indeed, the condition $\delta = 0$ satisfies by the values of energies of auxiliary bosons: $\omega_k \to \infty$, which induces the self-consistent values for the VPs $\eta_k = 0$. If the exchange interactions resulting from this limit transition are frustrated, then the sought ground state has chances to be an RVB state. The half-filling case, in principle, can be treated within the general formalism, described in the previous chapter. However, we will show here an alternative approach, which starts directly from the Heisenberg model and gives the same results (Figure 4.9).

Starting from the Heisenberg Hamiltonian with the interactions J_τ acting between the sites connected by lattice vectors τ, the Hamiltonian can be written:

$$H = \sum_{i,\tau} J_\tau \mathbf{S}_i \mathbf{S}_{i+\tau}. \tag{4.28}$$

We express each of the spin operators in (4.28) through the effective fermionic (above spinon) operators:

$$S_i^\gamma = \frac{1}{2} \sum_{\alpha,\beta} \sigma_{\alpha\beta}^\gamma f_{i\alpha}^\dagger f_{i\beta},$$

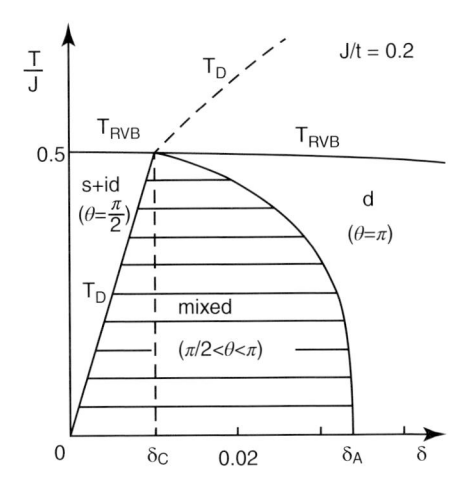

Figure 4.9 The phase diagram of the 2D t–J model on the δ–T plane. Here T_{RVB} and T_D are the onset temperatures of RVB and coherent motion of both spinons and holons, respectively. RVB order parameter is $s+id$, mixed or d type depending on δ and T. (Reprinted with permission from Ref. [62].) Reprinted with permission from [53], Japanese Chemical Society.

where $\sigma^\gamma_{\alpha\beta}$ are the Pauli matrices. This ensures that the wave function of the system is squeezed to the subspace of homeopolar configurations from very beginning. Substituting these operators into (4.28), we end up with the four-fermion interaction terms. This Hamiltonian is then solved in the mean-field approximation, assuming the decoupling of the four-fermion products with use of the following normal and anomalous averages:

$$\langle f^\dagger_{i,\alpha} f_{i+\tau,\beta}\rangle = \delta_{\alpha,\beta}\xi_\tau$$
$$\langle f^\dagger_{i,\alpha} f^\dagger_{i+\tau,\beta}\rangle = 2\alpha\delta_{\alpha,-\beta}\Delta^*_\tau.$$

With these notations, the average of (4.28) becomes

$$\langle H \rangle = 3N\sum_\tau J_\tau(\xi_\tau^2 + |\Delta_\tau|^2).$$

The corresponding mean-field Hamiltonian transformed to the k-space reads as follows:

$$H_{MF} = 3N\sum_{k,\tau} J_\tau \cos k\tau \left[\xi_\tau\left(f^\dagger_{k\uparrow}f_{k\uparrow} - f_{-k\uparrow}f^\dagger_{-k\uparrow}\right) + \Delta^*_\tau f_{-k\downarrow}f_{k\uparrow} + \Delta_\tau f^\dagger_{k\uparrow}f^\dagger_{-k\downarrow}\right] \quad (4.29)$$
$$+ 3N\sum_{k,\tau} J_\tau \xi_\tau \cos k\tau - \langle H\rangle.$$

It can be rewritten in a more compact way by introducing the Nambu spinors:

$$\Psi_k = \begin{pmatrix} f_{k,\uparrow} \\ f^\dagger_{-k,\downarrow} \end{pmatrix}; \quad \Psi^\dagger_k = \left(f^\dagger_{k,\uparrow}, f_{-k,\downarrow}\right).$$

Then in the Nambu notations, the first row of H_{MF} becomes

$$= \sum_k \Psi_k^\dagger H_k \Psi_k,$$

where H_k are 2×2 matrices defined as

$$H_k = \begin{pmatrix} \xi_k & \Delta_k \\ \Delta_k^* & -\xi_k \end{pmatrix},$$

with

$$\xi_k = 3\sum_\tau J_\tau \xi_\tau \cos k\tau,$$
$$\Delta_k = 3\sum_\tau J_\tau \Delta_\tau \cos k\tau.$$

H_k's are easily diagonalized, immediately giving the eigenvalues' spectrum: $\mp E_k$, with $E_k = \sqrt{\xi_k^2 + |\Delta_k|^2}$ and the eigenvectors' matrices:

$$S_k = \begin{pmatrix} u_k & v_k \\ -v_k^* & u_k \end{pmatrix},$$

where

$$u_k = \sqrt{\frac{E_k - \xi_k}{2E_k}},$$

$$v_k = \frac{\Delta_k}{\sqrt{2E_k(E_k - \xi_k)}}.$$

In terms of the eigenvectors of H_k (we denote them Φ_k: $\Phi_k = S_k \Psi_k$, $\Phi_k^\dagger = (\alpha_k^\dagger, \beta_k^\dagger)$), the first row of H_{MF} assumes the canonical form:

$$H_{\mathrm{MF}} = \sum_k \Phi_k^\dagger \begin{pmatrix} -E_k & 0 \\ 0 & E_k \end{pmatrix} \Phi_k = \sum_k E_k \left(\beta_k^\dagger \beta_k - \alpha_k^\dagger \alpha_k \right).$$

The population of the eigenstates $|\alpha_k\rangle$ and $|\beta_k\rangle$ obey the Fermi–Dirac statistics at temperature T and energies $\mp E_k$ respectively. To close the self-consistency conditions, we note that

$$\xi_\tau = \langle f_{i,\alpha}^\dagger f_{i+\tau,\alpha}\rangle = -\frac{1}{N}\sum_k (u_k^2 - v_k^2)\left\{ f\left(-\frac{E_k}{T}\right) - f\left(\frac{E_k}{T}\right)\right\}$$

$$2\alpha\delta_{\alpha,-\beta}\Delta_\tau^* = \langle f_{i,\alpha}^\dagger f_{i+\tau,\beta}^\dagger\rangle = \frac{1}{N}\sum_k u_k v_k \left\{ f\left(-\frac{E_k}{T}\right) - f\left(\frac{E_k}{T}\right)\right\}.$$

(f stands for the Fermi–Dirac distribution function). After further simplifications, the self-consistency equations becomes

$$\xi_\tau = \frac{1}{2N}\sum_k e^{ik\tau}\frac{\xi_k}{E_k}\tanh\frac{E_k}{2T}, \qquad (4.30)$$

$$\Delta_\tau = \frac{1}{2N}\sum_k e^{ik\tau}\frac{\Delta_k}{E_k}\tanh\frac{E_k}{2T}. \qquad (4.31)$$

At this point, a number of observations have to be made. First of all, it is surprising to see how a system of fermions having kinetic part and superconducting correlations emerges out from a purely insulating spin exchange model. The answer to this riddle is that the fermions meant by this theory are not the original physical charged electrons, rather the fictitious spinons bearing no charge and thus carrying no current. Second, if the number M of different couplings J_τ is less than or equal to 3, it is convenient to introduce the SU_2 invariant solution. Namely, such a solution that

$$\xi_\tau \xi_{\tau'} + \Delta_\tau \Delta_{\tau'}^* = 0, \qquad \forall \tau \neq \tau'.$$

In this way, E_k becomes

$$E_k = 3\sqrt{\sum_\tau J_\tau (\xi_\tau^2 + |\Delta_\tau|^2) \cos^2 k\tau}.$$

Third, if we have several Heisenberg couplings along different directions τ_i, $i = 1, \ldots, M$, how many independent parameters ξ_i, Δ_i can we have? Taking into account that ξ_i's are real, while Δ_i's may, in general, be complex, one would naively assume a total of $3M$ independent parameters. However, it can be easily demonstrated that at each τ_i, instead of three independent parameters there is only one independent parameter $\zeta_i = \sqrt{\xi_i^2 + |\Delta_i|^2}$. Indeed, if we set $\xi_i = \zeta_i \cos \theta_i$ and $\Delta_i = \zeta_i e^{\phi_i} \sin \theta_i$, we immediately see that two equations (4.30) and (4.31) reduce to a single one for ζ_τ not depending on ϕ_i, θ_i:

$$\zeta_\tau = \frac{3J_\tau}{2N} \sum_k \frac{\cos^2 k\tau}{E_k} \tanh \frac{E_k}{2T}. \tag{4.32}$$

Here, we used the orthogonality of $\cos k\tau$ and $\sin k\tau$ functions over the Brillouin zone and the positiveness and periodicity of $(1/E_k) \tanh(E_k/2T)$ with a period of at least 2π in units of inverse lattice spacing. Summarizing, we have demonstrated that the number of independent parameters is actually M instead of $3M$.

Different ζ_τ's are coupled through E_k so that Eq. (4.32) represents a system of nonlinear equations that admit the solutions with several nonvanishing ζ_τ's. Such an interoperability between ξ_τ and Δ_τ is due to the fact that the pure isotropic Heisenberg model produces identical contributions to the effective kinetic and superconducting coupling. If the true kinetic energy term (proportional to t) is retained in the Hamiltonian ξ_τ and Δ_τ also remain independent and a usual BCS-type solution recovers for (antiferromagnetic) $J_\tau > 0$ in the t–J model.

As already mentioned, Eq. Eq. (4.32) allows for the coexistence of resonating valence bonds in different directions. As was found in Ref. [63] and shown in Figures 4.10 and 4.11, a number of phases can be identified in a 2D Heisenberg model on rectangular lattice with couplings along the nearest neighbors (J_a and J_c) and along the diagonal, corresponding to the next-nearest ones (J_{ac}) – the c–a– ca-model. This model has been recently proposed to describe an important case of copper carbodiimide (CuNCN) – very probably the first material

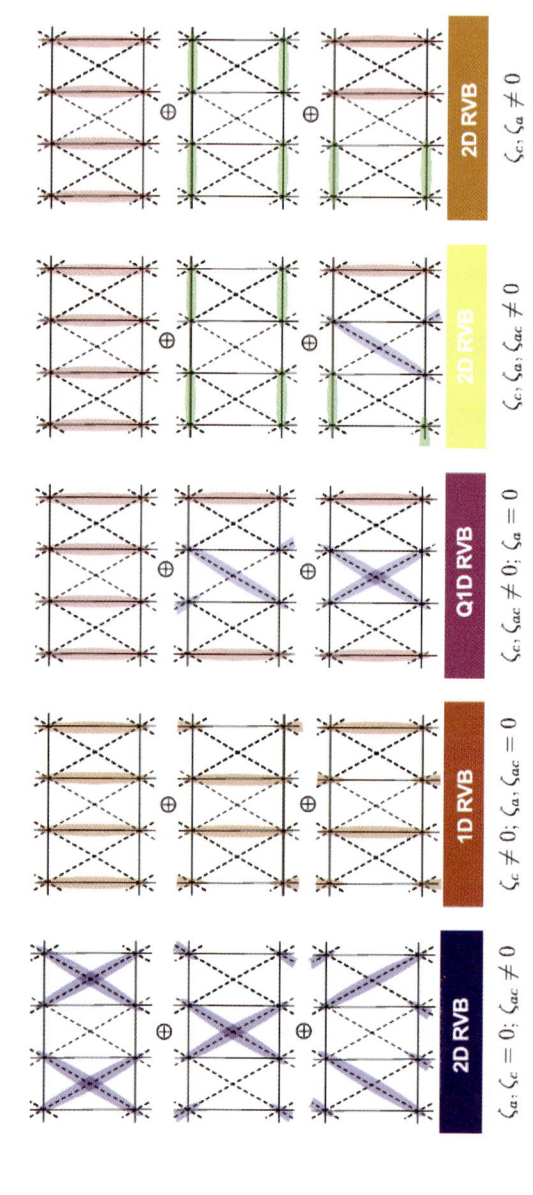

Figure 4.10 Characteristic spin-pairing (valence bond) contributing to produce various RVB states of the c-a-ca-model of CuNCN. (Reprinted with permission from Ref. [63].)

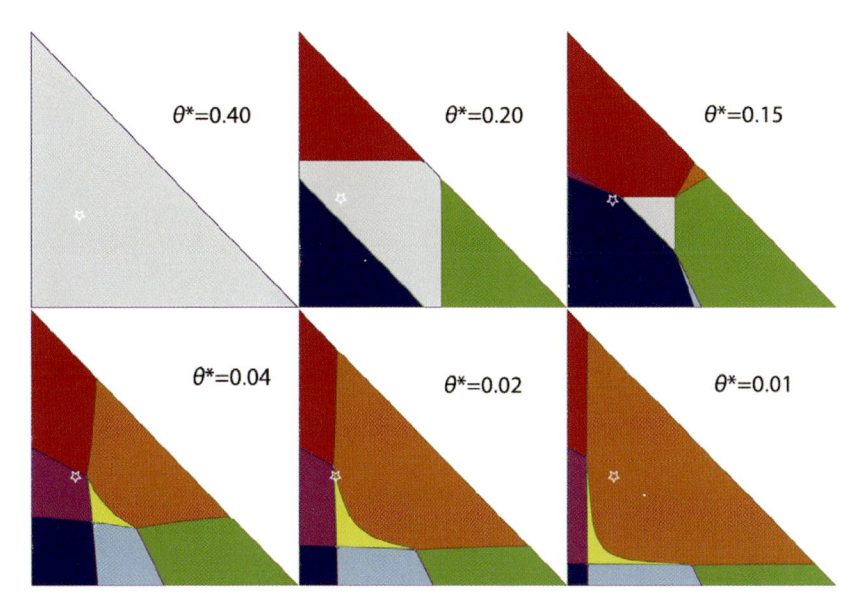

Figure 4.11 Parameter phase diagrams for the c– a– ca-RVB model in the high-temperature approximation. The abscissa and ordinate represent the reduced parameters J_a^* and J_c^*; these parameters, respectively, are equal to unity in the lower right and upper left corners; the entire set of reduced parameters is subject to the condition $J_a^* + J_c^* + J_{ac}^* = 1$. The reduced temperatures θ^* are fractions of $J_a^* + J_c^* + J_{ac}^*$. The color coding for phases is as follows: Curie paramagnetic, with all VPs zero = gray, only one nonvanishing VP = red or green (1D RVB, Pauli paramagnet), = blue for $\zeta_c, \zeta_a, \zeta_{ac} \neq 0$, respectively; one VP vanishing = magenta, cyan (Q1D-RVB, Pauli paramagnet), = orange for ζ_a, or ζ_c, or $\zeta_{ac} = 0$, respectively– observe the order of the list; yellow codes the phase with three nonvanishing VPs. Orange and yellow phases (2D RVB) feature a combination of the gapped Arrhenius-like temperature dependence of magnetic susceptibility and its linear dependence well below the pseudogap. The small star shows the tentative position of CuNCN on the parameter phase diagram. (Reprinted with permission from Ref. [63]).

featuring the RVB phases already at the room temperature and manifesting rich variety of RVB phases [64,65]. As shown in Figure 4.11, as the temperature lowers, a fixed point on the parameter's phase diagram undergoes a series of phase transitions. This is the case for parameters expected to be realistic for CuNCN (depicted in Figure 4.11 by a star symbol), which successfully describes the unusual physics of CuNCN.

4.5
Discussion and Conclusions

In this chapter, we review the applications of the theory of RVB in chemistry and condensed matter. The idea of constructing the ground-state wave function as a product of singlet states, distributed among the pairs of sites, revealed to be very fruitful [66–68]. However, a unique distribution of bonds (two-electron singlets) is not always possible, benzene molecule being the best-known example. In such cases the electronic ground state is assumed to be a superposition of numerous individual states with different (equivalent) distributions of the bonding singlets – the resonating valence bond (RV) state. The RVB *Ansatz* first appeared in *organic* chemistry and then spread to the field of frustrated magnets, high-T_c, and lightly doped Mott insulators. This versatility due to its general and intrinsically many-body nature being an advantage is at the same time one of the reasons why this wave function type cannot be treated analytically exactly. There is a fundamental difference between the approximations used in many-body and *ab initio* methods. Namely, the former uses an approximate (effective and simplified) Hamiltonian applied to a (strongly) correlated wave function, while the latter uses an, in principle, exact Hamiltonian (hence the name *ab initio*) at the expense of applying it to a much simpler and "uncorrelated" wave function of the Hartree–Fock (single determinant) type. Quantum chemical setting is somewhat misleading in this respect. The Hamiltonians used there tend to disguise as "exact" and only possible ones being predetermined by the composition and/or structure of the material. It is only partially true: Different *ab initio* basis sets and DFT functionals in reality define different Hamiltonians for the same composition and structure. More important, however, is the diversity of the solutions' types characteristic for different relations between Hamiltonian's matrix elements like Eq. (4.7) steering the exact solutions to the areas of the functional space inaccessible for the standard solid-state quantum chemistry codes because of the restricted forms of the trial wave functions implemented in these latter. Some more specific forms of effective model Hamiltonians like Hubbard, t–J, or similar ones looking out fairly arbitrary from the purist quantum chemical point of view, can nevertheless trace the key information about the type of the wave function relevant for the target class of compounds predetermined by the interrelations between matrix elements of the "exact" quantum chemical Hamiltonian. Among these general lines, two types of approaches have been developed in the RVB context and reviewed here: (i) approximate analytical (mean-field auxiliary- (slave)-boson, Gutzwiller approximation) and (ii) numerically exact Variational Monte Carlo one. The models studied are normally the effective t–J, single-band Hubbard, t–J–U, or three-band Hubbard Hamiltonians at different electron counts (fillings). Recently, a tendency to combine both approaches stepped forward, thanks to the variational principle renewing the interest to the RVB ideas in general. This was the possibility to estimate the energy of the RVB states and the relevant order parameters with use of the so-called mean-field RVB, actually a broken symmetry Slater determinants mimicking the structure of the RVB state through its variational parameters at least at half-filling.

Acknowledgment

This work is partially supported by the RFBR through the grant No. 14-03-00867.

References

1 Hoffmann, R., Shaik, S., and Hiberty, P.C. (2003) A conversation on VB vs MO theory: a never-ending rivalry? *Acc. Chem. Res.*, **36** (10), 750–756.
2 Frankland, E. (1854) *Lecture Notes for Chemical Students*, Vols. 1 and 2, Van Voorst, London.
3 Kekulé, A (1858) Ueber die constitution und die metamorphosen der chemischen verbindungen und ÅCEber die chemische natur des kohlenstoffs. *Ann. Chem. Pharm.*, **106** (2), 129–159.
4 Kekulé, A. (1865) Sur la constitution des substances aromatiques. *Bull. Soc. Chim. Paris*, **3** (2), 98–110.
5 Kekulé, A. (1866) Untersuchungen uber aromatische verbindungen. *Ann. Chem. Pharm.*, **137** (2), 129–196.
6 Kekulé, A. (1872) Ueber einige condensationsprodukte des aldehyds. *Ann. Chem. Pharm.*, **162** (1), 77–124, 309–320.
7 Ladenburg, A. (1869) Bemerkungen zur aromatischen theorie. *Ber. Dtsch. Chem. Ges.*, **2**, 140–142.
8 Heitler, W. and London, F. (1927) Wechselwirkung neutraler Atome und homöopolare Bindung nach der Quantenmechanik. *Z. Physik*, **44**, 455.
9 Heitler, W. and Rumer, G. (1930) Quantenchemie mehratomiger Moleküle. *Nachr. Ges. Wiss. Gött. Math. Phys.*, 277.
10 Weyl, H. (1930) Zur quantentheoretischen Berechnung molekularer Bindungsenergie. *Nachr. Ges. Wiss. Gött. Math. Phys.*, 285.
11 Weyl, H. (1931) Zur quantentheoretischen Berechnung molekularer Bindungsenergie II. *Nachr. Ges. Wiss. Gött. Math. Phys.*, 33.
12 Pauling, L. (1046) A resonating-valence-bond theory of metals and intermetallic compounds. *Proc. R. Soc. Lond. A*, **196**, 343. 1949.
13 Pauling, L. (1926) The dynamic model of the chemical bond and its application to the structure of benzene. *J. Am. Chem. Soc.*, **48** (5), 1132–1143.
14 Arndt, F., Scholz, E., and Nachtwey, P. (1924) Über dipyrylene und über die bindungsverhältnisse in pyron-ringsystemen. *Ber. Dtsch. Chem. Ges. (A and B Series)*, **57** (10), 1903–1911.
15 Linderberg, J. and Öhrn, Y. (1973) *Propagators in Quantum Chemistry*, Academic Press, London.
16 Surján, P.R. (1989) *Second Quantized Approach to Quantum Chemistry*, Springer, Heidelberg, Germany.
17 Hückel, E. (1931) Quantentheoretische beiträge zum benzolproblem. i. die elektrokonfiguration des benzols und verwandter verbindungen. *Z. Phys.*, **70**, 204.
18 Hückel, E. (1931) Quantentheoretische beiträge zum benzolproblem. ii. quantentheorie der induzierten polaritäten. *Z. Phys.*, **72**, 310.
19 Hückel, E. (1932) Quantentheoretische beiträge zum benzolproblem. iii. Quantentheoretische beiträge zum problem der aromatischen und ungesättigten verbindungen. *Z. Phys.*, **76**, 628.
20 Hückel., E. (1933) Quantentheoretische beiträge zum benzolproblem. iv. Die freien radikale der organischen chemie. *Z. Phys.*, **83**, 632.
21 Hückel, E. (1936) Die bedeutung der neuen quantentheorie für Die chemie. *Z. Elektrochem.*, **42**, 657.
22 Hückel, E. (1936) Über die c-c-bindung in hexaphenyläthan. kritische bemerkungen zu arbeiten von h. e. bent und mitarbeitern. *Z. Phys. Chem. B*, **34**, 335.
23 Hückel, E. (1936) Zur theorie des magnetismus sogenannter biradikale. *Z. Phys. Chem. B*, **34**, 339.
24 Hückel, E. (1937) Grundzüge der theorie ungesättigter und aromatischer verbindungen. *Z. Elektrochem.*, **43**, 827–849.
25 Hückel, E. (1937) Kritische betrachtungen zur theorie der substitutionsreaktionen an substituierten benzolen. *Z. Phys. Chem. B*, **35**, 163.
26 Hückel, E. (1957) Zur modernen theorie ungesä ttigter und aromatischer verbindungen. *Z. Elektrochem. Ber. Bunsenges*, **61**, 866–890.
27 Shaik, S.S. and Hiberty, P.C. (2008) *A Chemist's Guide to Valence Bond Theory*, John Wiley & Sons, Inc., Hoboken, NJ.

28 Tchougréeff, A.L. (2016) Several stories from theoretical chemistry with some Russian flavor and implications for theorems of chemistry, vagueness of its concepts, fuzziness of its definitions, iconicity of its language, and peculiarities of its nomenclature. *Int. J. Quantum Chem.*, **116** (3), 137–160.

29 Misurkin, I.A. (1967) Elektronnoe stroenie molekul s sopryazhennymi swjazjami (Electronic structure of molecules with conjugated bonds). PhD thesis, Karpov Institute, Moscow (in Russian).

30 Luzanov, A.V., Pedash, Yu.F., and Mokhamad, S. (1990) Interpretation of multiconfigurational states in the wave-operator method. *Theor. Exp. Chem.*, **26** (5), 485–495.

31 Gel'man, G. / Hellmann, H. *Kwantowaja Khimija (Einfuhrung in die Quantenchemie)*, 1st edn, Moscow/Leipzig; 2nd edn, Moscow/Berlin/Heidelberg, 1st edn onti/f. deuticke; 2nd edition binom/springer edition, 1937, 2012, 2015.

32 Anderson, P.W. (1959) New approach to the theory of superexchange interactions. *Phys. Rev.*, **115**, 2–13.

33 Liechtenstein, A. (2016) *DMFT* (chapter 5 of volume 5).

34 MP2 (2016) MP2 (chapter 5 of volume 5). Wiley.

35 Vonsovskii, S.V. (1974) *Magnetism*, John Wiley & Sons, Inc., New York.

36 Coldea, R., Tennant, D.A., and Tylczynski, Z. (2003) Extended scattering continua characteristic of spin fractionalization in the two-dimensional frustrated quantum magnet Cs2CuCl4 observed by neutron scattering. *Phys. Rev. B*, **68**, 134424.

37 Ono, T., Tanaka, H., Aruga Katori, H., Ishikawa, F., Mitamura, H., and Goto, T. (2003) Magnetization plateau in the frustrated quantum spin system Cs_2CuBr_4. *Phys. Rev. B*, **67**, 104431.

38 Nakatsuji, S., Nambu, Y., Tonomura, H., Sakai, O., Jonas, S., Broholm, C., Tsunetsugu, H., Qiu, Y., and Maeno. Y. (2005) Spin disorder on a triangular lattice. *Science*, **309**, 1697–1700.

39 Fazekas, P. and Anderson, P.W. (1974) On the ground state properties of the anisotropic triangular antiferromagnet. *Philos. Mag.*, **30** (2), 423.

40 Anderson, P.W. (1987) The resonating valence bond state in La_2CuO_4 and superconductivity. *Science*, **235** (4793), 1196–1198.

41 Baskaran, G., Zou, Z., and Anderson., P.W. (1987) The resonating valence bond state and high-T_c superconductivity: a mean field theory. *Solid State Commun.*, **63** (11), 973–976.

42 Faria, R.M. and Oliveira, O.N., Jr. (1999) Exploiting the electrical properties of thin films of semiconducting polymers. *Braz. J. Phys.*, **29**, 360–370.

43 Black-Schaffer, A.M. and Doniach, S. (2007) Resonating valence bonds and mean-field d-wave superconductivity in graphite. *Phys. Rev. B*, **75**, 134512.

44 Yokoyama, H. and Shiba, H. (1987) Variational Monte-Carlo studies of Hubbard model: I. *J. Phys. Soc. Jpn.*, **56** (4), 1490.

45 Yokoyama, H. and Shiba, H. (1987) Variational Monte-Carlo studies of Hubbard model. II. *J. Phys. Soc. Jpn.*, **56** (10), 3582.

46 Sorella, S. (2001) Generalized Lanczos algorithm for variational quantum Monte Carlo. *Phys., Rev.*, **64**, 024512.

47 Edegger, B., Muthukumar, V.N., and Gros, C. (2007) Gutzwiller–RVB theory of high-temperature superconductivity: results from renormalized mean-field theory and variational Monte Carlo calculations. *Adv. Phys.*, **56** (6), 927–1033.

48 Ogata, M. and Fukuyama, H. (2008) The t–J model for the oxide high-T_c superconductors. *Rep. Prog. Phys.*, **71** (3), 36501.

49 Plekhanov, E., Sorella, S., and Fabrizio, M. (2003) Increasing d-wave superconductivity by on-site repulsion. *Phys. Rev. Lett.*, **90**, 187004.

50 Plekhanov, E., Becca, F., and Sorella, S. (2005) d-Wave pairing in lightly doped Mott insulators. *Phys. Rev. B*, **71**, 064511.

51 Marchi, M., Azadi, S., and Sorella, S. (2011) Fate of the resonating valence bond in graphene. *Phys. Rev. Lett.*, **107**, 086807.

52 Kotliar, G. (2014) *Electronic Structure of Correlated Materials: Slave-Boson and Dynamical Mean-Field Theory*,

Forschungszentrum JÃCElich (Verlag), Jölich.

53 Wölfle, P. (1995) Slave boson theories of correlated electron systems. *J. Low Temp. Phys.*, **99** (3), 625–635.

54 Barnes, S.E. (1976) New method for the Anderson model. *J. Phys. F Met. Phys.*, **6** (7), 1375.

55 Barnes, S.E. (1977) New method for the Anderson model. II. The $u = 0$ limit. *J. Phys. F Met. Phys.*, **7** (12), 2637.

56 Coleman, P. (1984) New approach to the mixed-valence problem. *Phys. Rev. B*, **29**, 3035.

57 Read, N. and Newns, D.M. (1983) On the solution of the Coqblin–Schreiffer Hamiltonian by the large-N expansion technique. *J. Phys. C Solid State Phys.*, **16** (17), 3273.

58 Zou, Z. and Anderson, P.W. (1988) Neutral fermion, charge-e boson excitations in the resonating-valence-bond state and superconductivity in La_2CuO_4-based compounds. *Phys. Rev. B*, **37**, 627.

59 Kotliar, G. (1988) Resonating valence bonds and d-wave superconductivity. *Phys. Rev. B*, **37**, 3664.

60 Affleck, I. and Brad Marston, J. (1988) Large-n limit of the Heisenberg–Hubbard model: implications for high-T_c superconductors. *Phys. Rev. B*, **37**, 3774.

61 Tchougreeff, A.L. and Hoffmann, R. (1992) Charge and spin-density waves in the electronic-structure of graphite: application to analysis of STM images. *J. Phys. Chem.*, **96** (22), 8993–8998.

62 Suzumura, Y., Hasegawa, Y., and Fukuyama, H. (1988) Mean field theory of RVB and superconductivity. *J. Phys. Soc. Jpn.*, **57** (8), 2768–2778.

63 Tchougréeff, A.L. and Dronskowski, R. (2013) Low-temperature structure anomalies in CuNCN. Manifestations of RVB phase transitions? *J. Phys. Condens. Matter*, **25** (43), 435602.

64 Tchougréeff, A.L. and Dronskowski, R. (2014) Mean-field RVB ground states of lattice models of CuNCN. *Low Temp. Phys.*, **40** (1), 73.

65 Tchougréeff, A.L., Liu, X., Müller, P., van Beek, W., Ruschewitz, U., and Dronskowski, R. (2012) Structural study of CuNCN and its theoretical implications: a case of a resonating-valence-bond state? *J. Phys. Chem. Lett.*, **3** (22), 3360–3366.

66 Tokmachev, A.M. and Chugreev, A.L. (1999) Semiempirical electron pair correlation method for calculation of the electronic structure of molecules. *Zh. Fiz. Khim.*, **73** (2), 320–331.

67 Tokmachev, A.M. and Tchougreeff, A.L. (2001) Semiempirical implementation of strictly localized geminals for analysis of molecular electronic structure. *J. Comput. Chem.*, **22** (7), 752–764.

68 Tokmachev, A.M., Tchougreeff, A.L., and Misurkin, I.A. (2000) Effective electronic Hamiltonian for quantum subsystem in hybrid QM/MM methods as derived from APSLG description of electronic structure of classical part of molecular system. *Theochem*, **506** (1–3), 17–34.

5
Many Body Perturbation Theory, Dynamical Mean Field Theory and All That

Silke Biermann[1] and Alexander Lichtenstein[2]

[1]Ecole Polytechnique, Centre de Physique Théorique, CNRS UMR7644, 91128 Palaiseau, France
[2]Universität Hamburg, Institut für Theoretische Physik, Jungiusstraße 9, 20355 Hamburg, Germany

5.1
Introduction: Electronic Correlations from First Principles

Scientific progress of the last century in solid-state materials is closely related to the design of semiconducting materials for the semiconductor industry. Therefore, theoretical developments of the last 50 years associated with realistic electronic structure calculations largely focussed on such weakly correlated solids. Density functional theory (DFT) started with the seminal work by Walter Kohn, Pierre Hohenberg and Lu Sham [1,2]. It provides a first principles scheme based on an exact theorem, stating that the ground state of an interacting electron system can be found by minimizing an universal functional of the density in some additional external field. However, this functional is not known in general and can be calculated numerically with a reasonable accuracy only for the simple case of the homogeneous electron gas. These calculations, which have been proven to be very useful for the DFT scheme, have been done by Ceperley and Alder [3] using the two-step quantum Monte Carlo procedure starting from the "fixed-node" approximation followed by another complicated step called "nodal relaxation." Nevertheless, the accuracy of such a scheme is still limited and is very sensitive to the computational details [4]. The main restriction of density functional theory is however that it is limited to the description of ground-state properties; spectral properties can be found only within a time-dependent extension, the time-dependent density functional theory (TDDFT) scheme [5]. While the structural minimization of complex materials can be carried out very efficiently in the generalized gradient approximation to the DFT, due to the almost spherical exchange-correlation hole [6], the quality of estimates for spectral properties crucially depends on the system in question. TDDFT has more problems than the static DFT approach, since there are no suitable time-dependent reference systems to find the exchange correlation kernel.

Understanding the properties of transition metal systems is the key to getting insights into important materials questions such as high-temperature superconductivity, biological molecules like hemoglobin containing iron and many others. The enormous progress of the last three decades in synthesizing completely new materials for high-T_C superconductivity, giant and colossal magneto-resistance, or artificially creating two-dimensional lattices opens new views onto the theory of correlated fermionic systems. It turns out that even the ground-state properties of antiferromagnetic oxides or orbitally ordered compounds are not well described in the DFT scheme [7]. Accurate angle-resolved photoemission studies of cuprate superconductors show that the spectra of such systems, with strong electron–electron interactions in the 3d-shell of transition metals have well pronounced incoherent features [8].

The photoemission intensity is proportional to the so-called spectral function, which – for $\omega < 0$ – can be defined as $A(k,\omega) = \sum_e |\langle \psi_e | c_k | \psi_0 \rangle|^2 \delta\left(\omega + E_e^{(N-1)} - E_0^{(N)}\right)$. Here, $|\psi_0\rangle$ denotes the N-body ground state (of energy $E_0^{(N)}$) of an N-electron system, the $|\psi_e\rangle$ are excited states (of energies $E_e^{(N-1)}$) containing $N-1$ electrons, and the sum runs over all of them. The operator c_k destroys an electron with wave vector k, thus connecting N- and $N-1$-particle states in a way that simulates the photoemission process in which an electron of wave vector k and energy $\hbar\omega$ is ejected via Einstein's photoelectric effect (for a review, see Ref. [9]). For a noninteracting system, both the N-particle and $N-1$ particle states would be Slater determinants of one-particle states, and the spectral function would consist of delta-peaks at the band energies. The situation is however qualitatively different in the presence of interactions: according to Landau's theory of Fermi liquids, in normal metals the low-energy excitations can still be characterized by the quantum numbers of single-particle states, but these acquire a characteristic life time corresponding to their decay due to electron–electron interactions. We present in Figure 5.1 the qualitative

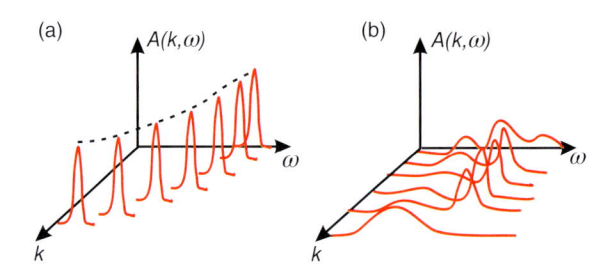

Figure 5.1 Schematic view of ARPES spectra for materials with weak (a) and strong (b) electronic Coulomb correlations. The material in (a) is well described by band theory; in the strongly correlated case in (b), peaks are broadened by lifetime effects and spectral weight transfers to satellite features take place.

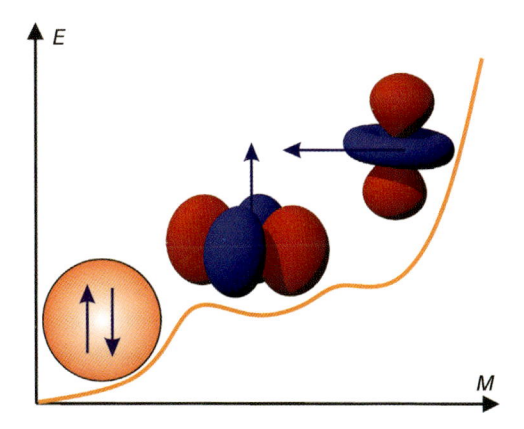

Figure 5.2 Schematic representation of magnetic and orbital fluctuations in correlated electron systems: the electronic energy landscape exhibits several (local) minima corresponding to states with different spin or orbital polarizations. The abscissa M denotes schematically some – possibly nonscalar – moment.

difference between the $\vec{k}$-resolved spectral function of normal metals with well-defined quasiparticle peaks at all momenta **k** and the strongly correlated case with an incoherent part and nonquasiparticle behavior in some parts of the Brillouin zone.

The main reason for complex behavior of correlated electronic systems is related to strong quantum fluctuations between different low-energy fermionic configurations (due to the tunneling), as shown schematically in Figure 5.2. For example, if the free energy of electronic systems has only one well-defined minimum at zero local moment (see Figure 5.2) then one can expect small electron fluctuations and normal diamagnetic quasiparticle behavior. In the case of two low-lying minima corresponding to singlet and triplet excitations (when the second minimum in Figure 5.2 will be as low as the first one) one can expect strong many body fluctuations and possible nonquasiparticle behavior related to local Hund's rule physics [9–12]. In order to treat the system with such effective energy profiles including low-energy singlet–triplet excitations, we need to use the path-integral approach and calculate the corresponding correlation functions using path-integral quantum Monte Carlo schemes that can handle many local minima in the free energy functional on an equal footing. In the quantum chemistry view, this corresponds to effects of low-energy transitions between almost degenerate energy states of individual shells or atoms with different total magnetic states (low-spin–high-spin transition) in large molecules.

The accuracy of functional approaches to the electronic structure of correlated solids is closely related to the choise of appropriate reference systems. In Figure 5.3 we show schematically three celebrated approaches to describe many-body effetcs in crystals: density functional theory is based on the simplest reference system, namely, the interacting homogeneous electron gas that was solved

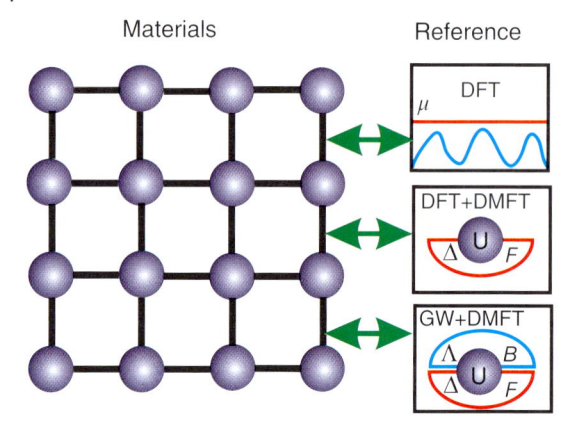

Figure 5.3 Schematic representation of reference systems in the DFT (constant external potential defined by μ), DFT + DMFT (fermionic bath defined by Δ) and GW + DMFT (fermionic and bosonic bath defined by Δ and Λ) methods.

for different densities within the fixed-node QMC scheme [3]. The DFT + DMFT scheme is based on the self-consistent solution of an effective impurity problem, a representative correlated atom embedded into a fermionic bath. Finally, the GW + DMFT approach introduces a more complex impurity reference system of a correlated atom in fermionic and bosonic baths, describing the combined effects of electronic hybridization and frequency-dependent screening effects of long-range Coulomb interactions in solids.

5.2
Effective Low-Energy Hamiltonians: "Downfolding" the Interaction

The electronic properties of solids are dominated by the electronic states in the immediate proximity to the Fermi level (for metals) or to the energy gap (for insulators). This observation – together with the insights from renormalization group techniques – is the motivation for the quest of low-energy effective models that describe the physical phenomena taking place in condensed matter. In the early days of correlated electron physics, models were most often phenomenologically motivated without the ambition of a microscopic derivation, let alone a quantitative description. Within the last decade, however, a new research field has developed at the interface of many-body theory and first prinicples electronic structure calculations. The aim is the construction of parameter-free many-body theories that preserve the *ab initio* nature of density functional-based electronic structure calculations, but incorporate at the same time a many-body description of Coulomb interactions beyond the independent-electron picture into the description of spectroscopic or finite-temperature properties.

Such "correlation" effects, that is, effects beyond the effective one-particle picture, are indeed most striking in spectroscopic probes, where they take the form of quasi-particle renormalizations (corresponding to an enhancement of the effective masses of the carriers) or broadening due to finite lifetimes, and give rise to satellite features or atomic multiplets. An intrinsic temperature dependence of the electronic structure of a metal, with a coherence–incoherence crossover delimiting Fermi liquid properties, or a strongly temperature-dependent gap – beyond what can be explained by a Fermi factor – are further hallmarks of electronic correlations [13].

Historically, the first nonperturbative electronic structure techniques for correlated materials evolved from many-body treatments of the multiorbital Hubbard Hamiltonian with realistic parameters. The general strategy of these so-called "DFT++" approaches [14,15] consists in the extraction of the parameters of a many-body Hamiltonian from first principles calculations and then solving the problem by many-body techniques. The procedure becomes conceptually involved, however, through the need of incorporating effects of higher energy degrees of freedom on the low-energy part, the so-called "downfolding."

For the one-particle part of the Hamiltonian, downfolding techniques have been the subject of a vast literature [16,17], and are by now well established. The task here is to define orbitals spanning the low-energy Hilbert space of the electronic degrees of freedom of a solid in such a way that a low-energy one-particle Hamiltonian can be constructed whose spectrum coincides with the low-energy part of the spectrum of the original one-particle Hamiltonian.[1] Downfolding of the interacting part of a many-body Hamiltonian is a less straightforward problem, which has attracted a lot of attention recently. The challenge is an accurate description of screening of low-energy interactions by high-energy degrees of freedom. Indeed, the net result of the rearrangement of the high-energy degrees of freedom as response to a perturbation of the system is an effective reduction of the perturbation strength in the low-energy space. It is for this reason that the effective Coulomb interaction in a low-energy effective model for a correlated system is in general up to an order of magnitude smaller than the matrix element of the bare Coulomb interaction.[2] Nevertheless, the latter is recovered in the limit of high frequencies of the perturbation, when screening becomes inefficient. The crossover – as a function of frequency – from the low-energy screened regime to the high-frequency bare matrix element of $e^2/|\mathbf{r}-\mathbf{r}'|$ takes place at a characteristic screening (plasma) frequency where the dielectric function exhibits a pole structure.

This frequency-dependence of the effective local Coulomb interactions, the *dynamical Hubbard* $\mathcal{U}(\omega)$ and its consequences on the electronic structure of

---

1) We do not enter here into details concerning the different strategies of achieving such a construction: various frameworks, such as muffin-tin orbitals methods [17], maximally localized Wannier functions [18], or projected atomic orbitals [19] have been employed.
2) For example, for elemental nickel, the matrix element of the screened interactions in Wannier functions for the 3d states is about 2–3 eV, while the unscreened interaction reaches a value of 20–30 eV [20].

correlated materials are at the center of the present review. Dealing with frequency-dependent interactions at the DMFT level has been a major bottleneck in the implementation of the combined "GW + DMFT" scheme since its proposal in 2003 [21]. The recent advances concerning this issue, both concerning Monte Carlo techniques and through the Bose factor ansatz, have now unblocked the situation: two calculations within GW + DMFT taking into account dynamical interactions have been achieved recently, for SrVO$_3$ [22,23] and for systems of adatoms on surfaces [24,25]. We review these calculations, together with systematic studies of an extended Hubbard model [26], which demonstrate how the GW + DMFT scheme enables an additional type of "downfolding": effects of long-range interactions can in fact be "backfolded" into a purely local effective quantity, a generalized Hubbard $\mathcal{U}(\omega)$, which acquires its frequency-dependence due to screening by nonlocal processes. The strength of these screening processes are shown to be strongly system-dependent when the true long-range nature of Coulomb interactions is taken into account, while simple rules of thumb work relatively well in the case of an extended Hubbard model with nearest-neighbor interactions only.

With the advent of the so-called "constrained random phase approximation" (cRPA) [20], explicit calculations of the effective Coulomb interactions after screening by degrees of freedom that lie further away from the Fermi level, have become possible. An important example are effective Coulomb interactions on transition metal ions screened by ligands. Detailed studies within the cRPA have been performed, for example, for transition metal oxides [27] or iron pnictides and chalcogenides [28]. Physically speaking, two factors determine the obtained values: first, the extension of the Wannier functions on the transition metal ion, which in turn is a measure of the overlap of transition metal and ligand states. Second, the efficiency of screening as expressed by the polarization of the ligand. At the cRPA level, these two effects can in fact be disentangled and analyzed separately by investigating the bare and screened effective interactions: the former only contains the first effect, while both contribute to the latter.

It is most interesting to connect these considerations with the nephelauxetic effect discussed in the chemistry literature [29], the reduction of the effective Coulomb interaction parameters (in the form of Slater or Racah parameters) on a transition metal ion through the presence of ligands. Also in this case, the extension of the orbital and the ligand screening play the determining roles. The prevailing picture is, however, that of a local cluster of ligands surrounding the transition metal ion, whereas in the cRPA the ligand states are supposed to be band-forming, and the polarization is determined by calculating electron–hole fluctuations involving these ligand bands. The solution of a finite cluster with transition metal and ligands can be investigated within the exact diagonalization scheme [30], or using the effective Hamiltonian crystal field method [31], while for self-consistent DMFT scheme we need a complicated solution of the transition metal in continuum of fermionic bath states. The classical approach to electronic structure of solids is based on the Hamiltonian formulation that can be solved for many-body problems only in approximate way or using exact diagonalization for small finite cluster. More advanced functional DMFT approach introduces an effective action formulation which is based on numerically exact solution of quantum impurity problem. The later plays role of a reference system similar to homogeneous electron gas in the DFT theory. It should be most interesting to elaborate further comparisons, based on the quantitative techniques available nowadays.

In this chapter a general functional approach to the strongly correlated electron systems will be discussed. We will separate the local and nonlocal correlations, and show that it is possible to solve the local correlation problem using the recently developed continuous time Monte Carlo (CT-QMC) scheme. Finally, we discuss an efficient way how to go from the simple model investigation of strongly correlated systems to realistic investigations of complex electronic materials. We end by defining current open questions and commenting on further perspectives in the field.

5.3
Functional Approach: From DFT to DMFT

We introduce a general functional approach which will cover DFT, dynamical mean-field theory (DMFT) and Baym–Kadanoff (BK) theories [32]. Let us start from the full many-body Hamiltonian describing electrons moving in the periodic external potential $V(\mathbf{r})$ of the ions with the chemical potential μ and interacting via the Coulomb law $U(\mathbf{r}-\mathbf{r}') = 1/|\mathbf{r}-\mathbf{r}'|$. We use atomic units $\hbar = m = e = 1$. In the field-operators ($\hat{\psi}^+$ and $\hat{\psi}$) representation, the Hamiltonian has the following form [1]:

$$H = \sum_\sigma \int d\mathbf{r}\, \hat{\psi}_\sigma^+(\mathbf{r}) \left[-\frac{1}{2}\nabla^2 + V(\mathbf{r}) - \mu \right] \hat{\psi}_\sigma(\mathbf{r}) \\ + \frac{1}{2} \sum_{\sigma\sigma'} \int d\mathbf{r} \int d\mathbf{r}'\, \hat{\psi}_\sigma^+(\mathbf{r}) \hat{\psi}_{\sigma'}^+(\mathbf{r}') U(\mathbf{r}-\mathbf{r}') \hat{\psi}_{\sigma'}(\mathbf{r}') \hat{\psi}_\sigma(\mathbf{r}).$$

(5.1)

On the way to more standard chemical description in terms of optimal orbitals in solids we will use a single-particle orthonormal basis set of functions $\phi_n(\mathbf{r})$ (e.g., Wannier orbitals indexed by a full set of quantum numbers, such as site, orbital, and spin index: $n = (i, m, \sigma)$. Then one can expand the quantum fields in terms of creation and annihilation operators ($\hat{c}_n^+$ and $\hat{c}_n$) for such basis orbitals

$$\hat{\psi}(\mathbf{r}) = \sum_n \phi_n(\mathbf{r}) \hat{c}_n \\ \hat{\psi}^+(\mathbf{r}) = \sum_n \phi_n^*(\mathbf{r}) \hat{c}_n^+ .$$

(5.2)

When many-electron Hamiltonian was written in this second quantization operator, it is easy to find exact energy spectrum of small finite systems that is equivalent to classical quantum chemical method, namely, full configuration interactions. On the other hand, for more general functional approach based on the Green's function that describe different electronic correlations in space and time and can be apply for open quantum system in the external bath, such operators technique is not very useful due to complicated algebra of noncommuting objects. The main idea of seminal works by Paul Dirac and Richard Feynman on the path integral related with introduction of additional dimension or imaginary time axes $\tau \subset [0, \beta)$, where fermionic operators in the Heisenberg representation can be ordered and transformed to more simple objects, namely, the Grassmann variables $\{c_n^*(\tau), c_n(\tau)\}$ (β is the inverse temperature). In this way one can develop a finite-temperature description of many-body systems and easily describe the frequency dependent screening effects. Using an efficient path integral representation of quantum statistical mechanics in $d+1$ dimension with an additional imaginary time axes, we can write the partition function of the many-body systems using the Euclidean action S

$$Z = \int \mathcal{D}[c^*, c] e^{-S}$$

$$S = \sum_{12} c_1^* (\partial_\tau + t_{12}) c_2 + \frac{1}{2} \sum_{1234} c_1^* c_2^* U_{1234} c_4 c_3.$$

(5.3)

where we use the following short definition of the sum: $\sum_1 \ldots \equiv \sum_{im\sigma} \int_0^\beta d\tau \ldots$ and the one- and two-electron matrix elements are defined as

$$t_{12} = \int d\mathbf{r} \phi_1^*(\mathbf{r}) \left[-\frac{1}{2} \nabla^2 + V(\mathbf{r}) - \mu \right] \phi_2(\mathbf{r})$$

$$U_{1234} = \int d\mathbf{r} \int d\mathbf{r}' \phi_1^*(\mathbf{r}) \phi_2^*(\mathbf{r}') U(\mathbf{r} - \mathbf{r}') \phi_3(\mathbf{r}) \phi_4(\mathbf{r}').$$

(5.4)

We note that the electronic free energy of the system is related to the partition function via the usual relation $F = -kT \log(Z)$ of statistical physics, stressing that the present formalism allows for a true finite temperature description. The one-electron matrix t_{12} can be related with celebrated in theoretical quantum chemistry Hückel molecular orbital method. If we ignore the many-body interactions ($U_{1234} = 0$), then the spectrum of one-dimensional Hückel chain related with diagonalization of t_{ij} matrix for a single s-orbital case with the periodic boundary condition is equal to $\epsilon_k = \alpha + 2\beta \cos(kx)$, where α and β are the standard notation for the diagonal and nearest neighbor elements of the electron hopping matrix. For such one-electron problem, one can analytically calculate corresponding path integral that will be of a Gaussian type and reproduce the well-known results for partition function $Z = \prod_k \left(1 + e^{-(\epsilon_k - \mu)/kT}\right)$, which related to the Fermi–Dirac distribution function. The one-electron Green's function is defined as the simplest nonzero correlation function for fermions

$$G_{12} = -\langle c_1 c_2^* \rangle_S = -\frac{1}{Z} \int \mathcal{D}[c^*, c] c_1 c_2^* \exp(-S).$$

(5.5)

The main problem of strongly interacting electronic systems is related to the fact that the higher order correlation functions do not separate into products of

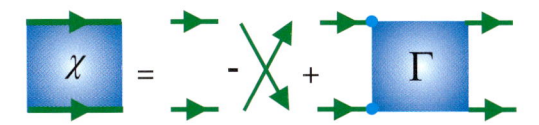

Figure 5.4 Representation of the full two-particle Green's function in terms of trivial products of single-particle Green's function and the full vertex function Γ. The figure shows a graphical way to express Eq. (5.7). Green lines denotes single-particle Green's functions, and the many-body vertex is represented as blue square.

lower order correlation function. For example, the two-particle Green's function or generalized susceptibility (X) is defined in the following form [33]:

$$X_{1234} = \langle c_1 c_2 c_3^* c_4^* \rangle_S = \frac{1}{Z} \int \mathcal{D}[c^*, c] c_1 c_2 c_3^* c_4^* \exp(-S), \quad (5.6)$$

and can be expressed graphically through the Green functions and the full vertex function Γ_{1234} [34] (see Figure 5.4)

$$X_{1234} = G_{14} G_{23} - G_{13} G_{24} + \sum_{1'2'3'4'} G_{11'} G_{22'} \Gamma_{1'2'3'4'} G_{3'3} G_{4'4} \quad (5.7)$$

In the case of noninteracting electron systems, the high-order correlations X are reduced to the antisymmetrized product of lower order correlations G, which would correspond to the first two terms (Hartree and Fock like) and a vertex function Γ in Eq. (5.7) equal to zero. In strongly correlated electron systems, the last part with the vertex is dominant and even diverges close to electronic phase transitions.

The Baym–Kadanoff functional [35] gives the one-particle Green's function and the total free energy at its stationary point. In order to construct the exact functional of the Green's function (Baym–Kadanoff) we modify the action by introducing the source term J in the following form:

$$S[J] = S + \sum_{12} c_1^* J_{12} c_2. \quad (5.8)$$

The partition function Z, or equivalently the free energy of the system F, becomes a functional of the auxiliary source field

$$Z[J] = e^{-F[J]} = \int \mathcal{D}[c^*, c] e^{-S'[J]}. \quad (5.9)$$

Variation of this source function gives all multiparticle correlation functions, for example, the single-electron Green's function

$$G_{12} = \left.\frac{\delta F[J]}{\delta J_{21}}\right|_{J=0}. \quad (5.10)$$

Alternatively, one could define the generalized susceptibility as the second variation of the $F[J]$ functional instead of spelling out the above integral. In that case, one would get only what is called the "connected part" of correlation function. In the case of the X-function, this is represented by the last term in Eq. (5.7) only.

The Baym–Kadanoff functional can be obtained by the Legendre transform from the variable J to G

$$F[G] = F[J] - \text{Tr}(JG). \tag{5.11}$$

We can use the standard decomposition of the free energy F into the single particle part and the correlated part

$$F[G] = \text{Tr} \ln G - \text{Tr}(\Sigma G) + \Phi[G], \tag{5.12}$$

where Σ_{12} is the single particle self-energy and $\Phi[G]$ is the correlated part of the Baym–Kadanoff functional and is equal to the sum of all two-particle irreducible diagrams. In the stationary point, this functional gives the free energy of the system. We note that this functional depends only on the full one-particle Green's function and bare interactions. The conditions for the existence of this functional for strongly correlated systems are the subject of ongoing work [36]. One can use different Legendre transforms and obtain different variants of functional constructions, for example, a functional of the self-energy Σ [37], a double Legendre transform leading to functionals of two variables G and Γ [38], or a more simple functional of G and screened Coulomb interactions W [39] useful in many-body perturbation theory. (The GW approximation is the lowest-order approximation to the latter.)

In practice, $\Phi[G]$ is not known for interacting electron systems, which is similar to the problem in the density functional theory. Moreover, this general functional approach reduces to the standard version of DFT theory, if one only uses the diagonal part in space-time of the Green function, which corresponds to the one-electron density:

$$n_1 = G_{12}\delta_{12} = \langle c_1^* c_1 \rangle_S, \tag{5.13}$$

with the Kohn–Sham potential $V_{\text{KS}} = V_{\text{ext}} + V_{\text{H}} + V_{\text{xc}}$ playing the role of the "constrained field" – J. Here, V_{ext} is the external potential and V_{H} is a Hartee potential. In practice, the exchange-correlation potential V_{xc} is known only for the homogeneous electron gas; therefore, in practical applications one uses approximations such as the so-called local density approximation to DFT. In this case, the DFT functional is defined in the following way:

$$F_{\text{DFT}}[n] = T_0[n] + V_{\text{ext}}[n] + V_{\text{H}}[n] + V_{\text{xc}}[n], \tag{5.14}$$

where T_0 is the kinetic energy of noninteracting systems. Finally, if we define the total electron density as

$$n(\mathbf{r}) = \sum_i \phi_i^*(\mathbf{r})\phi_i(\mathbf{r}), \tag{5.15}$$

the local density approximation to the DFT reads

$$T_0[n] + V_{\text{ext}}[n] = \sum_i \int d\mathbf{r} \phi_i^*(\mathbf{r}) \left[-\frac{1}{2}\nabla^2 + V_{\text{ext}}(\mathbf{r}) - \mu \right] \phi_i(\mathbf{r})$$

$$V_H[n] = \frac{1}{2} \int d\mathbf{r} n(\mathbf{r}) U(\mathbf{r} - \mathbf{r}') n(\mathbf{r}') \qquad (5.16)$$

$$V_{\text{xc}}[n] = \int d\mathbf{r} n(\mathbf{r}) \epsilon(n(\mathbf{r})),$$

where $\epsilon(n)$ is exchange correlation density for homogeneous electron gas that can be calculated with QMC-scheme [3].

In DFT, we loose information about the nonequal space-time Green's function, which gives the single particle excitation spectrum as well as the k-dependence of the spectral function, and restrict ourselves only to the ground-state energy of the many-electron system. Moreover, we also loose information about all collective excitations in solids, such as plasmons or magnons, contained in the generalized susceptibility or the second variation of the free energy.

One could in principle find the Baym–Kadanoff interacting potential $\Phi[G]$ for simple lattice models using the quantum Monte Carlo scheme (QMC). Unfortunately, in practice, due to the sign problem in lattice simulations such a numerically exact solution of the electronic correlation problem is not possible. On the other hand, one can obtain the solution of a local interacting quantum problem in a general fermionic bath, using the QMC scheme. For the single orbital case, there is no sign problem in the impurity QMC, since there is only one quantum fermion in the impurity action. Moreover, in the case of density–density interactions in multiorbital impurity problems that are diagonal in spin and orbital space there is also no sign problem in QMC. This observation is at the heart of the success of the DMFT approach to strongly correlated systems: in DMFT, the idea is to keep only the local part of the many-body fluctuations. One can then obtain the correlated part of the local functional numerically. In this scheme, we only use the local part of the many electron vertex obtained in a self-consistent way as an effective functional of the local Green's function. Combination of density functional theory with a local many-body DMFT scheme recently attracted attention in quantum chemistry [40]. In the following section we discuss a general dual fermion (DF) transformation [41], which helps us to separate the local fluctuations in many-body system and shows a perturbative way to go beyond the DMFT approximation.

5.4
DMFT and GW + DMFT from an Impurity Reference System

We start from the discussion of a general impurity model as a reference system for the DFT+DMFT and GW+DMFT schemes. Within a functional approach the general action of lattice fermions with nonlocal Coulomb interactions reads

$$S[c^*, c] = -\sum_{\mathbf{k}\omega}(i\omega + \mu - t_\mathbf{k})c^*_{\mathbf{k}\omega}c_{\mathbf{k}\omega} + \frac{1}{2}\sum_{\mathbf{q}\nu}V_\mathbf{q}n_{\mathbf{q}\nu}n_{\mathbf{q},-\nu}, \tag{5.17}$$

where $\omega = (2n+1)\pi/\beta$ and $\nu = 2n\pi/\beta, n = 0, \pm 1, \ldots$ are the fermionic and bosonic Matsubara frequencies, μ is a chemical potential. For simplicity with skip index for different orbitals (m) and spin projection (σ), while keeping the **k**-vectors of quasimomenta in solids.

We introduce the effective impurity reference system (see Figure 5.3) rewriting the general action of interacting lattice fernions in the folowing form [42]:

$$S[c^*, c] = S_{\text{imp}}[c^*, c] + \sum_{\mathbf{k}\omega}\tilde{t}_{\mathbf{k}\omega}c^*_{\mathbf{k}\omega}c_{\mathbf{k}\omega} + \frac{1}{2}\sum_{\mathbf{q}\nu}\tilde{V}_{\mathbf{q}\nu}n_{\mathbf{q}\nu}n_{\mathbf{q},-\nu}, \tag{5.18}$$

with

$$\begin{aligned}\tilde{t}_{\mathbf{k}\omega} &= t_\mathbf{k} - \Delta_\omega \\ \tilde{V}_{\mathbf{q}\nu} &= V_\mathbf{q} - \mathcal{U}_\nu,\end{aligned} \tag{5.19}$$

where the effective impurity action (the reference system in the Figure 5.3) reads

$$\begin{aligned}S_{\text{imp}}[c^*, c] &= -\sum_\omega \mathcal{G}_\omega^{-1} c^*_\omega c_\omega + \frac{1}{2}\sum_\nu \mathcal{U}_\nu n_\nu n_{-\nu} \\ \mathcal{G}_\omega &= (i\omega + \mu - \Delta_\omega)^{-1}.\end{aligned} \tag{5.20}$$

In the dual boson notation, one splits the dynamical Hubbard interaction into a local static part and a bosonic bath function in the following form: $\mathcal{U}_\omega = U + \Lambda_\omega$ (cf. Figure 5.3). For the remaining parts of the lattice fermion action in Eq. (5.18), we use a path-integral perturbation theory scheme, similar to the dual boson formalism [42–44]. We define new dual degrees of freedom f^*, f, ϕ via a Hubbard–Stratonovich transformation and integrate out the original degrees of freedom c^* and c [41]. We now decouple the QIMs by applying suitable Hubbard–Stratonovich transformations to the remainder of Eq. (5.18). The first term in Eq. (5.18) is decoupled through the following identity for fermionic Grassmann variables:

$$e^{c^*_i \tilde{t}_{ij} c_j} = \det[g\tilde{t}g]\int D[f^*, f] e^{-f^*_i g_i^{-1}[\tilde{t}^{-1}]_{ij}g_j^{-1}f_j + c^*_i g_i^{-1} f_i + f^*_i g_i^{-1} c_i}, \tag{5.21}$$

we introduce a scaling factor g (in fact an impurity Green function) to have the same dimention for transform (dual) and real variables. It is natural to decouple the density–density interaction term in the charge channel, in particular since we are interested in the charge fluctuations and screening effects induced by V. This is achieved through the following transformation based on real bosonic fields:

$$e^{\frac{1}{2}n_i \tilde{V}_{ij} n_j} = \sqrt{\det[\tilde{V}^{-1}]}\int D[\phi] e^{-\frac{1}{2}\phi_i [\tilde{V}^{-1}]_{ij}\phi_j + \phi_i n_i}. \tag{5.22}$$

Here the matrix $\tilde{V}$ is assumed to be positive definite, which does not change the final equations [41]. This allows us to integrate out the fermionic degrees of freedom locally, leading to a theory in terms of dual variables f, f^*, and ϕ only. Since g_ω is local, the sum over all states labeled by **k** could be replaced by the equivalent summation over all sites by a change of basis in the second term. The crucial point is that the coupling to the auxiliary fermions is purely local and S_{loc} decomposes into a sum of local terms. The lattice fermions can therefore be integrated out from S_{loc} for each site i separately. This completes the change of variables

$$Z_{\text{imp}}^{-1} \int D[c^*, c] \exp\left(-S[c^*, c, f^*, f, n]\right) = \exp \tilde{S}[f^*, f; \phi]. \tag{5.23}$$

The above equation may be viewed as the defining equation for the effective potential $V_{\text{imp}}[f^*, f; \phi]$. An explicit expression is found by expanding both sides of Eq. (5.23) and equating the resulting expressions order by order. In this way, the original action Eq. (5.17) is subdivided into two separate problems: the impurity action Eq. (5.21) and a dual action

$$\tilde{S}[f^*, f; \phi] = -\sum_{k\omega\sigma} f^*_{k\omega\sigma} \tilde{G}_{0,k\omega}^{-1} f_{k\omega\sigma} - \frac{1}{2} \sum_{q\nu} \phi_{q\nu} \tilde{W}_{0,q\nu}^{-1} \phi_{q\nu} + V_{\text{imp}}[f^*, f, \phi].$$

This dual action has the bare propagators of the following form

$$\begin{aligned}\tilde{G}_{0,k\omega}^{-1} &= \left(g_\omega \tilde{t}_{k\omega} g_\omega\right)^{-1} - g_\omega^{-1} \\ \tilde{W}_{0,q\nu}^{-1} &= \tilde{V}_{q\nu}^{-1} - \chi_\nu .\end{aligned} \tag{5.24}$$

One can prove that

$$\begin{aligned}\tilde{G}_{k\omega}^0 &= \left(g_\omega^{-1} - \tilde{t}_{k\omega}\right)^{-1} - g_\omega = G_{k\omega}^{\text{DMFT}} - g_\omega \\ \alpha_\nu \tilde{W}_{q\nu}^0 \alpha_\nu &= \left(w_\nu^{-1} - \mathcal{U}_\nu^{-1} + V_q^{-1}\right)^{-1} - w_\nu = W_{q\nu}^{\text{EDMFT}} - w_\nu .\end{aligned} \tag{5.25}$$

Here $\alpha_\omega = 1 + \mathcal{U}_\omega \chi_\omega$. The hybridization function Δ_ν and local effective interactions $\mathcal{U}_\omega$ can be found from DMFT or EDMFT-like self-consistency equations that cancel all local contributions to dual-perturbation theory

$$\begin{aligned}\sum_{\mathbf{k}} \tilde{G}_{k\omega} &= 0 \\ \sum_{\mathbf{q}} \tilde{W}_{q\nu} &= 0.\end{aligned} \tag{5.26}$$

Note, t_k and V_q are the Fourier transform of t_{ij} and V_{ij} and g and χ are the impurity Green's function and charge susceptibility, respectively,

$$\begin{aligned}g_\omega &= -\langle c_{\omega\sigma} c^*_{\omega\sigma}\rangle_{\text{imp}} \\ \chi_\nu &= -\langle \bar{n}_\nu \bar{n}_{-\nu}\rangle_{\text{imp}} \\ w_\nu &= -\langle \phi_\nu \phi_{-\nu}\rangle_{\text{imp}},\end{aligned} \tag{5.27}$$

where $\bar{n}_\nu = n_\nu - \langle n \rangle \delta_\omega$ and $\langle \ldots \rangle_{\text{imp}}$ denotes the impurity average with respect to the action (5.21). (Thereafter, we use combined frequency–spin subscripts, e.g.,

$1 \equiv \omega_1, \sigma_1$). Here, the impurity correlation function is defined as

$$g_{12} = -\langle c_1 c_2^* \rangle_{\text{imp}} = -\mathcal{Z}_{\text{imp}}^{-1} \int \mathcal{D}[c^*, c] c_1 c_2^* \exp(-S_{\text{imp}}[c^*, c]). \tag{5.28}$$

The impurity susceptibility and dual correlator are related via a HS-transformation [42]

$$w_\nu = \mathcal{U}_\nu + \mathcal{U}_\nu \chi_\nu \mathcal{U}_\nu. \tag{5.29}$$

The main advantage of introducing the present reference system is that the impurity action can be solved numerically exactly, so we should try to put as much of the physics as possible into the impurity model. The exact solution of the impurity model can be found using a continuous-time quantum Monte Carlo [45–47] solver. Only weaker correlations remain to be treated in the dual perturbation theory, which yields the fully renormalized fermionic and bosonic propagators $\tilde{G}_{\mathbf{k}\omega}$ and $\tilde{W}_{\mathbf{q}\nu}$.

The dual interaction $\tilde{V}_{\text{imp}}$ contains interaction terms of arbitrary order in f^*, f, and ϕ. Usually, the interactions are restricted to the two-particle level, where there is a fermion–fermion interaction γ and a fermion–boson interaction λ. The interaction vertices can be obtained from the numerically exact solution of the impurity problem. For simplicity, we assume that the impurity Green's function is diagonal in the spin–orbital indices

$$\gamma_{1234} = (g_1 g_2 g_3 g_4)^{-1} \left[\langle c_1 c_2 c_4^* c_3^* \rangle - \langle c_1 c_3^* \rangle \langle c_2 c_4^* \rangle + \langle c_1 c_4^* \rangle \langle c_2 c_3^* \rangle \right]$$
$$\lambda_{123} = (g_1 g_2)^{-1} \langle c_1 c_2 \bar{n}_3 \rangle. \tag{5.30}$$

In this case

$$\tilde{V}[f^*, f, \phi] = \sum_{123} \lambda_{123} f_1^* f_2 \phi_3 + \frac{1}{4} \sum_{1234} \gamma_{1234} f_1^* f_2^* f_4 f_3. \tag{5.31}$$

In Matsubara space, the local electron–electron vertex depends on two fermion and one boson frequencies $\gamma^\nu_{\omega\omega'}$, while the electron–boson vertex depends on one fermion and one boson frequency λ^ν_ω. These interaction vertices are related by [42,44]

$$\lambda^\nu_\omega = 1 + \sum_{\omega'} \gamma^\nu_{\omega\omega'} g_{\omega'} g_{\omega'+\nu}. \tag{5.32}$$

We note that in the limit of zero effective electron–electron interaction ($\gamma = 0$) the electron–boson vertex does not vanish ($\lambda = 1$). An important consequence of this property is the fact that the EDMFT approximation is not well grounded and there are always interactions between femionic and bosonic degrees of freedom. The simplest fix to this issue is the so-called GW + DMFT approximation. In this case we use the lowest-order correction in the electron–boson interactions with $\lambda = 1$. This corresponds to the second-order boson self-energy (fermionic Green's functions can in principle contain the self-energy) [44]

$$\tilde{P}^{(2)}_{q\nu} = -\sum_{k,\omega} \tilde{G}_{k\omega} \tilde{G}_{k+q,\omega+\nu}$$

$$\tilde{W}^{-1}_{GW,q\nu} = \tilde{W}^{-1}_{0,q\nu} - \tilde{P}^{(2)}_{q\nu}.$$
(5.33)

Finally, the nonlocal dual part of the GW + DMFT self-energy is equal to

$$\tilde{\Sigma}^{GW}_{k\omega} = \sum_{q\nu} \tilde{G}_{k-q,\omega-\nu} \tilde{W}^{GW}_{q\nu}$$

$$\tilde{W}^{GW}_{q\nu} = \tilde{V}_{q\nu} \left[1 - \left(\chi_\nu + \tilde{P}^{(2)}_{q\nu} \right) \tilde{V}_{q\nu} \right]^{-1}.$$
(5.34)

We point out two main properties of such a combined GW + DMFT scheme: (a) the effective susceptibility in the denominator of $\tilde{W}^{GW}_{q\nu}$ contains the exact local part χ_ω, which can be expected to increase the accuracy of this approximation; (b) the lattice interactions V_q need to be corrected for the impurity local part $\mathcal{U}_\nu$ to give effective bare interactions $\tilde{V}_{q\nu}$. In the first-order approximation, in the denominator of the RPA this effect vanishes since the summation over q-space gives the local dual green-function, which is zero due to the self-consistency condition in the DMFT loop.

Since our reference system is an effective impurity model, an additional self-energy $\tilde{\Sigma}^{GW}_{k\omega}$ plays the role of a T-matrix for the original lattice fermions and the renormalized Green's function is given by the following equation:

$$G^{-1}_{k\omega} = \left[g_\omega + g_\omega \tilde{\Sigma}^{GW}_{k\omega} g_\omega \right]^{-1} - \tilde{t}_{k\omega}.$$
(5.35)

If we take into account the formal definition of the DMFT self-energy,

$$g^{-1}_\omega = i\omega + \mu - \Delta_\omega - \Sigma^{DMFT}_{imp,\omega}.$$
(5.36)

In the limit of a small dual self-energy $\left| g_\omega \tilde{\Sigma}^{GW}_{k\omega} \right| = 1$ this construction reduces to the standard separation of the GW + DMFT self-energy into a local and a non-local part

$$G^{-1}_{k\omega} = i\omega + \mu - t_k - \Sigma^{GW+DMFT}_{k\omega}$$

$$\Sigma^{GW+DMFT}_{k\omega} = \Sigma^{DMFT}_{imp,\omega} + \tilde{\Sigma}^{GW}_{k\omega}.$$
(5.37)

5.5
Derivation of GW + DMFT from a Free Energy Functional

We now turn to a derivation of the GW + DMFT scheme from a functional point of view. The discussion follows closely the original derivation in Ref. [21] in the review [48,49]. As noted in Refs [39,50], the free energy of a solid can be viewed as a functional $\Gamma[G, W]$ of the Green's function G and the screened Coulomb interaction W. The functional Γ can trivially be split into a Hartree part Γ_H and a many body (exchange-correlations) correction Ψ, which contains all corrections beyond the Hartree approximation: $\Gamma = \Gamma_H + \Psi$. The Hartree part can be given in the form

$$\Gamma_H[G, W] = Tr \ln G - Tr\left[\left(G_H^{-1} - G^{-1}\right)G\right] \\ - \frac{1}{2} Tr \ln W + \frac{1}{2} Tr\left[\left(V_q^{-1} - W^{-1}\right)W\right]. \tag{5.38}$$

The Ψ-functional is the sum of all skeleton diagrams that are irreducible with respect to both, one-electron propagator and interaction lines. $\Psi[G, W]$ has the following properties:

$$\frac{\delta \Psi}{\delta G} = \Sigma^{xc} \\ \frac{\delta \Psi}{\delta W} = P. \tag{5.39}$$

The Ψ functional was first derived in Ref. [39]. A detailed discussion in the context of extended DMFT can be found in Ref. [50]. A detailed review of the derivation in a path-integral language is also found in Ref. [51].

The GW approximation consists in retaining the first-order term in the screened interaction W only, thus approximating the Ψ-functional by the first-order polarization loop with effective interaction

$$\Psi[G, W] = -\frac{1}{2} Tr(GWG). \tag{5.40}$$

We then find trivially

$$\Sigma = \frac{\delta \Psi}{\delta G} = -GW, \tag{5.41}$$

$$P = \frac{\delta \Psi}{\delta W} = GG. \tag{5.42}$$

Extended DMFT, on the other hand, would calculate all quantities derived from this function from a local impurity model, that is, one can formally write

$$\Psi = \Psi_{\text{imp}}\left[G^{RR}, W^{RR}\right], \tag{5.43}$$

where G^{RR} and W^{RR} are the local parts of the Green's function and screened interaction, respectively.

In Ref. [21], an approximation to the Ψ functional was constructed that corresponds to the combined GW+DMFT scheme. It approximates the Ψ functional as a direct combination of local and nonlocal parts from GW and extended DMFT, respectively

$$\Psi = \Psi_{GW}^{\text{nonloc}}\left[G^{RR'}, W^{RR'}\right] + \Psi_{\text{imp}}\left[G^{RR}, W^{RR}\right]. \tag{5.44}$$

More explicitly, the nonlocal part of the GW + DMFT Ψ-functional is given by

$$\Psi_{\text{GW}}^{\text{nonloc}}[G^{RR'}, W^{RR'}] = \Psi_{\text{GW}}[G^{RR'}, W^{RR'}] - \Psi_{\text{GW}}^{\text{loc}}[G^{RR'}, W^{RR'}] \quad (5.45)$$

while the local part is taken to be an impurity model Ψ functional. Following (extended) DMFT, this onsite part of the functional is generated from a local quantum impurity problem (defined on a single atomic site). The expression for its free energy functional $\Gamma_{\text{imp}}[G_{\text{imp}}, W_{\text{imp}}]$ is analogous to Eq. (5.38) with $\mathcal{G}$ replacing G_H and $\mathcal{U}$ replacing V

$$\begin{aligned}\Gamma_{\text{imp}}[G_{\text{imp}}, W_{\text{imp}}] &= Tr \ln G_{\text{imp}} - Tr\left[\left(\mathcal{G}^{-1} - G_{\text{imp}}^{-1}\right)G_{\text{imp}}\right] \\ &\quad - \frac{1}{2} Tr \ln W_{\text{imp}} + \frac{1}{2} Tr\left[\left(\mathcal{U}^{-1} - W_{\text{imp}}^{-1}\right)W_{\text{imp}}\right] \\ &\quad + \Psi_{\text{imp}} G_{\text{imp}}, W_{\text{imp}}.\end{aligned} \quad (5.46)$$

The impurity quantities $G_{\text{imp}}, W_{\text{imp}}$ can thus be calculated from the effective action

$$S = \int d\tau d\tau' \Big[-\sum c_L^*(\tau) \mathcal{G}_{LL'}^{-1}(\tau - \tau') c_{L'}(\tau') \\ + \frac{1}{2} \sum c_{L_1}^*(\tau) c_{L_3}^*(\tau') \mathcal{U}_{L_1 L_2 L_3 L_4}(\tau - \tau') c_{L_2}(\tau) c_{L_4}(\tau') \Big], \quad (5.47)$$

where the sums run over all orbital indices L. In this expression, c_L^* is a Grassmann number associated with orbital L on a given sphere. The Hartree terms are not included at the level of the impurity problem. Indeed, the Hartree potential is present already at the level of the effective "noninteracting" Green's functions.[3]

The construction (5.44) of the Ψ-functional is the only *ad hoc* assumption in the GW + DMFT approach. The explicit form of the GW + DMFT equations follows then directly from the functional relations between the free energy, the Green's function, the screened Coulomb interaction, and so on. Taking derivatives of Eq. (5.44) as in Eq. (5.39) it is seen that the complete self-energy and polarization operators read

$$\begin{aligned}\Sigma^{xc}(\mathbf{k}, i\omega_n)_{LL'} &= \Sigma_{\text{GW}}^{xc}(\mathbf{k}, i\omega_n)_{LL'} \\ &\quad - \sum_k \Sigma_{\text{GW}}^{xc}(\mathbf{k}, i\omega_n)_{LL'} + [\Sigma_{\text{imp}}^{xc}(i\omega_n)]_{LL'},\end{aligned} \quad (5.48)$$

$$\begin{aligned}P(\mathbf{q}, i\nu_n)_{\alpha\beta} &= P_{\text{GW}}(\mathbf{q}, i\nu_n)_{\alpha\beta} \\ &\quad - \sum_q P_{\text{GW}}(\mathbf{q}, i\nu_n)_{\alpha\beta} + P_{\text{imp}}(i\nu_n)_{\alpha\beta}.\end{aligned} \quad (5.49)$$

It is worthwhile to mention, that Σ_{imp}^{xc} and P_{imp} take into account local vertex corrections to all orders in perturbation theory.

---

3) The Fock exchange contribution is included through the GW terms.

The meaning of Eq. (5.48) is transparent: the off-site part of the self-energy is taken from the GW approximation, whereas the onsite part is calculated to all orders from the dynamical impurity model. This treatment thus goes beyond usual E-DMFT, where the lattice self-energy and polarization are just taken to be their impurity counterparts. The second term in Ref. (5.48) subtracts the onsite component of the GW self-energy thus avoiding double counting. At self-consistency this term can be rewritten as

$$\sum_{\mathbf{k}} \Sigma^{xc}_{GW}(\tau)_{LL'} = -\sum_{L_1 L_1'} W^{imp}_{LL_1 L' L_1'}(\tau) G_{L_1' L_1}(\tau), \quad (5.50)$$

so that it precisely subtracts the contribution of the GW diagram to the impurity self-energy. Similar considerations apply to the polarization operator.

We now outline the iterative loop that determines $\mathcal{G}$ and $\mathcal{U}$ self-consistently (and, eventually, the full self-energy and polarization operator):

- The impurity problem (5.47) is solved, for a given choice of $\mathcal{G}_{LL'}$ and $\mathcal{U}_{\alpha\beta}$: the "impurity" Green's function

$$G^{LL'}_{imp} \equiv -\langle c_L(\tau) c^*_{L'}(\tau') \rangle_S \quad (5.51)$$

is calculated, together with the impurity self-energy

$$\Sigma^{xc}_{imp} \equiv \delta \Psi_{imp}/\delta G_{imp} = \mathcal{G}^{-1} - G^{-1}_{imp}. \quad (5.52)$$

The two-particle correlation function (only connected part)

$$\chi_{L_1 L_2 L_3 L_4} = \langle c^*_{L_1}(\tau) c_{L_2}(\tau) c^*_{L_3}(\tau') c_{L_4}(\tau') \rangle_S \quad (5.53)$$

must also be evaluated.

- The impurity effective interaction is constructed as follows:

$$W^{\alpha\beta}_{imp} = \mathcal{U}_{\alpha\beta} - \sum_{L_1 \cdots L_4} \sum_{\gamma\delta} \mathcal{U}_{\alpha\gamma} O^{\gamma}_{L_1 L_2} \chi_{L_1 L_2 L_3 L_4} [O^{\delta}_{L_3 L_4}]^* \mathcal{U}_{\delta\beta}, \quad (5.54)$$

where $O^{\alpha}_{L_1 L_2} \equiv \langle \phi_{L_1} \phi_{L_2} | B^{\alpha} \rangle$ is the overlap matrix between two-particle states and products of one-particle basis functions. The polarization operator of the impurity problem is then obtained as

$$P_{imp} \equiv -2\delta \Psi_{imp}/\delta W_{imp} = \mathcal{U}^{-1} - W^{-1}_{imp}, \quad (5.55)$$

where all matrix inversions are performed in the two-particle basis B^{α} (see the discussion in Refs [52,53]).

- From Eqs. (5.48) and (5.49), the full **k**-dependent Green's function $G(\mathbf{k}, i\omega_n)$ and effective interaction $W(\mathbf{q}, i\nu_n)$ can be constructed. The self-consistency condition is obtained, as in the usual DMFT context, by requiring that the onsite components of these quantities coincide with G_{imp} and W_{imp}. In practice, this is done by computing the onsite quantities

$$G_{loc}(i\omega_n) = \sum_{\mathbf{k}} [G^{-1}_H(\mathbf{k}, i\omega_n) - \Sigma^{xc}(\mathbf{k}, i\omega_n)]^{-1}, \quad (5.56)$$

$$W_{\text{loc}}(i\nu_n) = \sum_q [V_{\mathbf{q}}^{-1} - P(\mathbf{q}, i\nu_n)]^{-1}, \tag{5.57}$$

and using them to update the Weiss dynamical mean field $\mathcal{G}$ and the impurity model interaction $\mathcal{U}$ according to

$$\mathcal{G}^{-1} = G_{\text{loc}}^{-1} + \Sigma_{\text{imp}} \tag{5.58}$$

$$\mathcal{U}^{-1} = W_{\text{loc}}^{-1} + P_{\text{imp}}. \tag{5.59}$$

This set of equations is iterated until self-consistency.

5.6 LDA + DMFT Scheme for Real Materials

Before turning to a more detailed discussion of how the GW + DMFT scheme works in practice, with illustrations on specific examples, we describe the current workhorse of correlated electronic structure theory, the combination of DFT and DMFT (the so-called "LDA + DMFT" scheme). First, we need to have an efficient scheme for partitioning the space and orbital degrees of freedom into active degrees of freedom included in the DMFT description and passive ones that are assumed to be well-described within DFT. For example, in the high-temperature superconducting oxide YBa$_2$Cu$_3$O$_7$ the strongly correlated electrons are the ones on the Cu 3d shell, and moreover there is only one per nonequivalent copper d$_{x^2-y^2}$ band which crosses the Fermi level with strong many-body fluctuations. Just a few percent of the total number of electronic states need to be included in the DMFT calculations. Therefore, the simplest realistic correlated scheme would be a DFT + DMFT approach [14,15] with partitioning of the orbital space to the normal band electrons $|K\rangle$ described by the DFT Bloch basis and the correlated local orbitals $|L\rangle$ described by some optimal Wannier basis.

The treatment of correlated electron systems requires the calculation of Green functions and hybridization functions in terms of local orbitals. This is readily achieved when using a basis set, which is localized in real space, such as linear (or Nth order) muffin-tin orbitals (NMTO) [17] or Gaussian basis sets [54,55]. However, many implementations of density functional theory use a delocalized plane wave basis set. This has the advantages that the basis set is simple and universal and its convergence is controlled in principle by a single parameter, the energy cutoff. The projector augmented wave (PAW) method [56], being a representative of a plane wave-based methods, can be used as a simple example of the general projection scheme from the Bloch to the local basis: $\langle K|L\rangle$. Following the general projection scheme of Ref. [54,55], the quantity needed in an implementation of DFT + DMFT is the projection $\mathcal{P}^C = \sum_L |L\rangle\langle L|$ of the full DFT Kohn–Sham Green function $G_{\text{KS}}(\omega)$ on a set of localized orbitals $|L\rangle$

$$G^C(\omega) = \mathcal{P}^C G^{\text{KS}}(\omega) \mathcal{P}^C. \tag{5.60}$$

The subspace $\mathcal{C}$ denotes here the correlated subspace, where many-body fluctuations play a major role and where the DMFT corrections to the DFT will be considered. In plane wave-based calculations, $G^{KS}(\omega)$ in Matsubara space is available in terms of an almost complete set of Bloch states $|K\rangle$ that are eigenstates of the Kohn–Sham DFT Hamiltonian $H^{KS}|K\rangle = \epsilon_K |K\rangle$

$$G^{KS}(\omega) = \sum_K \frac{|K\rangle\langle K|}{i\omega + \mu - \epsilon_K}. \tag{5.61}$$

Inserting Eq. (5.61) into Eq. (5.60) shows that one needs to evaluate projections of the type $\langle L|K\rangle$ in order to access the matrix elements $G^{\mathcal{C}}_{LL'}(\omega)$ of the local Green function. In most cases, the correlated orbitals are d or f orbitals, which are localized inside the PAW augmentation spheres to a good approximation (more than 90% of d(f) – density located inside the sphere). For $|L\rangle$ within these spheres and given the PAW decomposition [56] of a Bloch state $|K\rangle$, one obtains

$$\langle L|K\rangle = \sum_i \langle L|\phi_i\rangle \langle \tilde{p}_i|\tilde{K}\rangle. \tag{5.62}$$

The index i of the augmentation functions $|\phi_i\rangle$ includes site s, angular momentum l and m, and an index ν labeling the radial function: $i = (s, l, m, \nu)$ and $|\tilde{p}_i\rangle$ are projectors of the PAW scheme.

In the described projection scheme the $|L\rangle\langle L|$ matrix are not properly normalized for two reasons: (1) the Bloch basis is incomplete since only a limited number of Bloch bands is included and (2) the PAW augmentation functions are in general not orthonormal. The simplest way is to orthonormalize the projection matrices by the following Wannier type construction: by definition, the localized states $|L\rangle$ are labeled by site and angular momentum indices: $L = (s, l, m)$. We split the site index $s = \mathbf{R} + \mathbf{T}$ such that $\mathbf{R}$ labels the position within the unit cell and $\mathbf{T}$ is the Bravais lattice vector of the unit cell in which s is located. This allows us to construct the Bloch transform of the localized states,

$$|L_{\mathbf{k}}\rangle = \sum_T e^{i\mathbf{k}\mathbf{T}}|L_{\mathbf{T}}\rangle, \tag{5.63}$$

where $\mathbf{k}$ is from the first Brillouin zone and $|L_{\mathbf{T}}\rangle \equiv |L\rangle = |s, l, m\rangle$. The sum in Eq. (5.63) runs over the Bravais lattice. Labeling the Bloch states $|K\rangle = |\mathbf{k}, n\rangle$ by their crystal momentum $\mathbf{k}$, and band index n, we normalize our projection matrices $\mathcal{P}^{\mathcal{C}}_{Ln}(\mathbf{k}) = \langle L_{\mathbf{k}}|\mathbf{k}, n\rangle$ using the overlap operator

$$O_{LL'}(\mathbf{k}) = \sum_n \mathcal{P}^{\mathcal{C}}_{Ln}(\mathbf{k})\mathcal{P}^{*\mathcal{C}}_{L'n}(\mathbf{k}) \tag{5.64}$$

in

$$\overline{\mathcal{P}}^{\mathcal{C}}_{Ln}(\mathbf{k}) = \sum_{L'} O^{-1/2}_{LL'}(\mathbf{k})\mathcal{P}^{\mathcal{C}}_{L'n}(\mathbf{k}). \tag{5.65}$$

These orthonormalized projection matrices are calculated once at the beginning of any calculation and gives a good description of the local Green function of the correlated orbitals projected from the full Bloch Green function $G_{nn'}^B$

$$G_{LL'}^{\mathcal{C}}(\omega) = \sum_{\mathbf{k},nn'} \overline{\mathcal{P}}_{Ln}^{\mathcal{C}}(\mathbf{k}) G_{nn'}^B(\mathbf{k},\omega) \overline{\mathcal{P}}_{L'n'}^{*\mathcal{C}}(\mathbf{k}).$$

Similarly, the hybridization function, $\Delta(\omega)$, is available. It is related to the local Green function by

$$G^{-1}(\omega) = i\omega - \epsilon_d - \Delta(\omega), \tag{5.66}$$

where ϵ_d is the static crystal field. Equation (5.66) is a matrix equation with G, Δ, and ϵ_d being $(\dim \mathcal{C}) \times (\dim \mathcal{C})$ matrices, in general. To separate the hybridization from the static DFT crystal field, we numerically evaluate the limit $\omega \to \infty$, where $\omega - G^{-1}(\omega) \to \epsilon_d$.

In a DFT + DMFT calculation, the projection matrices $\overline{\mathcal{P}}_{Ln}^{\mathcal{C}}(k)$ are used for up- and downfolding quantities like the Green function and the self energy in the course of the iterative DMFT procedure in exactly the same way as shown for the local Green function above. For example, the self-energy obtained by an impurity solver for the effective impurity model $\Sigma_{LL'}^{\mathcal{C}}(\omega)$ can be upfolded to the Bloch basis as follows:

$$\Sigma_{nn'}^B(\mathbf{k},\omega) = \sum_{LL'} \overline{P}_{Ln}^{*\mathcal{C}}(\mathbf{k}) \Sigma_{LL'}^{\mathcal{C}}(\omega) \overline{P}_{L'n'}^{\mathcal{C}}(\mathbf{k}).$$

Since the self-energy in DMFT is a purely local quantity, the index $\mathbf{k}$ on $\Sigma_{nn'}^B(\mathbf{k},\omega)$ reflects the momentum dependence brought about by the projection matrices. The presented projection scheme allows for the inclusion of both correlated and uncorrelated states in the procedure. Therefore, information about the interplay of correlated orbitals with their uncorrelated ligands can be obtained.

As a first example, we show a realistic DFT + DMFT calculation of the $SrVO_3$ spectral function in Figure 5.5, where one can see the renormalization of the valence correlated $V\,t_{2g}$ states as well as broadening of the Bloch O 2p states [57]. It is important to note that the DFT + DMFT scheme have access to temperature effects which is completely new for solid-state quantum chemistry. Since the solution of an effective many-body DMFT impurity problem normally involve the finite-temperature Quantum Monte Carlo simulations [47] all local electronic excitations in the thermal bath were taken into account. One can study the temperature dependence of electronic structure and exchange splitting of ferromagnetic materials. As example, we show the finite-temperature spectral function of metallic iron for the temperature close to critical Curie point in Figure 5.6 compared to zero-temperature DFT density of states [58]. The spin splitting of magnetic d-states is drastically reduced compared to DFT results as well as band-width renormalization due to correlation effects. Estimation of the Curie temperature in DFT + DMFT scheme gives $T_C = 1900$ K in reasonable agreement with experimental values of $T_C^{\text{exp}} = 1043$ K. Calculations of the local spin–spin correlation function in high-temperature paramagnetic phase of iron (Figure 5.6) show that the magnetic moment is neither localized nor fully

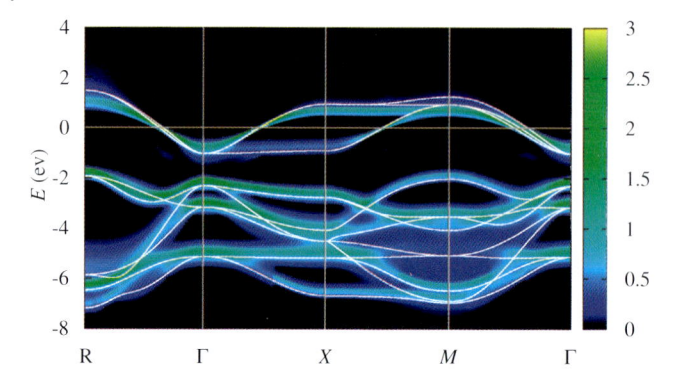

140 5 Many Body Perturbation Theory, Dynamical Mean Field Theory and All That

Figure 5.5 Momentum resolved electronic spectral function of SrVO₃ obtained within the DFT+DMFT scheme. The original DFT band structure of the V t_{2g} states near $E_F = 0$ and O 2p Bloch states (around −4 eV) is shown for comparison. Close to the Fermi level (origin of the energy scale) one distinguishes a threefold degenerate manifold of t_{2g} states, the filling of which corresponds to one electron. The oxygen states extend between −2 and −8 eV. Compared to the DFT band structure, the DFT+DMFT spectral function displays a renormalization of spectral features. The t_{2g} bandwidth is roughly reduced by a factor 0.6, corresponding to a quasi-particle mass enhancement of 1/0.6. Concomitantly, spectral weight is shifted to higher energy (nonband-like) features, see dark blue shadows. (From Ref. [57].)

itinerant. The classical problem of paramagnetic Mott insulator like NiO for temperature higher than antiferromagnetic Neel transition also have been successfully solved in the realistic DMFT approach [59] (see Figure 5.7). In this case the magnitude of the energy gap as well as the density of state are in good

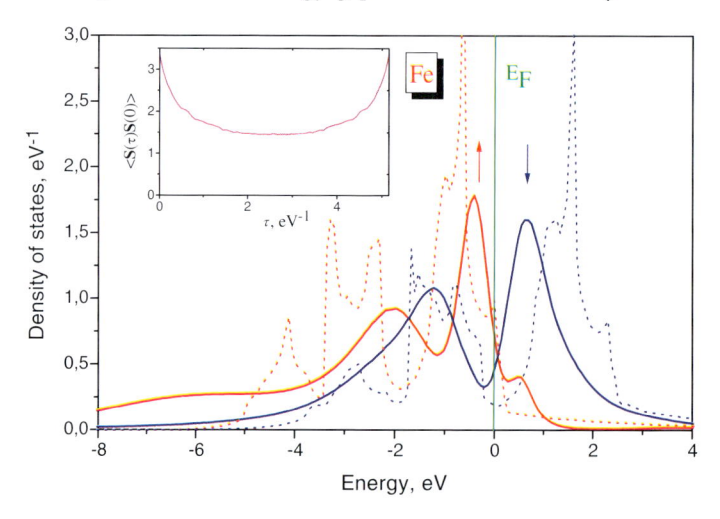

Figure 5.6 DFT+DMFT calculations of ferromagnetic iron at temperature $T = 0.8\, T_C$. The partial densities of d-states (full lines) are compared with the corresponding LSDA results at zero temperature (dashed lines) for the spin-up (red lines, arrow-up) and spin-down (blue lines, arrow-down) states. The insert shows the spin–spin correlation function for $T = 1.2\, T_C$. (From Ref. [58].)

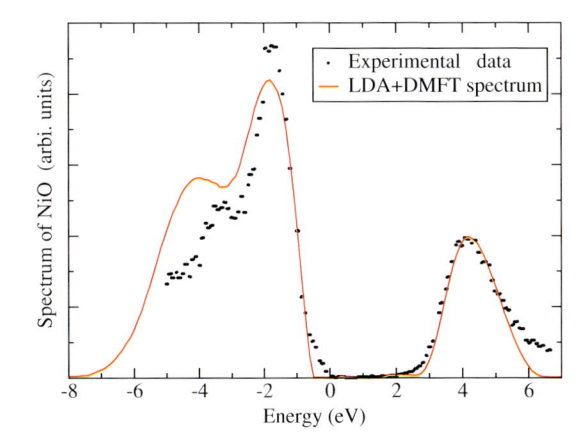

Figure 5.7 Theoretical spectrum of NiO obtained by the LDA + DMFT approach (red curve) at $T = 1160$ K, compared with experimental XPS + BIS data (Ref. [60]). (From Ref. [59].)

agreement with experimental data [60]. In principle, one can investigate also a superconducting transition for iron pnictides wthin a (cluster) DFT + DMFT scheme, but this is much more complicated problem due to the low critical temperature. Much more suitable problem for the DFT + DMFT related with a high-spin to low-spin transition as function of temperature for solids like LaCoO$_3$ [61] or local valence fluctuations in the kernel of hemoglobin for ligand binding problem [62]. The main success of the DFT + DMFT approach is related with investigation of metal-insulator transition in oxide compounds as function of d-states occupation, temperature, crystal distortions and stoichiometry [32].

5.7
The GW + DMFT Approach in Practice

With the development of efficient solvers for impurity models with dynamical interactions, the development of the GW + DMFT scheme described above has made a great leap foreward. In the realistic multiorbital context, fully dynamical calculations are now possible [22,23], and at the model level or for effective realistic systems, even fully self-consistent calculations have been performed [24,26,51,63,64].

As explained above, in GW + DMFT, the self-consistency requirement is generalized to encompass two-particle quantities as well, namely, the local projection of the screened interaction is required to equal the impurity screened interaction. This in principle promotes the Hubbard U from an adjustable parameter in DMFT techniques to a self-consistent auxiliary function that incorporates long-range screening effects in an *ab initio* fashion. Indeed, not only higher energy degrees of freedom can be downfolded into an effective dynamical interaction but also one can aim at incorporating nonlocal screening effects into an effective dynamical $\mathcal{U}(\omega)$. The theory is then free of any Hubbard *parameter*, and the interactions are directly determined from the full long-range Coulomb interactions in the continuum. Within the terminology of Hedin's equations, this means in particular that screening is assessed beyond the random-phase approximation or standard GW-scheme. Fully self-consistent GW+DMFT are rare, and have so far been performed only in the case of an extended Hubbard model [26,51,64] and for the system of adatoms on semiconductor surfaces [24] discussed below. In these studies, it was in particular shown that in the regime close to the metal-insulator transition, the RPA yields a poor estimate for screening which is strongly suppressed by correlation effects, see Ref. [26].

Interestingly, as a side product, these calculations demonstrated – at least within the model context – that GW+DMFT indeed cures the problems related to self-consistency in the GW approximation. In fact, it is well known that due to missing error cancellation effects, self-consistency at the pure GW level does not improve upon the one-shot estimate of spectral functions. The implicit presence of the local vertex in GW+DMFT, however, overcomes this shortcoming and allows for fully self-consistent calculations, see Figure 5.8.

From a formal point of view, the GW+DMFT method, as introduced in Ref. [21], corresponds to a specific approximation to the correlation part of the free energy of a solid, expressed as a functional of the Green's function G and the screened Coulomb interaction W: the nonlocal part is taken to be the first

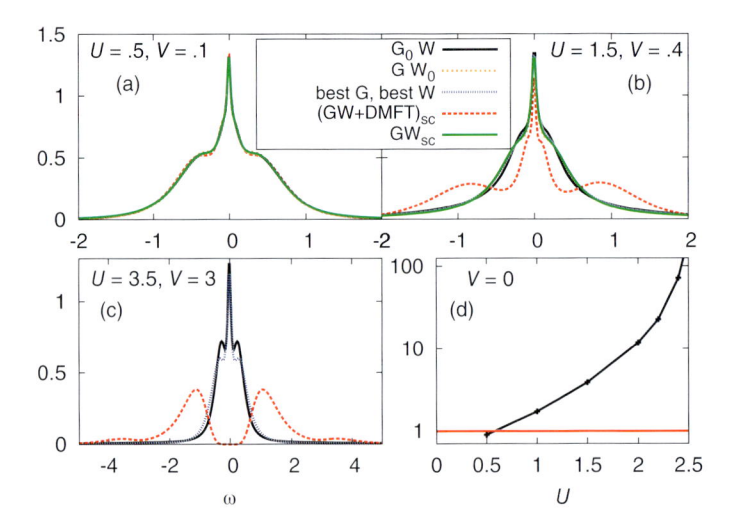

Figure 5.8 Benchmark of self-consistency within various flavours of GW and GW+DMFT for an extended Hubbard model with local interaction U and nearest neighbor V. Panels (a–c) present the spectral function $A_{loc}(\omega)$ obtained using different one-shot GW schemes, fully self-consistent GW and GW+DMFT. (d) Gives a crude estimate of the static part of the vertex $\Lambda(\omega=0)$ as a function of U. The estimate is obtained by first determining $\tilde{W}(\tau) = \Sigma_{imp}(\tau)/G_{imp}(\tau)$ from EDMFT, and then calculating $\Lambda(0) \approx \tilde{W}(i\nu_0)/W_{loc}(i\nu_0)$. (From Ref. [26].)

order term in W, while the local part is calculated from a local impurity model as in (extended) dynamical mean field theory. In practice, however, the calculation of a self-energy for (rather delocalized) s or p orbitals has never been performed within DMFT, and it appears to be more physical to approximate this part also by a GW-like expression. For these reasons, Tomczak *et al.* [22,23] proposed a practical scheme, in which only the local part of the self-energy of the " correlated" orbitals is calculated from the impurity model and all other local and nonlocal components are approximated by their first-order expressions in W. We review an example in Section 5.8.

5.8
GW + DMFT: the Example of SrVO$_3$

The very first dynamical implementation of GW + DMFT was achieved in 2012, and applied to the ternary transition metal oxide SrVO$_3$ [22,23]. Our target material, SrVO$_3$, has been thoroughly studied, both, experimentally and theoretically. It crystallizes in the cubic perovskite structure, splitting the V-d states into a threefold degenerate t_{2g} manifold, filled with one electron per V, and an empty e_g doublet. It has been characterized as a correlated metal with a quasiparticle renormalization of about 0.6 [65–67], and a photoemission (Hubbard-) satellite at around -1.6 eV binding energy [68]. Inverse photoemission has located the electron addition $d^1 \to d^2$ peak at an energy of about 2.7 eV [68].

In the implementation of Ref. [22,23], the GW + DMFT equations were solved self-consistently at the DMFT level, that is, for a fixed screened interaction (corresponding to the cRPA one) and fixed nonlocal GW self-energies. However, the fully dynamical interactions were retained, and the GW + DMFT equations solved within the BFA [69] reviewed above.

In Figure 5.9, we compare the momentum resolved GW + DMFT spectral function to pure GW results as well as to the most recent angle resolved

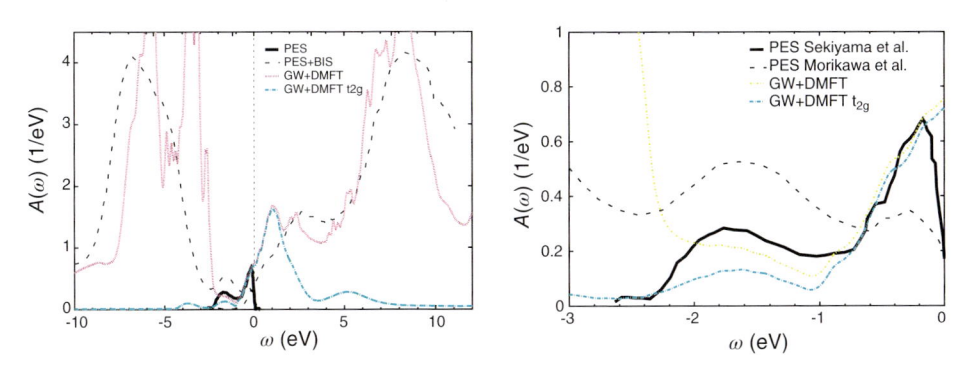

Figure 5.9 Momentum-resolved t_{2g} spectral function from GW + DMFT, compared to experimental angle resolved photoemission spectra of Refs [67,70]. (Adapted from Ref. [22].)

photoemission experiments (ARPES) [70]. The low-energy part of the spectral function is dominated by the tr2g contribution, which is profoundly modified within GW + DMFT compared to the pure GW spectrum. A renormalized quasi-particle band disperses around the Fermi level, with for example, at Γ a sharp peak in the occupied part of the spectrum at about -0.5 eV binding energy, corresponding to a strong renormalization of the LDA Kohn–Sham state that has, at this momentum, an energy of -1 eV. At the X-point, the three t_{2g} bands are no longer degenerate, and surprisingly weakly renormalized xz/yz states are observed at 0.9 eV, while the yz band is located at nearly the same energy as at the Γ point. This is in excellent agreement with experiments: in photoemission spectroscopy a renormalized quasi-particle band structure is dispersing between the Γ and X points. At binding energies of -1.6 eV a weakly dispersive Hubbard band forms, whose intensity varies significantly as a function of momentum [67]. In the GW + DMFT spectral function, the Hubbard band is observed at about -1.6 eV, and its k-dependent intensity variation is indeed quite strong. As anticipated from the total spectral function discussed above, the GW spectrum also displays a satellite structure, which is however of plasmonic origin, arising from the structure in Im W. This is a well-known failure of the GWA, which has been analyzed in detail in Ref. [71]: the simple form of the GW self-energy is not able to encode multiple satellites, and the single plasmon is then located at a too high energy. The lower Hubbard band is absent in GW, as expected.

As discussed above, the low-energy features in W are absent from the dynamical Hubbard interaction entering the combined GW and dynamical calculation, consistently with the fact that the full GW + DMFT calculation does not show spurious features at 3 eV as does the GW.

The overall picture of the GW + DMFT spectra results in an occupied band structure that resembles closely the dynamical mean field picture (see, e.g., Ref. [72]) though the lower Hubbard band is located slightly closer to the Fermi level, at about -1.6 eV, in agreement with experiments. This improvement is an effect of the relatively smaller zero-frequency U value (3.6 eV) compared to parameters commonly used in standard LDA + DMFT calculations (around 4–5 eV). In the latter, larger U values are required to match the experimentally observed mass enhancements, thus necessarily placing the lower Hubbard band at energies slightly too far away from the Fermi level. The dynamical screening, included in our calculations, results in additional spectral weight transfers [69,73], thus yielding at the same time a good description of the renormalization of the band near the Fermi level (corresponding to an enhancement of the effective masses) and the Hubbard band.

In the unoccupied part of the spectrum, nonlocal self-energy effects are larger. Interestingly, our total spectral function, right panel of Figure 5.10, does not display a clearly separated Hubbard band. The reason is visible from the k-resolved spectra: the upper Hubbard band is located at around 2 eV, as expected from the location of the lower Hubbard band and the fact that their separation is roughly given by the zero-frequency value of U (=3.6 eV). The peak around 2.7 eV that

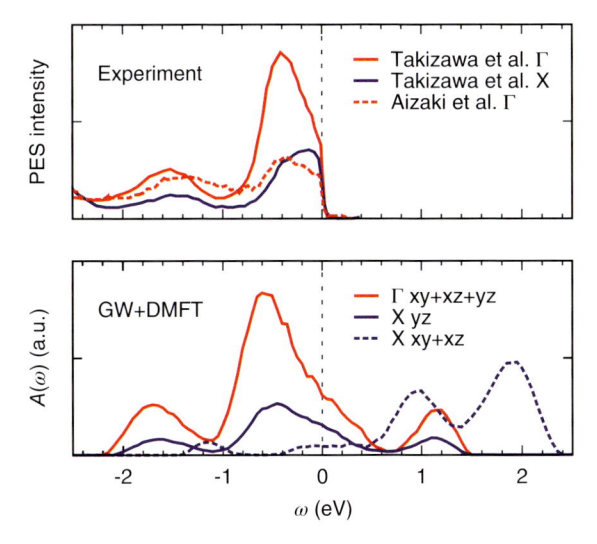

Figure 5.10 Spectral function from GW + DMFT in comparison to experiments. The tr2g orbital contribution is resolved as dashed line. The Fermi energy is set to zero. (Adapted from Ref. [23] (see also Ref. [22]).)

appears in the inverse photoemission spectrum [68] – commonly interpreted as the upper Hubbard band of tr2g character in the DMFT literature – arises from e_g states located in this energy range. The nonlocal self-energy effects lead, in the unoccupied part of the spectrum, to overlapping features from different k-points and the impression of an overall smearing out of the total spectral function. From the k-resolved spectra, it is also clear that while nonlocal self-energy effects stemming from the GW part have little influence on the occupied part of the spectrum, they widen the bands in the unoccupied part substantially. This leads to interesting effects beyond the previous DFT + DMFT results. A rough estimate of the various renormalization effects on the overall bandwidth leads to a – at first sight – astonishing conclusion: effects of the dynamical tail of the Hubbard $\mathcal{U}$ have been estimated to roughly lead to a band renormalization of $Z_B : 0.7$ [73], and the renormalization due to the static part $\mathcal{U}(\omega = 0)$ still adds to this. Nevertheless, the final position of the empty quasi-particle bands after the GW + DMFT calculation nearly coincides with the initial LDA energies. This gives an order of magnitude for the substantial widening of the band induced by nonlocal (Fock exchange) effects, which are not included in DFT + DMFT. The picture is consistent with the observation of Refs [74,75], that a purely local GW calculation in fact leads to much stronger renormalization effects than the full GW calculation, and the band structure within the latter results from subtle cancelations of band narrowing due to the local self-energy and widening due to its nonlocal parts. One could thus be tempted to conclude that renormalization effects due to local dynamical interactions and widening due to nonlocal self-energies cancel, giving new justifications to combined

LDA++ schemes with static interactions. There are several reasons why this conclusion would be too quick: first, the widening effects rather selectively act on the unoccupied band structure, since the exchange-correlation potential of LDA is a much better approximation to the many-body self-energy for occupied states than for empty ones. Second, the renormalization effect due to dynamical interactions goes hand in hand with a spectral weight loss at low-energies. These are barely observable in photoemission spectroscopy since spectra are generally not measured in absolute units, and even then would matrix element effects make a comparison of absolute normalizations intractable. Probes that can assess absolute units, such as optical spectroscopy, however, can be expected to be sensitive to such shifts of spectral weight. At present, unfortunately, available data for $SrVO_3$ do not have the necessary resolution to assess this question.

5.9
Dynamical Screening in the Iron Pnictide $BaFe_2As_2$

As a nontrivial example for a simplified GW + DMFT approach (retaining here only the dynamical nature of the effective local Coulomb interactions, without including nonlocal self-energy corrections), we review now calculations on the iron pnictide compound $BaFe_2As_2$. This material is the prototypical compound of the so-called "122-family" of iron pnictide superconductors. It exhibits superconductivity under pressure [76,77] or hole- as well as electron-doping [78,79]. Many experimental probes including angle-resolved and angle-integrated photoemission spectroscopy [80–86], optics and transport, Raman and neutron scattering, and nuclear magnetic resonance have been employed to characterize the electronic properties [87]. Experimental estimates of the (doping-dependent) mass enhancements vary substantially with doping; literature values range from about 1.4 [88] to 5, at least for the orbital pointing toward the As-sites [86]. The orbital character of the Fermi surface pockets are still subject to debate, but there seems to emerge a consensus about stronger correlation effects for holes than electrons.

The constrained RPA result for the average intra-orbital Coulomb repulsion $\mathcal{U}$ of $BaFe_2As_2$ is shown in Figure 5.11. Here, $\mathcal{U}(\omega)$ represents a partially screened Coulomb interaction for the Fe-d states, which accounts for screening by all degrees of freedom except the Fe-d states themselves. The real part ranges from the static value $U_0 \equiv \mathrm{Re}\, U(\omega = 0) = 3.6$ eV to the bare interaction V of about 20 eV at large ω. In this case, screening does not arise from a single well-defined plasmon excitation. Instead, $\mathrm{Im}\,\mathcal{U}(\omega)$ is characterized by a broad structure beginning from a peak at ~ 26 eV and extending down to a few eV, implying that any plasmon excitations overlap strongly with the one-particle excitations. We note moreover that the term "plasmon" is used here a bit abusively for any bosonic excitation mode that screens the Coulomb interactions, regardless of the precise nature (plasmon, particle-hole excitations or any other many-body satellite feature).

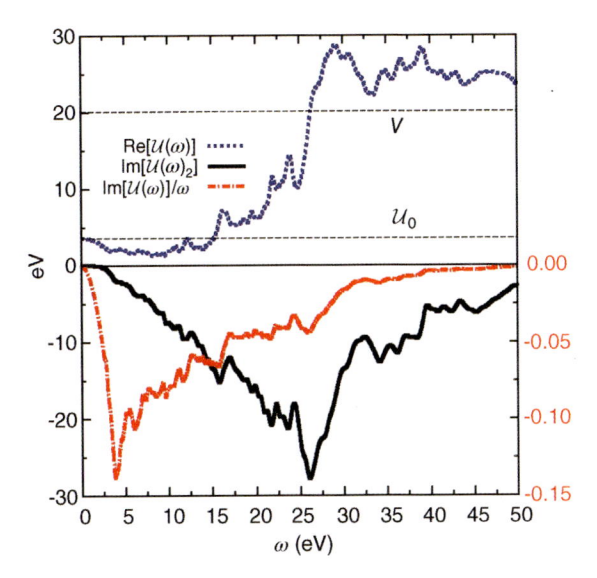

Figure 5.11 Frequency-dependent Hubbard interaction $\mathcal{U}(\omega)$ for BaFe$_2$As$_2$: real and imaginary parts as well as the mode distribution function Im $\mathcal{U}(\omega)/\omega^2$. (From Ref. [12].)

In a standard DFT + DMFT calculation without dynamical effects, the relatively small value of the interaction U_0 would result in a rather weakly correlated picture. This is demonstrated in Figure 5.12, which presents the Fe-d spectral function obtained by using the static U_0 in a standard DFT + DMFT calculation.

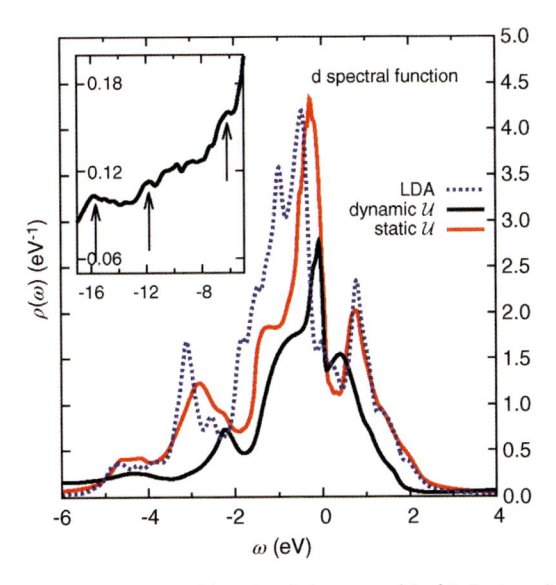

Figure 5.12 Spectral function (3d-states only) of BaFe$_2$As$_2$ within LDA + $\mathcal{U}(\omega)$ + DMFT, in comparison to standard LDA + DMFT and LDA, see text. (From Ref. [12].)

Interaction effects lead to a moderate renormalization of the Fe-d states, with a mass enhancement of 1.6. Comparison to the DFT density of states shows that the peaks at −3 eV and 1 eV are weakly renormalized band states. No Hubbard satellites or other correlation features appear, in agreement with previous studies [89,90].

The function $\frac{\text{Im}\, U(\omega)}{\omega^2}$ is plotted as the red dash-dotted line in Figure 5.11. Besides a first large peak at 3.8 eV, which comes from the rapid decay of Im $U(\omega)$ at small frequencies, there are smaller peaks at 6.1, 16, and 26 eV, as well as less pronounced features at 10 and 12 eV. Physically, the function Im $U(\omega)$ represents the density of screening modes weighted by their coupling strengths. It would be proportional to a single delta-function in the case of a system with a single well-defined plasmon excitation. Such a situation is a rather reasonable approximation to what happens in simple oxides, such as $SrVO_3$. The multitude of structures observed in the present case means that in the iron pnictides a variety of modes (corresponding to subplasmons, particle-hole transitions ...) results in a quasi-continuum of screening modes. For a detailed comparative discussion, see Ref. [91]. Coupling of these modes to the original fermions leads to a spectral weight transfer to higher energies which – in the present case – makes up for about 30% of the overall spectral weight [73]. The presence of a sharp low-energy peak in the bare (i.e., without coupling to bosonic modes) d-electron spectral function results in a whole series of weak, but well-defined satellites when the coupling to the screening modes is taken into account. The inset of Figure 5.12 shows the high energy tail of the occupied part of the spectrum, with arrows marking three satellites around −6.1, −12, and −16 eV. Experimentally, the observation of satellites at −6.5 eV and −12 eV was emphasized in the photoemission study of Ding *et al.* [82]. While de Jong *et al.* [92] confirms a hump in the d-electron spectral function around −6.5 eV, these authors suggest that the feature at −12 eV is an As-4s line. Our calculation suggests that a d-feature, originating from the structure in the frequency-dependent interaction, is superimposed to the As-4s spectral contribution. The −16 eV feature is probably not visible in experiments, because it overlaps with Ba-5p states, while a satellite that we predict at −3.8 eV is masked by structures arising from p–d hybridization.

A detailed analysis of the many-body self-energy revealed a further interesting aspect of this compound, namely, a pronounced "non-Fermi liquid" (incoherent) regime in the metallic phase near optimal doping, characterized by square-root behavior of the self-energy. We refer the interested reader to the original Ref. [12], where also implications for the doping and temperature dependence of the low energy electronic structure are discussed.

5.10
Screened Exchange Dynamical Mean Field Theory

Recently, a simple and extremely efficient simplified GW + DMFT was proposed, tested, and implemented [93,94]: "Screened Exchange Dynamical Mean Field

Theory" (SEx+DMFT) is based on the observation that – within the GW approximation – the nonlocal self-energy is to a good approximation frequency-independent. It therefore becomes a simple static (though k-dependent) correction to the LDA Hamiltonian. SEx+DMFT approximates this correction as a screened exchange term. The scheme thus retains the essential physical mechanisms of widening the bands by exchange and narrowing them (beyond the usual renormalizations induced by static Hubbard interactions) by the dynamical interactions, present in GW+DMFT. So far, the scheme has been tested for transition metal pnictides [93,95] and oxides [94] with very promising results, see Figure 5.13.

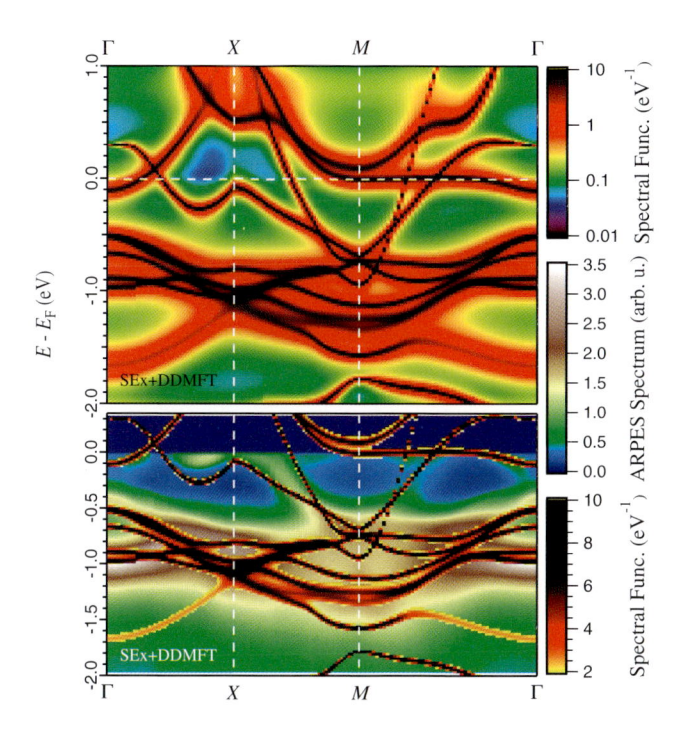

Figure 5.13 Spectral function of $BaCo_2As_2$ from SEx+DMFT (upper panel), in comparison to experiment (lower panel). This compound is a moderately correlated metal, with an overall quasi-particle renormalization factor as compared to LDA of about 0.7 and very modest lifetime effects. As a consequence, relatively well-defined single-particle excitations are present even on an eV energy scale. These can be distinguished in the theoretical spectral function (upper panel) as dark band-like features. The lower panel represents an angle-resolved photoemission spectrum where comparable features (brown-whitish) have been measured. Superimposed in orange–red–black color coding are the maxima of the theoretical spectral function (more precisely, all features present in the upper panel that exceed a cutoff value chosen to be $2\,eV^{-1}$). For further comparisons, in particular of the LDA band structure, LDA+DMFT, LDA+DMFT with frequency-dependent interactions, and the screened exchange band structure to the photoemission spectra we refer the reader to the original literature. (From Ref. [93].)

5.11
Perspectives

From a technical point of view, fully dynamical GW + DMFT calculations remain a challenge, even nowadays, and full self-consistency could at present be achieved only for the surface system, thanks to the simplifications that come along with a single orbital description. Self-consistent GW + DMFT calculations for realistic multiorbital systems are thus an important goal for future work. In this context, the orbital-separated scheme proposed in the application to $SrVO_3$ reviewed above, where only a subset of low-energy states is treated within DMFT presents appealing features. For late transition metal oxides, and in particular charge-transfer systems, interactions between the correlated shell and the ligand electrons are likely to be important for an accurate estimation of the charge transfer energy and thus the whole electronic structure, and exploring the performance of the GW + DMFT scheme in this context is an interesting open problem. One could still expect a perturbative treatment of the intershell interactions to be sufficient, so that explicitly including an intershell self-energy Σ_{pd} from GW in the orbital-separated scheme discussed above seems a promising way to describe charge transfer insulators.

At the same time, given the cost of full GW + DMFT, the design of schemes that can – at least approximately – reproduce the results of the full GW + DMFT calculations at lower computational cost remains an important task. Two routes are currently being pursued:

- At the GW level, is it possible to avoid full GW calculations, and to base the combination with DMFT on approximations to a full GW scheme? The SEx + DMFT scheme mentioned above is a promising option. Alternatively, is it possible to use a quasi-particle self-consistent GW (qsGW) scheme, for example, the one by Schilfgaarde *et al.* [96]?
- At the DMFT level, is it possible to avoid the solution of a local problem with frequency-dependent interaction, replacing it by an optimized impurity problem with effective hybridizations and static interactions?

Recent work provided important insights into the first question [75,97]. Tomczak *et al.* [75] argued that at least for certain classes of materials – the authors investigated iron pnictide compounds – nonlocal and dynamic contributions to the many-body self-energy are mostly separable, so that a purely local self-energy correction to an – in some sense – optimized one-particle band structure would be sufficient for an accurate description. The authors thus proposed a combination of quasi-particle self-consistent GW [96] with DMFT. Tests along these lines – although without attempting to avoid double counting of the local self-energies (see below) and based on a one-shot GW calculation – were performed in Ref. [97] for $SrVO_3$.

A subtle issue arises from the necessity of avoiding double counting of the local part of the GW self-energy, when "quasi-particlising" the GW calculations. Indeed, only the nonlocal part of the GW self-energy enters the combined GW + DMFT self-energy, which means that when a qsGW scheme is employed the local part $\Sigma_{GW}^{loc}(\omega) = \sum_k GW$ has to be explicitly substracted. This is not entirely trivial, since this term acquires both k- and state-dependence through the "quasi-particlization" of its original frequency-dependence. Indeed, the correction acting on a quasi-particle state ϵ_{kn} reads $\Sigma_{GW}^{loc}(\omega = \epsilon_{kn})$. This term is responsible for the widening of the unoccupied bands discussed above (see Section 5.6) and can thus by no means be considered as negligible. The implementation of efficient yet double-counting-free qsGW + DMFT schemes remains therefore an important open challenge.

An optimistic answer can be given to the second question, concerning the need of using frequency-dependent interactions, at least when only the very low-energy electronic structure is of interest. Indeed, as has been shown in Ref. [73], at least in the antiadiabatic limit where the characteristic energy scale of the variation of the interactions is the largest energy of the problem, one can reduce a dynamical Hubbard model to a static one at the price of a renormalization of the one-body part of the Hamiltonian. This trick can be applied also within a GW + DMFT calculation. On the other hand, the gain in computational cost obtained by this procedure is not tremendous in this case, since at least within the so-called "Bose factor approximation" of Refs [69,98] the solution of a dynamical local model can anyhow be achieved at the cost of the solution of the corresponding static model.

Finally, the GW + DMFT construction also introduces some nonlocality into the otherwise purely local self-energy (and polarization) of (extended) DMFT. We have reviewed the first realistic GW + DMFT calculations with fully dynamical interactions, which – for $SrVO_3$ – have evidenced a crucial effect of this nonlocal self-energy on the unoccupied part of the Kohn–Sham band structure. These findings call for further experimental investigations, in particular inverse photoemission experiments. A widening of the band due to the exchange interaction, which – for the empty states – is not well described by the Kohn–Sham potential leads to a substantial reinterpretation of the electronic structure. Nevertheless, this correction is essentially static and given by a (screened) Fock exchange term, the truly frequency-dependent nonlocal part of the self-energy correction is small. A recent systematic study on a simple two-dimensional extended Hubbard model [26] is consistent with those findings in the realistic context: there, it was shown that unless considering a system close to the charge-ordering transition, these nonlocal self-energy effects are tiny. To capture the strong k-dependence of the self-energies in the immediate vicinity of the Mott transition, experimentally observed, for example, in the form of highly k-dependent variations of the effective masses in doped Mott insulators, extending GW + DMFT to include not only fluctuations in the charge but also in the spin channel seems necessary. This is yet another promising topic for future works.

5.12
Conclusions

We have reviewed a series of recent papers dealing with dynamical screening of the Coulomb interactions in materials with correlated electrons. Quite generally, dynamically screened interactions arise as a "representative" within a restricted subspace of the full long-range Coulomb interactions in the original Hilbert space. In practice, it appears to be useful to disentangle two different mechanisms: (1) screening by high-energy degrees of freedom, when the original Coulomb Hamiltonian is downfolded to a low-energy effective Hamiltonian, (2) screening by nonlocal degrees of freedom, when a Hamiltonian with long-range interaction is backfolded into one with purely local interactions. The cRPA provides a simple and transparent description of the first mechanism, and the energetic separation of the different degrees of freedom can probably in many cases justify the neglect of vertex corrections. This is less obvious in the case of spatial downfolding, where both "backfolded" and retained degrees of freedom live on the same energy scales. The GW + DMFT scheme provides an elegant construction of effective local interactions, while including a local vertex in a nonperturbative manner.

This review has discussed the effects of dynamical interactions, specifically on the examples of the ternary transition metal compound $SrVO_3$ and the iron pnictide $BaFe_2As_2$. It was argued that the dynamical nature of the Hubbard interaction leads to many-body satellites in the spectral function (corresponding to plasmons, particle–hole excitations, or more general bosonic excitations), and that the corresponding transfer of spectral weight leads to additional renormalizations in the low-energy electronic structure.

For $SrVO_3$, which was the first compound treated within GW + DMFT in a fully dynamical manner (albeit selfconsistent only at the DMFT level), the nonlocal self-energy effects introduced by this scheme moreover lead to profound modifications of the unoccupied spectra, as compared to simple LDA + DMFT calculations. In particular, the peak at 2.7 eV seen in inverse photoemission experiments was identified as an e_g feature, while the energy scale expected for a potential upper Hubbard ($\sim$2 eV) would make this feature hard to separate from the quite dispersive unoccupied t_{2g} band states. These findings call for a reinterpretation of early LDA + DMFT calculations using a t_{2g}-model for this compound, where the peak at 2.7 eV was commonly interpreted as an upper Hubbard band of t_{2g} character. They do reconcile however DMFT-based electronic structure calculations with cluster model calculations for $SrVO_3$ [99]. The results are furthermore encouraging, since this first benchmark on $SrVO_3$ confirms the ability of the GW + DMFT approach to describe simultaneously Hubbard bands, higher energy satellite structures and corrected energy gaps, in an *ab initio* fashion. Quite generally, materials with a " double LDA failure," an inappropriate description of correlated states, and deficiencies of LDA for the more itinerant states, such as an underestimated "pd-gap," can be treated within this scheme.

5.12 Conclusions

These examples demonstrate the virtues of the GW + DMFT scheme

- GW + DMFT is entirely formulated in the Green's function language, even at the one-body level. In this way, the theory is conceptually double counting-free, since it avoids the mismatch between the Green's function-based description of DMFT and the density-based one of DFT that is inherent to any combined DFT + DMFT scheme.
- It deals directly with the long-range Coulomb interactions, and the effective local Hubbard interactions arise only as intermediate auxiliary quantities. This has several consequences: first, it allows for a truly *ab initio* description of the Coulomb interactions, including dynamical screening effects (and the associated transfers of spectral weight, plasmon replicae, etc.) as described above. Second, when intersite interactions become important, as for example, in the surface systems described above, new instabilities (here toward charge-ordered phases) can appear.
- It retains the nonperturbative character of dynamical mean field theory, thus avoiding limitations due to a truncation of the perturbation series. This latter point is essential to ensure the scheme to be equally appropriate in the weak, strong, and intermediate coupling regimes.
- Finally, the framework of the orbitally separated GW + DMFT scheme described above, where perturbative corrections to the LDA band structure are applied to ligand or conduction band states allows for an accurate description of the electronic structure on larger energy scales than in conventional methods. Indeed, within conventional LDA + DMFT techniques, there is no correction to the LDA for ligands or conduction band states (other than shifts resulting from suppressed hybridizations with the correlated shell). As demonstrated in the example of $SrVO_3$, such effects can be crucial even for a qualitative assessment of electronic states on energy scales of 2 or 3 eV.

We close this review by stressing again that the most important lesson to be learnt from the recent calculations is probably the finding that without addressing dynamical screening effects a "first principles" description is not even possible for the one-body part of the Hamiltonian: even in cases where a good separation of energy scales ($\omega_P \gg$ any other energy scale in the system) allows for an accurate mapping onto a static Hubbard model with effective local interaction, the bosonic renormalizations are crucial for assessing the one-particle Hamiltonian in a quantitative way. On the other hand, it is well known – and has been demonstrated again and again by numerous examples over the last years – that even tiny differences in the one-particle part of the Hamiltonian can lead, in a strongly correlated situation, to very different physical behaviors. Indeed, quantitative differences at the one-particle level can easily make the difference between a metal or an insulator [72,100]. Even on the level of simple one-orbital models, inclusion of dynamical screening changes the critical interaction for the Mott transition, thus shifting the metal-insulator transition line [73].

When are dynamical screening effects stemming from higher energy degrees of freedom large? The strength of the renormalizations induced by dynamical screening depends on the ratio of the electron–boson coupling strength over the plasma frequency, so that strong effects can be expected if the plasma frequency is small and the coupling large. Within the class of t_{2g} transition metal oxides, the plasma frequencies vary substantially less than the couplings. Indeed, the coupling strength is given by the difference of bare and (partially) screened interaction, $\Delta\mathcal{U} = \mathcal{U}(\infty) - \mathcal{U}(0)$. Typical values vary from $\Delta\mathcal{U}\sim 12$ eV for $SrVO_3$ to ~ 8 eV for 5d oxides, such as Sr_2IrO_4 [100]. The importance of the effects can thus be expected to decrease from 3d to 4d and 5d oxides. On the other hand, 4d and 5d orbitals are more extended than 3d orbitals, and intersite interactions are therefore larger. Backfolding of long-range interactions as in the case of the surface systems discussed above is therefore expected to generate larger corrections than in the 3d oxides.

For late transition metal oxides, where small charge-transfer energies and entangled d- and ligand states lead to additional complications, these questions are still largely unexplored. The role of intershell (correlated to ligand shell) interactions, of nonlocal exchange corrections to LDA, or of corrections to the LDA estimate of the d–p hybridizations are ingredients that are still awaiting systematic investigations.

We described how to treat electronic correlations in correlated materials within the local DMFT scheme. This knowledge can be used in realistic DFT + DMFT calculations for the strongly correlated transition metals and rare earth systems, where the spin, orbital, and charge fluctuations in the d- or f-shell play the crucial role in photoemission spectrum, magnetic and optical excitations. The numerically exact solution of the quantum impurity problem (using the available ALPS or TRIQS software) give us an effective local exchange-correlations functional for given correlation materials in specific external field. Combinations of the DMFT scheme with first-principles approaches (such as DFT or GW) are useful and promising tools for investigating correlated electronic materials.

Acknowledgments

We would like to acknowledge F. Aryasetiawan, T. Ayral, M. Casula, M. Ferrero, A. Georges, P. Hansmann, M. Imada, H. Jiang, M. I. Katsnelson, A. Millis, T. Miyake, O. Parcollet, L. Pourovskii, A. van Roekeghem, A. Rubtsov, J.M. Tomczak, L. Vaugier, P. Werner, and G. Sawatzky for fruitful collaborations and useful discussions on various aspects related to the topics of this review.

This work was supported by the European Research Council under its Consolidator Grant scheme (project 617196), IDRIS/GENCI under project t2016091393, and FOR1346 of the German DFG.

References

1. Hohenberg, P. and Kohn, W. (1964) *Phys. Rev.*, **136**, B864.
2. Kohn, W. and Sham, L.J. (1965) *Phys. Rev.*, **140**, A1133.
3. Ceperley, D.M. and Alder, B.J. (1980) *Phys. Rev. Lett.*, **45**, 566.
4. Ortiz, G. and Ballone, P. (1994) *Phys. Rev. B*, **50**, 1391.
5. Gross, E.K.U. and Kohn, W. (1990) *Adv. Quantum Chem.*, **21**, 255.
6. Jones, R.O. and Gunnarsson, O. (1989) *Rev. Mod. Phys.*, **61**, 689.
7. Anisimov, V.I., Aryasetiawan, F., and Lichtenstein, A.I. (1997) *J. Phys. Condens. Matter*, **9**, 767.
8. Damascelli, A., Hussain, Z., and Shen, Z.-X. (2003) *Rev. Mod. Phys.*, **75**, 473.
9. van Roekeghem, A., Richard, P., Ding, H., and Biermann, S. (2016) *C. R. Sci. Acad. Sci D*, **17**, 140.
10. Biermann, S., de' Medici, L., and Georges, A. (2005) *Phys. Rev. Lett.*, **95**, 206401.
11. Medici, L.de., Mravlje, J., and Georges, A. (2011) *Phys. Rev. Lett.*, **107**, 256401.
12. Werner, P., Casula, M., Miyake, T., Aryasetiawan, F., Millis, A.J., and Biermann, S. (2012) *Nat. Phys.*, **8**, 331.
13. Tomczak, J.M., Haule, K., and Kotliar, G. (2012) Signatures of electronic correlations in iron silicide. *Proc. Natl. Acad. Sci. USA*, **109**, 3243.
14. Anisimov, V.I., Poteryaev, A.I., Korotin, M.A., Anokhin, A.O., and Kotliar, G. (1997) *J. Phys. Condens. Matter*, **9**, 7359.
15. Lichtenstein, A.I. and Katsnelson, M.I. (1998) *Phys. Rev. B*, **57**, 6884.
16. Löwdin, P. (1951) *J. Chem. Phys.*, **19**, 1396.
17. Andersen, O.K. and Saha-Dasgupta, T. (2000) *Phys. Rev. B*, **62**, R16219.
18. Marzari, N. and Vanderbilt, D. (1997) *Phys. Rev. B*, **56**, 12847.
19. Aichhorn, M., Pourovskii, L., Vildosola, V., Ferrero, M., Parcollet, O., Miyake, T., Georges, A., and Biermann, S. (2009) *Phys. Rev. B*, **80**, 085101.
20. Aryasetiawan, F., Imada, M., Georges, A., Kotliar, G., Biermann, S., and Lichtenstein, A.I. (2004) *Phys. Rev. B*, **70**, 195104.
21. Biermann, S., Aryasetiawan, F., and Georges, A. (2003) *Phys. Rev. Lett.*, **90**, 086402.
22. Tomczak, J.M., Casula, M., Miyake, T., Aryasetiawan, F., and Biermann, S. (2012) *Europhys. Lett.*, **100**, 67001.
23. Tomczak, J.M., Casula, M., Miyake, T., and Biermann, S. (2014) *Phys. Rev. B*, **90**, 165138.
24. Hansmann, P., Ayral, T., Vaugier, L., Werner, P., and Biermann, S. (2013) *Phys. Rev. Lett.*, **110**, 166401.
25. Hansmann, P., Ayral, T., Tejeda, A., and Biermann, S. (2016) *Sci. Rep.*, **6**, 19728.
26. Th., Ayral., Ph., Werner., and Biermann, S. (2012) *Phys. Rev. Lett.*, **110**, 226401.
27. Vaugier, L., Jiang, H., and Biermann, S. (2012) *Phys. Rev. B*, **86**, 165105.
28. van Roekeghem, A., Vaugier, L., Jiang, H., and Biermann, S. (2016) *Phys. Rev. B*, **94**, 125147.
29. Tchougréeff, A.L. and Dronskowski, R. (2009) *Int. J. Quant. Chem.*, **109**, 2606.
30. Haverkort, M.W., Zwierzycki, M., and Andersen, O.K. (2012) *Phys. Rev. B*, **85**, 165113.
31. Tchougréeff, A.L., Soudackov, A.V., van Leusen, J., Krögerler, P., Becker, K.D., and Dronskowski, R. (2016) *Int. J. Quant. Chem.*, **116**, 282.
32. Kotliar, G., Savrasov, S.Y., Haule, K., Oudovenko, V.S., Parcollet, O., and Marianetti, C.A. (2006) *Rev. Mod. Phys.*, **78**, 865.
33. Migdal, A.B. (1967) *Theory of Finite Fermi Systems and Applications to Atomic Nuclei*, Interscience Publishers, New York.
34. Nozières, P. (1964) *Theory of Interacting Fermi Systems*, Benjamin, New York.
35. Baym, G. and Kadanoff, L.P. (1961) *Phys. Rev.*, **124**, 287.
36. Kozik, E., Ferrero, M., and Georges, A. (2015) *Phys. Rev. Lett.*, **114**, 156402.
37. Potthoff, M. (2003) *Eur. Phys. J. B*, **32**, 429.
38. van Leeuwen, R., Dahlen, N.E., and Stan, A. (2006) *Phys. Rev. B*, **74**, 195105.
39. Almbladh, C.O., von Barth, U., and van Leeuwen, R. (1999) *Int. J. Mod. Phys. B*, **13**, 535.

40 Lin, N., Marianetti, C.A., Millis, A.J., and Reichman, D.R. (2011) *Phys. Rev. Lett.*, **106**, 096402.

41 Rubtsov, A.N., Katsnelson, M.I., and Lichtenstein, A.I. (2008) *Phys. Rev. B*, **77**, 033101.

42 Rubtsov, A.N., Katsnelson, M.I., and Lichtenstein, A.I. (2012) *Annals Phys.*, **327**, 1320.

43 van Loon, E.G.C.P., Hafermann, H., Lichtenstein, A.I., Rubtsov, A.N., and Katsnelson, M.I. (2014) *Phys. Rev. Lett.*, **113**, 246407.

44 E. G. C. P., vanLoon, Lichtenstein, A.I., Katsnelson, M.I., Parcollet, O., and Hafermann, H. (2014) *Phys. Rev. B*, **90**, 235135.

45 Rubtsov, A.N., Savkin, V.V., and Lichtenstein, A.I. (2005) *Phys. Rev. B*, **72**, 035122.

46 Werner, P., Comanac, A., Medici, L.de., Troyer, M., and Millis, A.J. (2006) *Phys. Rev. Lett.*, **97**, 076405.

47 Gull, E., Millis, A.J., Lichtenstein, A.I., Rubtsov, A.N., Troyer, M., and Werner, P. (2011) *Rev. Mod. Phys.*, **83**, 349.

48 Biermann, S. (2014) *J. Phys. Condens. Matter*, **26**, 173202.

49 Biermann, S. and van Roekeghem, A. (2016) *J. Electron. Spectros. Relat. Phenomena*, **208**, 17.

50 Chitra, R. and Kotliar, G. (2001) *Phys. Rev. B*, **63**, 115110.

51 Ayral, T., Biermann, S., and Werner, Ph. (2013) *Phys. Rev. B*, **87**, 125149.

52 Biermann, S., Aryasetiawan, F., and Georges, A. (2004) *Proceedings of the NATO Advanced Research Workshop*, Kluwer Academic Publishers B.V..

53 Aryasetiawan, F., Biermann, S., and Georges, A. (2004) Proceedings of the conference "Coincidence Studies of Surfaces, Thin Films and Nanostructures," Ringberg Castle, September 2003.

54 Anisimov, V.I., Kondakov, D.E., Kozhevnikov, A.V. et al. (2005) *Phys. Rev. B*, **71**, 125119.

55 Lechermann, F., Georges, A., Poteryaev, A., Biermann, S., Posternak, M., Yamasaki, A., and Andersen, O.K. (2005) *Phys. Rev. B*, **74**, 125120.

56 Blöchl, P.E. (1994) *Phys. Rev. B*, **50**, 17953.

57 Karolak, M., Wehling, T.O., Lechermann, F., and Lichtenstein, A.I. (2011) *J. Phys. Condens. Matter*, **23**, 085601.

58 Lichtenstein, A.I., Katsnelson, M.I., and Kotliar, G. (2001) *Phys. Rev. Lett.*, **87**, 067205.

59 Ren, X., Leonov, I., Keller, G., Kollar, M., Nekrasov, I., and Vollhardt, D. (2006) *Phys. Rev. B*, **74**, 195114.

60 Sawatzky, G.A. and Allen, J.W. (1984) *Phys. Rev. Lett.*, **53**, 2339.

61 Karolak, M., Izquierdo, M., Molodtsov, S.L., and Lichtenstein, A.I. (2015) *Phys. Rev. Lett.*, **115**, 046401.

62 Weber, C., O'Regan, D.D., Hine, N.D.M., Littlewood, P.B., Kotliar, G., and Payne, M.C. (2013) *Phys. Rev. Lett.*, **110**, 106402.

63 Huang, L. and Wang, Y. (2012) *Europhys. Lett.*, **99**, 67003.

64 Sun, P. and Kotliar, G. (2004) *Phys. Rev. Lett.*, **29**, 196402.

65 Maiti, K., Manju, U., Ray, S., Mahadevan, P., Inoue, I.H., Carbone, C., and Sarma, D.D. (2006) *Phys. Rev. B*, **73**, 052508.

66 Sekiyama, A., Fujiwara, H., Imada, S. et al. (2004) *Phys. Rev. Lett.*, **93**, 156402.

67 Takizawa, M., Minohara, M., Kumigashira, H. et al. (2009) *Phys. Pev. B*, **80**, 235104.

68 Morikawa, K., Mizokawa, T., Kobayashi, K. et al. (1995) *Phys. Rev. B*, **52**, 13711.

69 Casula, M., Rubtsov, A., and Biermann, S. (2012) *Phys. Rev. B*, **85**, 035115.

70 Aizaki, S., Yoshida, T., Yoshimatsu, K. et al. (2012) *Phys. Rev. Lett.*, **109**, 056401.

71 Hedin, L. (1999) *J. Phys. Condens. Matter*, **11**, R489.

72 Pavarini, E., Biermann, S., Poteryaev, A., Lichtenstein, A.I., Georges, A., and Andersenr, O.K. (2004) *Phys. Rev. Lett.*, **92**, 176403.

73 Casula, M., Werner, Ph., Vaugier, L., Aryasetiawan, F., Miyake, T., Millis, A.J., and Biermann, S. (2012) *Phys. Rev. Lett.*, **92**, 126408.

74 Miyake, T., Martins, C., Sakuma, R., and Aryasetiawan, F. (2013) *Phys. Rev. B*, **87**, 115110.

75 Tomczak, J.M., van Schilfgaarde, M., and Kotliar, G. (2012) *Phys. Rev. Lett.*, **109**, 237010.

76 Alireza, P.M. et al. (2009) *J. Phys. Condens. Matter*, **21**, 012208.

77 Kimber, S.A.J. et al. (2009) *Nat. Mater.*, **8**, 471.
78 Rotter, M., Tegel, M., and Johrendt, D. (2008) *Phys. Rev. Lett.*, **101**, 107006.
79 Sefat, A.S., Jin, R., McGuire, M.A. et al. (2008) *Phys. Rev. Lett.*, **101**, 117004.
80 Liu, C., Samolyuk, G.D., Lee, Y. et al. (2008) *Phys. Rev. Lett.*, **101**, 177005.
81 Brouet, V., Rullier-Albenque, F., Marsi, M. et al. (2010) *Phys. Rev. Lett.*, **105**, 087001.
82 Ding, H., Nakayama, K., Richard, P. et al. (2011) *J. Phys. Condens. Matter*, **23**, 135701.
83 Koitzsch, A., Inosov, D.S., Evtushinsky, D.V. et al. (2009) *Phys. Rev. Lett.*, **102**, 167001.
84 Fink, J., Thirupathaiah, S., Ovsyannikov, R. et al. (2009) *Phys. Rev. B*, **79**, 155118.
85 Zhang, Y., Chen, F., He, C. et al. (2011) *Phys. Rev. B*, **83**, 054510.
86 Mansart, B., Brouet, V., Papalazarou, E. et al. (2011) *Phys. Rev. B*, **83**, 064516.
87 Wen, H.-H. and Li, S. (2011) *Annu. Rev. Condens. Matter Phys.*, **2**, 121.
88 Yi, M., Lu, D.H., Analytis, J.G. et al. (2009) *Phys. Rev. B*, **80**, 024515.
89 Skornyakov, S.L., Efremov, A.V., Skorikov, N.A. et al. (2009) *Phys. Rev. B*, **80**, 092501.
90 Yin, Z.P., Haule, K., and Kotliar, G. (2011) *Nat. Mater.*, **10**, 932.
91 van Roekeghem, A., Vaugier, L., Jiang, H., and Biermann, S. (2016) *Phys. Rev. B*, **94**, 125147.
92 de Jong, S., Huang, Y., Huisman, R. et al. (2009) *Phys. Rev. B*, **79**, 115125.
93 van Roekeghem, A., Ayral, T., Tomczak, J.M., Casula, M., Xu, N., Ding, H., Ferrero, M., Parcollet, O., Jiang, H., and Biermann, S. (2014) *Phys. Rev. Lett.*, **113**, 266403.
94 Roekeghem, A.van. and Biermann, S. (2014) *Europhys. Lett.*, **108**, 57003.
95 van Roekeghem, A., Richard, P., Shi, X., Wu, S., Zen, L., Saparov, B.I., Ohtsubo, Y., Su, S., Qian, T., Safa-Sefat, A., Biermann, S., and Ding, H. (2016) *Phys. Rev. B*, **93**, 245139.
96 van Schilfgaarde, M., Kotani, Takao., and Faleev, S. (2006) *Phys. Rev. Lett.*, **96**, 226402.
97 Taranto, C., Kaltak, M., Parragh, N., Sangiovanni, G., Kresse, G., Toschi, A., and Held, K. (2013) *Phys. Rev. B*, **88**, 165119.
98 Krivenko, I. and Biermann, S. (2015) *Phys. Rev. B*, **91**, 155149.
99 Mossanek, R.J.O., Abbate, M., Yoshida, T. et al. (2009) *Phys. Rev. B*, **79**, 033104.
100 Martins, C., Aichhorn, M., Vaugier, L., and Biermann, S. (2011) *Phys. Rev. Lett.*, **107**, 266404.

6
Semiempirical Molecular Orbital Methods

Thomas Bredow[1] and Karl Jug[2]

[1]*Universität Bonn, Institut für Physikalische und Theoretische Chemie, Mulliken Center for Theoretical Chemistry, Beringstr. 4, 53115 Bonn, Germany*
[2]*Leibniz University Hannover, Theoretical Chemistry, Callinstr. 3A, 30167 Hannover, Germany*

6.1
Early Approaches

Semiempirical methods date back almost to the beginning of quantum mechanics with the advent of the Schrödinger equation [1]. These and subsequent papers by other authors were also dedicated to problems of physics. Only a few years later, the physicist Erich Hückel addressed himself to a problem of chemistry, namely, electronic spectra of unsaturated molecules. In a monumental paper, which represented his second habilitation thesis for the venia legendi in theoretical physics in Stuttgart, he studied the electronic configuration of benzene and related molecules [2]. Starting from the Schrödinger equation, he introduced a number of simplifications to be able to solve it. Hückel chose the molecular orbital (MO) theory to approach the problem. In the MO theory, the molecular orbitals are approximated as a linear combination of atomic orbitals. Hückel restricted himself to the π electrons, in particular to the six π electrons of benzene. Since the resulting equations were still too complicated to be solved in this pre-computer age, further simplifications had to be made. The electronic repulsion was neglected and the remaining energy terms considered in a one-electron model. The atomic energy of a π electron in the field of all six carbon atoms was described by a single parameter later called α and the binding between two nearest-neighbor atoms characterized by the so-called resonance parameter β. In this way, the secular equations were greatly simplified and reduced to a 6×6 secular determinant with two parameters only. These were determined by experimental spectroscopic values. In particular, the first excitation energy of benzene was used to fix β. The parameter α remained unspecified. The scheme was easy to handle for hydrocarbons, but it did not work well for heteroatoms. Here the ω-technique [3] was used where the α parameter of carbon was modified by a charge-dependent term that contained a parameter ω to get better agreement with experiment. Another attempt to improve the Hückel method was the extension of the basis set from 2pπ to the inclusion of 3pπ [4].

The original ideas of Hückel lasted for over 20 years, before they were improved. The major drawback was the neglect of the electronic interaction with the consequence of singlet–triplet degeneracy. It was due to Pariser and Parr [5,6] and independently by Pople [7] to remedy the situation. Whereas Pariser and Parr focused on the excitations energies like Hückel, Pople related his approach to the ionization energies of unsaturated compounds. Both methods started from the self-consistent field (SCF) approach [8]. Hence, the parameters α and β had a different meaning and consequently different values from those in the Hückel method. Both methods used the zero-differential overlap (ZDO) approach for simplification. This means that the overlap between different orbitals was neglected. This neglect of differential overlap was efficient to simplify the procedure, but unsatisfactory in terms of an explanation. A good account of this work and other work at that time can be found in Parr's book [9], which contains not only the theory but also reprints of important original papers up to 1961.

It took more than 10 years to supply a convincing description of the theoretical background. McWeeny [10] already assumed orthogonal atomic orbitals as a basis for the neglect of overlap in the Hückel theory. The breakthrough came with the work of Fischer-Hjalmars [11]. She investigated the zero differential overlap approximation by means of a systematic expansion in terms of a parameter on the order of magnitude of a typical overlap integral between neighboring atoms. The terms of the Fock operator in a basis of orthogonal atomic orbitals can be arranged after different orders of the parameter. It was demonstrated that, by inclusion of second-order terms, the conditions imposed by the zero differential overlap approximation are automatically fulfilled. The orthogonalized atomic orbitals (OAOs) underlying the explanation by Fischer-Hjalmars were introduced by Löwdin [12]. The idea was to generate a symmetrically orthogonalized set of atomic valence orbitals for different atoms in a molecular system. In this way the atomic character of the orbitals was retained and the neglect of differential overlap between different atoms was justified. A focus of Löwdin's work was the treatment of solids, a topic that will also be treated more extensively in this chapter.

An early review [13] presented a unified notation to simplify the comparison of different methods on the development of semiempirical methods in the MO formalism. A most recent and quite extensive review on the development of quantum chemical methods starting from the 1920s and including a discussion of the various periods of progress and establishment with a particular focus on Germany was given by Gey [14]. The first period covers the time from 1927 to 1933, where the first quantum chemical investigations were made based on quantum mechanics and chemical valence structure theory. This period was followed by one from 1933 to 1950 with numerical and nonnumerical studies of selected molecules and the beginning of computer techniques. Subsequently, from 1950 to 1965, quantum chemical methods were improved and applied to a multitude of problems. The emphasis was on MO theory and computer programs were written. In the period from 1965 to 1980, quantum chemistry was established worldwide due to the development of computer techniques. After 1980, quantum chemistry was an indispensable field of chemistry. An interdisciplinary development of new models and computational procedures started.

6.2 All-Valence Electron Methods

6.2.1 Extended Hückel Theory

The restriction of the methods of the early approaches discussed in Section 6.1 was on planar unsaturated systems. This was a great limitation that was overcome in the 1960s. Also, these methods were based on the ZDO assumption. Before Fischer–Hjalmars delivered a satisfactory explanation of its use in terms of Löwdin orbitals, it was not completely clear whether this was a substantial drawback. In this sense an extension of the Hückel method to all valence electrons was proposed, which avoided the ZDO assumption by inclusion of overlap. It was called extended Hückel theory (EHT) by Hoffmann [15]. The idea was to use a generalized Mulliken approximation for the off-diagonal elements of the Hamiltonian proposed by Wolfsberg and Helmholz [16]. These elements are averages of the diagonal elements of the Hamiltonian multiplied with the overlap integral and with a constant factor empirically determined. However, the Hamiltonian is still unspecified and the electronic repulsion is not explicitly defined. Although the abstract gives the impression that the geometry is the most predictable property also in this method, no bond distances were optimized and it turned out that the EHT method is unsuitable to yield reliable bond lengths. An important achievement of this method was to show that there are σ electron levels that are higher in energy than the lowest π electron level.

The method was suitable and served mainly to support studies on organic systems. It was later extended to band structure calculations [17]. It was also the underlying basis for the development of theoretical increments and indices for reactivity, acidity, and basicity within solid-state materials [18]. The latter work was one approach for a theoretical way of aiding the design of solid-state syntheses [19] and goes back to the breakthrough of Parr and Pearson in modern acid–base theory [20].

6.2.2 NDO Methods

The π electron theories of the 1950s still suffered from a number of drawbacks. One was the wrong assumption that the π electron energy levels were higher than the σ electron levels. This was the basis for the separation of σ and π electrons. More important was the fact that π electron theories were unable to provide the geometrical structure of molecules. It was therefore not surprising that

Pople and his group worked on these problems. In the course of development of new all-valence electron methods, the question of consistency of semiempirical methods was raised. The SCF equations [8] are invariant under rotation of local axes. This rotational invariance must be preserved because otherwise unphysical results would occur, where the molecular properties would be dependent on the coordinate system. To achieve this goal, the ZDO assumption must be introduced with care.

6.2.2.1 CNDO

From 1965 Pople and his coworkers provided a series of methods that introduced various levels of the ZDO approximation observing invariance requirements. In the Complete Neglect of Differential Overlap (CNDO) method [21], the overlap between all different s- and p-valence orbitals was neglected. This meant that invariance was achieved not only under rotation but also under hybridization. The latter resulted in a neglect of one-center exchange integrals ($\mu\nu|\mu\nu$), where μ and ν are atomic orbitals on the same center. For example, the neglect of ($2px2py|2px2py$), ($2px2pz|2px2pz$), and ($2py2pz|2py2pz$) retains invariance under rotation and the neglect of ($2s2px|2s2px$), ($2s2py|2s2py$), and ($2s2pz|2s2pz$) leads to invariance under hybridization. However, these integrals are not small and the invariance under hybridization cannot be justified by physical reasons. The unnecessary restriction was criticized and a better procedure to retain the physically necessary invariance was proposed in terms of local axes [22]. Pople and his group adjusted the parameters, in particular the resonance integral β to *ab initio* calculations. Since the first results [23] with the original CNDO method were not very satisfactory, a new version called CNDO/2 was proposed [24]. It differed from the CNDO method only by the approximation of the two-center core potential by a two-center electronic repulsion integral multiplied by the charge of the effective nuclear charge of the second atom.

6.2.2.2 INDO

The CNDO approach was improved to the higher level of Intermediate Neglect of Differential Overlap (INDO) in the following year [25]. Here the ZDO assumption for one-center terms was removed and only retained for two-center terms. An even higher level of ZDO assumption was already suggested under the name Neglect of Diatomic Differential Overlap (NDDO) in the first paper [21], but not really pursued. In the NDDO method, integrals of the type ($\mu_A \nu_A | \mu_B \nu_B$), where A and B are different atomic centers, were retained. The essence of all these methods and the results for molecular properties, such as structures, force constants, binding energies, ionization potentials, and dipole moments, of small molecules and the computer program was collected a few years later in a book [26]. Although the results of these years were unsatisfactory and so Pople abandoned semiempirical theory, the standards of consistency and the levels of ZDO approximation survived and were a guideline for subsequent developments by other groups.

6.2.2.3 MINDO

The first who picked up the new ideas about the various levels of semiempirical methods and created his own methods was Dewar. In his book on the MO theory of organic chemistry [27], he discussed Pople's levels of CNDO, INDO, and NDDO, but explained also his philosophy, namely, to create a method that delivered chemical accuracy for heats of formation and geometries of molecules. This method was called MINDO, which stands for modified INDO. It took many years to achieve this goal. The first two versions MINDO and MINDO/2 fell short. In 1975, the MINDO/3 method [28] was presented. In the introduction it was stated that the purpose of their work was to develop a quantum mechanical treatment of molecular structure and reactivity simple enough, accurate enough, and reliable enough to serve as a practical chemical tool in areas at present inaccessible to experimental study. Dewar explained why he started at the INDO level and not at the CNDO level. He also made it clear that he considered the NDDO level as the only ultimately acceptable level. The number of parameters in MINDO/3 was excessively large, because the method needed atom pair parameters. Dewar gave up the relation to the underlying *ab initio* theory in various respects. For example, he adjusted the parameters to the experimental heats of formation at room temperature, that is, at 298 K. Also his empirical core repulsion function was theoretically unjustified. Nevertheless, he proved in an accompanying paper [29] that the results for heats of formation and geometries of a large variety of hydrocarbons were better than any previous semiempirical method. Subsequent papers in the same year extended the scope to compounds with nitrogen, oxygen [30], fluorine, and chlorine [31]. Unfortunately, some carbon–fluorine bonds were reproduced poorly with their heats of formation in error by more than 50 kcal/mol.

6.2.2.4 SINDO

After the comparative analysis of semiempirical methods [13], Jug worked on an alternative to the CNDO and INDO versions of Pople. The goal was to surpass the accuracy of these methods. It was soon recognized that the failure of the original CNDO method was due to a wrong approximation of the β parameter [32]. Pople had used the conventional form where β is the average of the atomic α parameters of the two related atoms multiplied with the overlap integral S. This Mulliken approximation is appropriate for nonorthogonal atomic orbitals (AOs). But the ZDO approximation implies that the underlying atomic orbitals are orthogonal, hence OAOs. A transformation from AOs to OAOs leads to a new form of the β parameter. A consequent implementation of the OAO basis enabled the creation of a new semiempirical method [33] that was superior to the CNDO and INDO methods in all calculated properties. With the success of this approach, the method was given the name SINDO [34], where S stands for symmetrically orthogonalized.

6.2.2.5 SINDO1

The continuous search for improvements of SINDO led to higher level of accuracy in 1980 when the SINDO1 version was published [35]. The label 1 was given, because in this version the CNDO/1 approximation was used instead of CNDO/2. SINDO1 covered all elements of the first row except the noble gas atoms [36,37]. The method was designed in such a way that it was able to compete not only with MINDO/3 in terms of accuracy of molecular properties but also with MNDO [38,39]. The latter method will be discussed in the next section. An advantage of SINDO1 was the introduction of a pseudopotential for the inner shells, an idea taken over from Zerner [40]. In SINDO1 binding energies were calculated in contrast to heats of formation $\Delta_f H^{298}$ in MINDO/3 and MNDO. To treat second-row elements, the inclusion of d-orbitals was considered. It was shown that the rigorous application of the ZDO assumption violated the rotational invariance of the Fock equations [41] and that a certain number of one-center hybrid integrals had to be retained, for example, $(3s3p_x|3p_x3d_{z^2})$. A list of 21 types of such hybrid integrals was given. The rotational invariance had been ignored by other authors in previous work. On this basis of the new insight, SINDO1 was extended to second-row elements [42,43]: sodium and chlorine [44], phosphorus [45], and silicon, aluminum, and magnesium [46]. The inclusion of d-orbitals in the basis set enabled a reliable calculation of the bonding in hypervalent compounds with P, S, and Cl atoms. It also opened the way to the treatment of clusters [47] and fragments of crystalline solids [48] with Al and Si atoms. To extend the scope of SINDO1, first-row transition elements from scandium to zinc were parametrized [49]. These developments paved the way for the simulation of solids by clusters and the study of adsorption of small molecules on crystal surfaces simulated by clusters. The crystal clusters that were studied were ionic. In the case of NaCl and MgO, the systematic extension of cluster size allowed to calculate the bond length and binding energy of the cluster in dependence of the relative average coordination number and thus extrapolate these properties to a cluster of infinite size [50]. For the adsorption of molecules, the surface of the clusters was systematically increased by increasing the clusters size in such a way that the surface size was increased, but the number of layers was kept small and constant. In this way the structure and adsorption energies of CO_2 on NaCl, H_2O on MgO, and H_2O on TiO_2 in the rutile and anatase structures could be determined [51]. To get a realistic bonding of hydrogen to the surface, 2p orbitals on hydrogen were introduced [52].

6.2.2.6 MSINDO

Up to that time the advancement of semiempirical methods had been approached in two different ways. Pople had emphasized the development of approximate formulas for integrals and the comparison with the underlying *ab initio* framework, whereas Dewar followed the idea of adjustment of parameters to empirical data. Since Pople failed to generate a theoretically reliable β parameter, Dewar produced better results for molecular properties. However, it was realized that Dewar's approach was hampered by inconsistencies. To remedy the situation, a third category of parametrization was proposed [53], namely, the consistency of parameters in rows of elements. This was an improved combination of Pople's and Dewar's philosophy. It opened the way for further improvements of the SINDO method. In this way a new version of SINDO1 was created a few years later [54] and applied to a large number of compounds containing first- and second-row elements [55]. It was called MSINDO, which means modified SINDO. Different from SINDO1, the new MSINDO contained orthogonalization corrections only to the first order, because the expansion of transformation matrix $\mathbf{S}^{-1/2}$ does not converge fast, if the overlap S is large. The second important modification concerns the orbital exponents. Different orbital exponents were used for the one- and two-center integrals. In this way MSINDO was no longer a minimal basis set method and better competitive with *ab initio* methods. It also calculated zero-point energies and produced heats of formation at 0 K. Soon afterward the method was extended to third-row main group elements [56] and third-row transition metals [57]. Later higher multipole terms were added [58]. In the INDO scheme only diatomic monopole–monopole interactions were included. This is different from the NDDO level where also higher multipole terms are considered in the one- and two-electron interactions. The neglect of these terms may be problematic if electrostatic contributions are important for the total energy. Thus dipole–monopole and dipole–dipole terms were added to the formalism [58]. This meant that MSINDO was now in between the INDO and NDDO level.

To bridge the gap between molecules and solids, the idea of clusters for the simulation of crystalline solids was improved by the cyclic cluster approach [59]. This model avoids the use of unit cells with infinite periodicity. Instead, it is periodic in a finite system. Different from the conventional approach to translational periodicity, the model is able to treat nonperiodic defects in a crystal in a simple way. It was also attempted to extend these ideas to the general Hartree–Fock scheme [60]. Since the finite cluster cannot describe long-range effects properly, the Madelung potential was included in free and cyclic clusters of metal oxides [61]. In this way the convergence of calculated properties with cluster size could be improved. In order to cope with the new challenge of quantum molecular dynamics, the Born–Oppenheimer molecular dynamics scheme was implemented in MSINDO and applied to silicon clusters [62]. This could be achieved because the reference of MSINDO was 0 K and not room temperature as in MINDO/3 and MNDO. The melting of small silicon clusters and the simulated annealing of clusters with up to 60 atoms was investigated. Most recently, the MSINDO-sCIS method was developed, a new method for the calculation of excited states of large molecules [63].

6.2.2.7 ZINDO

At the same time when SINDO was developed, Zerner worked on a program package that was later called ZINDO [64]. It contained a variety of semiempirical programs, including INDO/1 and INDO/S. Whereas INDO/1 can be used to obtain geometries, INDO/S was designed for the calculation of electronic spectra of organic molecules and complexes containing transition metals. The latter model was calibrated at the configuration interaction level with single excitations (CIS) level for energy differences with fixed geometries. The latter were taken from experiment. No attempts were made to optimize geometries or total ground-state energies in this model. An important feature was the introduction of a local pseudopotential for the inner shells [40], which was later implemented in SINDO1. Special attention was given to the calculation of one-center two-electron integrals from Slater–Condon factors F and G and the evaluation of $U_{\mu\mu}$ from experimental ionization energies [65]. The emphasis was now on the spectra of transition metal compounds. Zerner was also concerned with the efficiency of geometry optimization and introduced the Broyden–Fletcher–Goldfarb–Shanno (BFGS) quasi-Newton update [66]. One of the last publications was on a new parametrization of the resonance integral $H_{\mu\nu}$ in the INDO/1 scheme for main group elements [67], with which a considerable improvement in the geometries was reached. A special feature was the use of distance-dependent orbital exponents ζ, which mimic a multi-zeta basis set. Very recently, INDO/S was completely reparametrized and modified using Thiel's TBE-2 benchmark excitation energies and oscillator strengths [68]. The new method INDO/X reduces the mean deviation for singlet and triplet excitation energies by 50% compared to INDO/S.

6.2.3
NDDO Methods

6.2.3.1 MNDO

As already mentioned, Dewar considered the NDDO level as the only suitable level for semiempirical molecular orbital methods. He was not satisfied with the initial success of MINDO/3 in 1975 and despite a number of interesting applications worked on the implementation of a method on the NDDO level. This method was presented in 1977 under the name MNDO, which stands for Modified Neglect of Differential Overlap [38]. The first applications were on molecules containing hydrogen, carbon, nitrogen, and oxygen [39]. The advantage of MNDO over MINDO/3 was not only the inclusion of multipole–multipole interactions for two-center electronic repulsion integrals but also the abandoning of atomic pair parameters, which reduced the number of parameters considerably. The Mataga–Nishimoto approximation for the core–core repulsion was still abandoned and replaced by a more realistic potential, where the nucleus–nucleus interaction was replaced by an electron–electron interaction plus a combination of exponential terms containing two atomic parameters, one for each core. But the heats of formation were still calculated at room temperature. The results for heats of formation, geometries, ionization potentials, and dipole moments of small molecules containing H, C, N, and O atoms were quite improved compared to MINDO/3. A subsequent paper on molecules containing fluorine atoms failed with respect to heats of formation, showing errors over 20 kcal/mol for some compounds [69]. Parameters for second-row elements were also presented, but d orbitals were not included [70]. That this omission can lead to totally unrealistic heats of formation for hypervalent compounds can be seen in the case of chlorine [71]. For ClF_3 an error of 116.8 kcal/mol appeared from the calculation. Later the method was parametrized also for aluminum and boron [72], still without d functions.

6.2.3.2 AM1

It was therefore not surprising that Dewar and his group worked on an improvement. In 1985 the AM1 (Austin model 1) method was presented [73]. It differed from MNDO only in the parametrization and in the correction function for the core–core repulsion. Applications were made on molecules containing H, C, N, and O atoms. In particular, the new method managed to remove the problem of failure for the description of hydrogen bonding.

6.2.3.3 PM3

After his participation in the development of AM1, Stewart started to develop his own version of a semiempirical method, based again on a new parametrization of the original MNDO. This method was called PM3 (Parametric method 3) [74]. A comprehensive amount of applications to molecules with first- and second-row atoms was presented [75]. A comparison of calculated heats of formation with MNDO and AM1 results clearly showed significant improvements, in particular in systems with phosphorus atoms. The method was eventually made suitable for the calculation of large biomolecules with several thousand atoms. For this purpose, a technique based on localized molecular orbitals combined with pseudodiagonalization [76] was developed, which reduced the time for a single self-consistent-field (SCF) calculation in such way that it was almost linearly scaling with the size of the system [77]. Stewart continued to improve the accuracy of the original PM3 method by improved parametrization schemes and increased training sets, also extending the number of elements covered [78,79]. The latest prominent developments in this direction are PM6 [80] and PM7 [81]. The PMx methods – together with AM1 – are among the most popular semiempirical methods nowadays and represent the basis for many recent developments (see Section 6.2.4).

6.2.3.4 MNDOC

A closer look at the theoretical basis of semiempirical methods was taken by Thiel. He also started from the MNDO method, of which he was a codeveloper. His first idea was to include part of the correlation explicitly. The MNDOC model [82] treated correlation by second-order Brillouin–Wigner perturbation theory. The method was parametrized for the elements H, C, N, and O. It was found that the results for calculated heats of formation and geometries for closed-shell ground states were of similar accuracy in MNDOC and MNDO, which was rationalized by an analysis of the MNDOC correlation effects in these systems. Thiel's next step was the inclusion of d-orbitals in the basis set [83]. In this respect, MNDO had a deficiency that it shared with AM1 and PM3, but that was avoided in earlier in SINDO1 and later in MSINDO. It was not surprising that Thiel arrived at enormous improvements for heats of formation and geometries for hypervalent chlorine compounds, in particular compared to MNDO and

AM1. A few years later d orbitals were introduced for all second-row elements from sodium to chlorine and for the zinc group [84]. The method was called MNDO/d. It was shown that this method is clearly superior to MNDO, AM1, and PM3 also for hypervalent compounds with sulfur and phosphorus atoms.

6.2.3.5 OMx

Thiel also realized the importance of orthogonalization corrections, not present in MNDO, AM1, and PM3. He suggested to improve the MNDO model by the explicit inclusion of valence-shell orthogonalization corrections, penetration integrals, and effective core potentials (ECPs) in the one-center part of the core Hamiltonian matrix [85]. Later he introduced two new orthogonalization corrections in the MNDO model and called the new method OM2 [86]. He acknowledged that SINDO [33] and SINDO1 [35] had introduced orthogonalization corrections much earlier. The orthogonalization corrections in the latter methods were concerned with symmetrical orthogonalization and the resonance integrals β. In OM2 the orthogonalization between core and valence orbitals was also considered. Thiel's approach to improvements in MNDO was theoretically sounder than Dewar's and Stewart's. In 2010, the OM3 method was introduced and benchmarked for the calculation of vertical excitation energies of a benchmark set of 28 medium-sized organic molecules together with MNDO, AM1, PM3, OM1, OM2, OM3, INDO/S, and INDO/S2 [87]. The mean absolute deviations relative to the theoretical best estimates were lowest for OM3, and only slightly higher for OM1 and OM2, whereas INDO/S performs similar to OM2 for singlet states, but deteriorates considerably for triplet states.

6.2.3.6 Other Semiempirical Methods

A method with the same acronym as Jug's early SINDO version (see Section 6.2.2.4) was independently developed by Golebiewski *et al.* in 1977 [88]. It was used until the mid-1990s mainly for the study of vanadium pentoxide surfaces [89]. An INDO technique was developed for lanthanide complexes in 1987 [90] but was not frequently used afterward. In 1992, Zhidomirov and coworkers introduced the NDDO variant NDDO/MC that was particularly designed for transition metal complexes [91]. Similar to the earlier SINDO1 and later OMx methods, NDDO/MC also gives an explicit account of the orthogonality of the basis set and uses modified one-electron two-center terms compared to the original NDDO-type versions. It was in use until the early 2000s mainly to study catalytic processes in zeolites and adsorption processes on metal surfaces [92]. Only 1 year later, Dewar *et al.* modified their AM1 method by replacing the Ohno–Klopman approximation for the two-center two-electron integrals by a direct calculation using minimal STO-3G contracted Gaussian basis sets. Due to the similarity of this approach with *ab initio* methods, the new version was dubbed semi-*ab initio* model 1 (SAM1) [93]. The so-called sparkle model with focus on the electrostatic character of the metal–ligand interaction was introduced by De Andrade *et al.* in 1994 for the treatment of lanthanide complexes [94]. Its first parametrization was based on AM1 for europium(III), but it was later parametrized for all lanthanides, and also combined with PM3, PM6, PM7, and RM1 [95–103]. The Sparkle model is still under development, recently extensions for RM1, PM3, PM6, and PM7 methods and other lanthanides have been proposed [101,104]. A parametric Hamiltonian (H-pa) based on the integral formalism of MINDO/3 was suggested by Primera *et al.* in 1999 [105]. Unfortunately, this approach was not followed later on. Voityuk *et al.* combined the ideas of the NDDO approximation for two-center integrals with the CI ansatz for optical spectra calculation of INDO/S in their NDDO-G method [106]. Voityuk and Rösch extended the AM1 method to d-orbitals [107]. Originally their AM1/d method was parametrized only for molybdenum compounds, but was later extended to other transition metals. A pairwise distance-directed Gaussian modification was developed for PM3 and MNDO (PDDG/PM3 and PDDG/MNDO) in order to account for functional group information via pairwise atomic interactions [108]. The authors claimed that several specific problems of the original methods were overcome, including the relative stability of hydrocarbon isomers and energetics of small rings and molecules containing multiple heteroatoms. Later the PDDG/PM3 parametrization was extended to sulfur, silicon, and phosphorus [109]. It is frequently employed for studies of biomolecules (see Section 6.3.1). In 2006, Stewart and coworkers reparametrized the AM1 method for elements C, H, N, O, P, S, F, Cl, Br, and I for a large training set of molecules representative of organic and biochemistry. Their method called Recife Model 1 (RM1) reduced the average deviation of calculated heats of formation by 50% compared to AM1 [110]. The RM1 model was recently extended and parametrized for complexes of the trications of lanthanum, cerium, praseodymium, dysprosium, holmium, and erbium [111,112]. It was reported to describe bond distances with a higher accuracy than Sparkle/PMx.

6.2.4
Recent Developments

Reviews on semiempirical MO methods have been repeatedly given. Among the more recent ones are those by Clark [113], Bredow and Jug [114], and Thiel [115]. In 2015, Akimov and Prezhdo provided a comprehensive overview of *large-scale computations in chemistry* [116], covering semiempirical methods, density functional theory, and empirical force fields. In these reviews, a detailed view of the features of semiempirical methods are given and a comparison of their relative advantages is presented for various fields of application, covering biochemistry and materials science.

6.2.4.1 Analytic Second Derivatives

Many modern quantum chemical techniques available in *ab initio* programs have been specially designed for semiempirical methods in order to exploit their high computational efficiency. For example, analytical second derivatives of the energy are available since a long time [117], and have later been implemented in MNDO-type semiempirical methods for interfaces to ONIOM hybrid models [118]. Different from *ab initio* methods, analytic second derivatives were found to be much more efficient than numerical derivatives at low computational cost even for systems containing hundreds of atoms.

6.2.4.2 Optical Spectra

Another evolving area of semiempirical methods is the calculation of optical absorption spectra for large systems. Thiel and coworkers implemented a multireference configuration interaction approach, including analytic gradients based on the graphical unitary group method [119]. Later a surface hopping algorithm for nonadiabatic dynamics has been implemented in OM2 for excited-state geometry optimization and molecular dynamics simulations [120,121]. The importance of molecular orbital tracking and electronic state following for numerically stable optimizations was stressed. Semiempirical molecular dynamics with surface hopping was also reported for AM1 by Lischka and coworkers [122]. Multireference configuration interaction with single and double excitations (MRCISD) as well as its analytic CI gradients were later also incorporated in semiempirical methods by Lei *et al.* [123]. The shortcomings of localized minimal atomic orbital basis sets for linear response properties were partly remedied by using a density basis [124]. In an analogy to excited-state calculations at DFT level, a time-dependent semiempirical method to determine excitation energies (TD-PM3) was introduced by Bartell *et al.* [125]. It was shown that TD-PM3 gives results for excitation energies comparable to TD-DFT with large basis sets. Deviations from experiment were up to 20%.

The configuration interaction with singles (CIS) technique was implemented in MSINDO [63] and benchmarked against Thiel's TBE-2 (theoretical best estimate) data set. Different from INDO/S, the MSINDO-sCIS method takes excitations from all occupied and virtual orbitals into account and therefore does not require convergence tests of the active space. The computational effort was reduced by applying the Davidson–Liu block diagonalization method. Mean absolute deviations with respect to the TBEs for MSINDO-sCIS is 0.44 eV, comparable to the OM3 method but significantly smaller than for INDO/S. Soon the method was extended to analytical gradients for excited-state optimizations and molecular dynamics simulations, and also to open-shell systems using the unrestricted UCIS approach [126].

An alternative approach, unrestricted natural orbital complete active space configuration interaction (UNO-CAS) and UNO-CIS were implemented for NDDO methods by Dral and Clark [127]. Promising results were obtained for optical bandgaps of organic semiconductors such as polyynes and polyacenes, which are promising materials for nanoelectronics. In general, better agreement with experiment was obtained than with conventional semiempirical CI approaches.

Over the last 20 years, the effective Hamiltonian crystal field (EHCF) approach has been developed for the calculation of intra-atomic d–d transitions of third-row transition metal complexes [128–130]. The method combines the McWeeny's group function approximation and the Löwdin partition technique to construct strictly localized geminals [131,132]. The local nature of these geminals has been exploited to generate quasilinear scaling NDDO methods [133,134]. With only small modifications of the original parametrization, it was possible to increase the accuracy of semiempirical methods for the prediction of d–d transitions up to the 100 cm^{-1} regime [135]. With the most recent EHCF implementation in the MagAixTic program package, it is also possible to calculate magnetic interactions of polynuclear transition metal complexes.

6.2.4.3 Solvent Effects

For the simulation of biomacromolecules in an aqueous environment, the COnductor-like Screening MOdel (COSMO) was combined with AM1, PM3, PM5, and RM1 semiempirical Hamiltonians, and parametrized by reproducing experimental hydration free energies of a set consisting of 507 neutral and 99 ionic molecules [136]. A smooth version of COSMO with improved convergence behavior for geometry optimization was introduced for MNDO/d (MNDO/d-SCOSMO) [137,138]. In an approach to simulate the solid–liquid interphase region for heterogeneous photocatalysis, the COSMO was also implemented in MSINDO-sCIS with an extension for solvent relaxation in the excited state [139].

6.2.4.4 Dispersion Correction

As they are based on the Hartree–Fock approximation, semiempirical methods cannot describe long-range electron correlation effects such as London dispersion (also denoted as van der Waals interaction). This was recognized in several benchmark studies for geometries and energetics of biomolecules [140] where it was found that classical force field methods perform better than MNDO, AM1, PM3, RM1, PDDG/MNDO, and PDDG/PM3. Similar to modern DFT, post-SCF correction schemes have therefore been introduced in order to take dispersion into account. In 2006, Martin and Clark introduced additive "atomic orbital" polarizability tensors in NDDO-based methods and performed test calculations with MNDO [141]. In more popular correction schemes, the latest DFT-D developments by Grimme *et al.* (D3(BJ)) were combined with effective hydrogen bonding corrections by Rezac and Hobza [142]. The most recent development in this series, the so-called D3H4 correction, was parametrized for PM6, RM1, OM3, PM3, AM1, and SCC-DFTB methods on a large set of benchmark data (the S66 × 8 data set) including nonequilibrium geometries. and represents the state of the art for modern semiempirical quantum mechanics (SQM). Earlier versions such as PM6-DH2X with corrections for halogen bonding [143] were applied to complexes of CK2 protein kinase with halogenated inhibitors [144]. The complex geometries obtained with PM6-DH2X agree better with the X-ray crystal structures than those calculated with the popular Amber force field. PM6-DH2, corrected only for dispersion energy and hydrogen bonding, a fast and reliable rescoring scheme for docked complexes, was developed intended for rational drug discovery and *de novo* ligand design [145]. Later its accuracy for the benzene dimer was compared with *ab initio* approaches and force fields [146]. Semiempirical models were found to be less accurate than force field methods, PM6-DH2 performing better than others. The first schemes of this kind with corrections for dispersion (D) and hydrogen bond (H) interactions were introduced by Korth *et al.* in 2009 and 2010 [147], initially only for PM6 (PM6-DH) [148] and later also for AM1, OM3, and SCC-DFTB [149]. On the basis of several test sets for noncovalent interactions it was shown that these methods reach the quality of current DFT-D approaches for noncovalent intermolecular interaction. The performance of modern SQM methods for computer-aided drug design with focus on the post-SCF corrected methods was then given by Lepsik *et al.* [150]. It was concluded that SQM-based scoring is a promising modern efficient strategy to be exploited in computer-aided drug design, despite the enormous advance of DFT methods.

6.2.4.5 Computational Efficiency

The most promising field for future applications of semiempirical methods are very large systems with many thousands of atoms where even the most efficient DFT methods reach their limits. However, even semiempirical methods inevitably face computational bottlenecks when the valence basis set size is 10^4 and larger, mainly for matrix diagonalization and multiplication. In order to make AM1 applicable to very large systems, a Millam–Scuseria conjugate gradient density matrix search was applied to replace conventional matrix diagonalization. A proof-of-principle was given by the energy calculations on a 19 995 atom polymer of glycine and a 6304 atom RNA molecule [151]. Many linear scaling semiempirical methods make use of Stewart's local MO (MOZYME) and pseudodiagonalization approaches [77]. The performance of later MNDO, AM1, PM3, and PM5 implementations was evaluated for calculations of large biological systems [152]. A complementary approach is to modify program codes for modern computer architectures with graphical coprocessors. In 2012, MOPAC2009 code was modified for an efficient use of GPU coprocessors and multithreaded shared-memory CPUs [153]. A speedup of up to 14 was obtained, for example, for a methanol simulation box containing 2400 atoms and 4800 basis functions. Similar modifications were performed for MNDO, AM1, PM3, OM1, OM2, and OM3 for exploiting the hybrid multi-core CPU and GPU computing platform [154]. The overall computation times for energy calculations and geometry optimizations of large molecules were reduced by one order of magnitude for all methods compared to runs on a single CPU core. In an even more rigorous approach, Hennemann and Clark completely rewrote the code of the AM1, PM3, and MNDO programs, fully exploiting modern parallel computing techniques such as Message Passing Interface, MKL, ScaLAPACK, and BLACS [155]. Since – different from Stewart's MOZYME method – no further approximations were introduced, the EMPIRE results are directly comparable to the original AM1 or PM3 results. By using initial guesses for the density matrices derived from diagonalization of matrices constructed from the two-atom blocks of the NDDO one-electron matrix combined with diagonal one-atom blocks that consist of the orbital ionization energies taken from INDO/S, fast SCF convergence is achieved. Rigorous data distribution over the compute nodes, on-the-fly calculation of two-electron integrals, and pseudodiagonalization lead to a highly efficient parallel implementation. Perfectly linear scaling of the computer time was obtained up to 64 cores for a system with 52 000 atoms, and even larger calculations for 100 000 atoms on 1024 processors were reported. One year later, in 2014, EMPIRE was extended to the study of solids and surfaces by introducing periodic boundary conditions [156]. Due to the use of ideas from the large unit cell model and the cyclic cluster model [59,61,157–159], the AM1 and PM3 Hamiltonians had to be only slightly modified.

6.2.4.6 Parametrization Strategies

Improvement of semiempirical methods in terms of their accuracy and applicability is a continuous and apparently never-ending process. One of the most active developers in this area is Stewart. In 2004 he extended the NDDO methods MNDO, AM1, and PM3 to all remaining nonradioactive elements of the main group, excluding the noble gases (e.g., PM3 to B, Na, K, Ca, Rb, Sr, Cs, and Ba) [79]. As mentioned in Section 6.2.3.3, the PM3 method was frequently improved and extended leading to the modern versions PM6 [80] and PM7 [81]. But there are also alternative approaches. For example, the effect of reintroducing the overlap matrix into the secular equations for MNDO was investigated by Sattelmeyer *et al.* [160]. After an extension of the training sets by introducing MP2 results and reparametrization of the method, the accuracy of the so-called NO-MNDO method for heats of formation was nearly identical to that of AM1. In 2011 AM1 was reparametrized for palladium and silver (denoted as AM1*) [161]. A general strategy of reparametrizing semiempirical methods against *ab initio* methods for combined quantum mechanical and molecular mechanical (QM/MM) simulations of specific chemical reactions in condensed phases was presented in 2014 [162]. Reparametrization of existing methods is performed by introducing specific reaction parameters so that the atomic forces computed at the target QM/MM level are reproduced. Possible general improvements in the accuracy of semiempirical methods through the use of machine learning models for the parameters were explored by Thiel and coworkers in 2015 [163]. The so-called ML-SQC approach allows the automatic optimization of semiempirical parameters for individual molecules, improving the accuracy without deteriorating transferability to molecules with molecular descriptors different from those in the training set. For OM2, the mean error for atomization enthalpies for a test set comprised of 6095 constitutional $C_7H_{10}O_2$ isomers was decreased from 6.3 to 1.7 kcal/mol in this way. The authors conclude that this novel parametrization approach allows reasonably accurate high-throughput screening of materials and molecules.

6.2.4.7 Implementation in Solid-State Software Packages

Semiempirical Hamiltonians have been implemented in the widely used solid-state program cp2k [164]. It is expected that this development will lead to an increased number of applications of SQM in the field of condensed matter simulation. Direct comparison with DFT, Hartree–Fock, and post-Hartree–Fock methods is possible in this way.

6.3
Applications in Biochemistry and Materials Science

6.3.1
Biochemistry and Pharmacology

This section presents selected applications of the popular semiempirical methods INDO/S, MNDO, AM1, PMx, OMx, and their modifications to structure determination in biochemistry. Since the methods MNDO, AM1, and PM3 are available together in many program packages (e.g., MOPAC [165,166] and AMPAC)[1], they are frequently used together in combined and comparative studies (see, for example, Ref. [167]). For further details about the use of semiempirical methods in the field of organic chemistry, biochemistry, and pharmacology, see, for example, Refs [150,168–174]. In the latter review [174], it is shown that semiempirical quantum mechanics (SQM), in particular if corrected for hydrogen bonding and dispersion, can describe noncovalent intermolecular interactions similar to dispersion-corrected density functional theory. Bryce and Hillier [173] give a general survey on the use of quantum chemical methods to aid drug discovery with a focus on hybrid QM/MM and semiempirical MO methods for accurate predictions of macromolecular structure and reactivity.

6.3.1.1 Parametrization
Very often it was found that the standard parametrization of semiempirical methods was not sufficient for certain systems. Although this contradicts the original ideas of semiempirical theory, it has become popular to develop system-specific parameter sets. This has been done for molecular dynamics simulations of gas-phase reactions of small organic molecules [175], but also for biomolecules. For example, a specific reaction parameter (SRP) Hamiltonian based on the semiempirical AM1 method was developed and applied for model calculations of reactions in the gas phase that are relevant to the formate dehydrogenase (FDH)-catalyzed hydride transfer [176]. On this basis, the classical potential of mean force (PMF) was computed using a centroid Feynman path integral (PI) approach. The PM6 parameters had to be specially modified for these systems on the basis of the SRP strategy in order to match relative energies of optimized structures, as well as water binding energies obtained with DFT in a path integral molecular dynamics simulations for hydrated sulfuric acid clusters [177]. A modification of the AM1/d Hamiltonian called AM1/d-CB1 (AM1/d-Chemical Biology 1) was introduced to model chemical glycobiological systems [178]. The parameter training set included glycans and the chemical environment often found in glycoenzymes. It was found that AM1/d-CB1 is considerably more accurate than existing NDDO methods for modeling carbohydrates and amino acids.

---

1) AMPAC 10, 1992–2013 Semichem, Inc. 12456W 62nd Terrace – Suite D, Shawnee, KS 66216.

6.3.1.2 Conformational Analysis

Due to their computational efficiency, semiempirical methods allow structure optimizations of large biomolecules and have therefore applied for a long time. For example, MNDO was applied for a conformation analysis of cyclic ADP-ribose [179] and the binding of methylguanidinium to a methylphosphate entity [180]. A comparison between OM1 and OM2 with AM1 and PM3 for the description of the secondary structure in peptides and proteins has been performed [181] and it was shown that the description of the peptide conformers is considerably improved by OMx. Predictions of the three-dimensional structure of proteins from their amino acid sequence were made on the basis of semi-empirical methods and the AMBER force field [182]. Stewart reviewed the use of the PM6 method for modeling proteins [183]. It was reported that in many examples, PM6 gave structures in good agreement with the reported X-ray structures, in particular if the optimization procedure started with X-ray structures. NDDO and self-consistent density functional tight-binding (SCC-DFTB) semiempirical models were employed in structure determinations of deoxyribose and ribose sugars [184]. It was found that the semiempirical methods had to be corrected via B-spline interpolation of the potential energy difference surface relative to the *ab initio* benchmark data in order to correctly describe the sugar ring puckers. In an attempt to remove inaccurate structure data from protein databases, the semiempirical PM7 method was used for identifying anomalously short hydrogen bonds, C—H···O/C—H. N interactions, nonbonding close contacts, and unrealistic covalent bond lengths in recently published Protein Data Bank files [185]. Even the tertiary structure of the proteins was well conserved with PM7. The authors conclude that this approach may be useful to theoreticians, experimentalists, and crystallographers investigating the structure and function of proteins. Similarly, the PHENIX/DivCon package was introduced with the intention to improve macromolecular crystallographic refinement based on ligand stereochemical restraint files [186]. Semiempirical methods were used to refine the PDB structures. The authors found that their method results in reasonable ligand geometries and low ligand strains, even when the original refinement was inaccurate. They conclude that PHENIX/DivCon is applicable to both single-structure and high-throughput crystallography.

6.3.1.3 Spectroscopy

There are basically two strategies for calculations of optical properties for biomolecules: QM/MM with accurate QM methods for a small but representative core, or full QM calculations at semiempirical level of theory using the methods as described in Chapter 2 (see c.f. Ref. [187]). Among the others reported there, ZINDO (INDO/S) is still one of the most frequently used methods. It was, for example, employed to study electronic properties of the DNA base guanine [188], and was also found to be the most reliable method for charge transfer energies in π-stacks of nucleobases [189]. MNDO, AM1, PM3, RM1, PM6, and PM7 were used to calculate circular dichroism of organic model molecules such as helicenes [190]. In particular, RM1 and PM7 provide significantly improved spectral profiles compared to CNDO/S or INDO/S. OM2 is frequently used for the determination of optical properties. For example, the photoinduced relaxation dynamics of pyrimidine nucleobases uracil, thymine, and cytosine was studied using the surface hopping approach with OM2/MRCI [191]. Several semiempirical methods were compared for the prediction of two-photon absorption spectra for substituted olygophenylvinylenes of donor–acceptor type [192]. The cross sections were not accurately reproduced by PM6 due to inaccurate excitation energies and incorrect transition dipoles between the one-photon and two-photon states. INDO/S gave a better agreement with experiment, but only after empirical corrections to the excitation energies. Therefore, the authors conclude that semiempirical methods are not a suitable choice for the high-throughput screening technique. Not only excited states but also electron transport is studied with semiempirical methods. Nonequilibrium Green's function (NEGF) formalisms were combined with the EHT theory for computations of quantum transport properties for heterocyclic molecules such as pyrrole, furan, and thiophene [193]. Results obtained with the massively parallel EMPIRE code and linear scaling localized molecular orbital (LMO) calculations using MOPAC2009 were compared with electronic structure calculations of gas-phase proteins [194]. It was observed that the LMO procedure may lead to artificially overpolarized wavefunctions in gas-phase proteins. The full SCF iteration procedure may be slowly convergent due to problems with the overpolarization by inductive charge shifts.

6.3.1.4 Intermolecular Interactions and Reactions

An important issue in the theoretical treatment of biomolecules and the reactivity of enzymes is intra- and intermolecular hydrogen bonding interaction. The calculated H-bonding interactions calculated with AM1, PM3, and SAM1 were compared with accurate *ab initio* results and experiments [195]. It was found that AM1 performs better than the other two methods, but still is not satisfactory for O—H···O interactions. A similar comparison of the same methods was performed for the normal modes in several local anesthetics of amino ester type [196]. It is well known that dispersion effects are crucial for a correct theoretical description of intermolecular interactions. For this purpose, many correction schemes have been combined with semiempirical methods (see Section 6.2.4). AM1-D and PM3-D were tested against DFT-D for the treatment of noncovalent interactions between molecules involving sulfur atoms [197]. Intermolecular distances and binding energies obtained with DFT-D were better than the semiempirical results by factors of 8 and 2, respectively. Also, the accuracy of semiempirical methods to describe hydrogen bonding has been assessed many times. In many cases it was found that modern semiempirical quantum mechanics (SQM) approaches the accuracy of DFT methods. This has been shown, for example, in a combined PM6 and ONIOM2 hybrid study of structural aspects for the complexation of *ortho*-anisidine/beta-cyclodextrin [198]. The description of a broader variety of noncovalent interactions, including London dispersion, hydrogen bonds, and halogen bonds, was benchmarked for PM6 and PM7 [199].

It was found that post-SCF corrected PM6, denoted as PM6-D3H4X (see Section 6.2.4), yields better results than PM7. Without such corrections, semiempirical methods, for example, AM1, were found to be inferior to MM methods to describe interaction energies of halogen-bonded systems [200]. In this respect, Tuttle and Thiel showed that their dispersion-corrected OMx-D suite of methods describe biochemical systems more accurately than AM1 [201] without modifying the standard OMx parameters. Structural aspects for the complexation of *ortho*-anisidine/beta-cyclodextrin were explored by using PM6 and DFT methods [198]. The importance of intermolecular hydrogen bonding for the calculated stabilities is outlined. Reactants, transition states and products of the polar nucleophilic proton transfer, the water-assisted proton transfer, and the hydride transfer mechanisms were investigated for drugs used in the treatment of depression, Parkinson, Alzheimer, and other neurodegenerative disorders with semiempirical methods and DFT [202]. A multiscale QM/MM computational approach was used to provide atomic resolution electronic structure insights into possible catalytic scenarios of the G hairpin ribozyme [203]. A modified AM1 version (AM1/d-PhoT) and SCC-DFTB methods provided similar activation barriers as SCS-MP2 but predicted different reaction pathways, whereas DFT methods gave closer agreement. An important tool to identify chemical reactions of complex systems is molecular dynamics. Semiempirical methods allow for trajectories of much larger timescales compared to DFT. Still the computational effort, for example, for simulations of several nanoseconds is considerable. Therefore, several approaches to accelerate MD simulations were implemented in semiempirical methods. A recent example is the implementation of the atom-centered density matrix propagation (ACDMP) and the curvy steps (CURV) methods in MNDO, AM1, PM3, MNDO/d, and AM1/d using the extended Lagrangian molecular dynamics simulation approach [204]. It was shown that the two approaches – together with an efficient parallelization – reduce the computation time for BOMD simulations by a factor of 20. In order to improve the convergence behavior of free energy perturbation (FEP) methods at QM/MM level for enzyme reactions, the performance of MNDO, AM1, PM3, RM1, PDDG-MNDO, and PDDG-PM3 was compared with DFT methods [205]. The semiempirical methods have slightly better convergence properties, but very long simulations are still needed. However, convergence is obtained only if electrostatic interactions between the QM system and the surroundings are ignored. This casts some doubts on the common practice to base QM/MM FEPs on semiempirical simulations without any reweighting of the trajectories.

6.3.2
Solid-State Properties

Since the beginning of solid-state theory, the modeling of perfect crystalline solids has been the domain of theoretical physics, generally making use of DFT and plane wave basis sets. Chemists, who are more interested in imperfect crystals, developed different models that are based on finite cuts of the solids, so-called clusters, and atom-centered basis functions. Until now the cluster model is popular in theoretical studies of bulk properties [206–208]. However, it soon became clear that periodic boundary conditions (PBC) have to be introduced in order to account for translational symmetry. Furthermore, long-range electrostatic interactions, the lattice sums, have to be taken into account [209,210]. For these reasons, several extensions have been proposed for semiempirical theories, starting from the early 1970s. Basically there are two approaches for introducing PBC in clusters, the (quasimolecular) large unit cell (LUC), and the cyclic cluster model (CCM). The former was introduced by Evarestov *et al.* [211–213], first on tight-binding level, but soon extended to all existing semiempirical methods [214–221]. The latter was invented independently by Messmer and coworkers and Zunger, and first called *small periodic cluster* [222–228]. Both methods use the $k = 0$ approximation in the periodic wavefunction expansion over Bloch functions and are therefore real space methods. The **k**-point sampling of conventional periodic approaches is equivalently invoked by using large supercell models. The basic difference between the LUC model and the CCM is the definition of the interaction domains around the atoms. In the LUC model, they are defined as spheres centered at the nuclear positions, and the CCM employs Wigner–Seitz supercells (WSSCs). Only the WSSC domains guarantee the correct stoichiometry, therefore distance-dependent modulation functions had to be introduced in the LUC model [229]. Comparisons between the two types of models have been repeatedly given [214,230,231]. Only few semiempirical methods that are based on conventional k-point sampling have been reported [232], probably because this way of solid-state modeling requires only small unit cells and is therefore suitable for DFT.

6.3.2.1 Quasimolecular LUC Model

From the very beginning, the LUC model has attracted more attention, probably due to its slightly simpler formulation [229,233–240]. Stefanovich *et al.* presented a reparametrized INDO method for solids called CLUSTERD [241]. They tested their approach for a variety of ionic crystals, LiH, LiF, LiCl, KCl, MgO, and K_2SO_4, and later for triplet exciton self-trapping and Frenkel-defect pair creation in SiO_2 with unrestricted INDO [242]. A related approach, the quasicluster embedding scheme, was applied to neutral and charged point defects in KCl, SiO_2, and diamond crystals [243]. Optical absorption spectra of pure $KTaO_3$ and $KTaO_3$:Li were calculated with a modified INDO method [244]. Moliver combined Zerner's restricted open-shell HF methods with the LUC model in order to study neutral and charged C and Si vacancies in bulk diamond and silicon [245–247].

Perkins and Stewart introduced periodic boundary conditions to PM3 in 1980 [157]. He implemented the LUC model (here denoted as *cluster* approach [158]) for band-structure calculations of three-dimensional solids. In later NDDO versions, the point charge approximations of the two-center electron repulsion integrals had to be modified in order to avoid divergence of the lattice sums [159]. Most recently, the cluster method was implemented in PM6 [248]. Stewart presented a comprehensive statistics of the performance of his method to molecular crystals and inorganic solids, focusing on structural and energetic properties. Although the geometries of most systems were reproduced with good accuracy, large errors were found in the calculated structures of a small number of solids.

In a more recent LUC-INDO implementation, correlation corrections were added via second-order Møller–Plesset perturbation theory [249]. In this way a more consistent parametrization could be achieved, but at the cost of increased computational effort. The method was then applied to study the effect of pressure on the electronic structure of diamond [250].

The most recent applications of semiempirical LUC models to solid-state problems are a study of relaxation effects in La-doped Pb(Zr,Ti)O_3 [251] and a study of F centers in crystalline metal oxides [252]. The latter study presents a comparison of periodic DFT and semiempirical methods and models. A modified version of the extended Hückel theory denoted as EHT was applied to calculate the electronic structure of stoichiometric calcium and lead apatites [253]. The size convergence of clusters within the LUC model for cohesive energy, energy gap, valence band width, and hybridization of diamond-structured semiconductors C, Si, Ge, and Sn was studied with LUC–CNDO [254]. The solid-state structures of metal–organic frameworks (MOFs) were calculated with Sparkle/AM1, Sparkle/PM3 and Sparkle/PM6 [255]. Since the LUC naturally also allows one-dimensional periodicity, the piezoelectric effect in the ferroelectric polymer poly(vinylidene fluoride) could be studied with PM3 [256]. Optical properties of fluorinated liquid crystals were studied with molecular CNDO/S and INDO/S calculations [257]. The influence of the amount of dopant on crystal cell dimensions of CeF_3 doped with Tb^{3+} ions (CeF_3:Tb^{3+}) was studied with AM1, RM1, PM3, PM6, PM7, and DFT methods [104]. It was observed that Sparkle/PM3 and DFT methods provided best agreement with the experiment, the average deviations being only 1% in both cases.

6.3.2.2 Cyclic Cluster Model

After publishing a critical article on the early LUC formulations [258], Deák developed CCM-implementations for various semiempirical Hamiltonians and applied the methods to solid-state problems, with a focus on defect properties of silica and silicon [259–267]. Noga *et al.* developed the SOLID98 program that is based on quasirelativistic INDO-CCM [268–270]. Initial convergence tests for the cyclic cluster size were performed for diamond, silicon, germanium, boron nitride, gallium phosphide, gallium arsenide, and gallium antimonide. In their CCM version, **k**-point symmetry is used to generate the wavefunction of the supercell models, which is advantageous for perfect solids. Their WSSC interaction domains were not symmetric however, which lead to problems with SCF convergence and artificial dipole moments.

The problem of site symmetry conservation of the interaction domains was addressed more carefully in the MSINDO-CCM implementation [59]. Based on original ideas by Evjen, a weighting scheme was applied for symmetry-related atoms on the boundaries of the WSSCs. In this way artificial dipole moments were avoided and fast convergence of calculated properties with cluster size was achieved. Lattice sums were explicitly taken into account by an Ewald summation [61]. The method was applied to structural, energetic, and electronic properties calculations of aluminum-doped anatase and rutile [271], and isolated oxygen vacancies in alumina [272]. In these studies one advantage of the CCM due to the use of well-defined finite interaction domains was exploited. Since a WSSC excludes all images from the neighbor cells, no direct defect–defect interactions occur, and the model corresponds to the limit of isolated defects. MSINDO-CCM was therefore further applied to studies of anion substitution in zinc chalcogenides [273], the effects of oxygen vacancies and zinc interstitials on the structure and energy of zinc oxide [274], and unusual optical properties of Mn-doped ZnO [275]. For the latter study, it is interesting to note that almost in parallel a LUC–INDO study was published on electronic effects of O vacancies and F-centers in ZnO [276]. Mn, Fe, and Co doping of ZnO was studied with MSINDO–CCM in order to elucidate the effect of doping and defects on the electronic properties [277]. One of the first approaches to calculated optical absorption spectra for solids was introduced in 2012 [278]. The MSINDO–CCM–CIS method used a Davidson–Liu algorithm for calculations of excited state calculations based on a parametrized configuration with singles ansatz. First applications included the optical spectrum of stoichiometric and oxygen-deficient rutile and anatase. Analytical gradients were implemented and allowed to study the relaxation of the first excited state of an F center in NaCl and the subsequent fluorescence process.

6.3.3
Adsorption and Surface Reactions

The intuitive chemical approach to model crystalline surfaces was to cut a small portion out of the bulk and to study the terminating faces of the resulting cluster, whereas physicists always preferred models with periodic boundary conditions. The development of the CCM (or small periodic cluster) by Bennett *et al.* was the first step to bridge this gap at semiempirical (EHT and CNDO) level [222,223]. In the beginning structurally simple and electroneutral systems such as graphite were investigated where long-range Coulomb effects did not play a role.

The advantage of the CCM, the avoidance of direct defect–defect interactions, was exploited by Litinskii and Zakharov in their study of neutral and charged defects on the silica surface [279]. In this study, the problem of two-dimensional lattice sums was tackled, which is solved in conventional periodic approaches by the Ewald summation technique. If a finite cluster is used as a surface model, convergence of electronic and energetic properties with model size and shape is mandatory. This was done, for example, in studies of CO_2 and H_2O adsorption at NaCl(100), MgO(100), and TiO_2 surfaces with SINDO1 [50,51], and in hydrogen adsorption studies at the V_2O_5(010) surface [280]. For the water–titania system, the relative stability of molecular and dissociative adsorption was investigated for the first time [281], an issue that is still under debate.

In order to model surfaces of covalent systems, low-coordinated atoms that do not represent surface sites have to be saturated, most simply by hydrogen atoms. This was done in a study of relaxations of the Si(111)(1×1) surface [282].

Embedding techniques for the treatment of long-range electrostatic interactions in finite cluster models, either in point charge fields or in arrays of pseudoatoms, were developed in the framework of SINDO1 [209]. The latter technique was later also implemented in MSINDO. Further SINDO1 and MSINDO adsorption studies comprise water and other small molecules at clean and stepped MgO(001) surfaces [283–285], Cr_2O_3(001) [286], perfect and defective NaCl(001) [287,288]. Embedded cluster models have been used to calculate the surface structure of MgO(100) with MSINDO [289]. In general, qualitative or even semiquantitative agreement with experimental results has been obtained.

A combination of MSINDO and DFT techniques was applied in a study of Cu deposition on the MgO(001) surface [290]. Cu_n clusters with $n = 2-6$ were adsorbed on a free (8×8×3) $Mg_{96}O_{96}$ cluster.

A LUC–INDO method with readjusted parameters was employed to study structural, energetic, and electronic features of the submonolayer adsorption of bi-isonicotinic acid on the rutile(110) and anatase(101) surfaces [291–294]. NDDO/MC was used to study the atomic and molecular adsorption of oxygen on Ag(111) using a finite Ag_{38} cluster as surface model [295]. The treatment of metallic systems is in general difficult for semiempirical methods that are based on the Hartree–Fock approximation. Therefore, an equalization of the HOMO of the cluster to the experimental Fermi level had to be forced in the self-consistent procedure. In a subsequent work, correlation corrections based on double excitations were calculated for surface and subsurface diffusion and desorption barriers [296,297].

A series of investigations by Gerson et al. was devoted to the dissolution processes of oxides like MgO and NiO [298–302]. These studies were performed with MSINDO and employed pseudoatom-embedded cluster models. It was shown that defects, that is, single-atom vacancies, Schottky defects, and step edges, are the initial centers for water dissociation and subsequent acidic dissolution. The reaction of water with MgO cannot only lead to dissolution but also to the formation of brucite $Mg(OH)_2$. Various initial reaction steps were studied with MSINDO-CCM [303–305].

The effect of oxygen defects on the electronic structure of perovskite surfaces $SrTiO_3$(110) and $BaTiO_3$(001) was studied by Stashans and coworkers with LUC–INDO [306,307]. Formation of F centers with two unpaired electrons localized at the oxygen vacancy was found for various crystal phases.

The catalytic properties of vanadia and titania surfaces were studied by Jug and coworkers with MSINDO and hydrogen saturated, free, and cyclic cluster models [308–313]. Both static and molecular dynamics (MD) calculations were employed to reveal the catalytic mechanism of the selective reduction of NO with NH_3 to N_2 on vanadium pentoxide supported by anatase.

Later the Born–Oppenheimer MD techniques was also employed for the study of water adsorption at rutile surfaces and subsurface oxygen vacancy diffusion in rutile [314,315]. The vibration spectra of adsorbed water obtained from the velocity autocorrelation function were in excellent agreement with experimental spectra [314]. Metadynamics was used to accelerate oxygen diffusion processes and to calculate rate constants of diffusion processes for different temperatures [315].

The adsorption and subsequent photoreaction of oxalic acid on anatase and rutile surfaces was studied both experimentally and theoretically with MSINDO by Mendive et al. [316–319]. The adsorption mode of oxalic acid was identified by comparison of measured and calculated vibration spectra based on BOMD simulations. The first steps of the photoreaction was calculated with excited-state optimization using MSINDO-CCM-sCIS [319].

Molecular and dissociative adsorption of small organic molecules at the surfaces of titania particles was studied by Wahab et al. with MSINDO and hydrogen-saturated clusters [320–324].

The effect of Fe and Cu doping on the wettability of sphalerite ZnS(110) surfaces was studied with MSINDO-CCM [325]. It was found that iron doping leads to more hydrophilic surfaces, and copper-containing surfaces are more hydrophobic than the undoped surface.

PM6 was employed to study methanethiol adsorption on GaAs(100) surfaces [326]. Dissociative adsorption was found to be more stable than molecular adsorption.

The epitaxial growth of graphene on the C and Si faces of SiC was modeled with AM1 and PM3 [327,328]. Surface reconstructions were found to be an essential part of the growth mechanism.

A combined AM1 and DFT study of nonstoichiometric cadmium sulfide quantum dot models was performed by Clark and coworkers [329]. For the semiempirical calculations, the EMPIRE program package was used. Binding energies and optical and vibration spectroscopy were calculated and compared with experiment. EMPIRE was also used to study conductance properties of C_{60} containing self-assembled monolayers [330].

Dispersion-corrected PM6-DH2 was employed to study the adsorption energies of unsaturated molecules on graphene in the presence of water [331]. The results obtained with the semiempirical method were comparable to DFT values and were in good agreement with the experiment.

6.3.4
Clusters and Nanoparticles

One has to distinguish between cluster models for the simulation of bulk and surface properties, which are usually cutouts from the corresponding crystal structure as discussed in the previous sections, and gas-phase clusters that are real chemical entities.

This section is dedicated to semiempirical studies on geometric and electronic properties of the latter species. An intermediate case is a SINDO1 study on the effect of doping with Si, B, N, Al, and P on the electronic properties of silicon using the ROHF method [332]. Pseudoatoms were used to saturate the cluster boundary atoms. The melting behavior of free Si clusters was later studied with MSINDO-BOMD [62]. Simulated annealing of the Si_{45} and Si_{60} clusters resulted in previously unreported structures. SINDO1-CI was employed to study the photocatalytical formation of OH radicals at anatase nanoparticles [333]. MSINDO was applied to full structure optimizations for large siloxane clusters ($Si_{2n}O_{3n}H_{2n}$ compounds with n up to 120) [334] in order to examine the experimentally observed growth pattern in the gas-phase reaction of $SiCl_4$ with O_2. After reparametrization for third-row transition metal compounds [57], MSINDO was used in a study of Al doping of anatase and rutile TiO_2 particles [271]. The most stable defect structure for anatase was obtained by the replacement of a Ti–O–Ti unit by two Al atoms separated by an oxygen vacancy. In rutile the simple substitution of a Ti atom by an Al atom was the most favorable defect type. Very recently, the optical and electronic bandgaps of anatase nanoparticles were studied with PM6 and PM7 [335]. Nanoparticles of a different oxide, Eu_2O_3, were calculated with Sparkle/PM3 in a combined experimental and theoretical study [336].

One of the most frequently studied family of molecular clusters are the fullerenes. A considerable number of semiempirical studies is also devoted to this system. The structures, stabilities, and electronic spectra of the heterofullerenes $C_{59}N$ and $C_{69}N$ and the formation of dimers with N—N bonds were examined with INDO/S [337]. The relative stabilities of fullerene C_{74} isomers were calculated with AM1, taking into account electronically excited states [338]. MNDO/d and OMx have been used for the structure optimizations and bond analyses of large carbon clusters, for example, a C_{960} fullerene [339]. The performance of AM1 and PM3 for predicting geometries and energies of C_{20}–C_{86} fullerene isomers was compared with DFTB and DFT methods [340]. The proton conductivity of single-walled carbon nanotubes was studied by Zaporotskova et al. [341]. Possible mechanisms for migration of a proton over the surfaces of single-walled carbon nanotubes are proposed. The relative stabilities of C_{74} isomers was calculated by Slanina et al. [338]. Almost at the same time, the reliability of some traditional semiempirical methods for large fullerenes was questioned by Xu et al. [342]. They compared AM1, PM3, and MNDO for selected isomers of C_{116}–C_{120}. The semiempirical methods were not able to reproduce the relative energies obtained with B3LYP, contrary to a tight-binding method that was employed in the same study.

Also, other carbon-based particles were studied with semiempirical methods. Semiempirical and DFT studies of brown coal chars with iron species were reviewed by Domazetis et al. [343]. The focus was on structure optimizations and reaction mechanisms of CO formation in pyrolysis reaction at elevated temperatures. UNO-CI and DFT calculations were used to investigate the effect of heteroatom substitution on the optical properties for a model polycyclic aromatic hydrocarbon [344]. Carbon nanodots were characterized both experimentally and theoretically in terms of their photophysical properties [345]. The electronic structure of amorphous carbon nanoparticles that were initially cut out from the bulk structure was studied with EMPIRE [346].

The search for global minima for large clusters with low symmetry is a tremendous task. Many techniques have been developed over the years to solve this problem. A global minimum search via annealing based on semiempirical MD simulations was proposed for nanoscale gold clusters [347]. The computational efficiency of semiempirical methods is exploited for the extensive MD runs.

6.3.5
Simulation of Liquids

The simulation of liquids has come into focus relatively recently. It requires MD simulations at extremely large timescale and length scale in order to obtain well-equilibrated results. Historically, these simulations were the domain of classical force fields. Due to the rapid progress of software and hardware, *ab initio* MD based on DFT has become possible. Nevertheless, for some cases semiempirical methods were considered in order to bridge the gap between the computationally efficient but parameter-dependent classical methods and costly DFT.

The most famous system in this context is liquid water. There exists a plethora of theoretical studies on structural, energetic, and spectroscopic properties, depending on temperature and pressure. Also, semiempirical methods were examined to describe the structure of liquid water [348]. Standard AM1 and PM3 but also reparametrized PM3-PIF and PM3-MAIS methods were implemented in a linear scaling semiempirical code (DivCon) that is based on a divide-and-conquer strategy. The predicted liquid water structure obtained from AM1 or PM3 was very poor due to the limitations of these methods describing hydrogen bonds. PM3-PIF and PM3-MAIS calculations were in much better agreement with experimental data. The same failure of most standard NDDO methods to describe the structure of liquid water was reported by Welborn *et al.* [349]. The authors suggested a reparametrized NDDO model and found that it qualitatively reproduces the experimental radial distribution function of water much better than the original PM6 version. Therefore, they concluded that the deficiencies are due to poor parametrization and not due to the NDDO approximations themselves. More recently, the self-consistent polarization NDDO method, as implemented in CP2k and parametrized to reproduce properties of water clusters and cubic ice, was applied to the calculation of bulk water properties [350]. Good results for the density, radial distribution functions, enthalpy of vaporization, self-diffusion coefficient, molecular dipole moment distribution, and hydrogen bond populations, in comparison to experimental measurements, were obtained. Scaling factors for atomic charges derived from RM1 for the CM1 and CM3 charge models were optimized for a set of 40 molecules [351]. The aim was to find a reliable but efficient method for QM/MM simulations. AM1, PM3, and OMx were reparametrized to improve their description of bulk water and of proton transfer in water [352]. The new set of OM3 parameters gave the best overall results for the structural and dynamic properties of water and the hydrated proton. PDDG/PM3 was used in QM/MM simulations for organic and enzymatic reactions in solution [353]. Likewise, AM1 was modified and reparametrized for proton transfer reactions in water clusters [354]. The modifications were denoted as AM1-W and AM1PG-W and show good performance on hydrogen bonding energies and on proton transfer energy profiles.

Besides the structural properties the solvation of molecules is an important issue addressed by many studies. The ability of semiempirical methods to describe intermolecular interactions in water solution were reviewed by Marion et al. [355]. It was found that none of the available methods provides an entirely satisfactory description of both hydrophobic and hydrophilic interactions in water. A consistent reparametrization based on theoretical data sets obtained with MP2 was suggested to improve this situation. A generalized solvent boundary potential was implemented in the framework of SQM/MM methods by Benighaus and Thiel [356]. HCl dissociation in water clusters was studied using the PM3-MAIS model [357]. A water-mediated interaction between the ions was found for all investigated systems. PM3-MAIS was later used to calculate vibration spectra of HCl solvated in water clusters [358]. The theoretical results were in good agreement with the available experimental data. MD simulations were performed with PM7 in order to study electrolyte solutions based on propylene carbonate–dimethoxyethane mixtures [359]. A weak solvation of the anions was reported, leading to the formation of neutral ion pairs (LiBF4, LiPF6). With the same method, also solutions of Li^+ in dimethoxyethane and propylene carbonate were studied [360].

References

1 Schrödinger, E. (1926) Quantisierung als Eigenwertproblem. Erste Mitteilung. *Ann. Phys.*, **384** (4), 361–376.
2 Hückel, E. (1931) Quantentheoretische Beiträge zum Benzolproblem: I. Die Elektronenkonfiguration des Benzols und verwandter Verbindungen. *Z. Phys*, **70** (3–4), 204–286.
3 Streitwieser, A. (1961) *Molecular Orbital Theory for Organic Chemists*, John Wiley & Sons, Inc., New York.
4 Hartmann, H. (1960) Zur Theorie der π-Elektronensysteme. *Z. Naturforsch. A*, **15** (11), 993–1003.
5 Pariser, R. and Parr, R.G. (1953) A semi-empirical theory of the electronic spectra and electronic structure of complex unsaturated molecules 1. *J. Chem. Phys.*, **21** (3), 466–471.
6 Pariser, R. and Parr, R.G. (1953) A semi-empirical theory of the electronic spectra and electronic structure of complex unsaturated molecules 2. *J. Chem. Phys.*, **21** (5), 767–776.
7 Pople, J.A. (1953) Electron interaction in unsaturated hydrocarbons. *Trans. Faraday. Soc.*, **49** (12), 1375–1385.
8 Roothaan, C.C.J. (1951) New developments in molecular orbital theory. *Rev. Mod. Phys.*, **23** (2), 69–89.
9 Parr, R.G. (1964) *Quantum Theory of Molecular Electronic Structure*, W. A. Benjamin, New York.
10 McWeeny, R. (1964) Self-consistent generalization of Hückel theory, in *Molecular Orbitals in Chemistry, Physics and Biology* (eds P.O. Löwdin and B. Pullman), Academic Press, London, pp. 305–328.
11 Fischer-Hjalmars, I. (1965) Deduction of the zero differential overlap approximation from an orthogonal

atomic basis set. *J. Chem. Phys.*, **42** (6), 1962–1972.

12 Löwdin, P.O. (1950) On the non-orthogonality problem connected with the use of atomic wave functions in the theory of molecules and crystals. *J. Chem. Phys.*, **18** (3), 365–375.

13 Jug, K. (1969) On the development of semiempirical methods in the MO formalism. *Theor. Chim. Acta*, **14** (2), 91–135.

14 Gey, E. (2010) Interdisziplinäre Forschungssituation und Kooperationsformen der Quantenchemie, in *Interdisziplinarität und Institutionalisierung der Wissenschaft* (eds K. Fischer, H. Laitko, and H. Parthey), Wissenschaftlicher Verlag Berlin Olaf Gaudig & Peter Veit GbR, pp. 153–174.

15 Hoffmann, R. (1963) An extended Hückel theory: I. Hydrocarbons. *J. Chem. Phys.*, **39** (6), 1397–1412.

16 Wolfsberg, M. and Helmholz, L. (1952) The spectra and electronic structure of the tetrahedral ions MnO_4^-, CrO_4^-, and ClO_4^-. *J. Chem. Phys.*, **20** (5), 837–843.

17 Whangbo, M.H. and Hoffmann, R. (1978) The band structure of the tetracyanoplatinate chain. *J. Am. Chem. Soc.*, **100** (19), 6903–6908.

18 Dronskowski, R. (1992) Theoretical increments and indices for reactivity, acidity, and basicity within solid-state materials. *J. Am. Chem. Soc.*, **114** (18), 7230–7244.

19 Dronskowski, R. and Hoffmann, R. (1992) Theoretical way of aiding the design of solid-state materials. *Adv. Mater.*, **4** (7/8), 514–521.

20 Parr, R.G. and Pearson, R. (1983) Absolute hardness: companion parameter to absolute electronegativity. *J. Am. Chem. Soc.*, **105** (26), 7512–7516.

21 Pople, J.A., Santry, D.P., and Segal, G.A. (1965) Approximate self-consistent molecular orbital theory: I. Invariant procedures. *J. Chem. Phys.*, **43** (10), S129–S136.

22 Jug, K. (1969) On invariant procedures in approximate SCF MO theories. *Int. J. Quantum Chem.*, **3** (S3a), 241–249.

23 Pople, J.A. and Segal, G.A. (1965) Approximate self-consistent molecular orbital theory: II. Calculations with complete neglect of differential overlap. *J. Chem. Phys.*, **43** (10), S136–S151.

24 Pople, J.A. and Segal, G.A. (1966) Approximate self-consistent molecular orbital theory: III. CNDO results for AB_2 and AB_3 systems. *J. Chem. Phys.*, **44** (9), 3289–3296.

25 Pople, J.A., Beveridge, D.L., and Dobosh, P.A. (1967) Approximate self-consistent molecular orbital theory. V: intermediate neglect of differential overlap. *J. Chem. Phys.*, **47** (6), 2026–2033.

26 Pople, J.A. and Beveridge, D.L. (1970) *Approximate Molecular Orbital Theory*, McGraw-Hill, New York.

27 Dewar, M.J.S. (1969) *The Molecular Orbital Theory of Organic Chemistry*, McGraw-Hill, New York.

28 Bingham, R.C., Dewar, M.J.S., and Lo, D.H. (1975) Ground states of molecules: XXV. MINDO/3: an improved version of the MINDO semiempirical SCF-MO method. *J. Am. Chem. Soc.*, **97** (6), 1285–1293.

29 Bingham, R.C., Dewar, M.J.S., and Lo, D.H. (1975) Ground states of molecules: XXVI. MINDO/3 calculations for hydrocarbons. *J. Am. Chem. Soc.*, **97** (6), 1294–1301.

30 Bingham, R.C., Dewar, M.J.S., and Lo, D.H. (1975) Ground states of molecules: XXVII. MINDO/3 calculations for CHNO species. *J. Am. Chem. Soc.*, **97** (6), 1302–1306.

31 Bingham, R.C., Dewar, M.J.S., and Lo, D.H. (1975) Ground states of molecules: XXVIII. MINDO/3 calculations for compound containing carbon, hydrogen, fluorine, and chlorine. *J. Am. Chem. Soc.*, **97** (6), 1307–1311.

32 Jug, K. (1972) Operator equations in approximate molecular orbital theories: II. Integral approximations and charge distributions. *Theor. Chim. Acta*, **26** (3), 231–236.

33 Coffey, P. and Jug, K. (1973) Semiempirical molecular orbital calculations and molecular energies: a new formula for the β parameter. *J. Am. Chem. Soc.*, **95** (23), 7575–7580.

34 Jug, K. (1976) Mechanism of cyclopropane-propene isomerization. *Theor. Chim. Acta*, **42** (4), 303–310.

35 Nanda, D. and Jug, K. (1980) SINDO1. A semiempirical SCF MO method for molecular binding energy and geometry. I. Approximations and parametrization. *Theor. Chim. Acta*, **57** (2), 95–106.

36 Jug, K. and Nanda, D. (1980) SINDO1. II. Application to ground states of molecules containing carbon, nitrogen and oxygen atoms. *Theor. Chim. Acta*, **57** (2), 107–130.

37 Jug, K. and Nanda, D. (1980) SINDO1. III. Application to ground states of molecules containing fluorine, boron, beryllium and lithium atoms. *Theor. Chim. Acta*, **57** (2), 131–144.

38 Dewar, M.J.S. and Thiel, W. (1977) Ground states of molecules. 38. The MNDO method: approximations and parameters. *J. Am. Chem. Soc.*, **99** (15), 4899–4907.

39 Dewar, M.J.S. and Thiel, W. (1977) Ground states of molecules. 39. MNDO results for molecules containing hydrogen, carbon, nitrogen, and oxygen. *J. Am. Chem. Soc.*, **99** (15), 4907–4917.

40 Zerner, M. (1972) Removal of core orbitals in 'valence orbital only' calculations. *Mol. Phys.*, **23** (5), 963–978.

41 Schulz, J., Iffert, R., and K., J. (1985) On the rotational invariance of the Fock equations in INDO methods using d functions. *Int. J. Quantum Chem.*, **27** (4), 461–464.

42 Jug, K., Iffert, R., and Schulz, J. (1987) Development and parametrization of SINDO1 for second-row elements. *Int. J. Quantum Chem.*, **32** (2), 265–277.

43 Jug, K. and Iffert, R. (1987) Application of SINDO1 to sulphur compounds. *J. Comput. Chem.*, **8** (7), 1004–1015.

44 Jug, K. and Schulz, J. (1987) Application of SINDO1 to chlorine and sodium compounds. *J. Comput. Chem.*, **8** (7), 1040–1050.

45 Jug, K. and Schulz, J. (1988) Application of SINDO1 to phosphorus compounds. *J. Comput. Chem.*, **9** (1), 40–50.

46 Jug, K. and Iffert, R. (1988) Application of SINDO1 to silicon, aluminum, and magnesium compounds. *J. Comput. Chem.*, **9** (1), 51–62.

47 Jug, K., Schluff, H.P., Kupka, H., and Iffert, R. (1988) Structure and properties of small silicon and aluminum clusters. *J. Comput. Chem.*, **9** (8), 803–809.

48 Kupka, H. and Jug, K. (1989) Clusters as crystal fragments of silicon bulk. *Chem. Phys.*, **130** (1–3), 23–29.

49 Li, J., de Mello, P.C., and Jug, K. (1992) Extension of SINDO1 to transition metal compounds. *J. Comput. Chem.*, **13** (1), 85–92.

50 Jug, K. and Geudtner, G. (1993) Binding energies and bond distances of ion crystal clusters. *Chem. Phys. Lett.*, **208** (5–6), 537–540.

51 Jug, K., Geudtner, G., and Bredow, T. (1993) Theoretical investigations on adsorption at ion crystal surfaces. *J. Mol. Catal.*, **82** (2–3), 171–194.

52 Jug, K. and Geudtner, G. (1993) Treatment of hydrogen bonding in SINDO1. *J. Comput. Chem.*, **14** (6), 639–646.

53 Jug, K. and Krack, M. (1992) Consistent parametrization of semiempirical MO methods. *Int. J. Quantum Chem.*, **44** (4), 517–531.

54 Ahlswede, B. and Jug, K. (1999) Consistent modifications of SINDO1: I. Approximations and parameters. *J. Comput. Chem.*, **20** (6), 563–571.

55 Ahlswede, B. and Jug, K. (1999) Consistent modifications of SINDO1: II. Applications to first- and second-row elements. *J. Comput. Chem.*, **20** (6), 572–578.

56 Jug, K., Geudtner, G., and Homann, T. (2000) MSINDO parameterization for third-row main group elements. *J. Comput. Chem.*, **21** (11), 974–987.

57 Bredow, T., Geudtner, G., and Jug, K. (2001) MSINDO parameterization for third-row transition metals. *J. Comput. Chem.*, **22** (8), 861–887.

58 Bredow, T. and Jug, K. (2004) MSINDO, in *Electronic Encyclopedia of Computational Chemistry* (eds P.V.R. Schleyer, N.L. Allinger, T. Clark, J. Gasteiger, P.A. Kollmann, H.F. Schaefer, III, and P.R. Schreiner), John Wiley & Sons, Inc., Chicester, UK.

59 Bredow, T., Geudtner, G., and Jug, K. (2001) Development of the cyclic cluster

approach for ionic systems. *J. Comput. Chem.*, **22** (1), 89–101.

60 Bredow, T., Evarestov, R., and Jug, K. (2000) Implementation of the cyclic cluster model in Hartree–Fock LCAO calculations of crystalline systems. *Phys. Status Solidi B*, **222** (2), 495–516.

61 Janetzko, F., Bredow, T., and Jug, K. (2002) Effects of long-range interactions in cyclic cluster calculations of metal oxides. *J. Chem. Phys.*, **116** (20), 8994–9004.

62 Nair, N., Bredow, T., and Jug, K. (2004) Molecular dynamics implementation in MSINDO: study of silicon clusters. *J. Comput. Chem.*, **25** (10), 1255–1263.

63 Gadaczek, I., Krause, K., Hintze, K.J., and Bredow, T. (2011) MSINDO-sCIS: a new method for the calculation of excited states of large molecules. *J. Chem. Theory Comput.*, **7** (11), 3675–3685.

64 Zerner, M. and Ridley, J. (1973) Intermediate neglect of differential overlap technique for spectroscopy: pyrrole and azines. *Theor. Chim. Acta*, **32** (2), 111–134.

65 Bacon, A.D. and Zerner, M.C. (1979) Intermediate neglect of differential overlap theory for transition metal complexes: Fe, Co and Cu chlorides. *Theor. Chim. Acta*, **53** (1), 21–54.

66 Head, J.D. and Zerner, M.C. (1985) A Broyden–Fletcher–Goldfarb–Shanno optimization procedure for molecular geometries. *Chem. Phys. Lett.*, **122** (3), 264–270.

67 Da Motta Neto, J.D. and Zerner, M.C. (2001) New parametrization scheme for the resonance integrals $H_{\mu\nu}$ within the INDO/1 approximation. Main group elements. *Int. J. Quantum Chem.*, **81** (3), 187–201.

68 Voityuk, A.A. (2014) INDO/X: a new semiempirical method for excited states of organic and biological molecules. *J. Chem. Theory Comput.*, **10** (11), 4950–4958.

69 Dewar, M.J.S. and Rzepa, H.S. (1978) Ground states of molecules. 40. MNDO results for molecules containing fluorine. *J. Am. Chem. Soc.*, **100** (1), 58–67.

70 Dewar, M.J.S., McKee, M.L., and Rzepa, H.S. (1978) MNDO parameters for third period elements. *J. Am. Chem. Soc.*, **100** (11), 3607–3607.

71 Dewar, M.J.S. and Rzepa, H.S. (1983) Ground states of molecules. 53. MNDO results for molecules containing chlorine. *J. Comput. Chem.*, **4** (2), 158–169.

72 Davis, L.P., Guidry, R.M., Williams, J.R., Dewar, M.J.S., and Rzepa, H.S. (1981) MNDO calculations for compounds containing aluminum and boron. *J. Comput. Chem.*, **2** (4), 433–445.

73 Dewar, M.J.S., Zoebisch, E.G., Healy, E.F., and Stewart, J.J.P. (1985) AM1: a new general purpose quantum mechanical molecular model. *J. Am. Chem. Soc.*, **107** (13), 3902–3909.

74 Stewart, J.J.P. (1989) Optimization of parameters for semiempirical methods: I. Method. *J. Comput. Chem.*, **10** (2), 209–220.

75 Stewart, J.J.P. (1989) Optimization of parameters for semiempirical methods: II. Applications. *J. Comput. Chem.*, **10** (2), 221–264.

76 Stewart, J., Császár, P., and Pulay, P. (1982) Fast semi-empirical calculations. *J. Comput. Chem.*, **3** (2), 227–228.

77 Stewart, J.J.P. (1996) Application of localized molecular orbitals to the solution of semiempirical self-consistent field equations. *Int. J. Quantum Chem.*, **58** (2), 133–146.

78 Stewart, J.J. (1991) Optimization of parameters for semiempirical methods: III. Extension of PM3 to Be, Mg, Zn, Ga, Ge, As, Se, Cd, In, Sn, Sb, Te, Hg, Tl, Pb, and Bi. *J. Comput. Chem.*, **12**, 320–341.

79 Stewart, J. (2004) Optimization of parameters for semiempirical methods: IV. Extension of MNDO, AM1, and PM3 to more main group elements. *J. Mol. Model.*, **10** (2), 155–164.

80 Stewart, J.J.P. (2007) Optimization of parameters for semiempirical methods: V. Modification of NDDO approximations and application to 70 elements. *J. Mol. Model.*, **13** (12), 1173–1213.

81 Stewart, J.J.P. (2013) Optimization of parameters for semiempirical methods:

VI. More modifications to the NDDO approximations and re-optimization of parameters. *J. Mol. Model.*, **19** (1), 1–32.

82 Thiel, W. (1981) The MNDOC method, a correlated version of the MNDO model. *J. Am. Chem. Soc.*, **103** (6), 1413–1420.

83 Thiel, W. and Voityuk, A.A. (1992) Extension of the MNDO formalism to d orbitals: integral approximations and preliminary numerical results. *Theor. Chim. Acta*, **81** (6), 391–404.

84 Thiel, W. and Voityuk, A.A. (1996) Extension of MNDO to d orbitals: parameters and results for the second-row elements and for the zinc group. *J. Phys. Chem.*, **100** (2), 616–626.

85 Kolb, M. and Thiel, W. (1993) Beyond the MNDO model: methodical considerations and numerical results. *J. Comput. Chem.*, **14** (7), 775–789.

86 Weber, W. and Thiel, W. (2000) Orthogonalization corrections for semiempirical methods. *Theor. Chem. Acc.*, **103** (6), 495–506.

87 Silva-Junior, M.R. and Thiel, W. (2010) Benchmark of electronically excited states for semiempirical methods: MNDO, AM1, PM3, OM1, OM2, OM3, INDO/S, and INDO/S2. *J. Chem. Theory Comput.*, **6** (5), 1546–1564.

88 Golebiewski, A., Nalewajski, R., and Witko, M. (1977) SINDO theory of transition-metal compounds: 1. General analysis. *Acta Phys. Pol. A*, **51** (4), 617–627.

89 Witko, M., Hermann, K., and Tokarz, R. (1994) *Ab initio* and semiempirical cluster studies on the reactivity of the vanadium pentoxide (010)-surface. *J. Electron Spectros. Relat. Phenomena*, **69**, 89–98.

90 Culberson, J.C., Knappe, P., Rösch, N., and Zerner, M.C. (1987) An intermediate neglect of differential overlap (INDO) technique for lanthanide complexes: studies on lanthanide halides. *Theor. Chim. Acta*, **71**, 21–39.

91 Filatov, M., Zilberberg, I., and Zhidomirov, G. (1992) NDDO/MC: a new semiempirical SCF-MO method for transition-metal complexes. *Int. J. Quantum Chem.*, **44** (4), 565–585.

92 Filatov, M., Pelmenschikov, A., and Zhidomirov, G. (1993) On the nature of superactive centers in H-ZSM-5 zeolites: quantum chemical calculations. *J. Mol. Catal.*, **80** (2), 243–251.

93 Dewar, M., Jie, C., and Yu, J. (1993) SAM1: the first of a new series of general-purpose quantum-mechanical molecular-models. *Tetrahedron*, **49** (23), 5003–5038.

94 De Andrade, A., Da Costa, N., Simas, A., and De Sa, G. (1994) Sparkle model for the quantum-chemical AM1 calculation of europium complexes. *Chem. Phys. Lett.*, **227** (3), 349–353.

95 Rocha, G., Freire, R., Da Costa, N., De Sa, G., and Simas, A. (2004) Sparkle model for AM1 calculation of lanthanide complexes: improved parameters for europium. *Inorg. Chem.*, **43** (7), 2346–2354.

96 Freire, R., Rocha, G., Albuquerque, R., and Simas, A. (2005) Efficacy of the semiempirical sparkle model as compared to ECP *ab initio* calculations for the prediction of ligand field parameters of europium(III) complexes. *J. Lumin.*, **111** (1–2), 81–87.

97 Freire, R., Rocha, G., and Simas, A. (2005) Sparkle model for the calculation of lanthanide complexes: AM1 parameters for Eu(III), Gd(III), and Tb(III). *Inorg. Chem.*, **44** (9), 3299–3310.

98 Freire, R.O., da Costa, N.B., Rocha, G.B., and Simas, A.M. (2007) Sparkle/PM3 parameters for the modeling of neodymium(III), promethium(III), and samarium(III) complexes. *J. Chem. Theory Comput.*, **3** (4), 1588–1596.

99 Freire, R.O. and Simas, A.M. (2010) Sparkle/PM6 parameters for all lanthanide trications from La(III) to Lu(III). *J. Chem. Theory Comput.*, **6** (7), 2019–2023.

100 Freire, R.O. and Simas, A.M. (2011) The lanthanide contraction within the sparkle model. *Int. J. Quantum Chem.*, **111** (7–8), 1734–1739.

101 Filho, M.A.M., Dutra, J.D.L., Rocha, G.B., Freire, R.O., and Simas, A.M. (2013)

Sparkle/RM1 parameters for the semiempirical quantum chemical calculation of lanthanide complexes. *RSC Adv.*, **3** (37), 16747–16755.
102. Dutra, J.D.L., Filho, M.A.M., Rocha, G.B., Freire, R.O., Simas, A.M., and Stewart, J.J.P. (2013) Sparkle/PM7 lanthanide parameters for the modeling of complexes and materials. *J. Chem. Theory Comput.*, **9** (8), 3333–3341.
103. Filho, M.A.M., Dutra, J.D.L., Cavalcanti, H.L.B., Rocha, G.B., Simas, A.M., and Freire, R.O. (2014) RM1 model for the prediction of geometries of complexes of the trications of Eu, Gd, and Tb. *J. Chem. Theory Comput.*, **10** (8), 3031–3037.
104. Shyichuk, A., Runowski, M., Lis, S., Kaczkowski, J., and Jezierski, A. (2015) Semiempirical and DFT computations of the influence of Tb(III) dopant on unit cell dimensions of cerium(III) fluoride. *J. Comput. Chem.*, **36** (3), 193–199.
105. Primera, J., Sanchez, M., Romero, M., Sierraalta, A., and Ruette, F. (1999) Analysis of parametric functionals in semiempirical approaches using simulation techniques. *J. Mol. Struct.*, **469**, 177–190.
106. Voityuk, A., Zerner, M., and Rösch, N. (1999) Extension of the neglect of diatomic differential overlap method to spectroscopy: NDDO-G parametrization and results for organic molecules. *J. Phys. Chem. A*, **103** (23), 4553–4559.
107. Voityuk, A.A. and Rösch, N. (2000) AM1/d parameters for molybdenum. *J. Phys. Chem. A*, **104** (17), 4089–4094.
108. Repasky, M.P. and Jayaraman Chandrasekhar, W.L.J. (2002) PDDG/PM3 and PDDG/MNDO: improved semiempirical methods. *J. Comput. Chem.*, **23** (16), 1601–1622.
109. Tubert-Brohman, I., Guimaraes, C., and Jorgensen, W. (2005) Extension of the PDDG/PM3 semiempirical molecular orbital method to sulfur, silicon, and phosphorus. *J. Chem. Theory Comput.*, **1** (5), 817–823.
110. Rocha, G.B., Freire, R.O., Simas, A.M., and Stewart, J.J.P. (2006) RM1: a reparameterization of AM1 for H, C, N, O, P, S, F, Cl, Br, and I. *J. Comput. Chem.*, **27** (10), 1101–1111.
111. Filho, M.A.M., Dutra, J.D.L., Rocha, G.B., Simas, A.M., and Freire, R.O. (2014) Semiempirical quantum chemistry model for the lanthanides: RM1 (Recife Model 1) parameters for dysprosium, holmium and erbium. *PLoS One*, **9** (1)
112. Dutra, J.D.L., Filho, M.A.M., Rocha, G.B., Simas, A.M., and Freire, R.O. (2015) RM1 semiempirical quantum chemistry: parameters for trivalent lanthanum, cerium and praseodymium. *PLoS One*, **10** (7)
113. Clark, T. (2000) Quo vadis semiempirical MO-theory? *J. Mol. Struct.*, **530** (1–2), 1–10.
114. Bredow, T. and Jug, K. (2005) Theory and range of modern semiempirical molecular orbital methods. *Theor. Chem. Acc.*, **113** (1), 1–14.
115. Thiel, W. (2014) Semiempirical quantum-chemical methods. *Wiley Interdiscip. Rev. Comput. Mol. Sci.*, **4** (2), 145–157.
116. Akimov, A.V. and Prezhdo, O.V. (2015) Large-scale computations in chemistry: a bird's eye view of a vibrant field. *Chem. Rev.*, **115** (12), 5797–5890.
117. Patchkovskii, S. and Thiel, W. (1996) Analytical second derivatives of the energy in MNDO methods. *J. Comput. Chem.*, **17** (11), 1318–1327.
118. Frisch, M., Scalmani, G., Vreven, T., and Zheng, G. (2009) Analytic second derivatives for semiempirical models based on MNDO. *Mol. Phys.*, **107** (8–12), 881–887.
119. Koslowski, A., Beck, M.E., and Thiel, W. (2003) Implementation of a general multireference configuration interaction procedure with analytic gradients in a semiempirical context using the graphical unitary group approach. *J. Comput. Chem.*, **24** (6), 714–726.
120. Keal, T.W., Koslowski, A., and Thiel, W. (2007) Comparison of algorithms for conical intersection optimisation using semiempirical methods. *Theor. Chem. Acc.*, **118** (5–6), 837–844.
121. Fabiano, E., Keal, T.W., and Thiel, W. (2008) Implementation of surface hopping molecular dynamics using semiempirical methods. *Chem. Phys.*, **349** (1–3), 334–347.

122 Barbatti, M., Granucci, G., Persico, M., and Lischka, H. (2005) Semiempirical molecular dynamics investigation of the excited state lifetime of ethylene. *Chem. Phys. Lett.*, **401** (1–3), 276–281.

123 Lei, Y., Suo, B., Dou, Y., Wang, Y., and Wen, Z. (2010) New implementations of MRCI in semiempirical frameworks. *J. Comput. Chem.*, **31** (8), 1752–1758.

124 Giese, T. and York, D. (2005) Improvement of semiempirical response properties with charge-dependent response density. *J. Chem. Phys.*, **123** (16)

125 Bartell, L.A., Wall, M.R., and Neuhauser, D. (2010) A time-dependent semiempirical approach to determining excited states. *J. Chem. Phys.*, **132** (23), 234106.

126 Gadaczek, I., Krause, K., Hintze, K.J., and Bredow, T. (2012) Analytical gradients for the MSINDO-sCIS and MSINDO-UCIS method: theory, implementation, benchmarks, and examples. *J. Chem. Theory Comput.*, **8** (3), 986–996.

127 Dral, P.O. and Clark, T. (2011) Semiempirical UNO-CAS and UNO-CI: method and applications in nanoelectronics. *J. Phys. Chem. A*, **115** (41), 11303–11312.

128 Tchougréeff, A.L., Soudackov, A.V., van Leusen, J., Kägerler, P., Becker, K.D., and Dronskowski, R. (2015) Effective Hamiltonian crystal field: present status and applications to iron compounds. *Int. J. Quantum Chem*, **116** (4), 282–294.

129 Tchougréeff, A.L. (ed.) (2008) *Hybrid Methods of Molecular Modeling: Progress in Theoretical Chemistry and Physics*, vol. **17**, Springer.

130 Soudackov, A.V. and Jug, K. (1997) Effective Hamiltonian-crystal field on the INDO level: calculations of d-d spectra of some iron(II) compounds. *Int. J. Quantum Chem.*, **62** (4), 403–418.

131 Darkhovskii, M.B., Tokmachev, A.M., and Tchougréeff, A.L. (2006) MNDO parameterized hybrid SLG/SCF method as used for molecular modeling of Zn(II) complexes. *Int. J. Quantum Chem.*, **106** (10), 2268–2280.

132 Chugreev, A. (2007) New generation of semiempirical methods of molecular modeling based on the theory of group functions. *J. Struct. Chem.*, **48** (1), S32–S54.

133 Tokmachev, A.M. (2015) Perspectives of geminal methods for large molecular systems. *Int. J. Quantum Chem*, **116** (4), 265–269.

134 Tokmachev, A.M. and Tchougréeff, A.L. (2005) Efficient multipole model and linear scaling of NDDO-based methods. *J. Phys. Chem. A*, **109** (33), 7613–7620.

135 Tchougréeff, A.L. and Soudackov, A.V. (2014) Effective Hamiltonian crystal fields: present status and applicability to magnetic interactions in polynuclear transition metal complexes. *Russ. J. Phys. Chem. A*, **88** (11), 1904–1913.

136 Anisimov, V.M. and Cavasotto, C.N. (2011) Hydration free energies using semiempirical quantum mechanical Hamiltonians and a continuum solvent model with multiple atomic-type parameters. *J. Phys. Chem. B*, **115** (24), 7896–7905.

137 Khandogin, J., Gregersen, B., Thiel, W., and York, D. (2005) Smooth solvation method for d-orbital semiempirical calculations of biological reactions: 1. Implementation. *J. Phys. Chem. B*, **109** (19), 9799–9809.

138 Gregersen, B., Khandogin, J., Thiel, W., and York, D. (2005) Smooth solvation method for d-orbital semiempirical calculations of biological reactions: 2. Application to transphosphorylation thio effects in solution. *J. Phys. Chem. B*, **109** (19), 9810–9817.

139 Krause, K. and Bredow, T. (2014) Conductor-like screening model for relaxed excited states: implementation in the semiempirical method MSINDO. *J. Comput. Chem.*, **35** (7), 553–564.

140 Seabra, G.D.M., Walker, R.C., and Roitberg, A.E. (2009) Are current semiempirical methods better than force fields? A study from the thermodynamics perspective. *J. Phys. Chem. A*, **113** (43), 11938–11948.

141 Martin, B. and Clark, T. (2006) Dispersion treatment for NDDO-based semiempirical MO techniques. *Int. J. Quantum Chem*, **106** (5), 1208–1216.

142 Rezac, J. and Hobza, P. (2012) Advanced corrections of hydrogen bonding and

dispersion for semiempirical quantum mechanical methods. *J. Chem. Theory Comput.*, **8** (1), 141–151.

143 Rezac, J. and Hobza, P. (2011) A halogen-bonding correction for the semiempirical PM6 method. *Chem. Phys. Lett.*, **506** (4–6), 286–289.

144 Dobes, P., Rezac, J., Fanfrlik, J., Otyepka, M., and Hobza, P. (2011) Semiempirical quantum mechanical method PM6-DH2X describes the geometry and energetics of CK2-inhibitor complexes involving halogen bonds well, while the empirical potential fails. *J. Phys. Chem. B*, **115** (26), 8581–8589.

145 Fanfrlik, J., Bronowska, A.K., Rezac, J., Prenosil, O., Konvalinka, J., and Hobza, P. (2010) A reliable docking/scoring scheme based on the semiempirical quantum mechanical PM6-DH2 method accurately covering dispersion and H-bonding: HIV-1 protease with 22 ligands. *J. Phys. Chem. B*, **114** (39), 12666–12678.

146 Strutynski, K., Gomes, J.A.N.F., and Melle-Franco, M. (2014) Accuracy of dispersion interactions in semiempirical and molecular mechanics models: the benzene dimer case. *J. Phys. Chem. A*, **118** (40), 9561–9567.

147 Korth, M., Pitonak, M., Rezac, J., and Hobza, P. (2010) A transferable H-bonding correction for semiempirical quantum-chemical methods. *J. Chem. Theory Comput.*, **6** (1), 344–352.

148 Rezac, J., Fanfrlik, J., Salahub, D., and Hobza, P. (2009) Semiempirical quantum chemical PM6 method augmented by dispersion and h-bonding correction terms reliably describes various types of noncovalent complexes. *J. Chem. Theory Comput.*, **5** (7), 1749–1760.

149 Korth, M. (2010) Third-generation hydrogen-bonding corrections for semiempirical QM methods and force fields. *J. Chem. Theory Comput.*, **6** (12), 3808–3816.

150 Lepsik, M., Rezac, J., Kolar, M., Pecina, A., Hobza, P., and Fanfrlik, J. (2013) The semiempirical quantum mechanical scoring function for *in silico* drug design. *ChemPlusChem*, **78** (9), 921–931.

151 Daniels, A., Millam, J., and Scuseria, G. (1997) Semiempirical methods with conjugate gradient density matrix search to replace diagonalization for molecular systems containing thousands of atoms. *J. Chem. Phys.*, **107** (2), 425–431.

152 Anisimov, V.M., Bugaenko, V.L., and Bobrikov, V.V. (2006) Validation of linear scaling semiempirical LocalSCF method. *J. Chem. Theory Comput.*, **2** (6), 1685–1692.

153 Carvalho Maia, J.D., Urquiza Carvalho, G.A., Mangueira, C.P., Jr., Santana, S.R., Formiga Cabral, L.A., and Rocha, G.B. (2012) GPU linear algebra libraries and GPGPU programming for accelerating MOPAC semiempirical quantum chemistry calculations. *J. Chem. Theory Comput.*, **8** (9), 3072–3081.

154 Wu, X., Koslowski, A., and Thiel, W. (2012) Semiempirical quantum chemical calculations accelerated on a hybrid multicore CPU–GPU computing platform. *J. Chem. Theory Comput.*, **8** (7), 2272–2281.

155 Hennemann, M. and Clark, T. (2014) EMPIRE: a highly parallel semiempirical molecular orbital program: 1. Self-consistent field calculations. *J. Mol. Model.*, **20**, 2331.

156 Margraf, J.T., Hennemann, M., Meyer, B., and Clark, T. (2015) EMPIRE: a highly parallel semiempirical molecular orbital program: 2. Periodic boundary conditions. *J. Mol. Model.*, **21** (6), 144.

157 Perkins, P. and Stewart, J. (1980) Cluster model for solids. *J. Chem. Soc. Faraday Trans. II*, **76** (5), 520–533.

158 Perkins, P.G., Marwaha, A.K., and Stewart, J.J. (1980) An improved LCAO SCF method for three-dimensional solids and its application to polyethylene, graphite, diamond, and boron nitride. *Theor. Chim. Acta*, **57** (1), 1–23.

159 Stewart, J. (2000) A practical method for modeling solids using semiempirical methods. *J. Mol. Struct.*, **556** (1–3), 59–67.

160 Sattelmeyer, K.W., Tubert-Brohman, I., and Jorgensen, W.L. (2006) NO-MNDO: reintroduction of the overlap matrix into MNDO. *J. Chem. Theory Comput.*, **2** (2), 413–419.

161 Kayi, H. and Clark, T. (2011) AM1* parameters for palladium and silver. *J. Mol. Model.*, **17** (10), 2585–2600.

162 Zhou, Y. and Pu, J. (2014) Reaction path force matching: a new strategy of fitting specific reaction parameters for semiempirical methods in combined QM/MM simulations. *J. Chem. Theory Comput.*, **10** (8), 3038–3054.

163 Dral, P.O., von Lilienfeld, O.A., and Thiel, W. (2015) Machine learning of parameters for accurate semiempirical quantum chemical calculations. *J. Chem. Theory Comput.*, **11** (5), 2120–2125.

164 Hutter, J., Iannuzzi, M., Schiffmann, F., and VandeVondele, J. (2014) CP2K: atomistic simulations of condensed matter systems. *Wiley Interdiscip. Rev. Comput. Mol. Sci.*, **4** (1), 15–25.

165 Stewart, J. (1990) MOPAC: a semiempirical molecular-orbital program. *J. Comput.-Aided Mol. Design*, **4**, 1–45.

166 MOPAC2012 and Stewart, J.P. (2012) Stewart Computational Chemistry, Colorado Springs, CO. Available at http://HTTP://OpenMOPAC.net.

167 Sattelmeyer, K.W., Tirado-Rives, J., and Jorgensen, W.L. (2006) Comparison of SCC-DFTB and NDDO-based semiempirical molecular orbital methods for organic molecules. *J. Phys. Chem. A*, **110** (50), 13551–13559.

168 Thiel, W. (1996) Perspectives on semiempirical molecular orbital theory, in *Advances in Chemical Physics, Vol. XCIII: New Methods in Computational Quantum Mechanics*, vol. **93** (eds I. Prigogine and S.A. Rice), John Wiley & Sons, Inc., New York, pp. 703–757.

169 Thiel, W. (1997) Computational methods for large molecules. *J. Mol. Struct.*, **398–399**, 1–6.

170 Ford, G. (1997) Semiempirical molecular orbital theory in carcinogenesis research. *J. Mol. Struct.*, **401** (3), 253–266.

171 Gogonea, V., Suarez, D., van der Vaart, A., and Merz, K. (2001) New developments in applying quantum mechanics to proteins. *Curr. Opin. Struct. Biol.*, **11** (2), 217–223.

172 Merz, K.M., Jr. (2014) Using quantum mechanical approaches to study biological systems. *Acc. Chem. Res.*, **47**, 2804–2811.

173 Bryce, R.A. and Hillier, I.H. (2014) Quantum chemical approaches: semiempirical molecular orbital and hybrid quantum mechanical/molecular mechanical techniques. *Curr. Pharm. Des.*, **20** (20), 3293–3302.

174 Yilmazer, N.D. and Korth, M. (2015) Enhanced semiempirical QM methods for biomolecular interactions. *Comput. Struct. Biotechnol. J.*, **13**, 169–175.

175 Layfield, J.P., Owens, M.D., and Troya, D. (2008) Theoretical study of the dynamics of the H + CH_4 and H + C_2H_6 reactions using a specific-reaction-parameter semiempirical Hamiltonian. *J. Chem. Phys.*, **128** (19), 194302.

176 Vardi-Kilshtain, A., Major, D.T., Kohen, A., Engel, H., and Doron, D. (2012) Hybrid quantum and classical simulations of the formate dehydrogenase catalyzed hydride transfer reaction on an accurate semiempirical potential energy surface. *J. Chem. Theory Comput.*, **8** (11), 4786–4796.

177 Sugawara, S., Yoshikawa, T., Takayanagi, T., Shiga, M., and Tachikawa, M. (2011) Quantum proton transfer in hydrated sulfuric acid clusters: a perspective from semiempirical path integral simulations. *J. Phys. Chem. A*, **115** (42), 11486–11494.

178 Govender, K., Gao, J., and Naidoo, K.J. (2014) AM1/d-CB1: a semiempirical model for QM/MM simulations of chemical glycobiology systems. *J. Chem. Theory Comput.*, **10** (10), 4694–4707.

179 Rutherford, T., Wilkie, J., Vu, C., Schnackerz, K., Jacobson, M., and Gani, D. (2001) NMR studies and semi-empirical energy calculations for cyclic ADP-ribose. *Nucleosides Nucleotides Nucleic Acids*, **20** (8), 1485–1495.

180 Sapse, A., Capiaux, G., and Bertino, J. (2001) Examination of the reduced affinity of the thymidylate synthase G52S mutation for FdUMP by *ab initio* and semi-empirical studies. *Mol. Med.*, **7** (3), 200–204.

181 Möhle, K., Hofmann, H.J., and Thiel, W. (2001) Description of peptide and protein secondary structures employing semiempirical methods. *J. Comput. Chem.*, **22** (5), 509–520.

182 Wollacott, A.M. and Merz, K.M., Jr. (2007) Assessment of semiempirical quantum mechanical methods for the

183 Stewart, J.J.P. (2009) Application of the PM6 method to modeling proteins. *J. Mol. Model.*, **15** (7), 765–805.

184 Huang, M., Giese, T.J., Lee, T.S., and York, D.M. (2014) Improvement of DNA and RNA sugar pucker profiles from semiempirical quantum methods. *J. Chem. Theory Comput.*, **10** (4), 1538–1545.

185 Brandon, C.J., Martin, B.P., McGee, K.J., Stewart, J.J.P., and Braun-Sand, S.B. (2015) An approach to creating a more realistic working model from a protein data bank entry. *J. Mol. Model.*, **21** (1), 3.

186 Borbulevych, O.Y., Plumley, J.A., Martin, R.I., Merz, K.M., Jr., and Westerhoff, L.M. (2014) Accurate macromolecular crystallographic refinement: incorporation of the linear scaling, semiempirical quantum-mechanics program DivCon into the PHENIX refinement package. *Acta Crystallogr. D*, **70** (5), 1233–1247.

187 Wanko, M., Hoffmann, M., Strodel, P., Koslowski, A., Thiel, W., Neese, F., Frauenheim, T., and Elstner, M. (2005) Calculating absorption shifts for retinal proteins: computational challenges. *J. Phys. Chem. B*, **109** (8), 3606–3615.

188 Zhu, Q. and LeBreton, P. (2000) DNA photoionization and alkylation patterns in the interior of guanine runs. *J. Am. Chem. Soc.*, **122** (51), 12824–12834.

189 Voityuk, A.A. (2006) Assessment of semiempirical methods for the computation of charge transfer in DNA pi-stacks. *Chem. Phys. Lett.*, **427** (1–3), 177–180.

190 Hatanaka, M. (2013) Evaluation of optical activities by modern semi-empirical methods. *Int. J. Quantum Chem.*, **113** (22), 2447–2456.

191 Lan, Z., Fabiano, E., and Thiel, W. (2009) Photoinduced nonadiabatic dynamics of pyrimidine nucleobases: on-the-fly surface-hopping study with semiempirical methods. *J. Phys. Chem. B*, **113** (11), 3548–3555.

192 Nayyar, I.H. and Masunov, A.E. (2014) Two-photon absorption spectra predicted by semiempirical methods. *J. Comput. Theor. Nanoscience*, **11** (10), 2208–2220.

193 Kala, C.P., Priya, P.A., and Thiruvadigal, D.J. (2013) Semiempirical study of electron transport of heterocyclic molecule based molecular devices. *J. Comput. Theor. Nanosci.*, **10** (1), 213–217.

194 Wick, C.R., Hennemann, M., Stewart, J.J.P., and Clark, T. (2014) Self-consistent field convergence for proteins: a comparison of full and localized-molecular-orbital schemes. *J. Mol. Model.*, **20** (3), 2159.

195 Dannenberg, J. (1997) Hydrogen bonds: a comparison of semiempirical and *ab initio* treatments. *J. Mol. Struct.*, **401** (3), 279–286.

196 Palafox, M. and Melendez, F. (1999) A comparative study of the scaled vibrational frequencies in the local anesthetics procaine, tetracaine and propoxycaine by means of semiempirical methods: AM1, PM3 and SAM1. *J. Mol. Struct.*, **459** (1–3), 239–271.

197 Morgado, C.A., McNamara, J.P., Hillier, I.H., and Burton, N.A. (2007) Density functional and semiempirical molecular orbital methods including dispersion corrections for the accurate description of noncovalent interactions involving sulfur-containing molecules. *J. Chem. Theory Comput.*, **3** (5), 1656–1664.

198 Imene, D., Leila, N., Sakina, M.F.H., Abdelaziz, B., and DjamelEddine, K. (2013) Inclusion complexes of *ortho*-anisidine and β-cyclodextrin: a quantum mechanical calculation. *C. R. Chim.*, **16** (8), 696–703.

199 Hostas, J., Rezac, J., and Hobza, P. (2013) On the performance of the semiempirical quantum mechanical PM6 and PM7 methods for noncovalent interactions. *Chem. Phys. Lett.*, **568**, 161–166.

200 Ibrahim, M.A.A. (2011) Performance assessment of semiempirical molecular orbital methods in describing halogen bonding: quantum mechanical and quantum mechanical/molecular mechanical–molecular dynamics study. *J. Chem. Inform. Model.*, **51** (10), 2549–2559.

201 Tuttle, T. and Thiel, W. (2008) OMx-D: semiempirical methods with

orthogonalization and dispersion corrections. Implementation and biochemical application. *Phys. Chem. Chem. Phys.*, **10** (16), 2159–2166.
202 Atalay, V.E. and Erdem, S.S. (2013) A comparative computational investigation on the proton and hydride transfer mechanisms of monoamine oxidase using model molecules. *Comput. Biol. Chem.*, **47**, 181–191.
203 Mlynsky, V., Banas, P., Sponer, J., van der Kamp, M.W., Mulholland, A.J., and Otyepka, M. (2014) Comparison of *ab initio*, DFT, and semiempirical QM/MM approaches for description of catalytic mechanism of hairpin ribozyme. *J. Chem. Theory Comput.*, **10** (4), 1608–1622.
204 Nam, K. (2013) Acceleration of semiempirical quantum mechanical calculations by extended Lagrangian molecular dynamics approach. *J. Chem. Theory Comput.*, **9** (8), 3393–3403.
205 Heimdal, J. and Ryde, U. (2012) Convergence of QM/MM free-energy perturbations based on molecular-mechanics or semiempirical simulations. *Phys. Chem. Chem. Phys.*, **14** (36), 12592–12604.
206 Deák, P., Snyder, L.C., Singh, R.K., and Corbett, J.W. (1987) Evaluation of semiempirical quantum-chemical methods in solid-state applications: I. Molecular-cluster calculations of defects in silicon. *Phys. Rev. B*, **36** (18), 9612–9618.
207 Bredow, T. and Jug, K. (1994) Cluster simulation of bulk properties for stoichiometric and nonstoichiometric rutile. *Chem. Phys. Lett.*, **223** (1–2), 89–94.
208 Geudtner, G., Köster, A., and Jug, K. (1998) SINDO1 study of diffusion in β-LiAl. *Ber. Bunsenges. Phys. Chem.*, **102** (6), 833–836.
209 Bredow, T., Geudtner, G., and Jug, K. (1996) Embedding procedure for cluster calculations of ionic crystals. *J. Chem. Phys.*, **105** (15), 6395–6400.
210 Gale, J. (1997) Semi-empirical methods as a tool in solid-state chemistry. *Faraday Discuss.*, **106**, 219–232.
211 Dobrotvorskii, A.M. and Evarestov, E.A. (1974) The quasi-molecular large unit cell model in the theory of deep levels in imperfect crystals: point defects in graphitic boron nitride. *Phys. Status Solidi B*, **66** (1), 83–91.
212 Evarestov, R., Petrashen, M., and Ledovskaya, E. (1975) Translational symmetry in molecular models of solids. *Phys. Status Solidi B*, **68** (1), 453–461.
213 Evarestov, R. (1975) Use of representative points of the Brillouin zone for the self-consistent calculations of solids in the large unit cell approach. *Phys. Status Solidi B*, **72** (2), 569–578.
214 Evarestov, R., Petrashen, M., and Ledovskaya, E. (1976) Comparison of large unit-cell and small periodic cluster approaches to quasimolecular calculations of band spectra of crystals. *Phys. Status Solidi B*, **76** (1), 377–383.
215 Evarestov, R.A. and Lovchikov, V.A. (1977) Large unit cell calculations of solids in the CNDO approximation. *Phys. Status Solidi B*, **79** (2), 743–751.
216 Evarestov, R. and Lovchikov, V. (1977) Program for carrying out quantum-chemical calculations for quasi-molecules and crystals with the introduction of cyclic boundary conditions. *J. Struct. Chem.*, **18** (2), 335–337.
217 Evarestov, R.A. and Ermoshkin, A.N. (1978) Large unit cell calculations of the band structure of ionic crystals using the Mulliken–Ruedenberg approximation. *Phys. Status Solidi B*, **86** (1), 47–55.
218 Evarestov, R.A. and Lovchikov, V.A. (1979) Use of the symmetry-adapted atomic orbitals in the large unit cell approach to solids. *Phys. Status Solidi B*, **93** (2), 469–482.
219 Harker, A.H. and Larkins, F.P. (1979) A large unit cell semiempirical molecular orbital approach to the properties of solids: I. General theory. *J. Phys. C Solid State Phys.*, **12** (13), 2487–2495.
220 Harker, A. and Larkins, F. (1979) Large unit-cell semi-empirical molecular-orbital approach to the properties of solids. 2. Covalent materials: diamond and silicon. *J. Phys. C Solid State Phys.*, **12** (13), 2497–2508.
221 Harker, A. and Larkins, F. (1979) Large unit-cell semi-empirical molecular-orbital approach to the properties of

221 solids. 2. Covalent materials: lithium-fluoride and potassium chloride. *J. Phys. C Solid State Phys.*, **12** (13), 2509–2517.

222 Bennett, A., B, M., and Messmer, R. (1971) Molecular orbital approach to chemisorption: I. Atomic hydrogen on graphite. *Surf. Sci.*, **24** (1), 191–208.

223 Bennett, A.J., McCarroll, B., and Messmer, R.P. (1971) Molecular orbital approach to chemisorption: II. Atomic H, C, N, O, and F on graphite. *Phys. Rev. B*, **3** (4), 1397–1406.

224 Weigel, C., Messmer, R., and Corbett, J. (1973) Semi-empirical LCAO band structure calculation on the transition of diamond into the metallic state under pressure. *Solid State Commun.*, **13** (6), 723–726.

225 Zunger, A. (1974) A molecular calculation of electronic properties of layered crystals: I. Truncated crystal approach for hexagonal boron nitride. *J. Phys. C Solid State Phys.*, **7** (1), 76–95.

226 Zunger, A. (1974) Molecular calculation of electronic properties of layered crystals: II. Periodic small cluster calculation for graphite and boron nitride. *J. Phys. C Solid State Phys.*, **7** (1), 96–106.

227 Zunger, A. (1975) Small periodic cluster calculation on point defect problems in hexagonal layered solids. *J. Chem. Phys.*, **62** (5), 1861–1868.

228 Zunger, A. (1975) Band structure, crystal conformation, and hydrogen bond potentials for solid HF. *J. Chem. Phys.*, **63** (5), 1713–1731.

229 Smith, P.V. and Szymanski, J.E. (1986) Model calculations of the LUC-CNDO modulating function. *Phys. Status Solidi B*, **134** (1), 185–194.

230 Deak, P. (2000) Choosing models for solids. *Phys. Status Solidi B*, **217** (1), 9–21.

231 Jug, K. and Bredow, T. (2004) Models for the treatment of crystalline solids and surfaces. *J. Comput. Chem.*, **25** (13), 1551–1567.

232 Ramirez, R. and Böhm, M.C. (1988) A crystal orbital approach for two- and three-dimensional solids on the basis of CNDO/INDO Hamiltonians: basis equations. *Int. J. Quantum Chem.*, **34** (1), 47–71.

233 Evarestov, R.A., Sokolov, A.R., Leko, A.V., and Verjazov, V.A. (1989) Application of embedded-cluster models in the theory of local centres in graphite. *J. Phys Condens. Matter*, **1** (37), 6611–6616.

234 Smith, P. and Szymanski, J. (1983) Improved CNDO formalism for bulk properties and surface geometries. *Vacuum*, **33** (10–12), 625–626.

235 Szymanski, J.E., Smith, P.V., and Matthew, J.A.D. (1985) LUC–LCAO–CNDO calculations on covalent semiconductors. *Philos. Mag. B*, **51** (2), 193–198.

236 Shluger, A. (1985) Application of molecular models to electronic structure calculations of defects in oxide crystals: I. Parametrization of the modified INDO method. *Theor. Chim. Acta*, **66** (6), 355–363.

237 Smith, P.V., Szymanski, J.E., and Matthew, J.A.D. (1986) A reformulation of the LUC–CNDO approach to the properties of solids: application to diamond and silicon. *Phys. Status Solidi B*, **136** (1), 261–271.

238 Craig, B.I. and Smith, P.V. (1988) The periodic LUC–MINDO method and its application to diamond. *Phys. Status Solidi B*, **146** (1), 149–160.

239 Craig, B.I. and Smith, P.V. (1988) Determination of the bulk properties of silicon by the periodic LUC–MINDO method. *Phys. Status Solidi B*, **147** (1), K31–K35.

240 Craig, B.I. and Smith, P.V. (1989) Determination of the bulk properties of cubic SiC by the periodic LUC–MINDO method. *Phys. Status Solidi B*, **154** (2), K127–K130.

241 Stefanovich, E., Shidlovskaya, E., Shluger, A., and Zakharov, M. (1990) Modification of the INDO calculation scheme and parametrization for ionic crystals. *Phys. Status Solidi B*, **160** (2), 529–540.

242 Shluger, A. and Stefanovich, E. (1990) Models of the self-trapped exciton and nearest-neighbor defect pair in SiO_2. *Phys. Rev. B*, **42** (15), 9664–9673.

243 Titov, S. (1995) Quasicluster electronic-structure calculations of point-defects in crystals. *Phys. Status Solidi B*, **190** (1), 261–270.

244 Deineka, A., Trepakov, V., Tupicyn, I., Jastrabik, L., Kapphan, S., and Syrnikov, P. (1999) Optical spectra and Li-impurity effect in energy structure of KTaO$_3$ quantum paraelectric. *Radiat. Effect Defects Solid*, **149** (1–4), 113–117.

245 Moliver, S. (1999) The electronic structure of a silicon divacancy calculated by the open shell method. *Phys. Solid State*, **41** (3), 362–368.

246 Moliver, S. (2000) An open-shell method for neutral vacancy in silicon and diamond. *Phys. Solid State*, **42** (4), 673–682.

247 Moliver, S. (2003) Electronic structure of neutral silicon-vacancy complex in diamond. *Tech. Phys.*, **48** (11), 1449–1453.

248 Stewart, J.J.P. (2008) Application of the PM6 method to modeling the solid state. *J. Mol. Model.*, **14** (6), 499–535.

249 Abdulsattar, M.A. and Al-Bayati, K.H. (2007) Corrections and parametrization of semiempirical large unit cell method for covalent semiconductors. *Phys. Rev. B*, **75** (24), 245201.

250 Radi, I.O., Abdulsattar, M.A., and Abdul-Lettif, A.M. (2007) Semiempirical LUC–INDO calculations on the effect of pressure on the electronic structure of diamond. *Phys. Status Solidi B*, **244** (4), 1304–1317.

251 Stashans, A. and Maldonado, F. (2007) A quantum mechanical study of La-doped Pb(Zr,Ti)O$_3$. *Physica B*, **392** (1–2), 237–241.

252 Zhukovskii, Y.F., Kotomin, E.A., Evarestov, R.A., and Ellis, D.E. (2007) Periodic models in quantum chemical simulations of F centers in crystalline metal oxides. *Int. J. Quantum Chem.*, **107** (14), 2956–2985.

253 Matsos, M., Terra, J., and Ellis, D.E. (2009) Semiempirical electronic structure calculation on Ca and Pb apatites. *Int. J. Quantum Chem.*, **109** (4), 849–860.

254 Abdulsattar, M.A. (2009) Size effects of semiempirical large unit cell method in comparison with nanoclusters properties of diamond-structured covalent semiconductors. *Physica E*, **41** (9), 1679–1688.

255 de Oliveira, C.A.F., da Silva, F.F., Malvestiti, I., Malta, V.R.d.S., Dutra, J.D.L., da Costa, N.B., Jr., Freire, R.O., and Alves Severino, J. (2013) Synthesis, characterization, luminescent properties and theoretical study of two new coordination polymers containing lanthanide [Ce(III) or Yb(III)] and succinate ions. *J. Mol. Struct.*, **1041**, 61–67.

256 Bystrov, V., Paramonova, E., Bdikin, I., Bystrova, A., Pullar, R., and Kholkin, A. (2013) Molecular modeling of the piezoelectric effect in the ferroelectric polymer poly(vinylidene fluoride) (PVDF). *J. Mol. Model.*, **19** (9), 3591–3602.

257 Praveen, P.L. and Ojha, D.P. (2014) Optical absorption behavior and spectral shifts of fluorinated liquid crystals in ultraviolet region: a comparative study based on DFT and semiempirical approaches. *J. Mol. Liquids*, **194**, 8–12.

258 Deák, P. (1978) Notes on the semiempirical LUC approach of solids. *Phys. Lett. A*, **83** (1), 39–42.

259 Deák, P. (1981) Cyclic cluster-models with quantum-chemical Hamiltonians. *Acta Phys. Acad. Sci. Hung.*, **50** (3), 247–262.

260 Deák, P. and Giber, J. (1982) Cyclic cluster calculation on the electronic structure of α-quartz and of its vacancy. *Phys. Lett. A*, **88** (5), 237–240.

261 Deák, P. and Giber, J. (1983) Vacancy related structure defects in SiO$_2$: cyclic cluster calculations compared with experimental results, in *Defect Complexes in Semiconductor Structures*, Lecture Notes in Physics, vol. 175 (eds J. Giber, F. Beleznay, I. Szap, and J. Laszlo), Springer, Berlin, pp. 92–101.

262 Deák, P. (1983) CNDO/S parameters of C, O, and Si for solid state calculations. *Int. J. Quantum Chem.*, **23** (4), 1165–1177.

263 Deák, P. and Snyder, L.C. (1987) Evaluation of semiempirical quantum-chemical methods in solid-state applications: 2. Cyclic-cluster calculations of silicon. *Phys. Rev. B*, **36** (18), 9619–9627.

264 Deák, P., Snyder, L.C., Lindstrøm, J., Corbett, J., Pearton, S., and Tavendale, A. (1988) The self-trapping of hydrogen in

semiconductors. *Phys. Lett. A*, **126** (7), 427–430.
265 Deák, P., Snyder, L.C., and Corbett, J. (1991) Silicon-interstitial oxygen-interstitial complex as a model of the 450°C oxygen thermal donor in silicon. *Phys. Rev. Lett.*, **66** (6), 747–749.
266 Deák, P., Snyder, L.C., and Corbett, J. (1992) Theoretical studies on the core structure of the 450°C oxygen thermal donors in silicon. *Phys. Rev. B*, **45** (20), 11612–11626.
267 Deák, P., Rosenbauer, M., Stutzmann, M., Weber, J., and Brandt, M. (1992) Siloxene: chemical quantum confinement due to oxygen in a silicon matrix. *Phys. Rev. Lett.*, **69** (17), 2531–2534.
268 Noga, J., Banacký, P., Biskupic, S., Boca, R., Pelikán, P., Svrcek, M., and Zajac, A. (1999) Approaching bulk limit for three-dimensional solids via the cyclic cluster approximation: semiempirical INDO study. *J. Comput. Chem.*, **20** (2), 253–261.
269 Zajac, A., Pelikan, P., Noga, J., Banacky, P., Biskupic, S., and Svrcek, M. (2000) Cyclic cluster approach to three-dimensional solids: quasi-relativistic INDO treatment. *J. Phys. Chem. B*, **104** (8), 1708–1714.
270 Zajac, A., Pelikan, P., Minar, J., Noga, J., Straka, M., Banacky, P., and Biskupic, S. (2000) The structure and properties of graphite monofluoride using the three-dimensional cyclic cluster approach. *J. Solid State Chem.*, **150** (2), 286–293.
271 Steveson, M., Bredow, T., and Gerson, A. (2002) MSINDO quantum chemical modelling study of the structure of aluminium-doped anatase and rutile titanium dioxide. *Phys. Chem. Chem. Phys.*, **4** (2), 358–365.
272 Janetzko, F., Evarestov, R., Bredow, T., and Jug, K. (2004) First-principles periodic and semiempirical cyclic cluster calculations for single oxygen vacancies in crystalline Al_2O_3. *Phys. Status Solidi B*, **241**, 1032–1040.
273 Jug, K. and Tikhomirov, V. (2006) Anion substitution in zinc chalcogenides. *J. Comput. Chem.*, **27** (10), 1088–1092.
274 Jug, K. and Tikhomirov, V.A. (2008) Influence of intrinsic defects on the properties of zinc oxide. *J. Comput. Chem.*, **29** (13), 2250–2254.
275 Saal, H., Binnewies, M., Schrader, M., Boerger, A., Becker, K.D., Tikhomirov, V.A., and Jug, K. (2009) Unusual optical properties of Mn-doped ZnO: the search for a new red pigment – a combined experimental and theoretical study. *Chemistry*, **15** (26), 6408–6414.
276 Maldonado, F. and Stashans, A. (2009) Simulation of pure and defective wurtzite-type ZnO. *Phys. Scr.*, **80** (6), 065601.
277 Jug, K. and Tikhomirov, V.A. (2009) Comparative studies of cation doping of ZnO with Mn, Fe, and Co. *J. Phys. Chem. A*, **113** (43), 11651–11655.
278 Gadaczek, I., Hintze, K.J., and Bredow, T. (2012) Periodic calculations of excited state properties for solids using a semiempirical approach. *Phys. Chem. Chem. Phys.*, **14** (2)
279 Litinskii, A. and Zakharov, I. (1985) Electronic structure and energy spectrum of defects on catalyst surfaces:1. Electroneutral and charged sites on a silica-gel surface: investigation in the framework of the cyclic cluster model and the MINDO-3 approximation. *Kinet. Catal.*, **26** (2), 252–259.
280 Zhanpeisov, N., Bredow, T., and Jug, K. (1996) Quantum chemical SINDO1 study of vanadium pentoxide. *Catal. Lett.*, **39** (1–2), 111–118.
281 Bredow, T. and Jug, K. (1995) Theoretical investigation of water-adsorption at rutile and anatase surfaces. *Surf. Sci.*, **327** (3), 398–408.
282 Poredda, A. and Jug, K. (1994) Calculation of adsorption energies at silicon surfaces. *Surf. Sci.*, **313** (1–2), 168–178.
283 Tikhomirov, V., Geudtner, G., and Jug, K. (1997) SINDO1 study of water dissociation at the clean and stepped MgO(100) surface. *J. Phys. Chem. B*, **101** (49), 10398–10402.
284 Ahlswede, B., Homann, T., and Jug, K. (2000) MSINDO study of the adsorption of water molecules at defective MgO (100) surfaces. *Surf. Sci.*, **445** (1), 49–59.
285 Tikhomirov, V. and Jug, K. (2000) MSINDO study of water adsorption on

the clean MgO(100) surface. *J. Phys. Chem. B*, **104** (32), 7619–7622.

286 Bredow, T. (1998) Embedded cluster study of water adsorption at Cr_2O_3(0001). *Surf. Sci.*, **401** (1), 82–95.

287 Jug, K. and Geudtner, G. (1997) Quantum chemical study of water adsorption at the NaCl(100) surface. *Surf. Sci.*, **371** (1), 95–99.

288 Ahlswede, B. and Jug, K. (1999) MSINDO study of the adsorption of water molecules at defective NaCl(100) surfaces. *Surf. Sci.*, **439** (1–3), 86–94.

289 Gerson, A. and Bredow, T. (1999) MgO (100) surface relaxation and vacancy defects: a semi-empirical quantum-chemical study. *Phys. Chem. Chem. Phys.*, **1** (20), 4889–4896.

290 Geudtner, G., Jug, K., and Köster, A. (2000) Cu adsorption on the MgO(100) surface. *Surf. Sci.*, **467** (1–3), 98–106.

291 Persson, P., Stashans, A., Bergstrom, R., and Lunell, S. (1998) Periodic INDO calculations of organic adsorbates on a TiO_2 surface. *Int. J. Quantum Chem.*, **70** (4–5), 1055–1066.

292 Patthey, L., Rensmo, H., Persson, P., Westermark, K., Vayssieres, L., Stashans, A., Petersson, A., Bruhwiler, P., Siegbahn, H., Lunell, S., and Mårtensson, N. (1999) Adsorption of bi-isonicotinic acid on rutile TiO_2(110). *J. Chem. Phys.*, **110** (12), 5913–5918.

293 Persson, P. and Lunell, S. (2000) Binding of bi-isonicotinic acid to anatase TiO_2 (101) *Sol. Energy Mater. Sol. Cells*, **63** (2), 139–148.

294 Odelius, M., Persson, P., and Lunell, S. (2003) Bi-isonicotinic acid on rutile (110): calculated molecular and electronic structure. *Surf. Sci.*, **529** (1–2), 47–58.

295 Zilberberg, I., Milov, M., and Zhidomirov, G. (1999) Forms of chemisorbed oxygen on Ag(111): NDDO/MC semiempirical investigation. *J. Struct. Chem.*, **40** (3), 350–357.

296 Milov, M., Zilberberg, I., Ruzankin, S., and Zhidomirov, G. (2000) Oxygen adsorption on the Ag(111) surface: a quantum chemical study by the NDDO/MC method. *J. Mol. Catal. A*, **158** (1), 309–312.

297 Milov, M., Zilberberg, I., Ruzankin, S., and Zhidomirov, G. (2000) Oxygen diffusion through the Ag(111) surface: a quantum chemical study by the NDDO/MC method. *J. Struct. Chem.*, **41** (2), 200–205.

298 Gerson, A.R., Jones, R., Simpson, D., Pacchioni, G., and Bredow, T. (2001) Quantum chemical modelling case studies relevant to metal oxide dissolution and catalysis. *Ionics*, **7** (4–6), 290–309.

299 Simpson, D., Bredow, T., Smart, R., and Gerson, A. (2002) Mechanisms of acidic dissolution at the MgO(100) surface. *Surf. Sci.*, **516** (1–2), 134–146.

300 Simpson, D., Bredow, T., and Gerson, A. (2003) Contrasting dissolution mechanisms of MgO and NiO: a quantum-chemical study, in *Ceramic Interfaces: Properties and Applications V*, Key Engineering Materials, vol. **253** (eds K. Uematsu and H. Yokokawa), pp. 43–62.

301 Simpson, D., Bredow, T., and Gerson, A. (2005) MSINDO study of water adsorption on NiO surfaces. *Theor. Chem. Acc.*, **114** (1–3), 242–252.

302 Simpson, D.J., Bredow, T., and Gerson, A.R. (2009) MSINDO study of acid promoted dissolution of planar MgO and NiO surfaces. *J. Comput. Chem.*, **30** (4), 581–588.

303 Heidberg, B., Bredow, T., Littmann, K., and Jug, K. (2005) Ceramic hydration with expansion. The structure and reaction of water layers on magnesium oxide: a cyclic cluster study. *Mater. Sci.*, **23** (2), 501–508.

304 Jug, K., Heidberg, B., and Bredow, T. (2007) Cyclic cluster study on the formation of brucite from periclase and water. *J. Phys. Chem. C*, **111** (35), 13103–13108.

305 Jug, K., Heidberg, B., and Bredow, T. (2007) Molecular dynamics study of water adsorption structures on the MgO (100) surface. *J. Phys. Chem. C*, **111** (18), 6846–6851.

306 Stashans, A. and Serrano, S. (2002) Quantum-chemical study of polar $SrTiO_3$(110) surface and oxygen-vacancy defects therein. *Surf. Sci.*, **497** (1–3), 285–293.

307 Duque, C. and Stashans, A. (2003) Oxygen-vacancy defects on $BaTiO_3$ (001) surface: a quantum chemical study. *Physica B*, **336** (3–4), 227–235.

308 Homann, T., Bredow, T., and Jug, K. (2002) Adsorption of small molecules on the V_2O_5(001) surface. *Surf. Sci.*, **515** (1), 205–218.

309 Homann, T., Bredow, T., and Jug, K. (2004) Adsorption of small molecules on the anatase(100) surface. *Surf. Sci.*, **555** (1–3), 135–144.

310 Jug, K., Homann, T., and Bredow, T. (2004) Reaction mechanism of the selective catalytic reduction of NO with NH_3 and O_2 to N_2 and H_2O. *J. Phys. Chem. A*, **108** (15), 2966–2971.

311 Bredow, T., Homann, T., and Jug, K. (2004) Adsorption of NO, NH_3 and H_2O on V_2O_5/TiO_2 catalysts. *Res. Chem. Intermediates*, **30** (1), 65–73.

312 Jug, K., Nair, N., and Bredow, T. (2005) Reaction of surface hydroxyl groups with VO_4H_3 on anatase surfaces. *Surf. Sci.*, **596** (1–3), 108–116.

313 Nair, N., Bredow, T., and Jug, K. (2005) Toward an understanding of the formation of vanadia–titania catalysts. *J. Phys. Chem. B*, **109** (24), 12115–12123.

314 Jug, K., Nair, N., and Bredow, T. (2005) Molecular dynamics investigation of water adsorption on rutile surfaces. *Surf. Sci.*, **590** (1), 9–20.

315 Jug, K., Nair, N., and Bredow, T. (2005) Molecular dynamics investigation of oxygen vacancy diffusion in rutile. *Phys. Chem. Chem. Phys.*, **7** (13), 2616–2621.

316 Mendive, C.B., Bredow, T., Blesa, M.A., and Bahnemann, D.W. (2006) ATR-FTIR measurements and quantum chemical calculations concerning the adsorption and photoreaction of oxalic acid on TiO_2. *Phys. Chem. Chem. Phys.*, **8** (27), 3232–3247.

317 Mendive, C.B., Bredow, T., Feldhoff, A., Blesa, M., and Bahnemann, D. (2008) Adsorption of oxalate on rutile particles in aqueous solutions: a spectroscopic, electron-microscopic and theoretical study. *Phys. Chem. Chem. Phys.*, **10** (14), 1960–1974.

318 Mendive, C.B., Bredow, T., Feldhoff, A., Blesa, M.A., and Bahnemann, D. (2009) Adsorption of oxalate on anatase (100) and rutile (110) surfaces in aqueous systems: experimental results vs. theoretical predictions. *Phys. Chem. Chem. Phys.*, **11** (11), 1794–1808.

319 Mendive, C.B., Bredow, T., Schneider, J., Blesa, M., and Bahnemann, D. (2015) Oxalic acid at the TiO_2/water interface under UV(A) illumination: surface reaction mechanisms. *J. Catal.*, **322**, 60–72.

320 Wahab, H.S., Bredow, T., and Aliwi, S.M. (2008) MSINDO quantum chemical modeling study of water molecule adsorption at nano-sized anatase TiO_2 surfaces. *Chem. Phys.*, **354** (1–3), 50–57.

321 Wahab, H.S., Bredow, T., and Aliwi, S.M. (2008) Computational investigation of water and oxygen adsorption on the anatase TiO_2(100) surface. *J. Mol. Struct.*, **868** (1–3), 101–108.

322 Wahab, H.S., Bredow, T., and Aliwi, S.M. (2008) Computational investigation of the adsorption and photocleavage of chlorobenzene on anatase TiO_2 surfaces. *Chem. Phys.*, **353** (1–3), 93–103.

323 Wahab, H.S., Bredow, T., and Aliwi, S.M. (2008) Computational modeling of the adsorption and photodegradation of 4-chlorophenol on anatase TiO_2 particles. *J. Mol. Struct.*, **863** (1–3), 84–90.

324 Wahab, H.S., Bredow, T., and Aliwi, S.M. (2009) A computational study on the adsorption and ring cleavage of *para*-chlorophenol on anatase TiO_2 surface. *Surf. Sci.*, **603** (4), 664–669.

325 Simpson, D.J., Bredow, T., Chandra, A.P., Cavallaro, G.P., and Gerson, A.R. (2011) The effect of iron and copper impurities on the wettability of sphalerite (110) surface. *J. Comput. Chem.*, **32** (9), 2022–2030.

326 Saavedra, M., Buljan, A., and Munoz, M. (2009) Theoretical study of methanethiol adsorbed on GaAs (100) surface. *J. Mol. Struct.*, **906** (1–3), 72–77.

327 Alekseev, N.I., Luchinin, V.V., and Charykov, N.A. (2013) Evaporation of carbon atoms from the open surface of silicon carbide and through graphene cells: semiempirical quantum-chemical modeling. *Russ. J. Phys. Chem. A*, **87** (11), 1830–1837.

328 Alekseev, N.I., Luchinin, V.V., and Charykov, N.A. (2013) Initial stage of the epitaxial assembly of graphene from silicon carbide and its simulation by semiempirical quantum chemical methods: carbon face. *Russ. J. Phys. Chem. A*, **87** (10), 1709–1720.

329 Margraf, J.T., Ruland, A., Sgobba, V., Guldi, D.M., and Clark, T. (2013) Theoretical and experimental insights into the surface chemistry of semiconductor quantum dots. *Langmuir*, **29** (49), 15450–15456.

330 Leitherer, S., Jaeger, C.M., Halik, M., Clark, T., and Thoss, M. (2014) Modeling charge transport in C_{60}-based self-assembled monolayers for applications in field-effect transistors. *J. Chem. Phys.*, **140**, 204702.

331 Vincent, M.A. and Hillier, I.H. (2014) Accurate prediction of adsorption energies on graphene, using a dispersion-corrected semiempirical method including solvation. *J. Chem. Inform. Model.*, **54** (8), 2255–2260.

332 Krack, M., Bredow, T., and Jug, K. (1995) Doping of saturated silicon clusters. *Chem. Phys. Lett.*, **237** (3–4), 250–255.

333 Bredow, T. and Jug, K. (1995) SINDO1 study of photocatalytic formation and reactions of OH radicals at anatase particles. *J. Phys. Chem.*, **99** (1), 285–291.

334 Jug, K. and Wichmann, D. (2000) MSINDO study of large silsesquioxanes. *J. Comput. Chem.*, **21** (16), 1549–1553.

335 Vorontsov, A.V. (2015) Cluster models of photocatalytic anatase TiO_2 nanoparticles and their computational characterization. *Catal. Today*, **252**, 168–176.

336 Viana, R.d.S., Lago Falcao, E.H., Lisboa Dutra, J.D., da Costa, N.B., Freire, R.O., and Alves, S. (2014) New experimental and theoretical approach in Eu_2O_3 microspheres: from synthesis to a study of the energy transfer. *J. Photochem. Photobiol. A*, **281**, 1–7.

337 Ren, A., Feng, J., Sun, X., Li, W., Tian, W., Sun, C., Zheng, X., and Zerner, M. (2000) Theoretical investigation of the heterofullerenes $C_{59}N$ and $C_{69}N$ and their dimers. *Int. J. Quantum Chem.*, **78** (6), 422–436.

338 Slanina, Z., Uhlik, F., Zhao, X., Adamowicz, L., and Nagase, S. (2007) Relative stabilities of C_{74} isomers. *Fullerenes Nanotubes Carbon Nanostruct.*, **15** (3), 195–205.

339 Bakowies, D., Buhl, M., and Thiel, W. (1995) Can large fullerenes be spherical. *J. Am. Chem. Soc.*, **117** (40), 10113–10118.

340 Zheng, G., Irle, S., and Morokuma, K. (2005) Performance of the DFTB method in comparison to DFT and semiempirical methods for geometries and energies of C_{20}–C_{86} fullerene isomers. *Chem. Phys. Lett.*, **412** (1–3), 210–216.

341 Zaporotskova, I.V., Lebedev, N.G., and Zaporotskov, P.A. (2006) Proton conductivity of single-walled carbon nanotubes: a semiempirical study. *Phys. Solid State*, **48** (4), 806–811.

342 Xu, L., Cai, W., and Shao, X. (2007) Performance of the semiempirical AM1, PM3, MNDO, and tight-binding methods in comparison with DFT method for the large fullerenes C_{116}–C_{120}. *J. Mol. Struct.*, **817** (1–3), 35–41.

343 Domazetis, G., Raoarun, M., and James, B.D. (2007) Semiempirical and density functional theory molecular modeling of brown coal chars with iron species and H_2, CO formation. *Energy Fuels*, **21** (5), 2531–2542.

344 Dral, P.O., Kivala, M., and Clark, T. (2013) Doped polycyclic aromatic hydrocarbons as building blocks for nanoelectronics: a theoretical study. *J. Org. Chem.*, **78** (5), 1894–1902.

345 Strauss, V., Margraf, J.T., Dolle, C., Butz, B., Nacken, T.J., Walter, J., Bauer, W., Peukert, W., Spiecker, E., Clark, T., and Guldi, D.M. (2014) Carbon nanodots: toward a comprehensive understanding of their photoluminescence. *J. Am. Chem. Soc.*, **136** (49), 17308–17316.

346 Margraf, J.T., Strauss, V., Guldi, D.M., and Clark, T. (2015) The electronic structure of amorphous carbon nanodots. *J. Phys. Chem. B*, **119** (24), 7258–7265.

347 Andreeva, N.A. and Chaban, V.V. (2015) Global minimum search via annealing: nanoscale gold clusters. *Chem. Phys. Lett.*, **622**, 75–79.

348 Monard, G., Bernal-Uruchurtu, M., van der Vaart, A., Merz, K., and Ruiz-Lopez,

M. (2005) Simulation of liquid water using semiempirical Hamiltonians and the divide and conquer approach. *J. Phys. Chem. A*, **109** (15), 3425–3432.

349 Welborn, M., Chen, J., Wang, L.P., and Van Voorhis, T. (2015) Why many semiempirical molecular orbital theories fail for liquid water and how to fix them. *J. Comput. Chem.*, **36** (12), 934–939.

350 Murdachaew, G., Mundy, C., Schenter, G., Laino, T., and Hutter, J. (2011) Semiempirical self-consistent polarization description of bulk water, the liquid–vapor interface, and cubic ice. *J. Phys. Chem. A*, **115** (23), 6046–6053.

351 Vilseck, J.Z., Sambasivarao, S.V., and Acevedo, O. (2011) Optimal scaling factors for CM1 and CM3 atomic charges in RM1-based aqueous simulations. *J. Comput. Chem.*, **32** (13), 2836–2842.

352 Wu, X., Thiel, W., Pezeshki, S., and Lin, H. (2013) Specific reaction path Hamiltonian for proton transfer in water: reparameterized semiempirical models. *J. Chem. Theory Comput.*, **9** (6), 2672–2686.

353 Acevedo, O. and Jorgensen, W.L. (2014) Quantum and molecular mechanical Monte Carlo techniques for modeling condensed-phase reactions. *Wiley Interdiscip. Rev. Comput. Mol. Sci.*, **4** (5), 422–435.

354 Wang, S., MacKay, L., and Lamoureux, G. (2014) Development of semiempirical models for proton transfer reactions in water. *J. Chem. Theory Comput.*, **10** (8), 2881–2890.

355 Marion, A., Monard, G., Ruiz-Lopez, M.F., and Ingrosso, F. (2014) Water interactions with hydrophobic groups: assessment and recalibration of semiempirical molecular orbital methods. *J. Chem. Phys.*, **141**, 034106.

356 Benighaus, T. and Thiel, W. (2008) Efficiency and accuracy of the generalized solvent boundary potential for hybrid QM/MM simulations: implementation for semiempirical Hamiltonians. *J. Chem. Theory Comput.*, **4** (10), 1600–1609.

357 Arillo Flores, O.I. and Bernal-Uruchurtu, M.I. (2010) Charge separation process in water clusters containing HCl: molecular dynamics study using semiempirical Hamiltonians. *J. Phys. Chem. A*, **114** (34), 8975–8983.

358 Lin, W. and Paesani, F. (2015) Infrared spectra of $HCl(H_2O)_n$ clusters from semiempirical Born–Oppenheimer molecular dynamics simulations. *J. Phys. Chem. A*, **119** (19), 4450–4456.

359 Chaban, V. (2015) Solvation of the fluorine containing anions and their lithium salts in propylene carbonate and dimethoxyethane. *J. Mol. Model.*, **21** (7), 172.

360 Chaban, V. (2015) Solvation of lithium ion in dimethoxyethane and propylene carbonate. *Chem. Phys. Lett.*, **631**, 1–5.

7
Tight-Binding Density Functional Theory: DFTB

Gotthard Seifert

Technische Universität Dresden, Theoretische Chemie, Bergstrasse 66b, 01062 Dresden, Germany

7.1
Introduction

The density functional theory (DFT) within the Kohn–Sham (KS) realization is a very successful methodology for calculations of a variety of physical and chemical properties of molecules and the condensed phase. The development of efficient DFT–KS computer programs and the access to computers with rapidly increasing speed and node numbers have led to the possibility of studying systems with increasing size and complexity. But the size of systems to be studied is growing faster than the acceleration of DFT computer codes on corresponding computers can compensate. Therefore, there is and will be the need of even faster but approximate methods that allow studying systems as large as possible and large systems with small computational effort or for long simulation times. Classical force fields are often applied for this purpose. However, classical force fields possess a very limited transferability from one system to another, they cannot consider quantum effects and for many elements good parameter sets are missing. Traditional semiempirical quantum chemical methods have been widely used in chemistry especially for the calculations of large biomolecules. Originally, these methods were derived from the Hartree–Fock (HF) method. The missing correlation effects in the Hartree–Fock methods were "compensated" by a large set of empirical parameters with only little physical meaning. This limits their applicability strongly, mainly to molecules containing the elements C, O, H, and N. The corresponding methods in solid-state physics are empirical tight-binding (TB) methods. Tight-binding models have often been used in electronic structure calculations, although uncertainty about the foundations of the tight-binding method has severely limited progress. On the other hand, TB calculations are comparatively straightforward and yield a simple and appealing picture of bonding in terms of orbital occupations and bond orders. For reviews of semi-empirical tight-binding methods, see Refs [1,2]. Concerning the perspective of

---

Handbook of Solid State Chemistry, First Edition. Edited by Richard Dronskowski, Shinichi Kikkawa, and Andreas Stein.
© 2017 Wiley-VCH Verlag GmbH & Co. KGaA. Published 2017 by Wiley-VCH Verlag GmbH & Co. KGaA.

TB methods see Ref. [3], and for a recent overview about semiempirical methods for large-scale computations in Chemistry see, for example, Ref. [4].

It has been already shown [5] that a "tight-binding" scheme, that is, the application of an atom-specific limited basis set representation, can also be used for a more rigorous treatment based on the KS–DFT. The original ideas have been further developed during the past 20 years resulting in the density functional tight-binding (DFTB) method. The DFTB method is computationally as efficient as traditional quantum chemical or empirical "tight-binding" methods, but it is closely related to DFT and contains only a small number of quite transferable parameters. By now, it already has been applied for studies of large molecules, for example, biomolecules, as well as condensed matter systems with a broad range of elements.

This chapter comprises the description of the DFTB methodology with a specific emphasis of its applicability to large molecular systems and condensed matter problems. The treatment of weak (van der Waals) interactions is also described. This includes a compact description of its combination with "molecular mechanics" (MM) and the discussion of boundary conditions for periodic and open systems in a consistent way. The extension to "linear scaling" methods is briefly outlined. The capability for the calculation of optical properties and time-dependent processes is described. The practical realization of DFTB is sketched and the different computational implementations are referred as well as recent methodological developments are explained.

7.2
Basic Methodology of DFTB

7.2.1
Approximate Density Functional Theory

Hohenberg and Kohn [6] proved in 1964 that the ground state energy E_{tot} is uniquely defined by the electron density $n(\vec{r})$:

$$E_{tot}[n(\vec{r})] = T[n(\vec{r})] + E_{ext} + \overbrace{\frac{1}{2}\iint \frac{n(\vec{r})n(\vec{r}')}{|\vec{r}-\vec{r}'|}d\vec{r}\,d\vec{r}'}^{E_H} + E_{nn} + E_{xc}[n(r)]. \quad (7.1)$$

Here, T is the kinetic energy of the electrons, and E_{ext}, E_H, and E_{nn} are the electron nuclei, the mean field (Hartree), and the nuclear interaction energies, respectively. $E_{xc}[n]$ denotes the functional of the exchange and correlation energy. The practical applicability of the DFT was formulated in 1965 by Kohn and Sham [7]. They introduced a fictitious system of noninteracting particles with an orbital-based ansatz for the electron density distribution:

$$n(\vec{r}) = \sum_i^{occ} \psi_i^*(\vec{r})\psi_i(\vec{r}). \quad (7.2)$$

Applying a variation principle, a set of single particle equations is obtained:[1]

$$\left[-\frac{1}{2}\nabla^2 + V_{\text{eff}}(\vec{r})\right]\psi_i(\vec{r}) = \varepsilon_i \psi_i(\vec{r}),$$

$$V_{\text{eff}} = V_{\text{ext}} + V_{\text{H}} + V_{\text{xc}}, \quad (7.3)$$

$$V_{\text{H}} = \int \frac{n(\vec{r}\,')}{|\vec{r}-\vec{r}\,'|} d\vec{r}, \quad V_{\text{ext}} = -\sum_j \frac{Z_j}{r_j}, \quad V_{\text{xc}} = \frac{\delta E_{\text{xc}}}{\delta n}.$$

The highest challenge in applying the Kohn–Sham scheme is the nature of the exchange-correlation (E_{xc}) functional. Unfortunately, the exact exchange-correlation functional is unknown and has to be approximated by fitting it in different ways. These approximations result in either local, gradient-corrected, or hybrid functionals. For a long period, the most common exchange-correlation functional was the local-density approximation (LDA), in which the xc functional depends locally on the electron density, for example, in a homogeneous electron gas. For a more detailed description of the KS–DFT, see, for example, Chapter 1 of Volume 5.

Using the Kohn–Sham Eq. (7.3), the total energy can also be written as

$$E_{\text{tot}} = \sum_i^{\text{occ}} \langle \psi_i | \overbrace{-\frac{1}{2}\nabla^2 + V_{\text{eff}}}^{\hat{h}^0} | \psi_i \rangle - \frac{1}{2}\left[\int V_{\text{eff}} n\, d\vec{r} - \int V_{\text{ext}} n\, d\vec{r}\right]$$

$$+ E_{\text{xc}}[n] - \frac{1}{2}\int V_{\text{xc}} n\, d\vec{r} + E_{\text{nn}}. \quad (7.4)$$

Since E_{tot} is variational with respect to density variations, the total energy can be calculated from an approximate electronic charge density $n_0(\vec{r})$. This density should be close to the exact density $n(\vec{r})$. The exact density could be related to the approximate density by a density fluctuation: $n = n_0 + \delta n$. Foulkes and Haydock [8] showed that the energy at the reference density $n = n_0$ up to the second order in the density fluctuations δn, may be written as

$$E_{\text{tot}} = \sum_i^{\text{occ}} \langle \psi_i | \hat{h}^0 | \psi_i \rangle + \iint \left(\frac{1}{|\vec{r}-\vec{r}\,'|} + \frac{\delta^2 E_{\text{xc}}}{\delta n \delta n'}\bigg|_{n=n_0}\right) \delta n\, \delta n'\, d\vec{r}\, d\vec{r}\,'$$

$$-\frac{1}{2}\left[\int V_{\text{eff}} n_0 - \int V_{\text{ext}} n_0\right] + E_{\text{xc}}[n_0] - \frac{1}{2}\int V_{\text{xc}} n_0 + E_{\text{nn}}. \quad (7.5)$$

---

1) The following conventions will be used: The electron density will be denoted as $n(\vec{r})$. It is considered to be a positive definite quantity, so that, for example, the term of the electron-nuclear energy has to be considered negatively as $-\int \frac{Z}{r} n$. For simplification, we omit the mathematically complete and correct notations and leave out the dependences on $\vec{r}$ or the differential $d\vec{r}$ in cases, in which the context is explicit, for example, $\int V(\vec{r}) n(\vec{r}) d\vec{r} = \int V n$.

This approximate density may be obtained by the superposition of atomic densities, so that $n_0(\vec{r}) = \sum_j n_j^0(\vec{r})$. Likewise, the effective potential can be written as a superposition of atomic contributions: $V_{\text{eff}}(\vec{r}) = \sum_j V_{\text{eff},j}^0(\vec{r})$. The exchange-correlation energy may be decomposed into atomic contributions as well,

$$E_{xc} - \frac{1}{2}\int V_{xc} n = \frac{1}{2}\int V_{xc} n - \int n^2 \frac{\delta \varepsilon_{xc}}{\delta n} = \frac{1}{2}\sum_j \sum_l \left[\int V_{xc,j}^0 n_l^0 - 2 n_j^0 n_l^0 \frac{\delta \varepsilon_{xc}}{\delta n^0}\right], \tag{7.6}$$

which leads to the following expression for E_{tot}:

$$E_{\text{tot}}(n) \approx \sum_i^{\text{occ}} \langle \psi_i | \overbrace{-\frac{1}{2}\nabla^2 + \sum_j V_{\text{eff},j}^0}^{\hat{h}^0} | \psi_i \rangle$$

$$+ \frac{1}{2}\iint \left(\frac{1}{|\vec{r}-\vec{r}'|} + \frac{\delta^2 E_{xc}}{\delta n \delta n'}\bigg|_{n_0}\right) \delta n \delta n'$$

$$- \frac{1}{2}\sum_j \sum_l \int V_{\text{eff},j}^0 n_l^0 + \frac{1}{2}\sum_j \sum_l \left[\int V_{xc,j}^0 n_l^0 - 2 n_j^0 n_l^0 \frac{\delta \varepsilon_{xc}}{\delta n^0}\right]$$

$$- \frac{1}{2}\sum_j \sum_l \int \frac{Z_j}{r_j} n_l^0 + \frac{1}{2}\sum_j \sum_{l \neq j} \frac{Z_j Z_l}{R_{lj}}. \tag{7.7}$$

For large interatomic distances, the electron–nucleus energy can be approximated by a point-charge approximation:

$$\int \frac{Z_j}{r_j} n_l^0 \approx \frac{Z_j Q_l}{R_{jl}}, \quad \text{with} \quad j \neq l \quad \text{and} \quad Q_l^0 = \int n_l^0. \tag{7.8}$$

In the limit of large distances, $Q_l \to Z_l$. Therefore, for large distances the energy terms from nuclear–nuclear repulsion and electron–nuclear energy in Eq. (7.7) compensate each other. Due to the exponential decay of $V_{\text{eff},j}^0$ and n_j^0 in the large distance range, the two-center terms containing the potential vanish:

$$\int V_{\text{eff},j}^0 n_l^0 = 0, \quad \text{for} \quad l \neq j. \tag{7.9}$$

With the same argument we can assume that

$$\int V_{xc,j}^0 n_l^0 dx = 0 \quad \wedge \quad \int n_j^0 n_l^0 \frac{\delta \varepsilon_{xc}}{\delta n^0} = 0 \quad \text{for} \quad l \neq j. \tag{7.10}$$

This means that all many-center terms ($l \neq j$) in the third line of Eq. (7.7) vanish at large interatomic distances.

Assuming $\delta n \to 0$ and considering the corresponding expressions for the total energies of the isolated atoms,

$$E_j(n_j^0) = \sum_k^{N_j} \overbrace{\left\langle \psi_k^j \left| -\frac{1}{2}\nabla^2 + V_{\text{eff},j}^0 \right| \psi_k^j \right\rangle}^{\varepsilon_k^j}$$

$$-\frac{1}{2}\int V_{\text{eff},j}^0 n_j^0 + \frac{1}{2}\int\left[\int V_{\text{xc},j}^0 n_j^0 - 2\int n_j^0 n_j^0 \frac{\delta \varepsilon_{\text{xc}}}{\delta n^0}\right]$$

$$-\frac{1}{2}\int \frac{Z_j}{r_j} n_j^0, \tag{7.11}$$

the binding energy can be simply written as

$$E_B = E_{\text{tot}} - \sum_j E_j \approx \sum_i \varepsilon_i - \sum_j \sum_{k \in j} \varepsilon_k^j, \quad \text{with} \quad \varepsilon_i = \left\langle \psi_i \left| \hat{h}^0 \right| \psi_i \right\rangle. \tag{7.12}$$

With decreasing interatomic distances the assumptions that led to the compensation of terms in Eq. (7.7) are not valid anymore. For short interatomic distances, the nuclear repulsion dominates so that the energy becomes repulsive in this limit. Therefore, one can summarize the terms in lines 3 and 4 of Eq. (7.7) into a sum of pairwise repulsive energy terms:

$$-\frac{1}{2}\sum_j\sum_{l \neq j}\int V_{\text{eff},j}^0 n_l^0 + \frac{1}{2}\sum_j\sum_{l \neq j}\left[\int V_{\text{xc},j}^0 n_l^0 - 2\int n_j^0 n_l^0 \frac{\delta \varepsilon_{\text{xc}}}{\delta n^0}\right]$$

$$-\frac{1}{2}\sum_j\sum_{l \neq j}\int \frac{Z_j}{r_j} n_l^0 + \frac{1}{2}\sum_j\sum_{l \neq j}\frac{Z_j Z_l}{R_{lj}} \Rightarrow \sum_j\sum_{l \neq j} V_{\text{rep},lj}(R_{lj}). \tag{7.13}$$

Asymptotically, they vanish in the long-range limit as already mentioned.

The second-order term in the energy expression (Eq. (7.7))

$$E_{\text{2nd}} = \iint \left(\frac{1}{|\vec{r}-\vec{r}'|} + \left.\frac{\delta^2 E_{\text{xc}}}{\delta n \delta n'}\right|_{n_0}\right) \delta n \delta n', \tag{7.14}$$

can be approximated in a very simple way replacing the density fluctuations by "charge fluctuations" $\delta n \to \Delta q_j$:

$$E_{\text{2nd}} \approx \frac{1}{2}\sum_{jl} \gamma_{jl}(|\vec{R}_l - \vec{R}_j|)\Delta q_j \Delta q_l. \tag{7.15}$$

Expressions for γ_{lj} as abbreviations for the integral kernels of Eq. (7.11) are given and discussed in Ref. [9].

Considering the approximation for E_{2nd} (Eq. (7.12)) and using a pairwise repulsive energy term (Eq. (7.13)), the total energy can be written as

$$E_{tot}(n) \approx \sum_i^{occ} \varepsilon_i + \frac{1}{2}\sum_l\sum_j \gamma_{lj}(|\vec{R}_l - \vec{R}_j|)\Delta q_l \Delta q_j$$
$$+ \sum_j\sum_{l \neq j} V_{rep.lj}(R_{lj}) + \sum_j E_j - \sum_j\sum_{k \in j} \varepsilon_k^j. \quad (7.16)$$

The repulsive energy contributions can be calculated using Eq. (7.13), but in practice these pairwise contributions are fitted to full density functional theory calculations. In principle, any high-level method could be used here. The pairwise repulsive energy can be fitted to polynomials of the following type [10]:

$$V_{rep}(R) = \begin{cases} \sum_{n=2} d_n(R_c - R)^n, & \text{for } R < R_c, \\ 0, & \text{otherwise.} \end{cases} \quad (7.17)$$

Thereby, the repulsive potential has to be fitted for each pair of elements, not for each pair of atoms j and k. It can be tabulated as the Hamilton and overlap matrix elements.

As an illustrative example for the range of the repulsive potential, the binding energy curve for the Li dimer calculated using Eq. (7.7) is compared with that from Eq. (7.12), that is, neglecting $V_{rep.lj}$ (Figure 7.1).

From this example it can be clearly seen that the repulsive potential V_{rep} is rather short ranged. Generally, R_c is on the order of ~1.5 times of the typical interatomic bonding distance. This characteristic makes V_{rep} rather well transferable from one system to another.

The Kohn–Sham-like equations corresponding to the total energy expression of Eq. (7.7) and the density $n = n_0 + \delta n$ can be written as

$$\left(-\frac{1}{2}\nabla^2 + \sum_j V_{eff,j}^0 + V_{2nd}\right)\psi_i = \left(\hat{h}^0 + \hat{h}^1\right)\psi_i = \tilde{\varepsilon}_i\psi_i. \quad (7.18)$$

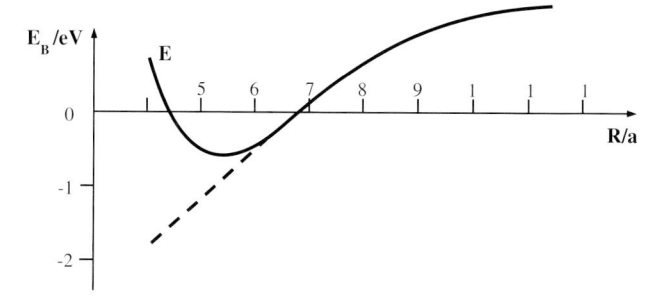

Figure 7.1 Energy (E) of Li$_2$ as a function of interatomic distance (R) calculated using Eq. (7.7) minus the atomic energies (Eq. (7.11)) and E_{2nd} approximated according Eq. (7.15). The dashed line is the energy according Eq. (7.12), that is, without the (repulsive) contributions (Eq. (7.13)). (The binding energy (E_B) is given with respect to the spin-polarized Li atoms.)

The effective potential V_{eff} consists of the sum of the atomic potentials $V^0_{\text{eff},j}$ (see Eq. (7.7)) and the corresponding contributions related to E_{2nd}.

Expressing the Kohn–Sham orbitals within an LCAO ("tight binding") approach

$$\psi_i(\vec{r}) = \sum_{\nu}^{N} C_{i\nu} \phi_\nu (\vec{r} - \vec{R}_j), \qquad (7.19)$$

leads to the corresponding secular equations

$$\sum_{\nu}^{N} C_{i\nu} (H_{\mu\nu} - \varepsilon_i S_{\mu\nu}) = 0. \qquad (7.20)$$

The matrix elements of the Hamilton operator as given in Eq. (7.18)

$$H_{\mu\nu} \equiv \langle \phi_\mu | \hat{h}^0 + \hat{h}^1 | \phi_\nu \rangle, \qquad (7.21)$$

can be written as

$$\begin{aligned} H_{\mu\nu} &= \langle \phi_\mu | \hat{h}^0 | \phi_\nu \rangle + \frac{1}{2} S_{\mu\nu} \sum_{k}^{N} (\gamma_{jk} + \gamma_{kl}) \Delta q_k \\ &= H^0_{\mu\nu} + H^1_{\mu\nu}, \quad \forall \mu \in l, \quad \nu \in j, \quad \text{with } S_{\mu\nu} = \langle \phi_\mu | \phi_\nu \rangle. \end{aligned} \qquad (7.22)$$

The orbitals in the Kohn–Sham equations (Eq. (7.18)) are represented in a minimal basis of optimized atomic orbitals φ_μ. These pseudoatomic basis functions φ_μ are obtained by solving the Kohn–Sham equations

$$\left[-\frac{1}{2} \nabla^2 + V^0_{\text{eff},j}(r_j) + \left(\frac{r_j}{r_0} \right)^{n_0} \right] \varphi_\mu = \varepsilon_\mu \varphi_\mu, \qquad (7.23)$$

for a spherical symmetric spin-unpolarized neutral atom self-consistently. The atomic orbital can be expressed in terms of Slater-type orbitals

$$\varphi_\mu(\vec{r}) = \sum_k \sum_\alpha A_{k\alpha} r^{l+k} e^{-\alpha r} Y_{lm}(\vec{r}/r), \qquad (7.24)$$

where l and m are the angular momentum and the magnetic quantum numbers associated with the orbital μ. Extensive tests have shown that five different values of α in a geometric series from ~0.8 to ~Z_j (Z_j – nuclear charge of atom j) and $k = 0,1,2,3,4$ form a sufficiently accurate basis for all elements of the periodic table [11]. For heavy elements ($Z_j \geq 50$), Eq. (7.18) can be replaced by the corresponding Kohn–Sham–Dirac equation and the relativistic effects can easily be considered within a scalar relativistic treatment [12]. In Eq. (7.23), a contraction potential $(r_j/r_0)^{n_0}$ has been added to form an efficient basis set for molecular and solid-state systems that avoids numerical instabilities due to the long-range tails of atomic orbitals of free atoms. It was first introduced by Eschrig [11,13] for LCAO–DFT band structure calculations. For solid-state structures and large molecules or clusters r_0 can be obtained unambiguously by a variational method [11]. Test calculations on molecules and solids proved that r_0 to be 1.85 of the atomic covalent radius is an optimum choice [11,13]. A value close to that was obtained by a variational procedure for close packed metal structures. Originally $n_0 = 2$ was chosen, but for band structure calculations $n_0 = 4$ seems to be more appropriate [11].

In order to restrict the basis to valence orbitals only, it is necessary to assure orthogonality of the basis functions to the atomic core orbitals of the system. The valence orbitals $|\phi_\mu\rangle$ at atom j, as solutions of Eq. (7.23), are orthogonal to the core orbitals $|\phi_c^j\rangle$ of the same atom j. However, the orthogonality to the core orbitals $|\phi_c^k\rangle$ of all other atoms of the system can be achieved by a Schmidt orthogonalization

$$|\phi_\mu\rangle = |\varphi_\mu\rangle - \sum_{k \neq j} \sum_c |\phi_c^k\rangle (\phi_c^k |\varphi_\mu\rangle), \quad \mu \in \{j\}. \quad (7.25)$$

Introducing the orthogonalization procedure into the calculation of the matrix elements $H_{\mu\nu}^0$ leads to

$$H_{\mu\nu}^0 = \langle \varphi_\mu | -\frac{1}{2}\nabla^2 + \sum_j V_{\text{eff},j}^0 |\varphi_\nu\rangle$$
$$- \sum_l \sum_{c_l} \langle \varphi_\nu | \varphi_{c_l}\rangle \varepsilon_{c_l} \langle \varphi_{c_l} | \varphi_\mu \rangle, \quad \mu, \nu \notin \{l\}. \quad (7.26)$$

In Eq. (7.26), ε_{c_l} is the energy of a core state at center l. Together with its core correction term in Eq. (7.26), the potential $V_{\text{eff},j}^0$ can be interpreted as a pseudopotential. For the atoms, on which φ_μ and φ_ν are centered, the potential does not become a pseudopotential. This means that the pseudopotential appears only in the three-center terms (φ_μ at j, φ_ν at k, $j \neq k \neq l$) and in the so-called crystal field-like terms (both φ_μ and φ_ν at j, $j \neq l$) of the matrix elements $H_{\mu\nu}^0$, whereas the "full" potential $V_{\text{eff},j}^0$ stands in all the other terms. Assuming the pseudopotentials are "weak," one may apply a two-center approximation for the Hamilton matrix elements:

$$H_{\mu\nu}^0 = \langle \phi_\mu | -\frac{1}{2}\nabla^2 + V_l + (1 - \delta_{lj})V_j |\phi_\nu\rangle, \quad \mu \in \{j\}, \quad \nu \in \{l\}. \quad (7.27)$$

The approximations formulated in Eq. (7.27), containing only two-center terms, lead to the same structure of the secular equations as in (nonorthogonal) TB schemes or the extended Hückel method (EHT), but the important advantage is that all matrix elements are calculated within density functional theory. This approximation was already introduced in Ref. [5] and its validity discussed in detail, for example, in Ref. [14].

The validity of the approximations discussed above is illustrated by the band structure of copper, as a representative example. The band structure calculated with the approximations of the matrix elements according to Eq. (7.27) agree very well with that from a "full" KS–DFT calculation, using the same basis set (Figure 7.2).

The second-order corrections ($H_{\mu\nu}^1$) to $H_{\mu\nu}$ (Eq. (7.22)) are within the approximation as atomic charge fluctuations (Δq_j) two-center terms anyway. That is, the

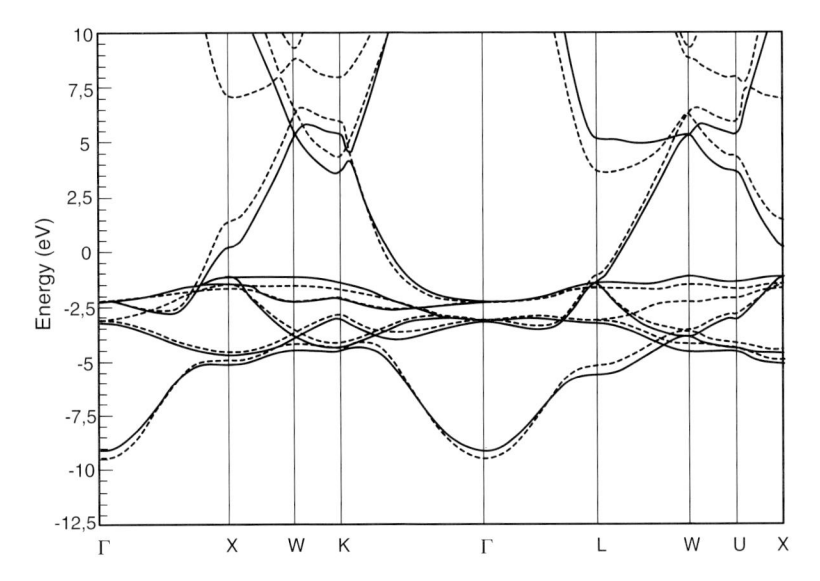

Figure 7.2 Band structure of copper (fcc, $a = 6.82\, a_B$), calculated with a 3d4s4p AO basis as already described ($r_0 = 3.5\, a_B$, $n_0 = 4$ – see Eq. (7.23)) in comparison with the results of a "full" KS–DFT calculation, using the same basis (FPLO (dashed line) [15]). The energy is given relative to E_F.

Hamilton matrices contain one- and two-center contributions only. This allows a tabulation of the corresponding integrals as functions of the distance between atomic pairs. These tables can be generated by DFT calculations for all combinations of elements of the periodic table. This means, it is not necessary to recalculate any integral during the actual calculation of a given structure. The pairwise tabulated integrals can be easily transformed to the specific coordinate system by transformations of the angular part of the basis functions (φ_μ) [16].

The Kohn–Sham equations with the Hamiltonian $H_{\mu\nu}$ (Eq. (7.19)) have to be solved self-consistently. The self-consistency has to be achieved for the charges Δq_j of the atoms in the system. Such a restricted self-consistency is called a self-consistent charge (SCC) treatment.

However, the charge of an atom in a molecule or a solid is not uniquely defined. But, in line with the ansatz for the Kohn–Sham orbitals as already discussed, the charges can be uniquely defined via the projection of the atomic orbitals to each atom. The partitioning of the overlapping contributions can simply be done with the Mulliken approach. The resulting charges then correspond to the widely used Mulliken charges $\Delta q_j = Q_j - Q_j^0$ with Q_j as the Mulliken (gross) populations of the atom j

$$Q_j = \frac{1}{2} \sum_i^{occ} n_i \sum_{\mu \in j} \sum_\nu^N \left(c_{\mu i}^* c_{\nu i} S_{\mu\nu} + c_{\nu i}^* c_{\mu i} S_{\nu\mu} \right),$$

with respect to the atomic population of the free atom Q_j^0.

Analytical expressions for the forces on the atoms can be derived as well:

$$\vec{F}_j = -\frac{\partial E}{\partial \vec{R}_j} = -\sum_i^M \sum_{\mu,\nu}^N c_{i\mu} c_{i\nu} \left[\frac{\partial H^0_{\mu\nu}}{\partial \vec{R}_j} - \left(\varepsilon_i - \frac{H^1_{\mu\nu}}{S_{\mu\nu}}\right) \frac{\partial S_{\mu\nu}}{\partial \vec{R}_j} \right] - \Delta q_j \sum_l^N \frac{\partial \gamma_{jk}}{\partial \vec{R}_j} \Delta q_l - \frac{\partial V_{\text{rep}}}{\partial \vec{R}_j}. \quad (7.28)$$

It should be noted that the computation of derivatives of the Hamilton and the overlap matrices can be performed efficiently, since the integrals are stored in tables. An efficient calculation of the forces is essential for the realization of molecular dynamics (MD) simulations with the DFTB approach. Corresponding formulations of the second derivate of the energy with respect to the nuclear coordinates $\partial^2 E/(\partial \vec{R}_j \partial \vec{R}_k)$ exist as well. These are needed, for example, for the calculation of vibrational spectra. However, a numerical differentiation of the forces turned out to be more practical.

The concept of density fluctuations (δn – see Eq. (7.12)) can be extended to a magnetization density $\mu(r) = n_\uparrow(r) - n_\downarrow(r)$ with the spin-up $n_\uparrow(r)$ and down $n_\downarrow(r)$ densities, corresponding to the electron density $n = n_\uparrow(r) + n_\downarrow(r)$. Expanding the magnetization density around $\mu=0$ up to second order, an additional term can be added to the expression of the total energy (Eq. (7.2)):

$$E_{\text{2nd},\mu} = \frac{1}{2} \iint \left. \frac{\delta^2 E_{\text{xc}}}{\delta \mu \delta \mu'} \right|_{\mu=0} \delta\mu \delta\mu' \, d^3r \, d^3r'. \quad (7.29)$$

Similar to the approximation density fluctuations by "charge fluctuations" $\delta n \to \Delta q_j$, the magnetization density can be approximated by spin populations at the atom j with the angular momentum l:

$$p_{jl} = q_{jl\uparrow} - q_{jl\downarrow}, \quad q_{jl\sigma} = \frac{1}{2} \sum_i^{\text{occ.}} n_{i\sigma} \sum_{\mu \in lj}\sum_\nu^N \left(c^*_{\mu i\sigma} c_{\nu i\sigma} S_{\mu\nu} + c^*_{\nu i\sigma} c_{\mu i\sigma} S_{\nu\mu} \right).$$

The second-order energy contribution can then be simplified to

$$E_{\text{2nd},\mu} \approx \frac{1}{2} \sum_j^N \sum_{l \in j} \sum_{l' \in j} p_{jl} p_{jl'} W_{jll'}. \quad (7.30)$$

The term $W_{jll'}$ to the spin polarization energy can be obtained from spin-polarized atomic DFT calculations applying Janak's theorem [17]:

$$W_{jll'} = \frac{1}{2} \left(\frac{\partial \varepsilon_{l\uparrow}}{\partial n_{l'\uparrow}} - \frac{\partial \varepsilon_{l\uparrow}}{\partial n_{l'\downarrow}} \right). \quad (7.31)$$

The Hamiltonian $H_{\mu\nu}$ (Eq. (7.19)) becomes correspondingly spin dependent:

$$H_{\mu\nu\sigma} = \langle \phi_\mu | \hat{h}^0 | \phi_\nu \rangle + \frac{1}{2} S_{\mu\nu} \sum_k^N \left(\gamma_{jk} + \gamma_{kl} \right) \Delta q_k$$
$$+ \delta_\sigma \delta_{\mu\nu} \sum_{l' \in j(\mu)} W_{jll'} p_{jl'}, \quad \forall \mu \in l, \quad \nu \in j'. \quad (7.32)$$

7.2.2
Weak Interactions – QM/MM Approach

The London dispersion interactions as part of the van der Waals (vdW) interaction that are not related to permanent electric mono- or multipoles cannot be properly described within DFT, using local or even semilocal functionals. Therefore, a first principle consideration of these interactions within DFT is still missing, although significant progress in an efficient approximate treatment of the vdW interactions has been achieved in recent years (see, for example, Ref. [18] and references therein). Also DFTB – as already described – does not take into account dispersion interactions. However, these interactions may be included into the DFTB approach *a posteriori* [19,20], that is, the dispersion energy (E_{disp}) is calculated in a semiempirical approach independently and added to the total energy. Any double counting of contributions is explicitly excluded, since the DFTB total energy does not contain any dispersion energy contribution. This is in fact an advantage over the *a posteriori* consideration of dispersion forces within LDA or GGA functionals, where spurious terms of the van der Waals interaction may be contained. Practically, the van der Waals interactions are considered as a sum of pairwise atomic interaction energies:

$$E_{\text{disp}} = \frac{1}{2} \sum_{i,j=1}^{N} U_{ij}(R). \tag{7.33}$$

For the U_{ij} Lennard-Jones potentials of the form

$$U_{ij}(R) = d_{ij} \left[-2 \left(\frac{R_{ij}}{R} \right)^6 + \left(\frac{R_{ij}}{R} \right)^{12} \right], \tag{7.34}$$

can be used, where d_{ij} and R_{ij} are the well depth and the van der Waals distance for the force between atoms i and j. This description of van der Waals interactions are considered in the DFTB and SCC–DFTB implementations in the deMon computer code [21], where the universal force field (UFF) parameterization [22] is used. The parameters d_{ij} and R_{ij} are available for nearly the entire periodic table [22]. Eq. (7.34) imposes, however, the problem that the R^{-12} term is repulsive for short-range distances, in which the energy is calculated by the quantum mechanical DFT or DFTB approach. Therefore, the function has to be modified, so that no repulsive terms do occur. Thus, in Ref. [19], the Lennard-Jones potential of Eq. (7.34) is used in the attractive, long-range part, and the potential is corrected for distances smaller than the node at $R_0 = R_{ij}/\sqrt[6]{2}$. Here, a polynomial of the form

$$U_{ij}^{\text{short-range}}(R) = U_0 - U_1 R^n - U_2 R^{2n}, \tag{7.35}$$

is used. The three parameters U_0, U_1, and U_2 are determined under the condition that the derivatives at R_0 match with those of equation. With the choice of

$n = 5$, they result in

$$U_0 = \frac{396}{25} d_{ij}, \quad U_1 = 2^{5/6} \frac{672\, d_{ij}}{25\, r_{ij}^5}, \quad \text{and } U_2 = -2^{2/3} \frac{552\, d_{ij}}{25\, r_{ij}^{10}}, \tag{7.36}$$

where $d_{ij} > 0$. Combining Eqs. (7.34)–(7.36), we can write the dispersion correction as

$$U_{ij}(R) = \begin{cases} d_{ij}\left[-2\left(\dfrac{R_{ij}}{R}\right)^6 + \left(\dfrac{R_{ij}}{R}\right)^{12}\right], & \text{if } R \geq R_{ij}/\sqrt[6]{2}, \\ \dfrac{396}{25} d_{ij} - 2^{5/6}\dfrac{672\, d_{ij}}{25\, R_{ij}^5} R^5 + 2^{2/3}\dfrac{552\, d_{ij}}{25\, R_{ij}^{10}} R^{10}, & \text{if } R < R_{ij}/\sqrt[6]{2}. \end{cases} \tag{7.37}$$

This treatment is similar to Grimme's approach [23] for the *a posteriori* consideration of the van der Waals interaction in DFT calculations.

The consideration of van der Waals interactions, as already described, allows the inclusion of further nonquantum mechanically calculated interactions in the spirit of a QM/MM methodology. In such a treatment, the total energy is written as:

$$E_{\text{tot}} = E_{\text{QM}} + E_{\text{MM}} + E_{\text{QM-MM}}, \tag{7.38}$$

where E_{QM} is the energy of the quantum mechanically described part of the system, and E_{MM} is the energy of the classically described part. For the QM part, the DFTB approach of Eq. (7.16) can be used, for the classical molecular mechanics part we can choose an empirical force field, for example, the dispersion energy (E_{disp}), as already described. The term $E_{\text{QM-MM}}$ describes the energy contribution resulting from the linkage of the QM to the MM region. The QM/MM approach is very useful for large systems in which a quantum mechanically treated small region (QM) of prominent interest can well be separated from the rest of the large system, the MM region. This is often the case in biomolecular systems. The DFTB method allows describing large parts of a QM–MM system quantum mechanically. Implementations are, for example, available in CHARMM [24] and AMBER [25,26]. Also the third-order extension of DFTB (see further) has been combined with a QM/MM approach [27]. For a comparison with the capabilities and shortcomings of recent force field-based methods, as the so-called "reactive force fields" (ReaXXFF), see, for example, Ref. [28].

The QM/MM treatment allows the consideration of the environment in a simple way because such calculations are much faster than a full quantum mechanical description. It allows formally studying systems that in many cases can even not be studied at all in a full quantum mechanical treatment. However, the QM/MM approach has also severe drawbacks. Bond breaking and bond formation cannot be properly described within a classical approach. Also, the separation into the MM and QM region and especially the linkage by $E_{\text{QM-MM}}$ is only possible in unique way in cases, where the QM and the MM part are only weakly "linked," for example, a molecule (QM) in a solvent (MM).

7.3
Extended Systems

7.3.1
Periodic Boundary Conditions – PBC

For the study of solid-state structures, the periodicity can be exploited in the same way as for other KS–DFT-based methods using the Bloch state representation for regular crystalline lattices:

$$\psi_{\vec{k}}(\vec{r}+\vec{R}) = \exp(i\vec{k}\vec{R})\psi_{\vec{k}}(\vec{r}). \tag{7.39}$$

The LCAO ansatz (Eq. (7.19)) for the Bloch states

$$\psi_{\vec{k}} = \sum_{\mu} C_{\mu}(\vec{k})\varphi_{\mu\vec{k}} \quad \varphi_{\mu\vec{k}} = \sum_{\vec{R}} \exp(i\vec{k}\vec{R})\Phi_{\mu}(\vec{r}-\vec{R}), \tag{7.40}$$

gives a corresponding eigenvalue problem

$$\sum_{\mu} \left(h_{\mu\nu}(\vec{k}) - \varepsilon_{\vec{k}} S_{\mu\nu}(\vec{k}) \right) C_{\mu}(\vec{k}) = 0, \tag{7.41}$$

with the matrix elements

$$h_{\mu\nu}(\vec{k}) = \left\langle \varphi^*_{\mu k} \middle| \hat{h} \middle| \varphi_{\mu k} \right\rangle \quad S_{\mu\nu}(\vec{k}) = \left\langle \varphi^*_{\mu k} \middle| \varphi_{\mu k} \right\rangle.$$

For the calculation of the electronic properties of periodic systems – for example, total energy forces and so on – the Brillouin zone has to be properly sampled by summations over the reciprocal lattice vectors $\vec{k}$.

This summation approximates then the accurate integration over the Brillouin zone. For large unit cells ("super cells"), the consideration of a single point ($\vec{k} = 0$) is often sufficient and simplifies the complex eigenvalue problem (Eq. (7.41)) to a real one, as for finite systems.

In some cases, for example, helical nanotubes, standard PBC would lead to too large super cells. In such cases, specific symmetry-adapted boundary conditions [29] are more appropriate than standard PBC. Such helical symmetry-adapted boundary conditions have been implemented to DFTB [30].

7.3.2
Open Boundary Conditions

Another group of applications also require approaches beyond the standard PBC. This holds especially for the quantum mechanical description of the charge transport through nanoscopic devices with an applied external potential, but also for a fully atomistic description of electrochemical processes. It requires open boundary conditions. These can be treated within the so-called nonequilibrium Green's function theory (NEGF) developed by Keldysh [31]. It has been first implemented into the DFTB formalism by di Carlo et al. [32] – gDFTB. In this

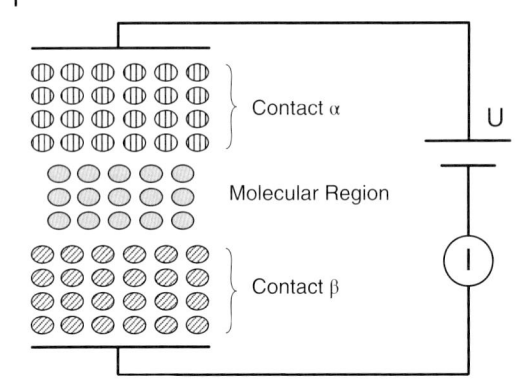

Figure 7.3 Schematic diagram of a molecular system embedded into two semi-infinite regions.

method, the system is divided into two semi-infinite lead or contact regions (Contact α, Contact β) and in between the finite molecular region (Molecule). At the contacts, an external bias can be applied – see Figure 7.3.

The (retarded) Green's function $G(E)$ of such system is defined as

$$G(E) = \lim_{\eta \to 0} [(E + i\eta)S - H]^{-1}. \quad (7.42)$$

An infinitesimal imaginary part $i\eta(\eta > 0)$ is added to eliminate the advanced Green's function as a nonphysical solution. H and S stand for the Hamilton and Overlap matrices within an LCAO representation of G. They can be separated into a block matrix form:

$$H = \begin{pmatrix} H_\pm & T_{\pm M} & 0 \\ T_{M\pm} & H_M & T_{M''} \\ 0 & T_{''M} & H_{''} \end{pmatrix}, \quad S = \begin{pmatrix} S_\pm & S_{\pm M} & 0 \\ S_{M\pm} & S_M & S_{M''} \\ 0 & S_{''M} & S_{''} \end{pmatrix}. \quad (7.43)$$

The transmission coefficient T for a given energy E can be calculated from the Green's function of the molecule (G_M) and the spectral functions of the self-energy operators ($\Gamma_{\alpha,\beta}$):

$$T_{\alpha\beta}(E) = Tr(\Gamma_\alpha G_M \Gamma_\beta G_M). \quad (7.44)$$

The self-energy terms $\Gamma_{\alpha,\beta}$ are determined by the surface Green's function of the contact region and describe the coupling to the contacts. The current I between the contacts α and β is related to the Green's function and the contact scattering functions:

$$\sum_\alpha^{in} = f_\alpha(E)\Gamma_\alpha, \quad \sum_\alpha^{out} = [1 - f_\alpha(E)]\Gamma_\alpha \quad (f_\alpha : \text{Fermi function for contact } \alpha),$$

$$I = \frac{e}{h}\int dE\, Tr\left(\sum_\alpha^{in} G_M - \sum_\alpha^{out} G_M\right). \quad (7.45)$$

The scattering functions can consider elastic as well as inelastic processes, as electron–electron or electron–phonon interactions. Even forces can be considered and a practical formulation has been developed for DFTB [33]. Accordingly, the gDFTB method is not only able to describe the electronic current through devices, but it can also handle in combination with molecular dynamics, for example, the formation of a Helmholtz double layer at electrode/electrolyte interfaces and the transport processes in electrochemical setups, as for example, batteries. For a detailed description of gDFTB see, for example, Ref. [34].

7.3.3
Order N Methods

DFTB-based methods are computationally very efficient in comparison with "standard" DFT methods. However, the "standard" implementation suffers from the $O(N^3)$ scaling behavior of the solution of the eigenvalue problem by diagonalization. Unfortunately, the breakthrough has not come with the use of parallel computers with optimized linear algebra packages (Scalapack, PBLAS, etc.) [35,36]. A sparse self-consistent field algorithm and its parallel implementation is described in Ref. [37].

More than a decade ago, linear scaling $O(N)$ methods were already adapted and implemented into DFTB program packages [38]. The divide-and-conquer (DAC) [38–40] is certainly an effective approach for the study of large biomolecular systems, but the method can hardly be generalized to solid-state or condensed systems in general. Also, other methods [41,42] have not yet brought the expected breakthrough. The exploitation of the sparsity of the Hamilton and the Overlap matrices are more useful for storage saving than for an $O(N)$ scaling in the computational time [43]. The concept of Car–Parrinello was also implemented into DFTB. It allows MD simulations in the picosecond range for systems with more than $\sim 10^4$ atoms [44], but can also not be viewed as an $O(N)$ method. That is, up to now the linear scaling behavior and accordingly the study of systems with $\sim 10^5$ or even more atoms is still an unsolved challenge. Also for the linear response time-dependent DFT (TD-DFTB), an $O(N)$ implementation was realized [45].

7.4
Time-Dependent DFTB (TD-DFTB) – Excited States

The DFTB approach can be extended within the spirit of the time-dependent density functional theory (TD-DFT) [46]. The linear response representation of the TD-DFT is a very successful and efficient approach for the calculation of electronic excitations. The corresponding DFTB approach has been put forward by Niehaus *et al.* [47].

In the framework of the linear response DFT (TD-DFRT), the excitation energies ω_I can be obtained by solving the following eigenvalue problem:

$$\sum_{i,j,\sigma}\left[\omega_{ij}^2\delta_{ik}\delta_{jl}\delta_{\sigma\tau} + 2\sqrt{\omega_{ij}}K_{ij\sigma,kl\tau}\sqrt{\omega_{kl}}\right]F_{ij\sigma}^I = \omega_I^2 F_{kl\tau}^I. \tag{7.46}$$

$K_{ij\sigma,kl\tau}$ are the coupling matrix elements:

$$K_{ij\sigma,kl\tau} = \iint \psi_i(\vec{r})\psi_j(\vec{r})\left(\frac{1}{|\vec{r}-\vec{r}\,'|} + \frac{\delta^2 E_{xc}}{\delta n_\sigma(\vec{r})\delta n_\tau(\vec{r}\,')}\right)\psi_k(\vec{r}\,')\psi_l(\vec{r}\,')\,\mathrm{d}\vec{r}\,\mathrm{d}\vec{r}\,'. \tag{7.47}$$

The $\psi_i(\vec{r})$ orbitals, σ and τ stand for the spin indices and $\omega_{ij} = \varepsilon_i - \varepsilon_j$ are the orbital energy differences between occupied (index i, k) and unoccupied orbitals (index j, l). The formal structure of the coupling matrix elements $K_{ij\sigma,kl\tau}$ in Eq. (7.47) looks very similar to the second-order energy (Eq. (7.14)). Therefore, a similar approach as for approximating the density fluctuation is used by decomposing the transition density between different orbitals in two atom-centered contributions: $p_{ij}(\vec{r}) = \psi_i(\vec{r})\psi_j(\vec{r}) = \sum_J p_{ij}^J(\vec{r})$. The atomic-centered contributions are similarly expanded in a multipole expansion truncated after the monopole term: $p_{ij}^J(\vec{r}) \approx q_{ij}^J F_J(\vec{r})$, with the Mulliken transition charges at atom J of

$$q_{ij}^J = \frac{1}{2}\sum_{\mu \in J}\sum_\nu \left(c_i^\mu c_j^\nu S_{\mu\nu} + c_i^\nu c_j^\mu S_{\nu\mu}\right). \tag{7.48}$$

To consider the spin dependence, a similar approximation as for the spin-dependent DFTB method is used – see Eq. (7.30). Then, the coupling matrix elements in Eq. (7.47) is approximated as

$$K_{ij\sigma,kl\tau} = \sum_{I,J} q_{ij}^I q_{kl}^J \left[\tilde{\gamma}_{IJ} + (\delta_{\sigma\tau} - 1)W_{IJ}\right]. \tag{7.49}$$

In Eq. (7.49) the $\tilde{\gamma}_{IJ}$ functional is nearly equivalent to the γ_{IJ} functional in Eq. (7.15). Similarly, the $\tilde{\gamma}_{IJ}$ functional is approximated by a Hubbard-like term, which in this case is calculated as the second derivative of the total energy with respect to the occupation number of the highest occupied atomic orbital of an atom with charge Δq_I. Niehaus et al. [47] found that the dependence of the Hubbard-like term on the charge is small so that the approximation $\tilde{\gamma}_{IJ} = \gamma_{IJ}$ is justified.

The magnetization term W_{IJ} in Eq. (7.49) is only considered for the onsite terms ($I = J$), since it is very short range. It contains the second functional derivative of E_{xc} with respect to the magnetization as defined in the context of Eq. (7.49) and can be approximated according to Eq. (7.31).

Finally, with the transition-dipole moment $\langle\psi_i|\vec{r}|\psi_i\rangle = \sum_I \vec{R}_I q_{ij}^I$, the oscillator strength of the respective transition ω_I can be calculated as

$$f^I = \frac{2}{3}\omega_I \sum_{k=x,y,z}\left|\sum_{i,j}\langle\psi_i|\vec{r}|\psi_i\rangle\sqrt{\frac{\omega_{ij}}{\omega_I}}\left(F_{ij\uparrow}^I + F_{ij\downarrow}^I\right)\right|^2. \tag{7.50}$$

7.5
Computational Implementations and Extensions

The implementation of the DFTB methodology, as described above, is summarized in Figure 7.4. The basis functions are represented in a minimal basis of optimized orthogonal atomic orbitals. These pseudoatomic basis functions are obtained by solving the Kohn–Sham equations for a spherical symmetric spin-unpolarized neutral atom self-consistently. The second-order terms are represented in a multipole expansion considering only the monopolar terms (γ, W). The pairwise repulsive energy terms (V_{rep} – Eq. (7.17)) can be obtained from "full" DFT calculations in appropriate reference systems. Since the whole Hamilton and overlap matrices as well as the V_{rep} contain only one- and two-center contributions, they can be calculated and tabulated in advance. A collection of DFTB integrals and parameters covering a large part of the periodic table has been published recently [48].

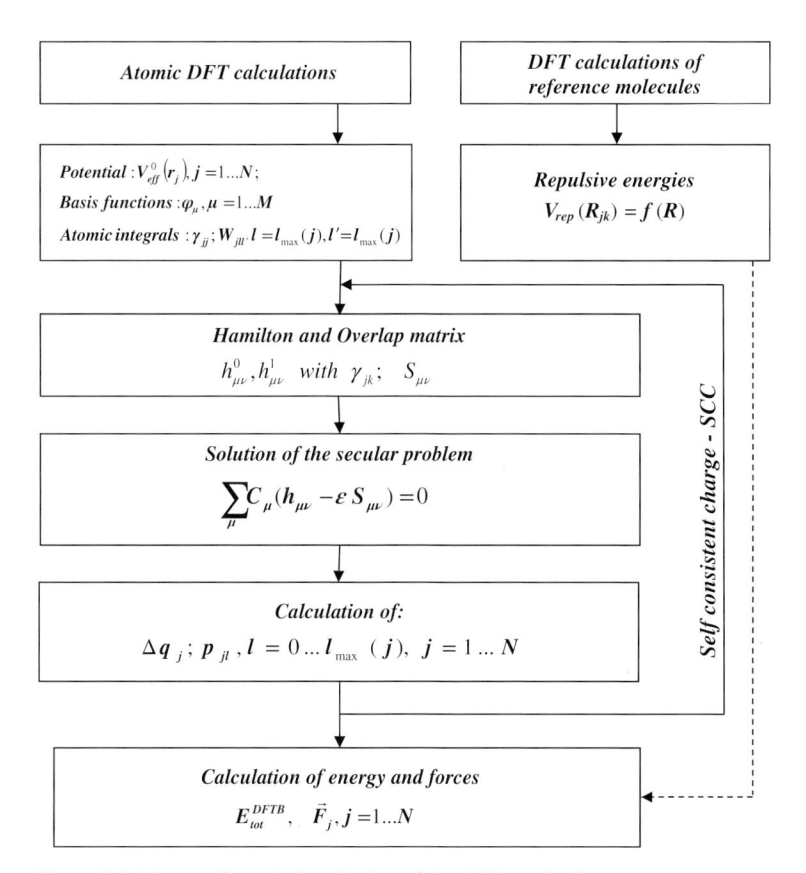

Figure 7.4 Scheme of practical realization of the DFTB method.

It has been realized in numerous computer codes by now. Among them are deMon [21], Gaussian [49], and ADF [50], which are originally standard DFT codes. A DFTB implementation is also available with the Accelrys Materials Studio package and as QM/MM and QM implementations in CHARMM and AMBER force fields. Additionally, there are a number of stand-alone DFTB implementations, for example, DFTB+ [43], dylax, or hotbit [27]. Many of them are free to use, not all are documented sufficiently.

Besides the different implementations, many extensions to the original DFTB approach have been put forward. The treatment of van der Waals forces and the linear-response formalism have been discussed above.

As DFT within the LDA or GGA, also DFTB suffers from a too strong stabilization of delocalized states. These effects have been discussed quite in detail for the DFTB method recently [51]. Extensions to correct this behavior have also been developed for DFTB. These are an efficient LDA+U approach [52], a combination of DFTB with configuration interaction [53], or a so called self-interaction correction (SIC) scheme [54].

The expansion of the energy (Eqs. (7.7) and (7.14)) can be extended to a third-order term [55–57]

$$E_{3rd} = \frac{1}{6} \iint \left(\frac{\delta^3 E}{\delta n \delta n' \delta n''} \bigg|_{n_0} \right) \delta n \delta n' \delta n''.$$

However, this extension introduces more parameters [58]. The third-order extension has been also implemented in to the (linear response) time-dependent DFTB method (TDDFTB) [59]. This extension was realized within the monopole approximation for δn ($\delta n \to \Delta q_j$). Probably the replacement of the monopole approximation by consideration of the spatial variation of δn will be more promising rather than going to higher orders in the Taylor expansion of the energy.

DFTB allows to calculate the electronic structure (electronic density distribution, density of states etc.), geometric structure and the energetics of systems with up to several 10^4 atoms. The efficient calculation of the atomic forces is also the basis for molecular-dynamics (MD) simulations of such large systems and time scales up to the nanosecond range within Born–Oppenheimer approximation. A Car–Parrinello approach has been implemented as well. An extended Lagrangian density functional tight-binding molecular dynamics for molecules and solids has been also developed [60].

Furthermore, the DFTB methodology has been extended to a time-dependent and nonadiabatic approach [46,61,62], considering also a correct long-range behavior [63].

Within DFTB, the conceptual and numerical difficult problem of screening the Coulombic repulsion between the nuclei is treated in a simple but efficient way by the replacement of corresponding compensating terms in Eq. (7.13) by a pairwise (repulsive) potential. However, the fitting procedure for the repulsive energy given in Eq. (7.17) is still a critical point. It is vital for receiving meaningful results that this procedure is done with greatest care. Since they depend on the fitting procedure, the tables are not generally transferable from one class of systems to another. This is the reason why these parameters are not available publicly for all elements of the periodic table. Attempts to automatize this procedure [64] have to be viewed with caution. Concerning other procedures to obtain repulsive potentials, see, for example, Refs [65,66]. Promising activities toward a nonempirical calculation of the repulsive energy terms in terms of Eq. (7.13) have been published recently [67].

7.6 Conclusions

In this chapter, the foundations of the DFTB method have been reviewed in detail. Based on density functional theory, the formalism introduces several approximations. The electronic charge density is written as a superposition of atomic densities. Likewise, the effective potential can be written as a superposition of atomic contributions, so that the exchange-correlation energy is also decomposed into atomic contributions.

For large interatomic distances, the integral describing the electron–nucleus interaction can be approximated by a point charge approximation. Correspondingly, the atomic population converges to the nuclear charge at large interatomic distances, and the energy terms from nuclear–nuclear repulsion and electron–nuclear energy compensate each other for large distances. Additionally, in the large-distance range, the two-center terms containing the potential vanish. In this limit – the so-called Wigner–Seitz limit – the energy can be approximated by the sum of the energies of the occupied orbitals. For small interatomic distances, these arguments are not valid anymore. Therefore, the compensating terms are calculated using the atomic densities and potentials and then summarized together with the nuclear repulsion energy terms into a sum of pairwise repulsive energy terms. For short interatomic distances, the nuclear repulsion dominates. In this limit, the pairwise energy terms are repulsive and they asymptotically approach zero in the long range.

The Kohn–Sham-like equations can be set up, in which the effective potential consists of the sum of the atomic potentials and the corresponding contributions related to the second-order terms. The corresponding Kohn–Sham orbitals are written with an LCAO approach. They are represented in a minimal basis of optimized orthogonal atomic orbitals. These pseudoatomic basis functions are obtained by solving the Kohn–Sham equations for a spherical symmetric spin-unpolarized neutral atom self-consistently. The approximations formulated above containing two-center terms only lead to the same structure of the secular equations as in other nonorthogonal tight-binding schemes. It is important to stress that all matrix elements are calculated within density functional theory.

As an extension of superposition of atomic densities, the total energy can be expanded in terms of density "fluctuations" around this reference density. This expansion is written in terms of a multipole expansion up to second order considering only the monopolar terms. The second-order expansion of the energy leads to Kohn–Sham equations that have to be solved self-consistently with respect to atomic charges – SCC–DFTB. Furthermore, forces can be calculated efficiently that allows large-scale molecular dynamics simulations. Weak interatomic interactions can be included by a dispersion energy correction through a Lennard-Jones-like potential describing the van der Waals interactions. Moreover, the combination of quantum mechanics and molecular mechanics (QM-MM) enables the consideration of the molecular environment in an efficient way. Extended systems can be treated using periodic but also open boundary conditions. For very large systems, $O(N)$ like methods have been briefly sketched. Finally, the description of excited states within time-dependent density functional response theory has been discussed, and some practical aspects have been given.

Further information on the DFTB method can be found in a number of review articles (see, for example, Refs [68–70]).

A list of implementations can be found, for example, at the Web site www.dftb.org. Information is also given about how to obtain sets of parameter files for performing the calculations. On the Web site www.dftb-plus.info, the DFTB+ code is presented in a similar way. DFTB+ is a stand-alone DFTB implementation developed at the Bremen Center for Computational Materials Science (BCCMS). As an example for the implementation into other software packages, see https://www.scm.com/DFTB/.

References

1 Bullett, D., Haydock, R., Heine, V., and Kelly, M.J. (1980) *Advances in Research and Applications* (eds H. Ehrenreich and D. Turnbull), Academic Press, New York.

2 Harrison, W.A. (1980) *Electronic Structure and the Properties of Solids*, Freeman, San Francisco, CA.

3 Horsfield, A. (2012) Where does tight binding go from here? *Phys. Status Solidi B*, **249** (2), 231–236.

4 Akimov, A.V. and Prezhdo, O.V. (2015) Large-scale computations in chemistry: a bird's eye view of a vibrant field. *Chem. Rev.*, **115** (12), 5797–5890.

5 Seifert, G., Eschrig, H., and Bieger, W. (1986) An approximation variant of LCAO-X-alpha methods. *Z. Phys. Chem.*, **267** (3), 529–539.

6 Hohenberg, P. and Kohn, W. (1964) Inhomogeneous electron gas. *Phys. Rev.*, **136**, B864.

7 Kohn, W. and Sham, L.J. (1965) Self-consistent equations including exchange and correlation effects. *Phys. Rev.*, **140**, A1133.

8 Foulkes, W.M.C. and Haydock, R. (1989) Tight-binding models and density-functional theory. *Phys. Rev. B*, **39** (17), 12520–12536.

9 Elstner, M., Porezag, D., Seifert, G., Frauenheim, T., and Suhai, S. (1999) Self-consistent charge–density functional tight-binding method for simulations of biological molecules. *Multiscale Model. Mater.*, **538**, 541–546.

10 Porezag, D., Frauenheim, T., Köhler, T., Seifert, G., and Kaschner, R. (1995) Construction of tight-binding-like potentials on the basis of density-functional theory – application to carbon. *Phys. Rev. B*, **51** (19), 12947–12957.

11 Eschrig, H. (1988) *Optimized LCAO Method and the Electronic Structure of*

Extended Systems, Akademie-Verlag, Berlin.

12 Heera, V., Seifert, G., and Ziesche, P. (1984) A semi-relativistic variant of the scattered-wave X-alpha method. *J. Phys. B At. Mol. Opt. Phys.*, **17** (4), 519–530.

13 Eschrig, H. and Bergert, I. (1978) An optimized LCAO version for band structure calculations application to copper. *Phys. Status Solidi B*, **90** (2), 621–628.

14 Frauenheim, T., Seifert, G., Elstner, M., Hajnal, Z., Jungnickel, G., Porezag, D., Suhai, S., and Scholz, R. (2000) A self-consistent charge density- functional based tight-binding method for predictive materials simulations in physics, chemistry, and biology. *Phys. Status Solidi B*, **217** (1), 41–62.

15 Koepernik, K. and Eschrig, H. (1999) Full-potential nonorthogonal local-orbital minimum-basis band-structure scheme. *Phys. Rev. B*, **59** (3), 1743–1757.

16 Slater, J.C. and Koster, G.F. (1954) Simplified LCAO method for the periodic potential problem. *Phys. Rev.*, **94**, 1498.

17 Janak, J.F. (1978) Proof that $\partial E \partial ni = \varepsilon$ in density-functional theory. *Phys. Rev. B*, **18** (12), 7165–7168.

18 Maurer, R.J., Ruiz, V.G., and Tkatchenko, A. (2015) Many-body dispersion effects in the binding of adsorbates on metal surfaces. *J. Chem. Phys.*, **143** (10), 102808.

19 Zhechkov, L., Heine, T., Patchkovskii, S., Seifert, G., and Duarte, H.A. (2005) An efficient *a posteriori* treatment for dispersion interaction in density-functional-based tight binding. *J Chem. Theory Comput.*, **1** (5), 841–847.

20 Elstner, M., Hobza, P., Frauenheim, T., Suhai, S., and Kaxiras, E. (2001) Hydrogen bonding and stacking interactions of nucleic acid base pairs: a density-functional-theory based treatment. *J. Chem. Phys.*, **114**, 5149.

21 Köster, A.M., Flores, R., Geudtner, G., Goursot, A., Heine, T., Patchkovskii, S., Reveles, J.U., Vela, A., and Salahub, D. (2003) deMon, NRC, Ottawa, Canada.

22 Rappé, A.K., Casewit, C.J., Colwell, K.S., Goddard, W.A., III, and Skiff, W.M. (1992) UFF, a full periodic table force field for molecular mechanics and molecular dynamics simulations. *J. Am. Chem. Soc.*, **114**, 10024–10035.

23 Grimme, S. (2004) Accurate description of van der Waals complexes by density functional theory including empirical corrections. *J. Comput. Chem.*, **25** (12), 1463–1473.

24 Cui, Q., Elstner, M., Kaxiras, E., Frauenheim, T., and Karplus, M. (2001) A QM/MM implementation of the self-consistent charge density functional tight binding (SCC-DFTB) method. *J. Phys. Chem. B*, **105** (2), 569–585.

25 Seabra, G.M., Walker, R.C., and Roitberg, A.E. (2006) Implementation of SCC-DFTB in AMBER. *Abstr. Pap. Am. Chem. Soc.*, **231**, 17-COMP.

26 Seabra, G.M., Walker, R.C., Elstner, M., Case, D.A., and Roitberg, A.E. (2007) Implementation of the SCC-DFTB method for hybrid QM/MM simulations within the amber molecular dynamics package. *J. Phys. Chem. A*, **111** (26), 5655–5664.

27 Kubar, T., Welke, K., and Groenhof, G. (2015) New QM/MM implementation of the DFTB3 method in the gromacs package. *J. Comput. Chem.*, **36** (26), 1978–1989.

28 Qi, T.T., Bauschlicher, C.W., Lawson, J.W., Desai, T.G., and Reed, E.J. (2013) Comparison of ReaxFF, DFTB, and DFT for phenolic pyrolysis. 1. molecular dynamics simulations. *J. Phys. Chem. A*, **117** (44), 11115–11125.

29 Milošević, I., Vuković, T., Damnjanović, M., and Nikolić, B. (2000) Symmetry based properties of the transition metal dichalcogenide nanotubes. *Eur. Phys. J. B*, **17** (4), 707–712.

30 Zhang, D.B., James, R.D., and Dumitrică, T. (2009) Electromechanical characterization of carbon nanotubes in torsion via symmetry adapted tight-binding objective molecular dynamics. *Phys. Rev. B*, **80** (11), 115418.

31 Keldysh, L.V. (1965) Diagram technique for nonequilibrium processes. *Sov. Phys. JETP*, **20** (4), 1018.

32 Di Carlo, A., Gheorghe, M., Lugli, P., Sternberg, M., Seifert, G., and Frauenheim, T. (2002) Theoretical tools for transport in molecular nanostructures. *Physica B Condens. Matter*, **314** (1–4), 86–90.

33 Di Carlo, A., Pecchia, A., Latessa, L., Frauenheim, T., and Seifert, G. (2005) *Tight-Binding DFT for Molecular Electronics*, Springer, Berlin, pp. 153–184.

34 Pecchia, A., Penazzi, G., Salvucci, L., and Di Carlo, A. (2008) Non-equilibrium Green's functions in density functional tight binding: method and applications. *New J. Phys.*, **10** (6), 065022.

35 Haugk, M., Elsner, J., Heine, T., Frauenheim, T., and Seifert, G. (1999) A parallel code for a self-consistent charge density functional based tight binding method: total energy calculations for extended systems. *Comp. Mater. Sci.*, **13** (4), 239–251.

36 Mniszewski, S.M., Cawkwell, M.J., Wall, M.E., Mohd-Yusof, J., Bock, N., Germann, T.C., and Niklasson, A.M.N. (2015) Efficient parallel linear scaling construction of the density matrix for born-oppenheimer molecular dynamics. *J. Chem. Theory Comput.*, **11** (10), 4644–4654.

37 Scemama, A., Renon, N., and Rapacioli, M. (2014) A sparse self-consistent field algorithm and its parallel implementation: application to density- functional-based tight binding. *J. Chem. Theory Comput.*, **10** (6), 2344–2354.

38 Liu, H., Elstner, M., Kaxiras, E., Frauenheim, T., Hermans, J., and Yang, W. (2001) Quantum mechanics simulation of protein dynamics on long timescale. *Proteins*, **44** (4), 484–489.

39 Hu, H., Lu, Z.Y., Elstner, M., Hermans, J., and Yang, W.T. (2007) Simulating water with the self-consistent-charge density functional tight binding method: from molecular clusters to the liquid state. *J. Phys. Chem. A*, **111** (26), 5685–5691.

40 Lee, T.S., Lewis, J.P., and Yang, W.T. (1998) Linear-scaling quantum mechanical calculations of biological molecules: the divide-and-conquer approach. *Comp. Mater. Sci.*, **12** (3), 259–277.

41 Sternberg, M., Galli, G., and Frauenheim, T. (1999) NOON – a non-orthogonal localised orbital order-N method. *Comput. Phys. Commun.*, **118** (2–3), 200–212.

42 Stephan, U. (2000) Comparison of the convergence properties of linear-scaling electronic-structure schemes for nonorthogonal bases. *Phys. Rev. B*, **62** (24), 16412–16424.

43 Aradi, B., Hourahine, B., and Frauenheim, T. (2007) DFTB+, a sparse matrix-based implementation of the DFTB method. *J. Phys. Chem. A*, **111** (26), 5678–5684.

44 Rapacioli, M., Barthel, R., Heine, T., and Seifert, G. (2007) Car-Parrinello treatment for an approximate density-functional theory method. *J. Chem. Phys.*, **126** (12), 124103.

45 Wang, F., Yam, C.Y., Chen, G., Wang, X., Fan, K., Niehaus, T.A., and Frauenheim, T. (2007) Linear scaling time-dependent density-functional tight-binding method for absorption spectra of large systems. *Phys. Rev. B*, **76** (4), 045114.

46 Frauenheim, T., Seifert, G., Elstner, M., Niehaus, T., Kohler, C., Amkreutz, M., Sternberg, M., Hajnal, Z., Di Carlo, A., and Suhai, S. (2002) Atomistic simulations of complex materials: ground-state and excited-state properties. *J. Phys.Condens. Matter*, **14** (11), 3015–3047.

47 Niehaus, T.A., Suhai, S., Della Sala, F., Lugli, P., Elstner, M., and Seifert, G., and Frauenheim, T. (2001) Tight-binding approach to time-dependent density-functional response theory. *Phys. Rev. B*, **63** (8), 085108.

48 Wahiduzzaman, M., Oliveira, A.F., Philipsen, P., Zhechkov, L., van Lenthe, E., and Witek, H.A., and Heine, T. (2013) DFTB parameters for the periodic table: part 1– electronic structure. *J. Chem. Theory Comput.*, **9** (9), 4006–4017.

49 Frisch, M.J., Trucks, G.W., Schlegel, H.B., Scuseria, G.E., Robb, M.A., Cheeseman, J.R., Scalmani, G., Barone, V., Mennucci, B., Petersson, G.A., Nakatsuji, H., Caricato, M., Li, X., Hratchian, H.P., Izmaylov, A.F., Bloino, J., Zheng, G., Sonnenberg, J.L., Hada, M., Ehara, M., Toyota, K., Fukuda, R., Hasegawa, J., Ishida, M., Nakajima, T., Honda, Y., Kitao, O., Nakai, H., Vreven, T., Montgomery, J.A., Jr., Peralta, J.E., Ogliaro, F., Bearpark, M., Heyd, J.J., Brothers, E., Kudin, K.N., Staroverov, V.N., Kobayashi,

R., Normand, J., Raghavachari, K., Rendell, A., Burant, J.C., Iyengar, S.S., Tomasi, J., Cossi, M., Rega, N., Millam, N.J., Klene, M., Knox, J.E., Cross, J.B., Bakken, V., Adamo, C., Jaramillo, J., Gomperts, R., Stratmann, R.E., Yazyev, O., Austin, A.J., Cammi, R., Pomelli, C., Ochterski, J.W., Martin, R.L., Morokuma, K., Zakrzewski, V.G., Voth, G.A., Salvador, P., Dannenberg, J.J., Dapprich, S., Daniels, A.D., Farkas, Ö., Foresman, J.B., Ortiz, J.V., Cioslowski, J., and Fox, D.J. (2009) Gaussian 09, Gaussian, Inc., Wallingford CT.

50 Fonseca Guerra, C., Snijders, J.G., te Velde, G., and Baerends, E.J. (1998) Towards an order-N DFT method. *Theor. Chem. Acc.*, **99**, 291.

51 Lundberg, M., Nishimoto, Y., and Irle, S. (2011) Delocalization errors in a hubbard-like model: consequences for density-functional tight-binding calculations of molecular systems. *Int. J. Quantum. Chem.* **112** (6), 1701–1711.

52 Sanna, S., Hourahine, B., Gallauner, T., and Frauenheim, T. (2007) An efficient LDA+U based tight binding approach. *J. Phys. Chem. A*, **111** (26), 5665–5670.

53 Rapacioli, M., Spiegelman, F., Scemama, A., and Mirtschink, A. (2011) Modeling charge resonance in cationic molecular clusters: combining DFT-tight binding with configuration interaction. *J. Chem. Theory Comput.*, **7**, 44.

54 Hourahine, B., Sanna, S., Aradi, B., Kohler, C., Niehaus, T., and Frauenheim, T. (2007) Self-interaction and strong correlation in DFTB. *J. Phys. Chem. A*, **111** (26), 5671–5677.

55 Christensen, A.S., Elstner, M., and Cui, Q. (2015) Improving intermolecular interactions in DFTB3 using extended polarization from chemical-potential equalization. *J. Chem. Phys.*, **143** (8), 084123.

56 Elstner, M. (2007) SCC-DFTB: what is the proper degree of self-consistency? *The J. Phys. Chem. A*, **111** (26), 5614–5621.

57 Yang, Y., Yu, H.B., York, D., Cui, Q., and Elstner, M. (2007) Extension of the self-consistent-charge density-functional tight-binding method: third-order expansion of the density functional theory total energy and introduction of a modified effective coulomb interaction. *J. Phys. Chem. A*, **111** (42), 10861–10873.

58 Kubillus, M., Kubar, T., Gaus, M., Rezac, J., and Elstner, M. (2015) Parameterization of the DFTB3 method for Br, Ca, Cl, F, I, K, and Na in organic and biological systems. *J. Chem. Theory Comput.*, **11** (1), 332–342.

59 Nishimoto, Y. (2015) Time-dependent density-functional tight-binding method with the third-order expansion of electron density. *J. Chem. Phys.*, **143** (9), 094108.

60 Aradi, B., Niklasson, A.M.N., and Frauenheim, T. (2015) Extended Lagrangian density functional tight-binding molecular dynamics for molecules and solids. *J. Chem. Theory Comput.*, **11** (7), 3357–3363.

61 Torralva, B., Niehaus, T.A., Elstner, M., Suhai, S., Frauenheim, T., and Allen, R.E. (2001) Response of C60 and Cn to ultrashort laser pulses. *Phys. Rev. B*, **64** (15), 153105.

62 Mitrić, R., Werner, U., Wohlgemuth, M., Seifert, G., and Bonačić-Koutecký, V. (2009) Nonadiabatic dynamics within time-dependent density functional tight binding method. *J. Phys. Chem. A*, **113** (45), 12700–12705.

63 Humeniuk, A. and Mitric, R. (2015) Long-range correction for tight-binding TD-DFT. *J. Chem. Phys.*, **143** (13), 134120.

64 Knaup, J.M., Hourahine, B., and Frauenheim, T. (2007) Initial steps toward automating the fitting of DFTB Erep(r). *J. Phys. Chem. A*, **111** (26), 5637–5641.

65 Doemer, M., Liberatore, E., Knaup, J.M., Tavernelli, I., and Rothlisberger, U. (2013) In situ parameterisation of SCC-DFTB repulsive potentials by iterative Boltzmann inversion. *Mol. Phys.*, **111** (22–23), 3595–3607.

66 Goldman, N., Fried, L.E., and Koziol, L. (2015) Using force-matched potentials to improve the accuracy of density functional tight binding for reactive conditions. *J. Chem. Theory Comput.*, **11** (10), 4530–4535.

67 Oliveira, A.F., Philipsen, P., and Heine, T. (2015) DFTB parameters for the periodic table, part 2: energies and energy gradients

from hydrogen to calcium. *J. Chem. Theory Comput.*, **11** (11), 5209–5218.
68 Elstner, M. and Seifert, G. (2014) Density functional tight binding. *Philos. Trans. R. Soc. A*, **372** (2011). doi: 10.1098/rsta.2012.0483.
69 Seifert, G. and Joswig, J.-O. (2012) Density functional tight binding – an approximate density functional theory method. *Comput. Mol. Sci.*, **2** (3), 456–465.
70 Oliveira, A.F., Seifert, G., Heine, T., and Duarte, H.A. (2009) Density-functional based tight-binding: an approximate DFT method. *J. Brazil Chem. Soc.*, **20** (7), 1193–1205.

8
DFT Calculations for Real Solids

Karlheinz Schwarz and Peter Blaha

Technical University Vienna, Institute for Materials Chemistry, Getreidemarkt 9-165-TC, A-1060 Vienna, Austria

8.1
Introduction

Nowadays, quantum mechanical calculations play an important role when studying the electronic structure of solids at the length scale of nanometers. In this context, different aspects need to be discussed, such as the details of the atomic structure, the quantum mechanical treatment of the electronic structure, which is mainly based on density functional theory (DFT), the computer codes for solving DFT, the parameters that determine the accuracy of the results, and various properties that can be derived from DFT calculations. This chapter introduces these simulation techniques and illustrates, for selected examples, which insights can be gained in structure–property relations.

8.2
Atomic Structure

For this chapter, we assume that the basic concepts of solids and their electronic structure are known and thus they will not be discussed. Such topics include the unit cell in direct space or reciprocal space (Brillouin zone), symmetry operations, translations, group theory (space group, point group), **k**-vectors, Bloch theorem, band structures, and density of states (DOS). The reader can find details about these points, for example, in Refs [1–3]. Take one example, namely, symmetry, which can provide useful information for interpreting properties or make calculations more efficient. However, a chosen (fixed) symmetry may also restrict or prevent changes such as a phase transition.

Handbook of Solid State Chemistry, First Edition. Edited by Richard Dronskowski, Shinichi Kikkawa, and Andreas Stein.
© 2017 Wiley-VCH Verlag GmbH & Co. KGaA. Published 2017 by Wiley-VCH Verlag GmbH & Co. KGaA.

Inorganic solid materials appear in nature as minerals, but many are synthesized by sophisticated methods for various applications and thus are of great interest for solid-state chemistry. Solid samples can vary from powder to single crystals with different length scales, where nanoparticles play a special role. Crystallography focuses on bulk properties and can determine the crystal structure in terms of a unit cell, inside which various atoms occupy atomic positions. When an atomic structure is determined, for example, by X-ray diffraction, one averages over a certain domain. There are other experimental techniques (e.g., scanning tunneling microscopy (STM)) that are surface sensitive and thus allow determination of surface structures that may deviate from the bulk structure. A real material is finite and may contain defects or impurities, which are disordered. Furthermore, in a compound, the composition may deviate from its ideal stoichiometry. In summary, one must keep in mind that there is a difference between the real and ideal atomic structures.

One often argues that nanoshaped crystals have different properties than their bulk counterparts. The obvious reason is that for a crystal a few nanometers in length, the ratio of the number of atoms on the surface to atoms in the bulk is relatively large. However, in a large crystal with a length of micrometer (or larger), this ratio is extremely small and thus the effects coming from the surface atoms can – to a good approximation – be ignored, at least for certain properties. We illustrate this principle in Figure 8.1 for a finite cubic particle, which we artificially construct by repeating a cube n times in all three directions, where

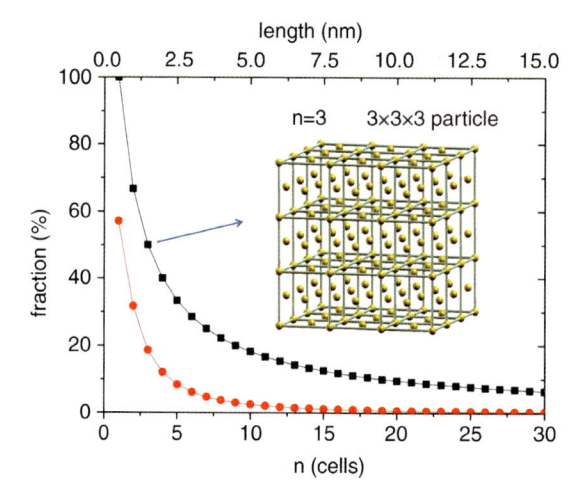

Figure 8.1 Fraction (in percent) of both the surface atoms (black line) and edge atoms including the corners (red line) is shown as function of n, the size of the finite particle (n is the number of cubic cells in one direction). This artificial particle is built by repeating the single cube (representing a face-centered cubic structure) n times in all three directions. The insert shows a finite particle for $n = 3$ for illustration. In order to provide a length scale (proportional to n), the size of the single cube is assumed to be 5 Å long.

each cube represents a face-centered cubic (fcc) structure. The insert shows the case for $n=3$, that is, a finite particle (composed of $3\times3\times3$ cubic cells), which contains 86 atoms on the surface (with 32 on the edges and corners) and 86 atoms in the bulk. In Figure 8.1, the fraction of atoms on the surface with respect to the total number of atoms is shown as a function of n. This fraction of surface atoms drastically reduces from 100% (for $n=1$) to about 3.3% for $n=60$, corresponding to a cell with a length of about 30 nm (containing already 885 781 atoms in total). Sometimes, for example, in catalysis, the atoms at the edges or corners of a particle play a crucial role. We also show in Figure 8.1 the fraction of these edge atoms, which is about 32% for $n=2$ but already near 1% for $n=16$.

The coordination number of the atoms at the surface differs from those in the bulk and thus these atoms may relax to slightly different positions. In addition, their electronic structure will differ and the lower coordination may substantially affect some surface-sensitive properties and functionalities (e.g., for heterogeneous catalysis). The large diversity of crystal facets in mineral crystals is another aspect recently described by Churakov [4] in more detail.

For any theoretical treatment, an atomic structure must be idealized. The standard assumption is that a solid is represented by a unit cell, which is repeated (in all three directions) by applying periodic boundary conditions. Accordingly, one assumes a solid is perfect, ordered, and infinite. As mentioned above, a sample of a real crystal differs from this ideal situation. Let us consider a defect as an example. If one would add a defect into the conventional unit cell, it would appear in all other cells and would be perfectly ordered and correspond to a high defect concentration. In reality, defects are distributed statistically and appear in small concentrations. One possibility for simulating such a situation is the concept of a "supercell," which is formed by combining a number of unit cells (e.g., $3\times3\times3$ cells) to form a new larger one. In such a supercell, one can make a substitution, for example, in the central (original) unit cell, but now the periodic images appear in all three dimensions but three lattice constants away from the original cell. The larger one makes such a supercell, the smaller will be the effects from the neighboring defects (caused by the periodic boundary condition). However, the computational effort, which approximately scales with the third power of the number of atoms per cell, increases substantially. Therefore, one must make a reasonable compromise. A similar concept can be used to simulate a surface of a crystal by constructing a "surface slab" that consists of several layers of a crystal, which are separated from their periodic images by an empty region (representing a vacuum). Such cases are mentioned in Section 8.6.

A real strength of theory is that it can calculate artificial model systems irrespective of their existence in nature. Atoms can be moved, added, or removed and the effects of such modifications can be investigated step-by-step. In a computer simulation, the atomic structure is well defined, since it is an input. However, in each specific case one must explore whether the assumed idealized structure is a good representation of a real material, whose atomic structure is known only approximately from experiment.

8.3
Quantum Mechanics for Treating the Electronic Structure

In order to describe the electronic structure of a system of interest, a quantum mechanical treatment is needed. The basic concepts are outlined in Chapter 1 of Volume 5. The authors start with the time-independent Schrödinger equation and describe the corresponding mathematical details and the formalism for many nuclei and electrons. In conventional calculations, the nuclei are treated as classical particles at $T = 0$. Since the mass of the nuclei is much higher than that of an electron, the Born–Oppenheimer approximation is assumed for most applications. This means that the nuclei can be taken at fixed positions when the electronic states are investigated by means of quantum mechanics, where the electronic states are described by an electronic wave function (WF) often labeled ψ. Its modulus $\psi * \psi$ defines the electron density $\rho(\mathbf{r})$, the probability of finding an electron at position $\mathbf{r}$. There are two classes of electronic structure methods: wave function based and density based. There are also schemes that combine both methods, and so they are considered here.

8.3.1
Hartree–Fock and Beyond

In the Hartree–Fock method, the total electronic wave function of an N-electron system can be expressed as a determinant composed of N single-electron WFs, known as the Slater determinant. This determinant has – by construction – the property that it changes sign when two electrons are interchanged, that is, the Pauli exclusion principle. This concept is called exchange interaction and is properly treated in the Hartree–Fock (HF) method. The corresponding single electronic states are orthogonal to each other and each electron feels the electrostatic field created by all other electrons (and nuclei) in the system. This problem can only be solved iteratively, since for the computation of the field, the other WFs are needed, but for finding each of these WFs the corresponding field is required. Therefore, the iterations must be carried out until self-consistency is reached (to be discussed below in connection with Figure 8.3), a procedure called self-consistent field (SCF).

The HF method is an approximation, since correlation effects are completely missing. They can be included with post-HF schemes such as configuration interaction (CI) or coupled cluster (CC) methods [5]. Such schemes are *ab inito* methods often used in quantum chemistry especially for studying molecules. They can become very accurate (almost exact) but quickly become computationally very expensive when the system size increases. Such schemes have recently also been applied to solids, but only for very simple ones [6]. The required computation time makes such calculations infeasible for realistic material science studies.

8.3.2
Density Functional Theory

At present, Kohn–Sham (KS) DFT is the state-of-the-art *ab initio* method [7,8] for studying the electronic properties of materials due to its balance between accuracy and computational efficiency. The fundamental statement by Hohenberg and Kohn [7] is that the total energy of an interacting electron system is uniquely defined by the electron density of the system, $\rho(\mathbf{r})$. The latter depends on $\mathbf{r}$, that is, three coordinates:

$$E = T_0[\rho] + \int V_{\text{ext}} \rho(\vec{r}) \, d\vec{r} + \frac{1}{2} \int \frac{\rho(\vec{r})\rho(\vec{r}\,')}{|\vec{r}\,' - \vec{r}|} \, d\vec{r} \, d\vec{r}\,' + E_{\text{xc}}[\rho]. \qquad (8.1)$$

Equation (8.1) is written (for an atom for simplicity) in atomic units, where the terms are the kinetic energy of noninteracting electrons T_0, the Coulomb interaction between the nucleus (with the external potential V_{ext}) and each electron, the Coulomb interaction between the electrons, and the important exchange–correlation energy E_{xc}, which includes all the quantum mechanical effects, such as the electron exchange, as well as static and dynamic electron–electron correlations. In Eq. (8.1) the Coulomb interaction contains the unphysical interaction of an electron with itself (the self-interaction), which is canceled in the exchange term (for example, exactly in Hartree–Fock). In DFT there often remains a self-interaction error due to the approximations used for E_{xc}. The real advantage of DFT is that one does not need to know the many-electron WF (which depends on $4N$ variables for N electrons). This is in principle an enormous simplification, since the density $\rho(\mathbf{r})$ even for very large systems depends only on three coordinates. Equation (8.1) is formally exact but E, the total energy of the system, can only be proven to be a functional of the density but its explicit mathematical form is unknown and thus needs to be approximated. In mathematics, such a case is called functional instead of function.

Kohn and Sham [8] proposed a scheme to make DFT calculations feasible. Their idea relies on mapping the interacting many-body system (the nuclei and electrons) onto a noninteracting system of KS electrons (quasiparticles) that lead to the true density. This idea introduces the KS equations (Eq. (8.2)) with a local one-body potential, the KS potential, mimicking all interelectronic interactions via Hartree and exchange–correlation (XC) contributions:

$$\left[-\frac{1}{2} \nabla^2 + V_{\text{ext}}(\vec{r}) + V_{\text{C}}[\rho(\vec{r})] + V_{\text{xc}}[\rho(\vec{r})] \right] \Phi_i(\vec{r}) = \varepsilon_i \Phi_i(\vec{r}). \qquad (8.2)$$

In Eq. (8.2) we have the terms of the kinetic energy, the external potential (from the nuclei), the Coulomb potential V_{C}, and the important exchange–correlation potential V_{xc}, which formally is the functional derivative of E_{xc} (in Eq. (8.1)) with respect to the density ρ. There is only one system for which these expressions for E_{xc} and V_{xc} can be accurately determined, namely, the homogeneous electron gas [9], which is a highly idealized case. For all practical

8 DFT Calculations for Real Solids

- DFT heaven
- unoccupied orbitals (ACFDT-RPA)
- occupied orbitals (hybrid-DFT)
- meta-GGAs ($\nabla^2\rho$, kinetic energy density)
- GGA ($\nabla\rho$)
- LDA (ρ)
- Hartree

(accuracy / computer time increases up the ladder)

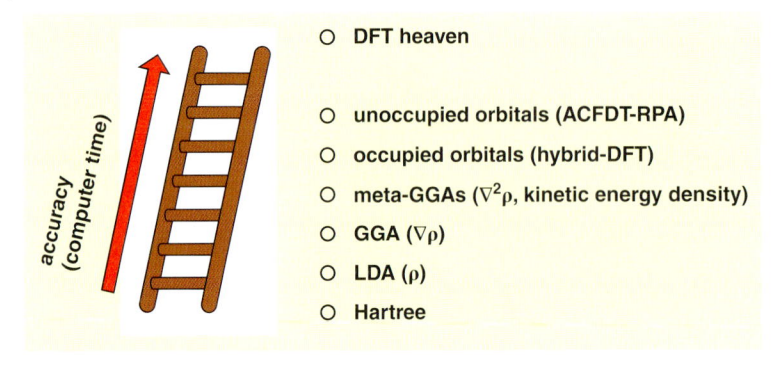

Figure 8.2 Jacob's ladder [10]: Find a functional that leads to DFT heaven.

applications, approximate functionals must be used, of which a large variety exist. Here we only mention the main categories, referring to Jacob's ladder shown in Figure 8.2 [10].

By going up this ladder, the exchange–correlation functional depends (at the various rungs) on ρ, $\nabla\rho$, $\nabla^2\rho$, and/or the kinetic energy density τ, and at the highest rungs on the occupied or unoccupied orbitals. The accuracy should improve for higher rungs, but the required computer time increases substantially. Therefore, DFT heaven cannot be reached. Electronic structure calculations of periodic systems, most commonly, use the local density approximations (LDA) [8,9] or the generalized gradient approximation (GGA) [11] parameterized in various forms, at the lower rungs of the ladder. On the next rung, one reaches the *meta*-GGAs to further improve the accuracy. In this scheme, E_{xc} depends also on the kinetic energy density τ. The next class of functionals consider an explicit dependence on occupied orbitals, which allow an exact treatment of exchange. Here the hybrid functionals [12] shall be mentioned that take a certain fraction of Hartree–Fock (with the exact treatment of exchange) and combine it with DFT to account for correlation. On the highest rung, unoccupied orbitals are also included, leading to schemes such as the adiabatic connection and fluctuation dissipation theory (ACFDT) and the random phase approximation (RPA), which is computationally very demanding [13]. We refer the reader to a recent paper [14] critically analyzing DFT functionals in terms of accuracy and addressing their differences for molecules and solids. This paper contains far too many details for this chapter.

For highly correlated systems, especially for those containing late transition elements (with localized 3d electrons) or rare earth atoms (with 4f electrons), a Hubbard U is often used, where the parameter U accounts for the energy penalty (repulsion) for occupying 3d or 4f electrons on the same atom (instead of delocalizing them). This leads to schemes such as LDA + U or GGA + U [15]. The U parameter can be estimated, but often it is adjusted to improve the agreement with experimental data.

Improving, optimizing, and validating various functionals is an active field of research, with many publications. Some conceptual aspects are presented in Refs [2,3] and some additional aspects will be mentioned in Section 8.5.

8.4 Solving DFT

In this chapter we first make some general remarks concerning DFT calculations, which will be rather similar for many computer codes, but then focus on our approach based on the augmented plane wave (APW) method described in Section 4.2. The general scheme for DFT-based electronic structure calculations is summarized in Figure 8.3.

We make a few comments in order to describe the various aspects sketched in Figure 8.3. Assuming periodic boundary conditions, we define the atomic structure by a unit cell (with the lattice constants a, b, c and the angles α, β, γ), its

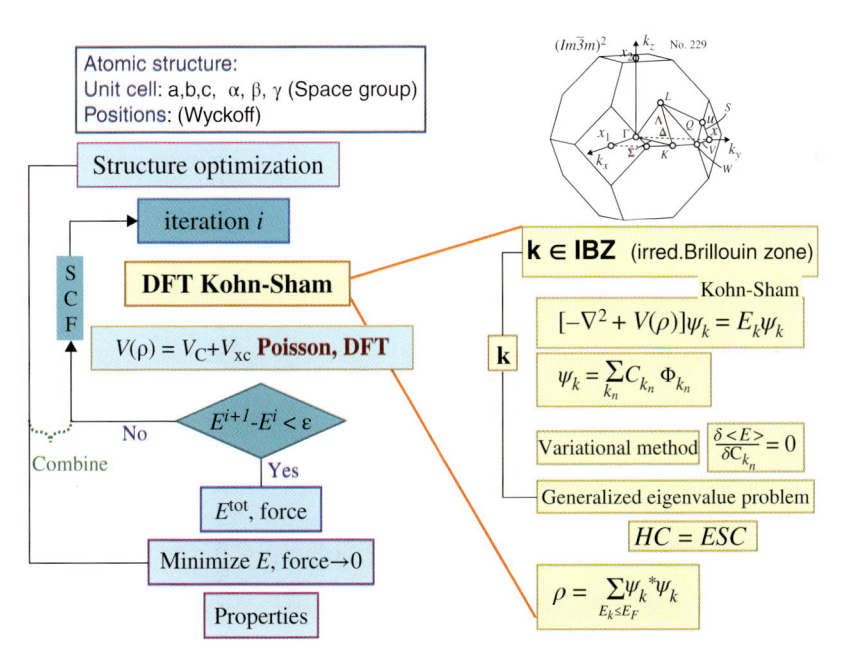

Figure 8.3 Major steps of typical DFT electronic structure calculations are schematically shown. For a given atomic structure, the self-consistent field (SCF) iterations must be carried out to convergence. In each loop, the Kohn–Sham (KS) equations must be solved in another loop, namely, the **k**-point loop. The irreducible wedge in the Brillouin zone is shown (top right) for a face-centered cubic (fcc) structure as an example. After the SCF cycle is converged, one has the total energy and forces (if present), which can be used to minimize the energy and thus find the equilibrium structure, for which properties can be computed.

space group, and the atomic (Wyckoff) positions [1,16].[1] For the chosen atomic structure, one determines the corresponding electronic structure via DFT in SCF cycles. In each cycle, the KS equations must be solved for a set of **k**-points. Symmetry allows one to reduce the number of **k**-points to the irreducible wedge of the first Brillouin zone (IBZ) (right top of Figure 8.3). This is illustrated for a cubic crystal (fcc or bcc), where the 48 symmetry operations lead to only 1/48th of the whole BZ being necessary (the IBZ). The wave function of each KS one-electron state (called a quasiparticle) is expanded in a basis set (labeled ϕ_{kn}) with weights (C_{kn}). The latter can be determined by the variational method leading to a (computationally heavy) generalized eigenvalue problem ($HC = ESC$), where H stands for the Hamiltonian, C for the basis set coefficients, S the overlap matrix, and E for the energy eigenvalues. The electron density $\rho(\mathbf{r})$ can be obtained from these KS orbitals by summing the moduli $\psi_i^*\psi_i$ of all occupied states, that is, those for which the KS one-electron energy E_k is smaller than the Fermi energy E_F. This density $\rho(\mathbf{r})$ together with the density from the inner electrons (core and semi-core) is needed in order to find the effective potential $V(\rho)$ for the next SCF cycle. At this step, Poisson's equation must be solved to find the Coulomb potential V_C to which the exchange–correlation potential V_{xc} (for the chosen DFT potential) is added. The SCF cycles are continued till convergence is reached. This is the case when, for example, the total energy of two successive iterations deviate less than a convergence criterion ε (e.g., 0.0001 Ry).

If symmetry allows, there can be forces acting on certain atoms in the unit cell. Such forces are the negative gradient of the total energy with respect to the corresponding position parameters. Take, for example, the rutile TiO_2 structure, in which oxygen sits on Wyckoff position 4f that has the coordinates $(x, x, 0)$, where x is not specified by symmetry. In this case, one can vary x to minimize the energy and thus a force can occur on this oxygen atom. This force vanishes at the equilibrium geometry. In the rutile example, Ti is located at the Wyckoff position 2a with the fixed coordinates $(0, 0, 0)$ and $(1/2, 1/2, 1/2)$, thus these positions are fixed by symmetry and no force will act on Ti. When all atoms are essentially at their equilibrium positions (with forces around zero), then one can change the volume of the unit cell and minimize the total energy E (the structure optimization loop in Figure 8.3). After all these geometry optimizations, a system (in the given structure) reaches its equilibrium geometry. As a last step in Figure 8.3, one can calculate various properties of this solid, which now should be in the optimized structure.

8.4.1
General Considerations for DFT Calculations

8.4.1.1 All-Electron Treatment

One can first ask the question which electrons shall be included carrying out a calculation? Often one argues that chemical bonding is essentially due to the

---

1) Inorganic Crystal Structure Database (ICSD), www.fiz-karlruhe.de.

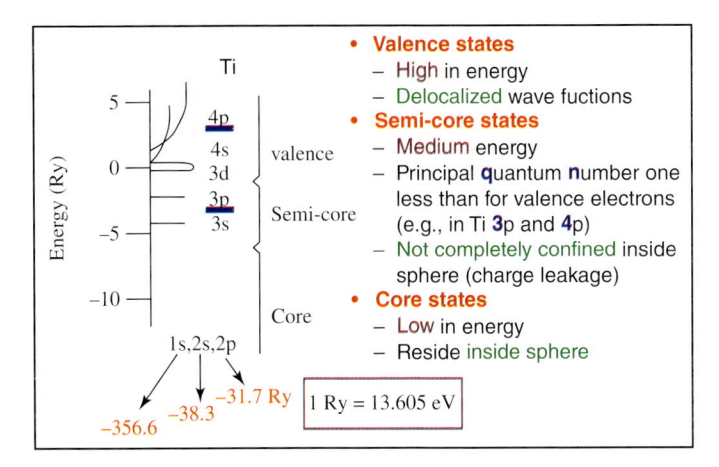

Figure 8.4 Three types of electronic states of titanium: core, semi-core, and valence electrons.

valence electrons and thus the core electrons remain frozen. In such a frozen-core approximation, one assumes that the core electron density does not change in the SCF cycle of DFT. Another choice is the pseudopotential scheme, in which the core electrons are only indirectly included (or excluded). To some extent, such schemes are justified, because they simplify the calculations. However, there are cases (for example, when a significant charge transfer between atoms occurs) where such approximations neglect changes due to the inner electrons (especially near the nuclei) that might be small but important for properties such as hyperfine interactions. The alternative to pseudopotentials is to include all electrons, which makes the corresponding calculations more demanding in terms of computational effort but leads to higher accuracy.

In the all-electron case, the atomic states can be characterized by three types as is illustrated in Figure 8.4 for the example Ti (with the atomic number $Z = 22$). Its electronic configuration is $1s^2 2s^2 2p^6 3s^2 3p^6 3d^2 4s^2$. We surround the Ti nucleus by an atomic sphere with a radius of about 1 Å. Then we have the three types of states:

- *Core states*, Ti 1s, 2s, 2p, which are very low in energy (around −357, −38, −32 Ryd, respectively), and have their wave functions (or electron densities) reside completely inside this sphere.
- *Valence states*, Ti 3d, 4s, 4p, which are high in energy and are delocalized.
- *Semi-core states*, Ti 3s, 3p, which are in between in energy (−2 to −4 Ryd) and have a charge leakage (a few percent of the charge density lies outside the sphere). They have a principal quantum number that is one less than the valence states (3s versus 4s, 3p versus 4p).

The so-called (closed-shell) semi-core states have wave functions, which are not completely confined in the atomic regions but weakly interact with neighbors. An all-electron treatment has the advantage of being able to explore the contribution of each electron to various properties such as the electric field gradients or hyperfine fields, which can have significant contributions from the core electrons.

8.4.1.2 Relativistic Effects

A material of interest may contain all kinds of elements. For a system with heavy elements, relativistic effects must be included. The inner electrons of these elements can reach a high velocity (approaching the velocity of light, c) which in simple terms is proportional to Z/n, where Z is the atomic number and n the principal quantum number. This leads to a mass enhancement of the electron (with a rest mass m_e) by the Lorentz factor γ. Let us take as an example the 1s orbital of Au (with a large $Z = 79$ and a small $n = 1$). This 1s electron reaches a speed of $0.58c$ and has a $\gamma = 1.22$ as Lorentz factor. Consequently, this mass enhancement contracts this orbital and thus causes a higher screening of the nuclear charge, which significantly affects the valence electrons. The latter must have orbitals that are orthogonal to all inner electrons and thus relativity affects all electrons (Figure 8.5). The 1s orbital contracts due to the relativistic mass

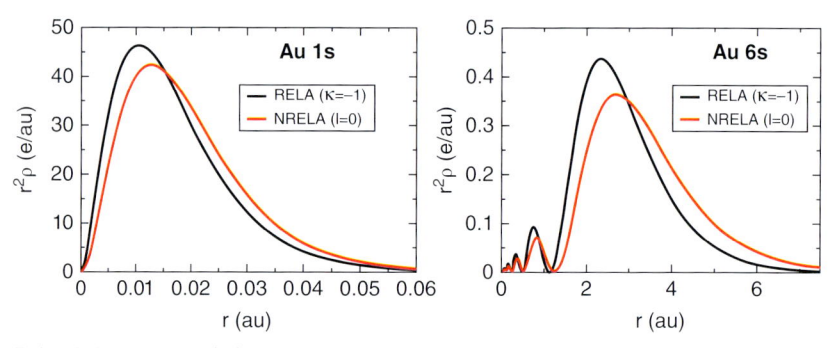

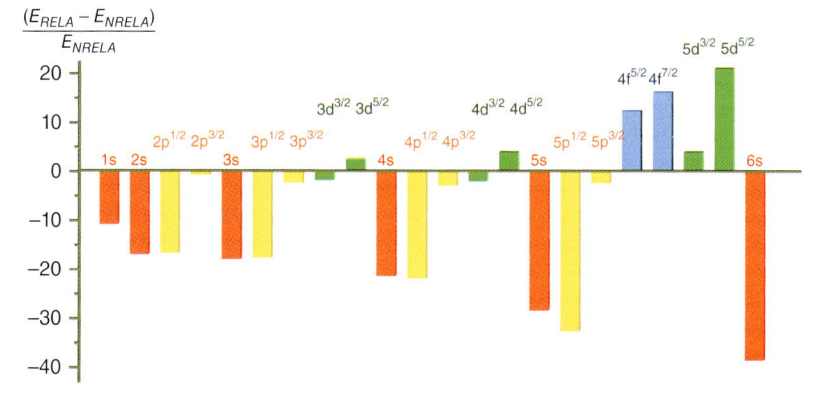

Figure 8.5 Relativistic effects on the orbitals of the Au atom: contraction, expansion, spin–orbital splitting. In the upper part, the radial probability is shown for the 1s and 6s states, which both show a contraction with respect to the nonrelativistic case. The lower panel shows the relative change in orbital energies due to relativity.

enhancement, while the 2s–6s orbitals contract mainly due to the orthogonality requirement to the lower states. In addition, the Darwin s-shift (Zitterbewegung) affects the orbitals, and spin–orbit coupling causes, for example, a splitting between the $p_{1/2}$ and $p_{3/2}$ states.

For the lighter elements (up to an atomic number of about 54 (Xe)), a scalar relativistic scheme [17] is often used, which is computationally very efficient. It can properly describe the main contraction or expansion of the various atomic orbitals (due to the Darwin s-shift and the mass enhancement) but it omits spin–orbit coupling. Inclusion of the latter couples spin-up and spin-down electrons and requires doubling the basis set, making the calculation much more costly. The spin–orbit coupling can be treated – to a good approximation – in a second variational treatment [18]. In the WIEN2k implementation (see Section 4.2), the core electrons are treated (fully relativistic) by solving Dirac's equation, whereas the semi-core and valence states are described with the scalar relativistic scheme. When needed, spin-orbital coupling can also be included.

8.4.1.3 Full Potential and Basis Sets

In DFT, the potential contains the effects from the nuclei, all-electrons, and the quantum mechanical treatment for exchange and correlation in the form of a mean field potential. In some quantum mechanical schemes, the potential can become orbital dependent (as, for example, in Hartree–Fock-related cases). In conventional DFT schemes (e.g., LDA or GGA), the potential is directly determined by the electron density of the system and is called "full potential" provided no shape approximation is made. In the old days (the 1970s and 1980s), the muffin-tin (MT) approximation was used, in which the potential (and the charge density) was assumed to be spherically symmetric inside each atomic sphere (centered on the corresponding nucleus) but constant in the interstitial region (outside all atomic spheres). The concept of the full potential versus muffin tin was illustrated in Figure 10 of Ref. [19]. In the case of a spherical symmetric potential, the wave functions inside an atomic sphere can be written – as for the hydrogen atom – as radial wave functions $u_{n\ell}(r)$ times spherical harmonics $Y_{\ell m}(\theta,\varphi)$. In the region where the potential is assumed to be constant (such as for a free electron), plane waves (PWs) are a proper basis set to expand the wave function. In a crystal, the wave function has an atomic-like character close to each nucleus and may have many nodes (for example, a 4s wave function of Ti). Such a wave function can hardly be expressed in PWs.

This situation has created the idea to replace the real wave function of the valence electrons by pseudowave functions that remain the same at large distances from the nucleus (where the chemical bonding occurs) but are smoothed (made nodeless) so that they can be well described using PWs. This concept requires replacing the true electrostatic potential (due to the nuclei and inner electrons) by an effective potential, called a pseudopotential. In such simulations, the core electrons are excluded and the valence electrons are replaced by pseudo-WFs, which are derived from the pseudopotential.

Closely related to the assumed form of the potential is the choice of basis set. The wave functions (in DFT, the Kohn–Sham orbitals) are expanded in basis functions. In quantum chemistry (mostly for molecules), Gaussians or Slater-type orbitals (STO) are often used as basis set to describe atomic orbitals (AOs). A linear combination of AOs (LCAO) is a scheme with the advantage that it can be adapted to interesting regions (active sites, catalysis) but its convergence depends on many parameters and thus is more difficult to control. In contrast to LCAO, the convergence of a PW basis set is easy to control (like a Fourier series) with one number, the PW cutoff. However, one disadvantage is that accuracy is uniform. For example, if an accurate representation of the wave function is reached in one region of space, it is the same everywhere, even in less important regions (e.g., the region without atoms, which represents vacuum in a surface simulation). Pseudopotentials or atomic basis sets must be properly chosen and adapted for each atom. A certain choice may be valid for certain atoms but not for all atoms in the periodic table.

There is another conceptual difference in basis sets, namely, (i) *atom-centered* basis functions (e.g., Gaussians) as used in LCAO schemes, or (ii) *spatially confined* basis sets, which will be described next. This has consequences for the interpretation of results, especially when a property (e.g., charge transfer, bonding analysis, or core-level spectrum) is decomposed into atomic contributions.

8.4.2
The Linearized Augmented Plane Wave (LAPW) Method and WIEN2k

Already in 1937 Slater [20] proposed the APW method. The development of APW and its linearized version, which led to the WIEN code [21] and later to its present version WIEN2k [22], has been described in detail in several reviews [19,23–28]. The LAPW method is our choice to solve the DFT Kohn–Sham equations, as illustrated in Figure 8.3. Detailed descriptions including many conceptual and mathematical details are given, for example, in Refs [28,29], and therefore we only summarize the main concepts.

In the APW method, the unit cell is partitioned into two types of regions: (i) (nonoverlapping) *atomic spheres* centered at the atomic sites and (ii) the remaining *interstitial region* (Figure 8.6). The size of the spheres (labeled R_{mt}, for muffin tin) can be adjusted for each of the various atoms. Inside each atomic sphere, the wave functions have nearly an atomic character and thus (assuming a muffin tin potential) can be written as a radial function times spherical harmonics. It should be stressed that the *muffin tin* approximation (MTA) is used only for the construction of the radial APW basis functions $u_\ell(r, E_\ell)$. The radial Schrödinger equation is solved numerically (and thus highly accurately), but the energy E_ℓ must be provided as input. In the interstitial region, the potential varies only slowly and thus the wave functions can be well expressed by a set of plane wave. Each plane wave is *augmented* by partial atomic waves inside each atomic sphere, as illustrated in Figure 8.6. This makes APW (in contrast to the LCAO approach) a spatially confined basis set, where, for example, an electron

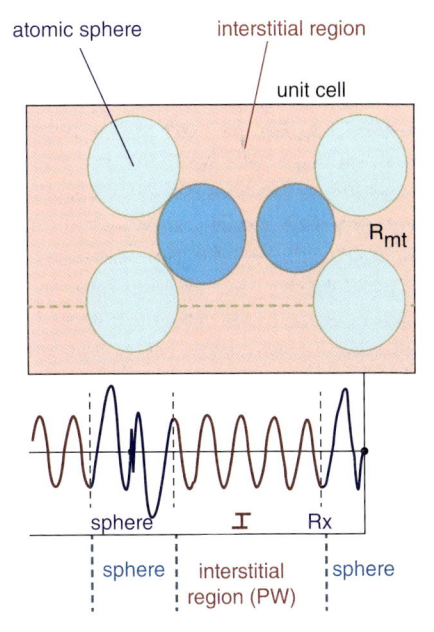

Figure 8.6 The different regions in the unit cell for the augmented plane wave scheme. (i) Atomic spheres (with a radius R_{mt}). (ii) Interstitial region between the spheres. The dashed line indicates the path for which the wave function is shown.

that is partially located in the interstitial region cannot be attributed to a specific atomic orbital.

The energy dependence of the atomic radial functions $u_\ell(r, E)$ can be treated in different ways. In Slater's original APW [20], this was done by varying the energy E, which leads to a nonlinear eigenvalue problem since the basis functions become energy dependent. At the sphere boundary, each plane wave must match the corresponding partial wave (its augmentation) as illustrated for a Fe-4p type wave function in Figure 8.7. In APW, this matching in value (at the sphere radius R_{mt}) determines the weight $A_{\ell m}$.

Three schemes of augmentation (APW, LAPW, APW+lo) have been suggested over the years. This illustrates how progress has been made in developing APW-type calculations, as has been discussed in several review articles [19,23,29]. In the linearized APW, called LAPW, Andersen [30] suggested linearizing (that is treat to linear order) this energy dependence. The radial Schrödinger equation is (numerically) solved for a fixed linearization energy E_ℓ (taken at the center of the corresponding energy bands) leading to $u_\ell(E_\ell, r)$. By adding an energy derivative of this function, $\dot{u}_\ell = \partial u_\ell / \partial \varepsilon$ (taken at the same energy), one can retain the variational flexibility. This linearization is a good approximation for a sufficiently small energy range around E_ℓ. In the LAPW scheme (see Figure 8.7), the atomic function inside the sphere α is given by a sum of partial waves, namely, (linearized) radial functions times spherical

- Atomic partial waves
  - LAPW

$$\Phi_{k_n} = \sum_{\ell m}[A_{\ell m}(k_n)u_\ell(E_\ell, r) + B_{\ell m}(k_n)\dot{u}_\ell(E_\ell, r)]Y_{\ell m}(\hat{r})$$

  - APW+lo

$$\Phi_{k_n} = \sum_{\ell m} A_{\ell m}(k_n)u_\ell(E_\ell, r)Y_{\ell m}(\hat{r})$$ Plus local orbitals (lo)

- Plane waves (PWs)

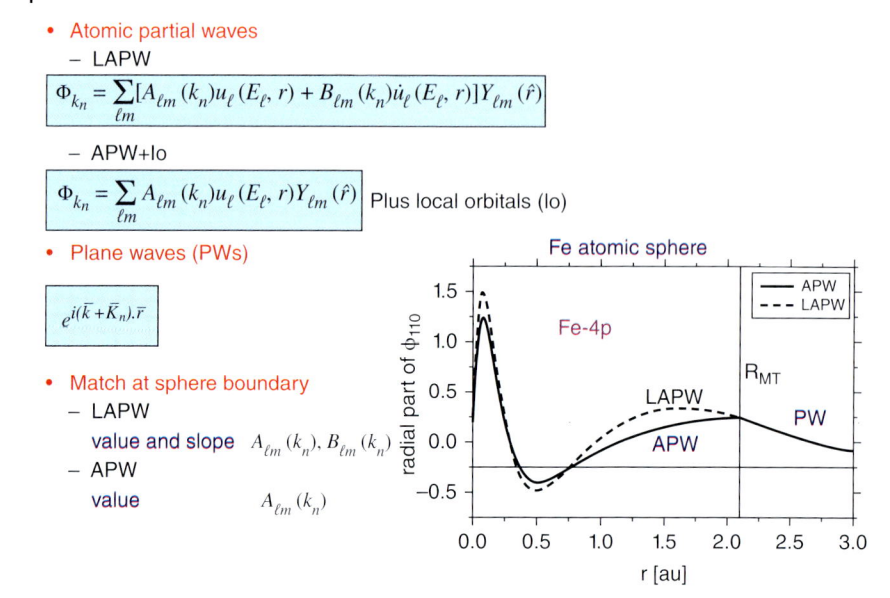

$$e^{i(\bar{k}+\bar{K}_n)\cdot\bar{r}}$$

- Match at sphere boundary
  - LAPW
    value and slope $A_{\ell m}(k_n), B_{\ell m}(k_n)$
  - APW
    value $A_{\ell m}(k_n)$

Figure 8.7 Basis sets and concepts used in the linearized augmented plane wave (LAPW) method as implemented in WIEN2k [22] are shown:

i) Inside the atomic sphere, different atomic partial waves are used in LAPW and APW + lo.
ii) In the interstitial region plane, waves are used.
iii) At the sphere boundary, a matching is done.
iv) Matching of LAPW versus APW is illustrated for an Fe-4p orbital.

harmonics $Y_{\ell m}$ labeled with the quantum numbers (ℓ,m). The two coefficients $A_{\ell m}$ and $B_{\ell m}$ (weights for each function and its derivative) – as given in Figure 8.7 – are determined to match each plane wave (characterized by the reciprocal vector K) continuously (in value and slope) to the one-center solution inside the atomic sphere at the sphere boundary (for details, see, for example, Refs [24,29]). The main advantage of the LAPW basis set is that it allows finding all needed eigenvalues with a single diagonalization, in contrast to APW, which presents a nonlinear eigenvalue problem. Historically, the stricter constraint (a matching in both, value and slope) had the disadvantage that LAPW required more PWs than APW to reach comparable convergence. The LAPW basis functions u and its derivative $\dot{u}$ are recalculated in each iteration cycle (see Figure 8.3) and thus can adjust to chemical changes. For example, charge transfer can change the oxidation state, requiring an expansion or contraction of the radial function. The LAPW method made it computationally attractive to go beyond the muffin tin approximation and treat both the crystal potential and the charge density without any approximation of their shape (called full potential), as pioneered by the Freeman group [31].

In Section 8.4.1.1, the partition of electronic states into core, semi-core, and valence states was described and illustrated for Ti in Figure 8.4. Let us focus on the p-type orbitals. The 2p state is treated fully relativistically as an atomic core state, while the valence 4p state is computed within LAPW using a higher linearization energy. The 3p semi-core states reside mostly inside the Ti sphere but have a "core-leakage" of a few percent. The 3p states are well separated in energy from the 4p states and thus linearization (with the linearization energy of the 4p state) would not work here. For such a case, Singh [32] proposed adding local orbitals (LO) to the LAPW basis set in order to accurately treat states with different principal quantum numbers (e.g., 3p and 4p) while retaining orthogonality. The 3p LO is similar to the 3p radial function (as an example) but is constrained to have zero value and slope at the sphere radius R_{mt}.

The concept of LOs fostered another idea, namely, the APW plus local orbitals (APW + lo) method by Sjöstedt et al. [33]. We label these local orbitals in lowercase to distinguish them from the semi-core LOs. In APW + lo, one returns to the APW basis but with the crucial difference that each radial function is solved only for a fixed energy E_ℓ. The PW matching at the atomic sphere radius is again (as in APW) only made between values but not in slope (see Figure 8.7). With this new scheme, calculations converge significantly faster with the number of plane waves while maintaining the convenience of LAPW [34].

The APW + lo scheme therefore combines the best features of all APW-based methods. It is well known that LAPW converges somewhat slower than APW due to the constraint of matching the PWs in both value and slope at the sphere boundary, and thus it is an improvement to return to APW, but only for the orbitals involved in chemical bonding. The energy-independent basis introduced in LAPW is crucial for avoiding the nonlinear eigenvalue problem of APW and thus is used for all higher ℓ components. The local orbitals provide the necessary variational flexibility to make this new scheme efficient, but they are added to the basis set only where needed (to avoid a further increase of the basis set). The crystalline wave functions (of Bloch type) are expanded in these APWs leading to a general eigenvalue problem. The size of the matrix is mainly given by the number of plane waves, but is increased slightly by the additional local orbitals that are used. As a rough estimate, one can say that about 50–100 PWs are needed for each atom in the unit cell in order to achieve good convergence.

8.4.3
Various Computer Codes

There are several computer codes available for solid-state simulations, either open source or commercial. A variety of codes is needed, where each code is often optimized for different materials or properties. Different codes offer different properties to be calculated, but they have advantages and disadvantages compared to others. Some codes are listed here: ABINIT (www.abinit.org), CASTEP (www.castep.org), CPMD (www.cpmd.org), QE (www.quantum-espresso.org), VASP (www.vasp.at), and WIEN2k (www.wien2k.at) are PW-based codes; CP2K (www.cp2k.org) is a hybrid code using both PW and Gaussian basis sets. CRYSTAL (www.crystal.unito.it) uses mostly Gaussian basis sets. For further details, see the corresponding web pages.

Different computer codes should give the same results provided they use the same first-principles formalism (e.g., the same DFT functional) and are carried out to full convergence. In practice, the results obtained with different codes depend on details such as the basis set, number of **k**-points, and choice of (pseudo) potential. Recently, error estimates have been derived [35] for solid-state DFT calculations, in which the WIEN2 k code plays the role of the standard (in terms of the most accurate results). These tests showed, for example, that several pseudopotentials had to be improved considerably in order to reach full agreement with the standard. A positive aspect of this analysis was that a good agreement between different accurate calculations was obtained. These validation tests have shown that the deviation between accurate codes is often significantly smaller than the typical difference between theory (based on different functionals) and experimental data.

There are different ways of distributing a code, which we list with representative examples:

- Open source with a free download (ABINIT, QE, CP2 K)
- Use for registered users (with a license fee), who have access to the source code (WIEN2 k)
- Limited access for registered users (for a certain number of users and/or a yearly license) (VASP)
- Software companies that distribute only executables (CASTEP).

A commercial company wants to have strict rules and – for obvious reasons – does not make the source code available. From a scientific perspective, which the WIEN2 k group favors, it is important that the source code is made available to the registered users, who pay a small license fee once. This policy has helped to generate a "WIEN2 k community" of researchers around the world. Many of them have contributed in several aspects, such as bug fixes, adding new features, which were made available to all the WIEN2 k users, improving the documentation, and suggesting new features. It has become a standard to help each other and thus users contribute to the development of computations of solids and surfaces. In total, this policy has had very positive impacts for WIEN2 k and the field. Other groups have set up different rules for jointly developing simulation codes so that many researchers can contribute.

The user-friendliness of WIEN2 k has been improved over the years. A graphical user interface *w2web* was mainly developed by Luitz and coworkers (see Ref. [22]) and is especially useful for novice users (see, for example, Figure 2 of Ref. [24] for the structure generation for CrO_2). *w2web* is also useful in cases that are not done routinely. In the meantime, many default options, based on the experience of previous calculations, have been implemented. This has made it much easier to set up a calculation. For novice users or experimentalists, this helps to get started without being an expert. However, there is also a drawback, namely, the danger that the code will be used as a black box: "push a button and receive the result." In the old version, the users were forced to think about how to run the calculation and thus had to look at details. This is a common problem, which all codes face.

8.5 How to Run DFT Calculations

Several aspects mentioned in this section are valid for many types of DFT calculations, but here we focus on the WIEN2k program package. Many details are available from the user's guide (see www.wien2k.at) that is continuously updated. Other computer codes may have equivalent or similar options and features, but some choices must be made for all of them. In modern times, the complexity of interesting systems requires the treatment of relatively large systems, so that surfaces, impurities, or defects can be studied. Sometimes fine details (such as the magnetic anisotropy energy, or phase transitions) shall be studied, for which very high precision (or a more sophisticated quantum mechanical treatment) is required that can only be achieved for small systems. One must consider which aspects can be neglected and which must be included. One often must make a compromise between a realistic (large) system and the high-accuracy calculation (on small systems) by choosing a model being as accurate as needed and as simple as possible. Such a choice of modeling is not a trivial task.

8.5.1 Idealize the Atomic Structure

In order to make calculations of a system like a solid (with very many atoms) feasible at all, one must obey periodic boundary condition, that is, one must define a unit cell that is repeated infinitely (as discussed in Section 8.2). This clearly is an idealization, since it excludes, for example, all kinds of surface effects.

For a bulk system with a known crystal structure, one can use either a CIF file to generate the *struct* file as input for WIEN2k or one can manually enter the space group and atomic positions (using the graphical interface *w2web*, see, for example, Figure 2 of Ref. [24] for the structure generation for CrO_2). Using such *struct* files one can build supercells with the WIEN2k tools *supercell* or the *structeditor* and create a model for defects, impurities, or disorder. The same procedure applies for the generation of surface slabs, where typically 5–13 atomic layers and vacuum regions of at least 10 Å should be used.

Another tool is *sgroup*, which allows determining the space group of structure models automatically. In cases where the space group has different settings (choice of origin), one should choose (if possible) the origin such that inversion symmetry is kept. This leads to real matrix elements instead of complex ones (without inversion symmetry) lowering the computational effort.

8.5.2
Choice of Quantum Mechanical Treatment (DFT)

Unfortunately, there is not an optimum DFT functional that works well for all cases and properties. DFT functionals are an active field of research and thus recent publications should be considered. WIEN2k can handle all atoms of the periodic table, but each of them may have a favorable functional for a given property (see, for example, Chapter 3.5 in Ref. [19]). Let us consider the equilibrium volume (lattice constant) of 3d, 4d, or 5d metals (such as Cu, Ag, Au). For Cu GGA PBE [11] is a good choice, for Ag the Wu Cohen functional [36] agrees better, while for Au LDA is best [37]. If one wants to study an intermetallic compound such as Cu_3Au one must make a compromise. A well-known example is the Fe case, comparing the fcc and bcc structure (including magnetism). Within the local spin density approximation (LSDA), one would predict that the fcc structure has a lower total energy than the bcc phase and it should be nonmagnetic, in contrast to experiment. With GGA (PBE) [11], the correct phase stability of the magnetic bcc structure is found with an equilibrium volume close to the experimental data.

For a material of interest, it is important to know whether it is a metallic or an insulator (semiconductor). In the latter case, a crucial property is the bandgap that involves the excitation from the occupied valence to the unoccupied conduction bands. It is well known and understood why conventional DFT approximations (such as LDA or GGA) often give only about half the bandgap with respect to the experiment. A more sophisticated treatment of this excitation process can be handled with the GW approximation [38], which computationally is much more demanding. A simple and pragmatic way to obtain good bandgaps is the choice of the modified Becke–Johnson (mBJ) potential (see Ref. [39]), which yields highly accurate gaps at a cost comparable to a GGA calculation. Alternatively, for semiconductors hybrid DFT [40] can also be used, although at significantly high computational cost. In cases where van der Waals interactions play a role, an approximate but efficient scheme from Grimme [41] is implemented into WIEN2k.

For correlated electrons, the $LDA + U$ or $GGA + U$ schemes [15] (as mentioned in Section 8.3.2) can be used. They are computationally efficient but have the drawback that in practice the U parameter must be properly chosen, which makes this approach no longer fully *ab initio*. Many-body physics can treat correlation effects, for example, by the dynamical mean field theory (DMFT) but traditionally such schemes require parameters, which nowadays can be extracted from DFT results. For example, one can start with an LDA calculation and transform the basis set from a Bloch picture to a Wannier description (see, for example, Ref. [42]). In the latter, the correlated electrons can be described by the DMFT, which can account for the local correlation effects using a Hubbard U and hopping parameters that were extracted from LDA results. Such combinations are called LDA + DMFT, as described in a recent review [43].

If the property of interest is a ground-state property such as the total energy, or of a form, which can be derived from the electron density, for example, the electric field gradient (EFG), then DFT is well suited. In cases where excitations are involved, such as K-edges in core-level spectroscopy, a proper treatment of core–hole electron interactions can sometime be simulated using a static core–hole supercell [44,45]. However, in particular for $L_{2,3}$ X-ray absorption spectra (XAS), one needs to solve the Bethe–Salpeter equations (BSE), leading to heavy computations that are feasible, but only for small systems. Such calculations made it possible to explain and resolve the deviation from the branching ratio between L_2 and L_3 edges of $L_{2,3}$ XAS in compounds such as CaF_2, TiO_2, or $SrTiO_3$ and reproduced all fine details of the experimental spectra [46].

8.5.3
Basis Set and Brillouin Zone Sampling

The atomic sphere radii should be set by the *setrmt* utility so that the ratio of the muffin tin radii (R_{mt}) of different atoms is properly set and optimal convergence is obtained. For instance, *setrmt* will set the R_{mt} of H about half the size of that of C or O, and R_{mt}s of transition metals will be larger than that of sp elements, allowing a more effective augmentation for elements with orbitals (e.g., the 3d states) that are difficult to converge. R_{mt}s should be kept fixed when comparing different calculations (for example, when comparing different phases or changing the unit cell volume).

The convergence of a calculation should be checked for the property of interest. In plane wave schemes (such as LAPW and WIEN2k), the main parameter is the number of plane waves. In WIEN2k the parameter RK_{max} has become the standard convergence parameter and is defined as the product of R, the smallest sphere radius, times K_{max}, the largest K-vector in the plane wave expansion. Reasonable values of RK_{max} depend on the type of atom with the smallest R_{mt} and range from about $RK_{max} = 3$ for H to $RK_{max} = 9$ for elements with 4f electrons (see http://www.wien2k.at/reg_user/faq).

The augmentation of the plane waves inside each atomic spheres can be done differently, namely, using LAPW or the APW + lo (plus local orbitals) scheme (as discussed in Section 8.4.2). Here also the partition of the electronic states into core, semi-core, and valence electrons enters. Especially in the case of high-pressure studies, it may be necessary to include some states as semi-core states instead of treating them as core states, because they will (weakly) interact with the neighboring atoms and thus form bands.

Since the computational effort for finding the energy eigenvalues (see Figure 8.3) is proportional to the number of **k**-points, one restricts the calculations to the irreducible Brillouin zone (IBZ) and uses symmetry to sample the full BZ. This can save a lot of computer time. For instance, for a cubic crystal (fcc or bcc) with 48 symmetry operations, only 1/48th of the whole BZ is needed (namely the IBZ). How fine the **k**-mesh should be taken depends on the size of the BZ, which is reciprocal to the unit cell in direct space. The larger the latter, the smaller the corresponding BZ and thus a cruder **k**-mesh is sufficient. Experience has shown that one can use a cruder **k**-mesh during the SCF cycle (Figure 8.3) than for calculating certain properties (especially for optics or magnetic anisotropy) where a very fine **k**-mesh is needed. In metals (especially magnetic ones) energy bands are partially occupied and thus a fine **k**-mesh is needed in order to define which of the band states $E_{n\mathbf{k}}$ are occupied or unoccupied. In this context each state (with energy $E_{n\mathbf{k}}$) can be broadened (e.g., by thermal smearing) in order to determine which fraction is occupied; this reduces charge fluctuations during the SCF cycles and accelerates convergence. The calculation of some properties (e.g., electron densities or density of states) requires an integration over the BZ. For such integrations, the tetrahedron method [47] is often used instead of a simple summation. For any property related to the Fermi surface, one must use a rather fine **k**-mesh. In insulators, a gap separates the occupied from the unoccupied states and consequently a cruder **k**-mesh is often sufficient.

We summarize the crucial points for efficiency and accuracy below:

- Number of plane waves, choice of augmentation, **k**-mesh (http://www.wien2k.at/reg_user/faq/).
  - Select a proper RK_{max} (compromise between accuracy and computational cost).
  - Select semi-core states described by local orbitals.
  - Choose a proper **k**-mesh (compromise between accuracy and computational cost).
- DFT functional
  - Choose the DFT functional (Section 3.2) according to Jacob's ladder (Figure 8.2). Often the "optimal" choice depends on your system and the calculated properties.
- Magnetism
  - Choose a nonmagnetic or magnetic (spin polarized) calculation.
  - Define a specific collinear magnetic order (ferro- or a particular ferri- or antiferromagnetic state) or use even a noncollinear magnetic structure.
  - If necessary, add spin–orbit coupling to the default scalar relativistic treatment.

Chemical effects are often determined by subtle differences and thus one must keep all computational parameters fixed if one wants to compare two similar situations. Such details are, for example, the DFT functional, atomic sphere sizes, basis sets, and plane wave cutoff. To answer the question, which of two phases of a compound is more stable, one must obeys these rules. The corresponding total energy can be a few hundred thousand Ryd but the difference can be in the mRyd range. This is a challenge in terms of numerical precision. It is important to mention that it is mandatory to check specifically the calculated properties for convergence with respect to RK_{max} and the **k**-mesh.

8.5.4
Hardware Considerations and Parallelization

WIEN2k runs on any LINUX operating system. At present (in our opinion) Intel processors of the I7 family (with a large memory bandwidth) provide the best and most economic hardware platform. One can do calculations for cells with 50–100 atoms on a modern PC (with 16 GB RAM and four cores). A few PCs such as this even allow an efficient k-point parallelization. For larger systems (up to about 1000 atoms), a cluster of PCs with fast communication (InfiniBand) is necessary, because on such supercomputers also MPI parallelization of the large eigenvalue problem is possible. An efficient implementation of a code running on a modern hardware requires optimized linear algebra libraries (Intel's MKL library) and algorithms adjusted to our applications. This is an important part of program development but not be covered here.

8.6
Examples of DFT Results for Bulk Materials and Surfaces

Here we focus on WIEN2k applications for which many references can be found on the web page (http://www.wien2k.at/papers). A summary of main results and properties obtained with WIEN2k was given, for example, in Chapter 5 of Ref. [19]. Here we provide a short, comprehensive list, and give for some cases a reference to a publication in which one can read more details:

- Atomic structure (see Chapter 2) by means of total energy and forces (see also Chapter 2 of Ref. [19]).
  - Accurate determination of unit cell (space group), equilibrium volume (Section 6.1), and lattice constants [37,48].
  - Optimizing atomic (equilibrium) positions using forces (see, for example, Chapter 6.5 of Ref. [26]) and phonon instabilities as discussed in Chapter 6.4 for the pyrochlore structure $Y_2Nb_2O_7$ [49].
  - Creating and optimizing models (supercells, surface slabs) for novel nanomaterials; for example, the h-BN "nanomesh," a hexagonal-BN layer on a Rh(111) substrate with a large lattice mismatch [50] causing corrugation of the BN layer.
  - Determine the phase stability under pressure [51].
  - Calculate phonons [3,52] and compare with IR, Raman and neutron scattering, and free energy calculations [53]
- Electronic structure
  - Band structure with band character (see, for example, Chapter 6.2 in Ref. [26] for the refractory metal compound TiC), and DOS, Fermi surfaces (see Section 6.2).
  - Bandgap [39].

- Chemical bonding and type of bonds: covalent, ionic, metallic bonds, van der Waals (see, for example, Chapter 6.2 in Ref. [26] or [2]).
- Electron density, Bader charges [54]: see, for example, Chapter 6.3 in Ref. [26] for the Verwey transition in the double perovskite $YBaFe_2O_5$, where one can distinguish between the charged ordered CO and valence-mixed VM state for Fe^{2+} and Fe^{3+} [49].
- Magnetic properties (see, for example, Chapter 4 in Ref. [24] for the half-metallic CrO_2)
  - Magnetic structure (FM, AFM [49], noncollinear in UO_2 [55]). Magnetization density, magnetic anisotropy.
- Spectroscopies and other properties (see user's guide: www.wien2k.at)
  - Core-level spectroscopy: XPS, XAS [49], XES, EELS [45].
  - Optical spectroscopy: UV-vis [56].
  - IR, Raman [3,52,54].
  - NMR: chemical shift and Knight shift [57,58], EFG; (see, for example, Chapter 6.4. in Ref. [26] for TiO_2) and [49].
  - Mössbauer spectroscopy: isomer shifts, hyperfine fields, EFG for $YBaFe_2O_5$ (see Section 6.3) [49].
  - Scanning tunneling microscopy (STM) in constant height and constant current mode [50,59].
  - Berry phases: polarizability, Born effective charges [60,61].

Now we present a short summary of selected examples for which details can be found in the given publications.

8.6.1
Equilibrium Volume Using Different DFT Functionals

Figure 8.8 illustrates how to find the equilibrium unit cell for elemental silicon using the LDA or GGA (PBE) as the DFT functional. LDA mainly overbinds leading to a volume less than experiment, whereas PBE leads to too large an equilibrium volume. Experiment is often in between. Other functionals can perform better than LDA or GGA-PBE, as was shown in Ref. [37] for many compounds. The recent analysis [14] extends the search for better DFT functionals to other properties than just the lattice constants and contains many details.

In more complex cases with lower symmetry, the optimization of the lattice parameters a,b,c and angles α,β,γ by total energy minimization only is not feasible anymore and only the stress tensor $\sigma_{\alpha\beta} = \frac{1}{\Omega}\frac{dE}{d\varepsilon_{\alpha\beta}}$ [3] can provide the necessary information to optimize all lattice parameters simultaneously in an efficient way. While the stress tensor has been implemented in many pseudopotential codes [3] and in the Soler–Williams APW method [62], it has not been implemented successfully in the regular LAPW method.

Another major problem concerns the prediction of experimentally unknown structures (such as for novel materials or high-pressure phases). This delicate and hard to solve problem has been tackled by various approaches such as

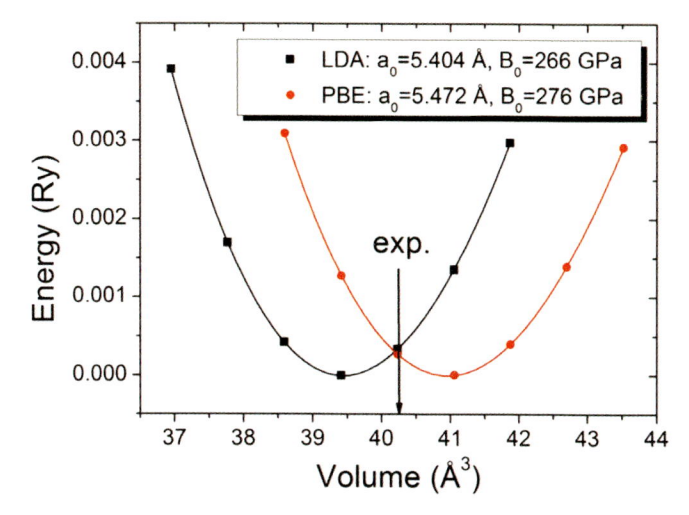

Figure 8.8 The total energy (with respect to the minimum) of silicon as a function of unit cell volume for LDA and PBE: For both functionals the equilibrium lattice constant a_0 (at the energy minimum) is given and the corresponding bulk modulus B_0.

brute force random search strategies, simulated annealing, or evolutionary schemes [48].

8.6.2
Band Structure and Chemical Bonding

The energy band structure of a material is a fundamental property of any solid-state calculation. Many experimentally relevant quantities, such as the direct or indirect energy bandgap, effective electronic masses and the optical absorption properties in semiconductors, or the Fermi surface and related properties in metals are directly connected to the momentum–energy dispersion relations in the solid. The eigenvalues in DFT are in principle not true excitation energies except the highest occupied state ("valence band maximum" (VBM) in semiconductors or Fermi energy in metals) that has a strict physical meaning. Therefore, one should not expect good agreement with experiment, although in particular for metals the situation is often quite satisfactory. Figure 8.9 shows as an example the calculated energy bands of VSe_2 compared to high-resolution angle-resolved photoemission (ARPES) spectra (for more details, see Ref. [63]). The agreement between theory and experiment is surprising even over a large energy range. Small discrepancies appear only at Γ close to E_F, where experimentally a larger splitting between the p_z and p_{xy} states appear compared to theory.

A 3D plot of those **k**-vectors for which the V-3d bands cut the Fermi energy yields the Fermi surface (FS) (Figure 8.10). The FS is important to understand the 3D charge density wave in VSe_2 due to nesting features (for more details, see Ref. [63]).

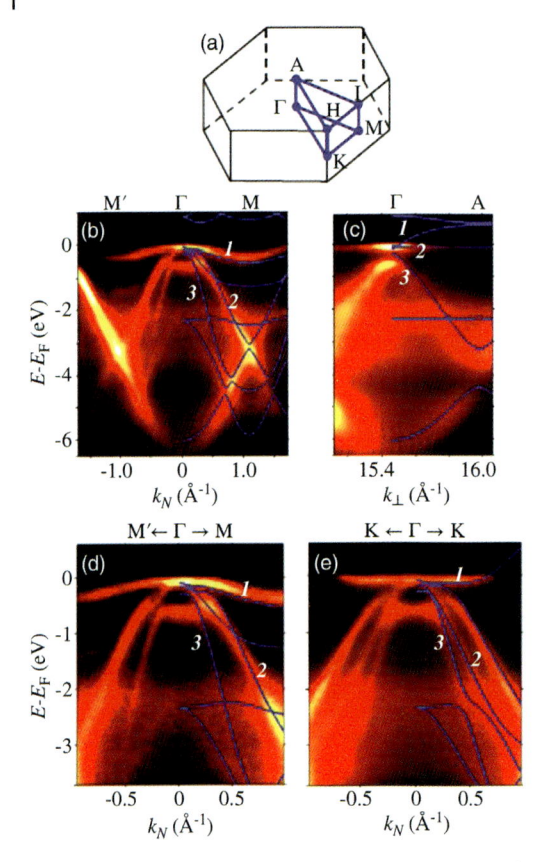

Figure 8.9 ARPES experimental band structure of VSe_2 along selected BZ directions together with DFT calculations (blue lines). The V-3d band (1) forms the Fermi surface. The Se-p_{xy} and Se-p_z bands are labeled with (2) and (3), respectively. (Taken from Ref. [63]).

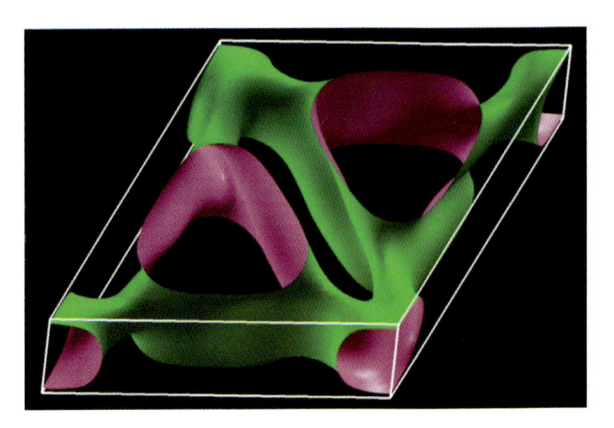

Figure 8.10 Calculated Fermi surface of VSe_2 (using XCrystDen [64]).

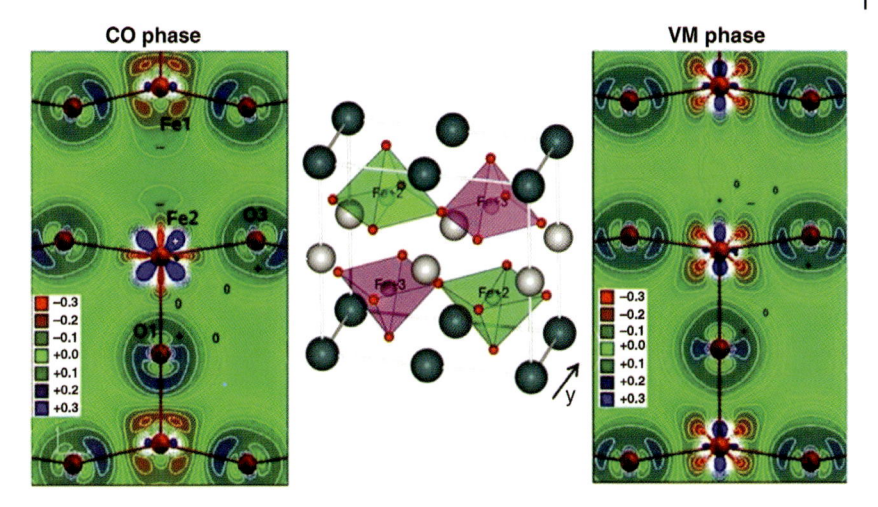

Figure 8.11 A charge-ordered (CO) phase at low temperature and a valence-mixed (VM) phase at high temperature exist in YBaFe$_2$O$_5$. The atomic structure of the CO phase is sketched in the center (partly taken from Ref. [49]). The Fe^{2+} and Fe^{3+} ions form chains along the y-axis. In addition, the Fe moments are antiferromagnetically aligned along this y-direction. The difference electron density $\Delta\rho = \rho_{cryst} - \rho_{at}^{sup}$ is shown for the CO (left) and VM (right) phase.

8.6.3
Verwey Transition in YBaFe$_2$O$_5$

The oxygen-deficient double-perovskite YBaFe$_2$O$_5$ shows a phase transition around 308 K between an orthorhombic (space group Pmma) and a tetragonal (space group P4/mmm) phase. In both phases, the magnetic moments from the Fe atoms are antiferromagnetically ordered. In the low temperature phase, a charge-ordered (CO) state appears in which Fe^{2+} and Fe^{3+} occupy different atomic positions (as sketched in the center of Figure 8.11). At high temperatures, a valence-mixed (VM) state exists in which all Fe atoms are equivalent and have the formal oxidation state Fe$^{2.5+}$. Such a situation is called a Verwey transition. This material has been studied in great detail. A DFT simulation [49] provides many results and references. A conventional GGA calculation leads to a metallic behavior in contrast to experiment and finds too small magnetic moments. GGA + U calculations, however, can explain all known experimental data, and there is a great deal of it. With GGA + U, this perovskite is found to be an insulator where the gap is smaller for the VM than for the CO state. The magnetic moments also agree well with experiment. A good quantity to analyze the chemical bonding in this system is the difference density $\Delta\rho$, taken between the crystal density ρ_{cryst} (from the SCF result) minus a superposition of (neutral) atomic densities ρ_{at}^{sup} placed at all crystal positions. We use this superposition density as a reference system without chemical interaction. The chemical binding can then induce charge transfer, covalent, or metallic bonds, the extent of which

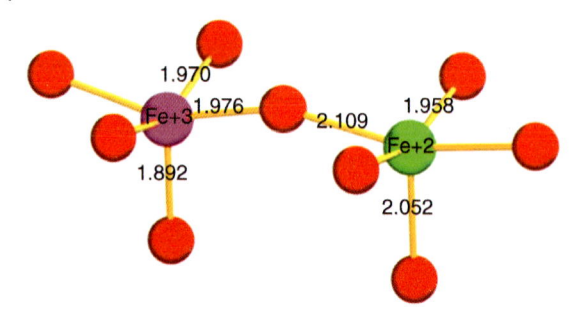

Figure 8.12 The nearest-neighbor distances (in a.u.) around Fe^{2+} and Fe^{3+} in the charge-ordered (CO) phase of $YBaFe_2O_5$.

shows up in $\Delta\rho$. For the CO phase, $\Delta\rho$ differs significantly between Fe1 and Fe2 (Figure 8.11 left), Fe^{2+} and Fe^{3+}, respectively. However, for the VM phase, $\Delta\rho$ is identical for all Fe positions (Figure 8.11 right) as to be expected.

In order to gain insight into structure–property relations, we analyzed the nearest-neighbor distances around the two types of iron atoms (Figure 8.12) in the CO phase. The experimental distances to the neighboring oxygen atoms agree well with our structure optimization.

The oxygen–iron distances vary significantly for Fe^{2+} and Fe^{3+}, whereas they are very similar (about 2 a.u.) in the VM phase. Consequently, the local distortion around the Fe atoms is the driving force for the Verwey transition causing a change in occupation of certain orbitals with a high Fe-3d character (to be seen in the partial density of states not shown here). Another quantity that is sensitive to this local structure is the EFG, which primarily arises from the nonspherical charge distribution around the measured nucleus (Fe in our case). The calculated EFGs agree well for GGA + U but not for LDA or GGA alone (as discussed in Ref. [49]). In addition, isomer shifts and the magnetic hyperfine fields were computed and agree with Mössbauer experiments. In this case, we can summarize that the GGA + U calculations yield a consistent picture of the structure–property situation. The U parameter had to be chosen as parameter making such a calculation less *ab initio* but the results only weakly depend on U. When U is chosen in a realistic range, many experimental results can be properly determined, making the results convincing.

8.6.4
Structure Relaxation of $Y_2Nb_2O_7$

From X-ray power diffraction data, Fukazawa and Maeno [65] proposed that $Y_2Nb_2O_7$ crystallizes in the pyrochlore structure (Fd-3 m), which contains 22 atoms per unit cell. They suggested that this compound may be an example of a geometrically frustrated 4d oxide with a novel ground state. The pyrochlore structure can be represented by corner-shared octahedra. In this structure, the two types of Nb atoms, Nb_1 and Nb_2, would have the same

Nb–Nb nearest-neighbor distance of 3.65 Å. An LDA calculation results in a nonmagnetic metallic ground state with a conventional t_{2g} and e_g splitting due to the octahedral crystal field of the oxygen ligands, but these results are in contrast to the semiconducting behavior in experiment (for further details, see Ref. [66]). In a simple ionic model, both types of Nb atoms would be Nb^{4+} with a $4d^1$ configuration. The bands at the Fermi level are mainly of t_{2g} character and are partially occupied, leading to a metallic behavior. Thus, the question arises: how can one get a nonmagnetic insulator. Can the electron–electron correlation cause the insulating nature of this system? Correlations should be less important for 4d than for 3d systems. Another possibility to explain the insulating nature of $Y_2Nb_2O_7$ is a charge (lattice) instability, for example, the tendency of Nb^{4+} ions to disproportionate into Nb^{5+} and Nb^{3+}. Such a case would be strongly frustrated and thus difficult to determine by X-ray diffraction (especially with power samples). Such a structural distortion would break the dominant octahedral crystal field and thus could cause the insulating behavior.

In order to study possible structural relaxations, the authors of Ref. [66] have calculated the phonon modes. This has been done using the so-called direct method [52], where forces are calculated in suitable supercells with atomic displacements in all directions and for all atoms. Using these calculated forces, one can derive the force constants in the harmonic approximation and setup and solve the dynamical matrices for the desired q-vectors, which yield the phonon frequencies at those points. This led to the surprising result that imaginary frequencies appear (Figure 8.13), indicating a strong phonon instability of the system in the pyrochlore structure. The authors have chosen the unstable phonons at the Γ, X, K, or L points and have frozen them in (with a certain amplitude) in order to build several starting structures for a search of lower symmetry structures.

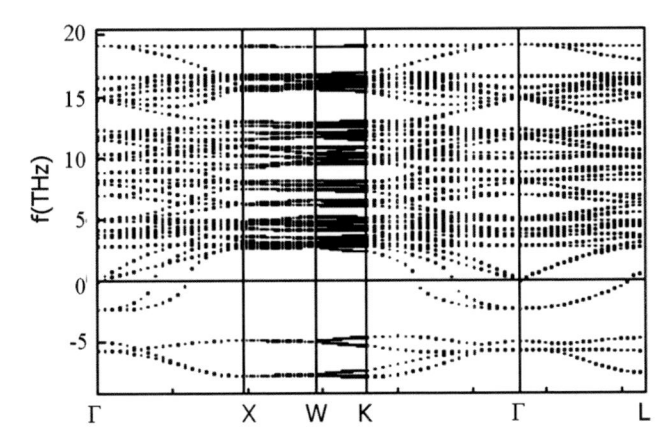

Figure 8.13 Phonon dispersions of $Y_2Nb_2O_7$ in the ideal pyrochlore structure show imaginary frequencies indicating an instability.

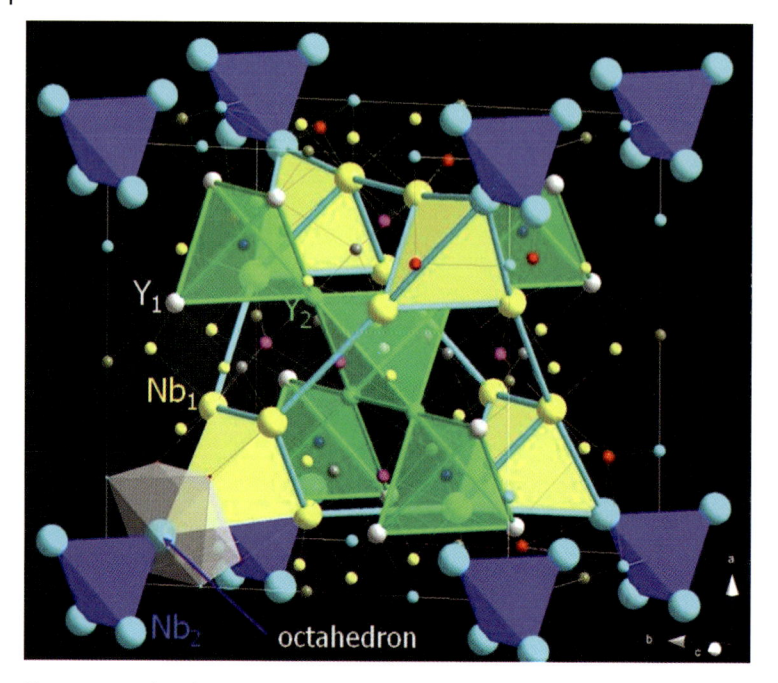

Figure 8.14 Relaxed structure of $Y_2Nb_2O_7$ (in space group P-43 m) is shown emphasizing the metal sublattices (Nb_1, Nb_2, Y_1, Y_2) forming tetrahedral networks. In addition, various oxygen positions and one octahedron around a Nb_2 site are shown.

Starting with these structures, the authors have performed full structural relaxations using supercells with up to 88 atoms without any symmetry constraint, so that the atoms can move in all directions. Several local minima were found but the relaxed structure with the lowest energy became an insulator and thus should be the ground state. As a next step, a search was made for a higher symmetry (with a given tolerance) using the program KPLOT [67]. This search led to the space group P-43 m. The structure was relaxed for this symmetry as shown in Figure 8.14. The main result (just focusing on the Nb sublattices) is (i) Nb_1^{3+} atoms form tetrahedra that are highly elongated along the (111) direction with a Nb_1–Nb_2 distance of 3.89 Å, (ii) Nb_2^{5+} atoms form perfect tetrahedra with short Nb–Nb bond distances of 2.91 Å. In the pyrochlore structure (originally proposed), the distance was 3.65 Å for all nearest-neighbor Nb–Nb bonds. The disproportionation into Nb^{5+} and Nb^{3+} therefore becomes evident.

8.6.5
Surfaces and Heterogeneous Catalysis

For many properties – in particular for heterogeneous catalysis, it is not the bulk but the surface geometrical and electronic structures that determine the behavior and catalytic activity of a material. The study of surfaces is therefore of vital

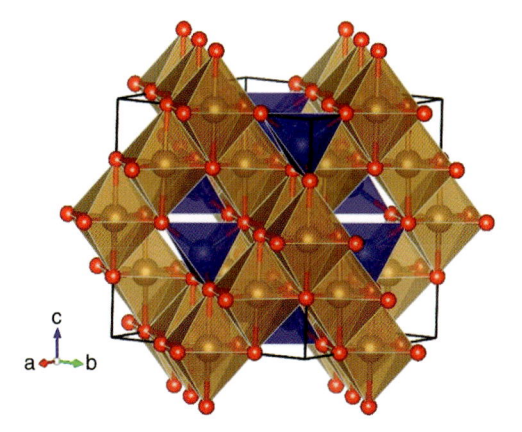

Figure 8.15 Unit cell of cubic Fe_3O_4 with Fe_A (blue, tetrahedral) and Fe_B (red, octahedral) coordination.

interest and they are usually modeled with a surface slab (as mentioned in Section 8.2). Surface atoms usually lose about half of their bonding neighbors, leading to many dangling bonds. In addition, a surface may not be charge neutral but possesses a finite charge, leading to the famous "polar catastrophe." In such cases one can expect at the surface a large structural reconstruction, which can even break the crystalline periodicity with a commensurate (e.g., a $\sqrt{2} \times \sqrt{2}$) or even incommensurate structure, which also affects the electronic structure.

Here we focus on an example for a specific surface, namely, the (001) surface of magnetite, Fe_3O_4. Already in ancient times, magnetite was a well-known and important material, which was used in compasses. Nowadays its interesting properties, such as the charge order–disorder Verwey transition, catches the interest of scientists (see also Section 6.3), as well as its applications as a catalyst. At room temperature, magnetite crystallizes in the cubic inverse spinel structure (AB_2O_3) where the Fe_A^{3+} atoms are tetrahedrally coordinated and the $Fe_B^{2.5+}$ atoms have octahedral coordination with oxygens (see Figure 8.15). In addition, the Fe_A and Fe_B atoms have a ferrimagnetic spin-alignment, but the cell has a net ferromagnetic moment. This mixed-valence state exists only above the Verwey temperature ($T_V = 117$ K), while below T_V, the Fe_B sublattice shows a charge-ordered phase with a complicated structure due to electron–lattice coupling [68].

The most stable (001) surface of Fe_3O_4 has a "B"-layer truncation and consists of alternating $Fe_{B2}O_4$ and Fe_A layers. The $(Fe_BO_2)_2$ layers consist of FeO_2 rows running alternatively along the [110] and [−110] directions in adjacent planes. This surface exhibits the well-known $\sqrt{2} \times \sqrt{2}(R45°)$ reconstruction [69], where the straight Fe-rows get modulated leading to "narrow" and "wide" sites. This surface was also studied theoretically by GGA and GGA + U calculations [70,71]. From them, such a reconstruction was found to be most stable, which has been explained by a charge order effect in the subsurface Fe_B layer. Later studies by the Diebold group revealed that ad-atoms such as Au, Ag, or Pd adsorb

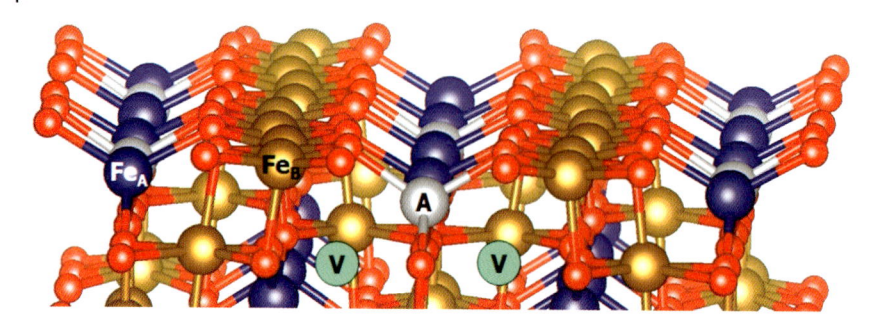

Figure 8.16 The (001) surface of magnetite Fe_3O_4. The image shows the strong undulation of the surface Fe_BO_2 rows, the Fe ad-atom (A) in the subsurface Fe_A-layer, and the double Fe_B-vacancy (V) in the Fe_BO_2 sub-subsurface layer. Au adsorption is solely possible at the geometrical "wide" site, because the A-atoms block the narrow site.

exclusively on the "narrow" site [72]. This single atom adsorption is stable up to high temperature and the magnetite surface effectively prevents any clustering. This site selectivity, however, could at first not be reproduced by DFT calculations, since both adsorption sites were found to be energetically almost degenerate. As a next step, this puzzle was solved by a new structural model [73], which takes into account two subsurface Fe_B vacancies and an additional Fe_A subsurface atom that selectively blocks one of the adsorption sites (Figure 8.16). This structural model sounds quite complicated, but it turns out to be the most thermodynamically stable and as confirmed by new LEED-IV measurements. The new structural model led to a significant better R-value than the previous attempts. However, there was another discrepancy between theory and experiment, namely, that Au atoms adsorb solely on the narrow site according to experimental STM investigations, whereas in theory it would be the geometrically wide site. Detailed simulations of STM images showed that for a small bias voltage, the geometrically wide site actually looks narrow due to a tilted Fe-3d electron density. For a larger bias voltage, the true geometry can be seen, both in the simulations and in experiment. The experimental STM was compared without the theoretical one, which was interpreted by the proper partial DOS (for further details and the corresponding discussion, see, for example, Figure 3 of Ref. [73]). This illustrates the importance of the close collaboration between theory and experiment.

8.7
Conclusions

Calculations are mainly done in order to gain insight for a material or to find trends for a class of compounds so that a systematic improvement can be proposed, in contrast to the trial-and-error scheme sometimes used experimentally. Theory can treat all systems irrespective of abundance, environmental aspects, or costs. Today, theory is often more efficient in optimizing materials than preparing all kinds of possible cases that then need to be characterized experimentally. The experiments can focus on the promising cases and do not need to study all possible cases. With computer experiments, one can study even unstable/artificial systems, which obviously cannot be measured. If one speculates that a relaxation of the atomic structure due to impurities or defects cause a certain special property, one can perform a computer experiment with or without relaxation and analyze the property. This can provide important insight.

In the comparison between theory and experiment, some general questions need to be kept in mind (as discussed in Section 7.1 of Ref. [19]), especially when deviations occur: Is the atomic structure properly described (and idealized) as discussed in Section 8.2? How accurate is the quantum mechanical treatment by DFT (Section 8.3)? Were the calculations carried to full convergence as summarized in Section 8.5 (in terms of **k**-points and basis sets)? Are there additional aspects, such as relativistic treatment, finite temperature, pressure, ground state versus metastable state, which may affect the results?

With all the possibilities mentioned in the previous sections, it is often useful to combine different theories according to their advantages but keeping in mind their disadvantages. About 20 years ago, the fields of quantum chemistry, DFT, and many-body theory were completely separated with hardly any cooperation between them: This has fortunately changed. The strength and weaknesses of the different approaches are recognized and mutually appreciated. One can solve complex problems only by close collaboration with the corresponding experts.

References

1 Schwarz, K. (2003) Electrons, in *International Tables for Crystallography, Volume D, Physical Properties of Crystals* (ed. A. Authier), Kluwer Academic Publishers, pp. 294–313.

2 Dronskowski, R. (2005) *Computational Chemistry of Solid State Materials: A Guide for Material Scientists, Chemists, Physicists and Others*, Wiley-VCH Verlag GmbH, Weinheim, Germany, ISBN 3-527-31410-5.

3 Martin, R.M. (2008) *Electronic Structure: Basic Theory and Practical Methods*, Cambridge University Press, ISBN-13: 978–0531534406.

4 Churakov, V.S.y (2015) Ab initio simulations of mineral surfaces: Recent advances in numerical methods and selected applications, in *Highlights in Mineralogical Crystallography* (eds T. Armbruster and R.M. Danisi), DeGruyter Verlag Berlin, ISBN 978-3-11-041710-4.

5 Bartlett, R.J. and Musial, M. (2007) *Rev. Mod. Phys.*, **79**, 291.

6 Booth, G.H., Grüneis, A., Kresse, G., and Alavi, A. (2013) *Nature*, **493**, 365.

7 Hohenberg, P. and Kohn, W. (1964) *Phys. Rev.*, **136**, 864.

8 Kohn, W. and Sham, L.S. (1965) *Phys. Rev.*, **140**, A1133.

9 Ceperley, C.M. and Alder, D.J. (1980) *Phys. Rev. Lett.*, **45**, 566.

10 Perdew, J.P. and Schmidt, K. (2001) Jacob's ladder of density functional approximations for the exchange-correlation energy, in *Density Functional Theory and Its Application to Materials*, vol. **577** (eds V. Van Doren, C. Van Alsenoy, and P. Geerlings), American Institute of Physics.

11 Perdew, J.P., Burke, K., and Ernzerhof, M. (1996) *Phys. Rev. Lett.*, **77**, 3865.
12 Perdew, J.P., Ernzerhof, M., and Burke, K. (1996) *J. Chem. Phys.*, **105**, 9982.
13 Harl, J., Schimka, L., and Kresse, G. (2010) *Phys. Rev. B*, **81**, 11526.
14 Tran, F., Blaha, P., and Schwarz, K. (2015) *J. Chem. Theory Comput.*, **11**, 4717.
15 Anisimov, V.I., Solovyev, I.V., Korotin, M.A., Czyżyk, M.T., and Sawatzky, G.A. (1993) *Phys. Rev. B*, **48**, 16929.
16 Hahn, T. (ed.) (1995) *International Tables for Crystallography, Volume A, Space-Group Symmetry*, Kluwer Academic Publishers, ISBN 0-7923-2950-3.
17 Koelling, D.D. and Harmon, B.N. (1977) *Solid State Phys.*, **10**, 3107.
18 MacDonnald, A.H., Picket, W.E., and Koelling, D.D. (1980) *J. Phys. C Solid State Phys.*, **13**, 2675.
19 Schwarz, K. (2015) Computation of materials properties at the atomic scale, in *Quantum Mechanics* (ed. M.R. Pahlavani), InTech, ISBN 978-953-51-4131-0.
20 Slater, J.C. (1937) *Phys. Rev.*, **51**, 846.
21 Blaha, P., Schwarz, K., Sorantin, P., and Trickey, S.B. (1990) *Comput. Phys. Commun.*, **59**, 399.
22 Blaha, P., Schwarz, K., Madsen, G.K.H., Kvasnicka, D., and Luitz, J. (2001) *An Augmented Plane Wave Plus Local Orbitals Program for Calculating Crystal Properties*, Vienna University of Technology, ISBN 3-9501031-1-2.
23 Schwarz, K., Blaha, P., and Madsen, G.K.H. (2002) *Comput. Phys. Commun.*, **147**, 71.
24 Schwarz, K. and Blaha, P. (2003) *Comput. Mater. Sci.*, **28**, 259.
25 Schwarz, K. (2003) *J. Solid State Chem.*, **176**, 319–328.
26 Schwarz, K., Blaha, P., and Trickey, S.B. (2010) *Mol. Phys.*, **108**, 3147–3166.
27 Schwarz, K. and Blaha., P. (2012) Electronic structure of solids and surfaces with WIEN2k, in *Practical Aspects of Computational Chemistry I: An Overview of the Last Two Decades and Current Trends* (eds J. Leszczynscki and M.K. Shukla), Springer Science+Business Media B.V., pp. 191–207, ISBN 978-94-007-0918-8.
28 Singh, D. and Nordström, L. (2006) *Plane Waves, Pseudopotentials and the LAPW Method*, 2nd edn, Springer, New York.
29 Cottenier, S. (2016) *Density Functional Theory and the Family of (L)APW-Methods: A Step-by-Step Introduction*, 2nd edn, Instituut voor Kern- en Stralingsfysica, KU Leuven, Belgium, pp. 2002–2013, ISBN 978-90-807215-1-7.
30 Andersen, O.K. (1975) *Phys. Rev. B*, **12**, 3060.
31 Weinert, M., Wimmer, E., and Freeman, A.J. (1982) *Phys. Rev. B*, **24**, 4571.
32 Singh, D.J. (1975) *Phys. Rev. B*, **43**, 6388.
33 Sjöstedt, E., Nordström, L., and Singh, D.J. (2000) *Solid State Commun.*, **114**, 15.
34 Madsen, G.H.K., Blaha, P., Schwarz, K., Sjöstedt, E., and Nordström, L. (2001) *Phys. Rev. B*, **64**, 195134.
35 Lejaeghere, K., Van Speaybroeck, V., Van Oost, G., and Cottentier, S. (2013) *Critical Reviews in Solid State and Materials Science*, **39** (1), 1–24.
36 Wu, Z. and Cohen, R. (2006) *Phys. Rev.*, **B73**, 235116.
37 Haas, P., Tran, F., and Blaha, P. (2009) *Phys. Rev. B*, **79**, 085104.
38 Jiang, H., Gomez-Abal, R.I., Li, X., Meisenbichler, C., Ambrosch-Draxl, C., and Scheffler, M. (2013) *Comput. Phys. Commun.*, **184**, 384.
39 Tran, F. and Blaha, P. (2009) *Phys. Rev. Lett.*, **102**, 226401.
40 Tran, F. and Blaha, P. (2011) *Phys. Rev. B*, **83**, 235118.
41 Grimme, S., Antony, J., Ehrlich, J., and Kieg, H. (2010) *J. Chem. Phys.*, **132**, 154104.
42 Kunes, J., Arita, R., Wissgott, P., Toschi, A., Ikeda, H., and Held, K. (2010) *Comput. Phys. Commun.*, **181**, 1888.
43 Held, K. (2007) *Adv. Phys.*, **65**, 829.926.
44 Strocov, V.N., Schmitt, T., Rubenson, J., Blaha, P., Paskova, T., and Nielsson, P.O. (2005) *Phys. Rev. B*, **72**, 085221.
45 Hetaba, W., Blaha, P., Tran, F., and Schattschneider, P. (2012) *Phys. Rev. B*, **85**, 205108.
46 Laskowski, R. and Blaha, P. (2010) *Phys. Rev. B*, **82**, 205104.
47 Blöchl, P.E., Jepsen, O., and Andersen, O.K. (1994) *Phys. Rev. B*, **49**, 16223.
48 Oganov, A.R. (ed.) (2010) *Modern Methods of Crystal Structure Prediction*, Wiley-VCH Verlag GmbH, Weinheim, Germany, ISBN 9783527409396.

49 Spiel, C., Blaha, P., and Schwarz, K. (2009) *Phys. Rev. B*, **79**, 115123.
50 Laskowski, R., Blaha, P., Gallauner, T., and Schwarz, K. (2007) *Phys. Rev. Lett.*, **98**, 106802.
51 Tröster, A., Schranz, W., Karsai, F., and Blaha, P. (2014) *Phys. Rev. X*, **4**, 031010.
52 Parlinski, K., Li, Z.Q., and Kawazoe, Y. (1997) *Phys. Rev. Lett.*, **78**, 4063.
53 Kahlenberg, V., Perfler, L., Konzett, J., and Blaha, P. (2013) *Inorg. Chem.*, **52**, 8941–8949.
54 Bader, R.W.F. (1994) *Atoms in Molecules: A Quantum Theory*, Oxford University Press, New York.
55 Laskowski, R., Madsen, G.K.H., Blaha, P., and Schwarz, K. (2004) *Phys. Rev. B*, **69**, 140408-1-4.
56 Ambrosch-Draxl, C. and Sofo, J. (2006) *Comput. Phys. Commun.*, **175**, 1.
57 Laskowski, R. and Blaha, P. (2015) *J. Phys. Chem.*, **119**, 19390–139396.
58 Laskowski, R. and Blaha, P. (2015) *J. Phys. Chem.*, **119**, 731–740.
59 Tersoff, J. and Hamann, D.R. (1985) *Phys. Rev. B*, **31**, 805.
60 Resta, R. (1994) *Rev. Mod. Phys.*, **66**, 899.
61 Ahmed, S.J., Kivinen, J., Zarborzan, W., Curiel, L., Pichardo, S., and Rubel, O. (2013) *Comput. Phys. Commun.*, **184**, 647651.
62 Nagasako, N. and Oguchi, T. (2011) *J. Phys. Soc. Jpn.*, **80**, 024701.
63 Strocov, V.N., Shi, M., Kobayashi, M., Monney, C., Wang, X., Krempasky, J., Schmitt, T., Patthey, L., Berger, H., and Blaha, P. (2012) *Phys. Rev. Lett.*, **109**, 086401.
64 Kokalj, A. (1999) *J. Mol. Graph. Model.*, **17**, 176.
65 Fukazawa, H. and Maeno, Y. (2003) *Phys. Rev. B*, **67**, 054410.
66 Blaha, P., Singh, D.J., and Schwarz, K. (2004) *Phys. Rev. Lett.*, **93**, 216403.
67 Hundt, R. *et al.* (1999) *J. Appl. Crystallogr.*, **32**, 413.
68 Senn, M.S., Wright, J.P., and Attfield, J. Paul. (1992) *Supramol. Sci.*, **255**, 583.
69 Wiesendanger, R., Shvets, I.V., Bürgler, D., Tarrach, G., Güntherodt, H.J., Coey, J.M.D., and Gräser, S. (1992) *Supramol. Sci.*, **255**, 583.
70 Pentcheva, R., Wendler, F., Meyerheim, H.L., Moritz, W., Jedrecy, N., and Scheffler, M. (2005) *Phys. Rev. Lett.*, **94**, 126101.
71 Lodziana, Z. (2007) *Phys. Rev. Lett.*, **99**, 206402.
72 Parkinson, G., Novotny, Z., Argentero, G., Schmid, M., Pavelec, J., Kosak, R., Blaha, P., and Diebold, U. (2013) *Nat. Mater.*, **12**, 724.
73 Bliem, R., McDermott, E., Ferstl, P., Setvin, M., Vasquez, O. Gamba., Pavelec, J., Schneider, M., Schmid, M., Diebold, U., Blaha, P., Hammer, L., and Parkinson, G. (2014) *Supramol. Sci.*, **346**, 1215.

9
Spin Polarization

Dong-Kyun Seo

Arizona State University, School of Molecular Sciences, Tempe, AZ 85287-1604, USA

9.1
Introduction

Spin polarization in electronic systems is a fundamental phenomenon associated with electron–electron interactions. An open-shell atom in its ground state exhibits unpaired electrons and hence carries a magnetic moment through spin polarization in valence orbitals. In a simplistic view, the magnetic moments would be aligned in parallel among neighboring atoms to maximize the lowering of the electron–electron repulsion by taking advantage of the Pauli exclusion principle. However, energy lowering by chemical bond formation or electron delocalization favors double electron occupancy of bonding (delocalized) orbitals, leading to an antiparallel spin alignment or even to a spin unpolarized electronic structure.

Figure 9.1a illustrates a situation where magnetic moments are highly localized on individual atomic sites in an ionic compound. The presence of the localized magnetic moments is a consequence of localization of d-electrons, as the interactions between d-orbitals and ligand orbitals are too weak to allow the former to spread out in the structure. In ionic transition metal compounds called Mott–Hubbard insulators, the weak d- and ligand orbital interactions are still significant enough to align the magnetic moments.

For rare earth metal compounds, however, f-electrons are severely localized regardless of their chemical environment because the f-orbitals do not mix substantially with ligand orbitals, unlike the upper d/s/p-orbitals on the same rare earth atom (ion). The magnetic exchange interactions among the localized magnetic centers are possible indirectly when the spin polarization in the f-orbitals induce an additional spin polarization in the upper d/s/p-orbitals that propagates in multiatomic distances. Such a long-range spin polarization would appear among the electrons that are well delocalized in the structure (Figure 9.1b).

Handbook of Solid State Chemistry, First Edition. Edited by Richard Dronskowski, Shinichi Kikkawa, and Andreas Stein.
© 2017 Wiley-VCH Verlag GmbH & Co. KGaA. Published 2017 by Wiley-VCH Verlag GmbH & Co. KGaA.

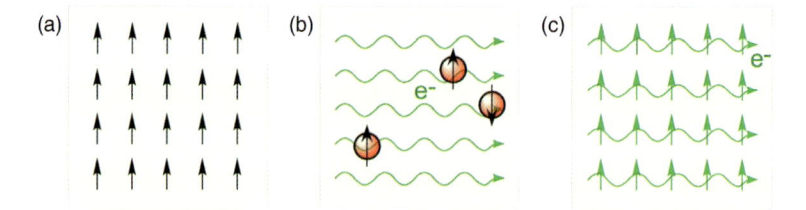

Figure 9.1 Schematic illustration of three different types of magnetism. In (a), localized magnetic moments, shown as black arrows, interact ferromagnetically and are ordered in parallel. In (b), localized magnetic moments interact through delocalized electrons, depicted by green waves, whose magnetic moments are negligible. In (c), delocalized electrons exhibit localized magnetic moments that interact with each other directly or indirectly.

Meanwhile, d-orbitals in transition metal elements, intermetallics, and alloys interact strongly with neighboring orbitals, forming metallic bonding, and the consequent orbital delocalization leads to less strong electron–electron repulsion among the d-electrons. In all transition metals except Cr, Fe, Co, and Ni, the reduction of electron–electron repulsion is so severe that spin polarization does not emerge. When spin polarization appears on transition metal atoms in the structure, their magnetic exchange interactions would be mediated through the delocalization of d-orbitals through neighboring transition metal or nontransition metal atoms (Figure 9.1c) [1].

Observation of various magnetic behaviors indicates that neither of two competing effects is always dominant in molecules and solids. Challenge in understanding the competition and its consequence, especially in solid-state compounds, is twofold. First, there exist diverse magnetic compounds which seemingly require different conceptual starting points to apprehend their properties. Second, different models for different types of the magnetism have been incoherent with disparate degrees of approximations and sophistications. This chapter is intended to examine those traditional models and to provide a unified conceptual model, under the name of "spin polarization perturbational orbital theory (SPPOT)" [2,3], in understanding the spin polarization phenomenon within the framework of Kohn–Sham spin-polarized density functional theory (KS-SDFT).

Over the years, the KS-SDFT has been a predominant tool for description of those spin-polarized systems by allowing electrons of different spins to have different spatial densities. In many cases, the predicted ground state and spin density from calculations are consistent with experimental observations. However, mere agreement between numerical calculations and experimental results is often not satisfactory, unless a comprehensible rationalization – an explanation couched in words and accessing trends – is given [2–4]. Underlying an estimation of total energies of various spin-polarized states are the one-electron orbital energies of Kohn–Sham spin orbitals, since the total electronic energy is obtained as the sum of the one-electron orbital energies of occupied Kohn–Sham spin orbitals after double-counting correction to the electronic repulsion terms. The summations of the one-electron densities given by the occupied Kohn–Sham spin orbitals provide total electron density and spin densities. While bearing no physical meaning, therefore, Kohn–Sham orbitals act as a vehicle that embodies and materializes the effect of electron interactions in numerical results. SPPOT is of a great value in understanding the spin polarization phenomenon, by providing approximate but analytical expressions that describe how Kohn–Sham spin orbitals change their energy and shape upon a given spin polarization pattern, how those changes result in the total energy differences among different spin arrangements, and how KS-SDFT results should be understood within the context of (spin-dependent) orbital interactions, no matter what DFT software programs are used for the calculations.

9.2
Spin Polarization Perturbational Orbital Theory (SPPOT)

In KS-SDFT [1,5], each magnetic state is described uniquely by a spin density,

$$\rho_S(\mathbf{r}) = \rho_\uparrow(\mathbf{r}) - \rho_\downarrow(\mathbf{r}), \tag{9.1}$$

as well as by a total electron density $\rho(\mathbf{r}) = \rho_\uparrow + \rho_\downarrow$ where

$$\rho_\sigma(\mathbf{r}) = \sum_i n_{i\sigma} \rho_{i\sigma}(\mathbf{r}) = \sum_i n_{i\sigma} |\psi_{i\sigma}(\mathbf{r})|^2, \tag{9.2}$$

with $\sigma = \uparrow$ or $\downarrow$. $\psi_{i\sigma}(\mathbf{r})$ and $n_{i\sigma}$ represent Kohn–Sham spin orbitals and their electron occupancies, respectively. Orbital density $\rho_{i\sigma}(\mathbf{r})$ is the electron density of the spin orbital $\psi_{i\sigma}(\mathbf{r})$. Combination of Eqs (9.1) and (9.2) leads to an alternative expression for $\rho_S(\mathbf{r})$

$$\rho_S(\mathbf{r}) = \sum_i \rho_{Si}(\mathbf{r}) = \sum_i \left(n_{i\uparrow} |\psi_{i\uparrow}(\mathbf{r})|^2 - n_{i\downarrow} |\psi_{i\downarrow}(\mathbf{r})|^2 \right), \tag{9.3}$$

where $\rho_{Si}(\mathbf{r})$ is the spin density from the orbital densities of a pair of Kohn–Sham spin orbitals, $\psi_{i\uparrow}(\mathbf{r})$ and $\psi_{i\downarrow}(\mathbf{r})$. Upon integration in space, $\rho_S(\mathbf{r})$ gives the net magnetic moment m as the number difference of $\uparrow$- and $\downarrow$-spin electrons in the system:

$$m = \int \rho_S(\mathbf{r}) d\mathbf{r} = \sum_i (n_{i\uparrow} - n_{i\downarrow}). \tag{9.4}$$

Broadly, Kohn–Sham spin orbitals act as a vehicle that embodies the spin polarization effect in numerical results. By following the underpinning of perturbational molecular orbital (PMO) theory [6], therefore, SPPOT focuses systematic analysis on how a spin polarization affects the energies and shapes of Kohn–Sham spin orbitals. All the discussions in this chapter will be limited to collinear magnetism, as the extension of the theory to noncollinear magnetism will be given elsewhere in the future.

9.2.1
Approximate Expressions of Exchange-Correlation Functional and Spin Polarization Energy

In SPPOT, an approximate form of exchange-correlation potential in Kohn–Sham Hamiltonian shows analytically how Kohn–Sham orbitals with different spins act differently upon a spin polarization in terms of their mixing and energy changes ("spin-dependent orbital mixing") in the case the spin polarization does not affect total electron density. Unlike PMO theory [6], SPPOT also allows us to examine quantitatively "total electronic energy changes" upon the spin polarization. This analysis is general for various types of wave functions regardless of the nature of basis functions including atomic-like orbitals and plane waves [2,3].

For collinear magnetic systems, the Kohn–Sham equation for a magnetic state **M** is given as

$$\mathcal{H}_\sigma \Psi_{i\sigma} = \varepsilon_{i\sigma} \Psi_{i\sigma}. \tag{9.5}$$

The SPPOT formalism employs an approximate Hamiltonian,

$$\mathcal{H}_\sigma = \mathcal{H}^{NM} + v^S_{xc\sigma}(\mathbf{r}), \tag{9.6}$$

where the first term $\mathcal{H}^{NM}$ is the Kohn–Sham Hamiltonian for the nonmagnetic state **NM** with the electron density $\rho(\mathbf{r})$, with its corresponding Kohn–Sham equation:

$$\mathcal{H}^{NM} \psi_i^0 = \varepsilon_i^0 \psi_i^0, \tag{9.7}$$

and the second term $v^S_{xc\sigma}(\mathbf{r})$ is the spin-polarized part of the exchange-correlation potential which is identified as the perturbation potential that leads to the magnetic state **M** defined by its spin density $\rho_S(\mathbf{r})$. The approximate Kohn–Sham Hamiltonian in Eq. (9.6) is obtained variationally from the expansion of the total electronic energy up to the second order with spin density as the expansion parameter [2].

Meanwhile, the exchange-correlation functional $E_{xc}[\rho(\mathbf{r}), \rho_S(\mathbf{r})]$ is decomposed into spin-independent and spin-dependent terms

$$E_{xc}[\rho(\mathbf{r}), \rho_S(\mathbf{r})] \approx E_{xc}[\rho(\mathbf{r}), 0] - \frac{1}{4} \iint \mathcal{K}_{xc}(\mathbf{r}, \mathbf{r}') \rho_S(\mathbf{r}') \rho_S(\mathbf{r}) d\mathbf{r}' d\mathbf{r}, \tag{9.8}$$

where

$$\mathcal{K}_{xc}(\mathbf{r}, \mathbf{r}') \equiv -2 \left[\frac{\partial^2 E_{xc}}{\partial \rho_S(\mathbf{r}) \partial \rho_S(\mathbf{r}')} \right]_0. \tag{9.9}$$

$\mathcal{K}_{xc}(\mathbf{r}, \mathbf{r}')$, a spin-independent positive function, is called "spin hardness function" [1,2]. The approximation in Eq. (9.8) is accurate up to the third order in $\rho_S(\mathbf{r})$ thanks to a time-reversal symmetry. Within the local spin density (LSD) approximation, the exchange correlation in Eq. (9.8) becomes

$$E_{xc}^{LSD}[\rho(\mathbf{r}), \rho_S(\mathbf{r})] \approx E_{xc}^{LSD}[\rho(\mathbf{r}), 0] - \frac{1}{4} \int \kappa_{xc}(\mathbf{r})[\rho_S(\mathbf{r})]^2 d\mathbf{r}, \tag{9.10}$$

where $\kappa_{xc}(\mathbf{r})$ is called "local spin hardness function":

$$\kappa_{xc}(\mathbf{r}) \equiv -2\left[\frac{\partial^2 E_{xc}^{LSD}}{\partial \rho_S(\mathbf{r})^2}\right]_0. \quad (9.11)$$

With Eqs (9.8) and (9.10), $v_{xc\sigma}^S(\mathbf{r})$ in Eq. (9.6) is approximately expressed as

$$v_{xc\sigma}^S(\mathbf{r}) \equiv \frac{\partial E_{xc}[\rho(\mathbf{r}), \rho_S(\mathbf{r})]}{\partial \rho_\sigma(\mathbf{r})} \approx \mp \frac{1}{2}\int \mathcal{K}_{xc}(\mathbf{r},\mathbf{r}')\rho_S(\mathbf{r}')d\mathbf{r}' \stackrel{LSD}{\approx} \mp \frac{1}{2}\kappa_{xc}(\mathbf{r})\rho_S(\mathbf{r}). \quad (9.12)$$

The − and + signs are for ↑- and ↓-spin orbitals, respectively. In other words, when a system becomes spin-polarized, the perturbational potential acts in opposite ways to ↑- and ↓-spin orbitals.

When the total electron densities are not significantly different between the magnetic state **M** and the nonmagnetic state **NM** (i.e., $\rho^M(\mathbf{r}) \approx \rho^{NM}(\mathbf{r}) \approx \rho(\mathbf{r})$) these results lead to a simple expression for the "spin polarization energy" as the total electronic energy change upon the spin polarization:

$$E_{sp}[\rho(\mathbf{r}), \rho_S(\mathbf{r})] \approx E[\rho(\mathbf{r}), \rho_S(\mathbf{r}) - E\rho(\mathbf{r}), 0]$$
$$\approx \sum_{i,\sigma}\left(n_{i\sigma}\varepsilon_{i\sigma} - n_{i\sigma}^0 \varepsilon_i^0\right) + E_{d.c.}^S[\rho(\mathbf{r}), \rho_S(\mathbf{r})], \quad (9.13)$$

where $E_{d.c.}^S[\rho, \rho_S]$ is the spin polarization part of the double-count correction for electron repulsion:

$$E_{d.c.}^S[\rho(\mathbf{r}), \rho_S(\mathbf{r})] = \frac{1}{4}\iint \mathcal{K}_{xc}(\mathbf{r},\mathbf{r}')\rho_S(\mathbf{r})\rho_S(\mathbf{r}')d\mathbf{r}'d\mathbf{r} \stackrel{LSD}{\approx} \frac{1}{4}\int \kappa_{xc}(\mathbf{r})[\rho_S(\mathbf{r})]^2 d\mathbf{r}. \quad (9.14)$$

9.2.2
Approximate Expressions of Spin Orbital shape and Energy Changes

SPPOT breaks down a spin polarization process into two sequential steps: (a) electron transfer among the zeroth-order spin orbitals to give an initial spin polarization; and (b) spin-dependent orbital mixing among those spin orbitals which results in a new electronic structure. Figure 9.2 depicts how SPPOT is conceptually related to KS-SDFT calculation procedures, as follows:

1) An iterative SCF procedure starts with an initially spin-unpolarized (unperturbed) electronic structure described by the Kohn–Sham equation in Eq. (9.7). ↑- and ↓-electrons equally occupy the unperturbed (zeroth-order) orbitals (i.e., $n_{i\uparrow}^0 = n_{i\downarrow}^0$ or all ψ_i^0's). Starting with this nonmagnetic state, a given magnetic state *M* arises with its spin density $\rho_S(\mathbf{r})$ and a practically unchanged total electron density $\rho(\mathbf{r})$. Active orbitals are defined as the zeroth-order orbitals among which the electron occupancies change to model different magnetic states. Preferably, suitable active orbitals are chosen as the ones that maximize the spin polarization on individual magnetic

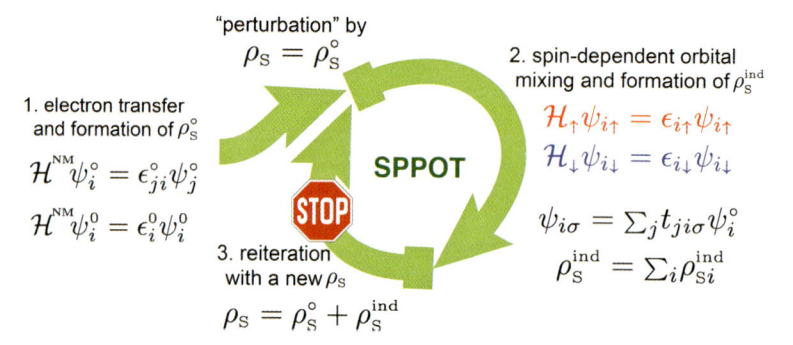

Figure 9.2 Schematic diagram of the analysis procedure of spin polarization perturbational orbital theory (SPPOT).

sites. Passive orbitals are the rest of the zeroth-order orbitals and do not contribute to ρ_S^o in Eq. (9.17). In the first step, $\rho_S(\mathbf{r})$ is approximately mimicked by $\rho_S^o(\mathbf{r})$ which is obtained by a suitable redistribution of the electrons among the active orbitals. In forming active orbitals, the initial orbitals $\psi_i^{o'}$'s may be hybridized to form an alternative set of orthonormal zeroth-order orbitals ψ_i^{o}'s:

$$\psi_i^o = \sum_j c_{ji}^o \psi_j^{o'}, \tag{9.15}$$

which gives

$$\mathcal{H}^{NM}\psi_i^o = \sum_j \varepsilon_{ji}^o \psi_j^o, \quad \text{where} \quad \varepsilon_{ji}^o \equiv \langle \psi_j^o | \mathcal{H}^{NM} | \psi_i^o \rangle = \varepsilon_{ij}^{o*}. \tag{9.16}$$

A model spin density is given as

$$\rho_S^o = \rho_\uparrow^o - \rho_\downarrow^o = \sum_i \rho_{Si}^o = \sum_i (n_{i\uparrow}^o - n_{i\downarrow}^o)|\psi_i^o|^2, \tag{9.17}$$

that is, the net spin density change among the active orbitals after the electron rearrangement.

2) Using the model spin polarization as a perturbation, new Kohn–Sham equations are obtained with different Hamiltonian expressions for different spins (i.e., $\mathcal{H}_\uparrow \neq \mathcal{H}_\downarrow$). By solving the new Kohn–Sham equation in Eq. (9.5), new Kohn–Sham spin orbitals are obtained with new energies and shapes. Like PMO theory, SPPOT expresses the new spin orbitals as a linear combination of the ψ_i^{o}'s in an approximated but analytical form

$$\psi_{i\sigma} = \sum_j t_{ji\sigma} \psi_j^o, \tag{9.18}$$

and their energies are given by correcting the initial orbital energies (ε_{ii}^o):

$$\varepsilon_{i\sigma} = \langle \psi_{i\sigma} | \mathcal{H}_\sigma | \psi_{i\sigma} \rangle \approx \varepsilon_{ii}^o + \text{energy correction}. \tag{9.19}$$

Meanwhile, Eq. (9.13) for the resulting magnetic (perturbed) state is rewritten as

$$E_{sp}[\rho, \rho_S] \approx \sum_{i,\sigma} \left(n_{i\sigma}^o \varepsilon_{i\sigma} - n_{i\sigma}^o \varepsilon_i^0\right) + \frac{1}{4}\iint \mathcal{K}_{xc}(\mathbf{r}, \mathbf{r}')\rho_S^o(\mathbf{r}')\rho_S^o(\mathbf{r})d\mathbf{r}'d\mathbf{r}$$
$$\overset{LSD}{\approx} \sum_{i,\sigma} \left(n_{i\sigma}^o \varepsilon_{i\sigma} - n_{i\sigma}^o \varepsilon_i^0\right) + \frac{1}{4}\int \mathcal{K}_{xc}(\mathbf{r})[\rho_S^o(\mathbf{r})]^2 d\mathbf{r}. \quad (9.20)$$

It is noted that the changes in the shape of the spin orbitals induce an additional spin polarization, because the zeroth-order orbitals no longer have the same shape for ↑- and ↓-electrons. Each pair of the perturbed spin orbitals carries a spin polarization which is decomposed to the initial spin polarization ρ_{Si}^o and the induced spin polarization ρ_{Si}^{ind}:

$$\rho_{Si}(\mathbf{r}) \equiv n_{i\uparrow}^o|\psi_{i\uparrow}(\mathbf{r})|^2 - n_{i\downarrow}^o|\psi_{i\downarrow}(\mathbf{r})|^2 = \rho_{Si}^o(\mathbf{r}) + \rho_{Si}^{ind}(\mathbf{r}). \quad (9.21)$$

The summation of the individual induced spin polarization over all the occupied spin orbitals give the total induced spin polarization, $\rho_S^{ind}(\mathbf{r})$, and thus

$$\rho_S(\mathbf{r}) = \sum_i \rho_{Si}(\mathbf{r}) = \sum_i \rho_{Si}^o(\mathbf{r}) + \sum_i \rho_{Si}^{ind}(\mathbf{r}) = \rho_S^o(\mathbf{r}) + \rho_S^{ind}(\mathbf{r}). \quad (9.22)$$

3) In KS-SDFT calculation procedure, the new ρ_S is used to start the next SCF cycle. However, SPPOT does not take this step but uses the analytical expressions of changes in the spin orbital energies/shapes and the approximate total energy change so as to understand the spin polarization effect in magnetic systems.

In carrying out these steps mathematically, the SPPOT formalism starts with the Hamiltonian matrix elements decomposed into an unperturbed term and a perturbation term:

$$H_{kj\sigma} = \langle \psi_k^o|\mathcal{H}^{NM}|\psi_j^o\rangle + \langle \psi_k^o|v_{xc\sigma}^S|\psi_j^o\rangle = \varepsilon_{kj}^o + \tilde{\Delta}_{kj\sigma}, \quad (9.23)$$

where $\tilde{\Delta}_{kj\sigma}$ is fully expressed as

$$\tilde{\Delta}_{kj\sigma} = \mp\frac{1}{2}\iint \mathcal{K}_{xc}(\mathbf{r}, \mathbf{r}')\rho_S^o(\mathbf{r}')\psi_k^{o*}(\mathbf{r})\psi_j^o(\mathbf{r})d\mathbf{r}'d\mathbf{r} = \tilde{\Delta}_{jk\sigma}^*. \quad (9.24)$$

Upon employing the PMO theory for nongenerate systems, we are interested in obtaining the orbital coefficients $t_{ji\sigma}$ and energy $\varepsilon_{i\sigma}$ in approximate but "analytical" forms as a finite sum of perturbation terms:

$$t_{ji\sigma} \approx \delta_{ji} + t_{ji\sigma}^{(1)}, \quad \text{and} \quad \varepsilon_{i\sigma} \approx \varepsilon_{ii}^o + \varepsilon_{i\sigma}^{(1)} + \varepsilon_{i\sigma}^{(2)}. \quad (9.25)$$

Without showing the derivation processes, the first-order spin orbital energy correction, $\varepsilon_{i\sigma}^{(1)}$, is given as

$$\varepsilon_{i\sigma}^{(1)} = \tilde{\Delta}_{ii\sigma} = \mp\frac{1}{2}\iint \mathcal{K}_{xc}(\mathbf{r}, \mathbf{r}')\rho_S^o(\mathbf{r}')|\psi_i^o(\mathbf{r})|^2 d\mathbf{r}'d\mathbf{r}, \quad (9.26)$$

where − and + sign in front of the double integral are for ↑- and ↓-spin orbitals, respectively. It describes the energy change of a zeroth-order spin orbital due to its own response to the spin polarization; that is, a spin orbital is stabilized (destabilized) when the electron density of the same spin increases (decreases).

The first-order correction in orbital coefficients is then

$$t_{ji\sigma}^{(1)} = -\frac{\varepsilon_{ji}^o + \tilde{\Delta}_{ji\sigma}}{\varepsilon_{j\sigma}^\oplus - \varepsilon_{i\sigma}^\oplus} = -t_{ij\sigma}^{(1)*} \quad \text{for} \quad i \neq j, \quad \text{and} \quad t_{ii\sigma}^{(1)} = 0, \tag{9.27}$$

where $\varepsilon_{i\sigma}^\oplus = \varepsilon_{ii}^o + \varepsilon_{i\sigma}^{(1)}$. Thus, the ψ_j^o mixes into ψ_i^o in forming the new $\psi_{i\sigma}$ as in Eq. (9.18), if either ε_{ji}^o or $\tilde{\Delta}_{ji\sigma}$ is not zero. This spin-dependent first-order orbital mixing results in the spin orbital energy correction in the second order:

$$\varepsilon_{i\sigma}^{(2)} = -\sum_{j \neq i} \frac{|\varepsilon_{ji}^o + \tilde{\Delta}_{ji\sigma}|^2}{\varepsilon_{j\sigma}^\oplus - \varepsilon_{i\sigma}^\oplus} \approx \varepsilon_{i\sigma}^{\text{rehyb}} \mp \frac{1}{4} \iint \mathcal{K}_{xc}(\mathbf{r}, \mathbf{r}') \rho_S^o(\mathbf{r}') \rho_{i\sigma}^{\text{ind}}(\mathbf{r}) d\mathbf{r}' d\mathbf{r}, \tag{9.28}$$

in which $\rho_{i\sigma}^{\text{ind}}$ is an approximate expression of the induced spin orbital density in transformation of ψ_i^o to $\psi_{i\sigma}$, after ignoring the second-order contribution

$$\rho_{i\sigma}^{\text{ind}} = |\psi_{i\sigma}|^2 - |\psi_i^o|^2 \approx \sum_{j \neq i} \left(t_{ji\sigma}^{(1)} \psi_i^{o*} \psi_j^o + t_{ji\sigma}^{(1)*} \psi_j^{o*} \psi_i^o \right), \tag{9.29}$$

and $\varepsilon_{i\sigma}^{\text{rehyb}}$ in Eq. (9.28) is due to the rehybridization of $\psi_{i\sigma}^o$'s that leads to more appreciable electron delocalization in the resulting $\psi_{i\sigma}$'s:

$$\varepsilon_{i\sigma}^{\text{rehyb}} \equiv \frac{1}{2} \sum_{j \neq i} \left(t_{ji\sigma}^{(1)} \varepsilon_{ji}^{o*} + t_{ji\sigma}^{(1)*} \varepsilon_{ji}^o \right) = -\sum_{j \neq i} \frac{|\varepsilon_{ji}^o|^2 + 1/2 \left(\varepsilon_{ji}^o \tilde{\Delta}_{ji\sigma}^{o*} + \varepsilon_{ji}^{o*} \tilde{\Delta}_{ji\sigma}^o \right)}{\varepsilon_{j\sigma}^\oplus - \varepsilon_{i\sigma}^\oplus}. \tag{9.30}$$

$\varepsilon_{i\sigma}^{\text{rehyb}}$ would be zero (no rehybridization), when modeling of ρ_S^o did not require the hybridization of ψ_i^{0}'s (i.e., $\varepsilon_{ji}^o = \varepsilon_i^o \delta_{ji}$). By using Eqs (9.1), (9.2), (9.21), (9.22), (9.25), (9.26) and (9.28), Eq. (9.20) is decomposed into the kinetic energy component ($E_{\text{sp}}^{\text{kin}}$) and the potential energy component ($E_{\text{sp}}^{\text{pot}}$) during the transition from the nonmagnetic state **NM** to the magnetic state **M**

$$E_{\text{sp}}[\rho, \rho_S] = E_{\text{sp}}^{\text{kin}} + E_{\text{sp}}^{\text{pot}}, \tag{9.31}$$

where

$$E_{\text{sp}}^{\text{kin}} \approx \sum_{i,\sigma} \left[n_{i\sigma}^o \left(\varepsilon_{ii}^o + \varepsilon_{i\sigma}^{\text{rehyb}} \right) - n_{i\sigma}^0 \varepsilon_i^0 \right], \tag{9.32}$$

and

$$E_{\text{sp}}^{\text{pot}} \approx -\frac{1}{4} \iint \mathcal{K}_{xc}(\mathbf{r}, \mathbf{r}') \rho_S^o(\mathbf{r}) \rho_S(\mathbf{r}') d\mathbf{r}' d\mathbf{r} \stackrel{\text{LSD}}{\approx} -\frac{1}{4} \int \kappa_{xc}(\mathbf{r}) \rho_S^o(\mathbf{r}) \rho_S(\mathbf{r}) d\mathbf{r}. \tag{9.33}$$

By using Eqs (9.31) and (9.32), the spin polarization energy difference between two magnetic states **M1** and **M2** is expressed as the following:

$$E_{sp}^{M2} - E_{sp}^{M1} \approx \sum_{i,\sigma} \left[n_{i\sigma}^{M2o} \left(\varepsilon_{ii}^{M2o} + \varepsilon_{i\sigma}^{M2\text{rehyb}} \right) - n_{i\sigma}^{M1o} \left(\varepsilon_{ii}^{M1o} + \varepsilon_{i\sigma}^{M1\text{rehyb}} \right) \right]$$
$$+ E_{sp}^{M2 \text{ pot}} - E_{sp}^{M1 \text{ pot}}.$$
(9.34)

It is important to note that the potential component E_{sp}^{pot} of the spin polarization energy is maximized when ρ_S overlaps with ρ_S^o most effectively. As shown in the following section, this effect favors a parallel spin alignment (i.e., the largest ρ_S^o among active orbitals, and it also plays a key role in understanding spin polarization induction from passive orbitals.

9.3
Applications to Simple Systems

In this section, we show how SPPOT can analyze spin polarization in various simplified systems and make connections to the known theories of spin polarization and magnetism. In some cases, the original nonmagnetic orbitals are good zeroth-order wave functions that can produce ρ_S^o by electron rearrangement among them (i.e., $\varepsilon_{i\sigma}^{\text{rehyb}} = 0$). Some systems can be described sufficiently only with ρ_S^o from highly localized active orbitals whose shape change contributes little to the total spin density (frozen orbital approximation; $\rho_{i\sigma}^{\text{ind}}(\mathbf{r}) \approx 0$).

9.3.1
Spin Polarization in Isolated Atoms and Ions

While the one-determinant nature of Kohn–Sham theory has its limitation for atoms, understanding of calculation results for spin-polarized atomic states provides valuable insights for the studies of magnetic systems from molecules to solids. Spin polarization in isolated atoms/ions is a case where the nonmagnetic orbitals do not require a hybridization for modeling ρ_S^o and where a frozen orbital approximation works well [2]. These approximations lead to no kinetic contribution to $E_{sp}(E_{sp}^{kin} = 0)$. The following process is applicable to both d- and f-electron systems but we will use a transition metal ion for illustration.

In a fictitious spin-unpolarized state (nonmagnetic state) of an isolated $Cr^{2+}(3d)^4$ ion, the 10 degenerate 3d spin orbitals share four electrons to become partially occupied with 4/10 per spin orbital (the left in Figure 9.3). Then, spin polarization occurs by electron transfer from the five ↓-spin 3d-orbitals to the corresponding ↑-spin orbitals (the center in Figure 9.3) to give an electron configuration $(xy_\uparrow^{4/5} xy_\downarrow^0)(yz_\uparrow^{4/5} yz_\downarrow^0)(xz_\uparrow^{4/5} xz_\downarrow^0)(x^2-y^2{}_\uparrow^{4/5} x^2-y^2{}_\downarrow^0)(z^2{}_\uparrow^{4/5} z^2{}_\downarrow^0)$ with $m = 4$.

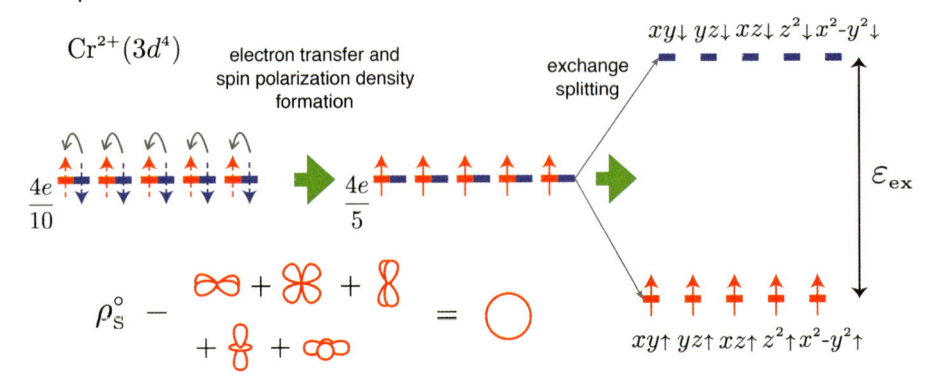

Figure 9.3 Spin polarization and exchange splitting of the 3d-orbitals from the nonmagnetic state of an isolated Cr^{2+} on.

From Eq. (9.17), this occupancy change among the 3d-orbitals (the active orbitals in this process) identifies a model spin density ρ_S^o:

$$\rho_S^o = \frac{m}{2\ell+1}\left(|\chi_{xy}|^2 + |\chi_{yz}|^2 + |\chi_{xz}|^2 + |\chi_{x^2-y^2}|^2 + |\chi_{z^2}|^2\right), \tag{9.35}$$

with $\ell = 2$ or d-orbitals. For any valence orbital χ_i, the first-order energy correction due to ρ_S^o is given from Eq. (9.26) as

$$\varepsilon_{i\sigma}^{(1)} \approx \mp \frac{1}{2}\iint \mathcal{K}_{xc}(\mathbf{r},\mathbf{r}')\rho_S^o(\mathbf{r}')|\chi_i(\mathbf{r})|^2 d\mathbf{r}' d\mathbf{r}. \tag{9.36}$$

The appearance in the net ↑-spin polarization stabilizes all the ↑-spin orbitals and destabilizes all the ↓-spin orbitals (the right in Figure 9.3). Meanwhile, $\varepsilon_{i\sigma}^{(2)} = 0$ under our approximations here.

For the d-orbitals, the energy difference between each pair of the ↑- and ↓-spin orbitals, exchange splitting energy (ε_{ex}), is given as

$$\varepsilon_{ex}^{3d} \approx \varepsilon_{3d\downarrow}^{(1)} - \varepsilon_{3d\uparrow}^{(1)} = U + 2\ell J, \tag{9.37}$$

where U and J are the average self-orbital and average inter-orbital interaction energies, respectively, for the 3d-orbitals:

$$U = \frac{1}{2\ell+1}\sum_i^{\ell} \iint \mathcal{K}_{xc}(\mathbf{r},\mathbf{r}')|\chi_i(\mathbf{r}')|^2|\chi_i(\mathbf{r})|^2 d\mathbf{r}' d\mathbf{r}, \tag{9.38a}$$

$$J = \frac{1}{2\ell(2\ell+1)}\sum_{i,j\neq i}^{\ell} \iint \mathcal{K}_{xc}(\mathbf{r},\mathbf{r}')|\chi_j(\mathbf{r}')|^2|\chi_i(\mathbf{r})|^2 d\mathbf{r}' d\mathbf{r}. \tag{9.38b}$$

Given the $(2\ell+1)$ orbital-degeneracy with a total m number of unpaired electrons, the general expression for the exchange splitting energy is

$$\varepsilon_{ex} \approx I_{ex} m, \tag{9.39}$$

where we define atomic exchange parameter (I_{ex}), as the average value of the self- and inter-orbital interaction energies within LSD approximation,

$$I_{ex} \equiv \frac{1}{(2\ell+1)^2} \iint \mathcal{K}_{xc}(\mathbf{r},\mathbf{r}') \sum_{i,j}^{\ell} |\chi_j(\mathbf{r}')|^2 |\chi_i(\mathbf{r})|^2 d\mathbf{r}' d\mathbf{r} = \frac{1}{2\ell+1}(U+2\ell J).$$

(9.40)

Therefore, the spin polarization energy in Eq. (9.31) is given as the following:

$$E_{sp}[\rho,\rho_S] = E_{sp}^{pot} \approx -\frac{1}{4} I_{ex} m^2.$$

(9.41)

Within LSD approximation, the I_{ex} is underestimated for atoms/ions and can be calculated correctly when self-interaction correction (SIC) is made properly [2]. These results have been utilized in analyzing the extent of SIC (or, a lack thereof) in a well-known LSD+U method [7].

9.3.2
magnetic coupling, in transition-metal dimer complexes

When metal-ligand bonds are relatively ionic like in Mott–Hubbard insulators depicted in Figure 9.1a, the active orbitals are obtained from d-orbitals localized on each metal ion and a frozen orbital approximation can be still valid in describing the interactions of those orbitals. Upon the assumption of pairwise interactions among magnetic sites, the smallest building units in such systems are magnetic transition-metal dimer complexes. The transition metal ions in those molecules may or may not be related by symmetry. Although a general description is available [3], for simplicity we consider a transition-metal dimer complex that contains only one unpaired electron on each of two transition metal sites, with $Cu_2(fsa)_2en$ molecule as an example (Figure 9.4) [8]. Magnetic exchange interactions between the unpaired electrons are assumed to be well described by considering only the interactions between the two d-orbitals containing those unpaired electrons (active electron approximation).

Our starting point is a hypothetical nonmagnetic state **NM** for which the zeroth-order orbitals are two delocalized bonding and antibonding molecular orbitals (MOs), ψ_1^0 and ψ_2^0, satisfying

$$\mathcal{H}^{NM}\psi_i^0 = \varepsilon_i^0 \psi_i^0, \quad (i=1,2).$$

(9.42)

A unitary transformation of the delocalized MOs can lead to a pair of orthonormal magnetic orbitals (OMOs) that are highly localized on each magnetic center **A** (the left Cu^{2+} on) or **B** the right) of the $Cu_2(fsa)_2en$ molecule, as shown in Figure 9.4a [8]:

$$\psi_a^o = \psi_1^0 \cos\gamma + \psi_2^0 \sin\gamma, \quad \text{and} \quad \psi_b^o = -\psi_1^0 \sin\gamma + \psi_2^0 \cos\gamma,$$

(9.43)

where $|\gamma| \leq \pi$. We define the following parameters neccessary for spin polarization energy expression: The off-diagonal energy matrix elements of the OMOs

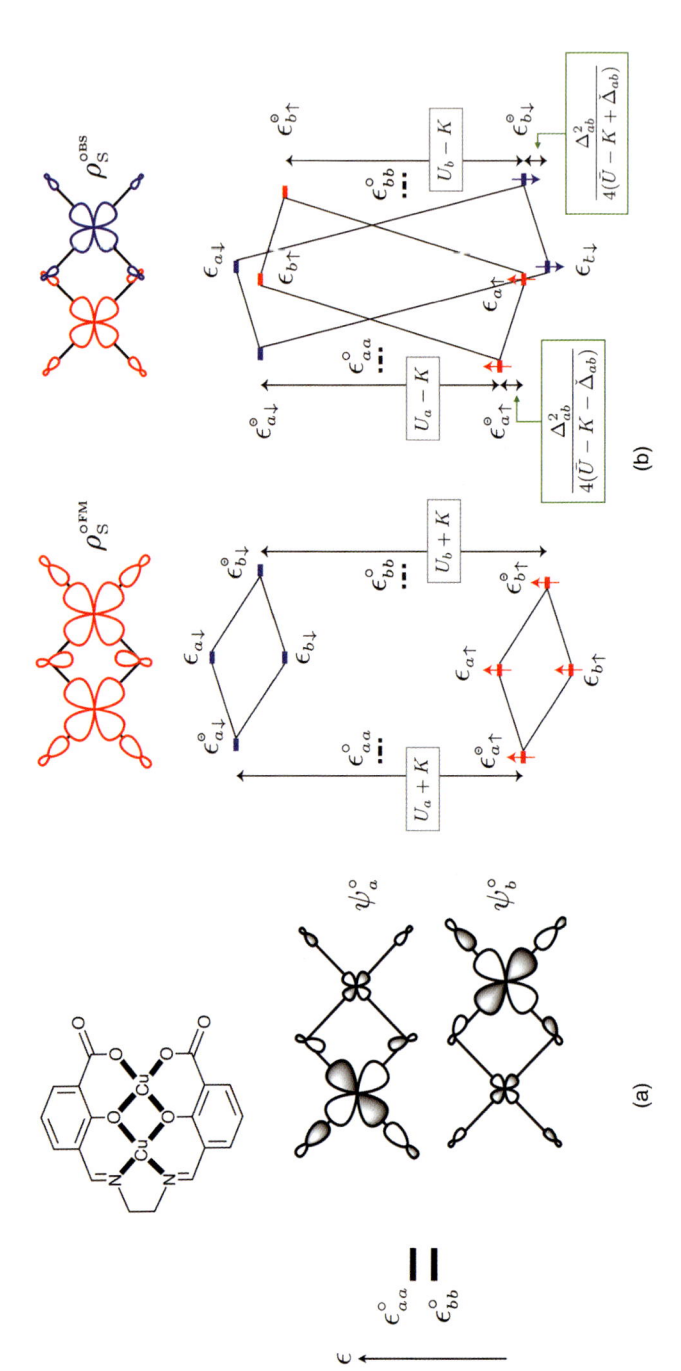

Figure 9.4 (a) Orthogonal magnetic orbitals (OMOs) as the active orbitals in the nonmagnetic state of $Cu_2(fsa)_2en$ and (b) the schematic energy diagrams of the OMOs for ferromagnetic and broken-symmetry (BS) spin polarization perturbations.

are

$$\Delta \equiv \varepsilon_2^0 - \varepsilon_1^0, \qquad (9.44a)$$

$$\Delta_{ab} \equiv 2\varepsilon_{ab}^o = 2\langle\psi_a^o|\mathcal{H}^{NM}|\psi_b^o\rangle = \Delta_{ba} = \Delta\sin 2\gamma, \quad \text{and} \qquad (9.44b)$$

$$\check{\Delta}_{ab} \equiv \varepsilon_{aa}^o - \varepsilon_{bb}^o = -\Delta\cos 2\gamma = -\check{\Delta}_{ba}. \qquad (9.44c)$$

The γ value for a given dimer complex can be guessed intuitively or determined using various orbital localization methods that provides a maximum orbital localization [3]. For a symmetric dimer, $|\gamma| = \pi/4$, hence $|\Delta_{ab}| = |\Delta|$ and $\varepsilon_{aa}^o = \varepsilon_{bb}^o$. The Cu$_2(fsa)_2$en molecule is almost symmetric and thus $|\gamma|$ is close to $\pi/4$.

The **FM** and **BS** (broken-symmetry) states of the molecule are then represented by their corresponding spin densities ρ_S^{oFM} and ρ_S^{oBS} when the electrons are redistributed among the two active orbitals, ψ_a^o and ψ_b^o, with the expected spin orientations at the magnetic sites, as shown at the top of Figure 9.4b:

$$\rho_S^{oFM} = |\psi_a^o|^2 + |\psi_b^o|^2 = \rho_a^o + \rho_b^o, \quad \text{and} \qquad (9.45a)$$

$$\rho_S^{oBS} = |\psi_a^o|^2 - |\psi_b^o|^2 = \rho_a^o - \rho_b^o. \qquad (9.45b)$$

Upon a frozen orbital approximation ($\rho_S \approx \rho_S^o$), insertion of Eq. (9.45) into Eq. (9.33) leads to

$$E_{sp}^{FM\,pot} \approx -\frac{1}{2}(\bar{U} + K), \quad \text{and} \quad E_{sp}^{BS\,pot} \approx -\frac{1}{2}(\bar{U} - K), \qquad (9.46)$$

where $\bar{U}$ is the average value between U_a and U_b, the self-orbital exchange-correlation energies for ψ_a^o and ψ_b^o, (i.e., $\bar{U} = (U_a + U_b)/2$) and K is the inter-orbital exchange-correlation energy between the orbital pair

$$U_i = \iint \mathcal{K}_{xc}(\mathbf{r},\mathbf{r}')\rho_i^o(\mathbf{r}')\rho_i^o(\mathbf{r})d\mathbf{r}'d\mathbf{r}, \quad \text{and} \quad K = \iint \mathcal{K}_{xc}(\mathbf{r},\mathbf{r}')\rho_a^o(\mathbf{r}')\rho_b^o(\mathbf{r})d\mathbf{r}'d\mathbf{r}, \qquad (9.47)$$

with $U \gg K$. From Eq. (9.46), the potential energy contribution to the total spin polarization energy between the **FM** and **BS** states is given as

$$E_{sp}^{BS\,pot} - E_{sp}^{FM\,pot} \approx K. \qquad (9.48)$$

Meanwhile, the approximate energies of the active spin orbitals $\psi_{a\sigma}$ and $\psi_{b\sigma}$ are summarized for an asymmetric dimer as

$$\varepsilon_{a\sigma} \approx \varepsilon_{a\sigma}^{\ominus} - \frac{|\varepsilon_{ba}^{\ominus}|^2}{\varepsilon_{b\sigma}^{\ominus} - \varepsilon_{a\sigma}^{\ominus}}, \quad \text{and} \quad \varepsilon_{b\sigma} \approx \varepsilon_{b\sigma}^{\ominus} - \frac{|\varepsilon_{ab}^{\ominus}|^2}{\varepsilon_{a\sigma}^{\ominus} - \varepsilon_{b\sigma}^{\ominus}}, \qquad (9.49)$$

with

$$\varepsilon_{a\sigma}^{\ominus} = \varepsilon_{aa}^o \mp \frac{U_a + cK}{2}, \quad \text{and} \quad \varepsilon_{b\sigma}^{\ominus} = \varepsilon_{bb}^o \mp \frac{c(U_b + cK)}{2}, \qquad (9.50)$$

where − is for ↑ and + or ↓, while c = +1 for the **FM** state and −1 for the **BS** state. The second terms in Eq. (9.49) are the second-order spin orbital energy corrections that are reduced to the orbital rehybridization energy $\varepsilon_{i\sigma}^{\text{rehyb}}$ under the frozen orbital approximation ($\tilde{\Delta}_{ab\sigma} \approx 0$). Figure 9.4b shows the spin orbital energy diagrams for both magnetic states when $\varepsilon_{aa}^o > \varepsilon_{bb}^o$. As expected from Figure 9.4b or from Eqs (9.30), 9.32, 9.44b and (9.50), the kinetic components contribute to the total spin polarization energy for the **BS** state, while they cancel out each other for the **FM** state [3]:

$$E_{\text{sp}}^{\text{FM kin}} = 0, \quad \text{and} \tag{9.51a}$$

$$E_{\text{sp}}^{\text{BS kin}} = -\frac{|\Delta_{ab}|^2}{4\left(\bar{U} - K - \check{\Delta}_{ab}\right)} - \frac{|\Delta_{ab}|^2}{4\left(\bar{U} - K + \check{\Delta}_{ab}\right)} \approx -\frac{|\Delta_{ab}|^2}{2(\bar{U} - K)}, \tag{9.51b}$$

in which the last expression is obtained since $\bar{U} \gg |\check{\Delta}_{ab}|$ for the molecule.

Combining Eqs (9.48) and (9.51), the electronic energy difference for the **FM** and **BS** states for the molecule is expressed as

$$E^{\text{BS}} - E^{\text{FM}} \approx E_{\text{sp}}^{\text{BS}} - E_{\text{sp}}^{\text{FM}} \approx K - \frac{|\Delta_{ab}|^2}{2(\bar{U} - K)}. \tag{9.52}$$

Based on the mapping analysis [9–11], the magnetic coupling constant J or the dimer is given as

$$J \approx 2K - \frac{|\Delta_{ab}|^2}{\bar{U} - K}, \tag{9.53}$$

in which the first and second terms favor ferromagnetic and antiferromagnetic interactions, respectively.

From the definition of U's and K in Eq. (9.47), it is clear that their numerical values depend significantly on the exchange-correlation functionals used in the KS-SDFT calculations. Since a proper correction of self interaction in the LSD approximation will give a larger value for both Us and K, we expect in calculated magnetic coupling constants a smaller antiferromagnetic contribution and a larger ferromagnetic contribution when a hybrid functional or SIC-LSD scheme is employed.

It is noted that the definition of K in Eq. (9.53) was given originally as the exchange integral k_{ab} between ψ_a^o and ψ_b^o in the theories based on UHF analysis by Anderson [12], or CI analysis by Hay et al. [13]. The definition of K in Eq. (9.47) implies that the ferromagnetic interaction is strong when the orbital densities of the magnetic orbitals overlap significantly. This is equivalent to the popular interpretation of k_{ab} by Kahn [14], in terms of overlap density between two magnetic orbitals when they are expressed as a real function in the studies of molecular magnetism:

$$k_{ab} = \iint \frac{\rho_{ab}^o(\mathbf{r})\rho_{ab}^o(\mathbf{r}')}{|\mathbf{r} - \mathbf{r}'|} d\mathbf{r}' d\mathbf{r}, \quad \text{where} \quad \rho_{ab}^o(\mathbf{r}) \equiv \psi_a^o(\mathbf{r})\psi_b^o(\mathbf{r}). \tag{9.54}$$

9.3.3
Stoner Condition for Spontaneous Spin Polarization

In the studies of metallic magnetism (itinerant electron magnetism) as shown in Figure 9.1c, the Stoner condition for spontaneous spin polarization provides a simple starting point [15]

$$I_S D_s(\varepsilon_F^0) > 1, \tag{9.55}$$

where I_S is called Stoner exchange parameter and $D_s(\varepsilon_F^0)$ is the density of states (DOS) at the Fermi energy ε_F^0 per unit volume per unit energy per spin of the "nonmagnetic" ("spin-unpolarized," to be precise) state of a metallic compound of interest. This expression is obtained when we consider that the spontaneous magnetization is stable when

$$E_{sp} = E_{sp}^{kin} + E_{sp}^{pot} < 0, \tag{9.56}$$

under a rather drastic assumption that the magnetization occurs only near the Fermi energy. The numerical values of I_S have been obtained by applying the linear response theory to the KS-SDFT calculation results for 32 elemental metals within the LSD approximation [16]. Here, we derive the Stoner condition by using the SPPOT formalism and examine the underlying assumptions and implications of the Stoner condition [15].

Consider a metal in its nonmagnetic state **NM** (the left in Figure 9.5) which tries to develop a spontaneous spin polarization with a small saturated magnetic moment m per unit cell volume. This is another case where the nonmagnetic orbitals are good for modeling ρ_S^0 and the frozen orbital approximation works well. Unlike the previous examples, however, the nonmagnetic orbitals are delocalized throughout the electronic system. We will assume that the spin density can be described by using the original nonmagnetic crystal orbitals

$$\psi_{i\mathbf{k}}^0(\mathbf{r}) = \psi_{i\mathbf{k}}^0(\mathbf{r}), \quad \text{and thus} \quad \varepsilon_{i\mathbf{k}\sigma}^{rehyb} = 0, \tag{9.57}$$

where i and $\mathbf{k}$ are the band index and wave vector for the crystal orbital $\psi_{i\mathbf{k}}^0$, respectively. For the spin polarization of our interest in this case, we assume that the crystal orbitals do not change their shape by the spin polarization perturbation (i.e., $\rho_S^{ind}(\mathbf{r}) = 0$; frozen orbital approximation). Then, the spin polarization process involves only the transfer of $m/2$ electrons from ↓-spin crystal orbitals to ↑-spin ones near the ε_F^0. The new highest-occupied energy levels are denoted as $\varepsilon_{F\uparrow}^0$ and $\varepsilon_{F\downarrow}^0$ for ↑- and ↓-spin channels, respectively. This electron transfer will increase the total electronic band structure energy as the electrons move to higher energy crystal orbitals. This increase corresponds to the kinetic component of the spin polarization energy E_{sp}^{kin} in Eq. (9.31) which is solely due to the electron occupancy changes during the electron transfer:

$$E_{sp}^{kin} = \sum_{i,\mathbf{k},\sigma} (n_{i\mathbf{k}\sigma}^0 - n_{i\mathbf{k}}^0)\varepsilon_{i\mathbf{k}}^0 = \int_{\varepsilon_F^0}^{\varepsilon_{F\uparrow}^0} \varepsilon D_s(\varepsilon) d\varepsilon + \int_{\varepsilon_F^0}^{\varepsilon_{F\downarrow}^0} \varepsilon D_s(\varepsilon) d\varepsilon, \tag{9.58}$$

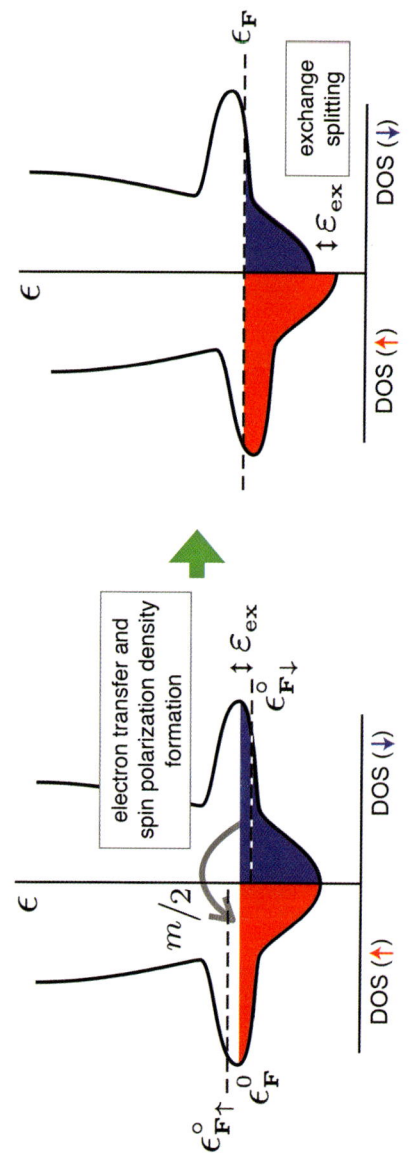

Figure 9.5 Spin polarization and exchange splitting of a metal that results in a net magnetic moment.

where ε_{ik}^0 is the energy of $\psi_{ik}^0(\mathbf{r})$'s and $D_s(\varepsilon)$ is the DOS per spin at ε in the nonmagnetic state. The same electron transfer leads to the spin density ρ_S^o as the sum of the orbital densities, $|\psi_{ik}^0(\mathbf{r})|^2$ in the energy window from $\varepsilon_{F\downarrow}^o$ to $\varepsilon_{F\uparrow}^o$:

$$\rho_S^o(\mathbf{r}) = \sum_{i,\mathbf{k}} (n_{i\mathbf{k}\uparrow}^o - n_{i\mathbf{k}\downarrow}^o)|\psi_{i\mathbf{k}}^0(\mathbf{r})|^2 = \sum_{i,\mathbf{k}} \int_{\varepsilon_{F\downarrow}^o}^{\varepsilon_{F\uparrow}^o} \delta(\varepsilon - \varepsilon_{i\mathbf{k}}^0)|\psi_{i\mathbf{k}}^0(\mathbf{r})|^2 d\varepsilon. \tag{9.59}$$

Integration of $\rho_S^o(\mathbf{r})$ in a unit cell gives the magnetic moment per unit cell, m. This value can be obtained alternatively as the number difference between ↑- and ↓-electrons per unit cell:

$$m = \int_{\varepsilon_{F\downarrow}^o}^{\varepsilon_{F\uparrow}^o} D_s(\varepsilon) d\varepsilon = 2\int_{\varepsilon_F^o}^{\varepsilon_{F\uparrow}^o} D_s(\varepsilon) d\varepsilon = 2\int_{\varepsilon_{F\downarrow}^o}^{\varepsilon_F^o} D_s(\varepsilon) d\varepsilon. \tag{9.60}$$

With the first-order spin orbital energy correction,

$$\varepsilon_{i\mathbf{k}\sigma}^{(1)} = \mp\frac{1}{2}\int \kappa_{xc}(\mathbf{r})\rho_S^o(\mathbf{r})|\psi_{i\mathbf{k}}^0(\mathbf{r})|^2 d\mathbf{r}, \tag{9.61}$$

SPPOT gives the exchange splitting energy for the $\psi_{i\mathbf{k}}^0(\mathbf{r})$ as

$$\varepsilon_{ex,i}(\mathbf{k}) = \varepsilon_{i\mathbf{k}\downarrow}^{(1)} - \varepsilon_{i\mathbf{k}\uparrow}^{(1)} \stackrel{\text{LSD}}{\approx} \int \kappa_{xc}\rho_S^o(\mathbf{r})|\psi_{i\mathbf{k}}^0(\mathbf{r})|^2 d\mathbf{r}. \tag{9.62}$$

The corresponding potential component of the spin polarization energy is given as

$$E_{sp}^{pot} \stackrel{\text{LSD}}{\approx} -\frac{1}{4}\int \kappa_{xc}(\mathbf{r})[\rho_S^o(\mathbf{r})]^2 d\mathbf{r}. \tag{9.63}$$

Derivation of the Stoner condition through the SPPOT formalism starts with a further simplification of the expression of E_{sp}^{pot}. When the magnetic moment is small (i.e., the spin polarization takes place only near ε_F^o), we may additionally assume that the exchange splitting energy is the same for all the crystal orbitals (rigid band approximation):

$$\varepsilon_{ex,i\mathbf{k}} \approx \varepsilon_{ex} \equiv \int \kappa_{xc}(\mathbf{r})\rho_S^o(\mathbf{r})\gamma(\mathbf{r}) d\mathbf{r} \quad \text{for all } i \text{ and } \mathbf{k}, \tag{9.64}$$

by employing an average orbital density $\gamma(\mathbf{r})$ among the crystal orbitals only at the Fermi energy in the spin-unpolarized state

$$\gamma(\mathbf{r}) \equiv \frac{\sum_{i,\mathbf{k}} \delta(\varepsilon_F^o - \varepsilon_{i\mathbf{k}}^0)|\psi_{i\mathbf{k}}^0(\mathbf{r})|^2}{D_s(\varepsilon_F^o)}. \tag{9.65}$$

This approximation allows us further to express $\rho_S^o(\mathbf{r})$ simply as $m\gamma(\mathbf{r})$ and hence Eqs (9.63) and (9.64) are simplified into

$$E_{sp}^{pot} \approx -\frac{1}{4}I_S m^2, \quad \text{and} \quad \varepsilon_{ex} = I_S m, \tag{9.66}$$

respectively, where I_S is the Stoner parameter given previously from the linear response theory [16]:

$$I_S \equiv \int \kappa_{xc}(\mathbf{r})[\gamma(\mathbf{r})]^2 d\mathbf{r}. \tag{9.67}$$

The $\gamma(\mathbf{r})$ in Eq. (9.65) corresponds to the normalized static linear spin-density response function in the original derivation of I_S in Eq. (9.67) [16]. From the SPPOT analysis, it is clear that the equations in Eq. (9.66) have a close resemblance to Eqs. (9.41) and (9.39) both analytically and numerically and that I_S is equivalent to I_{ex}, atomic exchange parameter, in its role. For the 3d-elements, the LSD calculation values are quite similar between I_S and I_{ex}, indicating that Stoner exchange parameters are indeed characteristic of atoms [2].

The kinetic energy term E_{sp}^{kin} in Eq. (9.58) can be simplified when we assume that at any ε, $D_s(\varepsilon)$ is $D_s(m)$ given as the average value of the DOS per spin in the small energy interval from $\varepsilon_{F\downarrow}^o$ to $\varepsilon_{F\uparrow}^o$. From Eq. (9.60), therefore,

$$D_s(m) = \frac{m}{\varepsilon_{F\uparrow}^o - \varepsilon_{F\downarrow}^o}, \quad \text{and} \quad \varepsilon_F^o = \frac{\varepsilon_{F\uparrow}^o + \varepsilon_{F\downarrow}^o}{2}, \tag{9.68}$$

and thus Eq. (9.58) is simplified as

$$E_{sp}^{kin} = D_s(m) \int_{\varepsilon_F^o}^{\varepsilon_{F\uparrow}^o} \varepsilon d\varepsilon + D_s(m) \int_{\varepsilon_F^o}^{\varepsilon_{F\downarrow}^o} \varepsilon d\varepsilon = \frac{m^2}{4D_s(m)}. \tag{9.69}$$

From Eqs (9.56), 9.66 and (9.69), the so-called extended Stoner condition is given as

$$I_S D_s(m) > 1, \tag{9.70}$$

for spontaneous magnetization [17].

Beyond the Stoner theory, however, it is critically important to recognize from Eq. (9.61) that the spin orbital energy corrections are different for different crystal orbitals (a "nonrigid band behavior"), as they depend on how the orbital density overlaps with spin density in space. This nonrigid band behavior is general and is observed in all KS-SDFT electronic band structures in the literature, as prominently recognized for half-metallic Heusler phases [18], Furthermore, the Stoner theory of metallic magnetism only predicts the formation of magnetic moments, rather than comparing spin polarization energies among different magnetic states, and it also does not count formation of induced spin polarization. Any elaboration of these aspects is beyond the scope of this chapter and yet it is remarked that SPPOT analysis provides an analytical tool to examine them in a tangible manner.

9.3.4
Ruderman–Kittel–Kasuya–Yosida (RKKY) Exchange Interactions

Unlike the cases above, the induced spin density plays a major role, when delocalized electrons mediate the magnetic exchange interactions among magnetic sites having localized electrons (Figure 9.1b) [19]. A full description of this RKKY exchange interaction requires expansion of SPPOT through a rigorous introduction of two-component spinor concept, which will appear in a separate publication. In this chapter, we will limit the discussions to elucidation of how an induced spin polarization develops around a localized magnetic center. As the simplest case, consider a hypothetical free electron system with a single magnetic atom (say, an f-electron atom) located at $\mathbf{R}$, with a magnetic moment $m_{\mathbf{R}}$. Its spin density $\rho^o_{S\mathbf{R}}$, along the spin orientation, is approximated as a point spin polarization in space and thus expressed by using the Dirac delta function $\delta(\mathbf{r} - \mathbf{R})$ centered at $\mathbf{R}$:

$$\rho^o_{S\mathbf{R}}(\mathbf{r}) \approx m_{\mathbf{R}}\delta(\mathbf{r} - \mathbf{R}). \tag{9.71}$$

The corresponding localized potential term is expressed as

$$v^{So}_{xc\mathbf{R}\sigma}(\mathbf{r}) \stackrel{LSD}{\approx} \mp \frac{m_{\mathbf{R}}}{2} \kappa_{xc\mathbf{R}}(\mathbf{r})\delta(\mathbf{r} - \mathbf{R}). \tag{9.72}$$

Although it is customary to use plane wave functions for description of RKKY interactions, they would not be correct zeroth-order wave functions if, as usual, we wanted to single out the spin polarization effect as the perturbation. This is because the magnetic site is atomistically different from the surrounding and the plane waves are strongly perturbed by the atomic charge of the site regardless of additional spin polarization perturbation. We include this nonmagnetic perturbation as another localized function $\delta v^{nsp}_{\mathbf{R}}(\mathbf{r})$ to give a total perturbational potential $\delta v_{\mathbf{R}}(\mathbf{r})$ for our analysis

$$\delta v_{\mathbf{R}}(\mathbf{r}) = \delta v^{nsp}_{\mathbf{R}}(\mathbf{r}) + v^{So}_{xc\mathbf{R}}(\mathbf{r}) \approx \left(Q_{\mathbf{R}} \mp \frac{m_{\mathbf{R}}}{2} \kappa_{xc\mathbf{R}}(\mathbf{r})\right)\delta(\mathbf{r} - \mathbf{R}), \tag{9.73}$$

where $Q_{\mathbf{R}}$ is a constant value that reflects the effect of the difference in effective nuclear charges between the magnetic atom and the surroundings.

Meanwhile, the delocalized electrons for the unperturbed state are described by the plane wave functions as the zeroth-order orbitals:

$$\psi^0_{\mathbf{k}}(\mathbf{r}) = \frac{1}{\sqrt{V}} e^{i\mathbf{k}\cdot\mathbf{r}}, \tag{9.74}$$

where V is the volume of the system, and by their orbital energy

$$\varepsilon^0_{\mathbf{k}} = -\frac{\hbar|\mathbf{k}|^2}{2m_e}. \tag{9.75}$$

The electron density given by a plane wave is a constant throughout the space:

$$\rho^0_{\mathbf{k}\sigma}(\mathbf{r}) = \psi^{0*}_{\mathbf{k}}(\mathbf{r})\psi^0_{\mathbf{k}}(\mathbf{r}) = \frac{1}{V}. \tag{9.76}$$

The first-order energy splitting is given as a small (constant) number, since V is very large for a bulk metal:

$$\varepsilon^{(1)}_{\mathbf{k}\sigma} = \langle \psi^0_{\mathbf{k}}|\delta v_{\mathbf{R}}|\psi^0_{\mathbf{k}}\rangle = \frac{1}{V}\left(Q_{\mathbf{R}} \mp \frac{\kappa_{xc\mathbf{R}}}{2} m_{\mathbf{R}}\right), \tag{9.77}$$

where we conveniently define a positive constant κ_{xcR}, which, like I_{ex} in Eq. (9.40), is an innate property of a magnetic site (atom):

$$\kappa_{xcR} \equiv \int \kappa_{xcR}(\mathbf{r})\delta(\mathbf{r}-\mathbf{R})d\mathbf{r} = \kappa_{xcR}(\mathbf{R}). \tag{9.78}$$

The plane wave spin orbitals then interact via the perturbational potential $\delta v_R(\mathbf{r})$ in Eq. (9.73) and the degree of their interactions is understood by the first-order correction in orbital coefficients

$$t^{(1)}_{\mathbf{k}'\mathbf{k}\sigma} = -\frac{\langle \psi^0_{\mathbf{k}'} | \delta v_R | \psi^0_{\mathbf{k}} \rangle}{\varepsilon^\sigma_{\mathbf{k}'\sigma} - \varepsilon^\sigma_{\mathbf{k}\sigma}} - -\frac{\langle \psi^0_{\mathbf{k}'} | \delta v_R | \psi^0_{\mathbf{k}} \rangle}{\varepsilon^0_{\mathbf{k}'} - \varepsilon^0_{\mathbf{k}}} \quad \text{for} \quad \mathbf{k}' \neq \mathbf{k}, \quad \text{and} \quad t^{(1)}_{\mathbf{k}\mathbf{k}\sigma} = 0. \tag{9.79}$$

By employing the Fourier transform of Dirac delta function,

$$\delta(\mathbf{r}-\mathbf{R}) = \frac{1}{V}\sum_{\mathbf{q}} e^{i\mathbf{q}\cdot(\mathbf{r}-\mathbf{R})}, \tag{9.80}$$

together with Eqs (9.29), (9.79), and (9.80), the approximate expression for the change in the electron density of the perturbed plane wave $\psi_{\mathbf{k}\sigma}(\mathbf{r})$ can be obtained

$$\rho^{ind}_{\mathbf{k}\sigma}(\mathbf{r}) \approx -\frac{1}{V^2}\left(Q_R \mp \frac{\kappa_{xcR}}{2}m_R\right)\left[\sum_{\mathbf{q}} \frac{e^{i\mathbf{q}\cdot(\mathbf{r}-\mathbf{R})} + e^{-i\mathbf{q}\cdot(\mathbf{r}-\mathbf{R})}}{\varepsilon^0_{\mathbf{k}+\mathbf{q}} - \varepsilon^0_{\mathbf{k}}}\right]. \tag{9.81}$$

The change in total electron density $\delta\rho_R(\mathbf{r})$ and induced spin density $\rho^{ind}_{SR}(\mathbf{r})$ caused by the magnetic site at $\mathbf{R}$ are calculated by summing the orbital density changes and induced spin densities, respectively, for all the perturbed wave functions up to the Fermi energy

$$\delta\rho_R(\mathbf{r}) = \sum_{\mathbf{k}} f(\varepsilon^0_{\mathbf{k}})\left(\rho^{ind}_{\mathbf{k}\uparrow}(\mathbf{r}) + \rho^{ind}_{\mathbf{k}\downarrow}(\mathbf{r})\right) \approx -2Q_R F(|\mathbf{r}-\mathbf{R}|), \quad \text{and} \tag{9.82a}$$

$$\rho^{ind}_{SR}(\mathbf{r}) = \sum_{\mathbf{k}} f(\varepsilon^0_{\mathbf{k}})\left(\rho^{ind}_{\mathbf{k}\uparrow}(\mathbf{r}) - \rho^{ind}_{\mathbf{k}\downarrow}(\mathbf{r})\right) \approx \kappa_{xcR} m_R F(|\mathbf{r}-\mathbf{R}|), \tag{9.82b}$$

where $f(\varepsilon^0_{\mathbf{k}})$ is the Fermi–Dirac distribution function, and $F(|\mathbf{r}-\mathbf{R}|)$ is Friedel function defined as

$$F(|\mathbf{r}-\mathbf{R}|) \equiv \frac{1}{V}\sum_{\mathbf{q}} \Pi(\mathbf{q}) e^{i\mathbf{q}\cdot(\mathbf{r}-\mathbf{R})}. \tag{9.83}$$

Although not shown here, this expression is obtained by utilizing the orthonormality of plane waves and the inversion symmetry in the reciprocal space for $\mathbf{q}$ so that $F(|\mathbf{r}-\mathbf{R}|)$ can be related to the static Lindhard fuction,

$$\Pi(\mathbf{q}) = \frac{1}{V}\sum_{\mathbf{k}} \frac{f(\varepsilon^0_{\mathbf{k}}) - f(\varepsilon^0_{\mathbf{k}+\mathbf{q}})}{\varepsilon^0_{\mathbf{k}+\mathbf{q}} - \varepsilon^0_{\mathbf{k}}}. \tag{9.84}$$

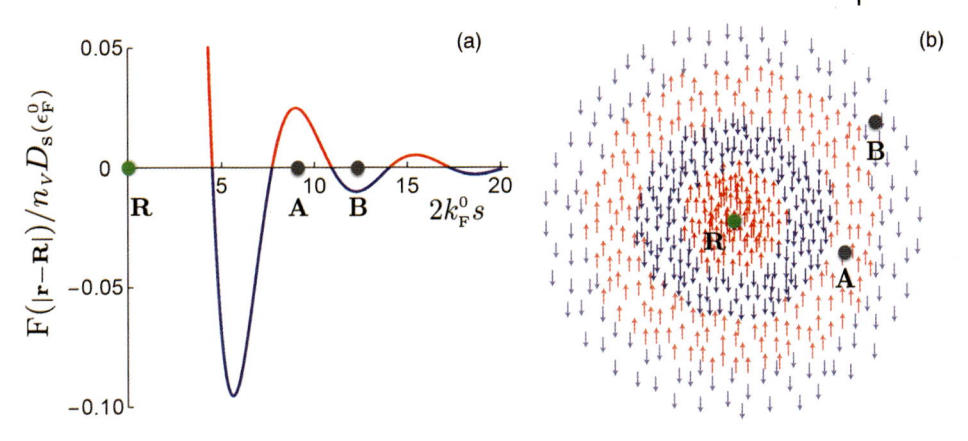

Figure 9.6 RKKY oscillation of a free electron metal depicted (a) in a graph as a function of $2k_F^0 s = 2k_F^0|\mathbf{r} - \mathbf{R}|$ and (b) in a schematic diagram where the induced spin density is shown with ↑ and ↓.

In essence, the induced spin density of a free electron system is described by Friedel function, as shown in Eq. (9.82b). At the absolute zero temperature, Friedel function can be expressed as an oscillatory function of $s = |\mathbf{r} - \mathbf{R}|$, the distance from the magnetic site, with an asymptotic periodicity of π/k_F^0 (RKKY oscillation):

$$F(|\mathbf{r} - \mathbf{R}|) = 6\pi n_V D_s\left(\varepsilon_F^0\right) \frac{\sin\left(2k_F^0 s\right) - \left(2k_F^0 s\right)\cos\left(2k_F^0 s\right)}{\left(2k_F^0 s\right)^4}, \quad (9.85)$$

where n_V is the electron concentration, and k_F^0 is the Fermi sphere radius determined for the unperturbed free electron state $\left(k_F^0 = [3\pi^2 n_V]^{\frac{1}{3}}\right)$ [20]. The same behavior for the electron density described by Eq. (9.82a) is called Friedel oscillation.

The result of Eq. (9.83) is shown in Figure 9.6, where the magnetic site $\mathbf{R}$ is at the origin, with a positive spin density. The radial dependence of the Friedel function indicates that the spin-dependent mixing of the plane waves leads to an accumulation of the spin density at the magnetic site at the expense of the opposite spin polarization off from the site. This is understandable from Eq. (9.33) in that the maximum energy stabilization occurs when an induced spin density has the same spin orientation as the initial spin density at the magnetic site. Likewise, Eq. (9.33) predicts that the magnetic sites located at the positions **A** and **B** will interact ferromagnetically and antiferromagnetically, respectively, with the magnetic site at the origin, judging from the induced spin density values at those sites. These findings are general even in the magnetic systems for which free electron model and point spin density approximation are not valid. For real electronic systems, however, we do not expect that the electron density and induced spin density show the same oscillatory behavior, which is a topic for future studies.

9.4
Prospect

The simple examples described above illustrate how SPPOT formalism can be used to analyze the spin polarization phenomena in seemingly different electronic systems under the same conceptual machinery. It is yet noted that SPPOT does not account for phase transition phenomena and for other electronic states such as superconductivity. The analytical expressions from SPPOT are universal for all the collinear magnetic systems whose ground states are well described by the one-determinant formalism of Kohn–Sham DFT. The mathematical exercises for individual simple cases shown here reveal how their underlying assumptions and approximations limit the scope of the existing magnetic exchange theories and yet they can provide an insight in developing our understanding of more complex magnetic systems that involve strong "spin-dependent orbital mixing" within the framework of chemical bonding concept. A similar approach has been adopted for transition intermetallics in physics community under a name of "covalent magnetism" [21], which unlike SPPOT considers a spin polarized band structure as a consequence of interactions of atomic d-spin orbitals that are already spin-polarized in energy. Meanwhile, SPPOT is related to density functional perturbation theory (DFPT) in which the sum-over-states representation of perturbational Green function for the linear density–density response leads to the mathematical expression of the electron density change equivalent to that of SPPOT [22], if the first-order orbital energy corrections are negligible.

However, further expansion of SPPOT formalism is necessary if one pursues comprehension of noncollinear magnetic systems with apparent magnetic anisotropy. While numerical successes have been reported based on two-component spinor DFT or multireference DFT, the interpretation of the calculation results will remain less satisfactory without a proper DFT-based model that encompasses the existing magnetic exchange theories, including the spin "vector" representation of Heisenberg model as the simplest example. It is mentioned in passing that explicit employment of two-component spinors as zeroth-order wave functions within the SPPOT formalism allows us to analyze magnetic coupling, spin–orbit interaction and Zeeman effect in the context of spin vector representation of the DFT calculation results for various systems including f-electron compounds.

References

1 Kübler, J.K. (2000) *Theory of Itinerant Electron Magnetism*, Clarendon Press/Oxford University Press, Oxford.

2 Seo, D.K. (2006) Density functional perturbational orbital theory of spin polarization in electronic systems. I. Formalism. *J. Chem. Phys.*, **125** (15), 154105.

3 Seo, D.K. (2007) Density functional perturbational orbital theory of spin

polarization in electronic systems. II. Transition metal dimer complexes. *J. Chem. Phys.*, **127** (18), 184103.

4 Landrum, G.A. and Dronskowski, R. (2000) The orbital origins of magnetism: From atoms to molecules to ferromagnetic alloys. *Angew. Chem., Int. Ed. Eng.*, **39** (9), 1560–1585.

5 Parr, R.G. and Yang, W. (1989) *Density-Functional Theory of Atoms and Molecules*, Oxford University Press.

6 Seo, D.K., Papoian, G., and Hoffmann, R. (2000) Generalized perturbational molecular orbital (PMO) theory. *Int. J. Quantum Chem.*, **77** (1), 408–420.

7 Seo, D.K. (2007) Self-interaction correction in the LDA+U method. *Phys. Rev. B*, **76** (3), 033102.

8 Kahn, O., Galy, J., Journaux, Y., Jaud, J., and Morgenstern-Badarau, I. (1982) Synthesis, crystal structure and molecular conformations, and magnetic properties of a copper-vanadyl (CuII-VOII) heterobinuclear complex: interaction between orthogonal magnetic orbitals. *J. Am. Chem. Soc.*, **104** (8), 2165–2176.

9 Noodleman, L. (1981) Valence bond description of antiferromagnetic coupling in transition metal dimers. *J. Chem. Phys.*, **74**, 5737–5743.

10 Illas, F., Moreira, I.de P.R., de Graaf, C., and Barone, V. (2000) Magnetic coupling in biradicals, binuclear complexes and wide-gap insulators: a survey of *ab initio* wave function and density functional theory approaches. *Theor. Chem. Acc.*, **104** (3), 265–272.

11 Dai, D. and Whangbo, M.H. (2003) Spin exchange interactions of a spin dimer: analysis of broken-symmetry spin states in terms of the eigenstates of Heisenberg and Ising spin hamiltonians. *J. Chem. Phys.*, **118** (1), 29–39.

12 Anderson, P.W. (1959) New approach to the theory of superexchange interactions. *Phys. Rev.*, **115**, 2–13.

13 Hay, P.J., Thibeault, J.C., and Hoffmann, R. (1975) Orbital interactions in metal dimer complexes. *J. Am. Chem. Soc.*, **97** (17), 4884–4899.

14 Kahn, O. (1993) *Molecular Magnetism*, Wiley-VCH Verlag GmbH, New York.

15 Seo, D.K. and Kim, S.H. (2008) Nature of Stoner condition for metallic ferromagnetism. *J. Comput. Chem.*, **29** (13), 2172–2176.

16 Janak, J.F. (1977) Uniform susceptibilities of metallic elements. *Phys. Rev. B*, **16** (1), 255–262.

17 Kulatov, E. and Mazin, I.I. (1990) Extended stoner factor calculations for the half-metallic ferromagnets NiMnSb and CrO_2. *J. Phys. Condens. Matter*, **2** (2), 343.

18 de Groot, R.A., Mueller, F.M., Engen, P.G. v., and Buschow, K.H.J. (1983) New class of materials: Half-metallic ferromagnets. *Phys. Rev. Lett.*, **50**, 2024–2027.

19 Yosida, K. (1996) *Theory of Magnetism*, Solid-State Sciences Series, Springer.

20 Kim, D. (2013) *New Perspectives in Magnetism of Metals*, Springer.

21 Mohn, P. (2006) *Magnetism in the Solid State*, Springer-Verlag, Berlin, Heidelberg.

22 Goncharov, V. (2014) *Non-Linear Optical Response in Atoms, Molecules and Clusters*, SpringerBriefs in Electrical and Magnetic Properties of Atoms, Molecules, and Clusters, Springer International Publishing.

10
Magnetic Properties from the Perspectives of Electronic Hamiltonian: Spin Exchange Parameters, Spin Orientation, and Spin-Half Misconception

Myung-Hwan Whangbo[1] and Hongjun Xiang[2,3]

[1]North Carolina State University, Department of Chemistry, 2620 Yarbrough Drive, Raleigh, NC 27695-8204, USA
[2]Fudan University, Department of Physics, Key Laboratory of Computational Physical Sciences (Ministry of Education), State Key Laboratory of Surface Physics, 220 Handan Rd., Shanghai 200433, P. R. China
[3]Nanjing University, Collaborative Innovation Center of Advanced Microstructures, Hankou Road 22, Nanjing 210093, P. R. China

10.1 Introduction

In this chapter, we examine how to think about and describe the magnetic properties of crystalline solids, which arise from their transition metal magnetic ions, from the perspectives of an electronic Hamiltonian. The latter represents the energy states of a magnetic system using both orbital and spin degrees of freedom, that is, the angular property of a magnetic ion is described by a set of orbital/spin states $|L, L_z\rangle|S, S_z\rangle$. Compared with the strength of chemical bonding (on the order of several electron volts), the unpaired electrons of a magnetic ion interact very weakly with those of neighboring magnetic ions so that the energy scale involved in magnetic states is very small, and the states responsible for the magnetic properties are closely packed in energy (Figure 10.1). (For example, at the magnetic field H of 1 T, $\mu_B H = 5.8 \times 10^{-2}$ meV = 0.67 K in k_B units. Other energy scales for discussing magnetic properties are 1 meV = 11.6 K = 8.06 cm^{-1}, and 1 cm^{-1} = 1.44 K.) To quantitatively describe the magnetic properties of such a system at any given temperature, it is necessary to obtain the spectrum of the energy states and subsequently Boltzmann average of the properties of these states. Since solving this problem on the basis of an electronic Hamiltonian is very difficult, one employs a spin Hamiltonian $\hat{H}_{\text{spin}}$ that represents each magnetic ion using only a set of spin states $|S, S_z\rangle$. This toy Hamiltonian allows one to generate the energy states without self-consistent field calculations thereby

---

Handbook of Solid State Chemistry, First Edition. Edited by Richard Dronskowski, Shinichi Kikkawa, and Andreas Stein.
© 2017 Wiley-VCH Verlag GmbH & Co. KGaA. Published 2017 by Wiley-VCH Verlag GmbH & Co. KGaA.

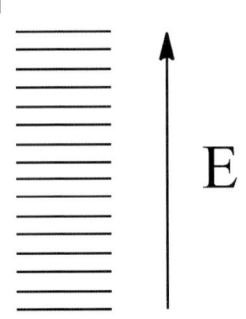

Figure 10.1 Close-packed energy states of a magnetic system that arise from weak interactions among the unpaired electrons of its magnetic ions.

greatly simplifying calculations, because it is specified by a few spin exchange interactions $J_{ij}\hat{S}_i \cdot \hat{S}_j$ between certain spin sites i and j,

$$\hat{H}_{\text{spin}} = \sum_{i<j} J_{ij}\, \hat{S}_i \cdot \hat{S}_j, \tag{10.1}$$

where the constants J_{ij} (i.e., spin exchange parameters) are the numerical parameters to be determined. The repeat pattern of the chosen spin exchange paths i–j forms the spin lattice (e.g., an isolated dimer, a uniform chain, an alternating chain, a two-leg ladder, etc.) of the magnetic ions (Figure 10.2). Once a spin lattice is selected, this model Hamiltonian greatly simplifies the generation of its energy states as a function of the numerical parameters J_{ij}, which are fixed as those that best simulate the experimental magnetic data (e.g., magnetic susceptibility, specific heat, and spin wave dispersion relations). The purpose of using such a toy Hamiltonian is to capture the essential physics of observed magnetic properties with a minimal number of adjustable parameters J_{ij}.

A general problem facing such a toy Hamiltonian analysis is that more than one spin lattice may equally well simulate the available experimental data. Since the novelty of a chosen spin lattice presents an opportunity to discover a new

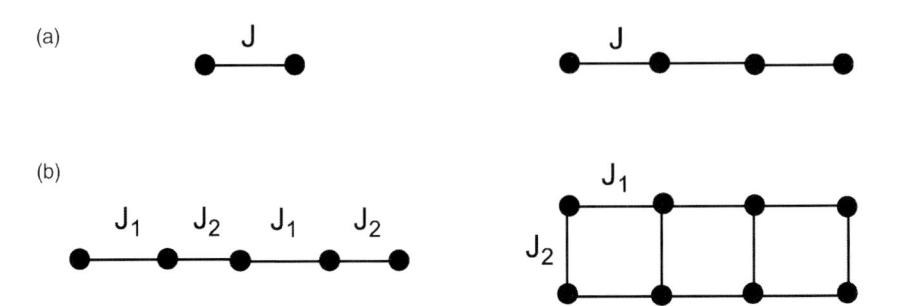

Figure 10.2 Examples of simple spin lattices. (a) An isolated spin dimer and a uniform chain requiring one spin exchange constant. (b) An alternating chain and a two-leg ladder requiring two spin exchange constants.

physics, the practitioners of spin Hamiltonian analyses tend to favor the interpretation of experimental data using a novel spin lattice without checking if the chosen spin lattice is consistent with the electronic structure of a magnetic system under examination. A chosen spin lattice often does not turn out to be relevant for the system under examination, thus generating "an answer in search of a problem." A bright side of such a regrettable situation would be that the generated physics could stimulate experimental interests to search for a system that fits the "predicted" physics. These days one can readily determine what spin exchange paths are relevant for any given magnetic system by performing energy-mapping analysis [1–4] on the basis of DFT electronic structure calculations. This theoretical/computational tool makes it possible to interpret experimental data in terms of the relevant spin lattice.

An implicit assumption behind using a spin Hamiltonian is that one can correctly describe all magnetic phenomena in terms of the energy states it generates. The strength of a spin Hamiltonian analysis is to simplify complex calculations as a result of using only the spin degree of freedom, but this strength is also the very cause for its failure at the fundamental level; a spin Hamiltonian description leads to the conceptual impasse recently termed the spin-half syndrome [5,6]. A classic example showing this spin-half misconception is the study of $CuCl_2 \cdot 2H_2O$ by Moriya and Yoshida more than six decades ago [7]; as the cause for the observed spin orientation of the $S=1/2$ ion Cu^{2+}, they dismissed outright the possibility that the $S=1/2$ ion has magnetic anisotropy induced by spin-orbit coupling (SOC), $\lambda \hat{S} \cdot \hat{L}$, which is a single-spin site interaction (i.e., a local interaction), and then proceeded to explain the observed spin orientation in terms of nonlocal interactions (e.g., anisotropic spin exchange and magnetic dipole–dipole interactions). Over the years, the spin-half misconception has been perpetuated in monographs and textbooks on magnetism [8]. However, this misconception contradicts not only the experimental observations that spin-half ions (e.g., Cu^{2+}, V^{4+}, Ir^{4+}) exhibit magnetic anisotropies [5,9], but also theoretical results based on electronic Hamiltonians in which the energy states of a magnetic system are described by using both orbital and spin degrees of freedom [5,6,9].

A transition metal magnetic ion of any spin ($S=1/2$–$5/2$) has magnetic anisotropy as a consequence of SOC, $\lambda \hat{S} \cdot \hat{L}$, because the latter induces interactions among its crystal field split d-states and because the energy lowering associated with these interactions depend on the spin orientation [2,5,6,9,10]. In an electronic Hamiltonian approach, the energy states of a magnetic system are discussed in terms of its magnetic orbitals (i.e., its singly occupied orbitals). Each magnetic orbital represents either the spin-up state $|\uparrow\rangle = |\frac{1}{2}, +\frac{1}{2}\rangle$ or the spin-down state $|\downarrow\rangle = |\frac{1}{2}, -\frac{1}{2}\rangle$, so the overall spin S of a magnetic ion is related to how many magnetic orbitals it generates. Thus, each magnetic ion of a magnetic orbital in spin state $|S, S_z\rangle$ ($= |\uparrow\rangle$ or $|\downarrow\rangle$) is described by the orbital/spin state $|L, L_z\rangle|S, S_z\rangle$. The magnetic states are modified by SOC, $\lambda \hat{S} \cdot \hat{L}$, due to the associated intermixing between them. For example, when there is no degeneracy in the magnetic orbitals, a given spin-up magnetic orbital $|L, L_z\rangle|\uparrow\rangle$ is modified by the intermixing as

$$\{(1-\gamma^2-\delta^2-\cdots)|L,L_z\rangle+\gamma|L',L'_z\rangle+\cdots\}|\uparrow\rangle+\{\delta|L'',L''_z\rangle+\cdots\}|\downarrow\rangle \quad (10.2)$$

where γ and δ are the mixing coefficients (see Section 10.7 for more details). Namely, due to SOC, the orbital/spin state $|L,L_z\rangle|\uparrow\rangle$ mixes not only with different orbital/spin states $|L',L'_z\rangle|\uparrow\rangle$ but also with different orbital/spin states $|L'',L''_z\rangle|\downarrow\rangle$. A given spin-down magnetic orbital $|L,L_z\rangle|\downarrow\rangle$ is similarly affected by SOC. This SOC-induced orbital mixing is independent of whether the overall spin S of the magnetic ion is 1/2 or greater because this mixing occurs in each individual magnetic orbital and hence, does not depend on how many magnetic orbitals a magnetic ion generates. This is why magnetic anisotropy is predicted for $S=1/2$ ions on an equal footing to $S>1/2$ ions in an electronic Hamiltonian approach. This fundamental result is not described by a spin Hamiltonian simply because it lacks the orbital degree of freedom; having completely suppressed the orbital $|L,L_z\rangle$ of a magnetic ion, a spin Hamiltonian does not allow one to discuss the SOC, $\lambda \hat{S}\cdot\hat{L}$, and hence, is unable to describe the preferred spin orientation of any magnetic ion. The spin-half misconception is a direct consequence from this deficiency of a spin Hamiltonian.

Anyone who attempted to publish the finding that the spin-half misconception is erroneous would have experienced eye-opening discourses with its proponents (mostly, practitioners of spin Hamiltonian analyses) to learn that they treat the attempt as an affront to their work and do their utmost to suppress its publication. For those schooled in the electronic structure description, it is only natural to describe the energy states of a magnetic system by using both orbital and spin degrees of freedom, because unpaired electrons responsible for magnetic properties must be accommodated in certain orbitals, and hence have no problem in finding that a spin Hamiltonian is a theoretically limited tool. However, most of those schooled in doing physics with spin Hamiltonian do not appear to realize that this toy Hamiltonian was born out of the necessity to simplify calculations. They tend to believe that the correct energy states of a magnetic system are those generated by using only the spin degree of freedom, and insist that an electronic Hamiltonian description should produce the same conclusion as does a spin Hamiltonian description even if it is an erroneous one resulting from its deficiency. To help break this conceptual impasse, it is necessary to expose the origin of the spin-half misconception by discussing how the properties of solid-state magnetic materials are described from the perspectives of an electronic Hamiltonian.

Analysis of magnetic properties on the basis of an electronic Hamiltonian deals with two competing issues: One is to produce accurate quantitative predictions and the other is to provide qualitative pictures with which to organize and think about. These two subjects are discussed by organizing our work as follows: In Section 10.2, we first discuss the angular properties of the atomic orbitals and then the crystal field split d-states of magnetic ions. Section 10.3 covers the energy-mapping analysis that allows one to relate the spin Hamiltonian analysis of a given magnetic system to its electronic structure by evaluating the spin exchange parameters this toy Hamiltonian needs. In Section 10.4, we discuss the qualitative features of spin exchange interactions in terms of orbital interactions involving magnetic orbitals. In Section 10.5, we describe indirect ways of incorporating SOC into a spin Hamiltonian and the associated energy-mapping analysis as well as the origin of the spin-half misconception. In order to prepare for our discussion of magnetic anisotropy, the condition leading to uniaxial magnetism is discussed in Section 10.6. Section 10.7 describes the qualitative rules that allow one to predict the preferred spin orientations of magnetic ions on the basis of the perturbation theory in which SOC is taken as perturbation with the crystal field split d-states as the unperturbed state. In Section 10.8, we discuss several issues concerning the magnetic properties of 5d magnetic ions. Our concluding remarks are summarized in Section 10.9.

10.2
Atomic Orbitals and Magnetic Orbitals

10.2.1
Angular Properties of Atomic Orbitals

The angular properties of atomic orbitals are specified by the spherical harmonics, $|L, L_z\rangle$, defined in terms of two quantum numbers; the orbital quantum number L ($= 0, 1, 2, \ldots$) and its z-axis component $L_z = -L, -L+1, \ldots, L-1, L$ for a given L. The angular behaviors of the atomic p- and d-orbitals are summarized in Table 10.1. In terms of the magnetic quantum numbers L_z, the d-orbitals are grouped into three sets:

$$L_z = 0 \quad \text{for} \quad 3z^2 - r^2,$$
$$L_z = \pm 1 \quad \text{for} \quad \{xz, yz\},$$
$$L_z = \pm 2 \quad \text{for} \quad \{xy, x^2 - y^2\}.$$

Similarly, the p-orbitals are expressed as linear combinations of the spherical harmonics $|L, L_z\rangle$, where $L=1$, and $L_z = 1, 0, -1$. Thus, the p-orbitals are grouped into two sets:

$$L_z = 0 \quad \text{for} \quad z,$$
$$L_z = \pm 1 \quad \text{for} \quad \{x, y\}.$$

Table 10.1 Angular properties of atomic p- and d-orbitals.

x	$(|1, -1\rangle - |1, +1\rangle)/\sqrt{2}$
y	$i(|1, -1\rangle + |1, +1\rangle)/\sqrt{2}$
z	$|1, 0\rangle$
$3z^2 - r^2$	$|2, 0\rangle$
xz	$(|2, -1\rangle - |2, +1\rangle)/\sqrt{2}$
yz	$i(|2, -1\rangle + |2, +1\rangle)/\sqrt{2}$
xy	$i(|2, -2\rangle - |2, +2\rangle)/\sqrt{2}$
$x^2 - y^2$	$(|2, -2\rangle + |2, +2\rangle)/\sqrt{2}$

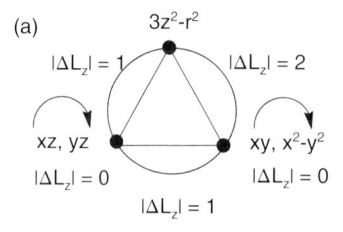

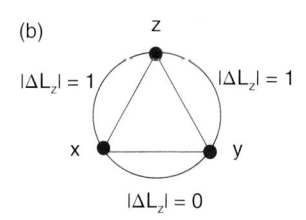

Figure 10.3 Minimum difference in the magnetic quantum numbers, $|\Delta L_z|$, between pairs of (a) d-orbitals and (b) p-orbitals.

Consequently, as depicted in Figure 10.3, the minimum difference $|\Delta L_z|$ in the magnetic quantum numbers between different atomic orbitals is summarized as follows:

$$|\Delta L_z| = 0 \quad \text{between} \begin{cases} xz \text{ and } yz, \\ xy \text{ and } x^2 - y^2, \\ x \text{ and } y. \end{cases}$$

$$|\Delta L_z| = 1 \quad \text{between} \begin{cases} 3z^2 - r^2 \text{ and } \{xz, yz\}, \\ \{xz, yz\} \text{ and } \{xy, x^2 - y^2\}, \\ z \text{ and } \{x, y\}. \end{cases}$$

$$|\Delta L_z| = 2 \quad \text{between } 3z^2 - r^2 \text{ and } \{xy, x^2 - y^2\}.$$

These $|\Delta L_z|$ values play a crucial role in understanding the preferred spin orientations of magnetic ions on the basis of the SOC-induced HOMO–LUMO interactions of their crystal field split d-states (see Section 10.7).

In quantum mechanics, the orbital angular momentum $\vec{L}$ is replaced by the orbital angular momentum operator $\hat{L}$ that has three components $\hat{L}_x$, $\hat{L}_y$, and $\hat{L}_z$ in a Cartesian coordinate system. Most calculations associated with orbital angular momentum make use of $\hat{L}_z$, $\hat{L}_+$, and $\hat{L}_-$, where $\hat{L}_+$ and $\hat{L}_-$ are the ladder operators defined by

$$\hat{L}_+ = \hat{L}_x + i\hat{L}_y,$$
$$\hat{L}_- = \hat{L}_x - i\hat{L}_y.$$

The orbitals $|L, L_z\rangle$ are affected by the operators $\hat{L}_z$, $\hat{L}_+$, and $\hat{L}_-$ as follows:

$$\hat{L}_z|L, L_z\rangle = L_z|L, L_z\rangle,$$
$$\hat{L}_+|L, L_z\rangle = \sqrt{L(L+1) - L_z(L_z+1)}|L, L_z+1\rangle, \quad (10.3)$$
$$\hat{L}_-|L, L_z\rangle = \sqrt{L(L+1) - L_z(L_z-1)}|L, L_z-1\rangle.$$

Here, we use the atomic unit in which the unit of angular momentum ħ is equal to 1. The $\hat{L}_+$ raises the L_z of $|L, L_z\rangle$ by 1 as long as $L_z + 1 \leq L$, while $\hat{L}_-$ lowers the L_z of $|L, L_z\rangle$ by 1 as long as $L_z - 1 \geq -L$. In our later discussion, we need to evaluate the integrals

$$\langle i|\hat{L}_x|j\rangle, \langle i|\hat{L}_y|j\rangle, \text{ and } \langle i|\hat{L}_z|j\rangle$$

involving atomic p-orbitals $(i, j = x, y, z)$ as well as those involving atomic d-orbitals $(i, j = 3z^2 - r^2, xz, yz, x^2 - y^2, xy)$. By using Eq. (10.3) and the expressions of the atomic orbitals listed in Table 10.1, we obtain the nonzero integrals listed in Table 10.2 [9].

10.2.2
Crystal Field Split d-States

In most cases, we are concerned with systems containing transition metal magnetic ions M in magnetic oxides. The preferred orientations of their spin moments are determined by their d-states split by their surrounding ligands L. It depends on the symmetry and composition of the ML$_n$ (typically, $n = 4$–6) polyhedron how the d-states of the ion M split. In a description of electronic structures using an effective one-electron Hamiltonian $\hat{H}^{\text{eff}}$, each split d-level of a ML$_n$ polyhedron does not change its energy and shape regardless of whether it is occupied by one or two electrons, because the presence of electron–electron

Table 10.2 Nonzero integrals of the angular momentum operators, $\langle i|\hat{L}_x|j\rangle$, $\langle i|\hat{L}_y|j\rangle$, and $\langle i|\hat{L}_z|j\rangle$, where $(i, j = x, y, z)$ or $(i, j = 3z^2 - r^2, xz, yz, x^2 - y^2, xy)$.

$\hat{L}_z$	$\langle y|\hat{L}_z|x\rangle = i$
$\hat{L}_x$	$\langle z|\hat{L}_x|y\rangle = i$
$\hat{L}_y$	$\langle x|\hat{L}_y|z\rangle = i$
$\hat{L}_z$	$\langle xy|\hat{L}_z|x^2 - y^2\rangle = 2i$
	$\langle xz|\hat{L}_z|yz\rangle = -i$
$\hat{L}_x$	$\langle 3z^2 - r^2|\hat{L}_x|yz\rangle = i\sqrt{3}$
	$\langle x^2 - y^2|\hat{L}_x|yz\rangle = i$
	$\langle xz|\hat{L}_x|xy\rangle = i$
$\hat{L}_y$	$\langle 3z^2 - r^2|\hat{L}_y|xz\rangle = -i\sqrt{3}$
	$\langle x^2 - y^2|\hat{L}_y|xz\rangle = i$
	$\langle yz|\hat{L}_y|xy\rangle = -i$

repulsion in a doubly occupied level is ignored. We discuss this simple picture first and then consider how to modify these one-electron levels by electron correlation.

10.2.2.1 One-Electron States without Electron Correlation

How strongly the d-orbitals of the transition metal M interact with the p-orbitals of the ligands L depends on the nature of the d-orbitals and the shape of the ML_n polyhedron [11]. In the split d-states that result from these interactions, the ligand p-orbitals are combined out of phase to the metal d-orbitals. Therefore, a given d-state lies high in energy if the M—L antibonding is strong in the state. Let us start from the d-states of an ML_6 octahedron (Figure 10.4a), which are split into the triply degenerate t_{2g} state lying below the doubly degenerate e_g state (Figure 10.4b). The three components of the t_{2g} state are each described by M—L π-antibonding, and the two components of the e_g state by M—L σ-antibonding (Figure 10.4c). Some ML_n (typically, $n = 4$–6) polyhedra can be regarded as derived from the ML_6 octahedron by lengthening and/or removing a few M—L bonds. The split d-states of such polyhedra can be readily predicted by considering how the extent of the σ-antibonding and/or π-antibonding of the M—L bonds varies under the geometrical changes (Figure 10.5).

For an axially elongated ML_6 octahedron with the z-axis taken along the elongated M—L bonds, the d-states are split as depicted in Figure 10.5b; the $3z^2 - r^2$ state (commonly, referred to as the z^2 state, for simplicity) is significantly lowered in energy because the σ-antibonding is reduced, while the xz and yz states are slightly lowered in energy because the π-antibonding is reduced. For a square planar ML_4 with the z-axis taken perpendicular to the plane, the d-states are split as shown in Figure 10.5c; the $3z^2 - r^2$ state is lowered to become the lowest

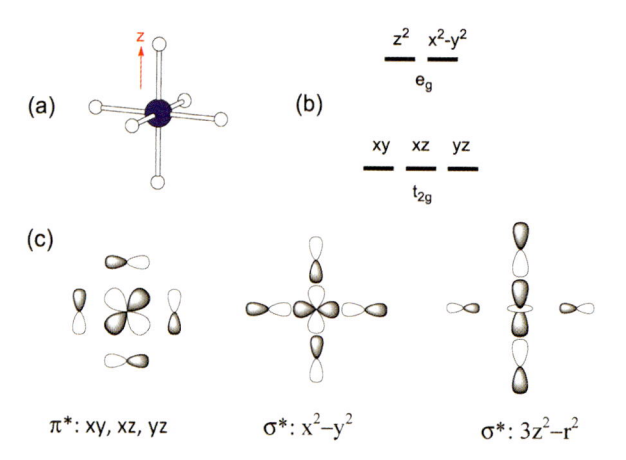

Figure 10.4 (a) An ideal ML_6 octahedron with the local z-axis taken along one M—O bond, that is, one fourfold rotational axis. (b) The orbital compositions of the t_{2g} and e_g states. (c) The π-antibonding in the xy, xz, and yz components of the t_{2g} state, and the σ-antibonding in the x^2-y^2 and the $3z^2-r^2$ components of the e_g state.

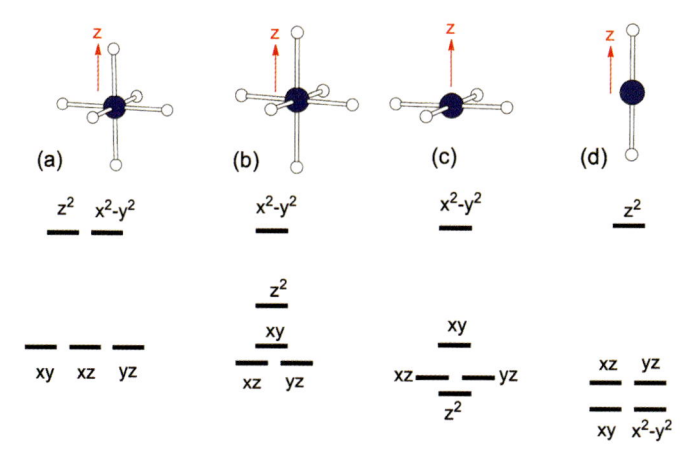

Figure 10.5 The split d-states of (a) an ideal ML$_6$ octahedron, (b) an axially elongated ML$_6$ octahedron, (c) a square planar ML$_4$, and (d) a linear ML$_2$.

in energy because the σ-antibonding along the z-direction is totally absent while that in the xy-plane is further reduced because the girdle of the $3z^2 - r^2$ state is diminished in size by the second-order orbital mixing of the upper s-orbital of M [11]. In addition, the xz and yz states of the ML$_4$ square plane are lower than those of the axially elongated ML$_6$ octahedron because the π-antibonding is absent along the z-direction. For a linear ML$_2$ with the z-axis taken along the M—L bonds, the d-states are split as depicted in Figure 10.5d; the xy and $x^2 - y^2$ states are lowered more in energy than are the xz and yz states because π-antibonding is absent in the xy and $x^2 - y^2$ states while it is present in the xz and yz states.

In discussing the t_{2g} and e_g states of an ML$_6$ octahedron, there occur cases when it is more convenient to take the z-axis along one threefold rotational axis of the octahedron (Figure 10.6a) [12] rather than along one M—L bond (i.e., along one fourfold rotational axis) (Figure 10.5a). Then their orbital character changes as summarized in Table 10.3; the $3z^2 - r^2$ state becomes one of the t_{2g} set, while the $(xy, x^2 - y^2)$ degenerate set mixes with the (xz, yz) degenerate set to give the $(1e_x, 1e_y)$ and $(2e_x, 2e_y)$ sets (Figure 10.6b). The $(xy, x^2 - y^2)$ set has a

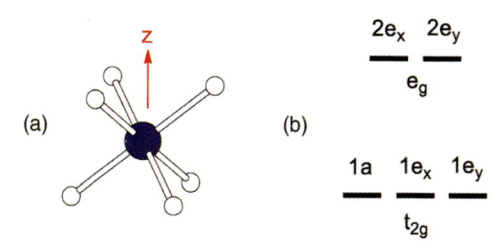

Figure 10.6 (a) An ideal ML$_6$ octahedron with the local z-axis taken along one threefold rotational axis. (b) The orbital compositions of the t_{2g} and e_g states as listed in Table 10.3.

Table 10.3 Orbital character of the d-states of an ML_6 octahedron in two different settings of the Cartesian coordinates.

z-axis direction	Along one M—L bond (Figure 10.4a)	Along one C_3-rotational axis (Figure 10.6a)
t_{2g}	xy	$1a \equiv 3z^2 - r^2$
	xz	$1e_x \equiv \sqrt{\frac{2}{3}}xy - \sqrt{\frac{1}{3}}xz$
	yz	$1e_y \equiv \sqrt{\frac{2}{3}}(x^2 - y^2) - \sqrt{\frac{1}{3}}yz$
e_g	$x^2 - y^2$	$2e_x \equiv \sqrt{\frac{1}{3}}xy + \sqrt{\frac{2}{3}}xz$
	$3z^2 - r^2$	$2e_y \equiv \sqrt{\frac{1}{3}}(x^2 - y^2) + \sqrt{\frac{2}{3}}yz$

larger contribution than does the (xz, yz) set in the $(1e_x, 1e_y)$ set, and the opposite is the case in the $(2e_x, 2e_y)$ set (Table 10.3). Such orbital representations as described in Figure 10.6 and Table 10.3 will be employed in Section 10.7.

10.2.2.2 One-Electron States with Electron Correlation

The essence of electron correlation is that when a given energy state is doubly occupied, its energy is raised by electron–electron repulsion. The latter is partly reduced in spin-polarized electronic structure calculations, in which spin-up states are allowed to differ in energy and shape from their spin-down counterparts. For strongly correlated systems, the energy split arising from spin-polarized electronic structure calculations is not strong enough to generate singly occupied states needed to describe their magnetic insulating states. In spin-polarized DFT calculations, this deficiency is corrected by adding the on-site repulsion U on magnetic ions to force a large split between their spin-up and spin-down states (Figure 10.7) [13]. Such calculations are referred to as DFT + U calculations.

An important consequence of spin-polarized DFT+U calculations is found when two adjacent spin sites interact [2]. If the two equivalent spin sites have a

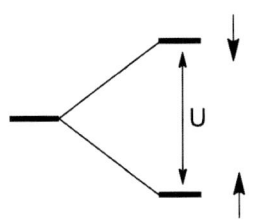

Figure 10.7 The split of the spin-up and spin-down states by an on-site repulsion U. These states are degenerate in the nonspin-polarized description (left), but are split in the spin-polarized description (right).

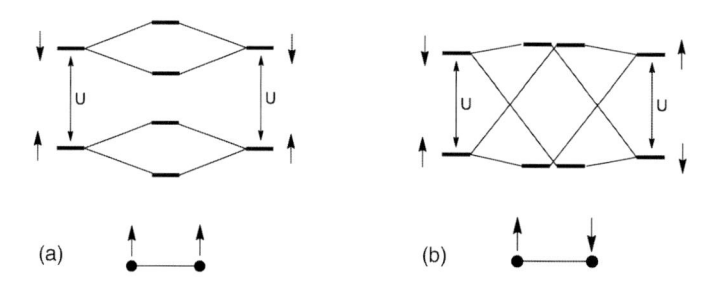

Figure 10.8 The orbital interactions between two equivalent spin sites for cases when they have (a) a FM arrangement and (b) an AFM arrangement.

ferromagnetic (FM) arrangement (Figure 10.8a), the spin-up states of the two sites are degenerate, and so are the spin-down states of the two sites. However, if the two equivalent spin sites have an antiferromagnetic (AFM) arrangement (Figure 10.8b), the spin-up states of the two sites are nondegenerate, and so are the spin-down states of the two sites. In general, orbital interactions between degenerate states are stronger than those between nondegenerate states [11]. Since orbital interactions between states require that their spins be identical, the AFM arrangement leads to a weaker orbital interaction between adjacent spin sites than does the FM arrangement [2].

From the viewpoint of the split d-states obtained from an effective one-electron Hamiltonian, the qualitative features of DFT + U calculations can be simulated by splitting the spin-up d-states from those of the spin-down d-states approximately by the amount of U, as illustrated in Figure 10.9, for a high-spin

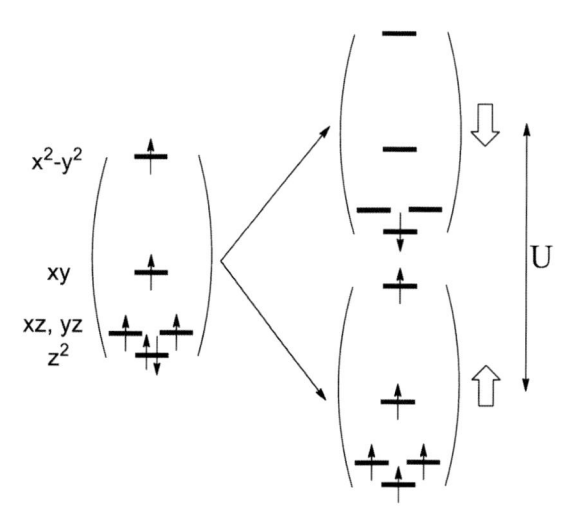

Figure 10.9 The simulation of the split d-states obtained from DFT + U calculations in terms of those obtained from an effective one-electron Hamiltonian for a high-spin ($S=2$) d^6 ion at a square planar site forming a FeL$_4$ square plane.

($S=2$) d^6 ion forming a FeL_4 square plane. For simplicity, the separation between the spin-up and spin-down d-states is exaggerated in Figure 10.9. What is important to note is that the HOMO and the LUMO levels occur within the spin-down states if the d-shell is more than half-filled, but within the spin-up states if the d-shell is less than half-filled. (This is due to the convention in which the majority and minority spin states are regarded as spin-up and spin-down states, respectively.) Only when the d-shell is half-filled in a high-spin manner, the HOMO and the LUMO levels occur between the spin-up and spin-down states.

An alternative way of correcting the deficiency of spin-polarized DFT calculations is the hybrid functional method [14], in which the exchange-correlation functional needed for calculations is obtained by mixing some amount, α (typically, 0.2), of the Hartree–Fock exchange potential into the DFT functional. The on-site repulsion U is an empirical parameter in DFT+U calculations, and so is the mixing parameter α in DFT+hybrid calculations. In general, DFT+U calculations are much less time-consuming than are DFT+hybrid calculations. It should be emphasized that density functional calculations are first principles calculations only after the value of U is fixed in DFT+U calculations, and only after the value of α is fixed in DFT+hybrid calculations.

Given computing resources, DFT calculations with or without including SOC effects [15] can be readily carried out by using user-friendly DFT program packages such as VASP [16], which considers only valence electrons using the frozen-core projector augmented waves, and WIEN2k [17], which considers all electrons. As the exchange-correlation functional needed for DFT calculations, the generalized gradient approximation [18] is commonly used for studying solid-state materials. In understanding results of DFT, DFT + U, and DFT + U + SOC calculations or predicting results prior to calculations, the concept of orbital interaction analysis [11] developed on the basis of one-electron picture is useful (see below).

10.3
Energy-Mapping Analysis

For two spins $\hat{S}_1$ and $\hat{S}_2$ at spin sites 1 and 2, respectively, the dot product $\hat{S}_1 \cdot \hat{S}_2$ has three Cartesian components, that is, $\hat{S}_1 \cdot \hat{S}_2 = \hat{S}_{1x}\hat{S}_{2x} + \hat{S}_{1y}\hat{S}_{2y} + \hat{S}_{1z}\hat{S}_{2z}$. Thus, a general expression for the spin exchange interaction energy between the two spin sites can be written as

$$\hat{H}_{spin} = J_x \hat{S}_{1x}\hat{S}_{2x} + J_y \hat{S}_{1y}\hat{S}_{2y} + J_z \hat{S}_{1z}\hat{S}_{2z},$$

where J_x, J_y, and J_z are anisotropic spin exchanges along the x-, y-, and z-directions, respectively. If $J_x = J_y = J_z = J$, namely, if the spin exchange is isotropic, the above expression is simplified as

$$\hat{H}_{spin} = J(\hat{S}_{1x}\hat{S}_{2x} + \hat{S}_{1y}\hat{S}_{2y} + \hat{S}_{1z}\hat{S}_{2z}),$$

which represents a Heisenberg spin Hamiltonian. Another extreme case is given by $J_x = J_y = 0$, for which we obtain an Ising spin Hamiltonian:

$$\hat{H}_{spin} = J_z \hat{S}_{1z} \hat{S}_{2z}.$$

This Hamiltonian describes a magnetic system made up of uniaxial magnetic ions (namely, those ions with a nonzero moment only in one direction, see Section 10.6). The deviation of spin exchange from the isotropic character is a consequence of SOC. In this section, we focus on how to determine isotropic spin exchanges that are often referred to as Heisenberg or symmetric spin exchanges. The evaluation of anisotropic spin exchanges will be discussed in Section 10.5.2.

10.3.1
Use of Eigenstates for an Isolated Spin Dimer [1,19]

To gain insight into the meaning of the spin exchange interaction, we consider a spin dimer consisting of two equivalent spin-1/2 sites, 1 and 2, with one electron at each spin site (Figure 10.10). The energy of the spin dimer arising from the spin exchange interaction between the spins $\hat{S}_1$ and $\hat{S}_2$ is given by the spin Hamiltonian:

$$\hat{H}_{spin} = J\hat{S}_1 \cdot \hat{S}_2, \tag{10.4a}$$

where J is the spin exchange parameter. If the spins are regarded as vectors $\vec{S}_1$ and $\vec{S}_2$, then the Hamiltonian is written as

$$\hat{H}_{spin} = J\vec{S}_1 \cdot \vec{S}_2. \tag{10.4b}$$

In the present work, we will use the operator and vector representations of spin interchangeably. Note the absence of the negative sign in this expression. With this definition, the AFM and FM spin exchange interactions are given by $J > 0$ and $J < 0$, respectively. Given the dot product between $\vec{S}_1$ and $\vec{S}_2$, the lowest energy for $J > 0$ occurs when the angle θ between the two spins is 180° (i.e., the spins are AFM), but that for $J < 0$ when $\theta = 0°$ (i.e., the spins are FM). In either case, the spin Hamiltonian leads to a collinear spin arrangement.

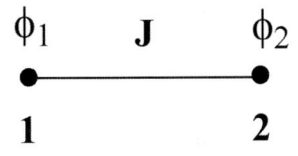

Figure 10.10 A spin dimer made up of two equivalent spin sites with an unpaired electron at each site. The unpaired electrons at the sites 1 and 2 are accommodated in the orbitals ϕ_1 and ϕ_2, respectively, and the spin exchange constant J describes the strength and sign of the interaction between the two unpaired electrons.

In principle, the spin at site i (= 1, 2) of the spin dimer can have either spin-up $|\uparrow\rangle$ or spin-down $|\downarrow\rangle$ state. For a single spin $S = 1/2$ and $S_z = \pm 1/2$, in terms of the $|S, S_z\rangle$ notations, these states are given by

$$|\uparrow\rangle = \left|\frac{1}{2}, +\frac{1}{2}\right\rangle,$$
$$|\downarrow\rangle = \left|\frac{1}{2}, -\frac{1}{2}\right\rangle.$$

These states obey the following general relationships:

$$\hat{S}_z|S, S_z\rangle = S_z|S, S_z\rangle,$$
$$\hat{S}_+|S, S_z\rangle = \sqrt{S(S+1) - S_z(S_z+1)}|S, S_z+1\rangle, \quad (10.5)$$
$$\hat{S}_-|S, S_z\rangle = \sqrt{S(S+1) - S_z(S_z-1)}|S, S_z-1\rangle,$$

where the ladder operators are given by

$$\hat{S}_+ = \hat{S}_x + i\hat{S}_y,$$
$$\hat{S}_- = \hat{S}_x - i\hat{S}_y.$$

Using these ladder operators, Eq. (10.4a) is rewritten as

$$\hat{H}_{spin} = J\hat{S}_{1z}\hat{S}_{2z} - J(\hat{S}_{1+}\hat{S}_{2-} + \hat{S}_{1-}\hat{S}_{2+})/2. \quad (10.4c)$$

The eigenstates of $\hat{H}_{spin}$ allowed for the spin dimer are the singlet state $|S\rangle$ and triplet state $|T\rangle$ that are given by

$$|S\rangle = (|\uparrow\rangle_1|\downarrow\rangle_2 - |\downarrow\rangle_1|\uparrow\rangle_2)/\sqrt{2},$$

$$|T\rangle = \begin{cases} |\uparrow\rangle_1|\uparrow\rangle_2 \\ |\downarrow\rangle_1|\downarrow\rangle_2 \\ (|\uparrow\rangle_1|\downarrow\rangle_2 + |\downarrow\rangle_1|\uparrow\rangle_2)/\sqrt{2}. \end{cases}$$

Note that the broken-symmetry (or Néel) states,

$$|\uparrow\rangle_1|\downarrow\rangle_2 \text{ and } |\downarrow\rangle_1|\uparrow\rangle_2,$$

interact through $\hat{H}_{spin}$ to give the symmetry-adapted states $|S\rangle$ and $|T\rangle$. We evaluate the energies of $|T\rangle$ and $|S\rangle$ by using Eq. (10.5) to find

$$E_{spin}(T) = \langle T|\hat{H}_{spin}|T\rangle = J/4,$$
$$E_{spin}(S) = \langle S|\hat{H}_{spin}|S\rangle = -3J/4.$$

Thus, the energy difference between the two states is given by

$$\Delta E_{spin} = E_{spin}(T) - E_{spin}(S) = J, \quad (10.6)$$

so the spin exchange constant J represents the energy difference between the singlet and triplet spin states of the spin dimer. The singlet state is lower in energy than the triplet state if the spin exchange J is AFM (i.e., $J > 0$), and the opposite is the case if the spin exchange J is FM (i.e., $J < 0$).

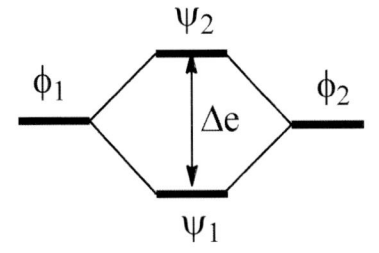

Figure 10.11 The interaction between the magnetic orbitals ϕ_1 and ϕ_2 of a spin dimer leading to the bonding and antibonding molecular orbitals ψ_1 and ψ_2 of the dimer, respectively, which are split by the energy Δe.

We now examine how the triplet and singlet states of the spin dimer are described in terms of electronic structure calculations. The electronic Hamiltonian $\hat{H}_{elec}$ for this two-electron system can be written as

$$\hat{H}_{elec} = \hat{h}(1) + \hat{h}(2) + 1/r_{12}, \tag{10.7}$$

where $\hat{h}(i)$ ($i = 1, 2$) is the one-electron energy (i.e., the kinetic and the electron–nuclear attraction energies) of the electron i ($= 1, 2$), and r_{12} is the distance between electrons 1 and 2. Assume that the unpaired electrons at sites 1 and 2 are accommodated in the orbitals ϕ_1 and ϕ_2, respectively, in the absence of interaction between them. Such singly occupied orbitals are referred to as magnetic orbitals. The weak interaction between ϕ_1 and ϕ_2 leads to the two levels ψ_1 and ψ_2 of the dimer separated by a small energy gap Δe (Figure 10.11), which are approximated by

$$\psi_1 = (\phi_1 + \phi_2)/\sqrt{2},$$
$$\psi_2 = (\phi_1 - \phi_2)/\sqrt{2}.$$

As depicted in Figure 10.12, one of the three triplet state wave functions is represented by the electron configuration Ψ_T. When Δe is very small (compared with that expected for chemical bonding), the singlet state electron

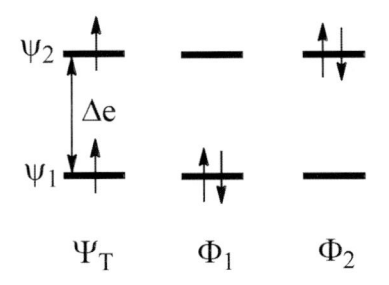

Figure 10.12 The occupation of the molecular orbitals ψ_1 and ψ_2 of the dimer with two electrons leading to the triplet configuration Ψ_T as well as two singlet configurations Φ_1 and Φ_2.

configurations Φ_1 and Φ_2 are very close in energy, and interact strongly under $\hat{H}_{\text{elec}}$ to give

$$\langle \Phi_1 | \hat{H}_{\text{elec}} | \Phi_2 \rangle = K_{12},$$

where K_{12} is the exchange repulsion between ϕ_1 and ϕ_2.

$$K_{12} = \langle \phi_1(1)\phi_2(2) | 1/r_{12} | \phi_2(1)\phi_1(2) \rangle,$$

which is the self-repulsion resulting from the overlap density $\phi_1\phi_2$. Thus, the true singlet state Ψ_S is described by the lower energy state of the configuration–interaction (CI) wave functions Ψ_i ($i = 1, 2$),

$$\Psi_i = C_{1i}\Phi_1 + C_{2i}\Phi_2, \quad (i = 1, 2),$$

namely, $\Psi_S = \Psi_1$. The energies of Ψ_S and Ψ_T, $E_{\text{CI}}(S)$, and $E_{\text{CI}}(T)$, respectively, can be evaluated in terms of $\hat{H}_{\text{elec}}$ by using the dimer orbitals ψ_1 and ψ_2 determined from the calculations for the triplet state Ψ_T. Then, after some manipulations, the electronic energy difference between the singlet and triplet state is written as [1,19]

$$\Delta E_{\text{CI}} = E_{\text{CI}}(S) - E_{\text{CI}}(T) = -2K_{12} + \frac{(\Delta e)^2}{U}. \tag{10.8}$$

The effective on-site repulsion U is given by

$$U = J_{11} - J_{12},$$

where J_{11} and J_{12} are the Coulomb repulsions:

$$J_{11} = \langle \phi_1(1)\phi_1(2) | 1/r_{12} | \phi_1(1)\phi_1(2) \rangle,$$
$$J_{12} = \langle \phi_1(1)\phi_2(2) | 1/r_{12} | \phi_1(1)\phi_2(2) \rangle.$$

Then, by mapping the energy spectrum of $\hat{H}_{\text{spin}}$ onto that of $\hat{H}_{\text{elec}}$, namely,

$$\Delta E_{\text{spin}} = \Delta E_{\text{CI}},$$

we obtain

$$J = \Delta E_{\text{CI}} = -2K_{12} + \frac{(\Delta e)^2}{U}. \tag{10.9}$$

It is important to note the qualitative aspect of the spin exchange J on the basis of the above expression. Since the repulsion terms K_{12} and U are always positive, the spin exchange J is divided into the FM and AFM components J_F (<0) and J_{AF} (>0), respectively. That is,

$$J = J_F + J_{AF},$$

where

$$J_F = -2K_{12}, \tag{10.10a}$$

$$J_{AF} = \frac{(\Delta e)^2}{U}. \tag{10.10b}$$

The FM term J_F becomes stronger with increasing the exchange integral K_{12} that in turn increases with increasing the overlap density, $\phi_1\phi_2$. The AFM term J_{AF} becomes stronger with increasing Δe that in turn becomes larger with increasing the overlap integral, $\langle\phi_1|\phi_2\rangle$. In addition, the J_{AF} term becomes weaker with increasing the on-site repulsion, U.

10.3.2
Use of Broken-Symmetry States for an Isolated Spin Dimer

For a general magnetic system, it is practically impossible to determine the eigenvalue spectrum of either $\hat{H}_{elec}$ or $\hat{H}_{spin}$. However, for broken-symmetry states, which are not eigenstates of $\hat{H}_{elec}$ and $\hat{H}_{spin}$, their relative energies can be readily determined in terms of both $\hat{H}_{elec}$ and $\hat{H}_{spin}$. With DFT calculations, the energy mapping for a spin dimer between the energy spectra of $\hat{H}_{elec}$ and $\hat{H}_{spin}$ is carried out by using high-spin and broken-symmetry states ($|HS\rangle$ and $|BS\rangle$, respectively) [1–5,20,21]. For example, let us reconsider the spin dimer shown in Figure 10.10 for which the pure-spin $|HS\rangle$ and $|BS\rangle$ states are given by

$$|HS\rangle = |\uparrow\rangle_1|\uparrow\rangle_2 \text{ or } |\downarrow\rangle_1|\downarrow\rangle_2,$$
$$|BS\rangle = |\uparrow\rangle_1|\downarrow\rangle_2 \text{ or } |\downarrow\rangle_1|\uparrow\rangle_2.$$

Here the $|HS\rangle$ state is an eigenstate of the spin Hamiltonian $\hat{H}_{spin}$ in Eq. (10.3a), but the $|BS\rangle$ state is not. In terms of this Hamiltonian, the energies of the collinear-spin states $|HS\rangle$ and $|BS\rangle$ are given by

$$E_{spin}(HS) = \langle HS|\hat{H}_{spin}|HS\rangle = J/4,$$
$$E_{spin}(BS) = \langle BS|\hat{H}_{spin}|BS\rangle = -J/4.$$

Thus,

$$\Delta E_{spin} = E_{spin}(HS) - E_{spin}(BS) = J/2.$$

In terms of DFT calculations, the electronic structures of the $|HS\rangle$ and $|BS\rangle$ states are readily evaluated to determine their energies, $E_{DFT}(HS)$ and $E_{DFT}(BS)$, respectively, and hence obtain the energy difference:

$$\Delta E_{DFT} = E_{DFT}(HS) - E_{DFT}(BS).$$

Consequently, by mapping ΔE_{spin} onto ΔE_{DFT}, we obtain

$$J/2 = \Delta E_{DFT}. \tag{10.11}$$

10.3.3
Use of Broken-Symmetry States for General Magnetic Solids

The energy-mapping analysis based on DFT calculations employs the broken-symmetry state that is not an eigenstate of the spin Hamiltonian. For a general spin Hamiltonian defined in terms of several spin exchange parameters (Eq. (10.1)), it is impossible to determine its eigenstates analytically in terms of the spin exchange parameters to be determined and is also difficult to determine them numerically even when their values are known. For any realistic magnetic system requiring a spin Hamiltonian defined in terms of various spin exchange parameters, the energy-mapping analysis based on DFT greatly facilitates the quantitative evaluation of the spin exchange parameters because it does not rely on the eigenstates but on the broken-symmetry states of the spin Hamiltonian. For broken-symmetry states, the energy expressions of the spin Hamiltonian can be readily written down (see below) and the corresponding electronic energies can be readily determined by DFT calculations as well.

In general, the magnetic energy levels of a magnetic system are described by employing a spin Hamiltonian $\hat{H}_{spin}$ defined in terms of several different spin exchange parameters (Eq. (10.1)). This model Hamiltonian generates a set of magnetic energy levels as the sum of pairwise interactions $J_{ij}\hat{S}_i \cdot \hat{S}_j$. It is interesting that the sum of such "two-body interactions" can reasonably well describe the magnetic energy spectrum. This is due to the fact that spin exchange interactions are determined primarily by the tails of magnetic orbitals (see Section 10.4) [1,2]. The spin exchange constants J_{ij} of a given magnetic system can be evaluated by employing the energy-mapping method as described below [2]:

a) Select a set of N spin exchange paths J_{ij} ($= J_1, J_2, \ldots, J_N$) for a given magnetic system on the basis of inspecting the geometrical arrangement of its magnetic ions and also considering the nature of its M—L—M and M—L$\cdots$L—M exchange paths.

b) Construct $N+1$ ordered spin states (i.e., broken-symmetry states) $i = 1, 2, \ldots, N+1$, in which all spins are collinear so that any given pair of spins has either FM or AFM arrangement. For a general spin dimer whose spin sites i and j possess N_i and N_j unpaired spins (hence, spins $S_i = N_i/2$ and $S_j = N_j/2$), respectively, the spin exchange energies of the FM and AFM arrangements (E_{FM} and E_{AFM}, respectively) are given by [3]

$$E_{FM} = +N_i N_j J_{ij}/4 = +S_i S_j J_{ij},$$

$$E_{AFM} = -N_i N_j J_{ij}/4 = -S_i S_j J_{ij}, \quad (10.12)$$

where J_{ij} ($= J_1, J_2, \ldots, J_N$) is the spin exchange parameter for the spin exchange path $ij = 1, 2, \ldots, N$. Thus, the total spin exchange energy of an ordered spin arrangement is readily obtained by summing up all pairwise interactions to find the energy expression $E_{spin}(i)$ ($i = 1, 2, \ldots, N+1$) in terms of the parameters to be determined and hence, the N relative energies:

$$\Delta E_{spin}(i-1) = E_{spin}(i) - E_{spin}(1), \quad i = 2, 3, \ldots, N+1.$$

c) Determine the electronic energies $E_{DFT}(i)$ of $N+1$ ordered spin states $i = 1, 2, \ldots, N+1$ by DFT calculations to obtain the N relative energies:

$$\Delta E_{DFT}(i-1) = E_{DFT}(i) - E_{DFT}(1), \quad j = 2, 3, \ldots, N+1.$$

As already mentioned, DFT calculations for a magnetic insulator tend to give a metallic electronic structure because the electron correlation of a magnetic ion leading to spin polarization is not well described. Since we deal with the energy spectrum of a magnetic insulator, it is necessary that the electronic structure of each ordered spin state obtained from DFT calculations be magnetic insulating. To ensure this aspect, it is necessary to perform DFT+U calculations [13] by adding on-site repulsion $U_{\text{eff}} = U - J$ with on-site repulsion U and on-site exchange interaction J on magnetic ions. Furthermore, as can be seen from Eq. (10.10b), the AFM component of a spin exchange decreases with increasing U_{eff} so that the magnitude and sign of a spin exchange constant may be affected by U_{eff}. It is therefore necessary to carry out DFT+U calculations with several different U_{eff} values.

d) Finally, determine the values of $J_1, J_2, \ldots, J_N$ by mapping the N relative energies ΔE_{DFT} onto the N relative energies ΔE_{spin},

$$\Delta E_{\text{DFT}}(i-1) = \Delta E_{\text{spin}}(i-1), \quad (i = 2 - N + 1). \tag{10.13}$$

In determining N spin exchanges $J_1, J_2, \ldots, J_N$, one may employ more than $N+1$ ordered spin states, hence obtaining more than N relative energies ΔE_{DFT} and ΔE_{spin} for the mapping. In this case, the N parameters $J_1, J_2, \ldots, J_N$ can be determined by performing least squares fitting analysis.

10.3.4
Energy-Mapping Based on Four Ordered Spin States [4]

For our calculations, we regard the spin operators $\hat{S}_i$ and $\hat{S}_j$ as the classical vectors of $\vec{S}_i$ and $\vec{S}_j$, respectively. Then, the spin Hamiltonian can be written as

$$\hat{H}_{\text{spin}} = \sum_{i<j} J_{ij} \hat{S}_i \cdot \hat{S}_j \rightarrow \sum_{i<j} J_{ij} \vec{S}_i \cdot \vec{S}_j. \tag{10.14}$$

Without loss of generality, the spin pair i–j will be regarded as 1–2. For simplicity, all spin sites are assumed to have an identical spin S. We carry out DFT+U calculations for the following four ordered spin states:

State	Spin 1	Spin 2	Other spin sites
1	(0, 0, S)	(0, 0, S)	Either (0, 0, S) or (0, 0, -S) according to the experimental (or a low-energy) spin state. Keep the same for the four spin states
2	(0, 0, S)	(0, 0, -S)	
3	(0, 0, -S)	(0, 0, S)	
4	(0, 0, -S)	(0, 0, -S)	

where the notations (0, 0, S) and (0, 0, -S), for example, mean that the spin vectors are pointed along the positive and negative z-directions, respectively. We represent the energies of the spin states 1–4 as E_1–E_4, respectively. Then, according to Eq. (10.14), the energy difference, $E_1 + E_4 - E_2 - E_3$, is related to the spin exchange J as

$$J_{12} = \frac{E_1 + E_4 - E_2 - E_3}{4S^2}. \tag{10.15}$$

Once the energies E_1–E_4 are obtained from DFT + U + SOC calculations, we can readily determine J_{12}.

10.3.5
General Features of Spin Exchanges Numerically Extracted

Common DFT functionals suffer from the self-interaction error, that is, a single electron interacts with itself, which is unphysical. This error results in a spurious delocalization of orbitals, including magnetic orbitals. Consequently, spin exchange interactions are overestimated by the usual DFT methods. This self-interaction error can be reduced by using the DFT + U method, in which the on-site Coulomb interaction is taken into consideration. This on-site interaction is parameterized by the effective on-site Coulomb interaction $U_{\text{eff}} = U - J$. By adding such Hartree–Fock-like terms, the DFT + U method makes the magnetic orbitals more localized and decreases the overlap between magnetic orbitals hence, reducing the magnitudes of spin exchange interactions. Currently, there is no reliable way of determining the U and J parameters needed for DFT + U calculations. A practical way of probing the magnetic properties of a given system is to carry out DFT + U calculations for several different U_{eff} values that provide several sets of the J_1, J_2, ..., J_N values. It is important to find trends common to these sets. What matters in finding a spin lattice are the relative magnitudes of the spin exchanges. As already pointed out, the purpose of using a spin Hamiltonian is to quantitatively describe the observed experimental data with a minimal set of J_{ij} values hence, capturing the essence of the chemistry and physics involved. Experimentally, such a set of J_{ij} values for a given magnetic system is deduced first by choosing a few spin exchange paths J_{ij} that one considers as important for the system and then by evaluating their signs and magnitudes such that the energy spectrum of the resulting spin Hamiltonian best simulates the observed experimental data. The numerical values of J_{ij} deduced from this fitting analysis depends on what spin lattice model one employs for the fitting, and hence, more than one spin lattice may fit the experimental data equally well. This nonuniqueness of the fitting analysis has been the source of controversies in the literature over the years. Ultimately, the spin lattice of a magnetic system deduced from experimental fitting analysis should be consistent with the one determined from the energy-mapping analysis based on DFT calculations, because the observed magnetic properties are a consequence of the electronic structure of the magnetic system.

10.4
Orbital Interactions Controlling Spin Exchanges

For a given magnetic system, one can determine the values of its various spin exchanges using the energy mapping analysis based on DFT + U calculations and hence, ultimately find the spin lattice appropriate for it. What the energy-mapping analysis cannot tell us is why a certain spin exchange is strong or weak. To answer this question, it is necessary to understand how the strength of a given spin exchange interaction between two magnetic ions is related to the orbital interaction between the magnetic orbitals representing the magnetic ions. In this section, we consider the qualitative aspects of the orbital interactions controlling spin exchange interactions.

Given a magnetic solid made up of ML_n polyhedra containing a magnetic transition cation M^{x+} (x = oxidation state), there may occur two types of spin exchange paths, namely, M—L—M exchange and/or M—L···L—M exchange paths. The qualitative factors governing the signs and magnitudes of M—L—M exchanges were well established many decades ago [22,23]. However, the importance of M—L···L—M exchange paths has been realized much later [1,2]. In leading to AFM interactions, M—L···L—M exchanges can be much stronger than M—L—M exchanges. What was not realized in the early studies of M—L—M exchanges is the importance of the magnetic orbitals of ML_n polyhedra, in which the M d-orbitals are combined out of phase with the L p-orbitals. In M—L···L—M spin exchanges, the magnetic orbitals of the two metal sites can interact strongly as long as their L p-orbital tails can interact through the L···L contact [1]. In what follows, we examine qualitatively the through-space and through-bond orbital interactions [2] that govern M—L···L—M spin exchanges.

As a representative example capturing the essence of spin exchange interactions, let us examine those of $LiCuVO_4$ [24–26] in which the CuO_2 ribbon chains, made up of edge-sharing CuO_4 square planes running along the b-direction are interconnected along the a-direction by sharing corners with VO_4 tetrahedra. This is shown in Figure 10.13. In $LiCuVO_4$ the Cu^{2+} ($S=1/2$, d^9) ions are magnetic, but the V^{5+} (d^0) ions are nonmagnetic. As for the spin exchange

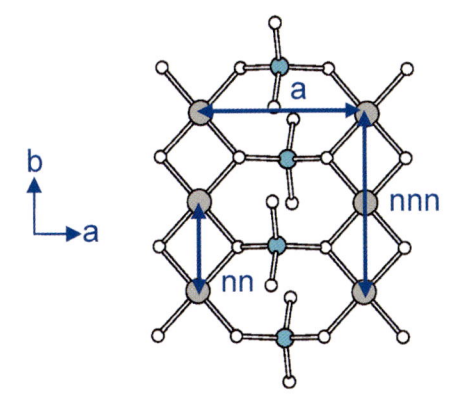

Figure 10.13 Two CuO_2 ribbon chains of $LiCuVO_4$ interconnected by VO_4 tetrahedra, where gray circle = Cu, cyan circle = V, and white circle = O. The intrachain spin exchange paths J_{nn} and J_{nnn} as well as the interchain spin exchange path J_a are indicated by the legends "nn," "nnn," and "a," respectively.

paths of LiCuVO$_4$, we consider the nearest-neighbor (nn) and next-nearest-neighbor (nnn) intrachain spin exchanges, J_{nn} and J_{nnn}, respectively, in each CuO$_2$ ribbon chain as well as the interchain spin exchange J_a along the a-direction (Figure 10.13).

The magnetic orbital of the Cu^{2+} ($S = 1/2$, d^9) ion is given by the x^2-y^2 σ-antibonding orbital contained in the CuO$_4$ square plane (Figure 10.14a), in which the Cu 3d x^2-y^2 orbital is combined out of phase with the 2p orbitals of the four surrounding O ligands. As already emphasized [1,2], it is not the "head" part (the Cu 3d x^2-y^2 orbital) but the "tail" part (the O 2p-orbitals) of the magnetic orbital that controls the magnitudes and signs of these spin exchange interactions. Let us first consider the Cu—O—Cu exchange J_{nn}. When the x^2-y^2 magnetic orbitals ϕ_1 and ϕ_2 of the two spin sites are brought together to form the Cu—O—Cu bridges, the O 2p-orbital tails at the bridging O atoms make a nearly orthogonal arrangement (Figure 10.14b). Thus, the overlap integral $\langle \phi_1 | \phi_2 \rangle$ between the two magnetic orbitals is almost zero that leads to $J_{AF} \approx 0$. In contrast, the overlap density $\phi_1 \phi_2$ of the magnetic orbitals is substantial that leads to nonzero J_F. As a consequence, the J_{nn} exchange becomes FM [25,26].

For the intrachain Cu—O···O—Cu exchange J_{nnn} (Figure 10.14c), the O 2p-orbital tails of the magnetic orbitals ϕ_1 and ϕ_2 at the terminal O atoms are well separated by the O···O contacts. Thus, the overlap density $\phi_1 \phi_2$ of the magnetic orbitals is negligible leading to $J_F \approx 0$. However, the overlap integral $\langle \phi_1 | \phi_2 \rangle$ is nonzero because the O 2p-tails of ϕ_1 and ϕ_2 overlap through the O···O contacts. This through-space interaction between ϕ_1 and ϕ_2 produces a large energy split Δe between ψ_+ and ψ_- that are in-phase and out-of-phase combinations of ϕ_1 and ϕ_2, respectively (Figure 10.15a), thereby leading to nonzero J_{AF}. Consequently, the J_{nnn} exchange becomes AFM [25,26].

In the interchain spin exchange path J_a, the two CuO$_4$ square planes are corner-shared with VO$_4$ tetrahedra. In the Cu—O···V^{5+}···O—Cu exchange paths,

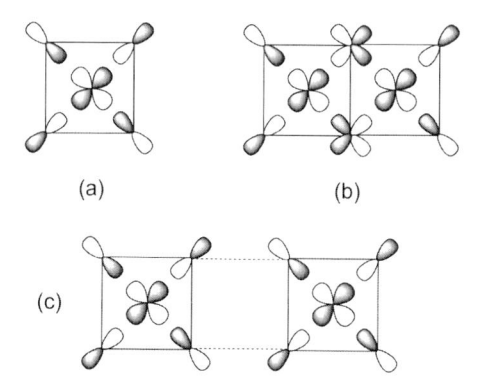

Figure 10.14 (a) The x^2-y^2 magnetic orbital of a CuO$_4$ square plane. (b) The Cu—O—Cu spin exchange interaction between nearest-neighbor CuO$_4$ square planes in a CuO$_2$ ribbon chain. (c) The Cu—O···O—Cu spin exchange interaction between next-nearest-neighbor CuO$_4$ square planes in a CuO$_2$ ribbon chain.

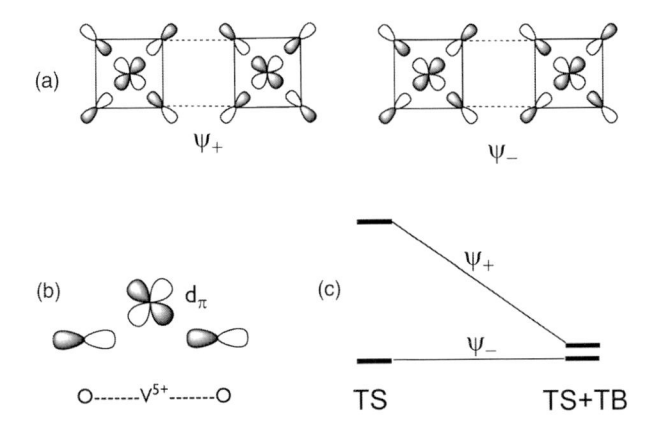

Figure 10.15 The through-space (TS) and the through-bond (TB) interactions between the two x^2-y^2 magnetic orbitals in the Cu−O···V^{5+}···O−Cu interchain spin exchange J_a in LiCuVO$_4$. (a) The energy split between ψ_+ and ψ_- due to the TS interaction. (b) The bonding interaction of the V d_π orbital with the O 2p tails of ψ_- in the O···V^{5+}···O bridge. (c) The energy split between ψ_+ and ψ_- due to the TS and TB interactions.

the empty V 3d-orbitals should interact in a bonding manner with the Cu x^2-y^2 orbitals. In the absence of the V 3d-orbitals, the energy split Δe between ψ_+ and ψ_- arising from the through-space interaction between ϕ_1 and ϕ_2 would be substantial, as expected from the intrachain exchange J_{nnn}, so that one might expect a strong AFM exchange for the interchain exchange J_a. However, in the Cu−O···V^{5+}···O−Cu exchange paths, the bridging VO$_4$ units provide a through-bond interaction between the empty V 3d$_\pi$-orbitals and the O 2p-tails of the magnetic orbitals on the O···O contacts. By symmetry, this through-bond interaction is possible only with ψ_- (Figure 10.15b and c). The V 3d$_\pi$-orbital being empty, the O 2p-tails of ψ_- on the O···O contacts interact in-phase with the empty V 3d$_\pi$-orbital hence, lowering the ψ_- level, whereas ψ_+ is unaffected by the V 3d$_\pi$-orbital, thereby reducing the energy split Δe between ψ_+ and ψ_- of the Cu−O···V^{5+}···O−Cu exchange paths and consequently weakening the interchain spin exchange J_a [25,26]. As a consequence, the magnetic properties are dominated by the one-dimensional character of the CuO$_2$ ribbon chain.

It is important to observe the corollary of the above observation for general M−L···A^{y+}···L−M spin exchange, where the cation A^{y+} (y = oxidation state) provides through-bond interactions. If the Δe between ψ_+ and ψ_- is negligible in terms of the through-space interaction, then the effect of the through-bond interaction would make Δe large leading to a strong AFM interaction [2].

When the ML$_n$ polyhedra containing M cations are condensed together by sharing a corner, an edge or a face, they give rise to M−L−M exchanges that are the subject of the Goodenough rules [22]. When these polyhedra are not condensed, they give rise to M−L···L−M and M−L···A^{y+}···L−M exchanges [1,2], where A^{y+} refers to the intervening d^0 metal cation. The importance of the latter spin exchanges, not covered by the Goodenough rules, was recognized [1,2] only after realizing that the magnetic orbitals of an M ion include both the M d-orbitals and the L p-orbitals of the ML_n polyhedron, and that the L p-orbital tails of the magnetic orbitals control the magnitudes and signs of such spin exchange interactions [1,2]. Concerning the M—L···L—M exchanges, there are several important consequences of this observation [1,2]:

a) The strength of a given M—L···L—M spin exchange is not determined by the shortness of the M···M distance, but rather by that of the L···L distance; it is strong when the L···L distance is in the vicinity of the van der Waals radii sum or shorter [1].

b) In a given magnetic system consisting of both M—L—M and M—L···L—M spin exchanges, the M—L···L—M spin exchanges are very often stronger than the M—L—M spin exchanges.

c) The strength of an M—L···L—M spin exchange determined by through-space interaction between the L np-tails on the L···L contact can be significantly modified when the L···L contact has a through-bond interaction with the intervening d^0 metal cation A^{y+} [25,27] or even the p^0 metal cation (e.g., Cs^+ as found for Cs_2CuCl_4 [28]). Such an M—L···A^{y+}···L—M spin exchange becomes strong if the corresponding M—L···—L—M through-space exchange is weak, but becomes weaker if the corresponding M—L···L—M through-space exchange is strong. This is so because the empty d_π-orbital of A^{y+} interacts only with the ψ_- orbital of the M—L···L—M exchange. In general, the empty d_π-orbital has a much stronger through-bond effect than does the empty p_π-orbital.

10.5
Incorporating the Effect of SOC Indirectly into Spin Hamiltonian

When a magnetic ion is present in molecules and solids to form a ML_n polyhedron with surrounding ligands L, its orbital momentum $\vec{L}$ is mostly quenched with a small momentum $\delta\vec{L}$ remaining unquenched [10]. Exceptional cases occur when the ML_n polyhedron has n-fold ($n \geq 3$) rotational symmetry so that it has doubly degenerate d-states and when the d-electron count of ML_n is such that a degenerate d-state is unevenly occupied. In this case, the orbital momentum $\vec{L}$ is not quenched so that the effect of the SOC, $\lambda \hat{S} \cdot \hat{L}$, becomes strong often leading to uniaxial magnetism (see Section 10.6). In this section, we consider the cases when the orbital quenching is not complete so a small orbital momentum $\delta\vec{L}$ remains at each magnetic ion. In the past, this situation has been discussed on the basis of the effective spin approximation [10,29], in which the need to explicitly describe the unquenched orbital momentum is circumvented by treating the system as a spin-only system. In this approximation, the effect of SOC arising from $\delta\vec{L}$ is absorbed into the coefficient for certain terms made up of only spin operators. This approximation deals with both single-spin site and two-spin site problems. The former includes the single-ion anisotropy, while the latter include the asymmetric spin exchange and the Dzyaloshinskii–Moriya (DM) exchange [30,31]. The DM exchange is often referred to as anti-symmetric exchange.

10.5.1
SOC Effect on a Single-Spin Site and Spin-Half Misconception

For a magnetic ion with nondegenerate magnetic orbital (e.g., Cu^{2+}), the SOC Hamiltonian $\hat{H}_{SO} = \lambda \hat{S} \cdot \hat{L}$ is transformed into the zero-field spin Hamiltonian $\hat{H}_{zf}$ [10]:

$$\hat{H}_{zf} = D(\hat{S}_z^2 - \frac{1}{3}\hat{S}^2) + E(\hat{S}_x^2 - \hat{S}_y^2)$$
$$= D(\hat{S}_z^2 - \frac{1}{3}\hat{S}^2) + \frac{1}{2}E(\hat{S}_+\hat{S}_+ + \hat{S}_-\hat{S}_-),$$
(10.16)

where the constants D and E originate from the SOC associated with the remnant orbital momentum $\delta\vec{L}$, that is,

$$D \propto \lambda^2(\delta L_\| - \delta L_\perp),$$
$$E \propto \lambda^2(\delta L_x - \delta L_y),$$

where $\delta L_\|$ and $\delta L_\perp$ are the $\|z$- and $\perp z$-components of $\delta\vec{L}$, respectively, while δL_x and δL_y are the x- and y-components of $\delta L_\perp$, respectively.

For $S > 1/2$ ions, Eq. (10.16) predicts magnetic anisotropy. For instance, a $S = 1$ ion is described by three spin states, $|1, +1\rangle$, $|1, 0\rangle$, and $|1, -1\rangle$. Thus,

$$D(\hat{S}_z^2 - \frac{1}{3}\hat{S}^2)|1, +1\rangle = D\left[\hat{S}_z^2 - \frac{1}{3}S(S+1)\right]|1, +1\rangle = +\frac{1}{3}D|1, +1\rangle,$$

$$D(\hat{S}_z^2 - \frac{1}{3}\hat{S}^2)|1, 0\rangle = D\left[\hat{S}_z^2 - \frac{1}{3}S(S+1)\right]|1, 0\rangle = -\frac{2}{3}D|1, 0\rangle,$$

$$D(\hat{S}_z^2 - \frac{1}{3}\hat{S}^2)|1, -1\rangle = D\left[\hat{S}_z^2 - \frac{1}{3}S(S+1)\right]|1, -1\rangle = +\frac{1}{3}D|1, -1\rangle,$$

and

$$E(\hat{S}_+\hat{S}_+ + \hat{S}_-\hat{S}_-)|1, +1\rangle = E|1, -1\rangle,$$
$$E(\hat{S}_+\hat{S}_+ + \hat{S}_-\hat{S}_-)|1, 0\rangle = 0,$$
$$E(\hat{S}_+\hat{S}_+ + \hat{S}_-\hat{S}_-)|1, -1\rangle = E|1, +1\rangle.$$

This shows that the $|1, \pm1\rangle$ states are separated in energy from the $|1, 0\rangle$ state by $|D|$. In addition, the $|1, +1\rangle$ and $|1, -1\rangle$ states interact and become split in energy by $|E|$. Due to this energy split, the thermal populations of the three states differ, hence leading to magnetic anisotropy. A similar conclusion is reached for $S > 1$ ions. For example, a $S = 3/2$ ion is described by the four states, $|\frac{3}{2}, +\frac{3}{2}\rangle$, $|\frac{3}{2}, +\frac{1}{2}\rangle$,

$\left|\frac{3}{2},-\frac{1}{2}\right\rangle$, and $\left|\frac{3}{2},-\frac{3}{2}\right\rangle$. Therefore,

$$D(\hat{S}_z^2 - \frac{1}{3}\hat{S}^2)\left|\frac{3}{2},+\frac{3}{2}\right\rangle = D\left[\hat{S}_z^2 - \frac{1}{3}S(S+1)\right]\left|\frac{3}{2},+\frac{3}{2}\right\rangle = D,$$

$$D(\hat{S}_z^2 - \frac{1}{3}\hat{S}^2)\left|\frac{3}{2},+\frac{1}{2}\right\rangle = D\left[\hat{S}_z^2 - \frac{1}{3}S(S+1)\right]\left|\frac{3}{2},+\frac{1}{2}\right\rangle = 0,$$

$$D(\hat{S}_z^2 - \frac{1}{3}\hat{S}^2)\left|\frac{3}{2},-\frac{1}{2}\right\rangle = D\left[\hat{S}_z^2 - \frac{1}{3}S(S+1)\right]\left|\frac{3}{2},+\frac{1}{2}\right\rangle = 0,$$

$$D(\hat{S}_z^2 - \frac{1}{3}\hat{S}^2)\left|\frac{3}{2},-\frac{3}{2}\right\rangle = D\left[\hat{S}_z^2 - \frac{1}{3}S(S+1)\right]\left|\frac{3}{2},+\frac{3}{2}\right\rangle = D,$$

and

$$E(\hat{S}_+\hat{S}_+ + \hat{S}_-\hat{S}_-)\left|\frac{3}{2},+\frac{3}{2}\right\rangle = E\left|\frac{3}{2},-\frac{1}{2}\right\rangle,$$

$$E(\hat{S}_+\hat{S}_+ + \hat{S}_-\hat{S}_-)\left|\frac{3}{2},+\frac{1}{2}\right\rangle = E\left|\frac{3}{2},-\frac{3}{2}\right\rangle,$$

$$E(\hat{S}_+\hat{S}_+ + \hat{S}_-\hat{S}_-)\left|\frac{3}{2},-\frac{1}{2}\right\rangle = E\left|\frac{3}{2},+\frac{3}{2}\right\rangle,$$

$$E(\hat{S}_+\hat{S}_+ + \hat{S}_-\hat{S}_-)\left|\frac{3}{2},-\frac{3}{2}\right\rangle = E\left|\frac{3}{2},+\frac{1}{2}\right\rangle.$$

Thus, the $\left|\frac{3}{2},\pm\frac{3}{2}\right\rangle$ states are separated in energy from the $\left|\frac{3}{2},\pm\frac{1}{2}\right\rangle$ states by $|D|$. Without loss of generality, it can be assumed that the $\left|\frac{3}{2},\pm\frac{3}{2}\right\rangle$ states lie higher than the $\left|\frac{3}{2},\pm\frac{1}{2}\right\rangle$ states. The $\left|\frac{3}{2},+\frac{3}{2}\right\rangle$ and $\left|\frac{3}{2},-\frac{1}{2}\right\rangle$ states interact with interaction energy E, and so the states $\left|\frac{3}{2},-\frac{3}{2}\right\rangle$ and $\left|\frac{3}{2},+\frac{1}{2}\right\rangle$. Then, according to perturbation theory, the $\left|\frac{3}{2},\pm\frac{3}{2}\right\rangle$ states are raised in energy by $E^2/|D|$, and the $\left|\frac{3}{2},\pm\frac{1}{2}\right\rangle$ states are lowered in energy by $E^2/|D|$. Consequently, the $\left|\frac{3}{2},\pm\frac{3}{2}\right\rangle$ states become separated in energy from the $\left|\frac{3}{2},\pm\frac{1}{2}\right\rangle$ states by $|D| + 2E^2/|D|$.

The aforementioned energy split for $S > 1/2$ ions, and the associated magnetic anisotropy is a consequence of SOC, albeit indirectly through the constants D and E. Since the information about the orbital $|L, L_z\rangle$ of the magnetic ion is completely hidden in these constants, it is not possible to predict the preferred spin orientation of a $S > 1/2$ ion on the basis of Eq. (10.16), although one can infer that such an ion has magnetic anisotropy as already described.

A rather different situation occurs for a $S = 1/2$ ion, which is described by two spin states, $|\uparrow\rangle = \left|\frac{1}{2},+\frac{1}{2}\right\rangle$ and $|\downarrow\rangle = \left|\frac{1}{2},-\frac{1}{2}\right\rangle$. We note that

$$D(\hat{S}_z^2 - \frac{1}{3}\hat{S}^2)\left|\frac{1}{2},+\frac{1}{2}\right\rangle = D\left[\hat{S}_z^2 - \frac{1}{3}S(S+1)\right]\left|\frac{1}{2},+\frac{1}{2}\right\rangle = 0,$$

$$D(\hat{S}_z^2 - \frac{1}{3}\hat{S}^2)\left|\frac{1}{2},-\frac{1}{2}\right\rangle = D\left[\hat{S}_z^2 - \frac{1}{3}S(S+1)\right]\left|\frac{1}{2},-\frac{1}{2}\right\rangle = 0,$$

and

$$E(\hat{S}_+\hat{S}_+ + \hat{S}_-\hat{S}_-)\left|\frac{1}{2},+\frac{1}{2}\right\rangle = 0,$$

$$E(\hat{S}_+\hat{S}_+ + \hat{S}_-\hat{S}_-)\left|\frac{1}{2},-\frac{1}{2}\right\rangle = 0.$$

Consequently, the spin-up and spin-down states do not interact under $\hat{H}_{zf}$, so their degeneracy is not split. (This result obeys the Kramers degeneracy theorem [32] which states that the degeneracy of an odd-spin system should not be split in the absence of an external magnetic field.) This is so even though the constants D and E are nonzero, that is, even though SOC effects are taken into consideration, though indirectly. Thus, the thermal populations of the two states $|\uparrow\rangle$ and $|\downarrow\rangle$ are identical, hence leading to the conclusion that an $S=1/2$ ion has no magnetic anisotropy that arise from SOC. This is the origin of the spin-half misconception.

Note that $\hat{H}_{SO} = \lambda \hat{S} \cdot \hat{L}$ and $\hat{H}_{zf}$ are local (i.e., single-spin site) operators, and do not describe interactions between different spin sites. The SOC-induced magnetic anisotropy for $S>1/2$ ions is commonly referred to as the single-ion anisotropy, to which practitioners of spin Hamiltonian analysis have no objection. However, most of them deny strenuously that $S=1/2$ ions have single-ion anisotropy and suggest the use of the term "magnetocrystalline anisotropy" to describe the experimentally observed magnetic anisotropy of $S=1/2$ ions. In the vernacular, this term is a red herring, because it means that the observed anisotropy is not caused by the single-spin site effect (i.e., SOC) but rather by nonlocal effects (i.e., anything other than SOC, for example, asymmetric spin exchange and magnetic dipole–dipole interactions), just as Moriya and Yoshida argued for the $S=1/2$ system $CuCl_2 \cdot 2H_2O$ more than six decades ago [7]. However, as recently shown [5,6,9] for various magnetic solids of $S=1/2$ ions (see Section 10.7), the spin-half misconception is erroneous. Unfortunately, this misconception remains unabated because it is perpetuated in monographs and textbooks on magnetism [8]. In defense of the spin-half misconception, one might argue that the true magnetic energy states are not those generated by an electronic Hamiltonian, but those generated by a spin Hamiltonian. However, this argument is even more fallacious than the spin-half misconception, because it amounts to arguing that there exists no orbital momentum. The magnetic properties of a magnetic ion are ultimately related to its moment $\vec{\mu}$ that is the derivative of its total electronic energy with respect to an applied magnetic field (see Section 10.6) [10]. The moment $\vec{\mu}$ consists of both orbital and spin components, that is, $\vec{\mu} = \vec{\mu}_L + \vec{\mu}_S$, and these components are related to the orbital and spin momenta as $\vec{\mu}_L = -\mu_B \vec{L}$ and $\vec{\mu}_S = -2\mu_B \vec{S}$, where μ_B is the Bohr magneton. Consequently, the magnetic energy states become identical to those generated by a spin Hamiltonian, only if $\vec{L} = 0$, that is, only if the quenching of orbital momentum is complete. The latter condition is hardly met for all magnetic ions in molecules and solids. It is satisfied for all magnetic ions in a spin Hamiltonian analysis by definition. In short, $S=1/2$ ions do possess single-ion anisotropy, but a spin Hamiltonian analysis predicts erroneously that they do not.

10.5.2
SOC Effect on Spin Exchange: Mapping Analysis for Anisotropic Spin Exchange [33]

In some cases, the spin exchange between two spin sites may not be isotropic. This is an indirect consequence of SOC because a spin at a given site has a preferred orientation due to SOC and because this orientation preference can influence the strength of the spin exchange. Given two spin sites, say, 1 and 2, one may take the z-axis along the exchange paths between 1 and 2. As already mentioned in Section 10.3, the anisotropic spin exchange interaction between two sites 1 and 2 is written as

$$\hat{H}_{spin} = J_x \hat{S}_{1x}\hat{S}_{2x} + J_y \hat{S}_{1y}\hat{S}_{2y} + J_z \hat{S}_{1z}\hat{S}_{2z}. \tag{10.17a}$$

To evaluate J_x, J_y, and J_z, we perform energy-mapping analysis by determining the energies of appropriate broken-symmetry spin states on the basis of DFT + U + SOC calculations. To determine the J_x component, we consider the following four ordered spin states:

State	Spin 1	Spin 2	Other spin site
1	(S, 0, 0)	(S, 0, 0)	Either (0, 0, S) or (0, 0, -S) according to the experimental (or a low-energy) spin state. Keep the same for the four spin states
2	(S, 0, 0)	(-S, 0, 0)	
3	(-S, 0, 0)	(S, 0, 0)	
4	(-S, 0, 0)	(-S, 0, 0)	

Then, the energy difference, $E_1 + E_4 - E_2 - E_3$, of the four states is related to the spin exchange J_x as,

$$J_x = \frac{E_1 + E_4 - E_2 - E_3}{4S^2}.$$

On the basis of DFT + U + SOC calculations for the four spin states, the value of J_x is readily determined. The values of J_y and J_z are obtained in a similar manner. To obtain J_y, we do DFT + U + SOC calculations for the following states:

State	Spin 1	Spin 2	Other spin site
1	(0, S, 0)	(0, S, 0)	Either (0, 0, S) or (0, 0, -S) according to the experimental (or a low-energy) spin state. Keep the same for the four spin states
2	(0, S, 0)	(0, -S, 0)	
3	(0, -S, 0)	(0, S, 0)	
4	(0, -S, 0)	(0, -S, 0)	

Then, we find

$$J_y = \frac{E_1 + E_4 - E_2 - E_3}{4S^2}.$$

To determine J_z, we perform DFT+U+SOC calculations for the following states:

State	Spin 1	Spin 2	Other spin sites
1	(0, 0, S)	(0, 0, S)	Either (S, 0, 0) or (-S, 0, 0) according to the experimental (or a low-energy) spin state. Keep the same for the four spin states
2	(0, 0, S)	(0, 0, -S)	
3	(0, 0, -S)	(0, 0, S)	
4	(0, 0, -S)	(0, 0, -S)	

Then, we find

$$J_z = \frac{E_1 + E_4 - E_2 - E_3}{4S^2}.$$

10.5.3
SOC Effect on Two Adjacent Spin Sites

Another important consequence of SOC is the DM interaction between two adjacent spin sites. Consider the SOC in a spin dimer made up of two spin sites 1 and 2, for which the SOC Hamiltonian is given by [10]

$$\hat{H}_{SO} = \lambda \hat{L} \cdot \hat{S} = \lambda(\hat{L}_1 + \hat{L}_2) \cdot (\hat{S}_1 + \hat{S}_2) \approx \lambda(\hat{L}_1 \cdot \hat{S}_1 + \hat{L}_2 \cdot \hat{S}_2), \quad (10.18)$$

where the last equality follows from the fact that the SOC is a local interaction. It is important to note that, although SOC describes a single-spin site interaction, the two spin sites can interact indirectly hence, influencing their relative spin orientations [2,31]. As illustrated in Figure 10.16, we suppose that an occupied orbital ϕ_i interacts with an unoccupied orbital ϕ_j at spin site 1 via SOC, and that the ϕ_i and ϕ_j of site 1 interact with an occupied orbital ϕ_k of site 2 via orbital interaction. The orbital mixing between ϕ_i and ϕ_k introduces the spin character of site 2 into ϕ_i of site 1, while that between ϕ_j and ϕ_k introduces the spin character of site 2 into ϕ_j of site 1. Namely,

$$\phi_i \rightarrow \phi'_i \approx (1 - \gamma^2)\phi_i + \gamma\phi_k,$$
$$\phi_j \rightarrow \phi'_j \approx (1 - \gamma^2)\phi_j + \gamma\phi_k,$$

where γ refers to a small mixing coefficient. Then, the SOC between such modified ϕ'_i and ϕ'_j at site 1 indirectly introduces the SOC-induced interaction between the spins at sites 1 and 2. For a spin dimer, there can be a number of interactions, like the one depicted in Figure 10.16 at both spin sites, so summing

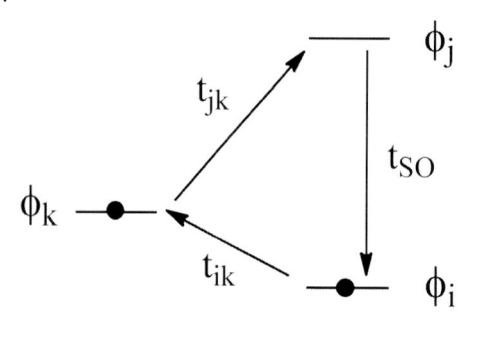

Figure 10.16 Three interactions controlling the strength of a DM interaction.

up all such contributions gives rise to the DM interaction energy E_{DM} between spin sites 1 and 2.

Suppose that $\delta\vec{L}_1$ and $\delta\vec{L}_2$ are the remnant orbital angular momenta at sites 1 and 2, respectively. Then, use of the $\hat{H}_{SO}$ (Eq. (10.18)) as perturbation leads to the DM interaction energy E_{DM} [10,31]:

$$E_{DM} = [\lambda J_{12}(\delta\vec{L}_1 - \delta\vec{L}_2)] \cdot (\vec{S}_1 \times \vec{S}_2) \equiv \vec{D}_{12} \cdot (\vec{S}_1 \times \vec{S}_2).$$

In this expression, the DM vector $\vec{D}_{12}$ is related to the difference in the unquenched orbital angular momenta on the two magnetic sites 1 and 2, namely,

$$\vec{D}_{12} \equiv \lambda J_{12}(\delta\vec{L}_1 - \delta\vec{L}_2).$$

For a spin dimer with spin exchange J_{12}, the strength of its DM exchange $\vec{D}_{12}$ is discussed by considering the ratio $|D_{12}/J_{12}|$, which is often approximated by

$$|D_{12}/J_{12}| \approx \Delta g/g,$$

where Δg is the contribution of the orbital moment to the g-factor g in the effective spin approximation. In general, the $\Delta g/g$ value is at most 0.1, so that the $|D_{12}/J_{12}|$ ratio is often expected to be 0.1 at most. However, it is important to recognize an implicit assumption behind this reasoning, namely, that the spin sites 1 and 2 have an identical chemical environment. When the two spin sites have different chemical environments, the $|D_{12}/J_{12}|$ ratio can be very large as found for a particular $Mn(2)^{3+}$—O—$Mn(3)^{4+}$ spin exchange path of $CaMn_7O_{12}$ (i.e., $|D_{12}/J_{12}| = 0.54$) [34]. As depicted in Figure 10.16, the magnitude of a DM vector D_{12} is determined by the three matrix elements:

$$t_{SO} = \langle \phi_i | \hat{H}_{SO} | \phi_j \rangle, \quad t_{ik} = \langle \phi_i | \hat{H}^{eff} | \phi_k \rangle, \text{ and } t_{jk} = \langle \phi_j | \hat{H}^{eff} | \phi_k \rangle.$$

When t_{SO}, t_{ik}, and t_{jk} are all strong, the magnitude of the DM vector D_{12} can be unusually large [35].

10.5.4
Mapping Analysis for the DM Vector of an Isolated Spin Dimer [2]

Let us consider how to determine the DM vector of an isolated spin dimer. So far, a spin dimer made up of spin sites 1 and 2 has been described by the spin Hamiltonian, $\hat{H}_{\text{spin}} = J_{12}\hat{S}_1 \cdot \hat{S}_2$, composed of only a Heisenberg spin exchange. This Hamiltonian leads to a collinear spin arrangement (either FM or AFM), as already mentioned. To allow for a canting of the spins $\vec{S}_1$ and $\vec{S}_2$ from the collinear arrangement (typically from the AFM arrangement), which is experimentally observed, it is necessary to include the DM exchange interaction $\vec{D}_{12} \cdot (\hat{S}_1 \times \hat{S}_2)$ into the spin Hamiltonian. That is,

$$\hat{H}_{\text{spin}} = J_{12}\hat{S}_1 \cdot \hat{S}_2 + \vec{D}_{12} \cdot (\hat{S}_1 \times \hat{S}_2). \tag{10.19}$$

The $\hat{S}_1 \times \hat{S}_2$ term, being proportional to $\sin\theta$, where θ is the angle between the two spin vectors $\vec{S}_1$ and $\vec{S}_2$, is nonzero only if the two spins are not collinear. Thus, the DM interaction $\vec{D}_{12} \cdot (\hat{S}_1 \times \hat{S}_2)$ induces spin canting. Even when a model Hamiltonian consists of only Heisenberg spin exchanges, a magnetic system with more than two spin sites can have a noncollinear spin arrangement so as to reduce the extent of spin frustration, if there exists substantial spin frustration.

As discussed in Section 10.3, the spin exchange J_{12} of Eq. (10.19) can be evaluated on the basis of energy-mapping analysis by considering two collinear spin states |HS⟩ and |BS⟩ (i.e., FM and AFM spin arrangements, respectively), because the DM exchange $\vec{D}_{12} \cdot (\hat{S}_1 \times \hat{S}_2)$ is zero for such collinear spin states. To evaluate the DM vector $\vec{D}_{12}$, we carry out energy-mapping analysis on the basis of DFT + U + SOC calculations. In terms of its Cartesian components, $\vec{D}_{12}$ is expressed as

$$\vec{D}_{12} = \hat{i}D_{12}^x + \hat{j}D_{12}^y + \hat{k}D_{12}^z.$$

Therefore, the DM interaction energy $\vec{D}_{12} \cdot (\hat{S}_1 \times \hat{S}_2)$ is rewritten as

$$\vec{D}_{12} \cdot (\hat{S}_1 \times \hat{S}_2) = (\hat{i}D_{12}^x + \hat{j}D_{12}^y + \hat{k}D_{12}^z) \cdot \begin{pmatrix} \hat{i} & \hat{j} & \hat{k} \\ \hat{S}_1^x & \hat{S}_1^y & \hat{S}_1^z \\ \hat{S}_2^x & \hat{S}_2^y & \hat{S}_2^z \end{pmatrix}$$

$$= D_{12}^x(\hat{S}_1^y\hat{S}_2^z - \hat{S}_1^z\hat{S}_2^y) - D_{12}^y(\hat{S}_1^x\hat{S}_2^z - \hat{S}_1^z\hat{S}_2^x) + D_{12}^z(\hat{S}_1^x\hat{S}_2^y - \hat{S}_1^y\hat{S}_2^x). \tag{10.20}$$

To determine the D_{12}^z component, we consider the following two orthogonally ordered spin states:

State	Spin 1	Spin 2
1	(S, 0, 0)	(0, S, 0)
2	(S, 0, 0,)	(0, -S, 0)

For these states, $\vec{S}_1 \cdot \vec{S}_2 = 0$ and $|\vec{S}_1 \times \vec{S}_2| = S^2$ so that, according to Eq. (10.20), the energies of the two states are given by

$$E_1 = S^2 D_{12}^z, \text{ and } E_2 = -S^2 D_{12}^z.$$

Consequently,

$$D_{12}^z = \frac{E_1 - E_2}{2S^2}. \tag{10.21a}$$

Thus, the D_{12}^z is determined by evaluating the energies E_1 and E_2 on the basis of DFT + U + SOC calculations.

The D_{12}^y and D_{12}^x components are determined in a similar manner. Using the following two orthogonal spin states,

State	Spin 1	Spin 2
3	(S, 0, 0)	(0, 0, S)
4	(S, 0, 0,)	(0, 0, -S)

the D_{12}^y component is obtained as

$$D_{12}^y = \frac{E_3 - E_4}{2S^2}. \tag{10.21b}$$

In terms of the following two orthogonal spin states:

State	Spin 1	Spin 2
5	(0, S, 0)	(0, 0, S)
6	(0, S, 0,)	(0, 0, -S)

the D_{12}^x term is given by

$$D_{12}^z = \frac{E_5 - E_6}{2S^2}. \tag{10.21c}$$

10.5.5
Mapping Analysis for the DM Vectors Using the Four-State Method for a General Magnetic Solid [4]

For a given pair of spins in a general magnetic solid, the D_{12}^x, D_{12}^y, and D_{12}^z components can be similarly extracted by performing DFT + U + SOC calculations for four noncollinearly ordered spin states in which all spin exchange interactions associated with the spin sites 1 and 2 vanish [4]. In such a case, the relative energies of the four states are related only to the energy differences in their DM interactions. To calculate the z-component of D_{12}, that is, D_{12}^z, we carry out DFT + U + SOC calculations for the following four ordered spin states:

State	Spin 1	Spin 2	Other spin sites
1	(S, 0, 0)	(0, S, 0)	Either (0, 0, S) or (0, 0, -S) according to the experimental (or a low-energy) spin state. Keep the same for the four spin states
2	(S, 0, 0)	(0, -S, 0)	
3	(-S, 0, 0)	(0, S, 0)	
4	(-S, 0, 0)	(0, -S, 0)	

Then, we obtain

$$D_{12}^z = \frac{E_1 + E_4 - E_2 - E_3}{4S^2}. \quad (10.22a)$$

To determine the y-component of D_{12}, that is, D_{12}^y, we perform DFT + U + SOC calculations for the following four ordered spin states:

State	Spin 1	Spin 2	Other spin site
1	(S, 0, 0)	(0, 0, S)	Either (0, S, 0) or (0, -S, 0) according to the experimental (or a low-energy) spin state. Keep the same for the four spin states
2	(S, 0, 0)	(0, 0, -S)	
3	(-S, 0, 0)	(0, 0, S)	
4	(-S, 0, 0)	(0, 0, -S)	

Then,

$$D_{12}^y = \frac{-E_1 - E_4 + E_2 + E_3}{4S^2}. \quad (10.22b)$$

To determine the x-component of D_{12}, that is, D_{12}^x, we carry out DFT + U + SOC calculations for the following four ordered spin states:

State	Spin 1	Spin 2	Other spin site
1	(0, S, 0)	(0, 0, S)	Either (S, 0, 0) or (-S, 0, 0) according to the experimental (or a low-energy) spin state. Keep the same for the four spin states
2	(0, S, 0)	(0, 0, -S)	
3	(0, -S, 0)	(0, 0, S)	
4	(0, -S, 0)	(0, 0, -S)	

Then,

$$D_{12}^x = \frac{E_1 + E_4 - E_2 - E_3}{4S^2}. \quad (10.22c)$$

10.6
Uniaxial Magnetism [10,36]

In classical mechanics, the magnetic moment $\vec{\mu}$ of a system refers to the change of its energy E with respect to the applied magnetic field $\vec{H}$,

$$\vec{\mu} = -\frac{\partial E}{\partial \vec{H}}. \tag{10.23}$$

A uniaxial magnetic ion has a nonzero magnetic moment only in one direction in coordinate space, while an isotropic magnetic ion has a nonzero moment in all directions with equal magnitude. An anisotropic magnetic ion, lying between these two cases, has a moment with magnitude depending on the spin direction. When a transition metal magnetic ion is located at a coordination site with threefold or higher rotational symmetry, its d-states have doubly degenerate levels, namely,

$$\{xz, yz\} \text{ and } \{xy, x^2 - y^2\},$$

if the z-axis is taken along the rotational axis. In terms of the $\{L_z, -L_z\}$ set of magnetic quantum numbers, the $\{xz, yz\}$ and $\{xy, x^2-y^2\}$ sets are equivalent to

$$\{xz, yz\} \leftrightarrow \{1, -1\},$$
$$\{xy, x^2 - y^2\} \leftrightarrow \{2, -2\}.$$

An uneven filling of such a degenerate level leading to configurations such as $(L_z, -L_z)^1$ and $(L_z, -L_z)^3$ generates an unquenched orbital angular momentum of magnitude L (in units of $\hbar$). Thus, an uneven filling of the $\{1, -1\}$ set leads to $L=1$, and that of the $\{2, -2\}$ set to $L=2$. Such an electron filling generates a Jahn–Teller (JT) instability, but the unquenched orbital momentum remains if the associated JT distortion is prevented by steric congestion around the magnetic ions. The orbital momentum $\vec{L}$ couples with the spin momentum $\vec{S}$ by the SOC, $\lambda \vec{S} \cdot \vec{L}$, leading to the total angular momentum $\vec{J} = \vec{L} + \vec{S}$. The resulting total angular momentum states $|J, J_z\rangle$ are doublets specified by the two quantum numbers J and $J_z = \pm J$, that is, $\{|J, +J\rangle, |J, -J\rangle\}$ [36]. In identifying the ground doublet state, it is important to notice [10] the following:

$\lambda < 0$ for an ion with more than half-filled d-shell
$\lambda > 0$ for an ion with less than half-filled d-shell.

If $\lambda < 0$, the lowest-energy doublet state of the $\lambda \vec{S} \cdot \vec{L}$ term results when $\vec{S}$ and $\vec{L}$ are in the same direction. If $\lambda > 0$, however, it results when $\vec{S}$ and $\vec{L}$ have the opposite directions. Consequently, for a magnetic ion with L and S, the total angular quantum number J for the spin–orbit-coupled ground state is given by

$$\text{Ground doublet}: J = \begin{cases} L + S & \text{if } \lambda < 0 \\ L - S & \text{if } \lambda > 0 \end{cases}.$$

For $\lambda < 0$, the energy of the J-state increases as J decreases. However, the opposite is the case for $\lambda > 0$ [36].

In quantum mechanical description, the moment is related to an energy split of a degenerate level by an applied magnetic field. The Zeeman interaction under magnetic field is given by [36]

$$\hat{H}_Z = \mu_B(\hat{L} + 2\hat{S}) \cdot \vec{H}. \tag{10.24}$$

If we take the z-axis along the rotational axis responsible for the degeneracy of the doublet state $\{|J, +J\rangle, |J, -J\rangle\}$, the Zeeman interaction for the field along the z-direction, $H_\parallel$, is written as

$$\hat{H}_{Z\parallel} = \mu_B(\hat{L}_z + 2\hat{S}_z)H_\parallel. \tag{10.25a}$$

This Hamiltonian always lifts the degeneracy of $\{|J, +J\rangle, |J, -J\rangle\}$, because

$$\langle J, +J|\hat{H}_\parallel|J, +J\rangle = (L + 2S)\mu_B H_\parallel,$$
$$\langle J, -J|\hat{H}_\parallel|J, -J\rangle = -\langle J, +J|\hat{H}_\parallel|J, +J\rangle,$$
$$\langle J, +J|\hat{H}_\parallel|J, -J\rangle = 0.$$

Therefore, the energy split $\Delta E_{J\parallel}$ is given by

$$\Delta E_{J\parallel} = 2(L + 2S)\mu_B H_\parallel, \tag{10.25b}$$

and the associated g-factor $g_\parallel$ by

$$g_\parallel = \Delta E_{J\parallel}/\mu_B H_\parallel = 2(L + 2S).$$

The Zeeman interaction for the field perpendicular to the z-direction (say, the x-direction), $H_\perp$, is written as

$$\hat{H}_{Z\perp} = \mu_B\left[\frac{1}{2}(\hat{L}_+ + \hat{L}_-) + (\hat{S}_+ + \hat{S}_-)\right]H_\perp, \tag{10.26a}$$

for which we find

$$\langle J, +J|\hat{H}_\perp|J, +J\rangle = \langle J, -J|\hat{H}_\perp|J, -J\rangle = 0,$$
$$\langle J, +J|\hat{H}_\perp|J, -J\rangle = \mu_B\langle J, +J|\frac{1}{2}(\hat{L}_+ + \hat{L}_-) + (\hat{S}_+ + \hat{S}_-)|J, -J\rangle \equiv \mu_B H_\perp \delta.$$

Then, the associated energy split $\Delta E_{J\perp}$ is given by

$$\Delta E_{J\perp} = 2\mu_B H_\perp|\delta|. \tag{10.26b}$$

The $|J, +J\rangle$ and $|J, -J\rangle$ states differ in their J_z values by $2J$, so $\Delta E_{J\perp} = 0$ unless $J = 1/2$ because $|J, -J\rangle$ state cannot become $|J, +J\rangle$ by the ladder operator $\hat{L}_+$ or $\hat{S}_+$ in such a case. Thus, for magnetic ions with unquenched orbital momentum $\vec{L}$, we find uniaxial magnetism if $J > 1/2$ [36].

It should be noted that a spin Hamiltonian does not allow one to predict whether or not a given magnetic ion in molecules and solids will exhibit uniaxial magnetism because it cannot describe SOC, $\lambda \hat{S} \cdot \hat{L}$, explicitly due to the lack of the orbital degree of freedom. Nevertheless, once a magnetic system is known to exhibit uniaxial magnetism, one might use an Ising spin Hamiltonian (Section 10.3) to discuss its magnetic property.

10.7
Describing SOC Effects with Both Orbital and Spin Degrees of Freedom: Magnetic Anisotropy [5]

In this section, we probe the effect of SOC by explicitly considering the orbital and spin degrees of freedom. This enables one to quantitatively determine the preferred spin orientation of a magnetic ion M with any spin (i.e., $S = 1/2$–$5/2$) by performing DFT+U+SOC calculations and qualitatively predict it on the basis of analyzing the HOMO–LUMO interactions of the ML$_n$ polyhedron induced by SOC, $\lambda \hat{S} \cdot \hat{L}$. For this purpose, the states of a magnetic ion are described by $|L, L_z\rangle |S, S_z\rangle$ instead of approximating it with $|S, S_z\rangle$. If a coordinate (x', y', z') is employed for the spin $\hat{S}$, and (x, y, z) for the orbital $\hat{L}$, the z' direction is the preferred spin orientation by convention. The latter is specified with respect to the (x, y, z) coordinate by defining the polar angles θ and ϕ as depicted in Figure 10.17. In evaluating whether or not the SOC-induced interactions between different electronic states vanish, one needs to recall that the orbital states $|L, L_z\rangle$ are orthonormal, and so are the spin states $|S, S_{z'}\rangle$. That is,

$$\langle L, L_z | L, L'_z \rangle = \begin{cases} 1, & \text{if } L_z = L'_z, \\ 0, & \text{otherwise}, \end{cases}$$

$$\langle S, S_z | S, S'_z \rangle = \begin{cases} 1, & \text{if } S_z = S'_z, \\ 0, & \text{otherwise}. \end{cases}$$

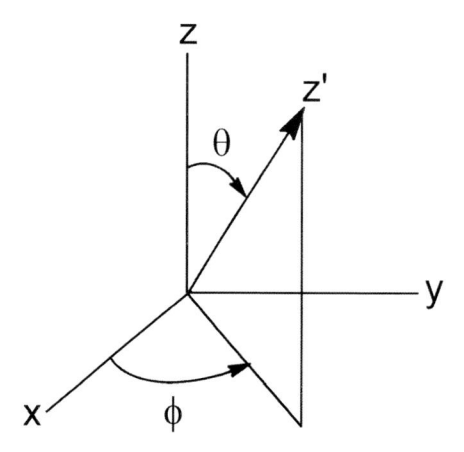

Figure 10.17 Polar angles θ and ϕ defining the preferred orientation of the spin (i.e., the z'-axis) with respect to the (x, y, z) coordinate used to describe the orbital.

10.7.1
Selection Rules for Preferred Spin-Orientation

Using the (x, y, z) and (x', y', z') coordinates for $\hat{L}$ and $\hat{S}$, respectively, the SOC Hamiltonian $\hat{H} = \lambda \hat{S} \cdot \hat{L}$ is rewritten as $\hat{H} = \hat{H}^{0}_{SO} + \hat{H}'_{SO}$ [2,10,37,38], where

$$\hat{H}^{0}_{SO} = \lambda \hat{S}_{z'} \left(\hat{L}_z \cos\theta + \frac{1}{2}\hat{L}_{+}e^{-i\varphi}\sin\theta + \frac{1}{2}\hat{L}_{-}e^{+i\varphi}\sin\theta \right) \tag{10.27a}$$

$$= \lambda \hat{S}_{z'} (\hat{L}_z \cos\theta + \hat{L}_x \sin\theta \cos\varphi + \hat{L}_y \sin\theta \sin\varphi). \tag{10.27b}$$

$$\hat{H}'_{SO} = \frac{\lambda}{2}(\hat{S}_{+'} + \hat{S}_{-'})(-\hat{L}_z \sin\theta + \hat{L}_x \cos\theta \cos\varphi + \hat{L}_y \cos\theta \sin\varphi). \tag{10.28}$$

We now consider if the preferred spin orientation is parallel to the local z-direction ($\|z$) (of the ML_n under consideration) or perpendicular to it ($\perp z$). The SOC-induced interaction between two d-states, ψ_i and ψ_j, involves the interaction energy $\langle \psi_i | \hat{H}_{SO} | \psi_j \rangle$. For our discussion, it is necessary to know whether this integral is zero or not. Since the angular part of a d- or p-orbital is expressed in terms of products $|L, L_z\rangle|S, S_{z'}\rangle$, the evaluation of $\langle \psi_i | \hat{H}_{SO} | \psi_j \rangle$ involves the spin integrals

$$\langle S, S'_{z'} | \hat{S}_{z'} | S, S_{z'} \rangle \text{ and } \langle S, S'_{z'} | \hat{S}_{\pm'} | S, S_{z'} \rangle,$$

as well as the orbital integrals

$$\langle L, L'_z | \hat{L}_z | L, L_z \rangle \text{ and } \langle L, L'_z | \hat{L}_{\pm} | L, L_z \rangle.$$

The SOC Hamiltonian $\hat{H}^{0}_{SO}$ allows interactions only between identical spin states, because $\langle \uparrow | \hat{S}_{z'} | \uparrow \rangle$ and $\langle \downarrow | \hat{S}_{z'} | \downarrow \rangle$ are nonzero. For two states, ψ_i and ψ_j, of identical spin, we consider the cases when $|\Delta L_z| = 0$ or 1. Then, we find

$$\langle \psi_i | \hat{H}^{0}_{SO} | \psi_j \rangle \propto \begin{cases} \cos\theta, & \text{if } |\Delta L_z| = 0 \\ \sin\theta, & \text{if } |\Delta L_z| = 1 \end{cases}. \tag{10.29a}$$

For the $|\Delta L_z| = 0$ case, $\langle \psi_i | \hat{H}^{0}_{SO} | \psi_j \rangle$ is maximum at $\theta = 0°$, that is, when the spin has the $\|z$ orientation. For the $|\Delta L_z| = 1$ case, $\langle \psi_i | \hat{H}^{0}_{SO} | \psi_j \rangle$ becomes maximum at $\theta = 90°$, that is, when the spin has the $\perp z$ orientation. Under SOC, ψ_i and ψ_j do not interact when $|\Delta L_z| > 1$, because $\langle \psi_i | \hat{H}^{0}_{SO} | \psi_j \rangle = 0$ in such a case.

The total energy of ML_n is lowered under SOC by the interactions of the filled d-states with the empty ones. Since the strength of SOC is very weak, these interactions can be described in terms of perturbation theory in which the SOC Hamiltonian is taken as perturbation with the split d-states of ML_n as unperturbed states. Then, the most important interaction of the occupied d-states with the unoccupied d-states is the one between the HOMO and the LUMO (with energies e_{HO} and e_{LU}, respectively), and the associated energy stabilization ΔE is given by [5]

Table 10.4 The preferred spin orientations of magnetic ions predicted using the $|\Delta L_z|$ values associated with the SOC-induced HOMO–LUMO interactions.

Spin orientation	Requirement	Interactions between
$\|z$	$\|\Delta L_z\| = 0$	xz and yz
		xy and $x^2 - y^2$
		x and y
$\perp z$	$\|\Delta L_z\| = 1$	$\{3z^2 - r^2\}$ and $\{xz, yz\}$
		$\{xz, yz\}$ and $\{xy, x^2 - y^2\}$
		z and $\{x, y\}$

$$\Delta E = \begin{cases} -\left|\langle HO|\hat{H}_{SO}^0|LU\rangle\right|, & \text{if } e_{HO} = e_{LU}, \\ -\dfrac{\left|\langle HO|\hat{H}_{SO}^0|LU\rangle\right|^2}{|e_{HO} - e_{LU}|}, & \text{if } e_{HO} < e_{LU}. \end{cases} \quad (10.29b)$$

Thus, we obtain the predictions for the preferred spin orientation as summarized in Table 10.4. In general, the effect of a degenerate interaction is stronger than that of a nondegenerate interaction. A system with degenerate HOMO and LUMO has JT instability, and the degeneracy would be lifted if the associated JT distortion were to take place [39].

According to Eq. (10.29), the preferred spin orientation is either $\|z$ or $\perp z$. For the preferred spin orientation to lie in between the $\|z$ and $\perp z$ directions, therefore, there must be two "HOMO–LUMO" interactions that predict different spin orientations (one for $\|z$, and the other for $\perp z$). Such a situation occurs for Na_2IrO_3, as will be discussed later.

10.7.2
Degenerate Perturbation and Uniaxial Magnetism

For a certain metal ion M, the electron configuration of ML_n has unevenly filled degenerate level. For example, the hexagonal perovskites Ca_3CoMnO_6 [40] consist of $CoMnO_6$ chains in which CoO_6 trigonal prisms containing high-spin Co^{2+} ($S = 3/2$, d^7) ions alternate with MnO_6 octahedra containing high-spin Mn^{4+} ($S = 3/2$, d^3) ions by sharing their triangular faces (Figure 10.18a). The d-states of the high-spin Co^{2+} ($S = 3/2$, d^7) ion in each CoO_6 trigonal prism (Figure 10.18b) can be described by the electron configuration, $(z^2)^2 < (xy, x^2 - y^2)^3 < (xz, yz)^2$, in the one-electron picture [36,39]. Thus, the spin-polarized d-states of the high-spin Co^{2+} is written as,

$$(z^2 \uparrow)^1 < (xy \uparrow, x^2 - y^2 \uparrow)^2 < (xz \uparrow, yz \uparrow)^2 < (z^2 \downarrow)^1 < (xy \downarrow, x^2 - y^2 \downarrow)^1 < (xz \downarrow, yz \downarrow)^0.$$

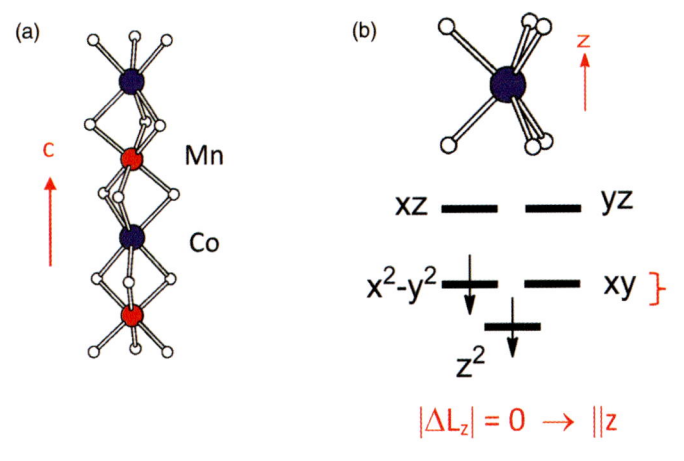

Figure 10.18 (a) A schematic view of an isolated CoMnO$_6$ chain of Ca$_3$CoMnO$_6$ that is made up of the CoO$_6$ trigonal prisms containing high-spin Co^{2+} (d^7, S = 3/2) ions and the MnO$_6$ octahedra containing high-spin Mn^{4+} (d^3, S = 3/2) ions. (b) The occupancy of the spin-down d-states for a high-spin Co^{2+} ion in an isolated CoO$_6$ trigonal prism.

Due to the half-filled configuration $(xy\downarrow, x^2 - y^2\downarrow)^1$, the HOMO and LUMO are degenerate with $|\Delta L_z| = 0$, so the preferred spin orientation is $\|z$, that is, along the threefold rotational axis of the trigonal prism. Furthermore, the configuration $(xy\downarrow, x^2 - y^2\downarrow)^1$ leads to an unquenched orbital momentum for $L = 2$. Since the d-shell of the high-spin Co^{2+} (d^7, S = 3/2) ion is more than half filled, $\lambda < 0$, so that $J = L + S = 2 + 3/2 = 7/2$ for the ground doublet state. Since $J > 1/2$, this ion has uniaxial magnetism, that is, it has a nonzero magnetic moment $\vec{\mu}$ only along the threefold rotational axis of the CoO$_6$ trigonal prism.

Each high-spin Fe^{2+} (S = 2, d^6) ion of Fe[C(Si(CH$_3$)$_3$)$_3$]$_2$ is located at a linear coordinate site (Figure 10.5d) [36,41], so that its spin-down d-states are filled as depicted in Figure 10.19 leading to the configuration $(xy\downarrow, x^2-y^2\downarrow)^1$. Thus, with $L = 2$ and $S = 2$, the spin–orbit-coupled ground doublet state is described by $J = L + S = 4$ with $J_z = \pm 4$. Since $J > 1/2$, this ion has uniaxial magnetism; it has a

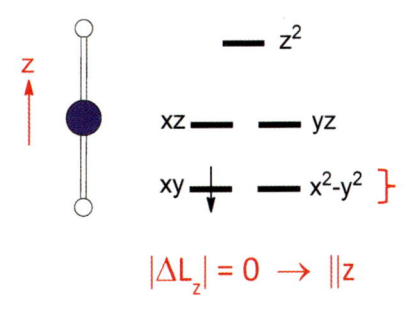

Figure 10.19 The spin-down electron configuration of a high-spin Fe^{2+} (d^6, S = 2) at a linear coordination site that induces uniaxial magnetism.

nonzero magnetic moment $\vec{\mu}$ only along the C–Fe–C axis (i.e., along the C_∞ rotational axis), and hence this Fe^{2+} ion has uniaxial magnetism.

We now examine the uniaxial magnetism that arises from metal ions at octahedral sites by considering the FeO_6 octahedra with high-spin Fe^{2+} (d^6, $S=2$) ions present in the oxide $BaFe_2(PO_4)_2$, the honeycomb layers of which are made up of edge-sharing FeO_6 octahedra. This oxide exhibits a uniaxial magnetism [42]. For our analysis of this observation, it is convenient to take the z-axis along one threefold rotational axis of an ML_6 octahedron (Figure 10.6a) [12]. The high-spin Fe^{2+} ion has the $(t_{2g})^4(e_g)^2$ configuration, the $(t_{2g})^4$ configuration of which can be described by $\Psi_{Fe,1}$ or $\Psi_{Fe,2}$ shown below:

$$\Psi_{Fe,1} = (1a)^1(1e_x, 1e_y)^3 = (1a\uparrow)^1(1e_x\uparrow, 1e_y\uparrow)^2(1e_x\downarrow, 1e_y\downarrow)^1,$$
$$\Psi_{Fe,2} = (1a)^2(1e_x, 1e_y)^2 = (1a\uparrow)^1(1e_x\uparrow, 1e_y\uparrow)^2(1a\downarrow)^1.$$

The occupancy of the spin-down d-states for $\Psi_{Fe,1}$ and $\Psi_{Fe,2}$ are presented in Figure 10.20a and b, respectively. An energy lowering through SOC is strong for $\Psi_{Fe,1}$ because it has an unevenly filled degenerate configuration $(1e_x\downarrow, 1e_y\downarrow)^1$, but not by $\Psi_{Fe,2}$ because the latter has an evenly filled degenerate configuration $(1e_x\downarrow, 1e_y\downarrow)^2$. According to Table 10.3, the spin-down configuration $(1e_x\downarrow, 1e_y\downarrow)^1$ of $\Psi_{Fe,1}$ is expressed as

$$(1e_x\downarrow, 1e_y\downarrow)^1 = \left(\sqrt{\frac{2}{3}}(xy\downarrow, x^2-y^2\downarrow)^1 - \sqrt{\frac{1}{3}}(xz\downarrow, yz\downarrow)^1\right). \tag{10.30}$$

The orbital-unquenched state $(xy\downarrow, x^2-y^2\downarrow)^1$ leads to $L=2$, but the state $(xz\downarrow, yz\downarrow)^1$ to $L=1$. The SOC constant $\lambda<0$ for the $\Psi_{Fe,1}$ configuration of Fe^{2+} ($S=2$, d^6) so that the ground doublet is $J=L+S=4$ from the component $(xy\downarrow, x^2-y^2\downarrow)^1$ ($L=2$), and $J=3$ from $(xz\downarrow, yz\downarrow)^1$ ($L=1$). In terms of the notation $\{J_z, -J_z\}$ representing a spin–orbit-coupled doublet set, the doublet $\{4, -4\}$ is more stable than $\{3, -3\}$ because $\lambda<0$, so the $(1e_x, 1e_y)^3$ configuration of Fe^{2+} is expressed as

$$Fe^{2+} : (1e_x, 1e_y)^3 = (1e_x\uparrow, 1e_y\uparrow)^2(1e_x\downarrow, 1e_y\downarrow)^1 \equiv \{4, -4\}^2\{3, -3\}^1.$$

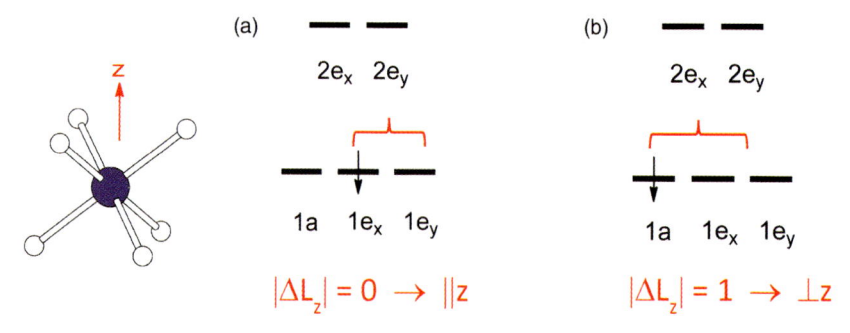

Figure 10.20 The spin-down electron configurations of a high-spin Fe^{2+} (d^6, $S=2$) at an octahedral site that induce (a) uniaxial magnetism and (b) no uniaxial magnetism.

With $J=3$ for the singly filled doublet, uniaxial magnetism is predicted for the high-spin Fe^{2+} ion at an octahedral site with $||z$ spin orientation. In support of this analysis, DFT calculations show the orbital moment of the Fe^{2+} ion to be $\sim 1\,\mu_B$ (i.e., $L \approx 1$) [43]. Note that the $\Psi_{Fe,2}$ configuration (Figure 10.20b) leads to $|\Delta L_z| = 1$ and hence, the preference for the $\perp z$ spin orientation.

10.7.3
Nondegenerate Perturbation and Weak Magnetic Anisotropy

We now examine the preferred spin orientations of magnetic ions with nondegenerate HOMO and LUMO. The layered compound $SrFeO_2$ consists of FeO_2 layers made up of corner-sharing FeO_4 square planes containing high-spin Fe^{2+} (d^6, $S=2$) ions [44]. Corner-sharing FeO_4 square planes are also found in $Sr_3Fe_2O_5$, in which they form two-leg ladder chains [45]. The d-states of a FeO_4 square plane are split as in Figure 10.5c [46,47], so that the spin-down d-states have only the $3z^2 - r^2 \downarrow$ level filled, with the empty $\{xz\downarrow, yz\downarrow\}$ set lying immediately above (Figure 10.9). Thus, between these HOMO and LUMO, with $|\Delta L_z| = 1$ so the preferred spin direction is $\perp z$, that is, parallel to the FeO_4 plane [46,47].

A regular MnO_6 octahedron containing a high-spin Mn^{3+} (d^4, $S=2$) ion has JT instability and hence adopts an axially elongated MnO_6 octahedron (Figure 10.5b). Such JT-distorted MnO_6 octahedra are found in $TbMnO_3$ [48] and Ag_2MnO_2 [49,50]. The neutron diffraction studies show that the spins of the Mn^{3+} ions are aligned along the elongated Mn—O bonds [48,50]. With four unpaired electrons to fill the split d-states, the LUMO is the $x^2 - y^2 \uparrow$ and the HOMO is the $3z^2 - r^2 \uparrow$ (Figure 10.21). Between these two states, $|\Delta L_z| = 2$ so

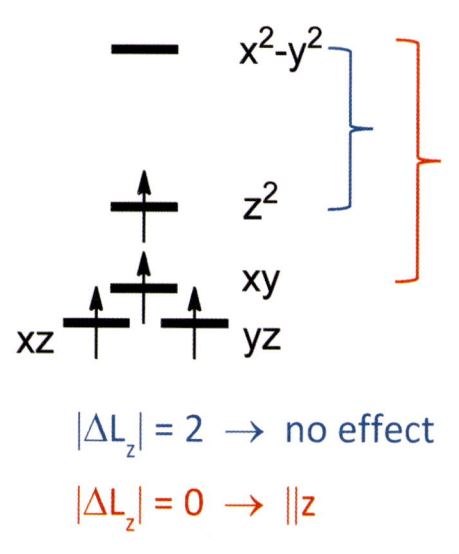

Figure 10.21 The high-spin configuration of an Mn^{3+} (d^4) ion in an axially elongated MnO_6 octahedron with the z-axis taken along the elongated Mn—O bonds.

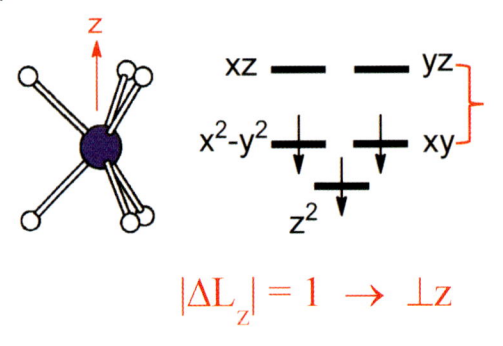

Figure 10.22 The spin-down electron configuration of a Ni^{2+} (d^8, $S=1$) ion at a trigonal prism site.

that they do not interact under SOC. The closest lying filled d-state that can interact with the LUMO is the $xy\uparrow$. Now, $|\Delta L_z|=0$ between the $x^2-y^2\uparrow$ and $xy\uparrow$ states, the preferred spin orientation is $||z$, that is, parallel to the elongated Mn—O bonds [50,51].

The NiO_6 trigonal prisms containing Ni^{2+} (d^8, $S=1$) ions are found in the $NiPtO_6$ chains of Sr_3NiPtO_6 [52] that is isostructural with Ca_3CoMnO_6. Each $NiPtO_6$ chain consists of face-sharing NiO_6 trigonal prisms and PtO_6 octahedra. The Pt^{4+} (d^6, $S=0$) ions are nonmagnetic. As depicted in Figure 10.22 for the spin-down d-states of Ni^{2+} (d^8, $S=1$), $|\Delta L_z|=1$ between the HOMO and LUMO. Consequently, the preferred spin orientation of the Ni^{2+} (d^8, $S=1$) ion is $\perp z$, that is, perpendicular to the $NiPtO_6$ chain. This is in agreement with DFT calculations [6].

10.7.4
Magnetic Anisotropy of S = 1/2 Systems and Spin-Half Misconception

In this section, we examine the experimentally observed magnetic anisotropies of various $S=1/2$ ions M. These observations are correctly reproduced by DFT + U + SOC calculations and also correctly explained by the SOC-induced HOMO–LUMO interactions of their ML_n polyhedra. The experimental and theoretical evidence against the spin-half misconception is overwhelming to say the least.

First, we consider the magnetic ions with $S=1/2$ in which the HOMO and LUMO of the crystal field d-states are not degenerate. An axially elongated IrO_6 octahedra containing low-spin Ir^{4+} (d^5, $S=1/2$) ions are found in the layered compound Sr_2IrO_4, in which the corner-sharing of the IrO_6 octahedra using the equatorial oxygen atoms forms the IrO_4 layers with the elongated Ir—O bonds perpendicular to the layer [53–55]. The neutron diffraction studies of Sr_2IrO_4 show that the Ir^{4+} spins are parallel to the IrO_4 layer [54,55]. With the z-axis chosen along the elongated Ir—O bond, the t_{2g} level of the IrO_6 octahedron is split into $\{xz, yz\} < xy$. With five d-electrons to fill the three levels, the spin-down states $xz\downarrow$ and $yz\downarrow$ are filled while the $xy\downarrow$ state is empty, as depicted in

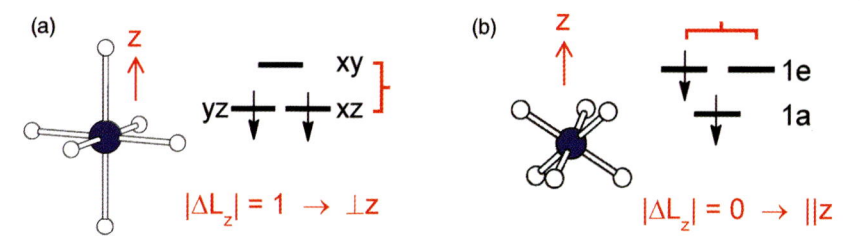

Figure 10.23 The spin-down states of the low-spin Ir^{4+} ($S = 1/2$, d^5) ion in (a) the axially elongated IrO$_6$ octahedron along the fourfold rotational axis in Sr$_2$IrO$_4$ and (b) the axially compressed IrO$_6$ octahedron along the threefold rotational axis in Na$_2$IrO$_3$.

Figure 10.23a. Consequently, $|\Delta L_z| = 1$ between the HOMO and LUMO, so that the preferred spin orientation is $\perp z$. This is in agreement with experiment and DFT calculations (see Section 10.8.1 for further discussion) [6,56].

Na$_2$IrO$_3$ consists of honeycomb layers made up of edge-sharing IrO$_6$ octahedra [57,58] that are substantially compressed along the direction perpendicular to the layer (lying in the *ab*-plane), that is, the c^* direction. Strictly speaking, each IrO$_6$ octahedron of Na$_2$IrO$_3$ has no threefold rotational symmetry but has a pseudo threefold rotational axis along the c^*-direction, which we take as the local z-axis. As for the preferred spin orientation of the Ir^{4+} ions of Na$_2$IrO$_3$, experimental studies have not been unequivocal, nor have been DFT studies, but it has become clear that the preferred spin orientation has components along the c^*- and *a*-directions (namely, $\|z$ and $\perp z$ components) [6,59,60]. Due to the compression of the IrO$_6$ octahedron along this axis, its t_{2g} state is split into $1a < (1e_x, 1e_y)$, where $1e_x$ and $1e_y$ are approximately degenerate, so that the spin-down d-states would be occupied as depicted in Figure 10.23b. For the Ir^{4+} ion of Na$_2$IrO$_3$, therefore, the HOMO and LUMO occur from the spin-down electron configuration close to $(1a\downarrow)^1(1e_x\downarrow, 1e_y\downarrow)^1$, so the preferred spin orientation would be the $\|z$ direction (namely, the c^*-direction) because $|\Delta L_z| = 0$. The electron configuration $(1a\downarrow)^1(1e_x\downarrow, 1e_y\downarrow)^1$ deduced from an isolated IrO$_6$ octahedron, explains the c^*-axis component, but cannot explain the presence of the *a*-axis component in the observed spin moment [59–61]. The perturbation theory analysis requires the split d-states of an IrO$_6$ octahedron present in Na$_2$IrO$_3$, not an isolated IrO$_6$ octahedron. The former have the effect of the intersite interactions, but the latter do not. Analysis of the intersite interaction showed [6] that they effectively reduce the energy split between $1a\downarrow$ and $(1e_x\downarrow, 1e_y\downarrow)$, so the $(1a\downarrow)^0(1e_x\downarrow, 1e_y\downarrow)^2$ configuration also participates substantially in controlling the spin orientation thereby giving rise to the *a*-axis component (see Section 10.8.1).

CuCl$_2$·2H$_2$O is a molecular crystal made up of CuCl$_2$(OH$_2$)$_2$ complexes containing Cu^{2+} (d^9, $S = 1/2$) ions, in which the linear O—Cu—O unit is perpendicular to the linear Cl—Cu—Cl unit (Figure 10.24a) [62]. The spins of the Cu^{2+} ions are aligned along the Cu—O direction [63], namely, the Cu^{2+} ions have easy plane anisotropy. The split spin-down d-states of CuCl$_2$·2H$_2$O show that the LUMO, $x^2 - y^2\downarrow$ has the smallest energy gap with the HOMO, $xz\downarrow$

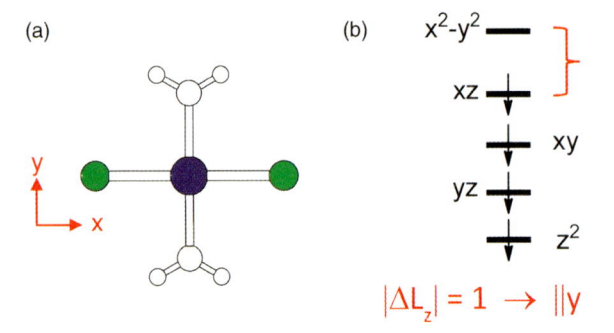

Figure 10.24 (a) The structure and the spin-down d-states of a CuCl$_2$(OH$_2$)$_2$ complex: blue circle = Cu, green circle = Cl, medium white circle = O, and small white circle = H. (b) The spin-down electron configuration of a Cu^{2+} (d^9, $S = 1/2$) ion.

(Figure 10.24b) [9]. Since $|\Delta L_z| = 1$, the preferred spin orientation is $\perp z$. To see if the spin prefers the x- or y-direction in the xy-plane, we use Eq. (10.27b). The matrix elements $\langle \psi_i | \hat{L}_\mu | \psi_j \rangle$ of the angular momentum operators $\hat{L}_\mu (\mu = x, y, z)$ are nonzero only for the following $\{\psi_i, \psi_j\}$ sets (see Table 10.2) [9]:

For $\hat{L}_z$: $\{xz, yz\}$, $\{xy, x^2 - y^2\}$.
For $\hat{L}_x$: $\{yz, 3z^2 - r^2\}$, $\{yz, x^2 - y^2\}$, $\{xz, xy\}$.
For $\hat{L}_y$: $\{xz, 3z^2 - r^2\}$, $\{xz, x^2 - y^2\}$, $\{yz, xy\}$.

The only nonzero interaction between the LUMO $x^2 - y^2 \downarrow$ and the HOMO $xz \downarrow$ under SOC is the term $\langle x^2 - y^2 | \hat{L}_y | xz \rangle$ involving $\hat{L}_y$. Eq. (10.27b) shows that this term comes with angular dependency of $\sin \theta \sin \varphi$ that is maximized when $\theta = 90°$ and $\phi = 90°$. Thus, the preferred spin orientation of CuCl$_2$(OH$_2$)$_2$ is along the y-direction, namely, along the Cu—O bonds [9].

In CuCl$_2$ [64,65], CuBr$_2$ [66], and LiCuVO$_4$ [24], the square planar CuL$_4$ units (L = Cl, Br, O) share their opposite edges to form CuL$_2$ ribbon chains (Figure 10.25a).

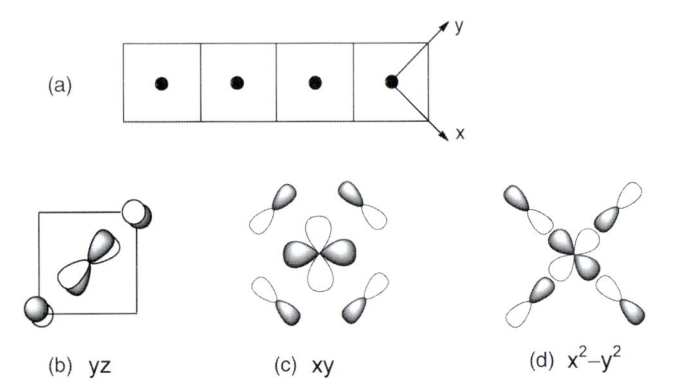

Figure 10.25 (a) The CuL$_2$ ribbon chain made up of edge-sharing CuL$_4$ square planes. The contributions of the metal d-orbitals and the ligand p-orbitals in the (b) yz, (c) xy, and (d) x^2-y^2 states of a CuL$_4$ square plane.

The split d-states in the CuL$_2$ ribbon chains of CuCl$_2$, CuBr$_2$, and LiCuVO$_4$ can be deduced by examining their projected density of states (PDOS) plots. Analyses of these plots can be best described by the effective sequence of the spin-down d-states shown in Eq. (10.31a) [9].

$$(3z^2 - r^2 \downarrow)^1 (xy \downarrow)^1 (xz \downarrow, yz \downarrow)^2 (x^2 - y^2 \downarrow)^0 \text{ for a CuL}_4 \text{ of a CuL}_2 \text{ ribbon chain,}$$
(10.31a)

$$(3z^2 - r^2 \downarrow)^1 (xz \downarrow, yz \downarrow)^2 (xy \downarrow)^1 (x^2 - y^2 \downarrow)^0 \text{ for an isolated CuL}_4 \text{ square plane.}$$
(10.31b)

Consequently, the interaction of the LUMO x^2-$y^2\downarrow$ with the HOMO $(xz\downarrow, yz\downarrow)$ will lead to the $\perp z$ spin orientation for the Cu^{2+} ions of the CuL$_2$ ribbon chains [9]. This spin-down d-state sequence is different from the corresponding one expected for an isolated CuL$_4$ square plane (shown in Eq. (10.31b)). This is due to the orbital interactions between adjacent CuL$_4$ square planes in the CuL$_2$ ribbon chain, in particular, the direct metal–metal interactions involving the xy orbitals through the shared edges between adjacent CuL$_4$ square planes.

Now we consider the magnetic ions with $S=1/2$ whose HOMO and LUMO are degenerate. Sr$_3$NiIrO$_6$ [67] is isostructural with Ca$_3$CoMnO$_6$, and its NiIrO$_6$ chains are made up of face-sharing IrO$_6$ octahedra and NiO$_6$ trigonal prisms. Each NiO$_6$ trigonal prism has a Ni^{2+} (d^8, $S=1$) ion, and each IrO$_6$ octahedron a low-spin Ir^{4+} (d^5, $S=1/2$) ion. Magnetic susceptibility and magnetization measurements [68][1] indicate that Sr$_3$NiIrO$_6$ has uniaxial magnetism with the spins of both Ni^{2+} and Ir^{4+} ions aligned along the chain direction. Neutron diffraction measurements show that in each chain the spins of adjacent Ni^{2+} and Ir^{4+} ions are antiferromagnetically coupled [68]. The low-spin Ir^{4+}(d^5, $S=1/2$) ion has the configuration $(t_{2g})^5$ that can be represented by $\Psi_{\text{Ir},1}$ or $\Psi_{\text{Ir},2}$:

$$\Psi_{\text{Ir},1} = (1a)^2 (1e_x, 1e_y)^3,$$
$$\Psi_{\text{Ir},2} = (1a)^1 (1e_x, 1e_y)^4.$$

The occupancies of the spin-down d-states for $\Psi_{\text{Ir},1}$ and $\Psi_{\text{Ir},2}$ are given as depicted in Figure 10.26a and b, respectively. It is $\Psi_{\text{Ir},1}$, not $\Psi_{\text{Ir},2}$, that can lower energy strongly under SOC. The spin-down part $(1e_x \downarrow, 1e_y \downarrow)^1$ of the configuration $(1e_x, 1e_y)^3$ in $\Psi_{\text{Ir},1}$ can be rewritten as in Eq. (10.30) so that $L=2$. For the low-spin Ir^{4+}, $\lambda<0$ because the t_{2g} shell is more than half-filled [10]. With $S=1/2$, we have $J=L+S=5/2$ from $(xy, x^2-y^2)^3$ and $3/2$ from $(xz, yz)^3$. Thus, the $(1e_x, 1e_y)^3$ configuration of Ir^{4+} is expressed as

$$\text{Ir}^{4+}: (1e_x, 1e_y)^3 = (1e_x \uparrow, 1e_y \uparrow)^2 (1e_x \downarrow, 1e_y \downarrow)^1$$
$$\equiv \{5/2, -5/2\}^2 \{3/2, -3/2\}^1.$$

The singly filled doublet has $J=3/2$, so uniaxial magnetism is predicted with the spin orientation along the $\|z$ direction. This explains why the $S=1/2$ ion Ir^{4+}

---

1) Kim, J.W. and Cheong, S.-W. unpublished results (private communication).

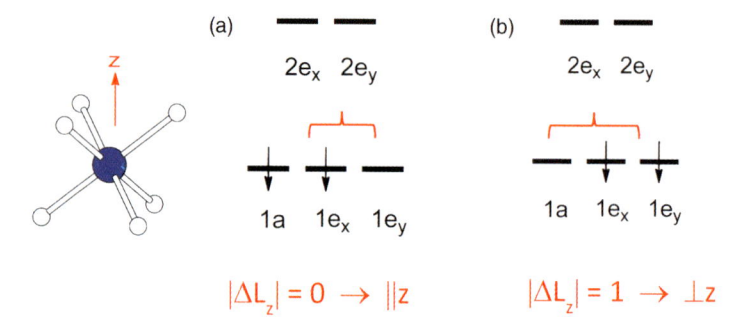

Figure 10.26 The spin-down electron configurations of a low-spin Ir^{4+} (d^5, $S = 1/2$) ion at an octahedral site that induce (a) uniaxial magnetism and (b) no uniaxial magnetism.

ion exhibits a strong magnetic anisotropy with the preferred spin direction along the z-axis. In contrast to the case of Sr_3NiPtO_6, the Ni^{2+} ions of Sr_3NiIrO_6 have the $\|z$ spin orientation. This is due to the combined effect of the uniaxial magnetism of the Ir^{4+} ions and the strong AFM spin exchange between adjacent Ir^{4+} and Ni^{2+} ions in each $NiIrO_6$ chain that overrides the weak preference for the $\perp z$ spin orientation for the Ni^{2+} ion in an "isolated NiO_6" trigonal prism (see Section 10.8.1 for further discussions) [5,6].

Let us consider the spin orientation of the $S = 1/2$ ions V^{4+} (d^1) in the VO_6 octahedra of $R_2V_2O_7$ (R = rare earth) [69], in which each VO_6 octahedron is axially compressed along the direction of its local threefold rotational axis (Figure 10.27a) so that its t_{2g} state is split into the $1a < 1e$ pattern (Figure 10.27b). With

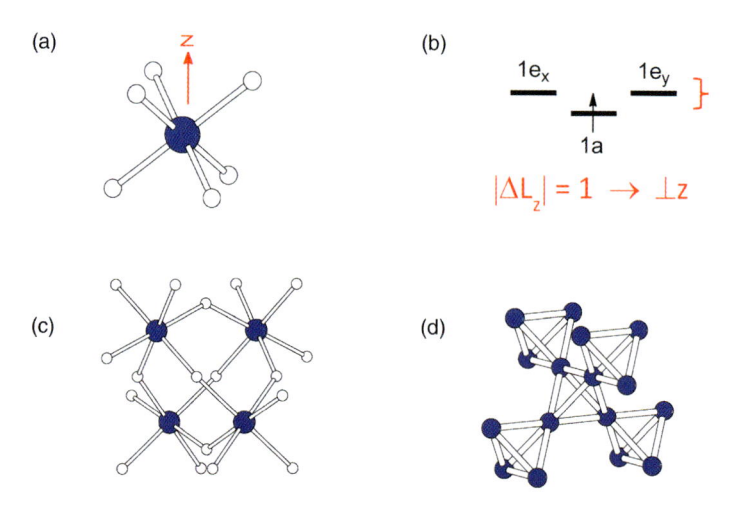

Figure 10.27 (a) An axially compressed VO_6 octahedron of $R_2V_2O_7$ (R = rare earth) along the local z-direction taken along a threefold rotational axis. (b) The split t_{2g} state of a V^{4+} (d^1, $S = 1/2$) ion at each VO_6 octahedron. (c) A tetrahedral cluster made up of four VO_6 octahedra. The local z-axes of the four VO_6 octahedra are all pointed to the center of the V_4 tetrahedron. (d) The pyrochlore lattice of the V^{4+} ions in $R_2V_2O_7$.

the local z-axis along the threefold rotational axis of VO_6, the HOMO is the $1a\uparrow$ state, which is represented by $3z^2 - r^2\uparrow$, that interacts with the LUMO $1e\uparrow = (1e_x\uparrow, 1e_y\uparrow)$ states under SOC through their $(xz\uparrow, yz\uparrow)$ components. Consequently, $|\Delta L_z| = 1$ and the preferred spin orientation would be $\perp z$. However, the observed spin orientation is $\|z$ [70] that has also been confirmed by DFT calculations [71]. This finding is explained if the V^{4+} ion has some uniaxial magnetic character despite that the HOMO and LUMO are not degenerate. For the latter to be true, the true ground state of each V^{4+} ion in $R_2V_2O_7$ should be a "contaminated state" $1a'$, which has some contributions of the $1e$ and $2e$ character of its isolated VO_6 octahedron, namely,

$$|1a'\rangle \propto |1a\rangle + \gamma|1e\rangle + \delta|2e\rangle,$$

where γ and δ are small mixing coefficients. This is possible because each VO_6 octahedron present in $R_2V_2O_7$ has a lower symmetry than does an isolated VO_6 octahedron. The VO_6 octahedra are corner-shared to form a tetrahedral cluster (Figure 10.27c), and such tetrahedral clusters further share their corners to form a pyroclore lattice (Figure 10.27d). Indeed, the PDOS plots for the spin-up d-states of the V^{4+} ions in $R_2V_2O_7$ show the presence of slight contributions of the $1e$ and $2e$ states to the occupied $1a$ state [5,71].

As reviewed above, both experimental and theoretical studies reveal that $S = 1/2$ ions do have magnetic anisotropy induced by SOC. The spin-half misconception is in clear contradiction to these experimental and theoretical observations. Due to the simplification it introduces for doing complex calculations, spin Hamiltonian has been a practical tool of choice in doing physics on magnetism and will remain so for some time to come. Nevertheless, this success does not justify the perpetuation of the spin-half misconception. This failure of a spin Hamiltonian should be considered as a small price to pay for the enormous gain it provides.

10.7.5
Ligand-Controlled Spin Orientation

For the $CuBr_4$ square planes of $CuBr_2$ ribbon chain [66], the $CuBr_5$ square pyramids of $(C_5H_{12}N)CuBr_3$ [72,73], and the CrI_6 octahedra of the layered compound CrI_3 [74,75], the ligand L is heavier than M, so the SOC between two d-states of ML_n results more from the SOC-induced interactions between the p-orbitals of the ligands L rather than from those between the d-orbitals of M. We clarify this point by considering a square planar ML_4 using the coordinate system of Figure 10.25a. The metal and ligand contributions in the yz, xy, and x^2-y^2 states of ML_4 are shown in Figure 10.25b–d, respectively. The SOC-induced interaction between different d-states can occur by the SOC of M, and also by that of each ligand L. The interaction between the z and $\{x, y\}$ orbitals at each L has $|\Delta L_z| = 1$, leading to the $\perp z$ spin orientation. In contrast, the interaction between the x and y orbitals at each L has $|\Delta L_z| = 0$, leading to the $\|z$ spin orientation (Table 10.2). When the ligand L is much heavier than the metal M, the SOC constant λ of L is greater than that of M. Furthermore, such ligands L possess diffuse and high-lying p-orbitals that makes the magnetic orbitals of ML_n dominated by the ligand p-orbitals and also makes the d-states of ML_n weakly split. This makes the SOC effect in ML_n dominated by the ligands.

10.7.6
High-Spin d^5 Systems

High-spin d^5 transition metal ions with $S=5/2$ possess a small nonzero orbital momentum $\delta\vec{L}$ and exhibit weakly preferred spin orientations. For such a magnetic ion, the SOC-induced HOMO–LUMO interaction should be based on the $\hat{H}'_{SO}$ term (Eq. (10.28)), because the HOMO and LUMO occur from different spin states. The comparison of Eq. (10.27b) with Eq. (10.28) reveals that the predictions concerning the $\|z$ versus $\perp z$ spin orientation from the term $\hat{H}'_{SO}$ are exactly opposite to those from the term $\hat{H}^0_{SO}$.

A similar situation occurs for a d^3 magnetic ion at octahedral sites, as found for the Os^{5+} ions in Ca_2ScOsO_6 [76] and the Ir^{6+} ions in Sr_2CuIrO_6 [77], because such an ion has the $(t_{2g})^3$ configuration and because the t_{2g} states are well separated in energy from the e_g states. Thus, the occupied spin-up t_{2g} states, $t_{2g}\uparrow$, become the HOMO, and the unoccupied spin-down t_{2g} states, $t_{2g}\downarrow$, the LUMO. It is known [78] that the orbital momentum of such a cation can be discussed by using the pseudo-orbital states $|L', L'_z\rangle$ with $L' = 1$ and $L'_z = 1, 0, -1$. To a first approximation, therefore, the orbital momentum of such a d^3 magnetic ion is zero. However, the quenching of the orbital momentum is not complete so that a $(t_{2g})^3$ ion has a small nonzero orbital momentum $\delta\vec{L}$. Thus, the preferred spin orientation of $(t_{2g})^3$ ions is governed by the SOC-induced HOMO–LUMO interaction based on the $\hat{H}'_{SO}$ term (Eq. (10.28)) [79].

10.8
Magnetic Properties of 5d Ion Oxides [6]

The d-orbitals of 5d ions are more diffuse than those of 3d ions, so that electron correlation is much weaker for 5d ions than for 3d ions. For a given MO_n polyhedron, the M 3d and O 2p orbitals do not differ strongly in their contractedness so that the associated crystal field splitting of an isolated MO_n polyhedron is strong. However, the M 5d orbitals are much more diffuse than O 2p orbitals so that the 5d-state splitting of an isolated MO_n polyhedron is weak. In addition, the interactions between adjacent metal ions M through the M—O—M bridges are stronger for 5d ions than for 3d ions. Thus, for 5d ion oxides, the relative ordering of their split d-states deduced from an isolated MO_n polyhedron might change by the interactions between adjacent metal ions (i.e., the intersite interactions). Furthermore, each of the crystal field split d-states can be split further by SOC [80], and this effect is much stronger for 5d ion oxides than for 3d ion oxides because the strength of SOC is much stronger for 5d ions than for 3d ions. The weak electron correlation and strong SOC in 5d ion oxides have important consequences, as discussed later.

10.8.1
Spin-Orbit Mott Insulating State and Madelung Potential

The combination of strong SOC and weak electron correlation creates a magnetic insulating state, as first reported for Ba_2NaOsO_6 containing Os^{7+} (d^1) ions [80]. This phenomenon, quite common in 5d ion oxides, was considered as a consequence of strong spin–orbital entanglement [81], and the resulting magnetic insulating state is described as a SOC-induced Mott insulating state [82] or spin–orbit Mott insulating state [83]. Both Sr_3NiIrO_6 and Sr_2IrO_4 are magnetic insulators, namely, they have a bandgap at all temperature [84–88].[2)] Na_2IrO_3 has been thought to be a magnetic insulator [89,90], but a recent DFT study suggested that it might be a Slater insulator [91]. The latter refers to a system with a partially filled bands and weak electron correlation that opens a bandgap when it undergoes a metal–insulator transition at a temperature below which an AFM ordering sets in [92]. In addition to the local factors affecting electron localization such as the oxidation state and the SOC constant λ of a metal ion M, the extent of electron localization is influenced by the Madelung potential acting at the M that is a nonlocal factor [6]. The Madelung potentials acting on the Ir^{4+} sites are less negative (i.e., less attractive) for Na_2IrO_3 than for Sr_3NiIrO_6 and Sr_2IrO_4, namely, the 5d electrons of an Ir^{4+} ion would be less strongly bound (i.e., less strongly localized) to the ion [6].

10.8.2
Influence of Intersite Interactions on Crystal Field Split *d*-States [6]

In predicting the preferred spin orientations of magnetic ions M in magnetic oxides on the basis of the SOC-induced HOMO–LUMO interactions, the split d-states of their local MO_n polyhedra are needed. As already mentioned, for oxides of 5d ions, the relative ordering of their split d-states deduced from an isolated MO_n polyhedron might change by the intersite interaction. In the following, we examine how the intersite interactions affect the split 5d-states of the Ir^{4+} ions in Sr_3NiIrO_6, Sr_2IrO_4, and Na_2IrO_3 and explore their consequences.

The ESR study of Sr_2IrO_4 showed [93] that the g-factors of the Ir^{4+} ion along the $\|c$ and $\perp c$ directions are explained if the t_{2g}-states are split as $xy < (xz, yz)$ rather than as $(xz, yz) < xy$ (discussed in Section 10.7.4). This finding, puzzling from the viewpoints of the split t_{2g} states of an isolated IrO_6 octahedron, reflects [6] that the split d-state patterns of Sr_2IrO_4 differ from those of an isolated IrO_6 octahedron due to the intersite interactions. In each IrO_4 layer of Sr_2IrO_4, the Ir—O—Ir linkages in the *ab*-plane are bent as shown in Figure 10.28a. This bending of the Ir—O—Ir linkages does not weaken the

2) Kim, J.W. and Cheong, S.-W. (2015) *unpublished work*.

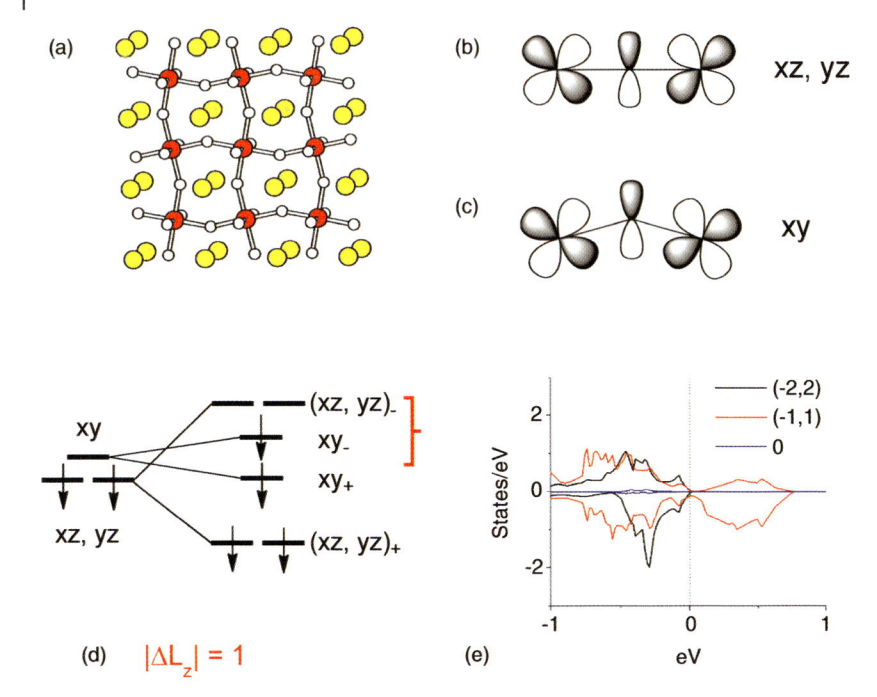

Figure 10.28 (a) A view of an isolated Sr$_2$IrO$_4$ layer made up of corner-sharing axially elongated IrO$_6$ octahedra approximately along the c-direction. (b) The interaction between adjacent xz orbitals (or adjacent yz orbitals) through the O 2p-orbitals through each bent Ir–O$_{eq}$–O bridge. (c) The interaction between adjacent xy orbitals through the O 2p-orbitals through each bent Ir–O$_{eq}$–O bridge. (d) The split d-states of a dimer made up of two adjacent Ir^{4+} ions after incorporating the effect of the intersite interactions for the cases of the axially elongated IrO$_6$ octahedra. (e) The PDOS plots for the d-states of Ir^{4+} in Sr$_2$IrO$_4$ in case when the IrO$_6$ octahedra are axially elongated, where the legends (2, -2), (1, -1), and 0 indicate the sets of orbitals (xy, x^2-y^2), (xz, yz), and $3z^2$-r^2, respectively.

π-antibonding interactions between adjacent xz/yz orbitals (Figure 10.28b), but does weaken those between adjacent xy orbitals (Figure 10.28c). Namely, the π-type interactions between adjacent xz/yz orbitals are stronger than those between adjacent xy orbitals. The split d-states of an IrO$_6$ octahedron embedded in Sr$_2$IrO$_4$ and hence, having the intersite interactions can be approximated by those of a dimer made up of two adjacent corner-sharing IrO$_6$ octahedra. Then, the interactions between two adjacent Ir^{4+} sites alter the crystal field split t_{2g} states as depicted in Figure 10.28d, so that the HOMO has the xy character, and the LUMO the xz/yz character. This picture explains the PDOS plots of Sr$_2$IrO$_4$ shown in Figure 10.28e, and predicts the ⊥c spin orientation as does the crystal field split t_{2g} states of an isolated IrO$_6$ octahedron (Figure 10.23a). In addition, this explains why the ESR results [93] of Sr$_2$IrO$_4$ are explained by the d-state ordering xy < (xz, yz), despite that it consists of axially elongated IrO$_6$ octahedra.

In Na$_2$IrO$_3$, edge-sharing IrO$_6$ octahedra form honeycomb layers (Figure 10.29a), and such layers are stacked along the c-direction (Figure 10.29b). DFT +

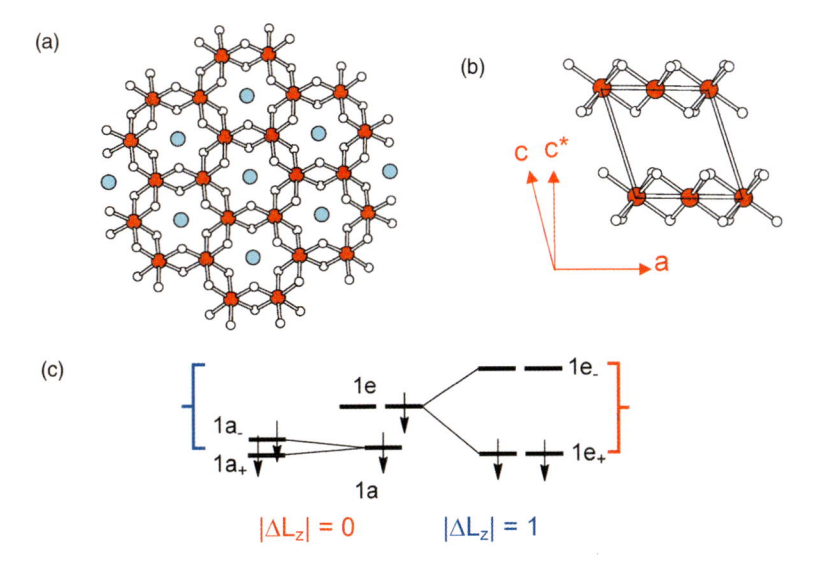

Figure 10.29 (a) A projection view of a NaIrO$_3$ honeycomb layer made up of edge-sharing IrO$_6$ octahedra with Na (light blue circle) at the center of each Ir$_6$ hexagon. (b) A perspective view of how the honeycomb NaIrO$_3$ layers repeat along the c-direction in Na$_2$IrO$_3$, where the layer of Na atoms lying in between the NaIrO$_3$ honeycomb layers is not shown for simplicity. (c) The split d-states of a dimer made up of two adjacent Ir^{4+} ions after incorporating the effect of the intersite interactions.

U + SOC calculations reveal that the preferred spin orientation of the Ir^{4+} ions in Na$_2$IrO$_3$ has both $||c^*$ and $||a$ components [6,94]. To examine the cause for this observation, we consider how the intersite interaction affects relative ordering of the spin-down $1a$ and $1e$ states of an Ir^{4+} ion (Figure 10.23b). Consider a dimer made up of two adjacent Ir^{4+} ions and recall that the d-orbital component of the $1a$ state is the $3z^2$-r^2 orbital, while those of the $1e$ state are the $(xy, x^2$-$y^2)$ and (xz, yz) orbitals (Table 10.3). As depicted in Figure 10.29c, the intersite interaction between the two $1a$ states leads to the $1a_+$ and $1a_-$ states, and that between the $1e$ states to the $1e_+$ and $1e_-$ states. The split between $1a_+$ and $1a_-$ states is weak because the lateral extension of the $3z^2$-r^2 orbitals within the plane of the honeycomb layer is small. In contrast, the split between the $1e_+$ and $1e_-$ states is large because the lateral extension of the $(xy, x^2$-$y^2)$ orbitals is large and so is that of the (xz, yz) orbitals. With four spin-down electrons in the dimer, the $1e_-$ states are empty while the remaining states are filled. The $|\Delta L_z| = 1$ interactions between the $1a_+/1a_-$ and $1e_-$ states predict the $\perp z$ spin orientation. The interactions between the $1e_+$ and $1e_-$ states give rise to the $|\Delta L_z| = 0$ interactions, between their (xz, yz) sets and between their $(xy, x^2$-$y^2)$ sets, predicting the $||z$ spin orientation. Consequently, if the $1a_+$ and $1a_-$ states are close in energy to the $1e_+$ states, then the preferred spin orientation of the Ir^{4+} ion would be the $(\perp z + ||z)$ direction. In essence, the $||a$ component of the Ir^{4+} spin orientation in Na$_2$IrO$_3$ is a consequence of the intersite interactions, because only the $||c^*$ direction is predicted in their absence.

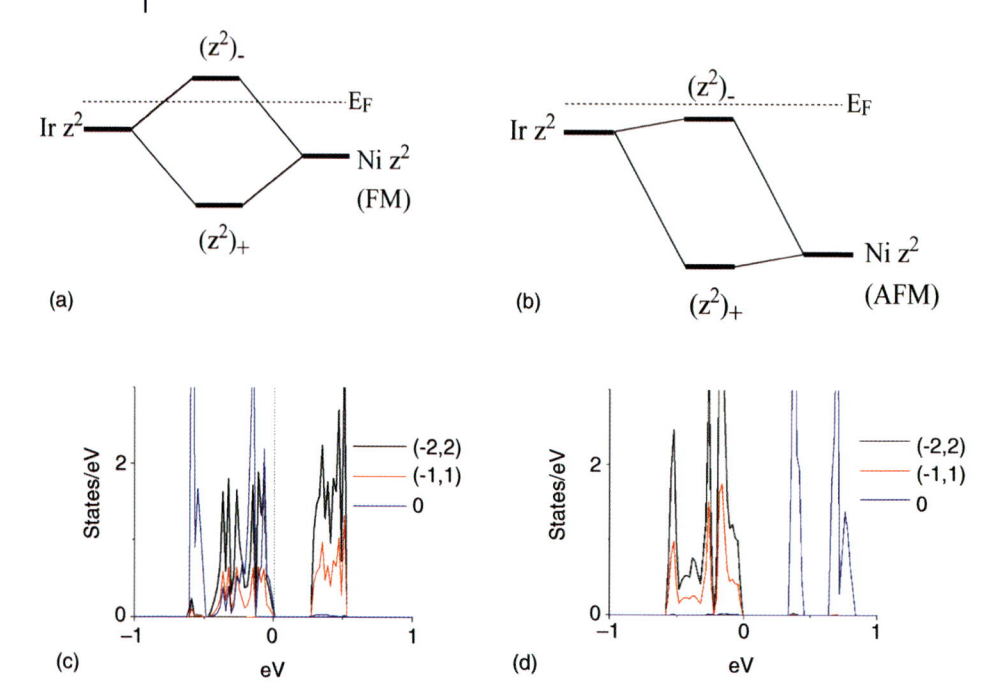

Figure 10.30 (a, b) Interactions between the Ir and Ni $3z^2-r^2$ states in each NiIrO$_6$ chain of Sr$_3$NiIrO$_6$ when the spins of adjacent Ir^{4+} and Ni^{2+} ions have a FM coupling in (a), and an AFM coupling in (b). (c, d) The PDOS plots for the spin-down d-states of Ir^{4+} in Sr$_3$NiIrO$_6$ in cases when adjacent Ir^{4+} and Ni^{2+} ions in each NiIrO$_6$ chain have a FM coupling in (c), and an AFM coupling in (d). The legends (2, -2), (1, -1), and 0 refer respectively to the (xy, x^2-y^2), (xz, yz), and $3z^2-r^2$ sets.

The magnetic insulating state of Sr$_3$NiIrO$_6$ is reproduced by DFT+U+SOC calculations only when adjacent Ni^{2+} and Ir^{4+} spins have an AFM coupling in each NiIrO$_6$ chain [6,95,96]. It is known experimentally [84,85] that the preferred orientation of the Ir^{4+} spins is the ||c-direction. DFT+U+SOC calculations showed that the preferred orientation of the Ir^{4+} spins is the ||c-direction if the Ni^{2+} and Ir^{4+} spins have an AFM coupling [6], but it is the ⊥c-direction if they have an FM coupling [6,97]. In each NiIrO$_6$ chain, the nearest-neighbor Ir...Ni distance is short due to the face-sharing between the IrO$_6$ and NiO$_6$ polyhedra so that the overlap between the Ir and Ni $3z^2-r^2$ orbitals can be strong. As illustrated in Figure 10.30a and b, the Ni $3z^2-r^2$ orbital is closer in energy to the Ir $3z^2-r^2$ orbital when adjacent Ni^{2+} and Ir^{4+} spins have an FM coupling than when they have an AFM coupling (see Section 10.2.2.2 and Figure 10.8). The latter makes the interaction between the Ir and Ni $3z^2-r^2$ states stronger for the FM than for the AFM spin arrangement [2,10,39b]. As a consequence, the resulting antibonding state $(3z^2-r^2)_-$ is unoccupied for the FM spin arrangement, but it is occupied for the AFM spin arrangement (Figure 10.30a and b), as found by DFT+U calculations for Sr$_3$NiIrO$_6$ [6]; the PDOS plots for the FM and AFM arrangements, presented in Figure 10.30c and d, respectively, reveal that the AFM arrangement is consistent with the local electron configuration $(1a\downarrow)^1(1e_x\downarrow, 1e_y\downarrow)^1$ (Figure 10.26a), predicting the $\parallel z$ spin orientation, while the FM arrangement is consistent with the local configuration $(1a\downarrow)^0(1e_x\downarrow, 1e_y\downarrow)^1$ (Figure 10.26b), predicting the $\perp z$ spin orientation.

10.8.3
Perturbation Theory Analysis of Preferred Spin Orientation [6]

The energy stabilization ΔE associated with the SOC-induced interaction between the HOMO and the LUMO (with energies e_{HO} and e_{LU}, respectively) is given by Eq. (10.29b). For the Ir^{4+} (low-spin d^5) ion systems Sr_3NiIrO_6, Sr_2IrO_4, and Na_2IrO_3, the overall widths of the t_{2g}-block bandwidths are of the order of 2 eV (i.e., 1.7, 2.6, and 2.4 eV, respectively, from our DFT + U calculations) and the HOMO–LUMO energy differences $|e_{HO} - e_{LU}|$ values are on the order of 0.2 eV (0.2, 0.2, and 0.3 eV, respectively) [6]. The SOC constant λ of Ir^{4+} is on the order of 0.5 eV [8e] so that λ^2 is comparable in magnitude to $|e_{HO} - e_{LU}|$ for the case of $e_{HO} < e_{LU}$. In such a case, use of perturbation theory does not lead to an accurate estimation of ΔE. However, this does not affect our qualitative predictions of the preferred spin orientations, because the latter do not require a quantitative evaluation of ΔE.

10.8.4
LS versus jj Coupling Scheme of SOC [6]

The effects of SOC are discussed in terms of either the LS or the jj coupling scheme depending on the strength of SOC. In the LS (or Russel–Saunders) scheme, the electron spin momenta are summed up to find the total spin momentum $\vec{S} = \sum \vec{s}_i$, and the orbital momenta of individual electrons to find the total orbital momentum $\vec{L} = \sum \vec{l}_i$. Then, the SOC is included to couple $\vec{S}$ and $\vec{L}$ to obtain the total angular momentum $\vec{J}$, leading to the SOC Hamiltonian, $\hat{H}_{SO} = \lambda \vec{S} \cdot \vec{L}$. The LS-coupling scheme is typically employed for elements with weak SOC (e.g., 3d- and 4d-elements). In this scheme the crystal field split d-states of a MO_n polyhedron are closely related to the orbital states $|L, L_z\rangle$ of M in the spin-up $|\uparrow\rangle$ or spin-down state $|\downarrow\rangle$ magnetic orbitals of MO_n. As found for Sr_3NiIrO_6, Sr_2IrO_4, and Na_2IrO_3 [6] and for Ba_2NaOsO_6 [80], our analyses based on the LS-coupling scheme explain the spin–orbit Mott insulating states of these 5d oxides as well as their observed magnetic anisotropies.

The jj-coupling scheme, appropriate for elements with strong SOC (e.g., 4f and 5f elements), has recently become popular in discussing the spin–orbit Mott insulating states of 5d oxides [81]. In this scheme, the spin and orbital momenta are added to obtain the total angular momentum $\vec{j}_i = \vec{l}_i + \vec{s}_i$ for each electron of a magnetic ion M, and the $\vec{j}_i$'s of the individual electrons are added to find the total angular momentum, $\vec{J} = \sum \vec{j}_i$, of M. In this approach, it is not readily obvious how to relate the $\vec{J}$ states to the crystal field split d-states of MO_n unless the corresponding analysis is done by using the LS-coupling scheme, because the crystal field split d-states of MO_n are determined by the interactions of the orbital states $|L, L_z\rangle$ of M with the 2p orbitals of the surrounding O ligands, and because the information about the orbital states $|L, L_z\rangle$ of M is completely hidden in the jj-coupling scheme. As a consequence, use of the jj scheme makes it difficult to predict such fundamental magnetic properties as the preferred spin orientation and the uniaxial magnetism of a magnetic ion M. The latter are readily predicted by the LS coupling scheme. As found for the Ir^{4+} ion of Sr_3NiIrO_6, the need to employ "J-states" in the LS scheme arises only when a magnetic ion has an unevenly filled degenerate d-state, leading to an unquenched orbital momentum $\vec{L}$ that combines with $\vec{S}$ to form $\vec{J} = \vec{S} + \vec{L}$. In the LS scheme, use of J-states is inappropriate for Sr_2IrO_4 and Na_2IrO_3 because they possess no unquenched orbital momentum $\vec{L}$ to combine with $\vec{S}$.

Studies on Sr_3NiIrO_6, Sr_2IrO_4, and Na_2IrO_3 [6] and on Ba_2NaOsO_6 [80] strongly suggest that the magnetic properties of the 5d oxides are better explained by the LS scheme than by the jj scheme. The latter implies that the spin–orbital entanglement in 5d elements is not as strong as has been assumed [81]. These conclusions are consistent with the view that SOC for 5d elements lies in between the LS- and jj-coupling schemes, but is closer to the LS scheme [98].

10.9
Concluding Remarks

In this chapter, we have reviewed how to think about magnetic properties of solid-state materials from the perspectives of an electronic Hamiltonian. On the quantitative level, use of this Hamiltonian enables one

a) to determine the relative stabilities of various spin arrangements on the basis of DFT + U or DFT + U + SOC calculations,
b) to evaluate the spin exchange and DM exchange parameters that a spin Hamiltonian requires by performing energy-mapping analysis based on DFT + U or DFT + U + SOC calculations, and
c) to characterize the magnetic anisotropy of a magnetic ion by performing DFT + U + SOC calculations.

On the qualitative level, use of an electronic Hamiltonian allows one

a) to examine spin lattices in terms of M—L—M as well as M—L⋯L—M spin exchanges,
b) to discuss how the strengths of M—L⋯L—M spin exchanges are modified by through-space and through-bond interactions, and
c) to predict/rationalize the preferred spin orientation of a magnetic ion on the basis of its SOC-induced HOMO–LUMO interactions.

The qualitative concepts governing these structure–property correlations help one organize/think about known experimental/theoretical observations, design new experiments to do and new calculations to perform, and predict/rationalize the outcomes of the new studies.

In the past, a spin lattice required for spin Hamiltonian analysis used to be chosen by inspecting the pattern of magnetic ion arrangement and employing the Goodenough rules [22] that cover only M—L—M spin exchanges. Use of Goodenough rules often led to spin lattices that are inconsistent with the electronic structures of the magnetic systems they are supposed to describe, to find that Goodenough rules are not adequate enough. The reason for this observation is that M—L—M spin exchanges are frequently much weaker than those spin exchanges not covered by Goodenough rules, namely, M—L$\cdots$L—M and/or M—L$\cdots$A^{y+}$\cdots$L—M spin exchanges. This is understandable, because Goodenough rules were formulated in the mid 1950s, when the magnetic orbitals of M ions were regarded as their singly occupied pure d-orbitals of M. The importance of M—L$\cdots$L—M and/or M—L$\cdots$A^{y+}$\cdots$L—M spin exchanges were recognized only in the late 1990s and the early 2000s, when it was realized [1,2] that the strengths of spin exchanges are not governed by the metal d-orbital components, but by the ligand p-orbital components of the magnetic orbitals of ML$_n$. Quantitative evaluations of M—L—M, M—L$\cdots$L—M, and M—L$\cdots$A^{y+}$\cdots$L—M spin exchanges became possible by the energy-mapping analysis [1–4] based on DFT + U calculations developed in the early 2000s. This quantitative analysis helps one find, for any magnetic system, the spin lattice consistent with its electronic structure.

The spin–orbit Mott insulating states of the 5d oxides Sr_3NiIrO_6, Sr_2IrO_4, and Na_2IrO_3 as well as Ba_2NaOsO_6 are well explained by analyses based on the LS-coupling scheme of SOC. Furthermore, their observed magnetic anisotropies are better explained by the LS scheme rather than by the jj scheme. Consequently, the spin–orbital entanglement invoked for 5d elements is not as strong as has been put forward [81]. These observations are in support of the view that SOC for 5d elements lies in between the LS- and jj-coupling schemes, but is closer to the LS-coupling scheme [98].

A magnetic ion has a preferred spin orientation because SOC induces interactions among its crystal field split d-states and because the associated energy lowering depends on the spin orientation. The preferred spin orientation of a magnetic ion is readily predicted on the basis of the selection rule involving the SOC-induced HOMO–LUMO interaction. In the electronic structure description of a magnetic ion, each of its states has both orbital and spin components, that is, each state is represented by a set of orbital/spin states $|L, L_z\rangle|S, S_z\rangle$. The states of a magnetic ion are modified by SOC, $\lambda \hat{S} \cdot \hat{L}$, because it induces intermixing between them, according to Eq. (10.2). This explains why a magnetic ion has magnetic anisotropy regardless of whether its spin is 1/2 or not. A spin Hamiltonian analysis fails to explain this fundamental result because it represents each magnetic state in terms of only spin states $|S, S_z\rangle$. The effects of SOC, $\lambda \hat{S} \cdot \hat{L}$, can be included into a spin Hamiltonian only indirectly by using the zero-field Hamiltonian $\hat{H}_{zf}$ (Eq. (10.16)). This Hamiltonian does not allow one to predict the preferred spin orientation for $S > 1/2$ ions, although it shows the presence of magnetic anisotropy arising from SOC for such ions in agreement with experiment. As for the $S=1/2$ ions, however, this Hamiltonian is downright incorrect because it predicts the absence of magnetic anisotropy induced by SOC, $\lambda \hat{S} \cdot \hat{L}$, not to mention that it cannot predict their preferred spin orientation.

It is high time for the proponents of the spin-half misconception to recognize this shortcoming of a spin Hamiltonian analysis. Nevertheless, we are not unaware of the astute observation by Max Planck: "A new scientific truth does not triumph by convincing its opponents and making them see the light, but rather because its opponents eventually die and a new generation grows up that is familiar with it [99]." This observation is more explicitly paraphrased as "Death is an essential element in the progress of science, since it takes care of conservative scientists of a previous generation reluctant to let go of an old, fallacious theory and embrace a new and accurate one [100]." The debate on the spin-half misconception, which has just begun [5,6,9], is certainly not as grand and epochal as that on the earth- versus sun-centered model of the planetary motion, the single galaxy versus multigalaxy universe, or the classical versus quantum theory in the past, but unmistakable parallels exist between them. It is our hope that the readers of this chapter will have an open-minded view on magnetism and avoid falling into such a conceptual trap as the spin-half misconception.

Acknowledgments

This research used resources of the National Energy Research Scientific Computing Center, a DOE Office of Science User Facility supported by the Office of Science of the US Department of Energy under Contract No. DE-AC02-05CH11231. Work at Fudan was supported by NSFC (11374056), the Special Funds for Major State Basic Research (2012CB921400, 2015CB921700), Program for Professor of Special Appointment (Eastern Scholar), Qing Nian Bo Jian Program, and Fok Ying Tung Education Foundation. MHW would like to thank Dr. Reinhard K. Kremer, Prof. Hyun-Joo Koo, Dr. Changhoon Lee, and Elijah E. Gordon for their invaluable discussions over the years.

References

1 Whangbo, M.-H., Koo, H.-J., and Dai, D. (2003) *J. Solid State Chem.*, **176**, 417.
2 Xiang, H.J., Lee, C., Koo, H.-J., Gong, X.G., and Whangbo, M.-H. (2013) *Dalton Trans.*, **42**, 823.
3 (a) Dai, D. and Whangbo, M.-H. (2001) *J. Chem. Phys.*, **114**, 2887; (b) Dai, D. and Whangbo, M.-H. (2003) *J. Chem. Phys.*, **118**, 29.
4 Xiang, H.J., Kan, E.J., Wei, S.-H., Whangbo, M.-H., and Gong, X.G. (2011) *Phys. Rev. B*, **84**, 224429.
5 Whangbo, M.-H., Gordon, E.E., Xiang, H.J., Koo, H.-J., Lee, C., and Whangbo, M.-H. (2015) *Acc. Chem. Res.*, **48**, 3080.

6 Gordon, E.E., Xiang, H.J., Köhler, J., and Whangbo, M.-H. (2016) *J. Chem. Phys.*, **144**, 114706.
7 Moriya, T. and Yoshida, K. (1953) *Prog. Theor. Phys.*, **9**, 663.
8 (a) Yoshida, K. (1996) *Theory of Magnetism*, Springer, New York, pp. 34–37; (b) Skomski, R. (2008) *Simple Models of Magnetism*, Oxford University Press, New York, pp. 84–104; (c) White, R.M. (2007) *Quantum Theory of Magnetism*, 3rd edn, Springer, Berlin, pp. 211–212 (d) Majlis, N. (2007) *The Quantum Theory of Magnetism*, 2nd edn, World Scientific, New Jersey, pp. 23–24; (e) Khomskii, D.I. (2014) *Transition Metal Compounds*, Cambridge University Press, Cambridge, Section 5.3.1.
9 Liu, J., Koo, H.-J., Xiang, H.J., Kremer, R.K., and Whangbo, M.-H. (2014) *J. Chem. Phys.*, **141**, 124113.
10 Dai, D., Xiang, H.J., and Whangbo, M.-H. (2008) *J. Comput. Chem.*, **29**, 2187.
11 Albright, T.A., Burdett, J.K., and Whangbo, M.-H. (2013) *Orbital Interactions in Chemistry*, 2nd edn, John Wiley & Sons, Ltd, New York.
12 Orgel, L.E. (1969) *An Introduction to Transition Metal Chemistry*, John Wiley & Sons, Ltd, New York, p. 174.
13 (a) Dudarev, S.L., Botton, G.A., Savrasov, S.Y., Humphreys, C.J., and Sutton, A.P. (1998) *Phys. Rev. B*, **57**, 1505; (b) Liechtenstein, A.I., Anisimov, V.I. and Zaanen, J. (1995) *Phys. Rev. B*, **52**, 5467.
14 Heyd, J., Scuseria, G.E., and Ernzerhof, M. (2003) *J. Chem. Phys.*, **118**, 8207.
15 Kuneš, J., Novák, P., Schmid, R., Blaha, P., and Schwarz, K. (2001) *Phys. Rev. B*, **64**, 153102.
16 (a) Kresse, G. and Hafner, J. (1993) *Phys. Rev. B*, **47**, 558; (b) Kresse, G. and Furthmüller, J. (1996) *J. Comput. Mater. Sci.*, **6**, 15; (c) Kresse, G. and Furthmüller, J. (1996) *Phys. Rev. B*, **54**, 11169.
17 Blaha, P., Schwarz, K., Madsen, G.K.H., Kvasnicka, D., and Luitz, J. (2001) *WIEN2k*, Vienna University of Technology, ISBN 3-9501031-1-2.
18 Perdew, J.P., Burke, K., and Ernzerhof, M. (1996) *Phys. Rev. Lett.*, **77**, 3865–3868.
19 Hay, P.J., Thibeault, J.C., and Hoffmann, R. (1975) *J. Am. Chem. Soc.*, **97**, 4884.
20 Noodleman, L. (1981) *J. Chem. Phys.*, **74**, 5737.
21 Noodleman, L. and Davidson, E.R. (1986) *Chem. Phys.*, **109**, 131.
22 Goodenough, J.B. (1963) *Magnetism and the Chemical Bond*, Wiley-Interscience, New York.
23 Anderson, P.W. (1950) *Phys. Rev.*, **79**, 350.
24 Gibson, B.J., Kremer, R.K., Prokofiev, A.V., Assmus, W., and McIntyre, G.J. (2004) *Physica B*, **350**, e253.
25 Dai, D., Koo, H.-J., and Whangbo, M.-H. (2004) *Inorg. Chem.*, **43**, 4026.
26 Koo, H.-J., Lee, C., Whangbo, M.-H., McIntyre, G.J., and Kremer, R.K. (2011) *Inorg. Chem.*, **50**, 3582.
27 Koo, H.-J. and Whangbo, M.-H. (2006) *Inorg. Chem.*, **45**, 4440.
28 Lee, C., Kang, J., Lee, K.H., and Whangbo, M.-H. (2009) *Inorg. Chem.*, **48**, 4185.
29 Pryce, M.H.L. (1950) *Proc. Phys. Soc (London) A*, **63**, 25.
30 Dzyaloshinskii, I. (1958) *J. Phys. Chem. Solids*, **4**, 241.
31 Moriya, T. (1960) *Phys. Rev. Lett.*, **4**, 228.
32 Kramers, H.A. (1930) *Proc. R. Acad. Sci. Amsterdam*, **33**, 959.
33 Li, Z.L., Yang, J.H., Chen, G.H., Whangbo, M.-H., Xiang, H.J., and Gong, X.G. (2012) *Phys. Rev. B*, **85**, 054426.
34 Lu, X.Z., Whangbo, M.-H., Dong, S.A., Gong, X.G., and Xiang, H.J. (2012) *Phys. Rev. Lett.*, **108**, 187204.
35 Yang, J.H., Li, Z.L., Lu, X.Z., Whangbo, M.-H., Wei, S.-H., Gong, X.G., and Xiang, H.J. (2012) *Phys. Rev. Lett.*, **109**, 107203.
36 Dai, D. and Whangbo, M.-H. (2005) *Inorg. Chem.*, **44**, 4407.
37 Wang, X., Wu, R., Wang, D.-S., and Freeman, A.J. (1996) *Phys. Rev. B*, **54**, 61.
38 Kim, H.H., Yu, I.H., Kim, H.S., Koo, H.-J., and Whangbo, M.-H. (2015) *Inorg. Chem.*, **54**, 4966.
39 (a) Zhang, Y., Xiang, H.J. and Whangbo, M.-H. (2009) *Phys. Rev. B*, **79**, 054432; (b) Zhang, Y., Kan, E.J., Xiang, H.J., Villesuzanne, A., and Whangbo, M.-H. (2011) *Inorg. Chem.*, **50**, 1758.
40 (a) Zubkov, V.G., Bazuev, G.V., Tyutyunnik, A.P., and Berger, I.F. (2001) *J. Solid State Chem.*, **160**, 293; (b) Wu,

H., Burnus, T., Hu, Z., Martin, C., Maignan, A., Cezar, J.C., Tanaka, A., Brookes, N.B., Khomskii, D.I., and Tjeng, L.H. (2009) *Phys. Rev. Lett.*, **102**, 026404.

41 Reiff, W.M., LaPointe, A.M., and Witten, E.H. (2004) *J. Am. Chem. Soc.*, **126**, 10206.

42 Kabbour, H., David, R., Pautrat, A., Koo, H.-J., Whangbo, M.-H., André, G., and Mentré, O. (2012) *Angew. Chem., Int. Ed.*, **51**, 11745.

43 Song, Y.-J., Lee, K.-W., and Pickett, W.E. (2015) *Phys. Rev. B*, **92**, 125109.

44 Sujimoto, Y., Tassel, C., Hayashi, N., Watanabe, T., Kageyama, H., Yoshimura, K., Takano, M., Ceretti, M., Ritter, C., and Paulus, W. (2007) *Nature*, **450**, 1062.

45 Kageyama, H., Watanabe, T., Tsujimoto, Y., Kitada, A., Sumida, Y., Kanamori, K., Yoshimura, K., Hayashi, N., Muranaka, S., Takano, M., Ceretti, M., Paulus, W., Ritter, C., and Gilles, A. (2008) *Angew. Chem., Int. Ed.*, **47**, 5740.

46 Xiang, H.J., Wei, S.-H., and Whangbo, M.-H. (2008) *Phys. Rev. Lett.*, **100**, 167207.

47 Koo, H.-J., Xiang, H.J., Lee, C., and Whangbo, M.-H. (2009) *Inorg. Chem.*, **48**, 9051.

48 Blasco, J., Ritter, C., Garcia, J., de Teresa, J.M., Perez-Cacho, J., and Ibarra, M.R. (2000) *Phys. Rev. B*, **62**, 5609.

49 Chang, F.M. and Jansen, M. (1983) *Z. Anorg. Allg. Chem.*, **507**, 59.

50 Ji, S., Kan, E.J., Whangbo, M.-H., Kim, J.-H., Qiu, Y., Matsuda, M., Yoshida, H., Hiroi, Z., Green, M.A., Ziman, T., and Lee, S.-H. (2010) *Phys. Rev. B*, **81**, 094421.

51 Xiang, H.J., Wei, S.-H., Whangbo, M.-H., and Da Silva, J.L.F. (2008) *Phys. Rev. Lett.*, **101**, 037209.

52 Claridge, J.B., Layland, R.C., Henley, W.H., and zur Loye, H.-C. (1999) *Chem. Mater.*, **11**, 1376.

53 Shimora, T., Inaguma, A., Nakamura, T., Itoh, M., and Morii, Y. (1995) *Phys. Rev. B*, **52**, 9143.

54 Ye, F., Chi, S.X., Chakoumakos, B.C., Fernandez-Baca, J.A., Qi, T.F., and Cao, G. (2013) *Phys. Rev. B*, **87**, 140406 (R).

55 Lovesey, S.W. and Khalyavin, D.D. (2014) *J. Phys. Condens. Matter*, **26**, 322201.

56 Liu, P., Khmelevskyi, S., Kim, B., Marsman, M., Li, D., Chen, X.-Q., Sarma, D.D., Kresse, G., and Franchini, C. (2015) *Phys. Rev. B*, **92**, 054428.

57 Choi, S.K., Coldea, R., Kolmogorov, A.N., Lancaster, T., Mazin, I.I., Blundell, S.J., Radaelli, P.G., Singh, Y., Gegenwart, P., Choi, K.R., Cheong, S.-W., Baker, P.J., Stock, C., and Taylor, J. (2012) *Phys. Rev. Lett.*, **108**, 127204.

58 Ye, F., Chi, S., Cao, H., Chakoumakos, B.C., Fernandez-Baca, J.A., Custelcean, R., Qi, T.F., Korneta, O.B., and Cao, G. (2012) *Phys. Rev. B*, **85**, 180403(R).

59 Kim, H.-J., Lee, J.-H., and Cho, J.-H. (2014) *Sci. Rep.*, **4**, 5253.

60 Chun, S.H., Kim, J.-W., Kim, J., Zheng, H., Stoumpos, C.C., Malliakas, C.D., Mitchell, J.F., Mehlawat, K., Singh, Y., Choi, Y., Gog, T., Al-Zein, A., Sala, M.M., Krisch, M., Chaloupka, J., Jackeli, G., Khaliullin, G., and Kim, B.J. (2015) *Nat. Phys.*, **11**, 3322.

61 Hu, K., Wang, F., and Feng, J. (2015) *Phys. Rev. Lett.*, **115**, 167204.

62 Brownstein, S., Han, N.F., Gabe, E.J., and le Page, Y. (1989) *Kristallogr.*, **189**, 13.

63 Poulis, N.J. and Haderman, G.E.G. (1952) *Physica*, **18**, 201.

64 Banks, M.G., Kremer, R.K., Hoch, C., Simon, A., Ouladdiaf, B., Broto, J.-M., Rakoto, H., Lee, C., and Whangbo, M.-H. (2009) *Phys. Rev. B*, **80**, 024404.

65 Zhao, L., Hung, T.-L., Li, C.-C., Chen, Y.-Y., Wu, M.-K., Kremer, R.K., Banks, M.G., Simon, A., Whangbo, M.-H., Lee, C., Kim, J.S., Kim, I.G., and Kim, K.H. (2012) *Adv. Mater.*, **24**, 2469.

66 Koo, H.-J., Lee, C., Whangbo, M.-H., McIntyre, G.J., and Kremer, R.K. (2011) *Inorg. Chem.*, **50**, 3582.

67 Nguyen, T.N. and zur Loye, H.-C. (1995) *J. Solid State Chem.*, **117**, 300.

68 Lefrançois, E., Chapon, L.C., Simonet, V., Lejay, P., Khalyavin, D., Rayaprol, S., Sampathkumaran, E.V., Ballou, R., and Adroja, D.T. (2014) *Phys. Rev. B*, **90**, 014408.

69 Haghighirad, A.A., Ritter, F., and Assmus, W. (2008) *Cryst. Growth Des.*, **8**, 1961.

70 Knoke, G.T., Niazi, A., Hill, J.M., and Johnston, D.C. (2007) *Phys. Rev. B*, **76**, 054439.

71 Xiang, H.J., Kan, E.J., Whangbo, M.-H., Lee, C., Wei, S.-H., and Gong, X.G. (2011) *Phys. Rev. B*, **83**, 174402.

72 Pan, B., Wang, Y., Zhang, L., and Li, S. (2014) *Inorg. Chem.*, **53**, 3606.

73 Lee, C., Hong, J., Son, W.J., Kan, E.J., Shim, J.H., and Whangbo, M.-H. (2016) *RSC Adv.*, **6**, 22722.

74 McGuire, M.A., Dixit, H., Cooper, V.R., and Sales, B.C. (2015) *Chem. Mater.*, **27**, 612.

75 Koo, H.-J., and Köhler, J., and Whangbo, M.-H., unpublished work.

76 Vasala, S., Yamauchi, H., and Karppinen, M. (2014) *J. Solid State Chem.*, **220**, 28.

77 Russell, D.D., Neer, A.J., Melot, B.C., and Derakhshan, S. (2016) *Inorg. Chem.*, **55**, 2240.

78 Lines, M.E. (1963) *Phys. Rev.*, **131**, 546.

79 Lee, C., and Koo, H.-J., and Whangbo, M.-H., unpublished work.

80 Xiang, H.J. and Whangbo, M.-H. (2007) *Phys. Rev. B*, **75**, 052407.

81 (a) Kim, B.J., Jin, H., Moon, S.J., Kim, J.-Y., Park, B.-G., Leem, C.S., Yu, J., Noh, T.W., Kim, C., Oh, S.-J., Park, J.-H., Durairaj, V., Cao, G., and Rotenberg, E. (2008) *Phys. Rev. Lett.*, **101**, 076402; (b) Kim, B.J., Ohsumi, H., Komesu, T., Sakai, S., Morita, T., Takagi, H., and Arima, T. (2009) *Science*, **323**, 1329.

82 Zhang, G.R., Zhang, X.L., Jia, T., Zeng, Z., and Lin, H.Q. (2010) *J. Appl. Phys.*, **107**, 09E120.

83 Song, Y.-J., Lee, K.-W., and Pickett, W.E. (2015) *Phys. Rev. B*, **92**, 125109.

84 Lefrançois, E., Chapon, L.C., Simonet, V., Lejay, P., Khalyavin, D., Rayaprol, S., Sampathkumaran, E.V., Ballou, R., and Adroja, D.T. (2014) *Phys. Rev. B*, **90**, 014408.

85 Wu, W., Adroja, D.T., Toth, S., Rayaprol, S., and Sampathkumaran, E.V. doi: arXiv:1501.05735.

86 Ye, F., Chi, S.X., Chakoumakos, B.C., Fernandez-Baca, J.A., Qi, T.F., and Cao, G. (2013) *Phys. Rev. B*, **87**, 140406(R).

87 Lovesey, S.W. and Khalyavin, D.D. (2014) *J. Phys. Condens. Matter*, **26**, 322201.

88 Liu, P., Khmelevskyi, S., Kim, B., Marsman, M., Li, D., Chen, X.-Q., Sarma, D.D., Kresse, G., and Franchini, C. (2015) *Phys. Rev. B*, **92**, 054428.

89 Comin, R., Levy, G., Ludbrook, B., Zhu, Z.-H., Veenstra, C.N., Rosen, J.A., Singh, Y., Gegenwart, P., Stricker, D., Hancock, J.N., van der Marel, D., Elfimov, I.S., and Damascelli, A. (2012) *Phys. Rev. Lett.*, **109**, 266406.

90 Gretarsson, H., Clancy, J.P., Liu, X., Hill, J.P., Bozin, E., Singh, Y., Manni, S., Gegenwart, P., Kim, J., Said, A.H., Casa, D., Gog, T., Upton, M.H., Kim, H.-S., Yu, J., Katukuri, V.M., Hozoi, L., van den Brink, J., and Kim, Y.-J. (2013) *Phys. Rev. Lett.*, **110**, 076402.

91 Kim, H.-J., Lee, J.-H., and Cho, J.-H. (2014) *Sci. Rep.*, **4**, 5253.

92 Slater, J.C. (1951) *Phys. Rev.*, **82**, 538.

93 Bogdanov, N.A., Katukuri, V.M., Romhányi, J., Yushankhai, V., Kataev, V., Büchner, B., van den Brink, J., and Hozoi, L. (2015) *Nat. Commun.*, **6**, 7306.

94 Hu, K., Wang, F., and Feng, J. (2015) *Phys. Rev. Lett.*, **115**, 167204.

95 Zhang, G.R., Zhang, X.L., Jia, T., Zeng, Z., and Lin, H.Q. (2010) *J. Appl. Phys.*, **107**, 09E120.

96 Ou, X.D. and Wu, H. (2014) *Sci. Rep.*, **4**, 4609.

97 Sarkar, S., Kanungo, S., and Saha-Dasgupta, T. (2010) *Phys. Rev. B*, **82**, 235122.

98 Khomskii, D.I. (2014) *Transition Metal Compounds*, Cambridge University Press, Cambridge, p. 34.

99 Planck, M.K. (1950) *Scientific Autobiography and Other Papers*, Philosophical Library, New York, p. 33.

100 Singh, S. (2004) *Big Bang*, HaperCollins Publishers, New York, p. 75.

11
Basic Properties of Well-Known Intermetallics and Some New Complex Magnetic Intermetallics

Peter Entel

Universität Duisburg-Essen, Fakultät für Physik, Lotharstrasse 1, 47048 Duisburg, Germany

11.1
Introduction and Hume-Rothery and Frank–Kasper Phases

Intermetallic compounds are chemical compounds of metals; they are materials of two or more metal atoms with predominantly metallic bonding, whereby transition metals with nonmetals may often be considered to border the intermetallic compounds or alloys. The crystal structure that forms is usually different from the crystal structures of the individual constituents and the bonding between the constituents is conventionally stronger than between the constituents themselves. The emerging crystal structures can be quite complex and may consist of a network of atomic clusters. By complex we mean that it may be a difficult task to determine the structure of an intermetallic compound experimentally or theoretically by accompanying *ab initio* band structure calculations (the β-phase in Mg–Al equilibrium system with many icosahedral subunits in the cubic unit cell of β-Al_3Mg_2 may be taken as an example, although Al and Mg are fairly simple metals). We will not pursue this in greater detail, nor do we accentuate here on compound behavior with stoichiometric ordering as distinguished from atomic disorder effects or alloy type of material, but, refer instead to the existing literature, for example, see Ref. [1,2].

Besides a collection of experimental and theoretical aspects in this section (Hume-Rothery and Frank–Kaspar phases), we summarize a few selected properties of Laves phase intermetallics in Section 11.2, of Ni–Al-based intermetallics in Section 11.3, and a few important new results of recent *ab initio* calculations in Section 11.4 regarding Ti–Ni alloys and also discuss martensitically transforming Ti–Ni–X alloys in Section 11.5 and their chemical bonding features in the subsequent Section 11.6. Finally, Sections 11.7 and 11.8 deal with the complex magnetic features of ternary magnetic Heusler intermetallics, which may be compared with some of the properties of the intermetallics discussed in the preceding sections. The term complex magnetic intermetallics refers to the many

Handbook of Solid State Chemistry, First Edition. Edited by Richard Dronskowski, Shinichi Kikkawa, and Andreas Stein.
© 2017 Wiley-VCH Verlag GmbH & Co. KGaA. Published 2017 by Wiley-VCH Verlag GmbH & Co. KGaA.

different magnetic states that due to composition, temperature, pressure, and external magnetic field can be realized in the ternary Heusler alloys such as metamagnetism, mictomagnetism, ferro-, ferri-, as well as antiferromagnetism, and noncollinear antiferromagnetism in conjunction with martensitic transformation tendencies. In particular, resulting functional properties such as the so-called inverse magnetocaloric effect are addressed, which relies on the strong interplay of the martensitic transformation and the metamagnetic phase transition (from the paramagnetic or antiferromagnetic low-temperature tetragonally distorted structure to the high-temperature ferromagnetic cubic structure).

So, important questions of materials science include the following: (i) How can we understand the driving forces of the atomic arrangements of the complex interesting systems? (ii) How do different elements distribute themselves in the structure? (iii) How can we predict the effects of valence electron concentration on structure and magnetic ordering and the influence of external magnetic fields? These cannot be answered in a straightforward manner for the intermetallics because their properties can be so different from their separate, elemental structures. All calculations presented in this text are based on highly precise up-to-date band structure calculations and comparison to experiment is made where possible as discussed in the text. In addition, regarding the calculation of functional properties such as shape memory and magnetocaloric properties, we have employed model Hamiltonians based on the Heisenberg and Potts model by using input parameters determined from *ab initio* calculations such as the magnetic coupling constants. This is defined further below.

So, in conclusion, a general classification scheme of intermetallics is not available; nevertheless, for historical reasons we mention that intermetallics involving Cu, Ag, and Au as well as transition metals with Be, Mg, Zn, Cd, Hg, Al, Ge, Sn, and Sb can be typified following the so-called Hume-Rothery rules [3]. Hume-Rothery showed in 1926 that the composition of these compounds is determined by the total number of valence electrons concentration e/a in the unit cell: At $e/a = 3/2$ β-phases with a bcc lattice, for $e/a = 21/13$ γ-phases with an fcc lattice, and for $e/a = 7/4$, hcp ε-phases are formed. (Note that in the discussion of ternary Heusler systems in Section 11.7, the valence electron concentration consists of the sum of s–, p–, and d-electrons, for example, see the phase diagrams shown in Figure 11.13.)

Hume-Rothery rules are still widely used, especially, when discussing (Frank–Kasper-type Al–Mg–Zn) approximants of quasicrystals using *ab initio* FLAPW (full-potential linearized augmented plane wave) electronic structure calculations [4]. (Here, d-electrons are several electronvolts below the Fermi energy and the stability of, for example, γ-brasses such as Cu_5Zn_8 (space group $I\bar{4}3m$) and Cu_9AQl_4 ($P\bar{4}3m$) with bcc and CsCl structure, respectively, can be discussed in terms of the pseudogap formation across the Fermi level ε_F, which originates from orbital hybridization and the Fermi surface–Brillouin zone interaction [4]. Frank–Kasper phases are topologically close-packed phases and are one of the largest groups of intermetallic compounds and because of their combination of periodic and aperiodic structures [5–7], it may be stated that Frank–

Kasper phases and the classes of quasicrystals and metallic glasses share some common features, when considering quasicrystals and so on as a "new class of ordered structures." Although the lattice contrasts with the conventional (crystalline) Frank–Kasper phases, it has been suggested that the glass transition occurs as a supercooled liquid approach to a Frank–Kasper phase [8–10].

The most common members of Frank–Kasper phases are the A15 compounds [11] (also known as β-W or A3B structure type where A is a transition metal and B any other element, such as Nb_3Sn showing superconductivity at 18.3 K) and Laves phases CN12–CN16. As incidental remark, we mention the intermetallic A15 phase superconductor Nb_3Sn, which is widely used in industry, as a type II superconductor around 4.2 K, the boiling point of liquid helium. It can withstand magnetic field intensity values up to 30 T and supports a record noncopper current of 2634 A/mm^2 at 4.2 K and has been proposed in 1986 for the Large Hadron Collider at CERN. Because of its brittleness, sometimes a combination of Nb_3Sn for the high-field sections and NbTi for the low-field sections is used. In 1990, the development work on Nb_3Sn was stopped in favor of the much more practical Nb–Ti material[1]. Probably high-temperature superconductors of type BSCOO and YBCO will be useful for next-generation accelerators. For a survey of applications of superconducting materials, see, for example, Refs [12,13]. For research associated with the LHC at CERN and some historical facts, see Ref. [14].

11.2
Laves Phase Intermetalllics and Ni– and Ti-Based Superalloys

We include in the discussion of intermetallic compounds a few remarks on Laves phase intermetallics that are mostly brittle and as is the case for the magnetic Heusler alloys to be discussed below, where stacking faults and kinetic arrest phenomena associated with martensitic formation hinder easy dislocation movement. This makes processing and workability of many binary and ternary intermetallics difficult. Remarks on Ti–Ni systems in this section and also in the next section, specifically their shape memory properties, find their corresponding magnetic counterparts in the magnetic Heusler alloys. Although magnetism in some Laves phases (e.g., Co–Cr–Nb or Fe–Nb) is a highly fascinating subject, this will not be pursued here; we refer the reader to the literature on first-principles calculations (and the semiempirical CAPHAD approach).

Laves showed in 1934 that atomic size-specific relationships are important: If the ratio of the atomic radii in AB_2 phase becomes $r_A/r_B = 1.1 - 1.3$, very compact structures can be formed with coordination numbers 12 and 16 and an ordered arrangement of atoms [15,16]. The Laves phases $MgCu_2$, $MgZn_2$, $MgNi_2$, and alike structural type form the most numerous group of intermetallics [1] and include about two-thirds of all intermetallic compounds. There are

---
1) CERN Courier (October 25, 2011).

several hundreds of Laves phases of AB_2 type reported so far, with three typical polytypes observed: cubic C15 ($MgCu_2$), hexagonal C14 ($MgZn_2$), and dihexagonal C36 ($MgNi_2$) (see images.google.de/imgres?imgurl=http) [17].

For the hexagonal Laves phase $MgZn_2$ (C14), recent *ab initio* calculations have been performed [18]. $MgZn_2$ crystallizes in the space group $P6_3/mmc$ (No. 194). The parameters of the primitive hexagonal unit cell are $a = 5.223$ and $c = 8.556$, $c/a = 1.62$ is almost identical to the ideal value of $c/a = (8/3)^{1/3} = 1.63$, it contains $Z = 4$ formula units. Andrae showed that among six different exchange–correlation energy density functionals employed in full optimization of the crystal structure, the GGA functional of Perdew, Burke, and Enzerhof (PBE) gave results closest to experiment (errors in cell parameters a and c below 0.6%).

Since Mg alloys find application where strength-to-weight ratio is advantageous, there is still much work going on regarding preaging deformation, precipitation, and structural transformation tendencies (see, for example, Ref. [19]), as well as corrosion chemistry and protection of zinc alloys. First-principles calculations underline the importance of this materials research [20,21]. From the calculation of elastic moduli of Laves phase $MgZn_2$ and related materials being larger than in Mg, it follows that directional bonding (larger shear modulus) and mechanical properties are improved after alloying.

Bonding features and glassy tendencies of binary Laves phases (search for structure induced minimum in the electronic density of states at the Fermi level) can best be described as a disordered tetrahedral close packing with a weak tendency to chemical short-range order in "amorphous" supercells of 60 atoms [22]. The minimum in the electronic states at ε_F exists not only in the amorphous phase but also for the crystalline structures and the stability of the crystalline structures is ensured by the fact that the interatomic distances to the first, second, and even more distant neighbors fall into the minima of the pair potentials, too. This means that at the composition of $MgZn_2$ the high stability of the crystalline compound suppresses the formation of a real metallic glass [22]. See also various aspects of metallic bonding discussed by Hafner and Pettifor [23,24]. Special hybridization in other alloys such as AlV_3 is associated with bandgap formation and covalent bond formation [25].

In the Laves phases but also in many other intermetallics, we encounter sometimes very complex bonding conditions because of the second (or third) element, which may lead to mixed bonding by reducing metal bonds and creating in part covalent and ionic bonds (if large electronegative differences between host and some impurity atoms occur), which is mostly responsible for thebrittleness of intermetallics, thereby reducing ductility, which hinders fabrication and easy deformation processes [26]. On the other hand, the higher toughness of intermetallics at elevated temperatures compared to elemental metals such as Fe is advantageous for applications, see Young's modulus of NiAl and TiAl compared to steel (Figure 11.1; also see Ref. [27]).

Nickel- and titanium-based superalloys are important in aircraft and gas turbine industry. Stoichiometric NiAl has a density of about $5.9\,g/cm^3$, which is about two-thirds of current Ni-based superalloys [28], and its high melting

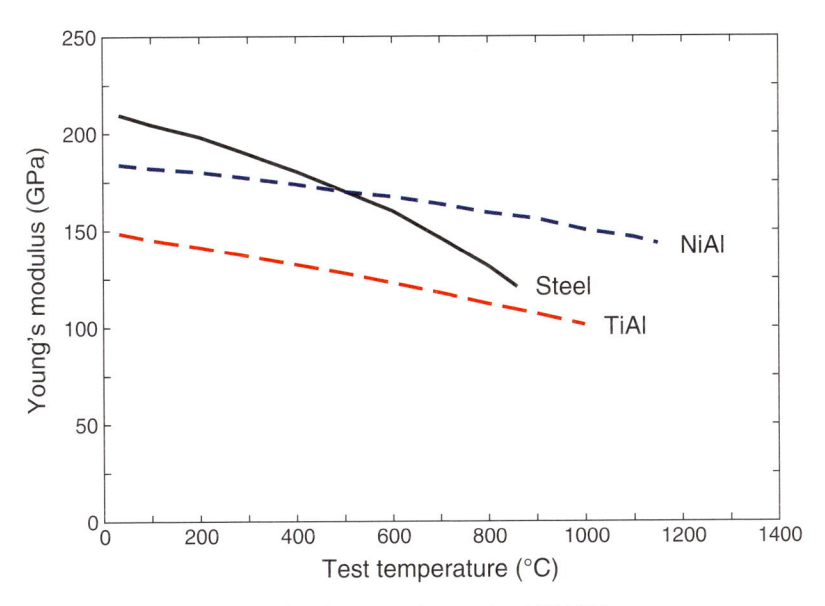

Figure 11.1 Elastic modulus of NiAl compared to steel and TiAl [27].

temperature of 1638 °C is important for technological applications. We will not pursue much further the discussion of superalloys here. Many structural, electromagnetic, chemical, and metallurgical applications can be found in Ref. [2].

11.3
Ni–Al-Based Intermetallics and Shape Memory Materials

NiAl- as well as CoAl- and FeAl-based intermetallics are well-known examples of B2 phase materials and form one of the largest groups of intermetallics [29]. NiAl with B2-type structure exhibits superior physical and high-temperature mechanical properties (for instance, high-temperature-resistant NiAl-based alloys with refractory metals Cr, Mo, and Re) and is therefore included here in a short discussion (Figure 11.1 shows the elastic modulus compared to steel and TiAl). Although very well known, we show in Figure 11.2 the crystal structure and phase diagram of NiAl with its rather extended compositional B2 phase range (adapted from Ref. [30]). The B2 CsCl-type crystal structure has a lattice constant of 2.887 Å at room temperature. If alloyed with a third element, we may well encounter Heusler-type structures that are discussed in Sections 11.7 and 11.8. Although it would be a good opportunity to discuss magnetism in B2-type intermetallics in general, we restrict the discussion of magnetism in this section to FeAl and pursue magnetism in relation to the functional properties of magnetic Heusler alloys in Sections 11.7 and 11.8.

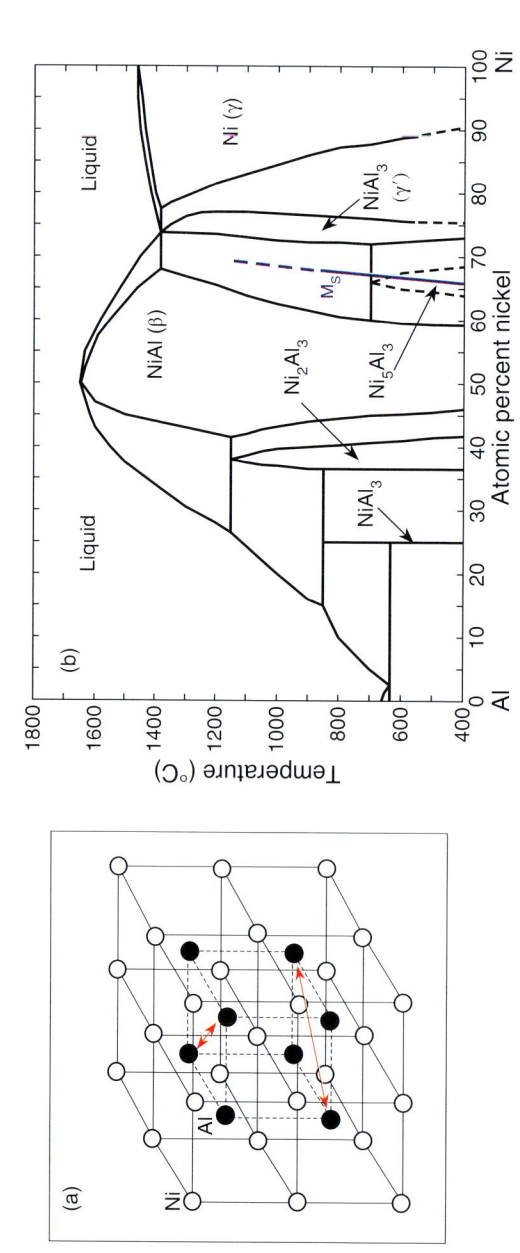

Figure 11.2 (a) The B2 CsCl-type crystal structure of NiAl with interpenetrating simple cubic Ni and Al sublattices. If the atoms marked by the arrows are replaced, for example, by Mn and Ga, respectively, the L2$_1$ Heusler structure of Ni$_2$MnGa is obtained (see Section 11.7). (b) Phase diagram of Ni–Al showing the dominant β-phase (around equiatomic composition) adapted from Ref. [30]. Quenching of Ni-rich NiAl alloys in a region centered around 64 at.% Ni from high temperatures (~1200 °C) leads to the formation of an ordered, fct (L1$_0$), twinned martensite with M_S data superimposed on the plot. The transformation in NiAl, satisfying prerequisites of shape memory behavior, is thermoelastic. Small additions of Ti, Cr, Mn, Fe, Nb, Mo, Ta, W, and Si to a binary 64at.%Ni36at.%Al alloy stabilize the parent B2 phase and lower M_S, while elements such as Co, Cu, and Ag increase the M_S. In particular, the Ni–Al–Mo system (Mo partially replacing Ni) has potential for applications at target temperatures of 1200–1300 °C (Mo has high melting temperature, room temperature toughness, and creep resistance). Alloys with Pd, Ir, and Rh would increase the melting temperature without forming detrimental intermetallic phases and would yield improved stability of the NiAl phase. The most common of the ternary intermetallic phases are the Heusler phases Ni$_2$AlX and the Laves phases NiAlX, which are, however, more brittle than NiAl.

NiAl with stoichiometric composition is highly ordered up to the melting temperature of 1649 °C, which is higher than those of the constituent elements. The high order is confirmed by *ab initio* calculations [31]. The authors also state that premartensitic phenomena observed in $Ni_{0.625}Al_{0.375}$ are Fermi surface driven. The lattice constant is a maximum for equiatomic composition [32]. For non-stoichiometric compositions constitutional disorder sets in, that is, antistructure atoms with excess Ni atoms on Al sites on the Ni-rich side and vacancies on the Al-rich side [1]. Magnetism in Ni–Al arises from magnetic clusters from antistructure Ni on the Al sites with $0.3\mu_B$ [33]. Likewise, experimental work shows that for CoAl and CoGa alloys the antistructure Co moment is $1\mu_B$ and the antistructure Fe in FeAl alloys has a moment of $2.2\mu_B$ [33]. Results of linearized muffin tin orbital (LMTO) and full potential linearized augmented plane wave (FLAPW) calculations for NiAl, Ni_3Al, FeAl, and Fe_3Al show that there exist a wide range of magnetic moments for which the changes in total energy are small (on the order of 100 meV [34]).

For samples with Ni compositions greater than ~62.8 at.%, binary $L1_0$ NiAl martensite with tetragonality $c/a = 0.86$ forms [35,36] (the M_S-data from Ref. [36] have been inserted in Figure 11.2), if decomposition and precipitation of Ni_3Al and/or Ni_5Al_3 can be avoided by annealing at high temperatures with rapid quenching to low temperatures [1]. In some cases, a monoclinic structure with rhombohedral 7R stacking was found depending on the Al content [37]. The 7R structure is intermediate in the sequence of the martensitic transformation and is derived from $L1_0$ by introducing periodic stacking faults with the energy barrier between B2 and $L1_0$ lowered by the formation of the intermediate 7R phase [38]. The martensitic transformation in Ni-rich NiAl can be induced by external applied stresses. It is thermoelastic and superelasticity has been observed, and the martensite deforms by twinning and detwinning and shows the shape memory effect [39]. The intriguing martensitic instability of the magnetic shape memory alloy Ni–Mn–Ga bears resemblance to the martensitic instability found in many β-phase materials (see, for example, Ref. [40]).

Regarding alloy design, Figure 11.3 shows various possibilities for alloying NiAl by adding elements with high solubility and little solubility in NiAl [41]. A short discussion of a hierarchical approach to alloy design involving Ni–Al as

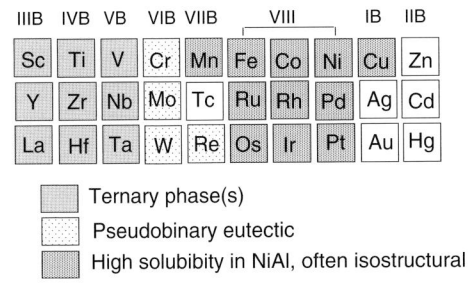

IIIB	IVB	VB	VIB	VIIB	VIII			IB	IIB
Sc	Ti	V	Cr	Mn	Fe	Co	Ni	Cu	Zn
Y	Zr	Nb	Mo	Tc	Ru	Rh	Pd	Ag	Cd
La	Hf	Ta	W	Re	Os	Ir	Pt	Au	Hg

▢ Ternary phase(s)
▢ Pseudobinary eutectic
▢ High solubility in NiAl, often isostructural

Figure 11.3 Portion of the periodic table of elements showing general alloying behavior of ternary additions to NiAl [41].

well as phase-partitioning and site substitution in Ni–Al–Mo superalloys can be found in Refs [42,43]. In particular, the Ni–Al–Mo system (Mo partially replacing Ni) has potential for applications at target temperatures of 1200–1300 °C. Mo has high melting temperature, room temperature toughness, and creep resistance). Alloys with Pd, Ir, and Rh would increase the melting temperature without forming detrimental intermetallic phases (strengthened mainly by dispersion of intermetallic particles, in the sense of steel strengthened by TiNi$_3$ η-phase precipitation) and would yield improved stability of the NiAl phase.

A large class of ternary intermetallic phases is formed by the Heusler phases Ni$_2$AlX and the Laves phases NiAlX, which are, however, more brittle than NiAl. Among the magnetic Heusler alloys showing the shape memory as well as the magnetocaloric effect, the pseudobinary Ni–Mn–(Al, Ga, In, Sn, Sb) alloys are the most well-known systems. [44–46]. The memory effect is related to the observation that, for example, stoichiometric Ni$_2$MnGa undergoes a structural transition from L2$_1$ parent phase to the tetragonal martensitic phase at around 200 K. The magnetic ordering in both parent and product phases is ferromagnetic and, in the martensitic phase, the application of an external magnetic field of about 1 T can induce a single-crystal reversible strain of about 9% [47] related to the field-induced martensitic variant rearrangement. This is known as the magnetic shape memory effect (MSME). The related strain has recently been pushed to 12% in a nonmodulated Ni$_{46}$Mn$_{24}$Ga$_{22}$Co$_4$Cu$_4$ martensite [48].

Besides Ni$_2$MnGa, Ni–Mn–Al also undergoes a martensitic transformation (at off-stoichiometric composition). However, the parent phase is either B2 or a mixture of B2 + L2$_1$ due to the lowered B2–L2$_1$ ordering temperature compared to Ni$_2$MnGa [45]. See the discussion of phase equilibria and stability of the B2 phase [44,49]. Also, the B2 phase shows mixed ferromagnetic (FM) and antiferromagnetic (AFM) nature with FM coupling in the L2$_1$ phase of Ni$_{50}$Mn$_{30}$Al$_{20}$ after annealing [45]. Ni$_2$MnAl shows similar coexisting FM and AFM ordering [50].

Another promising candidate for practical application using shape memory alloys is Cu–Al–Mn with excellent ductility, superelasticity, one- and two-way memory effect, respectively, and fatigue properties due to the addition of alloying elements, grain size, and texture [51]. The shape memory characteristics obtained in Table 11.1 are almost equal to those of Ni–Ti SMA (shape memory alloys).

Table 11.1 The shape memory characteristics of the Cu–Al–Mn alloys compared with those of Ni–Ti, Cu–Zn–Al and Cu–Al–Ni SMA.

Alloy	Transf. TT (°C)	CoW (%)	SE (%)	OWME (%)	TWME (%)
Cu–Al–Mn based	−200 to 150	>60	7.5	∼10	3.2
Ni–Ti	−40 to 100	∼30	8	8	5
Cu–Zn–Al	−200 to 120	∼30	2	5	2
Cu–Al–Ni	−200 to 170	∼10	2	5	—

TT: transformation temperature; CW: cold workability; SE: superelasticity; OWME and TWME are the one-way and two-shape memory effect, respectively.

Besides the many experimental and theoretical work on Ni–Al-based alloys using density functional theory, iron aluminides and related phases such as FeAl and Fe$_3$Al are of interest because of their outstanding mechanical properties (higher strength than comparable iron alloys and excellent oxidation and sulfidation properties). FeAl is closely related to NiAl, showing also the B2 structure and complete mutual miscibility [1]. The phase diagram of Fe–Al has been discussed by Massalski [52].

Regarding magnetism in the Fe-rich Fe–Al-based alloys, we find that small changes in the Fe content can induce large changes in its magnetic behavior, which can be linked to the intimate relationship between magnetism and microstructure. The magnetic exchange coupling constants for the ternary Fe–Mn–Al Heusler alloys have recently been investigated, which show strong nearest-neighbor ferromagnetic coupling in the cubic L2$_1$ phase [53]. Other transition metal elements give rise to exotic behavior (Ni–Co–Mn–Al [54]). Note that FeAl and its alloys show intriguing magnetic properties with ferromagnetism at room temperature (30.4 at.% Al alloy) and disappearance of spontaneous magnetization with cooling [55]. It was suggested that the transition is from ferromagnetism to antiferromagnetism resulting from indirect exchange interactions. However, no evidence for long-range antiferromagnetism was found and it was concluded that any magnetic moment that occurs in FeAl is associated with defects in the atomic ordering with "parasitic" antiferromagnetism [33], which was called "mictomagnetism" [55]. In summary, the mechanical and magnetic properties of Fe–Al intermetallic alloys strongly depend on the deviation from stoichiometry, whereby a third element can improve the ductility. Therefore, FeAl intermetallics can be viewed as a "test field" to test theories and hypothesis of fundamental magnetism and magnetostriction [56] and of the relation between magnetism and microstructure of B2; DO3 and disordered A2 phases on the Fe-rich side of the phase diagram of Fe–Al have been investigated in some detail [56]. We would like to add that metamagnetism, spin glass, cluster glass, mictomagnetism, and reentrant spin glass have also been assigned to characterize the strong disorder-influenced magnetism in FeAl-based alloys. Since metamagnetism is usually used to define a magnetically ordered state that shows zero magnetism under zero magnetic field but abrupt increasing magnetization at a critical magnetic field, mictomagnetism is better suited to define a spin system with various exchange interactions due to disorder as in Cu–Mn, Fe–Al, and Ni–Mn [57].

Beside *ab initio* calculations, extensive molecular dynamics simulations with adequate interatomic potentials have been used to describe the structural martensitic changes in intermetallics. We just cite here two prominent results of such simulations. In the first example, the embedded atom method (EAM) has been used to describe the martensitic transformation in Ni–Al alloys. In the EAM method, the total potential energy is discussed in Ref. [58] (which has later been improved to include the angular dependence of the host density to describe bond-bending forces in the modified embedded atom method (MEAM) [59]).

$$E_{\text{tot}} = \sum_i F_i(\rho_i) + \frac{1}{2} \sum_{i \neq j} \phi_{ij}(R_{ij}), \quad \rho_i = \sum_{j \neq i} \rho_j^{\text{at}}(r_{ij}). \tag{11.1}$$

Here, r_{ij} is the distance between atoms i and j. F_i is the so-called embedding function depending on the local electronic density ρ_i (as in density functional theory, one uses here that the energy of a system of arbitrarily distributed atoms can be expressed as a functional of the electronic density – in the approximation that the electronic density of the total system can be described by a superposition of local electronic densities). More precisely, the embedded charge density ρ_i between atoms i and j at the position of atom i is approximated by a superposition of element-dependent contributions $\rho_j^{at} = \rho_{Ni}^{at}, \rho_{Al}^{at}$ of the surrounding j atoms. ϕ_{ij} is the pair interaction potential function. The functions F_i and ϕ_{ij} are mostly determined by *ab initio* calculations and a fit to experimental data. For details see Refs [58,59].

In order to give an insight into the martensitic transformation in $Ni_{64}Al_{36}$ alloys, molecular dynamics simulations with 16^3 and 40^3 elementary cells were performed. Figure 11.4 shows resulting mean displacements of atomic positions from ideal positions of the atoms of the B2 lattice at 400 K. This is a typical precursor effect and gives rise to onset of twin boundary formation in the martensitic phase (see Figure 11.4) [60,61]. Impurities may destroy this long-range displacement texture leading to glassy behavior as shown in Figure 11.9.

Although magnetic degrees of freedom are difficult to retain when deriving many-body potential functions, the martensite–austenite transition and phonon dispersions of Fe–Ni intermetallics could successfully be simulated using EAM potentials [62,63] as well as phase field modeling going along with the molecular dynamics simulations [64]. The so-called extended Finnis–Sinclair potentials have been used for the simulation of bcc and fcc metals and alloys [65]. Since in the molecular dynamics simulation each translational degree of freedom contributes by $k_B/2$ to the total kinetic energy of the system, the temperature is fixed at $T = (2k_B/3N) \cdot E_{kin}$. Of interest is therefore the recent introduction of an effective interaction potential (EIP) where the temperature is included in the EIP model by

$$\phi(r, \Theta) = A \left\{ \exp\left[-2B\left(\frac{r}{\hat{r}(\Theta)} - 1\right)\right] - 2\exp\left[-B\left(\frac{r}{\hat{r}(\Theta)} - 1\right)\right] \right\}, \quad (11.2)$$

$$\hat{r}(\Theta) = [r_0 + r_\Theta(\Theta - 1)][t_0 + t_\Theta(\Theta - 1)], \quad (11.3)$$

where the Morse-type pair potential with constants A (which is related to the bond strength) and B (which is related to the bond stiffness) has been modified by introducing the absolute temperature T with respect to some reference temperature, $\Theta = T/T_{ref}$ [66], r is the distance between two atoms. In an even more general form, one has also tested the case where A and B have been used to depend on Θ [66]. Different forms for the function $\hat{r}(\Theta)$ have been tested. It seems that the quadratic form used in Eq. (11.3) provides the material model with enough flexibility to fit the thermal expansion coefficient of a given material. For biatomic crystals, the three bonds between the atoms require three pair

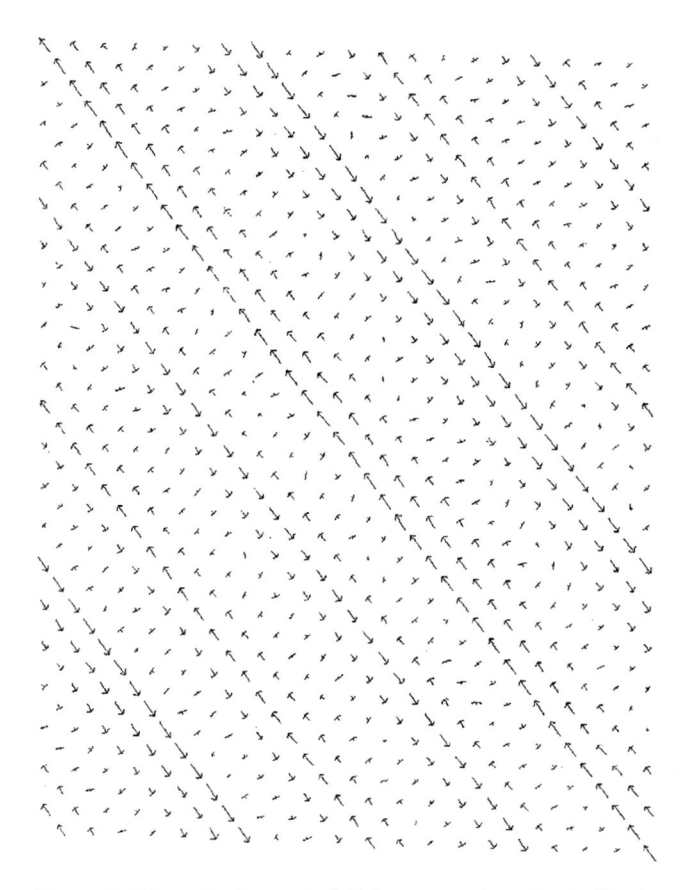

Figure 11.4 Mean displacements (×3) from ideal positions of the atoms of two (001) layers of $Ni_{64}Al_{36}$ at 400 K showing the appearance of static tweed structure in austenite B2 [60]. Since there are no defects present in the system, atomic disorder must be responsible for these deformations. Similar tweed structure have been observed in the simulations by Becquart et al. [61].

potentials with different fitting parameters, which allow an adequate description of the martensitic transformation [66]. For appropriate parameters, this model even allows a bifurcation type of description of B2 → αIrV → B19 transformation of Au–Cd [66]. Moreover, thermal expansion, bulk modulus, heat capacity, and entropy as a function of Θ for B2 $Au_{53}Cd_{47}$ could well be reproduced in agreement with experiment [66]. The advantage of such effective temperature-dependent pair interaction model is related to the possibility to capture the entire range of shape memory applications of microstructures. The method is computationally inexpensive compared to molecular dynamics, Monte Carlo, and density functional theory simulations.

Other recent developments include the derivation of bond order potentials, which comprise covalent bond formation, charge transfer, and may allow incorporating magnetism in future [67]. This would be a breakthrough in the simulation of magnetic materials.

Some 30 years ago, Car and Parrinello proposed a theory that unifies molecular dynamics method for the calculation of the atomic structure with local density approximation with the quantum mechanical many-body forces calculated using the Hellmann–Feynman theorem [68]. Assuming that the transfer of energy between the ionic and electronic subsystems is small, the electrons are kept in their ground-state configuration, while the trajectories of the ionic motion are subject to the Nosé–Hoover thermostat [69,70] controlling the ionic kinetic energy. While this transfer of energy is difficult to control in metallic systems, Kresse and Hafner have formulated an *ab initio* molecular dynamics method that allows treating in addition metallic systems [71]). The latter method would be really suited to treat the martensitic phase transformation fully quantum mechanically. Also, internal degrees of freedom such as magnetism could be automatically retained and their influence along the transformation path could be studied. However, the computational time to deal with large-enough system sizes and structural changes would still be too expensive.

The validity of such *ab initio* molecular dynamics procedures has been confined so far more to semiconductors and insulators with system sizes of about 100–150 atoms [72] than to intermetallics and structural phase transformations. There are other tools that have their own merits such as quantum Monte Carlo simulation, which is useful when dealing with growth processes and structural changes involving interfaces and surfaces. A further step to deal with structural changes would be to perform adiabatic *ab initio* molecular dynamics simulations in the linear response time-dependent density functional theory formalism (beyond standard *ab initio* molecular dynamics as used in the Car–Parrinello approach) [73].

11.4
Ti–Ni-Based Intermetallics and Shape Memory Materials

Besides a brief discussion of structure and phase diagram of Ni–Al alloys and molecular dynamics applied to this system in the previous section, we present in this and the following section breakthrough results of *ab initio* calculations for Ti–Ni alloys with B2 phase near equiatomic composition, which can explain behavior of details of the phase diagram of Ti–Ni the rapid decrease of the martensitic transformation with tiny changes in composition. Ti–Ni alloys have been used for more than three decades as shape memory material in different mechanical engineering and medical applications. Ti–Ni alloys are still the "best" of all shape memory alloys, because they show the one way (thermal memory) and pseudoelasticity (mechanical memory) [74]. They show good chemical resistance and cyclic stability [75]. Shape memory relies on the martensitic transformation from high-temperature austenite to the product phase and the reverse transformation. When the binary Ti–Ni alloys are quenched from high-temperature B2 phase (with lattice constant of 3.015 Å) to temperatures below the martensitic start temperature M_S, B2 → B19′ transformation takes place. B19′ refers to a monoclinic crystal structure with lattice parameters $a = 2.889$, $b = 2.4.120$, $c = 4.622$, and $\beta = 96.8°$.

There is an enormous amount of experimental and theoretical work on the various aspects of martensitic transformations, deformation, twinning, and grain boundary engineering in connection with the shape memory effect as well as new thermodynamic phases involving so-called strain glass phases, which cannot be discussed in detail here. Therefore, we refer only to the recent review and related work regarding the physical metallurgy aspects of Ti–Ni-based shape memory alloys by Otsuka and Ren [76]. In spite of the vast amount of literature, the phase diagram has been controversial for more than 30 years before it was established. The possible eutectoid reaction at 630 °C in the original phase diagram has been omitted [30]: Figure 11.5 shows the central region bounded by Ti_2Ni and $TiNi_3$ and the high temperature TiNi (B2) phase, to which the phase equilibrium between the B2 and T_3Ni_4 phases has been added and the onset of B19′ is sketched [77]. Ti_2Ni has a peculiar cubic fcc-like structure (space group Fd3 m) with 96 atoms per unit cell.

Trends of structural energy differences of $NiTi_2$ such as intermetallic compounds ($MnTi_2$, $FeTi_2$, $CoTi_2$, $NiTi_2$, $PdTi_2$, $PtTi_2$, and $CuTi_2h$) have been very early calculated [78]. The analysis has shown that the Ti sites with icosahedral symmetry determine the structural stability and a connection with the stability of quasicrystal models was established. Electronic structure calculations have shown that stoichiometric $TiNi_3$ displays a four-layer, hcp, ordered structure, which corresponds to the DO_{24} crystal structure [79].

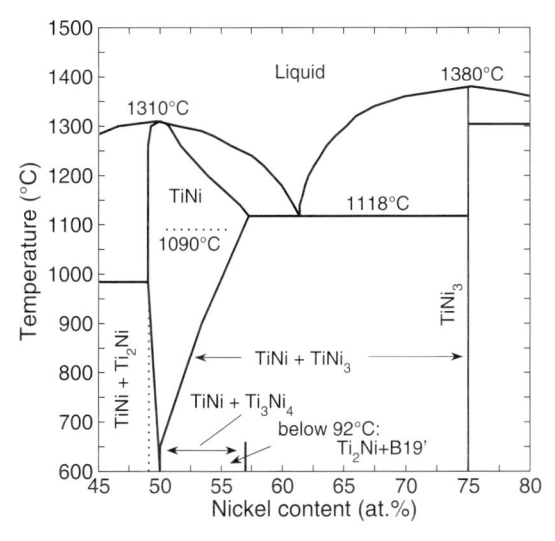

Figure 11.5 Phase diagram of Ti–Ni around its central part by Massalski et al. [30] to which the phase equilibrium between B2 and Ti_3Ni_4 phases has been added.

We have also added the possible order–disorder transition from B2 to bcc at 1090 °C [76], although the existence of such a transition has recently been questioned [80]. Previous data speaking in favor of an order–disorder transition were explained by a possible eutectic reaction due to oxidization at such high temperatures [80].

Characteristic of the phase diagram is the boundary of TiNi on Ni-rich side near the equiatomic composition that is nearly vertical and an explanation for this behavior has only recently been found on the basis of *ab initio* calculations involving large supercells [81]. In particular, the importance of relaxation around a Ni(Ti) antisite defect in $Ti_{95}Ni_{97}$ was highlighted, quite generally leading to a decrease of the latent heat when Ni is added to stoichiometric TiNi causing a rapid decrease of M_S [81].

Note that tight-binding (TB-LMTO) electronic structure calculations combined with a free-energy expression from the cluster variation method allowed studying phase equilibria in the Ti–Ni system successfully [79,82]. The calculations have shown that retaining the d-electron contribution to the TB-LMTO Hamiltonian is sufficient to explain the strong negative values of the alloying formation energies in the Ti–Ni system. The particular strong chemical interaction between Ti and Ni was already emphasized. This is briefly discussed below using density functional theory calculations to visualize the energy-resolved chemical bonding in TiNi using the crystal orbital Hamiltonian populations (COHP) procedure (see Section 11.6). For completeness, we mention that the formation energies of Ti–Ni alloys were already predicted earlier on Miedema's semiempirical model [83], which is built on the "macroscopic atom picture" where the reference is chosen as atoms embedded in a metal, instead of free atoms. An important parameter of the model is connected with the quantitative estimate of the effects of the change of the Wigner–Seitz cell boundary electron density. The Miedema model is very simple and presents useful rules for alloy formation, but it has been criticized because it has no justification based on quantum mechanics. The formation energies are more related to the bond order, since the electrons would be shared in the bonding state of the alloy [84]. Hence, *ab initio* calculation of formation energies and stability of intermetallic compounds are highly welcome. Another useful model has been established by Brewer by estimating enthalpy formation for equiatomic intermetallic compounds between elements at different ends of the transition metal series of the periodic table of elements, when stable alloy formation is related to multiple d-electron bonds formation [85].

The martensitic transformations were not included in these TB-LMTO calculations. Transformation temperatures, for example, upon heating and cooling of equiatomic compositions can be found in the literature [86]. This must now be discussed in the light of the recent *ab initio* calculations [81]. Figure 11.6a shows the dependence of the M_S-temperature on the Ni concentration around the equiatomic composition with data from five different samples [77,87]. Figure 11.6b shows the energy difference between B2 and B19′ phases using different computational methods [81]. Atomic relaxations have been performed by using the VASP code [88,89].

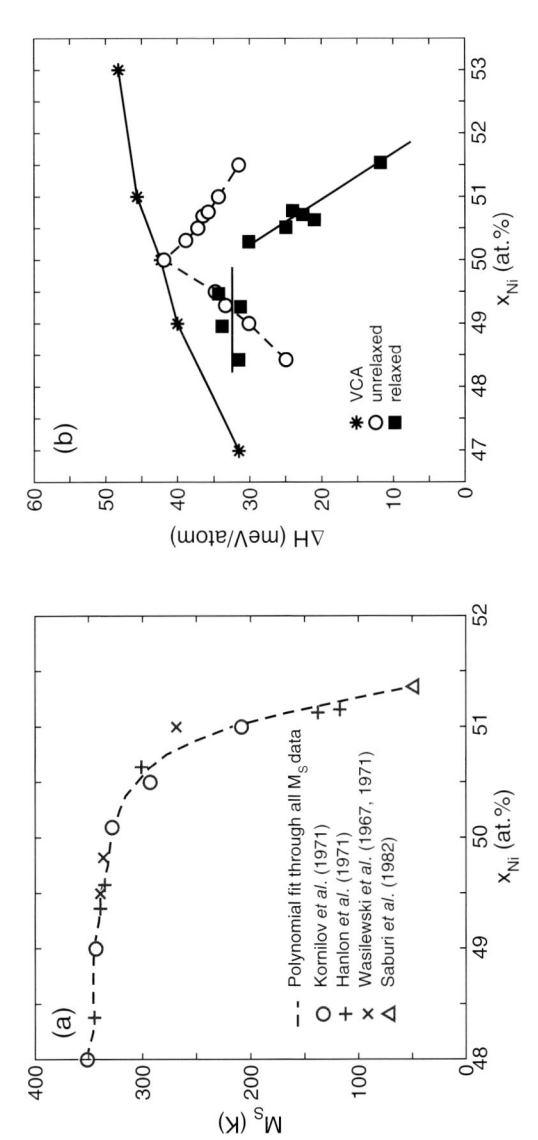

Figure 11.6 (a) Dependence of the martensitic start temperature from B2 austenite to B19′ martensite of solution annealed Ti–Ni samples. The data stem from five different sources of which the references are given in the original papers by Tang et al. and Khadil-Allafi et al., also a polynomial fit through all M_S data is shown in the figure [77,87]. (Reproduced with permission from Ref. [81]. Copyright 2015, Elsevier.) (b) Enthalpy difference between B2 and B19′ as a function of the Ni content for different computational approaches (unrelaxed and relaxed atomic position calculations have been performed with the VASP code [88,89] using projector augmented wave (PAW) pseudopotentials and an energy cutoff of 350 eV). In order to mimic alloy compositions, supercells of the relaxed stoichiometric B2 and B19′ martensite were constructed replacing Ti(Ni) by Ni(Ti) [81]. Results of VCA (virtual crystal approximation) calculations have been added to show the effect of valence electron concentration (band filling). (Reproduced with permission from Ref. [81]. Copyright 2015, Elsevier.)

In this section, we have discussed various Ti–Ni phases and have pointed out the importance of recent results of *ab initio* calculations of the B2 austenite to B19′ martensite [81]. The density functional theory results suggest that the strong decrease of ΔH is caused by the stabilization of the B2 phase by structural relaxation around Ni antisite atoms, together with the gradual destabilization of B19′. As discussed by Frenzel *et al.* [81], the martensite start temperatures and the latent heats of transformation for binary, ternary, and quaternary Ti–Ni-based shape memory alloys are closely related. Figure 11.6 shows some of the recent results. In the next section, we discuss some of the related pseudoelastic and giant elastocaloric effects of Ti–Ni alloys.

11.5
Martensitically Transforming Ti–Ni Intermetallics, Giant Elastocaloric Effects, and Strain Glass Formation in Ti–Ni–X Alloys

In many intermetallic alloys, we find a polymorphic transformation with a change of the relative arrangement of atoms, the displacements of neighboring atoms being small compared to the interatomic distances. The final phase of the martensitic transformation may be considered as a uniform deformation of the initial crystal structure [90]. It is a diffusionless transformation of the parent (austenitic) phase, which often shows incomplete phonon softening and so-called tweed precursors with spin glass nature [91]. We will see in Section 11.7 that the martensitic transformation in Heusler alloys is accompanied by a magnetic transition, which leads to the so-called magnetostructural transformation, which is responsible for the functional magnetocaloric effect in the Heusler alloys.

The β-phase Hume-Rothery alloys and the Ni-based martensitic shape memory alloys such as Ti–Ni–X (where, for example, X is some transition metal impurity such as Fe or Cu) are characterized by a moderate lattice instability at temperatures above the martensitic transformation temperature M_S such as tweed. Ti–Ni–X alloys display a variety of premartensitic behavior (including strain glass behavior [92]) and different paths of martensitic transformations (cubic B2 → monoclinic B19′, B2 → rhombohedral or R phase → B19′, B2 → orthorhombic B19 *to* B19′ [76]). Surprisingly, *ab initio* calculations have shown that the B19′ structure of TiNi is unstable relative to a base-centered orthorhombic structure that cannot store shape memory at the atomic level [93]. However, the B19′ structure can be stabilized by a wide range of applied or residual internal stresses and the memory is stored primarily at the microstructural level. Although the memory behavior of TiNi is well known, we add for completeness the behavior of the thermomechanical response of TiNi with 50.1 at.%Ni as schematically shown in Figure 11.7, where data have been taken from the corresponding literature [94–96]. Historically, the effect was first recognized by Buehler *et al.* [97]. For a recent review, we cite the work of Karaman [98].

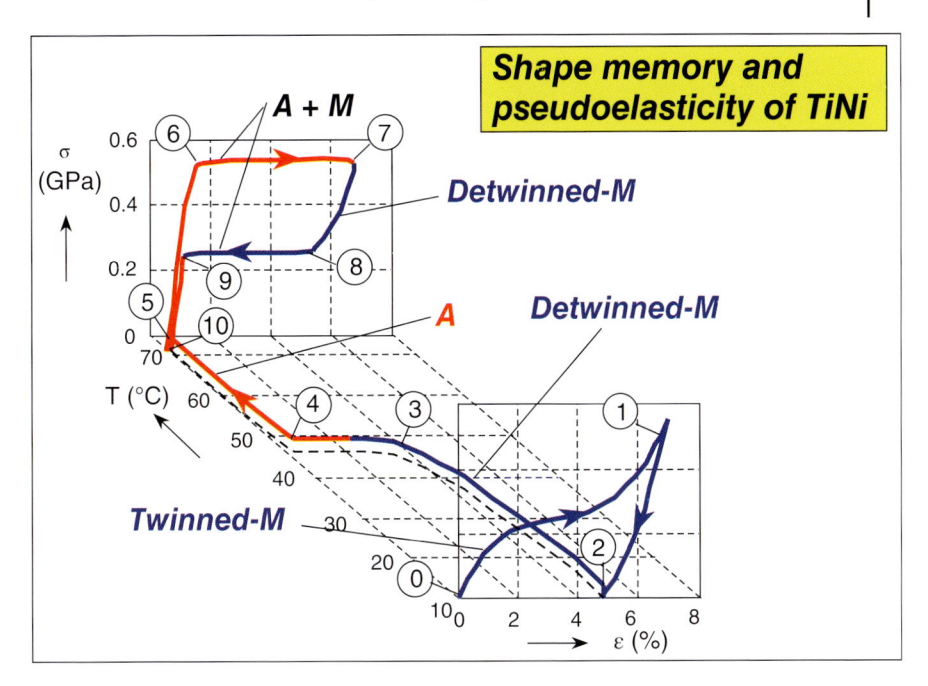

Figure 11.7 Thermomechanical response of a TiNi wire in water. The shape memory effect in steps 1–4 is the material's property to recover large mechanically induced strain (up to 8%) by moderate increase in temperature (≈ 10 – 20 °C). The pseudoelastic response in steps 5–10 accommodate strains of this magnitude during loading and then recover upon unloading (via hysteresis loop). The underlying mechanism is a reversible martensitic transformation. The transformation can be induced by changes in temperature or stress causing a strong thermomechanical coupling in the material behavior. The shape memory effect is seen as the strain is recovered between points 3 and 4 when the material transforms to austenite. At the beginning of the deformation at point 0, we find twinned martensite that transforms to detwinned monoclinic martensite (blue color) and then to cubic austenite (red color). Data have been adapted from the literature [94–96]. The cycles from austenite (B2) to martensite (B19′) and back include the intermediate martensitic R-phase.

The B2 → B19 martensite is, for example, found in Cu-dependent shape memory characteristics of Ti–Ni–Cu alloys (see Figure 11.8) [99]. It has recently been demonstrated that a Ti–Ni–Cu–Co thin film displays ultralow fatigue by next-to-identical 1st and 107th superelastic transformation cycles [100]. The transformation back and forth (B2 ⇌ B19) takes advantage of Ti_2Ni and Ti_2Cu precipitates (precursors), respectively and is 100%. That means that the structural transformation from austenite to martensite and vice versa is complete, respectively, whereby the precursors help to make the transformations from one structure to the other perfect. The associated giant elastocaloric effect allows large adiabatic temperature changes compared to other ferrocaloric materials. See also the recent comment by James pointing out the potential uses of multi-million cycle phase transformation materials [101].

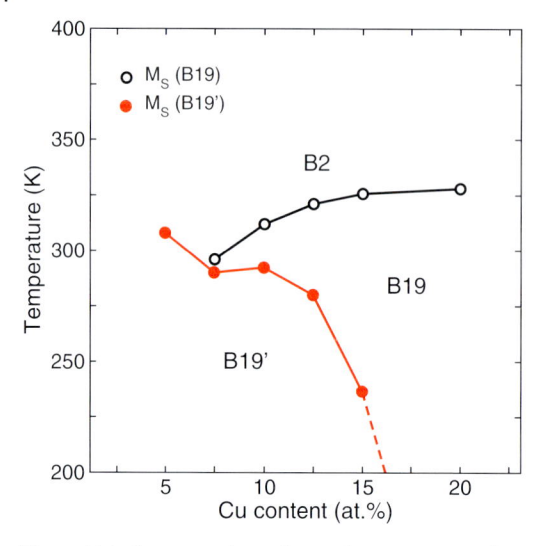

Figure 11.8 Cu content dependence of martensitic transformation temperatures of Ti–Ni–Cu alloys. (Reproduced with permission from Ref. [99]. Copyright 1990, American Physical Society.)

It has recently been suggested that like systems may exhibit a "glassy transition" from the premartensitic tweed phase (as unfrozen strain glass) to a frozen statically disordered "strain glass" state induced by a sufficient amount of doping of point defects, which breaks the long-range elastic interactions in the intermetallics [92]. Such glass transitions may not only occur in ferroelastic systems but may also occur in ferromagnetic and ferroelectric materials. Ti–Ni alloys display a variety of premartensitic behavior and different paths of martensitic transformations [76]. Note that the B2 → B19 martensite is, for example, found in Cu-dependent shape memory characteristics of Ti–Ni–Cu alloys [99].

There is evidence for strain glass transitions in ferroelastic $Ti_{50}Ni_{50-x}Fe_x$ [102], and in ferromagnetic $Ni_{55-x}Co_xMn_{20}Ga_{25}$ to name only two systems [103]. Figure 11.9 shows the complex phase diagram of the former system (adapted from Ref. [102]).

Evidence of such glassy and nonergodic behavior stem, for example, from DMA (dynamic mechanical analysis or dynamic mechanical spectroscopy, which is used to study and characterize materials) measurements, from which can be obtained elastic modulus and inverse mechanical susceptibility. This has recently been applied to Ti–Ni–Fe strain glass where both the storage modulus and internal friction curves show frequency dispersion that is characteristic of a strain glass transition [102]. The broken ergodicity is obvious from field-cooled (FC) and zero-field-cooled (ZFC) measurements where the field here is the tensile stress.

The shape memory effect and superelasticity in a Ti–Ni strain glass alloy has also been established recently [104]. The glassy behavior of ferroic intermetallics is an exciting new research field [105]. Although the relatively small transformation strains in most ferroelastic systems have limited values of observed strain so far, there is hope to find many more systems with strain glass phases by

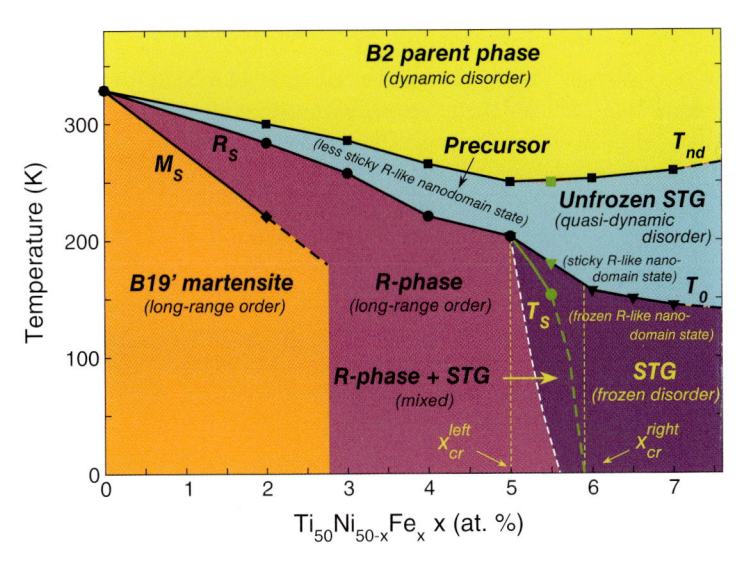

Figure 11.9 Phase diagram of $Ti_{50}Ni_{50-x}Fe_x$ showing B2 austenite and B19' and R-martensite as well as the frozen and unfrozen strain glass (STG) phases [102], with martensite start R_S and M_S and STG transition temperature T_0. T_S is the spontaneous STG-R transition start temperature of the compositional crossover regime ($x_{cr}^{left} < x < x_{cr}^{right}$) between R and STG phases. T_{nd} denotes the start temperature for appearance of precursory nanodomains. (Reproduced with permission from Ref. [102]. Copyright 2011, American Physical Society.)

exploring the interaction energy between martensitic variants, austenite–martensite interfaces, and between strain fields arising from localized defects, antisite defects, grain boundaries, and dislocation fields. If the strain fields arising from defect structures are larger than from the martensitic variants, there is a favorable condition to find spin glass features in the austenitic phase. In a series of articles by Lookman *et al.*, the glassy behavior in ferroic materials such as perovskite manganites has been investigated considering the interplay of long-range elastic interaction and disorder and the microstructure arising from ferroelastic transitions [106,107]. The importance of glassy nanodomain formation in intermetallics has also been discussed [108] as well as the interplay of elastic anisotropy and disorder [109,110] and the origin of magnetoelastic tweed such as precursor phenomena [111]. We conclude by citing work on a quite general aspect of ferroic glasses by Sherrington [112].

11.6
Bonding and Chemical Frustration in Ti–Ni–X Intermetallics and Chemical Pressure

At this stage a remark is necessary to distinguish in part Hume-Rothery alloys and intermetallic materials. Metals share common features that define them as a separate class of materials and that can largely be defined by their electronic structure (like the sp-bonded metals follow the Sommerfeld model, whereas the properties of transition metals depend strongly on the number of d-electrons and the localized f-orbitals define the properties of rare earth metals. Some alloy systems exhibit complete solid solubility (e.g., Cu–Ni), while others show limited solubility at any temperature. Several factors determine the limits of solubility that are expressed as a series of rules often called "Hume-Rothery Rules" [3], which are (i) the difference of atomic radii should not exceed 15%, (ii) the difference of electronegativity (chemical affinity) should be small, (iii) the crystal structure of the two elements must be identical, (iv) a metal will dissolve a metal of higher valency to a greater extent than of lower valency. Heusler alloys are intermetallics with particular composition and often face-centered cubic crystal structure. The structure of the magnetic Ni–Mn-based Heusler alloys considered in Section 11.7 consist of four interpenetrating fcc lattices and the complexity of the magnetic interactions that appear is mainly determined by the way of how the magnetic Ni and Mn atoms are distributed over these sublattices in the case of chemical disorder in the Heusler alloys. The Heusler alloys discussed below are all metallic "full Heusler alloys," that is, each lattice sites of the four interpenetrating sublattices is occupied by an atom. The so-called half-metallic Heusler alloys, where lattice sites of only three sublattices are occupied (and where a pseudogap may form in the spin-down electronic density of states), will not be discussed here, since the functional properties such as magnetic shape memory and mangetocaloric effects are more pronounced in the full Heusler alloys.

Intermetallics very often do not show complete solubility, but can have a very narrow stability range as in the rare earth transition metal (RETM) family. Structure of intermetallic compounds and constituting elements are different, which distinguishes them from Heusler materials. One aims at creating intermetallics that show precisely designed crystal structures with properties tuned for certain purposes. They also show smaller heats of formation compared to other alloys. So, the combination of metallic elements with different atomic sizes, electronic configuration, and electronegativities, in contrast to the Heusler phases, results in a large number of intermetallic compounds with known or novel crystal structure types, which often do not follow the Hume-Rothery rules or follow them only in part. Nonetheless, often the valence electron concentration per atom, (e/a), allows characterizing the intermetallic structures in the same manner as the Hume-Rothery alloys. For the case of magnetic Heusler alloys, the valence electron concentration includes also the number of d-electrons. For instance, the phase diagram of the four different Ni–Mn-based Heusler alloys shown in Figure 11.13 is conveniently plotted in the temperature versus electron concentration plane for each case showing the remarkable linear variation of the martensite start temperature with the valence electron concentration e/a. In case of Ni$_2$MnGa, we take into account 10 d-electrons for Ni, 7 d-electrons for Mn, and 3 (s, p)-electrons for Ga, that is, $e/a = 7.5$.

The chemical bonding properties of stoichiometric TiNi are shown in Figure 11.10 (calculations using the VASP package as well as the LMTO method

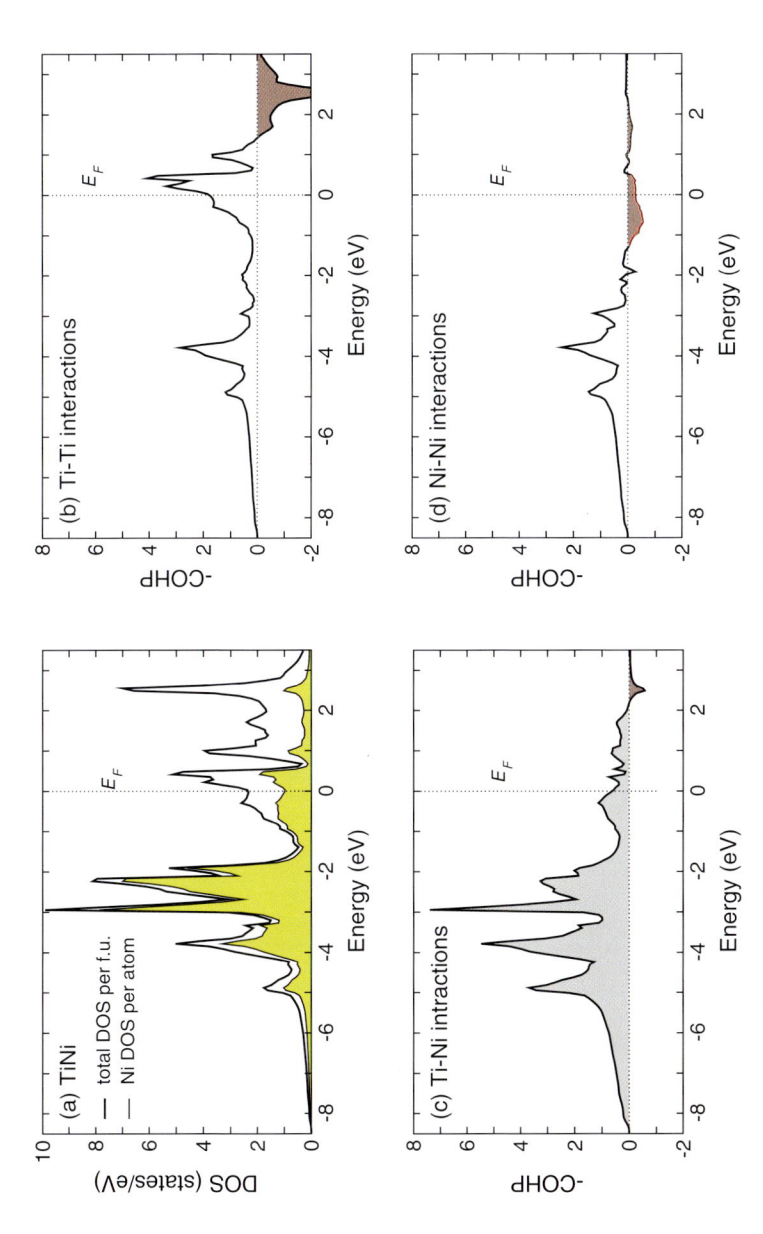

Figure 11.10 (a) Total and Ni-resolved electronic density of states of stoichiometric TiNi. (b–d) COHP diagrams of Ti–Ti, Ti–Ni, and Ni–Ni interactions, respectively. Light gray are bonding interactions, dark (brown, negative values) mark antibonding interactions. Ti–Ni bonding is strongest followed by Ti–Ti and then only by Ni–Ni. Non-spin-polarized and spin-polarized calculations yield comparable results. Austenitic CsCl structure (RG Pm3̄m with lattice parameter $a = 3.015$) [113].

yield comparable results). The experimental lattice constant is $a = 3.015$ [113], and the optimized lattice constant in the calculations is $a = 2.936$). It is interesting to compare the electronic density of states (DOS) of stoichiometric TiNi with bonding properties deduced from the crystal orbital Hamiltonian population (COHP) method as shown in Figure 11.10 [114–116]. The COHP is calculated from [115,116]

$$\text{COHP}_{\mu\vec{T},\nu\vec{T}'}(E) = H_{\mu\vec{T},\nu\vec{T}'} \times \sum_{jk} f_j(\vec{k}) C^*_{\mu\vec{T},j}(\vec{k}) C_{\nu\vec{T}',j}(\vec{k}) \delta(\varepsilon_j(\vec{k}) - E), \quad (11.4)$$

where $H_{\mu\vec{T},\nu\vec{T}'}$ are the elements of the Hamiltonian matrix **H** and $f_j(\vec{k})$ are the occupation numbers (which have been omitted in the COHP plots) [115], $C(\vec{k})$ contains the coefficients of linear combinations of atomic orbitals to represent crystal orbitals (LCAO-CO), building the wavefunctions of the jth band, and $\vec{T}$ is the unit cell lattice vector. The [-COHP] function gives an indication of the bonding or antibonding nature of the interaction of the orbitals defining the crystal orbitals. A positive value defines a bonding interaction, a negative value indicates an antibonding interaction (as is the case for Ti–Ti interactions in an energy range above the Fermi level and for Ni–Ni interactions around the Fermi energy, see Figure 11.10). Note that an analog of the COHP has already been applied earlier to the α-MoO$_3$ where its usefulness has been discussed by showing that the bonding of the oxygen to Mo in MoO$_3$ (010) has both ionic and covalent character, and, that differences in bonding between the various types of oxygen can be traced to differences in the amount of covalent bonding.

In summary, the COHP analysis can be considered as an energy-resolved local bonding analysis based on plane wave electronic structure calculations defined as "projected crystal orbital Hamiltonian population" (partitioned band structure energy into orbital pair interactions, that is, "bond-weighted" density of states between a pair of adjacent atoms indicating bonding and antibonding contributions to the band structure energy, the integrated COHP (ICOHP) hints toward the bond strength in electronvolt, which is shown in Table 11.2 for TiNi. The

Table 11.2 ICOHP values for Ti–Ti, Ti–Ti, and Ni–Ni interactions as well as the distances in the optimized structure, spin-up and spin-down contributions to IOCP are identical (f.u. = formula unit).

Interaction	Coordination number	$r_{A,B}$ (Å)	ICOHP (eV/bond)	ICOHP$_{f.u.}$ (eV/f.u.)
Ti–Ti	6	2.936	−0.732	−4.392
Ti–Ni	8	2.543	−1.468	−8.808
Ni–Ni	6	2.936	−0.506	−3.036

The optimized lattice constant is $a = 2.936$ Å. Note that the COHP data are the sum of all interactions between the species in the unit cell, that is, 6 Ti–Ti/Ni–Ni and 8 Ti–Ni interactions, which represent the total strength of one type of interaction. The averaged integrated (ICOHP) value for the individual interaction is also listed.

results of the COHP diagrams shown in Figure 11.10 and the corresponding bond strengths listed in Table 11.2 clearly show how dominant the Ti–Ni interaction strength is in this intermetallic compound compared to the other interactions. This may also explain in part the strong influence of antisite defects on the martensite start temperature discussed in Ref. [81], since any disorder by introducing antisite defects will weaken the bonding strengths in Ti–Ni alloy. However, calculations of the COHP diagrams for antisite defects have not been calculated so far. They are tedious and costly in computational time because of the large supercells with many atoms, which are required in the calculation.

With respect to chemical bonding issues in intermetallics, some kind of new aspect has been introduced. This addresses the concept of "chemical frustration," which can be regarded as an additional tool compared with the COHP method, which allows characterizing chemical bonding in connection with structural complexity, where a density functional theory also calibrated Hückel model is used [117–119]. This is important since any further ingredient to understand the appearance of the variety of complex structures of intermetallics and their bonding might help to design even better intermetallics with desired properties. One way to understand the structure of complex intermetallics is "chemical frustration," which shows light onto incompatible bonding or packing modes to induce complexity in ternary intermetallic alloys and compounds [120].

To give an example of chemical frustration, we cite recent DFT calculations of the intermetallic carbide system $Gd_{13}Fe_{10}C_{13}$ [121], where the unusual Fe–Fe bonding is likely a consequence of the conflict between metal–metal and metal–carbon bonding leading to complexity in this intermetallic carbide. The two emerging bonding types have preferences for tetrahedral and simple periodic packing, respectively – packing modes known empirically to be incompatible [122].

The role of "chemical pressure" in Ca_2Ag_7 was analyzed in a simple theoretical approach by calculating so-called binding surfaces (constant energy contour lines in the volume V versus tetragonal distortion c/a plane using VASP and the Hückel method, for further references, see Ref. [118]). Another convenient method to make connection between geometry and electronic structure is the method of moments of the electronic density of states $\rho(E)$ or by functions of the structure via sums of products of the Hamiltonian matrix elements [123,124],

$$\mu_n = \int dE E^n \rho(E),$$
$$\mu_n = \sum_{i_1}\sum_{i_2}\cdots\sum_{i_n} H_{i_1,i_2}H_{i_2,i_3}\cdots_{i_{n-1},i_n}. \quad (11.5)$$

The link between structure and $\rho(E)$ is achieved when $\rho(E)$ can be reconstructed from its moments by an inversion scheme. When the "Hückel Hamiltonian" is supplemented by a short-range repulsive interatomic potential, the method can reproduce the expected distance–dependence of bonding interactions, for example, in the μ_2-Hückel model with total energy [125], $E_{\mu_2-\text{Hückel}} = \gamma\mu_2 + E_{\text{Hückel}}$, where γ is usually chosen to reproduce the equilibrium volume of a structure.

The total energy can be split into a sum of onsite and pairwise interaction terms [125]:

$$E_{\mu_2\text{-Hückel}} = \sum_{ij} \left(\gamma H_{ij}^2 + \left(\sum_n o_n c_{ni}^* c_{nj} \right) H_{ij} \right), \tag{11.6}$$

where c_{nj} is the coefficient of atomic orbital j in crystal orbital n and o_n is the occupancy of that crystal orbital. The chemical pressure analysis can be undertaken in terms of

$$p = \frac{\partial E}{\partial V} = \frac{1}{N_{\text{orb}}} \sum_{ij} p_{ij}, \tag{11.7}$$

where N_{orb} is the number of orbitals in the crystal and p_{ij} is the pressure arising from the interaction between the orbitals i and j (the term $i = j$ is zero) [118]. Despite its simplicity, this method works and, for instance, allows discussing structural trends of Hume-Rothery phases in terms of the valence electron concentration [126]. Note that a chemical pressure analysis or a partial pressure analysis by decomposing the element-specific pressures into s, p, and d contributions has not yet been used to discuss magnetism in Ni–Mn-based Heusler alloys. This requires a further remark: A partial pressure analysis has been undertaken earlier by calculating the partial pressures associated with the spin polarized s, p, and d valence electrons in Invar-like transition metal alloys in order to calculate the influence of the partial pressures on the collapse of magnetism at a critical volume in alloys with so-called magnetovolume instabilities [127,128]. Here, the partial pressure analysis of Fe–Ni Invar alloys can be done separately for Fe and Ni, showing that the partial pressures from s- and p-electrons for both elements are positive in the whole volume range. Most of the bonding is done by the d-electrons, more precisely, by the Fe d-electrons. With onset of magnetism, majority-spin states lose part of their binding features; they even become antibonding at larger volumes. Minority-spin states behave just the other way round: Here, the bonding d-states become more bonding. This brief discussion shows, whatever pressure analysis is undertaken, that this is a useful concept to discuss bonding and magnetic features in systems that exhibit unusual magnetic behavior such as in intermetallic compounds with magnetovolume instabilities, where the systems are bound together dominantly by the minority d-electrons. Podgorny has discussed this in more detail for the ordered phases of Pt–Fe alloys [129], which show similar behavior of the interplay of magnetism and bonding features near a magnetovolume instability as the Fe–Ni alloys [127,128].

In conclusion, the structural complexity and competition of different types of chemical bonding trends show some resemblance to complex magnetic alloys and compounds of the Heusler type with competing, for example, ferromagnetic and antiferromagnetic interactions. Here, the COHP may also be a useful means to discuss the electronic density of states in terms of pairwise interactions. In frustrated magnets, geometry and magnetic interactions can suppress the

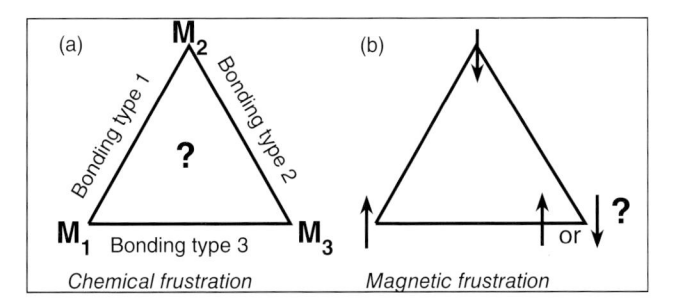

Figure 11.11 (a) Chemical frustration in a ternary system, where the interplay of mutually exclusive bonding types in the component binary subsystems (M_1–M_2, M_2–M_3, M_1–M_3) may lead to structural complexity in a very similar manner as (b) magnetic frustration due to antiferromagnetic interactions on triangular-like lattices (or competing ferromagnetic and antiferromagnetic interactions) may lead to complex magnetic configurations illustrated here for spins on a triangular lattice.

formation of long-range magnetic order, for instance, in systems where the atoms occupy a frustrated lattice, famous examples are fcc Fe iron [130] or Co-doped β-Mn [131]. This observation was already discussed in Ref. [118], but for convenience we briefly repeat this schematic illustration of competition and chemical frustration in ternary systems and the magnetic analog shown in Figure 11.11.

On a more general basis, discussion on families of intermetallic structure types involving chemical bonding characteristics can be found in Ref. [132]. The field of physics and chemistry of intermetallic compounds is an ever-growing field of research and it is not possible to give a comprehensive review of the great variety of intermetallic compounds in a short chapter. Therefore, in the following we will focus on a few examples of ternary intermetallics, in particular, on a few challenging ternary magnetic intermetallics, which have intriguing properties and which are of interest for fundamental research as well as new technological devices due to their functionality.

11.7
Magnetic and Functional Properties of Ternary Magnetic Heusler Intermetallics

The $L2_1$ structure consisting of four interpenetrating sublattices with unit cell of space group $Fm\bar{3}m$ (O_5^h) of the prototypical magnetic shape memory Heusler compound Ni_2MnGa is shown in Figure 11.12. The Ni sites have point symmetry m3m (octahedral site) and Ga and Mn sites have point symmetry 43m (tetrahedral site). The whole crystal has tetrahedral symmetry T_d. Preferential disorder type is B2 with Mn and Ga showing same site occupancy probability. For further structure details, see Ref. [133]. Instead of X_2YZ structure, the same class of materials may also form the so-called inverse XY)XZ crystal structure. A COHP bonding analysis of ferromagnetic Fe_2CuGa and further compounds

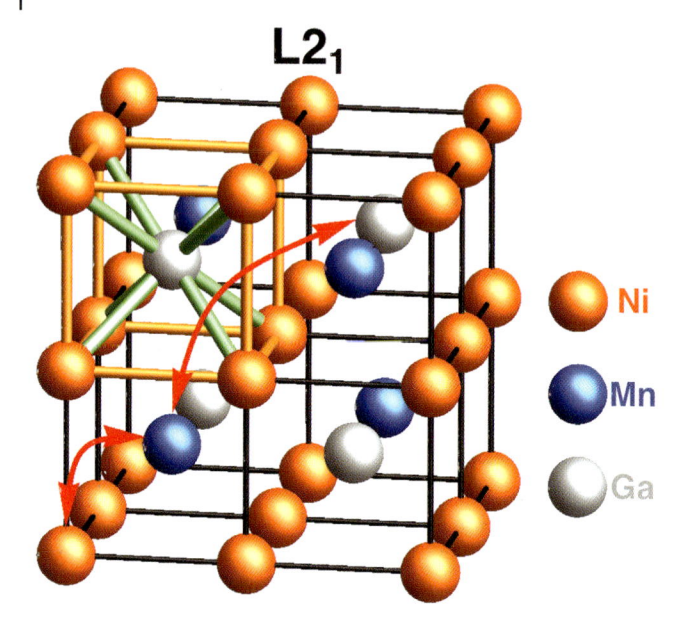

Figure 11.12 The cubic (L2$_1$) Heusler structure of Ni$_2$MnGa consists of four interpenetrating fcc lattices (full-Heusler alloys in contrast to semi-Heusler alloys when only three fcc lattices are occupied). The unmodulated martensitic phase is achieved for $c/a > 1$ (simple tetragonal distortion). Due to the specific bonding and distances between the atoms, any interchange of atoms, as indicated by the arrows, leads to antiferromagnetic interactions and spin flip of the exchanged Mn atom. Similarly, in case of excess Mn on the Ga sublattice in Ni$_2$Mn$_{1+x}$Ga$_{1-x}$, the excess Mn have their spins flipped. This is confirmed by *ab initio* calculation of magnetic exchange coupling constants and resulting magnetic configurations. (Reproduced with permission from Ref. [137]. Copyright 2013, Springer.)

shows an inherent tendency for structural distortion in a majority of these alloys [134] in much the same way as many magnetic Heusler materials with X$_2$YZ structure show a tendency for a martensitic transformation. In addition, a half-metal phase may appear for appropriate valence electron concentration e/a when the Fermi level falls into the gap between crystal-field split antibonding e_u and t_{1u} states [135]. Furthermore, many of the ferromagnetic shape memory Heusler alloys show crystal structure of modulated martensitic phases [136].

The particular interest in magnetic Heusler compounds and alloys is related to the functional properties of Mn excess Ni–Mn-based Ni–Mn–(Ga, In, Sn, Sb) intermetallics, which are promising for future technologies (various theoretical and experimental aspects have been highlighted in previous works [46,138–142]). A close look at the crystal structure in Figure 11.12 reveals that in case of Mn excess, the extra Mn start to occupy sites of the other sublattices such as Ni or Ga, In, Sn, and Sb with the appearance of new distances between some of the Mn atoms, which in turn leads to antiferromagnetic interactions between these

Mn atoms and an overall competition between ferromagnetic and antiferromagnetic interactions. The whole scenario of functional properties of nonstoichiometric magnetic Heusler alloys is based on this observation. In addition, there are strong magnetoelastic interactions, which lead to a coupled martensitic and magnetic phase transitions, which are called "magnetostructural" transition. We would like to emphasize that this has already been modeled some 40 years ago by using a degenerate tight-binding model for the e_g electrons, which has been supplemented by Coulomb interaction and charge and magnetization density coupled to the martensitic distortion of tetragonal nature. At the time when this model was established, it was applied to the interplay of magnetism and martensitic transformation but not to the magnetocaloric effect [143]. It turns out that this magnetostructural transition of first order is one of the significant magnetic responses of magnetic Heusler alloys that leads to giant magnetocaloric effects as discussed below. A prerequisite for this behavior is the large change of the magnetic moment across the martensitic transformation from a low-spin (LS) state of martensite to a high-spin (LS) state in austenite in an external magnetic field. This metamagnetic transition has its origin in the competition of antiferromagnetic and ferromagnetic interactions that are present in martensite and austenite.

The phase diagrams shown in Figure 11.13 [46,137,144] highlight the intersection of martensitic transformations and magnetic phase transitions. The phase diagrams have been calculated either by *ab initio* method [144] or approximated by energy differences between the phases [137].

Further complexity of structural phase diagrams due to coexisting intermartensitic phases is highlighted in Figure 11.14a for the case of Ni–Mn–Ga [145,146]. Associated magnetic ground states with different spin configurations, ferromagnetic (denoted by "ferro") and ferrimagnetic ("denoted by ferri", with spins flipped of Mn on the Ga sites, respective In sublattice sites in Ni–Mn–In alloys) are shown in Figure 11.14b and c. Figure 11.14b shows that $Ni_8Mn_5Ga_3$ with a valence electron concentration $e/a = 7.75$ has a ferrimagnetic ground state that is 20 meV (~200 K) lower in energy than the ferromagnetic state for the $L2_1$ phase. ("ferri" and "ferro" states have same energy for $c/a = 1$ if one Ni atom is replaced by Co as is the case for $Ni_8Co_1Mn_5Ga_3$ with $e/a = 7.6875$ in Figure 11.14c). Calculation of the temperature evolution of the "ferri" ($L1_0$) and "ferro" ($L2_1$) solutions in Figure 11.14b along c/a would finally lead to a crossing and a stable "ferro" solution at higher temperatures. Such a result could be obtained by fixed spin moment calculations fixing the total magnetic moment of the 16-atom supercell to low values. In this way, the calculations can describe the martensite-to-austenite transformation along $e/a = $ const. in Figure 11.14a. This shows in which complex way structural and magnetic degrees of freedom are involved in the martensitic transformation of magnetic Heusler alloys.

For alloy design, it is necessary to know the site preference of each element in the Heusler compound with generic formula X_2YZ. Alloy design by doping the

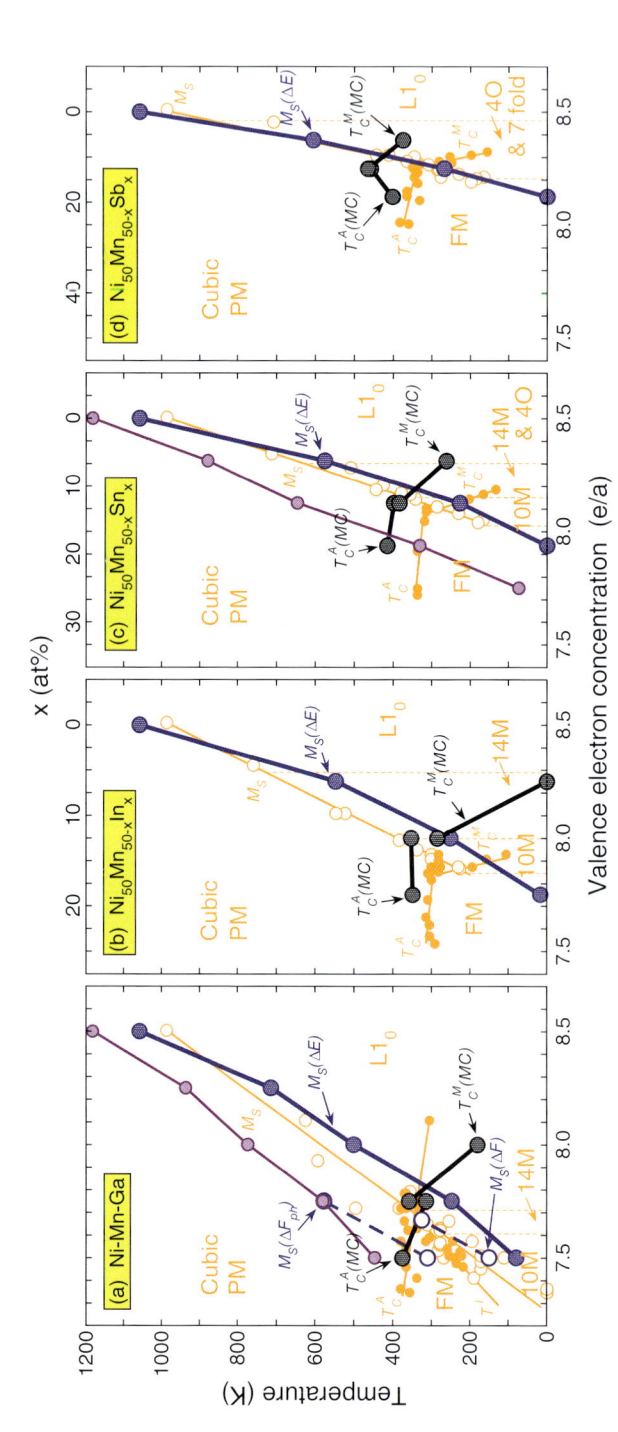

Figure 11.13 The calculated phase diagram of $Ni_2Mn_{1+x}Z_{1-x}$ ($Z = Ga$, In, Sn, Sb) (blue and black lines and blue circles in comparison to the experimental one [46] (orange data). Blue dots are from *ab initio* energy differences between structures, violet curves refer to $Ni_{25}Pt_{25}Mn_{50-x}(Ga, Sn)_x$, black curves are Monte Carlo results of Curie temperatures for austenite (A) and martensite (M). For Ni–Mn–Ga, $M_S(\Delta F_{ph})$ refers to harmonic lattice calculations, while $M_S(\Delta F)$ takes in addition the magnetic contribution to the free energy into account [137]. Limits of modulated 10M and 14M and 4O structures are also indicated. The characteristic feature of this phase diagram is that for each system martensite transformation and Curie temperature cross and that the martensite transformation temperature varies linearly with the valence electron concentration. (Reproduced with permission from Ref. [137]. Copyright, 2013, Springer.)

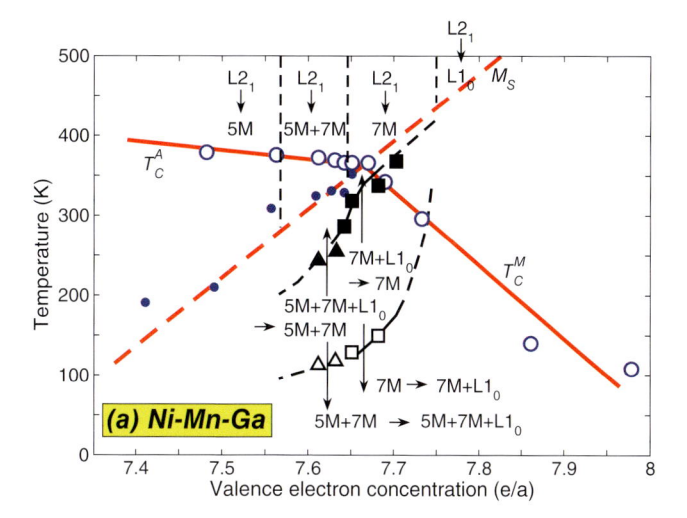

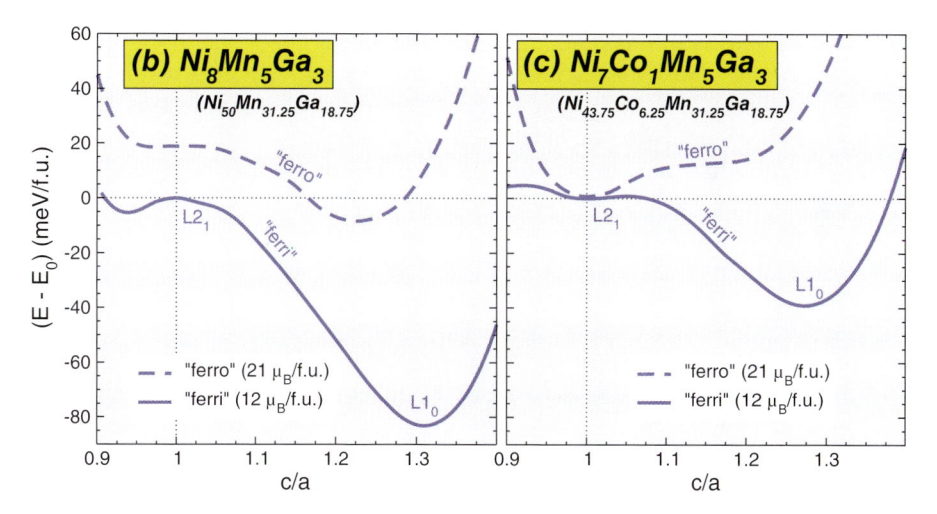

Figure 11.14 (a) More detailed structural phase diagram of Ni–Mn–Ga based on an alloy series with 50 at.% fixed Ni concentration (Figure adapted from Refs [145,146] with data provided by M. Acet). The forward and reverse martensitic transformation lines are shown with the arrows down- and up-pointing arrows, respectively. 5M and 7M denote modulated phases, $L1_0$ nonmodulated (tetragonal) martensite. Filled circles: M_S, open circles mark Curie temperatures of austenite and martensite. (Reproduced with permission from Ref. [146]. Copyright 2015, Elsevier.) Total energy variation of (b) $Ni_8Mn_5Ga_3$ $e/a = 7.75$) and (c) $Ni_7Co_1Mn_5Ga_3$ ($e/a = 7.6875$) (16-atom supercells) as a function of the tetragonal distortion c/a, where $c/a = 1$ is cubic austenite and $c/a > 1$ is tetragonal martensite. Energies in parts (b) and (c) are relative to the "ferri" solution corresponding to the spin reversal of Mn spins on the Ga sublattice, which is lowest in energy. "ferro" denotes the ferromagnetic solution with all spins parallel. Doping with Co weakens the stability of the martensitic structure as the minimum of $L1_0$ in (c) moves up in energy compared to (b). This shows that there is no uniform magnetic ground state because of frustration effects.

Table 11.3 Preferred sites of different atoms in known Heusler alloys X_2YZ (X and Y are transition metals and Z is from the group of sp elements). (Reproduced with permission from Ref. [153]. Copyright 1982, Elsevier.)

	Heusler alloy compositions													
	Y sites								X sites		Z sites			
	IIIB	IVB	VB	VIB	VIIB	VIIIB			IB	IIB	IIIA	IVA	VA	VIA
1														
2											B	C	N	O
3											Al	Si	P	S
4	Sc	Ti	V	Cr	Mn	Fe	Co	Ni	Cu	Zn	Ga	Ge	As	Se
5	Y	Zr	Nb	Mo	Tc	Ru	Rh	Pd	Ag	Cd	In	Sn	Sb	Te
6	La	Hf	Ta	W	Re	Os	Ir	Pt	Au	Hg	Tl	Pb	Bi	Po

ferromagnetic Heusler alloys by other transition metals such as Co or Cr may lead to systems with optimized magnetocaloric effect as in the case of Ni–Co–Mn–In [147–150]. (Results for Ni–Mn–In without Co are discussed by Barandiarán et al. [151].) The role of Co (with preference to occupy Ni sites [152]) is to enhance ferromagnetism in austenite; Co also leads to a larger change of magnetization across the austenite–martensite transformation, which in turn enhances the magnetocaloric effect, which is proportional to ΔM at the magnetostructrual transition.

Some site preferences of elements forming X_2YZ Heusler alloys are listed in Table 11.3 [153].

Before we discuss the magnetic and functional features such as magnetic shape memory and magnetocaloric properties in more detail, let us check the bonding properties of the ternary Heusler alloys since the third element brings in more bonding features compared to the binary shape memory Ti–Ni alloy. Figure 11.15 shows results of LMTO calculations of element-resolved density of states curves of Ni_2MnGa as well as the associated COHP curves. The cutoff distance in real space for the COHP curves has been set to 3.5 Å; therefore, the Mn–Mn interactions have not been included.

We have also listed in Figure 11.15 for each bonding case the total bonding strength as well as the spin-resolved values. Strong bonds exist between Ni–Mn and between Ga–Ni, which show approximately twice the bonding strength as compared to the strength of the Ni–Ni and Ga–Mn bonding, respectively. Hence, it is obvious that with change of the valence electron concentration e/a, and in specific with excess of Mn atoms, the strength of the strong bonds will decrease and the Heusler alloy will transform earlier (at higher temperatures to martensite). This explains in part the linear variation of M_S but also the linear variation of Curie temperature with e/a. The COHP curves for martensite with a tetragonal distortion $c/a = 1.25$ for Ni_2MnGa (not shown here) reveal that Ni–Ni and Ga–Mn bonds become

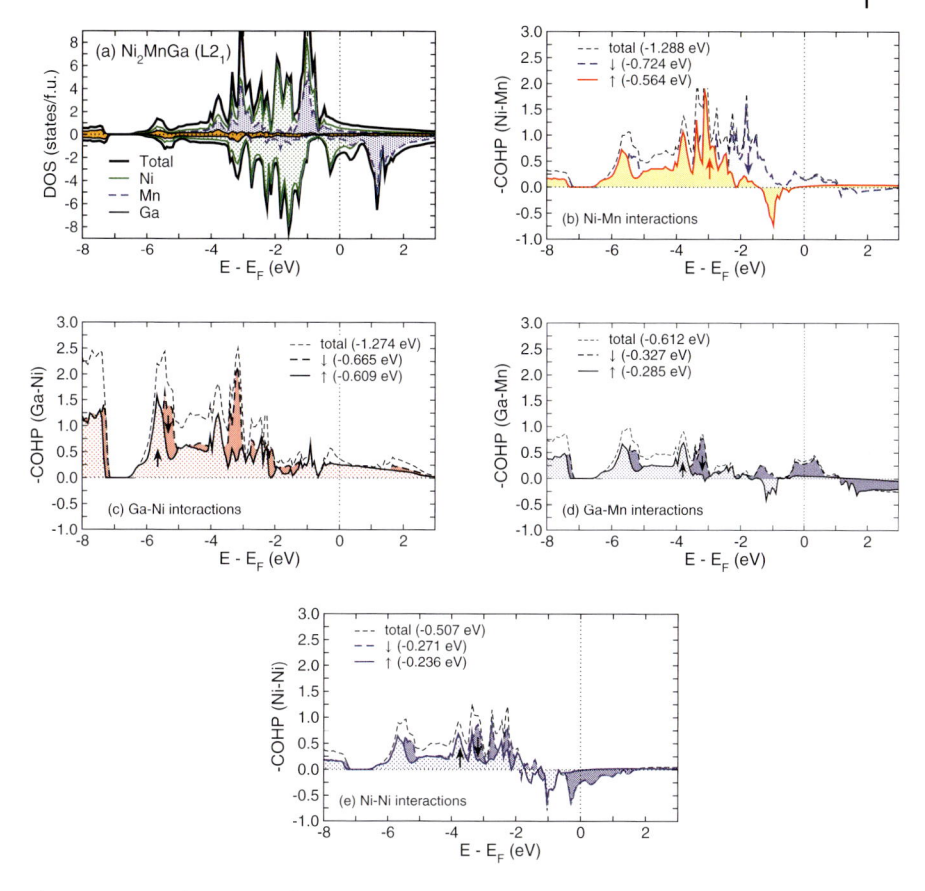

Figure 11.15 (a) The density of states of stoichiometric Ni_2MnGa (L2$_1$). For a tetragonal distortion of $c/a = 1.26$, the martensitic phase (L1$_0$) is stabilized by shifting the Ni-e_g peak from −0.3 eV in the minority-spin DOS to unoccupied states above the Fermi level that is pushed to the pseudogap region. (b–e) COHP diagrams of Ni–Mn (eight bonds), Ga–Ni (eight bonds), Ga–Mn (six bonds), and Ni–Ni (six bonds) interactions. The integrated COHP energies listed give the strength of the total and spin-down and spin-up-resolved interactions in electronvolts for a single bond. The element-distinct Ni–Mn and Ga–Ni interactions are the strongest ones.

stronger (because antibonding interactions are shifted in part across the Fermi level) and support the martensitic tendency, whereas Ga–Ni and Ni–Mn interactions favor more a $c/a = 0.94$ distortion.

A chemical bonding study has also recently been performed for Heusler compounds with the inverse (XY)XZ crystal structure (instead of full Heusler X$_2$YZ), some of which show an inherent tendency for a structural distortion [154]. Therefore, inverse magnetic Heusler alloys may also be of interest regarding magnetic functional properties such as the magnetic shape memory effect.

Hence, any change of the composition of Ni_2MnGa will lead to a weakening of the strong bonding features.

Heusler phases are intermetallics with specific compositions and fcc cubic crystal structure (the interpenetrating fcc lattices may be thought of as a bcc CsCl-type structure in which alternate positions on each sublattice may be occupied by different atoms).

The Heusler alloys exhibit ferromagnetic order, although the constituents of the materials may not be ferromagnetic themselves such as $Cu_2Mn(Sn, Al)$, which are ferromagnets at room temperature [155].

Different approaches have been used to discuss ferromagnetism in the full- and half-Heusler materials. Very early it has been thought to originate from the double-exchange (superexchange like) mechanism between the magnetic atoms (usually Mn in X_2MnZ) sitting at the bcc sites of the cubic structure, because direct exchange is small due to the large separation of neighboring Mn atoms, which is larger than 4 Å in $Cu_2Mn(Cu, Sn)$. This exchange is based on Zener's theory [156], and requires the two magnetic ions to have different oxidation states, which is then, for example, a very appropriate model to describe the magnetic coupling mechanism in mixed valence compounds. The physics of "double exchange" has been reviewed by Anderson and Gennes [157,158] with the observation that "double exchange" is completely different from the usual direct and indirect exchange coupling, since the coupling energy cannot be written as a sum of terms relating ionic spins by pairs and that the inverse of the high-temperature susceptibility $1/\chi$ versus T should follow a curved line [157]. Another model where the coupling between the spins arises from the s–d mixing in terms of the Anderson model [159] generalized to the case of two magnetic impurities has been called double resonance exchange coupling and is much larger than the coupling due to s–d exchange alone [160]. This model allowed for the first time to estimate Curie temperatures [161].

Experimental spin wave dispersions from neutron scattering analyzed using the Heisenberg Hamiltonian with long-range interactions [162,163] (as nowadays in theory [164]) can qualitatively be described by indirect exchange models such as Cyrot's double resonance respective Kasuya's virtual double exchange model [165] or the RKKY (Rudermann–Kittel–Kasuya–Yosida) interaction (see analysis by Caroli [160]). However, the sign of the exchange interaction cannot be estimated from the model Hamiltonians.

The first quantitative study of the exchange interactions in Heusler alloys was undertaken by Kübler et al. [166]. They confirmed that essentially the s,p-electrons of the Z atoms in X_2MnZ mediate the interaction between the d-electrons (type of superexchange with the group III–V electrons responsible for the Mn–Mn coupling). The *ab initio* results were projected onto the Heisenberg model to extract the Heisenberg exchange constants for the Heusler alloys.

Nowadays the magnetic moments and magnetic exchange coupling constants are successfully calculated by two different *ab initio* methods. Since magnetic exchange coupling constants are among the most important parameters in magnetic studies, we list here the computational methods that have been used in

11.7 Magnetic and Functional Properties of Ternary Magnetic Heusler Intermetallics

these calculations: One is a reciprocal space method using the so-called frozen-magnon approach [167–169] to obtain the exchange constants of a classical multisublattice Hamiltonian,

$$\mathcal{H}_{\text{eff}} = \sum_{\mu\nu} \sum_{\substack{ij \\ (\mu i \neq \nu j)}} J_{ij}^{\mu\nu} e_i^{\mu} e_j^{\nu}, \qquad (11.8)$$

where μ, ν denote different sublattices and i, j refer to lattice sites within the sublattices, e_i^{μ} is the unit vector in direction of the magnetic moment, and the $J_{ij}^{\mu\nu}$ are calculated from the frozen magnon energies [170]. The other method is a real-space approach that was introduced by Lichtenstein *et al.* [171]. This method is based on multiple scattering theory and employs the magnetic force theorem to calculate the energy change associated with small rotations of the magnetic moments at sites i and j (and frozen potentials). The energy change can be related to the pair exchange interaction parameter of the Heisenberg model $H_{\text{eff}} = -\sum_{ij} J_{ij} e_i e_j$ (where the values of the magnetic moments are absorbed in the exchange parameters J_{ij}),

$$J_{ij} = \frac{1}{4\pi} \int^{E_F} dE \, \Im \text{Tr}_L \left(\Delta_i \hat{T}_{\uparrow}^{ij} \Delta_j \hat{T}_{\downarrow}^{ji} \right), \qquad (11.9)$$

which involves the scattering path operators within the KKR-CPA (Korringa–Kohn–Rostoker coherent potential approximation) scheme [171].

Figure 11.16a shows the magnetic exchange parameters [172] calculated with the spin-polarized relativistic Korringa–Kohn–Rostoker (SPR KKR) code [173] as a function of the distance between the atoms in units of the lattice constants d/a_0 for Ni_2MnGa (which is the only magnetic Heusler alloy that shows a martensitic transformation at stoichiometric composition, that is, $e/a = 7.5$). The plot in Figure 11.16b demonstrates the existence of long-range RKKY-like oscillations, which confirms the experimental findings of the oscillating contribution from spin-wave analysis [162,163]. See also the computation of J_{ij} using the frozen magnon approach of Şaşıoğlu *et al.* and the RKKY-type analysis for half-Heusler CuMnIn and PdCuIn and full Heusler Cu_2MnIn and Pd_2CuIn [164]. Different mechanisms contributing to the exchange coupling have also been discussed by Dronskowski and coworkers [174]. Finally, in order to calculate the orbital decomposed exchange coupling constants for the case of a Heusler alloy with strong atomic disorder ($Ni_{45}Co_5Mn_{34}In_{16}$), which are shown in Figure 11.24, the code of Ogura *et al.* [175] has been used. In this code, also Lichtenstein's method is implemented.

The Heusler alloys exhibit ferromagnetic order and the shape memory effect, which is called magnetic shape memory effect (MSME) [46]. The memory effect is due to the rather large magnetic anisotropy of the martensitic phase due to finite tetragonality ($c/a \neq 1$) and large spin–orbit coupling, which is confirmed by the *ab initio* calculations of Enkovaara *et al.* on Ni_2MnGa [176] using the full potential linearized augmented plane wave (FLAPW) method as implemented in

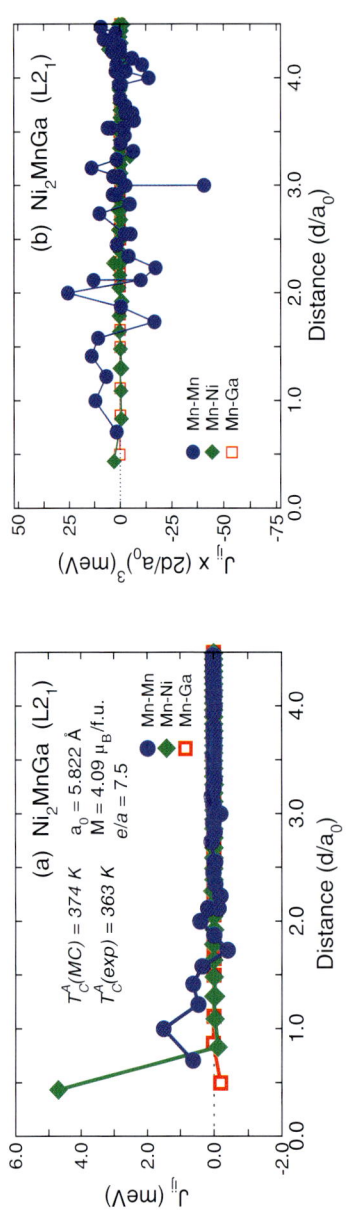

Figure 11.16 (a) Element-resolved magnetic exchange coupling constants of Ni$_2$MnGa as a function of the distance between the atoms with Mn at the origin. The Mn–Mn exchange is ferromagnetic and is the strongest interaction. (b) The magnetic exchange parameters of (a) multiplied by $(2d/a_0)^3$ which yields the characteristic oscillations of the RKKY interactions, which is finally responsible for the ferromagnetic order of Ni$_2$MnGa. Note that for disordered Mn-excess alloys antiferromagnetic interactions appear, which become stronger antiferromagnetic with $c/a \neq 1$ variation in martensite. The other alloys of the pseudobinary series Ni–Mn–(Al, Ga, In, Sn, Sb) show similar behavior compared to Ni–Mn–Ga. The RKKY-like oscillations become stronger with increasing number of s-, p-electrons of the Z element, which show the importance of the indirect interactions.

the code WIEN 2 k [177] and by experiment on Ni$_2$MnGa polycrystals [178,179]. For a detailed discussion of the linear change of the magnetic anisotropy energy with tetragonal distortion around $c/a = 1$, see Ref. [176]. Since in transition metals the spin–orbit coupling is about $\xi \approx 50$ meV, the magnetic anisotropy energy as a fourth-order effect in cubic symmetry is about μ eV. In the presence of tetragonal distortion as in the martensitic state of Heusler alloys, the symmetry is lowered and the magnetic anisotropy energy is a second-order effect and can be much larger, see Figure 11.17 for some prominent Heusler allys. The spin–orbit coupling induces also orbital moment anisotropy, which is not further pursued here, see Ref. [176] for discussions.

There are hardly measurements available on single crystals or other Heusler alloys. Figure 11.17 shows results of calculations provided by Gruner *et al.* using the VASP code [180] and by Koepernick and Eschrig using the FPLO code, which is a full-potential local-orbital minimum basis code [181].

At the starting point of Ullakko's investigations [182], the magnetic field-induced strain (MFIS) in the martensitic phase of Ni$_2$MnGa single crystal was about 0.2% at 265 K and the strain associated with martensitic transformation is accommodated by the formation of twin variants that minimize the elastic energy. Intensive research has led to increasing field induced strains of 6.5 and 9.5% [47] and finally of 12% [48] in nonmodulated structures of Ni$_2$MnGa due to considerably lowering the σ_{TW} value of Ni$_{46}$Mn$_{24}$Ga$_{22}$Co$_4$Cu$_4$ compared to the usual values of σ_{TW} in nonmodulated martensite of the ternary Ni–Mn–Ga intermetallics. In order to activate twin boundary motion in ferromagnetic martensite, the maximal magnetic stress σ_{MAG} must be lower than the twinning stress σ_{TW}, that is,

$$\sigma_{TW} < \sigma_{MAG} = K/\varepsilon_0, \ \varepsilon_0 = |c/a - 1|, \tag{11.10}$$

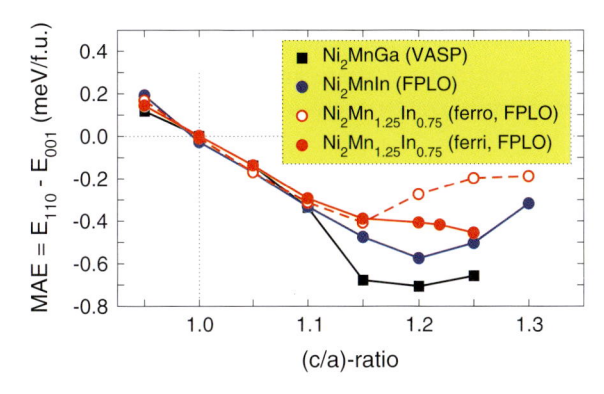

Figure 11.17 Second-order magnetocrystalline anisotropy energy per formula unit of Ni$_2$MnGa [180], Ni$_2$MnIn, and Ni$_2$Mn$_{1.25}$In$_{0.75}$ for ferromagnetic and ferrimagnetic spin configurations (for the latter, spins of excess Mn on the In sublattice have their spins flipped). For Ni–Mn–In, calculations were done with the full potential FPLO code [181] with up to 32.768 k points. The subscripts [110] and [001] refer to the magnetization direction. For $c/a = 1$, the anisotropy energy vanishes or becomes very small because of cubic symmetry.

where K is the magnetic anisotropy energy density. Its temperature dependence in the martensitic phase is deduced from $K = H_A M_S/2$ with anisotropy field H_A corrected for demagnetization effects (M_S is the saturation magnetization) and has been determined for polycrystalline Ni_2MnGa, $Ni_{2.15}Mn_{0.85}Ga$, and $Ni_2Mn_{1.1}Ga_{0.9}$ [179]. Note that the theoretical values shown in Figure 11.17 are all too large compared to the experimental values mainly because of the zero-temperature calculations, whereas the experimental values have been obtained at finite temperatures, which may lead to a reduction of about 30% [179].

Note that the modulated 10 and 14 M martensite structures of Ni_2MnGa exhibit a tetragonal distortion $c/a < 1$ with easy axis in c-direction (high-temperature $a(L2_1) = 5.822$ with martensitic transformation temperature $M_S \approx 276$, low-temperature $a = b = 5.90$, $c = 5.44$, $c/a = 0.922$) [182].

This easy axis becomes the hard axis for $c/a > 1$ for nonmodulated martensite (with energy minimum at approximately $c/a = 1.25$) and easy plane magnetocrystalline anisotropy perpendicular to the c-axis. The modulated martensite phases are not discussed in great detail here. For the interrelation of modulated and nonmodulated martensitic phases in terms of adaptive martensite, see the work of Fähler and coworkers [183,184]. Note that the origin of the martensitic transformation has been widely discussed in terms of an electronically driven instability due to Fermi surface nesting and the dramatic depletion of states (pseudogap), which vanishes below the transformation [185]. In addition, it was recently found that a peculiar nonharmonic lattice modulation of the adaptive martensite model goes hand in hand with the martensitic transition [186].

Figure 11.17 shows that with tetragonal distortion the martensitic phase develops magnetocrystalline anisotropy, and the strain associated with the martensitic transformation is accommodated by the formation of twin variants in configurations that minimize the elastic energy. With applied magnetic field, the twins can be reoriented, releasing the accommodated strain and heating up brings the crystal back to its cubic structure (this is at the heart of the magnetic shape memory effect).

There is another effect associated with the magnetic field applied to the martensitic phase: When the Heusler alloy becomes magnetic at the Curie temperature, there is practically no magnetic anisotropy present in the cubic phase and magnetization rises quickly with decreasing temperature. However, below M_S, magnetization may start to vanish in the polycrystalline samples because of the different magnetization directions in the polycrystalline sample and because of the competing ferromagnetic and antiferromagnetic interactions brought about by atomic disorder and cluster regions with Mn excess atoms. A sufficiently strong magnetic field when applied to martensite can lead to a first-order magnetic transition coexisting with the martensitic transformation, which has been named magnetostructural transition and which brings the sample back to the ferromagnetic austenitic phase. The magnetocaloric effect associated with this magnetostructural transition has been called inverse magnetocaloric effect in contrast to the conventional magnetocaloric effect around the Curie temperature because it leads to reverse signs of isothermal entropy and adiabatic temperature changes over the transition [54,187–189].

Here, *ab initio* calculations and Monte Carlo simulations of the finite-temperature magnetoelastic as well as the magnetocaloric effects have well reproduced the experimental observations. For discussion of theoretical work and further references we refer to the work of Comtesse and Sokolovskiy *et al.* [142,190,191]. The so-called metamagnetic behavior may lead to giant magnetocaloric effects in magnetic Heusler alloys doped additionally with Co or other transition metal elements in Mn excess systems such as Ni–Co–Mn–In [147]. However, the metamagnetic effect is not only present in magnetic Heusler alloys but is shown also by other intermetallics not having the Heusler crystal structure such as Fe–Rh [192], Mn_3GaC [193], and $CeFe_2$ [194–196], to name but a few.

There are large conventional magnetocaloric effects that are associated with a considerable adiabatic temperature change across the magnetic phase transition in an external magnetic field in systems such as $Gd_5(Si_2Ge_2)$ [197], $La(Fe,Si)_{13}$ [198], and Mn–Fe–P–As [199], which we will not discuss here. We limit the discussion here to the magnetocaloric effects of Heusler alloys, which may be of considerable importance for future refrigeration technologies, provided hysteresis effect, which hinder so far applications, can be mastered.

Figure 11.18 illustrates the magnetostructural effect in Ni–Mn–Z (Z = Ga, In, Sn, Sb) alloys in an external magnetic field of 5 T, showing the temperature dependence of the magnetization and the isothermal magnetic entropy change of Ni–Mn–In across the transition. The magnetic field-induced entropy change $\Delta S(T)$ has been determined by integrating numerically $\Delta S_{mag}(T,H) = \mu_0 \int_0^H dH\,(\partial M/\partial T)_H(T,H)$ [200]. Note that the corresponding adiabatic temperature change (not shown here) is obtained from $\Delta T_{ad} \approx -T(\Delta S(,T,H)/C(T,H)$, where the specific heat is the sum of magnetic and lattice contribution and, for simplicity, the field dependence of C is neglected during the integration. (For first-order magnetostructural phase transition, the Clausius–Clapeyron equation must be used, but note that for a sufficiently large magnetic field, the transition is always smoothened and Maxwell's relation may be used, see also the discussion of Porcari *et al.* [201].) The entropy change exhibits two peaks: one associated with the onset of magnetization at the Curie temperature, while the other peak is associated with the magnetostructural transition (inverse magnetocaloric effect) [46,200]. Note that the sign of ΔT_{ad} is opposite to ΔS and is negative for the inverse magnetocaloric effect, which leads to cooling. See also the discussion by Acet *et al.* [202].

Since the "nearly" first-order magnetostructural transition in an external magnetic field is a prerequisite to obtain a large magnetocaloric effect, we briefly discuss the origin of this coupled magnetic and structural transition. This was discussed earlier by Marcos *et al.*, observing that there are mainly two different

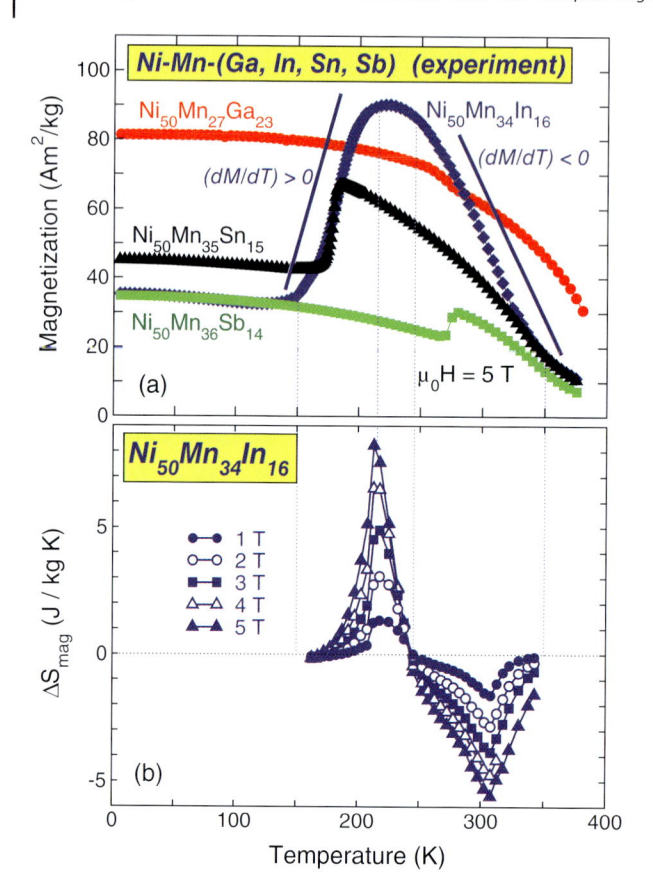

Figure 11.18 (a) The temperature dependence of the magnetization for Ni–Mn–Z martensitic Heusler alloys (Z = Ga, In, Sn, Sb). (b) Isothermal magnetic entropy change of Ni–Mn–In, which is negative around the Curie temperature and positive around the magnetostructural transition, which is reflected by the different slopes of the magnetization around the two transitions. (Data have been provided by Acet and coworkers [46,200]. Reproduced with permission from Ref. [46]. Copyright 2009, IOP Publishing Ltd.)

contributions: (i) magnetostructural coupling on the mesoscopic scale between the magnetic moments and the martensitic variants being simultaneously responsible for the magnetic shape memory effect as mentioned before and (ii) the microscopic spin–phonon coupling that gives rise to the shift of the transition temperatures with the applied magnetic field [203]. Both effects are there, magnetocrystalline anisotropy supporting the effect of magnetic field-induced reorientation of magnetic domains and variants and strong magnetoelastic coupling, although microscopic spin–phonon coupling still needs to be calculated. Note that (iii) a third effect contributes, which becomes dominant in all magnetic Heusler alloys with Mn excess, which leads to competing ferro- and antiferromagnetic interactions that are present already in the austenitic phase. But, the antiferromagnetic interactions become really dominant in the martensitic phase, which yields a very low magnetization of martensite and spin glass behavior.

Mechanisms (i) and (ii) are responsible for the behavior of the low-temperature magnetizations curves of near-stoichiometric polycrystalline and single crystal $Ni_{50}Mn_{25}Ga_{25}$. In Figure 11.19a and b we show corresponding experimental and calculated magnetizations curves, which have been discussed in detail by Sokolovskiy et al. [191,204]. Since martensite has a much larger anisotropy than austenite, it becomes much more difficult to magnetize the material below the structural transition in an external field. This leads to flat magnetization curves as observed in experiment [204]. When the Zeeman energy becomes much larger than the anisotropy energy, magnetic domains get destroyed and

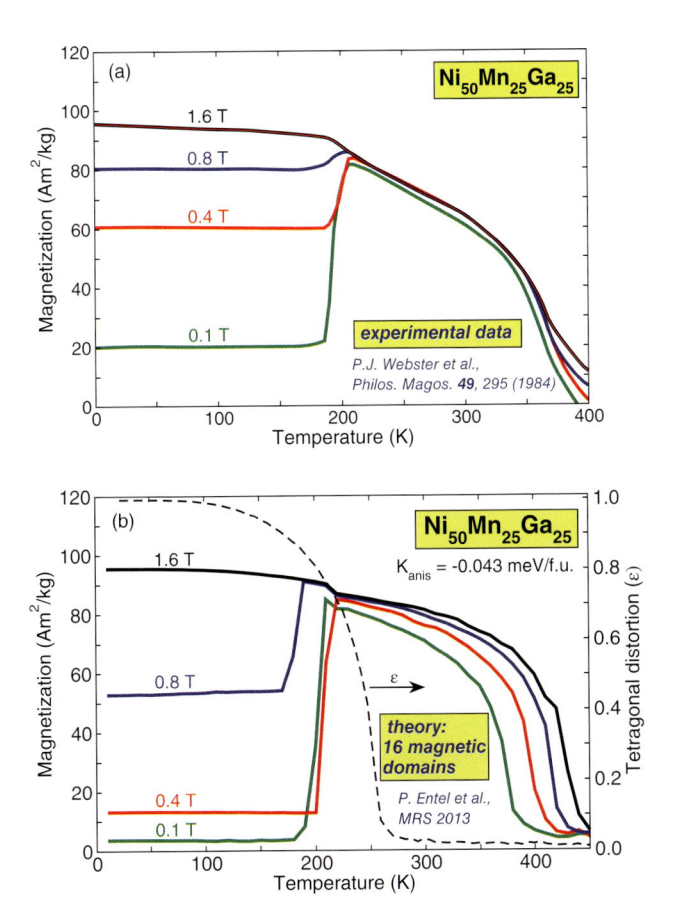

Figure 11.19 (a) Experimental and (b) theoretical isofield magnetization curves for polycrystalline $Ni_{50}Mn_{25}ga_{25}$ at various magnetic fields. The simulation cell used in the Monte Carlo simulations contained 16 magnetic domains. For details of simulations using an effective spin Hamiltonian, see Ref. [191].

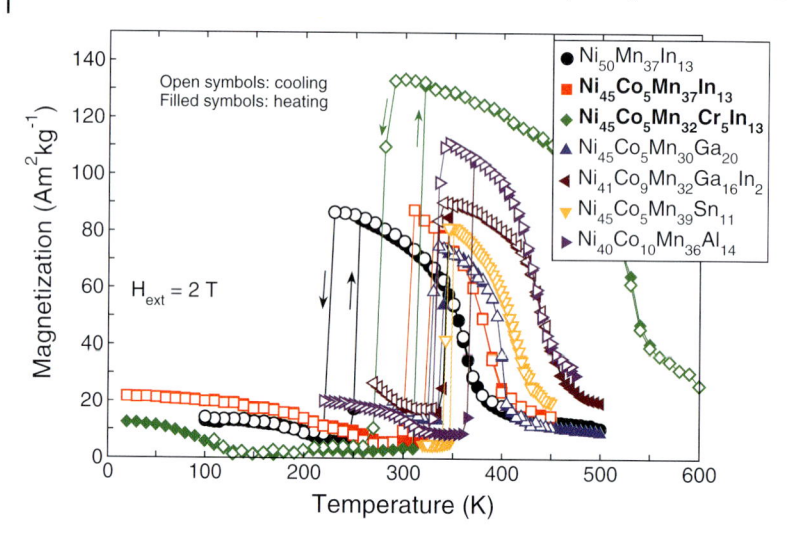

Figure 11.20 Magnetization curves upon cooling and heating across the magnetostructural phase transition (of "nearly" first order) for a series of Heusler alloys as listed in the figure. (Reproduced with permission from Ref. [190]. Copyright 2015, American Physical Society.)

the low-temperature magnetization curves of multidomain and single crystals show similar behavior [191].

For Mn-excess alloys, mechanism (iii) of the competing ferro and antiferromagnetic interactions destroys the long-range ferromagnetic order in the martensitic phase, which results in a large change of the magnetization at the magnetostructural transition. The magnitude of this magnetization jump mainly determines then the size of the magnetocaloric effect, which is shown in Figure 11.20 for a series of Heusler alloys [190].

Comparison of the Monte Carlo results for the large change of magnetization across the magnetostructural transition with experiment is made where possible. Agreement between theory and experiment with respect to magnetization changes, and large magnetic entropy and adiabatic temperature changes (where available) is observed in practically all cases. Nearly quantitative agreement of Monte Carlo simulations shown in Figure 11.21 with experiment is found for Ni–Co–Mn–In [147,149]. The simulations show the importance of optimal doping with Co, which enhances the Curie temperature T_C and martensitic transformation temperature M_S and leads to a giant magnetocaloric effect [147]. Further doping with Cr even enhances the cooling power [190].

The Monte Carlo simulations for the ferroic systems were done by using an appropriate model Hamiltonian taking into account $H = H_{\text{magnetic}} + H_{\text{elastic}} + H_{\text{magnetoelastic}}$ with parameters obtained from *ab initio* calculations. The heat capacity can be calculated from $C(T,H) = (1/k_B T^2)[\langle H^2 \rangle - \langle H \rangle^2]$, where the lattice contribution may be added by using for simplicity the Debye model. The integration of the heat capacity over the magnetostructural transition in an

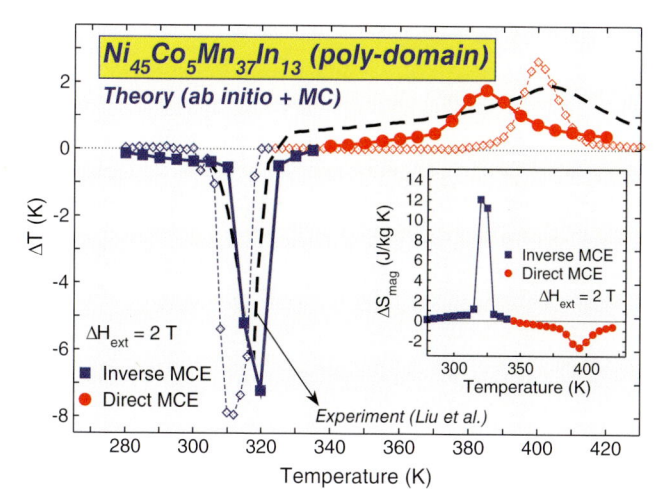

Figure 11.21 Calculated adiabatic temperature change ΔT_{ad} of the conventional or direct (red circles) and inverse (blue squares) magnetocaloric effect of polycrystalline $Ni_{45}Co_5Mn_{37}In_{13}$, that is, simulations of a magnetic polydomain structure. As spin model we have used the Potts model with spins coupled to the elastic degrees of freedom [142,190]. The parameters of this model have been fixed by first-principles calculations [142,190]. The thin curve with diamond symbols is the result for a single crystal without magnetic domains. The inset shows the calculated temperature variation of the entropy. The theoretical results agree well with the experimental data that are shown by the thick dashed curve [147]. (Reproduced with permission from Ref. [142]. Copyright 2014, APS.)

external field yields the isothermal entropy change and adiabatic temperature change. Since the external magnetic field will smooth the first-order magnet-structural transition, alternatively the Maxwell relation may be used to simulate the relations governing the magnetocaloric effect:

$$\Delta S_{mag}(T,H) = S_{mag}(T,H) - S_{mag}(T,0), \tag{11.11}$$

$$\Delta S_{mag}(T,H) = \mu_0 \int_0^H dH' \left(\frac{\partial M}{\partial T}\right)_{H'}, \tag{11.12}$$

$$\Delta T_{ad}(T,H) = -\mu_0 \int_0^H dH' \frac{T}{C(T,H')} \left(\frac{\partial M}{\partial T}\right)_{H'}, \tag{11.13}$$

$$\Delta T_{ad}(T,H) \approx -T \frac{\Delta S_{mag}(T,H)}{C(T,H)}. \tag{11.14}$$

All ferrocaloric and elastocaloric effects may be gathered together in a compact way by writing

$$\Delta S(T,Y) = \int_{\Delta Y} dY' \left(\frac{\partial X}{\partial T}\right)_{Y'}, \tag{11.15}$$

where $X = M$, $Y = H$ defines the magnetocaloric effect, $X = V$, $Y = -p$, the barocaloric effect, $X = \varepsilon$, $Y = \sigma$, the elastocaloric effect, and $X = P$, $Y = E$ is the electrocaloric effect.

For the Co-doped Ni–Mn–In alloys saturation magnetization is observed in very large magnetic fields of 10–14 T [149]. The magnetic phase diagram is furthermore enriched by the appearance of a spin glass phase at low temperatures with a spin glass freezing temperature of ≈ 60 K in the polycrystalline Ni–Co–Mn–In sample. The spin glass phase does not disappear in a single crystal but its freezing temperature T_f is lowered to ≈ 30 K [205]. This chapter also discusses the occurrence of multiple ferroic glasses in $Ni_{45}Co_5Mn_{36.6}In_{13.4}$, that is, besides the spin glass phase, the single crystal also exhibits a so-called strain glass phase, which is a frozen in type of atomic displacement fields that had earlier been denoted as a static frozen in tweed structure [92]. Conventionally, such glassy behavior is brought about by the introduction of point defects. In the present study, it has been shown that configurational disorder in a single ferromagnetic alloy can be used to stabilize both magnetic and strain glass phases. From the *ab initio* calculation of the long-range RKKY-type exchange coupling constants for the same material and from Monte Carlo simulations of the alloy we know that it is not a simple ferromagnet but exhibits strong competition of ferromagnetic, antiferromagnetic, and noncollinear interactions. The strength of all interactions change dramatically across the magnetostructural transition, in particular, the antiferromagnetic interactions become stronger, which is mainly responsible for the tiny displacements of the atoms and the appearance of the strain glass phase [142,206].

In addition, the same single crystal of $Ni_{45}Co_5Mn_{36.6}In_{13.4}$ exhibits a Kauzmann point [207], that is, the collapse of the entropy difference between a crystalline and a glassy phase, which is shown in Figure 11.23. The Kauzmann point for a magnetic system is obtained for vanishing entropy difference in the magnetic analog of the Clausius–Clapeyron equation of a martensitic system, which is given by $\mu_0 dH_0/dT = -\Delta S^{A \to M}/\Delta M^{A \to M}$ with $\Delta S \to 0$, and H_0 is the equilibrium magnetic field obtained from the H–M curves by using $H_0 = (H_{M_s} + H_{A_f})/2$. Alternatively, this point of vanishing entropy difference between crystalline and glassy phases, that is, where $dH_0/dT \approx 0$ holds, occurs below ≈ 150 K according to Ito *et al.* [208] and has been characterized as the kinetic arrest temperature below which no further martensitic transformation occurs because the driving force of the transformation vanishes upon cooling [208,209].

Data for the magnetic field–temperature phase diagram shown in Figure 11.23 can be obtained from the magnetization curves versus temperature in different magnetic fields shown in Figure 11.22 for $Ni_{45}Co_5Mn_{36.7}In_{13.3}$ [208], which is not much different from the composition $Ni_{45}Co_5Mn_{36.6}In_{13.4}$ for which Karaman and coworkers have established the phase diagram in Figure 11.23, which shows the Kauzmann point [205].

So, this magnetic intermetallic alloy system Ni–Co–Mn–In displays really complex behavior with four different solid–solid-type of phase transitions

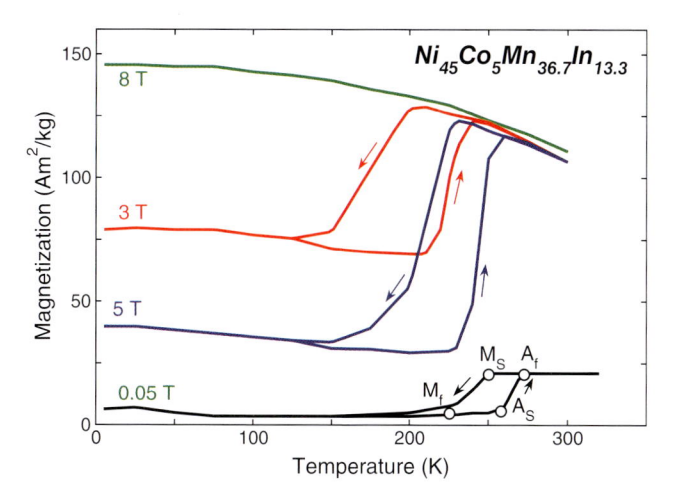

Figure 11.22 Isofield magnetization curves as a function of temperature for Ni$_{45}$Co$_5$Mn$_{36.7}$In$_{13.3}$ in external magnetic fields of 0.05, 3, and 5 T showing the decrease of the martensitic start temperature to lower temperatures as well as the decrease of the change of the magnetization ΔM at the magnetostructural transition with increasing magnetic field [208]. The isofield curves allow determining the characteristic temperatures M_S, M_f, A_S, A_f associated with the martensitic transformation that can be used to establish the magnetic field–temperature type of phase diagram as shown in Figure 11.23. (Reproduced with permission from Ref. [208]. Copyright @ AIP 2008 [211].)

(ferromagnetic–paramagnetic, martensitic transformation, magnetic, and strain glasses). Similar complexity is encountered when replacing In by Sn or Al [210–212].

In the case of Cr-doped Ni–Co–Mn–Cr–In, no experimental data are available; here, theory predicts a giant magnetocaloric effect provided the magnetic moments of Cr and Mn are parallel and the simultaneously spin flip at the magnetostructural transition gives rise to the huge change of magnetization, which is the largest one of all systems considered in Figure 11.20 (in a moderate external field of 2 T).

In order to discuss the microscopic origin of the metamagnetic behavior of Ni–Mn– (In, Sn), we show in Figure 11.24, besides the element-resolved exchange coupling constants of austenite ($c/a = 1$) and martensite ($c/a = 1.29$), some of the e_g and t_{2g} orbital-resolved contributions for Ni$_{45}$Co$_5$Mn$_{34}$In$_{16}$, which had been discussed for the Mn-rich Ni$_{50}$Mn$_{30}$Ga$_{20}$ alloys before [142]. The *ab initio* calculations of the exchange coupling constants have been done by the decomposition of the expression for the J_{ij} into orbital contributions using the code of Massako et al. [175] to decompose the J_{ij} and extract the contribution between L states ($L = (l, m)$ is the set of angular momentum and magnetic quantum numbers) at the ith site and L' states at jth site [142],

$$J_{ij}^{L-L'} = \frac{1}{4\pi} \int_{-\infty}^{\varepsilon_F} dE\, \text{Im}\, \Delta_{iL} \tau_{\uparrow LL'}^{ij} \Delta_{jL'} \tau_{\downarrow L'L}^{ji}. \tag{11.16}$$

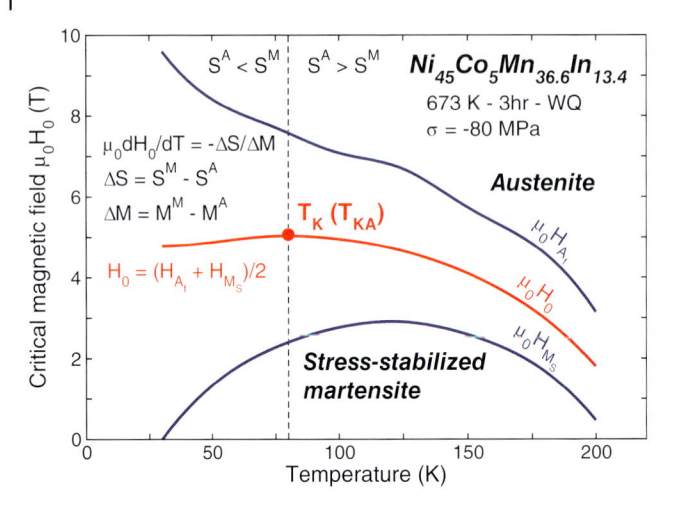

Figure 11.23 Magnetic (temperature field) phase diagram showing the martensitic start (H_{M_s}), austenite finish (H_{A_f}), and equilibrium (H_0) transformation magnetic fields as a function of temperature under −80 MPa stress. The vanishing of the entropy at T_K has been discussed in terms of a Kauzmann point T_K [205], respectively, in terms of a kinetic arrest temperature T_{KA} [209,210]. Chaddah and coworkers have discussed the kinetic arrest of the associated first-order ferromagnetic-to-antiferromagnetic transition in such materials (they had considered Ce(Fe$_{0.96}$Ru$_{0.04}$)$_2$ [195]) as a result of the formation of a nonergodic glass-like magnetic phase due to the slowing down of the magnetization dynamics, whereby the glassy state is formed with the help of an external magnetic field and this is distinctly different from a spin glass state. (Reproduced with permission from Ref. [205]. Copyright 2015, Elsevier.)

Fixing l, l' and summing over m, m', we arrive, for example, at a decomposition of exchange coupling constants in the form

$$J_{ij} = J_{ij}^{\text{d-d}} + J_{ij}^{\text{d-sp}} + J_{ij}^{\text{sp-d}} + J_{ij}^{\text{sp-sp}}, \tag{11.17}$$

$$J_{ij}^{\text{d-d}} = J_{ij}^{t_{2g}-t_{2g}} + J_{ij}^{t_{2g}-e_g} + J_{ij}^{e_g-t_{2g}} + J_{ij}^{e_g-e_g}. \tag{11.18}$$

When the elements at the ith and jth sites are the same, $J_{ij}^{\text{d-sp}} = J_{ij}^{\text{sp-d}}$ and $J_{ij}^{t_{2g}-e_g} = J_{ij}^{e_g-t_{2g}}$ we obtain

$$J_{ij} = J_{ij}^{\text{d-d}} + 2J_{ij}^{\text{d-sp}} + J_{ij}^{\text{sp-sp}}, \tag{11.19}$$

$$J_{ij}^{\text{d-d}} = J_{ij}^{t_{2g}-t_{2g}} + 2J_{ij}^{t_{2g}-e_g} + J_{ij}^{e_g-e_g}. \tag{11.20}$$

For example, Figure 11.24 shows some of the orbital-resolved magnetic coupling constants showing the remarkable feature that ferromagnetic t_{2g}–t_{2g} contributions are nearly compensated by the antiferromagnetic and e_g–e_g orbital exchange contributions, making the total exchange contributions rather small. Note that not all decomposed exchange contribution are shown in the Figure 11.24.

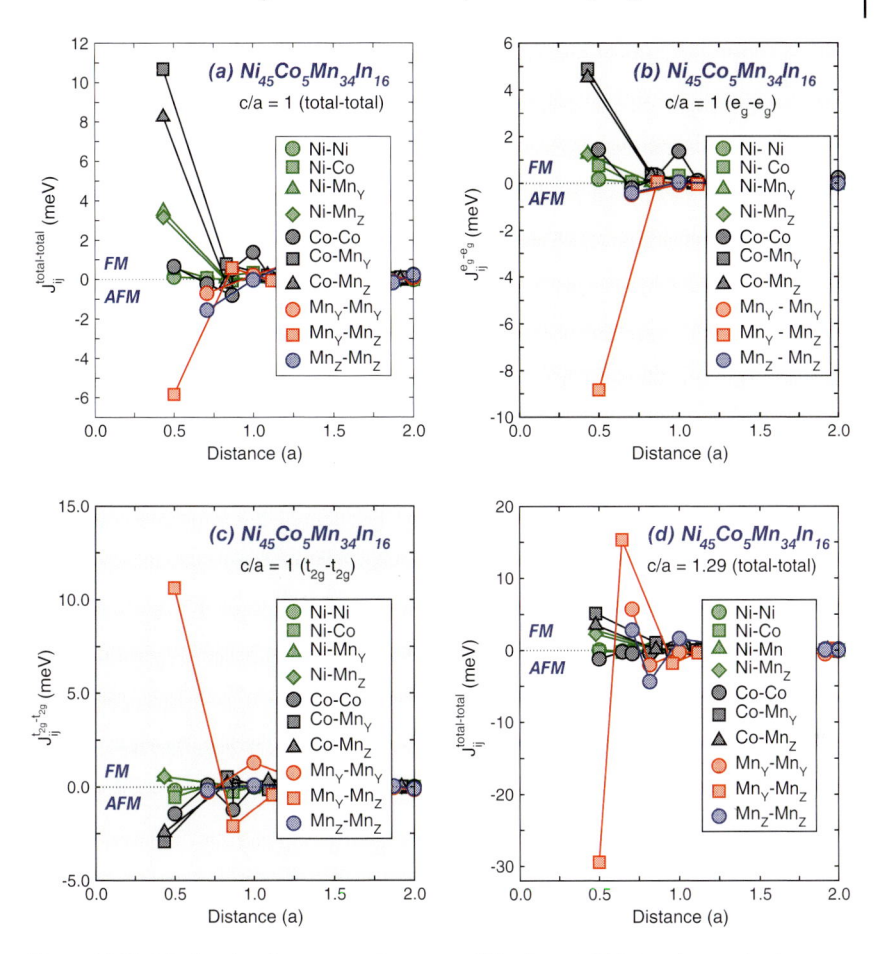

Figure 11.24 Calculated exchange coupling constants for the disordered Heusler alloy Ni$_{45}$Co$_5$Mn$_{34}$In$_{16}$ using the code of Ref. [175]. (a) and (d) show the element-resolved magnetic exchange coupling parameters for austenite and martensite of Ni$_{45}$Co$_5$Mn$_{34}$In$_{16}$ as a function of the distance between the atoms. Positive (negative) J_{ij} denote ferromagnetic (antiferromagnetic) interactions, respectively. In each case the first element listed is placed at the origin. In addition, parts (b) and (c) show the $e_g - e_g$ and $t_{2g} - t_{2g}$ orbital-resolved exchange contributions (not all orbitally decomposed contributions are shown). Remarkably, the large antiferromagnetic exchange coupling between e_g orbitals of Mn$_Y$ and Mn$_Z$ in part (b) is to a large extent compensated by the large ferromagnetic exchange coupling between t_{2g} orbitals of the same atoms in part (c). With the onset of the martensitic transformation, this particular exchange interaction becomes dominantly antiferromagnetic. Thus, austenite is dominantly ferromagnetic with competing ferromagnetic and antiferromagnetic interactions, while martensite is dominantly antiferromagnetic. It seems that this is the microscopic explanation of the metamagnetic behavior (i.e., transformation from antiferromagnetic martensite to ferromagnetic austenite in a sufficiently large external magnetic field [208]). Copyright @ AIP 2008 [211].

It is reasonable to assume that the increase of the strength of antiferromagnetic contributions in the martensitic phase allows an easy and sudden increase of ferromagnetism in a strong enough external magnetic field and a crossover from nearly antiferromagnetic martensite to ferromagnetic austenite, which is the first-order magnetostructural transition. Any doping of the magnetic Heusler alloys Ni–Mn– (Al, Ga, In, S, Sb) with transition metal elements such as Co strengthen the ferromagnetism of austenite and stabilizes the metamagnetic phase transition.

11.8
Concluding Remarks and Critical Assessment of Present Work

Intermetallic compounds and ordered intermetallic structures have attracted great interest as high temperature structural materials such as Fe–Al alloys with excellent oxidation resistance. In this chapter we have focused more on a few intermetallic shape memory alloys with emphasis on their magnetic properties, which show a change of their crystal structure and kinetic arrest features with a change of external parameters such as temperature, pressure, or magnetic field. The magnetic properties of Fe–Al alloys were briefly discussed. Here, the interest is on their mictomagnetic and metamagnetic properties, see, for example, Takahashi et al. regarding the spin distributions in plastically deformed Fe–Al intermetallic compounds [213] or Palstra et al. regarding La–Fe–Al compounds with cubic $NaZn_{13}$ structure [214]. Such mictomagnetic and metamagnetic trends appear also in the context of complex magnetic Heusler alloys. We like to remark that functional properties such as the giant magnetocaloric effect, which has been discussed here more closely in context of the complex magnetic Heusler alloys such as Ni–Mn– (Al, Ga, In, Sn, Sb), are not limited to these Mn-excess Heusler compounds, but have been discussed early for the rare earth compounds such as $Gd_5(Si_xGe_{1-x})$ near $x = 0.5$ [197] and subsequently for Tb–Si–Ge alloys [215], which also exhibit the giant magnetocaloric effect, metamagnetism, and a coupled magnetostructural phase transformation [216]. See also Ref. [217] and references therein. Another famous, classical metamagnetic magnetocaloric system with giant magnetocaloric features is Fe–Rh with cubic Cs–Cl structure that exhibits an isostructural ferromagnetic to antiferromagnetic phase transition [192], which is accompanied by a large magnetocaloric effect of $\Delta T_{ad} = -12.9$ K. A magnetostructural transition in this compound can be enforced by alloying Fe–Rh with Pd [218]. There have been attempts to tailor and to increase the cooling power of the Fe–Rh alloys and systematics of magnetostructural trends have been predicted [219,220]. Many other fascinating systems such as Mn_3GaC [193] with properties close to FeRh could not be addressed, here.

Most part of the discussion has been devoted to the complex magnetic intermetallics such as Ni–Mn– (Ga, In, Sn, Sb) showing that, in particular, the Ni–Mn–In alloys doped with Co exhibit extraordinary complex magnetic behavior accompanied by the magnetostructural phase transformation leading to the giant magnetocaloric effect, kinetic arrest or features in relation to a Kauzmann point, a cluster spin glass and a strain glass phase [189,205,208,210]. An interpretation of the kinetics of martensitic transformation in a Ni–Co–Mn–In alloy has recently been presented by Fukuda *et al.*, estimating the thermally activated cluster (nuclei of martensite) to be $(5\,\text{nm})^3$, which is not so different from what has been found by neutron scattering data [221,222].

The fascinating world of ferrocaloric material properties and underlying physical properties has recently been reviewed by Moya *et al.* [223] and many physical properties of magnetic intermetallics, which could not be discussed here, have been treated in the recent work by Felser (the book cited in [222]).

However, a critical remark needs to be added regarding our present understanding of the existence of different magnetic reference states ("ferri" and "ferro") as sketched in Figure 11.14b and c. Our *ab initio* treatment of relaxing atomic positions within the 16-atom supercell calculations suggests that there is no unique ground state, neither structurally nor magnetically, but, that, instead, the system is nonergodic at low temperatures, which may best be described as a mixture of a "strain glass" and a "magnetic glass" as mentioned before. "Ferri" ($c/a \neq 1$) and "ferro" ($c/a = 1$) solutions may smoothly be joined by a line of infinitely many noncollinear magnetic spin configuration solutions in the same way as for the Invar solution [224] where low-spin and high-spin solutions are smoothly joined by a line of noncollinear solutions.

Furthermore, the scenario of "kinetic arrest" in connection with the magnetostructural transition in magnetic Heusler alloys, which emerges naturally within this picture, has not yet satisfactorily been investigated up to day.

This "kinetic arrest" phenomenon, which may be regarded as the natural source of hysteretic effects, also requires further theoretical and experimental work. This could be achieved by simulating the glassy texture as has recently been proposed when discussing the formation of nanodomains in ferroelectrics [225] and ergodic-to-nonergodic state transformation. The formation of the complicated martensitic texture and glassy texture is also related to the experimental observation of intermartensitic phase boundaries shown in Figure 11.14a. Here, a recent successful *ab initio* simulation has been performed to describe the Ni–Mn–Ga Heusler alloy with magnetocaloric and shape memory properties and phase boundaries [226].

We would like to conclude that so far only successful work was highlighted, for example, the new explanation for the rapid decrease of the martensitic transformation temperature M_S in the shape memory intermetallic alloy Ti–Ni with small compositional changes due to Ti(Ni) and Ni(Ti) antisite defects [81] and the very successful simulation of the magnetocaloric effect of Ni–Co–Mn–In alloys using an *ab initio* based model Hamiltonian with magnetic, elastic, and magnetoelastic degrees of freedom [150]. Figures 11.6 and 11.21 show the most important results obtained by computation. The other diverse properties from structures to finite temperatures presented in this chapter are based on experimental and computational results where, for example, the VASP code [88] has proven to be very useful. So, the final question remains, of what are the challenges in understanding experimental and computational results for the compounds that are seemingly very different in structures and functional properties? Here, one major aspect is, if we restrict the discussion to the magnetic shape memory Heusler alloys such as Ni– (Co) –Mn–In, where the entropy $\Delta S(T)$ associated with the magnetostructural phase transition shows the "dilemma" of inverse magnetocaloric materials. On the one hand, a huge change in magnetization is necessary in order to drive the magnetocaloric effect. On the other hand, the large entropy change of the magnetic subsystem ΔS_{mag}, which is directly related to the magnetization, acts more and more against the lattice entropy change ΔS_{lat}. Hence, the transition may become arrested when both contributions compensate each other. For a discussion of the entropy change at the martensitic transformation and this dilemma, see Ref. [227,228].

Related to this problem are hysteresis effects that are associated with the first-order martensitic transformation whose temperature range becomes lowered by the application of high enough external magnetic field. The reversibility of the magnetocaloric effect strongly depends on the width of the thermal hysteresis and therefore it is necessary to find solutions how to minimize losses associated with thermal hysteresis in order to maximize the efficiency of magnetic refrigeration devices [229]. Here experiment and theory may in future successfully complement each other.

Acknowledgments

I would like to give special thanks to Richard Dronskowski and Dimitri Bogdanovski, who provided the COHP data, to Mehmet Acet and Asl Çakr for the phase diagram data, to Franca Albertini and Ingo Opahle for the magnetic anisotropy data, and to Ibrahim Karaman for the critical magnetic field data. Discussions with Raimundo Arróyave, Navdeep Singh, Vasiliy D. Buchelnikov, and Vladimir V. Sokolovskiy are gratefully acknowledged.

References

1 Sauthoff, G. (2008) *Intermetallics*, Wiley-VCH Verlag GmbH, Weinheim, Germany.
2 Westbrook, J.H. and Fleischer, R.L. (eds) (1955) *Intermetallic Compounds: Principles and Practice*, vols. 1 and 2, John Wiley & Sons, Inc., New York.
3 Hume-Rothery, W. (1936) *The Structure of Metals and Alloys*, The Institute of Metals, London, UK.
4 Asahi, R., Sato, H., Takeuchi, T., and Mizutami, U. (2005) Verification of Hume-Rothery electron concentration rule in Cu_5Zn_8 and Cu_9Al_4 γ brasses by *ab initio* FLAPW band calculations. *Phys. Rev. B*, **16**, 165103.
5 Frank, F.C. and Kasper, J.S. (1958) Complex alloy structures regarded as sphere packings: I. Definitions and classification of representative structures. *Acta Crystallogr.*, **11**, 184–190.

6 Frank, F.C. and Kasper, J.S. (1959) Complex alloy structures regarded as sphere packings: II. Analysis and basic principles. *Acta Crystallogr.*, **12**, 483–499.

7 Joubert, J.M. and Civello, J.C. (2012) Non-stoichiometric and Calphad modeling of Frank–Kasper phases. *Appl. Sci.*, **2**, 669–681.

8 Nelson, D.R. (1983) Liquids and glasses in spaces of incommensurate curvature. *Phys. Rev. Lett.*, **50**, 982–985.

9 Nelson, D.R. (1983) Order, frustration, and defects in liquids and glasses. *Phys. Rev. B*, **528**, 5515–5535.

10 Levine, D. and Steinhardt, P.J. (1984) Quasicrystals: a new class of ordered structures. *Phys. Rev. Lett.*, **53**, 2477–2480.

11 Hartmann, H., Ebert, F., and Bretschneider, O. (1931) Elektrolysen in Phosphatschmelzen: I. Die elektrolytische Gewinnung von α- und β-Wolfram. *Z. Anorg. Allg. Chem.*, **198**, 116–140.

12 Shi, D. (ed.) (1995) *High-Temperature Superconducting Materials, Science and Engineering*, Pergamon, Oxford, UK.

13 Bhattacharya, N. and Paranthaman, M.P. (eds) (2010) *High-Temperature Superconductors*, Wiley-VCH Verlag GmbH, Weinheim, Germany.

14 Krause, M. (2014) *CERN: How We Found the HIGGS BOSON*, World Scientific, Singapore.

15 Laves, F. and Löhberg, K. (1934) Die Kristallstruktur von intermetallischen Verbindungen der Formel AB_2. *Nachr. Ges. Wiss. Göttingen*, **1**, 59–66.

16 Murray, M.J. and Sanders, J.V. (2006) Close-packed structures of spheres of two different sizes: II. The packing densities of likely arrangements. *Philos. Mag.*, **42**, 721–740.

17 Sinha, A.K. (1972) Topologically close-packed structures in transition metal alloys. *Prog. Mater. Sci.*, **15**, 79–185.

18 Andrae, D., Paulus, B., Wedig, U., and Jansen, M. (2013) A first-principles study of electronic structure of the Laves phase $MgZn_2$. *Z. Anorg. Allg. Chem.*, **639**, 1963–1967.

19 Rosalie, J.M., Somekawa, H., Singh, A., and Mukai, T. (2013) Effect of precipitation on strength and ductility in a Mg–Zn–Y alloy. *J. Alloys Comp.*, **550**, 114–123.

20 Mao, P., Yu, B., Liu, Z., Wang, F., and Ju, Y. (2013) First-principles calculations of structural, elastic and electronic properties of a B2 type intermetallics in Mg–Zn–Ca–Cu alloy. *J. Magnes. Alloys*, **1**, 256–262.

21 Andersson, A., de Boissieu, M., Brüne, S., Drescher, C., Assmuss, W., Ohahshi, S., Tsai, A.P., Mihalkovič, M., Krajči, M., and Rapp, Ö. (2010) Electronic and structural properties of Laves-phase $MgZn_2$ varying chemical disorder. *Phys. Rev. B*, **82**, 024202.

22 Hafner, J., Jaswal, S.S., Tegze, M., Pflugi, A., Krieg, J., Oelhafen, P., and Günterodt, H.J. (1988) The atomic and electronic structure of metallic glasses: search for a structure-induced minimum in the density of states. *J. Phys. F*, **18**, 2583–2604.

23 Hafner, J. (2002) *From Hamiltonians to Phase Diagrams: The Electronic Mechanical and Statistical Mechanical Theory of s, p-Bonded Metals and Alloys*, Springer, Berlin, Germany.

24 Pettifor, D.G. (1982) Cohesion and decohesion in the metal bond. *Phys. Scr.*, **1**, 26–29.

25 Krajči, M. and Hafner, J. (2002) Covalent bonding and bandgap formation in intermetallic compounds: a case study for Al_3V. *J. Phys. Condens. Matter*, **14**, 1865–1879.

26 Russel, A. and Lee, K.L. (2005) *Structure–Property Relations in Nonferrous Metals*, John Wiley & Sons, Inc., Hoboken.

27 Brammer, T.M. (2012) Improving the phase stability and oxidation resistance of β-NiAl. M.Sc. thesis, Iowa State University, Ames, IA.

28 Donatchi, M.J., Jr., and Donatchi, S.J. (2002) *Superalloys: A Technical Guide*, 2nd edn, ASM International, Materials Park, OH.

29 Miracle, D.B. (1993) Overview No. 104: the physical and mechanical properties of NiAl. *Acta Metall. Mater.*, **41**, 649–684.

30 Massalski, T.B., Okamoto, H., Subramanian, P.R., and Kacprzak, L. (eds) (1990) *Binary Alloy Phase Diagrams*, 2nd

edn, vol. **3**, ASM, Materials Park, OH, p. 2874.

31 Stocks, G.M., Shelton, W.A., Nicholson, D.M., Pinski, F.J., Ginatempo, B., Barbieri, A., Györffy, B.L., Johnson, D.D., Staunton, J.B., Turchi, P.E.A., and Sluiter, M. (1992) First principles theory of alloy phase stability: ordering and pre-martensitic phenomena in β-phase NiAl, in *Ordered Intermetallics: Physical Metallurgy and Mechanical Behaviour* (eds C.T. Liu, R.W. Cahn, and G. Sauthoff), Kluwer, Dordrecht, The Netherlands, pp. 15–36.

32 Noebe, R.D., Bowman, R.R., and Nathal, M.V. (1996) The physical and mechanical metallurgy of NiAl, in *Physical Metallurgy and Processing of Intermetallic Compounds* (eds N.S. Stoloff and V.K. Sikka), Chapman & Hall, New York, NY.

33 Parthasarathi, A. and Beck, P.A. (1076) Magnetism in FeAl, CoAl and NiAl. *Solid State Commun.*, **18**, 211–216.

34 Min, B.T., Oguchi, T., and Jansen, H.J.F. (1986) Structural, electronic and magnetic properties of NiAl and FeAl alloys. *J. Magn. Magn. Mater.*, **54–57**, 1091–1092.

35 Au, Y.K. and Wayman, C.M. (1972) Thermoelastic behavior of the martensitic transformation in β′ NiAl alloys. *Scr. Metall.*, **6**, 1209–1214.

36 Thomson, R.J., Zhao, J.-C., and Hemker, K.J. (2010) Effect of ternary elements on a martensitic transformation in β-NiAl. *Intermetallics*, **18**, 796–802.

37 Martynov, V.V., Enami, K., Khandros, L.G., Nenno, S., and Tkachenko, A.V. (1983) Structure of martensitic phases formed in alloy 63.1 at%NiAl in tension. *Phys. Met. Metall.*, **55**, 136–143.

38 Khachaturyan, A.G., Shapiro, S.M., and Semenoskaya, S. (1992) Adaptive phase in martensitic transformation. *Phys. Rev. B Condens. Matter*, **33**, 278–281.

39 Kim, Y.D. and Wayman, C.M. (1992) Shape recovery and phase transformation behavior in Ni–Al alloys. *Metall. Trans.*, **23**, 2981–2986.

40 Chernenko, V.A. (1992) Compositional instability of β-phase in Ni–Mn–Ga alloys. *Scr. Mater.*, **40**, 523–527.

41 Cotton, J.D. (1991) The influence of chromium on structure and mechanical properties of B2 nickel aluminides alloys. Ph.D. thesis, the University of Florida, Gaineville, FL.

42 Ray, P.K., Brammer, T., Ye, Y.Y., Akine, M., and Kramer, M.J. (2010) A multi-stage hierarchical approach to alloy design. *JOM*, **62**, 25–29.

43 Tu, Y., Mao, Z., and Seidman, D.N. (2012) Phase-partitioning and site-substitution patterns of molybdenum in a model Ni–Al–Mo superalloy: an atom-probe tomographic and first-principles study. *Appl. Phys. Lett.*, **101**, 121910.

44 Kainuma, R., Geijima, F., Ohnuma, T., and Ishida, K. (2000) Ordering, martensitic ferromagnetic transformations in Ni–Al–Mn Heusler shape memory alloys. *Mater. Trans. JIM*, **41**, 943–949.

45 Mañosa, L., Planes, A., Acet, M., Duman, E., and Wassermann, E.F. (2004) Magnetic shape memory in Ni–Mn–Ga and Ni–Mn–Al. *J. Magn. Magn. Mater.*, **272–276**, 2090–2092.

46 Planes, A., Mañosa, L., and Acet, M. (2009) Magnetocaloric effect and its relation to shape-memory properties in ferromagnetic Heusler alloys. *J. Phys. Condens. Matter*, **21**, 233201.

47 Sozinov, A., Likhachev, A.A., Lanska, N., and Ullakko, K. (2002) Giant magnetic-field induced strain in NiMnGa seven-layered martensitic phase. *Appl. Phys. Lett.*, **80**, 1746–1748.

48 Sozinov, A., Lanska, L., Soroka, A., and Zou, W. (2013) 12% magnetic-field induced strain in Ni–Mn–Ga-based non-modulated martensite. *Appl. Phys. Lett.*, **102**, 021902.

49 Kainuma, R., Ise, M., Ishikawa, K., Ohnuma, I., and Ishida, K. (1998) Phase equilibria and stability of the B2 phase in the Ni–Mn–Al and Co–Mn–Al system. *J. Alloys Comp.*, **269**, 173–180.

50 Acet, M., Duman, E., Wassermann, E.F., Mañosa, L., and Planes, A. (2002) Coexisting ferro- and antiferromagnetism in Ni_2MnAl Heusler alloys. *J. Appl. Phys.*, **92**, 3867–3871.

51 Sutou, Y., Omori, T., Kainuma, R., and Ishida, K. (2008) Ductile Cu–Al–Mn

based shape memory alloys: general properties and applications. *Mater. Sci. Technol.*, **24**, 896–901.

52 Massalski, T.B., Murray, J.L., Bennett, L.H., and Baker, H. (1990) *Binary Alloy Phase Diagrams* (eds T.B. Massalski, J.L. Murray, L.H. Bennett, and H. Baker), ASM, Materials Park, OH.

53 Sokolovskiy, V.V., Buchelnikov, V.D., Zagrebin, M.A., Taskaev, S.V., Khovaylo, V.V., and Entel, P. (2013) *Ab initio* study of magnetic properties of Fe–Mn–Al Heusler alloys. *Mater. Res. Symp. Proc.*, **1581**. doi: 10.1557/opl.2013.888

54 Kainuma, R., Ito, W., Umetsu, R.Y., Oikawa, K., and Ishida, K. (2008) Magnetic field-induced reverse transformation in B2-type NiCoMnAl shape memory alloys. *Appl. Phys. Lett.*, **93**, 091906.

55 Shull, R.D., Okamoto, H., and Beck, P.A. (1976) Transition from ferromagnetism to mictomagnetism in Fe–Al alloys. *Solid State Commun.*, **20**, 863–868.

56 Plazaola, F., Apiñaniz, E., Rodriguez, D.M., Lagarra, E., and Garitaonandia, J.S. (2012) Fe–Al alloys' magnetism, in *Advance Magnetic Materials* (ed. L. Malkinski), InTech, Rijeka, Croatia, p. 133.

57 Nakai, Y. and Tsunoda, T. (1999) 3d elements with Cu, Ag or Au, *Magnetic Properties of Metals*, Springer, Berlin, Germany, p. 1.

58 Daw, M.S. and Baskes, M.I. (1984) Embedded-atom method: derivation and applications to impurities, surfaces, and other defects in metals. *Phys. Rev. B*, **29**, 6443–6453.

59 Baskes, M.I., Nelson, J.S., and Wright, A.F. (1989) Semiempirical modified embedded-atom potentials for silicon and germanium. *Phys. Rev. B*, **40**, 6085–6100.

60 Meyer, R. and Entel, P. (1998) Computer simulations of martensitic transformation in NiAl alloys. *Comput. Mater. Sci.*, **10**, 10–15.

61 Becquart, C.S., Clapp, P.C., and Rifkin, J.A. (1993) Molecular-dynamics simulation of tweed and the ω phase in Ni–Al. *Phys. Rev. B*, **48**, 6–13.

62 Meyer, R. and Entel, P. (1998) Martensite–austenite transition and phonon dispersion of $Fe_{1-x}Ni_x$ studied by molecular-dynamics simulations. *Phys. Rev. B*, **57**, 5140–5147.

63 Urbassek, H.M. and Sandoval, J.S. (2012) Molecular dynamics modeling of martensitic transformations is steels, in *Phase Transformations in Steels, Volume 2: Diffusionless Transformations, High Strength Steels, Modeling and Advanced Analytical Techniques* (eds E. Pereloma and D.V. Edmonds), Woodhead, Oxford, UK, p. 433.

64 Schmitt, U., Wang, B., Urbassek, H.M., and Müller, R. (2013) Modeling of martensitic transformations in pure iron by a phase field approach using information from atomistic simulations. *Tech. Mech.*, **33**, 119–130.

65 Dai, X.D., Kong, Y., Li, J.H., and Liu, B.X. (2006) Extended Finnis–Sinclair potential for bcc and fcc metals and alloys. *J. Phys. Condens. Matter*, **18**, 4527–4542.

66 Guthikonda, V.S. and Elliott, R.S. (2009) An effective interaction potential model for the shape memory alloy AuCd. *Continuum Mech. Thermodyn.*, **21**, 269–295.

67 Drautz, R., Hammerschmidt, T., Čák, M., and Pettifor, D.G. (2015) Bond-order potentials: derivation and parameterization for refractory elements. *Model. Simul. Mater. Sci. Eng.*, **23**, 074004.

68 Car, R. and Parinello, M. (1985) Unified approach for molecular dynamics and density-functional theory. *Phys. Rev. Lett.*, **55**, 2471–2474.

69 Nosé, S. (1984) A unified formulation of the constant temperature molecular-dynamics methods. *J. Chem. Phys.*, **81**, 511–519.

70 Hoover, W.G. (1985) Canonical dynamics: equilibrium phase-space distributions. *Phys. Rev. A*, **31**, 1695–1697.

71 Kresse, G. and Hafner, J. (1993) *Ab initio* molecular dynamics for liquid metals. *Phys. Rev. B*, **47**, 558–561.

72 Turchi, P.E.A. and Gonis, A. (eds) (1994) *Statics and Dynamics of Alloy Phase Transformations*, NATO Science Series B, Springer, New York, NY.

73 Marx, D. and Hutter, J. (2009) *Ab Initio Molecular Dynamics: Basic Theory and Advanced Methods*, Cambridge University Press, Cambridge, NY.

74 Otsuka, K. and Wayman, C.M. (1998) Mechanism of shape memory effect and superelasticity, in *Shape Memory Materials* (eds K. Otsuka and C.M. Wayman), Cambridge University Press, Cambridge, p. 27.

75 Miyazaki, S. (1990) Thermal and stress cycling effects and fatique properties of Ni–Ti alloys, in *Engineering Aspects of Shape Memory Alloys*, 1st edn (eds T.W. Duerig, W.K.N. Melton, D. Stöckel, and C.M. Wayman), Butterworth-Heinemann, London, UK, p. 394.

76 Otsuka, K. and Ren, X. (2005) Physical metallurgy of Ti–Ni-based shape memory alloys. *Prog. Mater. Sci.*, **50**, 511–678.

77 Tang, W., Sundman, S., Sandström, R., and Qiu, C. (1999) New modeling of the B2 phase and its associated martensitic transformation in the Ti–Ni system. *Acta Mater.*, **47**, 3457–3468.

78 Nguyen Manh, D., Pasturel, A.S., Paxton, A.T., and van Schilfgaarde, M. (1993) Structural stability of $NiTi_2$ intermetallic compounds. *J. Phys. Condens. Matter*, **5**, 9087.

79 Pasturel, A., Colinet, C., Nguyen Manh, D., Paxton, A.T., and van Schilgaarde, M. (1995) Electronic structure and phase stability study in the Ni–Ti system. *Phys. Rev. B*, **52**, 15176–15190.

80 Zhang, J., Fan, G., Zhou, Y., Ding, X., Ren, X., Sun, J., Nakamura, K., and Otsuka, K. (2006) The nonexistence of an order–disorder transition in near-stoichiometric TiNi alloy. *Mater. Sci. Eng. A*, **438–440**, 608–611.

81 Frenzel, J., Wieczorek, A., Opahle, I., Maaß, B., Drautz, R., and Eggeler, G. (2015) On the effect of alloy composition on martensite start temperatures and latent heats in Ni–Ti based shape memory alloys. *Acta Mater.*, **90**, 213–231.

82 Colinet, C. and Paturel, A. (1993) Thermodynamics of the nickel–titanium system. *Physica B*, **192**, 238–246.

83 Miedema, A.R., Châtel, P.F., and de Boer, F.R. (1980) Cohesion in alloys: fundamental of a semi-empirical model. *Physica B*, **100B**, 1–28.

84 Pettifor, D.G. (1987) A quantum-mechanical critique of the Miedema rules for alloy formation. *Solid State Physics*, **1987**, 43–92.

85 Brewer, L. (1968) Bonding and structures of transition metals. *Science*, **161**, 115–122.

86 Zheng, Y., Jiang, F., Li, L., Yong, H., and Liu, Y. (2008) Effect of ageing treatment on the transformation behavior of Ti-50.9 at.% Ni alloy. *Acta Mater.*, **56**, 736–745.

87 Khalil-Allafi, J., Dlouhy, A., and Eggeler, G. (2002) Ni_4Ti_3-precipitation during aging of NiTi shape memory alloys and its influence on martensitic phase transformations. *Acta Mater.*, **50**, 4255–4274.

88 Kresse, G. and Furthmüller, J. (1996) Efficient iterative schemes for *ab initio* total-energy calculations using a plane-wave basis set. *Phys. Rev. B*, **54**, 11169–11186.

89 Kresse, G. and Joubert, D. (1999) From ultrasoft pseudopotentials to the projector augmented-wave method. *Phys. Rev. B*, **59**, 1758–1775.

90 Kacharuryan, A.G. (1983) *Theory of Structural Transformation in Solids*;, Dover, New York.

91 Kartha, S.V., Castán, T., Krumhansl, J.A., and Sethna, J. (1991) Spin-glass nature of tweed precursors in martensitic transformations. *Phys. Rev. Lett.*, **67**, 3630–3633.

92 Ren, X., Wang, Y., Zhou, Y., Zhang, Z., Wang, D., Fan, G., Otsuka, K., Suzuke, T., Ji, Y., Zhang, J., Tian, Y., Hou, S., and Ding, X. (2010) Strain glass in ferroelastic systems: premartensitic tweed versus strain glass. *Philos. Mag.*, **90**, 141–157.

93 Huang, X., Ackland, G., and Rabe, K.M. (2003) Crystal structures and shape-memory behavior of NiTi. *Nat. Mater.*, **2**, 307–311.

94 Shaw, J.A. and Kyriakides, S. (1995) Thermomechanical aspects of NiTi. *J. Mech. Phys. Solids*, **43**, 1243–1281.

95 Shaw, J.A. and Kyyiakides, S. (1997) On the nucleation and propagation of phase

96 Shaw, J.A. (2000) Simulations of localized thermo-mechanical behavior in a NiTi shape memory alloy. *Int. J. Plast.*, **16**, 541–562.
  transformation fronts in a NiTi alloy. *Acta Mater.*, **45**, 683–700.
97 Buehler, W.J., Gilfrich, J.V., and Wiley, R.C. (1963) Effect of low-temperature phase changes on the mechanical properties of alloys near composition NiTi. *J. Appl. Phys.*, **3**, 1475–1477.
98 Ma, J., Karaman, I., and Noebe, R.D. (2010) High temperature shape memory alloy. *Int. Mater. Rev.*, **55**, 257–315.
99 Nam, T.H., Saburi, T., and Shimizu, K. (1990) Cu-content dependence of shape memory characteristics if Tu–Ni–Cu alloys. *Mater. Trans. JIM*, **31**, 959–967.
100 Chluba, C., Ge, W., de Miranda, R.L., Strobel, J., Kienle, L., Quandt, E., and Wuttig, M. (2015) Ultra-low fatigue shape memory alloys. *Science*, **384**, 1004–1007 (see also Supplementary Materials: http://www.sciencemag.org/content/348/6238/1004/ suppl/DCI Materials and Methods).
101 James, R.D. (2015) Taming the temperamental metal transformation. *Science*, **348**, 968–969.
102 Zhang, J., Wang, Y., Ding, X., Zhang, Z., Zhou, Y., Ren, X-., Wong, D., Ji, Y., Song, M., Otsuka, K., and Sun, J. (2011) Spontaneous strain glass to martensite transition in a $Ti_{50}Ni_{44.5}Fe_{5.5}$ strain glass. *Phys. Rev. B*, **84**, 214201.
103 Wang, Y., Huang, C., Gao, J., Yang, S., Ding, X., Song, X., and Ren, X. (2012) Evidence for ferromagnetic strain glass in Ni–Co–N–Ga Heusler alloy system. *Appl. Phys. Lett.*, **101**, 101913.
104 Wang, Y., Ren, X., and Otsuka, K. (2006) Shape memory effect and superelasticity in a strain glass alloy. *Phys. Rev. Lett.*, **97**, 225703.
105 Entel, P., Arróyave, R., Fähler, S., Kainuma, R., Planes, A., Ren, X., and Saxena, A. (eds) (2014) Ferroic glasses: magnetic, polar, and strain glass. *Phys. Status Solidi B*, **251** (Special Issue), 1957–2150.
106 Lookman, T., Xue, D., Vasseur, R., Zong, H., and Ding, X. (2014) On glassy behavior in ferroics. *Phys. Status Solidi B*, **251**, 2003–2009.
107 Vasseur, R., Lookman, T., and Shenoy, S. (2010) Microstructure from ferroelastic transitions using strain pseudospin clock models in two and three dimensions: a local mean-field analysis. *Phys. Rev. B*, **82**, 094118.
108 Ren, X. (2014) Strain glass and ferroic glass: unusual properties from glassy nano-domains. *Phys. Status Solidi B*, **251**, 1982–1992.
109 Lloveras, P., Castán, T., Porta, M., Planes, A., and Saxena, A. (2008) Influence of elastic anisotropy on structural nanoscale textures. *Phys. Rev. Lett.*, **100**, 165707.
110 Lloveras, P., Castán, T., Porta, M., Planes, A., and Saxena, A. (2009) Glassy behavior in martensites: interplay between elastic anisotropy and disorder in zero-field-cooling/field-cooling simulation experiments. *Phys. Rev. B*, **80**, 054107.
111 Saxena, A., Castán, T., Planes, A., Porta, M., Kishi, Y., Lograsso, T.A., Viehland, D., Wuttig, M., and De Graef, M. (2004) Origin of magnetic and magnetoelastic tweed like precursor modulations in ferroic materials. *Phys. Rev. Lett.*, **92**, 197203.
112 Sherrington, D. (2014) A spin glass perspective on ferroic glasses. *Phys. Status Solidi*, **251**, 1967–1981.
113 Schmidt, R., Schlereth, M., Wipf, H., Assmus, W., and Müllner, M. (1989) Hydrogen solubility and diffusion in the shape-memory alloy NiTi. *J. Phys. Condens. Matter*, **1**, 2473–2482.
114 Dronskowski, R. (1993) Crystal orbital populations (COHP): energy-resolved visualization of chemical bonding in solids based on density-functional calculations. *J. Phys. Chem.*, **97**, 8617–8624.
115 Dronskowski, R. (2005) *Computational Chemistry of Solid State Materials: A Guide for Materials Scientists, Chemists, Physicists and Others*, Wiley-VCH Verlag GmbH, Weinheim, Germany.
116 Maintz, S., Deringer, V.L., Tchougréeff, A.L., and Dronskowski, R. (2013) Analytic projection from plane-wave and PAW wavefunctions to chemical bonding

117 Harris, N.A., Hadler, A.B., and Fredrickson, D.C. (2011) In search of chemical frustration in the Ca–Cu–Cd system: chemical pressure relief in the crystal structures of Ca_5Cu_2Cd and $Ca_2Cu_2Cd_9$. *Z. Anorg. Allg. Chem.*, **637**, 1961–1974.

analysis in solids. *J. Comput. Chem.*, **34**, 2557–2567.

118 Fredrickson, D.C. (2011) Electronic packing frustration in complex intermetallic structure: the role of chemical pressure in Ca_2Ag_7. *J. Am. Chem. Soc.*, **133**, 10070–10073.

119 Yanello, V.J., Kilduff, B.J., and Fredrickson, D.C. (2014) Isolobal analogies in intermetallics: the reversed approximation MO approach and application to $CrGa_4$- and Ir_3Ge_7-type phases. *Inorg. Chem.*, **53**, 2730–2741.

120 Hadler, A.B. (2014) Bergman clusters, multiple bonds, and defect planes: synthetic outcomes of chemical frustration in ternary intermetallic systems. Ph.D. thesis, the University of Wisconsin, Madison, WI.

121 Hadler, A.B. and Fredrickson, D.C. (2012) $Gd_{13}Fe_{10}C_{13}$: indications of Fe–Fe multiple bonding emerging from chemical frustration. *J. Am. Chem. Soc.*, **134**, 10361–10364.

122 Bernal, J.D. (1959) A geometrical approach to the structure of liquids. *Nature*, **183**, 141–147.

123 Gaspard, J.P. and Cyrot-Lackmann, F. (1973) Density of states from moments: application to the impurity band. *J. Phys. C*, **6**, 3077–3096.

124 Burdett, J.K. and Lee, S.J. (1985) Moments and the energy of solids. *J. Am. Chem. Soc.*, **107**, 3050–3063.

125 Lee, S. (1996) Structural diversity in solid state chemistry. *Annu. Rev. Phys. Chem.*, **47**, 397–419.

126 Hoistad, L.M. and Lee, S. (1991) The Hume-Rothery electron concentration rules and second moment scaling. *J. Am. Chem. Soc.*, **113**, 8216–8220.

127 Schröter, M., Entel, P., and Mishra, S.G. (1990) Metallic magnetism and magnetic volume collapse. *J. Magn. Magn. Mater.*, **87**, 163–176.

128 Hoffmann, E., Entel, P., Schwarz, K., and Mohn, P. (1995) First-principles description of magnetovolume instabilities in intermetallic compounds. *Magn. Magn. Mater.*, **140–144**, 237–238.

129 Podgorny, M. (1991) Electronic structure of the ordered phases of Pt–Fe alloys. *Phys. Rev. B*, **43**, 11300–11318.

130 Sabiryanov, R.F. and Jaswal, A.S. (1999) Magnon and magnon–phonon interaction in iron. *Phys. Rev. Lett.*, **83**, 2062–2064.

131 Paddison, J.A.M., Stewart, J.R., Manuel, P., Courtois, P., McIntyre, G.J., Rainford, B.D., and Goodwin, A.L. (2013) Emergent frustration in Co-doped β-Mn. *Phys. Rev. Lett.*, **110**, 267207.

132 Ferro, R. and Saccone, A. (2008) *Intermetallic Chemistry*, vol. **13**, Pergamon Materials Series (R.W. Cahn), Elsevier, Amsterdam, NL.

133 Webster, P.J. and Ziebeck, K.R.A. (1988) Heusler alloys, in *Landolt–Börnstein New Series Group III: Crystal and Solid State Physics – Magnetic Properties of Metals* (ed. H.P.J. Wijn), Springer, Berlin, Germany, pp. 75–185.

134 Gilleben, M. and Dronskowski, R. (2009) A combinatorial study of inverse Heusler alloys by first-principles computational methods. *J. Comput. Chem.*, **31**, 612.

135 Galanakis, I. and Dederichs, P.H. (2002) Slaterr–Pauling behavior and origin of the Hals-metallicity of the full-Heusler alloys. *Phys. Rev. B*, **66**, 174429.

136 Righi, L., Albertini, F., Fabbrici, S., and Paoluzzi, A. (2011) Crystal structures of modulated martensitic phase of FSM Heusler alloys. *Mater. Sci. Forum*, **684**, 105–116.

137 Entel, P., Siewert, M., Gruner, M.E., Herper, H.C., Comtesse, D., Arróyave, R., Singh, N., Talapatra, A., Sokolovskiy, V.V., Buchelnikov, V.D., Albertini, F., Righi, L., and Chernenko, V.A. (2013) Complex magnetic ordering as a driving mechanism of multifunctional properties of Heusler alloys from first principles. *Eur. Phys. J. B*, **86**, 65–76.

138 Buchelnikov, V., Sokolovskiy, V., Taranenko, I., Taskaev, S., and Entel, P. (2011) Monte Carlo modeling of exchange bias effect in $Ni_{50}Mn_{25+x}Sb_{25-x}$

Heusler alloys. *J. Phys. Conf. Series*, **303**, 012084.

139 Buchelnikov, V.D., Sokolovskiy, V.V., Taskaev, S.V., Khovaylo, V.V., Aliev, A.A., Khanov, L.N., Batdalov, A.B., Entel, P., Miki, H., and Takagi, T. (2011) Monte Carlo simulations of the magnetocaloric effect in magnetic Ni–Mn–X (X=Ga, In) Heusler alloys. *J. Phys. D*, **44**, 064012.

140 Moya, X., Hueso, L.E., Maccherozzi, F., Tovstolytkin, A.I., Podyalovskii, D.I., Ducati, C., Phillips, L.C., Ghidini, M., Hovorka, O., Berger, A., Vickers, M.E., Defay, E., Dhesi, S.S., and Mathur, N.D. (2013) Giant and reversible extrinsic magnetocaloric effects in $La_{0.7}Ca_{0.3}MnO_3$ films due to strain. *Nat. Mater.*, **12**, 52–58.

141 Moya, X., Defay, E., Heine, V., and Mathur, N.D. (2015) Too cool to work. *Nat. Phys.*, **11**, 202–205.

142 Comtesse, D., Gruner, M.E., Ogura, M., Sokolovskiy, V.V., Buchelnikov, V.D., Grünebohm, A., Arróyave, R., Singh, N., Gottschall, T., Gutfleisch, O., Chernenko, V.A., Albertini, F., Fähler, S., and Entel, P. (2014) First-principles instability leading to giant inverse magnetocaloric effects. *Phys. Rev. B*, **89**, 184403.

143 Ray, D.K. and Jardin, J.P. (1986) Elastic and magnetic interactions in a narrow twofold-degenerate band. *Phys. Rev. B*, **33**, 5021–5027.

144 Uijttewaal, M.A., Hickel, T., Neugebauer, J., and Gruner, M.E., and Entel, P. (2009) Understanding the phase transitions of the Ni_2MnGa shape memory system from first principles. *Phys. Rev. Lett.*, **102**, 035702.

145 Çakır, A., Righi, L., Albrtini, F., Acet, M., and Farle, M. (2015) Extended investigation of intermediate transitions in Ni–Mn–Ga magnetic shape memory alloys: a detailed phase diagram determination. *J. Appl. Phys.*, **114**, 183912.

146 Çakır, A., Righi, L., Albrtini, F., Acet, M., and Farle, M. (2015) Intermartensitic transitions and phase stability in $Ni_{50}Mn_{50-x}Sn_x$ Heusler alloys. *Acta Mater.*, **99**, 140–149.

147 Liu, J., Gottschall, T., Skokov, K.P., Moore, J.D., and Gutfleisch, O. (2012) Giant magnetocaloric effect driven by structural transitions. *Nat. Mater.*, **11**, 620–626.

148 Bourgault, D., Tiller, J., Courtois, P., Chaud, X., Caillault, N., and Carbone, L. (2010) Large magneto-caloric effect in Ni–Co–Mn–In systems at room temperature. *Phys. Procedia*, **10**, 120–124.

149 Gottschall, T., Skokov, K.P., Frincu, B., and Gutfleisch, O. (2015) Large reversible magnetocaloric effect in Ni–Mn–In–Co. *Appl. Phys. Lett.*, **106**, 021901.

150 Sokolovskiy, V.V., Entel, P., Buchelnikov, V.D., and Gruner, M.E. (2015) Achieving large magnetocaloric effects in Co- and Cr-substituted Heusler alloys: predictions from first-principles and Monte Carlo studies. *Phys. Rev. B*, **91**, 220409.

151 Barandiarán, J.M., Chernenko, V.A., Cesari, E., Sala, D., Lazpita, P., Gutierrez, J., and Orue, I. (2013) Magnetic influence on the martensitic transformation entropy in Ni–Mn–In metamagnetic alloy. *Appl. Phys. Lett.*, **102**, 071904.

152 Sokolovskiy, V., Grünebohm, A., Buchelnikov, V., and Entel, P. (2014) *Ab initio* Monte Carlo approaches for magnetocaloric effect in Co- and In-doped Ni–Mn–Ga Heusler alloys. *Entropy*, **16**, 4992–5019.

153 Uhl, E. (1982) The ferromagnetic and paramagnetic properties of Heusler alloys $(Ni_{1-x}Co_xMnSn)$. *J. Solid State Chem.*, **43**, 354–358.

154 Gilleben, M. and Dronskowski, R. (2010) A combinatorial study of inverse Heusler alloys by first-principles computational methods. *J. Comput. Chem.*, **31**, 612–619.

155 Heusler, F. (1903) Über magnetische Manganlegierungen. *Verhandlungen Deutschen Phyikalischen Gesellschaft*, **12**, 219.

156 Zener, C. (1951) Interaction between the d-shells in transition metals: II. Ferromagnetic compounds of Mn with perovskite structure. *Phys. Rev.*, **82**, 403–405.

157 Anderson, P.W. and Hasegawa, H. (1955) Considerations on double exchange. *Phys. Rev.*, **100**, 675–681.

158 De Gennes, P.-G. (1960) Effects of double exchange in magnetic crystals. *Phys. Rev.*, **118**, 141–154.

159 Anderson, P.W. (1961) Localized magnetic states in metals. *Phys. Rev.*, **124**, 41–53.

160 Caroli, B. (1967) Interaction entre moments magnétiques localisés. *J. Phys. Chem. Solids*, **28**, 1427–1440.

161 Caroli, B. and Blandin, A. (1966) Champs hyperfins dans les alliages d'Heusler. *J. Phys. Chem. Solids*, **27**, 503–508.

162 Noda, Y. and Ishikawa, Y. (1976) Spin waves in Heusler alloys Pd_2MnSn and Ni_2MnSn. *J. Phys. Soc. Japan*, **40**, 690–698.

163 Tajima, K., Ishikawa, Y., Wester, P., Stringfellow, M.W., Tocchetti, D., and Ziebeck, K.R.A. (1977) Spin waves in Heusler alloy Cu_2MnAl. *J. Phys. Soc. Japan*, **43**, 483–489.

164 Şaşııõ, E., Sandratskii, L.M., and Bruno, P. (2008) Role of conduction electrons in mediating exchange interactions in Mn-based Heusler alloys. *Phys. Rev. B*, **77**, 064417.

165 Kasuya, T. (1974) Exchange mechanisms in Heusler alloys: virtual double exchange. *Solid State Commun.*, **15**, 1119–1122.

166 Kübler, K., William, A.R., and Sommers, C.B. (1983) Formation and coupling of magnetic moments in Heusler alloys. *Phys. Rev. B*, **28**, 1745–1755.

167 Rosengaard, N.M. and Johasson, B. (1997) Finite-temperature study of itinerant ferromagnetism in Fe, Co, and Ni. *Phys. Rev. B*, **55**, 14975–14986.

168 Halilov, S.V., Eschrig, H., Perlov, A.Y., and Oppeneer, P.M. (1998) Adiabatic spin dynamics from spin-density-functional theory: applications to Fe, Co, and Ni. *Phys. Rev. B*, **58**, 293–302.

169 Sandratskii, L.M. and Bruno, P. (2003) Electronic structure, exchange interactions, and Curie temperature in diluted III–V magnetic semiconductors: (GaCr)As, (GaMn)As, (GaFe)As. *Phys. Rev. B*, **67**, 214402.

170 Şaşıoğlu, E., Sandratskii, L.M., and Bruno, P. (2004) First-principles calculation of the intersublattice exchange interactions and Curie temperatures of the full Heusler alloys Ni_2MnX (X=Ga, In, Sn, Sb). *Phys. Rev. B*, **70**, 024427.

171 Liechtenstein, A.I., Katsnelson, M.I., Antropov, V.P., and Gubanov, V.A. (1987) Local spin density functional approach to the theory of exchange interactions in ferromagnetic metals and alloys. *J. Magn. Magn. Mater.*, **67**, 65–74.

172 Entel, P., Dannenberg, A., Siewert, M., Herper, H.C., Gruner, M.E., Buchelnikov, V.D., and Chernenko, V.A. (2011) Composition-Dependent Basics of Smart Heusler Materials from First- Principles Calculations. *Mater. Sci. Forum*, **684**, 1–29.

173 Ebert, H., Ködderitzsch, D., and Minàr, K. (2011) Calculating condensed matter properties using the KKR-Green's function method: recent developments and applications. *Rep. Prog. Phys.*, **74**, 096501.

174 Kurtulus, Y., Dronskowski, R., Samolyuk, G.D., and Antropov, V.P. (2005) Electronic structure and magnetic exchange coupling in ferromagnetic full Heusler alloys. *Phys. Rev. B*, **71**, 014425.

175 Ogura, M., Takahashi, C., and Akai, H. (2007) Calculated electronic structure and Néel temperatures of half-metallic diluted antiferromagnetic semiconductors. *J. Phys. Condens. Matter*, **19**, 365226.

176 Enkovaara, J., Ayuela, A., Nordström, L., and Nieminen, R.M. (2002) Magnetic anisotropy in Ni_2MnGa. *Phys. Rev. B*, **65**, 134422.

177 Blaha, P., Schwarz, K., Madson, G.K.H., Kvasnicka, D., and Luitz, J. (2001) *WIEN2k, an Augmented Plane Wave+Local Orbitals Program for Calculating Crystal Properties*, Karlheinz Schwarz, TU Wien, Vienna.

178 Albertini, F., Morellon, L., Algarabel, P.A., Ibarra, M.R., Pareti, L., Arnold, Z., and Calestani, G. (2001) Magnetoelastic effects and magnetic anisotropy in Ni_2MnGa polycrystals. *J. Appl. Phys.*, **89**, 5614–5617.

179 Albertini, F., Solzi, M., Paoluzi, A., and Righi, L. (2008) Magnetocaloric properties and magnetic anisotropy by tailoring phase transitions in NiMnGa alloys. *Mater. Sci. Forum*, **583**, 169–196.

180 Gruner, M.E., Entel, P., and Opahle, I. (2008) Ab initio investigation of twin boundary motion in the magnetic shape memory alloy Ni_2MnGa. J. Mater. Sci., **43**, 3825–3831.

181 Koepernick, K. and Eschrig, H. (1999) Full-potential nonorthogonal local-orbital minimum-basis band-structure scheme. Phys. Rev. B, **59**, 1743–1757 (FPLO-14: www.fplo.de/userspace).

182 Ullakko, K., Huang, J.K., Kantner, C., O'Handley, R.C., and Kokorin, V.V. (1996) Large magnetic-field induced strains in Ni_2MnGa single crystals. Appl. Phys. Lett., **69**, 1966–1968.

183 Kaufmann, S., Rößler, U.K., Heczko, O., Wuttig, M., Buschbeck, J., Schultz, L., and Fähler, S. (2010) Adaptive modulations of martensite. Phys. Rev. Lett., **104**, 145702.

184 Kaufmann, S., Niemann, R., Thersleff, T., Rößler, U.K., Heczko, O., Buschbeck, J., Holzapfel, B., Schultz, L., and Fähler, S. (2011) Modulated martensite: why it form and why it deforms easily. New J. Phys., **132**, 053029.

185 Opeil, C.P., Mihaila, B., Schulze, R.K., Mañosa, L., Planes, A., Hults, W.H., Fisger, R.A., Riseborough, P.S., Littlewood, P.B., Smith, J.L., and Lashley, J.C. (2008) Combined experimental and theoretical investigation of the premartensitc transition in Ni_2MnGa. Phys. Rev. Lett., **100**, 165703.

186 Schubert, M., Schaefer, H., Mayer, J., Hettich, M., Merklein, M., He, C., Rummel, C., Ristow, O., Großmann, M., Luo, Y., Gusev, V., Fonin, M., Dekorsy, T., and Demsar, J. (2015) Collective modes and structural modulation in Ni–Mn–Ga(Co) martensite thin films probed by femtosecond spectroscopy and scanning tunneling microscopy. Phys. Rev. Lett., **115**, 076402.

187 Krenke, T., Duman, E., Acet, M., Wassermann, E.F., Moya, X., Mañosa, L., and Planes, A. (2005) Inverse magnetocaloric effect in ferromagnetic Ni–Mn–Sn alloys. Nat. Mater., **4**, 450–454.

188 Kainuma, R., Imano, Y., Ito, W., Sutou, Y., Morito, H., Okamoto, S., Kitakami, O., Oikawa, K., Fujita, A., Kanomata, T., and Ishida, K. (2006) Magnetic-field-induced shape recovery by reverse phase transformation. Nature, **439**, 957–960.

189 Kihara, T., Xu, X., Ito, W., Kainuma, R., and Tolunaga, M. (2014) Direct measurement of inverse magnetocaloric effects in metamagnetic shape-memory alloys NiCoMnIn. Phys. Rev. B, **90**, 214409.

190 Sokolovskiy, V.V., Entel, P., Buchelnikov, V.D., and Gruner, M.E. (2015) Achieving large magnetocaloric effects in Co- and Cr-substituted Heusler alloys: predictions from first-principles and Monte Carlo studies. Phys. Rev. B, **91**, 220409.

191 Sokolovskiy, V.V., Pavlukhina, O., Buchelnikov, V.D., and Entel, P. (2014) Monte Carlo and first-principles approaches for single crystal and polycrystalline Ni_2MnGa Heusler alloys. J. Phys. D, **47**, 425002.

192 Kouvel, J.S. and Hartelius, C.C. (1962) Anomalous magnetic moments and transformations in the ordered alloy FeRh. J. Appl. Phys., **33**, 1343–1344.

193 Çakır, Ö. and Acet, M. (2012) Reversibility in the inverse magnetocaloric effect in Mn_3GaC studied by direct adiabatic temperature-change measurements. Appl. Phys. Lett., **100**, 202404.

194 Rajarajan, A.K., Roy, S.B., and Chaddah, P. (1997) Magnetic instability in $CeFGe_2$: effects of Re and Ir substitutions. Phys. Rev. B, **56**, 7808–7811.

195 Chattopadhyay, M.K., Roy, S.B., and Chaddah, P. (2005) Kinetic arrest of the first-order ferromagnetic-to-antiferromagnetic transition in Ce$(Fe_{096}Ru_{0.04})_2$: formation of a magnetic glass. Phys. Rev. B, **72**, 180401.

196 Chattopadhyay, M.K., Mamekar, M.A., and Roy, S.B. (2006) Magnetocaloric effect in $CeFe_2$ and Ru-doped $CeFe_2$ alloys. J. Phys. D, **39**, 1006–1011.

197 Pecharsky, V.K. and Gschneidner, K.A., Jr., (1997) Giant magnetocaloric effect in $Gd_5(Si_2Ge_2)$. Phys. Rev. Lett., **78**, 4494–4497.

198 Lyubina, L., Schäfer, R., Martin, N., Schultz, L., and Gutfleisch, O. (2010) Novel design of La(Fe,Si)$_{13}$ alloys towards high magnetic refrigeration performance. Adv. Mater., **22**, 3735–3739.

199 Tegus, O., Brück, E., Buschow, K.H.J., and de Boer, F.R. (2002) Transition-metal-based magnetic refrigerants for room-temperature applications. *Nature*, **145**, 150–152.

200 Aksoy, S., Krenke, T., Acet, M., Wassermann, E.F., Moya, X., Mañosa, L., and Planes, A. (2007) Tailoring magnetic and magnetocaloric properties of martensitic transitions in ferromagnetic Heusler alloys. *J. Appl. Phys.*, **91**, 241916.

201 Porcari, G., Fabbrici, S., Pernechele, C., Albertini, F., Paoluzi, A., Kamarad, J., Arnold, Z., and Solzi, M. (2012) Reverse magnetostructural transformation and adiabatic temperature change in Co- and In-substituted Ni–Mn–Ga alloys. *Phys. Rev. B*, **85**, 024414.

202 Acet, M., Mañosa, Ll., and Planes, A. (2011) Magnetic-field-induced effects in martensitic Heusler-based magnetic shape memory alloys, in *Handbook of Magnetic Materials*, vol. **19** (ed. K.H.J. Buschow), Elsevier, Amsterdam, The Netherlands, pp. 231–289.

203 Marcos, J., Mañosa, L., and Planes, A. (2003) Multiscale origin of the magnetocaloric effect in Ni–Mn–Ga shape memory alloys. *Phys. Rev. B*, **68**, 094401.

204 Webster, P.J., Ziebeck, K.R.A., Town, S.L., and Peak, S.M. (1984) Magnetic order and transformation in Ni_2MnGa. *Philos. Mag. B*, **49**, 295–310.

205 Monroe, J.A., Raymond, J.E., Xu, X., Nagasako, M., Kainuma, R., Chumlyakov, Y.I., Arróyave, R., and Karaman, I. (2015) Multiple ferroic glasses via ordering. *Acta Mater.*, **101**, 107–115.

206 Entel, P., Gruner, M.E., Ogura, M., Sokolovskiy, V.V., Buchelnikov, V.D., Grünebohm, A., Arróyave, R., Uebayashi, K., Singh, N., Talapatra, A., Acet, M., and Çakır, A. (2015) Large magnetocaloric effects in magnetic intermetallics: first-principles and Monte Carlo studies. *MATEC Web Conf.*, **33**, 02001.

207 Stillinger, F.H., Debenedetti, P.G., and Truskett, T. (2001) The Kauzmann paradox revisited. *J. Phys. Chem. B*, **105**, 11809–11816.

208 Ito, W., Ito, K., Umetsu, Y., Kainuma, R., Koyama, K., Watanabe, K., Fujita, A., Oikawa, K., Ishida, K., and Kanomata, T. (2008) Kinetic arrest of martensitic transformation in NiCoMnIn metamagnetic shape memory alloy. *Appl. Phys. Lett.*, **92**, 021908.

209 Xu, X., Kainuma, R., Kihara, T., Ito, W., Tokunaga, M., and Kanomata, T. (2015) Thermodynamics and kinetics of martensitic transformation in Ni–Mn-based magnetic shape memory alloys. *MATEC Web Conf.*, **33**, 01004.

210 Xu, X., Ito, W., Tokunaga, M., Umetsu, R.Y., Kainuma, R., and Ishida, K. (2010) Kinetic arrest of martensitic transformation in NiCoMnAl shape memory alloy. *Mater. Trans.*, **51**, 1357–1360.

211 Cong, D.Y., Roth, S., Luo, J., Pötschke, M., Hürrich, C., and Schultz, L. (2010) Superparamagnetic and superspin glass behaviors in the martensitic state of $Ni_{43.5}Co_{6.5}Mn_{39}Sn_{11}$ magnetic shape memory alloy. *Appl. Phys. Lett.*, **96**, 112504.

212 Cong, D.Y., Roth, S., and Schultz, L. (2012) Magnetic properties and structural transformations in Ni–Co–Mn–Sn multifunctional alloys. *Acta Mater.*, **60**, 5335–5351.

213 Takahashi, S., Li, X.G., and Chiba, A. (1996) Spin distribution in plastically deformed Fe–Al intermetallic compounds. *J. Phys. Condens. Matter*, **8**, 11243–11257.

214 Palstra, T.T.MN., Werij, H.G.C., Nieuwenhuys, G.J., Mydosh, J.A., de Boer, F.R., and Buschow, K.H.J. (1984) Metamagnetic transitions in cubic La$(Fe_xAl_{1-x})_{13}$ intermetallic compounds. *J. Phys. F*, **14**, 1961–1966.

215 Zou, M., Mudrykm, Ya., Pecharsky, V.K., Schneidner, K.A., Jr., Schlagerl, D.L., and Lograsso, T.A. (2007) Crystallographic, anisotropic metamagnetism, and magnetocaloric effect in $Tb_5Si_{2.2}Ge_{1.8}$. *Phys. Rev. B*, **75**, 024418.

216 Ritter, C., Morellon, L., Algarabel, P.A., Magen, C., and Ibarra, M.R. (2002) Magnetic and structural phase diagram of $Tb_5(Si_xGe_{1-x})_4$. *Phys. Rev. B*, **65**, 094405.

217 Oliveira, N.A. and van Ranke, P.J. (2010) Theoretical aspects of the magnetocaloric effect. *Phys. Rep.*, **489**, 89–159.

218 Uebayashi, K., Shimizui, H., and Yamada, H. (2006) Structure and magnetism of Fe (Rh,Pd) alloys. *Mater. Trans.*, **47**, 456–459.

219 Barua, R., Jiménez-Villacorta, F., and Lewis, L.H. (2013) Predicting magnetostructural trends in FeRh-based ternary alloys. *Appl. Phys. Lett.*, **103**, 102407.

220 Barua, R., Jiménez-Villacorta, F., and Lewis, L.H. (2014) Towards tailoring the magnetocaloric response in FeRh-based compounds. *J. Appl. Phys.*, **115**, 17A903.

221 Fukuda, T., Kakeshita, T., and Lee, Y. (2014) An interpretation of the kinetics of martensitic transformation in a $Ni_{45}Co_5Mn_{36.5}In_{13.5}$ alloy. *Acta Mater.*, **81**, 121–127.

222 Bhatti, K.P., Phelan, D.P., El-Khatib, S., James, R.D., and Leighton, C. (1915) Magnetic phase competition in off-stoichiometric martensitic Heusler alloys: the $Ni_{50}-xCoxMn25+ySn25-y$ system, in *Heusler Alloys*, Springer Series in Materials Science, vol. **222** (eds C. Felser and A. Hirohata), Springer.

223 Moya, X., Kar-Narayan, S., and Mathur, N.D. (2014) Caloric materials near ferroic phase transitions. *Nat. Mater.*, **13**, 439–450.

224 von Schilfgaarde, M., Abrikosov, A., and Johansson, B. (1999) Origin of the Invar effect in iron–nickel alloys. *Nature*, **400**, 46–49.

225 Ma, Y.-B., Albe, K., and Xu, B.-X. (2015) Lattice-based Monte Carlo simulations of the electrocaloric effect in ferroelectrics and relaxor ferroelectrics. *Phys. Rev. B*, **91**, 184108.

226 Dutta, B., Çakır, A., Giacobbe, C., Al-Zubi, A., Hickel, T., Acet, M., and Neugebauer, J. (2016) *Ab initio* prediction of martensitic and intermartensitic phase boundaries in Ni–Mn–Ga. *Phys. Rev. Lett.*, **116**, 025503.

227 Pérez-Landaábel, J.I., Recarte, V., Sánchez-Alarcos, V., Gómez-Polo, C., Kustov, S., and Cesari, E. (2011) Magnetic field induced martensitic transformation linked to the arrested austenite in a Ni–Mn–In–Co shape memory alloy. *J. Appl. Phys.*, **109**, 093515.

228 Gottschall, T., Skokov, K.P., Benke, D., Gruner, M.E., and Gutfleisch, O. (2016) Contradictory role of magnetic contribution in inverse magnetocaloric materials. *Phys. Rev. B*, **93**, 184431.

229 Gutfleisch, O., Gottschall, T., Fries, M., Benke, D., Radulov, I., Skokov, K.P., Wende, H., Gruner, M., Acet, M., Entel, P., and Farle, M. (2016) Mastering hysteresis in magnetocaloric materials. *Philos. Mag. A*, **374**. doi: 10.1098/rsta.2015.0308.

12
Chemical Bonding in Solids

Gordon J. Miller,[1] Yuemei Zhang,[1] and Frank R. Wagner[2]

[1]*Iowa State University, Department of Chemistry, 1605 Gilman Hall, Ames, IA 50011, USA*
[2]*Max-Planck-Institute for Chemical Physics of Solids, Chemical Metals Science, Nöthnitzer Straße 40, Dresden 01187, Germany*

Solids are typically identified by their physical properties, which originate from the nature of the forces holding atoms or molecules together in the condensed phase. As a result, an understanding of chemical bonding in solids lies at the heart of solid-state chemistry. The idealized cohesive forces lead to four main types of solids: *ionic*, *covalent*, *metallic*, and *molecular* [1]. The first three of these categories also describe their corresponding bonding limits, whereas molecular solids rely on van der Waals forces or anisotropic electrostatic interactions, such as dipole–dipole or hydrogen bonding forces, to condense individual atoms or molecules into the solid state. However, very few solids exclusively exhibit one kind of cohesive force. Most of them engender multiple bonding modes.

If solid-state structures are viewed as molecules with quasi-infinite numbers of atoms, then the same bonding concepts used in molecular chemistry should be appropriate for solid-state chemistry. The atomic scale of solids, however, compared to a single molecule, can bring about new collective features that lead to physical properties useful for technological applications. To reconcile these two extremes of atomic scale, quantum mechanics and the steady growth of computational capabilities have led to tremendous advances in how chemical bonding in solids, especially crystalline solids, can be analyzed and interpreted using localized and delocalized pictures of electronic structure. However, the experimental chemist relies significantly on simple valence electron counting rules to identify new synthetic targets, whereas the materials scientist and condensed matter physicist strive to relate the electronic structure of a solid to its measured properties. Both outlooks must be grounded in the computational results. With the goal of addressing these issues concerning chemical bonding in the solid state, this chapter is organized as follows: (1) a brief overview of fundamental characteristics and descriptions of chemical bonds in matter; (2) a summary of the types of solids and bonding motifs; (3) a description of how quantum

Handbook of Solid State Chemistry, First Edition. Edited by Richard Dronskowski, Shinichi Kikkawa, and Andreas Stein.
© 2017 Wiley-VCH Verlag GmbH & Co. KGaA. Published 2017 by Wiley-VCH Verlag GmbH & Co. KGaA.

mechanics is used to analyze chemical bonding in solids; (4) descriptions of important electron counting rules that are useful for experimentalists; and (5) selected examples of the synergies among chemical bonding, atomic structures, and properties.

12.1
Fundamental Characteristics of the Chemical Bond

In general, every chemical bond is characterized by the distance between a coupled pair of atoms and the energy of this pair with respect to the separate neutral atoms, which represents the reference state of the system. The energy versus distance profile $E(R)$ for this couple, illustrated in Figure 12.1a, indicates a long-range attraction and a short-range repulsion, the combination of which leads to three fundamental characteristics of the chemical bond: (i) the dissociation or binding energy E_0; (ii) the equilibrium bond distance R_0; and (iii) the vibrational (stretching) force constant. The dissociation energy is the difference between the minimum energy of the curve and that of the reference state. The bond distance is the interatomic separation at the energetic minimum, and the force constant corresponds to the curvature of the $E(R)$ profile near the equilibrium distance, that is, (d^2E/dR^2) at R_0. The shape of the $E(R)$ curve around R_0 can be approximated by a harmonic potential, but at longer distances the curve develops significant anharmonicity, which accounts for thermal expansion. For solids, there are numerous interatomic interactions, so the solid-state analog to $E(R)$ is the binding energy curve $U(V)$ with the following characteristics: (i) the cohesive energy U_0; (ii) the equilibrium volume V_0; and (iii) the bulk modulus B_0 [2]. The cohesive energy corresponds to the energy required to dissociate 1 mol of solid at volume V_0 completely into its constituent atoms [3]. The bulk

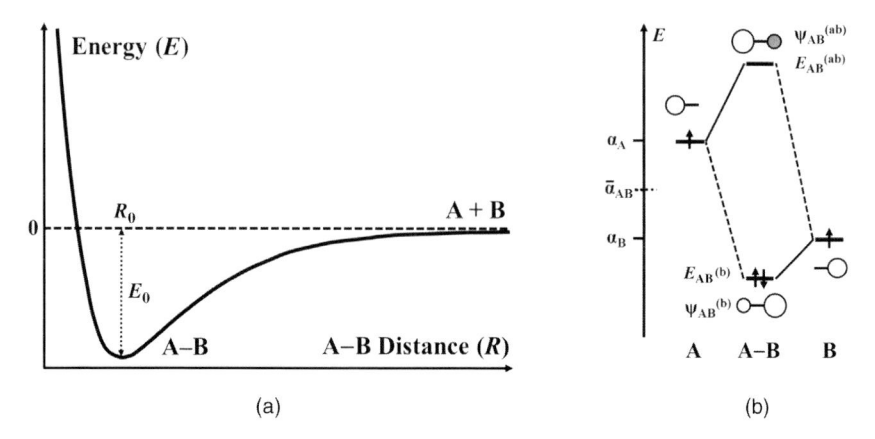

Figure 12.1 (a) Energy versus distance profile for an A—B bond. The dissociation energy E_0 and the equilibrium distance R_0 are marked. (b) MO energy diagram for a two-electron AB s-valent dimer.

modulus is the inverse of the isothermal compressibility κ_T, which is a measure of the curvature of the binding energy $U(V)$ near V_0:

$$B_0 = \frac{1}{\kappa_T} = -V_0 \left(\frac{\partial p}{\partial V}\right)_{T,V_0} = V_0 \left(\frac{\partial^2 U}{\partial V^2}\right)_{T,V_0}. \quad (12.1)$$

The short-range repulsive forces, which keep atoms from getting too close to each other, largely involve Pauli exclusion forces as well as interelectronic charge repulsions and internuclear repulsions, although the latter are somewhat screened by the electronic distribution around each nucleus. Primary attractive forces include electrostatic attractions and exchange, both of which may be accomplished by orbital overlap that encourages delocalization of electrons.

Chemical bonds are traditionally named according to the spatial distribution of valence electron densities relative to the atomic cores. The three major idealized bond types are (i) *ionic*, which exhibit valence electron transfer from electropositive (cationic) elements to electronegative (anionic) elements; (ii) *covalent*, which involve valence electron sharing among adjacent atoms; and (iii) *metallic*, which consists of cationic atom cores in a "sea" of valence electrons, the so-called "homogeneous electron gas." However, there are conflicting opinions about whether metallic bonds are, indeed, different in character than covalent interactions [4]. Nevertheless, all three bond categories span similar dissociation energies (no larger than ~10 eV) and equilibrium interatomic distances (no longer than ~3 Å) [3]. *Van der Waals* forces and *hydrogen bonds* can lead to condensed phases as well, but these interactions are weaker and generally involve longer distances than observed for the other three bond types. It is important, however, not to treat these idealized classifications too dogmatically, on account of the diverse solid-state phenomena arising from combinations of these bond types.

An important quantum mechanical approach to analyze chemical bonds relies on molecular orbitals (MOs) [5]. Consider the H_2 molecule, in which the two atoms H_A and H_B retain some atomic characteristics while simultaneously taking part in an electrostatic sharing of the valence electrons. Therefore, MOs may be represented by linear combinations of atomic orbitals (AOs), that is, the LCAO ansatz. The superposition of two AOs (φ_A and φ_B) creates two MOs: (i) a bonding MO that is the in-phase combination of AOs, $\psi_{AB}^{(b)} = c_A^{(b)} \varphi_A + c_B^{(b)} \varphi_B$; and (ii) an antibonding MO that is the out-of-phase combination, $\psi_{AB}^{(ab)} = c_A^{(ab)} \varphi_A - c_B^{(ab)} \varphi_B$, and has a node bisecting the $H_A - H_B$ contact. Solving the Schrödinger equation for H_2 using extended Hückel theory [6], which is one of the simplest quantum mechanical theories to study MOs, leads to the energies and wavefunctions of the bonding and antibonding MOs, respectively [7]:

$$E^{(b)} \sim (\alpha + \beta)(1 - S) = \alpha - |\beta| + |\beta|S + |\alpha|S, \quad \psi_{AB}^{(b)} \sim \left[\frac{1-S}{2}\right]^{1/2} (\varphi_A + \varphi_B), \quad (12.2a)$$

$$E^{(ab)} \sim (\alpha - \beta)(1 + S) = \alpha + |\beta| + |\beta|S - |\alpha|S, \quad \psi_{AB}^{(ab)} \sim \left[\frac{1+S}{2}\right]^{1/2} (\varphi_A - \varphi_B). \quad (12.2b)$$

In these equations α (<0) is the onsite energy term that approximates the energy of the H 1s AO, β (<0) is the intersite energy term, called a resonance or bond integral, that is a measure of the interaction energy between AOs on different sites, and S is the spatial overlap integral between the AOs on different sites (in this case $S > 0$). With respect to the AO energy α, the bonding MO is stabilized and the antibonding MO is destabilized. Also, the stabilization of the bonding MO is less than the destabilization of the antibonding MO because the AOs on different H atoms are not orthogonal, that is, $S \neq 0$. This latter outcome is called overlap repulsion, which is a positive energy shift of $|\beta|S$ for the energies of both the bonding and antibonding MOs [7]. In addition, the energy of the bonding MO includes the term $+|\alpha|S$, which measures the energy loss arising from a shift of electron density away from the atoms toward the bond region, whereas $-|\alpha|S$ is the energy gain of the antibonding MO by shifting electron density from the bond region toward the atomic regions. This shift in electron density is measured with respect to the superposition of neutral atom densities $(\varphi_A)^2 + (\varphi_B)^2$.

Within the Born–Oppenheimer approximation, which separates the electronic and nuclear motions, the energy of the chemical bond corresponds to electronic energy. So, what is the relative importance of electronic kinetic and potential energies that ultimately determines bonding? Consider the H_2^+ ion with one valence electron for two protons separated by a specific (bond) distance. Using a particle-in-a-box model, the one-electron bond of the H_2^+ molecule has lower kinetic energy than the separated $H^\bullet + H^+$ system because the electron moves in the larger "box" of molecular H_2^+, that is, the electronic motion becomes more delocalized. From the MO energy diagram for H_2^+, the electron occupies the bonding MO. However, in this picture, an interpretation of bonding as simply a lowering in kinetic energy ignores any changes in potential energy arising from the electrostatic attraction between the electron and the two nuclei. According to the Virial Theorem, the electronic kinetic (T) and potential (V) energies are related by $V = -2T$. Thus, any change in total electronic energy becomes $\Delta E = -\Delta T = \Delta V/2$. Since ΔE must be negative for the formation of a stable chemical bond, then ΔT must be positive and ΔV must be negative, that is, a change in potential energy is behind the energetics of bond formation [4,8].

To examine the chemical bond in H_2^+ using the Virial Theorem requires treating the exponent ζ of the 1s AO on each H atom, $\varphi \sim \exp(-\zeta r)$, as a variational parameter [9]. For the H atom, the average kinetic energy varies as ζ^2, whereas the average potential energy varies as -2ζ. The total energy achieves a minimum value at $\zeta = 1$ a.u.$^{-1}$ (1 a.u. = 0.529 Å), and this is the only value where the Virial Theorem is obeyed in the H atom. In the H_2^+ molecule, the bonding MO, which contains the single electron, is expressed as $\psi_{AB}^{(b)} \sim \exp(-\zeta r_A) + \exp(-\zeta r_B)$. The kinetic and potential energies follow similar functional forms as in the H atom with respect to ζ, but both are scaled differently because of the larger volume and two protons. The minimum energy for H_2^+ occurs for $\zeta \sim 1.228$, again where the Virial Theorem is obeyed, and shows that the AOs contract during bond formation. Variations of the kinetic and potential energies versus ζ indicate that kinetic energy favors delocalization because it decreases as the AO expands (decreasing ζ), whereas the potential energy favors localization because it decreases as the AO contracts (increasing ζ). Indeed, for fixed ζ, the average kinetic energy drops from the separated H$^{\bullet}$ + H$^+$ system to the H$_2^+$ molecule, but, when ζ is allowed to vary so as to achieve the overall minimum energy value, the drop in total energy arises from a decrease in potential energy via the increased buildup of electronic charge around the nuclei due to orbital contraction.

Orbital arguments are one-electron models and, therefore, suffer when trying to describe bond dissociation. Depending upon the extent of electron delocalization tolerated in the model, there are two limiting cases that can be derived for H$_2$. In the Hund–Mulliken *molecular orbital* approach, the ground-state electronic configuration is $(\psi_{AB}^{(b)})^2 \sim [\varphi_A(1) + \varphi_B(1)][\varphi_A(2) + \varphi_B(2)]$. Expanding this configuration yields four distinct terms, each of equal weight for the two electrons. Two of these terms, $\varphi_A(1)\varphi_A(2)$ and $\varphi_B(1)\varphi_B(2)$, represent *ionic* configurations (H$^+$ + H$^-$); the other two, $\varphi_A(1)\varphi_B(2)$ and $\varphi_B(1)\varphi_A(2)$, represent *covalent* configurations (H–H). Therefore, dissociation of H$_2$ according to the Hund–Mulliken model goes to 1/2(H$^{\bullet}$ + H$^{\bullet}$) + 1/2(H$^+$ + H$^-$) at an energy (+6.42 eV) that is the average of the ionization energy (+13.6 eV) and electron affinity (−0.86 eV) of the H atom. A second model, called the Heitler–London variant of the *valence bond* approach, includes just the covalent terms $\varphi_A(1)\varphi_B(2)$ and $\varphi_B(1)\varphi_A(2)$ while neglecting the ionic contribution for this normal, homopolar bond (in contrast to homopolar charge-shift bonds [10]) because the energetic effect of their mixing is less important. Dissociation of H$_2$ in this model occurs properly to (H$^{\bullet}$ + H$^{\bullet}$). Both models, however, underestimate the dissociation energy and do not accurately determine the equilibrium distance of H$_2$ [11]. In essence, the molecular orbital treatment underestimates electron–electron correlation and enhances electron delocalization throughout the molecule, whereas the valence bond treatment overestimates electron–electron correlation at equilibrium distances and electronic motion is too restricted [1].

An orbital-based, multielectron approach to reconcile these two treatments exists in the Hubbard model [12], which was developed to include electron–electron correlation in a simple way by including an on-site repulsion parameter (U) when two electrons occupy the same AO.

For H$_2$, six microstates arise from two electrons distributed among two AOs (see Figure 12.2). In the language of the Hubbard model, the electron is allowed to hop from one AO to the other, so the bond integral β is also called a hopping integral. Solution of this problem gives the ground-state energy of

$$E_{GS} = 2\alpha + \frac{1}{2}U - \left[4|\beta|^2 + \frac{1}{4}U^2\right]^{1/2} \tag{12.3}$$

for the singlet state for H$_2$ leading to a triplet–singlet energy difference of $\sim 4\beta^2/U$. Depending on the relationship between U and $|\beta|$, the two limiting cases of

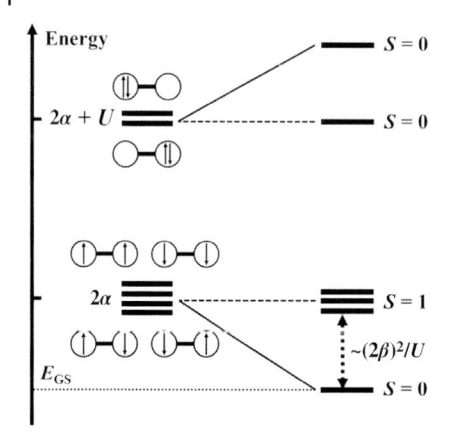

Figure 12.2 Application of the Hubbard model to H_2. Left: Energies and pictorial representation of the six microstates derived from two electrons and two AOs. Right: Result of allowing intersite interaction β, which couples singlet ($S = 0$) microstates. The energy difference between the singlet ground state and the triplet ($S = 1$) state is the direct interatomic exchange energy.

molecular orbital and localized valence bond behavior can be retrieved: As delocalization becomes more favorable ($U \ll |\beta|$), the ground-state approaches the Hund–Mulliken solution; as localization becomes more favorable ($U \gg |\beta|$), the ground-state approaches the Heitler–London solution.

Although MO theory has some shortcomings, as described above, it has been extremely useful to rationalize the structures of numerous molecules and solids as well as to interpret spectroscopic experiments and physical property measurements. Within the LCAO ansatz, the energies and spatial features of MOs depend on the relative energies between the component AOs and their corresponding spatial overlaps. For a heteroatomic, two-electron AB s-valent dimer with effective AO energies α_A and α_B ($\alpha_A > \alpha_B$), intersite integral β_{AB}, and overlap integral S_{AB}, the bonding and antibonding MOs and their corresponding energies are as follows (see Figure 12.1b) [7]:

$$E_{AB}^{(b)} \sim \bar{\alpha}_{AB}\left[1 - \frac{S_{AB}}{\sqrt{1+\delta^2}}\right]^{\frac{1}{2}} + |\beta_{AB}|S_{AB} - \left[\beta_{AB}^2 + \left(\frac{\Delta_{AB}}{2}\right)^2\right]^{1/2},$$

$$\psi_{AB}^{(b)} \sim \left[\frac{1}{2}\left(1 - \frac{\delta + S_{AB}}{\sqrt{1+\delta^2}}\right)\right]^{1/2}\varphi_A + \left[\frac{1}{2}\left(1 + \frac{\delta - S_{AB}}{\sqrt{1+\delta^2}}\right)\right]^{1/2}\varphi_B.$$
(12.4a)

$$E_{AB}^{(ab)} \sim \bar{\alpha}_{AB}\left[1 + \frac{S_{AB}}{\sqrt{1+\delta^2}}\right]^{1/2} + |\beta_{AB}|S_{AB} + \left[\beta_{AB}^2 + \left(\frac{\Delta_{AB}}{2}\right)^2\right]^{1/2},$$

$$\psi_{AB}^{(ab)} \sim \left[\frac{1}{2}\left(1 + \frac{\delta + S_{AB}}{\sqrt{1+\delta^2}}\right)\right]^{1/2}\varphi_A - \left[\frac{1}{2}\left(1 - \frac{\delta - S_{AB}}{\sqrt{1+\delta^2}}\right)\right]^{1/2}\varphi_B.$$
(12.4b)

In Eq. (12.3), $\bar{\alpha}_{AB} = (\alpha_A + \alpha_B)/2$, $\Delta_{AB} = \alpha_A - \alpha_B > 0$, and $\delta = \Delta_{AB}/(2|\beta_{AB}|)$. If atoms A and B are the same, then the dimer is homoatomic so that the energies and MOs are expressed by Eq. (12.2). From this simple model, the average MO energy is shifted upward by overlap repulsion $(+|\beta_{AB}|)S_{AB})$, and the energy difference between the two MOs involves terms relating to covalency (β_{AB}) and ionicity (Δ_{AB}). Furthermore, the bonding MO has greater contribution from the lower energy AO at atom B, whereas the antibonding MO has greater character from the higher energy AO at atom A (see also Figure 12.1b). For a given internuclear vector $\boldsymbol{R}$ between atoms A and B, the number density of valence electrons in the two-electron A—B dimer is

$$\rho_{AB}(\boldsymbol{r}) = 2[\psi_{AB}^{(b)}(\boldsymbol{r})]^2 = (1-q)\rho_A(\boldsymbol{r}) + (1+q)\rho_B(\boldsymbol{r}-\boldsymbol{R}) + p_{AB}\rho_{cov}(\boldsymbol{r}), \quad (12.5)$$

in which $q = \delta/[1+\delta^2]^{1/2}$ is the charge assigned to each AO, $p_{AB} = 1/[1+\delta^2]^{1/2} = [1-q^2]^{1/2}$ is the A—B bond order, and $\rho_A(\boldsymbol{r})$, $\rho_B(\boldsymbol{r})$, and $\rho_{cov}(\boldsymbol{r})$ are the number densities corresponding to atom A, atom B, and the covalent bond. Now, δ can be re-expressed as $q/[1-q^2]^{1/2}$, so that for the covalent limit ($\delta \to 0$) $q \to 0$ and for the ionic limit ($\delta \to \infty$) $q \to 1$. In the H$_2$ molecule, the atomic charge $q = 0$, the bond order $p_{HH} = 1$, and $\rho_{cov}(\boldsymbol{r})$ reveals that electron density flows from outside to inside the bond region.

According to this MO picture, which neglects explicit consideration of interelectronic interactions, all two-center, two-electron orbital interactions are stabilizing, regardless whether they are formed by two one-electron AOs, as in "2 H• → H$_2$" or "Li• + H• → LiH," or by one filled (two-electron) and one empty (zero-electron) AO, as in "H$^+$ + H$^-$ → H$_2$" or "Li$^+$ + H$^-$ → LiH." The energy changes of these interactions depend on the AO energy mismatch Δ_{AB}: (i) in the nearly degenerate case ($\Delta_{AB} \sim 0$), $\Delta E \propto -|\beta_{AB}|$; (ii) in the nondegenerate case, $\Delta E \propto -|\beta_{AB}|^2/\Delta_{AB}$ [4]. Therefore, chemical bonding interpretations using orbital arguments rely on evaluating intersite energy terms (β) and AO energy differences (Δ). β_{AB} values vary with distance, and can be estimated using the Wolfsberg–Helmholz relation $\beta_{AB} \propto \bar{\alpha}_{AB}S_{AB}$ [13]. Given this relationship, β_{AB} is also affected by the nodal character of the AOs involved, which lead to σ, π, or δ-type overlaps. The other important component, the AO energies (α), can be obtained from Hartree–Fock calculations or estimated from atomic spectra as valence-state ionization potentials, which are derived from the energy differences between the ground-state configurations of neutral atoms and the configurations of the corresponding ions obtained by removing one electron from the AO of interest [14]. In general, valence AO energies become more negative as the effective nuclear charge for that orbital increases. This variation in AO energies is reflected in the notion of *electronegativity* [15], which, according to Pauling, is the "power of an atom in a molecule or solid to attract elections to itself." Mulliken defined the electronegativity of an atom χ_A as the average of its first ionization energy and electron affinity [16]. From density functional theory, electronegativity is the negative derivative of the total electronic energy with respect to electron number [17]. Another related characteristic is the *chemical hardness* of an atom η_A, which is the curvature of the total energy versus electron number, and can be approximated by one-half the difference of the first ionization energy and the electron affinity [15,16]. Therefore, hardness gives a measure of the HOMO–LUMO gap in an atom (or molecule) or the on-site Coulomb repulsion, whereas electronegativity assigns the position of this gap on a common energy scale. Using these two parameters, the energy of a free ion A^{q+} with net charge $+q$ is [7,15].

$$U_A(q) \sim U_A(0) + \chi_A q + \eta_A q^2. \tag{12.6}$$

Thus, when two atoms come together to form a chemical bond, the electronegativity difference provides a measure of effective electron transfer from one atom to another, whereas the hardness values provide clues about the extent of electronic back donation so as to achieve a common electrochemical potential for the electrons in the chemical bond.

As implied from the previous discussion, applying orbital models to analyze chemical bonding requires recognizing the periodic trends of the "sizes" (extensions) and energies of the various core, valence, and virtual (unoccupied) AOs of the elements. The alkali and alkaline earth metals primarily use valence s AOs and, beginning with the fourth period, can involve virtual d-orbitals. The transition metals make use of valence d- and s-orbitals, and invoke increasing use of p-orbitals as the effective nuclear charge increases along a period. The lanthanide and actinide elements, like the early transition metals, rely significantly on valence s and d AOs, but can also involve valence f orbitals as well. However, the 4f AOs of the lanthanide elements are quite contracted and more core-like, whereas the 5f AOs of the actinides are more diffuse. Finally, the posttransition metals, metalloids, and nonmetals utilize valence s and p AOs for bonding. Now, the sizes of atoms and their corresponding AOs are not strictly defined because the radial parts of AOs decrease exponentially with increasing distance from the nucleus. Nevertheless, regardless of how atomic sizes are estimated, AO radii generally contract from left-to-right along a period as the effective nuclear charge increases, and they expand down a column as filled AOs are continually added to the core region. Likewise, the valence AO energies, as evaluated by Herman and Skillman, show the following trends [11]:

a) For main group elements, E_{ns} and E_{np} values decrease, that is, electrons assigned to these orbitals become more tightly bound, and $E_{np} - E_{ns}$ increase with increasing effective nuclear charge, whereas these energies increase down a group as the principal quantum number n increases. One notable exception is the 4s AO energy crossing below the 3s energy for elements to the left of group 15 (VB), like Ga and Al for which $E_{4s}(Ga) < E_{3s}(Al)$. This effect occurs because the 3d electrons do not completely screen the valence 4s electron(s) from the core.

b) For transition metals, E_{nd} values decrease gradually with effective nuclear charge, but then drop rapidly after the coinage metals (Cu, Ag, Au) once the valence d-shell is filled. However, on moving down a group, E_{4d} become more strongly bound than E_{3d} toward the middle of the series due to enhanced electron–electron repulsions for the smaller 3d AOs. Therefore $E_{(n+1)s} - E_{nd}$ tend to increase from 3d to 4d elements away from the left-hand side; relativistic effects reverse this trend for the corresponding 5d elements [18], decreasing $E_{(n+1)s} - E_{nd}$ by lowering E_{6s}.

In addition to these typical periodic trends, certain anomalous behavior can also affect chemical bonding. In particular, three regions of the periodic table deserve special emphasis: (i) the second period main group elements C, N, and O, which engage in σ- and π-bonding, whereas π-bonding is less extensive for their heavier analogs; (ii) the 3d elements Fe, Co, and Ni are ferromagnetic rather than Pauli paramagnetic as seen for most transition metals; and (iii) the sixth period elements, Pt, Au, Hg, Tl, Pb, and Bi, exhibit some structural features and properties that differ from their lighter group relatives. The phenomena observed for these subsets of elements arise from enhanced attractions or reduced repulsions that certain valence electrons experience from their corresponding atomic cores [11]. The rationale for the exceptional behavior of the 2p and 3d elements emerges from the concept of a pseudopotential, which is the effective interaction between the valence electrons and the atomic core, that is, nucleus plus core electrons. In short, the pseudo-orbitals of the valence electrons are orthogonal to all orbitals of different angular momentum quantum numbers l by virtue of their angular features, but not to core orbitals of the same l because the radial parts of these functions are not orthogonal to each other. Thus, since there are no "1p" and "2d" core orbitals, 2p and 3d valence electrons do not experience a typical Pauli repulsion from core electrons [19]. This outcome leads to exceptionally short interatomic distances and greater π-type orbital overlap for the 2p elements, and spontaneous magnetization for the 3d metals, arising from enhanced intra-atomic exchange that promotes unpaired electrons. This effect also influences the 4f AOs and electrons of the lanthanide metals, leading to their localized behavior in these elements and the effect on atomic sizes for the sixth period elements, the "lanthanide contraction." The unusual behavior of the later sixth period elements is due to a substantial relativistic contraction experienced by the 6s AO leading to unusually short interatomic distances, inertpair effects in Hg-Bi, and high electronegativities among these elements, especially for Pt and Au [18]. Thus, orbital-based models of chemical bonding significantly use the periodic trends of the elements, behavior that provides important perspectives for rationalizing different bonding modes in solids.

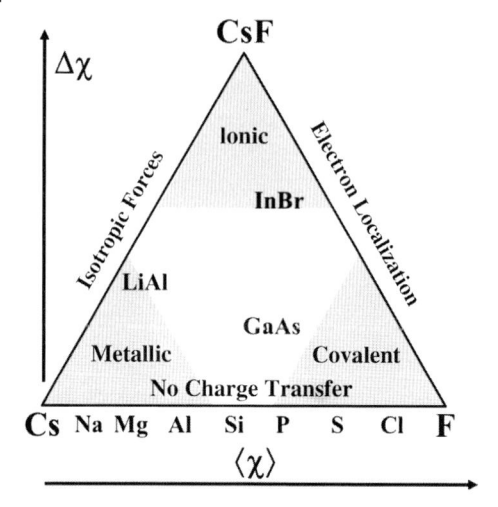

Figure 12.3 Van Arkel–Ketelaar diagram using as x-axis the average electronegativity χ and as y-axis the electronegativity difference between components. Major bonding regions and important characteristics of the edges are labeled.

12.2
Types of Solids and Bonding Motifs

The major types of solid-state compounds can be sorted effectively according to their electronic structures using a van Arkel–Ketelaar diagram (see Figure 12.3) [1], which is a structure map because each compound corresponds to a point on this diagram. There have been various manifestations of this diagram, using different qualitative or quantitative scales [20]. A pedagogically useful diagram sets the horizontal coordinate as the average electronegativity of the atoms and the vertical coordinate as the electronegativity difference between the atoms [20]. This scheme creates a triangular map in which the lower base is the sequence of elements from the least electronegative (Cs) to the most electronegative (F). The upper corner identifies the compound with the highest electronegativity difference (CsF). The upper region of this diagram corresponds to *ionic compounds*, held together by electrostatic forces with valence electrons localized on the anions and adopting structures that are usually characterized by close packings of anions and the generally smaller cations in interstitial sites; their crystals are considered to be brittle. Ionic compounds are electrically insulating as solids yet conductive in the liquid state. Most are diamagnetic, which is indicative of only electron pairs, unless they contain transition metals or rare earth elements, which can display magnetic features from any localized magnetic moments originating from unpaired 3d or 4f electrons. The lower left region is identified as *metallic compounds*, which are typically close-packed like ionic structures, but have no clear charge transfer between elements. Metallic solids are generally malleable and/or ductile, show good electrical conductivity, and are paramagnetic, although some late- and posttransition metals can be diamagnetic. Metallic elements have low electronegativities and ionization energies with their valence electrons delocalized throughout the atomic network. The lower right region is assigned to *covalent compounds*, whose structures are either isotropic but less efficiently packed than metals or anisotropic that show no clear distinction between cationic and anionic species. Like ionic compounds, they are electrical insulators and generally diamagnetic. Elements forming covalent compounds have high electronegativities and ionization energies, and their valence electrons localized into chemical bonds or lone pairs. Although these three bonding categories are idealizations and manifest significant differences, the diagram points out the possibilities for intermediacy and continuous variation among these bond types.

Whereas the corners of the van Arkel–Ketelaar diagram differentiate idealized bond types, the sides of the triangle indicate distinctions and relationships between pairs of bond categories. The base of the map from metals to covalent elements corresponds to the absence of charge transfer between components, that is, the elements themselves. The left-hand side between the metallic and ionic substances identifies isotropic, essentially electrostatic bonding forces between, respectively, free electrons and atom cores or cations and anions. Finally, the right-hand side connecting ionic and covalent compounds characterizes the highest degree of electron localization, which can occur either at atoms, to form ions, or in bonds. Thus, traversing this map from one side toward an opposite vertex leads to three trends in electronic structure: (i) increasing charge transfer (bottom side to top vertex); (ii) increasing anisotropic bonding (left-hand side to lower right); and (iii) increasing electronic delocalization (right-hand side to lower left). These qualitative transitions between different modes of electronic structure allow for additional compound types: for example, the metalloids found between metals and covalent compounds; Zintl phases, which behave like valence compounds, but exist between the metals and ionic compounds; and oligomeric, polar covalent oxides, sulfides, and halides, which are found between the ionic and covalent materials. Much scientific interest focuses on these intermediate regions because novel or interesting physical and chemical properties can originate from combining different bond types in a single material.

Since the van Arkel–Ketelaar diagram is able to portray a continuous variation of electronic behavior among different types of compounds, a unified approach to evaluate and understand electronic structure and to rationalize properties can emerge. As such, orbital-based schemes, although limited by being one-electron models, do provide a useful and relatively successful framework to address the three idealized bonding modes in solids. In general, the principal basis set includes the valence and important virtual or core AOs on each atom. Then, the Schrödinger equation is solved to obtain the desired results for chemical bonding analysis. In the next section, some of the quantum mechanical methods and philosophies to achieve these solutions will be discussed in more detail, but here simple models of electronic structure for typical ionic, covalent, and metallic solids will be discussed as fundamental starting points.

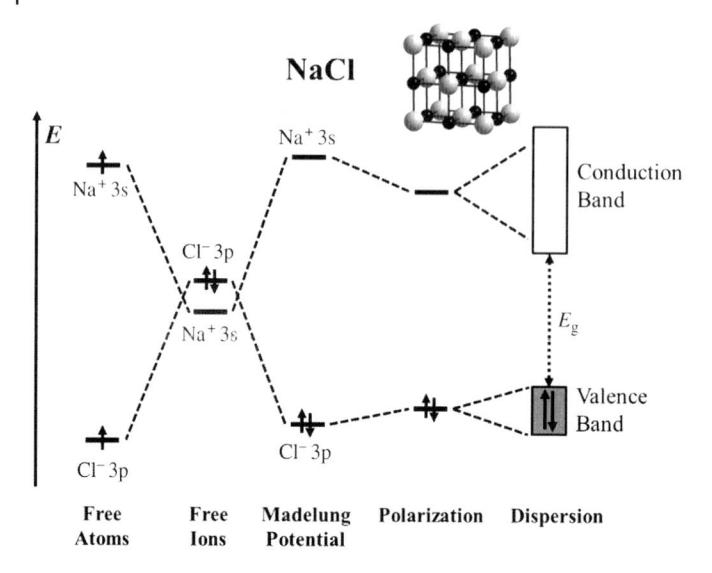

Figure 12.4 Evolution of the band diagram for a typical ionic solid NaCl(s) [1]. The free atom orbitals shift upon ionization to Na$^+$(g) and Cl$^-$(g). The Madelung potential and polarization shift the ion orbital energies in near alignment with the free atom orbital energies. Finally, after Cl$\cdots$Cl and Na$\cdots$Na dispersion, the final band diagram emerges. A bandgap E_g separates the occupied valence band from the empty conduction band.

12.2.1
Ionic Solids

For an "ionic" solid to be stable, the net electrostatic attraction between ions must exceed the energetic cost to exchange electrons between free atoms to produce gaseous ions. Therefore, ionic bonding will be most effective for elements with low ionization potentials to form cations and high electron affinities to form anions, such as for the alkali metal halides like NaCl(s). Furthermore, in the ideal ionic compound, the valence electrons are localized at the anions that achieve closed-shell electronic configurations. Therefore, the qualitative orbital energy diagram of an ionic solid like NaCl(s) [1,21], shown in Figure 12.4, has a filled valence band, originating from the Cl 3p AOs, that lies below the bottom of the empty conduction band, which is assumed to arise from the Na 3s AOs, according to the relative electronegativities of Cl and Na. Since the large energetic separation between Cl 3p and Na 3s AOs of ~8.8 eV does not give rise to significant stabilization or destabilization of bonding or antibonding orbitals, respectively, relative to the corresponding AOs, the valence bandwidth of 1.8(2) eV arises largely from Cl$\cdots$Cl orbital interactions with the band center being displaced to an energy slightly higher than E_{3p}(Cl) due to overlap repulsion. The conduction band would also be expected to have a bandwidth, arising from

Na···Na orbital interactions. To reconcile this simple orbital picture with an ionic model of Na$^+$ and Cl$^-$ ions that are held together by electrostatic forces, the Na 3s energy is shifted downward by the ionization potential and the Cl 3p energy is shifted upward by the electron affinity. Bringing Na$^+$ and Cl$^-$ ions together to form NaCl(s) stabilizes the anion states, which are surrounded by cations, and destabilizes the cation states, which are surrounded by anions, by the electrostatic Madelung potential. Once polarization and other relaxation processes are included, the energies of the ion states are nearly in line with those of the free atom states and broaden into bands via Na···Na and Cl···Cl dispersion. Thus, the two main observables associated with the electronic structure of ionic solids are the bandgaps [2] and valence bandwidths [22], which are listed for some alkali metal halides in Table 12.1.

The energy gap between the empty conduction band and the filled valence band of an ionic solid is typically greater than ~3 eV (less than ~400 nm), so these solids are electrical insulators and transparent to visible light. According to the model above, the bandgap is a charge transfer gap, and can be estimated by the difference in valence AO energies, for example, E_g(NaCl) = 8.5 eV and E_{3s}(Na) $- E_{3p}$(Cl) = 8.8 eV. Such estimates can be improved by including on-site Coulomb repulsions and the electrostatic potential at the anions, although the simple estimates already capture the periodic trends [2]. Among the alkali metal halides, bandgaps steadily decrease from MF(s) to MI(s) for a given metal M and specifically from LiF(s) to CsF(s) as the cation–anion distance increases, but these gaps are less sensitive to changes in the cations for chlorides, bromides, and iodides [1,2]. Calculations of the electronic structures of these heavier halides like NaCl(s) suggest that states at the bottom of the conduction band originate from the lowest virtual s AO of the anion (4s for Cl), rather than valence AOs of the cation [24].

The valence bandwidths for the alkali halides range from ~1.0–2.0 eV [21] and are derived primarily from nonbonding halide–halide interactions. Since the valence band is full, the valence electrons are considered to be localized on the anions. As cation–anion orbital interactions become stronger, the valence bandwidth will typically increase and the electronic distribution changes. Electron density distributions calculated for various NaCl-type compounds MX (M = Li-Rb, Ag, Mg, Ca; X = F-Br, O, S) show three distinct characteristics, depending upon the polarizabilities of the anions and cations: (i) for fluorides and oxides, all ions are spherical with saddle points in electron density found midway along the M···M (and X···X) contacts; (ii) for alkali and alkaline earth sulfides, chlorides, and bromides, cations are spherical and anions are distorted with local minima in electron density midway along the M···M contacts; and (iii) for silver halides, which have their highest electron density at the polarizable Ag ions due to the presence of filled 4d valence states, both cations and anions show distorted spherical distributions with density contours along the X···X contacts resembling those along the M···M contacts of sulfides, chlorides, and bromides [25].

Table 12.1 Chemical bonding characteristics for selected NaCl-type MX(s) solids using an ionic bonding model.

Compound	R_0 (Å) [2]	U_{coh} (eV)		B_0 (GPa)		E_g (eV) [2]	ΔE_{sp} (eV) [2]
		Experimental [2]	Calculated[a]	Experimental [23]	Calculated[b]		
LiF	2.01	8.85	8.38	67.1	52.5	13.6	14.5
LiCl	2.57	7.15	6.71	29.8	28.0	9.4	8.4
LiBr	2.75	6.45	5.97	23.8	23.4	7.6	7.1
NaF	2.32	7.91	7.49	46.5	36.5	11.6	14.9
NaCl	2.82	6.63	6.31	24.4	21.9	8.5	8.8
NaBr	2.99	6.01	5.68	19.9	18.7	7.5	7.5
KF	2.67	7.62	7.27	30.5	25.3	10.7	15.9
KCl	3.15	6.72	6.39	17.4	16.2	8.4	9.8
KBr	3.30	6.17	5.85	14.8	14.3	7.4	8.4

R_0 = equilibrium cation–anion distance; U_{coh} = cohesive energy; B_0 = bulk modulus; E_g = bandgap; and $\Delta E_{sp} = E_s(M) - E_p(X)$ for valence AOs of M and X.
a) Using Eq. (12.7).
b) Using Eq. (12.9).

Binding of ions in an ionic solid can be estimated by the lattice energy, which includes the attractive electrostatic Madelung energy and overlap repulsions between filled-shell ions, as well as smaller van der Waals and other multipole attractions, and a zero-point energy term [2,15,26,27]. By using just the electrostatic attraction and the Born–Mayer expression for the overlap repulsion, the lattice energy (eV) depends on the cation–anion distance R (Å) according to

$$U_{lat}(R) \sim -\frac{14.4AZ^+Z^-}{R} + Be^{-R/\rho} = -\frac{14.4AZ^+Z^-}{R}\left(1 - \frac{\rho R}{R_0^2}e^{(R_0-R)/\rho}\right), \quad (12.7)$$

where A is the structure-dependent Madelung constant, R_0 is the equilibrium distance, Z^+ and Z^- represent the ionic charges, and ρ is an empirical scaling parameter, frequently set to 0.345 Å. At R_0, the lattice energy expression becomes

$$U_{lat}(R_0) = -\frac{14.4AZ^+Z^-}{R_0}\left(1 - \frac{\rho}{R_0}\right), \quad (12.8)$$

which represents the energy change for dissociation of $M_mX_x(s)$ to the free ions $mM^{Z+}(g) + xX^{Z-}(g)$. This crude point-charge model of ions reasonably accounts for structural trends and bulk moduli of ionic solids. For example, among MX(s) salts, the CsCl structure type is favored when the cation and anion sizes are similar; the ZnS structure type is favored if there is extreme disparity between ionic sizes with generally larger anions; and the NaCl structure type is preferred for an intermediate size relationship. Since ionic forces are isotropic, these solids tend to be dense sphere packings of anions with essentially uniform coordination environments surrounding all atoms (ions).

The equilibrium lattice energy $U_{lat}(R_0)$ can be evaluated from experimental data using a Born–Haber cycle based upon the standard heat of formation of the salt $M_mX_x(s)$, heats of atomization of M and X, ionization potential(s) for the cation M^{Z+}, and electron affinity(ies) for the anion X^{Z-} [28]. Therefore, to obtain the cohesive energies of ionic solids, the ionization potential(s) and electron affinity(ies) must be subtracted from the equilibrium lattice energy. For example, the cohesive energies of the alkali metal halides MX(s) are $U_{coh} = |U_{lat}(R_0)| - [IP_1(M) - EA(X)]$, which are listed in Table 12.1 as evaluated from experiment [24] and calculated using Eq. (12.7). Since the volume dependence of the cohesive energy is found entirely in the lattice energy, estimates of bulk moduli [23] (GPa) for NaCl-type solids ($A = 1.7476$) may be derived from Eq. (12.6) to give [25]

$$B_0 = V_0 \left(\frac{d^2 U_{lat}}{dV^2}\right)_{V_0} = \frac{1}{18R_0}\left(\frac{d^2 U_{lat}}{dR^2}\right)_{R_0} = \frac{224 Z^+ Z^-}{R_0^4}\left(\frac{R_0}{\rho} - 2\right). \quad (12.9)$$

As seen in Table 12.1, the calculated cohesive energies and bulk moduli using Eqs. (12.7)–(12.9) typically underestimate the experimental values for NaCl-type alkali metal halides, but the variations with cation–anion distances are reproduced quite well. Once van der Waals and zero-point energy terms are included, all further differences between calculated and experimental values are attributed to polar covalent interactions between the cations and anions.

Overall, the point-charge model accounts well for bonding energetics and compressibilities of the alkali metal halides. However, extending this model to other systems is limited [15]. For example, many oxides are also considered to be ionic compounds because oxygen has a large electronegativity, but the doubly charged oxide ion, $O^{2-}(g)$, is unstable relative to $O^-(g) + e^-$. Thus, oxide lattice energies are related to a nonexistent reference state of ions, and it is unclear whether complete charge transfer occurs in oxides. Furthermore, the valence bandwidths in oxides, as determined from photoemission spectra, are larger than those in the alkali metal halides, for example, ~5.0 eV in MgO, a result that suggests stronger cation–anion orbital interactions in metal oxides than in alkali metal halides. Therefore, the oxide band is no longer purely anionic, but involves significant mixing of anion and cation wavefunctions [1]. According to the van Arkel–Ketelaar diagram, the variation of electronic structure from an ionic to covalent picture is essentially continuous. As electronegativity differences between "cations" and "anions" decrease, orbital interactions increase and the

Madelung energy will be reduced because the extent of charge transfer decreases. So, the nature of localized electronic states transforms from anions and cations in ionic solids to interatomic bonding and antibonding bonding in covalent solids.

12.2.2
Covalent Solids

Cohesion in the covalent bond occurs when the potential energy of the valence electrons decreases by increasing electron density between atom cores. Therefore, a simple orbital picture for a covalent compound assigns electron pairs into bonding orbitals and any nonbonding lone pairs. The covalent corner of the van Arkel–Ketelaar map consists of molecular solids formed by the most electronegative elements that have fewer bonding orbitals than lone pairs, which create weak intermolecular dispersion forces. On the other hand, a typical covalently bonded, extended solid is the diamond-type structure of silicon, which forms a three-dimensional tetrahedral network of two-center, two-electron bonds. This structure and other related covalent solids are not densely packed. To gain a qualitative understanding about the valence electronic structure of Si(s), as depicted in Figure 12.5a, the valence 3s (E_s) and 3p (E_p) AOs may be converted into four equivalent, isoenergetic sp^3 hybrid orbitals, $E(\mathrm{sp}^3) = (E_s + 3E_p)/4$, that are spatially orthogonal and directed toward the four nearest Si neighbors: $\varphi_1(\mathrm{sp}^3) = 1/2(s + p_x + p_y + p_z)$; $\varphi_2(\mathrm{sp}^3) = 1/2(s + p_x - p_y - p_z)$; $\varphi_3(\mathrm{sp}^3) = 1/2(s - p_x + p_y - p_z)$; and $\varphi_4(\mathrm{sp}^3) = 1/2(s - p_x - p_y + p_z)$.[2] Then, each sp^3 hybrid forms one bonding and one antibonding orbital with another sp^3 hybrid located on a neighboring atom directed along the interatomic axis. This interaction splits the four hybrid energy levels into two bonding and two antibonding levels per Si atom separated by twice the magnitude of the bond integral $2|\beta_{\mathrm{Si-Si}}|$ between two adjacent overlapping sp^3 hybrid orbitals. These hybrid orbitals broaden into bands via energetic coupling between different sp^3 hybrids on the same atom because they are not eigenfunctions of the Hamiltonian operator. The approximate bandwidths for both the occupied, bonding valence band and unoccupied, antibonding conduction band are $\Delta E_{\mathrm{sp}} = E_{\mathrm{np}} - E_{\mathrm{ns}}$, and a simple calculation shows that as long as $\Delta E_{\mathrm{sp}}(\mathrm{Si}) < 2|\beta_{\mathrm{Si-Si}}|$, a bandgap occurs. This condition is just satisfied for tetrahedral Sn, and is definitely observed for Ge and diamond. Therefore, the bandgap in tetrahedral semiconductors may be called a hybridization gap. For Si(s), Ge(s), and Sn(s), this gap is less than 3 eV (larger than ~400 nm), which makes these elements semiconducting by also showing decreasing electrical resistivity with increasing temperature. For compound tetrahedral semiconductors AX(s) like GaAs(s) and ZnSe(s), a similar qualitative orbital diagram may be constructed (see Figure 12.5b) in which the splitting between bonding and antibonding orbitals is $2\Delta_{\mathrm{A-X}} = 2[\beta^2 + (\Delta_{\mathrm{hyb}}/2)^2]^{1/2}$ ($\Delta_{\mathrm{hyb}} = E_A(\mathrm{sp}^3) - E_X(\mathrm{sp}^3)$) and the bandwidths differ because $\Delta E_{\mathrm{sp}}(A) \neq \Delta E_{\mathrm{sp}}(X)$.[2] Moreover, in all cases, there is energetic coupling between the bonding orbital of an A—X contact constructed from sp^3 hybrids and all other bonding and

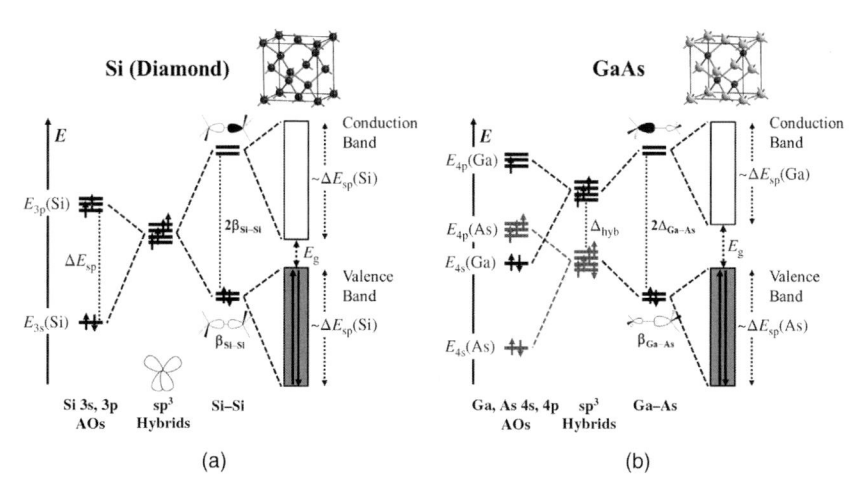

Figure 12.5 Evolution of the band diagrams for tetrahedral semiconductors (a) Si and (b) GaAs [2]. The free atom orbitals form sp³ hybrids, which interact according to the bond integral β_{Si-Si} or β_{Ga-As} to form bonding and antibonding orbitals. In the solids, these orbitals broaden into bands by intrasite and other intersite overlaps.

antibonding orbitals connected to the two atoms, a coupling that affects the bandwidths of and bandgap between the valence and conduction bands. Mixing of antibonding orbitals into bonding states also lowers the A—X bond order p_{AX} from unity, which would indicate a completely saturated, formally single bond [11]. The bond order in elemental tetrahedral semiconductors decreases as ΔE_{sp} increases, so the bonds in diamond, Si(s), and Ge(s) are, respectively, ~94, ~86, and ~81% saturated. Bond ionicity also affects bond order, as seen for GaAs (s) and ZnSe(s), which are ~79 and 75% saturated, respectively, arising, in part, from the energetic mismatch between sp³ hybrids on the different elements. The effect of bond ionicity is evident from plots of the calculated bonding densities along the nearest-neighbor contacts in Ge(s), GaAs(s), and ZnSe(s), plots that show that the maximum bonding density is midway along the Ge—Ge bond and steadily shifts toward the more electronegative component (As or Se) in GaAs and ZnSe [29].

The binding energy curve $U(R)$ of a tetrahedral semiconductor can be expressed as the sum of an attractive bond energy term $U_{bond}(R)$, which is related to $[\beta^2 + (\Delta_{hyb}/2)^2]^{1/2}$, and the overlap repulsion $U_{rep}(R)$, lowered by the promotion energy E_{prom} for the valence electrons [2]:

$$U(R) \sim U_{bond}(R) + U_{rep}(R) - E_{prom}. \tag{12.10}$$

Unfortunately, $U(R)$ is not a simple sum of two-center interactions, as seen for ionic solids, because multicenter interactions arise from the energetic coupling between bonding and antibonding orbitals on different sites [2,11]. Nevertheless, since the orbital model places electron pairs into bonding orbitals, the cohesive energy $U_{coh} = |U(R_0)|$, which is the magnitude of the binding energy at the equilibrium A—X distance R_0, provides a means to estimate effective pairwise bond dissociation energies $D(A—X)$ in covalent solids. For elemental tetrahedral semiconductors, $D(A—A) \sim U_{coh}(A)/2$; for compound semiconductors, $D(A—X) \sim U_{coh}(AX)/4$. To calculate the cohesive energies of covalent solids accurately requires elaborate electronic structure theory, but the values can be determined from experiment by using heats of formation of the solid and the heats of atomization of each component in a modified Born–Haber cycle [30]. For the tetrahedral semiconductors, this exercise yields

$$|U_A(R_0)| \sim 2D(A—A) = \Delta H_{atom}(A), \quad (12.11a)$$

$$|U_{AX}(R_0)| \sim 4D(A—X) = \Delta H_{atom}(A) + \Delta H_{atom}(X) - \Delta H_f^\circ(AX), \quad (12.11b)$$

and some results are listed in Table 12.2. The bond integral β between adjacent sp^3 hybrid orbitals can be estimated from the energetics of the pathway of first dissociating the solid into gaseous atoms using sp^3 hybrids and then relaxing the atoms to their ground-state configurations. For elemental semiconductors, $\beta_{A-A} \sim [U_{coh}(A) + E_{prom}(A)]/2$, and for compound semiconductors, $\beta_{A-X} \sim [U_{coh}(AX) + E_{prom}(A) + E_{prom}(X)]/4$, in which $E_{prom}(A)$, the promotion energy for element A with N_A valence electrons, is $[(8-N_A)/4]\Delta E_{sp}(A)$ [2]. These β values for selected tetrahedral semiconductors are also listed in Table 12.2. In general, bond dissociation energies $D(A—X)$ decrease as the sizes of the components increase, and the β values decrease as ionicity increases. From Tables 12.1 and 12.2, the cohesive energies of the tetrahedral semiconductors and NaCl-type alkali metal halides are similar in magnitude. However, experimental bulk moduli [31], which provide estimates for the curvatures of $U(R)$ near the equilibrium A—X distances, indicate stiffer bonds within the covalent, diamond-type solids than NaCl-type ionic salts by approximate factors of 3–5.

Although the orbital energy diagram of sp^3 hybrids accounts for the stability and bandgaps of the diamond-type structure for compounds with four valence electrons per atom, there is no simple rule that predicts which hybridization scheme and corresponding local structure is most preferred for each element. Other common hybrid orbitals for second period elements are the sp^2 hybrids, which, in addition to the single unhybridized p AO, can describe the bonding of carbon atoms in graphite and graphene. However, the particular set of hybrid orbitals chosen must represent the overall minimum total energy for the system by maximizing bond overlap with neighboring atoms while minimizing the promotion energy from the ground-state electronic configuration [11]. Moreover, the use of hybrid orbitals is valid only for any observable that includes all

Table 12.2 Chemical bonding characteristics for selected tetrahedral semiconductors AX(s).

Compound	R_0 (Å) [30]	U_{coh} (eV) [30]	$D(A-B)$ (eV)	β (eV)	B_0 (GPa) [31]	E_g (eV) [2]
C	1.54	7.42	3.71	7.87	433	5.47^{ind}
Si	2.35	4.68	2.34	5.94	98	1.12^{ind}
Ge	2.45	3.88	1.94	5.86	75	0.67^{ind}
Sn	2.81	3.13	1.57	4.71	53	~0.1^{ind}
AlP	2.36	8.59	2.15	5.53	86	2.45^{ind}
AlAs	2.43	7.76	1.94	5.37	77	2.16^{ind}
AlSb	2.66	7.15	1.79	4.83	58	1.6^{ind}
GaP	2.36	7.54	1.88	5.54	89	2.26^{ind}
GaAs	2.45	6.71	1.68	5.38	75	1.43^{dir}
GaSb	2.65	6.01	1.50	4.82	57	0.73^{dir}
ZnS	2.34	6.36	1.59	4.64	77	3.54^{dir}
ZnSe	2.46	5.40	1.35	4.37	62	2.7^{dir}
ZnTe	2.64	4.61	1.15	3.84	51	2.25^{dir}
CdS	2.52	5.71	1.43	4.23	62	2.42^{dir}
CdSe	2.62	4.93	1.23	4.00	53	1.74^{dir}
CdTe	2.81	4.15	1.04	3.48	42	1.49^{dir}

R_0 = equilibrium nearest-neighbor distance; U_{coh} = cohesive energy; $D(A-X)$ = A—X bond dissociation energy; β = estimate of intersite sp^3-sp^3 integral; B_0 = bulk modulus; and E_g = bandgap ("ind" = indirect; "dir" = direct).

electrons, that is, "collective" properties such as cohesive energy and electron density [1] Spectroscopic studies of solids, which provide insights about the different energy bands, demand a delocalized or MO-type approach.

This simple strategy to generate qualitative orbital energy diagrams of covalent solids can be easily extended to structures that do not consist of tetrahedrally bonded main group atoms, such as those with elements that contain lone pairs or transition elements, by taking into account electronegativities and using first- or second-order perturbation theory. Any lone pairs would occur as nonbonding bands intermediate in energy between bonding and antibonding bands with band centers close to the energies of the corresponding AOs or hybrids. Valence d AOs of transition metals would split according to the concepts of ligand field theory [13]. This qualitative approach becomes complicated, however, if the materials are metals. For example, in the case of graphite or graphene, the sp^2 hybrids are energetically separated into σ-bonding and σ-antibonding bands, but the unhybridized p AOs form a continuous π-band with no separation between bonding and antibonding character [1]. Thus, the orbital picture for metals develops some characteristics that distinguish them from covalent solids.

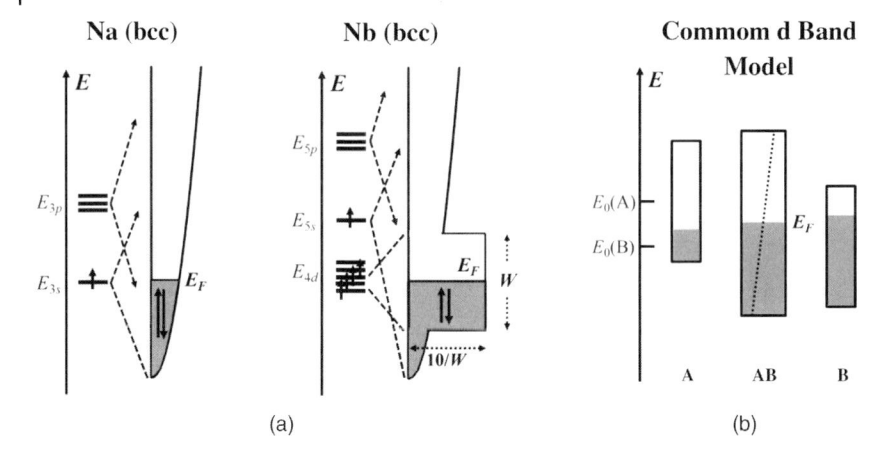

Figure 12.6 (a) Evolution of the band diagrams for main group metals, for example, Na, and transition metals, for example, Nb. The valence s and p form broad bands, whereas the d-band is narrow. Fermi levels, d bandwidth, and height are indicated. (b) Common d-band model for intermetallic alloy or compound. Dotted line indicate the contributions to the rectangular DOS from A (left) and B (right).

12.2.3
Metallic Solids

The electronic structure of a metal involves a broad, quasi-continuous energy band arising from the overlapping valence AOs of the metal atoms that form densely packed structures like fcc, hcp, or bcc packings with large coordination numbers for each atom [2]. The energy differences between adjacent levels within this band are small relative to kT, even at low temperatures, so the substance effectively conducts electric current because states above the highest occupied level are thermally accessible. The energy of the highest occupied orbital is called the Fermi level E_F, which measures the work function for a metal, that is, the energy needed to eject an electron to the vacuum. Orbital energy diagrams of main group metals like Na(s), Mg(s), and Al(s) differ from those of transition metals because of the valence d AOs (see Figure 12.6a). Both classes of metals have a broad conduction band consisting of valence s and p levels, whereas the transition metals include a narrower valence d-band, which generally shows features, that is, peaks and valleys, arising from the coordination at the transition metal atom. Nevertheless, important periodic trends can be elucidated by treating the valence d-band simply as rectangular with bandwidth W, height $n(E) = 10/W$, center E_0, and filled with N_d electrons, which depends on the transition metal [11]. In particular, the band center drops in energy from left-to-right across a transition series as N_d increases, and the bandwidth varies as $N_d \cdot (10 - N_d)$ such that W achieves maximum values near the half-filled d-band, that is, for Groups 5 and 6 (see Table 12.3 [32]). The corresponding picture for intermetallic compounds and alloys of different transition metals A and

Table 12.3 Chemical bonding characteristics for selected main group and fifth period transition metals.

Metal	R_{WS} (Å) [11]	R_C (Å) [11]	N	r_s (Å)	$|U_M|$ (eV) Experimental [11]	$|U_M|$ (eV) Calculated[a]	B_0 (GPa) Experimental [11]	B_0 (GPa) Calculated[b]
Li	1.73	0.70	1	1.73	1.63	2.12	11	8.76
Na	2.11	0.93	1	2.11	1.11	1.24	6.3	4.23
K	2.57	1.17	1	2.57	0.93	1.13	3.1	1.97
Rb	2.81	1.31	1	2.81	0.85	0.90	2.5	1.41
Be	1.25	0.40	2	0.99	3.32	5.55	130	99.88
Mg	1.77	0.69	2	1.40	1.51	2.26	45	27.95
Ca	2.18	0.92	2	1.73	1.84	2.92	17	12.97
Sr	2.38	1.02	2	1.89	1.72	2.67	12	9.39
Al	1.58	0.57	3	1.10	3.39	5.51	76	86.49
Ga	1.67	0.64	3	1.16	2.81	—	~60	—
In	1.84	0.72	3	1.28	2.52	—	~35	—
Cu	1.41	0.48	1	1.41	3.49	1.26	140	17.21
Ag	1.60	0.72	1	1.60	2.95	—	100	—
Zn	1.54	0.57	2	1.22	1.35	0.64	70	46.96
Cd	1.77	0.67	2	1.37	1.16	—	50	—

Metal (structure)	R_{WS} (Å) [32]	W (eV) [11]	$|U_T|$ (eV) [32]	B_T (GPa) [32]
Rb (bcc)	2.81	—	0.85	2.5
Sr (fcc)	2.38	—	1.72	12
Y (hcp)	1.99	6.28	4.39	41.0
Zr (hcp)	1.78	7.80	6.29	94.9
Nb (bcc)	1.63	9.34	7.44	169.0
Mo (bcc)	1.55	9.55	6.88	261.0
Tc (hcp)	1.52	9.10	7.13	297.0
Ru (hcp)	1.49	8.49	6.74	303.0
Rh (fcc)	1.40	7.59	5.72	282.0
Pd (fcc)	1.52	5.98	3.90	189.0
Ag (fcc)	1.60	3.86	2.94	98.8
Cd (hcp)	1.77	—	1.16	50

R_{WS} = Wigner–Seitz radius at equilibrium volume; R_C = radius of empty-core pseudopotential; N = number of valence electrons; r_s = effective radius of 1 valence electron; $|U_M|$, $|U_T|$ = cohesive energy; B_0 = bulk modulus; and W = 4d bandwidth.
a) Using Eq. (12.12b).
b) Using Eq. (12.13)

B is a common valence d-band with bandwidth W_{AB} that is somewhat larger than the bandwidths of the individual elements W_A and W_B due to the AO energy difference between the two metals (see Figure 12.6b) [11]. In addition, the total rectangular d-band height $n_{AB}(E)$ is the sum of two skewed distributions $n_A(E) + n_B(E)$. However, since there is no net charge transfer in a metal via effective screening from the conduction electrons, the effective band centers of each skewed distribution are adjusted from $E_0(A)$ and $E_0(B)$ to maintain local charge neutrality in the compound.

The internal energy of a close-packed, main group metal U_M can be estimated by treating it as empty-core pseudopotentials of radius R_C embedded in a delocalized "gas" of valence electrons with uniform number density $\rho = N/V$ in which N and V are the number of valence electrons and volume per atom in the solid [2,11]. Two important distance parameters are the effective sphere radius for one atom R_{WS}, called the Wigner–Seitz radius, and for one valence electron r_s, so that $4\pi R_{WS}^3/3 = V$ and $4\pi r_s^3/3 = V/N = 1/\rho$. U_M contains two main contributions: (i) the energy of the electron gas U_{eg}, which includes kinetic, exchange, and correlation; and (ii) an electrostatic term U_{es} arising from core–valence electron repulsions and the attractive Madelung energy of point ions in the uniform electron gas (U_M in eV; r_s, R_C, and R_{WS} in Å):

$$U_M(N, V, R_C) \sim N U_{eg}(r_s) + U_{es}(N, R_{WS}, R_C) \qquad (12.12a)$$

$$= N\left[\frac{8.416}{r_s^2} - \frac{6.59}{r_s} + U_{corr}(r_s)\right] + \frac{7.20 N^2}{R_{WS}}\left[3.00\left(\frac{R_C}{R_{WS}}\right)^2 - 1.80\right]. \qquad (12.12b)$$

These expressions explicitly omit structural effects which enter as a second-order correction to U_{es} and modify the electrostatic term by less than 0.5%. Although these differences are relatively insignificant regarding the cohesive energies, they are important for distinguishing different structures, such as exhibited by the trivalent elements Al (cubic), Ga (orthorhombic), and In (tetragonal). Also, there are various expressions for the correlation energy of the electron gas $U_{corr}(r_s)$, but the values typically constitute no more than ~10% of the attractive exchange term [15,33]. The pseudopotential core radii R_C are determined by setting $dU_M/dR_{WS} = 0$ at the equilibrium volume V_0. The cohesive energy for a main group metal may be estimated from Eq. (12.12) by subtracting the first N ionization potentials of the metallic element from $U_M(N,V_0,R_C)$. The corresponding bulk moduli B_0 (GPa) can also be estimated from Eq. (12.11b), showing contributions from both energy terms:

$$B_0 = \left(\frac{358.62}{r_s^5} - \frac{112.40}{r_s^4}\right) + \frac{122.58 N^2}{R_{WS}^4}\left(13.53\left(\frac{R_C}{R_{WS}}\right)^2 - 1.80\right). \qquad (12.13)$$

Reasonable agreement with experimental data for cohesive energies and bulk moduli are seen for the alkali and alkaline earth metals and even Al, but the agreement is poor for the coinage metals Cu, Ag, and Au and posttransition metals because the occupied d-orbitals are explicitly ignored (see Table 12.3).

For the transition metals, the rectangular model of the valence d-band provides a good rationale for the essential parabolic trend in cohesive energies with group number because the significant contributions come from the extent of d-band filling N_d. The cohesive energy $|U_T|$ turns out to be related to the bandwidth and d-band filling by [11]

$$|U_T| \propto W N_d(10 - N_d) \propto [N_d(10 - N_d)]^2. \tag{12.14}$$

At low N_d values, the cohesive energy increases as d–d bonding states are increasingly occupied. Likewise, equilibrium volumes and Wigner–Seitz radii drop. Once d–d antibonding states become accessible just beyond the half-filled point at $N_d = 5$, the cohesive energy decreases. However, the equilibrium volumes and Wigner–Seitz radii achieve minima beyond the half-filled band because the effective nuclear charge for the valence d-electrons increases as N_d increases. This model works best for the 4d elements, which show a maximum in cohesive energy for Nb but a minimum in equilibrium atomic volume and maximum in bulk modulus for Ru, as summarized in Table 12.3. The bulk moduli for metals in Groups 4–10 are also substantially larger than those observed for the covalently bonded tetrahedral semiconductors, except for diamond. Thus, occupation of the d-band provides substantial bond strength for these elements. The common, rectangular d-band model of transition metal alloys and intermetallics also accounts very well [11] for their trends in thermodynamic stabilities relative to the component metals, as found experimentally and predicted by Miedema *et al.* [34] and Brewer [35], that the most stable alloys would be comprised of elements at opposite ends of the transition metal series because their common d-band would be approximately half-filled and gain significant energetic stabilization via d–d bonding interactions.

The semiquantitative assessments of chemical bonding for ionic, covalent, and metallic solids using simple physical arguments to estimate cohesive energies and bulk moduli identify the important factors contributing to the stability and relative energetics of each bonding category. Each model involves an attractive term and a repulsive term to the energy profile in accord with the $E(R)$ curve of Figure 12.1. Since these simple descriptions are derived from the results of extensive quantum mechanical calculations on many different solid-state structures, it is now time to consider how these computational results are obtained.

12.3
Quantum Mechanical Calculations and Analyses

A thorough study of chemical bonding and electronic structure in molecules and solids requires a combination of experimental and computational analyses [36].

Calculating the electronic structure involves solving the time-independent Schrödinger equation [1,2,11], to which the Born–Oppenheimer approximation is applied because electronic velocities are much larger than nuclear velocities so that electronic motion can be separated from nuclear motion. Therefore, electronic energies and wavefunctions are evaluated using "static" nuclear structures, even ignoring the zero-point motion of nuclei at 0 K. As a result, temperature is not a direct factor in such calculations, but does influence the occupation of electronic states according to the Fermi–Dirac distribution, as well as the energetic distribution of states by any changes in atomic structure such as thermal expansion or solid–solid phase transitions. The resulting Schrödinger equation to be solved becomes

$$\hat{H}\Psi_n(r_1,\ldots,r_N) = (\hat{T} + \hat{V})\Psi_n(r_1,\ldots,r_N) = E_n\Psi_n(r_1,\ldots,r_N), \quad (12.15)$$

where $\Psi_n(r_1, \ldots, r_N)$ are the N-electron wavefunctions, E_n are the corresponding electronic energies, and the Hamiltonian operator $\hat{H}$ is the sum of kinetic ($\hat{T}$) and potential ($\hat{V}$) energy operators. In principle, Eq. (12.15) utilizes all electrons in the system, but typical results focus just on the valence and low-lying unoccupied states.

Since a solid is a chemical structure with a quasi-infinite number (about 10^{23}) of atoms and valence electrons, essential approximations to the atomic structure are also necessary to solve the Schrödinger equation. For this purpose, it helps to differentiate solids into molecular and extended structures. For molecular solids, calculations may be performed on an isolated molecule by ignoring "packing effects." However, these weak intermolecular interactions, especially when they are numerous, can influence the local structure of a molecule, so an effective medium could be included. Extended solids occur in three principle categories: (i) amorphous; (ii) incommensurately modulated or quasiperiodic; and (iii) crystalline or periodic. To examine aperiodic structures, appropriate fragments may be extracted from the solid and then terminated by s-type orbitals to reduce the effects of surface termination on the computational results. For a crystalline solid, the potential felt by an electron has the full periodicity of the lattice, that is, $V(r + T) = V(r)$, for a Bravais lattice vector T. To make the set of Bravais lattice vectors finite, Born–von Karman boundary conditions assign arbitrarily large lattice vectors equivalent to the null vector [36] Then, the wavefunction for a single electron $\psi_{nk}(r)$ contains the phase factor $e^{ik\cdot r}$ in which the wavevector k is a quantum number arising from these periodic boundary conditions. Furthermore, for the periodic potential, Bloch's theorem states that the wavefunctions take the form $\psi_{nk}(r) = e^{ik\cdot r}u_{nk}(r)$, with $u_{nk}(r+T) = u_{nk}(r)$, that is, $u_{nk}(r)$ also has the full periodicity of the lattice. There are three significant outcomes from Bloch's theorem [37]:

a) $\psi_{nk}(r+T) = e^{ik\cdot T}\psi_{nk}(r)$, which shows that to calculate electronic states in crystals, it is sufficent to evaluate the wavefunctions of one unit cell, because the wavefunctions for all other unit cells are given by a phase change with respect to the reference cell.

b) If K is a reciprocal lattice vector appropriate to the Bravais lattice, then $\psi_{nk}(r)$ and $\psi_{n\{k+K\}}(r)$ have the same symmetry properties with respect to lattice translations. Therefore, k and $k+K$ are equivalent symmetry labels and all information regarding electronic states in crystals is confined to one unit cell in reciprocal space surrounding a single K vector. The first Brillouin zone is the unit cell at $K=0$, called the Γ point, that is formed by the perpendicular bisectors to all neighboring reciprocal lattice vectors..

c) The wavefunction $\psi_{n,-k}(r)$ is the complex conjugate of $\psi_{nk}(r)$, so these wavefunctions will have the same energies $E(k)$.

Bloch's theorem allows accurate calculations of bulk electronic states and properties of periodic structures. So, the electronic structures of quasiperiodic and incommensurately modulated structures can also be estimated by using crystalline approximants, whose structures and compositions mimic the actual solids [38]. For crystals that display mixed site occupancies, using either periodic superstructures or ordered models with subgroups of the crystallographic space groups, or applying the coherent potential approximation (CPA) [39] to the electronic Hamiltonian become necessary to properly handle the disorder inherent in these solids.

Now, the major challenges for solving Eq. (12.14) involve setting up appropriate potential energy operators for the valence electrons and establishing useful expressions for the wavefunctions. For simple metals with delocalized electrons, the free-electron model using plane wave basis functions is reasonable [11,37]. On the other hand, for solids exhibiting more localized electrons, like transition metal compounds, semimetals, semiconductors, and insulators, the tight-binding method [1,11,37], which constructs wavefunctions by the LCAO ansatz, is preferred. These two philosophies will be compared shortly. Nevertheless, the energy operators may be separated into one-electron and two-electron contributions. The significant one-electron terms are the kinetic energies and the effective electrostatic electron–nuclear attractions, which are screened by the core electrons to create effective nuclear charges. The two-electron term involves interactions between pairs of valence electrons, leading to exchange and correlation. If there were no electron–electron interactions, then Eq. (12.15) could be easily solved and the wavefunctions would be products of one-electron spin orbitals, that is, $\Psi(r_1, \ldots, r_N) \sim \varphi_1(r_1) \cdots \varphi_N(r_N)$. Additional improvements to the Hamiltonian operator include relativistic effects of electronic motion and spin-orbit coupling, especially for the heavier elements.

Electron–electron interactions in the Schrödinger equation can be accounted for by the Hartree approximation [15], in which each electron moves in the average electrostatic field of all other electrons yielding a set of effective one-electron potentials and orbitals for each electron. From these one-electron orbitals, Slater determinants are constructed to account for the antisymmetry of electron exchange, a procedure giving the Hartree–Fock method. As a result, surrounding each electron is a region called a Fermi hole [11,15], which arises because other electrons of the same spin are prevented from closely approaching it via exchange repulsion. Therefore, like-spin electrons do not shield each other from the nucleus as well as electrons with opposing spins [15]. Consequently, given a system with unequal numbers of electron spins, there will be differential shielding and one set of spins will experience a higher effective nuclear charge than the other. This effect leads to exchange splitting of the corresponding spin orbitals and can have profound effects on chemical bonding, especially for high-spin versus low-spin transition metal complexes. For example, the five degenerate valence d AOs of a transition metal atom in an octahedral ligand field are split into three t_{2g} orbitals below two e_g orbitals. The e_g orbitals are metal–ligand σ-antibonding, whereas the t_{2g} orbitals are metal–ligand π-bonding or π-antibonding depending upon whether the ligands are, respectively, weak-field or strong-field ligands. The d-orbitals are also split by exchange into two sets of spin orbitals. If the exchange splitting exceeds the ligand field splitting, then high-spin complexes are favored; if not, low-spin complexes are favored. In the high-spin case, the σ-antibonding e_g orbital(s) will be occupied; in the low-spin case, the e_g orbital(s) will be unoccupied. In general, exchange effects are greatest for atoms with valence 2p, 3d, and 4f electrons because electrons assigned to these orbitals will experience especially high effective nuclear charges, leading to contracted AOs and large effective electron–electron interactions.

Correlation, on the other hand, reduces the electrostatic repulsion between electrons by mutual avoidance [15]. It can be included by applying perturbation theory or allowing electrons to occupy higher energy orbitals, which mixes other electronic configurations into the final ground-state wavefunction – a procedure called configuration interaction. In effective one-electron methods, introducing an empirical Coulomb repulsion parameter, sometimes called the Hubbard U parameter [12], is used to correct any underestimation of correlation by these computational procedures, especially for certain rare earth and transition metals [1,15].

Methods to solve Eq. (12.14) range from simple to complex arising from the two-electron terms, which necessitate approximations to achieve solutions that allow insights about chemical bonding. The simplest approximations, such as Hückel and extended Hückel [1,6], explicitly neglect all two-electron terms and treat the Hamiltonian operator empirically or semiempirically in which the resulting integrals are determined by simple algorithms using AO energies evaluated from atomic spectra or first-principles calculations. The electron–electron interactions are then "absorbed" by the numerical parameters. These strategies are most effective when angular overlap influences the energetics of structures or the two-electron energy terms are relatively insignificant compared to the one-electron terms. However, they poorly optimize interatomic distances and overestimate the energies of antibonding orbitals, thereby overestimating bandgaps in semiconductors and insulators. In first-principles methods, on the other hand, all integrals are explicitly calculated according to some physical model. Therefore, their accuracy and appropriateness depend on the system and the models involved to express the electronic potential energy. The Hartree–Fock method can give excellent results for insulators and semiconductors, but has not been thoroughly developed for metals because there is a singularity in the exchange energy for states at the Fermi level. Simplifications to Hartree–Fock methods include various "neglect of differential overlap" (NDO) approaches [15], which can treat large systems because many two-electron integrals are replaced by numerical parameters. Another effective way to solve the many-body problem of Eq. (12.14) uses density functional theory (DFT), in which the interacting system of electrons is described by its electronic probability density $\rho(r)$ rather than its many-body wavefunction because the ground-state electronic density uniquely determines the ground-state wavefunction and energy [40]. By introducing a reference system of noninteracting electrons with the same (spin) density, the functional of the total energy can be expressed as

$$E[\rho(r)] = T_0[\rho(r)] + \int \rho(r) v_{\text{ext}}(r) dr + \frac{1}{2} \int \int \frac{\rho(r)\rho(r')}{|r-r'|} dr dr' + E_{\text{XC}}[\rho(r)], \quad (12.16)$$

which includes the kinetic energy, the attractive Coulomb potential $v_{\text{ext}}(r)$ provided by the nuclei, a classical Coulomb repulsion term, and an energy functional for exchange and correlation. Although the exact functionals for exchange and correlation are not known except for the free-electron gas, various approximations like the local (spin) density (LDA/LSDA) [40] and generalized-gradient approximations (GGA) [40] permit calculation of certain physical quantities with reasonable success. The exchange–correlation functional in LDA/LSDA depends only on $\rho(r)$ at each point in space (LSDA decouples the two different electron spins), whereas GGA remains local but also takes into account the gradient $\nabla \rho(r)$ at the same position. Generally, LDA/LSDA is suitable for a nearly uniform electronic distribution, whereas GGA is necessary when larger density fluctuations are present. Since LDA/LSDA or GGA functionals often underestimate the on-site Coulomb interaction between electrons, especially for rare earth compounds with localized 4f electrons or transition metal oxides and halides with localized 3d electrons like NiO, one correction introduces a Hubbard on-site repulsion parameter U [41]. However, the choice of U is often made empirically rather than from first-principles techniques. Overall, DFT methods have been successful for evaluating ground-state properties, such as atomic structures, pressure-induced phase transitions, and isothermal bulk moduli, although equilibrium bond distances are often underestimated. On the other hand, DFT calculations typically underestimate bandgaps and overestimate cohesive energies [42].

Regardless of the mathematical method used to solve the Schrödinger equation, important results of these calculations on crystalline solids are depicted in electronic *band structure* (dispersion) diagrams $E(\mathbf{k})$ versus $\mathbf{k}$ and *density of states* (DOS) curves $n(E)$ versus E, which are the solid-state analogs of MO energy diagrams for discrete molecules. Band structures $E(\mathbf{k})$ are plotted along specified directions in the first Brillouin zone that exhibit special symmetry; the endpoints of these directions are given specific letters, for example, as already mentioned $\mathbf{k}=0$ is the Γ point. DOS curves $n(E)$, on the other hand, are obtained by sampling many wavevectors within the Brillouin zone. Each $n(E)$ value equals the number of spin-up and spin-down states found within an infinitesimally small energy window dE. Therefore, the total DOS curve, when integrated up to the Fermi level, gives the total number of valence electrons. The other important result is the total energy for a system of N valence electrons:

$$E_{\text{TOT}}(N) = E_{\text{Band}}(N) - E_{\text{DC}}(N) + E_{\text{core}} = \sum N_i E_i - E_{\text{DC}}(N) + E_{\text{core}}.$$
(12.17)

E_{Band} is the sum of energies of the occupied orbitals, and N_i are the occupation numbers (0, 1, or 2) for each orbital $\psi_i(\mathbf{r})$. Because this summation treats the two-electron terms twice for each pair of electrons, this double-counted energy E_{DC} must be withdrawn from E_{Band} [7,15]. The Coulomb repulsions between atomic cores E_{core} account for the final, repulsive energetic term in the total electronic energy.

12.3.1
Free-Electron versus Tight-Binding Approximations

As we have emphasized, important interpretations of chemical bonding emerge from the Schrödinger equation when orbital approximations are applied. Depending upon whether or not the system is metallic and the nature of the basis set, there are two limiting philosophies: the free-electron and the tight-binding approximations.

In the free-electron model the positive charge associated with the atom cores is smeared out so the valence electrons are not scattered by a lattice. The net attraction between the valence electron "gas" and this uniform positive background and the average repulsion within the valence electron gas exactly cancel each other, creating a constant potential. The solutions to the Schrödinger equation $\psi(\mathbf{r})$ after applying periodic boundary conditions $\psi(x+L, y, z) = \psi(x, y, z)$, and so on for a large cubic box of volume $V = L^3$ are as follows:

a) The orbitals are plane waves, $\psi_\mathbf{k}(\mathbf{r}) = (1/V)e^{i\mathbf{k}\cdot\mathbf{r}}$, periodic along a line parallel to the wavevector $\mathbf{k}$, which is also a quantum number arising from the boundary conditions, and has units in reciprocal space, that is, 1/distance such as Å^{-1}. Each plane wave can be assigned at most two electrons of opposite spins.

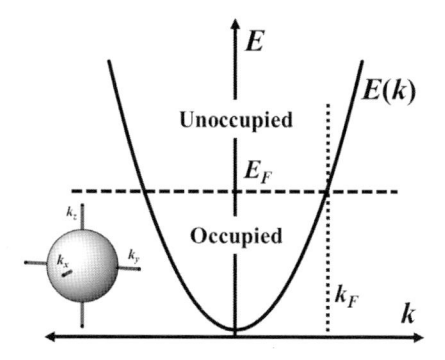

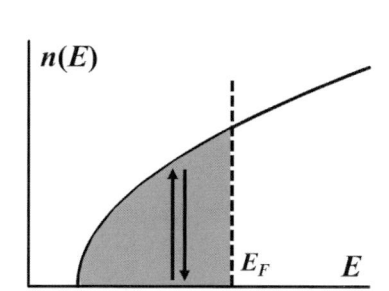

Figure 12.7 Energy band and DOS curves for the free-electron model. In the band diagram, the Fermi wavevector k_F and Fermi energy E_F are indicated. All states with $k \leq k_F$ at $T = 0\,K$ are occupied and form a sphere in k-space. The DOS states show that all states with $E \leq E_F$ are occupied.

b) The kinetic energy for the plane wave $\psi_k(r)$ is ($k = |\mathbf{k}|$):

$$E_{KE}(k) = \frac{\hbar^2 k^2}{2m} = (3.810\text{ eV\AA}^2)k^2. \quad (12.18)$$

At 0 K, all ψ_k are occupied such that $k \leq k_F$, the Fermi wavevector, and $E_{KE}(k) \leq E_{KE}(k_F) \equiv E_F$, the Fermi energy, so that the region of occupied states is a sphere in reciprocal space with radius k_F. This sphere contains N valence electrons, so k_F is a direct measure of electron density according to $k_F^3 = 3\pi^2\rho = 29.61\rho$. Also, $E_F = (36.47\text{ eV \AA}^2)\rho^{2/3}$. The band structure diagram, $E_{KE}(k)$ versus $\mathbf{k}$ is a parabola, and the resulting DOS curve $n(E)$ is proportional to $E_{KE}^{1/2}$, as illustrated in Figure 12.7.

c) The exchange energy associated with $\psi_k(r)$ is calculated by applying Hartree–Fock theory to the occupied plane waves, which gives [37]

$$E_{Exch}(k) = -\frac{e^2 k_F}{8\pi^2 \epsilon_0}\left(2 + \frac{k_F^2 - k^2}{k_F k}\ln\left|\frac{k_F + k}{k_F - k}\right|\right). \quad (12.19)$$

$E_{Exch}(k)$ is attractive, but there is a logarithmic singularity at k_F, so $E_{Exch}(k)$ is typically not included for plotting the band structure because it leads to a vanishing DOS at E_F, a result that is inconsistent with metallic behavior and arises by neglecting correlation in the electron gas. Nevertheless, the total exchange energy over all occupied plane waves can be evaluated.

The total energy of N free electrons follows from Eq. (12.17) by placing two electrons in each $\psi_k(r)$ with $k \leq k_F$ and treating the summation as an integration because $\psi_k(r)$ are quasi-continuous in reciprocal space, resulting in

$$NU_{eg}(\rho) \sim \int_0^{k_F} [2E_{KE}(k) + E_{Exch}(k)]dk \quad (12.20a)$$

$$= N\left(\frac{3}{5}E_F - \frac{3e^2 k_F}{16\pi^2\epsilon_0}\right) = N(21.87\rho^{2/3} - 10.63\rho^{1/3}). \quad (12.20b)$$

The first term is a repulsive, kinetic energy term; the second is an attractive exchange energy arising from the Fermi hole. An additional term may also take into account correlation $U_{\text{corr}}(\rho)$, included in Eq. (12.12b), for which there are approximate expressions [11,15,33,43].

Improvements to the free-electron model include incorporating a periodic pseudopotential to represent the valence electron–nuclear attraction from a periodic array of atomic cores, as well as to take into account the orthogonal relationship arising from the Pauli exclusion principle between orbitals for the valence and core electrons [2,11]. For metals, the pseudopotential also takes screening into account. The periodic screened pseudopotential $V_{\text{ps}}(r)$ can be expressed as a Fourier series over the reciprocal lattice vectors K:

$$V_{\text{ps}}(r) = V_{\text{ps}}(r + T) = \sum_K V_{\text{ps}}(K) e^{iK \cdot r}. \tag{12.21}$$

There are several kinds of pseudopotentials, but a simple approximation for an atom uses an empty-core pseudopotential [2,11], which is zero for $r < R_c$ and $-Z_{\text{eff}} e^2/4\pi\varepsilon_0 r$ for $r > R_c$. This approximation captures one essential aspect of pseudopotentials, that they are characterized by core radii. Using linear response theory, the Fourier transform of the screened empty-core pseudopotential is

$$V_{\text{ps}}(q) = -8\pi\rho \frac{\cos qR_c}{q^2 + 12\pi\rho/E_F}, \tag{12.22}$$

so that $V_{\text{ps}}(q=0) = -2E_F/3$ and $V_{\text{ps}}(q_0) = 0$ for $q_0 = \pi/2R_c$. For metals, $4R_c$ has a similar length as interatomic distances and lattice constants so that q_0 is close to the length of reciprocal lattice vectors connecting opposite faces of the Brillouin zone. Using Eq. (12.21), the integral of the Hamiltonian operator between two different plane wave states ψ_k and $\psi_{k'}$ is (using the Dirac braket notation)

$$\langle \psi_{k'} | \hat{T} + \sum_K V_{\text{ps}}(K) e^{iK \cdot r} | \psi_k \rangle = \left(\frac{\hbar^2 k^2}{2m} + V_{\text{ps}}(0) \right) (k - k')_\delta + V_{\text{ps}}(K) \delta(k + K - k'), \tag{12.23}$$

which yields the nearly free-electron approximation. With the periodic pseudopotential and Bloch's theorem, the energies of the plane waves are periodic in K, that is, $E(k + K) = E(k)$, and all information is found within the first Brillouin zone, which is bounded by a few vectors in reciprocal space located at $\pm K/2$. In the free-electron model, the wavefunctions $\psi_{+K/2}$ and $\psi_{-K/2}$ are degenerate; in the nearly free-electron model, they couple via the periodic potential according to Eq. (12.23). This coupling of degenerate wavefunctions opens an energy gap of $2|V_{\text{ps}}(K)|$ in the band structure at wavevector $K/2$. The resulting wavefunctions depend upon the sign of $V_{\text{ps}}(K)$: If $V_{\text{ps}}(K) < 0$, then the lower energy function builds up at the atomic cores; if $V_{\text{ps}}(K) > 0$, then the lower energy function builds up between atomic cores [44]. However, if $K \sim q_0$, then $V_{\text{ps}}(K) \sim 0$ and the energy bands closely resemble free-electron bands.

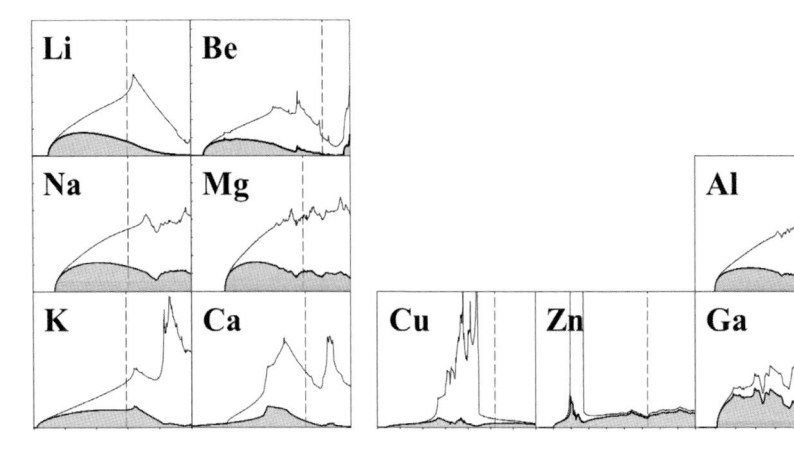

Figure 12.8 DOS curves for selected main group metals. The vertical axes are $n(E)$; the horizontal axes are energy E. The dashed lines indicate the corresponding Fermi levels. The lightly shaded regions identify contributions to the DOS from valence s AOs.

For 3d structures, the energy gaps at the various Brillouin zone boundaries do not occur at the same energies, so no gaps develop in the DOS curves, but there will be observable fluctuations relative to the free-electron DOS curve, as observed for various main group metals (see Figure 12.8). The elements best described by the nearly free-electron model include Li, Na, Mg, Al, and In. Most DOS curves follow the parabolic free-electron model near the lowest band energies, but significant deviations can occur close to the Fermi levels. Among the elements considered in Figure 12.8, Be, Ca, and Ga show the greatest discrepancies from the free-electron approximation. The deep minimum near E_F in the DOS for Be can be attributed to the absence of core p electrons, leading to negative $V_{ps}(K)$ values and contracted 2p AOs. Zn has a narrow 3d band superimposed over a broad valence 4s-4p band, which creates significant fluctuations near the Fermi level via 3d-4s-4p mixing. Ca has low-lying, virtual 3d orbitals that contribute to the features of its DOS curve just above the Fermi level. Ga, on the other hand, adopts a relatively complex structure, which creates significant covalent bonding effects that disrupt the nearly free-electron picture. The metals Na, Mg, and Al follow the free-electron model well because their significant $V_{ps}(K)$ values are rather close to 0.

In the *tight-binding* model of a crystalline solid [1,7,11,36,45], electrons move in a periodic array of atomic-like potentials, so the crystal orbitals are constructed via the LCAO ansatz. The overlap of these AOs slightly modifies the picture of isolated atoms, so that the Hamiltonian operator can be approximated as the sum of atomic Hamiltonians with a small correction to the potential energy arising from possible electron transfer and electron density accruing between atoms, that is,

$$\hat{H} = \left(\sum \hat{H}_A\right) + \hat{V}' = \sum \left(\hat{T}_A + \hat{V}_A\right) + \hat{V}'$$
$$= \left(\sum \hat{T}_A\right) + \left[\left(\sum \hat{V}_A\right) + \hat{V}'\right] = \hat{T} + \hat{V}. \qquad (12.24)$$

The basis functions are one-electron Bloch functions $\varphi_{\mu k}(r)$ delocalized throughout the entire structure and built from the AOs $\chi_\mu(r)$ in each unit cell:

$$\varphi_{\mu k}(r) = \frac{1}{\sqrt{N}} \sum_T e^{ik \cdot T} \chi_\mu(r - T). \tag{12.25}$$

By symmetry arguments, $\varphi_{\mu k}(r)$ and $\varphi_{\mu k'}(r)$ for two different wavevectors k and k' are orthogonal. The crystal orbitals for the solid are then constructed as linear combinations of these Bloch functions

$$\psi_{ik}(r) = \sum_\mu c_{i\mu}(k) \varphi_{\mu k}(r) = \frac{1}{\sqrt{N}} \sum_\mu \sum_T c_{i\mu}(k) e^{ik \cdot T} \chi_\mu(r - T). \tag{12.26}$$

Applying the variational theorem to the Schrödinger equation leads to solving a secular determinant for each selected wavevector k. The solutions are the effective one-electron energies $E_i(k)$ and coefficients $c_{i\mu}(k)$ [5,15,36,45]. The resulting Hamiltonian matrix elements $H_{ij}(k) = \langle \psi_{ik} | \hat{H} | \psi_{jk} \rangle$ contain two energetic contributions: (i) on-site integrals $\alpha_\mu = \langle \chi_\mu(r-T) | \hat{H} | \chi_\mu(r-T) \rangle$; and (ii) intersite integrals $\beta_{\mu\nu} = \langle \chi_\mu(r-T) | \hat{H} | \chi_\nu(r-T') \rangle$. Because of the interdependency among various energy terms and atomic charges, this problem should be solved self-consistently so that the output potential matches the input potential. Some simplified but useful approaches establish effective Hamiltonian matrix elements and achieve outcomes noniteratively. For example, in extended Hückel theory [6], the on-site integrals α_μ can be estimated by AO energies, and the intersite integrals $\beta_{\mu\nu}$ can be evaluated using the Wolfsberg–Helmholz relation.

There are no general analytical expressions for the electronic energy bands or DOS curves arising from tight-binding calculations, but some general outcomes include the following:

a) Energy bands $E_i(k)$ are characterized by their location and dispersion. Band centers are dictated largely by the respective energies of the AO basis functions and the strengths of intersite orbital overlaps. In some literature, these overlaps are called hybridizations, but they differ from on-site hybridizations that transform MOs into valence band orbitals, like sp^3 hybrids at tetrahedrally coordinated main group atoms. Bandwidths typically increase as the strengths of intersite orbital overlaps increase. Symmetry factors can disrupt band dispersion by introducing avoided crossings, which can lead to pseudogaps or actual band gaps in the DOS curves.

To demonstrate these concepts simply, consider a linear chain of main group atoms that have valence s and p AOs and are separated by a distance a. The first Brillouin zone contains the wavevectors $-\pi/a < k \leq \pi/a$, and the band structure is plotted from $k=0$ to π/a. Four energy bands, two σ-type and two degenerate π-type, occur in the band structure, shown in

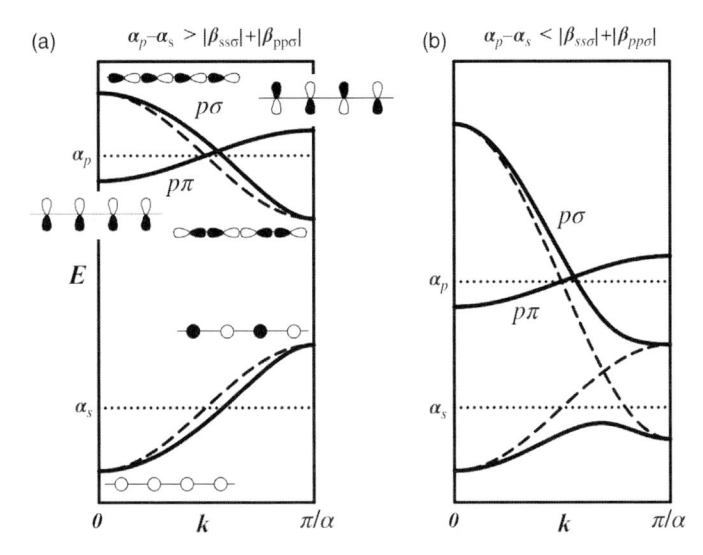

Figure 12.9 Tight-binding energy band diagrams for a linear chain of main group atoms. (a) $\alpha_p-\alpha_s$ is larger than the sum of intersite integrals $\beta_{ss\sigma}$ and $\beta_{pp\sigma}$ so that very little s–p mixing occurs. (b) $\alpha_p-\alpha_s$ is smaller than the sum of intersite integrals $\beta_{ss\sigma}$ and $\beta_{pp\sigma}$ so that an avoided crossing occurs along k. The dashed lines indicate the band dispersions without s–p mixing. The crystal orbitals for $k=0$ and π/a are shown; they are appropriate for both diagrams.

Figure 12.9, arising from the four valence s and p AOs. The key energy parameters are the on-site AO energies α_s and α_p, as well as the intersite integrals $\beta_{ss\sigma}$, $\beta_{sp\sigma}$, $\beta_{pp\sigma}$, and $\beta_{pp\pi}$. The dispersion of each band depends on the type of AO assigned to it: $dE/dk > 0$ for s(σ)-overlap and p (π)-overlap; $dE/dk < 0$ for p (σ)-overlap. Also, the bandwidths generally increase from π- to σ-overlap, and the band center of the s-band will be lower than the energy of the p-bands. Considering symmetry characteristics of the bands, inversion is a symmetry operation *only* at $k=0$ and π/a. Therefore, at these two wavevectors, the energy bands are either distinctly s (even) or p (odd) bands. For any intermediate k-point, s and p_z have identical symmetry characteristics because they are involved in σ-overlap between atoms and will mix, which means they may not cross along the k-axis. The p_π bands, however, have different rotational symmetry with respect to the chain than the s and p_σ bands, so p_π bands may cross p_σ bands. If $\alpha_p - \alpha_s$ is large with respect to $\beta_{ss\sigma}$ and $\beta_{pp\sigma}$, then the s- and p-bands will not intersect. However, if $\alpha_p - \alpha_s$ is small with respect to $\beta_{ss\sigma}$ and $\beta_{pp\sigma}$, then these bands will apparently cross – an avoided crossing occurs.

b) The DOS curve $n(E)$ is inversely proportional to $|\nabla_k E(k)|$ [37]. Therefore, when an energy band has high dispersion (large bandwidth), its contribution to $n(E)$ is low; when an energy band has low dispersion (narrow bandwidth), its contribution to $n(E)$ is large, creating a peak in the DOS curve.

c) The total number of valence electrons N can be partitioned by using the density matrix elements with the Bloch basis functions, $\rho_{\mu\nu} = \int \sum_i [N_i c_{i\mu}(\mathbf{k}) c_{i\nu}(\mathbf{k})] d\mathbf{k} = \int \rho_{\mu\nu}(\mathbf{k}) d\mathbf{k}$ (N_i = occupation numbers for crystal orbital $\psi_{i\mathbf{k}}$) [15]. Since each atom involves a subset of AOs for the entire problem, these density matrix elements can be divided among the atoms and atom pairs according to

$$N = \sum_A \sum_{\mu \in A} \rho_{\mu\mu} + \sum_A \sum_{B \neq A} \sum_{\mu \in A} \sum_{\nu \in B} \operatorname{Re}[\rho_{\mu\nu} S_{\mu\nu}]$$
$$= \sum_A \left[\sum_{\mu \in A} \left(\rho_{\mu\mu} + \sum_{B \neq A} \sum_{\nu \in B} \operatorname{Re}[\rho_{\mu\nu} S_{\mu\nu}] \right) \right], \quad (12.27)$$

in which "Re" identifies the real part of the possibly complex off-diagonal terms of the density matrix. The second term in the first part of Eq. (12.27) arises from interatomic orbital overlap and is called the overlap population [5,15], which is the average number of valence electrons shared between all pairs of atoms. The term in brackets in the second part of Eq. (12.27) represents the gross atomic population for each atom in the structure. This partitioning follows the Mulliken population analysis [5,13,15], which divides the overlap population between atom pairs strictly in half to obtain the gross atomic populations. Alternative population analyses exist providing different numerical results [15]. Likewise, the total DOS curve $n(E)$ can be separated into individual atom- or orbital-projected DOS curves,

$$n(E) = \sum_A \sum_{\mu \in A} n_{A\mu}(E). \quad (12.28)$$

d) The partitioning of valence electrons can be transformed to the energy scale because the density matrix elements $\rho_{\mu\nu}(\mathbf{k})$ can be transformed into energy-dependent values $\rho_{\mu\nu}(E)$. This leads to "crystal orbital overlap population" (COOP) curves [46], which are DOS curves weighted by the energy-dependent overlap population for an atom pair "A—B":

$$\operatorname{COOP}_{A-B}(E) = \sum_{\mu \in A} \sum_{\nu \in B} \operatorname{Re}[\rho_{\mu\nu}(E) S_{\mu\nu}]. \quad (12.29)$$

If $\operatorname{COOP}_{A-B}(E) > 0, = 0,$ or < 0, then the A—B interaction at energy E in the DOS is, respectively, bonding, nonbonding, or antibonding. Integrating the $\operatorname{COOP}_{A-B}(E)$ curve over all occupied states provides a measure of the A—B bond order. To bypass issues arising from basis set dependences on the overlap matrix, a partitioning based on the energies gives "crystal orbital Hamilton population" (COHP) curves [15,47]:

$$\operatorname{COHP}_{A-B}(E) = \sum_{\mu \in A} \sum_{\nu \in B} \rho_{\mu\nu}(E) H_{\mu\nu}(E), \quad (12.30)$$

which have become a valuable bond analysis procedure for DFT-type calculations. COHP curves resemble COOP curves, except that bonding interactions have negative COHP values and antibonding interactions have positive COHP values. The $COHP_{A-B}(E)$ curve integrated up to E_F yields an estimate of the A–B interaction strength, called ICOHP for the A–B contact.

Although COOP and COHP curves originated from tight-binding approaches to the electronic structure of solids because the crystal orbitals are constructed from local basis sets, it has become possible recently to extract similar chemical bonding information from computations based on various plane waves and pseudopotentials [48]. This energy-resolved bonding analysis, called "projected COHP" (pCOHP), involves evaluating a "transfer matrix" $T(\mathbf{k})$, which has matrix elements between the plane-wave-related functions $\psi_{i\mathbf{k}}$ and the local orbitals φ_μ, that is, $T_{j\mu}(\mathbf{k}) = \langle \psi_{i\mathbf{k}} | \varphi_\mu \rangle$. These transfer matrix elements are used to *transfer* the density and Hamiltonian matrices from the plane wave-type basis to a local orbital picture, and thus obtain interatomic COHP values [48].

12.3.2
Examples of Electronic Structures

Graphene (Graphite) versus Diamond

The results of tight-binding calculations for these two allotropes of carbon are depicted in Figure 12.10, which includes energy band diagrams plotted along high-symmetry directions of each Brillouin zone, DOS curves, and nearest-neighbor C—C COHP curves [1]. The DOS of graphene shows no bandgap at the Fermi level, although the curve goes to zero, so graphene is called a "zero-gap semiconductor." In the band structure, this feature of the DOS at E_F arises from the π-bands crossing exactly at the wavevector K of the Brillouin zone [49]. In addition, there are three σ-bonding bands overlapping the π-bands. According to the C—C COHP curve, the C—C interactions are bonding below E_F and antibonding above E_F. For diamond, four energy bands occur below an energy gap, which marks the position of the Fermi level. These four bands are C—C bonding, and above the gap, the bands are C—C antibonding. The valence bandwidth of diamond corresponds to the difference in energies of the two lowest valence states at the Γ point. A simplified tight-binding analysis provides some additional insights [50], as described below.

The graphene and diamond allotropes contain, respectively, trigonal planar and tetrahedral carbon atoms and primitive unit cells that contain two C atoms: C_A and C_B. The corresponding nearest-neighbor C—C distances are 1.418 [51] and 1.5446 [52] Å. If just 2s, $2p_x$, $2p_y$, and $2p_z$ valence AOs are assigned to each C atom and only nearest-neighbor interactions are considered, then the tight-binding equation involves an 8 × 8 Hamiltonian matrix $H(\mathbf{k})$, which consists of

Graphene

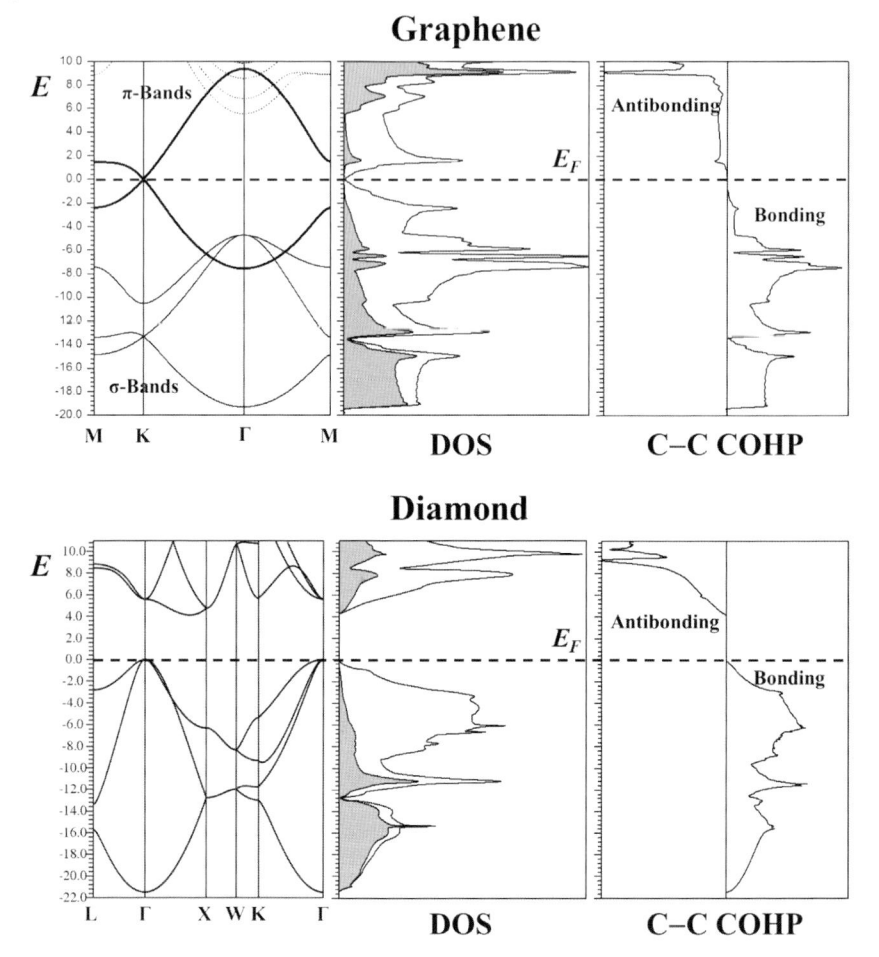

Diamond

Figure 12.10 Calculated electronic structures for graphene and diamond. Valence energy band diagrams plotted along high-symmetry directions in the corresponding Brillouin zones, electronic DOS curves with contributions from the 2s AO projected in gray, and nearest-neighbor C–C COHP curve are shown. The dashed line indicates the Fermi level for each system. Energies are in electrovolts.

two 4×4 intra-atomic blocks along the diagonal and two 4×4 interatomic off-diagonal blocks [2,50]:

$$H(k) = \begin{pmatrix} E_{As} & 0 & 0 & 0 & H_{As-Bs} & H_{As-Bx} & H_{As-By} & H_{As-Bz} \\ 0 & E_{Ap} & 0 & 0 & H_{Ax-Bs} & H_{Ax-Bx} & H_{Ax-By} & H_{Ax-Bz} \\ 0 & 0 & E_{Ap} & 0 & H_{Ay-Bs} & H_{Ay-Bx} & H_{Ay-By} & H_{Ay-Bz} \\ 0 & 0 & 0 & E_{Ap} & H_{Az-Bs} & H_{Az-Bx} & H_{Az-By} & H_{Az-Bz} \\ H^*_{As-Bs} & H^*_{Ax-Bs} & H^*_{Ay-Bs} & H^*_{Az-Bs} & E_{Bs} & 0 & 0 & 0 \\ H^*_{As-Bx} & H^*_{Ax-Bx} & H^*_{Ay-Bx} & H^*_{Az-Bx} & 0 & E_{Bp} & 0 & 0 \\ H^*_{As-By} & H^*_{Ax-By} & H^*_{Ay-By} & H^*_{Az-By} & 0 & 0 & E_{Bp} & 0 \\ H^*_{As-Bz} & H^*_{Ax-Bz} & H^*_{Ay-Bz} & H^*_{Az-Bz} & 0 & 0 & 0 & E_{Bp} \end{pmatrix}$$

(12.31)

Second nearest-neighbor interactions will add terms to the intra-atomic blocks, and provide improved agreement with the results of computations illustrated in Figure 12.10 [50]. The interatomic matrix elements H_{A-B} are $\mathbf{k}$-dependent and include β_{ss}, β_{sp}, $\beta_{pp\sigma}$, and $\beta_{pp\pi}$ intersite integrals that depend upon the local geometry at each C atom and the C—C distance. For graphene, since the node of the 2p$_z$ AOs is the plane of the structure, the only nonzero off-diagonal term involving 2p$_z$ AOs is H_{Az-Bz}. As a result, the 8×8 $H(\mathbf{k})$ matrix for graphene becomes block-diagonalized into submatrices that are 6×6 (σ-bands) and 2×2 (π-bands). A Hückel-type solution for the 2×2 submatrix leads to the eigenvalues ($\beta_{pp\pi} \equiv \beta_\pi$) [1]

$$E(k) = E(k_1, k_2) = E_p \pm [3\beta_\pi + 2\beta_\pi \cos 2\pi k_1 + 2\beta_\pi \cos 2\pi k_2 + 2\beta_\pi \cos 2\pi(k_1 + k_2)]^{1/2}, \quad (12.32)$$

which gives rise to π-bands that are symmetrical about E_p because the overlap repulsion term is explicitly ignored in the Hückel approximation. In the actual electronic structure, the overlap repulsion affects the positions of the antibonding levels significantly.

According to the results in Figure 12.10, the Fermi level of graphene falls amid the π-bands at wavevector K. From Eq. (12.32), in the absence of any second nearest-neighbor interactions or explicit two-electron interactions, the band energy $E(K) = E(1/3, 1/3) = E_p$, the energy of the 2p AOs. Furthermore, the slope of the π-bands are nonzero at K, so that the DOS is zero at $E(K)$. The bottom of the π-band occurs at Γ, and falls below the top of the σ-bonding bands, a result which is realized from the simple tight-binding model: $E_{min}(\pi\text{-bonding}) = E_p - 3|\beta_{pp\pi}|$ and $E_{max}(\sigma\text{-bonding}) = E_p - 3|1/2(\beta_{pp\pi} - \beta_{pp\sigma})|$; since $|\beta_{pp\pi}| < |\beta_{pp\sigma}|$, then $E_{min}(\pi\text{-bonding}) < E_{max}(\sigma\text{-bonding})$. Furthermore, the three σ-bonding bands and the corresponding three σ-antibonding bands are separated by an energy gap that depends on the magnitude of β_{sp}. For graphite, in which the honeycomb sheets of carbon atoms stack along the c-axis, weak $2p_z \cdots 2p_z$ interactions between sheets disrupt the band structure at the K-point, resulting in bands that overlap in energy by ~40 meV and lead to semimetallic behavior [53].

For diamond, the energy gap also arises from the inclusion of the interatomic β_{sp} term in the tight-binding Hamiltonian matrix. Through a unitary transformation of the 2s and 2p AOs into sp^3 hybrids, the bandgap can result from hybridization [2,11]. Without the β_{sp} term, the valence s- and p-bands would overlap in energy and diamond would be metallic [1,50]. According to the band structure, the bandgap is indirect because the highest occupied and lowest unoccupied crystal orbitals occur at different wavevectors $\mathbf{k}$, and the valence bandwidth is the energy difference between the two lowest energy bands at Γ, which are, respectively, 2s and 2p orbitals shifted by corresponding intersite integrals $W \sim (E_p - E_s) + 4(1/3|\beta_{pp\sigma} - 2\beta_{pp\pi}| - \beta_{ss})$ [51]. The three distinct peaked features in the DOS arise from states near the L and X points on the boundary of the Brillouin zone.

Binary derivatives of graphite and diamond include, respectively, hexagonal boron nitride BN and tetrahedral gallium arsenide GaAs or zinc blende ZnS. In these cases, the structures are alternant decorations of either the honeycomb (BN) or diamond (GaAs and ZnS) nets. For a 2d model of BN, the degeneracy of the π-bands at the K-point, as seen for graphene, is broken and a direct energy gap arises that can be approximated by $E_{2p}(B) - E_{2p}(N)$. The valence band orbitals achieve greater contributions from the N AOs, whereas the conduction band orbitals arise mostly from B AOs. For GaAs and ZnS, the energy gaps become enhanced from the electronegativity differences between the two components, and also become direct as ionicity increases. Likewise, the valence bands show greater AO contributions from the more electronegative elements, that is, As and S [50].

Rhenium Trioxide

ReO_3 is a network of vertex-sharing $ReO_{6/2}$ octahedra related to perovskite. The unit cell is a cube of side $a = 3.75$ Å with Re atoms at the corners and O atoms at the centers of every edge [54]. The valence AOs include 6s, 6p, and 5d orbitals at Re and 2s and 2p orbitals at each O, which follow the energetic sequence O 2s < O 2p < Re 5d < Re 6s < Re 6p. The tight-binding Hamiltonian matrix using this basis set is a 21×21 matrix with significant Re—O intersite integrals including $\beta_{dp\sigma}$ and $\beta_{dp\pi}$, and the resulting band structure, DOS and Re—O COHP curves are shown in Figure 12.11 with the Fermi level selected as the reference energy (0 eV) [2,55]. The energy range of these diagrams includes just the nine O 2p and some of the five Re 5d orbitals. The DOS curve and band structure diagram reveal that the O 2p and Re 5d bands fall in distinct energy regions arising from the electronegativity difference between Re and O as well as the Re—O

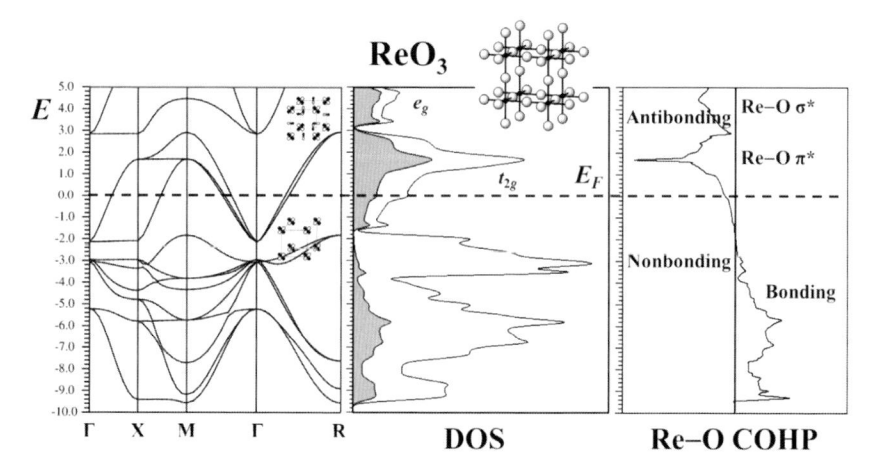

Figure 12.11 Calculated electronic structures for ReO_3. Valence energy band diagrams plotted along high-symmetry directions in the Brillouin zone, electronic DOS curves with contributions from Re valence AOs projected in gray, and nearest-neighbor Re—O COHP curve are shown. The dashed line indicates the Fermi level. Energies are in electrovolts.

interactions, with the Re 5d bands splitting into t_{2g} and e_g bands from the octahedral coordination by O atoms. The dispersion of the Re t_{2g} bands extend about 5 eV from the minimum at the Γ wavevector, $\boldsymbol{k} = (0, 0, 0)$, which is completely Re—O π-nonbonding, to the maximum at the R point, $\boldsymbol{k} = (1/2, 1/2, 1/2)$, which is maximally Re—O π-antibonding. In fact, the dispersion of the three Re—O π^* t_{2g} bands can be derived from the tight-binding matrix to be

$$E_{xy}(\boldsymbol{k}) = E_{Re5d} + 2\beta_{dp\pi}(\sin 2\pi k_x + \sin 2\pi k_y), \quad (12.33a)$$

$$E_{xz}(\boldsymbol{k}) = E_{Re5d} + 2\beta_{dp\pi}(\sin 2\pi k_x + \sin 2\pi k_z), \quad (12.33b)$$

$$E_{yz}(\boldsymbol{k}) = E_{Re5d} + 2\beta_{dp\pi}(\sin 2\pi k_y + \sin 2\pi k_z), \quad (12.33c)$$

for $\boldsymbol{k} = (k_x, k_y, k_z)$. Plots of these equations along the wavevector sequence Γ–X–M–Γ–R nicely reproduces essential dispersions of these bands. Weaker O$\cdots$O interactions break the accidental degeneracies derived from Eq. (12.33), which only includes nearest-neighbor Re—O interactions. From the DOS and Re—O COHP curves, the O 2p band shows some contributions of Re AOs that are clearly Re—O bonding below about -3.8 eV. The peaks between -3.8 and -1.6 eV are essentially Re—O nonbonding because there is little contribution from Re orbitals and the Re—O COHP is nearly zero in this region. The Fermi level falls within the Re—O π-antibonding t_{2g} band because, of the 25 valence electrons for ReO_3, 24 electrons fill the 12 O 2s and 2p bands, leaving one valence electron remaining to occupy the Re t_{2g} band. Therefore, ReO_3 is a metallic oxide, whereas nearly isostructural WO_3 is semiconducting but Na_xWO_3 and H_xWO_3 are metallic [55].

BCC Metals: Elements and Intermetallics

As a final example for this section, consider bcc Cr in its paramagnetic state and TiFe [56], which is similar but with Ti at the corners and Fe at the center of cubic unit cells. Their corresponding DOS and nearest-neighbor COHP curves are shown in Figure 12.12. Both DOS curves include a broad, nearly featureless valence 4s-4p band that extends around a narrower 3d band, which develops features arising from the cubic environment at each metal atom: the orbitals are split into e_g below t_{2g} bands. As a result of the intersite integrals, both electronic structures at the Fermi levels show pseudogaps, which occur because states below E_F are metal–metal bonding and states above E_F are metal–metal antibonding as seen in the Cr—Cr and Ti—Fe COHP curves. In TiFe, states below the pseudogap have greater contributions from Fe 3d AOs, and those above the pseudogap have greater contributions from Ti 3d AOs, a result in accord with the relative electronegativities of Ti and Fe. Body-centered cubic transition metals are preferred when the valence d-band is nearly half-filled, cases that occur for V, Nb, Ta, Cr, Mo, W, and these elements exhibit the largest cohesive energies among elements other than C [11]. Furthermore, binary intermetallic compounds formed by one early and one later transition metal, like TiFe, have among the largest standard enthalpies of formation of intermetallic

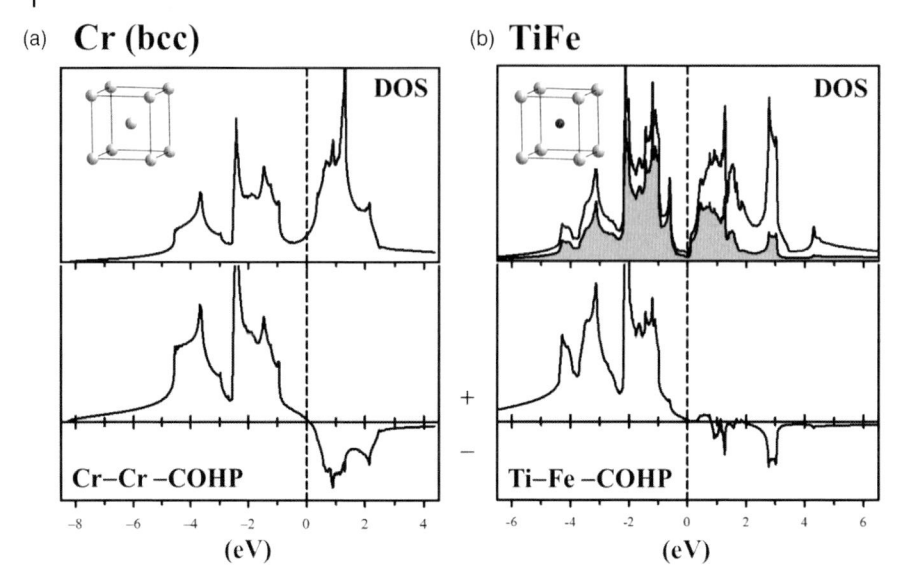

Figure 12.12 Electronic DOS and nearest-neighbor metal–metal COHP curves for (a) bcc Cr and (b) cubic TiFe. The Fermi levels are marked by dashed lines. In the DOS curve for TiFe, the gray shaded portion is the projection of Fe valence orbitals to the total DOS. COHP curves are plotted as "–COHP," so that + values are metal–metal bonding and – values are metal–metal antibonding.

compounds [34,35]. Both results are evident in the tight-binding analysis of the electronic structure for these two bcc-type examples.

The electronic structures of solids as described by the free-electron or tight-binding methods yield crystal orbitals that are delocalized throughout the entire solid. Although this spatial interpretation of bonding is reasonable for metals, an alternative picture for insulators and semiconductors considers electron pairs localized into bond and lone pair orbitals. This revision of the calculated electronic states can be accomplished by a unitary transformation of the delocalized orbitals, but these new functions are no longer solutions of the given form of the Schrödinger equation, that is, they are no longer Hamiltonian eigenstates. So, such functions are inappropriate to consider electronic transitions or ionization, but they are valid descriptors for collective properties such as the electron density and total energy, because they make use of all electrons in the ground state, whose total wavefunction has not physically changed by the transformation. Wannier functions are one type of "localized" functions for crystalline solids that are generated from the delocalized Bloch-type orbitals according to

$$W_\mu(r - T) = \frac{1}{\sqrt{N}} \sum_k e^{-ik \cdot T} \varphi_{\mu k}(r). \tag{12.34}$$

Localization of these functions is implied because $W_\mu(r - T)$ and $W_\mu(r - T')$ are orthogonal about the different lattice positions T and T'. However, their tails typically extend over a few unit cells. Nevertheless, the transformation between delocalized Bloch-type functions and localized Wannier functions is only strictly valid for filled or half-filled energy bands, such as occur for ferromagnetic insulators. Localization methods include Edmiston–Ruedenberg [57] and Foster–Boys [58] schemes, the latter being actually used to construct maximally localized Wannier functions in solids [59]. Unfortunately, different localization methods can yield different results, which can restrict their usefulness as a tool for analysis of chemical bonding.

A recent development to extract useful chemical bonding information from the calculated electronic structures of intermetallic solids is the reversed approximation molecular orbital (raMO) analysis [60], which creates localized MOs or Wannier functions with the goal of designing isolobal analogies [61] that have been especially useful for bonding insights in organometallic complexes and cluster compounds [1,5]. In the raMO procedure, the fully occupied wavefunctions of the system are used as an approximate basis set for the solution a simple chemical bonding question, for example, the interactions between a transition metal atom and its ligand σ orbitals. Underlying the raMO approach is the fact that the fully occupied wavefunctions will consist of linear combinations of any localized bonding features that may be present in the structure. For example, the raMO method could trace pseudogaps in the electronic DOS curves of $CrGa_4$, Os_3Sn_7 and Ir_3Sn_7 to 18-electron configurations at the transition metals [60].

12.3.3
Bonding Analyses in Position Space: QTAIM and ELF/ELI-D

Although solution of the electronic Schrödinger equation for molecules and solids yields wavefunctions and their energies, the key quantity of any chemical system is the total electron density distribution $\rho(r)$, which can be reconstructed to a certain degree from diffraction experiments. The atomic positions in a chemical structure are located at local maxima $\rho(r_A)$, where each atom type is identified from the electron density gradient $-\nabla\rho(r)$ close to the nucleus as an electron–nuclear cusp that is proportional to the core charge Z, and the total number of electrons is $\int\rho(r)dr$. Moreover, according to theorems of Hohenberg and Kohn [62], the ground-state wavefunction $\Psi_0(r_1, \ldots, r_N)$ is a unique functional of the electron density $\Psi_0\{\rho(r)\}$ and, likewise, the ground-state energy is the minimum of the functional $E\{\rho(r)\}$. This outcome implies that all bonding effects, in principle, are contained in $\rho(r)$. In fact, the electron density is one part of a hierarchy of density functions [14] $\rho_m(r_1, \ldots, r_N)$ for $1 \leq m < N$

$$\rho_m(r_1,\ldots,r_N) = \binom{N}{m} \int dr_{m+1} \cdots \int dr_N \Psi_0^*(r_1,\ldots,r_N)\Psi_0(r_1,\ldots,r_N),$$

(12.35)

which have probabilistic meanings and are useful for bonding analyses. Among these functions, the single-particle (electron) density $\rho(r) \equiv \rho_1(r_1)$, which represents N times the probability density of finding an electron at r_1, and the pair density $\rho_2(r_1, r_2)$, which is $N(N-1)/2$ times the probability density to find an electron at r_1 and a second electron at r_2, are widely used for chemical bonding studies. These densities can be decomposed into spin-dependent contributions (α = spin up; β = spin down), in particular,

$$\rho(r) = \rho^\alpha(r) + \rho^\beta(r), \tag{12.36a}$$

and

$$\rho_2(r_1, r_2) = \rho_2^{\alpha\alpha}(r_1, r_2) + \rho_2^{\beta\beta}(r_1, r_2) + \rho_2^{\alpha\beta}(r_1, r_2)\rho_2^{\beta\alpha} + \rho_2(r_1, r_2). \tag{12.36b}$$

Thus, once these density functions are obtained, chemical bonding can be comprehensively analyzed in position space. Furthermore, all kinds of quantum chemical codes delivering reasonable approximations to the wavefunctions can be utilized. Consequently, the different views of molecular orbital and valence bond theories become united in position space by extraction and analysis of $\rho_m(r_1, \ldots, r_N)$, which are the common physical kernels for any chemical system [63].

The notion of identifiable atoms within a chemical structure represents a central hypothesis in chemistry [64], and its precise definition has been achieved within the *Quantum Theory of Atoms in Molecules* (QTAIM) approach pioneered by Bader [65]. In this framework, atoms are defined as electron density basins associated with the local maxima at the nuclear positions $\rho(r_A)$. An electron density basin is the collection of all points in space whose density gradient paths $\nabla\rho(r)$ end at the same attractor r_A. Since, by construction, no gradient path crosses a basin boundary, these boundaries are zero-flux surfaces and space is exhaustively partitioned into non-overlapping electron density basins. Fulfillment of the zero-flux condition at the basin surfaces qualifies this scheme as a partitioning of position space into quantum mechanical subsystems because (a) there is a well-defined electronic kinetic energy for each QTAIM basin; (b) the total atomic energies of the QTAIM basins sum to the total electronic energy; and (c) for a system in a stationary state with vanishing net Hellmann–Feynman forces on the nuclei, the Virial Theorem $\langle V \rangle = -2\langle T \rangle$ is valid for the entire system as well as for each QTAIM basin. Although the QTAIM basins possess sharp boundaries, they behave like proper open quantum systems by freely exchanging electrons with their neighbors. Thus, QTAIM and qualitative LCAO–MO theory give consistent pictures of chemical bonding. In cases when orbitals are expanded using plane waves, an AO description can be regained by fitting an AO basis set to the converged molecular or crystal orbitals. In contrast, position–space bonding analysis is independent of the type of representation of the wavefunction.

Characterization of $\rho(r)$ for chemical bonding features is achieved by topological analysis, which provides the QTAIM basins as well as identifies critical points r_c of $\rho(r)$, that is, sites where $\nabla\rho(r_c) = 0$. Critical points are distinguished by the signs of their curvatures, which are the three eigenvalues of the diagonalized hessian matrix of $\rho(r_c)$, and they are referred to by "(*rank, signature*)." The rank

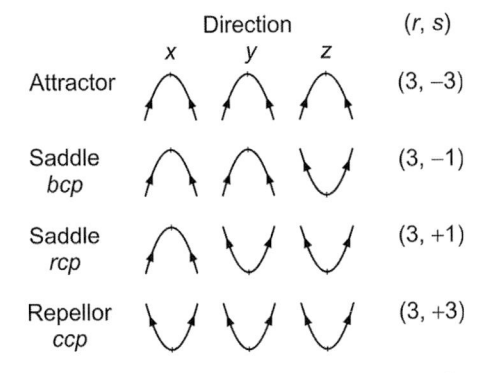

12.3 Quantum Mechanical Calculations and Analyses

Figure 12.13 Representation of the four types of structurally stable critical points occurring in the electron density distribution $\rho(r)$. The (r, s) designation refers to (rank, signature) as explained in the text.

is the number of nonzero curvatures and the signature sums the number of curvatures of each sign at the critical point r_c. Accordingly, there are four types of structurally stable critical points in $\rho(r)$ (see Figure 12.13):

- $(3, -3)$ is a local maximum (attractor), typically identified as an atomic critical point because maxima in $\rho(r)$ generally occur at nuclear positions. The nuclear maximum is strictly not a critical point because of the electron–nuclear cusp, but it may be considered as such in the topological analysis.
- $(3, -1)$ is a saddle point with one positive and two negative curvatures, and within QTAIM identifies a bond critical point (bcp), which occurs between two atomic critical points.
- $(3, +1)$, a ring critical point (rcp), is a saddle point with two positive and one negative curvatures, and is found at the middle of several bonds forming a loop.
- $(3, +3)$ is a local minimum (repellor), or a cage critical point (ccp), because it is located at the center of a region surrounded by several rings.

Thus, $\rho(r)$ typically displays a qualitatively simple topology with local maxima (attractors) at the nuclear positions, saddle points between close attractors, and local minima in cage-like environments.

In the framework of QTAIM, it is important to realize that the *total* electron density of the system is employed. Any approximations using just parts of $\rho(r)$, such as the valence electron density, can only be justified if they yield the same results as the total density. For example, the local valence charge density maxima found between nearest-neighbor Si atoms in diamond-type α-Si are insignificant within the strict QTAIM framework, because the total density only displays a $(3, -1)$ saddle point at the bond midpoint, a result which indicates some unspecified interatomic interaction. Using just $\rho(r)$, characterization of the interaction at each bcp can be accomplished from the Laplacian $\nabla^2\rho(\text{bcp})$, which is a measure of the average curvature of $\rho(\text{bcp})$. For shared interactions like covalent bonds, $\nabla^2\rho(\text{bcp}) < 0$, indicating local charge concentration, whereas for unshared interactions like ionic bonds or van der Waals interactions, $\nabla^2\rho(\text{bcp}) > 0$, indicating charge depletion. For the traditional covalent bond expected in α-Si, $\nabla^2\rho(\text{bcp}) < 0$ where the positive curvature in $\rho(\text{bcp})$ along the internuclear Si—Si contact is overcompensated by two negative curvatures orthogonal to the bond. In the valence electron density of α-Si, negative internuclear curvatures also occur at the local maxima along the Si—Si bonds caused by omitting the core contributions to the electron density. Thus, in this case accumulation of valence charge density along the Si—Si bond translates into charge concentration at the bcp within the QTAIM framework. Unfortunately, further examinations indicate that $\nabla^2\rho(\text{bcp})$ does not generally give a reliable classification of chemical bonding situations in terms of shared versus unshared interactions. Especially for bonds between transition metal atoms, no charge concentration is found, which is caused by the dominating positive contributions of the d AOs to $\nabla^2\rho(\text{bcp})$. To enhance the overall analysis, inclusion of additional quantities evaluated at the bcp like the energy density or the virial ratio $T(\text{bcp})/V(\text{bcp})$ are useful, but these values are not directly available from experimental data as is $\rho(r)$. Another recent effort on the basis of electron density alone employs the so-called source function to characterize covalent bonding between QTAIM atoms [66]. As discussed below, $\nabla^2\rho(r)$ must be supplemented by a two-particle quantity describing electron pairing, the pair-volume function, to display electron localizability information and serve as a more faithful indicator of covalent bonding.

Nevertheless, a valuable outcome of the QTAIM approach is that every property density defined in position space can be evaluated for each QTAIM atom by integration over its associated basin. Two important characteristics are the atomic basin volume V_A and its electron population N_A. For example, within certain groups of homodesmic compounds, that is, compounds exhibiting a single bond type, the QTAIM volumes exhibit clear correlations [67] with atomic volume increments developed by Biltz, who derived these increments from crystal chemical considerations [68]. In another case, QTAIM atomic volumes evaluated for seven different $Al_{1-x}Pt_x$ phases revealed a systematic variation of atomic volumes $V_{Al}(x)$ and $V_{Pt}(x)$ with changing composition and allowed an extended Vegard's law to be formulated, that is, $\overline{V}(x) = xV_{Pt}(x) + (1-x)V_{Al}(x)$, which accounts for the experimentally observed trend of the average atomic volumes and the cell volumes for all compounds in the binary phase diagram [67]. In yet another case, decomposition of the bulk modulus into contributions from QTAIM atoms clearly revealed transferability within compound classes such as oxygen atom compressibilities in different spinels [69].

The QTAIM basin electron population N_A is obtained by integrating $\rho(r)$ within the basin assigned to atom A and defines an effective atomic charge $Q_A^{\text{eff}} = Z_A - N_A$ for nuclear charge Z_A. Although these effective charges are useful for an ionic description of chemical bonding, it is important to recognize that N_A values are averaged populations because $\rho(r)$ is evaluated from the occupied orbitals, which are delocalized throughout the entire structure. Therefore, each N_A value also contains a dynamical part responsible for electron sharing and covalent bonding. The uncertainty of a basin's population, which can be expressed as the variance σ_A^2, is a measure of the population fluctuations in the wavefunction. For example, the ground-state MO wavefunction of H$_2$ contains the four terms $\varphi_A(1)\varphi_A(2)$, $\varphi_B(1)\varphi_B(2)$, $\varphi_A(1)\varphi_B(2)$, and $\varphi_B(1)\varphi_A(2)$, each with weight $w_i = 0.25$. For each atom, the average electron population is $N_A = N_B = 1.00$, and the variance is $\sigma_A^2 = \sigma_B^2 = \Sigma_i w_i (N_{iA} - N_A)^2 = 0.50$. The variance can be also calculated according to $\sigma_A^2 = \langle N_A^2 \rangle - (N_A)^2$, which requires knowledge of the electron pair density function $\rho_2(r_1, r_2)$ for the first term and the electron density $\rho(r)$ for the second term. $\rho_2(r_1, r_2)$ can be decomposed according to

$$\rho_2(r_1, r_2) = 1/2\rho(r_1)\rho(r_2) + 1/2\rho_{2,\text{XC}}(r_1, r_2), \tag{12.37a}$$

with

$$\rho_2^{\alpha\alpha}(r_1, r_2) = 1/2\rho^\alpha(r_1)\rho^\alpha(r_2) + 1/2\rho_{2,\text{XC}}^{\alpha\alpha}(r_1, r_2) \tag{12.37b}$$

and

$$\rho_2^{\alpha\beta}(r_1, r_2) = 1/2\rho^\alpha(r_1)\rho^\beta(r_2) + 1/2\rho_{2,\text{C}}^{\alpha\beta}(r_1, r_2). \tag{12.37c}$$

According to Eq. (12.37), the pair density function can be separated into a stochastically independent contribution, which is the product of two charge densities at r_1 and r_2, and an exchange–correlation hole contribution. The quantity $\rho_{2,\text{XC}}^{\alpha\alpha}(r_1, r_2)/\rho^\alpha(r_1)$ represents the exchange–correlation hole density of an electron at r_1, a value that measures the degree to which density is excluded at r_2 from the presence of an electron at r_1. For a single-determinant wavefunction, only Fermi correlation is included, which leads to the pure exchange part $\rho_{2,\text{X}}(r_1, r_2)$ of the pair density. Using $\rho_{2,\text{X}}(r_1, r_2)$ or $\rho_{2,\text{XC}}(r_1, r_2)$ to evaluate the electronic fluctuations in each basin allows derivation of an interatomic electron-sharing index called the "delocalization index" δ_{AB} between two basins A and B [70]:

$$\delta_{AB} = -2 \int_A dr_1 \int_B dr_2 \rho_{2,\text{XC}}(r_1, r_2). \tag{12.38}$$

The delocalization index δ_{AB} can be considered as the position–space analog of an effective covalent bond order, a concept that has a long history in quantum chemistry. Unlike the closely related Meyer–de Giambiagi–Wiberg bond index [71] defined for tight-binding basis functions, δ_{AB} can be directly obtained for any type of wavefunction construction and is not explicitly basis set dependent. In this framework, N_A consists of a sum of electrons localized inside the basin, λ_A, and the sum of all electrons delocalized (shared) between basins, that is,

$$N_A = \lambda_A + \sum_{B \neq A} \frac{1}{2}\delta_{AB} = \lambda_A + \sigma_A^2. \tag{12.39}$$

Therefore, the total fluctuation of N_A, given by the variance, is decomposed into a sum of two-center contributions δ_{AB}. Application of Eq. (12.38) to H$_2$ in the MO representation discussed above for which $\sigma_A^2 = 0.50$ yields $\delta_{H_A H_B} = 1.00$, that is, an effective H—H bond order of 1. Similarly, the expected C—C bond orders are fairly well reproduced for C$_2$H$_2$ ($\delta \sim 3$), C$_2$H$_4$ ($\delta \sim 2$), and C$_2$H$_6$ and crystalline diamond ($\delta \sim 1$). Highly polar bonds are characterized by strongly reduced effective covalent bond orders, for example, δ(Li,F) = 0.2 in molecular LiF [70].

This type of analysis has also been generalized to account for multicenter delocalization, because the two-center index δ_{AB} consists of a sum of three-center indices δ_{ABC} [72]. The ability to quantitatively assess multicenter bonding is very important for the study of numerous metal-rich and intermetallic phases as well as main group deltahedral cluster compounds, in which there are too few valence electrons to form two-center, two-electron bonds with respect to the number of nearest neighbors. For crystalline solids, research in this direction has recently begun [73].

Another important analysis of electronic distributions in position space involves the electron localization function (ELF) [74] and its subsequent developments, for example, the electron localizability indicator ELI-D. The ELF distribution $\eta(r)$ is a relative measure, with respect to the homogeneous electron gas, of the likelihood of finding an electron in the neighborhood of a reference electron with the same spin $\sigma = \alpha$ (up) or β (down) located at r, and, thus, has its origins in the study of the same-spin pair density functions of Eq. (12.37b). In the Taylor expansion of the σ-spin pair density spherically averaged around r of a Hartree–Fock type of wavefunction from spin orbitals $\varphi_i(r)$, Becke and Edgecombe rediscovered the function $D_\sigma(r)$ [75]:

$$D_\sigma(r) = \sum_{\text{occ. } i}^{\sigma} |\nabla \varphi_i(r)|^2 - \frac{1}{4} \frac{|\nabla \rho^\sigma(r)|^2}{\rho^\sigma(r)}, \tag{12.40}$$

which describes the probability of a σ-spin electron to approach another σ-spin electron located around the position r. Regions where this probability are small coincide with space where electrons are well localized and vice versa, which reveals $D_\sigma(r)$ to be an inverse measure of electron localizability. Also, $D_\sigma(r)$ is a monotonically decreasing function around each atom and does not display the interesting features of ELF. Therefore, the decisive step to create a proper electron localization function is to calibrate $D_\sigma(r)$ with the corresponding value of the homogeneous electron gas $D_\sigma^0(r) = 5.742 \rho(r)^{5/3}$ by forming the ratio $D_\sigma(r)/D_\sigma^0(r)$, which is called the kernel of ELF. To have a direct measure of electron localizability, the final function should contain the inverted kernel, and, for graphical convenience, its values are restricted between 0 and 1. To fulfill these demands, a Lorentzian scaling has been chosen out of several possibilities, which also importantly keeps the locations of the critical points of the scaled kernel:

$$\eta(r) = \left[1 + \left(\frac{D_\sigma(r)}{D_\sigma^0(r)}\right)^2\right]^{-1}. \tag{12.41}$$

According to this original definition of ELF for the closed-shell case, the values $\eta(r) = 1$ correspond to perfect localization and $\eta(r) = 0.5$ to "electron gas-like" delocalization. However, in the homogeneous electron gas, $\eta(r) = 0.5$ for all r

because $\nabla\eta(r)=0$. In real chemical systems, values of $\eta(r)\leq 0.5$ are typically connected with nonzero $\nabla\eta(r)$, which is a different scenario. Nevertheless, plots of $\eta(r)$ graphically describe regions of electron localization at each point in space, unlike Mulliken populations, which are centered at atoms.

Unfortunately, as a result of the physically unmotivated calibration step in Eq. (12.41), the physical meaning of ELF in this original form is obscured. Subsequently, several other interpretations of the ELF function have been proposed, most retaining the original name. As a result, there is no single interpretation of the physical meaning of ELF, but it should now be considered as a family of functions that essentially share a similar kernel $D_\sigma(r)/D_\sigma^0(r)$ for the special case of a single-determinant, closed-shell, time-independent wavefunction. However, the members of this family are all physically different, which becomes apparent when more general types of wavefunctions are employed, for example, multideterminant ones for correlated cases, time-dependent cases, or open-shell configurations [76]. One variant is the electron localizability indicator $\Upsilon_D^\sigma(r)$, called ELI-D, which is based on the same-spin pair density and can also be reduced to an ELF-like kernel for a single-determinant, closed-shell, time-independent wavefunction [76,77]. Among all these variants of ELF, the ELI-D approach has been developed most widely, covering consistent treatments of electron correlation, open-shell systems, momentum space, and relativistic wave functions. Moreover, ELI-D is a product of the electron density and the pair-volume function $\tilde{V}_D^\sigma(r)$, that is,

$$\Upsilon_D^\sigma(r) = \rho_\sigma(r)\cdot\tilde{V}_D^\sigma(r) \approx \rho_\sigma(r)\cdot\left(\frac{12}{\rho_\sigma(r)D_\sigma(r)}\right)^{3/8}, \qquad (12.42)$$

so that $\Upsilon_D^\sigma(r)$ can be regarded as a charge distribution. The second expression in Eq. (12.42) is valid only for a single-determinant wavefunction, and indicates the inverse relationship between $\Upsilon_D^\sigma(r)$ and the ELF kernel at this level of approximation. $\tilde{V}_D^\sigma(r)$ corresponds to a local volume required to encompass a pair of σ-spin electrons, so that the ELI-D function $\Upsilon_D^\sigma(r)$ is the position-resolved charge necessary to create a pair of σ-spin electrons. ELI-D analysis allows an exact orbital decomposition, which relates orbital-based and position–space approaches of chemical bonding analysis [73], and establishes a connection with the QTAIM approach as well [79]. The mechanism to create an ELI-D attractor results from a competitive interplay between $\rho(r)$ and $\tilde{V}_D^\sigma(r)$, which typically display inverse behavior. For example, to create an ELI-D attractor between two covalently bonded transition metal atoms as in $Re_2(CO)_{10}$, the positive value of $\nabla^2\rho(bcp)/\rho(bcp)$ is overcompensated by the negative value of $\nabla^2\tilde{V}_D^\sigma/\tilde{V}_D^\sigma$, indicating the decisive role of the pair-volume function to create electron localizability information.

Since the calculation of the kernel $D_\sigma(r)/D_\sigma^0(r)$ for any variant of ELF and ELI-D is very similar for a system whose wavefunction is given by a single closed-shell, time-independent Slater determinant, the locations of critical points, as well as the volumes and electron populations of corresponding basins are identical in $\eta(r)$ and $\Upsilon_D^\sigma(r)$. Therefore, the combined name ELI-D/ELF will be used for the remaining discussion.

ELI-D/ELF allows point-by-point analysis of electronic localizability in real space, which includes atomic shell structure, bonding electron pairs, and lone pairs leading to direct connections with the Lewis picture of chemical bonding and the VSEPR model [78]. Atomic shell structures are not faithfully displayed by the spherically averaged electron density $4\pi r^2 \rho(r)$ about any atom, although they are evident from the principal quantum numbers of the occupied AOs used in MO calculations. On the other hand, ELI-D/ELF displays the correct number of shells in position space and, upon integration of the electron density inside each shell, replicates the appropriate shell populations typically within about ±0.2e⁻ for the valence region [80]. ELI-D also reveals atomic shell structure even in momentum space [81]. The connection between ELI-D/ELF and localized orbital descriptions exists for the case of the C—C bonds in saturated hydrocarbons and diamond, in which a localized orbital for each bond can be found, and ELI-D/ELF displays an attractor in the valence region between carbon atoms. On the other hand, when π-bonding is involved, interpretations may differ. For example, two different localized orbital schemes for the C=C bond in ethylene yield either one σ- plus one π-bonding orbital (Edmiston–Ruedenberg) or two "banana bonds" (Foster–Boys), whereas ELI-D/ELF displays two attractors at the "banana bond" locations in support of the second localization scheme. On the other hand, in benzene, Foster–Boys localization retrieves one of the two degenerate Kekulé structures, but ELI-D/ELF yields a single attractor at the C—C internuclear line with a rather small curvature in the out-of-plane direction resulting from the π MOs. Since many interesting bonding situations like these will be challenging to assess using localized orbitals, physically motivated bonding indicators are preferred strategies. ELI-D/ELF functions follow the symmetry of the ground-state wavefunction and display the whole electronic localization scenario with one scalar distribution. Therefore, certain special bonding situations for which chemists traditionally depict mesomeric formulas based on the idea of resonating valence bond structures are shown in an averaged way by ELI-D/ELF, consistent with the total wavefunction.

The gradients of the ELI-D/ELF function can be used to construct attractor basins that exhaustively partition real space in a manner complementary to the QTAIM strategy. These regions correspond to inner atomic shells or, for the valence region, bonds and lone pairs. Since each partition scheme covers a different aspect of the chemical structure, a finer partitioning of position space can be achieved by employing both partitions simultaneously, which is called the ELI-D/QTAIM or ELF/QTAIM intersection technique [82]. For a typical two-center bond, an ELI-D/ELF attractor is located close to the internuclear line of the two atoms; it will be at the midpoint between two atoms for the nonpolar case. The corresponding ELI-D/ELF basin will exclusively touch the penultimate atomic shells of the two atoms, which classifies the basin as disynaptic (synaptic order defines the number of different penultimate shells that share a common surface with the reference basin). The QTAIM atomic basins intersect this ELI-

D/ELF basin at the attractor position, and the resulting ELI-D/ELF basin segments contain the same number of electrons. For polar two-center bonds, the ELI-D/ELF attractor is often located inside the basin of the more electronegative atom, which also shows the larger electronic basin population. This ELI-D/ELF/QTAIM technique has been directly employed to define a bond polarity index [82]. To relate this analysis to the concept of oxidation state, the electronic population of each diatomic ELI-D/ELF basin is completely assigned to the QTAIM atom basin that contains the majority of its charge. In this way, QTAIM rather than an electronegativity scale is used to assign ELI-based oxidation numbers on an electronegativity basis [83]. Examples of ELI-D/ELF and QTAIM space partitioning are illustrated in Figure 12.14 for graphene, bcc Na, and KCl.

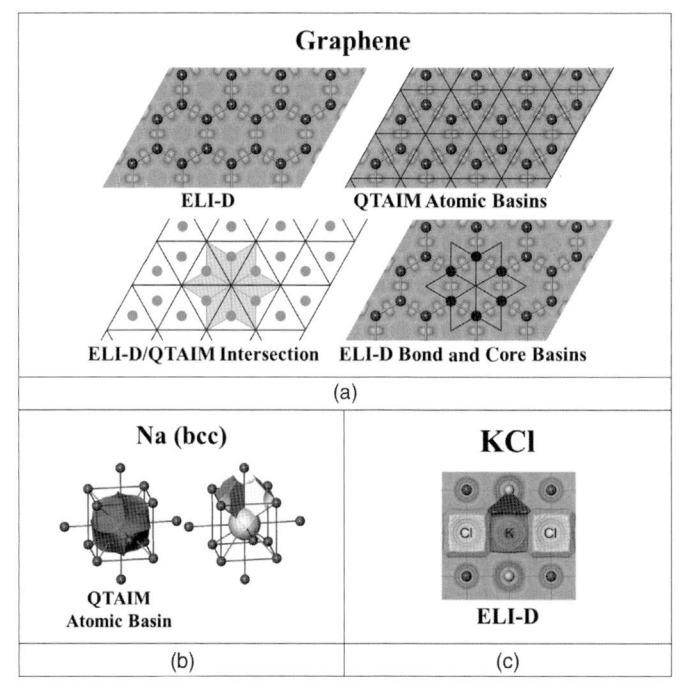

Figure 12.14 ELI-D/ELF and QTAIM space partitioning for (a) graphene; (001) slice, (b) bcc-Na; and (c) KCl. (a) Light gray regions in ELI-D diagram are at bond midpoints, signifying C—C bond attractors. The QTAIM atomic basins, indicated by the black lines are derived from the electron density. The ELI-D/QTAIM intersection occurs by superposition of both partitioning schemes. Each bond basin gets intersected by two QTAIM atoms; each QTAIM atom gets chemically decorated by 3/2 bond basins and one core basin. (b) QTAIM atom emphasizing 8 + 6 interatomic contact surfaces and, right, roughly spherical penultimate shell basin set of Na with specific surface structure caused by 24 bond basins surrounding each Na penultimate shell; each bond basin is attached to four Na penultimate shells (tetrasynaptic), indicating a multicenter bonding scenario. (c) ELI-D diagram of (100) slice with semitransparent QTAIM basins of K and Cl; ELI-D disynaptic basin (dark gray) between K and Cl is marginally intersected by K QTAIM atom, that is, it is effectively monoatomic representing a Cl lone pair-type feature.

Additional effects such as ring strain, multiple bonding, or multicenter bonding can lead to significant displacement of the ELI-D/ELF attractor from the internuclear line. For a nonpolar three-center bond, an ELI-D/ELF attractor exclusively touches the penultimate shells of three atoms (trisynaptic) and is expected to be equidistant from the three atoms involved. The corresponding intersection with the QTAIM atom basins yields equal charge contributions from each QTAIM atom. Originally, the synaptic order of an ELI-D/ELF basin was proposed to yield directly the multicenter character of a bond [84]. Although this prescription works well for molecules, experience shows that it should be considered as an upper bound for solid-state structures, in which a basin in the valence region is typically surrounded by a number of core basins in many directions, so that a high synaptic order is obtained even for classical cases. For example, in NaCl-type CaO, there are eight ELI-D/ELF valence basins around each oxygen core centering the O@Ca$_6$ octahedral faces and that are tetrasynaptic by one O and three Ca atoms. Therefore, determining the number of QTAIM atoms significantly intersecting a certain ELI-D/ELF basin is also important. From the relative electronic populations contained in each intersected part, the importance of each QTAIM atom contributing to an ELI-D/ELF basin can be assessed [85]. Again, for CaO, despite its tetrasynapticity with more than 90% of its population contained in the QTAIM oxygen atom, each ELI-D/ELF basin is effectively monoatomic, that is, it has oxygen lone pair character. In a further development, analysis of 51 binary main group compounds forming zinc blende or rock salt structures has revealed that the covalent bond numbers per anion, derived using the ELI-D/QTAIM basin intersection technique, can separate the more covalent compounds with zinc blende structure from the more ionic compounds with rock salt structure [85b].

This discussion of how to analyze chemical bonding from a position–space perspective would be incomplete if the energetic aspects were omitted. Once spatial partitioning is achieved via QTAIM, the total energy of the chemical system can be decomposed into intra-atomic terms summed over all atoms and pairwise, interatomic contributions summed over all atom pairs, that is,

$$E_{\text{TOT}} = \sum_A \left(T^{(A)} + V_{\text{en}}^{(A)} + V_{\text{ee}}^{(A)} \right) + \sum_{A<B} \left(V_{\text{en}}^{(AB)} + V_{\text{en}}^{(BA)} + V_{\text{ee}}^{(AB)} + V_{\text{nn}}^{(AB)} \right).$$

(12.43)

In Eq. (12.43), the symbols T, V_{en}, V_{ee}, and V_{nn} correspond, respectively, to kinetic, electron–nuclear potential, electron–electron potential, and nuclear–nuclear potential energies, and have clear physical meanings. To achieve a chemical interpretation of the total energy, the following decomposition of the total energy can be proposed [77]:

$$E_{\text{TOT}} = \sum_A E^{(A)} + \sum_{A<B} E^{(AB)} = \sum_A E_0^{(A)} + \left(\sum_A E_{\text{def}}^{(A)} + \sum_{A<B} E^{(AB)} \right)$$
$$\equiv \sum_A E_0^{(A)} + E_{\text{Bond}}.$$

(12.44)

According to Eq. (12.43), each intra-atomic energy term $E^{(A)}$ is given by the energy of a reference atomic state $E_0^{(A)}$, shifted by an atomic deformation energy $E_{\text{def}}^{(A)}$ for atom A in the chemical structure. This deformation energy can be further separated into charge transfer (CT) and charge reorganization (CR) terms that depend on the electron population N_A evaluated for the QTAIM basin A:

$$E_{\text{def}}^{(A)} = E_{\text{CT}}^{(A)}(N_A) + E_{\text{CR}}^{(A)}(N_A) = [E^{(A)}(N_A) - E_0^{(A)}] + [E^{(A)} - E^{(A)}(N_A)]. \tag{12.45}$$

Within the interatomic energy term $E^{(AB)}$ of Eq. (12.44), the electron–electron potential energy term $V_{\text{ee}}^{(AB)}$ can be decomposed into a repulsive Coulomb interaction between the charge clouds of atoms A and B and an attractive, non-classical exchange–correlation term that is closely related to the two-center delocalization index $\delta(A,B)$ defined in Eq. (12.38):

$$V_{\text{ee}}^{(AB)} = V_{\text{Coul}}^{(AB)} + V_{\text{XC}}^{(AB)} = V_{\text{Coul}}^{(AB)} + \int_A d\mathbf{r}_1 \int_B d\mathbf{r}_2 \frac{\rho_{2,\text{XC}}(\mathbf{r}_1, \mathbf{r}_2)}{|\mathbf{r}_1 - \mathbf{r}_2|}. \tag{12.46}$$

Numerical experience shows that larger A–B distances typically lead to lower effective covalent bond orders, and smaller covalent interaction energy terms $V_{\text{XC}}^{(AB)}$. According to this scheme, a chemical definition of the bonding energy E_{Bond} arises as the sum of generally positive atomic deformation energies and negative interaction energies, as given in Eq. (12.44).

The entire machinery of position–space bonding analysis is much more computationally demanding than qualitative MO theory, especially integrations over basin regions. For the energetic analysis, the precision of such integrations must be quite high because chemical bonding energies arise from the partial cancellation of much larger contributions to the total energy of a system. These challenges have been resolved for molecular systems and enlightening bonding analyses that provide quantitative links with useful qualitative concepts are rapidly emerging [86]. For crystalline solids, position–space analysis of chemical bonding using electron densities, ELI-D/ELF, and delocalization indices are well implemented, and development of the energetic decomposition is soon to be achieved.

In summary, the results of quantum mechanical calculations provide extensive measures of chemical bonding, whether in real space or in orbital space. The nature of the appropriate model and the level of approximation are important choices for assessing the outcome, and there are numerous codes in widespread use, some of which were used to produce Figure 12.8 (*TB–LMTO–ASA*), Figure 12.10 (*WIEN2k; TB–LMTO–ASA*), Figure 12.11 (*WIEN2k; VASP; LOBSTER* [48]), and Figure 12.12 (*TB–LMTO–ASA*) [87]. Of course, the next step becomes how to make the computational results useful for synthetic chemists or materials scientists, who want to make solids with targeted, high performance properties – one of the grand challenges of the twenty-first century for solid-state chemistry.

12.4
From Calculations to Electron Counting Rules

Chemists do not always use quantum chemical calculations to gain insights about chemical bonding. They also rely significantly on counting valence electrons and deriving seminal inferences from composition. Perhaps the most widely used tenet among chemists is the octet or $8-N$ rule, which accounts for many structures involving main group elements. Other important valence electron counting rules include the 18-electron rule for transition metal and organometallic complexes, the Wade–Mingos rules for cluster compounds, and Hückel's $4n+2\pi$-electron rule for aromatic rings, to name but a few [1,13]. All of them have been quite successful to enhance understanding of molecular structures and reactivities, but they are essentially guidelines. Exceptions occur, which do not diminish their importance, but simply emphasize that factors other than valence electron count can influence chemistry. These rules are often used diagnostically once a chemical structure has been characterized, but, in the hands of a creative synthetic chemist, they can also be predictive as a means toward proposing new compounds or reaction mechanisms. All of these valence rules can be rationalized using MO theory, typically by filling bonding and nonbonding MOs and relying on large energy separations between the HOMO and LUMO [1,5,15]. Many of these electron counting rules have been adapted to evaluate the stabilities and structures of extended solids, and some will be summarized in this section.

12.4.1
Octet or $8-N$ Rule [1,11]

The $8-N$ rule is widely used to rationalize the connectivities of covalently bonded s and p element compounds by identifying the number of near-neighbor contacts (coordination numbers) expected for an element with N valence electrons. The rule works for elements containing at least four valence electrons. So, C(diamond), Si, Ge are four-connected (tetrahedral); P, As, Sb are three-connected (trigonal pyramidal); and S, Se, Te are two-connected (bent). To establish the local coordination geometry requires inclusion of structure-dependent models, like VSEPR [13]. If multiple bonds occur, then the rule assigns the number of "localized" two-center, two-electron bonds to each atom in a chemical structure, such as three σ-bonds and one π-bond to each C atom in graphene. In any MO energy diagram for a main group element structure, the octet rule can be applied when all bonding and nonbonding MOs are occupied because the delocalized, occupied MOs can be transformed into localized orbitals, each containing two electrons [1,4,11]. Then, all bonds are treated as saturated covalent bonds so that each atom formally completes its octet using its valence s and p AOs. The underlying cause of the octet rule, however, relies both on the attractive forces and on the changes in overlap repulsion that arise by gradually filling antibonding MOs [11]. If just nearest-neighbor intersite integrals are considered and

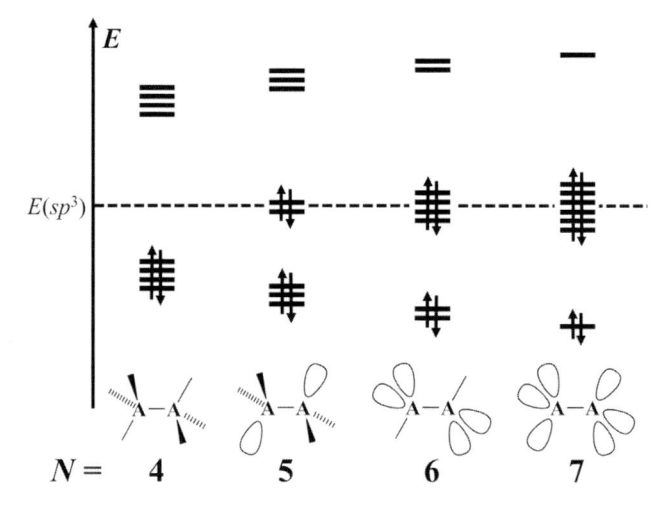

Figure 12.15 Qualitative σ-only MO scheme rationalizing the octet rule for main group bonding. For $N=4$, each atom pair has four bonding MOs and four antibonding MOs; for $N=5$, each atom pair has three bonding MOs, two nonbonding lone pairs, and three antibonding MOs; for $N=6$, each atom pair has two bonding MOs, 4 nonbonding lone pairs, and two antibonding MOs; for $N=7$, each atom pair has one bonding MO, six nonbonding lone pairs, and one antibonding MO. The energy scale reflects the overlap repulsion and the increasing intersite energy term as coordination number decreases.

overlap repulsion is ignored, then there would be no energy difference between equally filling a bonding and antibonding MO versus a pair of nonbonding MOs. However, since overlap repulsion destabilizes the antibonding orbital more than the bonding orbital is stabilized, lowering the coordination number by one for each fully occupied antibonding orbital stabilizes the system. Moreover, as the coordination number decreases for a given atom, the local overlap repulsions likewise decrease so that the interatomic distances may become smaller, and thereby enhance the intersite bonding integrals (see Figure 12.15) [11]. "Exceptions" such as three-bonded C rely on multiple bonding by invoking π-overlap, which is reasonable for the second period elements arising from the lack of core "1p" AOs. For the heavier elements, lack of π-bonding occurs because the intrinsically longer distances create weaker through-space orbital overlap as well-enhanced second-order mixing between filled π-antibonding and empty σ-antibonding orbitals, interactions that lower the $π^*$ MOs in energy and disrupt the purely π-type overlap.

A solid-state application of the octet rule is the Grimm–Sommerfeld valence rule [13,88]. Surveying compounds that adopt either the cubic zinc blende or hexagonal wurtzite structures, which consist of networks of four-bonded, tetrahedrally coordinated atoms, indicates that the average number of valence electrons per atom is 4. However, numerous examples related to these structures contain vacant positions, such as $CdIn_2Se_4$. The total number of valence electrons in $CdIn_2Se_4$ is $1(2e^-) + 2(3e^-) + 4(6e^-) = 32e^-$, which, when divided by 7 atoms in the chemical formula, yields 4.6 valence electrons per atom. Since this ratio exceeds 4, an intact network of tetrahedrally coordinated atoms would necessitate occupation of antibonding orbitals. To avoid this outcome in $CdIn_2Se_4$, 1/8 of the atomic sites in the tetrahedral framework are vacant, eliminating the occupation of antibonding orbitals and creating a semiconducting bandgap in the DOS. To accomplish this outcome via electron counting, a vacancy is treated as an "atomic site" with 0 valence electrons. Then, for $CdIn_2Se_4$, we obtain $32e^-/8$ atomic sites for an average of 4 valence electrons per atomic site; this is the Grimm–Sommerfeld valence rule for compounds related to tetrahedral semiconductors. A general formulation of such solids is "$0_a 1_b 2_c \cdots v_n^{(n)} \cdots 7_h$" in which n atoms in the formula contribute $v^{(n)}$ valence electrons, and the stoichiometric values a–h must satisfy the following equation:

$$\frac{\sum_{n=a}^{h} n v^{(n)}}{\sum_{n=a}^{h} n} = 4. \tag{12.47}$$

Selected examples following the Grimm–Sommerfeld rule are listed in Table 12.4 [13].

Another manifestation of the octet rule in solids is the Zintl–Klemm formalism [89]. Zintl compounds constitute a broad class of intermetallic and metal–metalloid compounds that generally (i) involve combinations of electropositive metals with posttransition metals or metalloids; (ii) exhibit narrow homogeneity widths; and (iii) are typically semiconducting and diamagnetic. The Zintl concept counts all valence electrons, but considers their structural influence only on the electronegative components, which are formally anions. Thus, the electropositive metals are treated simply as cations. For most Zintl phases $A_a X_x$ (A = electropositive metal; X = posttransition metal or metalloid), short interatomic X—X contacts are considered to be two-center, two-electron covalent bonds. If v_A and v_X are the numbers of valence electrons for atoms A and X, then the number of valence electrons per formula unit (N) $A_a X_x$ is

$$N = a v_A + x v_X = x\left(\frac{a}{x} v_A + v_X\right) = x N_X \tag{12.48}$$

in which N_X is the average number of valence electrons per anion. According to the octet rule, the average number of X—X contacts per X atom will be $8 - N_X$. Klemm proposed the pseudoatom concept for the negatively charged unit $X^{(av_A/x)-}$ that would show structural characteristics of the isoelectronic element [89]. If N_X is not an integer, then more than one pseudoatom are needed to account for the observed structure. Formal electron counting according to the $8 - N$ rule distinguishes between covalent bonds and lone pairs. Polar bonds, however, are ambiguous because they can be treated using either assignment. Within the pseudoatom concept, this outcome causes inconsistencies between the different formal descriptions: for example, As in GaAs can range from either 0-bonded "As^{3-}" with four lone pairs to 4-bonded "As^+" with four As—Ga

Table 12.4 Formulas and average number of valence electrons per atom for selected compounds related to tetrahedral semiconductors following the Grimm–Sommerfeld valence rule [13].

Formula "$0_a 1_b 2_c \ldots N_{(n)} \ldots 7_h$"	No. of valence e^- per atom	Examples
"4"	4.00	C, Si, Ge, Sn
"$4_1 4_1$"	4.00	SiC
"$3_1 5_1$"	4.00	BN, BP, AlN, AlP, AlAs, GaP, GaAs, GaSb
"$2_1 6_1$"	4.00	BeO, BeS, ZnO, ZnS, ZnSe
"$1_1 7_1$"	4.00	CuCl, CuBr
"$2_1 5_2$"	4.00	ZnP_2, $ZnAs_2$, CdP_2
"$1_1 3_1 6_2$"	4.00	$CuBSe_2$, $CuAlS_2$, $CuAlSe_2$, $CuGaS_2$, $CuGaSe_2$
"$2_1 4_1 5_2$"	4.00	$BeSiN_2$, $MgGeP_2$, $ZnSiP_2$, $ZnGeP_2$, $ZnSnAs_2$
"$1_3 5_1 6_4$"	4.00	Li_3PO_4, Cu_3PS_4, Cu_3AsSe_4, Cu_3SbS_4
"$1_1 4_2 5_3$"	4.00	$CuSi_2P_3$, $CuGe_2P_3$
"$1_2 4 6_3$"	4.00	Cu_2SiS_3, Cu_2GeS_3, Cu_2SnS_3, Cu_2SnSe_3
"$1_2 2_1 4_1 6_4$"	4.00	Cu_2ZnGeS_4, Cu_2ZnSnS_4
"$0_1 1_2 5_2 6_4$"	4.50	$CuSbS_2$
"$0_1 2_2 4_1 6_4$"	4.57	Be_2SiO_4, Zn_2SiO_4
"$0_1 2_1 3_2 6_4$"	4.57	$ZnAl_2S_4$, $ZnAl_2Se_4$, $ZnGa_2S_4$, $ZnGa_2Se_4$, $CdGa_2S_4$
"$0_1 1_2 2_1 7_4$"	4.57	Li_2BeF_4, Cu_2HgI_4
"$0_1 2_3 5_1 7_3$"	4.57	Zn_3PI_3, Zn_3AsI_3
"$0_1 3_2 6_3$"	4.80	Al_2S_3, Al_2Se_3, Ga_2S_3, Ga_2Se_3, In_2Se_3
"$0_1 5_4$"	5.00	P, As, Sb
"$0_1 4_1 6_2$"	5.33	SiO_2, SiS_2, GeS_2
"$0_1 2_1 7_2$"	5.33	BeF_2, $BeCl_2$, $ZnCl_2$, $ZnBr_2$, ZnI_2
"$0_2 3_1 5_1 6_4$"	5.33	BPO_4, BPS_4, $AlPO_4$, $AlPS_4$, $GaPO_4$

covalent bond pairs. Recently, using the ELI-D/QTAIM approach, a general formalism has been suggested to interpret polar bonding in a consistent way that unites both descriptions [90].

Further extensions of the Zintl–Klemm formalism include accounting for multiple bonds, vacancies, and clusters. As mentioned above, multiple bonds create lower coordination numbers, but the effective number of two-center, two-electron bonds remains intact, so that the electron counting rule in Eq. (12.48) does not change. If a Zintl-type structure contains a network with randomly distributed vacancies, the Zintl–Klemm concept can account for the fraction of vacancies in the polyanionic net. This effect typically occurs for $N_X > 4$.

For a compound formulated as $A_aX_x\square_y$, then

$$N = (x+y)N_X, \qquad (12.49a)$$

so that

$$y = (N - xN_X)/N_X = (N/N_X) - x. \qquad (12.49b)$$

These rules help assign vacancy fractions in tetrahedral and graphene-like nets of Zintl phases [91]. If $N_X < 4$, the electronegative metals or metalloids generally optimize bonding by forming clusters.

The Zintl–Klemm concept works best to rationalize structural behavior rather than to predict structures. Furthermore, the cations play important roles beyond just providing electrons to the polyanions, both by their sizes and by occasionally engaging in polar covalent interactions with the electronegative components [91,92]. Such influences are found in the isoelectronic series LiTl, NaTl, and KTl, which demonstrate significant energetic competition among metallic, ionic, and covalent interactions. This series can be formulated as A^+Tl^- but LiTl is cubic CsCl-type, NaTl is stuffed (double) diamond type, and $KTl \equiv K_6[Tl_6]$ contains distorted Tl_6^{6-} octahedra. Factors contributing to their structural differentiation include relativistic effects at Tl, electronegativity differences between A and Tl, and atomic size ratios between the constituent elements. Relativistic effects contract the Tl 6s AOs and hinder them from participating in effective interatomic overlap, rendering Tl—Tl covalent interactions relatively weaker than isoelectronic Al—Al or Ga—Ga contacts in related structures. Electronegativity differences determine the degree of valence electron transfer from the alkali metal to Tl; in these cases, greater transfer can strengthen the covalent Tl–Tl interactions. Finally, the increasing A:Tl size ratio from LiTl to KTl contributes to the formation of the CsCl-type LiTl, as well as the occurrence of Tl_6^{6-} clusters in KTl at ambient pressure rather than the double-diamond structure of NaTl. These findings demonstrate how even "simple" solid-state systems can display combinations of bonding modes.

12.4.2
18-Electron Rule [13]

Transition metal atoms have nine valence $(s+p+d)$ AOs that can contribute to chemical bonding. For a TL_n complex, in which each ligand L has a single σ-type orbital, there will generally be a collection of n symmetry-adapted orbitals that will find an appropriate match with one of the valence AOs on the metal T. These n ligand orbitals, which are usually lower in energy than the T valence AOs due to relative electronegativities, will be pushed to even lower energies via T—L σ-bonding overlap and n T AOs will be pushed to higher energy as the corresponding T—L σ-antibonding orbitals. This leaves $9-n$ orbitals at the T atom that remain σ-nonbonding, but that may engage in π-type overlap with other ligand orbitals. Therefore, by filling the n T—L bonding MOs and the $9-n$ σ-nonbonding MOs, there are nine filled σ-bonding and nonbonding MOs leading to the 18-electron rule. In general, this rule holds for molecules with large HOMO–LUMO gaps, when the ligands are good σ-donors by destabilizing the LUMO, and good π-acceptors by stabilizing the HOMO; such ligands are strong-field ligands on the spectrochemical series, such as CO, phosphines, CN^-, and ethylene. On the other hand, transition metals coordinated by weak-field ligands like H_2O and halides often do not follow the 18-electron rule, and lead to complexes with unpaired electrons. Other exceptions to this rule occur for special coordination geometries like square–planar complexes, which are 16-electron species because the valence T p_z AO does not overlap with ligand σ-orbitals and there is a large gap between the occupied z^2 and empty $x^2 - y^2$ orbitals. Thus, the 18-electron rule relies on using the valence d, s, and p AOs of a transition metal to form metal–ligand σ-bonds, and filling any remaining σ-nonbonding d-orbitals.

An important class of solids from the point of view of both structural and physical characteristics are the *Heusler* phases and its various relations [93]. This versatile collection of more than 800 compounds includes semiconductors, half-metallic ferromagnets, superconductors, and topological insulators. Heusler compounds $(Td)_2(Oh)X$ adopt the cubic Li_3Bi-type structure, which is a coloring of bcc packing with a fcc arrangement of main group metal/metalloid atoms X and transition metals Oh and Td, respectively, filling all octahedral and tetrahedral holes. Some important examples include ferromagnetic Cu_2MnAl, the ferromagnetic shape memory alloy Ni_2MnGa, and semiconducting Fe_2VAl. Defect derivatives of these Heusler phases have one-half of the tetrahedral holes occupied in an ordered manner "TdOhX," leading to the MgAgAs-type structure and numerous examples involving monovalent and divalent metals A, A', trivalent metals X', and late/early transition metals T/T', namely, AA'X (Nowotny–Juza phases: A = Li, Cu, Ag; A' = Be, Mg, Zn, Cd; and X = N-Bi), AX'X, ATX, T'TX (half-Heusler phases; T' includes rare earth metals). In these structures, one of the three components is surrounded by a cube of two interpenetrating tetrahedra of the other two elements, a position called the heterocubic (HC) site [90]. The other two components are tetrahedrally coordinated, so the MgAgAs-type also consists of two zinc blende networks intersecting at the HC sites. It seems that all half-Heusler phases T'TX have the late transition metal T in the heterocubic site, whereas all main group examples AA'X and AX'X have the main group metal/metalloid X in that site. ATX compounds occur either as $ATX^{(HC)}$ or as $AT^{(HC)}X$, depending on the constituents. Semiconducting representatives of MgAgAs-type compounds show either $8e^-$ (AA'X or AX'X) or $18e^-$ (ATX or T'TX) per formula unit and their bandgaps can be tuned from 0 to 4 eV by systematic changes in chemical composition [93,94]. On the other hand, $17e^-$ examples are metals and $19e^-$ or $22e^-$ cases are half-metallic ferromagnets, which originate when the nonmagnetic electronic DOS is split into majority and minority spin DOS curves and the Fermi level falls in a bandgap for the DOS of either spin [94]. Similarly, for the Heusler phases, $24e^-$ T'_2TX cases, like Fe_2VAl, are semiconducting and Co-rich cases exceeding $24e^-$ are half-metallic ferromagnets, displaying nearly fully polarized conduction electrons [95].

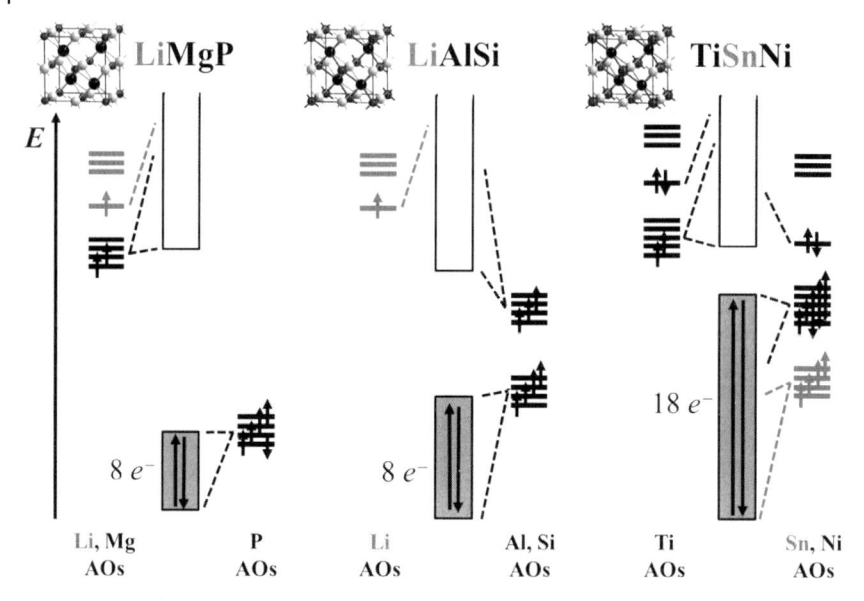

Figure 12.16 Schematic orbital energy diagrams of half-Heusler alloys AA'X (Nowotny–Juza phases; LiMgP), AX'X (LiAlSi), and T'TX (TiSnNi) to demonstrate the 8-electron and 18-electron rules followed for semiconducting examples of this structural class.

An orbital-based rationale for the 8e⁻ Nowotny–Juza and half-Heusler phases follows the Zintl–Klemm formalism, but there are some subtle differences between the AA'X (e.g., LiMgP) and AX'X (e.g., LiAlSi) arising from the electronegativity difference between A' and X' relative to elements A and X (see Figure 12.16), differences that affect the bandgap and valence electron density in the solid [90]. Nevertheless, the 8 valence electrons are assigned to A'—X or X'—X polar covalently bonding states that are derived mostly from the valence s and p AOs of element X. The electropositive element A donates its valence electrons to complete the octet, but the polar covalent interactions with the bonding orbitals are weak because of the large energy difference between the valence AOs of A and these bonding crystal orbitals. For the 18e⁻ semiconductors T'TX (e.g., TiSnNi), polar–covalent T—X interactions create bonding orbitals derived mostly from the valence s and p AOs of X. Then, T'—X (d–s,p) and T'—T d–d interactions further lower the X-centered bonding bands as well as separate the d-bands into T'—T bonding, which are centered more on the T atoms, and T'—T antibonding, which are centered more on the T' atoms. In these T'TX examples, the bandgap opens through d–p and d–d interactions between the two different transition metals. Recent electronic structure calculations also invoke the importance of σ-antibonding interactions between valence s-orbitals of the components to control the presence or absence of a bandgap in the DOS curve for half-Heusler compounds [96]. Nevertheless, the 18e⁻ semiconducting half-Heusler phases follow the 18-electron rule for transition metal complexes, albeit in a collective manner.

Table 12.5 Selected half-Heusler compounds ABC$^{(HC)}$ with element C in the heterocubic site and the results of a position–space analysis of chemical bonding.

ABC$^{(HC)}$	a_{exp} (Å)	a_{calc} (Å)	Q^{eff}(A)	Q^{eff}(B)	Q^{eff}(C)	E_{Mad}^{QTAIM} (eV)	δ(A,C)	δ(B,C)	ς_{nn}
— AX'X —									
LiAlSi	5.922	5.937	+0.82	+1.40	−2.22	−35.4	0.072	0.538	2.44
LiAlGe	5.977	6.020	+0.83	+1.34	−2.16	−33.3	0.070	0.536	2.42
LiInGe	6.304	6.404	+0.84	+0.18	−1.02	−7.7	0.048	0.688	2.94
LiInSn	6.676	6.820	+0.84	−0.00	−0.84	−5.6	0.046	0.681	2.91
BeAlB	4.93	4.962	+1.42	+2.08	−3.50	−104.5	0.208	0.352	2.24
— AA'X (Nowotny–Juza) —									
LiMgN	4.995	5.003	+0.81	+1.57	−2.38	−48.9	0.102	0.222	1.30
LiMgP	6.003	6.014	+0.83	+1.49	−2.32	−38.3	0.084	0.232	1.26
LiMgAs	6.181	6.212	+0.83	+1.46	−2.29	−36.4	0.081	0.240	1.28
LiMgSb	6.60	6.672	+0.84	+1.39	−2.23	−31.8	0.077	0.259	1.34
LiMgBi	6.73	6.856	+0.84	+1.31	−2.15	−28.6	0.074	0.278	1.41
— ATX —									
MgAgAs	6.20	6.341	+1.56	−0.16	−1.40	−18.9	0.150	0.612	3.05
MgSbCu	6.168	6.256	+1.54	−0.84	−0.70	−14.7	0.088	0.692	3.12
— T'TX (half-Heusler) —									
ScSbNi	6.065	6.118	+1.39	−0.60	−0.79	−12.4	0.354	0.674	4.11
TiSbCo	5.872	5.892	+1.21	−0.37	−0.84	−10.4	0.590	0.704	5.18
VSbFe	5.823	5.790	+0.87	−0.23	−0.64	−5.6	0.757	0.687	5.78
TiSnNi	5.927	5.949	+1.22	−0.20	−1.02	−11.6	0.510	0.616	4.50
ZrSbCo	6.068	6.109	+1.40	−0.53	−0.87	−12.9	0.576	0.668	4.98
TiSbRh	6.088	6.143	+1.30	−0.10	−1.20	−13.8	0.581	0.748	5.32
ZrSbRh	6.261	6.336	+1.53	−0.29	−1.24	−16.8	0.583	0.712	5.18
YSbPd	6.4973	6.619	+1.61	−0.58	−1.03	−15.9	0.336	0.674	4.04

Experimental and computationally optimized lattice constants, effective charges Q^{eff}, QTAIM Madelung energies, bond indices for the two contacts forming intersecting zinc blende networks, and a covalency index summing the nearest-neighbor bond indices are listed [85].

A position–space analysis of chemical bonding and electron density distributions also reveal some distinct differences among the semiconductors of the half-Heusler class formulated as AA'X, AX'X, and T'TX [90], especially by emphasizing the two intersecting zinc blende networks through the HC sites. Selected results are summarized in Table 12.5. In AA'X$^{(HC)}$ (LiMg*Pn*, *Pn* = N, P, As, Sb, Bi) both A—X$^{(HC)}$ and A'—X$^{(HC)}$ nets were found to be mainly ionic, whereas in AX'X$^{(HC)}$ (Li*TrTt*, *Tr* = Al, Ga, In; *Tt* = Ge, Sn) the Tr–Tt network becomes increasingly covalent with decreasing Tr–Tt electronegativity difference, while the A–Tt partial structure remains mainly ionic. In general, the effective charge distributions for these two cases are A$^{("+")}$A'$^{("+")}$X$^{("−")}$ and A$^{("+")}$X'$^{("+")}$X$^{("−")}$. For

T'T$^{(HC)}$X compounds, both T'—T$^{(HC)}$ and X—T$^{(HC)}$ tetrahedral substructures exhibit significant polar covalent bonding and smaller effective charges on the atoms than in AX'X cases. The effective charge distribution was found to be even qualitatively different than the others, namely, T'$^{("+")}$T$^{("-")}$X$^{("-")}$. For all "ABC$^{(HC)}$," ionicity and covalency were quantified, respectively, by the QTAIM Madelung energy $E_{\text{Mad}}^{\text{QTAIM}}$ and the index $\varsigma_{nn} = 4\delta_{AC} + 4\delta_{BC}$, involving the eight nearest-neighbor contacts at each HC site. The different bonding scenario in T'TX from the AA'X and AX'X cases was shown to be the reason for the different site preference of T$^{(HC)}$ instead of X$^{(HC)}$ in the T'TX compounds. Interestingly, BeAlB is unique among these half-Heusler compounds by displaying the effective charge distribution of AA'X compounds and the coexistence of intersecting, polar covalent tetrahedral partial structures like the T'TX representatives.

12.4.3
Wade–Mingos Rules: Polyhedral Skeletal Electron Pair Theory [5,13,97]

Unlike the network solids discussed above, main group element molecular clusters, especially those formed by Group 13 (B-Tl) and some Group 14 (C-Pb) and 15 elements, lack sufficient numbers of valence electrons to accommodate one electron pair for each edge, and so must be rationalized by electron counting schemes that invoke multicenter bonding. Many of these clusters are based upon n-vertex deltahedra, which are convex polyhedra with triangular faces and experiments show that the most common clusters range from 5 to 12 cluster atoms, such as the trigonal bipyramid ($n=5$), octahedron ($n=6$), and icosahedron ($n=12$). By using Euler's formula relating the numbers of vertices, edges, and faces of a convex polyhedron, an n-vertex deltahedron will have $2n-4$ faces and $3n-6$ edges. Clearly, such main group element clusters are deficient in the number of valence electrons if bonding is considered via either two-center, two-electron bonds for each edge, or three-center, two-electron bonds for each triangular face. The Wade–Mingos rules establish an effective electron counting scheme for such clusters and their derivatives by assigning a certain number of electrons to skeletal bonding, that is, multicenter interactions that hold the cluster geometry intact. For the three most commonly observed structural polyhedra, the Wade–Mingos rules state the following:

a) *Closo*-structures are intact n-vertex deltahedra for which there are $n+1$ skeletal electron pairs and n skeletal atoms, that is, vertices. The optimum valence electron concentration is $2(2n+1)/n = 4 + 2/n$ electrons per atom.
b) *Nido*-structures are $(n+1)$-vertex deltahedra missing one vertex for which there are $n+2$ skeletal electron pairs but only n skeletal atoms. The optimum valence electron concentration is $2(2n+2)/n = 4 + 4/n$ electrons per atom.
c) *Arachno*-structures are $(n+2)$-vertex deltahedra missing two vertices and have $n+3$ skeletal electron pairs for n skeletal atoms. The optimum valence electron concentration is $2(2n+3)/n = 4 + 6/n$ electrons per atom.

A solid-state application of the Wade–Mingos rules occurs for Rb_2In_3, which consists of two-dimensional networks of $[In_6]^{n-}$ octahedra connected by four *exo*-In—In bonds. Each $[In_6]^{n-}$ layer would require $14e^-$ (seven skeletal electron pairs for the six-vertex octahedron) + $4e^-$ (two *exo*-orbitals of terminal In atoms) + $4e^-$ (four *exo*-orbitals engaged in intercluster In—In bonding) for a total of 22 valence electrons. Therefore, $n=4$, so the formula $Rb_4In_6 \equiv Rb_2In_3$ accounts for the observed structure of interconnected clusters [98].

MO theory starting with a suitable AO basis set provides a decent rationale for the Wade–Mingos rules [5,13]. Each main group atom provides one valence s and three orthogonal p AOs, which can be oriented into one radial AO and two tangential AOs with respect to the cluster geometry. The s and radial p AOs can then be hybridized to form an outward-pointing (*exo*)orbital and an inward-pointing (skeletal) orbital. Empirical outcomes of solving the LCAO or tight-binding Schrödinger equation reveal the following general pattern of cluster orbitals for an intact, *n*-vertex deltahedron: (a) 1 totally bonding skeletal orbital from the inward-pointing radial orbitals; (b) *n* bonding tangential orbitals; (c) *n* radial *exo*-orbitals that are used either as lone pairs or to form bonds with ligand atoms or other clusters; (d) *n* antibonding tangential orbitals; and (e) $n-1$ antibonding skeletal orbitals from the inward-pointing radial orbitals. These five sets account for the total of $4n$ valence orbitals from the cluster atoms, and they are divided into $n+1$ bonding orbitals, *n exo*-directed, essentially nonbonding orbitals, and $2n-1$ antibonding orbitals. Thus, an *n*-vertex deltahedron consisting of *n* main group skeletal atoms will need $n+1$ skeletal bonding electron pairs and *n* additional pairs to fill the *exo*-orbitals. This is the Wade–Mingos rule for *closo*-deltahedra. One or two additional electron pairs would occupy the corresponding number of antibonding orbitals in this diagram, making the *closo*-deltahedron electronically unstable. These antibonding orbitals become stabilized into the bonding/nonbonding region of the MO diagram by adjusting the cluster to be based upon $(n+1)$ or $(n+2)$ vertex deltahedra, respectively. Thus, the rules for *nido*- and *arachno*-clusters emerge. From this orbital analysis, a striking analogy emerges between the Wade–Mingos rules for molecular clusters and the Grimm–Sommerfeld rule for extended tetrahedral structures concerning the occurrence of vacancies or defects in structures [13].

A solid-state application of Wade–Mingos rules applied to condensed clusters provides a simple rationale for the structural differences among the trialuminides of transition and rare earth metals TAl_3, all of which involve ordered, close-packed arrangements of T and Al atoms [93]. In $12e^-$ $LuAl_3$, the Al atoms form a simple cubic net of vertex-sharing $[Al_6]$ octahedra; in $14e^-$ $TaAl_3$, the Al atoms form a 3d net of vertex-sharing $[Al_5]$ square pyramids. Now, an isolated octahedral cluster requires $26e^-$ arising from $12e^-$ in six *exo*-orbitals plus $14e^-$ in skeletal bonding orbitals, which may be divided into one radial and six tangential bonding orbitals. The cubic network of $LuAl_3$ can be generated by $[Al_6]$ octahedra in a fcc array, which contains one octahedral hole per $[Al_6]$ octahedron. When the sizes of the $[Al_6]$ clusters and the octahedral holes are equal, the $[Al_{6/2}]$ net of $LuAl_3$ occurs. Upon forming the fcc array, the six *exo*-orbitals of the [Al$_6$] octahedra will create six radial orbitals in the octahedral holes, but only one of these will create a skeletal bonding orbital. Therefore, as the [Al$_{6/2}$] network forms, the two octahedral units have a total of $14+2=16$ skeletal bonding electrons, or eight skeletal bonding electrons per octahedron. The additional four electrons in LuAl$_3$ can be formally assigned to two of the valence 5d AOs of the rare earth metal because the coordination environment of Lu is a cuboctahedron, which splits the 5d AOs into e$_g$ below t$_{2g}$ sets. The calculated DOS and COHP curves of LuAl$_3$ reveal a pseudogap in the DOS for 12e$^-$ with optimized Al—Al orbital interactions [93]. For 14e$^-$ TaAl$_3$, the additional electron pair would occupy Al—Al antibonding orbitals for a network of [Al$_{6/2}$] octahedra, so creation of square pyramids, that is, *nido*-octahedra, accommodate the additional electron pair. Similar electronic structure calculations also indicate a pseudogap at the Fermi level in TaAl$_3$, with optimized Al—Al and Ta—Al interactions [99].

The Wade–Mingos rules have been extended into Jemmis *mno*-rules [100], which were formulated to account for structures derived from condensing clusters together via a face, an edge, a vertex, or four vertices. According to the Jemmis modification, the number of skeletal electron pairs required for a complete cluster is $m+n+o+p-q$, in which $m=$# of individual polyhedral fragments, $n=$# of vertices in the polyhedron, $o=$# of single-vertex shared condensations, $p=$# of missing vertices from the *closo*-skeleton, and $q=$# of capping atoms. Other schemes, such as Teo's topological electron counting rules [101], have been developed to rationalize the structures of a wide variety of clusters, typically containing transition metals.

12.4.4
Transition Metal Octahedral Clusters

An important class of solid-state structures that go back to the earliest days of inorganic cluster chemistry involve edge-capped T$_6$X$_{12}$ and face-capped T$_6$X$_8$ octahedral clusters of the early transition metals from Groups 3–6 and halides or chalcogenides X [102]. In crystals containing these clusters, each metal atom is terminated by an axial ligand, which completes a local square pyramidal coordination at each T atom. The T$_6$X$_{12}$ cluster requires 12–14 d-electrons for metal–metal bonding, whereas the T$_6$X$_8$ cluster needs 20–24 d-electrons [103]. As a result, T$_6$X$_{12}$ clusters are commonly observed for the Group 3–5 metals [96]. Among these, clusters of the more electron-deficient Groups 3 or 4 metals typically contain an interstitial element, which can range from main group elements to even later transition metals like Fe, Co, Ni, and Ru [104,105]. The chemistry of Zr, Nb, and Ta is especially rich in these clusters, as well as interstitial chemistry of Sc, Y, and other rare-earth metals,. On the other hand, T$_6$X$_8$ clusters, which require more d-electrons for metal–metal bonding, are found among the Groups 5–7 metals, especially for Mo, W, and Re [102]. These clusters seldom enclose interstitial atoms because they would create strong repulsions with the face-capping ligands. One notable exception occurs for

Nb$_6$I$_{11}$ [106], which may be formulated as [Nb$_6$I$_8$]I$_{6/2}$. The [Nb$_6$I$_8$]$^{3+}$ cluster, which has just 19 valence d-electrons for Nb–Nb cluster bonding, reacts with hydrogen to form the 20e$^-$ cluster [HNb$_6$I$_8$]I$_{6/2}$ = HNb$_6$I$_{11}$ [106]. Additional examples of T$_6$X$_8$ clusters occur in Chevrel phases, AMo$_6$S$_8$ and AMo$_6$Se$_8$ (A = divalent and trivalent cations) [107], which show interesting electronic properties and catalytic activities.

12.4.5
Shell Model

Alkali metal clusters observed in mass spectra can be rationalized using the free-electron model applied to a finite volume in which electrons are constrained to move under an attractive mean field potential due to atomic cores whose positions are not specified [108]. For a spherical well of radius R_0, the solutions of the Schrödinger equation are

$$E_{nl} = \frac{\hbar^2}{2mR_0^2}(\kappa_{nl}R_0)^2 \tag{12.50}$$

and

$$\varphi_{nlm}(r) \sim j_l(\kappa_{nl}r) \cdot Y_{lm}(\theta,\phi) = (-\kappa_{nl}r)^l \left(\frac{1}{\kappa_{nl}r}\frac{d}{d(\kappa_{nl}r)}\right)^l \frac{\sin \kappa_{nl}r}{\kappa_{nl}r} \cdot Y_{lm}(\theta,\phi). \tag{12.51}$$

The eigenfunctions $\varphi_{nlm}(r)$ are spherical waves creating shells whose energies increase according to their quantum numbers as follows: 1s < 1p < 1d < 2s < 1f < 2p < 1g < 2d < 1h < 3s < 2f and so on. The resulting closed-shell electron counts of 2, 8, 18, 20, 34, 40, 58, 68, 90, 92, 106, and so on provide excellent agreement with the enhanced peaks in the mass spectra of monovalent alkali metal and noble metal clusters, but do not satisfactorily explain all observables, such as the fine structure in mass spectra or the ionization potentials and electron affinities of clusters, to name but a few properties. This application of the free-electron model to clusters has provided some useful insights about the stabilities of certain Mg-rich complex metallic alloys [109].

12.4.6
Hume–Rothery Electron Compounds

The structures and stabilities of various noble metal alloys with main group metals can be rationalized by their valence electron-to-atom (e/a) ratios, as was shown by Hume–Rothery, who called them "electron compounds" [110]. The e/a ratios are determined by using just the valence s- and p-electrons from each component, irrespective of the specific chemical constituents making up the alloy [111]. Five structure types, found for binary or ternary mixtures of metals with e/a ratios ranging from 1 to 2, were identified: (1) fcc (α-brass); (2) bcc (β-brass); (3) hcp (ζ, c/a ~1.63; ε, c/a ~ 1.55–1.58; and η, c/a ~ 1.77-2); (4) β-Mn; and (5) γ-brass. Examination of several binary phase diagrams involving the noble metals Cu, Ag, or Au with sp-valent elements Zn, Al, Ga, In, Si, Ge, Sn, and so on reveal the following general trends: fcc (e/a = 1.00–1.35); ζ-hcp (e/a = 1.22–1.75); bcc (e/a = 1.36–1.59); β-Mn (e/a = 1.40–1.54); γ-brass (e/a = 1.54–1.70); ε-hcp (e/a = 1.66–1.89); and η-hcp (e/a = 1.96–2.00) [106]. For example, the bcc (β-brass) structure occurs for CuZn and Cu_3Al ($3e^-$/2 atoms or $6e^-$/4 atoms: e/a = 1.5) and the γ-brass structure occurs for Cu_5Zn_8 and Cu_9Al_4 ($21e^-$/13 atoms: e/a = 1.63). In this electron counting scheme, the noble metal valence d-band is formally filled so that the dominant orbital interactions will be sp–sp interactions. Transition metals with partially filled d-bands can be assigned negative valences, which have been useful to predict Al-based intermetallics and even new quasicrystals, but less effective for Cu-based intermetallics. However, electronic structure calculations reveal that the valence d-electrons have significant structural influences in all cases [112]. Thus, electronic structure calculations are important to provide insights into the basis for any Hume–Rothery electron-counting rule.

Original application of the free-electron model to rationalize these rules led to determining the number of electrons required to fill the free-electron Fermi sphere that makes first contact with the corresponding Brillouin zone boundary faces, an effect called Fermi surface–Brillouin zone (FsBz) interactions [111]. This construction in reciprocal space provides decent agreement with the upper bounds on e/a values for fcc, bcc, and ε-hcp phases. Application of pseudopotentials suggests that the sequence fcc–bcc–hcp among these Hume–Rothery phases is driven by the presence of van Hove singularities in the electron DOS curve, that is, places where the band structure has zero slope such as at Brillouin zone boundaries [11]. These singularities create pseudogaps in DOS curves, which are hallmarks of these electron phases and observed in the DOS and band structure of β-brass CuZn and arise in the screening function of electrons in a metal from states near the Fermi level. Furthermore, the valence d-bands are below the Fermi level, but, nonetheless, hybridize with the s- and p-bands. On the other hand, tight-binding calculations, using a computational technique called second moment scaling [107] to equate the overlap repulsion term in the total energy among different structure types, account very well for the sequence of observed structures [112]. These orbital-based conclusions yield the following observations: (a) a purely s and p valence orbital model gives enhanced stability to the bcc structure for e/a values below ~1.50, but reproduce stability trends for cases above 1.50, that is, β-Mn, γ-brass, and the various hcp structures; and (b) inclusion of contracted valence d AOs enhances the stability of the fcc structure for e/a below ~1.30. Therefore, short-ranged metal–metal interactions play an important role to stabilize various Hume–Rothery phases [44,112], and they are particularly useful to understand the atomic arrangements found for various γ-brasses. In general, maximizing the numbers of heteroatomic near-neighbor contacts is the major factor influencing the observed atomic distributions [111]. This outcome was especially important for the Cu-Zn γ-brasses, which occur for

57–68 atomic percent Zn, that is, $e/a = 1.58$–1.72 e$^-$/atom [114]. All of these structures contain a [(Zn$_4$)(Cu$_4$)] kernel that is surrounded by mixtures of Cu and Zn atoms. Analysis of the DOS and various metal–metal COHP curves of Cu$_5$Zn$_8$ indicates that optimizing Cu—Zn orbital interactions is responsible for the upper limit of the e/a ratio, that is, the Zn-rich border, whereas the lower limit arises from the competition in free energy with the β-brass CuZn phase [114].

So, there are numerous applications of valence electron counting rules for solids. All of them rely on filling a set of crystal orbitals, a result that often results in semiconducting or insulating behavior. Although these rules are powerful guidelines for electronic structure, and derive their justification from quantum chemical calculations, there are always subtleties that demand further analyses. This is especially true for assessing structure–property relationships in solids.

12.5
Selected Bonding Influences on Structure and Properties

There certainly exist synergistic relationships between atomic and electronic structures, so that chemical bonding features have profound influences on short-range and long-range structure and materials properties. The diversity of the chemical elements creates numerous trends and perplexing idiosyncrasies that attract the attention of experimentalists and theorists alike to manipulate existing solids and to predict the existence of new ones. In this final section, we summarize a few observations of how chemical bonding factors and features of electronic structure can influence chemical structure, electrical conductivity, itinerant magnetism, and unusual chemistry under pressure.

12.5.1
Structural Influences

Ground-state structures minimize the total electronic energy for a given configuration of energy states. Among the elements, there is a clear variation from close-packed arrangements for the metals with low-valence electron concentrations to more open networks for the metalloids and nonmetals with high-valence electron concentrations [1,11]. Within the transition metals, there is also a regular sequence of the close-packed structures hcp, fcc, and bcc depending on d-band filling [1,11]. In general, ground-state structures tend to generate electronic configurations that avoid occupation of antibonding orbitals, especially σ-antibonding orbitals, which destabilize chemical bonds. However, if a compound and a proposed structure allow antibonding orbitals to be occupied, then specific chemical and structural responses often occur.

To examine trends in structure as a function of valence electron count, the concept of second moment scaling [1,113] or the structural energy difference theorem [11] have provided useful guidelines to evaluate the relationship

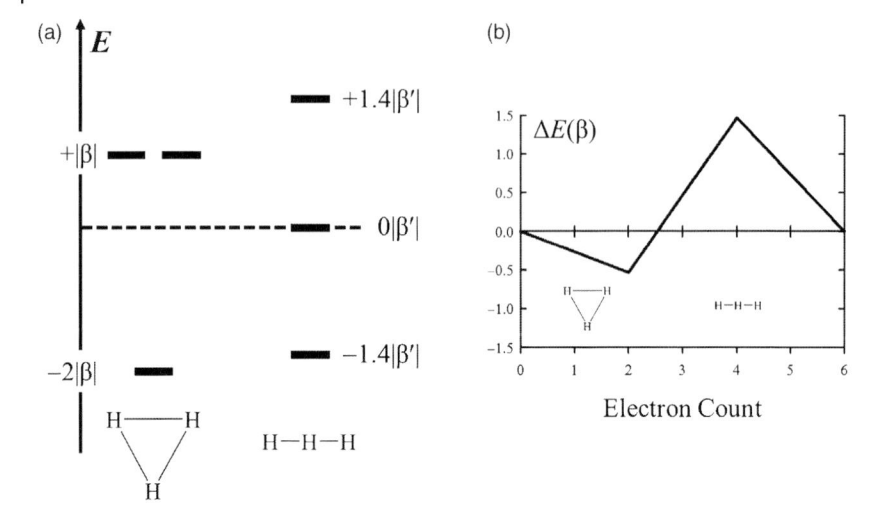

Figure 12.17 Demonstration of second moment scaling applied to the triangular and linear H_3 system. (a) MO energy diagrams for the two structures, indicating the MO energies using the intersite integrals β (for the triangle) and β' (for the linear molecule). (b) Energy difference curve over the full range of electron counts after applying second moment scaling. The energy scale is in units of β. For $\Delta E < 0$, the triangle structure is preferred; for $\Delta E > 0$, the linear structure is preferred.

between band occupations and chemical structure. The significant concept behind both approaches is to compare the energetics of structures under the condition of establishing equivalent effective repulsive energy terms to the total energy. Within tight-binding or LCAO–MO theory, this can be accomplished by setting the second moments of the electronic DOS, $\mu_2 = \int E^2 n(E) dE$, equal to each other for all structures to be compared. To illustrate how this procedure works, consider the case of three H atoms forming either a triangle, which is preferred for H_3^+ or a linear fragment preferred for H_3^-, shown in Figure 12.17. Using a simple MO approach and taking into account just nearest-neighbor intersite integrals β for the triangle and β' for the linear trimer, the second moments are, respectively, $\mu_2 = 6\beta^2$ and $\mu_2' = 4\beta'^2$. The concept of second moment scaling sets $\mu_2 = \mu_2'$, so that $\beta' = (1.5)^{1/2}\beta = 1.225\beta$ for comparing the total MO energies of the two cases as a function of electron count. A plot of $\Delta E(N) = E_{\text{linear}}(N) - E_{\text{triangle}}(N)$ versus N shows the triangle to be energetically favored at low electron counts whereas the linear trimer to be favored at higher electron counts. A thorough analysis of the DOS using moments provides valuable insights about the origins of the shape of the DOS curve and its relationship to structural components like rings and angular features [1,11].

For the main group elements, hybridization between valence s and p AOs significantly affects the electronic DOS curves and their corresponding structures [115]. DOS curves for these elements show the valence s-bands to drop steadily in energy relative to the p-bands as the effective nuclear charge increases

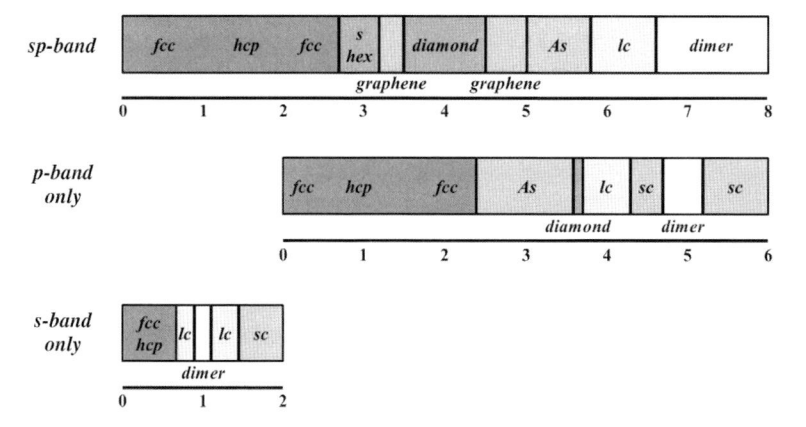

Figure 12.18 Results of second moment scaling calculations on various structural models versus band filling for main group elements shown for s-band only, p-band only, and sp-band cases [11]. Shading represents extent of sphere packing: light = molecular; dark = close-packed. lc = linear chain; sc = simple cubic; As = arsenic-type; s hex = simple hexagonal packing.

by moving left-to-right across a period. On the other hand, relativistic effects on the valence 6s AOs for Tl, Pb and Bi cause these 6s bands to split away below the 6p bands. The outcome of applying second moment scaling to various structural models ranging from close packed to molecular dimers is sketched in Figure 12.18 for the cases of filling (a) just an s-band; (b) just a p-band; and (c) an sp-band such that $\Delta E_{sp} = E_p - E_s \sim 0$. For just the single s- or p-band, close-packed structures are favored at low-band fillings, whereas more open, network structures prefer higher band fillings. This outcome is derived from the presence of three-membered rings in close-packed structures and not in open networks (see also Figure 12.17). Invoking s-p mixing largely accounts for the structural trends among the third–fifth period main group elements. The inconsistencies observed for second and sixth period elements, require, respectively, a softer or harder pairwise repulsive potential than modeled by the second moment scaling procedure. The softer core of the 2p elements leads to more effective π-bonding and lower coordination numbers. On the other hand, the larger ΔE_{sp} of the sixth period elements induced by relativistic effects and a harder core is mimicked by the p-only structural trends, which produces the sequence among the posttransition elements hcp (Tl) → fcc (Pb) → As-type (Bi) → simple cubic (Po). Finally, the bcc structures for K-Cs, rather than hcp, are attributed to the presence of d-bands just above the Fermi level for these elements (see the DOS for K in Figure 12.8).

For the transition metals, the rectangular d-band model discussed in Section 12.2 accounts for trends in cohesive energies but does not differentiate among structures because the actual DOS curves display structure arising from the crystal structures themselves (see DOS curves for Cr and TiFe in Figure 12.12) [11]. Therefore, a tight-binding model applying second moment scaling reproduces the variation in structure hcp → bcc → hcp → fcc across the 4d and 5d series, due in large part to the significant relative strength of d–d π-interactions as seen in the relationship among canonical d–d intersite integrals, β_σ: β_π: $\beta_\delta = 6:4:1$ [1]. When d–d π-overlap is ignored, the observed structural trend is not reproduced. That the transition metals adopt densely packed structures that contain numerous three-membered rings stems from the low occupancy of their s- and p-valence bands, with the d-band occupation influencing the finer structural details as well as contributing substantially to the cohesive energies. Structural differences within the 3d series arise from effects of spontaneous magnetism, which will be discussed shortly.

For all solids, bonding character and strengths depend on the relative energies and occupations of the valence orbitals created by the chemical structure. To summarize several salient points, the monosulfides of the fourth period elements "MS" provide a useful sequence that encompasses many effects of energy band filling and bonding characteristics. The features of these structures are summarized in Table 12.6. All of the fourth period elements except Br are more electropositive than S, so S becomes anionic in these examples. There are interesting trends in volumes, which suggest gradual filling of bonding, then antibonding orbitals, as well as variations in the occurrence of S—S and metal–metal bonding. For the monosulfides with the sp-elements, including Cu and Zn because their 3d AOs are formally filled, 7e⁻ "KS" and CuS have S—S dimers, 8e⁻ CaS and ZnS are rocksalt-type or tetrahedral, 9e⁻ GaS has 2d character with Ga—Ga bonds, 10e⁻ GeS is a distorted rocksalt-type, and then "AsS," "SeS," and "BrS" are molecular solids. Thus, when there are fewer than 8 valence electrons per "MS" formula unit, S—S bonds form, as rationalized by the octet rule, and when there are more than 8 valence electrons, M—M bonds form, unless the configuration of the metal is formally closed shell, as for Ge^{2+}, which is $[Ar](4s)^2$. Among the transition metal monosulfides, many show some degree of metal–metal bonding involving the valence 3d electrons except MnS, which is high-spin d^5 octahedral – its large volume indicates occupation of the e_g antibonding orbitals. In 9e⁻ ScS, there are vacancies associated with both the Sc and S sites and a broad homogeneity width, resulting in part from occupying Sc—S π-antibonding orbitals in stoichiometric, rocksalt-type ScS [116].

The examples discussed above exhibit structures in which σ-antibonding orbitals remain unoccupied. If a chemical structure would contain a sufficient number of valence electrons to occupy σ-antibonding orbitals, then Nature's typical "response" includes either (a) creating structural vacancies, which eliminate certain local orbital interactions, both bonding and antibonding, or (b) enacting structural distortion by bond stretching or bond angle bending to reduce certain interatomic overlaps and lower the energies of antibonding orbitals. An example of creating vacancies is cubic NbO, which can be viewed as a defect-NaCl structure with Nb and O vacancies ordered along the body diagonals of the unit cell [117]. The DOS and COOP curves for Nb—O and Nb—Nb contacts in NbO reveal [117]the following: (a) only Nb—O bonding and nonbonding orbitals are filled; (b) crystal orbitals below the Fermi level show strong Nb—Nb bonding;

12.5 Selected Bonding Influences on Structure and Properties | 473

Table 12.6 Ground-state structural features of the fourth period monosulfides from K to Br.

Compound	Pearson symbol	Voloume/formula unit ($Å^3$)	Distances (Å)	Coordination
KS (K_2S_2)	hP12	60.77	S—S: 2.11, 2.13 K—S: 3.14–3.29	S@K_6 (dist. octahedron; C_3, C_{3v}) K@S_6 (dist. tr.prism; C_{2v})
CaS	cF8	46.16	Ca—S: 2.85	S@Ca_6 (octahedron; O_h) Ca@S_6 (octahedron; O_h)
ScS	cF8	34.95	Sc—S: 2.60	S@Sc_6 (octahedron; O_h) Sc@S_6 (octahedron; O_h) (vacancies at Sc and S sites)
TiS	hP4	29.92	Ti—S: 2.48 Ti—Ti: 3.21, 3.28	S@Ti_6 (tr. prism; D_{3h}) Ti@S_6 (octahedron; D_{3d})
VS	oP8	28.11	V—S: 2.39–2.52 V—V: 2.95, 3.31	S@V_6 (dist. tr. prism; C_s) V@S_6 (dist. octahedron; C_s)
CrS	mS8	33.74	Cr—S: 2.43, 2.88 Cr—Cr: 3.04	S@Cr_6 (dist. tr. prism; C_2) Cr@S_6 (dist. octahedron; C_i)
MnS	cF8	35.97	Mn—S: 2.62	S@Mn_6 (octahedron; O_h) Mn@S_6 (octahedron; O_h)
FeS	hP24	30.19	Fe—S: 2.36–2.72 Fe—Fe: 2.93–2.99	S@Fe_6 (dist. tr. prism; C_s, C_3, D_3) Fe@S_6 (dist. octahedron; C_1)
CoS	hP4	25.77	Co—S: 2.35 Co—Co: 2.60, 3.38	S@Co_6 (tr. prism; D_{3h}) Co@S_6 (octahedron; D_{3d})
NiS	hR6	28.01	Ni—S: 2.26–2.38 Ni—Ni: 2.53, 3.15	S@Ni_5 (dist. sq. prism; C_s) Ni@S_5 (dist. sq. prism; C_s)
CuS	hP12	33.91	S—S: 2.00 Cu—S: 2.18 Cu—S: 2.32, 2.33	S@Cu_3 (trig. pyr.; C_{3v}) S@Cu_5 (trig. bipyr.; D_{3h}) Cu@S_4 (dist. tetrahedron; C_{3v}) Cu@S_3 (trig. pl.; D_{3h})
ZnS	cF8	39.39	Zn—S: 2.34	S@Zn_4 (tetrahedron; T_d) Zn@S_4 (tetrahedron; T_d)
GaS	hP8	43.20	Ga—S: 2.33 Ga—Ga: 2.45	S@Ga_3 (trig. pyr.; C_{3v}) Ga@S_3 (trig. pyr.; C_{3v})
GeS	oP8	41.07	Ge—S: 2.44–2.45; 3.28 Ge—Ge: 3.33	S@Ge_6 (dist. octahedron; C_s) Ge@S_6 (dist. octahedron; C_s)
AsS (As_4S_4)	mP32	49.98	As—S: 2.24 As—As: 2.57	S@As_2 (bent; C_1) As@S_2 (bent; C_s)
SeS (Se_4S_4)	mP32	56.77	Se/S—Se/S: 2.18–2.28	X_8 puckered rings
BrS (S_2Br_2)	oS16	60.04	S—Br: 2.24 S—S: 1.95	S_2Br_2 nonlinear molecules

The Pearson symbol [11] is a structural designator that summarizes the lattice symmetry and number of atoms per unit cell. Other relevant information includes volume per formula unit, shortest interatomic distances, and local coordination at each element.

and (iii) the Fermi level does not fall at a pseudogap, but is controlled by optimizing Nb—O bonding. The corresponding DOS and COOP curves for a hypothetical NaCl-type "NbO" indicate that Nb—O antibonding states would be occupied and, compared to actual NbO, Nb—Nb interactions are significantly weaker, although the O 2p band is slightly lower in energy with respect to the Nb 4d bands from having six neighboring Nb atoms in NaCl-type "NbO" rather than four Nb neighbors in the actual structure [117]. Thus, although vacancies lower the stabilizing Madelung potential at each O site, they contribute to strengthening Nb—O and Nb—Nb bonding. An example of structural distortion involves the metal trifluorides TF_3 [118], which are related to ReO_3, whose electronic structure is illustrated in Figure 12.11. ReO_3 and TF_3 are 3d networks of vertex-sharing $ReO_{6/2}$ or $TF_{6/2}$ octahedra; the O atoms in ReO_3 are linearly coordinated by Re, whereas the F atoms in TF_3 show various T—F—T bond angles, depending on d-electron count. The results of electronic structure calculations indicate that bent T—F—T bond angles are preferred for configurations beyond d^2 to alleviate T—F π-antibonding interactions. ReO_3 is a d^1 oxide, and as the Re—O COHP curve in Figure 12.11 shows, the states just below the Fermi level are essentially Re—O nonbonding. d^2-"NbF_3," which is most likely an oxyfluoride NbO_xF_{3-x}, is linear at "F," but d^3-MoF_3, d^5-RuF_3, and d^6-RhF_3 show decreasing T—F—T angles as the occupation of T—F π-antibonding increases. Further theoretical investigations suggest that the bond bending can be explained via a solid-state analog of a second-order Jahn–Teller distortion [118].

12.5.2
Metals versus Nonmetals

Metals are clearly distinguished from nonmetals by their electronic structures at the Fermi levels. In semiconductors or insulators, there is a bandgap largely relative to kT that separates the occupied valence band from the unoccupied conduction band, whereas in metals, the lowest unoccupied states fall well within kT of the occupied states. By applying an electric field, the degeneracy of electronic states ψ_k and ψ_{-k} is broken. If there is no energy gap at the Fermi level, then the occupation of states with $+k$ and $-k$ differs and a current develops. If the energy gap is sufficiently large, then there will be no net current. The existence of a bandgap in the electronic DOS curve for a solid can arise from various factors, including polar–covalent bonding or charge transfer induced by electronegativity differences. We have already pointed out the hybridization gap in elemental, tetrahedral semiconductors Si, Ge, and Sn, which arises via s–p mixing to optimize covalent bonding. The bandgap can be further adjusted in binary solids via differences in electronegativity. Another isoelectronic series, the NaCl-type CaO, ScN, and TiC, also demonstrate the effects of relative AO energies on the DOS curves [119]. In this series, anion electronegativities increase from C to N to O, whereas the effective nuclear charges increase among the cations from Ca to Sc to Ti. In their corresponding electronic DOS curves, (a) the centers between the anion 2p band and the metal 3d band steadily decrease along the series CaO–ScN–TiC; (b) the dispersion (bandwidth) of the anion 2p bands increases, an effect that can be attributed to stronger cation–anion orbital interactions from CaO to ScN to TiC; and (c) the contribution to the anion 2p band from the cation AOs steadily increases from CaO to ScN to TiC. The result is that CaO shows a clear energy gap between occupied and unoccupied levels (experimentally, ~7 eV); ScN shows a smaller energy gap, appropriate for a semiconductor (experimentally, ~2 eV); and TiC has a pseudogap, indicative of semimetallic character.

Solid-state compounds provide other mechanisms to create or eliminate bandgaps in DOS curves and that are related to chemical bonding features. These mechanisms are especially noteworthy in 1d systems: the Peierls distortion [1,36] and charge density waves [1], both involving creation of an energy gap in a partially occupied band. A classic example of a 1d system showing a Peierls distortion is polyacetylene $[CH]_\infty$ which is a planar, zigzag chain of CH units. Each C atom shows sp^2 hybridization to provide the σ-bonding backbone; the remaining $2p_z$ orbital forms a half-filled π-band because each C $2p_z$ AO contributes 1 valence electron. For a chain of equidistant C—C bonds, metallic behavior is expected because there is no energy gap at the Fermi level, as depicted in Figure 12.19a. This outcome can also be expressed via two resonance Lewis structures for the polymer chain. However, polyacetylene is a semiconductor implying a bandgap at the Fermi level. Indeed, an energy gap will open in the π-band at the half-filled point when C—C bond length alternation occurs along the polymer chain. A simple tight-binding model for the π-band of polyacetylene yields the following energy values:

$$E(k) = E_{2p} \pm \left[\beta_1^2 + \beta_2^2 + 2\beta_1\beta_2 \cos 2ka\right]^{1/2}, \tag{12.52}$$

where β_1 and β_2 represent the two intersite C—C π-integrals. The Fermi level is $E(k_F) = E(\pi/a) = E_{2p} - |\beta_1 - \beta_2|$ and $E_{gap} = 2|\beta_1 - \beta_2|$. If $\beta_1 = \beta_2$, that is, all C—C distances are equal, $E_{gap} = 0$. As soon as bond length alternation occurs, a bandgap opens, with the greatest changes to the energy bands occurring closest to the Fermi level. Therefore, the driving force for distortion of polyacetylene is closely related to the position of the Fermi level, which is controlled by the number of valence electrons. When polyacetylene is doped with an electron acceptor like bromine, the polymer chain becomes oxidized and metallic because some electrons are removed from the π-band, thereby eliminating the driving force for bond alternation and opening an energy gap at the Fermi level. Peierls distortions are not limited to metallic systems with half-filled energy bands, but any partially filled band for a 1d structure will be susceptible toward a distortion that opens a bandgap by adjusting the translational periodicity of the chain. Nevertheless, the largest driving force occurs for half-filled band cases.

Another way of opening energy gaps in DOS curves is by creating periodic variations in electron density at atomic sites rather than at bond sites. These are called *charge density waves*, which can be either static, as for a structural distortion, or dynamic, which is moving along a chain but elicits no static structural

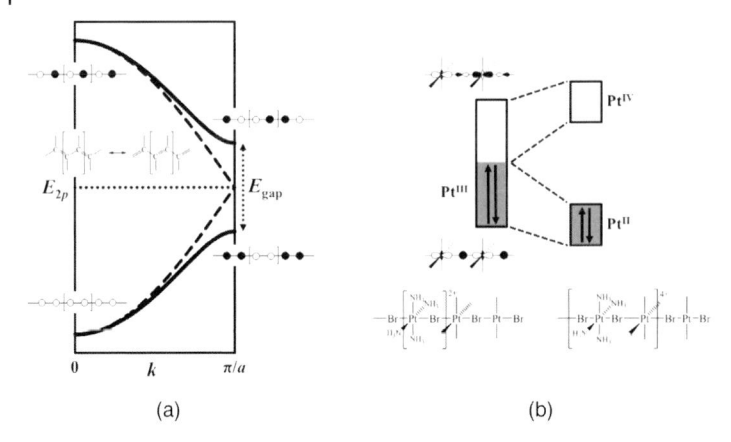

Figure 12.19 (a) Peierls distortion in the π-band of polyacetylene. (b) Charge density wave in the polycationic $[Pt(NH_3)_4Br]^{2+}$ chain.

effects. An example of a charge density wave in 1d occurs for the polycation of Wolfram's red salt $[Pt(NH_3)_4Br]^{2+}$, which contains square–planar $[Pt(NH_3)_4]^{3+}$ units alternating with Br^- ions [1]. A qualitative description of the band structure for an undistorted $[Pt(NH_3)_4Br]^{2+}$ chain focuses just on the half-filled Pt z^2 band, which gets its dispersion by being z^2-s antibonding at the bottom, at $k=0$, and z^2-p_z antibonding at the top, at $k=\pi/a$. For the half-filled band, Pt—Br bond alternation will not affect the band structure because all Pt atoms remain equivalent upon this distortion mode. However, the distortion that opens an energy gap leads to the formulation $[Pt(NH_3)_4Br_2]^{2+} \cdot [Pt(NH_3)_4]^{2+}$, so that there are two inequivalent Pt sites by formally doubling the periodicity along the chain (see Figure 12.19b). At one Pt, two neighboring Br^- ligands move closer, leading toward Pt^{4+}, d^6 pseudo-octahedral coordination; at the other Pt, the two Br^- ligands shift away, leading toward Pt^{2+}, d^8 pseudosquare planar coordination. Chemically, this is an example of disproportionation of Pt^{3+} into Pt^{4+} (d^6) and Pt^{2+} (d^8), creating a static charge density wave of alternating Pt^{4+} and Pt^{2+} sites along the chain and effecting a transition from metallic to semiconducting behavior. The z^2 orbital at the Pt site that has the shorter Pt—Br distances, the "Pt^{4+}" site, goes up in energy as the Pt—Br σ-antibonding interaction increases. The z^2 orbital at the Pt site that has the longer Pt—Br distances, the "Pt^{2+}" site, goes down in energy as the Pt—Br σ-antibonding interaction decreases.

Thus, the construction of energy bands for solids distinguishes metals from nonmetals by relying on partially filled, quasicontinuous bands. However, some systems benefit energetically by keeping electrons localized so as to minimize electron–electron repulsions. This approach was examined by Hubbard [12], who introduced an energy penalty U for occupying a localized orbital with two electrons, and applied this concept to a chain of H atoms, each with 1 AO and 1 valence electron leading to a half-filled energy band of bandwidth $W=4\beta$ ($\beta =$ intersite integral) [1,35]. In the "metallic state," the energy band is occupied by an equal number of spin-up and spin-down electrons, so that the total energy

E_{metal} is the sum of the band energy and the resulting energy penalty of having on average 1/2 spin-up and 1/2 spin-down electron per H atom. Therefore, $E_{metal} = -(W/\pi) + (U/4)$. In the "localized state," with one spin-up electron assigned to each H atom, the band energy is 0 because all crystal orbitals are half-occupied and the energy penalty is also 0 because no local orbital is fully occupied. Therefore, $E_{local} = 0$. The localized state will be preferred when $W < (\pi/4)U$ [120]. So, a half-filled band system with $W \sim U$ is susceptible to a metal–insulator transition induced by electron–electron repulsion, called a Mott–Hubbard transition. An example is NiO(s), which is a transparent green, insulating solid. According to tight-binding theory, NiO contains Ni^{2+} ($3d^8$) ions with a local electronic configuration $(t_{2g})^6(e_g)^2$. The partially filled e_g band would suggest metallic behavior for NiO. However, the Ni^{2+} ions have contracted 3d orbitals, so that the e_g bandwidth is small compared with the on-site electron repulsion term U, and NiO behaves as an antiferromagnetic insulator.

12.5.3
Magnetism

For the transition metals, the valence d-electrons contribute to cohesion, structure, conductivity and, for certain 3d metals, cooperative magnetization [11,15]. At 300 K, Fe, Co, and Ni are ferromagnetic, whereas Cr and Mn show antiferromagnetic faetures. Most transition metals are Pauli paramagnetic, except for the coinage metals Cu, Ag, and Au, which are diamagnetic. In a typical metal, the net numbers of spin-up ($N_\uparrow$) and spin-down ($N_\downarrow$) electrons are equal, $N_\uparrow = N_\downarrow$. When the metal is placed in a magnetic field, the degeneracy of the spin-up and spin-down states is broken, with the spin-up states moving to lower energies and the spin-down states moving to higher energies. Since the Fermi level remains essentially fixed, because it is determined by the total number of valence electrons $N = N_\uparrow + N_\downarrow$, a net magnetic moment is created that is proportional to the DOS at E_F, that is, $m = N_\uparrow - N_\downarrow \sim \mu_B n(E_F)$.

As shown in Table 12.3 and discussed earlier in this section, structures of the 4d and 5d metals correlate strikingly with d-band filling. Along the 3d series, however, the same trend does not hold entirely: Sc to Cr follow the sequence from hcp to bcc, as well as fcc Ni and Cu; but Mn adopts an exceedingly complex structure, Fe is bcc, and Co is hcp. Determination of the net magnetic moment per atom for the ferromagnetic metals also does not correspond to an integer number of electrons; they are $2.20\mu_B$ for Fe, $1.72\mu_B$ for Co, and $0.62\mu_B$ for Ni. However, the structures of the ferromagnetic metals Fe, Co, and Ni follow the same sequence as the 4d and 5d metals from Groups 5 to 10, that is, bcc–hcp–fcc. A chemical rationale for these characteristics of the ferromagnetic 3d metals emerges by considering the electronic structure of the "nonmagnetic" metals, that is, when the spin-up and spin-down states are degenerate [15,114]. For the case of nonmagnetic bcc-Fe, the Fermi level falls on a peak in the DOS curve and these states are Fe—Fe antibonding according to a COHP analysis. If the degeneracy between spin orbitals is removed, by applying LSDA, for example, then upon achieving self-consistency, the spin-up states are lowered compared to the spin-down states. Now, the Fermi level occurs nearly at the top of the spin-up 3d bands, so they are nearly filled, but in the midst of a deep pseudogap of the spin-down 3d bands. The magnetic moment is evaluated to be $2.18\mu_B$, which agrees very well with experiment. The COHP analysis shows that the Fe—Fe orbital interactions within the spin-down orbitals are nearly optimized. Thus, there are two important signatures of possible ferromagnetic behavior in metals arising from the nonmagnetic DOS: (a) a large DOS value at the Fermi level; and (b) the orbital interactions at the Fermi level are antibonding, in particular, those involving the magnetically active metals [15,121].

12.5.4
High-Pressure Effects

Studying solids under applied pressure provides salient insights about chemical bonding as revealed by possible structural changes and trends in electrical conductivity. On compression, nearest-neighbor distances usually decrease, which can lead to changes in coordination numbers. The contributions to the free energy have different volume dependencies: the electronic kinetic energy varies as $V^{-2/3}$; coulombic forces vary as $V^{-1/3}$; and the pV term can exceed many chemical bond energies at sufficiently high pressures. At lower applied pressures, the greatest effects occur for the energy bands derived from the valence AOs, bands that broaden as the intersite integrals increase upon decreasing interatomic separations. As a result, materials should generally become metallic under increased pressure. Although many solids do behave accordingly, there are some unexpected outcomes, both regarding structure and conductivity. Moreover, when the applied pressures become especially high, that is, in the gigapascal range, compound formation can be induced between elements that normally do not bond under ambient conditions, nonclassical compositions become energetically accessible, and unusual atomic and electronic structures can emerge from significant core–valence and core–core interactions.

The divalent hcp metals Mg and Zn remain metallic under pressure, but show different structural effects [122]. The c/a ratio of the unit cell lengths for Mg remains nearly ideally close-packed (1.624–1.673) for pressures up to ~300 kbar; the corresponding value for Zn is extended at ambient conditions (1.856), and steadily decreases to the ideal value of 1.63 at ~250 kbar. According to the results of first-principles calculations [122], the observed behavior arises from the extent of valence s–p orbital mixing, which exchanges s–s antibonding for p–p bonding states below the Fermi level, as well as from the opposing contributions to the total energy from occupying the energy bands and the electrostatic energy involving the atom cores with the screened electron density, which prefer, respectively, nonideal and ideal c/a ratios in these hcp structures. For Mg, the electrostatic energy overrides the band energy, and the extent of 3s–3p mixing is moderate such that small hybridization gaps occur in the band structure at

Brillouin zone boundaries, creating small deviations from free-electron behavior in the DOS curve at the Fermi level. In Zn, the band energy exerts stronger influence than the electrostatic term because the filled 3d subshell, which lies about 8 eV below the Fermi level, increases the effective nuclear charge primarily for 4s and partly for 4p electrons, so that these AOs have enhanced intersite integrals. The hybridization gaps are also larger in Zn than in Mg, so that the DOS of Zn shows a distinct pseudogap quite near the Fermi level. On compression, the occupied 3d states prevent extensive broadening of the 4d bands and the electrostatic energy term steadily dominates so that the c/a ratio of Zn becomes ideally close-packed.

Ca, on the other hand, exhibits nontrivial behavior under pressure [123]. Close to 4 GPa, fcc Ca undergoes a metal–nonmetal transition, and then changes to bcc at 19 GPa and becomes metallic again. At 32 GPa, it transforms to (distorted) simple cubic that exhibits superconductivity above 50 GPa. During this compression, the atomic volume of Ca shrinks by a factor of 2.4, which induces significant changes in its electronic structure. At even higher pressures, its structural complexity increases. The low-lying 3d AOs of Ca lead to pronounced s–p–d mixing, stabilizing the fcc structure over hcp observed for Mg. Upon compression, these 3d AOs become increasingly populated. Furthermore, the interatomic core–valence overlap increases, an effect that can expel valence electrons into the interstitial (empty space) regions of a structure, leading to strong, non-nuclear charge maxima and increased electron localization. First-principles calculations reveal a DOS curve with much structure, not at all resembling a parabolic, free-electron-like relation, and a 0.1 eV bandgap occurring at 18 GPa. An evaluation of ELF in fcc Ca shows maxima in the octahedral voids arising from multicenter bonding. These maxima become more pronounced on increasing pressures due to interatomic core–valence repulsions as the volume shrinks significantly. In the energy band structure, the bandgap opens via avoided band crossings involving s–p–d mixing [124].

The complex and non-close-packed structures exhibited by Ca above 32 GPa mimic behavior also observed for many other simple metals like Li, Rb, Cs, Sr, and Ba, which adopt close-packed crystal structures at low and high pressures, but complex networks at intermediate pressures [125]. In some cases, the participation of valence d AOs to bonding rationalizes this behavior. From the perspective of the nearly free-electron model, imperfect screening of the atom cores as the atoms approach more closely and Fermi surface–Brillouin zone interactions may also contribute, especially for Li, which has no low-lying valence d AOs. The complex structural candidates for these intermediate pressure phases will have strong diffraction peaks clustered around a narrow range of wavevectors close to $2k_F$.

In addition to unusual structures, highly compressed materials may achieve unprecedented compositions or show compound formation that are inaccessible under ambient conditions because the pV term in the free energy can exceed the energies of chemical bonds at sufficiently high pressures. For example, the enhanced involvement of 3d states in K under pressure may enable it to combine with Fe [126]. A recent computational study of the Li–Be system [127], which are immiscible under normal conditions, revealed stability ranges for "Li$_3$Be" (~100–120 GPa), "LiBe" (~50–100 GPa), "LiBe$_2$" (> ~20 GPa), and "LiBe$_4$" (> ~60 GPa). None of the predicted structures is close-packed, but all show aspects of a Hume–Rothery stabilization mechanism via Fermi surface–Brillouin zone interactions in their electronic structures. Also, somewhat peculiarly, the bottom of the valence DOS curve of "LiBe" at 82 GPa displays a step-like singularity that remains constant for ~4 eV until distinct features emerge. The Fermi level ultimately lies in a clearly broad pseudogap. The interpretation of the step-like feature in the DOS is as two-dimensional electron gas states associated with layers of Be atoms. This effect is attributed to larger size of the Li core over the Be core, such that as volume decreases, the Li cores begin to overlap, while the Be cores do not, and expel valence electrons into the vicinity of Be atoms where they form quasi-2d free-electron-like states. A high-pressure examination from 10 to 80 GPa of the "ionic" NaCl system also revealed unprecedented compounds [128]. NaCl transforms into the CsCl-type structure at 30 GPa with metallization expected at pressures at least 300 GPa, but will not spontaneously decompose to other compositions in the pressure range studied. Therefore, by varying composition and temperature, cubic and orthorhombic NaCl$_3$ and tetragonal Na$_3$Cl were obtained. At 20–48 GPa, NaCl$_3$ is semiconducting, orthorhombic builtup of Na$^+$ and Cl$_3^-$ linear trimers, as seen in I$_3^-$, in which the bonding may be described as a three-center, four-electron bond. At 48 GPa, NaCl$_3$ transforms into a metallic, cubic structure with linear chains of Cl atoms. The transition from the metallic Cl atom chains, which are formally $[Cl^{-1/3}]_\infty$, to Cl$_3^-$ units resembles a Peierls distortion for the 2/3-filled band in this case [128]. On the other hand, Na$_3$Cl is a layered superstructure of the CsCl-type structure, which can be formulated as insulating [NaCl] and metallic [Na$_2$] layers alternating along the c-axis.

Finally, at pressures exceeding 110 GPa hydrogen sulfide H$_2$S transforms into a metallic phase that becomes superconducting with a T_c of 203 K [129]. The likely superconducting phase involves a bcc lattice of S atoms but there are two differing interpretations [130]: (i) decomposition of H$_2$S into bcc "H$_3$S" and sulfur, but the amount of sulfur in the samples varies and is usually much less than expected; and (ii) dissociation of H$_2$S into H$_3$S$^+$ + SH$^-$, which together form a cubic perovskite-like structure based on the 3d network of vertex-sharing [SH$_{6/2}$]$^+$ octahedral with (SH)$^-$ units filling the voids with random orientations. DFT-based electronic structure calculations prefer the second choice [130]. The electronic band structure of [SH$_3$]$^+$(SH)$^-$ contains features that resemble the band structure for superconducting MgB$_2$ [131]. Furthermore, the occupied S 3p states from the [SH$_3$]$^+$ substructure lie below the S 3p states from the disordered (SH)$^-$ groups, which contribute substantially to states near the Fermi level of this phase. In effect, a description of the bcc-type [SH$_3$]$^+$(SH)$^-$ from its electronic structure is consistent with its formal charges. Nevertheless, further experimental and theoretical work on this remarkable phase transformation is warranted.

12.6
Summary

The solid state allows examination of all chemical bonding modes. A van Arkel–Ketelaar diagram shows that continuous variation among the three common bonding types, namely, covalent, metallic, and ionic, is possible. An orbital approach provides a unifying strategy to interpret the electronic structure of various types of solids, but the continual development of quantum chemical codes, especially via density functional theory, is providing increased accuracy and the ability to examine solids under extreme conditions of pressure. However, orbital methods do not provide clear real space characterization of chemical bonding, but through the development of important tools like QTAIM analysis of electron densities, delocalization index-based interatomic electron sharing, and ELF/ELI-D analysis of electron localizability provide direct access to atomic volumes, effective atomic charges, and assessments of oxidation states, effective bond orders and multicenter bonding, and evaluations of the effects of bond pairs and lone pairs. As proven for molecular systems, density matrix-based methods are well suited to be applied to correlated descriptions of solids as well, that is, density matrix-based or post-Hartree–Fock wavefunction-based methods, a perspective for solids that has already started [132]. Valence electron counting rules, which are useful for the experimentalists, still provide useful rationale for compound formation, their structures, and properties. However, the role of computations will continue to increase, and fundamental questions, like at which size does a nanoscale particle show chemical and physical properties that resemble those of the solid [133], are on the horizon for enriching our understanding of the chemical bond.

Acknowledgments

GJM wishes to thank his students, postdoctoral associates, colleagues, and mentors for all of their valuable contributions to the ideas presented in this chapter. Special appreciation goes to Prof. Michael Ruck, his research group, and colleagues at the Technical University Dresden for their kind hospitality and support via the Dresden Senior Fellowship Program, which allowed the topics for this contribution to be assembled. FRW wishes to thank Prof. Yuri Grin for support of this work. For additional financial support, the authors gratefully acknowledge the US National Science Foundation and Iowa State University.

References

1 Burdett, J.K. (1995) *Chemical Bonding in Solids*, Oxford University Press, New York.

2 Harrison, W.A. (2005) *Elementary Electronic Structure* (revised edition). World Scientific, Singapore.

3 M. Scheffler, A. Tkatchenko, P. Rinke, *Theoretical Material Science*, a lecture series at TU-Berlin, Chapter 6 ("Cohesion (Bonding) in Solids"), **2012**, pp. 139–177. (Available at http://th.fhi-berlin.mpg.de/sitesub/lectures/spring_2012/index.php?n=Lecture.Scripts.)

4 Burdett, J.K. (1997) *Chemical Bonds: A Dialog*, John Wiley & Sons, Inc., New York.

5 Albright, T.A., Burdett, J.K., and Whangbo, M. H. (1985) *Orbital Interactions in Chemistry*, John Wiley & Sons, Inc., New York.

6 Hoffmann, R. (1963) An extended Hückel theory: I. Hydrocarbons. *J. Chem. Phys.*, **39**, 1397–1412.

7 Pettifor, D.G. The tight-binding method: application to AB s-valent dimer. doi: arXiv:1112.4638 [cond-mat.mtrl-sci].

8 Ruedenberg, K. (1962) The physical nature of the chemical bond. *Rev. Mod. Phys.*, **34**, 326–376.

9 Schmidt, M.W., Ivanic, J., and Ruedenberg, K. (2014) The physical origin of covalent bonding, in *The Chemical Bond: Fundamental Aspects of Chemical Bonding* (eds G. Frenking and S. Shaik), Wiley-VCH Verlag GmbH, Weinheim, Germany, pp. 1–67.

10 Shaik, S., Danovich, D., Wu, W., and Hiberty, P.C. (2009) Charge-shift bonding and its manifestation in chemistry. *Nat. Chem.*, **1**, 443–449.

11 Pettifor, D. (1995) *Bonding and Structure of Molecules and Solids*, Clarendon Press, Oxford.

12 (a) Hubbard, J. (1963) Electron correlations in narrow energy bands. *Proc. R. Soc. Lond. A*, **276**, 238–257. (b) Hubbard, J. (1964) Electron correlatioins in narrow energy bands: II. The degenerate band case. *Proc. R. Soc. Lond. A*, **277**, 237–259. (c) Hubbard, J. (1964) Electron correlations in narrow energy bands: III. An improved solution. *Proc. R. Soc. Lond. A*, **281**, 401–419. (d) Hubbard, J. (1965) Electron correlations in narrow energy bands: IV. The atomic representation. *Proc. R. Soc. Lond. A*, **285**, 542–560. (e) Hubbard, J. (1967) Electron correlations in narrow energy bands: V. A perturbation expansion about the atomic limit. *Proc. R. Soc. Lond. A*, **296**, 82–99. (f) Hubbard, J. (1967) Electron correlations in narrow energy bands: VI. The connexion with many-body perturbation theory. *Proc. R. Soc. Lond. A*, **296**, 100–112.

13 Burdett, J.K. (1980) *Molecular Shapes*, John Wiley & Sons, Inc., New York.

14 McGlynn, S.P., Van Quickenborne, L., and Kinoshita, M. (1971) *Introduction to Applied Quantum Chemistry*, Holt, Rinehart, and Winston, New York.

15 Dronskowski, R. (2005) *Computational Chemistry of Solid State Materials*, Wiley-VCH Verlag GmbH, Weinheim, Germany.

16 Pearson, R.G. (1988) Absolute electronegativity and hardness: application to inorganic chemistry. *Inorg. Chem.*, **27**, 734–740.

17 Parr, R.G. and Yang, W. (1989) *Density-Functional Theory of Atoms and Molecules*, Oxford University Press, New York.

18 (a) Pyykkö, P. (1988) Relativistic effects in structural chemistry. *Chem. Rev.*, **88**, 563–594. (b) Pyykkö, P. (2012) Relativistic effects in chemistry: more common than you thought. *Annu. Rev. Phys. Chem.*, **63**, 45–64.

19 Kutzelnigg, W. (1984) Chemical bonding in higher main group metals. *Angew. Chem., Int. Ed. Engl.*, **23**, 272–295.

20 (a) Jensen, W.B. (1995) A quantitative van Arkel diagram. *J. Chem. Educ.*, **72**, 395–398. (b) Jensen, W.B. (1998) Logic, history, and the chemistry textbook II. Can we unmuddle the chemistry textbook? *J. Chem. Educ.*, **75**, 817–828.

21 Cox, P.A. (1987) *The Electronic Structure and Chemistry of Solids*, Oxford University Press, New York.

22 Wertheim, G.K., Rowe, J.E., Buchanan, D.N.E., and Citrin, P.H. (1995) Valence-band structure of alkali halides determined from photoemission data. *Phys. Rev. B*, **51**, 13675–13680.

23 Shanker, J. and Singh, K. (1981) The Anderson parameter and temperature dependence of bulk modulus for alkali halide crystals. *Phys. Status Solidi*, **108**, 151–157.

24 P. K., deBoer. and R.A., deGroot. (1999) With a grain of salt. *Phys. Lett. A*, **256**, 227–229.

25 Zhuravlev, Yu.N., Basalaev, Yu.M., and Poplavnoi, A.S. (2001) Electron density calculations for crystals with a NaCl lattice. *J. Struct. Chem.*, **42**, 172–176.

26 Doll, K. and Stoll, H. (1997) Cohesive properties of alkali halides. *Phys. Rev. B*, **56**, 10121–10127.

27 Thakur, K.P. (1976) The nature of the ionic bond. *Aust. J. Phys.*, **29**, 39–50.

28 Nasar, A. (2013) Correlation between standard enthalpy of formation, structural parameters and ionicity for alkali halides. *J. Serb. Chem. Soc.*, **78**, 241–253.

29 Cohen, M.L. (1973) Electronic charge densities in semiconductors. *Science*, **179**, 1189–1195.

30 Aresti, A., Garbato, L., and Rucci, A. (1984) Some cohesive energy features of tetrahedral semiconductors. *J. Phys. Chem. Solids*, **45**, 361–365.

31 Kumar, V., Shrivastava, A.K., and Jha, V. (2010) Bulk modulus and microhardness of tetrahedral semiconductors. *J. Chem. Phys. Solids*, **71**, 1513–1520.

32 Janthon, P., Luo, S., Kozlov, S.M., Viñes, F., Limtrakul, J., Truhlar, D.G., and Illas, F. (2014) Bulk properties of transition metals: a challenge for the design of universal density functionals. *J. Chem. Theory Comput.*, **10**, 3832–3839.

33 Giustino, F. (2014) *Materials Modelling Using Density Functional Theory*, Oxford University Press, Oxford.

34 Miedema, A.R., Chatel, P.F., and de Boer, F.R. (1980) Cohesion in alloys: fundamentals of a semi-empirical model. *Physica*, **B100**, 1–28.

35 Brewer, L. (1968) Bonding and structures of transition metals. *Science*, **161**, 115–122.

36 Alemany, P. and Canadell, E. (2014) Chemical bonding in solids, in *The Chemical Bond: Chemical Bonding across the Periodic Table*, 1st edn (eds G. Frenking and S. Shaik), Wiley-VCH Verlag GmbH, Weinheim, Germany, pp. 445–475.

37 Ashcroft, N.W. and Mermin, N.D. (1976) *Solid State Physics*, Saunders College, Philadelphia, PA.

38 Janot, C. (1994) *Quasicrystals: A Primer*, 2nd edn, Oxford University Press, Oxford.

39 Ziman, J.M. (1979) *Models of Disorder*, Cambridge University Press, London.

40 Levine, I.N. (2006) *Quantum Chemistry*, Prentice-Hall, India.

41 (a) Dudarev, S.L., Botton, G.A., Savrasov, S.Y., Humphreys, C.J., and Sutton, A.P. (1998) Electron-energy-loss spectra and the structural stability of nickel oxide: an LSDA+U study. *Phys. Rev. B*, **57**, 1505–1509. (b) Liechtenstein, A.I., Anisimov, V.I. and Zaane, J. (1995) Density-functional theory and strong interactions: orbital ordering in Mott–Hubbard insulators. *Phys. Rev. B*, **52**, R5467–R5470;

42 Cohen, A.J., Mori-Sánchez, P., and Yang, W. (2008) Insights into current limitations of density functional theory. *Science*, **321**, 792–794.

43 Perdew, J.P. and Zunger, A. (1981) Self-interaction correction to density functional approximations for many-electron systems. *Phys. Rev. B*, **23**, 5048–5078.

44 Berger, R.F., Walters, P.L., Lee, S., and Hoffmann, R. (2011) Connecting the chemical and physical viewpoints of what determines structure: from 1-D chains to γ-brasses. *Chem. Rev.*, **111**, 4522–4545.

45 Paxton, A.T. (2009) An introduction to the tight binding approximation – implementation by diagonalisation, in *Multiscale Simulation Methods in Molecular Sciences*, vol. **42**, NIC Series, Jülich Supercomputing Centre, Forschungszentrum Jülich, pp. 145–174.

46 (a) Hughbanks, T. and Hoffmann, R. (1983) Chains of *trans*-edge-sharing molybdenum octahedra: metal–metal bonding in extended systems. *J. Am. Chem. Soc.*, **105**, 3528–3537. (b) Wijeyesekera, S.D. and Hoffmann, R. (1984) Transition metal carbides: a comparison of bonding in extended and molecular interstitial carbides. *Organometallics*, **3**, 949–961.

47 Dronskowski, R. and Blöchl, P.E. (1993) Crystal orbital Hamilton populations

(COHP): energy-resolved visualization of chemical bonding in solids based on density-functional calculations. *J. Phys. Chem.*, **97**, 8617–8624.

48 (a) Deringer, V.L., Tchougréeff, A.L., and Dronskowski, R. (2011) Crystal orbital Hamilton population (COHP) analysis as projected from plane-wave basis sets. *J. Phys. Chem. A*, **115**, 5461–5466. (b) Maintz, S., Deringer, V.L., Tchougréeff, A.L., and Dronskowski, R. (2013) Analytic projection from plane-wave and PAW wavefunctions and application to chemical-bonding analysis in solids. *J. Comput. Chem.*, **34**, 2557–2567. (c) Maintz, S., Deringer, V.L., Tchougréeff, A.L., and Dronskowski, R. (2016) LOBSTER: a tool to extract chemical bonding from plane-wave based DFT. *J. Comput. Chem.*, **37**, 1030–1035.

49 Wegener, W. and Fritsche, L. (1976) Band structure of graphite based on the rigorous cellular method. *Phys. Status Solidi*, **78**, 585.

50 Chadi, D.J. and Cohen, M.L. (1975) Tight-binding calculations of the valence bands of diamond and zincblende crystals. *Phys. Status Solidi*, **68**, 405.

51 Lipson, H. and Stokes, A.R. (1942) The structure of graphite. *Proc. R. Soc. Lond. A*, **181**, 101–105.

52 Hom, T., Kiszenick, W., and Post, B. (1975) Accurate lattice constants from multiple reflection measurements: II. Lattice constants of germanium, silicon, and diamond. *J. Appl. Crystallogr.*, **8**, 457–458.

53 (a) Tatar, R.C., Holzwarth, N.A.W., and Rabii, S. (1981) Energy band structure of three dimensional graphite. *Synth. Met.*, **3**, 131–138. (b) Partoens, B. and Peeters, F.M. (2006) From graphene to graphite: electronic structure around the K point. *Phys. Rev. B*, **74**, 075404.

54 Colaitis, D., Lecaille, C., Lebas, D., and Axe, J.D. (1972) Precicions sur les parametres cristallines et sur L'indexation des raies debye and scherrer des oxydes ReO_2 et ReO_3. *Rev. Chim. Miner.*, **9**, 709–716.

55 (a) Corà, F., Stachiotti, M.G., Catlow, C.R.A., and Rodriguez, C.O. (1997) Transition metal oxide chemistry: electronic structure study of WO_3, ReO_3, and $NaWO_3$. *J. Phys. Chem. B*, **101**, 3945–3952. (b) Evarestov, R.A., Kalinko, A., Kuzmin, A., Losev, M., and Purans, J. (2009) First-principles LCAO calculations on 5d transition metal oxides: electronic and phonon properties. *Integr. Ferroelectr.*, **108**, 1–10.

56 Philip, T.V. and Beck, P.A. (1957) CsCl-type ordered structures in binary alloys of transition elements. *Trans. Am. Inst. Min. Metall. Petrol. Engr.*, **209**, 1269–1271.

57 Edmiston, C. and Ruedenberg, K. (1963) Localized atomic and molecular orbitals. *Rev. Mod. Phys.*, **35**, 457–465.

58 Foster, J.M. and Boys, S.F. (1960) Canonical configurational interaction procedure. *Rev. Mod. Phys.*, **32**, 300–302.

59 Marzari, N., Mostofi, A.A., Yates, J.R., Souza, I., and Vanderbilt, D. (2012) Maximally localized Wannier functions: theory and applications. *Rev. Mod. Phys.*, **84**, 1419–1475.

60 (a) Yannello, V.J., Kilduff, B.J., and Fredrickson, D.C. (2014) Isolobal analogies in intermetallics: the reversed approximation MO approach and applications to $CrGa_4$- and Ir_3Ge_7-type phases. *Inorg. Chem.*, **53**, 2730–2741. (b) Yannello, V.J. and Fredrickson, D.C. (2014) Orbital origins of helices and magic electron counts in the Nowotny chimney ladders: the $18-n$ rule and a path to incommensurability. *Inorg. Chem.*, **53**, 10627–10631.

61 Hoffmann, R. (1982) Building bridges between inorganic and organic chemistry (Nobel Lecture). *Angew. Chem., Int. Ed.*, **21**, 711–724.

62 Hohenberg, P. and Kohn, W. (1964) Inhomogeneous electron gas. *Phys. Rev.*, **136**, B864–B871.

63 Löwdin, P.O. (1955) Quantum theory of many-particle systems: I. Physical interpretations by means of density matrices, natural spin orbitals, and convergence problems in the method of configurational interaction. *Phys. Rev.*, **97**, 1474–1489.

64 (a) West, A.C., Schmidt, M.W., Gordon, M.S., and Ruedenberg, K. (2013) A comprehensive analysis of molecule-intrinsic quasi-atomic, bonding, and

correlating orbitals: I. Hartree–Fock wave functions. *J. Chem. Phys.*, **139**, 234107. (b) West, A.C., Schmidt, M.W., Gordon, M.S., and Ruedenberg, K. (2015) A comprehensive analysis in terms of molecule-intrinsic quasi-atomic orbitals: III. The covalent bonding structure of urea. *J. Phys. Chem. A*, **119**, 10368–10375.

65 Bader, R.F.W. (1990) *Atoms in Molecules: A Quantum Theory*, Clarendon Press, Oxford.

66 Gatti, C. (2012) The source function descriptor as a tool to extract chemical information from theoretical and experimental electron densities, in *Electron Density and Chemical Bonding: II. Theoretical Charge Density Studies, Structure and Bonding*, vol. **147** (ed. D. Stalke), Springer, Berlin, pp. 193–285.

67 Baranov, A., Kohout, M., Wagner, F.R., Grin, Yu., Kniep, R., and Bronger, W. (2008) On the volume chemistry of solid compounds: the legacy of Wilhelm Biltz. *Z. Anorg. Allg. Chem.*, **634**, 2747–2753.

68 Biltz, W. (1934) *Raumchemie Der Festen Stoffe*, Verlag Leopold Voss, Leipzig, Germany.

69 Pendás, A.M., Costales, A., Blanco, M.A., Recio, J.M., and Luaña, V. (2000) Local compressibilities in crystals. *Phys. Rev. B*, **62**, 13970–13978.

70 (a) Fradera, X., Austen, M., and Bader, R.F.W. (1999) The Lewis model and beyond. *J. Phys. Chem. A*, **103**, 304–314. (b) Angyan, J.G., Loos, M., and Mayer, I. (1994) Covalent bond orders and atomic valence indices in the topological theory of atoms in molecules. *J. Phys. Chem.*, **98**, 5244–5248. (c) Fradera, X., Poater, J., Simon, S., Duran, M., and Sola, M. (2002) Electron-pairing analysis from localization and delocalization indices in the framework of the atoms-in-molecules theory. *Theor. Chem. Acc.*, **108**, 214–224. (d) Bader, R.W.F. and Stephens, M.E. (1975) Spatial localization of the electronic pair and number distributions in molecules. *J. Am. Chem. Soc.*, **97**, 7391–7399.

71 (a) Wiberg, K.B. (1968) Application of the Pople–Santry–Segal CNDO method to the cyclopropylcarbinyl and cyclobutyl cation and to bicyclobutane. *Tetrahedron*, **24**, 1083–1096. (b) Giambiagi, M., de Giambiagi, M.S., Grempel, D.R., and Heymann, C.D. (1975) Definition of AVE bond index with monorthogonal basis: properties and applications. *J. Chim. Phys.*, **72**, 15–22. (c) Mayer, I. (1983) Charge, bond order and valence in the *ab initio* SCF theory. *Chem. Phys. Lett.*, **97**, 270–274.

72 Bochicchio, R., Ponec, R., Torre, A., and Lain, L. (2001) Multicenter bonding within the AIM theory. *Theor. Chem. Acc.*, **105**, 292–298.

73 Börrnert, C., Grin, Yu., and Wagner, F.R. (2013) Position–space bonding indicators for hexaborides of alkali, alkaline-earth, and rare-earth metals in comparison to the molecular crystal $K_2[B_6H_6]$. *Z. Anorg. Allg. Chem.*, **639**, 2013–2024.

74 Grin, Yu., Savin, A., and Silvi, B. (2014) The ELF perspective of chemical bonding, in *The Chemical Bond: Fundamental Aspects of Chemical Bonding*, 1st edn (eds G. Frenking and S. Shaik), Wiley-VCH Verlag GmbH, Weinheim, Germany, pp. 345–405.

75 (a) Becke, A.D. and Edgecombe, K.E. (1990) A simple measure of electron localization in atomic and molecular systems. *J. Chem. Phys.*, **92**, 5397–5403. (b) Deb, B.M. and Gosh, S.K. (1983) New method for the direct calculation of electron density in many-electron systems: I. Application to closed-shell atoms. *Int. J. Quantum Chem.*, **23**, 1–26.

76 Kohout, M. (2015) Electron pairs in position space. *Struct. Bond*, **170**, 119–168

77 Pendas, A.M., Kohout, M., Blanco, M.A., and Francisco, E. (2012) Beyond standard charge density topological analyses, in *Modern Charge Density Analysis* (eds C. Gatti and P. Macchi), Springer, Dordrecht, The Netherlands, pp. 303–358.

78 Wagner, F.R., Bezugly, V., Kohout, M., and Grin, Yu. (2007) Charge decomposition analysis of the electron localizability indicator: a bridge between the orbital and direct space representation of the chemical bond. *Chem. Eur. J.*, **13**, 5724–5741.

79 Wagner, F.R., Kohout, M., and Grin, Yu. (2008) Direct space decomposition of ELI-D: interplay of charge density and pair-volume function for different bonding situations. *J. Phys. Chem. A*, **112**, 9814–9828.

80 (a) Kohout, M. and Savin, A. (1996) Atomic shell structure and electron numbers. *Int. J. Quantum Chem.*, **60**, 875–882. (b) Baranov, A.I. (2014) Electron localizability indicators from spinor wavefunctions. *J. Comput. Chem.*, **35**, 565–585.

81 Kohout, M., Wagner, F.R., and Grin, Yu. (2006) Atomic shells from the electron localizability in momentum space. *Int. J. Quantum Chem.*, **106**, 1499–1507.

82 Raub, S. and Jansen, G. (2001) A quantitative measure of bond polarity from the electron localization function and the theory of atoms in molecules. *Theor. Chem. Acc.*, **106**, 223–232.

83 (a) Veremchuk, I., Mori, T., Prots, Yu., Leithe-Jasper, A., Kohout, M., and Grin, Yu. (2008) Synthesis, chemical bonding and physical properties of *RE*RhB$_4$ (*RE* = Y, Dy-Lu). *J. Solid State Chem.*, **181**, 1983–1991. (b) Höhn, P., Agrestini, S., Baranov, A., Hoffmann, S., Kohout, M., Nitsche, F., Wagner, F.R., and Kniep, R. (2011) Sr$_2$Ni$_3$: a strontium subnickelide? *Chem. Eur. J.*, **17**, 3347–3351.

84 Silvi, B. (2002) The synaptic order: a key concept to understand multicenter bonding. *J. Mol. Struct.*, **614**, 3–10.

85 (a) Bende, D., Wagner, F.R., and Grin, Yu. (2015) 8-*N* rule and chemical bonding in main-group MgAgAs-type compounds. *Inorg. Chem.*, **54**, 3970–3978. (b) Wagner, F.R., Bende, D., and Grin, Yu. (2016) *Dalton Trans.* doi: 10.1039/c5dt04140f

86 (a) Pendas, A.M., Blanco, M.A., and Francisco, E. (2009) Steric repulsions, rotation barriers, and stereoelectronic effects: a real space perspective. *J. Comput. Chem.*, **30**, 98–109. (b) Tiana, D., Francisco, E., Blanco, M.A., Macchi, P., Sironi, A., and Pendas, A.M. (2010) Bonding in classical and nonclassical transition metal carbonyls: the interacting quantum atoms perspective. *J. Chem. Theory Comput.*, **6**, 1064–1074.

87 (a) Blaha, P., Schwarz, K., Madsen, G.K.H., Kvasnicka, D., and Luitz, J. (2001) *WIEN2K, an Augmented Plane Wave + Local Orbitals Program for Calculating Crystal Properties*, Techn. Universität Wien, Austria; (b) Kresse, G. and Furthmüller, J. (1996) Efficiency of *ab-initio* total energy calculations for metals and semiconductors using a plane-wave basis set. *J. Comput. Mater. Sci.*, **6**, 15 (*Phys. Rev. B* **1996**, *54*, 11169). (c) Krier, G., Jepsen, O., Burkhardt, A., and Andersen, O.K. (2000) The TB-LMTO-ASA Program, version 4.7. (d) Andersen, O.K. and Jepsen, O. (1984) Explicit first-principles tight-binding theory. *Phys. Rev. Lett.*, **53**, 2571–2574. (e) Ebert, H., Ködderitzsch, D., and Minár, J. (2011) Calculating condensed matter properties using the KKR-Green's function method: recent developments and applications. *Rep. Prog. Phys.*, **74**, 096501.

88 Parthé, E. (1964) *Crystal Chemistry of Tetrahedral Structures*, Gordon and Breach, New York.

89 (a) (1996) *Chemistry, Structure, and Bonding of Zintl Phases and Ions* (ed. S.M. Kauzlarich), Wiley-VCH Verlag GmbH, Weinheim, Germany. (b) Nesper, R. (2014) The Zintl–Klemm concept: a historical survey. *Z. Anorg. Allg. Chem.*, **640**, 2639–2648.

90 (a) Bende, D., Wagner, F.R., and Grin, Y. (2015) 8−*N* rule and chemical bonding in main-group MgAgAs-type compounds. *Inorg. Chem.*, **54**, 3970–3978. (b) Bende, D., Grin, Y., and Wagner, F.R. (2014) Covalence and ionicity in MgAgAs-type compounds. *Chem. Eur. J.*, **20**, 1–8.

91 (a) Miller, G.J., Lee, C.-S., and Choe, W. (2002) Structure and bonding around the Zintl border, in *Inorganic Chemistry Highlights* (eds G. Meyer, D. Naumann, and L. Wesemann), Wiley-VCH Verlag GmbH, Weinheim, Germany, pp. 21–53. (b) Miller, G.J., Schmidt, M.W., Wang, F., and You, T.-S. (2011) Quantitative advances in the Zintl–Klemm formalism, in *Zintl Phases: Principles and Recent Developments*, *Structure and Bonding* (ed. T. Fässler), **139**, Springer, Berlin, pp. 1–55.

92 Wang, F. and Miller, G.J. (2011) Revisiting the Zintl–Klemm concept: alkali metal trielides. *Inorg. Chem.*, **50**, 7625–7636.

93 Graf, T., Felser, C., and Parkin, S.S.P. (2011) Simple rules for the understanding of Heusler compounds. *Prog. Solid State Chem.*, **39**, 1–50.

94 Jung, D., Koo, H.-J., and Whangbo, M.-H. (2000) Study of the 18-electron band gap and ferromagnetism in semi-Heusler compounds by non-spin-polarized electronic band structure calculations. *J. Mol. Struct.*, **527**, 113–119.

95 Felser, C., Fecher, G., and Balke, B. (2007) Spintronics: a challenge for materials science and solid-state chemistry. *Angew. Chem., Int. Ed.*, **46**, 668–699.

96 Köhler, J., Deng, S., Lee, C., and Whangbo, M.-H. (2007) On the origin of a band gap in compounds of diamond-like structures. *Inorg. Chem.*, **46**, 1957–1959.

97 (a) Wade, K. (1976) Structural and bonding patterns in cluster chemistry. *Adv. Inorg. Chem. Radiochem.*, **18**, 1–66. (b) Mingos, D.M.P. (1984) Polyhedral skeletal electron pair approach. *Acc. Chem. Res.*, **17**, 311–319.

98 Sevov, C.S. and Corbett, J.D. (1993) Synthesis, characterization, and bonding of indium clusters. Rb_2In_3, a Zintl phase with layers of *closo*-indium octahedra. *Z. Anorg. Allg. Chem.*, **619**, 128–132.

99 Condron, C.L., Miller, G.J., Strand, J.D., Bud'ko, S.L., and Canfield, P.C. (2003) A new look at bonding in trialuminides: reinvestigation of $TaAl_3$. *Inorg. Chem.*, **42**, 8371–8376.

100 Jemmis, E.D., Balakrishnarajan, M.M., and Pancharatna, P.D. (2002) Electronic requirements for macropolyhedral boranes. *Chem. Rev.*, **102**, 93–144.

101 Teo, B.K. and Zhang, H. (1990) Counting electrons and rationalizing structures of clusters and clusters of clusters. *Polyhedron*, **9**, 1985–1999.

102 (a) Schäfer, H. and Schnering, H.G. (1964) Metall-metall-bindungen bei niederen halogeniden, oxyden und oxydhalogeniden schwerer übergangsmetalle thermochemische und strukturelle prinzipien. *Angew. Chem.*, **76**, 833–849. (b) Simon, A. (1988) Clusters of valence electron poor metals: structure, bonding, and properties. *Angew. Chem., Int. Ed. Engl.*, **27**, 159–183. (c) Simon, A. (1994) Discrete and condensed transition metal clusters in solids, in *Clusters and Colloids: From Theory to Applications* (ed. G. Schmid), Wiley-VCH Verlag GmbH, Weinheim, Germany, pp. 371–458.

103 Cotton, F.A. and Haas, T.E. (1964) A molecular orbitals treatment of the bonding in certain metal atom clusters. *Inorg. Chem.*, **3**, 10–17.

104 (a) Simon, A., Mattausch, Hj., Miller, G.J., Bauhofer, W., and Kremer, R.K. (1991) Metal-rich halides of the rare earth elements: structure, bonding, and properties, in *Handbook on the Physics and Chemistry of Rare Earths* (eds K.A. Gschneidner and L. Eyring), North Holland, Amsterdam, pp. 191–285. (b) Corbett, J.D. (1992) Coordination chemistry in the solid state: cluster and condensed cluster halides of the early transition metals. *Pure Appl. Chem.*, **64**, 1395–1408.

105 Meyer, G. (2008) Cluster complexes as *anti-Werner* complexes. *Z. Anorg. Allg. Chem.*, **634**, 2729–2736.

106 (a) Simon, A., Schnering, H.G., and Schäfer, H. (1967) Beiträge zur chemie der elemente niob und tantal. LXIV. Nb_6J_{11} – eine verbindung mit $[Nb_6J_8]$-Gruppen. *Z. Anorg. Allg. Chem.*, **355**, 295–310. (b) Fitch, A.N., Barrett, S.A., Fender, B.E.F., and Simon, A. (1984) Nb_6I_{11}, and DNb_6I_{11}, and HNb_6I_{11}: a powder neutron diffraction and inelastic scattering study. *J. Chem. Soc., Dalton Trans.*, 501–505.

107 (a) Yvon, K. (1979) Bonding and relationships between structure and physical properties, in *Chevrel-Phase Compounds $M_xMo_6X_8$ (M = Metal, X = S, Se, Te)* (ed. E. Kaldis), vol. **3**, North-Holland Publishing Co., Amsterdam, pp. 53–129. (b) Chevrel, R., Hirrien, M., and Sergent, M. (1986) Superconducting Chevrel phases: prospects and perspectives. *Polyhedron*, **5**, 87–94.

108 Johnston, R.L. (2002) *Atomic and Molecular Clusters*, Taylor & Francis, London.

109 Makongo, J.P.A., Prots, Y., Burkhardt, U., Niewa, R., Kudla, C., and Kreiner, G. (2006) A case study of complex metallic alloy phases: structure and disorder phenomena of Mg-Pd compounds. *Philos. Mag. A*, **86**, 427–433.

110 Hume-Rothery, W. and Raynor, G.V. (1962) *The Structure of Metals and Alloys*, Institute of Metals, London.

111 Mizutani, U. (2011) *Hume-Rothery Rules for Structurally Complex Alloy Phases*, CRC Press, New York.

112 Hoistad, L.M. and Lee, S. (1991) The Hume–Rothery electron concentration rules and second moment scaling. *J. Am. Chem. Soc.*, **113**, 8216–8220.

113 Lee, S. (1991) Second-moment scaling and covalent crystal structures. *Acc. Chem. Res.*, **24**, 249–254.

114 Gourdon, O., Gout, D., Williams, D.J., Proffen, T., Hobbs, S., and Miller, G.J. (2007) Atomic distributions in the γ-brass structure of the Cu-Zn system: a structural and theoretical study. *Inorg. Chem.*, **46**, 251–260.

115 Cressoni, J.C. and Pettifor, D.G. (1991) Theory of structural trends with the sp bonded elements. *J. Phys. Condens. Matter*, **3**, 495–511.

116 (a) Burdett, J.K. and Mitchell, J.F. (1995) Nonstoichiometry in early transition metal compounds with the rock salt structure. *Prog. Solid State Chem.*, **23**, 131–170. (b) Burdett, J.K. and Mitchell, J.F. (1993) Origins of non-stoichiometry in early-transition-metal chalcogenides. *Chem. Mater.*, **5**, 1465–1473.

117 Burdett, J.K. and Hughbanks, T. (1984) NbO and TiO: a study of the structural and electronic stability of structures derived from rocksalt. *J. Am. Chem. Soc.*, **106**, 3101–3113.

118 Lin, J. and Miller, G.J. (1993) Dimensional diversity in transition metal trihalides. *Inorg. Chem.*, **32**, 1476–1487.

119 Neckel, A., Rastl, P., Eibler, R., Weinberger, P., and Schwarz, K. (1976) Results of self-consistent band-structure calculations for ScN, ScO, TiC, TiN, TiO, VC, VN, and VO. *J. Phys. C*, **9**, 579–592.

120 Whangbo, M.-H. (1979) Mott–Hubbard condition for electron localization in the Hartree-Fock band theory. *J. Chem. Phys.*, **70**, 4963–4966.

121 Landrun, G.A. and Dronskowski, R. (1999) Ferromagnetism in transition metals: a chemical bonding approach. *Angew. Chem., Int. Ed.*, **38**, 1389–1393.

122 Häussermann, U. and Simak, S.I. (2001) Origin of the *c*/*a* variation in hexagonal close-packed divalent metals. *Phys. Rev. B*, **64**, 245114.

123 Oganov, A.R., Ma, Y., Xu, Y., Errea, I., Bergara, A., and Lyakhov, A.O. (2010) Exotic behavior and crystal structures of calcium under pressure. *Proc. Natl. Acad. Sci. USA*, **107**, 7646–7651.

124 (a) Wang, G.M., Papaconstantopoulos, D.A., and Blaisten-Borojas, E. (2003) Pressure induced transitions in calcium: a tight-binding approach. *J. Phys. Chem. Sol.*, **64**, 185–192 (b) Brennan, T.D. and Burdett, J.K. (1993) Electronic structure of elemental calcium and zinc. *Inorg. Chem.*, **32**, 746–749.

125 Ackland, G.J. and Macleod, I.R. (2004) Origin of the complex crystal structures of elements at intermediate pressure. *New J. Phys.*, **6**, 138–146.

126 Parker, L.J., Atou, T., and Badding, J.V. (1996) Transition element-like chemistry for potassium under pressure. *Science*, **273**, 95–97.

127 Feng, J., Hennig, R.G., Ashcroft, N.W., and Hoffmann, R. (2008) Emergent reduction of electronic state dimensionality in dense ordered Li-Be alloys. *Nature*, **451**, 445–448.

128 Zhang, W., Oganov, A.R., Goncharov, A.F., Zhu, Q., Boulfelfel, S.E., Lyakhov, A.O., Stavrou, E., Somayazulu, M., Prakapenka, V.B., and Konôpkova, Z. (2013) Unexpected stable stoichiometries of sodium chlorides. *Science*, **342**, 1502–1505.

129 Drozdov, A.P., Eremets, M.I., Troyan, I.A., Ksenofontov, V., and Shylin, S.I. (2015) Conventional superconductivity at 203 Kelvin at high pressures in the sulfur-hydride system. *Nature*, **525**, 73–76.

130 Gordon, E.E., Xu, K., Xiang, H., Bussmann-Holder, A., Kremer, R.K., Simon, A., Köhler, J., and Whangbo, M.-H. (2016) Structure and composition of the 200 K-superconducting phase of H_2S at ultrahigh pressure: the perovskite $(SH^-)(H_3S^+)$. *Angew. Chem., Int. Ed.*, **55**, 3682–3684.

131 Deng, S., Simon, A., and Köhler, J. (2003) Superconductivity in MgB_2: a case study of the 'Flat Band-Steep Band' scenario. *J. Supercond.*, **16**, 477–481.

132 (a) Booth, G.H., Grüneis, A., Kresse, G., and Alavi, A. (2013) Towards an exact description of electronic wavefunctions in real solids. *Nature*, **493**, 365–370. (b) Shinohara, Y., Sharma, S., Dewhurst, J.K., Shallcross, S., Lathiotakis, N.N., and Gross, E.K.U. (2015) Doping induced meta-insulator phase transition in NiO: a reduced density matrix functional theory perspective. *New J. Phys.*, **17**, 093038-1-5.

133 Kleis, J., Greeley, J., Romero, N.A., Morozov, V.A., Falsig, H., Larsen, A.H., Lu, J., Mortensen, J.J., Dułak, M., Thygesen, K.S., Nørskov, J.K., and Jacobsen, K.W. (2011) Finite size effects in chemical bonding: from small clusters to solids. *Catal. Lett.*, **141**, 1067–1071.

13
Lattice Dynamics and Thermochemistry of Solid-State Materials from First-Principles Quantum-Chemical Calculations

Ralf Peter Stoffel and Richard Dronskowski

RWTH Aachen University, Chair of Solid-State and Quantum Chemistry, Institute of Inorganic Chemistry, Landoltweg 1, D-52056 Aachen, Germany

In this contribution, we first recall some basic pillars of classical thermodynamics, followed by an introduction to electronic-structure theory and first-principles lattice dynamics as a consequence of a well-defined electronic ground state. Then, these concepts are brought together to eventually yield thermodynamic state functions without any empirical bias, except from fundamental physical constants. Finally, the chapter closes with some chemistry-typical applications of the methods introduced before. Among these are theoretical vibrational spectroscopy, energetic polymorphism questions in terms of pressure and temperature, structural instabilities, anisotropic displacement parameters, free reaction energies, heat-capacity questions, and a critical comparison of the formalisms entering the electronic energies.

13.1
Classical Thermodynamics

Even in the modern, digital world of our twenty-first century, which considers itself radically enlightened, all of us are still dwarfs standing on the shoulders of giants, at least in scientific terms. And this holds especially true for thermochemistry [1], one of the pillars of our modern world because virtually everything related to changing matter (and that's what chemistry and, in particular, solid-state chemistry is all about) into other matter by energetic means is unthinkable without a solid thermochemical background. Interestingly enough, the fundamentals of thermodynamics and thermochemistry were laid in the nineteenth century already, the pre-Nobel era, and important names such as Carnot, Mayer, Joule, Clausius, Ostwald, and many others immediately come to mind [2–6]; all of that is covered in many textbooks of physical chemistry [7,8]. The final breakthrough came when science giants such as von Helmholtz [9] and Gibbs [10] eventually cast thermodynamics and thermochemistry into our

Handbook of Solid State Chemistry, First Edition. Edited by Richard Dronskowski, Shinichi Kikkawa, and Andreas Stein.
© 2017 Wiley-VCH Verlag GmbH & Co. KGaA. Published 2017 by Wiley-VCH Verlag GmbH & Co. KGaA.

present-day shape by a formulation that utilizes thermodynamic *potentials* such that we dwarfs may profit from that. The most fundamental thermodynamic quantity is given by the Gibbs free energy G, from which everything may be derived. Very often, G is expressed in the form of the Gibbs–Helmholtz equation in which the enthalpy H and the entropy S yield G at a given thermodynamic temperature T

$$G(T) = H(T) - T \cdot S(T).$$

In case our civilization fails and eventually blows up, virtually everything may be re-constructed (at least mathematically) from this very equation, at least if G or H and S have been tabulated and stored in a safe place. What a terrific intellectual framework that is, and it is *not* quantum mechanical in nature!

Having the Gibbs free energy at hand, related to a standardized zero point (for whatever pressure or temperature that looks convenient for us), the energy differences ΔG, which are also related to the chemical equilibrium constant K and the gas constant R, give access to phase stability

$$\Delta G(T) = \Delta H(T) - T \cdot \Delta S(T) = -R \cdot T \times \ln K.$$

By means of simple convention, the standardized zero point is set to the state of the elements at ambient conditions (say, room temperature and a pressure of 1 atm, for obvious reasons), which immediately translates into *gaseous* hydrogen, *liquid* mercury, or *solid* face-centered cubic copper. Needless to say, the same approach is extended from the elements to more complex chemical compounds that gives access to standard enthalpies or Gibbs energies of chemical reactions, that is, *formation* enthalpies or Gibbs energies. We also note that standardizing the pressure (but not the volume) simply reflects the typical conditions of an experimentalists in his or her laboratory, another consequence of the real world.

For obvious reasons, however, solid-state quantum-chemical *simulations* by means of a computer are conveniently carried out at constant *volume* since the unit cell does not change upon calculation (see below). Hence, it is important to reformulate the pressure-dependent Gibbs free energy by the constant-volume Helmholtz free energy A using pressure and volume

$$G(p, T) = A(T, V) + p(V) \cdot V.$$

Fortunately, the pressure can easily be derived from the volume-dependence of the Helmholtz free energy,

$$p(V) = -\frac{dA(V)}{dV},$$

as will be discussed in more detail below in Section 13.4.

As a rough approximation, the temperature dependence of the Gibbs energy can be approximated with satisfactory precision simply by taking the entropy and enthalpy as temperature independent and scaling the latter by the thermodynamic temperature. Nonetheless, the exact temperature dependence of G is determined by the heat capacity C_p because the temperature dependencies of the entropy and enthalpy are given by

$$S(T) = S_0 + \int_{T_0}^{T} \frac{C_p(T)}{T} dT \quad \text{and} \quad H(T) = H_0 + \int_{T_0}^{T} C_p(T) dT,$$

where S_0 and H_0 are standard entropies and enthalpies, respectively, related to the temperature T_0. Thus, for an accurate thermodynamic treatment, the knowledge of the temperature-dependent heat capacity C_p is mandatory. By far, the largest contribution to the heat capacity goes back to the vibrational properties of a crystal that are often modeled (at least in introductory textbooks of physical chemistry) by the famous Einstein [11] and Debye [12] models as the most popular approximations, respectively; here, the quantum-mechanical nature of macroscopic matter already pops up, opposed to the classical Dulong–Petit rule ($C_p = 3R$). While in the Einstein model only a single frequency per lattice site (=atom) is assumed, the Debye model comes up with an approximate function for the density of vibrational (=phonon) states. The real density of phonon states is accessible, for example, via inelastic neutron scattering, and it can also be obtained from quantum-chemical calculations as will be discussed below. Figure 13.1 shows a comparison between the Einstein and Debye models of the density of phonon states of face-centered cubic copper, together with a third one that has been calculated *ab initio*.

Let us bravely assume (correctly so, not an exceedingly bold claim) that accurately calculating the phonons of a solid-state material is possible, this will immediately pave the way toward correctly modeling its entire thermochemistry. If that material exists, one may probe the experimental data or the calculation itself in case the experiment is perfectly accurate. If the material is still nonexistent, one may think of *predicting* its thermodynamic properties ahead of its synthesis, a computational strategy of paramount importance at the beginning of the twenty-first century. The next sections will introduce the reader to the way this is done.

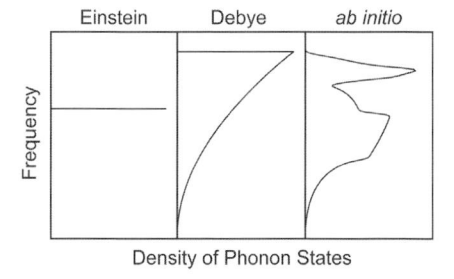

Figure 13.1 Schematic density of phonon states of fcc copper as approximated by the Einstein and Debye models together with the one resulting from *ab initio* calculations. (Reproduced with permission from Ref. [13]. Copyright 2010, John Wiley & Sons.)

13.2
Electronic Structure and Electronic Energy

Let us imagine a simple (or complex) crystalline solid-state material. In case we know the chemical composition (which atoms are involved?), the unit cell, its shape and symmetry (which metric? which space group?), and the position of all the atoms in three-dimensional space (how does the crystal structure look like?), we may try to understand its very existence and structure. To do so, an electronic-structure calculation by whatever method [14] is the perfect starting point, as briefly detailed below and more explicitly in the chapters by Springborg (Chapter 1), Bredow and Jug (Chapter 6), Seifert (Chapter 7), Schwarz (Chapter 8), and others. This mere calculation will yield the electronic ground-state energy E_{el}, an enormously important piece of information. Why?

The electronic ground-state energy E_{el} is the basis of all that follows, and at the same time it represents the largest energetic contribution within theoretical considerations of thermodynamic stability. E_{el} is so large that, in fact, simple differences in E_{el} are excellent approximations for internal energy differences without considering any temperature, hence $\Delta E_{el} \approx \Delta U$ (0 K). Not too long ago, when computer time was more precious than today, different materials or structures were compared based on the size of their corresponding electronic energies *only*. In the twenty-first century, however, we may also start from the volume-dependence of E_{el} and then derive the pressure-dependent zero-temperature enthalpy $H_{el}(p)$ as follows:

$$H_{el}(p) = E_{el}(V(p)) + pV(p) = E_{el}(V(p)) - \frac{dE_{el}(V(p))}{dV}V(p).$$

In computational practice, the pressure is determined by fitting the theoretically generated energy versus volume (*E*–*V*) data to a convenient equation of states (EOS) such as the one by Murnaghan [15], Birch [16], or Vinet et al. [17]. Such classical EOSs have been originally developed by geoscientists such as to describe the pressure–volume relationship of solid-state materials, for example, natural materials such as minerals. Integration of $p(V)$ gives a straightforward expression for $E(V)$.

As alluded to already, the most commonly used tool for the quantum-chemical calculation of a solid-state material's electronic energy E_{el} is given by density-functional theory (DFT) [18,19] in its various flavors, probably the most successful effective one-particle method ever invented. Here, the electronic energy is obtained as a functional of the electron density $\rho(\mathbf{r})$, which itself is a function of position $\mathbf{r}$. It can be separated into the kinetic energy T_0 of noninteraction pseudoelectrons, the attractive core-electron potential V, the repulsive coulomb (electron–electron) energy J and the energy contribution (a correction term, i.e., to account explicitly for the interactions of the electrons) which goes back to exchange and correlation, E_{XC}; all those terms are functionals of the electron density

$$E_{el} = E(\rho(\mathbf{r})) = T_0(\rho(\mathbf{r})) + V(\rho(\mathbf{r})) + J(\rho(\mathbf{r})) + E_{XC}(\rho(\mathbf{r})).$$

If the electron density $\rho(\mathbf{r})$ is known, the first three components can – in principle – be calculated exactly whereas E_{XC} has to be approximated. This critical approximation is carried out by using so-called exchange-correlation (XC) functionals, parametrized functions of the electron density. Commonly used XC functionals have been based on the local-density approximation (LDA) to start with Ref. [20–22] and the generalized gradient approximation (GGA) in the sequel Ref. [23–25], respectively. Nowadays, improved XC methods are available, such as hybrid [26] or meta-GGA [27] functionals, involving exact exchange or better-than-gradient terms. Also, van der Waals interactions by means of "dispersion corrections" have been incorporated into the framework of density-functional theory as has been extensively reviewed in Ref. [28].

The correct (or, better, fortunate) choice of the XC functional that depends on the class of materials under investigations (there goes first principles) does not only influence the electronic structure but also all other properties which are rooted on first-principles calculations, such as the vibrational and thermodynamic properties. Below we will also give some examples regarding this issue.

Before we do so, we reiterate that three-dimensional periodic systems (in other words, crystalline materials from an atomistic point of view) are most conveniently treated in the framework of Bloch's theorem [29] using periodic boundary conditions. In a periodic lattice with period $\mathbf{R}$, the periodic potential is defined as

$$V(\mathbf{r} + \mathbf{R}) = V(\mathbf{r}).$$

Bloch's theorem then ensures that the electronic wave function differs by a prefactor that depends on the electronic wave vector $\mathbf{k}$

$$\psi(\mathbf{r} + \mathbf{R}) = \exp(i\mathbf{k}\mathbf{R})\psi(\mathbf{r}).$$

The values of $\mathbf{k}$ are limited to the first Brillouin zone (the unit cell of reciprocal space) because larger values do not give any new information [30]. This circumstance is later discussed more in depth when introducing the phonon Brillouin zone. We emphasize that their mathematical apparatus is identical although the electronic and vibrational Brillouin zones describe something completely different in terms of their physical nature. Again, we refer to the chapter by Schwarz for a more detailed description of electronic-structure calculations of solid-state materials within this volume.

13.3 Lattice Dynamics

Atoms in crystalline materials, no matter if the bonding situation is described as "metallic," "covalent," or "ionic," vibrate around their equilibrium position already at 0 K, simply because of Heisenberg's famous uncertainty principle [31], and the amount of vibration increases with the thermodynamic temperature. Because the atoms are part of an incredibly large collective called crystal, these *collective* vibrations are characterized by the phononic wave vector **q**, eigenvectors **e**, and frequency ν (or angular frequency ω). If we move from the wave to the particle description, the excitation of the respective vibrational modes is related to their occupation by quasiparticles, the so-called *phonons*. Quite obviously, the occupation of the states is temperature dependent and needs to be described by Bose–Einstein statistics as phonons are bosonic quasiparticles. For a more detailed overview, we refer the reader to textbooks dedicated to phonons and related properties [32–34].

The wave vector of the phonon carries information of the propagation direction and the wavelength. Let us now assume that there are N atoms in the primitive unit cell. Then, at each wave vector there are $3N$ vibrational modes – which might be energetically degenerated – separated into $2N$ transversal and N longitudinal modes. Likewise, one may also separate between acoustic and optical phonons if more than one atomic species is involved in the lattice vibration, and there are three acoustic and $3N-3$ optical modes. The difference between acoustic and optical phonons is given by the phase of the vibration. For an acoustic mode, all atoms have the same phase; optical modes, however, are characterized by a phase *shift* between different kinds of atoms. Given a chemical difference between two atoms involved in the optical mode, there will be a (small or large) charge transfer and a dipole moment as a consequence of the phase shift such that this mode can be accessed by electromagnetic radiation, hence its name (optical). Figure 13.2 illustrates this circumstance using the grossly simplistic model of a one-dimensional two-atoms chain.

The eigenvectors $\mathbf{e}(\mathbf{q},\nu)$ characterize the displacements of the atoms building up a corresponding phonon mode. These consist of $3N$ components because they describe the displacements along the x, y, and z directions for each of the N atoms in the unit cell. At the Γ point ($\mathbf{q}=0$), the eigenvectors are real under all circumstances, which means that they can be directly applied to the unit cell. At other **q**-points away from the Γ point, they contain imaginary parts as well that characterize the phase of the respective vibrational modes.

To proceed further, we now assume a harmonic model [32] in which we imagine various atoms being part of a crystal, and these atoms are thought to be

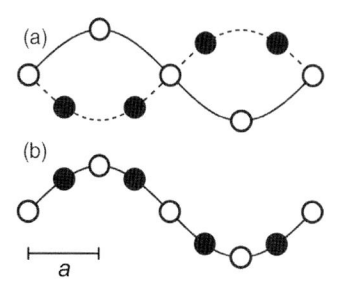

Figure 13.2 Optical (a) and acoustical (b) vibrational mode of a one-dimensional diatomic chain. (Figure taken from Ref. [35].)

connected with each other by virtual "springs" which we also assume to strictly obey Hooke's law. In addition, we stress that each atom is not only connected to its nearest neighbors but also to *every other* atom in the crystal (at least in principle; at certain distances, the force constants belonging to such two-atomic springs fall to zero, which allows for a practical application). Hence, the complexity of the ensemble of "springs" increases enormously and must be expressed by the so-called force constant matrix, $\Phi_{\alpha\beta ii'}$ [32]. Its elements are accessible from small displacements $d_{\beta i'}$ around the equilibrium positions of the atoms because these displacements create forces $F_{\alpha i}$ on all the other atoms

$$\Phi_{\alpha\beta ii'} = -\frac{F_{\alpha i}}{d_{\beta i'}}.$$

In the above equation, the indices α and β stand for the Cartesian directions x, y, and z while i and i' for the atoms involved. Hence, we witness the force of atom i along direction α that is introduced because atom i' is elongated by d along the direction β.

In the next step, the *dynamical matrix* $\mathbf{D}(\mathbf{q})$ connects the eigenvectors of a vibrational mode with its angular frequency ω:

$$\mathbf{D}(\mathbf{q})\mathbf{e}(\mathbf{q},\nu) = \omega^2(\mathbf{q},\nu)\mathbf{e}(\mathbf{q},\nu).$$

Its elements are given by

$$D_{\alpha\beta}(ii'\mathbf{q}) = \frac{1}{\sqrt{mm'}} \sum \Phi_{\alpha\beta}(ii') \exp(i\mathbf{q}[\mathbf{r}' - \mathbf{r}]),$$

where m and m' are the masses and $\mathbf{r}$ and $\mathbf{r}'$ are the positions of atoms i and i'.

The squared phonon frequencies ω^2 result from diagonalizing the dynamical matrix at each wave vector such as to solve the eigenvalue problem of the determinant

$$\left| D_{\alpha\beta}(ii'\mathbf{q}) - \delta_{\alpha\beta}\delta_{ii'}\omega^2(\mathbf{q}) \right| = 0,$$

with the Kronecker deltas $\delta_{\alpha\beta}$ and $\delta_{ii'}$. The phonon modes are usually given in units of wavenumbers (cm^{-1}), frequencies (THz), or energies (meV), respectively. Not too surprisingly, we arrive at vibrational energies which are *smaller by several orders of magnitude* than the electronic energies (eV); somewhat simplified, even in red-glowing iron with plenty of excited phonons the electrons still reside in the ground state. This is also the reason why we have been able to separate the nuclear degrees of freedom from the electronic ones at the very beginning without telling that to the reader.

Under normal circumstances, these ω^2 arrive at positive values, and the roots are also taken as positive frequencies; alternatively expressed, a dislocated atom will immediately swing back to its zero-force equilibrium position. If these ω^2 are negative, however, the frequencies will be imaginary, to be interpreted as an internal dynamical instability of the system under investigation because the structure is unstable against distortion along the eigenvectors of the respective imaginary frequencies such that a vibration destroys the structure; the atom never comes back to its spatial origin. Later, we will discuss the consequences of those instabilities in detail.

From what has been said in terms of the reciprocal-space approach of the crystalline state, it does not come as a surprise that the phonon frequencies – in perfect analogy to the electronic band structure – show a dispersion relation, the so-called *phonon dispersion* or *phonon band structure*. The latter is usually depicted as frequency versus wave vector along selected high-symmetry directions in reciprocal space, just like in the electronic case. Once again, we refer the reader to other chapters, say, the one by Springborg (Chapter 1).

As a simple example, let us now explain the phonon dispersion of a one-dimensional, infinitely long chain of hydrogen atoms connected by "springs" with the force constant f and bond distance a; both f and a result from the well-defined electronic ground state of that very one-dimensional hydrogen chain. The dependency of the frequency on the quantum number k (which is a one-dimensional wave vector) is given by the analytical expression

$$\omega = \sqrt{\frac{4f}{M}} \left| \sin \frac{1}{2} ka \right|,$$

with M being the mass of the respective atoms. All the resulting pairs of frequency ω and wave vector k are illustrated from the aforementioned phonon band structure as depicted in Figure 13.3. At the Γ point or zone center (wave vector = 0) all atoms move in the same direction and, therefore, the vibrational

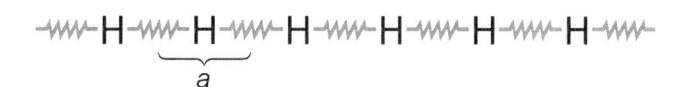

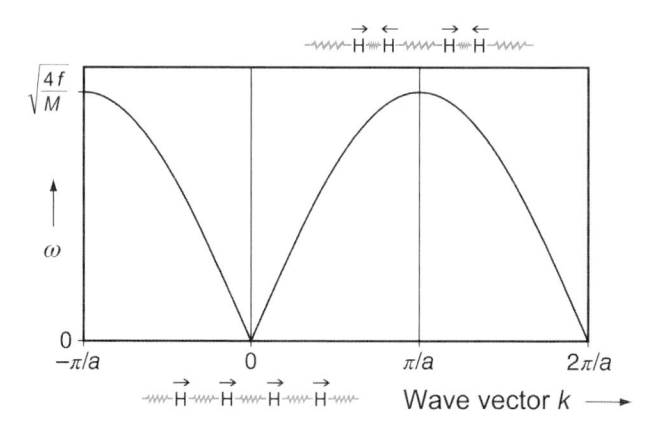

Figure 13.3 One-dimensional chain of hydrogen atoms connected by springs with force constant f and phonon dispersion relation of this chain along wave vector k. (Reproduced with permission from Ref. [13]. Copyright 2010, John Wiley & Sons.)

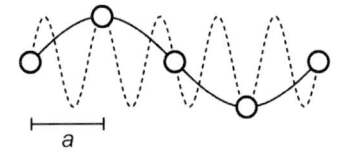

Figure 13.4 Vibrational mode of a one-dimensional monoatomic chain as described by two different wave vectors. The description using the larger wave vector (shorter wavelength; dashed line) does not give any new information. (Figure taken from Ref. [35].)

frequency equals zero. At the zone edge $k = \pi/a$ (or simply **X**), the highest frequency is obtained as ($\sqrt{4f/M}$), and the reason for the high energy is that neighboring H atoms vibrate against each other, thereby compressing the "spring." At this point, the zone boundary has been reached such that higher values of the wave vector do not lead to new vibrational modes. While the atomic motion can be described by different wave vectors with different wavelengths, larger wave vectors (shorter wavelengths) do not lead to any additional information, as illustrated in Figure 13.4. For that reason, only the knowledge of the vibrational modes within the first Brillouin zone is necessary, very much similar to what we know from the electronic structure.

Let us now turn to a slightly more chemical topic and study the product of a virtual reaction between two chemical species in one-dimensional space, namely, a one-dimensional infinite chain including two different atoms, lithium, Li and hydrogen, H. The treatment of such a periodic solid dubbed lithium hydride (with a charge transfer from Li to H) is not necessarily more complicated than that of the hydrogen chain, the main difference being the appearance of a so-called optical branch that does not start from the zero frequency at $k = 0$. This is a direct consequence of the unit cell including two atoms that can be phase-shifted at the Γ point. The phonon dispersion is depicted in Figure 13.5.

At Γ, the frequency of the optical branch is highest because Li and H are pushed against each other such that a lot of energy is needed to compress the Li–H "spring." At the zone edge, the lower-energy phonon corresponds to out-of-phase vibrating Li atoms while the H atoms are standing still; note that the atomic mass enters the denominator, hence the reciprocal relationship between frequency and mass. At higher energies, it is just the other way round such that only the H atoms vibrate against each other but the Li atoms are at complete rest.

We refrain from going to analytically solvable two- and three-dimensional examples but introduce, instead, the so-called *density of phonon states* (DPS) that gives an averaged picture (or just a counter) of the number of available states at each frequency ω.

$$\text{DPS}(\omega) = \sum_{\nu} g_{\nu}(\omega); \quad g_{\nu}(\omega) = \frac{Na}{\pi} \frac{d\mathbf{q}}{d\omega}.$$

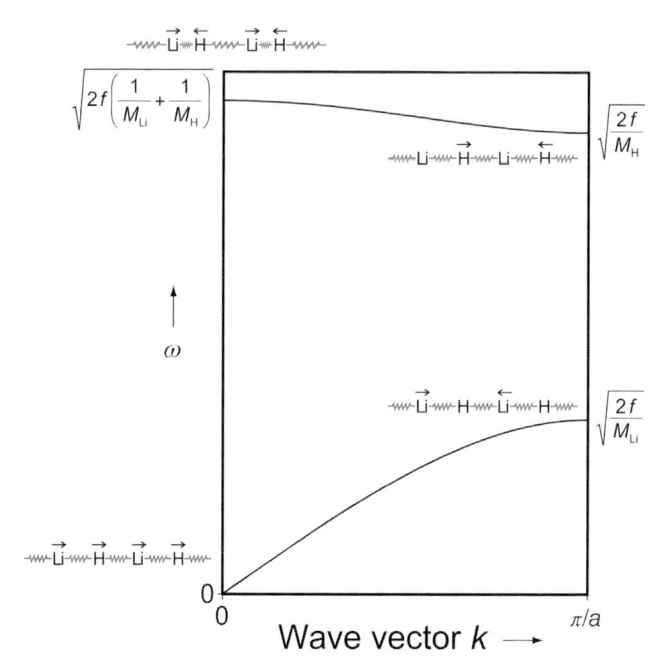

Figure 13.5 Phonon dispersion relation of a one-dimensional chain of lithium hydride along $0 < k < \pi/a$. (Reproduced with permission from Ref. [13]. Copyright 2010, John Wiley & Sons.)

Just like its electronic counterpart, the electronic density-of-states (DOS), the DPS is highly useful whenever we need to cover the entire reciprocal space for a real three-dimensional material. To do so, we offer a characteristic example of a more complex and truly three-dimensional compound, antimony telluride, Sb_2Te_3, whose phonon band structure (a true "spaghetti" structure due to its complexity) and the much simpler DPS (hence its usefulness) are shown in Figure 13.6. Needless to say, both result from sophisticated and numerical (instead of analytical) phonon computations based on density-functional electronic-structure theory and, in this mere case, the simple, yet transparent LDA. We will later come back to this compound and its enormous importance for nowadays technology and present several other applications. The band structure is depicted along several high-symmetry directions in the first Brillouin zone that are related to the hexagonal setup of space group $R\bar{3}m$. Also, long-range Coulomb interactions are included, nicely visible at the Γ point where the frequencies depend on the direction from which they reach this very point. For example,

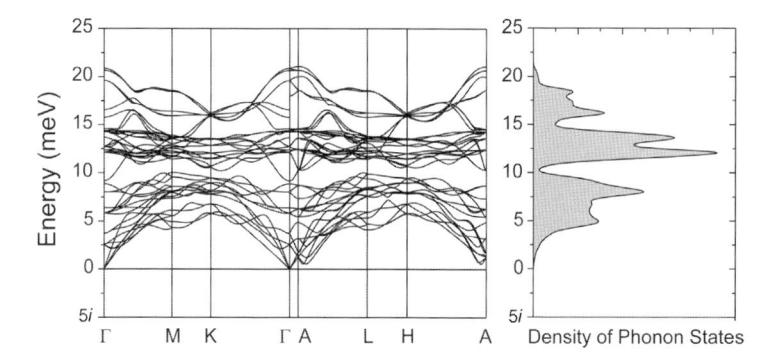

Figure 13.6 Phonon dispersion and density of phonon states of antimony telluride based on LDA electronic-structure calculations; all phonons are real, thereby reflecting the stability of the structure type. (Modified figure based on the one originally published in Ref. [36] with kind permission of IOP Publishing Ltd.)

this can be observed in the range between 15 and 20 meV; depending on the direction from which the bands arrive at Γ, say, from **K** and **A**, the frequencies at Γ differ significantly. Note also that for generating the DPS, the entire Brillouin zone has been sampled, including those points that do *not* lie on the high-symmetry directions shown within the band structure. Once again, this is known from electronic-structure theory just as well.

The simulation of phonons based on DFT can conveniently be carried out making use of the small-displacement method [37,38] in which all the atoms are dislocated just a little (about 0.01 Å, see above) to derive the forces and then the frequencies from the forces. Alternatively, this can be accomplished by density-functional perturbation theory [39,40]. We refrain from going into the details and refer to the original literature; instead, let us concentrate on the various applications.

13.4
Ab Initio Thermochemistry

Before going into the computational details, let's first discuss the sheer necessity of having such a theoretical tool dubbed *ab initio* thermochemistry. After all, whenever the solid-state compound is available and of reasonable interest, there will certainly be ingenious experimentalists aiming at measuring its thermochemical properties; why then do theory when good experimental data are at hand? First, not all measurements are necessarily accurate or easy to perform, for example, whenever phase-purity cannot be guaranteed. Second, we may know the Gibbs energy G of that exciting phase A at a pressure of 50 MPa and a temperature of 500 K; but what is its value at 10 GPa and 200 K? Redo the entire experiment? Before doing that, resynthesize A? And finally, how about synthesizing *hypothetical* compounds whose likewise hypothetical but reasonably accurate thermochemical data would be very much needed to arrive at more reasonable conditions for the synthesis? Clearly, there is an enormous potential in *ab initio* thermochemistry.

So let us proceed. Based on the laws of statistical thermodynamics, the microscopic (i.e., on the atomic scale) vibrational properties can indeed be expanded to the real macroscopic world. If we stick to the harmonic approximation for the vibrational part, the vibrational energy of any crystal is simply the sum of the energies of a set of noninteracting harmonic oscillators

$$E_{ph}(T) = \sum \left(n(T) + \frac{1}{2}\right) h\nu,$$

where n is the temperature-dependent occupation number of the phonon modes as given by Bose–Einstein statistics because phonons are bosonic quasiparticles, as said before. If we neglect (at least for the moment) other, even smaller energetic contributions, E_{ph} is the dominating temperature-dependent term of the internal energy U. Hence, the internal energy U is then given – in the simplified model presented here – as the sum of electronic ground-state energy and all vibrational energy contributions, both of which can be calculated *ab initio*.

$$U(V, T) = E_{el}(V, T = 0) + E_{ph}(V, T).$$

It is straightforward to obtain the constant-volume heat capacity C_V – the derivative of U with respect to the temperature – from the relation

$$C_V(T) = \left(\frac{dE_{ph}(T)}{dT}\right)_{V=\text{const.}}.$$

In addition, via the partition function, the harmonic phonon free energy A_{ph} can be expressed as

$$A_{ph}(T) = \frac{1}{2}\sum_{\mathbf{q},\nu} \hbar\omega(\mathbf{q},\nu) + k_B T \sum_{\mathbf{q},\nu} \ln\left[1 - \exp\left(\frac{-\hbar\omega(\mathbf{q},\nu)}{k_B T}\right)\right].$$

Clearly, this is the most important temperature-dependent contribution to the Helmholtz free energy. As can be seen from the last equation, only the knowledge of the phonon properties in the first Brillouin zone is necessary for its calculation, nothing else. Nonetheless, one may also proceed a bit more formally.

The full *ab initio* thermochemical treatment of any crystalline system starts with setting up an expression for the temperature-dependent constant-volume Helmholtz free energy. This very ansatz is a sum (not a Taylor series) of several independent contributions

$$A(T) = E_0(T = 0) \\ + A_{ph}(T) \left(+\Delta A_{el}(T) + T \cdot S_{conf} + A_{vib.\ ah}(T) + A_{mag}(T) + \ldots\right),$$

where the bracketed terms are rather small compared to the electronic and vibrational contributions as already discussed above. ΔA_{el} is the temperature-dependent electronic energy which goes back to the thermal excitation of electronic degrees of freedom, for example, those arising for the (almost) free electrons in metals. It can be safely neglected for insulators or semiconductors, and even metallic compounds only have a relatively small contribution. Nevertheless, if one were to calculate the phase transition from insulating or semiconducting phases to metallic ones or vice versa, including these temperature-dependent electronic contributions would certainly improve the numerical results. The configurational entropy, S_{conf}, plays an important role for systems with mixed occupation of atomic sites and can be included in a classical way. The explicit calculation of anharmonic vibrational and also magnetic contributions involves an enormous computational effort and will not be further discussed here; luckily, a significant part of the crystal's anharmonicity is already incorporated by the underlying electronic-structure theory (see below).

Given a sufficient quality of the Helmholtz free energy, the Gibbs free energy then simply results from the volume-dependent change of the Helmholtz free energy as

$$G(p, T) = A(T, V(p)) + p \cdot V(p) = A(T, V(p)) - \frac{dA(T, V(p))}{dV(p)} \cdot V(p).$$

We reiterate that accessing the constant-pressure Gibbs free energy directly from constant-volume electronic-structure theory is impossible, thus there must be a detour via the constant-volume Helmholtz free energy. In order to obtain the volume dependence of the constant-temperature (and constant-volume) Helmholtz free energy, the latter is calculated at several well-chosen volumes. These energy-volume data are then fitted to a parametrized equation of states. The fitting parameters at each temperature are the bulk modulus $B(T)$, its pressure-derivative $B'(T)$, the equilibrium Helmholtz energy $A_0(T)$, and the equilibrium volume $V_0(T)$. For each minimum of the A versus V curves (with the coordinates A_0 and V_0), the pressure is exactly zero ($p = -dA/dV = 0$), such that the value $A_0(T)$ in this minimum is equal to $G(T, p=0)$; we have eventually arrived at the *ab initio* Gibbs free energy. From this procedure, together with the pressure-dependent Gibbs free energy at each temperature, also the molar volume and the bulk modulus are obtained as functions of pressure. As a caveat, one needs to be careful and make sure that the pressure range is only valid within the range that was originally used for the fitting procedure. Figure 13.7 schematically shows the Helmholtz free energy as a function of the volume at several ascending temperatures. Interestingly enough (and somewhat surprising, but only at first sight), there is a considerable volume *expansion* while the temperature increases. This cannot be attributed to the harmonic approximation for the phonons; in harmonic theory, there is no such volume expansion at all up to *any* temperature (and there will never be melting). Nonetheless, the energetic part going back to electronic-structure theory assures that the underlying interatomic potentials are *per se* anharmonic because they have been calculated by, for example, density-functional theory. This lucky combination of anharmonic DFT

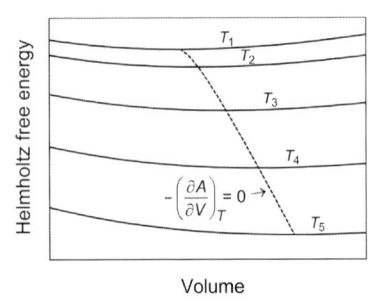

Figure 13.7 Schematic Helmholtz free energy A as a function of the volume V with increasing temperature ($T_1 < T_2 < \ldots < T_5$). The dashed line connects the minima of A. (Figure taken from Ref. [35].)

potentials plus harmonic phonons goes under the name *quasi-harmonic approximation*, and it does include anharmonic effects to some degree, such as the volume expansion. Other effects, such as heat conductivity, are not accounted for by this methodology.

The numerical ingredients *of ab initio* thermochemistry are practically complete. Once there is a reliable Gibbs energy function $G(p, T)$ from first principles, all other thermodynamic potentials such as the entropy S, the enthalpy H and the constant-pressure heat capacity C_p can be obtained from standard thermodynamic relations, such as:

$$S(T) = -\left(\frac{dG(T)}{dT}\right)_p, \quad H(T) = G(T) + TS(T), \quad \text{and} \quad C_p(T) = \left(\frac{dH(T)}{dT}\right)_p.$$

Figure 13.8 gives a schematic overview of the procedure of *ab initio* thermochemistry, beginning with the crystallographic input and finally leading to the Gibbs free energy.

13.5
Examples

13.5.1
Ag[Ag(C$_3$H$_3$)$_2$]: Infrared Spectroscopy

Without being too harsh with other methods of structural characterization, it is probably fair to say that most solid-state chemists more or less worship diffraction methods, in particular (single-crystal) X-ray crystallography, for very good reasons. Countless solid-state structures have already been clarified, and the atomistic accuracy in terms of bond lengths (usually better than 0.01 Å) and angles (often better than 0.05°) is nothing but spectacular and unmatched by any other method; indeed, solid-state chemists usually can live pretty well without spectroscopy, unlike their molecular colleagues.

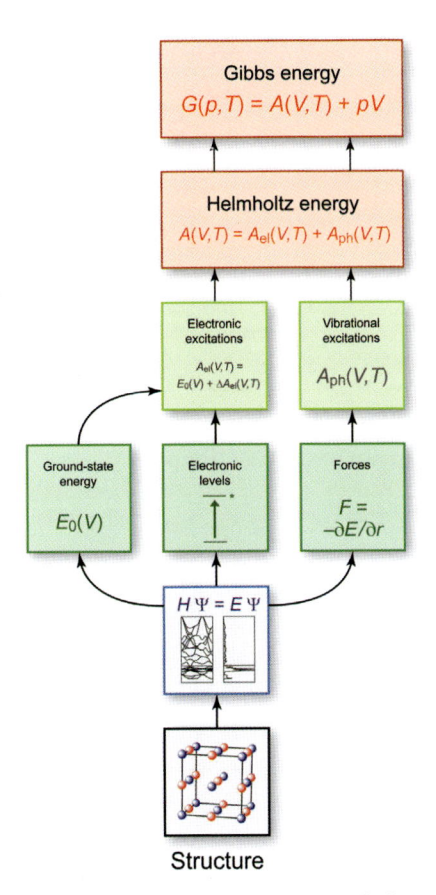

Figure 13.8 *ab initio* thermochemical calculation scheme finally leading to the Gibbs free energy. (Original figure from Ref. [41] reprinted with kind permission of Elsevier.)

On the other hand, there is the ubiquitous characterization method dubbed infrared (IR) spectroscopy, a technique that is widely used for molecules containing infrared-active vibrations. We will hence start with a first example in which infrared spectroscopy gives a helping hand to X-ray crystallography, a rare but nonetheless important experience. The crystal structure under investigation, namely, silver bis(propinyl)argentate, a somewhat weird metal alkynyl compound, contains silver cations and molecular anions. This class of solid-state compounds is not only difficult to make but also some of its representatives are also dangerously explosive. In addition, it is practically *impossible* to grow single crystals such that structure solution and refinement must be carried out using complex powder-X-ray-diffraction data. We show the eventually obtained structure model in Figure 13.9 a.

Puzzlingly enough, due to the large difference in the scattering power of the atoms the structure could *not* be solved by only relying on highest-resolution

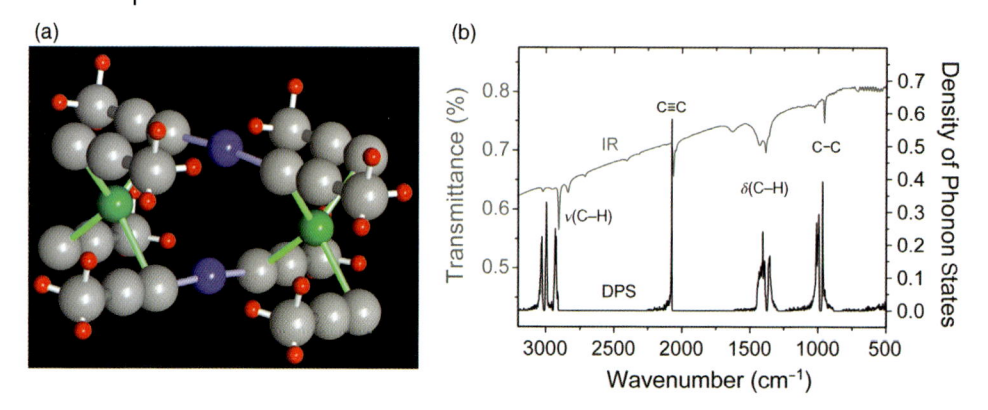

Figure 13.9 Crystal structure (a) and density of phonon states of Ag[Ag(C$_3$H$_3$)$_2$] (b) as compared to an experimentally measured infrared spectrum. (Reproduced with permission from Ref. [13]. Copyright 2010, John Wiley & Sons.)

(synchrotron) powder X-ray diffraction, simply because grossly different structural models (in centric and acentric space groups) led to identical residual values and standard deviations. Therefore, additionally infrared spectroscopy together with vibrational spectra calculated *ab initio* (alternatively expressed: the density of phonon states from first principles) were utilized. Figure 13.9b shows the comparison between experimental IR data and the DPS. Because this comparison results in good agreement between experiment and theory *only* for the acentric structure model (Figure 13.9a), the latter was fully verified. This structure contains two different silver cations, one (in blue) almost linearly coordinated by C$_3$H$_3$ anions while the other one (in green) was found to be coordinated by the C—C triple bond. For a more detailed description of the crystal structure and its solution, see Ref. [42]. The electronic reason for the found crystal structure is given by the Ag—C bonding because only in the acentric structure the amount of antibonding levels is significantly reduced. We also note that the DPS includes more vibrational modes than IR simply because not all the modes are infrared (but Raman) active. Nonetheless, by visually inspecting the theoretically calculated vibrational structure one may determine the nature of the vibrational modes because they are described by the respective phonon mode eigenvectors.

13.5.2
Sb$_2$Te$_3$: Partial Densities of Phonon States versus Nuclear Inelastic Scattering

As the next example, we present results of calculations on antimony telluride [36], Sb$_2$Te$_3$, a binary chalcogenide compound, which is of enormous scientific and also technological importance. Sb$_2$Te$_3$ is known to be a topological insulator [43,44] (that is, electrically conducting on the surface but insulating in the bulk) and has outstanding thermoelectric properties [45,46] as well. Also, it

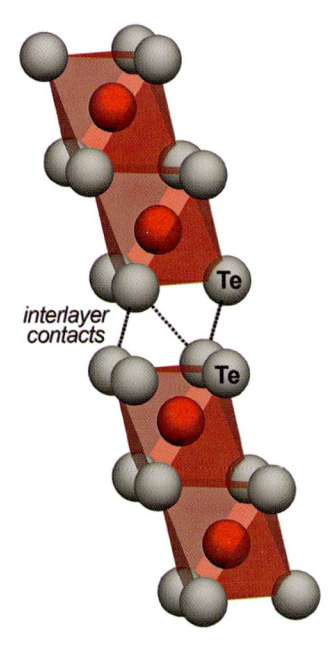

Figure 13.10 Detail of the crystal structure of antimony telluride. (Figure reprinted from Ref. [36] with kind permission of IOP Publishing Ltd.)

is structurally related to the Ge–Sb–Te phase-change materials [47] that are located on the pseudobinary tie-line GeTe-Sb_2Te_3 in the Ge–Sb–Te phase diagram. We will also come back to Sb_2Te_3 and, generally, phase-change materials in the remaining part of this chapter.

Antimony telluride adopts the rhombohedral structure type of its heavier homologue Bi_2Te_3 with space group $R\bar{3}m$. By doing so, the binary phase forms a layered structure with covalently bonded quintuple layers (Te—Sb—Te—Sb—Te). Between the Te atoms of two layers, there are weaker, van der Waals (vdW)-like interactions. This very structural motif of two quintuple layers connected by weak inter-layer contacts is depicted in Figure 13.10. To better describe and understand the physical properties of Sb_2Te_3, in particular for finite temperatures, deep insight into its vibrational properties are mandatory. At the same time, one would like to know how well density-functional parametrizations do perform for this very case.

The atom-resolved DPS can experimentally be measured by *nuclear inelastic scattering* (NIS) [48]. This ingenious method does allow for a local probe (in addition to what diffraction may provide) but one drawback of this method is that only Mössbauer-active species are detectable. Hence, some elements cannot be detected at all. Nonetheless, Figure 13.11 shows the DPS of Sb_2Te_3 as obtained from NIS measurements [49] because both elements, antimony and telluride, are Mössbauer-active. These measurements are shown together with

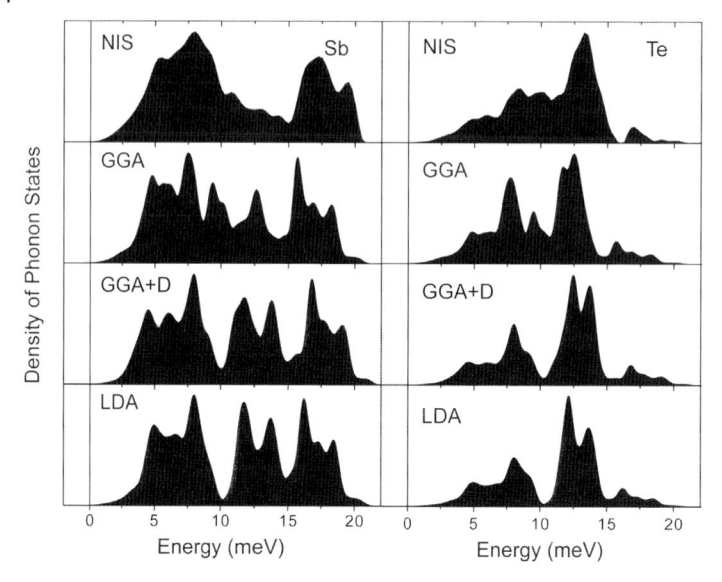

Figure 13.11 Partial DPS of antimony telluride (Sb$_2$Te$_3$) measured with NIS together with DFT results. Left: Sb DPS; right: Te DPS. (Figure reprinted from Ref. [36] with kind permission of IOP Publishing Ltd.)

theoretical results using different formulations of DFT, namely, LDA, GGA, and dispersion-corrected GGA, dubbed as GGA+D due to the application of Grimme's semi-empirical "+D" correction [28]. These different methods were investigated to judge their performance regarding the vdW-like inter-layer Te—Te bonds.

The theoretically calculated DPS is in good agreement with the experiment, albeit some features are overestimated. To start with, the DPS range can be divided into three main regimes. The first one, with vibrational energies up to 10 meV, is mostly dominated by Sb contributions. The same holds for energies above 15 meV while between 10 and 15 meV most vibrational contributions are related to Te.

The partial Sb densities of phonon states calculated with the LDA and GGA+D formalisms agree well with the experiment, although the features at 12 and 13.5 meV are overestimated. In the case of Te, LDA and GGA+D again show good accordance, whereas the frequencies calculated with the pure GGA are slightly underestimated. Hence, one may conclude that the simple LDA performs astonishingly well for the case of Sb$_2$Te$_3$, which eventually allows for calculations of other (macroscopic) physical properties such as the so-called Grüneisen parameter (describing how the phonon frequencies change with increasing volume), the bulk modulus or the volume expansion with temperature. We will come back to the latter two phenomena below.

13.5.3
Barium Cerate (BaCeO$_3$)

We now move on to the phenomenon of structural polymorphism and show how the methods introduced so far can be applied to this problem. Within solid-state chemistry, polymorphism means a compound's ability to exist in alternative extended structures. As will become obvious, being able to theoretically predict structural changes depends on the ability to properly treat imaginary phonon modes and understand the consequences for the respective structures and their stabilities. Also, considerations based on the *ab initio* thermochemically calculated Gibbs free energies are presented.

As an example for the aforementioned polymorphism, we show representative calculations on barium cerate [50], BaCeO$_3$, a ternary oxide adopting the perovskite motif. BaCeO$_3$ finds diverse applications, for example, as a proton conductor [51]. As expected for the ubiquitous perovskite type, several structural changes as a function of the temperature are likely and, indeed, barium cerate exhibits polymorphism with several structures *all of which* resemble the perovskite motif; hence, the structural changes are relatively small, even tiny, a true challenge for numerical theory. At high temperatures above 1200 K, BaCeO$_3$ has been described to adopt the perfect cubic perovskite structure. Below that temperature, there are at least three polymorphs, all related to perovskite, as said before; the lower the temperature, the lower is their symmetry. While the nearest-neighbor connectivity does not change at all, the polymorphs only differ by the tilting of the CeO$_6$ octahedra as well as the Ba positions. Figure 13.12 depicts the crystal structures as described in the literature [52,53] with ascending temperature. At "low" temperatures up to 563 K, BaCeO$_3$ adopts an orthorhombic crystal structure of space group *Pnma*, and above that temperature, up to 673 K, the structure can be described as being body-centered orthorhombic with space group *Imma*. Above 673 K, BaCeO$_3$ transforms into the trigonal space group $R\bar{3}c$. Admittedly, the accuracy of the experimental temperatures (563 and 673 K) is anything but clear, and the equality of the last temperature to *exactly* 400 °C causes a little discern for the theorist.

Taking a look on the calculated densities of phonon states (Figure 13.13) one immediately realizes the appearance of imaginary phonon modes for the three

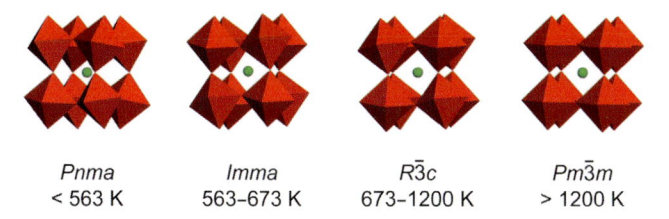

Pnma	*Imma*	$R\bar{3}c$	$Pm\bar{3}m$
< 563 K	563–673 K	673–1200 K	> 1200 K

Figure 13.12 Structural motifs of BaCeO$_3$ and the experimentally determined temperature regimes of their stability [52]. (Reprinted modified figure from Ref. [50] with kind permission of Wiley-VCH Verlag GmbH.)

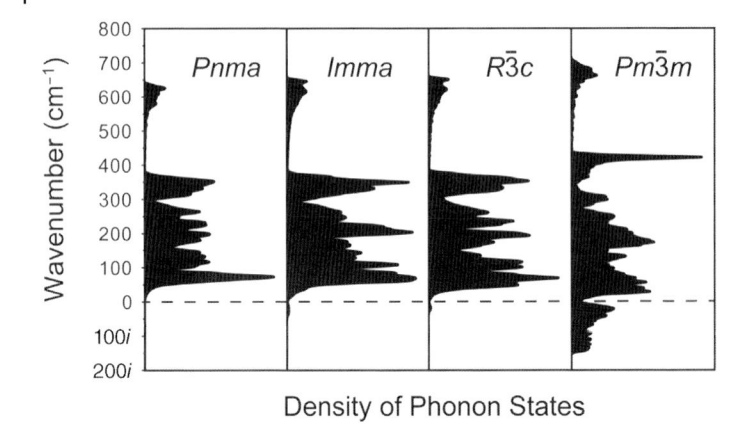

Figure 13.13 Densities of phonon states of BaCeO$_3$ polymorphs. The high temperature polymorphs (*Imma*, *R$\bar{3}$c*, *Pm$\bar{3}$m*) show imaginary frequencies. (Reproduced with permission from Ref. [50]. Copyright 2013, John Wiley & Sons.)

high-temperature polymorphs while the low-temperature *Pnma* polymorph turns out to be stable, with only positive wave numbers. This is a clear sign of the *Pnma* polymorph being stable at 0 K and (at least) somewhat above. The densities of phonon states of the *Imma* and *R$\bar{3}$c* polymorphs exhibit only small imaginary parts, whereas the structure that is forced to keep the ideal cubic perovskite structure (*Pm$\bar{3}$m*) has large imaginary contributions that indicates the structure to be highly unstable for zero temperature. Due to this enormous instability, the structure type has been excluded from the forthcoming considerations.

The phonon structures of the *Imma* and *R$\bar{3}$c* polymorphs have therefore been further analyzed in order to detect at which wave vectors the largest imaginary modes are located. Figure 13.14 shows the phonon dispersion along high-symmetry directions including the highest imaginary frequencies of the *Imma* and *R$\bar{3}$c* polymorphs. The highest imaginary mode of the latter is located at the Γ point, with $\mathbf{q} = (0\ 0\ 0)$. Therefore, the eigenvectors can directly be applied to the unit cell. The *Imma* polymorph shows the largest imaginary mode at a wave vector of $\mathbf{q} = (½\ 0\ ½)$, such that the unit cell is insufficiently sized for the application of the eigenvectors belonging to this mode. Instead, a $2 \times 1 \times 2$ supercell has to be constructed. After applying the eigenvectors to the respective simulation cells, structural optimizations lead to the formation of the lower-lying polymorphs related to symmetry and experimental temperature range.

Based on the quasi-harmonic approximation, the Gibbs energies of the three polymorphs under investigation (neglecting the cubic phase due to its large imaginary DPS contributions) were calculated and the thus obtained transition temperatures compared to the ones from the literature. Figure 13.15 shows the temperature-dependent Gibbs free energies of the three polymorphs relative to G of the *Pnma* structure. At each temperature, the lowest energy indicates the

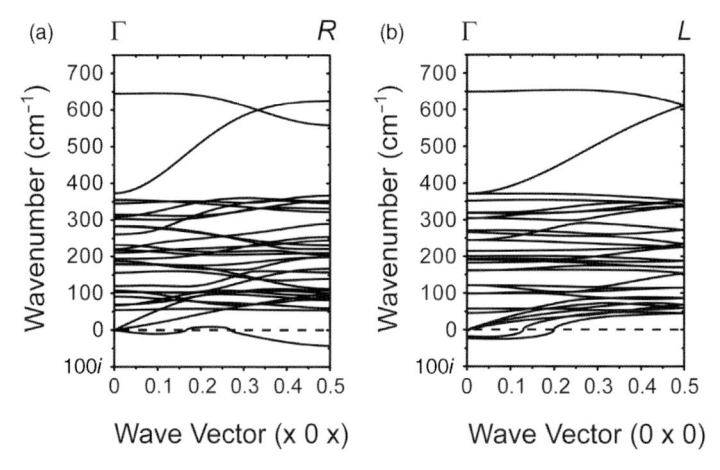

Figure 13.14 Phonon dispersion relations of the *Imma* (a) and $R\bar{3}c$ (b) polymorphs of BaCeO$_3$ along selected high-symmetry directions in the first Brillouin zone. (Reproduced with permission from Ref. [50]. Copyright 2013, John Wiley & Sons.)

most stable polymorph. The theoretical values reproduce the correct tendency, but the transition temperatures are largely underestimated. For the *Pnma–Imma* transition, the calculated value is 250 K compared to 563 K from experiment. The theoretical stability regime of the *Imma* polymorph reaches only up to 262 K, which is also a large underestimation of the experimental 673 K. This underestimation results from the extremely similar temperature dependencies of the respective Gibbs energies and the minute energy differences between the polymorphs (as alluded to already), all lying under 1 kJ/mol within the temperature range under consideration.

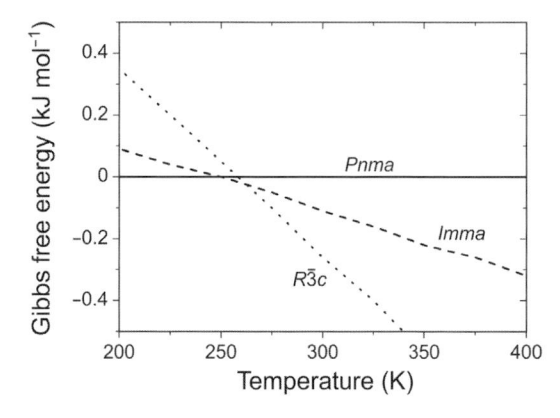

Figure 13.15 Temperature-dependent Gibbs free energies of the *Pnma*, *Imma*, and $R\bar{3}c$ polymorphs of BaCeO$_3$ relative to the *Pnma* phase. (Reproduced with permission from Ref. [50]. Copyright 2013, John Wiley & Sons.)

13.5.4
GeSe

Chalcogenides of group IV elements, such as GeTe, PbTe, or GeSe, are an important class of functional materials with various interesting physical properties. It is puzzling to note that phases such as GeTe, yet another fundamental phase-change material, date back to the year 1934 in which GeTe's technological application could not have been foreseen at all [54]; and yet, synthetic solid-state chemistry was far ahead of its time. Let us now, however, deal with the lighter homologue GeSe, which is structurally similar to GeS. All three phases are prominent IV–VI semiconductors, and a thorough understanding of GeSe's physical properties is mandatory because of its fascinating applications, for example, as a possible nanoscale photodetector [55,56].

Under ambient conditions, germanium selenide crystallizes in the orthorhombic space group *Pnma* (Figure 13.16), in which Ge is coordinated by three nearest Se neighbors around 2.56 Å and four second-nearest Se neighbors around 3.3 Å. Under pressure, however, the phase adopts the rock-salt structure with space group $Fm\bar{3}m$ and six equal Ge–Se bonds. These two polymorphs have been investigated by means of phonon calculations including the effect of pressure [57]. The phonon band structures of the two polymorphs at $p=0$ corroborate this experimental finding (Figure 13.17), not too surprisingly. While the rock-salt structure exhibits imaginary phonon modes, the *Pnma* structure does not. With increasing pressure, computationally arrived at by introducing smaller unit cells, the imaginary phonon modes of the rock-salt structure disappear, as can be seen in Figure 13.18, which presents the phonon dispersion curves along X–Γ–L of cubic GeSe. This disappearance of imaginary phonon frequencies indicates the stabilization of rock-salt type GeSe as a function of applied pressure, eventually arrived at, approximately, 7 GPa. This result goes in nice

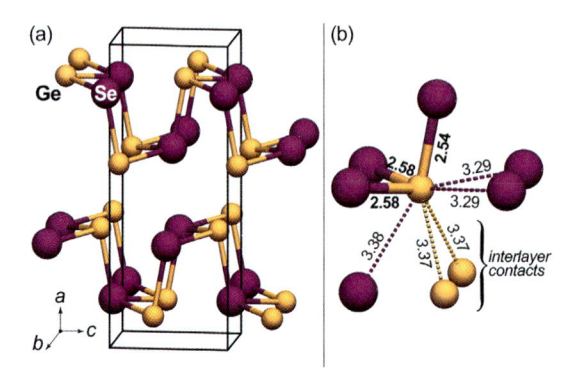

Figure 13.16 Crystal structure of the *Pnma* polymorph of germanium selenide, showing (a) the orthorhombic unit cell and providing (b) an insight into the Ge coordination. (Reproduced with permission from Ref. [57]. Copyright 2014, American Physical Society.)

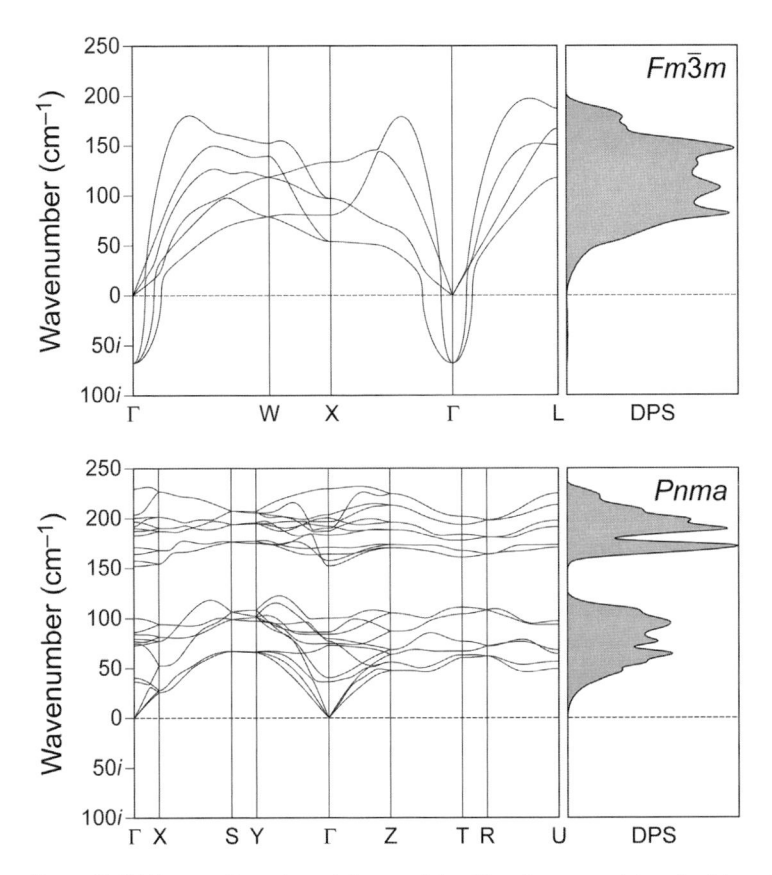

Figure 13.17 Phonon dispersion relations and densities of phonon states of cubic and orthorhombic GeSe at $p = 0$. (Reproduced with permission from Ref. [57]. Copyright 2014, American Physical Society.)

accordance with the experimental findings and once more shows the reliability of the methodology.

13.5.5
Ab Initio ORTEP

The thermal motion of atoms and functional groups are conveniently described by the so-called anisotropic displacement parameters (ADPs) whenever they result from a diffraction experiment; as such, they serve as key characteristics of any atom in a crystal. We already stressed out that in any given crystalline solid, all the atoms are moving all the time, and the amount of movement increases with the temperature. Even at absolute zero there is atomic movement due to Heisenberg's famous uncertainty principle. The fact that such atomic movement also affects the intensities in an X-ray (and neutron) diffraction experiment was

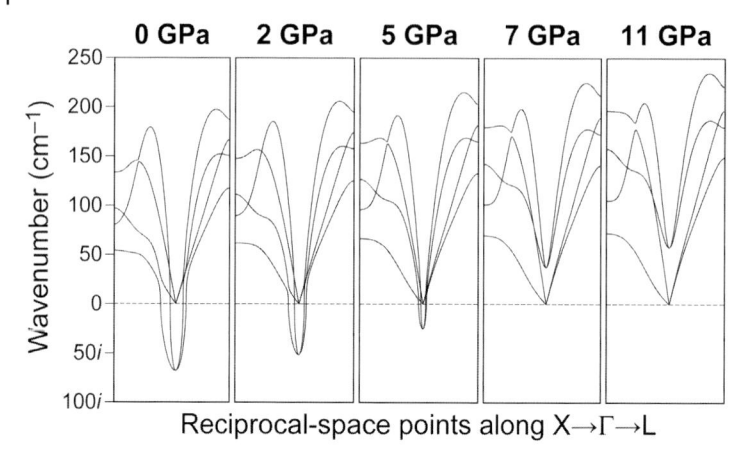

Figure 13.18 Pressure-dependent phonon dispersion of rock-salt GeSe along selected high-symmetry directions. (Reproduced with permission from Ref. [57]. Copyright 2014, American Physical Society.)

first realized by Debye [58] and further studied by Waller [59] – hence the historic "Debye–Waller factor" for correcting an atom's scattering power by a temperature-dependent term.

With the advent of digital computing in the 1960s, refining a crystallographic model based on X-ray (and neutron) intensities became rather straightforward, at least in principle (not necessarily in practice), and people soon also modeled the amount of atomic movement of an atom or a functional group (in a molecule contained in a crystal) by a refinable "atomic displacement parameter" (ADP) U as it is called using modern nomenclature. Because atomic movement is not necessarily isotropic, refinable anisotropic ADPs were the next step, typically expressed by a symmetric second-rank tensor (i.e., a symmetric 3×3 matrix composed of real numbers) with entries U_{ij}. Because the latter is pretty hard to imagine, a landmark contribution was Ref. [60] which managed to visualize the aforementioned numerical entries. Zillions of Oak Ridge Thermal Ellipsoid Plots (ORTEPs) drawn (and published) up to the present day with whatever kind of software tells us something about the ingenuity of Johnson's approach.

The fascinating finding is that, starting from the knowledge of the phonon frequencies and eigenvectors, it is possible to obtain those ADPs (or ORTEPs) of each atom involved in the respective structure as the 3×3 thermal displacement matrix. As an example, we present the comparison between theoretical and experimental ADPs of crystalline urea [61]. The latter have been experimentally extracted from (highly expensive) single-crystal neutron-diffraction data. As one can see from Figure 13.19, the theoretical values agree perfectly with the experimentally measured ones. As an agreement measure, we introduce the "similarity index" S (see numbers next to the atoms in Figure 13.19) whose values describe the residual between theory and experiment. Note that a value of zero means

(a) Neutron diffraction   (b) Computed

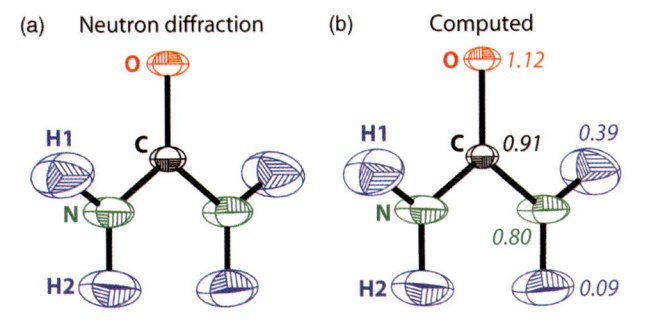

Figure 13.19 Thermal ellipsoids of urea at a temperature of 123 K from first-principles calculations (a), in excellent agreement with neutron diffraction data (b). (Reprinted from Ref. [61] with kind permission by The Royal Society of Chemistry.)

perfect agreement whereas the maximum value of 100 indicates absolutely no match between the ADPs.

Closely related, structurally and also chemically, to urea is the molecule guanidine whose solid-state crystal structure was unknown until as late as 2009 [62] despite the fact that generations of organic chemists had been busily dealing with guanidine. Here, also neutron diffraction data are now available that allow comparing the experimental ADPs with the theoretically computed values [61]. There is still good agreement between the two ORTEP drawings (Figure 13.20) but temperature plays a more decisive role. As can be seen from the scatterplot below which displays a 1:1 comparison between computed and measured U_{ij} values, the data at 100 K indicate very good agreement, whereas at higher temperature, namely, 273 K, the theoretical calculations underestimate the thermal motion of the atoms. The interpretation is clear: in the theoretical calculations, a harmonic approach has been used (and must be used in order to have a computational demand that can be accomplished in a reasonable time), but it is the anharmonicity that gets increasingly more important while the crystal reaches its melting temperature; for guanidine, 273 K is just 50 K below that very temperature. Such comparisons would be an ideal test ground for the development of methods explicitly including anharmonicity.

13.5.6
Tin Pest

Let us again turn to the macroscopic thermodynamic properties resulting from quantum-chemical calculations, as briefly touched upon for BaCeO$_3$ already. We recall that the enthalpy H and the Gibbs free energy G give access to phase stability or phase transitions as a function of various thermodynamic variables such as pressure p and temperature T. In any case, the phase with the lowest energy is the thermodynamically stable one at a given point of the energy landscape.

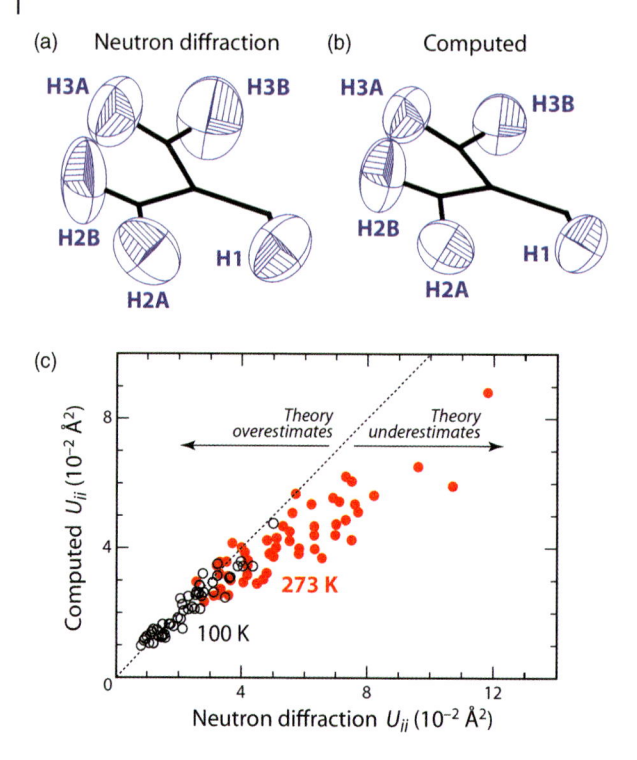

Figure 13.20 ORTEP plots of guanidine based on neutron diffraction data (a) and DFT (b), respectively, at a temperature of 273 K. For clarity, only the hydrogen atoms are shown. (c) Scatterplot of on-site displacement matrix elements U_{11}, U_{22}, and U_{33} for all C, N, and H atoms in crystalline guanidine at 100 K (open circles) and 273 K (red bullets). (Reprinted from Ref. [61] with kind permission by The Royal Society of Chemistry.)

One of the most well-known structural polymorphisms is the transition between gray α-tin and white β-tin. Since antiquity it has been known that upon increasing the temperature above 13.2 °C, the α-allotrope (no longer called polymorph because we deal with an element, not with a compound) turns into the much denser β-allotrope [63]. Figure 13.21 shows the unit cells of the two respective crystal structures. As taught in most textbooks, α-tin crystallizes in the cubic diamond-structure where each Sn atom is tetrahedrally coordinated by four nearest Sn atoms. The crystal structure of β-tin is slightly more complex; it can be described as a strongly compressed diamond structure only along one direction that results in a 4+2 coordination and, thus, a more "metallic" nature; in fact, β-tin is a metal, opposed to semiconducting α-tin which resembles its lighter homologues C, Si, and Ge. We also note that the atomic distances in β-tin are longer than the ones in α-tin despite the fact that β-tin is the denser polymorph, the notorious pressure–distance paradox [64].

This very circumstance, namely, the fact that there is a volume expansion by decreasing temperature below 13.2 °C, is known as the "tin pest" and stands for a

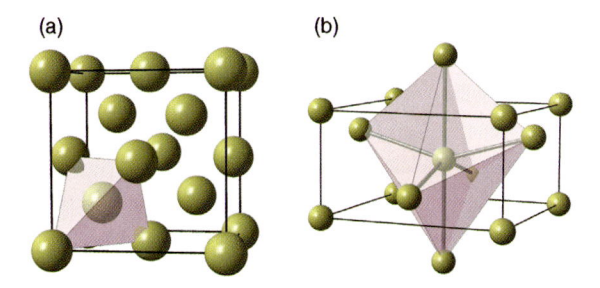

Figure 13.21 Crystal structures of α-tin (a) and β-tin (b). (Figure taken from Ref. [35].)

catastrophic material failure (in organ pipes). Likewise, the structural change may be induced by varying the pressure. At room temperature, for example, the pressure-caused phase transition occurs at about 1 GPa [65,66]. From the point of view of electronic-structure theory, the pressure-induced transformation is far easier to access. Figure 13.22 presents the calculated enthalpy difference ΔH without considering temperature at all; the differences in H just result as differences in electronic energy while the volume changes. The theoretical data are in good harmony with the experimental transition pressure that gets only slightly underestimated. Of course, calculating such data is much easier and a lot faster than performing the high-pressure experiment.

Moving forward, the temperature-dependent phase transition can be obtained by the Gibbs energy difference ΔG. In analogy to Figure 13.22, Figure 13.23 shows the temperature-dependent Gibbs free energy difference between the two

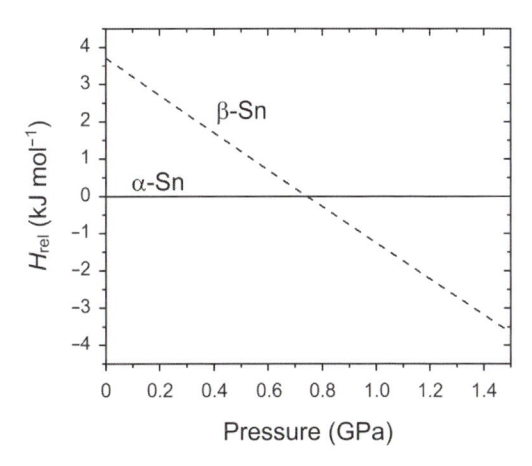

Figure 13.22 Relative enthalpies H_{rel} of α- and β-tin, respectively, calculated from density functional theory. The enthalpy of α-tin is set to zero and the phase transition occurs at the point where H_{rel} (β-Sn) gets negative. (Reproduced with permission from Ref. [13]. Copyright 2010, John Wiley & Sons.)

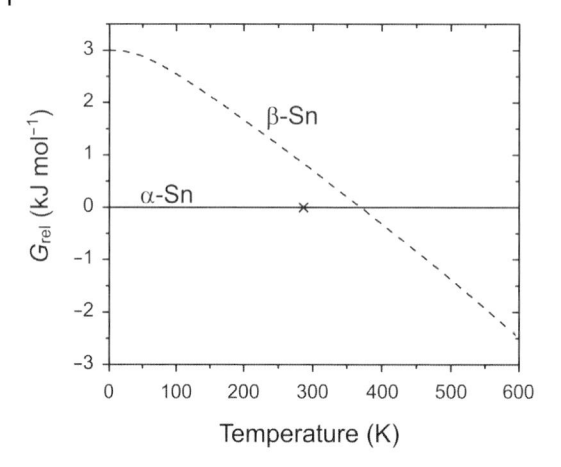

Figure 13.23 Relative Gibbs energies of α- and β-Sn as functions of the temperature. The cross marks the phase transition temperature from experiment. (Reproduced with permission from Ref. [13]. Copyright 2010, John Wiley & Sons.)

tin polymorphs, again setting the energy of α-tin to zero over the entire temperature range. The phase transition occurs at approximately 380 K when the Gibbs energy of β-tin becomes smaller than zero. While this is an overestimation of about 90 K, it nonetheless corresponds to a *minute* error in energy. In fact, the energetic error is smaller than 0.8 kJ/mol at the transition temperature (about 286 K or 13 °C), thereby reaching "chemical accuracy" as phrased by our molecular quantum-chemistry colleagues.

13.5.7
Formation Energies

In a handbook series on solid-state chemistry, there is probably no need to emphasize the high-temperature superconductor revolution that occurred already three decades ago. The oxocuprates as ingeniously invented by Bednorz and Müller [67] (and many other researchers following this discovery) simply redefined parts of solid-state research. Superconducting phases such as $YBa_2Cu_3O_{7-x}$ [68] are nowadays taught in freshmen chemistry classes because that weird macroscopic quantum phenomenon, so hard to comprehend, could finally be witnessed at the temperature of liquid nitrogen and became an everyday experience.

Likewise, the crystal structure of $YBa_2Cu_3O_{7-x}$ and its relatives are so fundamental that we do not have to redisplay it at this point; the structural science is clearly settled, not necessarily its interpretation. On the other hand, *manufacturing* considerable amounts of superconducting ceramics is anything but trivial, and industrial applications for high-temperature superconductivity had to wait until large-scale superconductor production using wet chemical precipitation on textured metal bands was eventually introduced [69] but not necessarily mastered; this is needed because brittle ceramics need to fit into any shape, a lesser problem for a metallic superconductor. And this process engineering of a ceramic is precisely the starting point for *ab initio* thermochemistry which could play an important (if not decisive) role, the reason being that the available thermochemical databases proved to be too fragmentary or unreliable for fully modeling the industrial process.

Among the less reliable data is the Gibbs energy of formation of orthorhombic $YBa_2Cu_3O_{6.5}$, a kind of reduced "superstructure" compound of $YBa_2Cu_3O_{7-x}$. Although its Gibbs formation energy had been measured over and over again using a multitude of differently prepared samples, the corresponding IUPAC report [70] provides only a disillusioningly inaccurate numerical value for this very thermochemical state function, easily seen from the enormously wide error bars in Figure 13.24. In fact, the upper and lower accuracy limits (dotted) for a certainty of 99% are so far away from the averaged value that any $YBa_2Cu_3O_{6.5}$ Gibbs formation energy between −30 and −130 kJ/mol at 1000 K seems fine; choose any value you like – and terribly fail during process optimization.

The *ab initio* alternative, however, consists of calculating the Gibbs free energies of all starting materials and the one of the product, so the requested Gibbs free formation energy from the oxides, ΔG_{ox}, simply follows the formal reaction

$$\tfrac{1}{2}\,Y_2O_{3(s)} + 2\,BaO_{(s)} + 3\,CuO_{(s)} \rightarrow YBa_2Cu_3O_{6.5(s)}$$

The theoretical result is also found in Figure 13.24. For $YBa_2Cu_3O_{6.5(s)}$, the (small) additional influence of the configurational entropy for the oxygen-atom positions in the one-dimensional Cu—O chains along the crystallographic *b* direction was included as well. We note that the first-principles curve runs

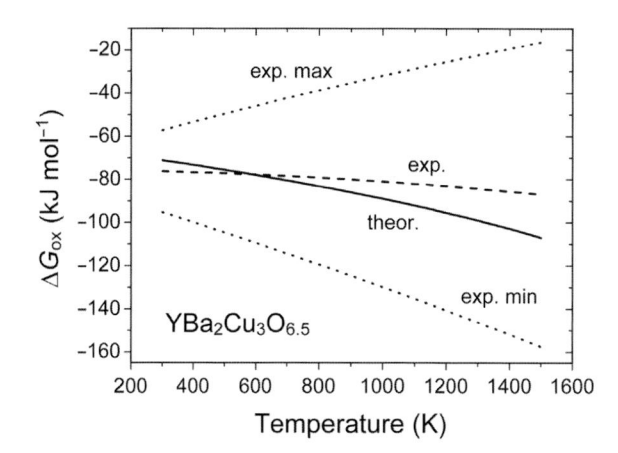

Figure 13.24 The Gibbs energy of formation of $YBa_2Cu_3O_{6.5}$ from the binary oxides, ΔG_{ox}, as a function of temperature as calculated *ab initio* compared to experimental data and their error bars [70]. (Reproduced with permission from Ref. [13]. Copyright 2010, John Wiley & Sons.)

relatively close to the experimental average, so experiment and theory provide the same qualitative message. Nonetheless, there was good reason to believe in a higher reliability of the *ab initio* G function which was originally estimated to be more accurate than 10 kJ/mol over the entire illustrated temperature range. In fact, the accuracy proved to be so good that it allowed for the world's first generation of a self-consistent thermochemical data set for the entire Y–Ba–Cu–O system based on 24 phases (i.e., 11 metallic ones, 12 oxidic matter plus the melt) which became part of a commercially used data base; unpublished but to be paid for. Your high-temperature superconductor is likely to have been manufactured using *ab initio* thermochemical data.

13.5.8
Heat Capacities

As mentioned before, the temperature dependencies of the Gibbs energy, the enthalpy and the entropy all go back to the constant-pressure heat capacity of the system under investigation. In fact, the paramount importance of the heat capacity was already recognized by Einstein in the very early days because, according to him, measuring the heat capacity of some material will provide more information about that material than any other physical property. It then does not come as a surprise that the quality of the theoretically obtained thermodynamic properties can therefore be nicely judged by just looking at the heat capacities. Thus, let us present various heat capacities and compare them to available experimental data. The first example (Figure 13.25) shows the data of copper(II)oxide, CuO, and cerium(IV)oxide, CeO_2, both compared to data included in the SGTE (*Scientific Group Thermodata Europe*) data base [1,71].

The deviation of the theoretical data from the experimental ones lies in the range of only a few J/(mol K). For the case of density-functional theory-based

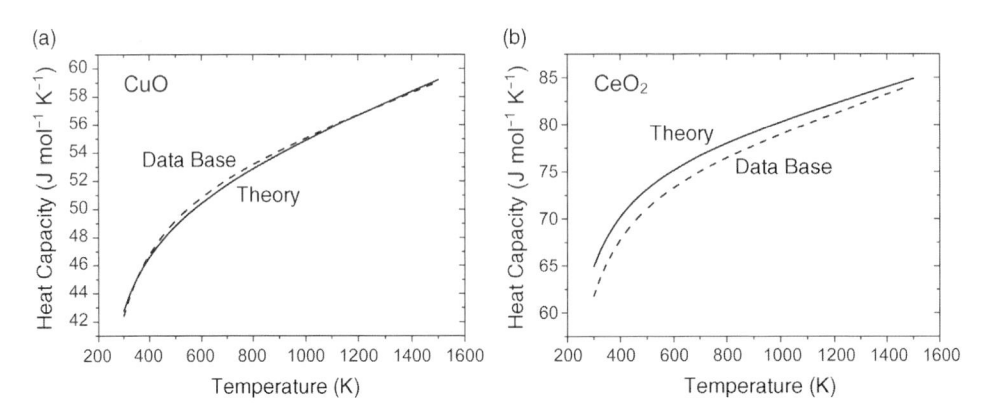

Figure 13.25 Constant-pressure heat capacities C_p of copper(II)oxide, CuO, and cerium(IV) oxide, CeO_2, compared to compiled experimental data from the SGTE data base [71]. (Figure taken from Ref. [35].)

calculations, this is a clear sign for rock-solid accuracy, as already stated within the section regarding the phase transition of tin; equally important is the fact that theory and experiment match each other over the entire temperature range, at least for the CuO data which show nearly perfect agreement. When it comes to the tabulated heat capacity of CeO_2, it is slightly overestimated by less than 3 J/(mol K). This deviation can be seen as uncritical since the error of a single heat-capacity measurement is about 3% [72] and the SGTE data is a compilation of various experiments.

It does not come as a surprise that different polymorphs (i.e., materials with identical chemical composition) will differ in their heat capacities because different structures will also result in different phonon states, the vibrational fingerprint of the material. This is especially true whenever different coordination polyhedra are involved, that is, a *connectivity* difference of a central atom with its neighboring atoms. As a characteristic example, let us present such thermochemical data on jadeite, an important high-pressure clinopyroxene mineral with the chemical composition $NaAlSi_2O_6$ and octahedral Al—O coordination besides the typical tetrahedral Si—O units. In addition to this crystalline system, recent theoretical research also arrived at a hypothetical low-density polymorph of jadeite that has not yet been experimentally established. This hypothetical phase was theoretically identified *ab initio* by density-functional theory at a larger-than-normal volume. It is fascinating to realize that the molar volume of the hypothetical low-density jadeite polymorph *exactly* matches the one of jadeite *glass* whose atomistic structure is still unknown. Why, then, should not we use the low-density jadeite model as a crystalline approximant for the glassy state [73] and see how various thermochemical data match each other? To do so, the data of the low-density-polymorph were compared to the data available for a glass phase with the jadeite composition [74].

The heat capacities of jadeite and its hypothetical low-density polymorph are shown together with experimental data of crystalline jadeite and jadeite glass in Figure 13.26. Not too surprisingly, the theoretical and experimental heat capacities of crystalline jadeite (lower two curves) are almost superimposable, a reassuring result for *ab initio* theory. It is more astonishing, however, that the experimental heat capacity of glassy jadeite agrees extraordinarily well with the theoretical one based on the hypothetical low-density polymorph, so the vibrational states of the glass are correctly reproduced. In the low-temperature regime, the C_p data for the low-density polymorph and the glass are significantly higher than for the crystalline counterparts (either measured experimentally or calculated theoretically), and here is an explanation: in the low-density crystalline approximant of the glass, one finds low-frequency vibrational modes caused by the appearance of Al—O tetrahedra, which are not present in the jadeite crystal where Al is octahedrally coordinated by oxygen atoms (see above). The unusual vibrational properties of the glass are most evident from a C_p/T^3 plot (Figure 13.26b) that clearly shows a much stronger deviation from the Debye-type low-temperature behavior only for the low-density polymorph. Hence, *ab*

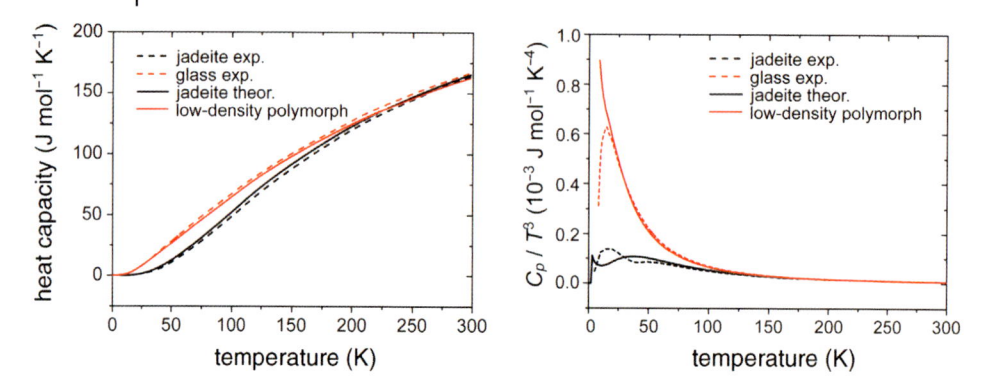

Figure 13.26 (a) Heat capacity of jadeite and its low-density polymorph compared to (b) experimental data for jadeite and glass with jadeite composition. (Reproduced with permission from Ref. [13]. Copyright 2010, John Wiley & Sons.)

initio theory and its hypothetical low-density jadeite polymorph turns out to be a reasonable crystalline model for the amorphous glass.

13.5.9
Temperature-Dependence of Volume and Bulk Modulus

Let us recall that the equation-of-state fits as introduced in the context of the quasi-harmonic approximation give access to the molar volume and bulk modulus at each temperature under consideration. Because of the underlying anharmonic DFT potentials, volumes expand while the temperature increases despite purely harmonic phonons. To look at this in greater detail, we return to the well-known antimony telluride, Sb_2Te_3, and present the respective data for this compound compared to experimental data. As already mentioned, the initial intention of the investigation was to choose an appropriate description of the van der Waals interactions occurring between sheets of tellurium atoms. Therefore, the data were calculated using different formulations of DFT also including dispersion corrections (+D).

As shown in Figure 13.27, the LDA calculations give good agreement with experimental data both for the molar volume (which increases with rising temperature) and the bulk modulus (which lowers with increasing temperature). A closer look might reveal that this friendly behavior of the LDA goes back to some fortunate error cancellation but we will not go into the electronic-structure details. For the more advanced GGA, it is shocking to realize that it totally fails for temperature-dependent properties. This GGA failure goes back to a deficient description of the electronic structure, as expected, because it does not (in fact, it cannot) properly account for the weak van der Waals-like interactions between the Te-terminated layers of the Sb_2Te_3 structure. The LDA does not include van der Waals attractions either but its typical overbinding fortunately "covers" the Te–Te attractions, leading to the aforementioned error cancellation.

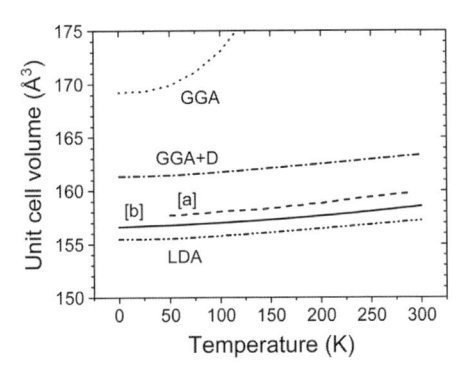

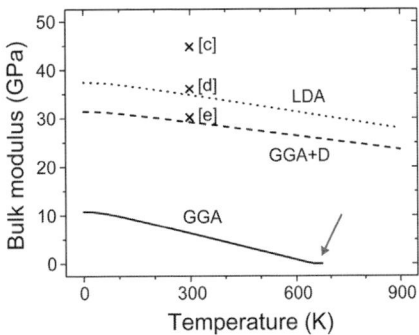

Figure 13.27 Ab initio thermochemically calculated volume expansion and temperature-dependent bulk modulus of Sb$_2$Te$_3$ probing different formulations of DFT compared with experimental data. Experimental data are denoted as [a] [49], [b] [75], [c] [76], [d] [77], and [e] [78]. The red arrow marks the breakdown of the bulk modulus calculated with pure GGA. (Modified figures reprinted from Ref. [36] with kind permission of IOP Publishing Ltd.)

Coming back to the GGA, by adding the "+D" correction this explicitly includes the wanted interactions in a semi-empirical way and, as could be expected, the results improve significantly. Note that the results presented here should not serve as a general recommendation regarding the hierarchy of DFT functionals as in many cases GGA works better than LDA [14].

13.6 Summary

In the twenty-first century, accurate first-principles methods are available to deal with the lattice dynamics and the thermochemistry of any solid-state material. To calculate its thermodynamic potentials, *ab initio* quantum-mechanical methods must first be utilized to establish the well-defined electronic ground-state energy. In most cases, such calculations will be carried out using density-functional theory but other methods (such as wave-function-based Hartree–Fock theory) are also possible. Second, the simulation of the vibrational states and the entire lattice dynamics is accomplished using a harmonic model and leads to the phonon energy via the laws of statistical thermodynamics. In combination with the anharmonic interatomic potentials, one eventually arrives at a quasi-harmonic representation of the volume- and temperature-dependent energy of the entire crystal from which everything – yes, everything, at least in terms of G – may be derived.

By following this train of thought, the vibrational states of solid-state materials may be calculated, visualized, and compared with various experiments, thereby greatly improving our analytical tools. Likewise, theoretical enthalpy or Gibbs energy data eventually explain the existence of solid-state materials and their mutual relationships in quantitative terms, in particular with respect to polymorphism as a function of the temperature. For hypothetical matter, only structural information (which is also available from electronic-structure theory) is needed to predict any thermochemical property ahead of the synthesis; only a few would have been brave enough to predict that around the year 2000. Also, structural changes on the atomic scale as a function of the temperature are accessible from analyzing the corresponding phonons in reciprocal space. Because of the quantitative nature of the underlying theories, atomic vibrations derived from experimental studies such as neutron diffraction are subject to theoretical analysis. Likewise, heat capacities and other thermochemical data, even energies of formation, may be thoroughly analyzed and explained. As such, we are entering an era where *ab initio* theory becomes an essential tool for the restless synthetic solid-state chemist.

And yet, there are challenges. Sometimes electronic-structure theory does fail, with severe consequences for the free energies derived from an imperfect electronic ground state; nonetheless, there are great improvements in, say, density-functional theory, and *ab initio* thermochemistry will automatically profit from that. Likewise, important research is going on to move beyond the limits of the quasi-harmonic approximation, and such contributions will not only better describe the thermochemistry close to the melting temperature but also will make the latter theoretically accessible. As predicted in the year 2010 by us [13], routine calculations of the Gibbs energy and pressure–temperature diagrams have become common practice in a couple of leading laboratories around the globe, not only because knowing G automatically makes synthetic attempts more rational, but also the necessary computational ressources are available. Also, industrial companies have started to use *ab initio* data for modeling their processes, in particular for economically relevant questions. We also remind the reader that there are those geoscientific (or geochemical?) questions that can only be solved using *ab initio* theory as an independent method because there are those harsh natural conditions (say, in the earth's mantle) that can hardly be experimentally reproduced by man.

References

1 Hack, K. (2008) *The SGTE Casebook*, 2nd edn, Woodhead, Cambridge.

2 Carnot, S. (1872) Réflexions sur la puissance motrice du feu et sur les machines propres à développer cette puissance. *Ann. Sci. Ecole Norm. S., Sér. 2*, **1**, 393–457.

3 Mayer, J.R. (1851) *Bemerkungen über das Mechanische Aequivalent der Wärme*, Verlag von Johann Ulrich Landherr, Heilbronn.

4 Joule, J.P. (1850) On the mechanical equivalent of heat. *Phil. Trans. R. Soc. Lond.*, **140**, 61–82.

5 Clausius, R. (1867) *The Mechanical Theory of Heat, with its Applications to the Steam-Engine and to the Physical Properties of Bodies*, J. Van Voorst.

6 Ostwald, W. (1885) *J. Prakt. Chem.*, **31**, 433–462.
7 Atkins, P.W. and de Paula, J. (2001) Atkins, in *Physical Chemistry*, 7th edn, Oxford University Press.
8 Wedler, G. (1997) *Lehrbuch der Physikalischen Chemie*, 4th edn, Wiley-VCH Verlag GmbH, Weinheim.
9 von Helmholtz, H. (1882) *Die Thermodynamik Chemischer Vorgänge, Sitzungsberichte der Preussischen Akademie der Wissenschaften*, Reichsdruckerei, Berlin.
10 Gibbs, J.W. (1876) *Trans. Conn. Acad.*, **3**, 108.
11 Einstein, A. (1907) *Ann. Physik*, **327**, 180.
12 Debye, P. (1912) *Ann. Physik*, **344**, 789.
13 (a) Stoffel, R.P., Wessel, C., Lumey, M.-W., and Dronskowski, R. (2010) *Angew. Chem.*, **122**, 5370; (b) Stoffel, R.P., Wessel, C., Lumey, M.-W., and Dronskowski, R. (2010) *Angew. Chem., Int. Ed.*, **49**, 5242.
14 Dronskowski, R. (2005) *Computational Chemistry of Solid State Materials*, Wiley-VCH Verlag GmbH, Weinheim.
15 Murnaghan, F.D. (1937) *Am. J. Math.*, **59**, 235.
16 Birch, F. (1947) *Phys. Rev.*, **71**, 809.
17 Vinet, P., Smith, J.R., Ferrante, J., and Rose, J.H. (1987) *Phys. Rev. B*, **35**, 1945–1953.
18 Hohenberg, P. and Kohn, W. (1964) *Phys. Rev.*, **136**, 864.
19 Kohn, W. and Sham, L. (1965) *Phys. Rev.*, **140**, 1133.
20 Ceperley, D.M. and Alder, B.J. (1980) *Phys. Rev. Lett.*, **45**, 566–569.
21 Perdew, J.P. and Zunger, A. (1981) *Phys. Rev. B*, **23**, 5048–5079.
22 Perdew, J.P. and Wang, Y. (1992) *Phys. Rev. B*, **45**, 13244.
23 Perdew, J.P. and Wang, Y. (1986) *Phys. Rev. B*, **33**, 8800.
24 Becke, A.D. (1988) *Phys. Rev. A*, **38**, 3098.
25 Perdew, J.P., Burke, K., and Ernzerhof, M. (1996) *Phys. Rev. Lett.*, **77**, 3865–3868.
26 Heyd, J., Scuseria, G.E., and Ernzerhof, M. (2003) *J. Chem. Phys.*, **118**, 8207.
27 Sun, J., Marsman, M., Csonka, G.I., Ruzsinszky, A., Hao, P., Kim, Y.-S., Kresse, G., and Perdew, J.P. (2011) *Phys. Rev. B*, **84**, 035117.
28 Grimme, S., Hansen, A., Brandenburg, J.G., and Bannwarth, C. (2016) *Chem. Rev.*, **116**, 5105–5154.
29 Bloch, F. (1928) *Z. Phys.*, **52**, 555.
30 Brillouin, L. (1930) *Comptes Rendus*, **191**, 292.
31 Heisenberg, W. (1927) *Z. Phys.*, **43**, 172–198.
32 Dove, M.T. (1993) *Introduction to Lattice Dynamics*, Cambridge University Press, Cambridge.
33 Brüesch, P. (1982) *Phonons: Theory and Experiments I*, Springer Verlag, Berlin.
34 Wallace, D.C. (1972) *Thermodynamics of Crystals*, John Wiley & Sons, Inc., New York.
35 Stoffel, R.P. (2012) Voraussetzungsfreie Thermochemie kristalliner Festkörper. Dissertation. RWTH Aachen University.
36 Stoffel, R.P., Deringer, V.L., Simon, R.E., Hermann, R.P., and Dronskowski, R. (2015) *J. Phys. Condens. Matter*, **27**, 085402.
37 Parlinski, K., Li, Z.Q., and Kawazoe, Y. (1997) *Phys. Rev. Lett.*, **78**, 4063.
38 (a) Togo, A., Oba, F., and Tanaka, I. (2008) *Phys. Rev. B*, **78**, 134106; (b) Togo, A. and Tanaka, I. (2015) *Scr. Mater.*, **108**, 1–5.
39 Baroni, S., Gironcoli, S.de., Corso, A.D., and Giannozzi, P. (2001) *Rev. Mod. Phys.*, **73**, 515.
40 Gonze, X. and Lee, C. (1997) *Phys. Rev. B*, **55**, 10355.
41 Deringer, V.L. and Dronskowski, R. (2013) Computational methods for Solids, in *Comprehensive Inorganic Chemistry II*, vol. 9 (eds J. Reedijk and K. Poeppelmeier), Elsevier, Oxford, p. 59–87.
42 Pak, L., Zibrowius, B., Lumeij, M.-W., Dronskowski, R., and Ruschewitz, U. (2014) *Z. Anorg. Allg. Chem.*, **640**, 3127–3134.
43 Zhang, H., Liu, C.-X., Qi, X.-L., Dai, X., Fang, Z., and Zhang, S.-C. (2009) *Nat. Phys.*, **5**, 438–442.
44 Hsieh, D., Xia, Y., Qian, D., Wray, L., Meier, F., Dil, J.H., Osterwalder, J., Patthey, L., Fedorov, A.V., Lin, H., Bansil, A., Grauer, D., Hor, Y.S., Cava, R.J., and Hasan, M.Z. (2009) *Phys. Rev. Lett.*, **103**, 146401.
45 Snyder, G.J. and Toberer, E.S. (2008) *Nat. Mater.*, **7**, 105–114.

46 Fujimori, S., Yagi, S., Yamazaki, H., and Funakoshi, N. (1988) *J. Appl. Phys.*, **64**, 1000–1004.
47 Wuttig, M. and Yamada, N. (2007) *Nat. Mater.*, **6**, 824–832.
48 Chumakov, A. and Rüffer, R. (1998) *Hyperfine Interact.*, **113**, 59–79.
49 Bessas, D., Sergueev, I., Wille, H.-C., Perßon, J., Ebling, D., and Hermann, R.P. (2012) *Phys. Rev. B*, **86**, 224301.
50 Stoffel, R.P. and Dronskowski, R. (2013) *Z. Anorg. Allg. Chem.*, **639**, 1227–1231.
51 Kreuer, K.D. (2003) *Annu. Rev. Mater. Res.*, **33**, 333–359.
52 Knight, K.S. (1994) *Solid State Ion.*, **74**, 109.
53 Genet, F., Loridant, S., Ritter, C., and Lucazeau, G. (1999) *J. Phys. Chem. Solids*, **60**, 2009.
54 Klemm, W. and Frischmuth, G. (1934) *Z. Anorg. Allg. Chem.*, **218**, 249.
55 Mukherjee, B., Cai, Y., Tan, H.R., Feng, Y.P., Tok, E.S., and Sow, C.H. (2013) *ACS Appl. Mater. Interfaces*, **5**, 9594.
56 Hsueh, H.C., Vass, H., Clark, S.J., Ackland, G.J., and Crain, J. (1995) *Phys. Rev. B*, **51**, 16750.
57 Deringer, V.L., Stoffel, R.P., and Dronskowski, R. (2014) *Phys. Rev. B*, **89**, 094303.
58 Debye, P. (1913) *Ann. Phys.*, **348**, 49.
59 Waller, I. (1923) *Z. Phys.*, **17**, 398.
60 Johnson, C.K. (June 1965) *OR TEP: A FORTRAN Thermal-Ellipsoid Plot Program for Crystal Structure Illustrations*, ORNL-3794 (Rev.), Union Carbide Corp., Oak Ridge Natl. Lab.
61 Deringer, V.L., Stoffel, R.P., Togo, A., Eck, B., Meven, M., and Dronskowski, R. (2014) *CrystEngComm*, **16**, 10907.
62 Yamada, T., Liu, X., Englert, U., Yamane, H., and Dronskowski, R. (2009) *Chem. Eur. J.*, **15**, 5651–5655.
63 Holleman, A.F. (1995) *Lehrbuch der Anorganischen Chemie*, 101, Auflage. de Gruyter, Berlin, New York.
64 Kleber, W. (1967) *Krist. Techn.*, **2**, 13–14.
65 Nikolaev, I.N., Marin, V.P., Panyushkin, V.W., and Pavlyukov, L.S. (1973) *Sov. Phys. Solid State*, **14**, 2022.
66 Appen, J.von., Hack, K., and Dronskowski, R. (2004) *J. Alloys Compd.*, **379**, 110.
67 Bednorz, J.G. and Müller, K.A. (1986) *Z. Phys. B*, **64**, 189.
68 Wu, M.K., Ashburn, R.J., Torng, C.J., Hor, P.H., Meng, R.L., Gao, L., Huang, Z.J., Wang, Y.Q., and Chu, C.W. (1987) *Phys. Rev. Lett.*, **58**, 908.
69 Bäcker, M. and Schneller, T. (2007) *Nachrichten aus der Chemie*, **55**, 1202.
70 Voronin, G.F. (2000) *Pure Appl. Chem.*, **72**, 463.
71 Bale, C., Chartrand, P., Degterov, S., Eriksson, G., Hack, K., Mahfoud, R.B., Melançon, J., Pelton, A., and Petersen, S. (2002) *Calphad*, **26**, 189–228.
72 Leitner, J., Chuchvalec, P., Sedmidubský, D., Strejc, A., and Abrman, P. (2002) *Thermochimica Acta*, **395**, 27.
73 Stoffel, R.P., Philipps, K., Conradt, R., and Dronskowski, R. (2016) *Z. Anorg. Allg. Chem.*, **642**, 590–596.
74 Jindal, R., Jayaganthan, R., Singh, I.V., and Conradt, R. (2011) *Ceram. Int.*, **37**, 741–748.
75 Kim, Y., Chen, X., Wang, Z., Shi, J., Miotkowski, I., Chen, Y.P., Sharma, P.A., Lima Sharma, A.L., Hekmaty, M.A., Jiang, Z., and Smirnov, D. (2012) *Appl. Phys. Lett.*, **100**, 071907.
76 Chen, X., Zhou, H.D., Kiswandhi, A., Miotkowski, I., Chen, Y.P., Sharma, P.A., Lima Sharma, A.L., Hekmaty, M.A., Smirnov, D., and Jiang, Z. (2011) *Appl. Phys. Lett.*, **99**, 261912.
77 Souza, S.M., Poffo, C.M., Trichês, D.M., de Lima, J.C., Grandi, T.A., Polian, A., and Gauthier, M. (2012) *Physica B*, **407**, 3781–3789.
78 Jacobsen, M.K., Kumar, R.S., Cornelius, A.L., Sinogeiken, S.V., and Nico, M.F. (2007) *AIP Conf. Proc.*, **955**, 171–174.

14
Predicting the Structure and Chemistry of Low-Dimensional Materials

Xiaohu Yu,[1] Artem R. Oganov,[2,1,3,4] Zhenhai Wang,[3] Gabriele Saleh,[1] Vinit Sharma,[5] Qiang Zhu,[3] Qinggao Wang,[1] Xiang-Feng Zhou,[3,6] Ivan A. Popov,[7] Alexander I. Boldyrev,[7] Vladimir S. Baturin,[1,8] and Sergey V. Lepeshkin[1,8]

[1] *Moscow Institute of Physics and Technology, Department of Problems of Physics and Energetics, 9 Institutskiy Lane, Dolgoprudny city, Moscow Region, 141700, Russia*
[2] *Skolkovo Institute of Science and Technology, Skolkovo Innovation Center, 3 Nobel St., Moscow 143026, Russia*
[3] *Stony Brook University, Department of Geosciences, Stony Brook, NY 11794, USA*
[4] *Northwestern Polytechnical University, International Center for Materials Discovery, Xi'an 720072, China*
[5] *Oak Ridge National Laboratory, Materials Science and Technology Division, Oak Ridge, TN 37831, USA*
[6] *Nankai University, School of Physics, Tianjin 300071, China*
[7] *Utah State University, Department of Chemistry and Biochemistry, Logan, UT 84322, USA*
[8] *Russian Academy of Sciences, P.N. Lebedev Physical Institute, Leninskiipr., 53, 119991 Moscow, Russia*

14.1
Introduction

Properties of a material are inextricably linked to its atomic structure, which makes a precise structural model a vital prerequisite to understanding and modeling most of its properties, as well as the physical phenomena that can occur in that system. However, the exact atomic structure is difficult to predict from just the chemical composition. This situation was described a few decades ago as a "continuing scandal in physical sciences" [1]. Now, with the advent of evolutionary and other methods, this problem can be considered as solved, at least for the most part. Such predictions are especially valuable where experiment is difficult or limited – extreme conditions (e.g., high pressure), low-dimensional materials, and the "holy grail" problem of designing novel materials entirely on the computer. Low-dimensional systems in our view rank among the most exciting subjects of modern scientific research, due to their novel physical and chemical properties tunable by dimensionality and particle size. It is well known that the structure, properties, and even chemical composition of

Handbook of Solid State Chemistry, First Edition. Edited by Richard Dronskowski, Shinichi Kikkawa, and Andreas Stein.
© 2017 Wiley-VCH Verlag GmbH & Co. KGaA. Published 2017 by Wiley-VCH Verlag GmbH & Co. KGaA.

low-dimensional systems can be very different from those of the bulk crystalline material. This makes their study particularly difficult – and particularly rewarding, especially in view of possible applications and guiding experimental efforts.

This chapter addresses modern challenges in computational science, within the context of the rapidly evolving field of nanotechnology. Originally, our evolutionary algorithm USPEX was developed for 3D crystals [2] (for some of its applications, see high-pressure phases of boron [3], transparent dense sodium [4], unexpected stable stoichiometries of Na_xCl_y [5]), but later extended to low-dimensional systems [6,7]. While one of our aims is to provide a clear understanding of how structure prediction algorithms work, and make the reader appreciate their strengths, this chapter is by no means a review of applications. Instead, this chapter summarizes the lessons and insight that we have learnt in our quest to understand the world of low-dimensional nanomaterials.

14.2
Methods

The first successful evolutionary algorithm for the prediction of cluster structures was developed by Deaven and Ho [8] and the first *ab initio* gradient embedded genetic algorithm (GEGA) method was developed by Alexandrova *et al.* [9,10]. There are a number of other major papers, in particular those written by Johnston on cluster structure prediction [11–13]. The adaptation of our crystal structure prediction method USPEX to clusters and other low-dimensional systems shares many similarities with the method of Deaven and Ho, but takes advantage of many special techniques for initialization of the population, constraint techniques, and selection mechanism, and involves many additional variation operators. Recent development of the USPEX algorithm focused on improving its efficiency for large systems, enabling automatic prediction of stable stoichiometries, and extending its applicability to different types of systems, for example, the clusters [7], polymers [14], 2D crystals [15], surface reconstructions [6], and so on. According to tests [7], performance of USPEX favorably compares to other methods for cluster structure prediction, while for inorganic crystals this method has unmatched efficiency and reliability, as evidenced, for example, by the blind test for inorganic crystal structure prediction [7].

14.2.1
Evolutionary Structure Prediction

Any evolutionary algorithm starts with a set of structures (population), and evolves them using selection (giving preference to better candidates to become parents) and specially designed variation operators in order to create offspring from parent structures. Variation operations, such as heredity (produces a child structure from two or more parents) and mutation (produces a child from a single parent), must retain some essential memory of parent structures in the

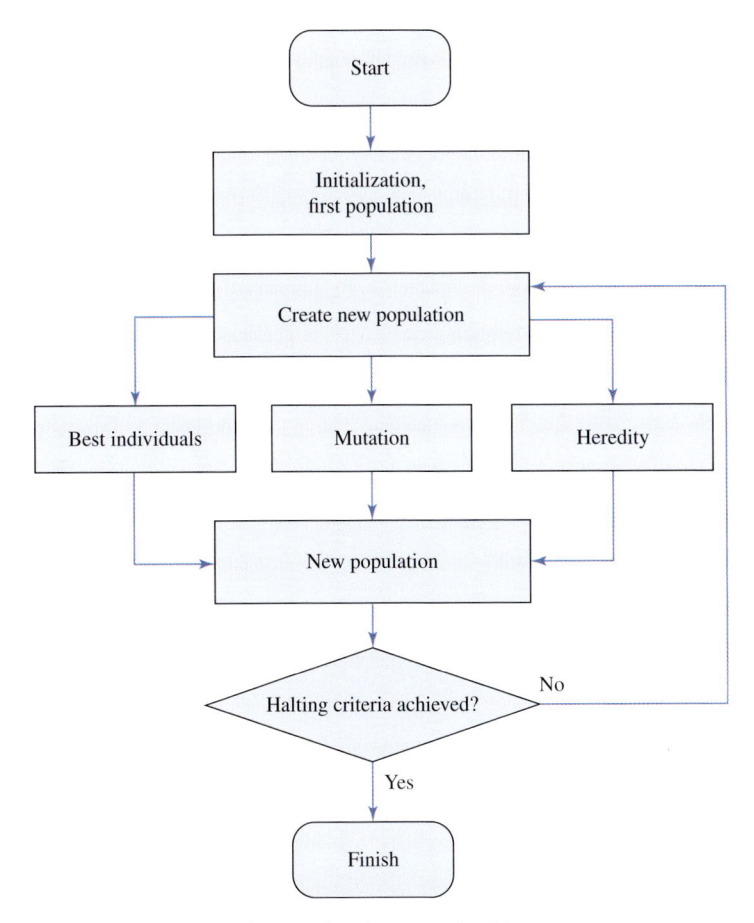

Figure 14.1 Flowchart of a typical evolutionary algorithm.

offspring, while also producing new structural features and retaining diversity of the population. In the following, we first review the main principles of evolutionary structure prediction, and then discuss how to predict materials of different dimensionality: clusters, polymers, surface reconstructions, and 2D materials.

Figure 14.1 shows a flowchart of a typical evolutionary algorithm. Usually, the initial population is produced randomly, unless some information (such as likely candidate structures, or lattice parameters, structural building blocks, symmetry, etc.) is available – this information can be used either for creating the initial population or for imposing constraints. All structures produced by the algorithm have to be locally optimized (i.e., relaxed) at a chosen level of theory; for example, empirical force fields, density functional theory, hybrid functionals, or quantum Monte Carlo. The new generation of individuals is then produced from the best individuals of the previous one by heredity and mutations. The calculation is stopped when the same best structure is found, without improvement, in sufficiently many generations.

14.2.2
New Developments

As the system size increases, certain difficulties arise. First, for most population-based global optimization methods, there is a danger of being trapped in some local minimum instead of the global one. The number of local minima rises exponentially with the system size and so does the danger of trapping, and therefore, one has to develop techniques to avoid this. In evolutionary algorithms, the second problem is "genetic drift" caused by the fact that the current best solution tends to create child structures similar to it, and after some generations these may come to dominate the population. There are several effective ways to cope with this problem (use of symmetries and other ways to reduce effective dimensionality of the problem, fingerprint functions for calculating structure similarity, antiseed technique for avoiding getting stuck in a local minimum, better design of variation operators), but it is not clear how well they work for very large systems (with thousands of degrees of freedom).

Increase of the number of degrees of freedom leads to the decline in the quality and diversity of random structures. A new initialization scheme was proposed [7,16] to deal with this problem, the idea of the scheme being to reduce the number of degrees of freedom by randomly applying symmetry to randomly generated atomic coordinates. For crystals, additional translational symmetries and pseudosymmetries are used, as well as the full apparatus of 230 space groups. For 2D systems, we use 17 plane groups or 80 layer groups. For clusters, an interesting dilemma appears – the number of possible symmetry groups (point groups) is infinite. We take a pragmatic solution, in actual calculations using 32 crystallographic point groups and the most important noncrystallographic groups (pentagonal, decagonal, icosahedral), although the program allows any additional point groups to be specified. Using symmetry in this fashion has the advantage of being unbiased, providing very diverse structures and covering the search space better. In this procedure, during relaxation, the initial symmetry of the structures can increase. Building in small random distortions in the generated symmetric structures, we also allow symmetry to be broken during relaxation. Variation operators also break symmetry and enable totally new structures with very different symmetries to emerge. It is also beneficial to add random symmetric structures in every generation. This development alone does not fully solve the problems of "curse of dimensionality," and variation operators also must be improved, so that not only good structures are given preference for creating offspring, but also good fragments of structures are identified and nurtured. These ideas are applied both to 3D crystal and low-dimensional systems.

14.2.2.1 Smart Variation Operations and Clustering

Variation operators are the heart of an evolutionary algorithm. They produce new structures out of those that were already sampled and drive the search toward the globally optimal structure. Stochastic nature of variation operators is both a blessing and a curse. On the one hand, randomness is required for ensuring diversity of the population, for avoiding any bias in exploring the search space, and for slowing down the genetic drift. This is why random selection and random mutations of spatially coherent slabs for heredity are preferred. On the other hand, completely random variation operators may be inefficient when applied to large systems. Therefore, "smart" operators that retain a degree of randomness should be designed, in order for the operators to be directed by the system itself to choices that have a higher probability to improve the fitness.

a) **Improved heredity using order–fitness correlation**

Heredity operator takes spatially coherent slabs from two parent structures and combines them to form a single child structure. It would be desirable to select defect-free slabs, because once a defect is passed from a parent to a child, it may take many generations to get rid of it. The concept of local order parameter [17] allows selection of less defective slabs from the parents. Order parameter quantifies the degree of symmetry of the environment of a given atom, and one can define an average order parameter in a structural fragment, and when cutting slabs, preference can be given to those slabs where this value is higher. However, it is wiser to first calculate the correlation between the average atomic order of structures and their energy (or the physical property of interest) and, depending on the strength of (anti)correlation, the extent to which the order parameter will influence variation operators.

b) **Softmutation**

The basic idea is to move the atoms along the eigenvectors of the softest modes, since low-frequency phonon modes correspond to directions of low curvature of the energy surface and such displacements will likely encounter low-energy barriers, beyond which new low-energy structures will be discovered. Physical idea behind such mutation is to move the system into a new structure across the lowest-energy barrier. Such mutations enable a very effective local and semilocal exploration of the energy landscape.

c) **Clustering**

It is very important to keep the population diverse, while enhancing sampling of the low-energy part of configuration space. To that aim, we allow part of the current generation (typically, 10%) to survive into the next generation, and these structures are selected based on a compromise between lowest energies and maximum diversity. This can be done by combining fingerprint theory with an appropriate clustering algorithm. After full relaxation, the lowest energy part of the population is clustered into groups using some value of the fingerprint distance threshold (d_{min}, for more details, see Ref. [7]): (1) First, a group is formed with structures differing by less than d_{min} from the lowest-energy structure. (2) This process is repeated with all ungrouped structures until they all are assigned into some groups. From each group, only one structure (with the lowest energy among that group) survives into the next generation.

Now, we move on to explain specific details of different methods for the prediction of structures of clusters, polymers, and surface reconstructions.

14.2.3
Cluster Structure Prediction

Cluster (nanoparticle) structure prediction has similarities to crystal structure prediction, but there are some critical differences. For consistency with most widely used codes, our implementation uses periodic boundary conditions. This produces reliable results if there is enough "vacuum" around the cluster to eliminate its interaction with its periodic images. We introduce two rectangular parallelepipeds – "big cell" that includes sufficiently large vacuum, and "small cell," that includes the smallest rectangular built around the cluster, as shown in Figure 14.2. The size of "small cell" is adjusted as the cluster geometry changes during relaxation. The thickness of the vacuum region around the cluster is a user-defined parameter; more vacuum means more accurate results, but greater computational costs (in particular, for plane wave methods).

When the cluster is generated or relaxed, its geometric center is placed at the center of the unit cell and it will be rotated to align the principal moment of

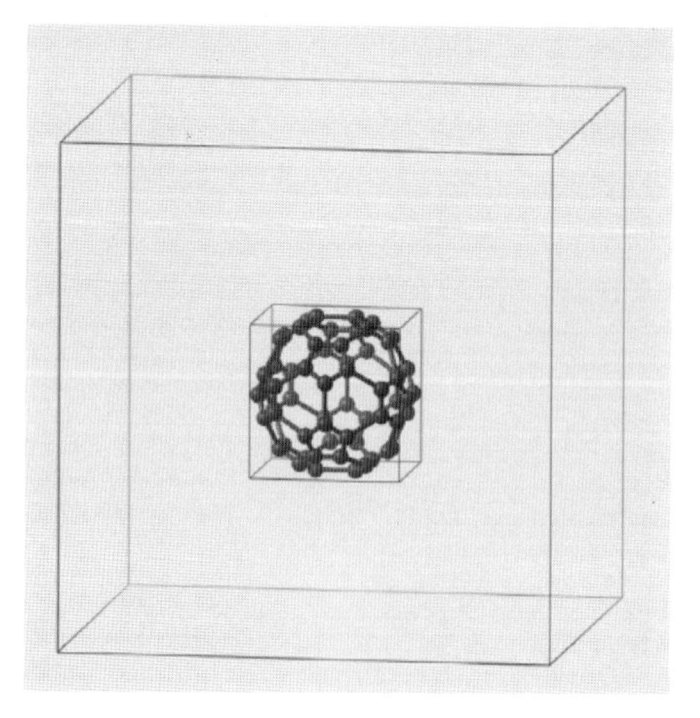

Figure 14.2 Cluster in a "small cell" surrounded by vacuum. During relaxation, atoms are allowed to get outside of "small cell" and the large cell is adjusted accordingly, to maintain enough vacuum. (Reprinted with permission from Ref. [7].)

inertia axis with the highest moment along the z-direction, and the other principal components along the x- and y-axes. Variation operators are quite similar to those applied in 3D crystals, but there are some differences. For example, to cultivate diversity, clusters are randomly rotated around a random axis that goes through their center of mass before performing a "cut-and-splice" heredity.

The standard test system, Lennard-Jones clusters, which have been thoroughly studied in the past two decades [18–27], are used for comparing the performance of our cluster structure prediction algorithm with other algorithms. Statistics were compared with the state-of-the art minima hopping (MH) algorithm [19,24], other evolutionary algorithms (EA) [19,20], and PSO algorithm [18], and are summarized in Table 14.1. Comparison clearly shows a much stronger performance of USPEX among these methods. For more details, see Ref. [7].

14.2.4
Polymer Structure Prediction

Now we introduce the very recent development of evolutionary algorithms for predicting the structure of polymers [14]. The procedure is illustrated in Figure 14.3.

a) *Define monomeric blocks:* Z-matrix is used to represent the monomeric units – polymers can be made of one or several types of blocks. The atom's bond

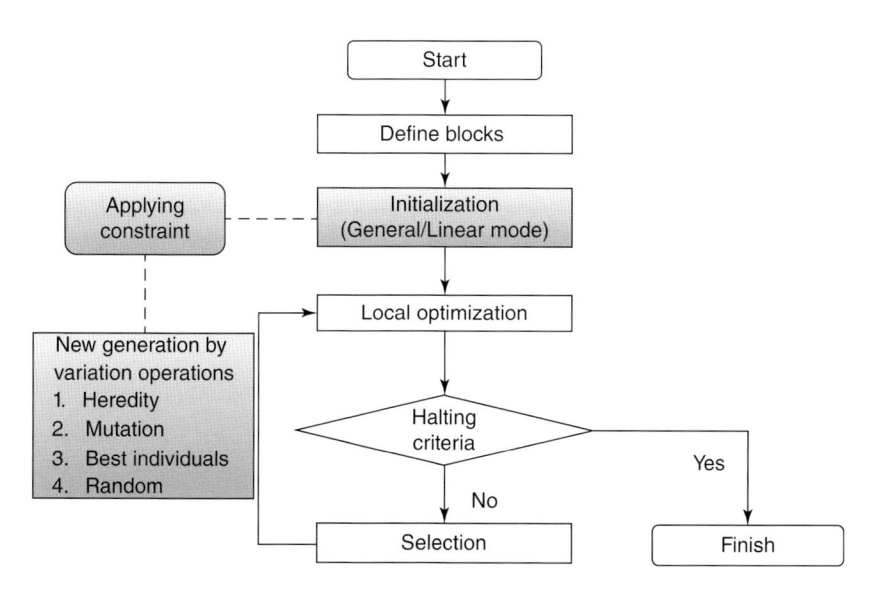

Figure 14.3 Flowchart of the constrained evolutionary algorithm. The key feature is that all generated structures before local optimization are constructed based on the prespecified building blocks (highlighted in gray box). "Linear mode" means that the polymeric chain runs parallel to the crystallographic c-axis. (Reprinted with permission from Ref. [14].)

Table 14.1 Performance of different algorithms for Lennard-Jones clusters.

	Symmetric initialization	Antiseeds	Order	Success rate (%)	Average number of structures until global minimum is found	Dispersion	Number of calculations
LJ_{38}(PSO [18])	+	−	−	100	605	N/a	100
LJ_{38}(USPEX)	+	+	−	**100**	**35**	**58**	**183**
LJ_{38}(USPEX)	−	−	−	67	2291	1443	100
LJ_{38}(USPEX)	−	+	−	98	3080	2119	100
LJ_{38}(EA [19])[a]	−	−	−	N/a	1265	N/a	100
LJ_{38}(MH [19])[a]	−	−	−	100	1190	N/a	100
LJ_{38}(EA [20])[a]	−	−	−	N/a	~2000[b]	N/a	N/a
LJ_{38}(PSO [18])	−	−	−	100	1649	N/a	20
LJ_{44}(USPEX)	−	−	−	100	1510	1079	35
LJ_{44}(USPEX)	−	+	+	**100**	**859**	**524**	**37**
LJ_{44}(USPEX)	+	+	+	100	1129	765	41
LJ_{44}(USPEX)	+	−	−	86	1551	1020	35
LJ_{44}(USPEX)	+	+	−	94	1423	867	35
LJ_{55}(PSO [18])	+	−	−	100	159	N/a	100
LJ_{55}(USPEX)	+	−	−	**100**	**11**	**30**	**60**
LJ_{55}(USPEX)	−	−	−	100	717	407	103
LJ_{55}(EA [19])[a]	−	−	−	100	100	N/a	100
LJ_{55}(MH [19])[a]	−	−	−	100	190	N/a	100
LJ_{75}(PSO [18])	+	−	−	98	2858	N/a	50
LJ_{75}(USPEX)	+	+	+	**100**	**2145**	**2024**	**53**
LJ_{75}(USPEX)	+	+	−	100	5419	4513	47

Best algorithms are highlighted in bold.
Source: Reprinted with permission from Ref. [7].
a) Results for algorithms [19,20] are given for optimized algorithm parameters. We did not attempt to optimize USPEX parameters and chose them based on intuition and experience. We could achieve even greater advantage over other approaches with parameter optimization, but did not do this because in realistic calculations (using quantum-mechanical energy evaluations) one cannot afford parameter tuning.
b) Depending on algorithm parameters. 2000–12 000 structures were required to reach the global minimum in Ref. [20].

connectivity in a given molecule can be defined by internal coordinates (i.e., bond lengths, bond angles, and torsion angles).

b) *Initialization:* Random symmetric structures are used for producing the first generation of structures. A fully random initialization is usually a poor choice for large systems, as it always leads to nearly identical glassy structures that have similar (and high) energies and low degree of order. It is difficult to obtain ordered ground states from such a starting population. In order to achieve both high structural diversity and quality, a better way is to create symmetric structures for the initial population using randomly selected symmetry groups with randomly generated initial positions of particles.

c) *Local optimization (structure relaxation):* This is done stepwise with increasing precision from low to high, to achieve greater efficiency.

d) *Selection:* At the end of each generation, all relaxed structures in the generation are compared using their fingerprints, and all nonidentical structures are ranked by their energies.

e) *Variation operators:* Child structures (new generation) are produced from parent structures (old generation) using two general types of variation operators: heredity and several kinds of mutations.

Heredity: This is a basic variation operator that cuts planar slices from two selected individuals and combines them to produce a child structure. During this process, each monomer is represented by its geometric center and orientation matrix, and the entirety of each monomer is retained during the operation of heredity, as well as all other variation operators.

Rotational mutation: A certain number of randomly selected monomers are rotated by random angles around the main axis of the polymer.

Softmutation: This operator involves atomic displacements along the softest mode eigenvectors, or a random linear combination of softest eigenvectors.

Addition of random structures: Although the search space has been effectively decreased by applying geometric constraints, we are still facing a high-dimensional configuration space. A general challenge for evolutionary algorithms (and many other global optimization methods) is how to avoid getting stuck in a local minimum when dealing with multidimensional search spaces – In other words, avoiding decrease of population diversity during evolutionary run. A key to maintain the population diversity is to add "new blood." Therefore, some fraction of each generation is produced using the random symmetric algorithm. Fingerprint niching also helps to maintain diversity.

f) *Stopping the simulation and postprocessing:* When the lowest-energy structure remains unchanged for a certain number of generations, the calculation automatically stops and the lowest-energy structures found in USPEX are then carefully relaxed with higher precision.

14.2.5
Prediction of Surface Reconstructions

In this part, we discuss the newly developed method for predicting surface reconstructions [6], implemented in USPEX. Figure 14.4 shows the standard representation of surfaces in periodic calculations, with three main regions: substrate, surface, and vacuum. Thickness of vacuum region and geometry and thickness of substrate are prespecified, while the surface region is fully variational and optimized by the evolutionary algorithm. The number of surface atoms varies up to a predefined maximum number for a given thickness of

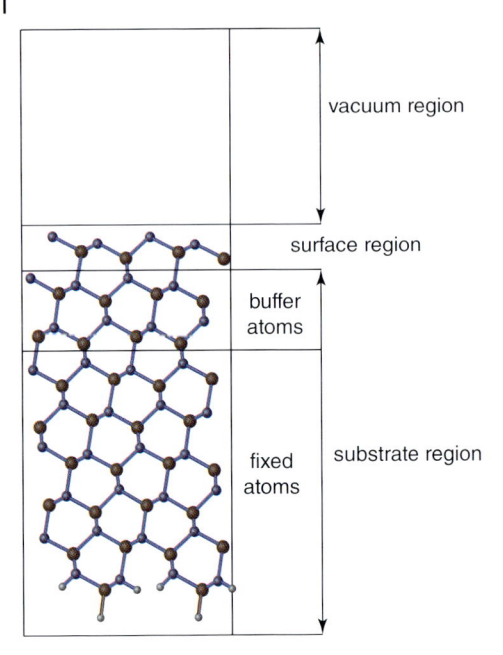

Figure 14.4 Representation of surfaces. (Reprinted with permission from Ref [6].)

surface. To allow automatic exploration of reconstructions with different surface unit cells, the cell size is also variable. It is advisable to divide substrate into two parts – fixed substrate representing the bulk, and "buffer" region that has structural topology of the substrate, but with atoms allowed to relax.

For predicting surface reconstructions, we create an initial population by placing a variable number of atoms of each type in the surface region on top of fixed substrate. The initial structures are then relaxed and ranked based on their fitness. Structures with better fitness are more likely to be selected as parents to generate new child structures. Three ways are used to produce offspring.

1) *Heredity:* Two structures are chosen from the previous generation. They are randomly sliced at the same position in the surface unit cell. Then pieces from both parent structures are combined to generate the offspring.
2) *Softmutation:* One structure is chosen as parent. Surface atoms are displaced along the eigenvectors of the softest surface modes.
3) *Transmutation:* One structure is chosen as parent. Randomly selected atoms are transmuted into another chemical species.

The offspring, together with a few best structures from the previous generation, comprise the next generation. This process is repeated until no lower energy structures are produced for sufficiently many generations. For more details about surface reconstruction prediction, see Ref. [6].

14.3
Results for Low-Dimensional Systems

14.3.1
0D Systems: Nanoclusters

During the last decades, nanoparticles, conventionally defined as atomic aggregates of size in the range of 1–100 nm, have attracted a great deal of attention from the scientific community. This is due to their peculiar properties, which qualitatively differ from their bulk counterparts [28]. Nanoparticles can be thought of as entities bridging the gap between small molecules and bulk materials. They can exhibit a range of fascinating chemical, optical, electronic, and magnetic properties, which are not observed in the corresponding bulk material. This distinct behavior is due to a variety of factors, including the high ratio of surface to bulk atoms, geometric shell closing, whole-cluster electronic orbitals and shell closing (e.g., as in the jellium model) and superatomic character, local chemical bonding, and some of these are reflected in the Hückel rule for aromatic clusters, and the Wade–Mingos rule for boranes. Accordingly, they exhibit a unique, heavily size-dependent behavior, of which reactivity of gold [29], paramagnetism of noble metals [30], and tunability of semiconductor bandgap with particle dimension [31] are but a few examples.

Much effort has been spent in developing models capable of predicting and explaining why clusters with certain compositions, that is, the so-called magic clusters, display an enhanced stability. The "islands of stability" that appear in clusters can be due to structural (filling geometric "shells," as in icosahedral clusters with 13, 55, 147, . . . atoms) or electronic (filling maximum number of bonding orbitals) reasons. Indeed, when nanoclusters are synthesized by cooling down a hot gas of atoms, mass spectra show that clusters with such magic numbers of atoms are far more abundantly present in the final product [31]. These particles are important for practical applications and may serve as nuclei or building block for crystallization.

To explain how certain electronic configurations confer increased stability, two concepts are usually invoked: the "superatom" model and the aromaticity of metal clusters. The concept of "superatom" was initially put forward to rationalize stability of certain simple spherical clusters of metals such as Li and Na. The main idea, demonstrated in 1984 through the use of "jellium model" [32], was that when electrons are delocalized throughout the cluster, its electronic structure develops whole-cluster orbitals that resemble atomic orbitals, making the whole cluster a kind of "superatom." One of the main consequences is that clusters become particularly stable when the number of electrons is just enough for the completion of an electronic shell, in a fashion that is reminiscent of noble gases. Further studies in the successive decades have shown that, in fact, a wide variety of metal-containing clusters display an atom-like behavior [33]. Aromaticity is a well-known paradigm in organic chemistry. In the late 1970s, it was extended to metal clusters displaying properties akin to benzene [34]. However, it was not until much later that this concept started to be widespread in the field of nanocluster research. In fact, the seminal papers investigating Al_4^{4-} and related clusters [35,36] triggered a plethora of new studies dealing with aromaticity of metal clusters. Nowadays, a number of different flavors of the aromaticity concept are available to rationalize stability of metal clusters, for example, σ- π- and δ-aromaticity and antiaromaticity, sometimes simultaneously present in the same cluster and competing with each other. For more details, we refer the reader to Refs [37,38].

In this section, we review some recent results on oxide and hydride clusters. The focus is primarily on their stability and peculiar chemical bonding features.

Ce_mO_n [39]

Cerium oxide is a material with a widespread range of technological applications, for example, in gas sensors, fuel cells, water splitting devices, catalytic converters of exhaust gases, and in catalysis in general. All of them are directly related to the ease with which the Ce^{3+}/Ce^{4+} redox process, and the associated release/uptake of oxygen, can take place. As most of the aforementioned applications occur on the surface, the efficiency of cerium oxide greatly improves when it is employed in the nanoparticle form. Consequently, understanding stability/reactivity of cerium oxide nanoclusters through electronic structure analysis is a fundamental step for improving its technological performance. As an example of such investigation, we discuss structural evolution and chemical bonding in Ce_6O_n ($n=1-14$) nanoclusters along their progressive oxidation, carried out by means of the evolutionary structure prediction algorithm USPEX [7,40,41] coupled with DFT+U calculations. The most stable isomers of each cluster are

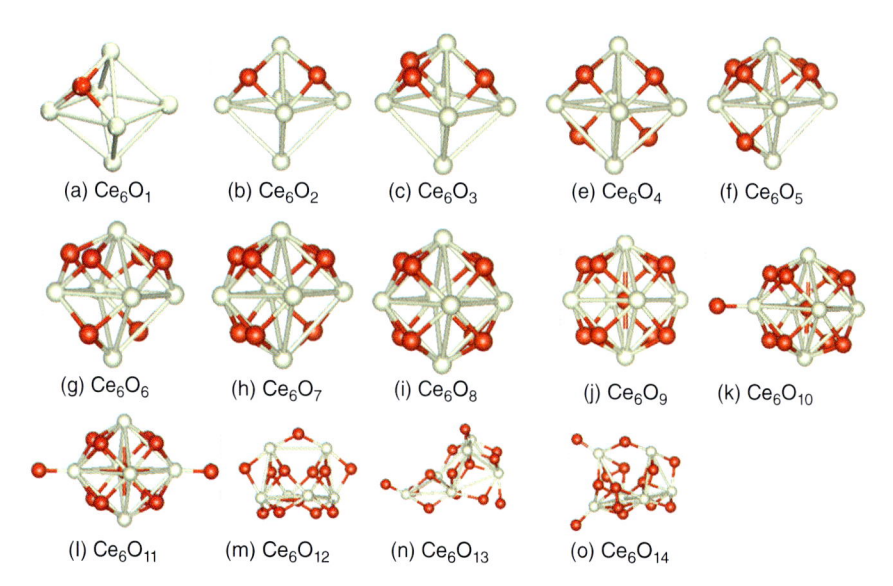

Figure 14.5 Ground states of Ce_6O_n ($n=1-14$) clusters. (Reprinted with permission from Ref. [39].)

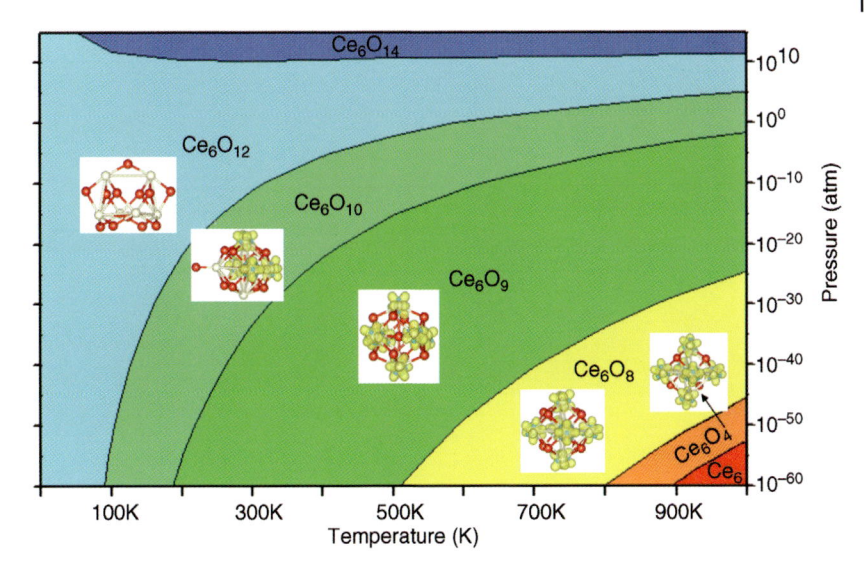

Figure 14.6 Phase diagram of Ce_6O_n in oxygen atmosphere. The spin density distribution (yellow) of the corresponding clusters is shown in insets. (Reprinted with permission from Ref. [39].)

shown in Figure 14.5. It is interesting to note that in clusters with $n = 1–11$, the Ce_6 octahedral core is invariably present. Oxygen atoms occupy hollow sites ($n = 1–8$), the interior of the cluster ($n = 9$), and outside positions ($n = 10, 11$).

Stating from $n = 12$, the structure of the octahedral Ce_6 core is destroyed. More interesting is that the six cerium atoms keep the octahedral Ce_6 core for lower concentration oxygen cerium oxide cluster, and both Ce_6O_8 and Ce_6O_9 have O_h symmetry.

Figure 14.6 shows the phase diagram of Ce_6O_n clusters in oxygen atmosphere; To obtain this figure, we performed analysis similar to the one in Ref. [42]. One can see that cerium oxide clusters with low oxygen content exist at high temperatures and lower oxygen pressures, while higher oxygen content clusters mainly exist at low temperatures and higher oxygen pressures. In particular, Ce_6O_8 cluster has a rather wide stability field.

Electron localization function (ELF) [43] of Ce_6O_8 shows concentration of charge density in the center of bare Ce_6O_8 cluster that is indicative of a multicentre covalent bond (Figure 14.7).

To interpret this, we use the natural bond orbital (NBO) [44] analysis and its extension adaptive natural density partitioning analysis (AdNDP) [45]. NBO allows determination of Lewis elements of localized bonding, such as lone pairs and 2c–2e bonds (classical two-center two-electron bonds), while AdNDP can find delocalized bonding (nc–2e bonds, $n > 2$) in clusters [46] and solid-state systems [47,48]. Ce atom has four valence electrons and O atom takes two electrons, so the total balance is $6*4(Ce^{4+}) - 8*2(O^{2-}) = 2$ electrons. From

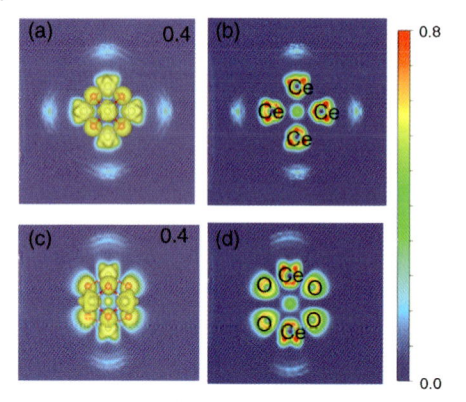

Figure 14.7 (a) 3D and (b) 2D images of the electron localization function (ELF) of Ce_6O_8 cluster; (c) 3D and (d) 2D ELF of Ce_6O_8 cluster. (Reprinted with permission from Ref. [39].)

Figure 14.8d, we can clearly see that these two electrons are shared by six Ce atoms. Thus Ce_6O_8 cluster represents the first case of spherical d-orbital σ-bond aromaticity in chemistry.

This picture of chemical bonding in Ce_6O_8 explains the wide stability field of another magic cluster, Ce_6O_9. The delocalized bonding orbital has a low ionization energy (similar to low work function of metals and, especially, electrides), and Ce_6O_8 cluster can be easily oxidized, donating its delocalized electron pair to an oxygen atom placed at the center of the cluster to form Ce_6O_9.

Ti_mO_n [49]

TiO_2 has been studied extensively because of its photocatalytic properties and numerous applications. For example, nanoparticulate TiO_2 is used in dye-sensitized solar cells. In order to maximize performance of the TiO_2 anode in Li–ion batteries, different TiO_2 nanostructures, such as nanoparticles, nanotubes, nanowires, and mesoporous structures, are studied. As the work on nanostructured TiO_2 gains importance, computational studies of TiO_2 clusters could provide valuable insights into their structures and properties.

The USPEX algorithm, extended to study magnetic clusters and augmented with a new variation operator (spin mutation, which changes the direction and magnitude of atomic spin moments), was used to look for stable Ti_8O_n ($n = 1$–18) clusters. In the calculation, 10% of new structures were produced by spin mutation, 40% by heredity, 20% by softmutation, 10% by permutation, and the remaining 20% by the random symmetric algorithm. To account for strong electron correlations in Ti_8O_n clusters, DFT + U ($U_{eff} = 3$ eV) method was used. In order to get accurate electronic structure of the Ti_8O_{12} cluster, accounting for its multiconfigurational nature, we used CASSCF calculations with 8×10 active space.

Figure 14.9 shows the lowest-energy structures of Ti_8O_n ($n = 1$–18) clusters obtained with USPEX at the GGA + U level of theory. One of magic clusters,

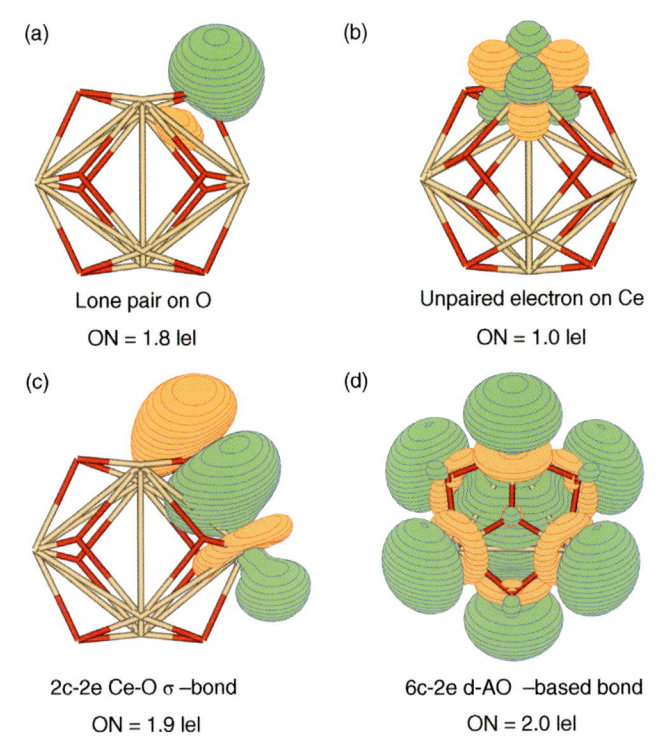

Figure 14.8 Representative elements of chemical bonding of the Ce$_6$O$_8$ cluster revealed by the NBO and AdNDP methods. (a) Lone pair on oxygen. (b) Unpaired electron on Ce. (c) 2c–2e Ce—O σ-bond. (d) 6c–2e σ-bond. "ON" denotes occupation number here and elsewhere. (Reprinted with permission from Ref. [39].)

Ti$_8$O$_{12}$, with wide stability, has eight Ti atoms in corner of a cube and 12 O atoms near midpoints of all edges of that cube.

Why Ti$_8$O$_{12}$ has the high stability can be understood through analysis of chemical bonding. Figure 14.10 shows partial charge plots. We can clearly see electron density localization inside the Ti$_8$ fragment from both HOMO and HOMO-1.

In order to get more insight into the electronic structure of the Ti$_8$O$_{12}$ cluster, NBO [44] and AdNDP [45] methods are used to further interpret the results described above. In total, there are eight valence electrons responsible for direct Ti–Ti bonding in the cubic Ti$_8$O$_{12}$ cluster: 8 titanium atoms give a total of $8 \times 4 = 32$ electrons for bonding and 12 oxygen atoms can get in total 12×2 (O^{2-}) = 24 electrons. Ti$_8$O$_{12}$ cluster may have cubic d-orbital σ-aromaticity according to the 6n + 2 or 2(n + 1) [2] electron counting rule in the case of nonmagnetic cluster. However, the electronic ground state of cubic Ti$_8$O$_{12}$ cluster is antiferromagnetic: its antiferromagnetic state is 0.74 eV lower in energy than the ferromagnetic state in DFT+U calculations, while CASSCF calculations show an even greater value of 1.60 eV. The AdNDP method was used to analyze the wave

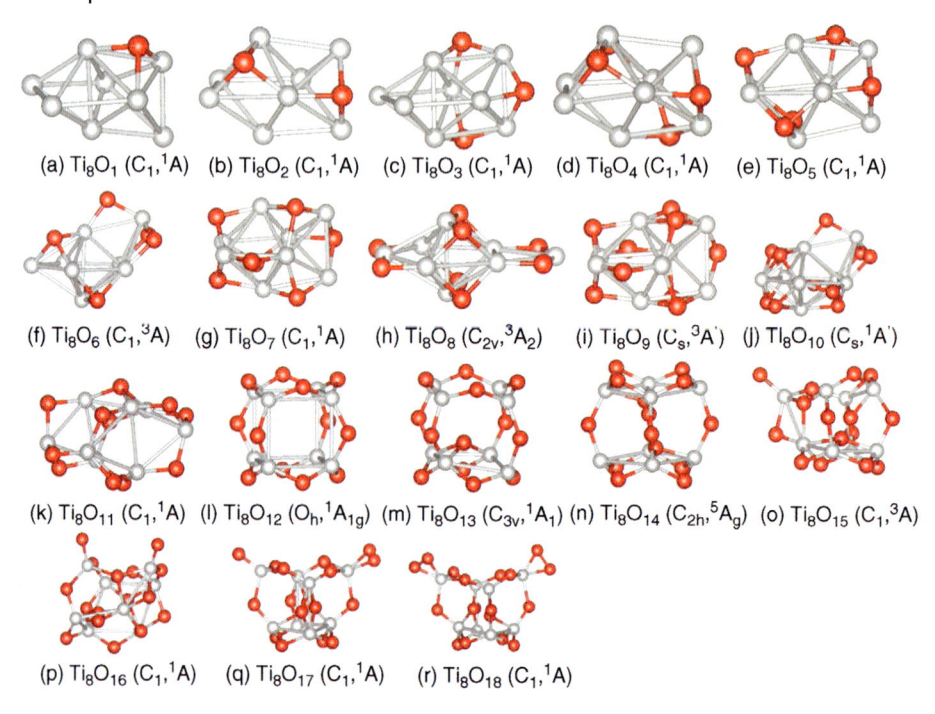

(a) Ti$_8$O$_1$ (C$_1$,^{1}A) (b) Ti$_8$O$_2$ (C$_1$,^{1}A) (c) Ti$_8$O$_3$ (C$_1$,^{1}A) (d) Ti$_8$O$_4$ (C$_1$,^{1}A) (e) Ti$_8$O$_5$ (C$_1$,^{1}A)

(f) Ti$_8$O$_6$ (C$_1$,^{3}A) (g) Ti$_8$O$_7$ (C$_1$,^{1}A) (h) Ti$_8$O$_8$ (C$_{2v}$,^{3}A$_2$) (i) Ti$_8$O$_9$ (C$_s$,^{3}A') (j) Ti$_8$O$_{10}$ (C$_s$,^{1}A')

(k) Ti$_8$O$_{11}$ (C$_1$,^{1}A) (l) Ti$_8$O$_{12}$ (O$_h$,^{1}A$_{1g}$) (m) Ti$_8$O$_{13}$ (C$_{3v}$,^{1}A$_1$) (n) Ti$_8$O$_{14}$ (C$_{2h}$,^{5}A$_g$) (o) Ti$_8$O$_{15}$ (C$_1$,^{3}A)

(p) Ti$_8$O$_{16}$ (C$_1$,^{1}A) (q) Ti$_8$O$_{17}$ (C$_1$,^{1}A) (r) Ti$_8$O$_{18}$ (C$_1$,^{1}A)

Figure 14.9 Ground states of Ti$_8$O$_n$ ($n = 1$–18) clusters. (Reprinted with permission from Ref. [49].)

function of the antiferromagnetic Ti$_8$O$_{12}$ obtained from the CASSCF calculations to understand how the chemical bonding picture of antiferromagnetic cluster compares to the nonmagnetic case. The s-type and p-type lone pairs on oxygen atoms as well as all the 2c–2e Ti–O σ-bonds were found to be exactly the same as for the nonmagnetic cluster with just slight differences in the occupation number (ON) values. The only difference is that the totally symmetric 8c–2e σ-bond (Figure 14.11) (originating from the bonding HOMO-1 (a$_{1g}$) orbital) was found to have lower ON value of 1.53 |e| instead of 1.89 |e|. Furthermore, three 8c–2e σ-bonds (Figure 14.11d–f) originating from the triply degenerate HOMO t$_{1u}$ orbital have appreciably lower ON values: 1.14 |e| instead of 1.87 |e|. There is indeed a certain stabilization coming from some d-AO overlap of eight Ti electrons, but it cannot be called aromatic as in the case of the hypothetical nonmagnetic structure with four bonding orbitals, each occupied by ∼2 electrons. In fact, superexchange plays an important role as eight Ti atoms separated from each other by 12 O atoms are forced to adopt an antiferromagnetic configuration to maximize d-orbital interactions. This is the first finding of antiferromagnetism in a small cluster. We see unique chemical bonding with eight electrons of Ti atoms interacting with each other in antiferromagnetic fashion to lower the total energy of the system.

(a) HOMO (isosurface: 0.001 e/Å³)

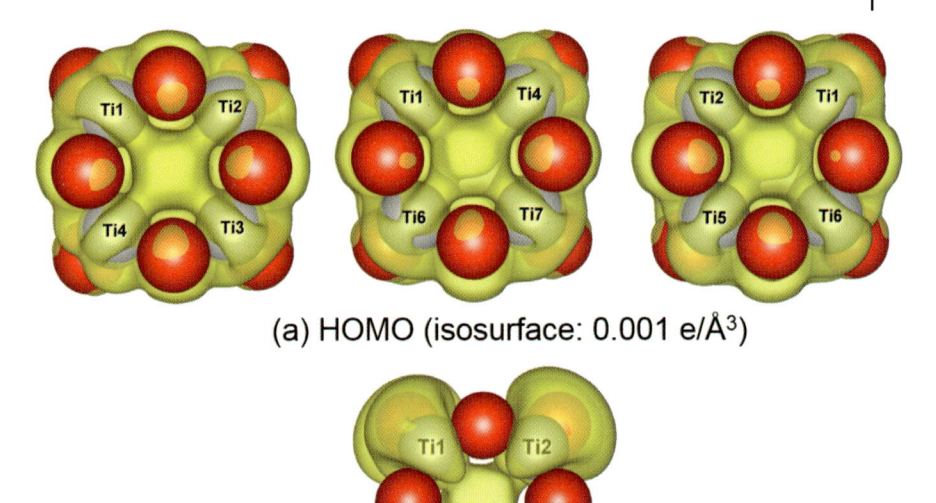

(b) HOMO-1 (isosurface: 0.0005 e/Å³)

Figure 14.10 Partial charge density of Ti_8O_{12} cluster. (a) HOMO. (b) HOMO-1. (Reprinted with permission from Ref. [49].)

14.3.1.3 $Si_{10}H_{2m}$

Silicon nanoclusters are in the center of intense theoretical and experimental studies for over two decades. The reasons of such a great attention come from several different fields. These systems are promising objects for further miniaturization of silicon-based microelectronic devices via the use of transistors on quantum dots [50]. Silicon clusters with different dopant atoms exhibit photoluminescence that distinguishes them from bulk silicon in a fundamental way (see, for example, Ref. [51]). This phenomenon, having a number of possible applications, has no clear explanation yet, although it is obvious that surface atoms and size-induced rearrangements of atoms play a crucial role here. Thus, knowledge of the atomic structure is of great importance. Another problem relevant for applications is the determination of conditions for mass production of identical particles. Most of the widely used cluster synthesis methods, such as laser ablation, produce ensembles of atoms, which are, in general case, nonuniform [52].

Here we present results of structure analysis and thermodynamic consideration of small silicon clusters, passivated by hydrogen, with formula $Si_{10}H_{2m}$, ($m = 0 \cdots 11$) [53].

We start with ground-state structures of clusters, obtained from *ab initio* USPEX searches, and presented in Figure 14.12.

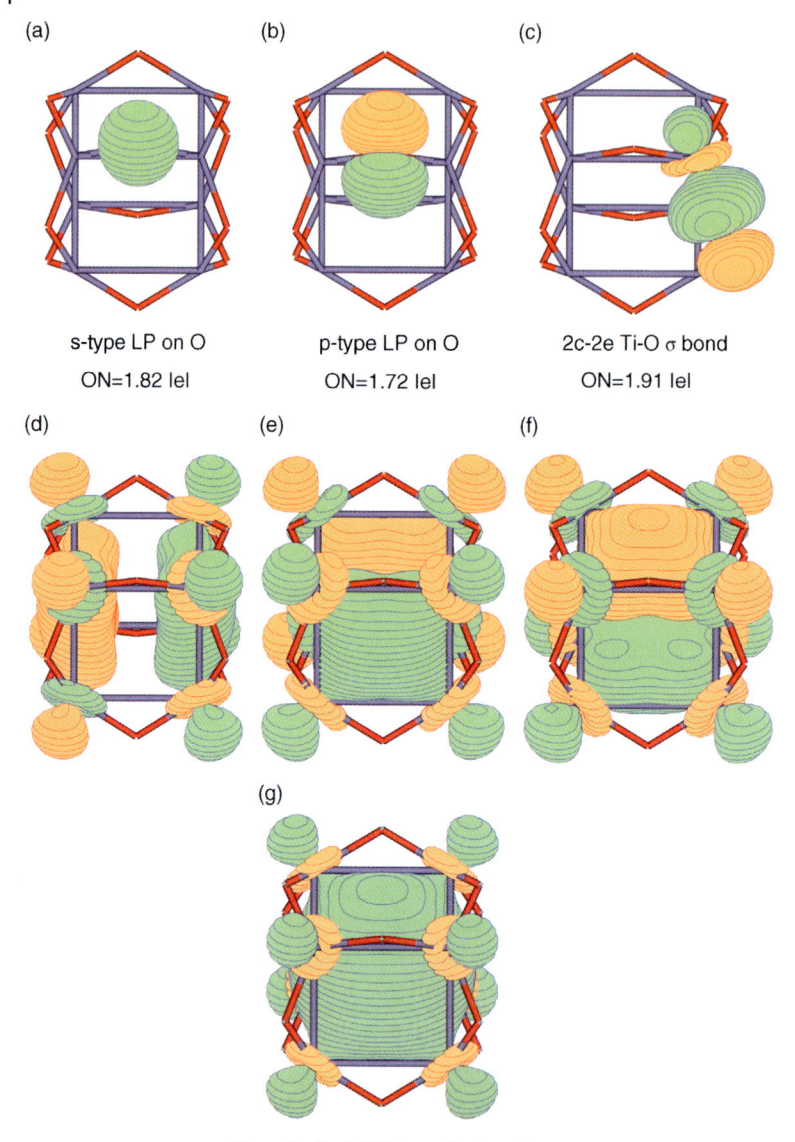

Figure 14.11 Representative chemical bonding elements of the Ti_8O_{12} cluster revealed via AdNDP method. (a) One s-type lone pair on oxygen. (b) One p-type lone pair on oxygen. (c) One 2c–2e Ti—O σ-bond. (d–g) Four 8c–2e d–AO-based σ-bonds. The occupation numbers (ON) shown were obtained for the nonmagnetic electronic configuration; as discussed in the text, these numbers remain virtually unchanged in the true antiferromagnetic configuration – except the major decrease of the ON of the 8c–2e bonds. (Reprinted with permission from Ref. [49].)

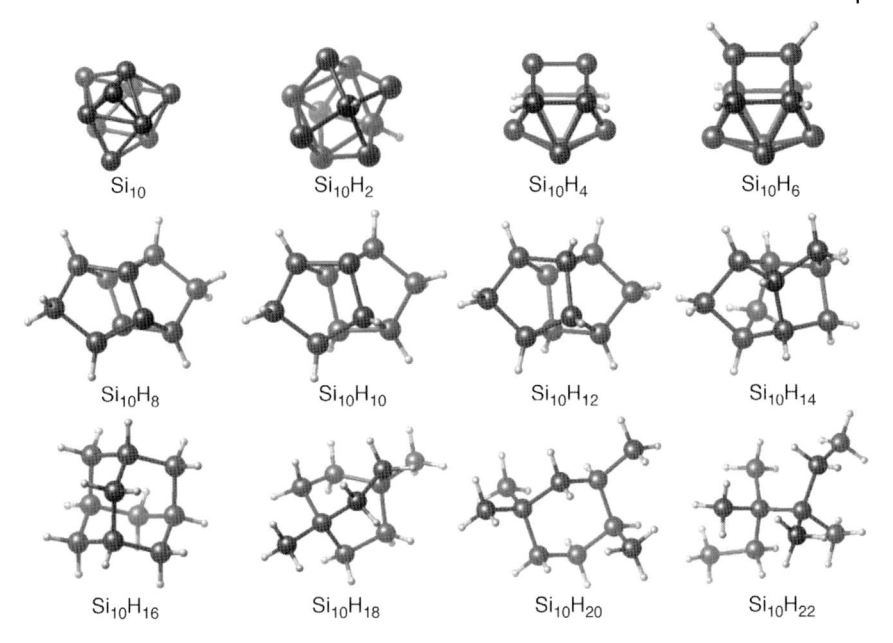

Figure 14.12 Ground-state structures of $Si_{10}H_{2m}$ clusters ($0 \leq m \leq 11$). (Reprinted with permission from Ref. [53].)

First of all, it is clear that when the size of clusters is small, they show great variability of structures depending on the number of passivating hydrogen atoms. Second, it is only in the case of $Si_{10}H_{16}$, that the atomic arrangement resembles the diamond-like structure of bulk silicon. As the degree of passivation grows, clusters become more open and branchy, as hydrogen atoms, forming only one bond, are always situated on the surface of clusters. Clusters with $m > 11$ are unstable.

Looking at low-passivated structures, $Si_{10} - Si_{10}H_{10}$, we see that unlike in the usual silicon compounds, silicon atoms have coordination numbers different from the standard valence of silicon, that is, four. This can be explained by means of graph theory. On the one hand, silicon atoms tend to have four neighbors. On the other hand, bonds of four-coordinate silicon atoms, for example, in bulk silicon, form tetrahedral angles 109°47'. We build the graph, corresponding to the cluster, in the following way. Hydrogen atoms are neglected, and the edges of graph represent the nonbranching chains of Si—Si bonds, so that Si atoms participating in three or more Si—Si bonds are represented as graph's vertices (Figure 14.11).

It is clear from Figure 14.13, that a silicon atom forms a vertex if it is bound to one or zero hydrogen atoms. Assuming that graphs constructed in such way are planar, which is true for all $Si_{10}H_{2m}$ clusters under consideration, we can say that it satisfies cycle double cover problem, meaning that it can be drawn on a sphere as a superposition of minimal cycles, with each graph edge participating only in

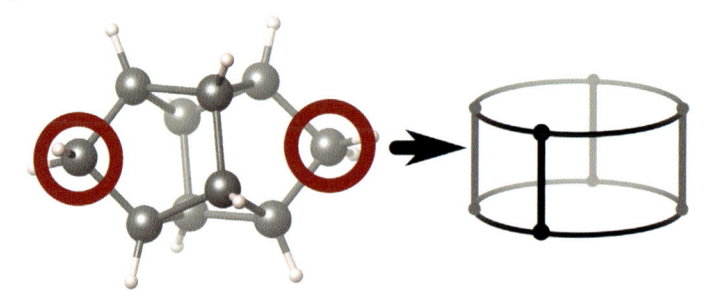

Figure 14.13 Cluster and corresponding graph. Atoms highlighted by circles do not form vertices of the graph.

two cycles (all further assumptions concerning graphs do not affect the generality of results). Such cycles are called faces (by analogy with convex polyhedra). If a graph has v vertices, e edges, and f faces, it satisfies Euler formula:

$$f = e - v + 2. \tag{14.1}$$

If we demand fourfold coordination of Si atoms in a Si_nH_{2m} cluster, then its graph has $v = 2(n-m)$ vertices, $e = (3/2)v = 2(n-m)$ edges, and $f = n-m+2$ faces. The average number of atoms per face is $n_{av} = (4n-2m)/f = (4n-2m)/(n-m+2)$. The numerator here comes from the fact that each vertex atom is related to three cycles, and each on-edge atom to two cycles. In case of $Si_{10}H_{2m}$, we have $n_{av} = 2 + 16/(12-m)$. The important consequence is that the average angle between adjacent atoms is $\alpha_{av} = 180° - 360°/n_{av}$. If a cycle has five or more atoms ($m > 5$), the mean angle can be greater than 103°, which is close to tetrahedral angle. If a cycle is big enough (i.e., $m > 7$), it has degrees of freedom to bend, so the angles will be equal to energetically favorable tetrahedral angles. On the other hand, if $m < 5$ ($Si_{10} - Si_{10}H_{10}$) average angle $\alpha_{av} < 96°$, so the structure would experience the so-called "ring strain." The structure "avoids" it by changing the coordination number of silicon atoms away from the ideal number of 4 and/or by following a delocalized bonding. This simple consideration is one of many examples when graph theory gives useful insights for the understanding of structure formation, as is also the case in automatic distinguishing of structural isomers or classification of crystal structures.

The thermodynamic approach allows one to find the most abundant clusters in hydrogen atmosphere at a given temperature and hydrogen pressure. The results, obtained from minimization of the Gibbs free energy, are presented in Figure 14.14.

We can see that at normal pressure only Si_{10}, $Si_{10}H_{14}$, $Si_{10}H_{16}$, $Si_{10}H_{20}$, and $Si_{10}H_{22}$ clusters can exist, provided the ensemble of clusters is uniform.

In experiments, researchers often investigate the properties of cluster ensembles consisting of different types of clusters. These properties substantially depend on concentrations of different clusters. The concentrations, in turn, are determined by the energies of different clusters that we have found as a result of global structure optimization. We solve the following model problem: Consider

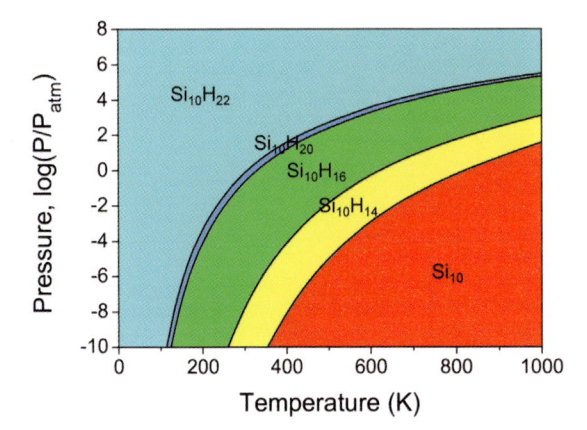

Figure 14.14 Phase diagram of $Si_{10}H_{2n}$ ($n=0-11$) clusters in hydrogen atmosphere.

the gas of bare Si_{10} clusters and fixed mean number of bound H_2 molecules per cluster ($N(H_2)$). The question is: what is the equilibrium distribution of $Si_{10}H_{2m}$ concentrations (C_m) at given temperature T? The results of free energy optimization with taking isomers into account are given in Figure 14.15a ($T=0$ K) and Figure 14.15b ($T=500$ K).

At zero temperature, the ensemble is a binary mixture with the exceptions of $N(H_2) = 0, 7, 10,$ and 11, when it is uniform. When the temperature is finite, the uniformity is lost even at hydrogen concentrations mentioned above. It is interesting to note that the clusters absent in the ensemble are those with non-four-coordinate silicon atoms.

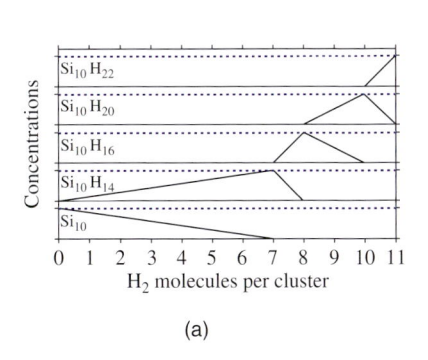

Figure 14.15 Equilibrium ensemble constituents C_m at $T=0$ K (a) and $T=500$ K (b). The height indicates the values of C_m (ranging from 0 (solid horizontal lines) to 1 (dotted lines). The filled area shows the respective share of isomeric structures. Only clusters with nonzero concentrations are shown. (Reprinted with permission from Ref. [53].)

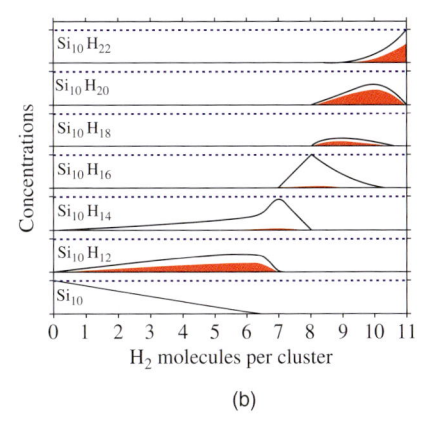

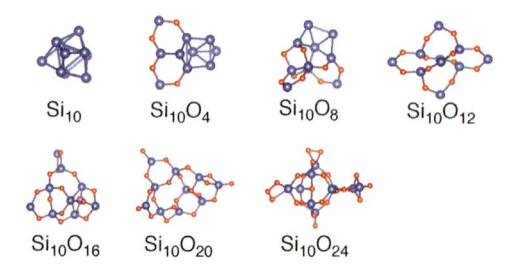

Figure 14.16 Ground-state structures of $Si_{10}O_{4n}$ ($n = 0$–6) clusters.

14.3.1.4 $Si_{10}O_{4n}$

In this part we present ground-state atomic structures of silicon nanoclusters $Si_{10}O_{4n}$ ($n = 0$–6) predicted using the evolutionary algorithm USPEX and density functional theory at the GGA level.

The atomic structures of $Si_{10}O_{4n}$ change drastically with increasing number of oxygen atoms. At low concentrations, oxygen atoms form Si–O–Si bridges, which results in the emergence of quartz-like regions in the cluster. The fact that low-oxygen Si–O clusters consist of chemically active pure-Si part and less-active quartz-like region can explain the process of forming silicon nanowires [54]. It is experimentally known that these wires have silicon core surrounded by a quartz shell. When oxygen concentration exceeds the SiO_2 composition, extra O atoms occupy the cluster surface and will presumably enhance the cluster's reactivity.

Figure 14.16 shows phase diagram of $Si_{10}O_{4n}$ clusters in oxygen atmosphere. In a wide range of oxygen pressures and temperatures, the most stable clusters are $Si_{10}O_{20}$ and $Si_{10}O_{24}$. Clusters with small numbers of oxygen atoms appear only at very low oxygen pressures and high temperatures.

Silicon–oxygen clusters have been previously studied mostly within $(SiO)_n$ or $(SiO_2)_n$ stoichiometries [55–57]. These compositions are quite obvious, as these compositions correspond to well-known bulk systems. Nevertheless, the phase diagram on Figure 14.17 shows that at ambient conditions, the dominant one is beyond-$(SiO_2)_n$ cluster $Si_{10}O_{24}$. The explanation of this fact comes from the great share of surface silicon atoms. These atoms have unsaturated bonds that react with the most abundant and chemically active gas in the atmosphere, that is, oxygen. The existence of such chemically active clusters at ambient conditions can have many interesting implications and is currently under detailed investigation.

14.3.2
1D Systems

Polymers [58]

Polymers are among the most diverse and widely used materials. Recently, we extended the USPEX method to enable prediction of polymeric structures made of prespecified monomeric blocks and this work has been considered as an important case of computational discovery of energy materials [59]. Here we

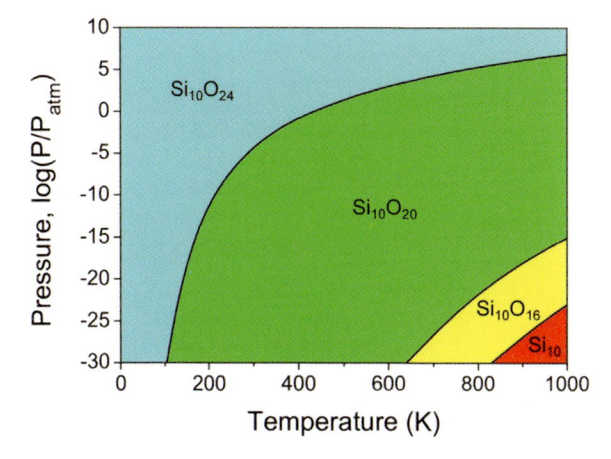

Figure 14.17 Phase diagram of $Si_{10}O_{4n}$ ($n = 0$–6) clusters in oxygen atmosphere.

illustrate its application to finding polymers with high static dielectric constant, for the use in flexible capacitors.

Polymeric dielectric materials are pervasive in modern electronics and electrical systems. The selection and design of a polymer depend on the requirements specific to the application, which, in the case of dielectric applications, can be stated in terms of a subset of the following properties: dielectric constant, bandgap, dielectric breakdown field, dielectric loss, morphology, glass transition temperature, mechanical strength, cost, and so on.

Figure 14.18 shows schematically the step-by-step computational search strategy to arrive at promising polymeric dielectrics. Figure 14.19 illustrates the pre-screening step (1), that is, the search for the most promising monomers.

Figure 14.19 shows the relationship between the dielectric constant (electronic, ionic, and total) and bandgap for 267 polymers, including the prototypical system, polyethylene, which has the largest bandgap and the smallest dielectric constant of all systems studied. From Figure 14.19a, we can see that the upper bound of the electronic part of the dielectric constant versus the bandgap displays a nearly perfect inverse dependence. This imposes a theoretical limit on the achievable electronic part of the dielectric constant. On the other hand, the ionic part of the dielectric constant, which is determined by the infrared (IR)-active zone-center phonon modes, is not correlated with the bandgap, as shown in Figure 14.19b. Figure 14.19c shows the variation of the total dielectric constant with the bandgap, and it can be seen as a "map" of achievable combinations of these properties within the chemical space explored. From the thus identified most promising monomers, we predicted the most stable polymeric structures.

Figure 14.20 shows the computed bandgap values of all the identified low-energy structures. We can see that except the case of $[-NH-CS-NH-C_6H_4-]_n$, the band gap (E_g) values are over 3.5 eV for all identified structures. The dielectric constants were determined using density-functional perturbation theory, with results for the orientational average ranging from 4 to 6, double that of

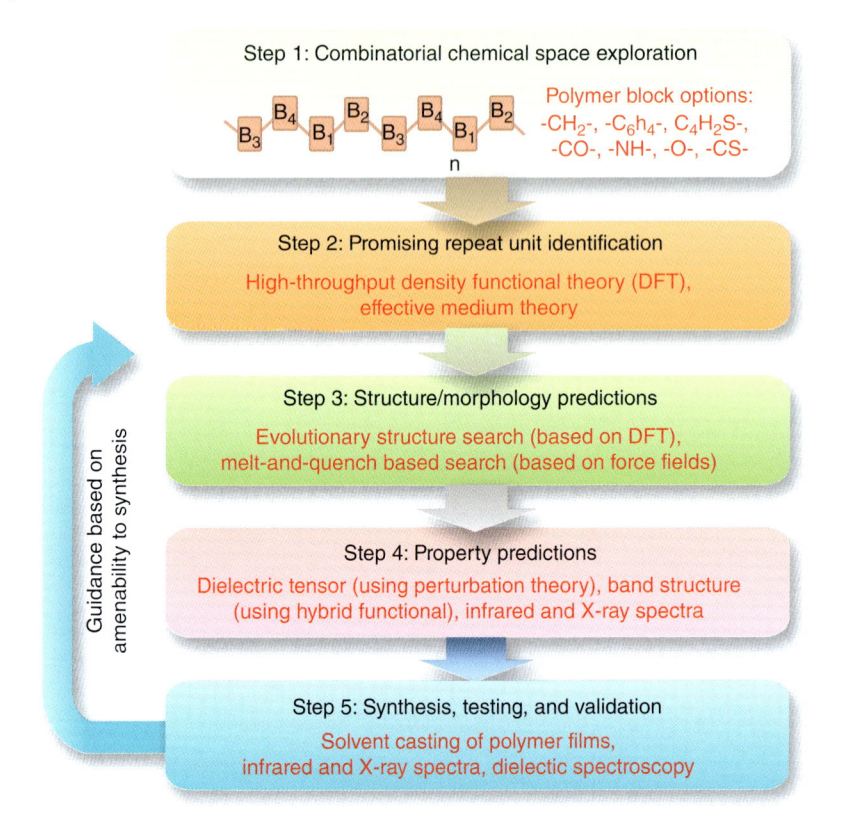

Figure 14.18 Schematic illustration of our rational dielectric polymer design strategy. The strategy involves five consecutive steps. (1) Combinatorial chemical space exploration, using 1D polymer chains containing four independent blocks with periodic boundary conditions along the chain axis. (2) Promising repeat unit identification, by screening based on bandgap and dielectric constant. (3) Three-dimensional structure predictions of polymers composed of the repeat units. (4) Property predictions for 3D systems. (5) Synthesis, testing, and validation. (Reprinted with permission from Ref. [58].)

polyethylene or polypropylene. Assuming that the predicted ground-state structures remain stable at room temperature, the predicted bandgaps and dielectric constants at room temperature are expected to be valid at those elevated temperatures as well. The predicted values of the total (ε_t) and electronic (ε_e) parts of dielectric constants for all three polymers considered are listed in Figure 14.20. In the same work [58], experimental synthesis of the predicted polymer was successfully attempted. Experiments have confirmed the predicted crystal structures and dielectric constants. Moreover, this computation-guided study also provides guidelines for exploration of chain derivatives. For example, aromatic, aliphatic, and oligoether with various additional dipoles were incorporated into the basic polythiourea backbone; which was successfully used in experimental investigations in order to design materials with improved solubility and processability [60].

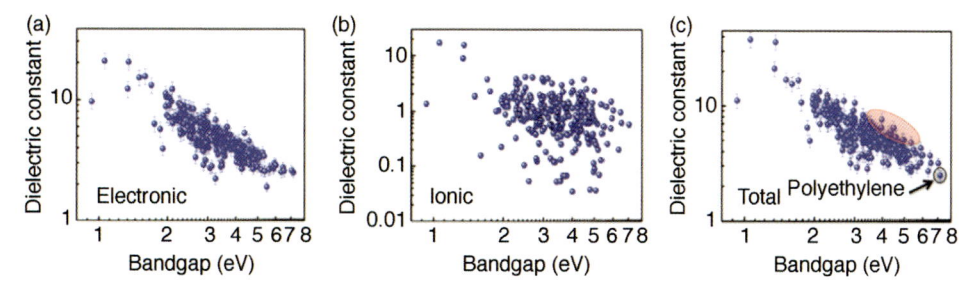

Figure 14.19 The dielectric constant versus bandgap relationship for 1D polymers. Computed (a) electronic, (b) ionic, and (c) total dielectric constant (along the polymer chain axis) versus bandgap. The associated errors in the dielectric constant computed using density functional perturbation theory for single chains, and subsequently estimated using effective medium theory for a bulk environment are also shown. The highlighted region corresponds to the most promising repeat unit composed of at least one of —NH—, —CO— and —O—, and at least one of —C_6H_4— and —C_6H_2S— blocks. Bandgap was computed using the hybrid HSE06 electronic exchange–correlation functional. (Reprinted with permission from Ref. [58].)

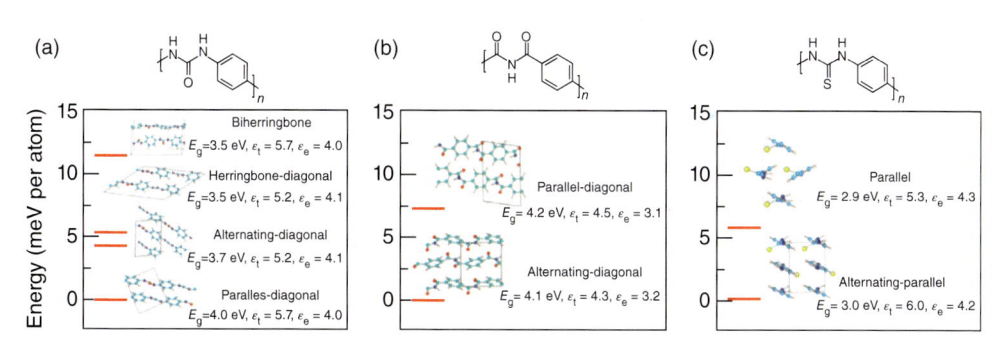

Figure 14.20 Predicted properties and structures of the lowest-energy polymers. The repeat units of the three identified polymers are (a) [—NH—CO—NH—C_6H_4—] (b) [—CO—NH—CO—C_6H_4—], and (c) [—NH—CS—NH—C_6H_4—]. The crystal structures were predicted by USPEX (using DFT) (for part (a), the structure was also found by melt-and-quench calculations using a force field). The zero of the energy scale corresponds to the most stable structures. For each predicted structure, the calculated values of the bandgap (E_g), total (ε_t), and electronic (ε_e) part of dielectric constants are also listed. (Reprinted with permission from Ref. [58].)

14.3.3
2D Systems

14.3.3.1 2D Crystals

Two-dimensional materials, due to their distinct properties and high specific surface areas, are important for various applications, for example, in optoelectronics, spintronics, catalysis, chemical and biological sensors, supercapacitors, solar cells, and lithium-ion batteries. In this part, we introduce the 2D materials discovered by evolutionary methods.

2D Carbon [61]

Graphene, was the first example of a true 2D atomic monolayer material. After it was experimentally realized in 2004, it rose as the hottest star in materials science.

For 2D carbon materials, systematic 2D structure searches were performed using the *ab initio* evolutionary algorithm USPEX with 6, 8, 10, 12, 14, 16, 18, 20, 22, and 24 carbon atoms per unit cell. The initial thickness was set to zero, since sp- and sp^2-carbon allotropes prefer planar structures at ambient pressure, and thickness was allowed to change during the evolution.

During structure searches, graphene always occupied the lowest position in the enthalpy evolution. With different atom numbers in unit cell, many 2D carbon structures with both sp and sp^2 carbon atoms were generated, but these structures usually have higher energies than sp^2 allotropes. For sp^2 carbon structures, many are composed of penta-, hexa-, hepta-, and octa- (5-, 6-, 7-, and 8-) carbon rings. Most of the planar carbon allotropes proposed in pervious works can be found in our systematic searches. Their energies and planar atomic densities are plotted in Figure 14.21. It is noted that Dirac allotropes are mostly at much higher energies compared with graphene (denoted as 1), and structures with higher energies usually have lower planar atomic densities.

Phagraphene: Systematic *ab initio* evolutionary structure searches for 2D carbon networks identified a new low-energy 2D form of carbon, which is composed of 5–6–7 carbon rings and we named it phagraphene. The planar atomic density of phagraphene is 0.37 atoms/Å^2, only slightly smaller than 0.38 atoms/Å^2 for graphene. Thanks to its sp^2 bonding and dense atomic packing comparable to graphene, this 2D carbon structure with *pmgm* layer group is lower in energy than most of the predicted 2D carbon allotropes. As shown in Figure 14.22, both DFT and tight-binding calculations confirm its distorted Dirac cones in the first Brillouin zone. The direction-dependent Dirac cones are further proved to be robust against the external strain - and strain allows one to tune Fermi velocities. This exotic structure is not only promising for fully investigating massless Dirac fermions in 2D carbon electronic systems, but also helpful for understanding topological and correlated phases in the corresponding photonic artificial lattices.

2D Boron [15]

Boron possesses unique richness and complexity of chemistry. It has been widely accepted that planar boron structures, composed of triangular and hexagonal motifs, are the most stable two-dimensional (2D) phases. In order to explore other potentially stable 2D B structures with novel electronic properties, we performed structure searches using the *ab initio* evolutionary algorithm USPEX. Initial structures were randomly produced using plane group symmetries with a user-defined initial thickness. The energetic stability is sensitive to the thickness;

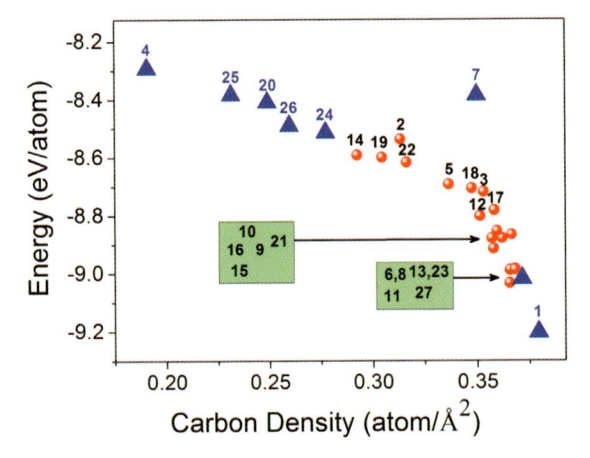

Figure 14.21 Energies and carbon densities for most of 2D carbon allotropes obtained from our evolutionary structure searches. Each structure is encoded by a number, and here we give the corresponding details for each one of them – the number within parentheses here denotes the number of atoms in the primitive cell. 1: graphene (2) [62]; 2: C31-sheet (4) [63,64]; 3: biphenylene (6); 4: α-graphyne (8) [65]; 5: planar T-graphene (8) [66]; 6: pentaheptite (8) [67]; 7: S-graphene (8) [68]; 8: rectangular heackelite (8) [69]; 9: pentahexoctite (8) [70]; 10: Net-W (8) [71]; 11: HOP graphene (10) [72]; 12: C65-sheet (10) [64]; 13: Pza-C10 (10) [73]; 14: γ-graphyne (12) [74]; 15: OPG-L (12) [75]; 16: OPG-Z (12) [75]; 17: C41-sheet (12) [64]; 18: C63-sheet (12) [64]; 19: C64-sheet or BPC (12); 20: structrure-9 (12) [76]; 21: oblique heackelite (12) [69]; 22: fused pentagon network (14) [77]; 23: hexagonal heackelite (16) [69]; 24: 6,6,12-graphyne (18) [65]; 25: β-graphyne (18) [65]; 26: δ-graphyne (20) [78]; 27: phagraphene (20) [61]. Allotropes with Dirac cones are labeled as blue triangles. (Reprinted with permission from Ref. [61].)

here, we studied monolayer and bilayer 2D boron structures. According to the experimental evidence for the spacing (~3.2 Å) between two adjacent layers of multiwalled boron nanotubes, the initial thickness was set to 3 Å and allowed to change during relaxation; all newly produced structures were fully relaxed, and relaxed lowest-energy structures were allowed to become parents for the new generation of structures. As with most of the other cases considered in this chapter, structure relaxations and total energy calculations were done at the DFT–GGA level of theory, using the all-electron PAW method, as implemented in the VASP code [79–81].

In the searches, we used 6, 8, 10, 12, 14, 16, and 18 atoms per unit cell in 2D structure searches. Two special structures (designated as 2D-B_{14} and 2D-B_{16}, which contain 14 and 16 atoms per unit cell, respectively) that are much lower in energy than the α-sheet were found. Their large thickness (the distance between the two planes that include the highest and lowest atomic positions) of 4.544 and 6.361 Å, respectively, is responsible for their superior energetic stability. In general, 2D structures become more stable with increasing their thickness as they approach the bulk state. There are many low-energy structures found by USPEX, and only two that possess special electronic structure will be discussed

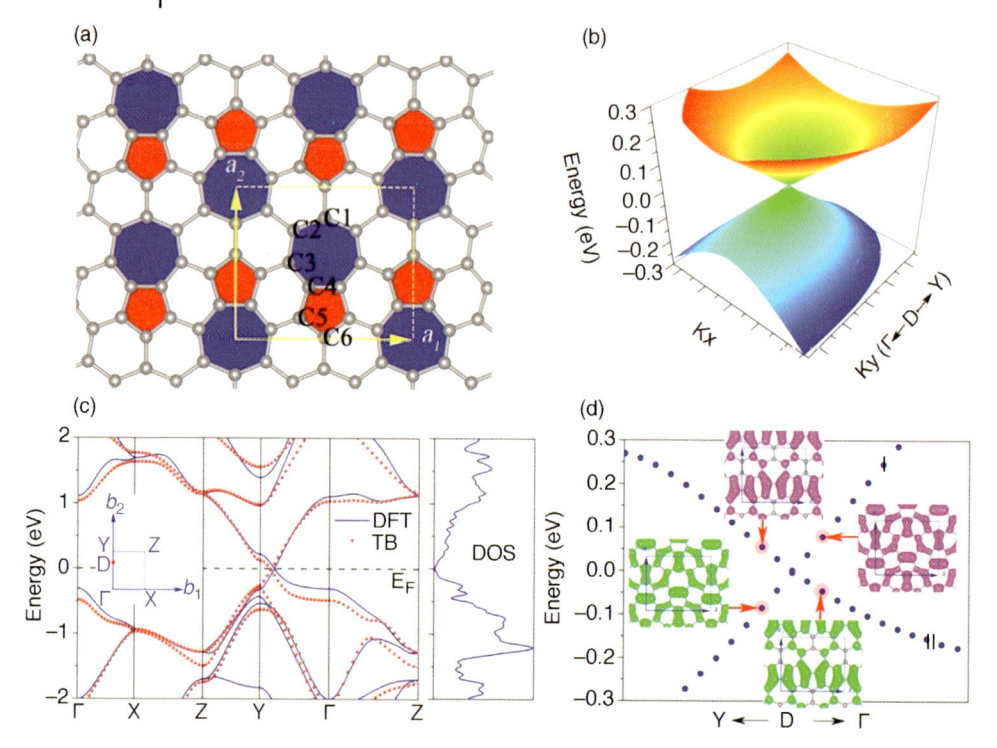

Figure 14.22 (a) Structure of phagraphene, C1–C6 are inequivalent carbon atoms in its unit cell. (b) Distorted Dirac cone formed by valence and conduction bands in the vicinity of the Dirac point. (c) Comparison of band structures from density functional theory (DFT) (blue lines) and tight-binding (red circles) model. The corresponding density of states (DOS) is zero at the Fermi level. Inset shows the first Brillouin zone (BZ) with high-symmetry k-points: Γ (0,0,0), X (0.5,0,0), Z (0.5,0.5,0), Y (0,0.5,0), and Dirac point D (0, 0.377,0). (d) Charge density distributions corresponding to crystal orbitals near the distorted Dirac cone, both Dirac bands (denoted as I and II) are from p_z orbitals of sp^2 carbon atoms. Fermi level has been set to zero. (Reprinted with permission from Ref. [61].)

further. We would like to focus on low-energy monolayer and bilayer structures that possess interesting electronic structures and have considerable chances of being prepared experimentally on a suitably chosen substrate.

Figure 14.23 shows a typical example (8 atoms/cell), which can serve as a benchmark case because the α-sheet also has eight atoms in the unit cell. This allows one to test whether the α-sheet or a better 8-atom structure is found in the search. The α-sheet structure is indeed reproduced during the search, as shown in Figure 14.23a. A more stable *pmmn* structure has two nonequivalent atomic positions (i.e., two sublattices), as illustrated in Figures 14.23b and c by different colors. Figure 14.23d and e shows an even more stable *pmmm* structure, which is made of buckled triangular layers, and has three nonequivalent atomic positions. There are similarities (buckling and coupling) between the

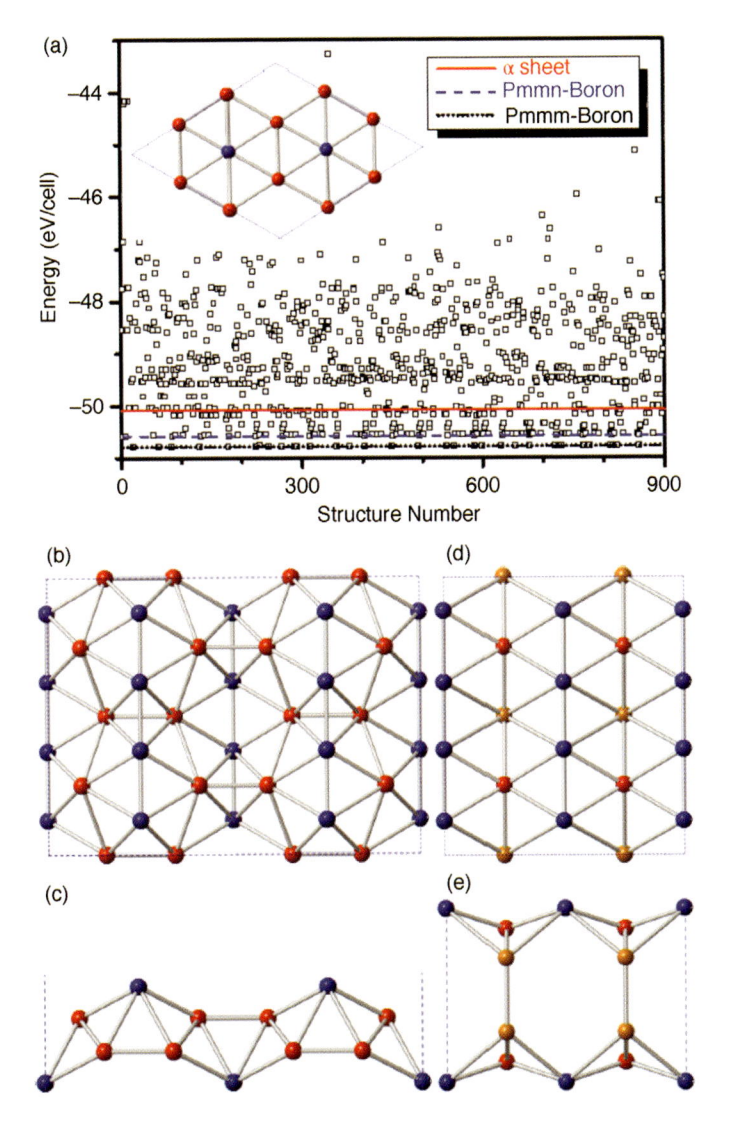

Figure 14.23 (a) Typical enthalpy evolution for an 8-atom 2D boron system during an evolutionary search. The inset shows the structure of the α-sheet. (b and c) Projections of 2×2×1 supercell of *pmmn*–boron structure along [001] and [100] directions. (d and e) Projections of 2×2×1 supercell of *pmmm*–boron structure along [001] and [100] directions. The nonequivalent atomic positions are shown by different colors. (Reprinted with permission from Ref. [15].)

geometric structures of *pmmn* and *pmmm*, as opposed to the planar α-sheet. Therefore, contrary to the general construction rules for flat monolayer boron sheets (mixing of triangular and hexagonal patterns), great complexity is uncovered with multilayer structures. The nonflatness of 2D boron sheets enhances its

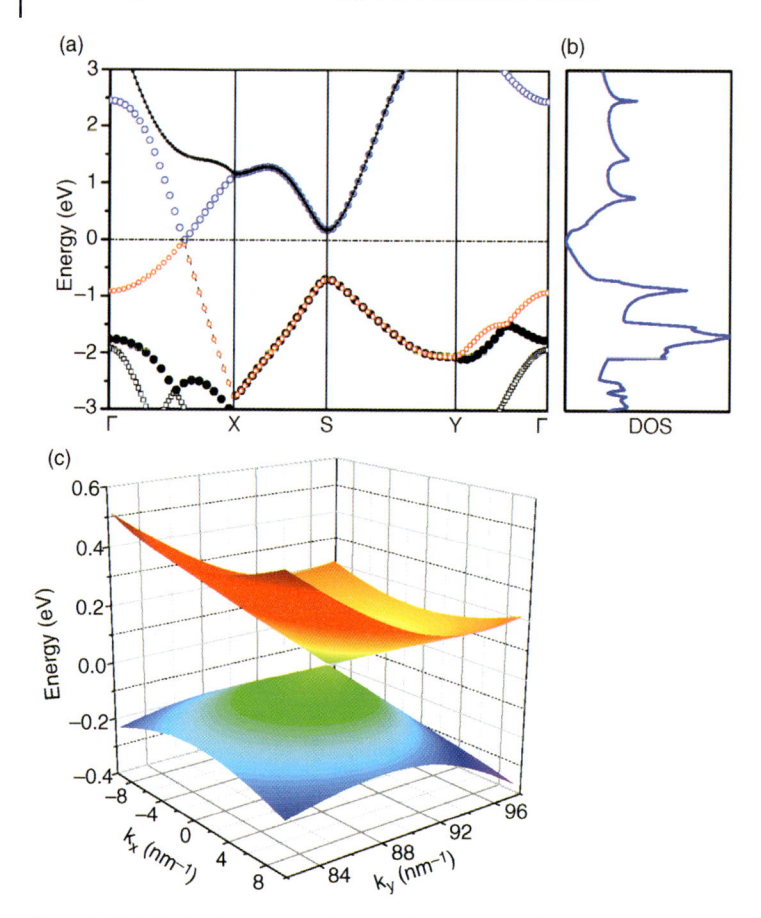

Figure 14.24 Electronic structure of pmmn–boron. (a) Band structure. (b) DOS. (c) Dirac cone formed by the valence and conduction band in the vicinity of the Dirac point. (Reprinted with permission from Ref. [15].)

energetic stability and creates novel electronic properties. In particular, as shown in Figure 14.24, 2D *pmmn*–boron can exhibit massless Dirac fermions with anisotropic Dirac cones. The slope of the bands in the k_x direction is ± 23 eV* Å, equivalent to a Fermi velocity 0.56×10^6 m/s. In the k_y direction, the slope of the bands equal -48 eV* Å (1.16×10^6 m/s) and 19 eV* Å (0.46×10^6 m/s), compared to ± 34 eV* Å (0.82×10^6 m/s) in graphene when approaching a Dirac point along the $\Gamma - K$ line. This gives *pmmn*–boron an advantage with respect to graphene: transport properties of these Dirac fermions will depend on direction, which gives an additional degree of freedom (with faster-than-graphene and slower-than-graphene directions) for electronic applications.

Recently, 2D boron monolayers were synthesized on Ag and Cu substrates [82,83]. It seems that the substrate has a considerable influence on the atomic and electronic structure of the B-layer, and a number of different

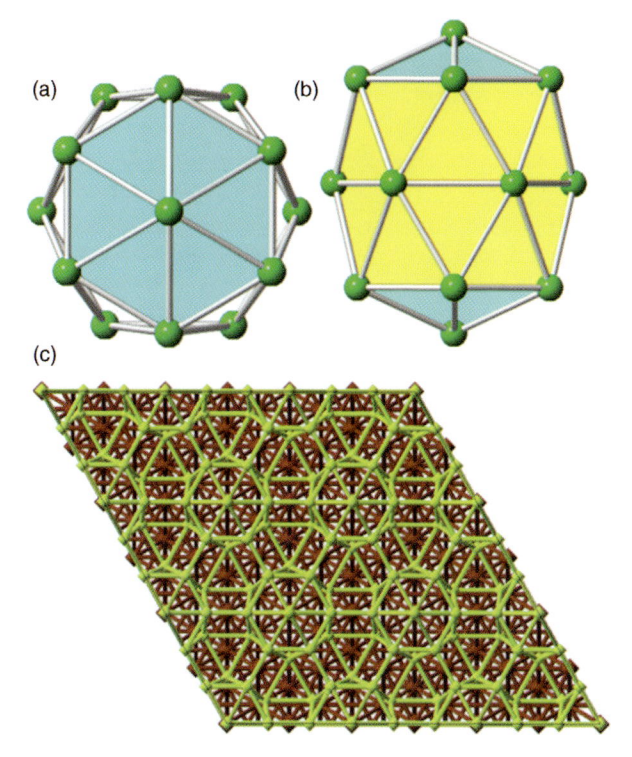

Figure 14.25 (a and b) Top and side views of B_{20} cluster, the building block of M-boron. (c) M-boron structure overlaid on Cu(111) substrate. (Reprinted with permission from Ref. [15].)

monolayer structures can be prepared. Another recent paper, to great surprise, predicted a low-energy magnetic 2D boron structure (M-boron) [84] with lattice dimensions compatible with Cu(111) surface and potentially synthesizable (see Figure 14.25.).

Cu_2Si [47]

The ground-state structure of Cu_2Si was predicted by other techniques in Ref. [47]. A new flat 2D structure was predicted, with each Si atom coordinated by six Cu atoms, and each Cu atom coordinated by three Si atoms and three Cu atoms, as shown in Figure 14.26. The most interesting aspect is chemical bonding of this 2D material.

The periodic NBO analysis was utilized to understand chemical bonding in Cu_2Si monolayer. In total, there are 22 valence electrons from two Cu atoms and four valence electrons from each Si atom per unit cell. This analysis found 10 d-type lone pairs on the two Cu atoms (five on each atom) with occupation numbers 1.9–2.0 $|e|$ (see Figure 14.27). Where the remaining six electrons are is crucial for understanding this compound.

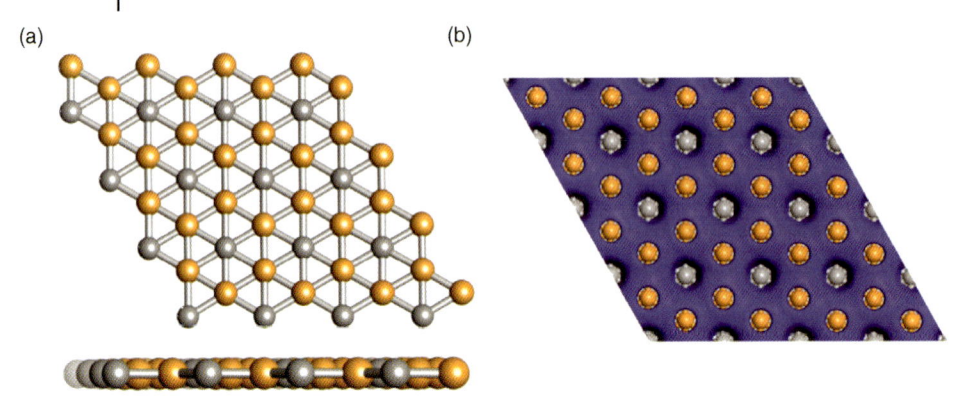

Figure 14.26 (a) Top and side views of the ball and stick model of the 2D Cu$_2$Si monolayer: 4×4 section of the extended monolayer. (b) Deformation charge density of the 2D Cu$_2$Si monolayer. Blue color corresponds to electron accumulation. Copper atoms are yellow; silicon atoms are gray. (Reprinted with permission from Ref. [47].)

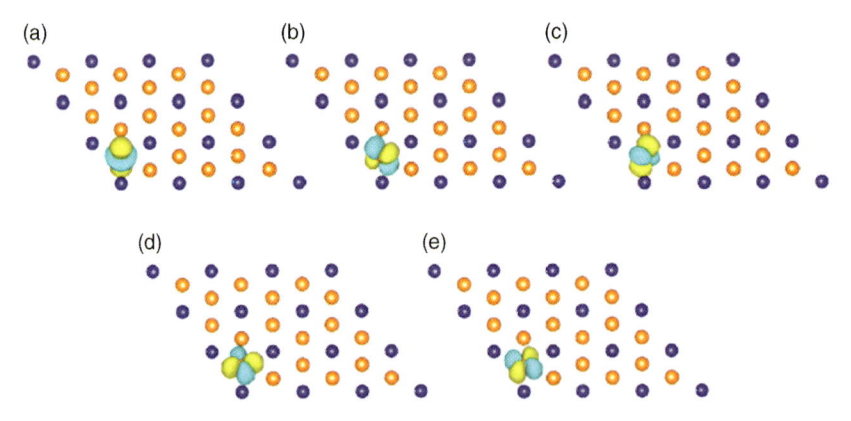

Figure 14.27 Bonding in Cu$_2$Si monolayer. Five d-type lone pairs on Cu atoms. (Reprinted with permission from Ref. [47].)

Both NBO and SSAdNDP methods have shown the absence of any classical two-center two-electron (2c–2e) σ-bonds. Using SSAdNDP analysis, the same 10 lone pairs on two Cu atoms with occupation number 1.9–2.0 |e| as well as three 4c–2e σ-bonds with occupation number 1.9 |e| are found, thus accounting for 26 electrons per unit cell. Figure 14.28a–c shows the individual 4c–2e σ-bonds. From Figure 14.28a–c, we can clearly see that the delocalized bonding governs the planar geometry of the Cu$_2$Si sheet. We can observe that the deformation charge density of Figure 14.26b is similar to the superposition of 4c–2e σ-bonds shown in Figure 14.28d. Recently, a few more atomically thin sheets (Ni$_2$Si and Ni$_2$Ge [85], Cu$_2$Ge [86]) have been computationally predicted. They all have atomic structure and chemical bonding similar to those of Cu$_2$Si sheet.

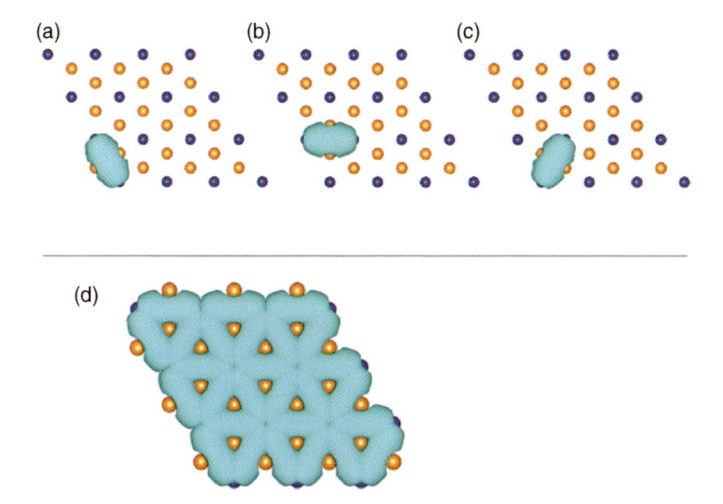

Figure 14.28 Bonding in the Cu$_2$Si monolayer. (a–c) Individual 4c–2e σ-bonds. (d) Superposition of 4c–2e σ-bonds. Copper is orange, and silicon is blue. (Reprinted with permission from Ref. [47].)

14.3.3.2 Surface Reconstructions

Surface structures play an important role in reactivity and catalytic activity of solids. In the following, we discuss some recent predictions of surface reconstructions.

B(111) Surface [87]

The element boron had long attracted enormous attention owing to its fascinating properties, such as exceptional structural complexity [88], superhardness, unusual partially ionic bonding in the γ-phase [3], and superconductivity at high pressures [89]. The reconstruction of the (111) surface of α-boron was first studied using the minima hopping method and a complex metallic surface reconstruction was predicted [90]. Subsequent study using USPEX found a simpler and much more stable semiconducting reconstruction of this surface [87].

In our predictions, the number of surface atoms within one run was allowed to vary from 1 to 20, constrained within a surface layer of thickness 4 Å, and a 10 Å thick vacuum region was used. Given that the thickness of B$_{12}$ icosahedron is about 3.7 Å, this gives enough space to fully explore the relevant chemical landscape (more than 2500 structures were sampled, to achieve full confidence in the final result). These calculations were performed at the GGA level of theory, using the exchange–correlation functional of Perdew, Burke, and Ernzerhof (PBE).

Amsler et al. [90] proposed a very complex metallic (111)-I$_{R,(a)}$ reconstruction involving three outer atomic layers above the icosahedral B$_{12}$ units. In contrast to (111)-I$_{R,(a)}$, new reconstruction [designated as (111)-I$_{R,(z)}$], proposed by Zhou et al. [87], has an unexpectedly simple structure, shown in Figure 14.29a, where a single boron atom occupies an interstitial position (named as B$_i$, colored in red), connects the B$_p$ atoms (colored in blue), and forms bridges with bond

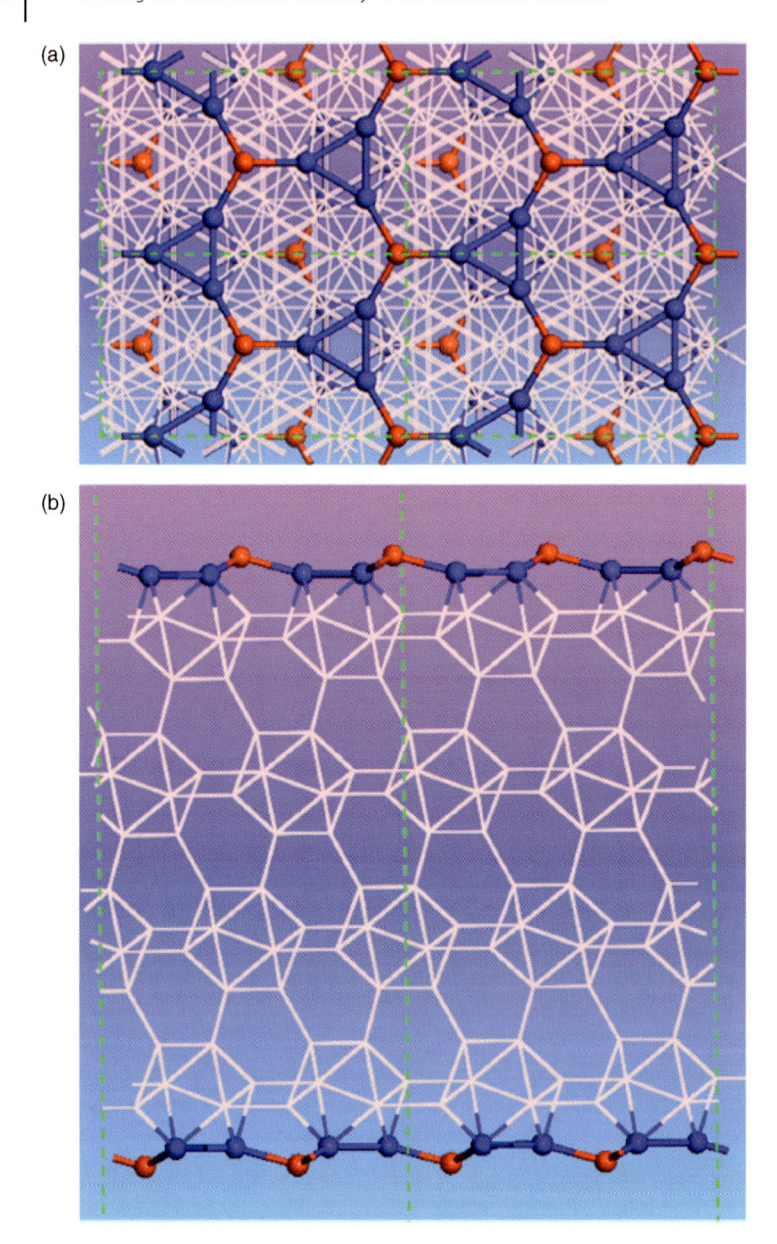

Figure 14.29 (a) Projection of the (111)-$I_{R,(z)}$ structure along the [111] direction. (b) Projection of the (111)-$I_{R,(z)}$ structure along the [$\bar{1}\bar{1}2$] direction. The inequivalent surface atoms are shown by different colors. (Reprinted with permission from Ref. [87].)

lengths of 1.793 Å and bond angles of 113.06°. It is clearly seen that the B_i atoms are slightly above the topmost icosahedral atom (B_p), as shown in Figure 14.29b, and together they form modulated "3 + 9" membered rings on the topmost surface. According to *ab initio* calculations, the surface energy of (111)-$I_{R,(z)}$ is 128.23 meV/Å^2 using GGA-PBE and 136.79 meV/Å^2 using HSE06 functional, which is considerably, by 42 and 60 meV/Å^2, respectively, lower than the energy of the (111)-$I_{R,(a)}$ surface reconstruction. The newly found stable surface reconstruction is semiconducting, just like bulk α-boron.

It is interesting to study charge transfer on the (111)-$I_{R,(z)}$ surface. Chemical effects of the B_i adsorption can be seen in Figure 14.30, which shows charge

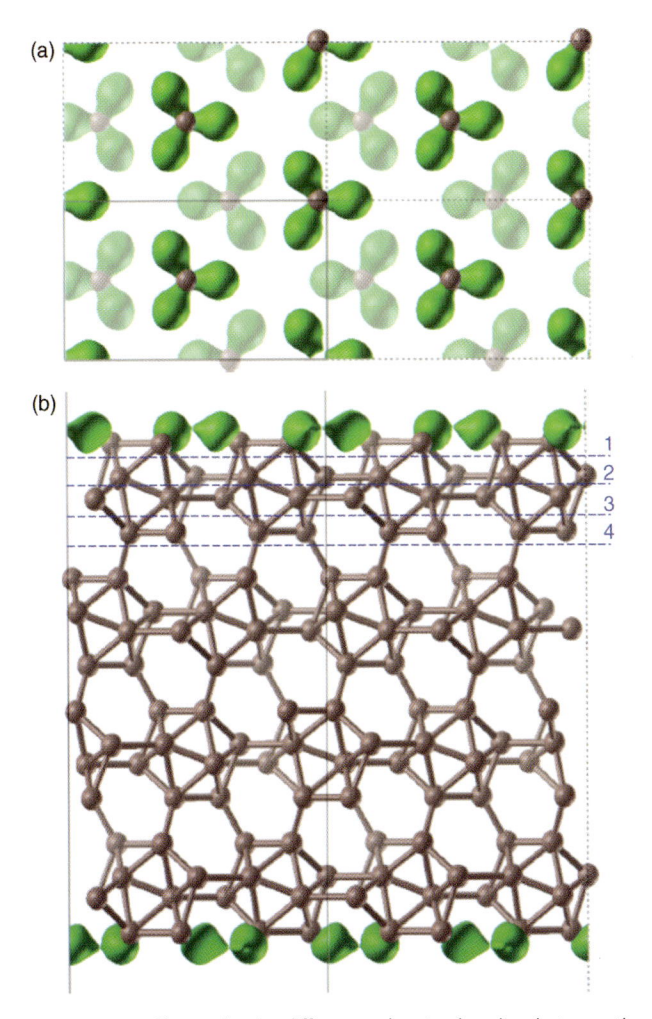

Figure 14.30 Charge density difference, showing bonding between the adsorbed boron atoms and the substrate. (a) Top view – for clarity, only the adsorbed atoms are shown). (b) Side view, showing the substrate. (Reprinted with permission from Ref. [87].)

density difference. There is notable charge transfer from B_i to the neighboring B_{12} icosahedra through B_p atoms. Bader charges for B_i are +0.17 e, and +0.04 e for B_p, which is close to the value in the bulk (~+0.05 e). As seen in Figure 14.30b, each B_{12} icosahedron comprises four atomic layers (labeled from 1 to 4); the net charges of these layers (per unit cell) are in the "+--+" order with the values of +0.56, −0.46, −0.46, and +0.36 e, much larger than the corresponding values of +0.20, −0.22, −0.22, and +0.24 e in α-boron.

New Reconstructions of Rutile TiO$_2$ (110) Surface [91]

Surfaces of semiconductors are crucially important for electronics, especially when the devices are reduced to the nanoscale. The determination of a material's surface structure is an important but a challenging step toward understanding its effects on the performance of devices. The exact atomic configurations of rutile TiO$_2$(110) surface reconstruction are under debate, that is, (1×1) [92], single-linked (1×2) [93], cross-linked (1×2), and pseudo-hexagonal rosette structures [94]. Knowing the exact surface structures of rutile TiO$_2$(110) will deepen our understanding of related photocatalytic reactions. Here, we discuss how surface reconstructions of rutile TiO$_2$(110) surfaces were resolved using USPEX.

In our USPEX calculations, surface cells with index-four multiplications of the single surface cell (from 1×1 to 1×2 to 2×2 and 1×4) were considered, and maximally, two Ti and four O atoms were allowed for per surface unit cell. We used a vacuum layer of 13 Å and a substrate slab of 4–5 TiO$_2$ layers with the upper 3.5 Å used as buffer. In the whole structure search, 1791 structures, or 40 generations, were explored. Low-energy structures were selected for detailed analysis. In order to obtain more accurate surface energies, slabs with six and seven TiO$_2$ layers (with only the bottom layer being kept fixed) and vacuum layer of 15 Å were used. All total energy calculations were done at the GGA level of theory, with the addition of a Hubbard U-term ($U_{eff} = 4.1$ eV) and inclusion of spin polarization, using the VASP package [79–81].

Surface energy determines stability of a surface, and is defined as

$$\Delta G = 1/N[G_{sur} - \sum n_i \mu_i], \tag{14.2}$$

where G_{sur} is the Gibbs free energy of a surface structure, n_i and μ_i are the number of atoms and chemical potential for each atomic species, respectively, and N is equal to $(m \times n)$ for an $(m \times n)$ surface cell, and serves as a normalization factor.

Since the contributions of temperature and ambient pressure are negligible, the Gibbs free energy of a solid phase can be replaced by its internal energy. Chemical potentials (μ_O and μ_{Ti}) should satisfy the following boundary conditions: (a) $\mu_O \leq 1/2\mu_{O2}$, (b) $\mu_{Ti} \leq \mu_{Ti}$ (bulk), and (c) $\mu_{Ti} + 2\mu_O = E_{TiO2}$, where E_{TiO2} is the internal energy of bulk rutile TiO$_2$. Accordingly, Eq. (14.2) can be rewritten as follows:

$$\Delta G = 1/N[E_{sur} - n_{Ti}E_{TiO_2} - m_O(n_O - 2n_{Ti})], \quad (14.3)$$

where E_{sur} is the internal energy of the model surface.

In equilibrium with O_2 gas, μ_O is expressed as follows:

$$\begin{aligned}\mu_O &= 1/2[E_{O2} + \Delta H_{O_2}(T, P^0) - T\Delta S_{O_2}(T, P^0) + k_B T \ln(P/P^0)] \\ &= 1/2E_{O_2} + \Delta\mu_o(T, P),\end{aligned} \quad (14.4)$$

where k_B, P^0, and P are the Boltzmann constant, standard atmospheric pressure, and oxygen partial pressure, respectively. E_{O2} was obtained from a spin-polarized GGA calculation, ΔH_{O2} (T, P^0) and $T\Delta S_{O2}$ (T, P^0) were taken from a thermodynamic database.[1] The fourth term is the contribution coming from the partial pressure of oxygen. Figure 14.31a shows dependence of surface energies on the chemical potential of oxygen. Structure with the lowest surface energy at a certain range of chemical potentials is stable at these conditions. With $E_O = 1/N[E_{sur} - n_{Ti}E_{TiO2}]$ and $\Delta n = 1/N(n_O - 2n_{Ti})$, Figure 14.31a can be recast in another insightful form as shown in Figure 14.31b, where points above the convex hull represent metastable structures.

Figure 14.31 shows that there are four stable structures: unreconstructed rutile $TiO_2(110)$ (-3.56 eV $\leq \Delta\mu_O \leq 0$ eV), "V-(4×1)" (-3.98 eV $\leq \Delta\mu_O \leq -3.56$ eV), Ti_2O_3-(1×2) (-4.11 eV $\leq \Delta\mu_O \leq -3.98$ eV), and "Ti_3O_3-(2×1)" (-4.96 eV $\leq \Delta\mu_O \leq -4.11$ eV). A number of interesting metastable structures were also predicted, for example, "$Ti_3O_2(1×2)$" and Ti_2O-(1×2) reconstructions. Figure 14.32 shows the four stable structures. From both Figure 14.31 and Figure 14.32, recently proposed "Ti_2O_3-(1×2)" and "Ti_2O-(1×2)" reconstructions are confirmed, and two new reconstructions "Ti_3O_2-(1×2)" and Ti_3O_3-(2×1)" are predicted. In addition, surface electronic states are found to be sensitive to reconstructions and, therefore, depend on the thermodynamics conditions. The predicted reconstructions clearly agree with experimental STM patterns. Ideally, these reconstructions should be further confirmed by surface X-ray diffraction and electron diffraction methods.

In conclusion, a complete and renewed understanding of stable reconstructions of rutile $TiO_2(110)$ surface was obtained. The predicted surface structures agree well with the experimental results, resolve previous controversies, and demonstrate the power of USPEX for predicting surface reconstructions. It is worth noting that the predicted Ti_2O_3-(1×2) reconstruction has been confirmed in recent experiments [95]. These works showed the great advantage and merit of USPEX package toward the understanding of surface structure at the atomic level.

---

1) *NIST Webbook*, webbook.nist.gov/.

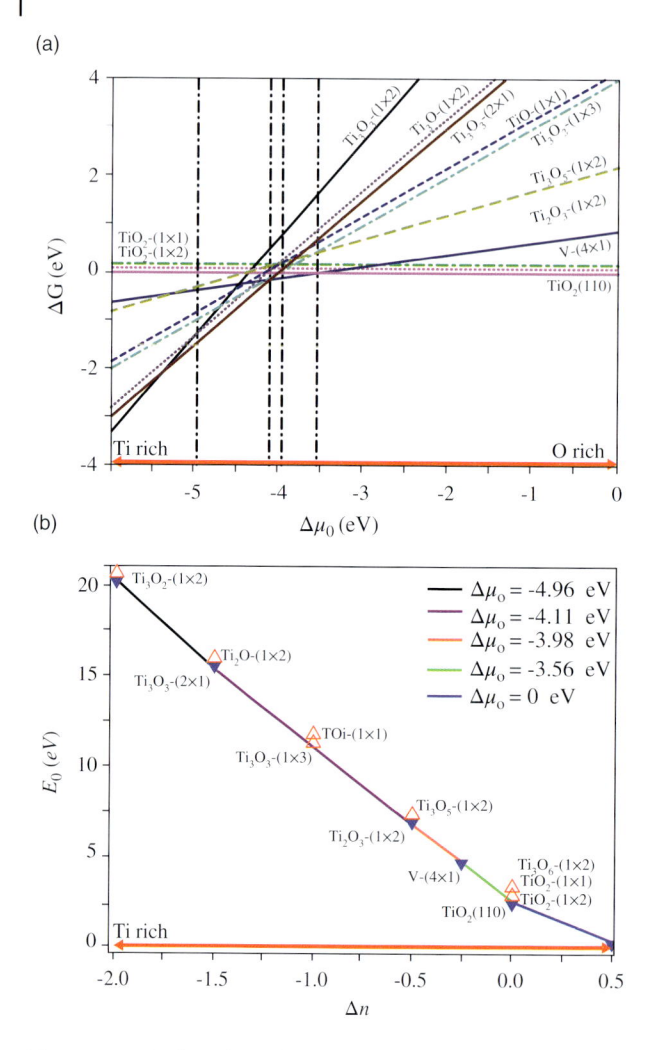

Figure 14.31 Surface phase diagram of rutile TiO$_2$(110): stability of different structures represented in terms of (a) oxygen chemical potential $\Delta\mu_O$ and (b) stoichiometry deviation Δn. Surface energies in part (a) are relative to the unreconstructed (110) surface. Stable and metastable structures in part (b) are represented by solid and hollow triangles, respectively. (Reprinted with permission from Ref. [91].)

14.4
Future Prospects

In this chapter, we briefly discussed various methodological aspects and several recent examples of the prediction of low-dimensional materials. Our understanding of chemical bonding in low-dimensional materials is gaining

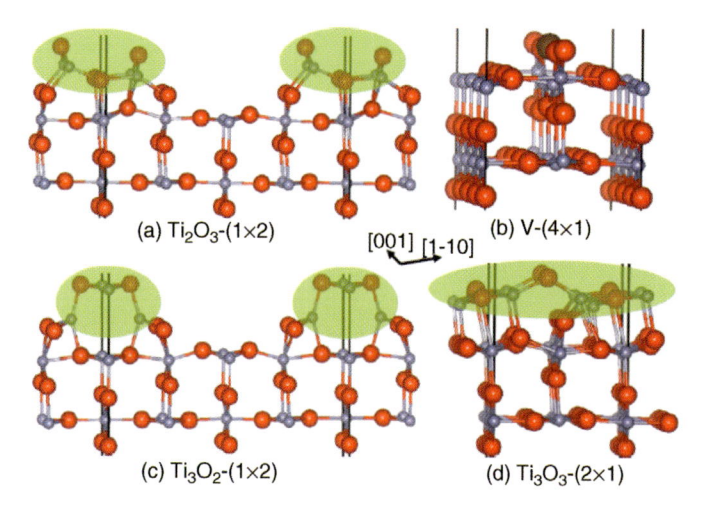

Figure 14.32 Side views of stable reconstructions of rutile TiO$_2$(110). (a) Ti$_2$O$_3$-(1 × 2). (b) V-(4 × 1). (c) Metastable Ti$_3$O$_2$-(1 × 2). (d) Ti$_3$O$_3$-(2 × 1). Ti and O atoms are shown as small gray and big red balls, respectively. The O vacancy in V-(4 × 1) is represented by the dark gray ball. Structural features are highlighted by yellow areas. (Reprinted with permission from Ref. [91].)

momentum, but still has a long way to go before these concepts become fully instrumental to transform various fields of application of low-dimensional materials. Structure prediction and chemical bonding analysis should be the main tools in design of systems with specific desired properties. Slight (and allowed) changes of the chemical composition can induce major changes in chemical bonding and properties of nanomaterials, and this relation needs to be further explored to allow new paths of designing functional nanomaterials. Some of the most exciting applications of low-dimensional materials include unique composite materials, catalysis, light-driven motor systems, and more applications are waiting to be discovered.

While this chapter was under review, we published a detailed study of Si–O nanoparticles [96]. Consistent with results shown here, it was found that for Si$_7$O$_n$ clusters, the most abundant cluster in equilibrium with ambient atmosphere at room temperature is Si$_7$O$_{19}$. This oxygen-rich cluster is magnetic and has ozonide O$_3$-groups, carrying magnetic moments, at the surface of the cluster. Presence of such reactive oxygen species as ozonide may explain the well-known carcinogenicity of silica dust.

Acknowledgments

This work was supported by Russian Science Foundation (grant 16-13-10459).

References

1 Maddox, J. (1988) Crystals from first principles. *Nature*, **335**, 201.
2 Oganov, A.R. (2011) *Modern Methods of Crystal Structure Prediction*, John Wiley & Sons, Inc., New York.
3 Oganov, A.R., Chen, J., Gatti, C., Ma, Y., Ma, Y., Glass, C.W., Liu, Z., Yu, T., Kurakevych, O.O., and Solozhenko, V.L. (2009) Ionic high-pressure form of elemental boron. *Nature*, **457**, 863–867.
4 Ma, Y., Eremets, M., Oganov, A.R., Xie, Y., Trojan, I., Medvedev, S., Lyakhov, A.O., Valle, M., and Prakapenka, V. (2009) Transparent dense sodium. *Nature*, **458**, 182–185.
5 Zhang, W., Oganov, A.R., Goncharov, A.F., Zhu, Q., Boulfelfel, S.E., Lyakhov, A.O., Stavrou, E., Somayazulu, M., Prakapenka, V.B., and Konôpková, Z. (2013) Unexpected stable stoichiometries of sodium chlorides. *Supramol. Sci.*, **342**, 1502–1505.
6 Zhu, Q., Li, L., Oganov, A.R., and Allen, P.B. (2013) Evolutionary method for predicting surface reconstructions with variable stoichiometry. *Phys. Rev. B*, **87**, 195317.
7 Lyakhov, A.O., Oganov, A.R., Stokes, H.T., and Zhu, Q. (2013) New developments in evolutionary structure prediction algorithm USPEX. *Comput. Phys. Commun.*, **184**, 1172–1182.
8 Deaven, D.M. and Ho, K.M. (1995) Molecular geometry optimization with a genetic algorithm. *Phys. Rev. Lett.*, **75**, 288–291.
9 Alexandrova, A.N. and Boldyrev, A.I. (2005) Search for the $Li_n^{0/+1/-1}$ ($n = 5-7$) lowest-energy structures using the *ab initio* gradient embedded genetic algorithm (GEGA) elucidation of the chemical bonding in the lithium clusters. *J. Chem. Theory Comput.*, **1**, 566–580.
10 Alexandrova, A.N., Boldyrev, A.I., Fu, Y.-J., Yang, X., Wang, X.-B., and Wang, L.-S. (2004) Structure of the $Na_xCl_{x+1}^-$ ($x = 1-4$) clusters via *ab initio* genetic algorithm and photoelectron spectroscopy. *J. Chem. Phys.*, **121**, 5709–5719.
11 Johnston, R.L. (2003) Evolving better nanoparticles: genetic algorithms for optimising cluster geometries. *Dalton Trans.*, 4193–4207.
12 Heiles, S. and Johnston, R.L. (2013) Global optimization of clusters using electronic structure methods. *Int. J. Quantum Chem.*, **113**, 2091–2109.
13 Ferrando, R., Jellinek, J., and Johnston, R.L. (2008) Nanoalloys: from theory to applications of alloy clusters and nanoparticles. *Chem. Rev.*, **108**, 845–910.
14 Zhu, Q., Sharma, V., Oganov, A.R., and Ramprasad, R. (2014) Predicting polymeric crystal structures by evolutionary algorithms. *J. Chem. Phys.*, **141**, 154102.
15 Zhou, X.-F., Dong, X., Oganov, A.R., Zhu, Q., Tian, Y., and Wang, H.-T. (2014) Semimetallic two-dimensional boron allotrope with massless Dirac fermions. *Phys. Rev. Lett.*, **112**, 085502.
16 Lyakhov, A.O., Oganov, A.R., and Valle, M. (2010) How to predict very large and complex crystal structures. *Comput. Phys. Commun.*, **181**, 1623–1632.
17 Oganov, A.R. and Valle, M. (2009) How to quantify energy landscapes of solids. *J. Chem. Phys.*, **130**, 104504.
18 Lv, J., Wang, Y., Zhu, L., and Ma, Y. (2012) Particle-swarm structure prediction on clusters. *J. Chem. Phys.*, **137**, 084104.
19 Schönborn, S.E., Goedecker, S., Roy, S., and Oganov, A.R. (2009) The performance of minima hopping and evolutionary algorithms for cluster structure prediction. *J. Chem. Phys.*, **130**, 144108.
20 Froltsov, V.A. and Reuter, K. (2009) Robustness of 'cut and splice' genetic algorithms in the structural optimization of atomic clusters. *Chem. Phys. Lett.*, **473**, 363–366.
21 Daven, D.M., Tit, N., Morris, J.R., and Ho, K.M. (1996) Structural optimization of Lennard-Jones clusters by a genetic algorithm. *Chem. Phys. Lett.*, **256**, 195–200.
22 Wales, D.J. and Doye, J.P.K. (1997) Global optimization by basin-hopping and the lowest energy structures of Lennard-Jones clusters containing up to

110 atoms. *J. Phys. Chem. A*, **101**, 5111–5116.
23 Doye, J.P.K. and Wales, D.J. (1998) Thermodynamics of global optimization. *Phys. Rev. Lett.*, **80**, 1357–1360.
24 Goedecker, S. (2004) Minima hopping: an efficient search method for the global minimum of the potential energy surface of complex molecular systems. *J. Chem. Phys.*, **120**, 9911–9917.
25 Hartke, B. (1999) Global cluster geometry optimization by a phenotype algorithm with Niches: location of elusive minima, and low-order scaling with cluster size. *J. Comput. Chem.*, **20**, 1752–1759.
26 Pullan, W. (2005) An unbiased population-based search for the geometry optimization of Lennard–Jones clusters: $2 \leq N \leq 372$. *J. Comput. Chem.*, **26**, 899–906.
27 Doye, J.P.K., Wales, D.J., and Berry, R.S. (1995) The effect of the range of the potential on the structures of clusters. *J. Chem. Phys.*, **103**, 4234.
28 Schmid, G., Baumle, M., Geerkens, M., Heim, I., Osemann, C., and Sawitowski, T. (1999) Current and future applications of nanoclusters. *Chem. Soc. Rev.*, **28**, 179–185.
29 Pyykkö, P. (2004) Theoretical chemistry of gold. *Angew. Chem., Int. Ed.*, **43**, 4412–4456.
30 Luo, W., Pennycook, S.J., and Pantelides, S.T. (2007) s-Electron ferromagnetism in gold and silver nanoclusters. *Nano. Lett.*, **7**, 3134–3137.
31 Baletto, F. and Ferrando, R. (2005) Structural properties of nanoclusters: energetic, thermodynamic, and kinetic effects. *Rev. Mod. Phys.*, **77**, 371–423.
32 Knight, W.-D., Clemenger, K., de Heer, W.A., Saunders, W.A., Chou, M., and Cohen, M.L. (1984) Electronic shell structure and abundances of sodium clusters. *Phys. Rev. Lett.*, **52**, 2141–2143.
33 Luo, Z. and Castleman, A.W. (2014) Special and general superatoms. *Acc. Chem. Res.*, **47**, 2931–2940.
34 Thorn, D.L. and Hoffmann, R. (1979) Delocalization in metallocycles. *Nouv. J. Chim.*, **3**, 39–45.
35 Li, X., Kuznetsov, A.E., Zhang, H.-F., Boldyrev, A.I., and Wang, L.-S. (2001) Observation of all-metal aromatic molecules. *Supramol. Sci.*, **291**, 859–861.
36 Kuznetsov, A.E., Birch, K.A., Boldyrev, A.I., Li, X., Zhai, H.-J., and Wang, L.-S. (2003) All-metal antiaromatic molecule: rectangular Al_4^{4-} in the $Li_3Al_4^-$ anion. *Supramol. Sci.*, **300**, 622–625.
37 Chattaraj, P.K. (2010) *Aromaticity and Metal Clusters*, CRC Press.
38 Zubarev, D.Y., Averkiev, B.B., Zhai, H.-J., Wang, L.-S., and Boldyrev, A.I. (2008) Aromaticity and antiaromaticity in transition-metal systems. *Phys. Chem. Chem. Phys.*, **10**, 257–267.
39 Yu, X., Oganov, A.R., Popov, I.A., and Boldyrev, A.I. (2016) d-AO spherical aromaticity in Ce_6O_8. *J. Comput. Chem.*, **37**, 103–109.
40 Oganov, A.R. and Glass, C.W. (2006) Crystal structure prediction using *ab initio* evolutionary techniques: principles and applications. *J. Chem. Phys.*, **124**, 244704–244715.
41 Oganov, A.R., Lyakhov, A.O., and Valle, M. (2011) How evolutionary crystal structure prediction works – and why. *Acc. Chem. Res.*, **44**, 227–237.
42 Bhattacharya, S., Levchenko, S.V., Ghiringhelli, L.M., and Scheffler, M. (2013) Stability and metastability of clusters in a reactive atmosphere: theoretical evidence for unexpected stoichiometries of Mg_MO_x. *Phys. Rev. Lett.*, **111**, 135501.
43 Silvi, B. and Savin, A. (1994) Classification of chemical bonds based on topological analysis of electron localization functions. *Nature*, **371**, 683–686.
44 Glendening, E.D., Landis, C.R., and Weinhold, F. (2012) Natural bond orbital methods. *Wiley Interdiscip. Rev. Comput. Mol. Sci.*, **2**, 1–42.
45 Zubarev, D.Y. and Boldyrev, A.I. (2008) Developing paradigms of chemical bonding: adaptive natural density partitioning. *Phys. Chem. Chem. Phys.*, **10**, 5207–5217.
46 Popov, I.A., Popov, V.F., Bozhenko, K.V., Černušák, I., and Boldyrev, A.I. (2013) Structural changes in the series of boron–carbon mixed clusters $C_xB_{10-x}^-$

47 Yang, L.-M., Bačić, V., Popov, I.A., Boldyrev, A.I., Heine, T., Frauenheim, T., and Ganz, E. (2015) Two-dimensional Cu_2Si monolayer with planar hexacoordinate copper and silicon bonding. *J. Am. Chem. Soc.*, **137**, 2757–2762.

48 Popov, I.A., Averkiev, B.B., Starikova, A.A., Boldyrev, A.I., Minyaev, R.M., and Minkin, V.I. (2015) Assessing the viability of extended nonmetal atom chains in M_nF_{4n+2} (M = S and Se). *Angew. Chem., Int. Ed.*, **54**, 1476–1480.

49 Yu, X., Oganov, A.R., Popov, I.A., Qian, G., and Boldyrev, A.I. (2016) Antiferromagnetic stabilization in Ti_8O_{12} cluster. *Angew. Chem., Int. Ed.*, **55**, 1699–1703.

50 Wilson, W.L., Szajowski, P.F., and Brus, L.E. (1993) Quantum confinement in size-selected, surface-oxidized silicon nanocrystals. *Supramol. Sci.*, **262**, 1242–1244.

51 Belomoin, G., Therrien, J., Smith, A., Rao, S., Twesten, R., Chaieb, S., Nayfeh, M.H., Wagner, L., and Mitas, L. (2002) Observation of a magic discrete family of ultrabright Si nanoparticles. *Appl. Phys. Lett.*, **80**, 841–843.

52 Okano, A. and Takayanagi, K. (1998) Neutral silicon clusters produced by laser ablation in vacuum. *Appl. Surf. Sci.*, **127–129**, 362–367.

53 Baturin, V.S., Lepeshkin, S.V., Matsko, N.L., Oganov, A.R., and Uspenskii, Y.A. (2014) Prediction of the atomic structure and stability for the ensemble of silicon nanoclusters passivated by hydrogen. *Europhys. Lett.*, **106**, 37002.

54 Zhang, Y.F., Tang, Y.H., Wang, N., Yu, D.P., Lee, C.S., Bello, I., and Lee, S.T. (1998) Silicon nanowires prepared by laser ablation at high temperature. *Appl. Phys. Lett.*, **72**, 1835–1837.

55 Flikkema, E. and Bromley, S.T. (2004) Dedicated global optimization search for ground state silica nanoclusters: $(SiO_2)N$ (N = 6–12). *J. Phys. Chem. B*, **108**, 9638–9645.

56 Hu, S.-X., Yu, J.-G., and Zeng, E.Y. (2010) A density functional study of the structural and electronic properties of silicon monoxide clusters. *J. Phys. Chem. A*, **114**, 10769–10774.

57 Catlow, C.R.A., Bromley, S.T., Hamad, S., Mora-Fonz, M., Sokol, A.A., and Woodley, S.M. (2010) Modelling nano-clusters and nucleation. *Phys. Chem. Chem. Phys.*, **12**, 786–811.

58 Sharma, V., Wang, C., Lorenzini, R.G., Ma, R., Zhu, Q., Sinkovits, D.W., Pilania, G., Oganov, A.R., Kumar, S., Sotzing, G.A., Boggs, S.A., and Ramprasad, R. (2014) Rational design of all organic polymer dielectrics. *Nat. Commun.*, **5**, 4845.

59 Jain, A., Shin, Y., and Persson, K.A. (2016) Computational predictions of energy materials using density functional theory. *Nat. Rev. Mater.*, **1**, 15004.

60 Ma, R., Sharma, V., Baldwin, A.F., Tefferi, M., Offenbach, I., Cakmak, M., Weiss, R., Cao, Y., Ramprasad, R., and Sotzing, G.A. (2015) Rational design and synthesis of polythioureas as capacitor dielectrics. *J. Mater. Chem. A*, **3**, 14845–14852.

61 Wang, Z., Zhou, X.-F., Zhang, X., Zhu, Q., Dong, H., Zhao, M., and Oganov, A.R. (2015) Phagraphene: a low-energy graphene allotrope composed of 5–6–7 carbon rings with distorted Dirac cones. *Nano Lett.*, **15**, 6182–6186.

62 Novoselov, K.S., Geim, A.K., Morozov, S.V., Jiang, D., Zhang, Y., Dubonos, S.V., Grigorieva, I.V., and Firsov, A.A. (2004) Electric field effect in atomically thin carbon films. *Supramol. Sci.*, **306**, 666–669.

63 Hudspeth, M.A., Whitman, B.W., Barone, V., and Peralta, J.E. (2010) Electronic properties of the biphenylene sheet and its one-dimensional derivatives. *ACS Nano*, **4**, 4565–4570.

64 Lu, H. and Li, S.-D. (2013) Two-dimensional carbon allotropes from graphene to graphyne. *J. Mater. Chem. C*, **1**, 3677–3680.

65 Malko, D., Neiss, C., Viñes, F., and Görling, A. (2012) Competition for graphene: graphynes with direction-dependent Dirac cones. *Phys. Rev. Lett.*, **108**, 086804.

66 Liu, Y., Wang, G., Huang, Q., Guo, L., and Chen, X. (2012) Structural and electronic properties of T graphene: a two-dimensional carbon allotrope with tetrarings. *Phys. Rev. Lett.*, **108**, 225505.

67 Crespi, V.H., Benedict, L.X., Cohen, M.L., and Louie, S.G. (1996) Prediction of a pure-carbon planar covalent metal. *Phys. Rev. B*, **53**, R13303–R13305.

68 Xu, L.-C., Wang, R.-Z., Miao, M.-S., Wei, X.-L., Chen, Y.-P., Yan, H., Lau, W.-M., Liu, L.-M., and Ma, Y.-M. (2014) Two-dimensional Dirac carbon allotropes from graphene. *Nanoscale*, **6**, 1113–1118.

69 Terrones, H., Terrones, M., Hernández, E., Grobert, N., Charlier, J.C., and Ajayan, P.M. (2000) New metallic allotropes of planar and tubular carbon. *Phys. Rev. Lett.*, **84**, 1716–1719.

70 Sharma, B.R., Manjanath, A., and Singh, A.K. (2014) Pentahexoctite: a new two-dimensional allotrope of carbon. *Sci. Rep.*, **4**, 7164.

71 Wang, X.-Q., Li, H.-D., and Wang, J.-T. (2013) Prediction of a new two-dimensional metallic carbon allotrope. *Phys. Chem. Chem. Phys.*, **15**, 2024–2030.

72 Mandal, B., Sarkar, S., Pramanik, A., and Sarkar, P. (2013) Theoretical prediction of a new two-dimensional carbon allotrope and NDR behaviour of its one-dimensional derivatives. *Phys. Chem. Chem. Phys.*, **15**, 21001–21006.

73 Luo, X., Liu, L.-M., Hu, Z., Wang, W.-H., Song, W.-X., Li, F., Zhao, S.-J., Liu, H., Wang, H.-T., and Tian, Y. (2012) Two-dimensional superlattice: modulation of band gaps in graphene-based monolayer carbon superlattices. *J. Phys. Chem. Lett.*, **3**, 3373–3378.

74 Narita, N., Nagai, S., Suzuki, S., and Nakao, K. (1998) Optimized geometries and electronic structures of graphyne and its family. *Phys. Rev. B*, **58**, 11009–11014.

75 Su, C., Jiang, H., and Feng, J. (2013) Two-dimensional carbon allotrope with strong electronic anisotropy. *Phys. Rev. B*, **87**, 075453.

76 Enyashin, A.N. and Ivanovskii, A.L. (2011) Graphene allotropes. *Phys. Status Solidi B*, **248**, 1879–1883.

77 Mina, M. and Susumu, O. (2013) Two-dimensional sp^2 carbon network of fused pentagons: all carbon ferromagnetic sheet. *Appl. Phys. Express*, **6**, 095101.

78 Zhao, M., Dong, W., and Wang, A. (2013) Two-dimensional carbon topological insulators superior to graphene. *Sci. Rep.*, **3**, 3532.

79 Kresse, G. and Joubert, D. (1999) From ultrasoft pseudopotentials to the projector augmented-wave method. *Phys. Rev. B*, **59**, 1758–1775.

80 Kresse, G. and Furthmüller, J. (1996) Efficiency of *ab-initio* total energy calculations for metals and semiconductors using a plane-wave basis set. *Comput. Mater. Sci.*, **6**, 15–50.

81 Kresse, G. and Furthmüller, J. (1996) Efficient iterative schemes for *ab initio* total-energy calculations using a plane-wave basis set. *Phys. Rev. B*, **54**, 11169–11186.

82 Tai, G., Hu, T., Zhou, Y., Wang, X., Kong, J., Zeng, T., You, Y., and Wang, Q. (2015) Synthesis of atomically thin boron films on copper foils. *Angew. Chem.*, **127**, 15693–15697.

83 Mannix, A.J., Zhou, X.F., Kiraly, B., Wood, J.D., Myers, B., Liu, X.L., Fisher, B.L., Santiago, U., Guest, J.R., Yacaman, M.J., Ponce-Pedraza, A., Oganov, A.R., Hersam, M.C., and Guisinger, N.P. (2015) Synthesis of borophene: an anisotropic, two-dimensional boron allotrope. *Supramol. Sci.*, **350**, 1513–1516.

84 Zhou, X.-F., Oganov, A.R., Wang, Z., Popov, I.A., Boldyrev, A.I., and Wang, H.-T. (2016) Two-dimensional magnetic boron. *Phys. Rev. B*, **93**, 085406.

85 Yang, L.-M., Popov, I.A., Frauenheim, T., Boldyrev, A.I., Heine, T., Bacic, V., and Ganz, E. (2015) Revealing unusual chemical bonding in planar hypercoordinate Ni_2Ge and quasi-planar Ni_2Si two-dimensional crystals. *Phys. Chem. Chem. Phys.*, **17**, 26043–26048.

86 Yang, L.-M., Popov, I.A., Boldyrev, A.I., Heine, T., Frauenheim, T., and Ganz, E. (2015) Post-anti-van't Hoff-Le Bel motif in atomically thin germanium–copper alloy film. *Phys. Chem. Chem. Phys.*, **17**, 17545–17551.

87 Zhou, X.-F., Oganov, A.R., Shao, X., Zhu, Q., and Wang, H.-T. (2014) Unexpected reconstruction of the α-boron (111) surface. *Phys. Rev. Lett.*, **113**, 176101.

88 Albert, B. and Hillebrecht, H. (2009) Boron: elementary challenge for experimenters and theoreticians. *Angew. Chem., Int. Ed.*, **48**, 8640–8668.

89 Eremets, M.I., Struzhkin, V.V., Mao, H.-K., and Hemley, R.J. (2001)

Superconductivity in boron. *Supramol. Sci.*, **293**, 272–274.

90 Amsler, M., Botti, S., Marques, M.A.L., and Goedecker, S. (2013) Conducting boron sheets formed by the reconstruction of the α-boron (111) surface. *Phys. Rev. Lett.*, **111**, 136101.

91 Wang, Q., Oganov, A.R., Zhu, Q., and Zhou, X.-F. (2014) New reconstructions of the (110) surface of rutile TiO_2 predicted by an evolutionary method. *Phys. Rev. Lett.*, **113**, 266101.

92 Park, K.T., Pan, M.H., Meunier, V., and Plummer, E.W. (2006) Surface reconstructions of TiO_2(110) driven by suboxides. *Phys. Rev. Lett.*, **96**, 226105.

93 Blanco-Rey, M., Abad, J., Rogero, C., Mendez, J., Lopez, M.F., Martin-Gago, J.A., and de Andres, P.L. (2006) Structure of rutile TiO_2(110)-(1∗2): formation of Ti_2O_3 quasi-1D metallic chains. *Phys. Rev. Lett.*, **96**, 055502.

94 Li, M., Hebenstreit, W., and Diebold, U. (2000) Morphology change of oxygen-restructured TiO_2(110) surfaces by UHV annealing: formation of a low-temperature (1∗2) structure. *Phys. Rev. B*, **61**, 4926–4933.

95 Mochizuki, I., Ariga, H., Fukaya, Y., Wada, K., Maekawa, M., Kawasuso, A., Shidara, T., Asakura, K., and Hyodo, T. (2016) Structure determination of the rutile-TiO_2(110)-(1 × 2) surface using total-reflection high-energy positron diffraction (TRHEPD). *Phys. Chem. Chem. Phys.*, **18**, 7085–7092.

96 Lepeshkin, S., Baturin, V., Tikhonov, E., Matsko, N., Uspenskii, Y., Naumova, A., Feya, O., Schoonen, M.A., and Oganov, A.R. (2016) Super-oxidation of silicon nanoclusters: magnetism and reactive oxygen species at the surface. *Nanoscale*, **8**, 1816–1820.

15
The Pressing Role of Theory in Studies of Compressed Matter

Eva Zurek

State University of New York, University at Buffalo, Department of Chemistry, 331 Natural Sciences Complex, Buffalo, NY 14260-3000, USA

15.1 Introduction

The periodic table as we know is based upon our experience at 1 atm. However, the pressure in our universe spans many orders of magnitude, from intergalactic space ($P = 10^{-32}$ atm) to the center of a neutron star ($P = 10^{32}$ atm). The earth's core is subject to 3.5 million atm (~350 GPa), and the cores of giant planets may experience pressures in the terapascal regime. Pressure is not only an important variable in earth and planetary sciences but is also relevant for extreme situations such as nuclear explosions, and it can be utilized to synthesize compounds with unique properties, such as superconductors or superhard materials, which remain metastable when quenched to atmospheric conditions. Until recently, static compression experiments in diamond anvil cells (DACs) were limited to below 400 GPa. Improvements to DAC technology, via micro-ball nanodiamond anvils, have now made it possible to reach pressures of 600 GPa or higher reliably [1]. Larger dynamic pressures can be created with shock waves, precompressed pressure cells have been placed inside of DACs, and DACs have been shocked. There are many ways experimentalists have interrogated compressed matter [2].

However, carrying out experiments at high pressures and analyzing the results is fraught with difficulties [3]. Compounds containing light elements are particularly challenging since these nuclei are prone to diffuse into the diamonds within a DAC, causing them to break, leading to an expensive and premature end of the experiment. The pressure range currently accessible by neutron diffraction is only a fraction of what can be applied by a DAC. This means that for many light element systems, it may not be possible to directly obtain structural information about a pressurized phase, so it is necessary to employ indirect probes such as

Raman or IR spectroscopy instead. In addition, the experimentally observed signals may result from a mixture of phases, decomposition products of the starting materials, or even new (unintended!) species that form from chemical reactions of the sample with the experimental apparatus. Furthermore, measuring the electrical properties of a material compressed by a DAC can be very challenging. These are some of the reasons why theoretical calculations are an integral part of high-pressure research. The experimental high-pressure community relies on theory to aid in structural characterization, confirm the properties of a compound once its structure is known, and predict novel compounds as targets for synthesis. There is substantial synergy between experiment and theory, and this feedback loop between the two communities is of utmost importance in order to advance the field.

The structures of solids under pressure – the stoichiometries in which they crystallize and the lattices they assume – are substantially different from what is found at atmospheric conditions [3–14]. As a result, typically one cannot use the knowledge gained via one's experience at 1 atm to reliably predict the pressure-dependent phase diagrams of compressed solids. This means that neither data mining [15] nor intuition-driven guesses such as phenomenological structure maps [16] are likely to be useful in predicting which compounds may form under pressure. As a result, many theoretical groups working in this field must first use crystal structure prediction (CSP) techniques to find the structures of the global minimum and important local minima for a given set of P, T conditions. Once these are known, their properties can be calculated, and compared with experimental results (if available). In this field, where computations can be much quicker and cheaper than experiments, many studies have been inspired via theoretical predictions. In addition, computations allow one to study well-defined compounds and stoichiometries, and they are able to probe pressures that are not yet currently accessible to experiment.

In this chapter, we describe the algorithms most often used for *a priori* CSP, followed by exciting examples from the recent literature showcasing just how different the behavior of matter under conditions of extreme pressure can be. We illustrate how pressure perturbs the electronic structure of condensed phases, and describe ways in which to think about chemical bonding in compressed matter. Next, we provide examples of unique stoichiometries that can be synthesized under pressure and their emergent properties. Finally, we briefly discuss how pressure influences melting. Many of these effects are interrelated. For example, some of the unique properties that emerge are due to pressure-induced changes of the electronic structure, or they may be found in a compound with a novel stoichiometry that only becomes stable under pressure. As a result, some of the examples provided in one section of this chapter could have fit equally well into other sections. We hope this chapter inspires the reader to speculate as to what the periodic table may have looked like if Mendeleev had developed it given access to experimental results for compressed matter [17].

15.2
Techniques for Crystal Structure Prediction

CSP is a global optimization problem in which the atomic positions and the unit cell parameters are the variables and the (free) energy is the function to be minimized [18–29]. Many of the algorithms used for CSP are well-known metaheuristics used to find good solutions for optimization problems. None of them are guaranteed to find the globally optimal solution; the only way in which this can be done is by locating every local minimum. The number of local minima increases exponentially with system size [30], so that an exhaustive search is only possible for the simplest systems. Some of the first applications of global optimization algorithms to find the structures of solids and molecules include the Li_3RuO_4 crystal lattice [31] and fullerene clusters [32]. Because of the large computational cost of CSP searches, the energies were evaluated using empirical potentials in these early works.

CSP methods are oftentimes employed to locate the global minimum, although the identification of important low-lying kinetically stable local minima may also be important. Whereas some of these methods are better designed to perform a thorough search of a particular region of the potential energy surface (PES), other techniques have been designed to explore the whole (chemically relevant) PES. Basin hopping [33], minima hopping [34], metadynamics [35], and simulated annealing [36] fall into the first category of algorithms. They perform well given a starting structure that lies within the same region of the PES as the target structure. As a result, one may need to carry out runs with different starting structures to enhance sufficient sampling of the PES. These techniques explore the PES by overcoming energy barriers. Random structure searches [23,37], evolutionary/genetic algorithms [38–45], and the particle swarm optimization technique [22,46] are the methods of choice when the structures are wholly unknown. These methods can traverse the entire potential energy landscape and hone in on the most promising regions. Hybrid methods that use a mixture of computational strategies have also been proposed [47,48]. The CSP techniques used in high-pressure research are typically employed in conjunction with programs that carry out density functional theory (DFT) optimizations of the geometries of the crystalline systems. This is because reliable interatomic potential parameters are typically not available for compressed matter. However, approaches that fit parameters on-the-fly to DFT data generated during a structure search and employ the custom parameter set in further optimizations have been developed [49].

In the following, we describe some of the most commonly used algorithms for CSP of compressed matter. We focus on techniques developed primarily for inorganic systems where each atom can be treated as an independent entity and intermolecular connectivity constraints are not required. It is of course possible to modify the algorithms discussed herein so that they may be used to predict the structures of molecular solids [50], but this requires special considerations resulting from the large degree of conformational flexibility, near degenerate energies of many crystalline polymorphs, and typically large unit cell sizes in molecular crystals [51]. The CSP algorithms discussed in this chapter can also be applied to inorganic systems at ambient conditions. However, approaches that incorporate experimental information, data mining, and machine learning may be more robust at predicting the structures of large complex crystals at 1 atm [15,52–54].

15.2.1
Potential Energy Surfaces

In order to describe a unit cell and the positions of the N atoms within it, $3N + 3$ parameters are required: three unit cell vectors and three angles, as well as $3N - 3$ degrees of freedom for the atomic coordinates. Locating the global minimum in the PES amounts to finding the values for these $3N + 3$ parameters that minimize the (free) energy. A schematic diagram of a one-dimensional PES is given in Figure 15.1. This PES possesses a number of local minima that may be metastable (kinetically stable) provided the barriers between them are sufficiently high. Each minimum can be reached via local optimization from a given set of atomic configurations; the area in the PES within which this set falls is called a basin. A superbasin is a region of the PES that contains many neighboring basins. If the barriers within a superbasin are small, it is called a funnel.

But just how difficult is it to locate the global minimum within a given PES? Stillinger has shown that the number of local minima increases exponentially with the number of atoms that make up the system [30]. In addition, it has been illustrated that finding the global minimum of homogeneous [55] and heterogeneous [56] clusters is a NP-hard (nondeterministic polynomial-time-hard) problem. This means that there is no algorithm that scales as a polynomial in the number of degrees of freedom. Moreover, the "no free lunch" theorems state that all of the algorithms employed for CSP will give equivalent success rates

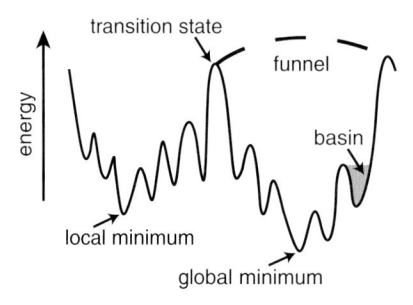

Figure 15.1 A schematic of a 1D PES. It contains a large number of local minima, one of which is pointed out, and a global minimum. A basin (one is shaded in gray) contains all of the geometries that will optimize to the same local minimum. This PES contains two funnels that are separated by a transition state.

when averaged over all PESs [57], implying that there is no single algorithm that will perform well for every system. Together, these properties do not bode well for the success of CSP, especially for systems with large and complex unit cells. However, one should not lose all hope: There are still a number of encouraging properties of PESs that can be harnessed to find good solutions to the CSP problem.

First of all, only certain areas of the PES correspond to structures that are chemically sensible. The other regions, which are likely to have atoms that are too close or too far apart from each other, will be high in energy and possess very few minima. This means that chemically irrelevant configurations can be removed from the structure search prior to carrying out computationally time-consuming relaxations. Second, it has been proposed that the basins containing low-energy minima take up the largest amount of "space" in the PES [58]. Thus, it is more likely that a randomly generated structure will fall in a low-energy basin than in a basin with a high energy. Finally, the Bell Evans Polanyi Principle [59,60] suggests that low-energy basins are located close to each other on the PES and the barriers separating them are expected to be small. This means that small perturbations of a structure will result in the exploration of the basins within a given funnel, ultimately locating the lowest energy point in this funnel. The simplest situation for CSP is if a PES contains a single funnel. If many funnels that are far apart from each other are present, or if the PES does not have any regions of attraction, the search will be more difficult. Now, let us get a quick overview of some of the most popular techniques for CSP.

15.2.2
Following Imaginary Phonons

The simplest method for CSP begins with optimizing the geometries of a few high-symmetry lattices followed by calculation of their phonon spectra. Next, the atoms are displaced along the eigenvectors of the softest phonon mode (the imaginary mode with the largest magnitude) detected and optimized to the nearest local minimum. The perturbation typically lowers the symmetry of the system. If the mode is located at the zone center, Γ, the unit cell of the distorted structure will be the same as that of the original lattice. It is more likely, however, that the soft mode is off the zone center so that an appropriate supercell must be made prior to the distortion. The distorted structure may also possess a number of soft modes, in which case this procedure needs to be iterated until a local minimum is reached.

A disadvantage of this method is that it requires calculation of the entire phonon spectrum, which is computationally expensive within DFT. Moreover, it only works well provided that the starting structure resembles the target, that is, if the two lattices lie close to each other in the PES. An advantage of this approach is that it reveals how complex lattices can be derived from electronically driven structural distortions of prototypical systems, thereby enhancing our chemical understanding and forging relationships between structures. The method of following imaginary modes has been employed extensively to predict interesting materials, a few examples include perovskites that can be derived from low-symmetry distortions of the prototypical cubic structure [61–63], complex crystal structures of late transition metal monoxides [64], and the structure of compressed silane [65], which was posited could be a high-temperature superconductor [66,67].

In the following sections, we discuss CSP methods that are based on techniques used to solve global optimization problems. These algorithms may result in structures that have instabilities off the zone center, if the unit cells employed in the simulations were too small. Thus, the method of following soft modes can be combined with automated global optimization schemes to locate minima.

15.2.3
Random (Sensible) Structure Searches

The simplest CSP method randomly generates the atomic coordinates and lattice vectors of a user-defined number of crystals, which are relaxed to the nearest stationary point, as illustrated in Figure 15.2a. In order to prevent such a search from exploring chemically unreasonable regions of the PES, which are bound to be high in energy, constraints that are based on chemical principles must be employed. Two of the most exceedingly useful constraints, appropriate unit cell volume ranges and minimum interatomic distances, ensure that atoms are neither too close nor too far apart. Constraining the Bravais lattices or space groups that are generated can also be particularly useful, especially if they can be obtained from experiment. And, imposing connectivity constraints so that the building blocks employed are molecules or clusters, when appropriate, can dramatically increase the success of an otherwise fully random search. Pickard and Needs have referred to this procedure as "generating random sensible structures" [23].

Because in a purely random search the structures that are created are independent of one another, this technique does not learn where the valleys in the PES are likely to lie as the search progresses. To overcome this deficiency, Pickard and Needs developed the *ab initio* random structure searching (AIRSS) method [23,37], wherein the geometries are relaxed using first-principles calculations. In addition to allowing the enforcement of a number of constraints, particularly stable structures are subject to random mutations such as atomic displacements and unit cell deformations. This so-called "shaking" procedure incorporates learning into the search. Shaking also helps to overcome barriers between basins, thereby enabling a thorough exploration of a particular region of the PES. The success of the AIRSS method has been rationalized by noting that because the funnel that contains the global minimum takes up the largest amount of "space" in the PES, the probability that a randomly generated structure will fall into this funnel is high. And, the shaking procedure encourages exploration of the basins within this funnel, inevitably leading toward the global minimum.

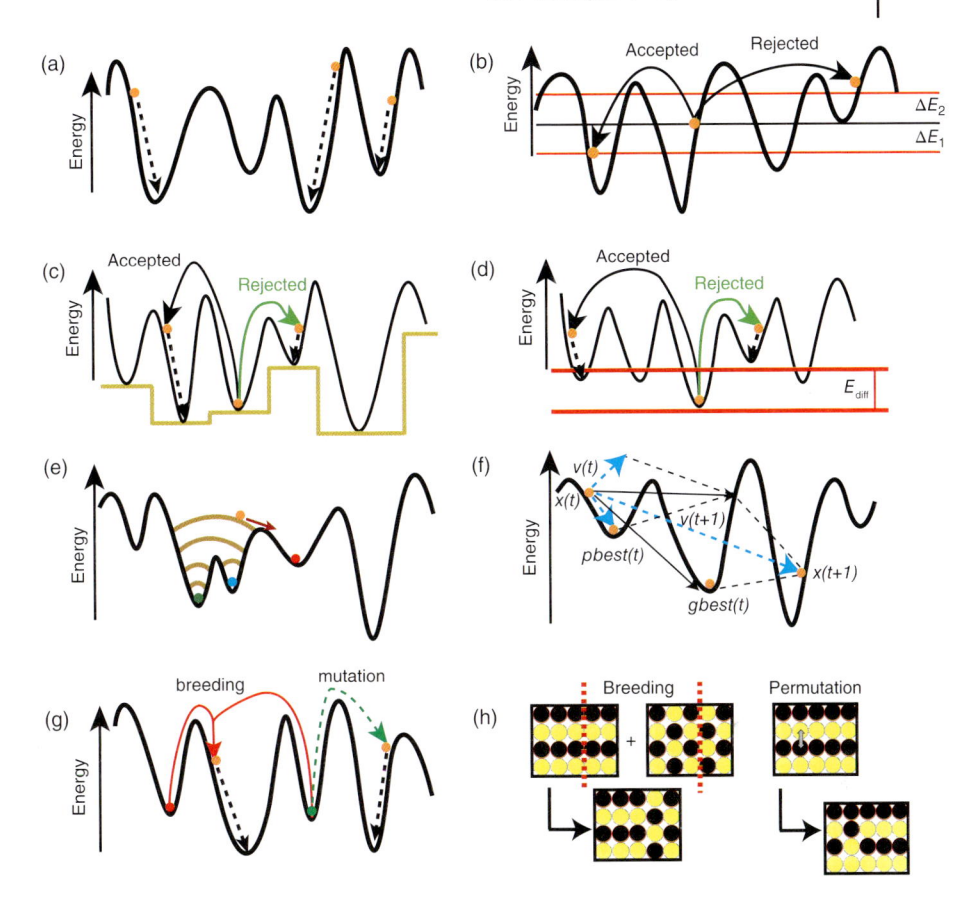

Figure 15.2 A schematic illustration of how a (a) random search, (b) simulated annealing, (c) basin hopping, (d) minima hopping, (e) metadynamics, (f) particle swarm optimization, or (g) evolutionary algorithm simulation traverses a 1D PES. The dots represent configurations and the dashed arrows optimization to the nearest local minimum (h). An illustration of how evolutionary operators act in real space on a 2D lattice. The breeding (or cut and splice) operator makes a single child from two parents. Permutation (of atomic positions) is an example of a mutation.

Random searches have been employed to predict the structures of a plethora of compressed solids. Just a few of these include alloys of Li and Be [68], hydrogen [69,70], lithium [71,72], and TiO_2 [73], as well as hydrogen–oxygen [74,75] and hydrogen–silane mixtures [76].

15.2.4
Simulated Annealing

The simulated annealing algorithm was inspired by the metallurgical process of heating a substance followed by controlled cooling until crystallization occurs [36]. A run begins with evaluating the energy of a (typically randomly generated) configuration of atoms. A new structure is created by perturbing the original system via randomly displacing the atoms, permuting atoms of different types, or by modifying the unit cell parameters. The Monte Carlo method of Metropolis is used to decide if this new configuration should be accepted or rejected. The probability of acceptance is calculated via $P = \exp(-\Delta E/k_B T)$, where ΔE is the energy difference between the two structures, k_B is the Boltzmann constant, and T is the simulation temperature. A random number ε that ranges between 0 and 1 is chosen, and if $\varepsilon < P$, this structure is accepted and the cycle repeats. Otherwise, new geometries are made via perturbations of the original configuration until one of them is accepted. This procedure is illustrated in Figure 15.2b. The simulation temperature is not a physical temperature, but rather a variable that is employed to control the rejection rate. The run begins with large temperatures so that most structures are accepted. During the search, the temperature is slowly decreased so that fewer and fewer high-energy structures are accepted, thereby mimicking the physical annealing procedure. At the end of the simulation, a quench run is performed wherein $T = 0$ K so that only downhill steps are accepted, and the final configuration is relaxed to the nearest local minimum. Unfortunately, because the high-temperature PES does not always coincide with the low-temperature PES, a simulated annealing run may become stuck in a local minimum upon cooling. To help prevent this problem from occurring, more complicated annealing schemes, wherein the temperature is sometimes raised, can be performed. Another drawback of this method is that it may be necessary to carry out a number of runs if a good starting structure is not known. The extent of the random mutation is also important – If it is too small, the method will not be able to escape local minima; but if it is too large, the run essentially becomes a random search. Another way to move the atoms in a simulated annealing simulation is by using molecular dynamics.

Doll et al. have extensively employed simulated annealing in conjunction with Hartree–Fock and DFT calculations to predict the crystal structures of a wide range of inorganic systems [20,77], including lead sulfide [78], calcium carbide [79], boron nitride [80], GeF_2 [81], and metal pernitrides [82].

15.2.5
Basin and Minima Hopping

Similar to simulated annealing, basin hopping [33] uses a Monte Carlo procedure to determine if new configurations should be accepted or rejected. Each configuration is optimized to the nearest local minimum so that every geometry that falls within a given basin optimizes to the same structure. It is said that this procedure transforms the PES into a series of interpenetrating step functions that connect the local minima, as shown in Figure 15.2c. Because the energies of the optimized species are employed to determine the acceptance probability, it is possible to maintain the same temperature, which must be chosen judiciously, during the course of the entire run. As in simulated annealing, the closeness of the starting structure to the target and the extent of the random mutation are important factors for a successful search. A drawback of this technique is that there is no penalty associated with visiting an already explored basin. Basin hopping has been extensively applied to predict the structure of finite clusters [33,83–86], but we do not know of examples where it was used in CSP of solids under pressure.

The related minima hopping method of Gödecker [34] differs from basin hopping since it does not explore the PES with a Monte Carlo procedure, but via molecular dynamics instead. This means that it is not necessary to choose a temperature that controls the acceptance probability. As in basin hopping, each structure is locally optimized. The algorithm compares the energy of a new configuration with that of its predecessor and the energy difference, E_{diff}, is continuously adjusted such that half of the structures are accepted (see Figure 15.2d). The kinetic energy used in the molecular dynamics simulation is modified so that there is a 50% chance that the system will overcome a barrier to enter a new basin. Moreover, if the algorithm revisits an already known area of the PES, the kinetic energy may be raised so as to promote enhanced exploration and escape from funnels that do not contain the local minimum. Minima hopping has been employed to predict the structures of a number of systems under high-pressure conditions, including disilane [87], carbon [88,89] as well as hydrides of phosphorus [90], and hydrides of sulfur and selenium [91].

15.2.6
Metadynamics

The metadynamics technique is a molecular dynamics-based method that overcomes barriers in the PES [35]. In metadynamics, the forces acting on the atoms are modified by a history-dependent term that discourages revisitation of known areas of the PES. The potential in the already visited regions of the PES is continuously lifted as these areas are filled with Gaussian functions; this is illustrated in Figure 15.2e. This so-called "basin-flooding" procedure allows the method to overcome a barrier and access a new local minimum when the potential within the original basin is lifted high enough. Typically, the flooding is carried out in a space whose dimensionality is smaller than the $3N + 3$ degrees of freedom used to describe the full crystal lattice. A space defined by a few collective variables, such as the unit cell parameters, may be used instead. A disadvantage of metadynamics is that it may flood a transition barrier between two or more basins, which could prevent the global minimum from being found in a single search. Therefore, it may be necessary to carry out a number of searches using different starting configurations to find the global minimum. Metadynamics has been used extensively to predict new structural modifications under pressure for a plethora of systems, including germanium [92], calcium [71], and CO_2 [93]. The metadynamics technique has also been applied to a wide range of dynamics simulations of rare events such as structural phase transitions (including those at high pressures [94,95]), conformational changes in solution, and calculating chemical reaction mechanisms [35].

15.2.7
Particle Swarm Optimization

The particle swarm optimization (PSO) technique is a global optimization method that was inspired by the migratory behavior of large groups of animals, such as a flock of birds or a school of fish [96] (the swarm). Wang et al. were the first to apply this algorithm toward CSP for solids [46] within the CALYPSO (Crystal structure AnaLYsis by Particle Swarm Optimization) code [22]. Call et al. previously applied the PSO method to predict the structures of various clusters [97].

In this method, the trajectory a single individual traverses within the PES depends upon its own path as well as on the path of the other members of the swarm. If the position of an individual at some instant is given as $x(t)$ and its velocity is $v(t)$, then

$$x(t+1) = x(t) + v(t+1), \tag{15.1}$$

where

$$v(t+1) = \omega v(t) + c_1 r_1 (pbest(t) - x(t)) + c_2 r_2 (gbest(t) - x(t)). \tag{15.2}$$

In Eq. (15.2) $pbest(t)$ is the position of the individual once it has been optimized to the nearest local minimum, $gbest(t)$ is the position of the global minimum of the swarm, ω is an inertia weight ranging from 0.4 to 0.9 that is modified during the search, r_1 and r_2 are random numbers ranging from 0 to 1, and the coefficients c_1 and c_2 indicate how much the individual trusts its own position versus that of the global minimum. Large values of ω promote global exploration, whereas small values lead to a more local search. The PSO procedure is illustrated schematically in Figure 15.2f.

The PSO technique begins by creating random sensible structures (subject to interatomic distance, volume, and symmetry constraints), which are locally optimized. A large fraction of the subsequent structures are created via Eqs. (15.1) and (15.2). However, random structures are continually injected into the swarm during the search to ensure diversity and prevent the algorithm from getting stuck in a particular region of the PES. All of these factors combined mean that the PSO algorithm is able to learn from its history, as well as being able to broadly explore the whole PES while simultaneously zooming in on the most promising areas.

The CALYPSO code has been used extensively to predict the structures of novel materials under pressure, including hydrides that may be high-temperature superconductors [98–101], cesium polyfluorides [102], and chlorine polyhydrides [103].

15.2.8
Evolutionary Algorithms

An evolutionary algorithm (EA) (sometimes referred to as a genetic algorithm (GA)) adapts concepts from evolutionary biology, such as selection, reproduction, and mutation, to the CSP problem. An EA begins by generating random sensible structures (seeding with known structure types is also possible) that are locally optimized. A thermodynamic quantity such as the energy or enthalpy is used to determine the fitness of each configuration, and the fittest individuals are chosen to be parents of a new generation of structures. This procedure ensures that traits promoting stability propagate into the next generation. Child structures can be made via a combination of the unit cells of two parents, thereby mimicking breeding, or perturbations of a single individual that are analogous to mutations, as illustrated in Figure 15.2g and h. The cut-and-splice operation that combines two parents into a single child in real space was introduced by Deaven and Ho for clusters [32], and was adapted toward crystals by Glass *et al.* in the USPEX (Universal Structure Predictor: Evolutionary Xtallography) algorithm [40]. Mutations include perturbations of the unit cell shape, as well as the atomic positions and atomic orderings. Whereas breeding typically leads to large structural changes that allow broad exploration of the PES, mutations result in a more local search. Because EAs learn about the PES during the course of a run, they are sometimes provided as examples of artificial intelligence.

In the last decade, a number of EAs that are interfaced with periodic program packages employing first-principles calculations or interatomic potentials have been released. Some of these include XTALOPT [38,104,105], GASP [39], USPEX [40,106,107], MAISE [41], EVO [42], as well as algorithms by Trimarchi and Zunger [43,108], Abraham and Probert [44], Fadda and Fadda [45], and the "adaptive-GA" of Wentzcovitch and coworkers [49].

Evolutionary algorithms have been extensively applied toward predicting the structures of compressed phases. Just a few examples include hydrides with stoichiometries that would not be stable at atmospheric conditions [109–121], lithium hydroxide (LiOH) [122], and xenon suboxides [122], as well as binary and ternary phases containing the light elements Li, Be, and B [123–127].

15.3
Chemistry under Pressure

When a material is subject to conditions of a few hundred gigapascal, the pV contribution to the free energy can be up to 10 eV – which is larger than the strength of any bond in a molecule at atmospheric conditions. At the same time, the volume of the substance decreases substantially. This brings atoms that were previously too far apart to interact close enough to each other so that new bonds can form. And those atoms that were already bonded at 1 atm are pushed closer together into the repulsive region of their interatomic potentials. Clearly, new chemistry emerges as the electronic structure of a compound responds to the change in density.

Prewitt and Downs proposed a number of rules of thumb that could be used to predict how the structure and bonding of a solid are influenced by pressure [128]. Actually, these rules, which arose from a tremendous body of experimental data, were inspired by Pauling's crystal chemistry rules [129]. About two decades later, Grochala et al. [4] proposed a new set of rules, some of which encompass those of Prewitt and Downs. The rules of Grochala and coworkers took into account astonishing new experimental results, and helped to further develop the scientific community's chemical intuition under pressure. In this section, we summarize a number of exciting research results showcasing the amazing effects of pressure on chemistry.

15.3.1
Unique Electronic Structures

It has long been recognized that pressure can profoundly impact the electronic structure of a condensed system. In 1926, J.D. Bernal suggested that under sufficient compression all substances should eventually become metallic; this was pointed out in the literature for the first time in 1935 by Wigner and Huntington [130]. To understand why metallization should occur, consider how the electronic structure of a gas of H_2 molecules, each with a large gap of >13 eV between their highest-occupied molecular orbitals (HOMOs) and lowest-unoccupied molecular orbitals (LUMOs), changes upon condensation. As the molecules begin to interact with each other, energy bands will form. Increased pressure leads to a larger orbital overlap so that the bonding combinations go down, and the antibonding combinations go up in energy. The pressure-induced broadening of both the filled and empty bands leads to the eventual overlap of the valence and the conduction bands; this is illustrated schematically in Figure 15.3a. In fact, because of pressure's propensity to induce multicentered bonding (see Section 15.3.2) as the intermolecular distances in a model 1D chain of H_2 molecules decrease, their inter- and intramolecular distances are expected to equalize. And, a 1D atomistic chain of hydrogen atoms must be metallic

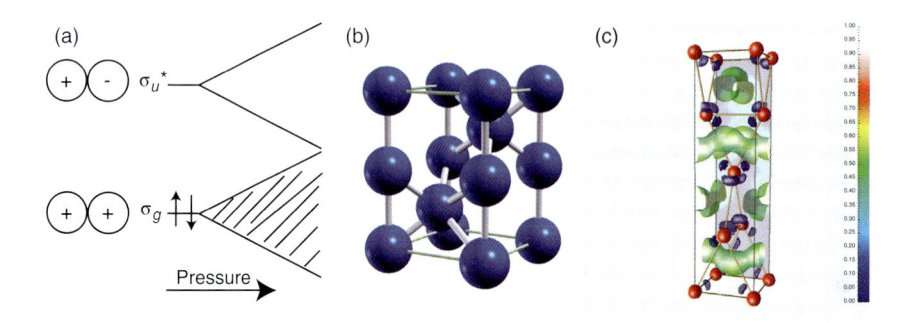

Figure 15.3 (a) Illustration showing how pressure-induced band broadening and overlap will lead to the eventual metallization of hydrogen. (b) Unit cell of the insulating structure of Na-*hp*4 at 320 GPa [142]. (c) The valence charge density of Cs-IV colored by the electron localization function (color bar shown) [143].

because the top of the bonding and the bottom of the antibonding bands are degenerate in energy. Wigner and Huntington predicted that molecular hydrogen would dissociate giving rise to atomistic metallic hydrogen above 25 GPa [130]. However, static pressures an order of magnitude greater than these have been unable to metallize hydrogen. This elusive state of matter is considered to be the "holy grail" of high-pressure research toward which substantial experimental [131–135] and theoretical [14,136–139] work has been devoted. Researchers have been pressing the limits of what is experimentally possible, and hydrogen has been subject to center-of-the-earth pressures at different temperatures in a DAC. Notably, two recent reports [140,141] provide evidence for new phases of hydrogen, one of which may be metallic [141].

Another surprising effect of pressure is on the so-called "simple" metals, which can become semiconducting or even insulating when squeezed. Neaton and Ashcroft were the first to predict that the pressure-induced overlap of the 2p cores will push the valence electrons of sodium into the interstitial regions rendering this element insulating [144]. They also conjectured that pressure will lead to the dimerization of atoms in compressed lithium, whose metallicity will decrease as the valence 2s electrons become localized in the interstitial regions [145]. About a decade later, these predictions were confirmed. Experiments showed that lithium becomes semiconducting above 70 GPa [146], and numerous DFT studies have examined the geometric and electronic structures of compressed lithium [71,72,147,148]. Groundbreaking studies revealed that at ∼200 GPa sodium becomes insulating with a bandgap of ∼1.3 eV [142]. *Ab initio* calculations found that the most likely candidate for the insulating phase of sodium is the double-hexagonal close-packed structure with a particularly small c/a ratio illustrated in Figure 15.3b. Because the valence electrons become confined in the interstitial regions in compressed sodium and lithium, both of these systems can be thought of as "electrides" (ionic compounds in which the electrons act as the anions). Electride behavior has been predicted in other compressed metals, and its electronic origins have been analyzed [149].

However, under pressure, valence orbitals can enter the picture as well. For example, cesium undergoes a pressure-induced s → d transition [150], wherein it essentially becomes a d^1 metal by 15 GPa [151]. This electronic transition occurs because of the energy reordering of atomic orbitals under pressure. It leads to an intriguing sequence of phase transitions, culminating in a nonclose-packed tetragonal Cs-IV phase that is stable between 4.3 and 12 GPa [152]. Because the maximum of electron density in this structure is found in the interstitial regions, as illustrated in Figure 15.3c, Cs-IV can also be thought of as an electride, Cs^+e^- [143,153]. At 70 GPa, a close-packed structure is assumed again, in which the 5p semi-core levels hybridize with the 6d [154]. The mixing of np or $(n-1)d$ character into the valence bands of the alkali metals and alkaline earth metals is quite common under pressure.

Finally, we note that the perturbation of the electronic structure under pressure can lead to quite unusual oxidation states. For example, recent calculations showed that mercury, which commonly assumes an oxidation state of +2, can attain oxidation states of +4 in HgF_4 and +3 in HgF_3. These phases were predicted to become stable around 40 and 75 GPa, respectively [155]. And, above 100 GPa, the energy reordering that occurs to atomic orbitals under pressure results in a Li 2s → Cs 5d electron transfer in phases with the stoichiometry Li_nCs, $n = 1 – 5$. As a result, cesium can assume a negative oxidation state going beyond −1 [156].

15.3.2
Structure and Bonding

With increasing pressure, the electron densities of atoms that were not interacting at atmospheric conditions begin to overlap and coordination numbers often increase (this is not always the case, for example, the fcc phase of Cs is stable between 2.3 and 4.2 GPa [157], whereas at higher pressures, 4.3–12 GPa, a non-close-packed tetragonal Cs-IV phase is assumed [152]). In order to adapt to such changes in the geometry, atoms must adopt new bonding strategies. This behavior has been suspected long before it was observed; for example, Pauling speculated that when water is sufficiently compressed, the distance between the H–O atoms that are hydrogen bonded would decrease and the intramolecular H–O distances would increase until the two bond lengths became equivalent [129]. In such a structure, the electrons that belonged to oxygen's lone pairs end up being shared with those hydrogen atoms to which they were hydrogen bonded to at 1 atm. Indeed, experiments have shown that above 60 GPa water forms ice X, whose unit cell is illustrated in Figure 15.4a. This phase is composed of interpenetrating diamond networks wherein each O—H bond is of the same length, and one can no longer distinguish between inter- and intramolecular hydrogen bonds [158,159]. Pressure-induced bond symmetrization in other hydrogen-bonded systems, including the hydrogen halides (HF, HCl and HBr), has been experimentally observed and studied theoretically [160]. The symmetric *Cmcm* phase of HBr, illustrated in Figure 15.4b, is assumed around 39 GPa [161]. Such bond equalization occurs in many compounds under pressure.

In fact, the electronic rearrangement that occurs upon increasing coordination can often be understood in terms of multicentered bonding schemes – whether they be electron-rich or electron-poor [4]. For example, recent theoretical calculations have predicted a number of pressure-stabilized hydrides with novel stoichiometries wherein the bonding can be explained using such schemes [103,111,112,115]. The simplest example of a three-centered two-electron bond (3c–2e) is H_3^+, which is one of the most abundant ions in interstellar space. This molecule adopts the geometry of an equilateral triangle because it maximizes the overlap in the bonding MO. Recently, a phase with the H_5Cl stoichiometry was predicted to become stable with respect to decomposition into H_2 and HCl between 60 and 300 GPa [103]. The H_3^+ motif, which approached the ideal equilateral triangle configuration by 300 GPa, was incorporated within H_5Cl, as illustrated in Figure 15.4c. If this phase is ever made, it would be the first extended compound to contain this iconic unit.

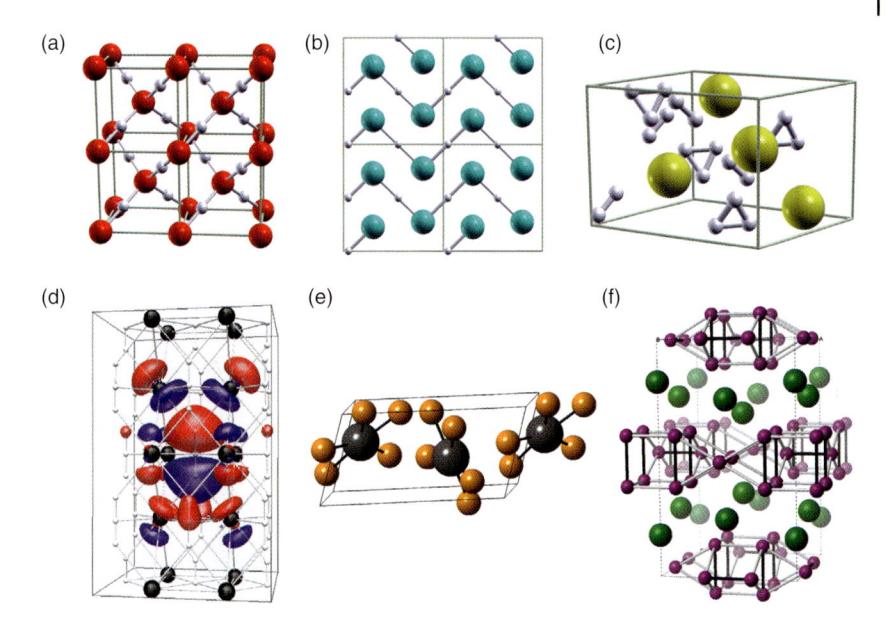

Figure 15.4 Crystal structures of (a) phase X of H$_2$O, (b) *Cmcm*-HBr [160], (c) H$_5$Cl at 300 GPa [103], (d) *Cmmm*-CsH$_3$ at 80 GPa [113], (e) CsF$_5$ at 150 GPa [102], and (f) *I4/mmm*-CaGe$_3$ [162]. In part (d) we also illustrate the H$_3^-$-centered Wannier function (isovalue ±0.03 au) obtained from a basis that yielded the two bands directly below the Fermi level. Oxygen atoms are red, hydrogen atoms white, bromine atoms turquoise, chlorine atoms yellow, cesium atoms black, fluorine atoms orange, calcium atoms green, and germanium atoms purple.

H$_3^-$, on the other hand – the simplest example of a three-centered four-electron bond (3c–4e) bond – is predicted to be linear. In the gas phase, however, this molecule would not have equidistant H—H bonds. *Ab initio* calculations have shown that an asymmetric H$_3^-$ unit whose H—H distances measure 0.75 and 2.84 Å [163] is a minimum on the PES. This species exhibits a double–well minimum, with the lowest-energy barrier being a symmetric configuration with H—H lengths of 1.06 Å. Similar H$_3^-$ motifs are predicted to be present in pressure-stabilized hydrides containing a heavy group I or group II element, such as K [112], Rb [111], Cs [113], and Ba [115]. Consider the CsH$_3$ stoichiometry that was found to be the lowest point on the convex hull above 30 GPa [113]. A number of nearly isoenthalpic CsH$_3$ phases were predicted to be stable, all of them consisting of Cs$^+$ and symmetric H$_3^-$ units; one such phase is illustrated in Figure 15.4d. Computations have shown that both pressure and the softness of the not-so-innocent bystander cations are key factors affecting the symmetrization of the H—H bonds in the H$_3^-$ motif. Equalization occurs via the interaction of a soft base, H$^-$, with an extremely weak acid, H$_2$. Because of the M$^+$ – H$^-$ interaction, the basicity of the hydride is decreased in the presence of the alkali metal cation, and model calculations revealed that symmetric bonds would form most easily in the gas phase [112].

The changes in electronic structure under pressure outlined in Section 15.3.1 (energy reordering of atomic orbitals, accessing unusual oxidation states) can also cause orbitals that do not participate in bonding at atmospheric conditions to become involved in bonding. For example, because the cesium in compressed CsH_3 assumes a +1 oxidation state and the energy of the semi-core Cs 5p bands increases with pressure, they are able to hybridize with the hydrogen s-orbitals [113]. In Figure 15.4d, we illustrate the Wannier function that describes the two bands below the Fermi level in one of the CsH_3 phases found to be stable under pressure. It shows clear signatures of hybridization between the H_3^- nonbonding molecular orbital with the Cs 5p states. A more dramatic involvement of the Cs 5p semi-core orbitals is predicted to occur in the cesium polyfluorides [102,164], CsF_n, $n = 2 - 6$, which are computed to become stable with respect to decomposition into CsF and F_2 around 10 GPa [102]. The geometries and bonding within these compounds are reminiscent of the isoelectronic XeF_n molecules [102]. For example, the pentagonal planar units comprising CsF_5 illustrated in Figure 15.4e resemble the XeF_5^- ion [102]. Remarkably, for $n \geq 2$, the cesium atoms were found to be oxidized past the +1 state. The Bader charges calculated for the CsF_n phases at 100 GPa showed an almost linear increase with the number of fluorine atoms comprising the phase. Moreover, the crystal orbital Hamiltonian populations (COHPs) [165] provided evidence of covalent bonding between the Cs 5p and F 2p states in CsF_2 and CsF_3. This study shows how pressure can cause so-called inner shell electrons to become involved in bonding.

As a final example of how pressure can be utilized to access phases that would not be stable otherwise, let us consider a certain set of polar intermetallic compounds comprised of group 14 elements and alkaline earth or rare earth metals. What differentiates them from the other phases discussed in this section is that they can be synthesized using high-temperature high-pressure (HTHP) techniques and quenched to ambient conditions. The reason why these phases are metastable at 1 atm is because large barriers are required to break the bonds between the group 14 elements that are formed under pressure. HTHP methods can be used to synthesize phases with unusually high-coordination environments, wherein the bonding cannot be explained using electron-precise counting schemes such as the Zintl concept [166]. For example, a $CaGe_3$ stoichiometry with $I4/mmm$ symmetry, whose constituents include Ge_2 dimers that condense to form square prisms that are capped on either side by another set of Ge_2 dimers (see Figure 15.4f), has been synthesized and its electronic structure has been theoretically studied [162]. Other binary silicides, $CaSi_3$, YSi_3, and $LuSi_3$ [167] assuming this structure were later synthesized, and computations predicted that an isotypic $BaGe_3$ phase could be accessed at $\sim$25 GPa [168]. First-principles calculations have shown that these systems exhibit delocalized metallic bonding and some of them behave as Bardeen–Cooper–Schrieffer (BCS)-type electron–phonon mediated superconductors below 10 K [162,167,169].

15.3.3
Novel Combinations

The pressure variable can be used to access chemical compositions that are not stable at 1 atm. For example, elements that would not normally form compounds may be coerced to do so under pressure, or they may crystallize in unique stoichiometries. This consequence of pressure is not only important in chemistry and materials science but it also affects the chemical species and minerals that may be prevalent on the surfaces of giant planets and in planetary interiors. In the following, we outline the results of only a couple of the plethora of theoretical predictions highlighting how unique chemical compositions may be attained under pressure. Some of these have been verified by experiments in DACs.

Consider the two light elements lithium and beryllium, which do not form binary phases at atmospheric conditions. Theoretical work predicted that Li–Be alloys with stoichiometries $LiBe_2$, $LiBe$, $LiBe_4$, and Li_3Be, would become thermodynamically stable at pressures ranging from ∼20 to 200 GPa [68]. A LiBe structure that assumed the $P2_1/m$ space group and was composed of alternating puckered square nets of lithium atoms and puckered triangular nets of boron atoms, see Figure 15.5a, was singled out for further analysis. Its density of states

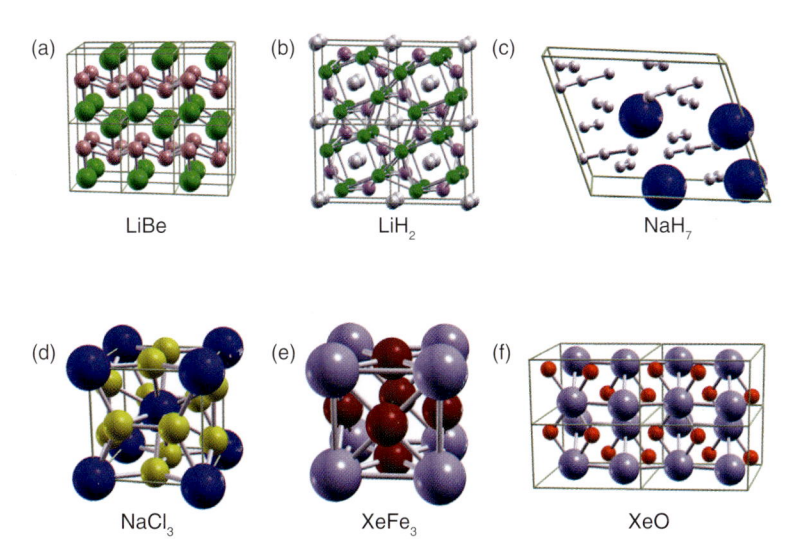

Figure 15.5 Illustrations of (a) $P2_1/m$-LiBe at 82 GPa [68], (b), $P4/mbm$-LiH_2 [109], (c) Cc-NaH_7 at 100 GPa [170] (note the not-yet-symmetrized H_3^- anion), (d) $Pm3n$-$NaCl_3$ at 200 GPa [171], (e) $Pm\bar{3}m$-XeF_3 at 250 GPa [172], and (f) $P2_1/m$-XeO [173] that were predicted to be stable under pressure. Phases with the XeO stoichiometry were found to be metastable in Ref. [122]. Lithium atoms are green, beryllium atoms pink, hydrogens in H_2 white, hydridic hydrogens mauve, chlorine atoms yellow, sodium atoms blue, xenon atoms gray, iron atoms dark red, and oxygen atoms bright red.

at 82 GPa contained a sharp step-like feature indicative of a 2D electronic structure that turned out to be associated with the beryllium layers. Because of the large size difference between lithium and beryllium, the 1s cores of the lithium atoms were found to overlap at this pressure, whereas the beryllium cores did not. As a result, the lithium valence electrons moved into the 2D beryllium layers, so that the bonding within this phase could be rationalized using the Zintl concept [166]. This is just one example of how pressure can induce two elements that are normally immiscible to form novel phases with unique electronic structures.

Examples of compounds with unusual stoichiometries can be found in the hydrogen/lithium phase diagram under pressure [109,117]. In the lithium-rich region, two nearly isoenthalpic Li_5H phases were calculated to have the most negative enthalpies of formation for the reaction $LiH + (m-1)Li \rightarrow Li_mH$ between 50 and 120 GPa [117]. The two polymorphs, one with $Cmc2_1$ and one with $Abm2$ symmetry, were both composed of face-linked Li_8H clusters. The Li_8H building block is best described as a distorted bicapped trigonal antiprism wherein the lithium atoms comprise the vertices, and the hydrogen atom lies within the cluster. The structure and electronic structure of the two Li_5H phases are reminiscent of the well-known alkali metal suboxides, Rb_9O_2 and $Cs_{11}O_3$, that are formed from fused M_6O octahedra [174]. Because of the large electronegativity difference between the lithium and the hydrogen atoms, the subhydrides can be thought of as $(Li^+)_5 (H^-)(e^-)_4$. DFT calculations showed ionic bonding on the inside of the clusters and metallic bonding in the interstitial regions between them, in analogy with the alkali metal suboxides [174]. At 90 GPa, both Li_5H phases were semimetallic, with the $Cmc2_1$ symmetry structure exhibiting two conical-like bands with near linear dispersion crossing the Fermi level. In the hydrogen-rich region, a number of LiH_n, $n > 1$, phases were found to be stable with respect to decomposition into LiH and H_2 above ~100 GPa [37,109]. Let us focus on describing LiH_2 and LiH_6 because they expound the main features of all of the stable polyhydrides of lithium, and because they fell on convex hulls spanning 150–300 GPa. LiH_2 was composed of a host Li^+/H^- lattice containing a 1D chain of H_2 guests, as illustrated in Figure 15.5b. This phase was predicted to become semimetallic as a result of the overlap of the "donor impurity" H^- band with the H_2 σ^*-bands. In LiH_6, on the other hand, the electropositive element lithium donated its valence electron to the hydrogen molecules so they could be thought of as $H_2^{-1/3}$. The partial filling of the antibonding MOs led to a weakening and lengthening of the H—H bond. DFT calculations showed that LiH_6 is likely to be superconducting [175]. The synthesis of the lithium polyhydrides proved to be quite difficult, but recently Pépin et al. succeeded by subjecting LiH to pressures above 130 GPa at 300 K in a DAC [176]. Synchrotron infrared absorption spectroscopy yielded peaks that differed from those seen in pure compressed hydrogen, but they were in reasonable agreement with the H_2 vibron frequencies calculated for LiH_2 and LiH_6. Pépin et al. suggested that the polyhydrides form via a mechanism where pressure causes lithium to diffuse into the DAC yielding a LiH_6 layer at the diamond/sample interface, and a LiH$_2$ layer at the LiH$_6$/LiH interface. Whereas the LiH$_6$ phase predicted via *a priori* CSP techniques is metallic by virtue of the fact that it contains 1 formula unit in the primitive cell, experiments did not show any evidence of metallicity up to 215 GPa. So, the exact stoichiometry and structure of the polyhydride phases formed in Ref. [176] is still open to deliberation.

Continuing in this trajectory, a number of theoretical studies have used DFT in conjunction with CSP techniques to predict the structures of MH$_n$ (M = Na, K, Rb, Cs and $n \geq 2$, or Mg, Ca, Sr, Ba and $n > 2$) [37,98,110–116,177,178]. The computational exploration of these phases was inspired by the prediction that the metallization of hydrogen could be hastened via chemical doping [179], and that metallic hydrogen-rich phases could potentially behave as high-temperature superconductors [66]. Notably, a recent study appears to have succeeded in synthesizing sodium polyhydrides above 40 GPa and 2000 K in a DAC [170]. The experimental X-ray diffraction patterns and Raman spectra were compared with those obtained for phases whose geometries were obtained via CSP. Structures with NaH$_3$ and NaH$_7$ stoichiometries gave the best match with experiment and provided evidence for the formation of an H$_3^-$ anion within NaH$_7$, as illustrated in Figure 15.5c. The general features of these structures are in-line with those found in the first computational study of sodium polyhydrides, but Ref. [110] did not consider the NaH$_3$ stoichiometry. Both theoretical studies agree that the thermodynamically stable sodium polyhydrides are unlikely to be high-temperature superconductors at pressures attainable in DACs [110,170]. However, within nonhybrid DFT, $R\bar{3}m$-SrH$_6$ at 250 GPa [116,178], and $Im\bar{3}m$-CaH$_6$ [98] at 150 GPa were found to have a high density of states at the Fermi level and the latter phase was calculated to be superconducting at high temperatures. Even though superconductivity in the hydrides of an alkali metal or alkaline earth has not been experimentally observed, evidence is beginning to mount that other hydrides with unique stoichiometries may be good superconductors, as discussed in Section 15.3.4. We also note that a recent mixed experimental–theoretical study of the Na/Cl phase diagram under pressure found at least two stable novel stoichiometries, Na$_3$Cl and NaCl$_3$ [171]. Evolutionary algorithms were used to predict stable crystal structures whose computed X-ray powder patterns could be compared with experimental data. An NaCl$_3$ phase with $Pm3n$ symmetry that was found to be stable above 48 GPa is shown in Figure 15.5d.

One example of how pressure's effect on chemical reactivity may be key to unraveling a mystery in earth and planetary science research manifests itself in the so-called "missing xenon paradox." Studies have shown that the earth's atmosphere contains an order of magnitude less xenon than expected based upon the abundance ratios of the elements found in interstellar space [180]. A number of hypotheses have emerged to explain how the xenon went missing [181], with one class of models focusing on pressure-induced compound formation between xenon and minerals or elements within the earth. For example, *a priori* CSP techniques have found a number of Xe–Fe, Xe–Ni, and Xe–O compounds that become thermodynamically stable with respect to the elemental phases under pressure [122,172,173]. High-pressure experiments have shown that xenon and iron do not react with each other below 155 GPa. However, DFT calculations found that at 0 K and above 200 and 250 GPa, respectively, stable Xe–Fe and Xe–Ni phases emerged [172]. The inclusion of temperatures up to those believed to be in the Earth's inner core, 6000 K, revealed that a number of metal-rich Xe-containing phases remained stable at these conditions, notably $XeNi_3$ and $XeFe_3$ (see Figure 15.5e). Xenon oxides are also calculated to be unstable with respect to their elemental constituents below 75–83 GPa [122,173]. Above these pressures however, a number of phases, with stoichiometries including Xe_7O_2, Xe_3O_2, Xe_2O, XeO_2, and XeO_3, were found to become thermodynamically stable. A phase with XeO stoichiometry is shown in Figure 15.5f. The propensity of bond formation between xenon and oxygen under pressure was thought to suggest that xenon may be retained in oxygen-containing minerals that comprise the earth's lower mantle [173]. Another explanation for the less than expected amount of xenon could be that it reacts with the magnesium comprising silicate and oxide phases, since MgXe compounds are predicted to be stable at center-of-the-earth conditions [182]. Moreover, experiments have shown that xenon becomes reactive with water at P/T conditions that are found in the interiors of giant planets such as Uranus and Neptune [183], and hydrogen-rich xenon-containing compounds such as $Xe(H_2)_7$ can be stabilized under pressure [184].

Above we have provided just a few examples of interesting compounds that may become stable when squeezed. High-pressure experiments can be difficult to carry out, and the data complicated to interpret, but evidence is starting to mount that pressure can be used to access solid-state compounds with unique compositions such as the lithium polyhydrides [176], sodium polyhydrides [170], and $Na_3Cl/NaCl_3$ [171]. It is unlikely that any of the aforementioned lithium–beryllium phases, or the hydrides of the alkali metals or alkaline earths could be quenched to atmospheric conditions because the barriers to their decomposition are likely to be low and they may not correspond to minima at these conditions. However, a metastable XeO_2 phase has been synthesized at 1 atm and characterized via Raman spectroscopy, but its structure is still a matter of debate [122].

15.3.4
Emergent Properties

Perhaps one of the most sought-after properties that is affected by pressure is superconductivity. At least 23 elements, including oxygen, lithium, iodine, and iron, become superconducting under pressure [185]. And the superconducting transition temperature, T_c, of 30 elements[1], and numerous compounds, is pressure dependent. Until recently, the material with the highest known T_c was $HgBa_2Ca_2Cu_3O_8$, whose 1 atm T_c of 133 K rises to 164 K by 31 GPa [186].

1) http://physics.wustl.edu/~jss/NewPeriodicTable.pdf.

Tremendous effort has been directed toward the computational prediction of hydrogen-rich materials that behave as high-temperature superconductors under pressure. The two examples we present below are significant because a feedback loop between theory and experiment has led to their discovery, and it appears that their high T_c can be explained within BCS theory.

Theoretical work [187] was carried out to predict the structures the classic H_2S hydride adopts under pressure and estimate their superconducting critical temperatures using the Allen–Dynes-modified McMillan equation [188]. It was concluded that above 130 GPa, hydrogen sulfide would transform to a phase whose T_c was 80 K above 160 GPa [187]. This study inspired Drozdov et al. to investigate the superconducting behavior of hydrogen sulfide under pressure [189]. The superconducting transition temperatures measured for samples prepared at low temperatures, $T \leq 100$ K, were in-line with the theoretical predictions for H_2S. However, when the samples were prepared at elevated temperatures, $T \geq 300$ K, a remarkably high T_c of 203 K at 150 GPa [189] was observed, and this finding could not be reconciled with the theoretical results obtained for H_2S. Therefore, it was suggested that the high preparation temperatures facilitated the decomposition of H_2S into a hydride with a stoichiometry that is unknown at atmospheric conditions, and this phase gave rise to the unusually large T_c [189].

Serendipitously, a theoretical study [190], which was inspired by the synthesis of a $(H_2S)_2H_2$ phase above 3.5 GPa [191], was published around the same time when the experimental results of Drozdov et al. were first communicated. DFT calculations were employed to predict the most stable geometries with the H_3S stoichiometry as a function of pressure, and estimate the T_c of the metallic phases. At 130 GPa, the T_c of a structure with $R3m$ symmetry was calculated as being 155–166 K, and at 200 GPa $Im\bar{3}m$-H_3S (illustrated in Figure 15.6a) was computed to have a T_c of 191–204 K. Comparison of the theoretical and experimental work suggested that the observed superconductivity was due to H_3S. Indeed, recently it has been shown that the experimentally observed XRD pattern of the superconducting phase can be explained by a mixture of the $R3m$

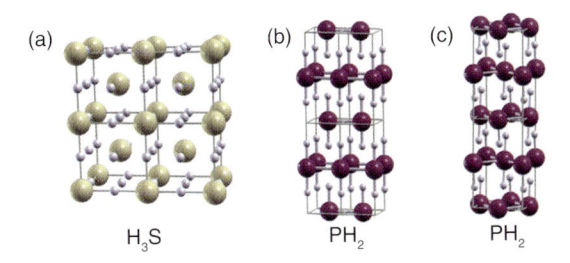

(a) H_3S (b) PH_2 (c) PH_2

Figure 15.6 Supercells of the (a) $Im\bar{3}m$-H_3S [190], (b) $I4/mmm$-PH_2 and (c) $C2/m$-PH_2 [118] phases that are likely to be contributors to the superconductivity observed when hydrogen sulfide [189] and phosphine [193] were compressed in a DAC. Hydrogen atoms are white, sulfur atoms are yellow, and phosphorus atoms are purple.

and $Im\bar{3}m$-H_3S structures along with the β-polonium structure of elemental sulfur [192].

The record-breaking T_c of compressed H_3S has resulted in a flurry of theoretical activity that has verified the stability of this phase, analyzed the factors yielding the remarkable superconducting behavior, examined anharmonicity as well as the isotope effect, and computed T_c for the related compounds selenium hydride and tellurium hydride [91,99,194–200]. In addition, a combined theoretical and experimental work showed that only H_2S can describe the superconductivity for the samples prepared at low temperatures, and that H_2S can be kinetically protected to very high pressures [200].

Recently, pressure-induced high-temperature superconductivity was observed in another hydride of a p-block element [193]. Resistance measurements showed that when phosphine, PH_3, was compressed, it became superconducting below 30 K at 83 GPa and 103 K at 207 GPa. Theoretical studies soon became available, which showed that none of the hydrides of phosphorus are thermodynamically stable with respect to decomposition into solid phosphorus and hydrogen above 100 GPa [90,118,201,202]. It was therefore concluded that like hydrogen sulfide, phosphine undergoes pressure-induced decomposition, and the superconductivity measured likely arises from PH, PH_2, PH_3, and other hydrides of phosphorus that are dynamically stable at these pressures [90,118,202]. For example, the two PH_2 phases illustrated in Figure 15.6b were computed to have a Tc of 70-75 K at 200 GPa [90] [118]. Thus, the theoretical and experimental results suggest that phosphine and its decomposition products may also be hydrogen-rich BCS-type superconductors.

15.3.5
Melting Behavior

Solid–liquid equilibria are affected by pressure – sometimes in counterintuitive ways. Typically, the melting temperatures increase under pressure because the bonds between the atoms become stiffer as the material is compressed. However, the slopes of the melting curves of some substances can be negative; one well-known example is water. It is not easy to unambiguously pinpoint exactly when a solid melts and different criteria – including an abrupt change in the resistivity, disappearance of X-ray diffraction lines, and visual observations – can be employed. Theoretical calculations of the melting lines face their own sets of problems, including a large computational expense, potential hysteresis, and overheating [203]. However, comparison of the melting lines obtained using different experimental procedures coupled with those calculated from theory help to more accurately determine the boundary between the solid and liquid phases as a function of pressure. Theoretical studies of melting are exceptionally important because they can be used to determine which experimental results are the most reliable, and also unveil the factors affecting the melting temperatures.

Below, we describe a few examples of anomalous melting behavior in some seemingly simple systems to give the reader a flavor of how pressure-induced quantum and electronic effects can impact melting.

Because the zero point energy (ZPE) increases as atoms are brought closer together, nuclear quantum effects (NQE) can be key factors in dictating the state of a compressed phase. For example, it has been postulated that under pressure hydrogen might undergo "cold melting," for example, it may adopt a metallic liquid ground state with exotic behavior at $T = 0$ K [204], provided that the difference in energy between various atomic configurations is smaller than the ZPE. Experiments have shown that the melting line of hydrogen peaks near 65 GPa [205], and first-principles calculations revealed that the slope in the melting curve remains negative at least to 200 GPa [206]. This decrease in the melting temperature was found to be a result of the softening of the intermolecular interactions in the liquid, and not due to NQE [206]. But, a number of theoretical studies have shown that NQE dramatically affect the melting temperatures at pressures approaching the TPa regime [207,208]. Even though liquid hydrogen was computed to melt below room temperature in these simulations, it is still a matter of debate as to whether the melting temperature ever reaches 0 K even at very high pressures [207].

Due to the difficulties inherent in studying the behavior of lithium in a DAC, theoretical predictions [209,210] of its melting line preceded experiments [211,212]. The melting temperature of lithium peaks near 10 GPa after which a sharp decrease is observed until ~40 GPa. Between 40 and 60 GPa, the melting line appears to be essentially flat, and above 60 GPa it recovers its positive slope. First-principles calculations suggested that the decrease in the melting temperature is caused by a Peierls distortion in the liquid that opens up a gap at the Fermi level, thereby lowering the total energy of the liquid with respect to that of a crystalline lattice [210]. An experimental study found the minimum in the melting temperature to be 190 K [211], whereas the calculations, which treated the protons classically, predicted a minimum of 280 K instead [210]. This discrepancy led to the suggestion that NQE may be responsible for the significantly lower melting temperature measured experimentally [211]. A later experimental study [212] postulated that because lithium undergoes several structural changes to low-symmetry phases above 40 GPa, the results obtained with the X-ray diffraction method adopted in Ref. [211] may not be a proof of melting. Electrical resistivity and visual appearance was employed to measure melting instead [212]. The minimum melting temperature obtained using these methods, 306 K at 44 GPa, agreed well with the first-principles results, calling into question the importance of NQE. Indeed, a recent theoretical study found that even though NQE do change the free energies of the liquid and the solid, they have little effect on the melting temperature of lithium [213]. Thus, the minimum melting temperatures obtained theoretically using both classic and quantum nuclei [210,213] are in good agreement with the experimental results from Ref. [212].

The melting line of sodium is also quite remarkable. A dramatic drop from 1000 K at 30 GPa to 300 K near 120 GPa has been observed [214], and a large number of stable low-symmetry phases have been experimentally identified near the melting minimum [215]. First-principles calculations with classical nuclei have reproduced this behavior [216]. In analogy with compressed lithium, it was proposed that a Peierls distortion in the liquid that opens up a gap at the Fermi level is responsible for the pronounced decrease in the melting temperature [216], although other explanations have also been put forward [217].

15.4
Conclusions

Because high-pressure experiments can be difficult to carry out, and the results may be complicated to analyze, first-principles calculations have steadily gained importance in this field. The emergence of global optimization techniques for crystal structure prediction has opened the door for the theoretical discovery of compounds with unique geometries, stoichiometries, and whose electronic structure, bonding, and properties could not have been easily foreseen using chemical rules of thumb derived at 1 atm. The increase in computational power and development of new algorithms has made it possible to calculate, with reasonable accuracy, the properties and melting behavior of matter under pressure. As a result, computational experiments are helping us to build chemical intuition about the behavior of compressed matter. Such knowledge is important for designing new materials and states of matter, as well as for earth and planetary science. A synergistic feedback loop between theory and experiments has proven to be key to inspiring experimental explorations and analyzing their results. However, a number of challenges for theory remain, we outline these briefly in the paragraphs below.

Despite the dramatic advances in CSP, it is still not possible to reliably predict the structures of solids with many degrees of freedom without any experimental information. This is because the number of local minima is an exponential function of the number of atoms in the unit cell, and because the time required for each local optimization increases as the unit cell size increases. We look forward to the continued development of ever-smarter algorithms that are able to find the global minimum and important local minima for large complex cells. Even though the most stable structure is temperature dependent, at the moment most CSP searches are carried out at 0 K because determining the free energies from first-principles can be extremely time consuming. In some cases the free energies of the most stable set of structures found in a CSP search are calculated in order to deduce which one will be preferred at a particular temperature. The drawback of this procedure is that it assumes the 0 K and finite T PES resemble each other. In principle, metadynamics or molecular dynamics could be employed to obtain free energies, but in practice very long simulation times are necessary to yield accurate results.

Due to the good balance between accuracy and computational expense typically provided by nonhybrid DFT calculations (of the GGA or LDA type), they are often the methods of choice for predicting structures and calculating the properties of compressed phases. But, an increasing number of studies are appearing that show the deficiency of these functionals for solids [218,219]. Nontrivial exchange–correlation effects can be important when valence electrons become localized, dispersion forces become important [220,221], semicore electrons overlap [222], and for strongly correlated systems or those where multireference methods are required [223] to accurately describe the electronic structure. It has been shown that inclusion of Hartree–Fock exchange can significantly impact the calculated transition pressures between phases and demixing transitions [222,224,225]. Moreover, geometry optimizations with hybrid functionals can lead to small structural distortions that open up a bandgap [226]. Hybrid DFT and screened exchange functionals can overcome some of these problems, but because they are extremely computationally expensive, they are not widely used. In particular, at the present time, it is computationally not feasible to optimize every geometry in a CSP search with these functionals. In addition, standard functionals lead to a severe underestimation of electronic bandgaps, and concomitantly of metalization pressures [227]. This can be overcome by using the DFT+U [228] or the GW [229] methods, but these too have their limitations. The Hubbard-like U term in the former is often chosen empirically, whereas the first-principles GW method is very time consuming. We are looking forward to the development of new functionals and algorithms that can be employed to calculate the electronic structure of solids that will remedy these problems. Theoretical advances are bound to keep "pressing-on" to enable ever more accurate calculations for compressed matter.

Acknowledgments

EZ acknowledges the NSF (DMR-1505817) for financial support, and thanks the Alfred P. Sloan Foundation for a research fellowship (2013–2015).

References

1 Dubrovinsky, L., Dubrovinskaia, N., Prakapenka, V.B., and Abajumov, A. (2012) Implementation of micro-ball nanodiamond anvils for high-pressure studies above 6 Mbar. *Nat. Commun.*, **3**, 1163.

2 Hemley, R.J. (2006) A pressing matter. *Physics World*, **19**, 26–30.

3 Zurek, E. and Grochala, W. (2015) Predicting crystal structures and properties of matter under extreme conditions via quantum mechanics: the pressure is on. *Phys. Chem. Chem. Phys.*, **17**, 2917–2934.

4 Grochala, W., Hoffmann, R., Feng, J., and Ashcroft, N.W. (2007) The chemical imagination at work in very tight places. *Angew. Chem. Int. Ed.*, **46**, 3620–3642.

5 Machon, D., Meersman, F., Wilding, M.C., Wilson, M., and McMillan, P.F.

(2014) Pressure-induced amorphization and polyamorphism: inorganic and biochemical systems. *Prog. Mater. Sci.*, **61**, 216–282.

6 Lee, R., Howard, J.A.K., Probert, M.R., and Steed, J.W. (2014) Structure of organic solids at low temperature and high pressure. *Chem. Soc. Rev.*, **43**, 4300–4311.

7 Manaa, M.R. and Fried, L.E. (2014) The reactivity of energetic materials under high pressure and temperature, in *Advances in Quantum Chemistry*, vol. 69 (ed. J.R. Sabin), Academic Press, pp. 221–252.

8 Boffa Ballaran, T., Kurnosov, A., and Trots, D. (2013) Single-crystal X-ray diffraction at extreme conditions: a review. *High Pressure Res.*, **33**, 453–465.

9 Goncharov, A.F., Howie, R.T., and Gregoryanz, E. (2013) Hydrogen at extreme pressures. *Low Temp. Phys.*, **39**, 402–408.

10 Bhardwaj, P. and Singh, S. (2012) Pressure induced structural transitions-a review. *Cent. Eur. J. Chem.*, **10**, 1391–1422.

11 Dubrovinsky, L. and Dubrovinskaia, N. (2013) Transition metal oxides under extreme conditions, in *Comprehensive Inorganic Chemistry II* (eds J. Reedijk and K.R. Poeppelmeier), vol. 2, Elsevier, pp. 223–239.

12 McMillan, P.F. (2013) Materials from extreme conditions, in *Comprehensive Inorganic Chemistry II* (eds J. Reedijk and K.R. Poeppelmeier), vol. 2, Elsevier, pp. 17–46.

13 Klug, D.D. and Yao, Y. (2011) Metallization of solid hydrogen: the challenge and possible solutions. *Phys. Chem. Chem. Phys.*, **13**, 16999–17006.

14 Naumov, I.I. and Hemley, R.J. (2014) Aromaticity, closed-shell effects, and metallization of hydrogen. *Acc. Chem. Res.*, **47**, 3551–3559.

15 Fischer, C.C., Tibbetts, K.J., Morgan, D., and Ceder, G. (2006) Predicting crystal structure by merging data mining with quantum mechanics. *Nat. Mater.*, **5**, 641–646.

16 Pettifor, D.G. (1986) The structures of binary compounds-I. Phenomenological structure maps. *J. Phys. C*, **19**, 285–313.

17 Ashcroft, N.W. (2009) Pressing some boundaries in Mendeleev's chart. *Physics*, **2**, 65.

18 Zurek, E. (2016) Discovering new materials via a priori crystal structure prediction, in *Reviews in Computational Chemistry*, vol. 29 (eds A.L. Parrill and K.B. Lipkowitz), John Wiley & Sons, Inc., Hoboken, NJ, 274.

19 Woodley, S.M. and Catlow, R. (2008) Crystal structure prediction from first principles. *Nat. Mater.*, **7**, 937–946.

20 Schön, J.C., Doll, K., and Jansen, M. (2010) Predicting solid compounds via global exploration of the energy landscape of solids on the *ab initio* level without recourse to experimental information. *Phys. Status Solidi B*, **247**, 23–29.

21 Rossi, G. and Ferrando, R. (2009) Searching for low-energy structures of nanoparticles: a comparison of different methods and algorithms. *J. Phys. Condens. Matter*, **21**, 084208.

22 Wang, Y. and Ma, Y. (2014) Perspective: crystal structure prediction at high pressures. *J. Chem. Phys.*, **140**, 040901.

23 Pickard, C.J. and Needs, R.J. (2009) Structures at high pressure from random searching. *Phys. Status Solidi B*, **246**, 536–540.

24 Revard, B.C., Tipton, W.W., and Hennig, R.G. (2014) Structure and stability prediction of compounds with evolutionary algorithms. *Top. Curr. Chem.*, **345**, 181–222.

25 Oganov, A.R. (ed.) (2011) *Modern Methods of Crystal Structure Prediction*, Wiley-VCH Verlag GmbH, Weinheim, Germany.

26 Wales, D.J. (2003) *Energy Landscapes: Applications to Clusters, Biomolecules and Glasses*, Cambridge University Press, Cambridge.

27 Jansen, M. (2015) Conceptual inorganic materials discovery-a road map. *Adv. Mater.*, **27**, 3229–3242.

28 Johnston, R.L. (2003) Evolving better nanoparticles: genetic algorithms for optimising cluster geometries. *Dalton. Trans.*, 4193–4207.

29 Zhu, Q., Oganov, A.R., Zeng, Q., and Zhou, X. (2015) Structure prediction and

its applications in computational materials design, *in Chemical Modelling*, vol 12 (eds M. Springborg and J.O. Joswig), The Royal Society of Chemistry, pp. 219–248.

30 Stillinger, F.H. (1999) Exponential multiplicity of inherent structures. *Phys. Rev. E*, **59**, 48–51.

31 Bush, T.S., Catlow, C.R.A., and Battle, P.D. (1995) Evolutionary programming techniques for predicting inorganic crystal structures. *J. Mater. Chem.*, **5**, 1269–1272.

32 Deaven, D.M. and Ho, K.M. (1995) Molecular geometry optimization with a genetic algorithm. *Phys. Rev. Lett.*, **75**, 288–291.

33 Wales, D. J. and Doye, J. P. K. (1997) Global optimization by basin-hopping and the lowest energy structures of Lennard-Jones clusters containing up to 110 atoms. *J. Phys. Chem. A*, **101**, 5111–5116.

34 Gödecker, S. (2004) Minima hopping: an efficient search method for the global minimum of the potential energy surface of complex molecular systems. *J. Chem. Phys.*, **120**, 9911–9917.

35 Laio, A. and Parrinello, M. (2002) Escaping free-energy minima. *Proc. Natl. Acad. Sci. USA*, **99**, 12562–12566.

36 Kirkpatrick, S., Gelatt, C.D., and Vecchi, M.P. (1983) Optimization by simulated annealing. *Science*, **220**, 671–680.

37 Pickard, C.J. and Needs, R.J. (2011) *Ab initio* random structure searching. *J. Phys. Condens. Matter*, **23**, 053201.

38 Lonie, D.C. and Zurek, E. (2011) Xtalopt: an open-source evolutionary algorithm for crystal structure prediction. *Comput. Phys. Commun.*, **182**, 372–387.

39 Tipton, W.W., Bealing, C.R., Mathew, K., and Hennig, R.G. (2013) Structures, phase stabilities, and electrical potentials of Li–Si battery anode materials. *Phys. Rev. B*, **87**, 184114.

40 Glass, C.W., Oganov, A.R., and Hansen, N. (2006) USPEX-evolutionary crystal structure prediction. *Comput. Phys. Commun.*, **175**, 713–720.

41 Kolmogorov, A.N., Shah, S., Margine, E.R., Bialon, A.F., Hammerschmidt, T., and Drautz, R. (2010) New superconducting and semiconducting Fe-B compounds predicted with an *ab initio* evolutionary search. *Phys. Rev. Lett.*, **105**, 217003.

42 Bahmann, S. and Kortus, J. (2013) EVO-evolutionary algorithm for crystal structure prediction. *Comput. Phys. Commun.*, **184**, 1618–1625.

43 Trimarchi, G. and Zunger, A. (2007) Global space-group optimization problem: finding the stablest crystal structure without constraints. *Phys. Rev. B.*, **75**, 104113.

44 Abraham, N.L. and Probert, M.I.J. (2006) A periodic genetic algorithm with real-space representation for crystal structure and polymorph prediction. *Phys. Rev. B.*, **73**, 224104.

45 Fadda, A. and Fadda, G. (2010) An evolutionary algorithm for the prediction of crystal structures. *Phys. Rev. B*, **82**, 104105.

46 Wang, Y., Lv, J., Zhu, L., and Ma, Y. (2010) Crystal structure prediction via particle-swarm optimization. *Phys. Rev. B*, **82**, 094116.

47 Zhu, Q., Oganov, A.R., and Lyakhov, A.O. (2012) Evolutionary metadynamics: a novel method to predict crystal structures. *CrystEngComm*, **14**, 3596–3601.

48 Liu, Z.L. (2014) Muse: multi-algorithm collaborative crystal structure prediction. *Comput. Phys. Commun.*, **185**, 1893–1900.

49 Wu, S.Q., Ji, M., Wang, C.Z., Nguyen, M.C., Zhao, X., Umemoto, K., Wentzcovitch, R.M., and Ho, K.M. (2014) An adaptive genetic algorithm for crystal structure prediction. *J. Phys. Condens. Matter*, **26**, 035402.

50 Zhu, Q., Oganov, A.R., Glass, C.W., and Stokes, H.T. (2012) Constrained evolutionary algorithm for structure prediction of molecular crystals: methodology and applications. *Acta. Crystall. B-Stru*, **68**, 215–226.

51 Bardwell, D.A., Adjiman, C.S., Arnautova, Y.A., Bartashevich, E., Boerrigter, S.X.M., Braun, D.E., Cruz-Cabeza, A.J., Day, G.M., Della Valle, R.G., Desiraju, G.R., van Eijck, B.P., Facelli, J.C., Ferraro, M.B., Grillo, D., Habgood, M., Hofmann,

D.W.M., Hofmann, F., Jose, K.V.J., Karamertzanis, P.G., Kazantsev, A.V., Kendrick, J., Kuleshova, L.N., Leusen, F.J.J., Maleev, A.V., Misquitta, A.J., Mohamed, S., Needs, R.J., Neumann, M.A., Nikylov, D., Orendt, A.M., Pal, R., Pantelides, C.C., Pickard, C.J., Price, L.S., Price, S.L., Scheraga, H.A., van de Streek, J., Thakur, T.S., Tiwari, S., Venuti, E., and Zhitkov, I.K. (2011) Towards crystal structure prediction of complex organic compounds: a report on the fifth blind test. *Acta. Crystall. B-Stru*, **67**, 535–551.

52 Meredig, B. and Wolverton, C. (2013) A hybrid computational–experimental approach for automated crystal structure solution. *Nat. Mater.*, **12**, 123–127.

53 Curtarolo, S., Hart, G.L.W., Nardelli, M.B., Mingo, N., Sanvito, S., and Levy, O. (2013) The high-throughput highway to computational materials design. *Nat. Mater.*, **12**, 191–201.

54 Meredig, B., Agrawal, A., Kirklin, S., Saal, J.E., Doak, J.W., Thompson, A., Zhang, K., Choudhary, A., and Wolverton, C. (2014) Combinatorial screening for new materials in unconstrained composition space with machine learning. *Phys. Rev. B*, **89**, 094104.

55 Wille, L.T. and Vennik, J. (1985) Computational complexity of the ground-state determination of atomic clusters. *J. Phys. A.*, **18**, L419–L422.

56 Greenwood, G.W. (1999) Revisiting the complexity of finding globally minimum energy configurations in atomic clusters. *Z. Phys. Chem.*, **211**, 105–114.

57 Wolpert, D.H. and Macready, W.G. (1997) No free lunch theorems for optimization. *IEEE Trans. Evol. Comput.*, **1**, 67–82.

58 Doye, J.P.K. and Massen, C.P. (2005) Characterizing the network topology of the energy landscapes of atomic clusters. *J. Chem. Phys.*, **122**, 084105.

59 Bell, R.P. (1936) The theory of reactions involving proton transfers. *Proc. R. Soc. Lond. A Math. Phys. Sci.*, **154**, 414–429.

60 Evans, M.G. and Polanyi, M. (1935) Some applications of the transition state method to the calculation of reaction velocities, especially in solution. *Trans. Faraday Soc.*, **31**, 875–894.

61 Baettig, P., Schelle, C.F., LeSar, R., Waghmare, U.V., and Spaldin, N.A. (2005) Theoretical prediction of new high-performance lead-free piezoelectrics. *Chem. Mater.*, **17**, 1376–1380.

62 Ghosez, P., Cockayne, E., Waghmare, U.V., and Rabe, K.M. (1999) Lattice dynamics of $BaTiO_3$, $PbTiO_3$, and $PbZrO_3$: a comparative first-principles study. *Phys. Rev. B*, **60**, 836–843.

63 Parlinski, K. and Kawazoe, Y. (2000) *Ab initio* study of phonons and structural stabilities of the perovskite-type $MgSiO_3$. *Eur. Phys. J. B*, **16**, 49–58.

64 Derzsi, M., Piekarz, P., and Grochala, W. (2014) Structures of late transition metal monoxides from Jahn–Teller instabilities in the rock salt lattice. *Phys. Rev. Lett.*, **113**, 025505.

65 Kurzydłoski, D. and Grochala, W. (2011) Phonon dispersion analysis as an indispensable tool for predictions of solid state polymorphism and dynamic metastability: case of compressed silane. *Acta. Phys. Pol. A*, **119**, 895–900.

66 Ashcroft, N.W. (2004) Hydrogen dominant metallic alloys: high temperature superconductors? *Phys. Rev. Lett.*, **92**, 187002.

67 Feng, J., Grochala, W., Jaron, T., Hoffmann, R., Bergara, A., and Ashcroft, N.W. (2006) Structures and potential superconductivity in SiH_4 at high pressure: en route to "metallic hydrogen." *Phys. Rev. Lett.*, **96**, 017006.

68 Feng, J., Hennig, R.G., Ashcroft, N.W., and Hoffmann, R. (2008) Emergent reduction of electronic state dimensionality in dense ordered Li–Be alloys. *Nature*, **451**, 445–448.

69 Pickard, C.J. and Needs, R.J. (2007) Structure of phase III of solid hydrogen. *Nat. Phys.*, **3**, 473–476.

70 McMahon, J.M. and Ceperley, D.M. (2011) Ground-state structures of atomic metallic hydrogen. *Phys. Rev. Lett.*, **106**, 165302.

71 Yao, Y., Tse, J.S., and Klug, D.D. (2009) Structures of insulating phases of dense lithium. *Phys. Rev. Lett.*, **102**, 115503.

72 Marques, M., McMahon, M.I., Gregoryanz, E., Hanfland, M., Guillaume, C.L., Pickard, C.J., Ackland, G.J., and Nelmes, R.J. (2011) Crystal structures of dense lithium: a metal–semiconductor–metal transition. *Phys. Rev. Lett.*, **106**, 095502.

73 Lyle, M.J., Pickard, C.J., and Needs, R.J. (2015) Prediction of 10-fold coordinated TiO_2 and SiO_2 structures at multimegabar pressures. *Proc. Natl. Acad. Sci. USA*, **112**, 6898–6901.

74 Zhang, S., Wilson, H.F., Driver, K.P., and Militzer, B. (2013) H_4O and other hydrogen–oxygen compounds at giant-planet core pressures. *Phys. Rev. B*, **87**, 024112.

75 Pickard, C.J., Martinez-Canales, M., and Needs, R.J. (2013) Decomposition and terapascal phases of water ice. *Phys. Rev. Lett.*, **110**, 245701.

76 Yao, Y. and Klug, D.D. (2010) Silane plus molecular hydrogen as a possible pathway to metallic hydrogen. *Proc. Natl. Acad. Sci. USA*, **107**, 20893–20898.

77 Doll, K., Schön, J.C., and Jansen, M. (2007) Global exploration of the energy landscape of solids on the *ab initio* level. *Phys. Chem. Chem. Phys.*, **9**, 6128–6133.

78 Zagorac, D., Doll, K., Schön, J.C., and Jansen, M. (2011) *Ab initio* structure prediction for lead sulfide at standard and elevated pressures. *Phys. Rev. B*, **84**, 045206.

79 Kulkarni, A., Doll, K., Schön, J.C., and Jansen, M. (2010) Global exploration of the enthalpy landscape of calcium carbide. *J. Phys. Chem. B*, **114**, 15573–15581.

80 Doll, K., Schön, J.C., and Jansen, M. (2008) Structure prediction based on *ab initio* simulated annealing for boron nitride. *Phys. Rev. B*, **78**, 144110.

81 Doll, K. and Jansen, M. (2011) *Ab initio* energy landscape of GeF_2: a system featuring lone pair structure candidates. *Angew. Chem., Int. Ed.*, **50**, 4627–4632.

82 Kulkarni, A., Schön, J.C., Doll, K., and Jansen, M. (2013) Structure prediction of binary pernitride MN_2 compounds (M = Ca, Sr, Ba, La, and Ti). *Chem. Asian J.*, **8**, 743–754.

83 Costales, A., Blanco, M.A., Francisco, E., Pandey, R., and Pendas, A.M. (2005) Evolution of the properties of Al_nn_n clusters with size. *J. Phys. Chem. B*, **109**, 24352–24360.

84 Harding, D.J., Kerpal, C., Meijer, G., and Fielicke, A. (2013) Unusual bonding in platinum carbido clusters. *J. Phys. Chem. Lett.*, **4**, 892–896.

85 Woodley, S.M. (2013) Knowledge led master code search for atomic and electronic structures of LaF_3 nanoclusters on hybrid rigid ion-shell model-DFT landscapes. *J. Phys. Chem. C*, **117**, 24003–24014.

86 Gao, Y., Shao, N., Zhou, R., Zhang, G., and Zeng, X.C. (2012) $[CTi_7^{2+}]$: heptacoordinate carbon motif? *J. Phys. Chem. Lett.*, **3**, 2264–2268.

87 Flores-Livas, J.A., Amsler, M., Lenosky, T.J., Lehtovaara, L., Botti, S., Marques, M.A.L., and Goedecker, S. (2012) High-pressure structures of disilane and their superconducting properties. *Phys. Rev. Lett.*, **108**, 117004.

88 Amsler, M., Flores-Livas, J.A., Lehtovaara, L., Balima, F., Ghasemi, S.A., Machon, D., Pailhes, S., Willand, A., Caliste, D., Botti, S., San Miguel, A., Goedecker, S., and Marques, M.A.L. (2012) Crystal structure of cold compressed graphite. *Phys. Rev. Lett.*, **108**, 065501.

89 Botti, S., Amsler, M., Flores-Livas, J.A., Ceria, P., Goedecker, S., and Marques, M.A.L. (2013) Carbon structures and defect planes in diamond at high pressure. *Phys. Rev. B*, **88**, 014102.

90 Flores-Livas, J.A., Amsler, M., Heil, C., Sanna, A., Boeri, L., Profeta, G., Wolverton, C., Goedecker, S., and Gross, E.K.U. (2016) Superconductivity in metastable phases of phosphorus–hydride compounds under high pressure. *Phys. Rev. B*, **93**, 020508.

91 Flores-Livas, J.A., Sanna, A., and Gross, E.K.U. (2015) High temperature superconductivity in sulfur and selenium hydrides at high pressure. *Eur. Phys. J.B*, **89**, 1–6.

92 Selli, D., Baburin, I.A., Martoňak, R., and Leoni, S. (2013) Novel metastable metallic and semiconducting germaniums. *Sci. Rep.*, **3**, 1466.

93 Sun, J., Klug, D.D., Martonak, R., Montoya, J.A., Lee, M.S., Scandolo, S., and Tosatti, E. (2009) High-pressure polymeric phases of carbon dioxide. *Proc. Natl. Acad. Sci. USA*, **106**, 6077–6081.

94 Behler, J., Martonak, R., Donadio, D., and Parrinello, M. (2008) Metadynamics simulations of the high-pressure phases of silicon employing a high-dimensional neural network potential. *Phys. Rev. Lett.*, **100**, 185501.

95 Plasienka, D. and Martonak, R. (2015) Transformation pathways in high-pressure solid nitrogen: from molecular N_2 to polymeric cg-N. *J. Chem. Phys.*, **142**, 094505.

96 Kennedy, J. and Eberhart, R. (1995) Particle swarm optimization. Proceedings of IEEE International Conference on Neural Networks IV, pp. 1942–1948.

97 Call, S.T., Zubarev, D.Y., and Boldyrev, A.I. (2007) Global minimum structure searches via particle swarm optimization. *J. Comput. Chem.*, **28**, 1177–1186.

98 Wang, H., Tse, J.S., Tanaka, K., Iitaka, T., and Ma, Y. (2012) Superconductive sodalite-like clathrate calcium hydride at high pressures. *Proc. Natl. Acad. Sci. USA*, **109**, 6463–6466.

99 Zhang, S., Wang, Y., Zhang, J., Liu, H., Zhong, X., Song, H.F., Yang, G., Zhang, L., and Ma, Y. (2015) Phase diagram and high-temperature superconductivity of compressed selenium hydrides. *Sci. Rep.*, **5**, 15433.

100 Li, Y., Hao, J., Liu, H., Tse, J.S., Wang, Y., and Ma, Y. (2015) Pressure-stabilized superconductive yttrium hydrides. *Sci. Rep.*, **5**, 09948.

101 Li, Y., Hao, J., Liu, H., Li, Y., and Ma, Y. (2014) The metallization and superconductivity of dense hydrogen sulfide. *J. Chem. Phys.*, **140**, 174712.

102 Miao, M.S. (2013) Caesium in high oxidation states and as a p-block element. *Nat. Chem.*, **5**, 846–852.

103 Wang, Z., Wang, H., Tse, J.S., Iitaka, T., and Ma, Y. (2015) Stabilization of H^{3+} in the high pressure crystalline structure of H_nCl ($n = 2$–7). *Chem. Sci.*, **6**, 522–526.

104 Lonie, D.C. and Zurek, E. (2011) XtalOpt version r7: an open-source evolutionary algorithm for crystal structure prediction. *Comput. Phys. Commun.*, **182**, 2305–2306.

105 Falls, Z., Lonie, D.C., Avery, P., Shamp, A., and Zurek, E. (2016) XtalOpt version r9: an open-source evolutionary algorithm for crystal structure prediction. *Comput. Phys. Commun.*, **199**, 178–179.

106 Oganov, A.R., Lyakhov, A.O., and Valle, M. (2011) How evolutionary crystal structure prediction works – and why. *Acc. Chem. Res.*, **44**, 227–237.

107 Lyakhov, A.O., Oganov, A.R., and Valle, M. (2010) How to predict very large and complex crystal structures. *Comput. Phys. Commun.*, **181**, 1623–1632.

108 d'Avezac, M. and Zunger, A. (2008) Identifying the minimum-energy atomic configuration on a lattice: Lamarckian twist on Darwinian evolution. *Phys. Rev. B.*, **78**, 064102.

109 Zurek, E., Hoffmann, R., Ashcroft, N.W., Oganov, A.R., and Lyakhov, A.O. (2009) A little bit of lithium does a lot for hydrogen. *Proc. Natl. Acad. Sci. USA*, **106**, 17640–17643.

110 Baettig, P. and Zurek, E. (2011) Pressure-stabilized sodium polyhydrides, NaH_n ($n > 1$). *Phys. Rev. Lett.*, **106**, 237002.

111 Hooper, J. and Zurek, E. (2012) Rubidium polyhydrides under pressure: emergence of the linear H_3^- species. *Chem. Eur. J.*, **18**, 5013–5021.

112 Hooper, J. and Zurek, E. (2012) High pressure potassium polyhydrides: a chemical perspective. *J. Phys. Chem. C*, **116**, 13322–13328.

113 Shamp, A., Hooper, J., and Zurek, E. (2012) Compressed cesium polyhydrides: Cs^+ sublattices and H_3^- three-connected nets. *Inorg. Chem.*, **51**, 9333–9342.

114 Lonie, D.C., Hooper, J., Altintas, B., and Zurek, E. (2013) Metallization of magnesium polyhydrides under pressure. *Phys. Rev. B*, **87**, 054107.

115 Hooper, J., Altintas, B., Shamp, A., and Zurek, E. (2013) Polyhydrides of the alkaline earth metals: a look at the extremes under pressure. *J. Phys. Chem. C*, **117**, 2982–2992.

116 Hooper, J., Terpstra, T., Shamp, A., and Zurek, E. (2014) Composition and constitution of compressed strontium

117 Hooper, J. and Zurek, E. (2012) Lithium subhydrides under pressure and their superatom-like building blocks. *ChemPlusChem*, **77**, 969–972.

118 Shamp, A., Terpstra, T., Bi, T., Falls, Z., Avery, P., and Zurek, E. (2016) Decomposition products of phosphine under pressure: PH2 stable and superconducting. *J. Am. Chem. Soc.*, **138**, 1884–1892.

119 Shamp, A. and Zurek, E. (2015) Superconducting high-pressure phases composed of hydrogen and iodine. *J. Phys. Chem. Lett.*, **6**, 4067–4072.

120 Zhou, X.F., Oganov, A.R., Dong, X., Zhang, L., Tian, Y., and Wang, H.T. (2011) Superconducting high-pressure phase of platinum hydride from first principles. *Phys. Rev. B.*, **84**, 054543.

121 Hu, C.H., Oganov, A.R., Zhu, Q., Qian, G.R., Frapper, G., Lyakhov, A.O., and Zhou, H.Y. (2013) Pressure-induced stabilization and insulator–superconductor transition of BH. *Phys. Rev. Lett.*, **110**, 165504..

122 Hermann, A. and Schwerdtfeger, P. (2014) Xenon suboxides stable under pressure. *J. Phys. Chem. Lett.*, **5**, 4336–4342.

123 Hermann, A., Suarez-Alcubilla, A., Gurtubay, I.G., Yang, L.M., Bergara, A., Ashcroft, N.W., and Hoffmann, R. (2012) LiB and its boron-deficient variants under pressure. *Phys. Rev. B.*, **86**, 144110.

124 Hermann, A., Ashcroft, N.W., and Hoffmann, R. (2012) Making sense of boron-rich binary Be-B phases. *Inorg. Chem.*, **51**, 9066–9075.

125 Hermann, A., Ashcroft, N.W., and Hoffmann, R. (2013) Binary compounds of boron and beryllium: a rich structural arena with space for predictions. *Chem. Eur. J.*, **19**, 4184–4197.

126 Hermann, A., McSorley, A., Ashcroft, N.W., and Hoffmann, R. (2012) From Wade–Mingos to Zintl–Klemm at 100 GPa: binary compounds of boron and lithium. *J. Am. Chem. Soc.*, **134**, 18606–18618.

127 Hermann, A., Ivanov, B.L., Ashcroft, N.W., and Hoffmann, R. (2012) LiBeB: a predicted phase with structural and electronic peculiarities. *Phys. Rev. B*, **86**, 014104.

128 Prewitt, C.T. and Downs, R.T. (1998) High-pressure crystal chemistry. *Rev. Mineral.*, **37**, 284–318.

129 Pauling, L. (1960) *The Nature of the Chemical Bond*, 3rd edn, Cornell University Press.

130 Wigner, E. and Huntington, H.B. (1935) On the possibility of a metallic modification of hydrogen. *J. Chem. Phys.*, **3**, 764–770.

131 Narayana, C., Luo, H., Orloff, J., and Ruoff, A.L. (1998) Solid hydrogen at 342 GPa: no evidence for an alkali metal. *Nature*, **393**, 46–49.

132 Loubeyre, P., Occelli, F., and LeToullec, R. (2002) Optical studies of solid hydrogen to 320 GPa and evidence for black hydrogen. *Nature*, **416**, 613–617.

133 Zha, C.S., Liu, Z., Ahart, M., Boehler, R., and Hemley, R.J. (2013) High-pressure measurements of hydrogen phase IV using synchrotron infrared spectroscopy. *Phys. Rev. Lett.*, **110**, 217402.

134 Eremets, M.I. and Troyan, I.A. (2011) Conductive dense hydrogen. *Nat. Mater.*, **10**, 927–931.

135 Howie, R.T., Guillaume, C.L., Scheler, T., Goncharov, A.F., and Gregoryanz, E. (2012) Mixed molecular and atomic phase of dense hydrogen. *Phys. Rev. Lett.*, **108**, 125501.

136 Ashcroft, N.W. (1968) Metallic hydrogen: a high-temperature superconductor? *Phys. Rev. Lett.*, **21**, 1748–1749.

137 Pickard, C.J. and Needs, R.J. (2007) Structure of phase III of solid hydrogen. *Nat. Phys.*, **3**, 473–476.

138 Pickard, C.J., Martinez-Canales, M., and Needs, R.J. (2012) Density functional theory study of phase IV of solid hydrogen. *Phys. Rev. B*, **85**, 214114.

139 McMinis, J., Clay III, R.C., Lee, D., and Morales, M.A. (2015) Molecular to atomic phase transition in hydrogen under high pressure. *Phys. Rev. Lett.*, **114**, 105305.

140 Dalladay-Simpson, P., Howie, R.T., and Gregoryanz, E. (2016) Evidence for a new phase of dense hydrogen above 325 gigapascals. *Nature*, **529**, 63–67.

141 Eremets, M.I., Troyan, I.A., and Drozdov, A.P. Low temperature phase diagram of hydrogen at pressures up to 380 GPa. A possible metallic phase at 360 GPa and 200 K. doi: ArXiv:1601.04479.

142 Ma, Y., Eremets, M., Oganov, A.R., Xie, Y., Trojan, I., Medvedev, S., Lyakhov, A.O., Valle, M., and Prakapenka, V. (2009) Transparent dense sodium. *Nature*, **458**, 182–185.

143 Zurek, E., Jepsen, O., and Andersen, O.K. (2005) Muffin-tin orbital Wannier-like functions for insulators and metals. *ChemPhysChem*, **6**, 1934–1942.

144 Neaton, J.B. and Ashcroft, N.W. (2001) On the constitution of sodium at higher densities. *Phys. Rev. Lett.*, **86**, 2830–2833.

145 Neaton, J.B. and Ashcroft, N.W. (1999) Pairing in dense lithium. *Nature*, **400**, 141–144.

146 Matsuoka, T. and Shimizu, K. (2009) Direct observation of a pressure-induced metal-to-semiconductor transition in lithium. *Nature*, **458**, 186–189.

147 Pickard, C.J. and Needs, R.J. (2009) Dense low-coordination phases of lithium. *Phys. Rev. Lett.*, **102**, 146401.

148 Lv, J., Wang, Y., Zhu, L., and Ma, Y. (2011) Predicted novel high-pressure phase of lithium. *Phys. Rev. Lett.*, **106**, 015503.

149 Miao, M.S. and Hoffmann, R. (2014) High pressure electrides: a predictive chemical and physical theory. *Acc. Chem. Res.*, **47**, 1311–1317.

150 Sternheimer, R. (1950) On the compressibility of metallic cesium. *Phys. Rev.*, **78**, 235–243.

151 McMahan, A.K. (1984) Alkali-metal structures above the s–d transition. *Phys. Rev. B*, **29**, 5982.

152 Takemura, K., Minomura, S., and Shimomura, O. (1982) X-ray diffraction study of electronic transitions in cesium under high pressure. *Phys. Rev. Lett.*, **49**, 1772–1775.

153 von Schnering, H.G. and Nesper, R. (1987) How nature adapts chemical structures to curved surfaces. *Angew. Chem., Int. Ed.*, **26**, 1059–1080.

154 Takemura, K., Christensen, N.E., Novikov, D.L., Syassen, K., Schwarz, U., and Hanfland, M. (2000) Phase stability of highly compressed cesium. *Phys. Rev. B*, **61**, 14399–14404.

155 Botana, J., Wang, X., Hou, C., Yan, D., Lin, H., Ma, Y., and Miao, M. (2015) Mercury under pressure acts as a transition metal: calculated from first principles. *Angew. Chem., Int. Ed.*, **127**, 9412–9415.

156 Botana, J. and Miao, M. (2014) Pressure-stabilized lithium caesides with caesium anions beyond the −1 state. *Nat. Commun.*, **5**, 4861.

157 Hall, H.T., Merrill, L., and Barnett, J.D. (1964) High pressure polymorphism in cesium. *Science*, **146**, 1297–1299.

158 Bernasconi, M., Silvestrelli, P.L., and Parrinello, M. (1998) *Ab initio* infrared absorption study of the hydrogen-bond symmetrization in ice. *Phys. Rev. Lett.*, **81**, 1235–1238.

159 Goncharov, A.F., Struzhkin, V.V., Somayazulu, M.S., Hemley, R.J., and Mao, H.K. (1996) Compression of ice to 210 gigapascals: infrared evidence for a symmetric hydrogen-bonded phase. *Science*, **273**, 218–220.

160 Zhang, L., Wang, Y., Zhang, X., and Ma, Y. (2010) High-pressure phase transitions of solid HF, HCl and HBr: an *ab initio* evolutionary study. *Phys. Rev. B*, **82**, 014108.

161 Katoh, E., Yamawaki, H., Fujihisa, H., Sakashita, M., and Aoki, K. (1999) Raman and infrared study of phase transitions in solid HBr under pressure. *Phys. Rev. B*, **59**, 11244–11250.

162 Schnelle, W., Ormeci, A., Wosylus, A., Meier, K., Grin, Y., and Schwarz, U. (2012) Dumbbells of five-connected Ge atoms and superconductivity in CaGe3. *Inorg. Chem.*, **51**, 5509–5511.

163 Ayouz, M., Dulieu, O., Guerout, R., Robert, J., and Kokoouline, V. (2010) Potential energy and dipole moment surfaces of H_3^- molecule. *J. Chem. Phys.*, **132**, 194309.

164 Zhu, Q., Oganov, A.R., and Zeng, Q. (2015) Formation of stoichiometric CsFn compounds. *Sci. Rep.*, **5**, 7875.

165 Dronskowski, R. and Blöchl, P.E. (1993) Crystal orbital Hamilton population (COHP): energy-resolved visualization of chemical bonding in solids based on

166 Kauzlarich, S.M. (ed.) (1996) *Chemistry, Structure, and Bonding of Zintl Phases and Ions*, Wiley-VCH Verlag GmbH, Weinheim, Germany.

167 Schwarz, U., Wosylus, A., Rosner, H., Schnelle, W., Ormeci, A., Meier, K., Baranov, A., Nicklas, M., Leipe, S., Müller, C.J., and Grin, Y. (2012) Dumbbells of five-connected silicon atoms and superconductivity in the binary silicides MSi_3 (m = Ca, Y, Lu). *J. Am. Chem. Soc.*, **134**, 13558–13561.

168 Zurek, E. and Yao, Y. (2015) Theoretical predictions of novel superconducting phases of $BaGe_3$ stable at atmospheric and high pressures. *Inorg. Chem.*, **54**, 2875–2884.

169 Fukuoka, H., Tomomitsu, Y., and Inumaru, K. (2011) High-pressure synthesis and superconductivity of a new binary barium germanide $BaGe_3$. *Inorg. Chem.*, **50**, 6372–6377.

170 Struzhkin, V.V., Kim, D.Y., Stavrou, E., Muramatsu, T., Mao, H.K., Pickard, C.J., Needs, R.J., Prakapenka, V.B., and Goncharov, A.F. (2016) Synthesis of sodium polyhydrides at high pressures. *Nat. Commun.*, **7**, 12267

171 Zhang, W., Oganov, A.R., Goncharov, A.F., Zhu, Q., Boulfelfel, S.E., Lyakhov, A.O., Stavrou, E., Somayazulu, M., Prakapenka, V.B., and Konopkova, Z. (2013) Unexpected stable stoichiometries of sodium chlorides. *Science*, **342**, 1502–1505.

172 Zhu, L., Liu, H., Pickard, C.J., Zou, G., and Ma, Y. (2014) Reactions of xenon with iron and nickel are predicted in the earth's inner core. *Nat. Chem.*, **6**, 644–648.

173 Zhu, Q., Jung, D.Y., Oganov, A.R., Glass, C.W., Gatti, C., and Lyakhov, A.O. (2013) Stability of xenon oxides at high pressures. *Nat. Chem.*, **5**, 61–65.

174 Simon, A. (2010) Metal clusters inside out. *Philos. Trans. R. Soc. A*, **368**, 1285–1299.

175 Xie, Y., Li, Q., Oganov, A.R., and Wang, H. (2014) Superconductivity of lithium-doped hydrogen under high pressure. *Acta. Crystallogr. C*, **70**, 104–111.

176 Pépin, C., Loubeyre, P., Occelli, F., and Dumas, P. (2015) Synthesis of lithium polyhydrides above 130 GPa at 300 K. *Proc. Natl. Acad. Sci. USA*, **112**, 7673–7676.

177 Zhou, D., Jin, X., Meng, X., Bao, G., Ma, Y., Liu, B., and Cui, T. (2012) Ab initio study revealing a layered structure in hydrogen-rich KH_6 under high pressure. *Phys. Rev. B*, **86**, 014118.

178 Wang, Y., Wang, H., Tse, J.S., Iitaka, T., and Ma, Y. (2015) Structural morphologies of high-pressure polymorphs of strontium hydrides. *Phys. Chem. Chem. Phys.*, **17**, 19379–19385.

179 Carlsson, A.E. and Ashcroft, N.W. (1983) Approaches for reducing the insulator–metal transition pressure in hydrogen. *Phys. Rev. Lett.*, **50**, 1305–1308.

180 Pepin, R.O. and Porcelli, D. (2002) Origin of noble gases in the terrestrial planets. *Rev. Mineral. Geochem.*, **47**, 191–246.

181 Shcheka, S.S. and Keppler, H. (2012) The origin of the terrestrial noble-gas signature. *Nature*, **490**, 531–534.

182 Miao, M., Wang, X., Brgoch, J., Spera, F., Jackson, M.G., Kresse, G., and Lin, H. (2015) Anionic chemistry of noble gases: formation of Mg–NG (NG = Xe, Kr, Ar) compounds under pressure. *J. Am. Chem. Soc.*, **137**, 14122–14128.

183 Sanloup, C., Bonev, S.A., Hochlaf, M., and Maynard-Casely, H.E. (2013) Reactivity of xenon with ice at planetary conditions. *Phys. Rev. Lett.*, **110**, 265501.

184 Somayazulu, M., Dera, P., Goncharov, A.F., Gramsch, S.A., Liermann, P., Yang, W., Liu, Z., Mao, H.K., and Hemley, R.J. (2010) Pressure-induced bonding and compound formation in xenon–hydrogen solids. *Nat. Chem.*, **2**, 50–53.

185 Buzea, C. and Robbie, K. (2004) Assembling the puzzle of superconducting elements: a review. *Supercond. Sci. Technol.*, **18**, R1–R8.

186 Gao, L., Xue, Y.Y., Chen, F., Xiong, Q., Meng, R.L., Ramirez, D., Chu, C.W., Eggert, J.H., and Mao, H.K. (1994) Superconductivity up to 164 K in $HgBa_2Ca_{m-1}Cu_mO_{2m+2+\delta}$ (m = 1, 2, and 3) under quasihydrostatic pressures. *Phys. Rev. B*, **50**, 4260–4263.

187 Li, Y., Hao, J., Liu, H., Li, Y., and Ma, Y. (2014) The metallization and superconductivity of dense hydrogen sulfide. *J. Chem. Phys.*, **140**, 174712.

188 Allen, P.B. and Dynes, R.C. (1975) Transition temperature of strong-coupled superconductors reanalyzed. *Phys. Rev. B*, **12**, 905–922.

189 Drozdov, A.P., Eremets, M.I., Troyan, I.A., Ksenofontov, V., and Shylin, S.I. (2015) Conventional superconductivity at 203 Kelvin at high pressures in the sulfur hydride system. *Nature*, **525**, 73–76.

190 Duan, D., Liu, Y., Tian, F., Li, D., Huang, X., Zhao, Z., Yu, H., Liu, B., Tian, W., and Cui, T. (2014) Pressure-induced metallization of dense $(H_2S)_2H_2$ with high-T_c superconductivity. *Sci. Rep.*, **4**, 6968.

191 Strobel, T.A., Ganesh, P., Somayazulu, M., Kent, P.R.C., and Hemley, R.J. (2011) Novel cooperative interactions and structural ordering in H_2S-H_2. *Phys. Rev. Lett.*, **107**, 255503.

192 Einaga, M., Sakata, M., Ishikawa, T., Shimizu, K., Eremets, M. I., Drozdov, A. P., Troyan, I. A., Hirao, N., Ohishi, Y. (2016) Crystal structure of the superconducting phase of sulfur hydride. *Nat. Phys.* **12**, 835–838

193 Drozdov, A.P., Eremets, M.I., and Troyan, I.A. (2015) Superconductivity above 100 K in PH_3 at high pressures. doi: ArXiv:1508.06224.

194 Papaconstantopoulos, D.A., Klein, B.M., Mehl, M.J., and Pickett, W.E. (2015) Cubic H_3S around 200 GPa: an atomic hydrogen superconductor stabilized by sulfur. *Phys. Rev. B*, **91**, 184511.

195 Bernstein, N., Hellberg, C.S., Johannes, M.D., Mazin, I.I., and Mehl, M.J. (2015) What superconducts in sulfur hydrides under pressure and why. *Phys. Rev. B*, **91**, 060511.

196 Duan, D., Huang, X., Tian, F., Li, D., Yu, H., Liu, Y., Ma, Y., Liu, B., and Cui, T. (2015) Pressure-induced decomposition of solid hydrogen sulfide. *Phys. Rev. B*, **91**, 180502.

197 Errea, I., Calandra, M., Pickard, C.J., Nelson, J., Needs, R.J., Li, Y., Liu, H., Zhang, Y., Ma, Y., and Mauri, F. (2015) High-pressure hydrogen sulfide from first principles: a strongly anharmonic phonon-mediated superconductor. *Phys. Rev. Lett.*, **114**, 157004.

198 Akashi, R., Kawamura, M., Tsuneyuki, S., Nomura, Y., and Arita, R. (2015) First-principles study of the pressure and crystal-structure dependences of the superconducting transition temperature in compressed sulfur hydrides. *Phys. Rev. B*, **91**, 224513.

199 Zhong, X., Wang, H., Zhang, J., Liu, H., Zhang, S., Song, H., Yang, G., Zhang, L., and Ma, Y. (2015) Tellurium hydrides at high pressures: high-temperature superconductors. *Phys. Rev. Lett.*, **116**, 057002.

200 Li, Y., Wang, L., Liu, H., Zhang, Y., Hao, J., Pickard, C.J., Nelson, J.R., Needs, R.J., Li, W., Huang, Y., Errea, I., Calandra, M., Mauri, F., and Ma, Y. (2016) Dissociation products and structures of solid H_2S at strong compression. *Phys. Rev. B*, **93**, 020103.

201 Fu, Y., Du, X., Zhang, L., Peng, F., Zhang, M., Pickard, C.J., Needs, R.J., Singh, D.J., Zheng, W., and Ma, Y. (2016) High-pressure phase stability and superconductivity of pnictogen hydrides and chemical trends for compressed hydrides. *Chem. Mater.*, **28**, 1746–1755.

202 Liu, H., Li, Y., Gao, G., Tse, J.S., and Naumov, I.I. (2016) Crystal structure and superconductivity of PH_3 at high pressures. *J. Phys. Chem. C*, **120**, 3458–3461.

203 Hong, Q.J. and van de Walle, A. (2013) Solid–liquid coexistence in small systems: a statistical method to calculate melting temperatures. *J. Chem. Phys.*, **139**, 094114.

204 Babaev, E., Sudbo, A., and Ashcroft, N.W. (2004) A superconductor to superfluid phase transition in liquid metallic hydrogen. *Nature*, **431**, 666–668.

205 Deemyad, S. and Silvera, I.F. (2008) Melting line of hydrogen at high pressures. *Phys. Rev. Lett.*, **100**, 155701.

206 Bonev, S.A., Schwegler, E., Ogitsu, T., and Galli, G. (2004) A quantum fluid of metallic hydrogen suggested by first-principles calculations. *Nature*, **431**, 669–672.

207 Geng, H.Y., Hoffmann, R., and Wu, Q. (2015) Lattice stability and high-pressure

208 Chen, J., Li, X., Zhang, Q., Probert, M.I.J., Pickard, C.J., Needs, R.J., Michaelides, A., and Wang, E. (2013) Quantum simulation of low-temperature metallic liquid hydrogen. *Nat. Commun.*, **4**, 2064.

melting mechanism of dense hydrogen up to 1.5 TPa. *Phys. Rev. B*, **92**, 104103.

209 Hernández, E.R., Rodriquez-Prieto, A., Bergara, A., and Alfé, D. (2010) First-principles simulations of lithium melting: stability of the bcc phase close to melting. *Phys. Rev. Lett.*, **104**, 185701.

210 Tamblyn, I., Raty, J.Y., and Bonev, S.A. (2008) Tetrahedral clustering in molten lithium under pressure. *Phys. Rev. Lett.*, **101**, 075703.

211 Guillaume, C.L., Gregoryanz, E., Degtyareva, O., McMahon, M.I., Hanfland, M., Evans, S., Guthrie, M., Sinogeikin, S.V., and Mao, H.K. (2011) Cold melting and solid structures of dense lithium. *Nat. Phys.*, **7**, 211–214.

212 Schaeffer, A.M.J., Talmadge, W.B., Temple, S.R., and Deemyad, S. (2012) High pressure melting of lithium. *Phys. Rev. Lett.*, **109**, 185702.

213 Feng, Y., Chen, J., Alfé, D., Li, X.Z., and Wang, E. (2015) Nuclear quantum effects on the high pressure melting of dense lithium. *J. Chem. Phys.*, **142**, 064506.

214 Gregoryanz, E., Degtyareva, O., Somayazulu, M., Hemley, R.J., and Mao, H.K. (2005) Melting of dense sodium. *Phys. Rev. Lett.*, **94**, 185502.

215 Gregoryanz, E., Lundegaard, L.F., McMahon, M.I., Guillaume, C., Nelmes, R.J., and Mezouar, M. (2008) Structural diversity of sodium. *Science*, **320**, 1054–1057.

216 Raty, J.Y., Schwegler, E., and Bonev, S.A. (2007) Electronic and structural transitions in dense liquid sodium. *Nature*, **449**, 448–451.

217 Eshet, H., Khaliullin, R.Z., Kühne, T.D., Behler, J., and Parrinello, M. (2012) Microscopic origins of the anomalous melting behavior of sodium under high pressure. *Phys. Rev. Lett.*, **108**, 115701.

218 George, J., Reimann, C., Deringer, V.L., Bredow, T., and Dronskowski, R. (2015) On the DFT ground state of crystalline bromine and iodine. *ChemPhysChem*, **16**, 728–732.

219 Grochala, W. (2014) Diamond: electronic ground state of carbon at temperatures approaching 0 K. *Angew. Chem., Int. Ed.*, **53**, 3680–3683.

220 Santra, B., Klimes, J., Alfe, D., Tkatchenko, A., Slater, B., Michaelides, A., Car, R., and Scheffler, M. (2011) Hydrogen bonds and van der Waals forces in ice at ambient and high pressures. *Phys. Rev. Lett.*, **107**, 185701.

221 Morales, M.A., McMahon, J.M., Pierleoni, C., and Ceperley, D.M. (2013) Nuclear quantum effects and nonlocal exchange–correlation functionals applied to liquid hydrogen at high pressure. *Phys. Rev. Lett.*, **110**, 065702.

222 Teweldeberhan, A.M., DuBois, J.L., and Bonev, S.A. (2012) Stability of the high-pressure phases of $CaTiO_3$ perovskite at finite temperatures. *Phys. Rev. B*, **86**, 064104.

223 Bartolomei, M., Carmona-Novillo, E., Hernandez, M.I., Perez-Rios, J., Campos-Martinez, J., and Hernandez-Lamoneda, R. (2011) Molecular oxygen tetramer $(O_2)_4$: intermolecular interactions and implications for the ε solid phase. *Phys. Rev. B*, **84**, 092105.

224 Liu, H., Cui, W., and Ma, Y. (2012) Hybrid functional study rationalizes the simple cubic phase of calcium at high pressures. *J. Chem. Phys.*, **137**, 184502.

225 Boates, B. and Bonev, S.A. (2013) Demixing instability in dense molten $MgSiO_3$ and the phase diagram of MgO. *Phys. Rev. Lett.*, **110**, 135504.

226 Boates, B. and Bonev, S.A. (2011) Electronic and structural properties of dense liquid and amorphous nitrogen. *Phys. Rev. B*, **83**, 174114.

227 Krukau, A.V., Vydrov, O.A., Izmaylov, A.F., and Scuseria, G.E. (2006) Influence of the exchange screening parameter on the performance of screened hybrid functionals. *J. Chem. Phys.*, **125**, 224106.

228 Anisimov, V.I., Zaanen, J., and Andersen, O.K. (1991) Band theory and Mott insulators: Hubbard U instead of stoner I. *Phys. Rev. B*, **44**, 943–954.

229 Aryasetiawan, F. and Gunnarsson, O. (1998) The GW method. *Rep. Prog. Phys.*, **61**, 237–312.

16
First-Principles Computation of NMR Parameters in Solid-State Chemistry

Jérôme Cuny,[1] *Régis Gautier,*[2] *and Jean-François Halet*[2]

[1]*Laboratoire de Chimie et Physique Quantiques (LCPQ), Université de Toulouse III – Paul Sabatier and CNRS, 118 Route de Narbonne, F-31062 Toulouse, France*
[2]*Institut des Sciences Chimiques de Rennes, UMR 6226 CNRS – Université de Rennes 1 – Ecole Nationale Supérieure de Chimie de Rennes, Avenue du Général Leclerc, F-35042 Rennes, France*

16.1
Introduction

Since the simultaneous experimental evidence of the NMR phenomenon in the mid-1940s by Purcell *et al.* [1], and Bloch *et al.* [2], nuclear magnetic resonance (NMR) spectroscopy is a technique that has advanced rapidly to become nowadays an essential tool to study the structure and dynamics of a variety of systems in chemistry, physics, biology, or medicine [3,4]. In molecular organic and organometallic chemistry for instance, NMR is one of the most popular techniques used to characterize the structure of chemical compounds in solution. In biology, NMR is essential to gain access to information about the secondary and ternary structure of proteins and large macromolecules. In contrast, NMR is generally less widely used in the solid state than in the liquid phase due to its higher complexity in terms of instrumentation and theoretical foundations. However, as NMR scrutinizes individually each nucleus in a sample, solid-state NMR (SSNMR) constitutes a local probe that provides useful additional information to other experimental techniques requiring long-range atomic order such as X-ray or neutron diffraction [5]. SSNMR is nowadays an important addition to crystallographers' toolboxes to characterize crystalline solids, and the term NMR crystallography is now in common usage [6–10]. SSNMR is also a method of choice to characterize the structure of amorphous materials such as glasses, surfaces, support materials, and nanosystems [11–13]. Indeed, SSNMR is a powerful but complex experimental technique that can be helpful, if not essential, to study a large gamut of parameters (internuclear distances, anisotropy, torsion angles, atomic orientations, spin diffusion, molecular dynamics, etc.) in inorganic solid-state chemistry [6,14,15].

Handbook of Solid State Chemistry, First Edition. Edited by Richard Dronskowski, Shinichi Kikkawa, and Andreas Stein.
© 2017 Wiley-VCH Verlag GmbH & Co. KGaA. Published 2017 by Wiley-VCH Verlag GmbH & Co. KGaA.

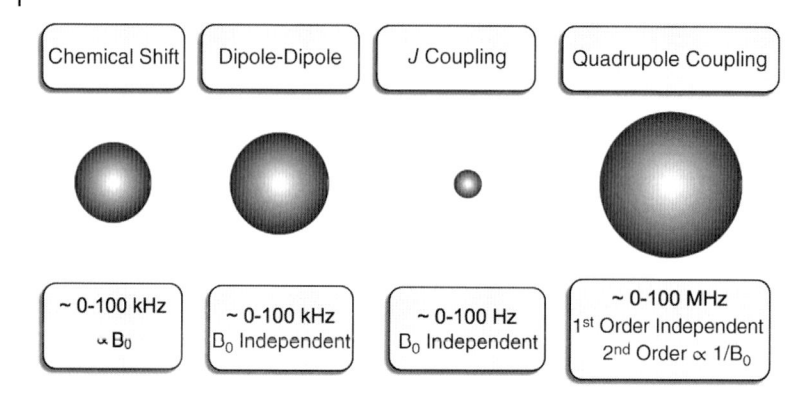

Figure 16.1 Relative magnitudes of various components of the internal spin interactions encountered in insulating nonmagnetic solid-state compounds. (Adapted from Ref. [16].)

Chemical Shift: ~0-100 kHz, $\propto B_0$
Dipole-Dipole: ~0-100 kHz, B_0 Independent
J Coupling: ~0-100 Hz, B_0 Independent
Quadrupole Coupling: ~0-100 MHz, 1st Order Independent, 2nd Order $\propto 1/B_0$

Specifically, NMR spectroscopy focuses on the behavior of the nuclear spin magnetic moments and more explicitly on their interactions between themselves and with their chemical environment under the action of various magnetic fields. These interactions are key quantities as they carry the chemical information from the probed nuclei. For insulating nonmagnetic materials, these interactions that are summarized in Figure 16.1 can be magnetic as for instance the chemical shift, the dipole–dipole, and the J-coupling interactions, or electrostatic in the case of the quadrupolar coupling interaction. Each one carries some specific information about the structure, dynamics, and distribution of chemical environments and asymmetries in the vicinity of the probed nuclei. Once collected together, they can provide a very accurate picture of a given material at the atomic level.

All this information makes SSNMR a powerful and convenient experimental technique but with one major limitation: the difficulty in interpreting the experimental data. To illustrate this point, Figure 16.2a displays the ^{23}Na SSNMR spectrum of $In_2Na_2Mo_5O_{16}$, a solid-state compound the structure of which is based on Mo_3O_{13} and MoO_4 motifs [17]. Although this compound is made up of only three crystallography independent sodium atoms, the spectrum of Figure 16.2a cannot be interpreted as it is without the use of additional tools such as fitting or numerical simulation programs as shown in Figure 16.2b. Furthermore, even after the decomposition of the spectrum in terms of the independent sodium contributions, it is still difficult to extract quantitative chemical information from them. This strongly contrasts with liquid-state NMR in which a direct qualitative interpretation can be made even for molecules of intermediate complexity. Thus, line broadening and band overlapping resulting from local asymmetries and the distribution of chemical environments, respectively, are common reasons that render difficult data interpretation in SSNMR. In a number of cases, this complexity is not only due to the specificity of the studied compounds but also to the intrinsic properties of the probed nuclei.

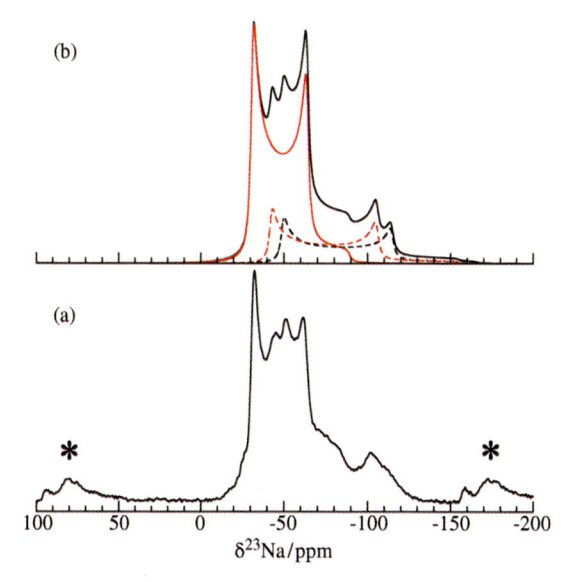

Figure 16.2 (a) Experimental ^{23}Na SSNMR spectrum of In$_2$Na$_2$Mo$_5$O$_{16}$ recorded at 7.05 T using a Hahn echo pulse sequence under MAS conditions at 10 kHz. The spectrum is referenced with respect to the ^{23}Na SSNMR signal of a NaCl powder. * symbols indicate spinning side-bands. (b) Best-fit simulation (black plain line) of the corresponding central transition obtained using three sodium contributions: Na1 (red plain line), Na2 (black dashed line), and Na3 (red dashed line). The fit was performed using the Dmfit program [18].

A number of experimental approaches have been developed to tackle these difficulties, among others one can note the development of high-resolution NMR techniques such as magic-angle spinning (MAS) [19,20], multiquantum magic-angle spinning (MQMAS) [21–24] and dynamic-angle spinning (DAS) [25,26], the development of very high magnetic field spectrometers and the possibility of isotopic enrichment. However, despite these developments, a large part of the information contained in the spectra can still be only partially exploited. Consequently, in order to get a better insight into the atomic structure and behavior of solid-state compounds, it was desirable to develop additional tools to complement SSNMR measurements.

To this end, significant efforts have been made in establishing theoretical tools to estimate NMR parameters from first-principles calculations, in particular chemical shifts. Such methods have been a long standing interest in the quantum chemistry community. Indeed, the first calculation of chemical shift parameters performed at the *ab initio* (Hartree–Fock) level of theory was proposed by Lipscomb and coworkers in 1963 [27]. The following decades have seen a number of theoretical improvements that have allowed for the development of efficient and accurate formalisms that are nowadays widely used by both experimentalist and theoretician communities to characterize molecular systems [28–35]. Numerous excellent review articles have appeared in the past dealing with the theoretical aspects of these methods and their applications [36–42].

Until the end of the 1990s, all the available approaches were limited to molecular systems, simply because methods to include the long-range effects of the crystal lattice were not developed. Those who were interested in solid-state compounds had thus to resort to the so-called *cluster approach* in which the local environment in a solid is modeled by a finite "molecular" entity constructed such that atoms in the center experience an environment similar to that in true extended solid. An illustration of such an approach is given in Figure 16.3 for the description of amorphous silica. Although being applicable to some particular cases such as molecular solids or surfaces, its limitations to accurately model inorganic compounds is obvious.

The main advances in the development of the formalisms devoted to solid-state systems were made in the mid 1990s when Mauri, Pfrommer, and Luiz proposed a first approach based on density functional theory (DFT), later referred as the MPL method [43,44]. Gregor *et al.* explicitly demonstrated later that, when correctly dealing with the gauge-origin problem, core and valence electrons can be treated separately for the calculation of chemical shifts [45]. Another approach was further proposed by Sebastiani and Parrinello [46]. However, the real breakthrough occurred the same year when the gauge-including projector augmented wave (GIPAW) approach was proposed by Pickard and Mauri [47]. In this method, the authors combined the MPL approach and the projector augmented wave (PAW) formalism to achieve all-electron accuracy [48,49], as well as the idea from the molecular gauge-including atomic orbitals (GIAO) [28,29] and independent gauge for localized orbitals (IGLO) [30,31] methods to ensure the gauge independence to ultimately provide a consistent and general formalism. Its efficiency was further demonstrated allowing to tackle a large range of solid-state topics [50,51] and to develop other theoretical formalisms to compute metallic-shift and *J*-coupling tensors in extended systems [52,53].

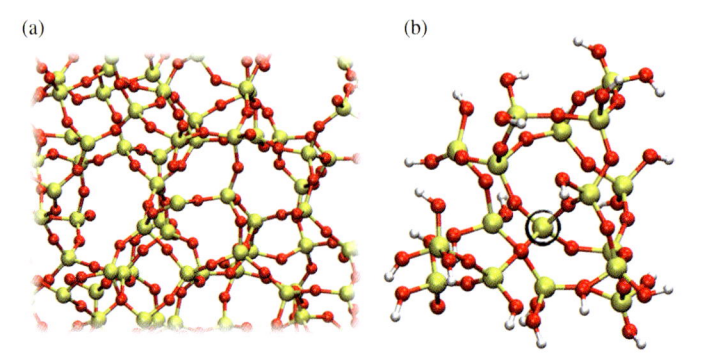

(a) (b)

Figure 16.3 Tridimensional structural arrangement of amorphous silica (a) and "molecular" cluster constructed to reproduce the correct chemical environment of the highlighted central silicon atom (b). External oxygen atoms are replaced by hydroxyl groups.

The three aforementioned approaches, that is, MPL, Sebastiani–Parrinello, and GIPAW were developed within the framework of DFT under periodic boundary conditions. More precisely, they use the pseudopotential plane wave approach for solving the DFT Kohn–Sham equations [54,55]. Another approach was later developed, still in the context of the DFT pseudopotential plane wave framework, namely, the converse approach [56,57]. Finally, another formalism, developed in the framework of the augmented plane wave electronic structure method [58–60], was recently proposed showing great potential for the study of solid-state compounds [61,62]. However, the GIPAW method largely dominates the others nowadays in terms of dissemination and applications [50,51]. In particular, it has proven to be powerful in the development of NMR crystallography and the structural characterization of many organic and inorganic solid-state compounds [6,63,64].

In the present chapter, our goal is to give the reader an overview of the computational approaches devoted to the theoretical evaluation of NMR parameters in solid-state chemistry and show how they can be used to complement other experimental studies for routine chemical analysis of solid compounds. After a brief recall of the basics of the NMR and SSNMR interactions that are intended to be computed, these approaches are described highlighting their strengths and weaknesses. Special attention is paid to the important theoretical considerations necessary to perform meaningful simulations. Finally, a selection of educational examples from different areas of materials science are discussed to illustrate the ability of these approaches, in particular the GIPAW one, to provide valuable insight into the understanding of both crystalline and amorphous solid-state compounds.

16.2
Theory and Methods

16.2.1
Physical Basis of NMR

The NMR response of a nucleus depends not only on its intrinsic magnetic properties but also on its surrounding environment, which encompasses other nuclear spins and electrons. Thus, modeling the NMR response of a given nucleus, that is, its nuclear spin dynamics, requires the time propagation of a time-dependent electronic and nuclear Schrödinger equation [16]. In practice, this is too complex to be solved and therefore one resorts to an effective spin Hamiltonian in which the influence of the surrounding nuclei and electrons is averaged. This Hamiltonian consists of a sum of terms, each one characteristic of a specific NMR interaction that are defined by a small number of second-rank tensors encompassing the averaged influence of the surrounding electrons and nuclei. The elements of these tensors are directly related to the shape of the corresponding spectra and can thus be extracted from experimental measurements. Being averaged quantities, these tensors depend on the atomic positions and electronic structure of the probed chemical system; they are thus accessible from electronic structure calculations as it will be discussed in Section 16.2.2.

Figure 16.1 displays the main interactions occurring in a SSNMR experiment with their relative intensities in common insulating nonmagnetic solid-state compounds. These relative contributions depend on the compound of interest and on the probed nuclei. Nevertheless, the following hierarchy can be drawn: When dealing with quadrupolar nuclei (see below for definition), the main features of a SSNMR spectrum are generally dominated by the chemical shift and quadrupolar coupling interactions. The direct dipole–dipole interaction becomes significant in compounds containing spin-1/2 nuclei that display a high gyromagnetic ratio, that is, ^{1}H and ^{19}F (sometimes ^{31}P). In these cases, the direct dipole–dipole interaction can contribute as much as the chemical shift to the spectra. Finally, the J-coupling interaction is often weak, that is, a few tens of hertz (although some exceptions exist as discussed below) [65], and its contribution to the spectra is generally hidden by the line broadening induced by the other interactions and by the distribution of chemical environments. Hence, in the following, we focus our discussion on the chemical shift and quadrupolar coupling interactions due to their predominance in most materials' spectra and to a lesser extent discuss the metallic shift and dipolar interactions.

Chemical Shift Interaction

The interaction between a nucleus and an externally applied magnetic field $\mathbf{B_0}$ is the so-called Zeeman interaction. For a set of isolated noninteracting nuclei, the corresponding Hamiltonian is given by:

$$H = -\sum_k \gamma_k \mathbf{I_k} \cdot \mathbf{B_0} \tag{16.1}$$

where $\mathbf{I_k}$ is the spin angular momentum and γ_k is the gyromagnetic ratio of nucleus k. In an insulating nonmagnetic compound, the chemical shift interaction results from the diamagnetic response of the electrons to $\mathbf{B_0}$. This response generates an induced magnetic field $\mathbf{B_{in}(r_k)}$ that shields the nucleus k from B_0. The nucleus effectively is subjected to a magnetic field $\mathbf{B(r_k)}$ defined as

$$\mathbf{B(r_k)} = \mathbf{B_0} + \mathbf{B_{in}(r_k)} = (1 - \overleftrightarrow{\sigma}_k)\mathbf{B_0} \tag{16.2}$$

where $\overleftrightarrow{\sigma}_k$ is the nuclear shielding tensor at the k nucleus position that describes the shielding of k by the surrounding electronic environment. $\overleftrightarrow{\sigma}_k$ is then defined by

$$\mathbf{B_{in}(r_k)} = -\overleftrightarrow{\sigma}_k \mathbf{B_0} \tag{16.3}$$

The chemical shift interaction is a perturbation of the Zeeman interaction that causes a slight modification of the nuclear resonance frequency with respect to the Larmor frequency of the bare nucleus. These modifications bring chemical information in NMR spectra.

Using the Haeberlen convention [66], and assuming $|\sigma_{zz} - \sigma_{iso}| \geq |\sigma_{xx} - \sigma_{iso}| \geq |\sigma_{yy} - \sigma_{iso}|$, it is possible to express the three eigenvalues σ_{xx}, σ_{yy}, and σ_{zz} of $\overleftrightarrow{\sigma}$ using the following three parameters:

$$\sigma_{iso} = \frac{1}{3}\left(\sigma_{xx} + \sigma_{yy} + \sigma_{zz}\right) \tag{16.4}$$

$$\sigma_{aniso} = \sigma_{zz} - \sigma_{iso} \tag{16.5}$$

$$\eta_\sigma = \frac{\sigma_{yy} - \sigma_{xx}}{\sigma_{aniso}} \tag{16.6}$$

that are the isotropic nuclear shielding, the anisotropy parameter and the asymmetry parameter, respectively. In liquid state, due to the fast Brownian motion of molecules, or under some specific experimental conditions in SSNMR, only the isotropic part of $\overleftrightarrow{\sigma}$ influences the NMR signal. In other cases, band broadening occurs leading to line shapes that are characteristic of the asymmetry of the chemical environment around the nuclei.

Experimentally, this interaction is expressed in terms of chemical shift δ instead of nuclear shielding σ. By introducing the isotropic nuclear shielding of a reference nucleus σ_{ref}, the three eigenvalues of the chemical shift tensor can be expressed as a function of the nuclear shielding tensor ones using $\delta = -(\sigma - \sigma_{ref})$. The aforementioned nuclear shielding parameters can thus be expressed in terms of three chemical shift parameters: $\delta_{iso} = -(\sigma_{iso} - \sigma_{ref})$, $\delta_{aniso} = -\sigma_{aniso}$, and $\eta_\delta = \eta_\sigma$.

Quadrupolar Coupling Interaction
The electric charge distribution in a nucleus is not necessarily spherical. When it is not, it can be written as an electric multipole development where the second-order term, that is, the electric quadrupolar moment, can interact with the electric field gradient (EFG) generated by its surrounding environment. This interaction modifies the spin state of the nucleus and consequently its NMR signal: This is the quadrupolar coupling or quadrupolar interaction. It is worth mentioning that spin-1/2 nuclei behave like spherical charges and are consequently not affected by the quadrupolar interaction. On the other hand, all the nuclei with a nuclear spin larger than 1/2 are quadrupolar nuclei, that is, they are affected by the quadrupolar coupling interaction. In practice, SSNMR of these nuclei requires specific experimental techniques due to the large band broadening they display [67,68]. Quadrupolar nuclei are characterized by their quadrupolar moment eQ that is characteristic of the strength of the quadrupolar interaction they undergo whereas the EFG is characterized by a tensor, $\overleftrightarrow{V}$, that is symmetric and traceless. It turns out that the quadrupolar coupling interaction can be described by two parameters related to the EFG tensor eigenvalues (V_{xx}, V_{yy}, V_{zz}) by

$$C_Q = \frac{eQV_{zz}}{h} \tag{16.7}$$

$$\eta_Q = \frac{V_{xx} - V_{yy}}{V_{zz}} \tag{16.8}$$

with $|V_{zz}| \geq |V_{xx}| \geq |V_{yy}|$. C_Q is the quadrupolar coupling constant and η_Q the asymmetry parameter.

Dipolar and Metallic Shift Interactions

As previously mentioned, other interactions can also affect SSNMR spectra. The J-coupling (or indirect dipole–dipole) and direct dipole–dipole interactions correspond to interactions between nuclear spin magnetic moments, mediated by the bonding electrons for the former and through space for the latter. The corresponding Hamiltonian can be written as

$$H = \sum_{k<l}{}' \mathbf{I}_k \cdot \left(\overleftrightarrow{D}_{kl} + \overleftrightarrow{J}_{kl} \right) \cdot \mathbf{I}_l \qquad (16.9)$$

where $\overleftrightarrow{D}_{kl}$ and $\overleftrightarrow{J}_{kl}$ are the direct dipole–dipole and J-coupling tensors, respectively. The J-coupling interaction is often rather weak, a few tens of hertz, compared to the line broadening induced, for instance, by the quadrupolar coupling interaction and the chemical shift anisotropy. Unfortunately, common MAS experiments are not sufficient to average to zero second-order quadrupolar effects, second-order cross-terms between interactions or residual dipolar coupling terms which make an accurate determination of J-coupling tensors very challenging. Consequently, the fine structure of NMR spectra observed in the liquid phase is generally obscured in the solid state [65,69]. Nevertheless, in the case of spin-1/2 nuclei (without quadrupolar coupling interactions) combining high-frequency spinning, accurate magic angle setting and applying J-resolved pulse sequences allow to characterize the J coupling in the solid state to access information about bond strength and connectivity [65,69–79]. J couplings have also been characterized between spin-1/2 and quadrupolar nuclei and, to a lesser extent, between quadrupolar nuclei (see Refs [65,69,80–82], and references therein). Finally, it is worth pointing out that in some particular cases, generally involving spin-1/2 heavy nuclei, the J-coupling interaction can reach several hundreds or thousands hertz [65]. The direct dipole–dipole coupling is of the order of magnitude of the chemical shift and can lead to strong broadening of the line shapes in samples containing spin-1/2 nuclei with high gyromagnetic ratio, in particular hydrogen and/or fluorine atoms. For low gyromagnetic ratio and quadrupolar nuclei, this interaction is generally negligible or easily averaged to zero using high resolution techniques such as MAS.

In metallic systems, the concept of chemical shift disappears to become the metallic shift. As for the chemical shift, the metallic shift results from the response of the electronic structure to an externally applied magnetic field. However, in contrast to insulating materials, it consists of two contributions: an orbital contribution which is similar to the chemical shift in insulating nonmagnetic materials, and a spin contribution, also called Knight shift. Generally, the magnitude of the Knight shift is larger than the orbital contribution, and it can be used to extract a wealth of information from SSNMR spectra of, for instance, intermetallic and superconducting materials but can induce difficulties in the spectra interpretation [83–89].

16.2.2
Computation of NMR Parameters

Accurate determination of the electronic structure of the studied compounds via a quantum chemical treatment is a prerequisite for a rigorous theoretical computation of the nuclear shielding and electric field gradient tensors of a given nucleus. In conventional calculations, the Born–Oppenheimer approximation (nuclei at fixed positions) is assumed, leading to solve the nonrelativistic time-independent electronic Schrödinger equation as follows:

$$H^R \Psi^R(\mathbf{r}) = E^R \Psi^R(\mathbf{r}) \tag{16.10}$$

where the nuclear coordinates $\{\mathbf{R}\}$ are fixed parameters and the many-body wave function $\Psi^R(\mathbf{r})$ is a function of the electron coordinates $\{\mathbf{r}\}$ at fixed $\{\mathbf{R}\}$. There are two classes of electronic structure methods to solve this equation: one is based on the wave function (Hartree–Fock and post-Hartree–Fock) whereas the other is based on the electron density (DFT) [90]. Nowadays, the latter is certainly the method that offers the best compromise between accuracy and computational efficiency for the description of solid-state compounds made of several tens or hundreds of atoms. In the following section, a brief overview of the main DFT approaches aiming at computing electric field gradient and nuclear shielding tensors, is given. For the benefit of providing the reader with a more complete introduction to DFT methods, we refer to the chapters of Springborg and Dong, and Schwarz and Blaha in this volume.

16.2.2.1 Density Functional Theory for Extended Systems

General Basis of DFT
In contrast to wave function-based methods, DFT aims at solving the electronic Schrödinger equation using the electronic density as the fundamental variable. Hohenberg and Kohn proposed in 1964 a formulation of DFT in which they demonstrated a unique one-to-one mapping between the external potential $V_{\text{ext}}(\mathbf{r})$ applied to a nondegenerated electronic system and its ground-state density $n_0(\mathbf{r})$ [91]. This implies that the ground-state energy of the system is a functional of its ground-state density $E[n_0(\mathbf{r})]$.

Kohn and Sham later introduced the notion of fictitious noninteracting electronic system of electrons that exhibits the same electronic density and energy as the real interacting electronic system [92]. For such a fictitious system, the total wave function can be written as a Slater determinant of monoelectronic wave functions $\psi_n(\mathbf{r})$, called the Kohn–Sham orbitals, that obey

$$n_0(\mathbf{r}) = \sum_n f_n |\psi_n(\mathbf{r})|^2 \tag{16.11}$$

where f_n is the occupation of $\psi_n(\mathbf{r})$. The $\psi_n(\mathbf{r})$'s are the eigenvectors of monoelectronic Schrödinger-like equations, the Kohn–Sham equations, obtained from the variational minimization of the total energy of the system with respect to the

$\psi_n(\mathbf{r})$'s under the constraint of a constant number of particles

$$\left\{-\frac{\hbar^2}{2m_e}\nabla^2 + V_{ext}(\mathbf{r}) + V^H[n_0(\mathbf{r})] + V_{xc}[n_0(\mathbf{r})]\right\}\psi_n(\mathbf{r}) = \varepsilon_n\psi_n(\mathbf{r}) \quad (16.12)$$

The first term of Eq. (16.12) is the kinetic energy operator of the noninteracting electrons, $V_{ext}(\mathbf{r})$ is an external potential that generally describes the nucleus–electron electrostatic interactions and $V^H[n_0(\mathbf{r})]$ is the Hartree potential. $V_{xc}[n_0(\mathbf{r})]$ is the exchange-correlation potential that incorporates exchange and correlation contributions to the total energy and also the difference in kinetic energy between the noninteracting and interacting electronic systems. These operators have well-defined analytic forms but for $V_{xc}[n_0(\mathbf{r})]$ the exact form of which is unknown. Intensive studies have been made to tentatively provide accurate approximated exchange-correlation functionals, going from the simplest local density (LDA) [93–97] and generalized-gradient (GGA) [98–103] approximations to meta-GGA and hybrid functionals [104–115]. Despite continuous progress in the development of functionals, functional quality still depends on the property of interest and on the studied system. A careful choice of functional is then necessary to obtain meaningful results. This is further discussed in Section 24.2.3 in the specific case of the computation of NMR parameters in solid-state compounds.

Modeling of Solids

Once a proper exchange-correlation functional is set, a major difficulty that arises is related to the size of the system. Indeed, any realistic crystal is composed of a large, quasi infinite number of electrons leading to an equivalently large number of Kohn–Sham orbitals, $\psi_n(\mathbf{r})$, that need to be determined. The Bloch's theorem allows to circumvent this issue. Indeed, as the Kohn–Sham potential displays the periodicity of the crystal lattice, the eigenvectors of Eq. (16.12) can be written as Bloch's functions as follows:

$$\psi_{n,\mathbf{k}}(\mathbf{r}) = e^{i\mathbf{k}\cdot\mathbf{r}}u_{n,\mathbf{k}}(\mathbf{r}) \quad (16.13)$$

where $\mathbf{k}$ are the crystal momentum values lying within the first Brillouin zone of the reciprocal lattice and the $u_{n,\mathbf{k}}(\mathbf{r})$'s are functions having the unit cell periodicity (i.e., $u_{n,\mathbf{k}}(\mathbf{r}) = u_{n,\mathbf{k}}(\mathbf{r}+\mathbf{R})$ for all lattice vectors $\mathbf{R}$). The problem is now reduced to solve Eq. (16.12) for a limited number of electrons at an infinite number of $\mathbf{k}$ values. The eigenvalues of Eq. (16.12) are now written $\varepsilon_n^\mathbf{k}$ and all the $\varepsilon_n^\mathbf{k}$ associated to the same n value form a *band*. In practice, using a limited number of $\mathbf{k}$ values (often called $\mathbf{k}$-points), that is discrete bands, is sufficient to compute most properties of physical interest [116].

The Pseudopotential Plane Wave Approach

In order to solve the Kohn–Sham equations numerically, the $\psi_{n,\mathbf{k}}(\mathbf{r})$'s must be expended on a basis set. There are many possible choices, each one with advantages and drawbacks. When dealing with extended systems, plane wave basis sets are well adapted due to their mathematical simplicity and computational efficiency. In this case, the $\psi_{n,\mathbf{k}}(\mathbf{r})$'s are written as follows:

$$\psi_{n,\mathbf{k}}(\mathbf{r}) = \sum_{\mathbf{G}} c_n^{\mathbf{k}}(\mathbf{G}) e^{i(\mathbf{k}+\mathbf{G}) \cdot \mathbf{r}} \qquad (16.14)$$

where the $c_n^{\mathbf{k}}(\mathbf{G})$ are the coefficients of the expansion that are to be determined when solving Eq. (16.12). Only plane waves for which $\mathbf{G}$ is a reciprocal lattice vector have a nonzero $c_n^{\mathbf{k}}(\mathbf{G})$. Equation (16.14) is exact in the limit of an infinite number of $\mathbf{G}$ vectors, but in practice, the basis set is truncated to a finite number of functions using the following criterion:

$$\frac{\hbar^2 |\mathbf{k}+\mathbf{G}|^2}{2m_e} \leq E_{\text{cut}} \qquad (16.15)$$

The so-called cutoff energy, E_{cut}, represents an upper limit for the plane wave kinetic energy beyond which plane waves are not included in the sum of Eq. (16.14). Increasing E_{cut} augments the number of basis functions and variationally improves the convergence of the studied physical quantities.

The major drawback of a plane wave basis set is its inefficiency to reproduce strongly oscillating functions such as core wave functions and valence wave functions close to the nucleus (later referred as the core region). Indeed, an accurate description of these oscillations requires a large number of plane waves at the expense of computational efficiency. However, from a chemical point of view, these oscillations are very close to the nucleus and are thus of little interest as they only slightly influence the electronic properties of a solid-state compound. In a pseudopotential approach, the core electrons, considered as frozen, are integrated within an effective ionic potential that acts on the valence electrons. Without explicit core wave functions, the valence wave functions need no longer to oscillate as the orthogonality constraint with the core states disappears. Consequently, using a properly constructed pseudopotential leads to smooth valence pseudowave functions that require far fewer plane waves but gives equivalently accurate results. An example of a valence wave function and its pseudowave function is shown in Figure 16.4. Several schemes have been proposed to generate optimal pseudopotentials [117–123]. Among them, schemes proposed by Troullier and Martins (leading to norm-conserving pseudopotentials) [123] and Vanderbilt (leading to ultrasoft pseudopotentials) [122] are widely used.

These ingredients constitute the so-called pseudopotential plane wave (PPW) electronic structure method [54,55]. Its high computational efficiency lies in the fact that numerically efficient fast-Fourier transforms can be used to convert the Kohn–Sham orbitals back and forth between real and Fourier (or reciprocal) space representations to take advantage from their respective properties.

The Augmented Plane Wave Approach

In the augmented plane wave (APW) approaches [58–60], the unit cell is partitioned into two regions: The interstitial region (I) and the augmentation region that is composed of nonoverlapping spheres (S_α) centered at each atomic

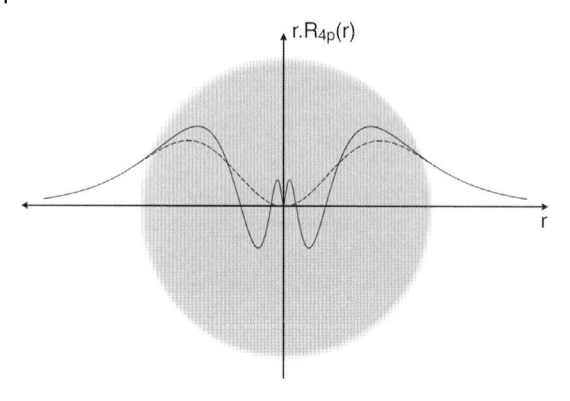

Figure 16.4 Representation of the radial part of a 4p monoelectronic wave function (solid line) and the corresponding pseudowave function (dashed line) as used in the PPW method. The gray area represents the core region in which the monoelectronic wave function is simplified.

position. In contrast to the PPW approach, the basis set is composed of plane waves in the interstitial region that are augmented by a linear combination of spherical harmonics times radial functions inside each atomic sphere. Within the linearized augmented plane wave (LAPW) [58] approach each radial function is taken as a linear combination of the solution of the scalar relativistic Schrödinger equation in a spherical potential evaluated at energy E_l and its energy derivative at the same E_l. Coefficients of this combination are defined such that the basis functions are continuous in value and slope at the S_α boundary of each atom. Further development supplied additional radial basis functions, the so-called local orbitals (LO), within the interstitial region that vanish at the sphere boundary. These LOs allow for the description of a larger energy spectrum of the compounds. A third derivation of the APW approach, the APW+lo method, has been proposed in which the linearization is insured by local orbitals [59,60]. In all cases, the Kohn–Sham orbitals can be written as follows:

$$\psi_{n,\mathbf{k}}(\mathbf{r}) = \begin{cases} \dfrac{1}{\sqrt{\Omega}} \sum_{\mathbf{G}} c_n^{\mathbf{k}}(\mathbf{G}) e^{i(\mathbf{k}+\mathbf{G})\cdot\mathbf{r}} & \mathbf{r} \in I \\ \sum_{l,m} W_{lm}^{n,\alpha,\mathbf{k}}(\mathbf{r}) Y_l^m(\hat{\mathbf{r}}) & \mathbf{r} \in S_\alpha \end{cases} \quad (16.16)$$

where Ω is the unit cell volume and $W_{lm}^{n,\alpha,\mathbf{k}}(\mathbf{r})$ is a double sum over the reciprocal lattice vectors $\mathbf{G}$ and the various radial functions used to define augmented plane waves and local orbitals. $Y_l^m(\hat{\mathbf{r}})$ are spherical harmonics and $\hat{\mathbf{r}}$ is the angular vector.

It is worth mentioning that the construction of the basis sets allows dealing with real wave functions instead of pseudowave functions and take explicitly into account the core states. Thus, the LAPW and APW+lo approaches are generally considered as full-potential all-electron methods. They are implemented in the widely used WIEN2k code for instance [124] (see the chapter by Schwarz and Blaha in this volume).

16.2.2.2 Electric Field Gradient Tensor

Theoretical evaluation of the EFG tensor from periodic DFT calculations is rather straightforward. Indeed, it is a ground state property that can be expressed as the derivative of the electric field at the nucleus position with respect to its position. A recent formulation in terms of the local electric field was proposed by Profeta et al. as follows:

$$V_{i,j}(\mathbf{r}) = \int \frac{\rho(\mathbf{r})}{|\mathbf{r}-\mathbf{r}'|^3} \left[\delta_{ij} - 3\frac{(r_i - r_i')(r_j - r_j')}{|\mathbf{r}-\mathbf{r}'|^2} \right] d^3\mathbf{r}' \tag{16.17}$$

where i and j represent Cartesian coordinates and $\rho(\mathbf{r})$ is the total ground state charge density [125].

NMR properties are strongly sensitive to the shape of the wave functions and density close to the nuclei. Consequently, within the PPW approach, the non-physical form of the pseudowave functions in the core region would inevitably lead to meaningless results if Eq. (16.17) is applied without care. To overcome this drawback, the PAW method can be employed to reconstruct all-electron quality wave functions and density from PPW calculations [48,49]. In this approach, a linear transformation operator $\mathcal{T}$ is defined to map pseudowave functions to all-electron wave functions $|\psi_n\rangle = \mathcal{T}|\tilde{\psi}_n\rangle$. To construct $\mathcal{T}$, one defines augmentation regions Ω_α centered around the atoms α. Outside the Ω_α's, the $|\psi_n\rangle$'s and $|\tilde{\psi}_n\rangle$'s are equal and then $\mathcal{T}$ is the unity operator, whereas inside the Ω_α's, $\mathcal{T}$ is defined as a sum of operators affecting only a specific augmentation region as follows:

$$\mathcal{T} = 1 + \sum_\alpha \sum_j \left[|\phi_{\alpha,j}\rangle - |\tilde{\phi}_{\alpha,j}\rangle \right] \langle \tilde{p}_{\alpha,j}| \tag{16.18}$$

where j runs over all the projectors defining the pseudopotential of atom α. The $|\phi_{\alpha,j}\rangle$'s and $|\tilde{\phi}_{\alpha,j}\rangle$'s are all-electron and pseudopartial wave functions, respectively, constructed during the pseudopotential generation. The projectors $\langle \tilde{p}_{\alpha,j}|$ are defined such that $\langle \tilde{p}_{\beta,i}|\tilde{\phi}_{\alpha,j}\rangle = \delta_{i,j}\delta_{\alpha,\beta}$ in each augmentation region. These functions are constructed from isolated atomic calculations in order to make the reconstruction operator exact for the isolated atom. However, the reconstruction is expected to be also accurate when the atom is placed in a chemical system.

Consequently, using a properly constructed $\mathcal{T}$ operator, the PAW approach allows to achieve an all-electron accuracy similar in quality to the LAPW and APW+lo methods (vide supra). It was applied to the EFG tensor calculation by Petrilli et al. [126] by decomposing $\rho(\mathbf{r})$ given in Eq. (16.17) as a sum of three terms: the charge density of the nuclei and core electrons ($n^{ion}(\mathbf{r})$), the pseudoelectronic charge density ($\tilde{n}^{el}(\mathbf{r})$), and the correction of the electronic charge density in the augmentation regions ($n^{el,\Omega_\alpha}(\mathbf{r})$). The PAW reconstruction approach has demonstrated its efficiency and accuracy for the calculation of EFG tensors in a large variety of compounds [50,51,126,127], and is thus implemented in various computer codes such as CASTEP, for instance [128].

Calculation of EFG tensors within the LAPW and APW+lo formalisms was proposed about a decade ago before the PAW approach for insulating and metallic systems [129,130]. Being all-electron methods, there is no need to resort to any reconstruction scheme and therefore, the acquisition of EFG tensors is more straightforward than for the PPW approach. These two methods have been in turn largely applied to the study of a large number of solid-state compounds and, as for the PAW approach, have demonstrated high accuracy [131–137].

16.2.2.3 Nuclear Shielding Tensor

Computation of the nuclear shielding tensor from first-principles in periodic systems is much less straightforward than the EFG tensor. Indeed, a number of technical and theoretical issues arise. Several approaches have been proposed that are briefly reviewed below.

Gauge-Including Projector Augmented Wave Method

The GIPAW approach was the first method that provided a consistent and general formalism to accurately calculate nuclear shielding tensors in periodic systems within DFT. Developed in the framework of the PPW electronic structure method, GIPAW aims at calculating the induced magnetic field $\mathbf{B}_{in}$ (see Section 16.2.1) at the nucleus position from the Biot–Savart law as follows:

$$\mathbf{B}_{in}(\mathbf{r}) = \frac{\mu_0}{4\pi} \int \mathbf{j}(\mathbf{r}') \times \frac{(\mathbf{r} - \mathbf{r}')}{|\mathbf{r} - \mathbf{r}'|^3} d^3\mathbf{r}' \tag{16.19}$$

where $\mathbf{j}(\mathbf{r})$ is the induced electronic current density that arises from the application of $\mathbf{B}_0$ to the sample. Consequently, computing the nuclear shielding tensor amounts to computing $\mathbf{j}(\mathbf{r})$. Considering that the effect of $\mathbf{B}_0$ is a small perturbation of the system, linear-response theory is applied to evaluate $\mathbf{B}_{in}$ and $\mathbf{j}(\mathbf{r})$ as first-order quantities: $\mathbf{B}_{in}^{(1)}$ and $\mathbf{j}^{(1)}(\mathbf{r})$. In this framework, the first-order induced current density $\mathbf{j}^{(1)}(\mathbf{r})$ can be written as follows:

$$\mathbf{j}^{(1)}(\mathbf{r}') = 2\sum_n^{occ.} \left[\langle \psi_n^{(0)} | \mathbf{J}^p(\mathbf{r}') | \psi_n^{(1)} \rangle + \langle \psi_n^{(1)} | \mathbf{J}^p(\mathbf{r}') | \psi_n^{(0)} \rangle \right]$$
$$+ 2\sum_n^{occ.} \langle \psi_n^{(0)} | \mathbf{J}^d(\mathbf{r}') | \psi_n^{(0)} \rangle \tag{16.20}$$

where the factor 2 accounts for spin degeneracy and the sum runs over all the occupied orbitals. $\mathbf{J}^p(\mathbf{r}')$ and $\mathbf{J}^d(\mathbf{r}')$ are the paramagnetic and diamagnetic components of the current density operator, respectively. $\psi_n^{(0)}(\mathbf{r})$ (previously, simply written $\psi_n(\mathbf{r})$) and $\psi_n^{(1)}(\mathbf{r})$ are the ground-state and first-order Kohn–Sham orbitals, respectively. Introducing the Green-function operator $\mathcal{G}(\varepsilon_n^{(0)})$ to write the $\psi_n^{(1)}(\mathbf{r})$'s as functions of the $\psi_n^{(0)}(\mathbf{r})$'s and the symmetric gauge for the vector potential $\mathbf{A}(\mathbf{r}) = (\mathbf{B} \times \mathbf{r})/2$, Eq. (16.20) can be rewritten as follows:

$$\mathbf{j}^{(1)}(\mathbf{r}') = \frac{4e}{m_e} \sum_n^{occ.} Re\left[\langle \psi_n^{(0)} | \mathbf{J}^p(\mathbf{r}') \mathcal{G}(\varepsilon_n^{(0)}) (\mathbf{r} - \mathbf{r}') \times \mathbf{p} | \psi_n^{(0)} \rangle \right].\mathbf{B} \tag{16.21}$$

Calculation of $\mathbf{j}^{(1)}(\mathbf{r})$ from Eq. (16.21) and its integration through Eq. (16.19) lead to the nuclear shielding tensor at the nucleus positions. However, this is made complex by several technical issues that are solved within the GIPAW approach. First, the position operator $\mathbf{r}$ does not display the periodicity of the unit cell. This prevents, *a priori*, the use of the $u_{n,\mathbf{k}}(\mathbf{r})$ part of the Bloch functions. One solution to this problem was proposed by Mauri and coworkers within the MPL approach [43,44], the so-called finite wavelength magnetic field, that was further applied to the GIPAW method. Other approaches to solve this issue were proposed, in particular through the use of maximally localized Wannier orbitals as proposed by Sebastiani and Parrinello [46]. The other issues in the evaluation of Eq. (16.21) are specific to the PPW method. Indeed, similar to the EFG tensor, the use of pseudowave functions that have a nonphysical form in the core region has a strong impact on the accuracy of calculated nuclear shielding tensors [138]. In addition, Gregor, Mauri and Car demonstrated that the shielding of the core electrons is chemically insensitive only if the core and valence states are partitioned in a properly gauge-invariant way [45]. Furthermore, using a PAW reconstruction of the pseudowave functions would require an infinitely large number of projectors in order to insure gauge-invariance. In the GIPAW approach, these issues are addressed by introducing a PAW-like field-dependent transformation operator $\mathcal{T}_B$:

$$\mathcal{T}_B = 1 + \sum_{\alpha,j} e^{\frac{ie}{2\hbar}\mathbf{r}\cdot(\mathbf{R}_\alpha \times \mathbf{B}_0)} \left[|\phi_{\alpha,j}\rangle - |\tilde{\phi}_{\alpha,j}\rangle \right] \langle \tilde{p}_{\alpha,j}| e^{-\frac{ie}{2\hbar}\mathbf{r}\cdot(\mathbf{R}_\alpha \times \mathbf{B}_0)} \quad (16.22)$$

These different points represent the main ingredients of the GIPAW approach. For more details about its theoretical developments, the reader can refer to the original and further publications [47,138,139].

Converse Approach

The converse approach developed by Thonhauser and coworkers provides a significantly different methodology to compute nuclear shielding tensors [56,57]. Indeed, when GIPAW aims at computing the current response to an externally applied uniform magnetic field at all nucleus positions, the converse approach derives nuclear shielding tensors from the orbital magnetization induced at a given nucleus by a magnetic point dipole sitting at this particular nucleus and all its periodic image positions, following:

$$\sigma_{i,j} = -\Omega \frac{\partial M_j^{ind}}{\partial m_i} \quad (16.23)$$

where Ω is the volume of the considered unit cell, i and j represent Cartesian coordinates, M_j^{ind} is the induced magnetization along j and m_i the magnetic dipole component along i. All these quantities are defined at a given nucleus position.

Evaluation of the induced magnetization for extended systems needed in Eq. (16.23) is achieved by using recent formulations of the Berry-phase theory of magnetization [140–144]. In contrast to GIPAW, the converse approach does not suffer from gauge-origin problems and, more importantly, does not rely upon linear-response theory which makes it considerably simpler. However, although being general with respect to the electronic-structure formalism it uses, it has only been implemented within PPW DFT computer codes and has thus required a PAW-like reconstruction of the density and wave functions.

In practice, the main difference between these two formalisms is that the converse approach can only be applied to one atom per calculation whereas, from the GIPAW induced magnetic field in reciprocal space, one only needs to perform a Fourier transform at a given nucleus position to evaluate its nuclear shielding tensor. Nevertheless, the simplicity of the converse approach makes it more likely to be used in conjunction with more advanced exchange-correlation functionals such as hybrid functionals which are rather difficult to handle using GIPAW.

Augmented Plane Wave Approach

Laskowski and Blaha have recently developed another formalism to compute nuclear shielding tensors within the LAPW and APW+lo frameworks [61,62]. Following the GIPAW formalism, it aims at calculating $\mathbf{j}^{(1)}(\mathbf{r})$ and the corresponding induced magnetic field at the nucleus positions following Eqs (16.19) and (16.21). In doing so, it resorts to linear-response theory and to the finite wavelength magnetic field approach to overcome the position operator problem in extended systems. However, the difference in basis set between the PPW and the APW methods leads to differences in the numerical evaluation of $\mathbf{j}^{(1)}(\mathbf{r})$ as well as in its integration. In particular, the approach requires to add extra basis functions in the form of local orbitals, the so-called NMR LOs, in order to insure a proper representation of the first-order perturbed wave functions. Although recent, the few applications of this formalism have demonstrated a result quality almost equivalent to that of GIPAW for various nuclei [61,62,145].

16.2.2.4 Other Interactions: J Coupling and Metallic Shift

J Coupling

J coupling, that arises from the interaction of two nuclei k and l can be physically interpreted in terms of a responding nucleus and a perturbing nucleus. One can then consider that nucleus l generates, at the first order, an induced magnetic field $\mathbf{B}_{in}^{(1)}$ at nucleus k (and reciprocally) that is related to the $\overleftrightarrow{\mathbf{J}}_{kl}$ tensor by the following:

$$\mathbf{B}_{in}^{(1)}(\mathbf{r}_k) = \frac{2\pi}{\hbar \gamma_k \gamma_l} \overleftrightarrow{\mathbf{J}}_{kl} \cdot \mu_l \qquad (16.24)$$

Therefore, computing the $\overleftrightarrow{\mathbf{J}}_{kl}$ tensor is equivalent to compute $\mathbf{B}_{in}^{(1)}(\mathbf{r}_k)$ [146,147]. When spin-orbit coupling is negligible, $\mathbf{B}_{in}^{(1)}(\mathbf{r}_k)$ can be expressed as the sum of two independent contributions. First, nucleus l induces an electronic current density $j^{(1)}(r)$ that, in turn, generates an induced magnetic field at nucleus k. This mechanism is somewhat similar to the mechanism of nuclear shielding. Second, nucleus l also generates an induced spin polarization $m^{(1)}(r)$ which also contributes to the induced magnetic field at nucleus k. Joyce and coworkers recently developed an approach that aims at calculating $j^{(1)}(r)$ and $m^{(1)}(r)$ to ultimately calculate $\vec{J}_{kl}$ in periodic systems [52,148]. It is based on the PPW framework and, as GIPAW, also applies a PAW reconstruction to the peudowave functions to reach all-electron accuracy. The method has already been applied to various inorganic compounds such as borates and silicates [75,149,150], as well as to organic solids [151,152]. In both cases the approach has provided unique and valuable information about the atomic origin of the J coupling which is hardly reachable experimentally.

Metallic Shift

D'Avezac and coworkers were the first to propose a general and consistent formalism to compute metallic shifts for periodic systems [53]. This approach developed within the PPW DFT framework consists in splitting the interaction into two contributions. On one hand, there is the orbital shift that results from the interaction of the external magnetic field with the electrons behaving as electric charges. This mechanism is similar to the chemical shift in insulating nonmagnetic compounds. From a theoretical point of view, it amounts to estimate an induced current density $j^{(1)}(r)$ using the same ingredients as those of the GIPAW formalism while taking into account the partial occupation of the electronic bands within first-order perturbation theory [153]. The formalism proposed by d'Avezac et al. describes the Fermi surface using an accurate and efficient smearing technique. Nevertheless, they showed that the quality of the orbital shift description strongly depends on the Fermi surface description, that is, it requires a very large k-point mesh and a careful check of the convergence with respect to the smearing. On the other hand, the spin contribution, or Knight shift, originates from the interaction of the external magnetic field with the electrons behaving as spins. As for the J coupling, its evaluation amounts to estimate a first-order induced magnetization $m^{(1)}(r)$. In a large number of cases, this contribution prevails over the orbital shift.

Very recently, Laskowski and Blaha proposed a similar formalism within the APW framework [154]. Although displaying the same convergence difficulties as the aforementioned method, it has the advantage of explicitly treating the core states, and thus to take into account the core polarization effects that can dominate in some cases. This approach was applied to both alkali, alkaline-earth, transition and noble metals and showed excellent agreement with experimental results.

In the near future, these two approaches should allow routine calculations of NMR shifts in metallic systems to be performed, hopefully providing valuable information to overcome the difficulty of SSNMR of metals.

16.2.3
Considerations about Computational Details

Owing to the development of user-friendly interfaces for quantum chemical softwares and the increase of available computational power and efficiency of algorithms, it becomes easier and easier to perform DFT calculations to compute physical properties of solid-state compounds. Nowadays, if the unit cell of the considered system is not too large, calculations can even be amenable on a desktop computer. However, these tools cannot be considered as black boxes as several requirements have to be fulfilled to obtain meaningful results. The main parameters influencing the quality of NMR parameter calculations and the way to improve this quality are briefly reviewed in the following section.

Basis Set and First Brillouin Zone Sampling

In any numerical quantum chemical simulation, a first requirement is to use a basis set that is extended enough to get reliable results. If localized basis sets such as Gaussian or Slater functions are used, one must increase the number of such functions per atomic orbital till convergence of the computed properties, be they energetic or magnetic. Adding diffuse and polarization functions provides some additional flexibility that can be required depending on the nature of the chemical system. In a PPW approach, the truncation of the plane wave basis set is controlled by a unique cutoff parameter (see Section 16.2.2.1) that represents an upper limit of the kinetic energy of the included plane waves. So by varying this unique parameter it is possible to test the convergence of the results with respect to the basis set extension, which has to be done carefully for each new calculation. The same requirement holds for the **k**-point sampling of the Brillouin zone that has to be fine enough to properly converge any calculated property. In this case, it is the density of this grid that has to be varied to ensure convergence.

Treatment of Relativistic Effects

It is well known from quantum calculations on molecules that taking into account relativistic effects on the treatment of the electronic structure can have significant influence on NMR parameters, in particular for heavy elements [38,155,156]. Similar influence is expected in solid-state systems. Unfortunately, it has only been marginally looked at so far. Actually, in the PAW and GIPAW approaches, scalar relativistic effects are taken into account only during the pseudopotential generation by solving the all-electron scalar-relativistic Koelling–Harmon radial equation instead of the all-electron radial Schrödinger equation [157]. Spin-orbit contribution is not taken into account. Nevertheless, this rather crude approximation provides satisfactory results as one generally looks at differences of chemical shifts which allows error compensation. To further discuss this question, Yates and coworkers proposed a scheme, the ZORA GIPAW method [158], based on the zeroth-order regular approximation (ZORA) [159–161] to explicitly take into account scalar relativistic effects in the augmentation regions without resorting to an explicit relativistic Hamiltonian. Although this approach provided improvements on the calculation of chemical shifts of some heavy nuclei, it has not been further used yet. A similar approach for the calculation of *J*-coupling tensors has been recently proposed by Green and Yates [162]. With the development of experimental NMR characterization of heavy elements, there is little doubt that in the coming years, the ZORA GIPAW method will experience an increasing interest. One can also assume that the APW-derived approach will also help in providing new insights about the influence of relativistic effects in solid-state compounds.

Exchange-Correlation Functional
The effect of electronic correlation on the calculations of NMR parameters has been largely investigated in molecules [37]. As mentioned in Section 16.2.2.1, the choice of the exchange-correlation functional is critical in DFT since an accurate description of most NMR parameters requires inclusion of the electron correlation. Therefore, computations of NMR parameters should benefit from the use of the most advanced functionals, that not only depend on the density, such as self-interaction corrected and hybrid functionals [163,164]. However, it is noteworthy to mention that, as for the calculations of the electronic structure of chemical systems, no universal recipe for the best XC functional can be given. A glance at the literature shows that PAW and GIPAW calculations of NMR parameters have exclusively been performed using GGA functionals as all their actual implementations in computer codes do not allow to use more elaborate ones [50,51]. In particular, the PBE functional [102] has extensively been used due to its well-recognized good performances for a wide range of solid-state systems. However, the GGA weaknesses have been evidenced for the calculation of NMR parameters of solid-state compounds when looking at the linear fit of the correlation between calculated isotropic nuclear shieldings and experimental chemical shifts. Indeed, the $\delta_{iso} = -(\sigma_{iso} - \sigma_{ref})$ relation leads, in an ideal case, to $\delta_{iso}^{exp.} = -\sigma_{iso}^{calc.} + \sigma_{ref}$ where the slope of the linear fit should be -1. It turns out that studies have shown some deviations from this value, its magnitude depending on the nucleus and also on the considered chemical shift range [165–168]. Large deviations are generally obtained for transition metals displaying a variety of oxidation states and a large chemical shift range as shown in Figure 16.5 for ^{95}Mo-containing cluster compounds [167]. Other studies have also evidenced more pathological deficiencies of the GGA functionals as for instance in the case of lanthanum and calcium compounds [169,170]. However, no extensive test has further been performed on this question using GIPAW. We only noted that a study by Laskowski and coworkers performed with the APW+lo method has provided new insights about these deficiencies [171]. By comparing GGA, GGA+U (U = Hubbard parameter which accounts for the energy penalty (repulsion) for occupying 3d or 4f electrons on the same atom) and hybrid functionals, the authors showed for a series of oxide and halide compounds that deviations from a -1 slope originate mostly from the computed GGA band-gap error. This demonstrates that nuclear shielding of solid-state compounds are very sensitive to the accuracy of the computed electronic band structure which indeed strongly depends upon the choice of the exchange-correlation functional.

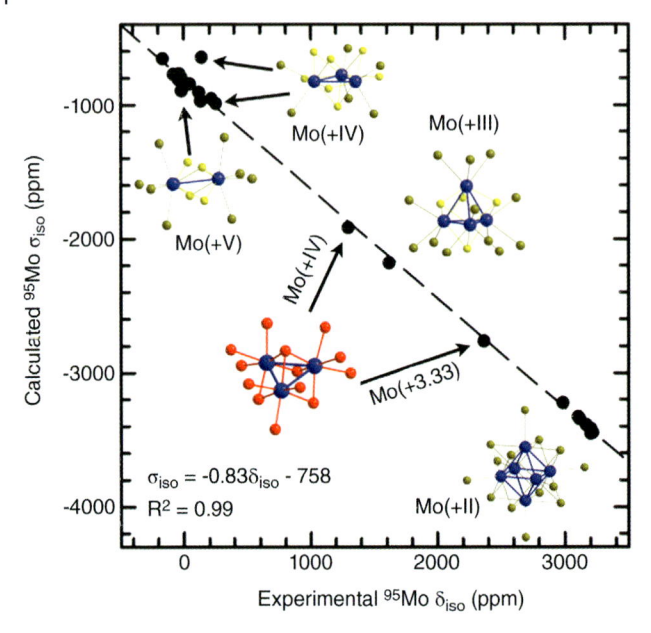

Figure 16.5 Experimental ^{95}Mo isotropic chemical shift versus calculated ^{95}Mo nuclear shielding for various molybdenum cluster compounds in which different molybdenum oxidation states. (Adapted from Ref. [167].)

Geometry

Another key point controlling the quality of the calculated NMR parameters in solid-state, or at least the agreement between theory and experiments, is the geometrical arrangement used for the calculations [50,51]. Indeed, it has been shown that in a large number of cases, even high quality X-ray or neutron diffraction data do not lead to accurate correlation between experimental and calculated results for numerous nuclei [166–168,170,172–177]. Indeed, this is not surprising since experimental structures do not necessarily correspond to energy minima in the computed potential energy surfaces. It is then generally recommended to first accurately optimize the geometry of the system, that is, all atomic positions and/or cell-parameters, prior to performing NMR calculations. This is illustrated in Figure 16.6 for a series of molybdates $AMoO_4$ (A = alkaline earth, Zn, Cd, Pb) and A'_2MoO_4 (A' = alkali, ammonium) [173].

16.3
Illustrative Examples

The approaches described above have been applied to investigate a large variety of topics in chemical crystalline and amorphous systems. The purpose of the following section is to illustrate the use of the theoretical NMR methods, in

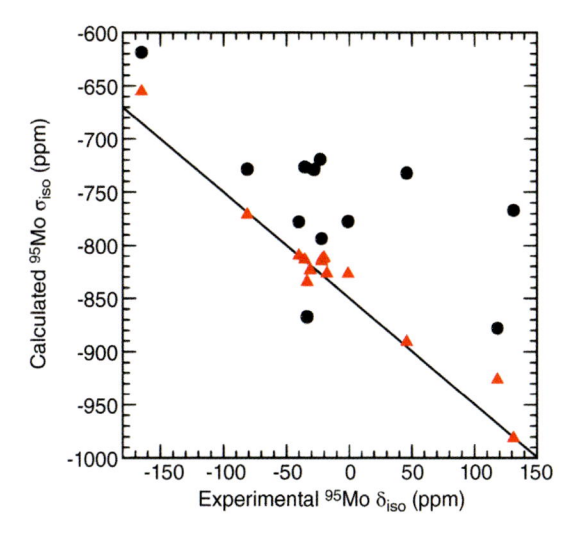

Figure 16.6 Correlation between ^{95}Mo experimental chemical shifts δ_{iso} and calculated isotropic nuclear shieldings σ_{iso} for nonoptimized (black circles) and fully optimized (red triangles) structures of a series of molybdates AMoO$_4$ (A = alkaline-earth, Zn, Cd, Pb) and A'$_2$MoO$_4$ (A' = alkali, ammonium). The solid line represents the perfect agreement regardless of the intercept of the linear model. (Adapted from Ref. [173].)

particular the GIPAW one with experimental examples. There are many solid systems that could serve this purpose and the selection is somewhat personal but that we find particularly educational.

16.3.1
Combining Theory and Experiments for Structural Study of Crystalline Compounds

16.3.1.1 The Coloring Problem in Rare-Earth Metal Boron Carbides

Thirty year ago, Burdett *et al.* introduced the "coloring problem" with the following question: "Given a molecular or extended network and two different types of atoms, we may wonder what is the best way to distribute them in the network for a fixed stoichiometry?" [178,179]. Indeed, the way the atoms are distributed can influence not only to what extent different atoms occur in a compound, that is, stoichiometry, but also the nature of the electronic structure, and therefore physical and chemical properties. This question about the atom site preference termed the "coloring problem" occurs in some classic molecular examples: The simplest case being the linear triatomic carbon dioxide molecule. The explanation for the arrangement OCO and not OOC is simply that using simple theoretical tools such as the Lewis concept the site with the lowest coordination number is preferentially occupied by the more electronegative atom, that is, the site with the higher electron density [180]. This is supported by quantum chemical calculations and comparison of the total energies of both arrangements.

This is not so straightforward in the case of extended networks with symmetry equivalent positions. This is where SSNMR can be useful. Let us approach this problem by looking at the structural arrangement of some solid-state boron carbide compounds for instance. All the MB_2C_2-layered structures (M = alkaline-earth or rare-earth metal) consist of a series of sheets alternately containing M atoms and boron–carbon networks [181,182]. The regain of interest for these compounds is due to their similarity with graphene and carbon derivatives [183]. Several different two-dimensional (2D) B/C topologies have been reported: Fused six-membered rings in MgB_2C_2 [184] and BeB_2C_2 [185], fused five- and seven-membered rings in ScB_2C_2 [186], and two different colorings of fused four- and eight-membered rings in MB_2C_2 (M = Ca or rare-earth metal) (Figure 16.7) [187]. When the metal atom is a heavy element, X-ray diffraction studies are unable to differentiate B (Z = 5) from C (Z = 6) for the different crystallographic sites inside the sheets. For instance, LaB_2C_2 was proposed by Bauer and Bars to adopt Coloring I [188] (Figure 16.7), although this distribution seems to be somewhat in conflict with well-known conclusions from molecular organic chemistry [189]. This was also in contradiction with an earlier study of Smith *et al.* who suggested Coloring II [190]. Later, neutron and X-ray diffraction structures of various lanthanide derivatives were solved assuming the model of Smith *et al.* [190]. Albert *et al.* have shown that CaB_2C_2 also adopts Coloring II, but with a different stacking of the 2D layers [191].

In 1986, Burdett *et al.* carried out a theoretical study at the extended Hückel level [192], to demonstrate that an isolated B/C sheet based on Coloring II is largely preferred than that adopting Coloring I for a formal dianionic charge per B_2C_2. Almost 20 years later, the question of the coloring in these compounds being still open, especially with one more electron per formula unit, that is with a trivalent metal, some of us investigated and revisited the structural and electronic properties of MB_2C_2 (M = Mg, Sc, Ca, Y, Ln) phases with the aid of first-principles calculations [187]. Full geometry optimizations showed that Coloring II is found to be the most stable arrangement regardless of the metal, whether it is divalent or trivalent. They also confirmed that, in the case of a divalent metal,

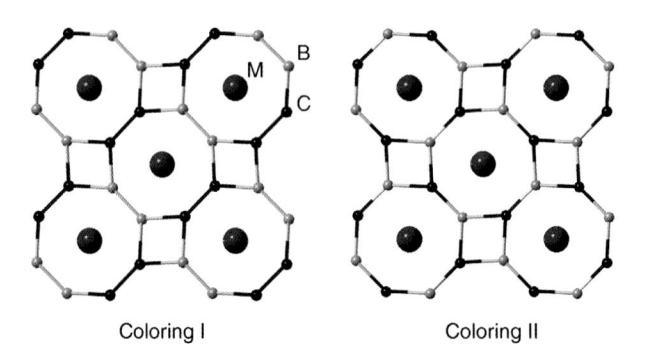

Figure 16.7 Possible arrangements of the MB_2C_2 structures proposed for M = Ca, Y, La, and Ce [187].

Coloring II allows the opening of an energy band gap at the Fermi level. In order to support these results, computed quadrupolar parameters of ^{11}B (I = 3/2) nuclei for both geometry-optimized structural hypotheses were computed at the APW + lo level and compared with those experimentally measured by Magishi et al. [193] who determined the EFG parameters and the components of the hyperfine coupling tensors at ^{11}B nucleus site of CeB$_2$C$_2$ from ^{11}B NMR experiments in order to elucidate its magnetic properties. Computed eigenvalues of the EFG tensor for Coloring II almost perfectly agree with the experimental data. As an illustration, ^{11}B η_Q computed for Coloring II compares well to the measured one (0.28 versus 0.22) whereas the one computed for Coloring I is equal to 0.89 [187].

YB$_2$C is another example of layered compound where boron and carbon atoms form planar networks based this time on fused four- and seven-membered rings with the metal atoms located above and below the heptagons [188,194]. Two of the boron versus carbon distributions among those which may be envisaged are shown in Figure 16.8. Arrangement of the boron atoms in Coloring I is described with two crystallographically independent positions, whereas only one

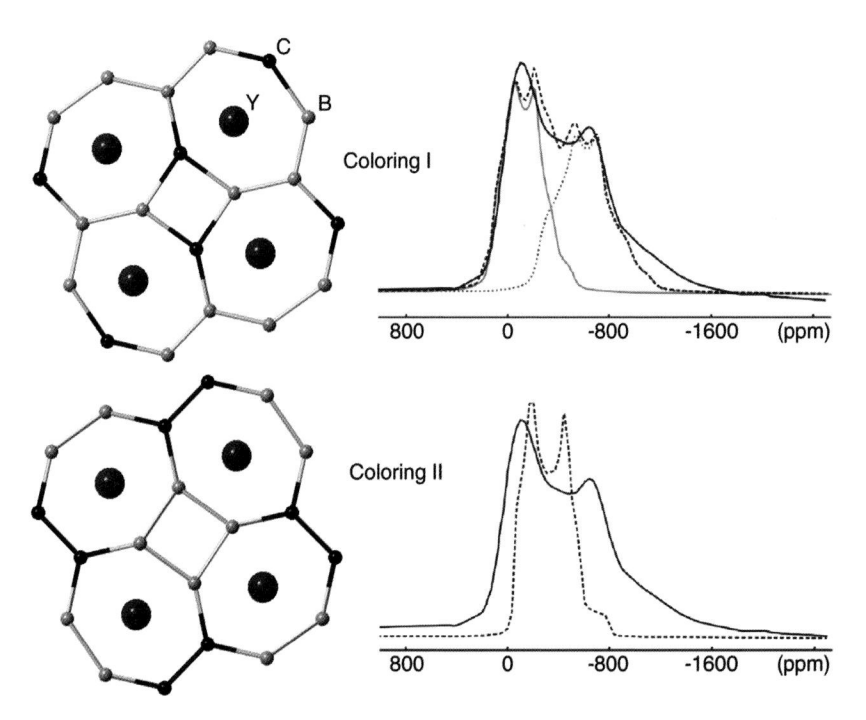

Figure 16.8 Structural arrangements of the experimental (Coloring I, top) and hypothetical (Coloring II, bottom) of YB$_2$C and corresponding simulated (dotted line) ^{11}B NMR spectra. The simulated spectrum for the Coloring I arrangement results from the addition of simulated spectra (light solid and light dotted lines) for the two different crystallographic positions of boron. For comparison, the experimental ^{11}B NMR spectrum is given (solid line) [195].

crystallographic position is required for Coloring II. Although with yttrium in the compound, it is not easy to differentiate between boron and carbon atoms around using X-ray or neutron-diffraction techniques, Coloring I was initially proposed. ^{11}B solid-state NMR experiments were combined to DFT calculations to solve this coloring problem. NMR experiments were performed on YB$_2$C crystalline powder samples under MAS conditions [195]. The quadrupolar coupling constant C_Q and asymmetry parameter η_Q were computed for YB$_2$C for DFT-optimized arrangements with Colorings I and II at the PBE level of theory using the PAW method. Values obtained were used to simulate the ^{11}B NMR spectrum, using the Dmfit program [18] that could be compared to the experimental one, as shown in Figure 16.8. Unambiguously, the simulated spectrum for Coloring I fits better than that simulated for Coloring II. Obviously, the two independent boron atoms in Coloring I arrangement fits well with the two ^{11}B NMR signals of the experimental spectrum, each of them being associated with small quadrupolar coupling constants and forming the major contribution to one of the two discontinuities of the signal. On the other hand, it is not possible to simulate satisfactorily the experimental spectrum using the DFT computed quadrupolar parameters of the unique boron site in the Coloring II structure. This supports X-ray diffraction experiments [196]. Similar investigations were carried out with the same method for LuB$_2$C which adopts the same arrangement [197].

These calculations provide a clear theoretical answer to the coloring problem in these layered boron carbide phases, showing that NMR parameters are very sensitive to atomic local environment, and therefore can easily bring crucial information about the "coloring problem" in solids.

16.3.1.2 Characterization of ^{93}Nb Solid-State NMR Parameters

Niobium can form chemical compounds with many elements from the periodic table. For instance, many inorganic niobates such as LiNbO$_3$, the high temperature form of PbNb$_2$O$_6$, and Pb(Mg$_{2/3}$Nb$_{1/3}$)O$_3$ exhibit ferroelectric or piezoelectric properties and are used in actuators, attenuators, or sensors. Niobium-based oxides also display catalytic properties and are frequently used as a support for catalytically active metals or other metal-oxide catalysts [198]. Metal clusters compounds such as Nb$_6$Cl$_{12}$Cl$_2$(H$_2$O)$_4$ [199], are also largely studied for their catalytic activity for the hydrogenation of alkynes, for instance [200]. Polyoxoniobates have attracted much attention during the past few years due to their unique structures and properties [201]. It is not surprising then that ^{93}Nb SSNMR shows a growing interest for the characterization of the local structure of niobium sites in niobium-containing systems [202–206] even if it has been much less used than some other transition metal nuclei despite the large gyromagnetic ratio and complete natural abundance of ^{93}Nb. This can be attributed to the lack of studies on Nb-containing model compounds which makes challenging the interpretation of experimental data and somewhat limits the routine application of ^{93}Nb SSNMR.

Aiming at establishing an accurate basis for the measurement of ^{93}Nb NMR interaction parameters, Hanna et al. theoretically and experimentally estimated the quadrupolar coupling and chemical shift tensors in some niobate systems where Nb atoms are tetra- or hexa-coordinated [207]. They showed in particular that the chemical shift anisotropy is usually an important and nonnegligible contribution to ^{93}Nb NMR line shapes, and that combined static/MAS approaches with variable applied magnetic field studies must be introduced. They observed a general trend between the ^{93}Nb isotropic chemical shift values and the coordination number of the niobium atoms. Experimental measurements were augmented with DFT calculations using the periodic WIEN2k and GIPAW NMR-CASTEP computer codes as well as the molecular Gaussian computer code to assess the real predictive capability of DFT calculations to estimate quadrupolar interaction parameters, especially the quadrupolar coupling constant for heavier elements such as niobium [207]. An excellent correlation between the computed and measured values of the chemical shift anisotropic parameters is also achieved whereas a modest one is obtained for δ_{iso}. It is worth mentioning that, as expected for crystalline materials, calculations carried out within a cluster approach using the Gaussian code with B3LYP hybrid functional do not perform as well as their periodic counterparts. Surprisingly, the ^{93}Nb NMR interaction parameters computed using the X-ray crystal structures show a better agreement with experimental values than those obtained with CASTEP-optimized crystal structures.

More recently, Papulovskiy et al. also applied the PAW and GIPAW approaches, in conjunction with some spectroscopic measurements, to revisit the ^{93}Nb NMR properties for a series of over fifty ternary niobates where the coordination number of niobium can vary from four to eight [208]. Indeed, these compounds often demonstrate multiple Nb sites often with very large quadrupolar coupling constants, which is difficult to deal with in practice by ^{93}Nb SSNMR, justifying complementary computations. The large set of data obtained allows not only to confirm the relation between the ^{93}Nb isotropic chemical shift values and the niobium coordination number in these oxides, but also to reveal that the magnitude of the chemical shift anisotropy also correlates (not linearly) with the niobium coordination number in NbO$_x$ polyhedra. It turns out that, while four-coordinated niobium sites typically have a small chemical shift span of about 400 ppm, it increases substantially with coordination to 900, 1100, and 1900 ppm for six-, seven-, and eight-coordinated niobium sites, respectively. These observations are of primary interest for future measurements on complex niobium materials. Using this joint experimental and theoretical approach, Papulovskiy et al. also revisited, for instance, the interpretation of the ^{93}Nb spectrum of PNb$_9$O$_{25}$ [208].

The same authors also determined experimental and theoretical ^{93}Nb NMR parameters in polyoxoniobates such as K$_8$Nb$_6$O$_{19}$ and H$_6$Nb$_{10}$O$_{28}$ which contain [Nb$_6$O$_{19}$]$^{8-}$ and [Nb$_{10}$O$_{28}$]$^{6-}$ cluster anions, respectively [208]. These compounds may serve as interesting examples of how an apparent idealized O$_h$ symmetry of Nb sites being disturbed by the presence of terminal oxygen atoms, resulting in niobium species being displaced from the center of the octahedral polyhedron thus severely affecting their ^{93}Nb NMR spectra. Indeed, ^{93}Nb NMR parameters computed with the GIPAW approach for the optimized structure of $K_8Nb_6O_{19}$ are very similar for the three nonequivalent NbO_6 sites, making their ^{93}Nb NMR signals to almost completely overlap. Similarly, the same approach for $H_6Nb_{10}O_{28}$ confirms five nonequivalent NbO_6 sites with all sites characterized by large quadrupolar coupling constants and the nonaxial symmetry of the electric field gradient tensors. This is a striking example of how first-principles calculations can help in understanding complex experimental spectra.

A recent experimental and theoretical study of the ^{93}Nb SSNMR parameters of two niobium halide cluster compounds, namely $[Nb_6Br_{12}(H_2O)_6][HgBr_4].12H_2O$ [209] and $[Nb_6Cl_{12}(OH)_2(H_2O)_4].4H_2O$ [210], both containing the $[Nb_6X_{12}]^{2+}$ diamagnetic cluster core (X = halogen atom), also demonstrates that a three-method combined approach that uses crystal structure, SSNMR measurements and first-principles calculations can provide important chemical information on complex niobium-containing compounds [176].

Examples discussed above clearly show that the PAW and GIPAW computational NMR results can be beneficial for correct interpretation and analysis of complex ^{93}Nb NMR spectra due to several niobium sites in the crystal structures. Of course, such an approach is also valid and useful for other solid-state systems containing NMR receptive heavy nuclei [167,173,211–215].

16.3.2
Structural Information in Amorphous Compounds

As amorphous systems do not display any long-range structural order, the number of experimental techniques to characterize their local arrangement is rather limited. This is why tremendous efforts have been devoted recently to the application of SSNMR to this kind of compounds and, in parallel, to the theoretical estimation of their NMR parameters. As stated by Charpentier and coworkers, "GIPAW represented a major step forward for the investigation of glass structure when combined to molecular dynamics (MD) simulations" [216]. Within this methodology, referred as the MD-GIPAW approach, MD simulations, classical or *ab initio*, are first performed on a crystalline structure, which undergoes a first strong increase of temperature to generate a liquid-like system and then a certain number of decreases of pressure and temperature applying a cooling protocol. This protocol generates a structural model that is representative of a glass atomic structure on which GIPAW or other theoretical calculations can be performed. Among applications of the MD-GIPAW approach, one has to mention numerous studies by Charpentier and coworkers not only on oxide glasses [216] but also on more complex glass systems containing boron, sodium, calcium, aluminum, and so on [217–226], in which a number of empirical correlations between NMR parameters, mainly chemical shifts and quadrupolar couplings, and structural quantities such as bond lengths and angles are provided. An example of such a correlation is given in Figure 16.9 for the study of amorphous silica [227].

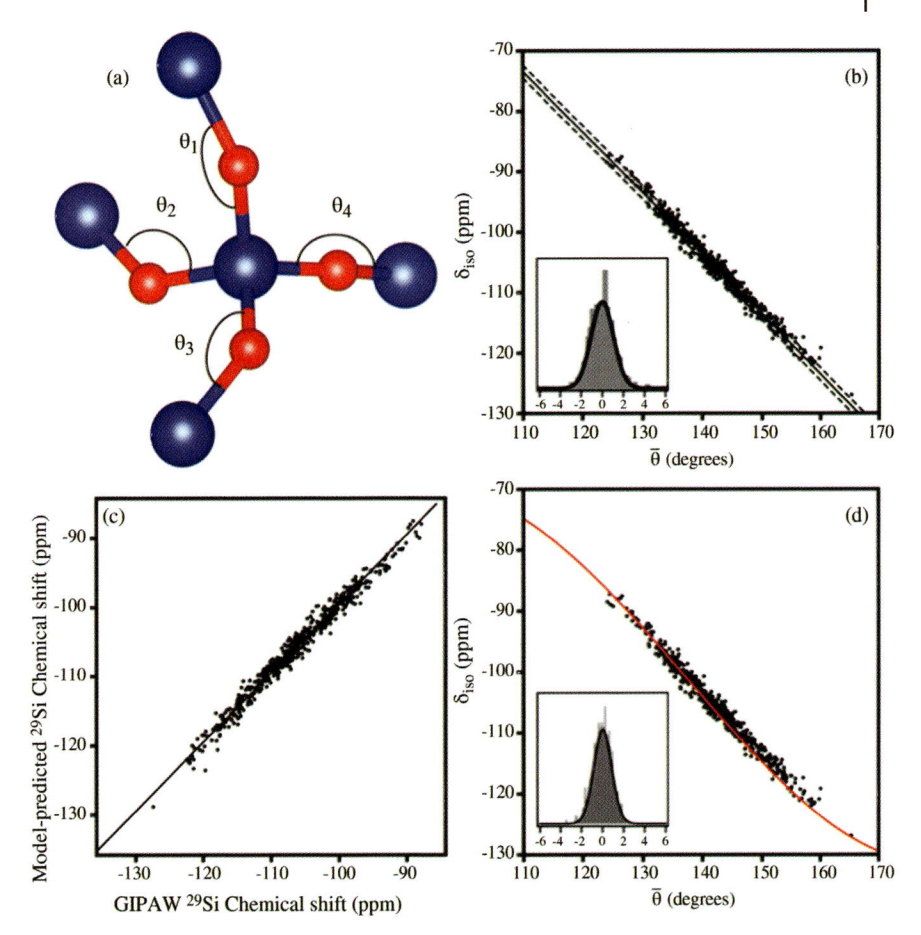

Figure 16.9 A central silicon atom surrounded by four oxygen atoms indicating four different Si–O–Si angles in silica (a). Computed ^{29}Si chemical shifts versus average of the adjacent Si–O–Si angles fitted using a linear correlation function. The inset shows a histogram of the δ_{iso} residuals fitted to a normal distribution with a standard deviation of 1.1 ppm (b). Comparison of the ^{29}Si chemical shifts predicted by the linear model versus values computed by GIPAW (c). The same data used in (b) fitted with a trigonometric function (d). The inset shows a histogram of the δ_{iso} residuals fitted to a normal distribution with standard deviation of 0.9 ppm. (Reprinted with permission from Ref. [227]. Copyright 2014, American Chemical Society.)

In addition to oxides, the field of binary chalcogenide glasses (GeSe$_x$, AsSe$_x$, GeSe$_x$Te$_y$) has also largely benefited from the possibilities given by the MD-GIPAW approach [228–231]. Indeed, the atomic structural arrangement in these systems of interest for their magnificent optical properties, is rather complex as it displays a subtle mixing of chemical environments involving a random network of corner-sharing GeSe$_4$ (or AsSe$_4$) tetrahedral arrangements (similar to the silica network), tetrahedra in edge-sharing configurations, homopolar bonds and most likely also undercoordinated and overcoordinated atoms. This becomes even more complex when adding a third element into the structure such as tellurium. Recently, the MD-GIPAW approach has helped in providing a clearer picture of the structure as well as of the shape of the ^{77}Se NMR signal of such glasses. For example, Bouessel du Bourg *et al.* showed that increasing a fraction of Te atoms in a GeSe$_4$ glass composition leads to a strong modification of the ^{77}Se SSNMR signal, even for a low Te amount [230]. Results indicate that the Te atoms that have a strong tendency to be dispersed within the Se chains, rather than to bind to germanium atoms, are responsible for these changes.

These examples show how this method combined to experimental SSNMR is a powerful tool for structural analysis of disordered and complex compounds, providing insight into the local chemical environment of nuclei complementary to structural information provided by other spectroscopic methods (e.g., IR and Raman). It is worth mentioning however, that the MD-GIPAW approach shows one major limitation which is the size of the systems that can computationally be handled. Recently, a new approach based on a high-dimensional neural-network representation of NMR parameters calculated using an *ab initio* formalism has been proposed [232]. It allows to efficiently and accurately predict the NMR parameters of very large systems and thus provides the opportunity to extend the MD-GIPAW approach to more complex chemical systems.

16.4
Conclusions

Still in infancy 20 years ago, first-principles NMR-DFT methods can nowadays routinely be employed for the determination of local structure, disorder, and even dynamics in the solid state. This is achieved thanks to the constant advance of computing hardware and improvements of codes suitable for studying (large) solids. Computed chemical shift, quadrupolar coupling and *J*-coupling parameters can in turn be exquisitely precise, not only allowing direct and meaningful comparison with experimental data, but also providing valuable and complementary chemical information as demonstrated with the illustrative examples briefly discussed above. Among the different theoretical methods which were developed, the GIPAW approach seems to be nowadays the best compromise between accuracy and computational efficiency for simulating materials made of several up to a few hundreds of atoms per unit cell.

Further developments must comprise the improvement of efficiency of the DFT codes in order to calculate SSNMR parameters from very large molecular dynamics calculations. This would allow to accurately take into account finite temperature effects and to simulate more realistic amorphous materials. In addition, the increasing use of dispersion correction schemes and the introduction of full relativistic corrections, already present in "molecular" codes, should improve the accuracy of SSNMR calculations, especially for compounds displaying weak-interactions and/or containing heavy elements.

We hope that this review has been convincing enough in showing that highly accurate computed chemical information is accessible through the use of relatively affordable computational SSNMR methods and that it will aid the solid-state chemistry community in adding SSNMR computations to their repertoire of techniques for structure elucidation.

References

1. Purcell, E.M., Torrey, H.C., and Pound, R.V. (1946) Resonance absorption by nuclear magnetic moments in a solid. *Phys. Rev.*, **69** (1–2), 37–38.
2. Bloch, F., Hansen, W.W., and Packard, M. (1946) Nuclear induction. *Phys. Rev.*, **69** (3–4), 127.
3. Emsley, J. and Feeney, J. (1995) Milestones in the first fifty years of NMR. *Prog. Nucl. Magn. Reson. Spectrosc.*, **28** (1), 1–9.
4. Emsley, J.W. and Feeney, J. (2007) Forty years of progress in nuclear magnetic resonance spectroscopy. *Prog. Nucl. Magn. Reson. Spectrosc.*, **50** (4), 179–198.
5. Laws, D.D., Bitter, H.M.L., and Jerschow, A. (2002) Solid-state NMR spectroscopic methods in chemistry. *Angew. Chem., Int. Ed.*, **41** (17), 3096–3129.
6. Harris, R.K., Wasylishen, R.E., and Duer, M.J. (eds.) (2009) *NMR Crystallography*, John Wiley & Sons, Ltd., Chichester.
7. Martineau, C. (2014) NMR crystallography: Applications to inorganic materials. *Solid State Nucl. Magn. Reson.*, **63–64**, 1–12.
8. Harris, R.K. (2004) NMR crystallography: the use of chemical shifts. *Solid State Sci.*, **6** (10), 1025–1037.
9. Elena, B., Pintacuda, G., Mifsud, N., and Emsley, L. (2006) Molecular structure determination in powders by NMR crystallography from proton spin diffusion. *J. Am. Chem. Soc.*, **128** (29), 9555–9560.
10. Ripmeester, J.A. and Wasylishen, R.E. (2013) Special issue devoted to NMR crystallography. *CrystEngComm*, **15** (43), 8598.
11. Dybowski, C. and Bai, S. (2008) Solid-state NMR spectroscopy. *Anal. Chem.*, **80** (12), 4295–4300.
12. Massiot, D., Messinger, R.J., Cadars, S., Deschamps, M., Montouillout, V., Pellerin, N., Veron, E., Allix, M., Florian, P., and Fayon, F. (2013) Topological, geometric, and chemical order in materials: Insights from solid-state NMR. *Acc. Chem. Res.*, **46** (9), 1975–1984.
13. Ashbrook, S.E., Dawson, D.M., and Seymour, V.R. (2014) Recent developments in solid-state NMR spectroscopy of crystalline microporous materials. *Phys. Chem. Chem. Phys.*, **16** (18), 8223–8242.
14. MacKenzie, K.J.D. and Smith, M.E. (2002) *Multinuclear Solid-State NMR of Inorganic Materials*, Pergamon, New York.
15. Bakhmutov, V.I. (2011) *Solid-State NMR in Materials Science: Principles and Applications*, CRC Press, Boca Raton.
16. Levitt, M.H. (2009) *Spin Dynamics: Basics of Nuclear Magnetic Resonance*, John Wiley & Sons, Chichester, 2nd edn.
17. Collins, B.T., Fine, S.M., Potenza, J.A., Greenblatt, M., and Tsai, P.P. (1989) Crystallographic and magnetic studies of an unusual Mo_3O_{13} cluster compound, sodium molybdoindate $Na_2In_2Mo_5O_{16}$. *Inorg. Chem.*, **28** (12), 2444–2447.
18. Massiot, D., Fayon, F., Capron, M., King, I., Le Calvé, S., Alonso, B., Durand, J.O., Bujoli, B., Gan, Z., and Hoatson, G. (2002) Modelling one- and two-dimensional solid-state NMR spectra. *Magn. Reson. Chem.*, **40** (1), 70–76.
19. Andrew, E.R. (1971) The narrowing of NMR spectra of solids by high-speed specimen rotation and the resolution of chemical shift and spin multiplet structures for solids. *Prog. Nucl. Magn. Reson. Spectrosc.*, **8** (1), 1–39.

20 Andrew, E.R., Bradbury, A., and Eades, R.G. (1959) Removal of dipolar broadening of nuclear magnetic resonance spectra of solids by specimen rotation. *Nature*, **183** (4678), 1802–1803.

21 Frydman, L. and Harwood, J.S. (1995) Isotropic spectra of half-integer quadrupolar spins from bidimensional magic-angle spinning NMR. *J. Am. Chem. Soc.*, **117** (19), 5367–5368.

22 Medek, A., Harwood, J.S., and Frydman, L. (1995) Multiple-quantum magic-angle spinning NMR: a new method for the study of quadrupolar nuclei in solids. *J. Am. Chem. Soc.*, **117** (51), 12779–12787.

23 Fernandez, C. and Amoureux, J.-P. (1996) Triple-quantum MAS-NMR of quadrupolar nuclei. *Solid State Nucl. Magn. Reson.*, **5** (4), 315–321.

24 Amoureux, J.-P., Fernandez, C., and Steuernagel, S. (1996) Z filtering in MQMAS NMR. *J. Magn. Reson. A*, **123** (1), 116–118.

25 Chmelka, B.F., Mueller, K.T., Pines, A., Stebbins, J., Wu, Y., and Zwanziger, J.W. (1989) Oxygen-17 NMR in solids by dynamic-angle spinning and double rotation. *Nature*, **339** (6219), 42–43.

26 Farnan, I., Grandinetti, P.J., Baltisberger, J.H., Stebbins, J.F., Werner, U., Eastman, M.A., and Pines, A. (1992) Quantification of the disorder in network-modified silicate glasses. *Nature*, **358** (6381), 31–35.

27 Stevens, R.M., Pitzer, R.M., and Lipscomb, W.N. (1963) Perturbed Hartree–Fock calculations. I. Magnetic susceptibility and shielding in the LiH molecule. *J. Chem. Phys.*, **38** (2), 550–560.

28 Ditchfield, R. (1972) Molecular orbital theory of magnetic shielding and magnetic susceptibility. *J. Chem. Phys.*, **56** (11), 5688–5691.

29 Wolinski, K., Hinton, J.F., and Pulay, P. (1990) Efficient implementation of the gauge-independent atomic orbital method for NMR chemical shift calculations. *J. Am. Chem. Soc.*, **112** (23), 8251–8260.

30 Van Wullen, C. and Kutzelnigg, W. (1996) Calculation of nuclear magnetic resonance shieldings and magnetic susceptibilities using multiconfiguration Hartree–Fock wave functions and local gauge origins. *J. Chem. Phys.*, **104** (6), 2330–2340.

31 Schindler, M. and Kutzelnigg, W. (1982) Theory of magnetic susceptibilities and NMR chemical shifts in terms of localized quantities. II. Application to some simple molecules. *J. Chem. Phys.*, **76** (4), 1919–1933.

32 Malkin, V.G., Malkina, O.L., and Salahub, D.R. (1993) Calculations of NMR shielding constants beyond uncoupled density functional theory. IGLO approach. *Chem. Phys. Lett.*, **204** (1–2), 87–95.

33 Malkin, V., Malkina, O., and Salahub, D.R. (1993) Calculations of NMR shielding constants by uncoupled density functional theory. *Chem. Phys. Lett.*, **204** (1–2), 80–86.

34 Keith, T.A. and Bader, R.F.W. (1992) Calculation of magnetic response properties using atoms in molecules. *Chem. Phys. Lett.*, **194** (1–2), 1–8.

35 Keith, T.A. and Bader, R.F.W. (1993) Calculation of magnetic response properties using a continuous set of gauge transformations. *Chem. Phys. Lett.*, **210** (1–3), 223–231.

36 Helgaker, T., Jaszuński, M., and Ruud, K. (1999) Ab initio methods for the calculation of NMR shielding and indirect spin-spin coupling constants. *Chem. Rev.*, **99** (1), 293–352.

37 Kaupp, M., Bühl, M., and Malkin, V.G. (2004) *Calculation of NMR and EPR Parameters: Theory and Applications*, Wiley-VCH Verlag GmbH, Weinheim.

38 Autschbach, J. (2004) The calculation of NMR parameters in transition metal complexes. *Struct. Bond.*, **112**, 1–48.

39 Bühl, M. and van Mourik, T. (2011) NMR spectroscopy: quantum-chemical calculations. *WIREs Comput. Mol. Sci.*, **1** (4), 634–647.

40 Lodewyk, M.W., Siebert, M.R., and Tantillo, D.J. (2012) Computational prediction of ^{1}H and ^{13}C chemical shifts: a useful tool for natural product, mechanistic, and synthetic organic chemistry. *Chem. Rev.*, **112** (3), 1839–1862.

41 Helgaker, T., Coriani, S., Jorgensen, P., Kristensen, K., Olsen, J., and Ruud, K. (2012) Recent advances in wave function-based methods of molecular-property calculations. *Chem. Rev.*, **112** (1), 543–631.

42 de Dios, A.C. and Jameson, C.J. (2012) Recent advances in nuclear shielding calculations. *Annual Reports on NMR Spectroscopy*, Elsevier, vol. 77, pp. 1–80.

43 Mauri, F. and Louie, S.G. (1996) Magnetic susceptibility of insulators from first principles. *Phys. Rev. Lett.*, **76** (22), 4246–4249.

44 Mauri, F., Pfrommer, B.G., and Louie, S.G. (1996) Ab initio theory of NMR chemical shifts in solids and liquids. *Phys. Rev. Lett.*, **77** (26), 5300–5303.

45 Gregor, T., Mauri, F., and Car, R. (1999) A comparison of methods for the calculation of NMR chemical shifts. *J. Chem. Phys.*, **111** (5), 1815–1822.

46 Sebastiani, D. and Parrinello, M. (2001) A new ab initio approach for NMR chemical shifts in periodic systems. *J. Phys. Chem. A*, **105** (10), 1951–1958.

47 Pickard, C.J. and Mauri, F. (2001) All-electron magnetic response with pseudopotentials: NMR chemical shifts. *Phys. Rev. B*, **63** (24), 245101.

48 Van de Walle, C.G. and Blöchl, P.E. (1993) First-principles calculations of hyperfine parameters. *Phys. Rev. B*, **47** (8), 4244–4255.

49 Blöchl, P.E. (1994) Projector augmented-wave method. *Phys. Rev. B*, **50** (24), 17953–17979.

50 Thibault, C. (2011) The PAW-GIPAW approach for computing NMR parameters: a new dimension added to NMR study of solids. *Solid State Nucl. Magn. Reson.*, **40** (1), 1–20.

51 Bonhomme, C., Gervais, C., Babonneau, F., Coelho, C., Pourpoint, F., Azaï, T., Ashbrook, S.E., Griffin, J.M., Yates, J.R., Mauri, F., and Pickard, C.J. (2012) First-principles calculation of NMR parameters using the gauge including projector augmented wave method: a chemist's point of view. *Chem. Rev.*, **112** (11), 5733–5779.

52 Joyce, S.A., Yates, J.R., Pickard, C.J., and Mauri, F. (2007) A first principles theory of nuclear magnetic resonance J-coupling in solid-state systems. *J. Chem. Phys.*, **127** (20), 204107.

53 d'Avezac, M., Marzari, N., and Mauri, F. (2007) Spin and orbital magnetic response in metals: susceptibility and NMR shifts. *Phys. Rev. B*, **76** (16), 165122.

54 Payne, M.C., Teter, M.P., Allan, D.C., Arias, T.A., and Joannopoulos, J.D. (1992) Iterative minimization techniques for ab initio total-energy calculations: molecular dynamics and conjugate gradients. *Rev. Mod. Phys.*, **64** (4), 1045–1097.

55 Martin, R.M. (2004) *Electronic Structure: Basic Theory and Practical Methods*, Cambridge University Press, Cambridge.

56 Thonhauser, T., Ceresoli, D., Mostofi, A.A., Marzari, N., Resta, R., and Vanderbilt, D. (2009) A converse approach to the calculation of NMR shielding tensors. *J. Chem. Phys.*, **131** (10), 101101.

57 Ceresoli, D., Marzari, N., Lopez, M.G., and Thonhauser, T. (2010) Ab initio converse NMR approach for pseudopotentials. *Phys. Rev. B*, **81** (18), 184424.

58 Singh, D.J. and Nordström, L. (2006) *Planewaves, Pseudopotentials and the LAPW Methods*, Springer, New York, 2nd edn.

59 Sjöstedt, E., Nordström, L., and Singh, D. (2000) An alternative way of linearizing the augmented plane-wave method. *Solid State Commun.*, **114** (1), 15–20.

60 Madsen, G.K.H., Blaha, P., Schwarz, K., Sjöstedt, E., and Nordström, L. (2001) Efficient linearization of the augmented plane-wave method. *Phys. Rev. B*, **64** (19), 195134.

61 Laskowski, R. and Blaha, P. (2012) Calculations of NMR chemical shifts with APW-based methods. *Phys. Rev. B*, **85** (3), 035132.

62 Laskowski, R. and Blaha, P. (2014) Calculating NMR chemical shifts using the augmented plane-wave method. *Phys. Rev. B*, **89** (1), 014402.

63 Dudenko, D.V., Williams, P.A., Hughes, C.E., Antzutkin, O.N., Velaga, S.P., Brown, S.P., and Harris, K.D.M. (2013) Exploiting the synergy of powder X-ray diffraction and solid-state NMR

spectroscopy in structure determination of organic molecular solids. *J. Phys. Chem. C*, **117** (23), 12258–12265.

64 Reddy, G.M., Cook, D.S., Iuga, D., Walton, R.I., Marsh, A., and Brown, S.P. (2015) An NMR crystallography study of the hemihydrate of 2′,3′-O-isopropylidineguanosine. *Solid State Nucl. Magn. Reson.*, **65**, 41–48.

65 Vaara, J., Jokisaari, J., Wasylishen, R.E., and Bryce, D.L. (2002) Spin-spin coupling tensors as determined by experiment and computational chemistry. *Prog. Nucl. Magn. Reson. Spectrosc.*, **41** (3–4), 233–304.

66 Haeberlen, U. (1976) *High Resolution NMR in Solids. Selective Averaging*, Academic Press, New York.

67 Wasylishen, R.E., Ashbrook, S.E., and Wimperis, S. (eds.). (2012) *NMR of Quadrupolar Nuclei in Solid Materials*, John Wiley & Sons, Ltd., Chichester.

68 Ashbrook, S.E. and Sneddon, S. (2014) New methods and applications in solid-state NMR spectroscopy of quadrupolar nuclei. *J. Am. Chem. Soc.*, **136** (44), 15440–15456.

69 Massiot, D., Fayon, F., Deschamps, M., Cadars, S., Florian, P., Montouillout, V., Pellerin, N., Hiet, J., Rakhmatullin, A., and Bessada, C. (2010) Detection and use of small J-couplings in solid-state NMR experiments. *C. R. Chim.*, **13** (1–2), 117–129.

70 Fayon, F., King, I.J., Harris, R.K., Gover, R.K.B., Evans, J.S.O., and Massiot, D. (2003) Characterization of the room-temperature structure of SnP_2O_7 by ^{31}P through-space and through-bond NMR correlation spectroscopy. *Chem. Mater.*, **15** (11), 2234–2239.

71 Cadars, S., Lesage, A., Hedin, N., Chmelka, B.F., and Emsley, L. (2006) Selective NMR measurements of homonuclear scalar couplings in isotopically enriched solids. *J. Phys. Chem. B*, **110** (34), 16982–16991.

72 Coelho, C., Azaïs, T., Bonhomme-Coury, L., Maquet, J., Massiot, D., and Bonhomme, C. (2006) Application of the MAS-J-HMQC experiment to a new pair of nuclei ^{29}Si, ^{31}P: $Si_5O(PO_4)_6$ and SiP_2O_7 polymorphs. *J. Magn. Reson.*, **179** (1), 114–119.

73 Coelho, C., Azaïs, T., Bonhomme-Coury, L., Laurent, G., and Bonhomme, C. (2007) Efficiency of the refocused ^{31}P-^{29}Si MAS-J-INEPT NMR experiment for the characterization of silicophosphate crystalline phases and amorphous gels. *Inorg. Chem.*, **46** (4), 1379–1387.

74 Cadars, S., Lesage, A., Trierweiler, M., Heux, L., and Emsley, L. (2007) NMR measurements of scalar-coupling distributions in disordered solids. *Phys. Chem. Chem. Phys.*, **9** (1), 92–103.

75 Florian, P., Fayon, F., and Massiot, D. (2009) 2J Si-O-Si scalar spin-spin coupling in the solid state: Crystalline and glassy wollastonite $CaSiO_3$. *J. Phys. Chem. C*, **113** (6), 2562–2572.

76 Bonhomme, C., Gervais, C., Coelho, C., Pourpoint, F., Azaï, T., Bonhomme-Coury, L., Babonneau, F., Jacob, G., Ferrari, M., Canet, D., Yates, J.R., Pickard, C.J., Joyce, S.A., Mauri, F., and Massiot, D. (2010) New perspectives in the PAW/GIPAW approach: J_{P-O-Si} coupling constants, antisymmetric parts of shift tensors and NQR predictions. *Magn. Reson. Chem.*, **48** (S1), S86–S102.

77 Thureau, P., Mollica, G., Ziarelli, F., and Viel, S. (2013) Selective measurements of long-range homonuclear J-couplings in solid-state NMR. *J. Magn. Reson.*, **231**, 90–94.

78 Viger-Gravel, J., Meyer, J.E., Korobkov, I., and Bryce, D.L. (2014) Probing halogen bonds with solid-state NMR spectroscopy: observation and interpretation of $J(^{77}Se,^{31}P)$ coupling in halogen-bonded P = Se I motifs. *CrystEngComm*, **16** (31), 7285–7297.

79 Sanz Camacho, P., Athukorala Arachchige, K.S., Slawin, A.M.Z., Green, T.F.G., Yates, J.R., Dawson, D.M., Woollins, J.D., and Ashbrook, S.E. (2015) Unusual intermolecular "through-space" J couplings in P–Se heterocycles. *J. Am. Chem. Soc.*, **137** (19), 6172–6175.

80 Perras, F.A. and Bryce, D.L. (2013) Symmetry-amplified J splittings for quadrupolar spin pairs: A solid-state NMR probe of homoatomic covalent

81 Perras, F.A. and Bryce, D.L. (2014) Boron–boron *J* coupling constants are unique probes of electronic structure: A solid-state NMR and molecular orbital study. *Chem. Sci.*, **5** (6), 2428–2437.

82 Garaga, M.N., Hsieh, M.F., Nour, Z., Deschamps, M., Massiot, D., Chmelka, B.F., and Cadars, S. (2015) Local environments of boron heteroatoms in non-crystalline layered borosilicates. *Phys. Chem. Chem. Phys.*, **17** (33), 21664–21682.

83 Ba, Y. and Ripmeester, J.A. (2002) ^{27}Al and ^{7}Li spin-echo and spin-echo double resonance NMR applications to the study of lithium depletion and surface oxidation in Al-Li alloys. *Magn. Reson. Chem.*, **40** (1), 81–86.

84 Walstedt, R.E. (2008) *The NMR Probe of High-Tc Materials*, 228, Springer, Berlin.

85 Langer, T., Dupke, S., Eckert, H., Matar, S.F., Winter, M., and Pöttgen, R. (2012) ^{7}Li and ^{29}Si solid state NMR and chemical bonding of $La_2Li_2Si_3$. *Solid State Sci.*, **14** (3), 367–374.

86 Hoting, C., Eckert, H., Haarmann, F., Winter, F., and Pöttgen, R. (2014) Equiatomic intermetallic compounds YTX (T = Ni, Ir; X = Si, Ge, Sn, Pb): a systematic study by ^{89}Y solid state NMR and ^{119}Sn Mössbauer spectroscopy. *Dalton Trans.*, **43** (21), 7860–7867.

87 Yang, J., Tang, Z.T., Cao, G.H., and Zheng, G.Q. (2015) Ferromagnetic spin fluctuation and unconventional superconductivity in $Rb_2Cr_3As_3$ revealed by ^{75}As NMR and NQR. *Phys. Rev. Lett.*, **115** (14), 147002.

88 Sirusi, A.A., Ballikaya, S., Uher, C., and Ross, J.H. (2015) Low-temperature structure and dynamics in Cu_2Se. *J. Phys. Chem. C*, **119** (35), 20293–20298.

89 Sirusi, A.A., Ross, J.H., Yan, X., and Paschen, S. (2015) NMR study of $Ba_8Cu_5Si_xGe_{41-x}$ clathrate semiconductors. *Phys. Chem. Chem. Phys.*, **17** (26), 16991–16996.

90 Parr, R.G. and Yang, W. (1994) *Density-Functional Theory of Atoms and Molecules*, Oxford University Press, New York.

91 Hohenberg, P. and Kohn, W. (1964) Inhomogeneous electron gas. *Phys. Rev.*, **136** (3B), B864–B871.

92 Kohn, W. and Sham, L.J. (1965) Self-consistent equations including exchange and correlation effects. *Phys. Rev.*, **140** (4A), A1133–A1138.

93 von Barth, U. and Hedin, L. (1972) A local exchange-correlation potential for the spin polarized case. *J. Phys. C Solid State Phys.*, **5** (13), 1629–1642.

94 Vosko, S.H., Wilk, L., and Nusair, M. (1980) Accurate spin-dependent electron liquid correlation energies for local spin density calculations: A critical analysis. *Can. J. Phys.*, **58** (8), 1200–1211.

95 Ceperley, D.M. and Alder, B.J. (1980) Ground state of the electron gas by a stochastic method. *Phys. Rev. Lett.*, **45** (7), 566–569.

96 Perdew, J.P. and Zunger, A. (1981) Self-interaction correction to density-functional approximations for many-electron systems. *Phys. Rev. B*, **23** (10), 5048–5079.

97 von Barth, U. (2004) Basic density-functional theory-an overview. *Phys. Scr.*, **T109**, 9–39.

98 Langreth, D.C. and Mehl, M.J. (1983) Beyond the local-density approximation in calculations of ground-state electronic properties. *Phys. Rev. B*, **28** (4), 1809–1834.

99 Perdew, J.P. and Yue, W. (1986) Accurate and simple density functional for the electronic exchange energy: Generalized gradient approximation. *Phys. Rev. B*, **33** (12), 8800–8802.

100 Lee, C., Yang, W., and Parr, R.G. (1988) Development of the Colle-Salvetti correlation-energy formula into a functional of the electron density. *Phys. Rev. B*, **37** (2), 785–789.

101 Becke, A.D. (1988) Density-functional exchange-energy approximation with correct asymptotic behavior. *Phys. Rev. A At. Mol. Opt. Phys.*, **38** (6), 3098–3100.

102 Perdew, J.P., Burke, K., and Ernzerhof, M. (1996) Generalized gradient approximation made simple. *Phys. Rev. Lett.*, **77** (18), 3865–3868.

103 Perdew, J.P. and Burke, K. (1996) Comparison shopping for a gradient-corrected density functional. *Int. J. Quantum Chem.*, **57** (3), 309–319.

104 Perdew, J.P., Kurth, S., Zupan, A., and Blaha, P. (1999) Accurate density functional with correct formal properties: A step beyond the generalized gradient approximation. *Phys. Rev. Lett.*, **82** (12), 2544–2547.

105 Tao, J., Perdew, J.P., Staroverov, V.N., and Scuseria, G.E. (2003) Climbing the density functional ladder: Nonempirical meta-generalized gradient approximation designed for molecules and solids. *Phys. Rev. Lett.*, **91** (14), 146401.

106 Perdew, J.P., Tao, J., Staroverov, V.N., and Scuseria, G.E. (2004) Meta-generalized gradient approximation: Explanation of a realistic nonempirical density functional. *J. Chem. Phys.*, **120** (15), 6898–6911.

107 Perdew, J.P., Ruzsinszky, A., Csonka, G.I., Constantin, L.A., and Sun, J. (2009) Workhorse semilocal density functional for condensed matter physics and quantum chemistry. *Phys. Rev. Lett.*, **103** (2), 026403.

108 Zhao, Y., Schultz, N.E., and Truhlar, D.G. (2005) Exchange-correlation functional with broad accuracy for metallic and nonmetallic compounds, kinetics, and noncovalent interactions. *J. Chem. Phys.*, **123** (16), 161103.

109 Zhao, Y. and Truhlar, D.G. (2008) Density functionals with broad applicability in chemistry. *Acc. Chem. Res.*, **41** (2), 157–167.

110 Zhao, Y. and Truhlar, D.G. (2008) Construction of a generalized gradient approximation by restoring the density-gradient expansion and enforcing a tight Lieb-Oxford bound. *J. Chem. Phys.*, **128** (18), 184109.

111 Zahariev, F., Leang, S.S., and Gordon, M.S. (2013) Functional derivatives of meta-generalized gradient approximation (meta-GGA) type exchange-correlation density functionals. *J. Chem. Phys.*, **138** (24), 244108.

112 Becke, A.D. (1993) A new mixing of Hartree–Fock and local density-functional theories. *J. Chem. Phys.*, **98** (2), 1372–1377.

113 Becke, A.D. (1993) Density functional thermochemistry. III. The role of exact exchange. *J. Chem. Phys.*, **98** (7), 5648–5652.

114 Adamo, C. and Barone, V. (1999) Toward reliable density functional methods without adjustable parameters: The PBE0 model. *J. Chem. Phys.*, **110** (13), 6158–6170.

115 Jaramillo, J., Scuseria, G.E., and Ernzerhof, M. (2003) Local hybrid functionals. *J. Chem. Phys.*, **118** (3), 1068–1073.

116 Monkhorst, H.J. and Pack, J.D. (1976) Special points for Brillouin-zone integrations. *Phys. Rev. B*, **13** (12), 5188–5192.

117 Zunger, A. and Cohen, M.L. (1978) First-principles nonlocal-pseudopotential approach in the density-functional formalism: Development and application to atoms. *Phys. Rev. B*, **18** (10), 5449.

118 Hamann, D.R., Schlüter, M., and Chiang, C. (1979) Norm-conserving pseudopotentials. *Phys. Rev. Lett.*, **43** (20), 1494–1497.

119 Kleinman, L. and Bylander, D.M. (1982) Efficacious form for model pseudopotentials. *Phys. Rev. Lett.*, **48** (20), 1425–1428.

120 Vanderbilt, D. (1985) Optimally smooth norm-conserving pseudopotentials. *Phys. Rev. B*, **32** (12), 8412–8416.

121 Hamann, D.R. (1989) Generalized norm-conserving pseudopotentials. *Phys. Rev. B*, **40** (5), 2980–2987.

122 Vanderbilt, D. (1990) Soft self-consistent pseudopotentials in a generalized eigen value formalism. *Phys. Rev. B*, **41** (11), 7892–7895.

123 Troullier, N. and Martins, J.L. (1991) Efficient pseudopotentials for plane-wave calculations. *Phys. Rev. B*, **43** (3), 1993–2006.

124 Blaha, P., Schwarz, K., Madsen, G.K.H., Kvasnicka, D., and Luitz, J. (2001) *WIEN2k, An Augmented Plane Wave Plus Local Orbitals Program for Calculating Crystal Properties*, Vienna University of Technology, Austria.

125 Profeta, M., Mauri, F., and Pickard, C.J. (2003) Accurate first principles prediction of ^{17}O NMR parameters in

SiO$_2$: Assignment of the zeolite ferrierite spectrum. *J. Am. Chem. Soc.*, **125** (2), 541–548.

126 Petrilli, H.M., Blöchl, P.E., Blaha, P., and Schwarz, K. (1998) Electric-field-gradient calculations using the projector augmented wave method. *Phys. Rev. B*, **57** (23), 14690–14697.

127 Zwanziger, J.W. and Torrent, M. (2008) First-principles calculation of electric field gradients in metals, semiconductors, and insulators. *Appl. Magn. Reson.*, **33** (4), 447–456.

128 Clark, S.J., Segall, M.D., Pickard, C.J., Hasnip, P.J., Probert, M.J., Refson, K., and Payne, M.C. (2005) First principles methods using CASTEP. *Z. Kristallogr.*, **220**, 567–570.

129 Blaha, P., Schwarz, K., and Herzig, P. (1985) First-principles calculation of the electric field gradient of Li$_3$N. *Phys. Rev. Lett.*, **54** (11), 1192–1195.

130 Blaha, P., Schwarz, K., and Dederichs, P.H. (1988) First-principles calculation of the electric-field gradient in hcp metals. *Phys. Rev. B*, **37** (6), 2792–2796.

131 Singh, D.J., Schwarz, K., and Blaha, P. (1992) Electric-field gradients in YBa$_2$Cu$_3$O$_7$: discrepancy between experimental and local-density-approximation charge distributions. *Phys. Rev. B*, **46** (9), 5849–5852.

132 Bryant, P.L., Harwell, C.R., Wu, K., Fronczek, F.R., Hall, R.W., and Butler, L.G. (1999) Single-crystal ^{27}Al NMR of andalusite and calculated electric field gradients: the first complete NMR assignment for a 5-coordinate aluminum site. *J. Phys. Chem. A*, **103** (27), 5246–5252.

133 Jäger, B., Paluch, S., Wolf, W., Herzig, P., Zogal, O.J., Shitsevalova, N., and Paderno, Y. (2004) Characterization of the electronic properties of YB$_4$ and YB$_6$ using ^{11}B NMR and first-principles calculations. *J. Alloys Compd.*, **383** (1–2), 232–238.

134 Laskowski, R., Madsen, G.K.H., Blaha, P., and Schwarz, K. (2004) Magnetic structure and electric-field gradients of uranium dioxide: An *ab initio* study. *Phys. Rev. B*, **69** (14), 140408.

135 Blaha, P., Schwarz, K., and Novák, P. (2005) Electric field gradients in cuprates: Does LDA+U give the correct charge distribution? *Int. J. Quantum Chem.*, **101** (5), 550–556.

136 Hansen, M.R., Madsen, G.K.H., Jakobsen, H.J., and Skibsted, J. (2005) Refinement of borate structures from ^{11}B MAS NMR spectroscopy and density functional theory calculations of ^{11}B electric field gradients. *J. Phys. Chem. A*, **109** (9), 1989–1997.

137 Lo, A.Y.H., Hanna, J.V., and Schurko, R.W. (2007) A theoretical study of ^{51}V electric field gradient tensors in pyrovanadates and metavanadates. *Appl. Magn. Reson.*, **32** (4), 691–708.

138 Yates, J.R., Pickard, C.J., and Mauri, F. (2007) Calculation of NMR chemical shifts for extended systems using ultrasoft pseudopotentials. *Phys. Rev. B*, **76** (2), 024401.

139 Yates, J.R. and Pickard, C.J. (2009) *NMR Crystallography*, John Wiley & Sons, Ltd., Chichester, chap. 3, pp. 29–40.

140 Resta, R., Ceresoli, D., Thonhauser, T., and Vanderbilt, D. (2005) Orbital magnetization in extended systems. *ChemPhysChem*, **6** (9), 1815–1819.

141 Thonhauser, T., Ceresoli, D., Vanderbilt, D., and Resta, R. (2005) Orbital magnetization in periodic insulators. *Phys. Rev. Lett.*, **95** (13), 137205.

142 Xiao, D., Shi, J., and Niu, Q. (2005) Berry phase correction to electron density of states in solids. *Phys. Rev. Lett.*, **95** (13), 137204.

143 Ceresoli, D., Thonhauser, T., Vanderbilt, D., and Resta, R. (2006) Orbital magnetization in crystalline solids: Multi-band insulators, chern insulators, and metals. *Phys. Rev. B*, **74** (2), 024408.

144 Ceresoli, D., Gerstmann, U., Seitsonen, A.P., and Mauri, F. (2010) First-principles theory of orbital magnetization. *Phys. Rev. B*, **81** (6), 060409.

145 Zhou, B., Michaelis, V.K., Yao, Y., Sherriff, B.L., Kroeker, S., and Pan, Y. (2014) Density functional theory study of the magnetic shielding mechanism for ^{11}B in pentaborate minerals: Ulexite and probertite. *CrystEngComm*, **16** (45), 10418–10427.

146 Ramsey, N.F. and Purcell, E.M. (1952) Interactions between nuclear spins in molecules. *Phys. Rev.*, **85** (1), 143–144.

147 Ramsey, N.F. (1953) Electron coupled interactions between nuclear spins in molecules. *Phys. Rev.*, **91** (2), 303–307.

148 Yates, J.R. (2010) Prediction of NMR *J*-coupling in solids with the planewave pseudopotential approach. *Magn. Reson. Chem.*, **48** (S1), S23–S31.

149 Griffin, J.M., Yates, J.R., Berry, A.J., Wimperis, S., and Ashbrook, S.E. (2010) High-resolution ^{19}F MAS NMR spectroscopy: Structural disorder and unusual *J* couplings in a fluorinated hydroxy-silicate. *J. Am. Chem. Soc.*, **132** (44), 15651–15660.

150 Barrow, N.S., Yates, J.R., Feller, S.A., Holland, D., Ashbrook, S.E., Hodgkinson, P., and Brown, S.P. (2011) Towards homonuclear *J* solid-state NMR correlation experiments for half-integer quadrupolar nuclei: Experimental and simulated ^{11}B MAS spin-echo dephasing and calculated $^2J_{BB}$ coupling constants for lithium diborate. *Phys. Chem. Chem. Phys.*, **13** (13), 5778–5789.

151 Joyce, S.A., Yates, J.R., Pickard, C.J., and Brown, S.P. (2008) Density functional theory calculations of hydrogen-bond-mediated NMR *J* coupling in the solid state. *J. Am. Chem. Soc.*, **130** (38), 12663–12670.

152 Hung, I., Uldry, A.C., Becker-Baldus, J., Webber, A.L., Wong, A., Smith, M.E., Joyce, S.A., Yates, J.R., Pickard, C.J., Dupree, R., and Brown, S.P. (2009) Probing heteronuclear ^{15}N-^{17}O and ^{13}C-^{17}O connectivities and proximities by solid-state NMR spectroscopy. *J. Am. Chem. Soc.*, **131** (5), 1820–1834.

153 de Gironcoli, S. (1995) Lattice dynamics of metals from density-functional perturbation theory. *Phys. Rev. B*, **51** (10), 6773–6776.

154 Laskowski, R. and Blaha, P. (2015) NMR shielding in metals using the augmented plane wave method. *J. Phys. Chem. C*, **119** (33), 19390–19396.

155 Autschbach, J. (2004) Calculation of heavy-nucleus chemical shifts. Relativistic all-electron methods, in *Calculation of NMR and EPR Parameters: Theory and Applications* (eds M. Kaupp, M. Bühl, and V.G. Malkin), Wiley-VCH Verlag GmbH, Weinheim, Chap. 14, pp. 227–247.

156 Autschbach, J. and Ziegler, T. (2004) Relativistic calculations of spin-spin coupling constants of heavy nuclei, in *Calculation of NMR and EPR Parameters: Theory and Applications* (eds M. Kaupp, M. Bühl, and V.G. Malkin), Wiley-VCH Verlag GmbH, Weinheim, Chap. 15, pp. 249–264.

157 Koelling, D.D. and Harmon, B.N. (1977) A technique for relativistic spin-polarised calculations. *J. Phys. C Solid State Phys.*, **10** (16), 3107–3114.

158 Yates, J.R., Pickard, C.J., Payne, M.C., and Mauri, F. (2003) Relativistic nuclear magnetic resonance chemical shifts of heavy nuclei with pseudopotentials and the zeroth-order regular approximation. *J. Chem. Phys.*, **118** (13), 5746–5753.

159 van Lenthe, E., Baerends, E.J., and Snijders, J.G. (1993) Relativistic regular two-component hamiltonians. *J. Chem. Phys.*, **99** (6), 4597–4610.

160 van Leeuwen, R., van Lenthe, E., Baerends, E.J., and Snijders, J.G. (1994) Exact solutions of regular approximate relativistic wave equations for hydrogen-like atoms. *J. Chem. Phys.*, **101** (2), 1272–1281.

161 van Lenthe, E., Baerends, E.J., and Snijders, J.G. (1994) Relativistic total energy using regular approximations. *J. Chem. Phys.*, **101** (11), 9783–9792.

162 Green, T.F.G. and Yates, J.R. (2014) Relativistic nuclear magnetic resonance J-coupling with ultrasoft pseudopotentials and the zeroth-order regular approximation. *J. Chem. Phys.*, **140** (23), 234106.

163 Perdew, J.P. and Zunger, A. (1981) Self-interaction correction to density-functional approximations for many-electron systems. *Phys. Rev. B*, **23**, 5048–5079.

164 Sousa, S.F., Fernandes, P.A., and Ramos, M.J. (2007) General performance of density functionals. *J. Phys. Chem. A*, **111** (42), 10439–10452.

165 Chapman, R.P. and Bryce, D.L. (2009) Application of multinuclear magnetic

resonance and gauge-including projector-augmented-wave calculations to the study of solid group 13 chlorides. *Phys. Chem. Chem. Phys.*, **11** (32), 6987–6998.

166 Widdifield, C.M. and Bryce, D.L. (2010) Solid-state ^{127}I NMR and GIPAW DFT study of metal iodides and their hydrates: Structure, symmetry, and higher-order quadrupole-induced effects. *J. Phys. Chem. A*, **114** (40), 10810–10823.

167 Cuny, J., Cordier, S., Perrin, C., Pickard, C.J., Delevoye, L., Trébosc, J., Gan, Z., Le Pollès, L., and Gautier, R. (2012) ^{95}Mo solid-state nuclear magnetic resonance spectroscopy and quantum simulations: synergetic tools for the study of molybdenum cluster materials. *Inorg. Chem.*, **52** (2), 617–627.

168 Laurencin, D., Gervais, C., Stork, H., Krämer, S., Massiot, D., and Fayon, F. (2012) ^{25}Mg solid-state NMR of magnesium phosphates: High magnetic field experiments and density functional theory calculations. *J. Phys. Chem. C*, **116** (37), 19984–19995.

169 Profeta, M., Benoit, M., Mauri, F., and Pickard, C.J. (2004) First-principles calculation of the ^{17}O NMR parameters in Ca oxide and Ca aluminosilicates: The partially covalent nature of the Ca–O bond, a challenge for density functional theory. *J. Am. Chem. Soc.*, **126** (39), 12628–12635.

170 Sadoc, A., Body, M., Legein, C., Biswal, M., Fayon, F., Rocquefelte, X., and Boucher, F. (2011) NMR parameters in alkali, alkaline earth and rare earth fluorides from first principle calculations. *Phys. Chem. Chem. Phys.*, **13** (41), 18539–18550.

171 Laskowski, R., Blaha, P., and Tran, F. (2013) Assessment of DFT functionals with NMR chemical shifts. *Phys. Rev. B*, **87** (11), 195130.

172 Laurencin, D., Gervais, C., Wong, A., Coelho, C., Mauri, F., Massiot, D., Smith, M.E., and Bonhomme, C. (2009) Implementation of high resolution ^{43}Ca solid-state NMR spectroscopy: Toward the elucidation of calcium sites in biological materials. *J. Am. Chem. Soc.*, **131** (37), 13430–13440.

173 Cuny, J., Furet, E., Gautier, R., Le Polles, L., Pickard, C.J., and d'Espinose de Lacaillerie, J.B. (2009) Density functional theory calculations of ^{95}Mo NMR parameters in solid-state compounds. *ChemPhysChem*, **10** (18), 3320–3329.

174 Widdifield, C.M. and Bryce, D.L. (2010) Solid-state $^{79/81}$Br NMR and gauge-including projector-augmented wave study of structure, symmetry, and hydration state in alkaline earth metal bromides. *J. Phys. Chem. A*, **114** (5), 2102–2116.

175 Griffin, J.M., Knight, F.R., Hua, G., Ferrara, J.S., Hogan, S.W.L., Woollins, J.D., and Ashbrook, S.E. (2011) ^{77}Se solid-state NMR of inorganic and organoselenium systems: A combined experimental and computational study. *J. Phys. Chem. C*, **115** (21), 10859–10872.

176 Perić, B., Gautier, R., Pickard, C.J., Bosiočić, M., Grbić, M.S., and Požek, M. (2014) Solid-state NMR/NQR and first-principles study of two niobium halide cluster compounds. *Solid State Nucl. Magn. Reson.*, **59–60**, 20–30.

177 Sadoc, A., Biswal, M., Body, M., Legein, C., Boucher, F., Massiot, D., and Fayon, F. (2014) NMR parameters in column 13 metal fluoride compounds (AlF$_3$, GaF$_3$, InF$_3$ and TlF) from first principle calculations. *Solid State Nucl. Magn. Reson.*, **59–60**, 1–7.

178 Burdett, J.K., Lee, S., and Mclarnan, T.J. (1985) The coloring problem. *J. Am. Chem. Soc.*, **107** (11), 3083–3089.

179 Miller, G.J. (1998) The "coloring problem" in solids: How it affects structure, composition and properties. *Eur. J. Inorg. Chem.*, **1998** (5), 523–536.

180 Albright, T.A., Burdett, J.K., and Whangbo, M.-H. (2013) *Orbital Interactions in Chemistry*, John Wiley & Sons, Inc, Hoboken, 2nd ed.

181 Bauer, J., Halet, J.-F., and Saillard, J.-Y. (1998) Rare earth metal borocarbides: Examples of coordination compounds in solid-state chemistry. *Coord. Chem. Rev.*, **178–180**, 723–753.

182 Ben Yahia, M., Roger, J., Rocquefelte, X., Gautier, R., Bauer, J., Guérin, R., Saillard, J.-Y., and Halet, J.-F. (2006) Portraits of some representatives of metal boride

183 Ewels, C.P., Rocquefelte, X., Kroto, H.W., Rayson, M.J., Briddon, P.R., and Heggie, M.I. (2015) Predicting experimentally stable allotropes: Instability of penta-graphene. *Proc. Natl. Acad. Sci.*, **112** (51), 201520402.

184 Wörle, M. and Nesper, R. (1994) MgB_2C_2, a new graphite-related refractory compound. *J. Alloys Compd.*, **216**, 75–83.

185 Hofmann, K., Rocquefelte, X., Halet, J.-F., Bähtz, C., and Albert, B. (2008) The eta (6),eta(1)-coordination of beryllium atoms in the graphite analogue BeB_2C_2. *Angew. Chem., Int. Ed.*, **47** (12), 2301–2303.

186 Smith, G.S., Johnson, Q., and Nordine, N.C. (1965) The crystal structure of ScB_2C_2. *Acta Crystallogr.*, **19** (4), 668–673.

187 Rocquefelte, X., Boulfelfel, S.E., Ben Yahia, M., Bauer, J., Saillard, J.-Y., and Halet, J.-F. (2005) Structural preference versus metal within the MB_2C_2 (M = Mg, Sc, Ca, Y, Ln) phases: The coloring problem revisited by DFT calculations. *Angew. Chem., Int. Ed.*, **44** (46), 7542–7545.

188 Bauer, J. and Bars, O. (1980) The ordering of boron and carbon-atoms in the LaB_2C_2 structure. *Acta Crystallogr., Sect. B Struct. Sci.*, **36**, 1540–1544.

189 Hoffmann, R. (1969) Symmetry requirements for the stabilization of one class of diradicals. *J. Chem. Soc. D, Chem. Commun.*, (5), 240–241.

190 Smith, P.K. and Gilles, P.W. (1967) High temperature rare earth-boron–carbon studies—III LnB_2C_2 and the ternary phase diagram. *J. Inorg. Nucl. Chem.*, **29** (2), 375–382.

191 Hofmann, K. and Albert, B. (2002) Near-edge fine structures in electron energy loss spectra: Are CaB_2C_2 and LaB_2C_2 isotypic? *ChemPhysChem*, **10** (6), 896–898.

192 Burdett, J.K., Canadell, E., and Hughbanks, T. (1986) Symmetry control of the coloring problem: the electronic structure of MB_2C_2 (M = calcium, lanthanum, . . .). *J. Am. Chem. Soc.*, **108** (14), 3971–3976.

193 Magishi, K., Koyama, K., Saito, T., Mizuno, K., Watanuki, R., and Suzuki, K. (2003) ^{11}B NMR study in the tetragonal compound CeB_2C_2. *Physica B: Cond. Matter.*, **329–333** (2), 649–650.

194 Bauer, J. (1982) New ternary diboride carbides: ScB_2C and LuB_2C. *J. Less Common Met.*, **87**, (1)45–52.

195 Cuny, J., Messaoudi, S., Alonzo, V., Furet, E., Halet, J.-F., Le Fur, E., Ashbrook, S.E., Pickard, C.J., Gautier, R., and Le Pollès, L. (2008) DFT calculations of quadrupolar solid-state NMR properties: Some examples in solid-state inorganic chemistry. *J. Comput. Chem*, **29** (13), 2279–2287.

196 Bauer, J. and Nowotny, H. (1971) Der Dreistoff Yttrium-Bohr-Kohlenstoff. *Monatsh. Chem.*, **102**, 1129–1145.

197 Babizhetskyy, V., Simon, A., Hoch, C., Hiebl, K., Le Pollès, L., Gautier, R., and Halet, J.-F. (2012) Structural, electronic and magnetic properties of layered REB_2C compounds (RE = Dy, Tm, Lu). *J. Solid State Chem.*, **191**, 121–128.

198 Tanabe, K. (2003) Catalytic application of niobium compounds. *Catal. Today*, **78**, (1-4) 65–77.

199 Koknat, F., Parsons, J.A., and Vongvusharintra, A. (1974) Metal cluster halide complexes. I. Efficient synthesis of hydrated hexanuclear niobium and tantalum cluster halides $M_6X_{14}\cdot 8H_2O$. *Inorg. Chem.*, **13** (7), 1699–1702.

200 Kamiguchi, S., Takaku, S., Kodomori, M., and Chihara, T. (2006) Variable catalytic behavior of Nb, Mo, Ta, W, and Re halide clusters: Isomerization of alkynes to conjugated dienes under nitrogen and hydrogenation to alkenes under hydrogen. *J. Mol. Catal. A: Chem.*, **260** (1-2), 43–48.

201 Wang, S.-S. and Yang, G.-U. (2015) Recent advances in polyoxometalate-catalysed reactions. *Chem. Rev.*, **115** (11), 4893–4962.

202 Du, L.S., Schurko, R.W., Lim, K.H., and Grey, C.P. (2001) A solid-state ^{93}Nb and ^{19}F NMR spectroscopy and X-ray diffraction study of potassium heptafluoroniobate(V): Characterization of ^{93}Nb, ^{19}F coupling, and fluorine motion. *J. Phys. Chem. A*, **105** (4), 760–768.

203 Du, L.S., Schurko, R.W., Kim, N., and Grey, C.P. (2002) Solid-state ^{93}Nb, ^{19}F, and ^{113}Cd nuclear magnetic resonance study of niobium oxyfluorides: Characterization of local distortions and oxygen/fluorine ordering. *J. Phys. Chem. A*, **106** (34), 7876–7886.

204 Lapina, O.B., Khabibulin, D.F., Romanenko, K.V., Gan, Z., Zuev, M.G., Krasil'nikov, V.N., and Fedorov, V.E. (2005) ^{93}Nb NMR chemical shift scale for niobia systems. *Solid State Nucl. Magn. Reson.*, **28** (2-4), 204–224.

205 Prasad, S., Zhao, P., Huang, J., Fitzgerald, J.J., and Shore, J.S. (2001) Niobium-93 MQMAS NMR spectroscopic study of alkali and lead niobates. *Solid State Nucl. Magn. Reson.*, **19** (1–2), 45–62.

206 Flambard, A., Montagne, L., Delevoye, L., and Steuernagel, S. (2007) ^{93}Nb and ^{17}O NMR chemical shifts of niobiophosphate compounds. *Solid State Nucl. Magn. Reson.*, **32** (2), 34–43.

207 Hanna, J.V., Pike, K.J., Charpentier, T., Kemp, T.F., Smith, M.E., Lucier, B.E.G., Schurko, R.W., and Cahill, L.S. (2010) A ^{93}Nb solid-state NMR and density functional theory study of four- and six-coordinate niobate systems. *Chem. Eur. J.*, **16** (10), 3222–3239.

208 Papulovskiy, E., Shubin, A., Terskikh, V.V., Pickard, C.J., and Lapina, O.B. (2013) Theoretical and experimental insights into applicability of solid-state ^{93}Nb NMR in catalysis. *Phys. Chem. Chem. Phys.*, **15** (14), 5115–5131.

209 Vojnovic, M., Antolic, S., Kojik-Prodit, B., Brnicevic, N., Miljak, M., and Aviani, I. (1997) Reactions of hexanuclear niobium and tantalum halide clusters with mercury(II) halides. I Synthesis and structures of the semiconducting compounds [M$_6$Br$_{12}$(H$_2$O)$_6$][HgBr$_4$] 12H$_2$O,M=Nb,Ta. *Z. Anorg. Allg. Chem.*, **623** (8), 1247–1254.

210 Brnicevic, N., Planinic, P., McCarley, R.E., Antolic, S., Luic, M., and Kojic-Prodic, B. (1995) Synthesis and structural characterization of three new compounds containing [Nb$_6$Cl$_{12}$]$^{2+}$ units as hydrated hydroxidests. *J. Chem. Soc., Dalton Trans.*, (9), 1441–1446.

211 Chandran, C.V., Cuny, J., Gautier, R., Le Pollès, L., Pickard, C.J., and Bräuniger, T. (2010) Improving sensitivity and resolution of mqmas spectra: A ^{45}Sc-NMR case study of scandium sulphate pentahydrate. *J. Magn. Reson.*, **203** (2), 226–235.

212 Bräuniger, T., Hofmann, A.J., Moudrakovski, I.L., Hoch, C., and Schnick, W. (2016) A ^{45}Sc-NMR and DFT calculation study of crystalline scandium compounds. *Solid State Sci.*, **51**, 1–7.

213 Cuny, J., Sykina, K., Fontaine, B., Le Pollès, L., Pickard, C.J., and Gautier, R. (2011) ^{95}Mo nuclear magnetic resonance parameters of molybdenum hexacarbonyl from density functional theory: Appraisal of computational and geometrical parameters. *Phys. Chem. Chem. Phys.*, **13** (43), 19471–19479.

214 Truflandier, L., Paris, M., Payen, C., and Boucher, F. (2006) First-principles calculations within periodic boundary conditions of the NMR shielding tensor for a transition metal nucleus in a solid state system: The example of ^{51}V in AlVO$_4$. *J. Phys. Chem. B*, **110** (43), 21403–21407.

215 Truflandier, L., Paris, M., and Boucher, F. (2007) Density functional theory investigation of 3d transition metal NMR shielding tensors in diamagnetic systems using the gauge-including projector augmented-wave methods. *Phys. Rev. B*, **76** (3), 035102.

216 Charpentier, T., Menziani, M.C., and Pedone, A. (2013) Computational simulations of solid state NMR spectra: A new era in structure determination of oxide glasses. *RSC Adv.*, **3** (27), 10550–10578.

217 Charpentier, T., Kroll, P., and Mauri, F. (2009) First-principles nuclear magnetic resonance structural analysis of vitreous silica. *J. Phys. Chem. C*, **113** (18), 7917–7929.

218 Soleilhavoup, A., Delaye, J.M., Angeli, F., Caurant, D., and Charpentier, T. (2010) Contribution of first-principles calculations to multinuclear NMR analysis of borosilicate glasses. *Magn. Reson. Chem.*, **48** (S1), S159–S170.

219 Angeli, F., Villain, O., Schuller, S., Charpentier, T., de Ligny, D., Bressel, L.,

and Wondraczek, L. (2012) Effect of temperature and thermal history on borosilicate glass structure. *Phys. Rev. B*, **85** (5), 054110.

220 Benoit, M., Profeta, M., Mauri, F., Pickard, C.J., and Tuckerman, M.E. (2005) First-principles calculation of the ^{17}O NMR parameters of a calcium aluminosilicate glass. *J. Phys. Chem. B*, **109** (13), 6052–6060.

221 Pedone, A., Charpentier, T., and Menziani, M.C. (2010) Multinuclear NMR of CaSiO$_3$ glass: Simulation from first-principles. *Phys. Chem. Chem. Phys.*, **12** (23), 6054–6066.

222 Gambuzzi, E., Pedone, A., Menziani, M.C., Angeli, F., Florian, P., and Charpentier, T. (2015) Calcium environment in silicate and aluminosilicate glasses probed by ^{43}Ca MQMAS NMR experiments and MD-GIPAW calculations. *Solid State Nucl. Magn. Reson.*, **68–69**, 31–36.

223 Charpentier, T., Ispas, S., Profeta, M., Mauri, F., and Pickard, C.J. (2004) First-principles calculation of ^{17}O, ^{29}Si, and ^{23}Na NMR spectra of sodium silicate crystals and glasses. *J. Phys. Chem. B*, **108** (13), 4147–4161.

224 Angeli, F., Villain, O., Schuller, S., Ispas, S., and Charpentier, T. (2011) Insight into sodium silicate glass structural organization by multinuclear NMR combined with first-principles calculations. *Geochim. Cosmochim. Acta*, **75** (9), 2453–2469.

225 Ispas, S., Charpentier, T., Mauri, F., and Neuville, D. (2010) Structural properties of lithium and sodium tetrasilicate glasses: Molecular dynamics simulations versus NMR experimental and first-principles data. *Solid State Sci.*, **12** (2), 183–192.

226 Gambuzzi, E., Pedone, A., Menziani, M.C., Angeli, F., Caurant, D., and Charpentier, T. (2014) Probing silicon and aluminium chemical environments in silicate and aluminosilicate glasses by solid state NMR spectroscopy and accurate first-principles calculations. *Geochim. Cosmochim. Acta*, **125**, 170–185.

227 Nimmo, J.P. and Kroll, P. (2014) First-principles calculations and analysis of ^{29}Si nuclear magnetic resonance chemical shifts in silicon oxycarbide ceramics. *J. Phys. Chem. C*, **118** (51), 29952–29961.

228 Kibalchenko, M., Yates, J.R., Massobrio, C., and Pasquarello, A. (2011) Structural composition of first-neighbor shells in GeSe$_2$ and GeSe$_4$ glasses from a first-principles analysis of NMR chemical shifts. *J. Phys. Chem. C*, **115** (15), 7755–7759.

229 Sykina, K., Bureau, B., Le Pollès, L., Roiland, C., Deschamps, M., Pickard, C.J., and Furet, E. (2014) A combined ^{77}Se NMR and molecular dynamics contribution to the structural understanding of the chalcogenide glasses. *Phys. Chem. Chem. Phys.*, **16** (33), 17975–17982.

230 Bouessel du Bourg, L., Roiland, C., le Pollès, L., Deschamps, M., Boussard-Plédel, C., Bureau, B., Pickard, C.J., and Furet, E. (2015) Impact of te on the structure and ^{77}Se NMR spectra of se-rich Ge–Te–Se glasses: A combined experimental and computational investigation. *Phys. Chem. Chem. Phys.*, **17** (43), 29020–29026.

231 Bouzid, A., Massobrio, C., Boero, M., Ori, G., Sykina, K., and Furet, E. (2015) Role of the van der waals interactions and impact of the exchange-correlation functional in determining the structure of glassy GeTe$_4$. *Phys. Rev. B*, **92** (13), 134208.

232 Cuny, J., Xie, Y., Pickard, C.J., and Hassanali, A.A. (2016) *Ab initio* quality NMR parameters in solid-state materials using a high-dimensional neural-network representation. *J. Chem. Theory Comput.*, **12** (2), 765–773.

17
Quantum Mechanical/Molecular Mechanical (QM/MM) Approaches

C. Richard A. Catlow,[1,2] John Buckeridge,[1] Matthew R. Farrow,[1] Andrew J. Logsdail,[1,2] and Alexey A. Sokol[1]

[1]*University College London, Kathleen Lonsdale Materials Chemistry, Department of Chemistry, 20 Gordon Street, London WC1H 0AJ, UK*
[2]*Cardiff University, School of Chemistry, Park Place, Cardiff CF10 3AT, UK*

17.1
Introduction

Computational modeling techniques are now standard tools in solid-state science. They are used routinely to model and predict structures, to investigate defect, transport, and spectroscopic properties of solids, to simulate sorption and diffusion, to develop models for nucleation and growth of solids, and increasingly to model and predict reaction mechanisms. They are applied to bulk solids, surfaces, and nanostructures, and successful applications are reported for all major classes of solid: metals, semiconductors, inorganic and ceramic materials, and molecular crystals. Modeling methods are now indeed tools that are used to guide, interpret, and predict experiment [1–3].

The solid-state modeler now has a rich and powerful range of computational tools. Methods based on interatomic potentials have a long and successful history in the field and have been used to great effect in structure modeling and prediction in modeling defects and in simulating molecular and ionic diffusion. These methods still have an important rôle in the field, but they are inevitably limited by their exclusion of any explicit information on electronic structure. Electronic structure methods, which solve the Schrödinger equation at some level of approximation, give such information but need substantially greater computational resources, limiting them to smaller systems, and, in dynamic simulations, shorter time scales. However, with the continuing exponential growth in computer power and with theoretical, technical, and algorithmic developments, the range and scope of such methods has expanded enormously in recent decades. In particular, there has been an explosion in the use of density functional theory (DFT)-based methods – see Chapters 1 and 8 of the current volume, which evaluate exchange and correlation energies by using an approximate

Handbook of Solid State Chemistry, First Edition. Edited by Richard Dronskowski, Shinichi Kikkawa, and Andreas Stein.
© 2017 Wiley-VCH Verlag GmbH & Co. KGaA. Published 2017 by Wiley-VCH Verlag GmbH & Co. KGaA.

functional of the electron density, and scale with the system size less drastically than the Hartree–Fock (HF) methods. DFT has become an increasingly standard tool in computational solid-state science.

In modeling crystalline solids, the most obvious procedure is to use three-dimensional periodic boundary conditions, by repeating the crystallographic unit cell (or a supercell of the latter) infinitely in three dimensions. Such approaches may also be adapted to model surfaces, where a two-dimensional slab is used with periodicity applied in the third dimension perpendicular to the slab leaving a large gap between slabs in neighboring cells. Three-dimensional periodicity has many substantial technical advantages, including the ability to represent the wave function with plane-wave basis sets.

Historically, a number of inventive schemes had been advanced, introducing the two-dimensional analogues of this approach and various central cell corrections [4–8]. With the advent of high-power modern computers, however, the methodological complexity of these techniques has mostly been replaced by less efficient but more flexible, easy-to-implement schemes. Modeling using 3D periodicity is facilitated by a number of excellent and versatile computer codes. Examples include the VASP [9] and CASTEP codes [10], which employ plane-wave basis sets, and CRYSTAL [11], which uses atomic-centered basis sets in common with molecular quantum chemistry. These and other computer codes have been very widely used and have given very extensive insight into the structure and bonding, as well as physical and chemical properties, of a wide range of solids and their surfaces [12–16]. There are, however, inherent limitations in this approach, the most obvious being the treatment of localized states such as defects, polarons, and adsorbed molecules. Here, the only strategy within the context of 3D periodicity is to embed these states within a supercell, to which the periodicity is applied. This approach, while in widespread usage, inevitably includes interactions between the defects or adsorbed species that may give uncertain and in some case spurious contributions to the calculated energy and properties. A related practical problem is that, to minimize interactions between the species of interest in neighboring cells, it may be necessary to use very large supercells with a consequent increase in the computational requirements. A second and subtler (and less well documented) problem is that, with an infinite 3D periodic system employing atom-centered pseudopotentials, there is no clear definition of the vacuum energy level. This difficulty is a consequence of the strategy adopted to resolve singularities that arise in the Fourier transforms of the periodic Coulomb potential and atomic pseudopotentials (specifically the so-called $\mathbf{G}=0$ term), which involves the introduction of a problem- and implementation-specific constant [17]. As a common reference is therefore not available, calculating ionization potentials or defects in different charge states consistently is problematic.

For these reasons, alternative methods are needed and the solid-state modeler can have recourse to a long standing approach widely used in molecular quantum chemistry and in biomolecular modeling, that is to represent the region of interest – the defect, localized electronic state, adsorbed molecule, and the surrounding species – by a molecular cluster. In earlier work, a simple cluster was often used, but it became apparent that such a strategy was severely limited as, in the majority of cases, the effects of long-range interactions could not be ignored, and the description of the species at the outermost parts of the cluster would be particularly problematic. For this reason, the "QM/MM" approach was developed in which the quantum mechanical (QM) cluster is embedded in an approximate representation of the surrounding lattice that is modeled using a "molecular mechanics" (MM) or interatomic-potential-based technique. Calculations involving interatomic potentials are computationally far cheaper than QM calculations, so the embedding region can be large and can effectively represent the electrostatic effects and steric constraints of the surrounding lattice. A number of methods are used for describing the interactions between the two regions and for handling the crucial issue of the interface between them, as will be discussed in later sections of this chapter. It is, however, clear that, given care in the choice of cluster size and embedding methodology and potentials, these methods can yield accurate results and are indeed the method of choice for several problems in condensed matter physics and chemistry. This chapter will, therefore, first outline the methodological aspects of the field before giving illustrations of the power of the techniques in modeling defects, electronic states, and the reactivity of solids. The examples given demonstrate the capability of the hybrid QM/MM (Quantum Mechanical/Molecular Mechanical) method to treat problems extending to the boundaries of contemporary solid-state chemistry and physics.

17.2
The Hybrid Quantum Mechanical/Molecular Mechanical Approach

As outlined above, the basic idea of a hybrid QM/MM method is that an extended system, such as a crystal, surface, nanowire, or large nanoparticle, can be split into two or more regions, each described with different levels of theory. A site of interest can then be modeled with high accuracy using quantum mechanical methods at amenable computational costs, whereas the surroundings will be included in a more approximate manner employing cheaper MM techniques. Development of an appropriate hybrid QM/MM model poses the following key questions:

- What is the nature of *chemical bonding* in the system of interest (see Chapter 12 of this volume)? The answer should tell us how electrons are distributed between the QM and MM regions.
- What is the *embedding potential* experienced by both nuclei and electrons in the QM region?
- What is the *polarizing potential* experienced by the surroundings?
- What is the master equation that will control the behavior of the model's components?

Below we will briefly outline the commonly used approaches to address these questions.

17.2.1
Background

17.2.1.1 One-Electron Problem

In solid-state theory, quantum mechanical/molecular mechanical models have been developed, to understand and predict defect properties of bulk materials, commonly in their crystalline form, and the chemical reactivity of their surfaces. Ideal extended systems (without defects) are most often described using one-electron wave functions, or orbitals, that are used as the basis of many-electron wave functions, charge densities, or Green's functions in HF, DFT, or many-body theories (see Chapters 1, 3–5, 7, 6, 9 of this volume). Such orbitals are solutions of corresponding one-electron Schrödinger equations with a periodic potential and, therefore, are normally taken as fully delocalized Bloch waves, which, however, can be recast as localized Wannier functions [18]; both types of functions are orthonormal and complete by design and can be transformed into each other by a unitary transformation. For metals, the transformation can present certain difficulties due to the fractionally occupied bands at the Fermi level [19], and it has not been proven mathematically that the resultant set is fully localized (although this result has been frequently obtained in practice, e.g., [20]). A local perturbation in the potential, due to a defect in bulk or an adsorbate on the surface, destroys the periodicity. The resulting orbitals for an extended system with a perturbation can either localize around the defect, with their energies below a certain threshold (or within a certain range, supporting bound states), or retain their delocalized character, while experiencing only a small fluctuation around the defect (scattered states).

Both bound and scattered states can be expanded using the perfect periodic solutions as a basis within perturbation theory or semiclassical approaches [21]. In either case, the long-range character of realistic perturbations requires further approximations in how (a) perturbed orbitals are combined in the many-electron solutions self consistently; (b) the perturbed nuclear potential energy is calculated around the defect; and (c) the whole electron-nuclear system is brought into equilibrium. Following Ref. [22], "perturbed crystal" techniques consider solutions for the whole crystal, describing the effect of a local perturbation in an approximate manner. Alternatively, for well-localized perturbations, one can adopt a "perturbed cluster" approach that focuses on the bound states perturbed by the crystalline environment in which the cluster is embedded. This environment can be represented as an external potential, continuous medium, or by special boundary conditions. These perturbed cluster approaches can be constructed in an *ad hoc* manner, or derived following certain approximations from the full defected crystal solutions, with a net effect that the external potential acting on the cluster orbitals is nonlocal and, in its simplest form, energy dependent [22–24]. The majority of contemporary quantum mechanical/molecular mechanical approaches make a further approximation, replacing the fully nonlocal potentials with atom-centered local or semilocal energy-independent approximate forms. Such embedding pseudopotentials take the same form and are derived in a similar spirit to effective core and/or model pseudopotentials employed in molecular and periodic calculations (see Chapter 2 of this volume).

Perturbed, embedded clusters present a significant improvement on simple molecular clusters. With these methods one can make use of numerous quantum chemical techniques that account for local and semilocal electron correlation effects on the bound, ground, and excited states, spectroscopic properties of the defect, and are implemented in widely available molecular quantum mechanical codes. Despite obvious limitations, a molecular cluster approach has been successfully used to calculate binding energies and frequency shifts of adsorbed molecules on the surfaces of ionic materials [25,26]. This success has, however, been clearly limited to the cases of neutral and weakly dipolar species such as CO, which coordinate to a local adsorption site. The long-range Coulomb interactions and electron exchange prove to be essential for the treatment of charged defects, charge transfer, and chemical reactivity, which are sensitive to the electrostatic field – see, for example, Refs [27,28].

17.2.1.2 Many-Electron Problem, Electron Groups, Localization, and Structural Elements

When we define the central QM region of an extended system, it is obvious how we select nuclei and the associated core electrons but not so for the valence electrons. In general, we would like to reduce the difficult problem of a very large number of N electrons (assumed infinite in periodic models) to explicit treatment of only a small number of n electrons (the QM group) and an approximate implicit tackling of the (still very large) $N-n$ number of electrons in the surroundings (the MM group). In the context of quantum mechanics, we should account for the interaction between electrons in these two groups. Such interactions can be split into the classical Coulomb term and quantum mechanical exchange and correlation terms. Whereas the former is straightforward, although the subject of many inventive optimization schemes, and can be described using the charge densities of the two regions [29–31], the latter is nontrivial. The exchange term can be formally set by choosing an antisymmetric ansatz for the wave function of the whole [32–34], and often the correlation term is neglected or included in an *a posteriori* fashion using an MM level of theory, which has underpinned a number of theoretical developments. This approach has been the basis of many embedding concepts for systems with well-localized electrons such as molecular solids. This class of material, however, forms a minority within the broad range of systems in the solid state. Usually, in materials of interest for chemical and physical applications, one should employ strong electron localization procedures to make a link from the standard molecular orbital type of theories, either based on the HF approximation or DFT, to such useful objects as ionic closed shells, lone pairs, or bonds, suggested by chemical intuition, and widely employed from early hybrid quantum mechanical/molecular mechanical work on solvents, organic macromolecules, enzymatic reactions, and so on [35–38]. To perform such a strong electron localization, one should remove the orthonormal constraint on the wave functions, which necessitates the switch from canonical Hartree–Fock (and correspondingly Kohn–Sham in DFT) equations and their orbital solutions to noncanonical orbitals, subject to Adams–Gilbert equations [39,40]. Much effort has, therefore, concentrated on the derivation of appropriate embedding schemes [29,39–56]. In this respect, the external potential can originate from the surroundings, represented by their charge density, giving rise to the family of so-called density functional embedding methods – see Refs [56–60] and references therein, which are however beyond the scope of the current review.

Following Refs [48,49,61,62], in a construction of a quantum mechanical/molecular mechanical model, it is useful to consider the whole extended system as a set of structural elements, each of which is associated with an electronic group, which, as mentioned, can be realized as an ion or a chemical bond. Such structural elements may be polarized or locally excited (transferred into an excited state) by external fields but remain intact, if the fields are sufficiently weak. In the core defect region, however, the structural elements may be destroyed by charge transfer, chemical bonding, or atoms (ions) added or removed from the system. Ideally, therefore, separation of the whole system into the QM cluster of interest and the surroundings described with an MM method should avoid cutting through structural elements, which allows one to associate with the QM cluster a given integral number of electrons, subject to treatments with standard methods of molecular quantum chemistry.

17.2.1.3 Electronic Structure of the Defect Containing System and the Choice of the Method

Considering the electronic structure, materials are classed as insulators or semiconductors – where all electrons can be localized in a relatively small volume – or metals, semimetals, and degenerate semiconductors – where localized electronic states still extend over many hundreds atoms and beyond. In practice, quantum mechanical/molecular mechanical methods are immediately applicable to a particular material, if the characteristic Bohr radius of the most extended occupied localized state is below that of a computationally tractable QM cluster. This approach is therefore suitable and practical for all strongly ionic (e.g., NaCl, Al_2O_3, or GaN) and covalent compounds (e.g., diamond C, Si, Ge, saturated carbon chain, and silicone polymers), but also for intermediate heteropolar materials (e.g., SiO_2, SiC, and MoS_2) and molecular solids (e.g., elemental arsenic, selenium or sulfur, fullerenes, ice, and hydrocarbons). Materials with naturally extended states would be more appropriately modeled with alternative approaches – where the electron separation into groups is performed, for example in reciprocal rather than real space, which is, however, beyond the scope of this chapter. Alternatively, a real-space approach to embedding, which allows for nonlocal perturbed solutions to be constructed, usually within a Green's function formalism, may be employed, which, however, poses the problem of how to account for atomic relaxations – see, for example, Refs [22,63–68] and the perturbed crystal approaches mentioned above. If the perturbation of an extended system is assumed to be localized within a small region, then one of the above approaches can still be applied, and the relaxation performed within the cluster [69–71]. The problem of self-consistent mutual polarization of the surroundings and the inner QM cluster has in turn been addressed in Refs [48,72–74].

A number of alternative approaches have been developed and implemented in computer codes to treat point defects in ionic solids, with ICECAP [72,75,76], CLUSTER [49,77,78], GUESS [79–81], and AIMP [52,73,82,83], and with extensions to covalent and other materials and types of application reported, for example, in Refs [25–28,35,54,84–88]. These implementations differ in

- the boundary conditions on the cluster – cf. the perturbed cluster approach [24,74,89] with the embedded and cyclic cluster approaches [6,85] and the localization procedure – cf. Refs [47,52,73] employing the strong orthogonality approximation with the more general approach in Ref [53];
- the cluster termination employed for semicovalent materials – cf. the popular hydrogen termination with the semiempirical link atom and *ab initio* separable potential operator approaches [41–43,81,87] and the use of the antisymmetrized product of strictly localized geminals [54]; and
- the level of coupling between the cluster and its environment, which is particularly important in the context of semicovalent and biological materials – cf. Refs [84,86,88,90].

Hybrid quantum mechanical/molecular mechanical techniques have a close relationship with fragmentation methods that simulate a large system of interest as a set of small interacting fragments, which, in fact, may even overlap with each other, reminiscent of divide-and-conquer geometry optimization strategies [44,45,57,91–93].

Now we will focus on the hybrid quantum mechanical/molecular mechanical approaches that have been implemented in the computational chemistry software ChemShell [94] with exemplar applications highlighted in Section 17.3.

17.2.2
Chemshell Quantum Mechanical/Molecular Mechanical Coupling and Master Equations

ChemShell, a particularly effective package for quantum mechanical/molecular mechanical modeling that implements a number of alternative additive and subtractive approaches, has been developed by Sherwood *et al.* [94–96]. Indeed, this package has been widely used in both biomolecular and solid-state modeling.

Quantum mechanical/molecular mechanical schemes implemented in ChemShell can be classified by the nature of the total energy expression [94,96,97]. The most physically intuitive, "additive" approach is to construct a total QM/MM energy expression as a sum of the QM, MM, and coupling terms:

$$E = E(\text{Inner, QM}) + E(\text{Outer, MM}) + E(\text{Couple, QM/MM}), \qquad (17.1)$$

where E(Couple,QM/MM) accounts for all interactions between the two regions, for example, classically handled bonding and van der Waals interactions and modifications to the QM Hamiltonian to reflect the electrostatic influence of some or all of the atoms in the outer region. Link atoms, used to saturate cluster terminating bonds in homo- or heteropolar covalent materials, need to be tackled carefully in this approach, to avoid unwanted QM/MM interactions. Indeed, in the additive zeolite model of Ref. [30], employed in the application work highlighted below, the link atoms are the intrinsic part of the Inner region. Both QM and MM centers enter bending and torsional, 3- and 4-body MM terms, but due to the electrostatic coupling between MM and QM regions they are also accounted for by the QM term. To exclude such unphysical double counting, the corresponding MM terms are omitted in the calculation of the total energy and its derivatives (forces).

An alternative "subtractive" approach applies the MM scheme to the whole (Inner + Outer) system and then corrects for the double counting by subtracting the energy (computed at the classical level) of the Inner system:

$$E = E(\text{Inner} + \text{Link}, QM) + E(\text{Inner} + \text{Outer}, MM) \\ - E(\text{Inner} + \text{Link}, MM). \tag{17.2}$$

Thus, the double counting of the Inner region energy and the link atom contributions (that are a model artifact) are removed. This approach relies on the MM energy closely reproducing the QM energy landscape for the Inner and Link regions. Many subtractive approaches, perhaps most notably of the multishell ONIOM type – cf. [44,98], include all coupling between the inner and outer regions at the MM level of theory, and the QM calculation is performed effectively as a gas-phase cluster calculation.

The additive and subtractive schemes are two extreme cases, and a number of intermediate approaches are possible. If both of the MM terms in Eq. (17.2) are redefined to include only an outer shell of the Inner region with link atoms, the requirement of a close match between MM and QM landscapes for the core Inner region is lifted. The subtractive approach can still be exploited to delete the terms associated with the link atoms. In the limit of excluding all Inner atoms, one arrives at an additive expression with a link atom correction:

$$E = E(\text{Inner} + \text{Link}, QM) + E(\text{Outer}, MM) - E(\text{Couple}, QM/MM) \\ - E(\text{Link}, MM). \tag{17.3}$$

A careful analysis of these approaches using such artificial objects as link atoms and their foundations can be found in Ref. [99].

Further, the quantum mechanical/molecular mechanical coupling can be classified depending on how the MM centers affect the electrons that are described with the QM terms and vice versa. Following Refs [94,97,100], we distinguish (a) mechanical embedding, where MM centers are not present in the QM calculation, and affect the QM centers only via interatomic potentials (a force field); (b) electrostatic embedding, where the charge distribution of the surroundings that are represented via MM centers is included in the QM Hamiltonian; (c) polarized embedding, in which MM centers respond to the electrostatic field of the QM region but their response in not fed back into the QM Hamiltonian; and (d) self-consistent polarized embedding, where both the electrostatic embedding and polarization effects are treated self-consistently. Only self-consistent schemes truly minimize the total QM/MM energy (Eq. (17.1)) with respect to both electronic and nuclear degrees of freedom in the entire system. As the response by MM centers in most interesting cases of charged and strongly dipolar defects, reactive centers, or adsorbates, is nonlinear (the system is essentially anharmonic), the calculation of the total energy and forces has to be iterative, with a number of improvements already implemented or under development in ChemShell [48,96,101–103].

In our work on ionic systems, including insulating metal oxides and wide-gap semiconductors, we have employed a so-called solid-state embedding scheme, which places semilocal pseudopotentials around the QM cluster, typically on cations, which constitutes an Interface region of the QM/MM model. The primary rôle of such pseudopotentials is to provide an appropriate account for short-range exchange and correlation embedding potential on the QM electron group from the surroundings. In the case of cations in low oxidations states, standard quantum-chemical effective large-core potential parameterizations are commonly employed, which, however, proves often too crude, and, typically, for cations in high oxidation states and/or transition metal ions, a customary parameterization is necessary.

For zeolites, a hetero-polar covalent binding model is assumed for the siliceous framework of the material, where each pair of Si and O atoms is linked by a σ bond, which is a linear combination of sp^3 hybrids residing on both atoms. The preferential occupation of O-centered states makes it anionic with two lone pairs and two σ bonds per center, whereas the occupation of Si-centered states renders it cationic supporting four σ bonds in a nearly perfect tetrahedral coordination. Using the link atom terminating scheme, at the boundary of the QM region any terminating dangling bond is saturated by hydrogen placed at an appropriate O—H or Si—H distance away from the QM bonded center along the corresponding QM—MM bond. The counterpart MM atom in this case is removed from the MM model and its charge is withdrawn onto further nearest MM neighbors. The resulting local bond dipoles induced by this termination procedure are compensated by adding to the model one pair of point charges forming opposite dipoles as described in Ref. [30].

As mentioned in the Introduction, a particular strength of quantum mechanical/molecular mechanical approaches is their ability to describe systems in different charge states with respect to the same reference level. For charged species, polarization outside the core defect region can be efficiently and accurately described in the continuous medium (dielectric continuum) approximation, which we employ to describe the contribution from the system surroundings beyond a certain cutoff radius, R. In this case, the total energy of the system can be shown to include the energy of the charged defect in the Madelung potential of the whole system, the self-energy of the defect, both of which are calculated via Eq. (17.1), and the polarization energy included *a posteriori*. Following Ref. [104], this correction can be obtained from a continuum approximation to a system of interest beyond a certain cutoff radius; it has, in fact, been first derived for bulk systems in Ref. [105] as the heat of hydration of an ion, and reads as follows:

$$E^{Born} = -\frac{Q^2}{2R}\left(1 - \frac{1}{\epsilon}\right), \tag{17.4}$$

where Q is the charge, and ϵ is the effective isotropic dielectric constant. This formula was later complemented with a similar result for the solvation of a dipole, μ [106]:

$$E^{Onsager} = -\frac{2\mu^2}{R^3}\frac{\epsilon - 1}{2\epsilon + 1}, \tag{17.5}$$

which can be readily extended to higher multipoles, with the leading quadrupolar term decaying as R^{-5} and effectively vanishing at typical cutoff radii in modern simulations. In vertical (optical) processes, the ionized system should be described using the high-frequency constant and in adiabatic processes the static constant should be employed instead.

17.2.3
Practicalities of Setting up a Quantum Mechanical/Molecular Mechanical Calculation

A quantum mechanical/molecular mechanical calculation should be preceded by the setup procedure, a nontrivial process that typically includes the following steps:

- Geometry optimization of the extended system, which may need to be preceded by the system preparation; for example, solid solutions may require selection of a representative sample and polar surfaces have to be reconstructed.
- An active, or defect site of interest is chosen in the extended system and a cluster is carved, for example, based on a cutoff radius – spherically shaped clusters are generated for bulk systems and interfaces while hemispherical for surfaces.
- Atomic centers within the cut cluster are designated for QM, MM, and QM/MM calculations according to the hybrid QM/MM coupling scheme. In particular, it is important to identify atoms around the active site, which should be treated at a QM level of theory, and introduce relevant chemical modifications in the model, that is, point defects and adsorbates.
- In the case of electrostatically coupled models, the terminating centers have their charge modified, and extra point charges are placed around the carved

cluster to simulate the Madelung field of the extended system within the cluster inner core.

These steps are illustrated in Figure 17.1 for an ionic system. We highlight regions in the model within which atomic centers play different rôles (five in total): (a) QM region – all centers are treated quantum mechanically; (b) Interface region – all atoms bear pseudopotentials, which provide a localizing embedding potential for the QM region, but interact with each other and MM centers by classical interatomic potentials; (c) active MM region – all centers are fully relaxed; (d) inactive, or frozen MM region, – all centers are fixed; and (e) terminating point charges maintaining the Madelung potential within the active region.

This procedure is implemented in the code ChemShell [30,96,94,103]; other software packages also implement this method or other closely related procedures.

In the next section, we present applications of this methodology to physical and chemical problems arising in the study of catalytic and energy materials.

17.3
Recent Applications of Quantum Mechanical/Molecular Mechanical Approaches

Quantum mechanical/molecular mechanical techniques are increasingly becoming the methods of choice in modeling localized states in solids. We now illustrate the range of applications by examining a number of key areas relating to electronic structure, to defect structures energies and properties, including processes of importance in energy materials, and to reactivity and catalysis.

First we focus on the calculation of ionization and redox potentials that, as argued in the Introduction, raises problems with 3D periodic methods that are absent within the quantum mechanical/molecular mechanical.

17.3.1
Ionization Potentials

In many applications, and to provide a fundamental view of materials, the absolute position of electronic energy bands is of primary interest. Measurements of the ionization potentials, work functions, and Fermi levels on various prepared surfaces and interfaces require computational input for interpretation, analysis, and general prediction and design. As mentioned in the Introduction, energies of processes that involve a change in the charge of a local center are exceptionally problematic for supercell approaches. In contrast, the additive hybrid QM/MM techniques with electrostatic coupling between the QM cluster and its surroundings, including polarization effects, are well suited to tackle this challenge [94]. Explicit simulations of charged models with a common reference, in bulk or on surfaces, provide a unified approach to the modeling of both vertical

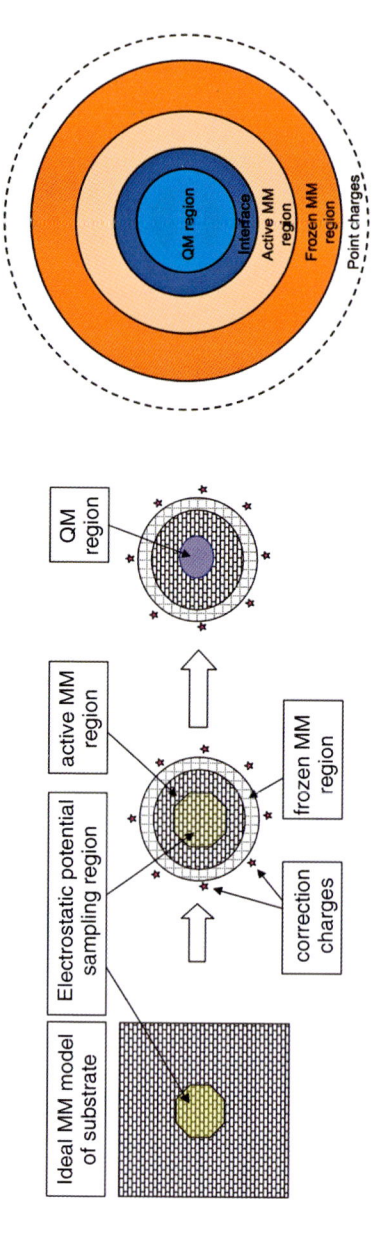

Figure 17.1 A sketch of a quantum mechanical/molecular mechanical model. A model generation is illustrated on the left with QM and MM regions highlighted, rendered from Figure 17.2 of [94]. A representative five-region model employed in the study of defects in ionic compounds is shown on the right.

(or optical, i.e., involving relaxation only of the electronic subsystem) and adiabatic (or thermodynamic, involving relaxation of ions and electrons) processes that are characterized by ionization potentials and/or electron affinities. We note that such simulations are also key for modeling chemical processes with charged intermediates, such as deprotonation.

The vertical ionization potential (or vertical ionization energy, I_v), can be computed as

$$I_v = E(\text{system}^+) - E(\text{system}^0) + E(e^-), \quad (17.6)$$

where $E(\text{system}^0)$ is the total energy of the relaxed neutral system of interest, $E(\text{system}^+)$ is the total energy of the singly (positively) charged system with the same atomic configuration, and $E(e^-)$ is the energy of an isolated electron (at the reference vacuum level). The equation for I_v can be applied to a model with any dimensionality, making it suitable for application to bulk and surface models. This approach has been adopted, for example, by Scanlon et al. to determine the energy band alignment between the two most abundant and technologically important TiO$_2$ polymorphs, rutile and anatase [107]. The band alignment was shown to be of Type II – staggered – with the anatase bands lying below those of rutile, in contrast to the previous, widely cited work of [108] and in agreement with accompanying experiments. This result has rationalized the improved charge carrier separation in mixed phase TiO$_2$ crystals. Buckeridge et al. then determined the band alignment of a wide range of TiO$_2$ polymorphs (illustrated in Figure 17.2), showing how the photocatalytic performance of mixed-phase

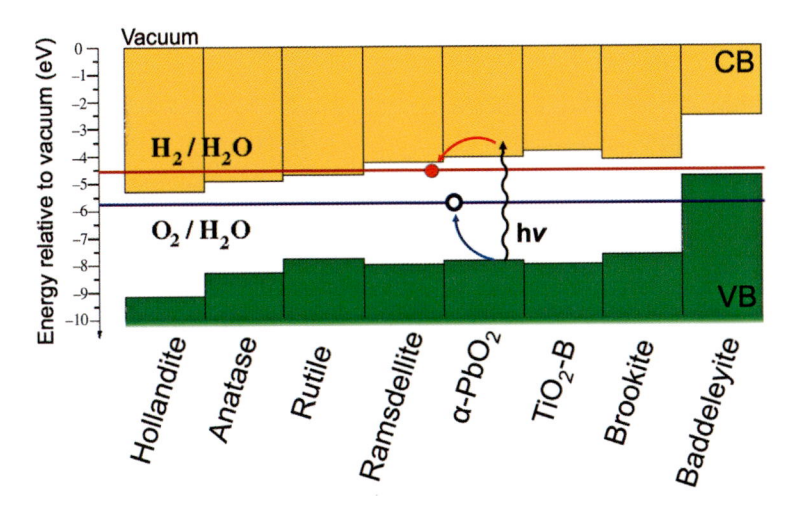

Figure 17.2 Quantum mechanical/molecular mechanical calculations using ChemShell showing the conduction band (CB) and valence band (VB) positions relative to the vacuum level for the various TiO$_2$ polymorphs, shown in comparison with the H$_2$ and O$_2$ redox potentials.

samples could be improved with novel combinations of different crystal structures [109].

The study of the polymorphs of titania followed the initial investigation into the electron binding energies of other binary wide-gap semiconducting compounds, including ZnO and GaN, which allowed the authors to determine their band offset [110]. Bulk studies are, however, rare due to the poor mathematical definition and inaccessibility of the bulk ionization potential to experiment. A step toward linking the bulk and surface ionization potentials has been reported in the investigation of the electrostatic potential in SnO_2 [111]. The direct measurability of surface ionization potentials and electron work functions has motivated several quantum mechanical/molecular mechanical studies, a representative sample of which we will now discuss.

The most tested material in this regard is MgO, the rock-salt structure of which exposes a very stable (100) surface. Initial calculations gave an ionization potential of 6.7 eV [81,112], which was consistent with the experimental value of 6.7 ± 0.4 eV [112]. In that study, however, the authors did not consider the long-range surface polarization. Subsequent work including this effect has resulted in $I_v = 6.46$ eV, still in very good agreement with experiment [113]. Complementary to I_v, one can compute the adiabatic ionization potential, I_a, which also allows the surrounding ions to respond to a charged defect. Downing et al. [113] have determined $I_a = 5.52$ eV, that is, roughly 1 eV lower than I_v.

While pristine surfaces are fundamentally important, more commonly it is intrinsic surface defects that offer reactivity for chemical applications including catalysis. Following the standard theory of chemical (acid, base, and redox) reactivity, the active sites are characterized by their electronegativity and hardness (or softness) that are, in turn, determined by the electron (or proton) ionization potential and affinity. For metal oxides, a well-studied example of such a site is the neutral F^0 center, the formation of which leaves two electrons trapped in the vacant oxygen lattice site. On reaction with electron-deficient species, this site can be ionized, resulting in an F^+ center. Using embedded-cluster quantum mechanical/molecular mechanical techniques, one can calculate such ionization energies (often referred to as defect levels) accurately: Ferrari and Pacchioni [114] reported the I_v of F^0 and F^+ center on (100) MgO as 4.71 and 9.69 eV, respectively. A subsequent study that included long-range polarization [113] reduced these levels to 3.05 and 4.55 eV, respectively, which indicated that both defect levels sit in the middle of the band gap. Pacchioni et al. demonstrated that I_v depends on the surface coordination of the defect site: for an edge vacancy (i.e., lower coordinated), their calculated I_v for the F^0 center reduced from 4.71 to 3.81 eV, a result confirmed by further studies [114–117]. The strong dependence of the defect levels on the surrounding electrostatic environment has further been highlighted in a quantum mechanical/molecular mechanical study of reconstructed ZnO surface sites [103], the authors of which found a significant destablization of electron levels in isovalent vacant oxygen interstitial surface sites compared to "true" surface vacancies by about 1.57 eV.

17.3.2
Modeling Point Defects with Hybrid Quantum Mechanical/Molecular Mechanical Methods

The hybrid quantum mechanical/molecular mechanical embedded cluster technique is particularly suited to modeling point defects and defect clusters in ionic solids. The methodology is based on the two-region approach of Mott and Littleton [118–121], whereby the system is divided into a central region that includes the defect, in which ionic and electronic relaxations are treated explicitly, and a surrounding region where the dielectric response of the infinite solid is calculated using a more simplified relaxation technique combined with a continuum picture, allowing one to model defects at the true dilute limit. With hybrid quantum mechanical/molecular mechanical methods, the local relaxations around a defect can be determined to a high degree of accuracy using density functional theory or beyond within the QM part of the calculation, while the polarization of the surrounding infinite system can be computed as precisely as the interatomic force field used in the MM part of the calculation allows (which can be very accurate if the force field reproduces well the structural, elastic and dielectric properties of the bulk material). This approach builds upon the earlier work of Ref. [115], who modeled a vacancy in binary metal oxides using a set of Gaussian basis functions to describe the wave function, embedded within a lattice of polarizable ions represented by point charges. The boundary conditions allow one to access an absolute reference energy, from which ionization energies can be calculated consistently (see Section 17.3.1), while two-dimensional [60,95,103,113,122–128] and one-dimensional [129,130] systems pose no significant problems to the technique.

The advantages of this approach, we reiterate, over the commonly used supercell method employing a plane-wave basis set are clear: no spurious interactions between periodic images of charged defects; access to a common reference; and ease of treatment of charged centers on surfaces and in wires. A drawback is the difficulty in treating defects with associated diffuse electron charge densities, such as shallow centers, where the density extends far beyond the confines of the QM region, although it is noteworthy that this difficulty also arises in plane-wave calculations, due to the limited size of the supercell used.

In the rest of this section, we will first present an application of hybrid quantum mechanical/molecular mechanical methods to treat localized defects in wide-gap semiconductors, where we demonstrate the power of the technique, then discuss an approach to treating diffuse defect-related electron states. The interested reader can find alternative examples of the theory and application of hybrid quantum mechanical/molecular mechanical approaches to treat defects in solids in Refs [59,131–139].

17.3.2.1 Defect Chemistry
Access to a common reference allows one to calculate defect ionizations, both vertical and adiabatic, in a consistent manner [140–142]. In predominantly ionic

systems, such as wide-gap semiconductors, most point defects of interest are localized in nature, and hence can be modeled quite well using hybrid quantum mechanical/molecular mechanical methods. In particular, reactions involving the formation and ionization of point defects balanced by charge carriers can be determined accurately, as one can calculate the ionization potential, giving the formation energy of a hole, and use a scissors operator approach, including the experimental band gap, to determine the electron affinity and therefore electron formation energies.

Such reactions and ionizations are key to understanding the defect properties of a material. In conventional semiconductors, such as Si or GaAs, where the band gap is of the order of 1 eV, electronic disorder governs the defect chemistry allowing the materials to be easily n- or p-doped with concentrations of free carriers controlled by the doping level. At the other extreme, in insulators, ionic disorder dominates via the formation of compensating sets of defects such as Frenkel pairs or stoichiometric vacancies (Schottky defects). Wide-gap semiconductors fall somewhere in between these two extremes (see Figure 17.3). Whether ionic or electronic disorder dominates is not immediately clear and needs to be carefully worked out using the relevant defect reactions and ionizations.

In the following, we take ZnO and GaN as representative wide-gap semiconductors [110,143,144]. ZnO (band gap 3.44 eV) is a widely used transparent conducting oxide that is easily n-doped but cannot be reliably p-doped [145]. GaN (band gap 3.5 eV) is an important material for high-powered microelectronics

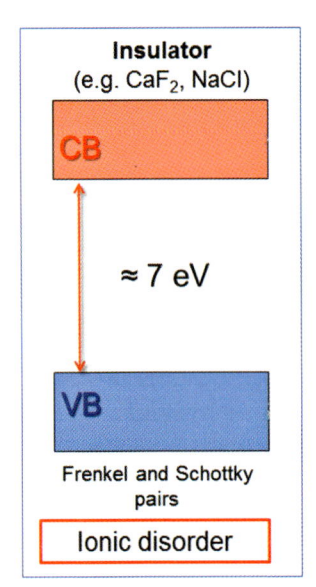

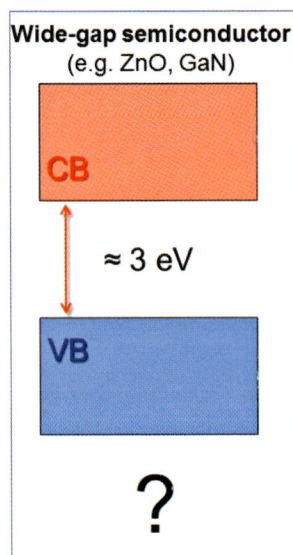

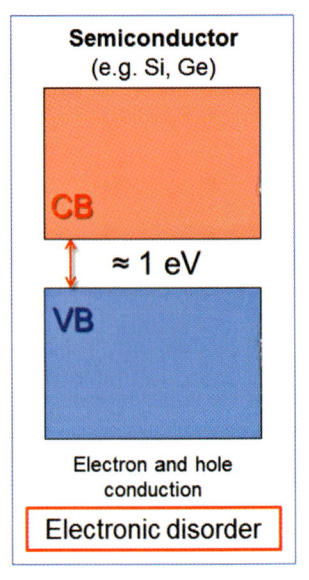

Figure 17.3 Cartoon depicting ionic versus electronic disorder in conventional semiconductors, wide-gap semiconductors and insulators. Conduction bands (CB) are in red, and valence bands (VB) are in blue.

and solid-state lighting that is intrinsically n-type, but can only be p-doped by introducing very high concentrations of Mg close to the solubility limit [146]. Modifying concentrations of electrons and holes in such systems can be achieved by changing the stoichiometry or by introducing aliovalent dopants, compensated by charge carriers. For example, one can introduce a monovalent cation dopant on the Zn site in ZnO, such as Li$^+$, which has an effective negative charge and can be compensated by the formation of a hole, while a monovalent anion dopant on the O can be compensated by an electron. Compensation by charged point defects is, however, always in principle possible. In the case of p-doping in ZnO, an acceptor-introduced hole can be compensated by the formation of a positively-charged O vacancy, V$_O$, using the standard Kröger–Vink notation [147]:

$$h^\bullet \rightleftharpoons \frac{1}{2}V_O^{\bullet\bullet} + \frac{1}{4}O_2 \qquad (17.7)$$

or by a Zn interstitial (Zn$_i$):

$$h^\bullet + \frac{1}{2}Zn(s) \rightleftharpoons \frac{1}{2}Zn_i^{\bullet\bullet} \qquad (17.8)$$

while for n-doping the relevant equations are (V$_{Zn}$ is a Zn vacancy and O$_i$ is an O interstitial):

$$e' + \frac{1}{4}O_2 \rightleftharpoons \frac{1}{2}O_i''; \; e' \rightleftharpoons \frac{1}{2}V_{Zn}'' + Zn(s) \qquad (17.9)$$

If the energies of these reactions are appreciatively negative, then the charge carrier in question will be thermodynamically unstable with respect to the formation of the corresponding point defect.

The defect formation energies can be calculated using the relevant defect reaction. For example, the formation of a doubly ionized oxygen vacancy in ZnO is governed by the equation:

$$O_O^\times \rightleftharpoons V_O^{\bullet\bullet} + 2e' + \frac{1}{2}O_2 \qquad (17.10)$$

where the reaction occurs in equilibrium with oxygen gas (termed "anion-rich" conditions) and exchange of electrons at the Fermi level. Equilibrium with ZnO and zinc metal may also be assumed:

$$O_O^\times + Zn(s) \rightleftharpoons V_O^{\bullet\bullet} + 2e' + ZnO(s) \qquad (17.11)$$

which corresponds to "anion-poor" conditions. Such environmental conditions may also be applied to the defect reactions above governing the stability of electrons and holes. The corresponding equations for oxygen interstitials and metal defects are obvious, as is the application of this analysis to GaN.

The calculated energies of these reactions, determined using hybrid quantum mechanical/molecular mechanical methods, are presented in Table 17.1. In both materials, electron carriers are thermodynamically stable, while hole carriers are

Table 17.1 Calculated reaction energies (ΔE_f) for processes in which charge carriers are compensated by point defect formation.

ZnO	ΔE_f (kJ/mol)		GaN	ΔE_f (kJ/mol)	
	Anion poor	Anion rich		Anion poor	Anion rich
$h^\bullet \to \frac{1}{2} V_O^{\bullet\bullet}$	−250	−71	$h^\bullet \to \frac{1}{3} V_N^{\bullet\bullet\bullet}$	−114	−74
$h^\bullet \to \frac{1}{2} Zn_i^{\bullet\bullet}$	−227	−48	$h^\bullet \to \frac{1}{3} Ga_i^{\bullet\bullet\bullet}$	−50	−11
$e' \to \frac{1}{2} O_i''$	251	167	$e' \to N_i'$	567	450
$e' \to \frac{1}{2} V_{Zn}''$	345	72	$e' \to \frac{1}{3} V_{Ga}''''$	173	134

Source: Results are taken from Ref. [144].

not, with respect to point defect formation. These results explain the observed ease with which both GaN and ZnO are *n*-doped and difficulty with which they are *p*-doped. Such analysis is much more challenging when using plane-wave supercell methods, due to the lack of a common reference and the problem of image charge interaction (we note that several schemes have been proposed to treat this image charge interaction problem, see, e.g., Refs [14,148–158]).

Of course, these reactions assume the condition of thermodynamic equilibrium with reservoirs of the relevant metals, gases, and metal oxides/nitrides. One can always use experimental techniques to overcome these doping limits in wide-gap semiconductors, such as applying capping layers or epitaxially growing materials far from equilibrium to "freeze in" defects with kinetic barriers. In doing so, however, one is constantly battling against thermodynamics. The ease with which such analysis can be carried out demonstrates the power of the hybrid quantum mechanical/molecular mechanical embedded cluster technique when applied to study defects in wide-gap systems. First-principles calculations of thermochemical properties are discussed in detail in Chapter 13.

17.3.2.2 Diffuse States

As mentioned above, the major limitation of hybrid quantum mechanical/molecular mechanical methods when studying defect structures is in treating defects with diffuse charge densities. A method to overcome this difficulty takes advantage of the two region approach: using hydrogenic eigenfunctions as the basis set, within the QM region one can determine a perturbing potential, or more simply apply a scissors operator approach to determine the defect energy, accounting for the quantum confinement effect, while outside a hydrogenic donor potential is assumed. The extent of the diffuse charge carrier within the two regions can be calculated semiclassically using effective mass theory. With this approach, more accurate defect levels can be calculated, as has been shown for the case of Si and O doping in GaN [159]. The resulting thermal ionization energies for the +/0 transition for both impurities, determined using two different exchange and correlation functionals, B97-2 [160] and BB1k [161], are

Table 17.2 Thermal ionization energies of the +/0 transition for Si and O dopants in GaN, determined using the B97-2 and BB1k hybrid density functionals and compared with experiment [162,163].

	B97-2	BB1k	Experiment
Si_{Ga}	0.046	0.037	0.022 ± 0.004
O_N	0.056	0.041	0.029

Energies are in eV.
Source: Results are taken from Ref. [159].

shown in Table 17.2, compared with the experimentally determined values [162,163]). The agreement between theory and experiment is unprecedented for first-principles calculations of such ionization levels. Further developments of this approach are underway within the hybrid QM/MM community.

17.3.3
Active Sites in Zeolites

Having discussed the application of quantum mechanical/molecular mechanical approaches to study the electronic properties of bulk, surface, and defect structures in wide-gap systems, we turn our attention to quantum mechanical/molecular mechanical investigations of chemical reactivity in nanoporous materials, a topic of fundamental importance for catalytic science.

17.3.3.1 Deprotonation Energies as a Measure of Site Reactivity

While electron ionization energies are key indices of Lewis acidity or basicity, the index of Brønsted acidity or basicity, that is, the ability to donate protons, is of primary interest in catalysis over porous siliceous and aluminum silicate materials (zeolites). When SiO_2 frameworks are doped with cationic species, such as Al^{3+} replacing Si^{4+}, counterions must be introduced to ensure charge neutrality. In basic zeolites, extra framework Li^+ or Na^+ ions (Lewis bases) are typically used as counterions, while in acidic, H^+ ions – forming framework hydroxyl groups – are found (Brønsted acids). The energy, E_{DP}, required to separate these protons from the framework, a key index of the acidity of the material, is given by

$$E_{DP}(Z-OH) = E(Z-O^-) - E(Z-OH^0) + E(H^+) \quad (17.12)$$

here Z—OH represents the zeolite material and Z—O$^-$ is the deprotonated model with an overall charge of minus one, where an electron has been left on the framework oxygen species. $E(H^+)$ is the energy of a proton at the reference vacuum level.

Early investigations of deprotonation energies for zeolites using a quantum mechanical/molecular mechanical approach were carried out by the groups of Sauer [88,164–169] and Sherwood [30,86,170]. Extensive investigations were made to show how the type of the zeolitic framework influenced the deprotonation energy as well as the crystallographic position of the counterion within each framework [170]. Initial calculations [164] suggested that the deprotonation energies for the frameworks Y, CHA, MOR, and ZSM-5 were ranked as Y (1171 kJ/mol) > CHA (1190 kJ/mol) > MOR (1195 kJ/mol) > ZSM-5 (1200 kJ/mol), with further tests of increasing concentrations of aluminum strongly correlated to increases in $E_{DP}(Z-OH)$ [169]. Following these studies, improvements in the description of long-range polarization effects within the embedding environment [171,172] have resulted in a reduction in deprotonation energies by ~10–20 kJ/mol. A recent study investigated two-dimensional zeolitic materials, showing that the deprotonation energy is reduced, relative to the three-dimensional case, due to the alteration of the long-electrostatic environment [173]; affected by the smaller dielectric constant. However, lower interaction energies between protonated molecules and negatively charged surfaces compensate for the reduced dimensionality of the substrate. As a consequence, overall adsorption energies for reaction intermediates are similar to those in bulk [174].

A key application of QM/MM-calculated deprotonation energies can be found in an extensive study of the zeolite-catalyzed conversion of methanol to hydrocarbons (MTH) [172], a process that has been researched significantly since its initial discovery by Chang and Silvestri [175] and first commercialized in 1985 [176]. The authors presented a comparison of the deprotonation, methanol adsorption (E_{ads}), and methoxylation energies (E_{methox}) in H-ZSM-5 and H-Y zeolite frameworks, in an attempt to investigate the effect of framework topology and active site location on reactive sorbate behavior. The main results are summarized in Table 17.3, and an example of their computed adsorption structures is shown in Figure 17.4.

The order of methoxylation energy proved to be H-Y < H-ZSM-5 (Z6) < H-ZSM-5 (M7) < H-ZSM5-I2 (I2), which was in contrast to the order of methanol adsorption energies (H-ZSM-5 (M7) < H-Y < H-ZSM-5 (I2) < H-ZSM-5 (Z6)), at least when only one methanol molecule is present in the immediate environment. The study highlighted the importance of highly coordinated structures in stabilizing the system when considering the interaction of sorbates such as methanol with the active sites of acidic zeolite catalysts.

Table 17.3 Methoxylation energies in kJ/mol of the four zeolite systems and energy difference between the adsorbed state and the methoxylated state.

System	E_{methox}	$E_{ads} - E_{methox}$
Me-Y + H_2O	−162.0	+17.6
Me-ZSM-5 (I2) + H_2O	−150.4	+27.1
Me-ZSM-5 (M7) + H_2O	−138.6	+56.2
Me-ZSM-5 (Z6) + H_2O	−118.0	+18.6

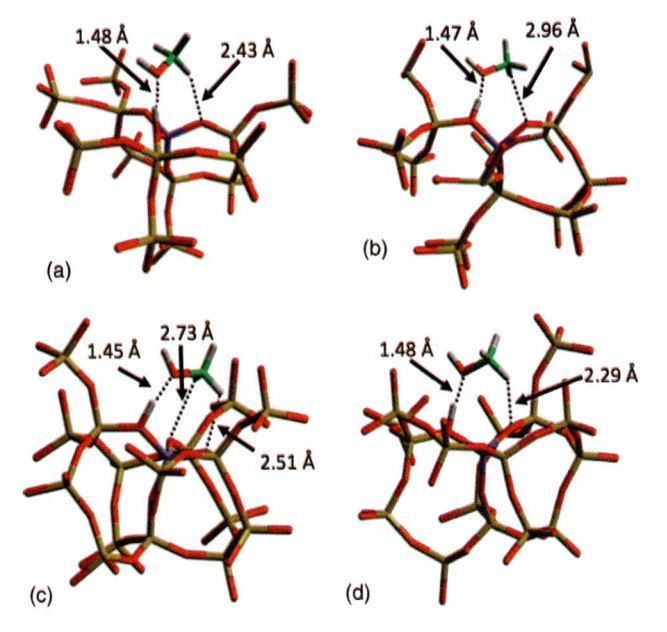

Figure 17.4 Optimized configurations of methanol adsorbed in (a) H-Y, (b) H-ZSM-5 (I2), (c) H-ZSM-5 (M7), and (d) the H-ZSM-5 (Z6). The shortest distances between methanol and host framework atoms are highlighted by dotted lines.

17.3.3.2 REDOX Potentials

An alternative measure of chemical reactivity is a material's redox potential that is determined as the energy of (de)hydrogenation, given by

$$E_{DH}(Z-OH) = E(Z-O) - E(Z-OH) + \frac{1}{2}E(H_2) \qquad (17.13)$$

in analogy with the deprotonation energy described by Eq. (17.12). In this case, the system of interest is neutral, which implies that, in principle, it would be possible to calculate E_{DH} using more common supercell methods; however, the typical size of siliceous unit cells, which can include hundreds of atoms, means that higher levels of theory are only affordable employing quantum mechanical/ molecular mechanical models. Rather than the deprotonation energy considered above, for transition-metal-doped zeolites (and other related materials) that exhibit high sustainable redox activity, the dehydrogenation energy provides an effective tool for characterization, as the process results in the oxidation of the corresponding cation.

Rather than investigating full reactions with quantum mechanical/molecular mechanical methods, redox potentials can be calculated to provide a useful index for the reactivity of a system. Berger *et al.* [122] showed that Fe(II) is more stable than the Fe(III) when doped into the MFI siliceous framework, although the computed dehydrogenation energy varied with the level of theory used: with standard density functional theory, different sites were found to have dehydrogenation energies of 0.16–0.39 eV; this range decreased to −0.04 to 0.09 eV when hybrid density functionals were employed, that is, the energy difference between the Fe(II) and Fe(III) states was insignificant. However, more sophisticated many body wave function theory calculations that include unoccupied orbitals in their description of electron correlation stabilized the Fe(II), resulting in oxidation energies of 0.18–0.34 eV, but with different ordering to that of standard DFT.

In contrast, Ti(IV) has been shown to be the most stable oxidation state for TS-1, an MFI framework doped with Ti. To *et al.* [177] studied a range of oxidation states and geometries for such Ti dopants, concluding that reduction from Ti(IV) to Ti(III) is easiest on a low-coordinated site. The reduction of a tripodal (three-coordinate) Ti(IV) cation to Ti(III) was found to be endothermic by 120–175 kJ/mol, while for a four-coordinate species, that is, part of the standard MFI framework, the Ti(III) species was an additional 290 kJ/mol higher in energy. These results showed that an inverted configuration, with the Ti dopant extending into the framework's porous channels, was considerably more favorable for reactivity than sites within the framework, as expected. This point was further confirmed when considering two-coordinated Ti cations, whereby the oxidation of Ti(III) to Ti(IV) was reduced by 70 kJ/mol, compared with three-coordinated sites. The results of this study are discussed in another context in the next section.

17.3.3.3 Coordination of Metal Centers

Incorporation of metal sites in zeolitic frameworks has been the subject of numerous investigations, some of which we discuss here. The material we introduced above, a Ti-doped silicalite-1 (TS-1) [178] is a siliceous counterpart of ZSM-5, which has been of great interest, since it is used with aqueous H_2O_2 as the primary oxidant in a large number of important reactions, such as the epoxidation of linear olefins, hydroxylation of aromatics, and oxidation of alcohols, amines, linear alkenes, sulfur compounds, and ethers [179].

French *et al.* and To *et al.* [177,180,181] have proposed a number of structural models for a Ti species that would result from the framework hydrolysis and site inversion that takes place during synthesis, postsynthetic treatment, and/or, depending on the catalytic process, during the catalytic transformation. The structures of interest are shown in Figure 17.5.

As the calculated geometries of the tripodal sites were in excellent agreement with experimental extended X-ray absorption fine structure (EXAFS) data [182], they helped rationalize the difference in the observed spectroscopic signature of metal centers in differently treated samples. The analysis of site transformation in the presence of water [40] resulted in the crucial conclusion that the catalytically active sites in TS-1 are hydrated tripodal rather than tetrapodal, a strikingly similar situation to that of other Ti-doped mesoporous materials previously explored using simple cluster models [183].

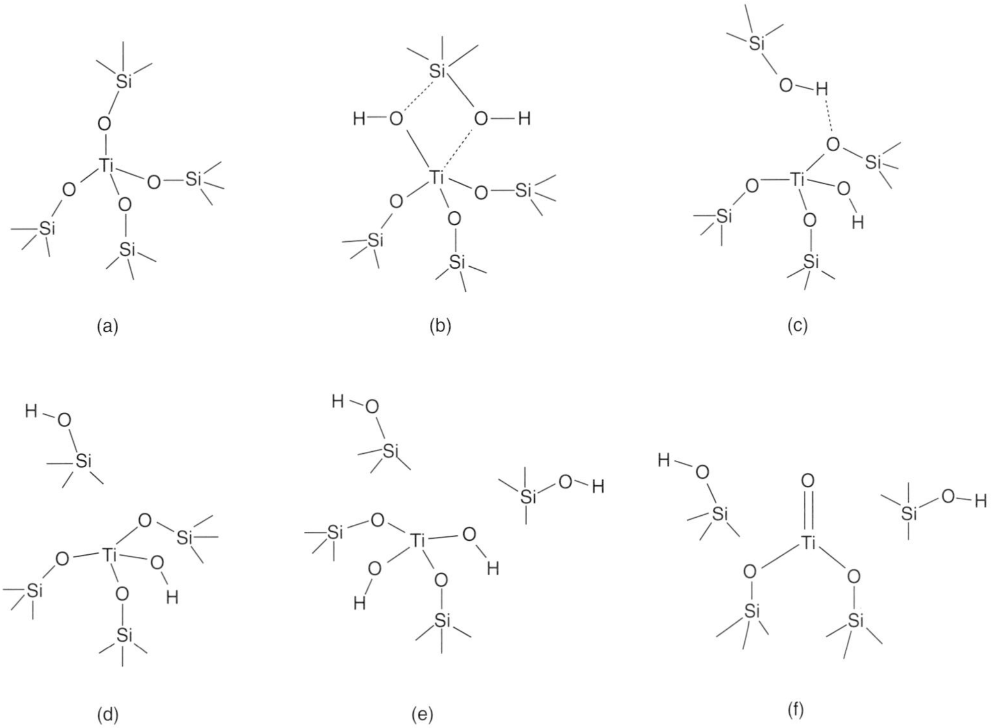

Figure 17.5 The six models used to represent the active Ti sites in TS-1: (a) tetrapodal, (b) 2MR, (c) tripodal (1l), (d) tripodal (2l), (e) bipodal, and (f) titanyl.

Consideration of the site reactivity toward H_2O_2 in anhydrous and hydrous conditions [184], confirmed the close competition between η^1 and η^2 complexes as candidates for the oxygen donating species in TS-1. Curiously, from electron paramagnetic resonance (EPR) studies, a much higher energy (by about 400–450 kJ mol^{-1}) superoxide radical species had been reported to emerge as a result of interaction of aqueous H_2O_2 and nonaqueous urea H_2O_2 [185]. Calculation of the related spectroscopic signature [184], unambiguously identified the superoxide supported on tripodal Ti as the magnetically active center (see Table 17.4, where the calculated components of the **g** tensor are compared to experimental values, which are expected to be accurate to at least the third decimal place), which allowed them to argue in support of the tetrahedral site inversion mechanism as essential for promoting the catalytic activity of these systems. In the absence of definitive structural experimental characterization, this combination of spectroscopic evidence and supporting computational results is key to unraveling the mechanisms of catalysis in atomistic and electronic detail (for more examples of such combinations see Chapters 9, 10, and 16 of this volume).

Table 17.4 EPR spin Hamiltonian parameters of the Ti(OO·) radical species. (Reproduced with permission from Ref. [185]. Copyright 2003, Elsevier.)

g tensor	Calculated (B3LYP) at BB1K geometry	Experiment[a]	
		aqueous H_2O_2	urea–H_2O_2
g_{xx}	2.003	2.002	2.004
g_{yy}	2.010	2.009	2.010
g_{zz}	2.025	2.026	2.028

a) From Ref. [185].

Clearly, it can be seen that the quantum mechanical/molecular mechanical embedded cluster methodology is beneficial in the above cases due to its ability to (a) deal with charged systems and calculate energies with a common reference, and (b) reduce the size of the system of interest to that of a finite fragment, while ensuring high accuracy is maintained. As a consequence, complex reactions using the substrate material as a heterogeneous catalyst can be studied effectively.

17.4
Final Remarks

The applications discussed above clearly show the range and power of the hybrid quantum mechanical/molecular mechanical technique. Furthermore, as we have argued, for many problems in the science of defects and in catalysis, this technique is increasingly the method of choice. The approach offers the possibility of achieving high levels of precision by the use of highly accurate QM techniques in the inner region as elegantly illustrated by the recent work of Piccini et al. [186] on catalytic reactions in zeolites (see also Ref. [187] for a discussion of the implications of this work). Applications in catalytic science are expected to be particularly fruitful in future work. The method also offers the possibility of enhanced accuracy in modeling defects in key materials such as those used as transparent conducting oxides. The biggest practical challenges relate to the development of viable interatomic potentials for new materials and to improvements in the usability of the techniques. However, with the growing sophistication of the software for quantum mechanical/molecular mechanical modeling, the range and impact of the applications of the techniques in solid-state science will rapidly grow.

Acknowledgments

We are grateful to Paul Sherwood and Tom Keal for many discussions and collaborations relating to QM/MM techniques. We acknowledge funding from

EPSRC (EP/K016288/1, EP/K038419/1). AJL is grateful to the Ramsay Trust and the Department of Chemistry at UCL for providing a Ramsay Fellowship. We thank P. V. Sushko, S. M. Woodley, S. A. Shevlin, D. O. Scanlon, T. Lazauskas, D. P. Mora-Fonz, C. A. Downing, M. Miskufova, G. Dutta, Z. Xie, A. G. H. Smith, A. Walsh, R. Deka, A. J. O'Malley, A. Chutia, and A. Thetford for input on quantum mechanical/molecular mechanical studies.

References

1 Butler, K.T., Frost, J.M., Skelton, J.M., Svane, K.L., and Walsh, A. (2016) Computational materials design of crystalline solids. *Chem. Soc. Rev.*, **45** (22), pp. 6138–6146.

2 Moniz, S.J.A., Shevlin, S.A., Martin, D.J., Guo, Z.-X., and Tang, J. (2015) Visible-light driven heterojunction photocatalysts for water splitting - a critical review. *Energy Environ. Sci.*, **8**, 731–759.

3 Woodley, S.M. (2016) Nanoclusters and nanoparticles, in *Computational Modeling of Inorganic Nanomaterials* (eds S.T. Bromley and M.A. Zwijnenburg), CRC Press, pp. 3–46.

4 Bennett, A.J., McCarroll, B., and Messmer, R.P. (1971) Molecular orbital approach to chemisorption. II. Atomic H, C, N, O, and F on graphite. *Phys. Rev. B*, **3**, 1397–1406.

5 Evarestov, R.A. (1975) Use of representative points of the Brillouin zone for the self-consistent calculations of solids in the large unit cell approach. *Phys. Status Solidi B*, **72**, 569–578.

6 Janetzko, F., Bredow, T., and Jug, K. (2002) Effects of long-range interactions in cyclic cluster calculations of metal oxides. *J. Chem. Phys.*, **116**, 8994–9004.

7 Stoneham, A.M. (1975) *Theory of Defects in Solids: Electronic Structure of Defects in Insulators and Semiconductors*, Oxford University Press, Oxford, UK.

8 Zhang, G., Canning, A., Grønbech-Jensen, N., Derenzo, S., and Wang, L.-W. (2013) Shallow impurity level calculations in semiconductors using *ab initio* methods. *Phys. Rev. Lett.*, **110**, 166404.

9 Kresse, G. and Furthmüller, J. (1996) Efficient iterative schemes for *ab initio* total-energy calculations using a plane-wave basis set. *Phys. Rev. B*, **54**, 11169–11186.

10 Clark, S.J., Segall, M.D., Pickard, C.J., Hasnip, P.J., Probert, M.J., Refson, K., and Payne, M.C. (2005) First principles methods using CASTEP. *Z. Kristallogr.*, **220**, 567–570.

11 Dovesi, R., Orlando, R., Erba, A., Zicovich-Wilson, C.M., Civalleri, B., Casassa, S., Maschio, L., Ferrabone, M., De La Pierre, M., D'Arco, P. et al. (2014) CRYSTAL14: A program for the *ab initio* investigation of crystalline solids. *Int. J. Quantum. Chem.*, **114**, 1287–1317.

12 Freysoldt, C., Grabowski, B., Hickel, T., Neugebauer, J., Kresse, G., Janotti, A., and Van de Walle, C.G. (2014) First-principles calculations for point defects in solids. *Rev. Mod. Phys.*, **86**, 253–305.

13 Groß, A. (2008) Adsorption at nanostructured surfaces from first principles. *J. Comput. Theor. Nanosci.*, **5**, 894–922.

14 Lany, S. and Zunger, A. (2008) Assessment of correction methods for the band-gap problem and for finite-size effects in supercell defect calculations: case studies for ZnO and GaAs. *Phys. Rev. B*, **78**, 235104.

15 Nieminen, R.M. (2007) Supercell methods for defect calculations, in *Theory of Defects in Semiconductors* (eds D.A. Drabold and S.K. Estreicher), Springer, Berlin, pp. 29–68.

16 Pacchioni, G. (2015) First principles calculations on oxide-based heterogeneous catalysts and photocatalysts: problems and advances. *Catal. Lett.*, **145**, 80–94.

17 Zunger, A. and Cohen, M.L. (1979) Self-consistent pseudopotential calculation of

the bulk properties of Mo and W. *Phys. Rev. B*, **19**, 568–582.
18 Brouder, C., Panati, G., Calandra, M., Mourougane, C., and Marzari, N. (2007) Exponential localization of Wannier functions in insulators. *Phys. Rev. Lett.*, **98**, 046402.
19 Souza, I., Marzari, N., and Vanderbilt, D. (2001) Maximally localized Wannier functions for entangled energy bands. *Phys. Rev. B*, **65**, 035109.
20 Sporkmann, B. and Bross, H. (1994) Calculation of Wannier functions for fcc transition metals by Fourier transformation of Bloch functions. *Phys. Rev. B*, **49**, 10869–10876.
21 Koster, G.F. and Slater, J.C. (1954) Wave functions for impurity levels. *Phys. Rev.*, **95**, 1167–1176.
22 Pisani, C., Dovesi, R., and Ugliengo, P. (1983) Comparison of different approaches to the study of local defects in crystals. I. Theoretical considerations and computational schemes. *Phys. Status Solidi B*, **116**, 249–259.
23 Pisani, C., Dovesi, R., Nada, R., and Kantorovich, L.N. (1990) Abinitio Hartree–Fock perturbed-cluster treatment of local defects in crystals. *J. Chem. Phys.*, **92**, 7448–7460.
24 Pisani, C., Dovesi, R., Roetti, C., Causà, M., Orlando, R., Casassa, S., and Saunders, V.R. (2000) CRYSTAL and EMBED, two computational tools for the ab initio study of electronic properties of crystals. *Int. J. Quantum Chem.*, **77**, 1032–1048.
25 Pelmenschikov, A.G., Morosi, G., Gamba, A., and Coluccia, S. (1995) A check of quantum chemical molecular models of adsorption on oxides against experimental infrared data. *J. Phys. Chem.*, **99**, 15018–15022.
26 Pelmenschikov, A.G., Morosi, G., Gamba, A., and Coluccia, S. (1998) Unimportance of the surrounding lattice in the adsorption of CO on low-coordinated Mg sites of MgO. *J. Phys. Chem. B*, **102**, 2226–2231.
27 Johnson, M.A. and Truong, T.N. (1999) Importance of polarization in simulations of condensed phase energetic materials. *J. Phys. Chem. B*, **103**, 9392–9393.
28 Kachurovskaya, N.A., Mikheeva, E.P., and Zhidomirov, G.M. (2002) Cluster molecular modeling of strong interaction for VO_x/TiO_2 supported catalyst. *J. Mol. Catal. A Chem.*, **178**, 191–198.
29 Laino, T., Mohamed, F., Laio, A., and Parrinello, M. (2006) An efficient linear-scaling electrostatic coupling for treating periodic boundary conditions in QM/MM simulations. *J. Chem. Theory Comput.*, **2**, 1370–1378.
30 Sherwood, P., de Vries, A.H., Collins, S.J., Greatbanks, S.P., Burton, N.A., Vincent, M.A., and Hillier, I.H. (1997) Computer simulation of zeolite structure and reactivity using embedded cluster methods. *Faraday Discuss.*, **106**, 79–92.
31 Sushko, P.V. and Abarenkov, I.V. (2010) General purpose electrostatic embedding potential. *J. Chem. Theory Comput.*, **6**, 1323–1333.
32 McWeeny, R. (1959) The density matrix in many-electron quantum mechanics. I. Generalized product functions. Factorization and physical interpretation of the density matrices. *Proc. Roy. Soc. Lond. A Mat.*, **253**, 242.
33 McWeeny, R. (1960) Some recent advances in density matrix theory. *Rev. Mod. Phys.*, **32**, 335–369.
34 Tolpygo, K.B. (1950) Physical properties of a rock salt lattice made up of deformable ions. *Zhurnal Eksperimentalnoi I Teoreticheskoi Fiziki*, **20**, 497–509.
35 Åqvist, J. and Warshel, A. (1993) Simulation of enzyme reactions using valence bond force fields and other hybrid quantum/classical approaches. *Chem. Rev.*, **93**, 2523–2544.
36 Lennard-Jones, J.E. (1937) The electronic structure of some polyenes and aromatic molecules. I. The nature of the links by the method of molecular orbitals. *Proc. R. Soc. Lond. A Math Phys. Sci.*, **158**, 280–296.
37 Warshel, A. and Karplus, M. (1972) Calculation of ground and excited state potential surfaces of conjugated molecules. I. Formulation and parametrization. *J. Am. Chem. Soc.*, **94**, 5612–5625.

38 Warshel, A. and Levitt, M. (1976) Theoretical studies of enzymic reactions: Dielectric, electrostatic and steric stabilization of the carbonium ion in the reaction of lysozyme. *J. Mol. Biol.*, **103**, 227–249.

39 Danyliv, O. and Kantorovich, L. (2004) Comparison of localization procedures for applications in crystal embedding. *Phys. Rev. B*, **70**, 075113.

40 Danyliv, O., Kantorovich, L., and Corá, F. (2007) Treating periodic systems using embedding: Adams-Gilbert approach. *Phys. Rev. B*, **76**, 045107.

41 Abarenkov, I.V., Boyko, M.A., and Sushko, P.V. (2011) Embedding and atomic orbitals hybridization. *Int. J. Quantum Chem.*, **111**, 2602–2619.

42 Abarenkov, I.V. and Tupitsyn, I.I. (2001) A new separable potential operator for representing a chemical bond and other applications. *J. Chem. Phys.*, **115**, 1650–1660.

43 Antes, I. and Thiel, W. (1999) Adjusted connection atoms for combined quantum mechanical and molecular mechanical methods. *J. Phys. Chem. A*, **103**, 9290–9295.

44 Guo, W., Wu, A., Zhang, I.Y., and Xu, X. (2012) XO: An extended ONIOM method for accurate and efficient modeling of large systems. *J. Comput. Chem.*, **33**, 2142–2160.

45 He, J., Di Paola, C., and Kantorovich, L. (2009) Partitioning scheme for density functional calculations of extended systems. *J. Chem. Phys.*, **130**, 144104.

46 Hégely, B., Nagy, P.R., Ferenczy, G.G., and Kállay, M. (2016) Exact density functional and wave function embedding schemes based on orbital localization. *J. Chem. Phys.*, **145**, 064107.

47 Huzinaga, S., Seijo, L., Barandiarán, Z., and Klobukowski, M. (1987) The ab initio model potential method. Main group elements. *J. Chem. Phys.*, **86**, 2132–2145.

48 Kantorovich, L.N. (1983) Multipole theory of the polarization of solids by point defects. I. Dipole approximation. *Phys. Status Solidi B*, **120**, 77–86.

49 Kantorovich, L.N. (1988) An embedded-molecular-cluster method for calculating the electronic structure of point defects in non-metallic crystals. I. General theory. *J. Phys. C Solid State*, **21**, 5041.

50 Manby, F.R., Stella, M., Goodpaster, J.D., and Miller, T.F. (2012) A simple, exact density-functional-theory embedding scheme. *J. Chem. Theory Comput.*, **8**, 2564–2568.

51 Milov, I.D., Abarenkov, I.V., and Tupitsyn, I.I. (2015) Application of the embedding potential method in calculations of the electronic structure and X-ray emission spectra of crystal MgO clusters. *Opt. Spectrosc.*, **118**, 519–528.

52 Seijo, L. and Barandiarán, Z. (1996) Applications of the group-function theory to the field of materials science. *Int. J. Quantum Chem.*, **60**, 617–634.

53 Shidlovskaya, E.K. (2002) Improved embedded molecular cluster model. *Int. J. Quantum Chem.*, **89**, 349–370.

54 Tchougréeff, A.L. (1999) Group functions, Lowdin partition, and hybrid QC/MM methods for large molecular systems. *Phys. Chem. Chem. Phys.*, **1**, 1051–1060.

55 Tchougréeff, A.L. (2016) Several stories from theoretical chemistry with some Russian flavor and implications for theorems of chemistry, vagueness of its concepts, fuzziness of its definitions, iconicity of its language, and peculiarities of its nomenclature. *Int. J. Quantum Chem.*, **116**, 137–160.

56 Wesolowski, T.A., Shedge, S., and Zhou, X. (2015) Frozen-density embedding strategy for multilevel simulations of electronic structure. *Chem. Rev.*, **115**, 5891–5928.

57 Fabiano, E., Laricchia, S., and Sala, F.D. (2014) Frozen density embedding with non-integer subsystems' particle numbers. *J. Chem. Phys.*, **140**, 114101.

58 Huang, C., Pavone, M., and Carter, E.A. (2011) Quantum mechanical embedding theory based on a unique embedding potential. *J. Chem. Phys.*, **134**, 154110.

59 Peng, Q., Zhang, X., Hung, L., Carter, E.A., and Lu, G. (2008) Quantum simulation of materials at micron scales and beyond. *Phys. Rev. B*, **78**, 054118.

60 Stefanovich, E.V. and Truong, T.N. (1997) A theoretical approach for

modeling reactivity at solid–liquid interfaces. *J. Chem. Phys.*, **106**, 7700–7705.

61 Kantorovich, L.N. (2000a) Application of the group function theory to infinite systems. *Int. J. Quantum Chem.*, **76**, 511–534.

62 Kantorovich, L.N. (2000b) Derivation of atomistic models for lattices consisting of weakly overlapping structural elements. *Int. J. Quantum Chem.*, **78**, 306–330.

63 Grimley, T.B. and Pisani, C. (1974) Chemisorption theory in the Hartree–Fock approximation. *J. Phys. C Solid State*, **7**, 2831.

64 Inglesfield, J.E. (1981) A method of embedding. *J. Phys. C Solid State*, **14**, 3795.

65 James, R. and Woodley, S. (1996) Extraction of Green's functions from total energy plane wave pseudopotential calculations. *Solid State Commun.*, **97**, 935–939.

66 Pisani, C. (1978) Approach to the embedding problem in chemisorption in a self-consistent-field-molecular-orbital formalism. *Phys. Rev. B*, **17**, 3143–3153.

67 Pisani, C., Dovesi, R., and Carosso, P. (1979) Moderately-large-embedded-cluster approach to the study of local defects in solids. Vacancy and substitutional impurities in graphite. *Phys. Rev. B*, **20**, 5345–5357.

68 Woodley, S.M. (1997) A real-space approach to surface and defect states, University of Bath, UK.

69 Evarestov, R.A., Sokolov, A.R., Leko, A.V., and Verjazov, V.A. (1989) Application of embedded-cluster models in the theory of local centres in graphite. *J. Phys. Condens. Matter*, **1**, 6611.

70 Gao, S. and Wang, D. (1990) An embedding scheme for calculating charge transfer between adsorbates and metal substrates. *J. Phys. Condens. Matter*, **2**, 2187.

71 Kruger, S. and Rösch, N. (1994) The moderately-large-embedded-cluster method for metal surfaces; a density-functional study of atomic adsorption. *J. Phys. Condens. Matter*, **6**, 8149.

72 Harding, J.H., Harker, A.H., Keegstra, P.B., Pandey, R., Vail, J.M., and Woodward, C. (1985) Hartree–Fock cluster computations of defect and perfect ionic crystal properties. *Physica B+C*, **131**, 151–156.

73 Pascual, J.L. and Seijo, L. (1995) Ab initio model potential embedded cluster calculations including lattice relaxation and polarization: Local distortions on Mn^{2+}-doped CaF_2. *J. Chem. Phys.*, **102**, 5368–5376.

74 Pisani, C., Corà, F., Nada, R., and Orlando, R. (1994) Hartree–Fock perturbed-cluster treatment of local defects in crystals: I. The EMBED program: general features. *Comput. Phys. Commun.*, **82**, 139–156.

75 Kunz, A.B. and Vail, J.M. (1988) Quantum-mechanical cluster-lattice interaction in crystal simulation: Hartree–Fock method. *Phys. Rev. B*, **38**, 1058–1063.

76 Vail, J.M. (1990) Theory of electronic defects: applications to MgO and alkali halides. *J. Phys. Chem. Solids*, **51**, 589–607.

77 Shluger, A.L., Harker, A.H., Puchin, V.E., Itoh, N., and Catlow, C.R.A. (1993) Simulation of defect processes: experiences with the self-trapped exciton. *Model. Simul. Mat. Sci. Eng.*, **1**, 673.

78 Shluger, A.L., Kotomin, E.A., and Kantorovich, L.N. (1986) Quantum-chemical simulation of impurity-induced trapping of a hole: (Li) 0 centre in MgO. *J. Phys. C Solid State*, **19**, 4183.

79 Mukhopadhyay, S., Sushko, P.V., Stoneham, A.M., and Shluger, A.L. (2004) Modeling of the structure and properties of oxygen vacancies in amorphous silica. *Phys. Rev. B*, **70**, 195203.

80 Sulimov, V.B., Sushko, P.V., Edwards, A.H., Shluger, A.L., and Stoneham, A.M. (2002) Asymmetry and long-range character of lattice deformation by neutral oxygen vacancy in α-quartz. *Phys. Rev. B*, **66**, 024108.

81 Sushko, P.V., Shluger, A.L., and Catlow, C.R.A. (2000) Relative energies of surface and defect states: *ab initio* calculations for the MgO (001) surface. *Surf. Sci.*, **450**, 153–170.

82 Pascual, J.L., Barros, N., Barandiarán, Z., and Seijo, L. (2009) Improved embedding

ab initio model potentials for embedded cluster calculations†. *J. Phys. Chem. A*, **113**, 12454–12460.

83 Seijo, L. and Barandiarán, Z. (1999) The ab initio model potential method: a common strategy for effective core potential and embedded cluster calculations, in *Computational Chemistry: Reviews of Current Trends*, World Scientific, pp. 55–152.

84 Bakowies, D. and Thiel, W. (1996b) Semiempirical treatment of electrostatic potentials and partial charges in combined quantum mechanical and molecular mechanical approaches. *J. Comput. Chem.*, **17**, 87–108.

85 Bredow, T., Geudtner, G., and Jug, K. (1996) Embedding procedure for cluster calculations of ionic crystals. *J. Chem. Phys.*, **105**, 6395–6400.

86 Greatbanks, S.P., Sherwood, P., and Hillier, I.H. (1994) Embedded cluster model for the *ab initio* study of broensted acidity in zeolites. *J. Phys. Chem.*, **98**, 8134–8139.

87 Nasluzov, V.A., Ivanova, E.A., Shor, A.M., Vayssilov, G.N., Birkenheuer, U., and Rösch, N. (2003) Elastic polarizable environment cluster embedding approach for covalent oxides and zeolites based on a density functional method. *J. Phys. Chem. B*, **107**, 2228–2241.

88 Sierka, M. and Joachim Sauer, A. (1997) Structure and reactivity of silica and zeolite catalysts by a combined quantum mechanics–shell-model potential approach based on DFT. *Faraday Discuss.*, **106**, 41–62.

89 Pisani, C. (1993) Embedded-cluster techniques for the quantum-mechanical study of surface reactivity. *J. Mol. Catal.*, **82**, 229–252.

90 Field, M.J., Bash, P.A., and Karplus, M. (1990) A combined quantum mechanical and molecular mechanical potential for molecular dynamics simulations. *J. Comput. Chem.*, **11**, 700–733.

91 Brinkmann, L., Heifets, E., and Kantorovich, L. (2014) Density functional calculations of extended, periodic systems using Coulomb corrected molecular fractionation with conjugated caps method (CC-MFCC). *Phys. Chem. Chem. Phys.*, **16**, 21252–21270.

92 Gordon, M.S., Fedorov, D.G., Pruitt, S.R., and Slipchenko, L.V. (2012) Fragmentation methods: a route to accurate calculations on large systems. *Chem. Rev.*, **112**, 632–672.

93 Yang, W. (1991) Direct calculation of electron density in density-functional theory. *Phys. Rev. Lett.*, **66**, 1438–1441.

94 Sherwood, P., de Vries, A.H., Guest, M.F., Schreckenbach, G., Catlow, C.R.A., French, S.A., Sokol, A.A., Bromley, S.T., Thiel, W., Turner, A.J. et al. (2003) QUASI: A general purpose implementation of the QM/MM approach and its application to problems in catalysis. *J. Mol. Struct.*, **632**, 1–28.

95 Keal, T.W., Sherwood, P., Dutta, G., Sokol, A.A., and Catlow, C.R.A. (2011) Characterization of hydrogen dissociation over aluminium-doped zinc oxide using an efficient massively parallel framework for QM/MM calculations. *Proc. Math. Phys. Eng. Sci.*, **467**, 1900–1924.

96 Metz, S., Kästner, J., Sokol, A.A., Keal, T.W., and Sherwood, P. (2014) ChemShell – a modular software package for QM/MM simulations. *Wiley Interdiscip. Rev. Comput. Mol. Sci.*, **4**, 101–110.

97 Sherwood, P. (2000) Hybrid quantum mechanics/molecular mechanics approaches, in *Modern Methods and Algorithms of Quantum Chemistry* (ed. J. Grotendorst), Neuman Institute for Computing, Julich, Germany, p. 285.

98 Svensson, M., Humbel, S., Froese, R.D.J., Matsubara, T., Sieber, S., and Morokuma, K. (1996) ONIOM: A multilayered integrated MO + MM method for geometry optimizations and single point energy predictions. A test for Diels–Alder reactions and Pt(P(t-Bu)$_3$)$_2$ + H$_2$ oxidative addition. *J. Phys. Chem.*, **100**, 19357–19363.

99 Tchougréeff, A.L. (2008) *Hybrid Methods of Molecular Modeling*, 1st edn, Springer Netherlands, Netherlands.

100 Bakowies, D. and Thiel, W. (1996a) Hybrid models for combined quantum mechanical and molecular mechanical

approaches. *J. Phys. Chem.*, **100**, 10580–10594.

101 Kästner, J., Carr, J.M., Keal, T.W., Thiel, W., Wander, A., and Sherwood, P. (2009) DL-FIND: an open-source geometry optimizer for atomistic simulations. *J. Phys. Chem. A*, **113**, 11856–11865.

102 Kästner, J., Thiel, S., Senn, H.M., Sherwood, P., and Thiel, W. (2007) Exploiting QM/MM capabilities in geometry optimization: a microiterative approach using electrostatic embedding. *J. Chem. Theory Comput.*, **3**, 1064–1072.

103 Sokol, A.A., Bromley, S.T., French, S.A., Catlow, C.R.A., and Sherwood, P. (2004) Hybrid QM/MM embedding approach for the treatment of localized surface states in ionic materials. *Int. J. Quantum Chem.*, **99**, 695–712.

104 Jost, W. (1933) Diffusion and electrolytic conduction in crystals (ionic semiconductors). *J. Chem. Phys.*, **1**, 466–475.

105 Born, M. (1920) Volumen und Hydratationswärme der Ionen. *Zeitschrift für Physik*, **1**, 45–48.

106 Onsager, L. (1936) Electric moments of molecules in liquids. *J. Am. Chem. Soc.*, **58**, 1486–1493.

107 Scanlon, D.O., Dunnill, C.W., Buckeridge, J., Shevlin, S.A., Logsdail, A.J., Woodley, S.M., Catlow, C.R.A., Powell, M.J., Palgrave, R.G., and Parkin, I.P. (2013) Band alignment of rutile and anatase TiO_2. *Nat. Mater.*, **12**, 798–801.

108 Kavan, L., Grätzel, M., Gilbert, S.E., Klemenz, C., and Scheel, H.J. (1996) Electrochemical and photoelectrochemical investigation of single-crystal anatase. *J. Am. Chem. Soc.*, **118**, 6716–6723.

109 Buckeridge, J., Butler, K.T., Catlow, C.R.A., Logsdail, A.J., Scanlon, D.O., Shevlin, S.A., Woodley, S.M., Sokol, A.A., and Walsh, A. (2015a) Polymorph engineering of TiO_2: demonstrating how absolute reference potentials are determined by local coordination. *Chem. Mater.*, **27**, 3844–3851.

110 Walsh, A., Catlow, C.R.A., Miskufova, M., and Sokol, A.A. (2011) Electron and hole stability in GaN and ZnO. *J. Phys. Condens. Matter*, **23**, 334217.

111 Butler, K.T., Buckeridge, J., Catlow, C.R.A., and Walsh, A. (2014) Crystal electron binding energy and surface work function control of tin dioxide. *Phys. Rev. B*, **89**, 115320.

112 Kantorovich, L.N., Shluger, A.L., Sushko, P.V., Gunster, J., Stracke, P., Goodman, D.W., and Kempter, V. (1999) Mg clusters on MgO surfaces: study of the nucleation mechanism with MIES and abinitio calculations. *Faraday Discuss.*, **114**, 173–194.

113 Downing, C.A., Sokol, A.A., and Catlow, C.R.A. (2014b) The reactivity of CO2 on the MgO(100) surface. *Phys. Chem. Chem. Phys.*, **16**, 184–195.

114 Ferrari, A.M. and Pacchioni, G. (1995) Electronic structure of F and V centers on the MgO surface. *J. Phys. Chem.*, **99**, 17010–17018.

115 Norgett, M.J. (1971) The electronic levels of the Ft centre in MgO and CaO. *J. Phys. C Solid State*, **4**, 1289.

116 Paganini, M.C., Chiesa, M., Giamello, E., Coluccia, S., Martra, G., Murphy, D.M., and Pacchioni, G. (1999) Colour centres at the surface of alkali-earth oxides. A new hypothesis on the location of surface electron traps. *Surf. Sci.*, **421**, 246–262.

117 Sousa, C., Pacchioni, G., and Illas, F. (1999) *Ab initio* study of the optical transitions of F centers at low-coordinated sites of the MgO surface. *Surf. Sci.*, **429**, 217–228.

118 Catlow, C.R.A. and Mackrodt, W.C. (1982) Theory of simulation methods for lattice and defect energy calculations in crystals, in *Computer Simulation of Solids* (eds C.R.A. Catlow and W.C. Mackrodt), Springer, Berlin, pp. 1–20.

119 Lidiard, A.B. (1989) The Mott–Littleton method: an introductory survey. *J. Chem. Soc., Farad. Trans.*, **85**, 341–349.

120 Mott, N.F. and Littleton, M.J. (1938) Conduction in polar crystals. I. Electrolytic conduction in solid salts. *Trans. Faraday Soc.*, **34**, 485–499.

121 Norgett, M.J., Establishment, A.E.R. Authority, U.K.A.E., and H.M.S.O. (1974) *A General Formulation of the Problem of Calculating the Energies of Lattice Defects in Ionic Crystals*, H.M. Stationery Office.

122 Berger, D., Logsdail, A.J., Oberhofer, H., Farrow, M.R., Catlow, C.R.A., Sherwood, P., Sokol, A.A., Blum, V., and Reuter, K. (2014) Embedded-cluster calculations in a numeric atomic orbital density-functional theory framework. *J. Chem. Phys.*, **141**, 024105.

123 Berger, D., Oberhofer, H., and Reuter, K. (2015) First-principles embedded-cluster calculations of the neutral and charged oxygen vacancy at the rutile $TiO_2(110)$ surface. *Phys. Rev. B*, **92**, 075308.

124 Downing, C.A., Sokol, A.A., and Catlow, C.R.A. (2014a) The reactivity of CO_2 and H_2 at trapped electron sites at an oxide surface. *Phys. Chem. Chem. Phys.*, **16**, 21153–21156.

125 Dutta, G., Sokol, A.A., Catlow, C.R.A., Keal, T.W., and Sherwood, P. (2012) Activation of carbon dioxide over zinc oxide by localised electrons. *ChemPhysChem*, **13**, 3453–3456.

126 Goumans, T.P.M., Catlow, C.R.A., and Brown, W.A. (2008) Hydrogenation of CO on a silica surface: An embedded cluster approach. *J. Chem. Phys.*, **128**, 134709.

127 Herschend, B., Baudin, M., and Hermansson, K. (2004) A combined molecular dynamics+quantum mechanics method for investigation of dynamic effects on local surface structures. *J. Chem. Phys.*, **120**, 4939–4948.

128 Logsdail, A.J., Downing, C.A., Keal, T.W., Sherwood, P., Sokol, A.A., and Catlow, C.R.A. (2016) Modelling the chemistry of Mn-doped MgO for bulk and (100) surfaces. *Phys. Chem. Chem. Phys.*, **18**, 28648.

129 Buckeridge, J., Bromley, S.T., Walsh, A., Woodley, S.M., Catlow, C.R.A., and Sokol, A.A. (2013) One-dimensional embedded cluster approach to modeling CdS nanowires. *J. Chem. Phys.*, **139**, 124101.

130 Oliva, C., Strodel, P., Goldbeck-Wood, G., and Maiti, A. (2008) Understanding the interaction of ammonia with carbon nanotubes.

131 Chen, H. and Ortner, C. (2015a) QM/MM methods for crystalline defects. Part 1: Consistent Energy and Force-Mixing. ArXiv e-prints ArXiv:150906627.

132 Chen, H. and Ortner, C. (2015b) QM/MM methods for crystalline defects. Part 1: Locality of the tight binding model. ArXiv e-prints ArXiv:150505541.

133 Fang, Z., Bueken, B., D.E., DeVos., and Fischer, R.A. (2015) Defect-engineered metal–organic frameworks. *Angew. Chem., Int. Ed.*, **54**, 7234–7254.

134 Fang, Z., Dürholt, J.P., Kauer, M., Zhang, W., Lochenie, C., Jee, B., Albada, B., Metzler-Nolte, N., Pöppl, A., Weber, B. et al. (2014) Structural complexity in metal–organic frameworks: simultaneous modification of open metal sites and hierarchical porosity by systematic doping with defective linkers. *J. Am. Chem. Soc.*, **136**, 9627–9636.

135 Peng, Q., Zhang, X., and Lu, G. (2010) Quantum mechanical simulations of nanoindentation of Al thin film. *Comp. Mater. Sci.*, **47**, 769–774.

136 Vörös, M., Gali, A., Kaxiras, E., Frauenheim, T., and Knaup, J.M. (2012) Identification of defects at the interface between 3C-SiC quantum dots and a SiO2 embedding matrix. *Phys. Status Solidi B*, **249**, 360–367.

137 Zhang, X. and Lu, G. (2008) Electronic origin of void formation in fcc metals. *Phys. Rev. B*, **77**, 174102.

138 Zhang, X. and Lu, G. (2010) Calculation of fast pipe diffusion along a dislocation stacking fault ribbon. *Phys. Rev. B*, **82**, 012101.

139 Zhang, X., Zhao, Y., and Lu, G. (2012) Recent development in quantum mechanics/molecular mechanics modeling for materials. *Int. J. Multiscale Comput. Eng.*, **10**, 65–82.

140 Buckeridge, J., Catlow, C.R.A., Scanlon, D.O., Keal, T.W., Sherwood, P., Miskufova, M., Walsh, A., Woodley, S.M., and Sokol, A.A. (2015b) Buckeridge et al. reply. *Phys. Rev. Lett.*, **115**, 029702.

141 Buckeridge, J., Catlow, C.R.A., Scanlon, D.O., Keal, T.W., Sherwood, P., Miskufova, M., Walsh, A., Woodley, S.M., and Sokol, A.A. (2015c) Determination of the nitrogen vacancy as a shallow compensating center in gan doped with divalent metals. *Phys. Rev. Lett.*, **114**, 016405.

142 Sokol, A.A., French, S.A., Bromley, S.T., Catlow, C.R.A., H.J.J., vanDam., and Sherwood, P. (2007) Point defects in ZnO. *Faraday Discuss.*, **134**, 267–282.

143 Catlow, C.R.A., Sokol, A.A., and Walsh, A. (2011) Microscopic origins of electron and hole stability in ZnO. *Chem. Commun.*, **47**, 3386–3388.

144 Walsh, A., Buckeridge, J., Catlow, C.R.A., Jackson, A.J., Keal, T.W., Miskufova, M., Sherwood, P., Shevlin, S.A., Watkins, M.B., Woodley, S.M. et al. (2013) Limits to doping of wide band gap semiconductors. *Chem. Mater.*, **25**, 2924–2926.

145 Ellmer, K. (2008) Electrical properties, in *Transparent Conductive Zinc Oxide: Basics and Applications in Thin Film Solar Cells* (eds K. Ellmer, A. Klein, and B. Rech), Springer, Berlin, pp. 35–78.

146 Morkoç, H. (2009) *Handbook of Nitride Semiconductors and Devices: Materials Properties, Physics and Growth*, vol. **1**, Wiley-VCH Verlag GmbH & Co. KGaA, Weinheim.

147 Kröger, F.A. and Vink, H.J. (1956) Relations between the concentrations of imperfections in crystalline solids, in *Solid State Physics* (eds S. Frederick and T. David), Academic Press, pp. 307–435.

148 Buckeridge, J. and Sokol, A.A. (2016) One-dimensional nanosystems, in *Computational Modeling of Inorganic Nanomaterials*, CRC Press, pp. 47–81.

149 Farrow, M.R. and Probert, M.I.J. (2011) Atomistic molecular dynamics simulations of shock compressed quartz. *J. Chem. Phys.*, **135**, 044508.

150 Freysoldt, C., Neugebauer, J., and Van de Walle, C.G. (2009) Fully *ab initio* finite-size corrections for charged-defect supercell calculations. *Phys. Rev. Lett.*, **102**, 016402.

151 Gavartin, J.L., Shluger, A.L., Foster, A.S., and Bersuker, G.I. (2005) The role of nitrogen-related defects in high-k dielectric oxides: density-functional studies. *J. Appl. Phys.*, **97**, 053704.

152 Kantorovich, L.N. (1999) Elimination of the long-range dipole interaction in calculations with periodic boundary conditions. *Phys. Rev. B*, **60**, 15476–15479.

153 Lento, J., Mozos, J.-L., and Nieminen, R.M. (2002) Charged point defects in semiconductors and the supercell approximation. *J. Phys. Condens. Matter*, **14**, 2637.

154 Leslie, M. and Gillan, N.J. (1985) The energy and elastic dipole tensor of defects in ionic crystals calculated by the supercell method. *J. Phys. C Solid State*, **18**, 973.

155 Makov, G. and Payne, M.C. (1995) Periodic boundary conditions in *ab initio* calculations. *Phys. Rev. B*, **51**, 4014–4022.

156 Murphy, S.T. and Hine, N.D.M. (2013) Anisotropic charge screening and supercell size convergence of defect formation energies. *Phys. Rev. B*, **87**, 094111.

157 Schultz, P.A. (2000) Charged local defects in extended systems. *Phys. Rev. Lett.*, **84**, 1942–1945.

158 Wilson, D.J., Sokol, A.A., French, S.A., and Catlow, C.R.A. (2008) Defect structures in the silver halides. *Phys. Rev. B*, **77**, 064115.

159 Xie, Z., Sui, Y., Buckeridge, J., Catlow, C.R.A., Keal, T.W., Sherwood, P., Walsh, A., Scanlon, D.O., Woodley, S.M., and Sokol, A.A. (2016) Demonstration of the donor characteristics of Si and O defects in GaN using hybrid QM/MM. *Phys. Status Solidi A*, 1–5.

160 Wilson, P.J., Bradley, T.J., and Tozer, D.J. (2001) Hybrid exchange-correlation functional determined from thermochemical data and *ab initio* potentials. *J. Chem. Phys.*, **115**, 9233–9242.

161 Zhao, Y., Lynch, B.J., and Truhlar, D.G. (2004) Development and assessment of a new hybrid density functional model for thermochemical kinetics. *J. Phys. Chem. A*, **108**, 2715–2719.

162 Götz, W., Johnson, N.M., Chen, C., Liu, H., Kuo, C., and Imler, W. (1996) Activation energies of Si donors in GaN. *Appl. Phys. Lett.*, **68**, 3144–3146.

163 Zolper, J.C., Wilson, R.G., Pearton, S.J., and Stall, R.A. (1996) Ca and O ion implantation doping of GaN. *Appl. Phys. Lett.*, **68**, 1945–1947.

164 Brändle, M. and Sauer, J. (1998) Acidity differences between inorganic solids

induced by their framework structure. A combined quantum mechanics/molecular mechanics ab initio study on zeolites. *J. Am. Chem. Soc.*, **120**, 1556–1570.

165 Brändle, M., Sauer, J., Dovesi, R., and Harrison, N.M. (1998) Comparison of a combined quantum mechanics/interatomic potential function approach with its periodic quantum-mechanical limit: proton siting and ammonia adsorption in zeolite chabazite. *J. Chem. Phys.*, **109**, 10379–10389.

166 Eichler, U., Brändle, M., and Sauer, J. (1997) Predicting absolute and site specific acidities for zeolite catalysts by a combined quantum mechanics/interatomic potential function approach. *J. Phys. Chem. B*, **101**, 10035–10050.

167 Haase, F. and Sauer, J. (2000) Ab initio molecular dynamics simulation of methanol interacting with acidic zeolites of different framework structure. *Microporous Mesoporous Mater.*, **35–36**, 379–385.

168 Haase, F., Sauer, J., and Hutter, J. (1997) Ab initio molecular dynamics simulation of methanol adsorbed in chabazite. *Chem. Phys. Lett.*, **266**, 397–402.

169 Sierka, M., Eichler, U., Datka, J., and Sauer, J. (1998) Heterogeneity of brønsted acidic sites in faujasite type zeolites due to aluminum content and framework structure. *J. Phys. Chem. B*, **102**, 6397–6404.

170 de Vries, A.H., Sherwood, P., Collins, S.J., Rigby, A.M., Rigutto, M., and Kramer, G.J. (1999) Zeolite structure and reactivity by combined quantum-chemical–classical calculations. *J. Phys. Chem. B*, **103**, 6133–6141.

171 Ivanova Shor, E.A., Shor, A.M., Nasluzov, V.A., Vayssilov, G.N., and Rösch, N. (2005) Effects of the aluminum content of a zeolite framework: a DFT/MM hybrid approach based on cluster models embedded in an elastic polarizable environment. *J. Chem. Theory Comput.*, **1**, 459–471.

172 O'Malley, A.J., Logsdail, A.J., Sokol, A.A., and Catlow, C.R.A. (2016) Modelling metal centres, acid sites and reaction mechanisms in microporous catalysts. *Faraday Discuss.*, **188**, 235–255.

173 Rybicki, M. and Sauer, J. (2015) Acidity of two-dimensional zeolites. *Phys. Chem. Chem. Phys.*, **17**, 27873–27882.

174 Sauer, J. (2016) Brønsted activity of two-dimensional zeolites compared to bulk materials. *Faraday Discuss.*, **188**, 227–234.

175 Chang, C.D. and Silvestri, A.J. (1977) The conversion of methanol and other O-compounds to hydrocarbons over zeolite catalysts. *J. Catal.*, **47**, 249–259.

176 Maiden, C. (1988) The New Zealand gas-to-gasoline project. *Stud. Surf. Sci. Catal.*, **36**, 1–16.

177 To, J., Sokol, A.A., French, S.A., Catlow, C.R.A., Sherwood, P., and van Dam, H.J.J. (2006b) Formation of heteroatom active sites in zeolites by hydrolysis and inversion. *Angew. Chem., Int. Ed.*, **45**, 1633–1638.

178 Taramasso, M., Perego, G., and Notari, B. (1983) Preparation of porous crystalline synthetic material comprised of silicon and titanium oxides (Google Patents).

179 Bordiga, S., Damin, A., Berlier, G., Bonino, F., Ricchiardi, G., Zecchina, A., and Lamberti, C. (2001) The role of isolated sites in heterogeneous catalysis: characterization and modeling. *Int. J. Mol. Sci.*, **2**, 167.

180 French, S.A., Sokol, A.A., To, J., Catlow, C.R.A., Phala, N.S., and Klatt, G., and van Steen, E. (2004) Active sites for heterogeneous catalysis by functionalisation of internal and external surfaces. *Catal. Today*, **93–5**, 535–540.

181 To, J., Sherwood, P., Sokol, A.A., Bush, I.J., Catlow, C.R.A., van Dam, H.J.J., French, S.A., and Guest, M.F. (2006a) QM/MM modelling of the TS-1 catalyst using HPCx. *J. Mater. Chem.*, **16**, 1919–1926.

182 Gleeson, D., Sankar, G., Catlow, C.R.A., Thomas, J.M., Spano, G., Bordiga, S., Zecchina, A., and Lamberti, C. (2000) The architecture of catalytically active centers in titanosilicate (TS-1) and related selective-oxidation catalysts. *Phys. Chem. Chem. Phys.*, **2**, 4812–4817.

183 Barker, C.M., Gleeson, D., Kaltsoyannis, N., Catlow, C.R.A., Sankar, G., and Thomas, J.M. (2002) On the structure and coordination of the oxygen-donating

species in Ti up arrow MCM-41/TBHP oxidation catalysts: a density functional theory and EXAFS study. *Phys. Chem. Chem. Phys.*, **4**, 1228–1240.

184 To, J., Sokol, A.A., French, S.A., and Catlow, C.R.A. (2008) Hybrid QM/MM investigations into the structure and properties of oxygen-donating species in TS-1. *J. Phys. Chem. C*, **112**, 7173–7185.

185 Shetti, V.N., Manikandan, P., Srinivas, D., and Ratnasamy, P. (2003) Reactive oxygen species in epoxidation reactions over titanosilicate molecular sieves. *J. Catal.*, **216**, 461–467.

186 Piccini, G., Alessio, M., and Sauer, J. (2016) Ab initio calculation of rate constants for molecule–surface reactions with chemical accuracy. *Angew. Chem., Int. Ed.*, **55**, 5235–5237.

187 Catlow, C.R.A. (2016) Prediction of rate constants for catalytic reactions with chemical accuracy. *Angew. Chem., Int. Ed.*, **55**, 9132–9133.

18
Modeling Crystal Nucleation and Growth and Polymorphic Transitions

Dirk Zahn

Friedrich-Alexander Universität Erlangen-Nürnberg, Lehrstuhl für Theoretische Chemie/Computer Chemie Centrum, Nägelsbachstraße 25, 91052 Erlangen, Germany

18.1
Introduction

The aim of this chapter is to bring together molecular simulation techniques and recent progress in exploring crystal formation. The latter involves solute association processes prior to nucleation, the evolution of forming aggregates, and later stages of crystal growth of mature nuclei. For each of these stages, recent observations of prenucleation species and multistage crystal nucleation processes have challenged the classical view on the thermodynamics and kinetics of crystal formation. Section 18.2 is largely adopted from a recent extension to classical nucleation theory as reported in Ref. [1]. To provide a general mainframe to characterizing crystal nucleation and growth, we discuss classical nucleation theory with a particular emphasis on going beyond the conventional implementation as used for more than a century by now. Apart from the classical concept "solution → nucleus → crystal," also nucleation inhibitors, precursor clusters, and multistep (nonclassical) nucleation processes are rationalized by the competition of interface and bulk energy terms. To a large extent, modern views on crystal formation are based on molecular simulation, in particular molecular dynamics simulations combined with advanced sampling techniques to bridge the timescale–length scale problem inherent to nucleation studies. In Section 18.3 we briefly summarize the molecular dynamics simulation method and describe special techniques for modeling both nucleation-controlled and diffusion-controlled aggregate formation. Moreover, we discuss approaches to study structural transitions as required for understanding secondary nucleation processes that may occur during mature stages of aggregate growth. Size-induced solid–solid transformations of forming aggregates are particularly relevant to molecular crystals comprising a considerable variety of molecular conformation, intermolecular binding, and packing into ordered motifs. A series of illustrations is provided in Section 18.4. Clearly, the manifold of studies characterizing prenucleation

Handbook of Solid State Chemistry, First Edition. Edited by Richard Dronskowski, Shinichi Kikkawa, and Andreas Stein.
© 2017 Wiley-VCH Verlag GmbH & Co. KGaA. Published 2017 by Wiley-VCH Verlag GmbH & Co. KGaA.

events, solute aggregation, crystal nucleation, and nuclei evolution thereafter is immense. A full account drastically exceeds the scope of this chapter. We therefore focus on a selection of case studies elaborated for zinc oxide – a popular, yet arbitrarily chosen example that has been intensely studied in our group. In Section 18.5, we review past success and future perspectives of molecular simulations in gradually evolving toward a reliable tool for the rational design of solid materials.

18.2
Thermodynamics and Kinetics of Classical and Noncrystal Nucleation

Since the seminar work of Ostwald in the late nineteenth century, crystal nucleation and growth remained of constant (if not even increasing) interest [2]. Today, applications of crystal engineering range from quantum dots and nanomaterials to biomimetic composites and pharmaceutical formulation. In parallel to pushing the limits of creating new materials by brute-force methods such as high-throughput screening robots, we are experiencing considerable improvements of experimental investigation techniques [3] and simulation methods [4]. The combination of both types of characterization approaches is currently paving the way to an increasingly rational design of crystalline compounds. The key to this highly desirable, but still rather far goal is in-depth understanding of the molecular scale processes involved.

The need for molecular scale understanding recently increased sharply, as the observation of multistage nucleation processes and the identification of prenucleation clusters challenge our current theoretical mainframe for rationalizing the underlying thermodynamics. For such phenomena, the traditional concept of classical nucleation theory faces serious shortcomings [5,6] – and eventually motivated the term "nonclassical nucleation." To exploit new perspectives of solid-state syntheses by designing precursor solutions and/or triggering secondary nucleation events, extensions to classical nucleation theory are required. This chapter provides a survey on the thermodynamics and kinetics of nucleation, be it multistep or straight (e.g., "classical"), also considering the recently identified prenucleation clusters and polymorphic transitions during crystal growth.

18.2.1
Classical and Nonclassical Nucleation

A simple and intuitive rationalization of crystal formation was provided by Gibbs contrasting two key driving forces of which one promotes and the other one disfavors the formation of a crystal [7,8]. From a purely thermodynamic point of view, crystallization should occur when supersaturation, undercooling, or pressure gives rise to the crystalline phase being more stable than the corresponding solution or melt, respectively. However, immediate phase transfer is typically observed only upon rather drastic favoring of the crystalline state, as, for example, by vaporization of the solvent. Unlike crystal formation by spinodal decomposition of the pristine phase, the more common scenario is that of crystal formation being hindered by a barrier in free energy. Classical nucleation theory (CNT) relates this barrier to the need of a forming crystal nucleus to establish an interface to the surrounding melt, vapor, or solution. Surface tension and unfavorable interactions at the interface give rise to an increase in free energy that scales with the surface area A of the forming nucleus. On the other hand, favorable interactions within the inner core of the forming nucleus lead to a prospective gain in free energy – which scales with the volume V of the nucleus. The central merit of CNT is to describe the competition of both aspects by two simple terms in order to provide an energy profile as a function of nucleus size. In what follows, $G(N)$ is considered as the free energy difference in comparing the dispersed solution comprising N solutes with an analogous system in which all N solutes form an aggregate and are embedded by equal amounts of solvent molecules and modeled at identical temperature and pressure as considered for the dispersed solution.

$$G(N) = g_{surface}(A) - g_{bulk}(V) \tag{18.1}$$

For spherical nuclei, the two energy terms may be written as a function of the radius, although a shape-independent formulation of CNT may be obtained by considering the surface and the bulk energy terms as a function of the number of solutes N in the nucleus. Assuming the inner structure of the nucleus as identical to that of the final crystal, the gain in free energy upon crystallization reads $-\mu \cdot N$ with μ being the change in free energy per solute. A further assumption often used relies on the nuclei to maintain a constant shape (such as spheres, cubes, prisms, polyhedra, etc.) during the whole formation process. In this case the surface area is given by $f \cdot N^{2/3}$ with f being a constant depending on the habit of the nucleus. For example, a spherical shape would lead to

$$N = \rho \cdot \frac{4}{3}\pi r^3 \quad \text{and} \quad A_{surface} = 4\pi r^2 = \left(\sqrt[3]{36\pi} \cdot \rho^{-2/3}\right) \cdot N^{2/3}, \tag{18.2}$$

where ρ refers to the particle density of the forming crystal. This allows describing nucleation free energy as a function of the number of precipitated solutes:

$$G(N) = c_{surface} \cdot N^{2/3} - \mu \cdot N \tag{18.3}$$

from which the nucleation barrier and the critical nucleus size is deduced as

$$0 = \left.\frac{\partial G(N)}{\partial N}\right|_{N_{crit}} \to N_{crit} = \left(\frac{2c_{surf}}{3\mu}\right)^3; \quad \Delta G = \frac{4}{27} \frac{c_{surf}^3}{\mu^2}. \tag{18.4}$$

It is important to point out that both the surface and the volume terms are related to constant prefactors $c_{surface}$ and μ, respectively. On this basis, CNT provides a simple rationale of the free energy profile as a function of size (Figure 18.1a). Results should however be regarded from a more qualitative viewpoint as there are a number of examples that showed the shortcomings of assuming constant shape and inner structure of the nuclei.

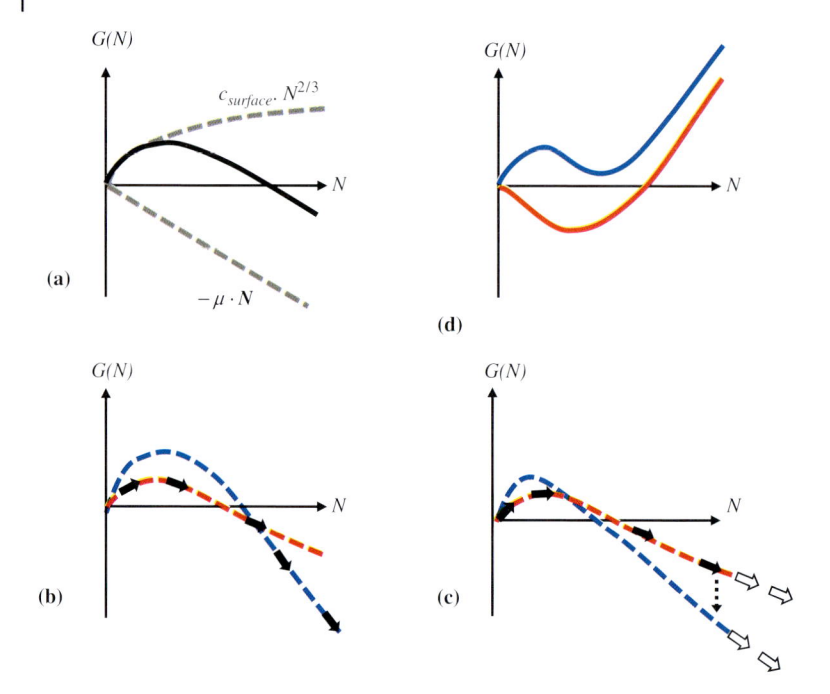

Figure 18.1 (a) Classical nucleation pathway. Small nuclei are dominated by interface/surface domains accounting for an increase in free energy, while sufficiently large nuclei are stabilized from favorable packing in the bulk leading to a net gain in free energy. (b) Two-step nucleation mechanism with low secondary nucleation barrier. The initially formed phase A (red curve) is that of lowest nucleation barrier, that is, nuclei of comparably low surface tension/interface energy are observed. Upon later stages of nuclei growth, the core domain in the aggregate becomes dominant and transformation to phase B (blue curve) is driven by more favorable packing in the bulk. (c) Competition of crystal structures with large barriers to polymorphic transitions. Same as part (b), but for weaker thermodynamic preference of the $A \to B$ transition and/or larger barrier to the secondary nucleation event. The size-induced transformation may be subject to large hysteresis effects or even inhibit structural reorganization at all. (d) Prenucleation clusters. Nonconstant bulk and surface/interface energy terms may lead to (local) minima in the energy profile and give rise to metastable (blue curve) or even stable (red curve) prenucleation clusters. (Figure adapted with permission from Ref. [1].)

Molecular simulation studies proved particularly valuable for extending our mechanistic understanding beyond classical nucleation theory. On the one side, this applies to the surfaces of forming nuclei: To minimize interfacial free energy, diffusive and dynamically changing nucleus–melt or nucleus–solvent interfaces appear more favorable [9]. An even more important issue is however the need to consider structural transitions within the inner core of a forming crystal. First evidence for this was collected by Ostwald leading to the famous Ostwald's step rule that suggests a series of structural transitions during crystal nucleation [2]. While the original argument was based on preferential nucleation of a phase that is structurally similar to the preceding one, a more thermodynamic rationale of multistep nucleation processes is given by following the pathway encompassing the lowest nucleation barrier [2,10]. The latter criterion can be considered as a qualification of the former, if "structural similarity" of two phases is interpreted in terms of the easiness to transfer one in the other.

18.2.2
Multistep Nucleation and Crystal Polymorphism

An important example of two-step nucleation is given by solute segregation in terms of a single disordered clusters or a partially ordered agglomerate of clusters, followed by aggregate ordering at a later stage of the precipitation process [11–14]. An explanation of such nonclassical nucleation may indeed be given by considering the competition of (at least) two phases [5]. Here, the final crystal structure implies most favorable solute packing in the bulk and is thus the predominant phase for large crystal nuclei. On the other hand, the formation of disordered clusters could give rise to roughly spherical aggregates of particularly favorable surface tension and/or interface energy. In this case, the disordered structure is thermodynamically preferred for small aggregates, while the crystalline structure is only stable for larger aggregates. Illustrations for such two-step nucleation are given in Figures 18.1b and c.

It is noteworthy that the secondary nucleation step requires a size-induced phase transition of the forming aggregate. Depending on the solute, such solid–solid or liquid–solid transformation may be subject to a considerable energy barrier. As a consequence, hysteresis effects might shift the secondary nucleation step to late stages of aggregate growth or even fully inhibit transformation to the thermodynamically preferred structure. A prominent example of the latter case is given by molecular crystals that are particularly often found to nucleate in terms of different polymorphic structures depending on the specific syntheses conditions (and thus different rankings of the corresponding nucleation barriers).

18.2.3
Prenucleation Clusters and Nucleation from Building Blocks

Similar to the size-dependent phase stability discussed earlier for nuclei of competing crystal polymorphs, also the structures and energetics of noncrystalline cluster need to be considered as functions of size. This gives rise to a variety of challenges to conventional CNT and the required extensions of the theory mainframe are still under development. Schematically, prenucleation clusters may occur in terms of (i) relatively favorable, yet thermodynamically unstable, intermediates to crystal nucleation (Figure 18.1d). Long-standing examples for discontinuous size dependence in cluster energy are "magic number" clusters observed during metal crystallization from the vapor [15,16]. Molecular dynamics simulations showed that the evolution of forming metal nuclei involves structural transitions from nuclei with crystalline bulk to compact polyhedra of particularly favorable surface tension and vice versa [16].

On the other hand, (ii) prenucleation clusters may also occur as the thermodynamically favored species with respect to the dispersed ions. The term "stable" is then used to describe prenucleation clusters that coexist with dispersed solutes in solutions below the saturation limit. A simple example of such a solution is given by tenside molecules in water. Here, the formation of micelles may be rationalized by a favorable interface energy and an unfavorable bulk energy term, thus flipping the chart characteristic to conventional CNT (compare Figures. 18.1a and d). Comparing micelles with dispersed tenside molecules, favorable interface energy arises from the segregation of hydrophobic moieties. On the other hand, unfavorable bulk (free) energy results from insufficient solute–solute interactions compared to the entropy change needed for solute aggregation. A similar argument might apply to the most popular type of prenucleation clusters, that is, clusters that are coordinated by surfactants – a prominent example being Zn_4O (acetate)$_6$ clusters observed in ethanolic solutions of zinc acetate dihydrate [17–19]. In such systems, the coordination by surfactants obviously lowers the interfacial energy of the clusters. We are however not aware of a rigorous proof of thermodynamic stability (which would refer to case (ii) and kinetic hindering could also account for preventing ripening to ZnO).

While the (relative) stability of the cluster types discussed above is intuitive and well established for many decades, the recent discovery of an unexpected type of prenucleation species considerably extended our picture of solutions prior to nucleation. Using ultracentrifugation, Coelfen and coworkers identified $CaCO_3$ prenucleation clusters in aqueous solution [20]. In the absence of surfactants, the rationale based on micelles does not apply. On the other hand, arguments based on specific structures of preferential energy such as the "magic number" clusters observed for metals also appear unreasonable as $CaCO_3$ is known to (initially) nucleate as amorphous aggregates. On the basis of molecular simulations, Wallace et al. instead suggest the formation of liquid-like droplets of high ion concentration [21]. While the interfacial energy between the ion-rich and the ion-poor solutions appears as practically zero [22,23], polyionic chains account for the interactions in the bulk. The latter appear to change from favorable to unfavorable with increasing droplet size [1 and references therein].

It is tempting to interpret crystal nucleation via prenucleation clusters as a special case of the nonclassical nucleation pathways discussed earlier. In this sense, the first nucleation step would be the agglomeration of prenucleation clusters, forming (partially) organized structures that could also be interpreted as mesocrystals [24]. The secondary nucleation step then refers to the ripening of such agglomerates into the final crystal structure [25]. An exciting perspective of such nucleation pathways is to manipulate crystal nucleation by selection of the prenucleation species acting as building blocks. This would allow promoting specific structural motifs and ideally directing crystal nucleation into specific polymorphs or compositions.

Using cyro-TEM, Sommerdijk and coworkers recently provided strong evidence for this concept by capturing different stages of iron oxide/hydroxide agglomeration from solution, followed by a secondary nucleation event leading to magnetite crystals [26]. As thermodynamic rationale, these authors suggested to use classical nucleation theory as developed for nucleation from disperse solution (Figure 18.1a), but introduce an offset in terms of size and free energy to describe nucleation from precursor clusters. This offset directly corresponds to the (local) minima in the free energy profiles illustrated in Figure 18.1d.

The key question is, whether such offsets to the free energy level of the solution generally lead to lower effective nucleation barriers (as evidently the case for the above study on magnetite). Nucleation from particularly favorable solute clusters could instead imply an increase in free energy barrier as compared to the dispersed solution. In this case, the putative precursor clusters actually reflect nucleation inhibitors. We argue that this depends on the *crossing of the free energy profiles* related to the prenucleation clusters with the energy curve corresponding to conventional nucleation from solution. Figure 18.2 illustrates the three possible scenarios depending on the degree of thermodynamical stability and the size distribution of the prenucleation clusters compared to the critical nucleus estimated from classical nucleation theory. Metastable prenucleation cluster may be interpreted as a relatively favorable intermediates to crystal nucleation and are thus well suited as building blocks to nonclassical crystal nucleation. For clusters that are thermodynamically preferred over solutions of dispersed solutes, the picture is more complex.

The free energy diagrams shown in Figure 18.2 illustrate two different types of such clusters. Here, only the example of large prenucleation clusters of broad size distribution exhibits a low effective barrier to crystal nucleation, thus qualifying the clusters as building blocks to crystal formation. (Note that the crossing

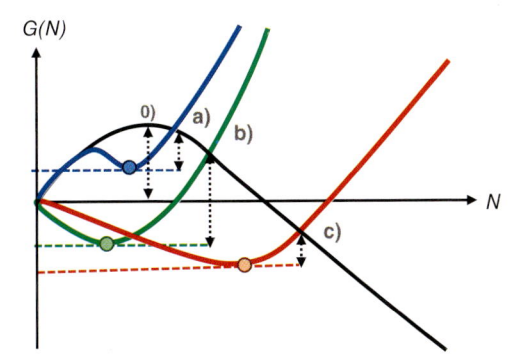

Figure 18.2 Crystal nucleation from prenucleation clusters. While the black curve refers to classical nucleation of a crystal from a solution of dispersed ions, the colored curves illustrate the possible energy profiles for prenucleation clusters. The blue curve corresponds to a metastable prenucleation cluster that may be interpreted as a relatively favorable intermediate to crystal nucleation. The effective barrier to crystal formation (a) is lower than that of conventional nucleation from the ionic solution (0). For stable prenucleation clusters, two scenarios may apply: large prenucleation clusters of broad size distribution also imply low effective barriers to crystal nucleation (c) and may thus serve as precursors to crystal formation. On the other hand, small prenucleation clusters of sharp size distribution would imply an increase in effective nucleation barrier (b). In this case, crystal nucleation appears more favorable in regions of the ionic solution, initially not affecting the prenucleation clusters. (Figure adapted with permission from Ref. [1].)

of the free energy curves only reflects a rough estimate of the transition barrier, actually ignoring the possibly quite large increase in energy barrier arising from cluster reorganization.) Contrary to this, the illustrated curve for small prenucleation clusters of sharp size distribution indicates a considerable barrier for the cluster → nucleus transition (even using the lower estimates as obtained from Figure 18.2). For the given example, this barrier is actually larger than the barrier to nucleation from bulk solution. Consequently, nucleation is expected to take place within the bulk ionic solution without affecting the previously formed clusters and the term "prenucleation cluster" is misleading. Indeed, this type of clusters formed prior to nucleation would persist until later stages of crystal growth as described in the following.

Clusters that are the most stable species present in solution prior to crystal nucleation may still be outperformed thermodynamically once the nuclei reach mature stages of crystal growth. The critical point is the intersection of the free energy gain from solute uptake in postcritical crystal nuclei compared to the free energy of the same number of solutes within the previously formed clusters (Figure 18.3, green curve). The fate of the beforehand stable clusters in solution then is to either (i) collide with a forming nucleus and merge into it; or alternatively (ii) the clusters might dissociate into solution to compensate for the depletion of dispersed ions in the solution arising from crystal precipitation. The choice of mechanisms also depends on cluster mobility compared to the diffusion of dispersed ions. Interestingly, the mechanistic picture (ii) is similar to the Ostwald ripening [27] of differently sized crystal nuclei. In particular, for this

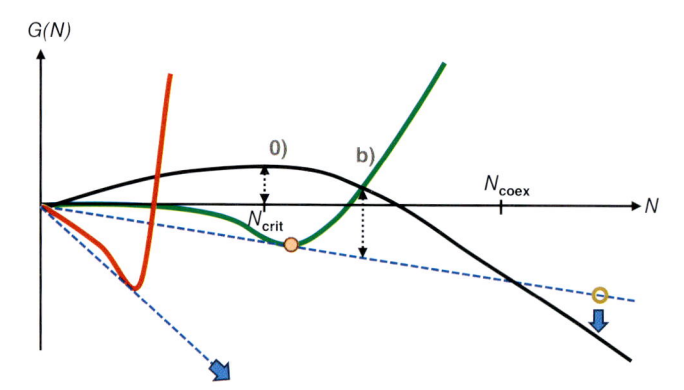

Figure 18.3 Free energy diagram for crystal nucleation from solution of dispersed solutes (black curve, barrier denoted as (0) initially ignoring the clusters formed prior to nucleation (green curve, barrier to direct transformation denoted as (b). Such clusters would coexist with (postcritical) crystal nuclei of small size ($N_{crit} < N < N_{coex}$). However, at more mature stages of crystal growth, the free energy gain in solute association with the growing nucleus may outperform thermodynamic preference of solute assembly in a single or multiple (here shown for 2) clusters. On the other hand (red curve), excessive stabilization of small clusters in solution may actually inhibit crystal nucleation. The dashed blue lines indicate the free energy of multiples of clusters taken as a simple linear extrapolation of the cluster of most favorable size. (Figure adapted with permission from Ref. [1].)

scenario we argue that the clusters formed prior to nucleation should be regarded as "buffers" to ion concentrations in solution rather than precursors to crystal nucleation. On the other hand, route (i) implies that the clusters are candidate building blocks to crystal growth, but not to nucleation.

From a practical point of view, crystal "design" from precursor solutions is probably most promising for two-step processes. In the first step, particularly stable, well-defined clusters are formed prior to nucleation. By changing the solution (e.g., adding a further component etc.), the cluster stability is then reduced to provide a more "reactive" species that agglomerate and turn into crystal nuclei. The beforehand mentioned $Zn_4O(acetate)_6$ clusters in ethanolic solution are prominent examples for this strategy. In the initial solution, these clusters are quite stable and the nucleation of ZnO is typically triggered by adding hydroxide [17].

An extreme case of promoting the stability of small clusters in solution is reflected by the free energy profile shown as the red curve in Figure 18.3. In this case, the cluster species is thermodynamically preferred over the bulk crystal and the solution would rather form multiples of clusters than a crystal. An illustrative example for this scenario are polyacrylate additives in aqueous solution used to hinder $CaCO_3$ nucleation. Combing molecular dynamics simulations with an extensive structural sampling technique, Parrinello and coworkers demonstrated the peculiar binding of calcium and carbonate ions to the additive [28,29]. Despite the local accumulation of ions, crystal nucleation is still disfavored as the association to the polyacrylate additive leads to Ca–Ca distances that mismatch with the packing in any of the known solid forms of calcium carbonate.

18.2.4
Nucleation Kinetics versus Diffusion-Controlled Crystal Formation

The almost 150-year-old concept of classical nucleation theory only requires small extensions to account for the manifold of nucleation and prenucleation phenomena known today. Crystal nucleation processes called "nonclassical" elude the concept of classical nucleation theory only in its conventional implementation, but multistep nucleation processes may be rationalized by considering multiple free energy profiles. Each of these is derived in a "classical" manner, that is, by contrasting unfavorable surface/interface tension to favorable bulk energy. The nucleation process will follow the route of lower barrier, which may arise from low interface tension rather than optimal bulk energy. In this case, size-induced changes in polymorph stability may be predicted from the crossing of the free energy profiles of competing crystal structures. The actual solid–solid transition is subject to a kinetic barrier that implies hysteresis or even prevents the direct transformation at all.

For "classical" crystal nucleation pathways corresponding to the energy profile illustrated in Figure 18.1a, the kinetics may be estimated from the rate of critical nucleus formation. Rate theory [30] implies

$$r_{\text{nucleation}} = r_0 \cdot \exp\left(-\frac{\Delta G}{k_B T}\right), \tag{18.5}$$

where T is the temperature, k_B is the Boltzmann constant, and r_0 is the kinetic prefactor. Processes that determine the kinetic prefactor are solute diffusion in solution and solute desolvation in order to incorporate the solute in the forming nucleus. Both of these processes are typically related to an activation energy stemming from the need to rearrange solvation shells or replacing solvent–solute by solute–solute contacts. Formally, the kinetic prefactor may thus be written as

$$\begin{aligned} r_0 &= r_{\text{diff}} \cdot r_{\text{desolv}}, \\ r_{\text{diff}} &= r_{\text{diff}}^0 \cdot \exp\left(-\frac{\Delta E^{\text{diff}}}{k_B T}\right), \\ r_{\text{desolv}} &= r_{\text{desolv}}^0 \cdot \exp\left(-\frac{\Delta E^{\text{desolv}}}{k_B T}\right). \end{aligned} \quad (18.6)$$

When combining Eqs. (18.6) and (18.5), it is tempting to combine the three exponential terms into a putative effective barrier $\Delta G + \Delta E^{\text{diff}} + \Delta E^{\text{desolv}}$ – which is however quite misleading as diffusion, desolvation, and nucleation are separate steps. This is best seen from considering the rate of critical nucleus dissociation during a failed attempt to nucleation:

$$\begin{aligned} r_{\text{crit}\rightarrow\text{desolv}} &= r_0 \cdot \exp\left(+\frac{\Delta G}{k_B T}\right) \\ &= r_{\text{desolv}}^0 \cdot r_{\text{diff}}^0 \cdot \exp\left(-\frac{\Delta E^{\text{diff}}}{k_B T}\right) \exp\left(-\frac{\Delta E^{\text{desolv}}}{k_B T}\right) \exp\left(+\frac{\Delta G}{k_B T}\right). \end{aligned} \quad (18.7)$$

Indeed, solute diffusion and de/re solvation processes always impose a barrier that slows the kinetics, be it the forward or the backward reaction. This motivated the interpretation of ΔE^{diff} and ΔE^{desolv} as kinetic barriers, while ΔG is called the thermodynamic barrier [31].

In view of the different process steps and their corresponding activation barriers, it is useful to discriminate the two limiting cases of diffusion-controlled and nucleation-controlled crystal formation. The latter type of processes is characterized by $\Delta G \gg \Delta E^{\text{diff}} + \Delta E^{\text{desolv}}$. On the other hand, diffusion-controlled crystal formation implies no or only low barriers to desolvation and nucleation, but not necessarily $\Delta E^{\text{diff}} \gg \Delta G + \Delta E^{\text{desolv}}$. Indeed, diffusion will always become rate-determining if solubility is low and nucleation occurs from very sparse solution. The two limiting scenarios – crystal growth limited by mass transport to the nucleus and nucleation triggered by crossing an activation barrier – give rise to rather different challenges to both experiment and molecular simulation.

18.2.5

Crystal Growth at Nonconstant Solute Concentration

The thermodynamic considerations discussed so far assumed at least roughly constant solute concentrations. This fully applies to crystal formation from the melt, but only to the early stages of nucleation from solution. Indeed, at late stages of crystal precipitation, we need to consider the gradual decrease of solute concentration that affects the gain in free energy upon crystallization which reads $-\mu \cdot N$ with μ being the change in free energy per solute (see also Eq. (18.3)). The free energy term μ may be related to the change in potential energy upon crystallization and an entropy contribution that reflects the binding of solvent molecules and, more importantly, the manifold of configuration space accessible to the solutes:

$$\mu = \Delta E_{\text{pot}} - \Delta S_{\text{solvent}} \cdot T - \Delta S_{\text{solutes}} \cdot T$$

$$\approx \Delta E_{\text{pot}} - \Delta S_{\text{solvent}} \cdot T - k_{\text{B}} \cdot T \cdot \ln\left(\frac{V_{\text{mol}}^{\text{crystal}}}{c_{\text{mol}}^{-1}}\right). \tag{18.8}$$

The latter term relates the average volume per molecule in the crystal to the average volume per solute in a solution of (molecular) concentration c_{mol}.

In the course of crystal precipitation, solute concentration will thus decrease until the gain in free energy per molecule μ reaches zero – which represents the thermodynamics definition of the saturation concentration and marks the end of the precipitation process. In-between nucleation and termination of crystal growth, the free energy profile as a function of nucleus size experiences rather drastic changes as illustrated in Figure 18.3. It is noteworthy that the depletion of solute concentration also shifts the size of the critical nuclei as estimated from CNT toward larger dimensions. This implies that postcritical nuclei may become precritical and actually get consumed by larger (and thus more stable) nuclei as a consequence of gradually decreasing solute concentration. The combination of Eqs. (18.3) and (18.8) hence allows for a (simplistic) thermodynamic description of Ostwald ripening (Figure 18.4) [27].

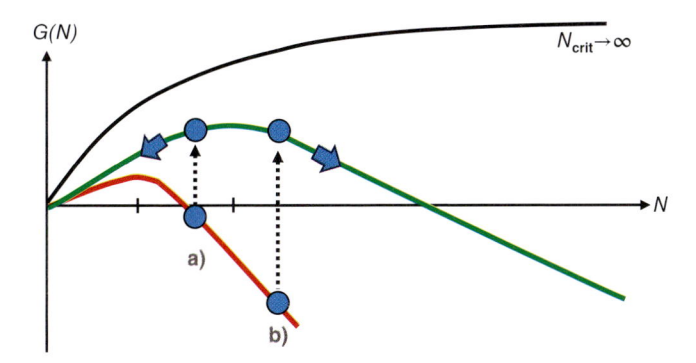

Figure 18.4 Free energy diagram for crystal formation from solution of nonconstant solute concentration. The red curve reflects the very early stage of crystal precipitation at highest degree of supersaturation. Nuclei (a) and (b), which are both postcritical at high solute concentration (red curve), may show dissimilar stability at lower concentration (green curve). In this simplified scheme of concentration depletion by crystal precipitation, (a) would redissolve, while (b) continues to grow until the saturation concentration is reached (black curve).

18.3
Molecular Dynamics Simulation Approaches to Understanding Crystal Nucleation

A broad overview of experimental techniques for characterizing nucleation processes was recently provided by Pienack and Bensch and the reader is directed to Ref. [3] for a detailed account. Moreover, a topical survey of experimental approaches to on clusters formed prior to nucleation and nonclassical nucleation was provided by Gebauer and Coelfen [6]. In the following, we give a brief summary of molecular simulation methods that – in combination with experimental characterization – reflect the state of the art in unraveling clusters, precursors, and nuclei in solution.

18.3.1
Molecular Mechanics: Quantum, Classical, and Hybrid Models

Molecular simulations are based on atomic interactions that may be calculated at best accuracy using quantum chemical approaches. The downside of calculating potential energy and interatomic forces at high levels of theory is however the immense computational demand. The latter implies limited sampling of the manifold of different configurations – which is of crucial importance for characterizing complex systems. In the worst case, insufficient statistical sampling may lead to divergent crystal nucleation pathways and thus conceptually wrong result, irrespective of how accurately the underlying interaction potentials were calculated. On the other hand, molecular mechanics models such as harmonic springs mimicking covalent bonds allow for much more extended evaluation of the configurational manifold. Despite the loss of accuracy in calculating the interaction potential of a specific atomic arrangement, the use of classical mechanics models may therefore lead to more reliable sampling of possible nucleation pathways compared to approaches based on quantum calculations [4].

To combine the advantages of both approaches, it is often useful to employ quantum calculations for small subsystems of particular interest, while the remaining part of the system is modeled by inexpensive classical mechanics. Such hybrid approaches qualify for the study of proton transfer reactions during crystal formation as described in Section 18.4 by the example of zinc oxide precipitation from hydroxide solution. Therein, only the reacting species are described quantum mechanically while the vast majority of the system is modeled at good accuracy by classical mechanics. Similarly, quantum or semi-empirical approaches have been employed for considering charge transfer reactions such as metal ion reduction leading to metallic nanoparticles [32,33].

18.3.2
Molecular Dynamics Simulations I: Equations of Motion and Numerical Solutions

With reasonable models describing the atomic interactions at hand, we can now address the simulation of molecular dynamics. Typically, this is implemented via classical dynamics, that is, by employing Newton's equations of motion for the atoms. Depending on the treatment of the atomic interactions by force fields or electronic quantum calculations, we use the terms "classical" and "quantum" molecular dynamics, respectively. In addition to the quantum treatment of the electrons, particularly accurate modeling of proton mobility may also motivate quantum mechanical approaches for the nuclei [34,35]. Here, we will focus on the former two types of molecular dynamics simulations and refer to Ref. [36] for an overview of full quantum modeling of molecular systems. Accordingly, the motion of the individual atoms i is determined by

$$m_i \cdot \ddot{\vec{r}}_i = -\nabla_{\vec{r}} \langle \hat{H} \rangle \approx -\nabla_{\vec{r}} \epsilon_0 \approx -\nabla_{\vec{r}} V_{FF}. \qquad (18.9)$$

Here the first part refers to the Hellmann–Feynman theorem and the right-hand side denotes approximations to the Hamiltonian H by the electronic ground state ϵ_0 or a molecular mechanics force field V_{FF}. In either case, the calculation of the interatomic forces of an N particles system leads to a function depending on the coordinates of all atomic positions $\vec{r}_1..\vec{r}_N$ and thus to N coupled differential equations of second order. In lack of analytic solutions beyond $N=2$, a numerical approach may be derived from approximating the second derivative in Eq. (18.9):

$$\ddot{\vec{r}}_i = \lim_{\Delta t \to 0} \frac{\frac{\vec{r}_i(t+\Delta t) - \vec{r}_i(t)}{\Delta t} - \frac{\vec{r}_i(t) - \vec{r}_i(t-\Delta t)}{\Delta t}}{\Delta t} \approx \frac{\vec{r}_i(t+\Delta t) - 2\vec{r}_i(t) + \vec{r}_i(t-\Delta t)}{\Delta t^2}. \qquad (18.10)$$

Using sufficiently small Δt (typically ~1 fs), the combination of Eqs. (18.9) and (18.10) allows predicting future atomic positions at time $t + \Delta t$:

$$\vec{r}_i(t+\Delta t) \approx 2\vec{r}_i(t) - \vec{r}_i(t-\Delta t) + \frac{\vec{F}_i(\vec{r}_1,..,\vec{r}_N)}{m_i} \Delta t^2 \qquad (18.11)$$

To assess longer trajectories, we need to employ the above equation iteratively to calculate the evolution of atomic positions and interaction forces in small time steps Δt.

With the time steps Δt typically being limited to 1 fs, the propagation of a molecular model system by 1 ns thus requires a million iterations, including the same number of calculations for the updates of the interaction forces. As a consequence, molecular dynamics simulations offer *in situ* insights at the atomic level of detail at the price of considerable computational costs. The latter limit the accessible timescale and length scale drastically, and the study of crystal nucleation and related processes calls for special techniques as described in the following.

18.3.3
Molecular Dynamics Simulations II: Bridging Timescale and Length Scale

Molecular dynamics simulations thus reflect iterative solutions to the Newtons equations of motion and calculate molecular trajectories as small increments of time. While this time step (and hence the maximum time resolution) is typically around 1 fs, the overall simulation time is usually chosen within the nano- to microsecond regime. These timescales are substantially lower than that of most nucleation process of experimental or industrial relevance. Using specialized simulation techniques for bridging the timescale/length scale problem, crystal nucleation may still be assessed by molecular simulation, as briefly summarized in the following (for a more detailed account, see Ref. [4]). While the discussed techniques apply to all types of molecular dynamics simulations, we note that the atomic interaction forces are most accurately calculated from quantum treatment of the electrons, and specifically developed molecular mechanics models are needed to obtain similar accuracy. The latter are usually preferred as force fields allow for assessing larger systems and longer timescales and thus provide drastically lower margins of the statistical error. Indeed, when studying complex processes like crystal nucleation, we should always consider two sources of errors – the beforehand mentioned accuracy in evaluating the interaction forces of specific configurations *and* the relevance of these specific configurations for actually describing the process of interest.

The key to studying nucleation-controlled processes is to provoke the model system to overcome the nucleation barrier that arises from ordering a domain within the melt or from desolvation and rearranging nearby solutes within a solution. To make such processes accessible to molecular simulation, the strategies developed – in one way or the other – are all based on first implementing an artificial boost of solute–solute interactions, and then correcting the biasing of the results. Auer and Frenkel pioneered this field by driving nucleation from the melt via a predefined order parameter that reflects nearest-neighbor distances and angles [37–39]. To avoid bias from excessively boosting nucleation kinetics, the process is described by a series of setups each mimicking steady state within a small interval of the order parameter. Using the umbrella sampling technique, artificial potentials were implemented to restrain the order parameter within a certain range, and sketches of the energy profile (potential-of-mean-force) were collected from Boltzmann statistics [37–39].

While umbrella sampling uses additional potentials to create attraction toward a desired state to the model system, it is also possible to induce nucleation processes by artificially creating "repulsion" from an unwanted configuration. The metadynamics technique reflects a systematic scan of configuration space by continuously disfavoring configurations that have been characterized before. This approach does not require prejudicing a reaction coordinate (as in umbrella sampling), but relies on a broader a set of predefined variables to which the biasing potential is applied. Within this choice of descriptors, metadynamics samples configuration space free of prejudicing [40].

More recently, transition path sampling molecular dynamics was employed to sample time-dependent pathways of nucleation from solution [41,42]. Here, an initial nucleation pathway is prepared from imposing high temperature, pressure, or manipulation of the interaction potentials [43]. Increasingly realistic pathways are then collected from performing a Monte Carlo sampling within trajectory space confined to liquid–solid transition routes. By the example of NaCl aggregation in water, this approach was also tested for investigating nucleation from solution [44].

To study diffusion-controlled crystal formation processes, an alternative class of simulation methods proved more performing. The general concept is to treat solute diffusion and solute aggregation by different methods. Gavezzoti pioneered this field by essentially ignoring long-ranged diffusion process and exploring the manifold of solute–solute contacts from small model systems comprising a few solutes, only [45,46]. This allows focusing molecular simulation to the critical issue of solute–solvent bond dissociation and replacement by solute–solute bonds. Possible crystal structures are then predicted from expanding the manifold of solute–solute contacts to periodic arrangements [46].

The Kawska–Zahn method describes solute diffusion to a forming nucleus by an inexpensive docking procedure implemented as a Monte Carlo step, while solute association and the reorganization of the aggregate is studied from explicit molecular dynamics simulations [47]. Depending on the model system, aggregate relaxation after each growth step can be modeled in different ways. (i) From direct molecular dynamics simulation of a given period of time. This leads to a kinetic Monte Carlo algorithm and is suited for fast precipitation processes, only. Alternatively (ii), Wallace *et al.* employed parallel replica simulations for an extensive sampling of configurations, thus mimicking infinite relaxation times [21]. In-between these extremes (iii), we suggest a simulated annealing-type procedure to allow aggregate relaxation to a degree that both energy profiles and aggregate structures evolve continuously as functions of nucleus size [16].

18.4
Case Studies on ZnO: Understanding Ion Aggregation, Ripening Reactions, Nucleation and Growth, Polymorphic Transitions and Nanoparticle Stabilization from Molecular Simulation

While zinc oxide belongs to the most widely studied nanomaterials, we are yet at the beginning of understanding the molecular scale mechanisms of ZnO crystal nucleation and growth [17,48–53]. Much of the recent advances were elaborated from molecular simulations, complementing the manifold of experiments, but allowing for particularly detailed insights. Indeed, the molecular dynamics simulations described in the following can be interpreted as a tool of *in situ* investigation offering femtosecond time resolution and <0.1 nm spatial resolution at the same time. This combination is unique compared to current experimental approaches that would allow for high resolution in either space or time (cf. HR-TEM and fs spectroscopy for example).

In 2008, we transferred the Kawska–Zahn approach to model crystal nucleation from solution to the investigation of zinc and hydroxide ion association in ethanol solution [54]. While the vast majority of atomic interactions are modeled by inexpensive force fields, a quantum picture is used to consider possible

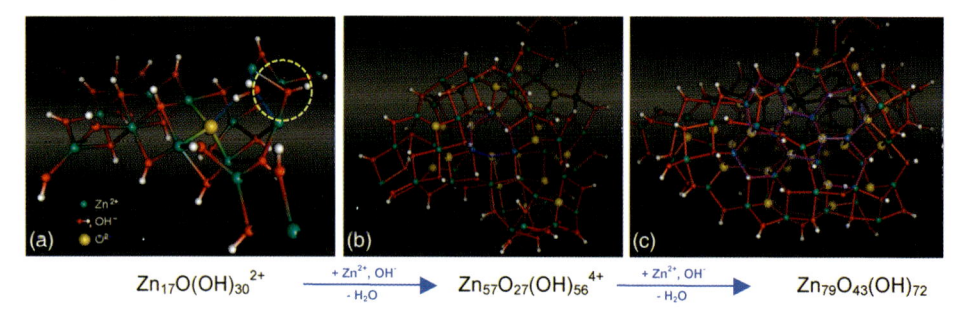

$$\text{Zn}_{17}\text{O}(\text{OH})_{30}{}^{2+} \xrightarrow[-\text{H}_2\text{O}]{+\text{Zn}^{2+},\text{OH}^-} \text{Zn}_{57}\text{O}_{27}(\text{OH})_{56}{}^{4+} \xrightarrow[-\text{H}_2\text{O}]{+\text{Zn}^{2+},\text{OH}^-} \text{Zn}_{79}\text{O}_{43}(\text{OH})_{72}$$

Figure 18.5 Association of zinc and hydroxide ions modeled in ethanol solution. Ripening reactions lead to the deprotonation of OH$^-$ ions of particularly strong coordination by Zn^{2+}. Structural relaxation upon the ongoing formation of O^{2-} ions leads to the nucleation and growth of a wurtzite ZnO domain in the aggregate core. (Adapted with permission from Ref. [54].)

proton transfer reactions among the hydroxide ions. The latter issue is of crucial relevance to explore the "ripening" of zinc hydroxide aggregates leading to zinc oxide nuclei. Starting from a [Zn $\cdots$ OH]$^+$ dimer, we explored ion-by-ion docking and subsequent relaxation of the forming aggregate in solution. On this basis, an early (and transient) stage of self-organization was identified in terms of octahedral coordination of Zn^{2+} by OH$^-$ ions. If persistent, such motifs would lead to a Zn(OH)$_2$ crystal, which is however not observed, as OH$^-\cdots$OH$^- \rightarrow$ O^{2-} + H$_2$O reactions lead to the formation of ZnO/Zn(OH)$_2$ core–shell nanocrystals. Figure 18.5a illustrates the first of such ripening reactions as it was observed for a zinc hydroxide aggregate of about 50 ions. A common feature of this kind of proton transfer reactions is the very dissimilar coordination of the two reacting OH$^-$ ions by Zn^{2+}. While the reaction in the gas phase or in plain solution is extremely endothermic, strong coordination by zinc ions favors hydroxide deprotonation, whereas weak coordination of the neighboring hydroxide facilitates proton association. Using a combined quantum/classical model, all possible proton transfer reactions were probed and, if exothermic, implemented as a ripening step during aggregate relaxation.

The formation of O^{2-} ions was found to provoke drastic structural changes, which lead to the nucleation of a ZnO core (of wurtzite structure) embedded by an otherwise weakly ordered Zn(OH)$_2$-type shell (Figure 18.5b). Upon association of further zinc and hydroxide ions, ongoing proton transfer reactions at the aggregate surface give rise to a roughly constant thickness of the Zn(OH)$_2$ shell while the ZnO core grows in favor of a well-defined ZnO/Zn(OH)$_2$ core–shell nanocrystal (Figure 18.5c).

Mimicking ion-by-ion association with model simulations, we can hence provide in-depth insights into the early stages of the evolution from dimers, trimers, and so on up to aggregates comprising hundreds of solutes. For the case study on zinc oxide, the key steps of self-organization were observed already for aggregates counting less than 200 ions. This provided the grounds for deriving

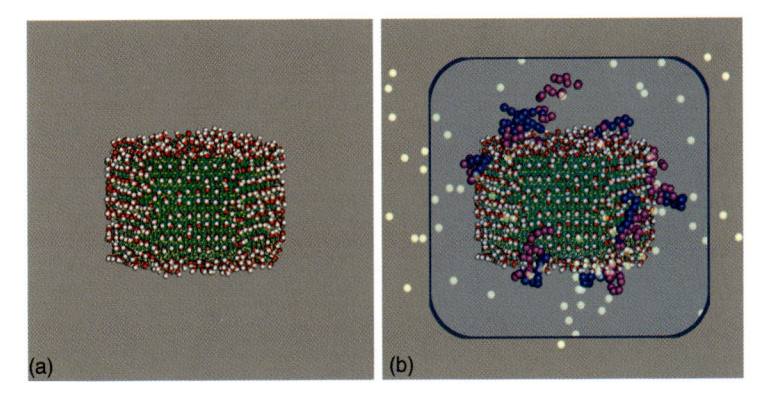

Figure 18.6 Scale-up model (a) of a ZnO/Zn(OH)$_2$ core–shell nanoparticle comprising ~10 000 atoms. Same model after (b) association of surfactant molecules (polyacyrlate–polyethylenoxide blockpolymers) and a diffuse halo of Na$^+$ ions (pale yellow) illustrating hydrodynamic radius and colloid charge. (Adapted with permission from Ref. [55].)

building rules for how to construct much larger ZnO/Zn(OH)$_2$ core–shell nanocrystal models without explicit simulation of crystal growth. Figure 18.6a illustrates such a scale-up model that was essentially prepared from cutting a hexagonal prism from a ZnO single crystal. To account for surface layers comprising only Zn^{2+} and OH$^-$ ions, exposed O^{2-} ions were avoided by (at equal occurrence) protonation to OH$^-$ or removing the whole ion.

On this basis, realistic models of nanoparticles comprising 10 000 of solutes can be subjected to molecular simulation. Such "extended" models are needed to explore processes that are connected to larger length scales. Along this line, we explored nanoparticle stabilization by surfactant association [55]. Figure 18.6b shows the association of negatively charged surfactant molecules to a hexagonal a ZnO/Zn(OH)$_2$ core–shell nanoparticle of $\sim$ 5 nm diameter. In solution (solvent is omitted for clarity in Figure 18.6b) the effective charge of the colloid is reduced by a halo of Na$^+$ counterions. Molecular dynamics simulations of several 100 ns allow assessing solvent and counter ion mobility in proximity of the nanoparticle and thus provide molecular scale insights into the foundations of effective particle dimensions and net charge in solution [55].

Moreover, we used scale-up models of ZnO nanorods to investigate polymorphic transitions, namely, the pressure-induced wurtzite to rock salt transformation [56]. Here, dimensions of a few nanometers proved sufficient to investigate nucleation and growth mechanisms for the solid–solid transformation. While brute-force molecular dynamics simulations at high pressure leads to drastic hysteresis effects (and misleading information on the transformation mechanisms), the transition path sampling molecular dynamics simulation technique allows studying the process at conditions directly mimicking the anvil experiments. Figure 18.7 shows snapshots from such a molecular dynamics trajectory to illustrate the subsequent nucleation of ionic displacements and the interplay of layer shifting and the final shape of the nanoparticle.

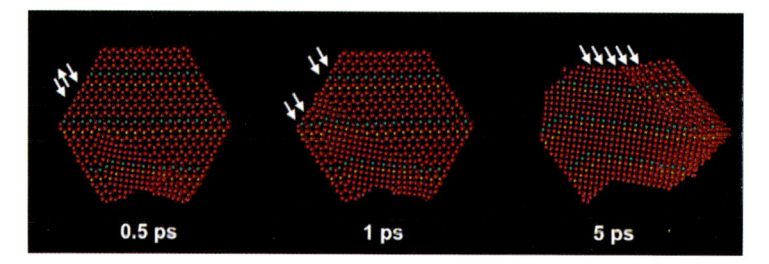

Figure 18.7 Wurtzite → rocksalt transition route illustrating the nucleation and growth of the high-pressure phase (6 GPa) in a ZnO nanorod. Along the horizontal direction, several atomic rows are highlighted to help the visualization of the observed (100) layer-shifting mechanism and its interplay with deforming the overall nanoparticle shape. Antiparallel layer shuffling as observed for the beginning of the nucleation process is disfavored at later stages of phase growth to avoid the formation of notched surfaces. (Adapted with permission from Ref. [56].)

18.5
Conclusion

Several factors have influenced the ongoing success of molecular simulations in many fields of research ranging from biology to physics and from gas-phase chemistry to materials science. The most obvious development might be the continuous improvement of computer technology. However, for application in solid-state chemistry and materials science, even more important progress was achieved by the evolution of simulation models from simplistic hard sphere particles to realistic accounts of how atoms and molecules interact. While there is of course still space for improvements, quantum chemistry achieved the required accuracy for most problems already a few decades ago. This provides robust reference data for creating semiempirical and molecular mechanics models. The effectiveness of such approaches allows studying materials properties in increasingly large systems that may account for defects, dislocations, grains, surfaces, and interfaces.

Assessing large length scales within realistic models is not only confined to the investigation of structures and static properties but also allows exploring dynamic properties and processes. Specialized molecular dynamics simulation techniques have been developed to study the association of solutes, nucleation, and growth of ordered structures and transformation processes that might occur during or after particle formation. The obtained mechanistic insight qualifies for inspiring new syntheses routes – the key perspective of molecular simulations in materials design. The tremendous potential at reach is reflected by the more established use of molecular modeling and simulation in current pharmaceutical research – for which the combination of experiments, informatics, and computational chemistry became indispensable already decades ago.

References

1. Zahn, D. (2015) *Chem. Phys. Chem.*, **16**, 2069–2075.
2. Ostwald, W. (1897) *Z. Phys. Chem.*, **22**, 289–330.
3. Pienack, N. and Bensch, W. (2011) *Angew. Chem., Int. Ed.*, **50**, 2014–2034.
4. Anwar, J. and Zahn, D. (2011) *Angew. Chem., Int. Ed.*, **50**, 1996–2014.
5. Vekilov, P. (2010) *Nanoscale*, **2**, 2346–2357.
6. Gebauer, D. and Coelfen, H. (2011) *Nanotoday*, **6**, 564–584.
7. Gibbs, J.W. (1876) *Trans. Conn. Acad. Sci.*, **3**, 108–248.
8. Gibbs, J.W. (1876) *Trans.Conn. Acad. Sci.*, **16**, 343–524.
9. Auer, S. and Frenkel, D. (2003) *Nature*, **413**, 711–713.
10. Stranski, I.N. and Totomanow, D. (1933) *Z. Phys. Chem.*, **163**, 399–408.
11. Bonnett, P.E., Carpenter, K.J., Dawson, S., and Davey, R.J. (2003) *Chem. Comm.*, 698–699.
12. Zhang, T.H. and Liu, X.Y. (2009) *Angew. Chem., Int. Ed.*, **48**, 1–6.
13. Vekilov, P.G. (2004) *Cryst. Growth Des.*, **4**, 671–685.
14. Anwar, J. and Boateng, P.K. (1998) *J. Am. Chem. Soc.*, **120**, 9600–9604.
15. Alonso, J.A. (2005) *Structure and Properties of Atomic Nanoclusters*, Imperial College Press.
16. Milek, T., Duchstein, P., Seifert, G., and Zahn, D. (2010) *Chem. Phys. Chem.*, **11**, 847–852.
17. Spanhel, L. (2006) *J. Sol-Gel Sci. Technol.*, **39**, 7–24.
18. Schmidt, T., Müller, G., Spanhel, L., Kerkel, K., and Forchel, A. (1998) *Chem. Mater.*, **10**, 65–71.
19. Segets, D., Martin, A., Hartig, J., Gradl, J., and Peukert, W. (2012) *Chem. Eng. Sci.*, **70**, 4–13.
20. Gebauer, D., Völkel, A., and Coelfen, H. (2008) *Science*, **322**, 1819–1822.
21. Wallace, A.F., Hedges, L.O., Fernandez-Martinez, A., Raiteri, P., Gale, J.D., Waychunas, G.A., Whitelam, S., Banfield, J.F., and De Yoreo, J.J. (2013) *Science*, **341**, 885–889.
22. Raiteri, P. and Gale, J.D. (2010) *J. Am. Chem. Soc.*, **132**, 17623–17634.
23. Demichelis, R., Raiteri, P., Gale, J.D., Quigley, D., and Gebauer, D. (2011) *Nat. Commun.*, **2**, 590.
24. Erdemir, D., Lee, A.Y., and Myerson, A.S. (2009) *Acc. Chem. Res.*, **42**, 621–629.
25. Gebauer, D., Kellermeier, M., Gale, J.D., Bergström, L., and Coelfen, H. (2014) *Chem. Rev. Soc.*, **43**, 2348–2371.
26. Baumgartner, J., Drey, A., Bomans, P.H.H., LeCoadou, C., Fratzl, P., Sommerdijk, N.A.J.M., and Faivre, D. (2013) *Nat. Mater.*, **12**, 310–314.
27. Ostwald, W. (1897) *Z. Phys. Chem.*, **22**, 289–330.
28. Bruneval, F., Donaldio, D., and Parrinello, M. (2007) *J. Phys. Chem. B*, **111**, 12219.
29. Tribello, G.A., Liew, C., and Parrinello, M. (2009) *J. Phys. Chem. B*, **113**, 7081.
30. Eyring, H. (1935) *Chem. Rev.*, **17**, 65–77.
31. Hu, Q., Nielsen, M.H., Freeman, C.L., Hamm, L.M., Tao, J., Lee, J.R.I., Han, T.Y.J., Becker, U., Harding, J.H., Dove, P.M., and De Yoreo, J.J. (2012) *Faraday Discuss.*, **129**, 164701.
32. Milek, T., Doepper, T., Neiss, C., Goerling, A., and Zahn, D. (2014) *J. Mol. Model.*, **20**, 2111.
33. Milek, T. and Zahn, D. (2014) *Nanoletters*, **14**, 4913–4917.
34. Marx, D., Tuckerman, M.E., Hutter, J., and Parrinello, M. (1999) *Nature (London)*, **397**, 601.
35. Zahn, D. and Brickmann, J. (1999) *Israel J. Chem.*, **39**, 469–482.
36. Agmon, N. (1999) *Israel J. Chemistry*, **39**, 493–502.
37. Auer, S. and Frenkel, D. (2004) *J. Chem. Phys.*, **120**, 3015–3029.
38. Auer, S. and Frenkel, D. (2004) *Annu. Rev. Phys. Chem.*, **55**, 333–361.
39. Auer, S. and Frenkel, D. (2001) *Nature*, **409**, 1020–1023.
40. Giberti, F., Salvalaglio, M., and Parrinello, M. (2015) *IUCrJ*, **2**, 256–266.
41. Moroni, D., ten Wolde, P.R., and Bolhuis, P.G. (2005) *Phys. Rev. Lett.*, **94**, 235703.
42. Zahn, D. (2007) *J. Phys. Chem. B*, **111**, 5249.

43 Zahn, D. (2006) *J. Chem. Theory Comput.*, **2**, 107.
44 Zahn, D. (2004) *Phys. Rev. Lett.*, **92**, 040801.
45 Gavezzotti, A., Filippini, G., Kroon, J., vanEijck, B.P., and Klewinghaus, P. (1997) *Chem. Eur. J.*, **3**, 893.
46 Gavezzotti, A. (1999) *Chem. Eur. J.*, **5**, 567–576.
47 Kawska, A., Brickmann, J., Kniep, R., Hochrein, O., and Zahn, D. (2006) *J. Chem. Phys*, **124**, 24513.
48 Wang, Z.L. (2004) *J. Phys. Condens Mater.*, **16**, R829–R858.
49 Özgür, Ü., Alivov, Ya.I., Liu, C., Teke, A., Reshchikov, M.A., Doğan, S., Avrutin, V., Cho, S.J., and Morkoç, H. (2005) *J. Appl. Phys.*, **98**, 041301.
50 Schmidt-Mende, L. and MacManus-Driscoll, J.L. (2007) *Mater. Today*, **10**, 40–48.
51 Klingshirn, C. (2007) *Phys. Status Solidi B*, **244**, 3027–3073.
52 Klingshirn, C., Fallert, J., Zhou, H., Sartor, J., Thiele, C., Maier-Flaig, F., Schneider, D., and Kalt, H. (2010) *Phys. Status Solidi B*, **247**, 1424–1447.
53 Kołodziejczak-Radzimska, A. and Jesionowski, T. (2014) *Materials*, **7**, 2833–2881.
54 Kawska, A., Duchstein, P., Hochrein, O., and Zahn, D. (2008) *Nanoletters*, **8**, 2336–2340.
55 Duchstein, P., Milek, T., and Zahn, D. (2015) *PLoS One*, **10**, e0125872.
56 Frenzel, J., Seifert, G., and Zahn, D. (2009) *Z. Anorg. Allg. Chem.*, **635**, 1773–1776.

Index

a

ab initio
- thermochemical calculation 505
- thermochemically calculated volume expansion 523

ab initio FLAPW (full-potential linearized augmented plane wave) 346
ab initio method 231
ab initio molecular dynamics method 356
ab initio ORTEP 513
ab initio random structure searching (AIRSS) 576
ab initio theory 163, 521
ab initio thermochemistry 501–504
A—B bond
- energy *vs.* distance profile for 406

ACDMP. *See* atom centered density matrix propagation (ACDMP)
acentric structure model 506
active Ti sites 669
Adams–Gilbert equations 652
adaptive-GA 581
adaptive natural density partitioning analysis (AdNDP) 539
- method 544

additive approach 653
additive zeolite 654
adiabatic connection and fluctuation dissipation theory (ACFDT) 232
adiabatic local-density approximation (ALDA) 17
AdNDP. *See* adaptive natural density partitioning analysis (AdNDP)
adsorption and surface reactions 180
AFM. *See* antiferromagnetic (AFM)
Ag[Ag(C$_3$H$_3$)$_2$]
- infrared spectroscopy 504
- crystal structure 506
- density of phonon states 506

Ag—C bonding 506
alkali metal clusters 468
Allen–Dynes-modified McMillan equation 591
all-electron calculations 31, 34
all-electron reference calculations 41
all-electron wave functions 39, 45
α-boron
- reconstruction of 559
- – complex metallic surface reconstruction 559
- – minima hopping method 559

α-tin
- crystal structures 517
- relative enthalpies 517
- relative Gibbs energies 518

AMBER force fields 175, 220
amino ester type local anesthetics 176
amorphous carbon nanoparticles 183
amorphous silica
- tridimensional structural arrangement 610

analytic second derivatives 169
anatase, defect structure 183
anatase nanoparticles 183
Anderson model 376
angle resolved photoemission experiment (ARPES) 144, 249
angle resolved photoemission experiments (ARPES)
- experimental band structure 250
- spectra, schematic view of 120

angular integrations 43
angular momenta 33
angular momentum operators
- nonzero integrals 291

anharmonicity 592
anion-poor conditions 663

Handbook of Solid State Chemistry, First Edition. Edited by Richard Dronskowski, Shinichi Kikkawa, and Andreas Stein.
© 2017 Wiley-VCH Verlag GmbH & Co. KGaA. Published 2017 by Wiley-VCH Verlag GmbH & Co. KGaA.

anisotropic displacement parameter (ADP) 513
anisotropic spin exchange
– mapping analysis 312
annihilation operators 125
antibonding orbitals 421
antiferromagnet 96
antiferromagnetic (AFM) 295, 352
– arrangement 295
– coupling 336
antiferromagnetic cluster 542
antiferromagnetic Heisenberg 94
antiferromagnetism 346
antimony telluride 507, 522
– crystal structure 507
– partial DPS measurement 508
– phonon dispersion 501
– phonon states, density of 501
AO. See atomic orbital (AO)
Arachno-structures 466
aromaticity 1
artificial dipole moments 179
artificial lattices 552
asymmetric spin exchange 309, 311
asymptotic periodicity 281
atom-centered density matrix propagation 177
atomic coordinates 576
atomic displacements 535, 576
atomic eigenvalues 35, 47
atomic exchange parameter 271
atomic motion 499
atomic orbital 171
atomic orbital (AO) 66, 163, 238, 289
– angular properties 288, 289
– coefficients 68
– energy 426
– expands 409
atomic pseudopotentials 648
atomic reference configurations 47
– wave functions 37
atomic shell structures 453
atomic structure 227
atomic valence orbitals 160
atomization energies 50
– chemical accuracy 50
– errors, maximum absolute deviation (maxAD) 52
atomization enthalpies 173
atomization/formation energies 49
atom pair parameters 163
atom terminating scheme 655
augmented plane wave (APW) 233

Austin model 167
auxiliary boson approximate (ABA) 106
– self-consistency conditions 107
auxiliary functions 61
auxiliary-boson theory (ABT) 105
average intra-orbital Coulomb repulsion 148
average pseudopotential 43

b
$BaCeO_3$
– polymorphs
  phonon states densities 510
– structural motifs 509
$BaCo_2As_2$
– spectral function 149
Bader charges 562, 586
$BaFe_2As_2$
– frequency-dependent Hubbard interaction 147
– spectral function 147
banana bond 454
band diagrams
– evolution of 424
band filling 471
bandgap 244, 551
– dielectric constant, relationship between 549
band structures 227, 432
Bardeen–Cooper–Schrieffer (BCS) function 101
– Slater determinant 101
barium cerate ($BaCeO_3$), 509
basin hopping 573, 577, 578
basin-flooding procedure 579
basis set superposition error (BSSE) 79
Baym–Kadanoff (BK) theories 125
B_{20} cluster 557
B2 CsCl-type crystal structure 349
Bell Evans Polanyi Principle 575
benzene
– electronic configuration 159
– first excitation energy 159
– resonating valence bonds 88
– RVB wave function
– – Rumer basis states 102
Berry-phase theory, of magnetization 622
beryllium 587
β-phase Hume-Rothery alloys 360
β-polonium 592
β-tin
– crystal structures 517
– relative enthalpies 517
– relative Gibbs energies 518

Bethe–Salpeter equation (BSE) 245
BFGS. *See* Broyden–Fletcher–Goldfarb–Shanno (BFGS)
B-spline interpolation 175
binary chalcogenide glasses 634
binary Laves phases
– bonding features and glassy tendencies 348
binary phase diagrams 469
binding energies 182, 407, 422
Biot–Savart law 620
black box 81
black-box character 61
Bloch's theorem 429, 435, 495, 616
Bloch basis 139, 141
Bloch functions 436, 621
Bloch Green function 139
Bloch state 138, 139
Bloch theorem 93, 227
Bloch waves 650
Bohr magneton 311
Bohr radius 652
Boltzmann average 285
Boltzmann constant 563, 578, 690
Boltzmann statistics 694
bond effect 308
bonding rationalizes 481
Born–Haber cycle 422
Born–Mayer expression 418
Born–Oppenheimer approximation 230, 408, 428, 615
Born–Oppenheimer molecular dynamics scheme 165
Born–von Karman boundary conditions 429
boron atoms
– substrate and, bonding between 561
boron monolayers
– 2D 556
boron nanotubes 553
boron structures
– planar 552
Bose–Einstein statistics 496, 502
bosonic baths 122
bosonic fields 131
bosonic quasiparticles 496
Boughton–Pulay procedure 69
B2 → B19 martensite 361
Bravais lattices 139, 576
– vector 139
Brillouin–Wigner perturbation theory 167
Brillouin zone (BZ) 62, 99, 111, 121, 139, 215, 234, 432, 435, 441, 552
Brillouin zone boundaries 469

broken-symmetry states 302
Brownian motion 613
Broyden–Fletcher–Goldfarb–Shanno (BFGS) 166
brute-force methods 682
brute-force molecular dynamics simulations 697
bulk ionic solution 688
bulk modulus 522
– temperature-dependence 522

c

Ca_2Ag_7
– chemical pressure 367
$CaCO_3$ prenucleation clusters
– in aqueous solution 686
calcium carbonate 689
CALYPSO 580
Ca–Ca distances 689
carbon allotropes 552, 553
– planar 552
– – atomic densities 552
carbon densities 553
carbon nanodots 183
– photophysical properties 183
carbon–fluorine bonds 163
Car–Parrinello approach 356
Cartesian components 315
Cartesian coordinates 619
Cartesian directions 497
Cartoon depicting ionic *vs.* electronic disorder 662
catalyzed hydride transfer 174
C–C antibonding 441
CCM. *See* cyclic cluster model (CCM)
$CdIn_2Se_4$ 459
Ce_6O_8 cluster
– chemical bonding elements of 541
– electron localization function of 540
Ce_6O_n
– clusters
– – ground states of 538
– – phase diagram of 539
– phase diagram of 539
cerium(IV)oxide
– constant-pressure heat capacities 520
cesium polyfluorides 580, 586
chalcogenide compound 506
chalcogenides 512
charge density 50
– cutoff 51
– difference 561, 562
– waves 477

charge fluctuations 207, 212
charge-ordered (CO) phase 251
charge relaxation 37
charge reorganization (CR) 456
charge transfer 561
chemical bond formation 261
chemical bonding 29, 411, 413
– antibonding MOs 408
– atomic orbital (AO) 407
– bonding influences on structure/properties 471
– – high-pressure effects 479
– – magnetism 478
– – metals *vs.* nonmetals 475
– – structural influences 471
– covalent solids 420
– energy *vs.* distance profile 406
– equilibrium lattice energy 419
– fundamental characteristics 406
– ionic/covalent configurations 409
– ionic solids 416
– metallic solids 426
– molecular orbitals (MO) 407
– NaCl-type solids 419
– orbital-based models 414
– position–space analysis 456, 465
– quantum mechanical calculations/analyses
– – BCC metals 444
– – bonding analyses in position space 446
– – electron counting rules 457
– – free-electron *vs.* tight-binding approximations 433
– – graphene (graphite) *vs.* diamond 441
– – Hume–Rothery electron compounds 469
– – 18-electron rule 461
– – Octet/8–N rule 458
– – polyhedral skeletal electron pair theory 465
– – rhenium trioxide 443
– – shell model 468
– – transition metal octahedral clusters 467
– – Wade–Mingos rules 465, 467
– selected half-Heusler compounds $ABC^{(HC)}$ 463
– semiquantitative assessments 428
– shift in electron density 408
– σ- and π-bonding 413
– solid-state structures 405
– solids /bonding motifs 414
– van der Waals and zero-point energy 419
chemical bonds 407
chemical doping 589
chemical frustration 367

chemical glycobiological systems 174
chemical hardness 412
chemical (Mulliken) notation 62
chemical shift tensor 613
chemical space exploration 550
chemical valence structure theory 160
chemistry, under pressure 581
ChemShell 653, 659
chemshell quantum mechanical/molecular mechanical coupling
– master equations 653
child structures 535, 536
– generation
– – heredity 536
– – soft mutation 536
– – transmutation 536
chlorine polyhydrides 580
chosen exchange-correlation functional 31
CK2 protein kinase 171
classical crystal nucleation pathways 689
classical nucleation pathway 684
classical nucleation theory (CNT) 683, 689
classical thermodynamics 491–493
Clausius–Clapeyron equation 381, 386
clinopyroxene 521
Closo-structures 466
CLUSTERD 178
cluster graph 546
cluster-based technique 60
cluster-based treatment 60
clusters 182
– size 60
– surrounded by vacuum 532
– synthesis methods 543
– – laser ablation 543
C–Fe–C axis 324
coherent potential approximation (CPA) 429
Co-doped β-Mn 369
Co-doped Ni–Mn–In alloys saturation magnetization 386
cold melting 593
collective vibrations 496
collinear magnetic systems 264
collinear magnetism 263
coloring problem 631
common sense explanation 92
competing crystal structures
– free energy profiles 689
complete neglect of differential overlap (CNDO) method 162
complex magnetic systems 282
complex powder-X-ray-diffraction data 505
computational box 35

computational cost 60, 80
computational efficiency 172
computational modeling techniques 647
computational resources 1
computational science
– challenges in 528
computational volume
– subdivision 32
computer codes 241
– advantages and disadvantages 241
– ways of distributing 242
computer-aided drug design 172
condensed matter systems 29
conduction band 556
configurational entropy 503
configuration interaction (CI) 230
conformational analysis 175
conformations 70
constant-pressure Gibbs free energy 503
constant-volume electronic-structure
 theory 503
constant-volume Helmholtz free energy 492, 503
– temperature-dependence 502
constrained random phase approximation
 (cRPA) 124
continuous time Monte Carlo (CT-QMC) 125
conventional matrix diagonalization 172
conventional unit cell 229
Cooper channel 103
coordination numbers 229, 584
copper carbodiimide (CuNCN) 111
copper(II) carbodiimide (CuNCN) 113
copper(II)oxide
– constant-pressure heat capacities 520
core electrons 29
core Hamiltonian matrix 168
core-leakage 241
core–core repulsion 166
core–hole supercell 245
core orbitals 30
core states 235
corner-sharing $GeSe_4$ 634
correct integer spin polarization 51
correlation contribution 60
Coulomb effects 180
Coulomb hole 71
Coulomb integrals 64, 73, 75
Coulomb interactions 76, 122, 123, 124, 128, 130, 136, 143, 144, 148, 231, 304, 432
Coulomb kernel 75
Coulomb law 125
Coulomb potential 5, 42, 231, 234

Coulomb repulsions 95, 300, 432, 433
– parameter 431
coupled cluster (CC) 230
– calculations 52
– methods 46
covalent compounds 415
covalent magnetism 282
Cr-doped Ni–Co–Mn–Cr–In 387
CRYSTAL code 65
crystal field split d-states 291, 337
– influence of intersite interactions 333
crystal formation 690
crystalline cell 49
crystalline metal oxides 179
crystalline polymorphs 574
crystalline solids
– electronic energy 494
– electronic structure 494
– lattice dynamics 495–501
– magnetic properties 285
– tight-binding model 436
crystalline state
– reciprocal-space approach 498
crystalline wave functions 241
crystallographic point groups 530
crystal nucleation
– free energy diagram 688
– molecular dynamics simulation
 approaches 692
– – bridging timescale and length scale 693
– – equations of motion and numerical
 solutions 692
– – quantum, classical, and hybrid models 692
– pathways 692
– processes 689
crystal orbital Hamiltonian populations
 (COHP) 358, 366, 439, 586
– curves 444
– diagrams 367
– – for antisite defects 367
– method 367
crystal orbital overlap population (COOP)
 curves 439
crystal precipitation 691
crystal structure prediction (CSP) 1, 532, 551, 572
– inorganic
– – prediction 528
– prediction method 528
– techniques for 573
crystal, vibrational properties 493
CsCl structure 419
cubic Fe_3O_4
– unit cell of 255

cubic perovskite structure 144
cubic scaling term 48
Cu–O–Cu bridges 306
Cu–O–Cu spin exchange interaction 306
$CuCl_2(OH_2)_2$ complex
– structure and the down-spin-down d-states 328
Cu doping 182
Cu-dependent shape memory 362
CuL_2 ribbon chain 328
CuL_4 square planes 329
Cu–Al–Mn alloys
– shape memory characteristics 352
CuO_2 ribbon chains 305, 307
Curie temperatures 142, 373, 376, 381
curvy steps (CURV) methods 177
Cu_2Si monolayer
– bonding in 558, 559
– chemical bonding in 557
– deformation charge density 558
Cu_2Si sheet
– planar geometry of 558
cutoff energy 617
cyclic ADP-ribose 175
cyclic cluster approach 165
cyclic cluster model (CCM) 173, 178, 179
cyro-TEM 686

d

Davidson–Liu algorithm 180
Davidson–Liu block diagonalization method 170
3D crystal structures 530, 533
Debye-type low-temperature behavior 521
Debye model 384, 493
Debye–Waller factor 514
degenerate perturbation 322
degrees of freedom 574
delocalization index 450
delocalized bonding 558
– orbital 540
delta test 49, 50
deMon computer code 213
density fitting 73
– approximation 74
– coefficients 74
density functional embedding methods 652
density functional-based electronic structure
– ab initio nature of 122
density functional perturbation theory 282, 551
density functional theory (DFT) 2, 29, 46, 59, 119, 129, 169, 227, 231, 353, 431, 494, 520, 529, 554, 573, 610, 647

– all-electron treatment 234
– atom-centered basis functions 238
– based electronic structure 482
– based methods 60
– basic principles 2
– Bloch basis 138
– calculations 578
– crystal field 141
– density 148
– descriptors 12
– electronic structure
– – calculations 233
– examples of applications 21
– excitations 15
– extended systems 8
– full potential and basis sets 237
– functionals 5
– Hubbard corrections 7
– Kohn–Sham equations 238
– open issues 21
– – charge (de)localization:, 23
– – functionals 21
– – response properties:, 24
– – weak interactions 22
– relativistic effects 236
– relaxation energy 79
– spatially confined basis sets 238
– surfaces and heterogeneous catalysis 254
– theory 129
– total energy 39
density-functional electronic-structure theory 500
density-functional methods 2
density-functional perturbation theory 501
density of phonon states (DPS) 499
density of states (DOS) 227, 275, 432, 554
– band structure and chemical bonding 249
– calculations 243
– – atomic structure 243
– – basis set and Brillouin zone sampling 245
– – choice of quantum mechanical treatment 244
– – hardware considerations and parallelization 247
– equilibrium volume 248
– examples of 247
– schematic representation 122
deprotonation 659
Dewar benzenes 90
Dewar structures 94
DFT++ approaches 123
D3H4 correction 171
d-electrons 261

d-states of Ir^{4+}
– PDOS plots 334
diagonal pair amplitudes 67
diamond 443
diamond allotropes 441
diamond anvil cell (DAC) 571
diatomic ELI-D/ELF basin 454
Diebold group 255
dielectric constant 549, 551
– bandgap vs 551
dielectric materials
– polymeric 549
dielectric matrix 19
dielectric polymer
– design strategy 550
diffusion-controlled crystal formation 690
– processes 695
digital computing 514
dipole moments 23
dipole–dipole interaction 71, 612
Dirac's equation 237
Dirac bra–ket notation 435
Dirac cones 552, 556
Dirac point 554, 556
direct defect–defect interactions 180
direct space 75
dispersion correction 79, 171
dissociation energy 406
distorted bicapped trigonal antiprism 588
divalent hcp metals Mg and Zn 480
divide-and-conquer (DAC) 217
DNA base guanine 175
DO_{24} crystal structure 357
domain approximation 69
domain error 69
donor impurity 588
doped Mott insulators 152
DOS. See density of states (DOS)
DOS curves 435, 436, 437, 469, 474
double electron occupancy 261
double-counting-free qsGW + DMFT schemes 152
downfolding 122, 123, 124
downfolding techniques 123
drug–surface interactions 81
dual fermion (DF) 130
dual-perturbation theory 132
dual-space Gaussian pseudopotentials 47, 49
Dulong–Petit rule 493
dynamical Hubbard 124
dynamical mean field theory (DMFT) 125, 130, 135, 244

– energy 134
– literature 146
– loop
– – self-consistency condition 134
– scheme 125
– self-energy 141
– techniques 143
dynamic electron–electron correlations 231
dynamic-angle spinning (DAS) 609
dynamic mechanical analysis (DMA) 362
dynamic mechanical spectroscopy 362
Dzyaloshinskii–Moriya (DM)
– exchange 309
– interaction energy 314
– interaction strength
– – interactions controlling 314
– vector of isolated spin dimer
– – mapping analysis 315

e
EA. See evolutionary algorithm (EA)
ECP. See effective core potential (ECP)
edge atoms 228
EDMFT approximation 133
effective core potential (ECP) 168
effective Hamiltonian crystal field (EHCF) 170
effective hydrogen bonding corrections 171
effective interaction potential (EIP) 354
effective nuclear charge 162
effective one-electron Hamiltonian 291
EHCF. See effective Hamiltonian crystal field (EHCF)
EHT. See extended Hückel method (EHT)
eigenfunctions 9
eigenvectors 496, 531, 535
eigenvectors' matrices 110
Einstein and Debye models 493
Einstein's photoelectric effect 120
electrical conductivity 415
electrical resistivity 593
electric field gradient (EFG) 245, 613
– tensors 613
electron compounds 469
electron configuration 299
electron correlation 292
– contribution 80
electron delocalization 261
electron density 13, 231, 263
– distributions 204, 418, 447
– localization 541
electronegativity 412, 414
electronic density of states (DOS) 444, 500

electronic distribution 19
electronic energy 274
electronic ground-state energy 494
electronic Hamiltonian approach 287
electronic-structure calculations 1
electronic-structure methods 1, 2
electronic-structure theory 491, 501, 503, 517
electronic repulsion 159, 161, 162
electronic Schrödinger equation 446
electronic states, of titanium 235
electronic structure calculations 29, 30, 31, 49
electronic structure codes 45
electronic structure methods 647
electronic systems 281
electronic transition 583
electron (or proton) ionization potential 660
electron localization function (ELF) 451, 539, 540
– definition 452
– kernel 452
– QTAIM intersection technique and 454
electron localization procedures 651
electron–boson vertex 133
electron–electron interactions 2, 120, 261, 430, 431
electron–electron repulsion 262, 292, 294
electron–lattice coupling 255
electron–nucleus cusp 31
electron–nucleus energy 206
electron–nucleus interactions 2
electron occupancies 265
electron paramagnetic resonance (EPR)
– spin Hamiltonian parameters 670
– studies 669
electron repulsion 265
electron theories 161
electrostatic electron–nuclear attractions 430
electrostatic embedding 655
electrostatic Hartree and exchange-correlation potentials 45
electrostatic Madelung potential 417
ELF. *See* electron localization function (ELF)
ELI-D attractor 452
ELI-D/ELF attractor 454
ELI-D/ELF basin 455
ELI-D/ELF function 454
ELI-D/ELF functions 454
embedded atom method (EAM) 353
embedding techniques 181
emergent properties 590
EMPIRE program package 182

empirical force fields 169, 529
energy bands 433, 438
energy consistent 46
energy differences 263
energy gap
– empty conduction band and filled valence band 417
energy-consistent pseudopotentials 46, 47
energy-mapping
– analysis 287, 296–304, 301, 304
– based on four ordered spin states 303
energy of (de)hydrogenation 667
energy split arising 294
energy, *vs.* logarithmic derivative curve 35
enthalpy 492, 520
– evolution 555
– vaporization 184
entropy 492, 520
equation of states (EOS) 494
equilibrium Helmholtz energy 503
Euclidean action 126
Euler formula 546
evolutionary algorithm (EA) 528, 530, 533, 535, 548, 552, 580
– challenge for 535
– constrained, flowchart 533
– flowchart 529
– procedure
– – heredity 535
– – initialization 534
– – local optimization 535
– – monomeric blocks 533
– – random structures, addition of 535
– – rotational mutation 535
– – selection 535
– – softmutation 535
– – stopping simulation and postprocessing 535
– – variation operators 535
– simulation traverses 577
– smart variation operations and clustering 530
Ewald summation technique 180
exchange energy 434
exchange-correlation (XC) 231, 495
exchange-correlation effects 2
exchange-correlation energy 6, 39
exchange-correlation functional and spin polarization energy, approximate expressions of 264
exchange-correlation potential 264
exchange interaction 230

exchange–correlation interaction 40
exchange–correlation potential 231
exchange splitting energy 270
excited-state geometry optimization 170
excited-state optimization 182
extended Hückel method (EHT) 210
extended Hückel theory (EHT) 161
extended Lagrangian molecular dynamics simulation approach 177
extended X-ray absorption fine structure (EXAFS) data 668

f

face-centered cubic (fcc) 229
– copper, schematic phonon density 493
– Fe iron 369
fast-Fourier transforms 618
FDH. See formate dehydrogenase (FDH)
FeAl-based intermetallics 349
Fe–Fe antibonding 479
Fe-rich Fe–Al-based alloys 353
FeL_4 square plane 295
Fe–Al intermetallic alloys
– mechanical and magnetic properties 353
Fe–Ni alloys 368
Fe_3O_4
– surface of magnetite 256
FEP. See free energy perturbation (FEP)
Fermi energy 249, 275
Fermi hole 430, 434
Fermi level 122, 366, 445, 476, 477, 479, 480, 481, 482, 586, 650
– graphene 442
Fermi liquids 123
– Landau's theory 120
Fermi–Dirac distribution 110, 127, 428
fermionic bath 122
fermionic Grassmann variables 131
fermionic Green's functions 133
fermion–fermion interaction 133
Fermi operators 105
Fermi surface 249, 624
Fermi surface–Brillouin zone interactions 481
Fermi surface pockets 148
Fermi velocities 552
ferroelastic $Ti_{50}Ni_{50-x}Fe_x$ 362
ferroic materials 363
ferromagnetic (FM) 295, 352
– arrangement 295
– coupling 336
ferromagnetic Fe_2CuGa
– COHP bonding analysis 369

ferromagnetic iron
– DFT + DMFT calculations 140
ferromagnetic shape memory Heusler alloys 370
ferromagnetic (FM) spin arrangement 336
FHI pseudopotentials 45
FHI Troullier–Martins pseudopotentials 50
field-cooled (FC) 362
fingerprint distance threshold (d_{min}) 531
fingerprint niching 535
finite-temperature magnetoelastic
– Monte Carlo simulations 381
Finnis–Sinclair potentials 354
first Brillouin zone 495, 500
– $BaCeO_3$ polymorphs of 511
– vibrational modes 499
first-order wave function 63
first-principles
– calculations 593
– lattice dynamics 491
fluorescence process 180
Fock matrix 62, 63
Fock operator 63
force constant matrix 497
formate dehydrogenase (FDH) 174
formation energies 518
formation enthalpies 492
fourfold rotational axis 292
four-index electron repulsion integral distribution 80
four-index integrals 64
four-index two-electron integral 74
Fourier series 434
Fourier space 48
Fourier transform 132
Fourier transformations 65
fragment-based models 61
Frank–Kasper phases 345, 346, 347
free energy optimization 547
free energy per solute 691
free energy perturbation (FEP) 177
free energy profile 683
free-electron model 433, 434
– DOS curves 433
Friedel oscillation 281
frozen-magnon approach 377
fullerene clusters 573
full Hamiltonian matrix 43
full Heusler alloys 364
full potential 237
full potential linearized augmented plane wave (FLAPW) 351

g

gauge-including atomic orbitals (GIAO) 610
Ga–Mn bonding 374
gauge-including projector augmented wave (GIPAW) 610
– approach 621
– ^{29}Si chemical shifts 633
Gaussian code
– with B3LYP hybrid function 632
Gaussian codes 30
Gaussian functions 579
Gaussian-type function 43
Gaussian-type orbital (GTO) 30, 31, 75
Gaussian/Slater functions 624
Gaussian2 test set
– errors in atomization energies 51
GBRV pseudopotentials 50
generalized gradient approximation (GGA) 6, 232, 432, 495
general magnetic solids
– broken-symmetry states, use of 301
generating random sensible structures 576
genetic algorithm (GA) 580
genetic drift 530
geometry optimizations 172
germanium selenide 512
– crystal structure 512
GeSe 512
– phonon dispersion relations 513
GGA band-gap error 626
G hairpin ribozyme 177
ghost states 46
giant magnetocaloric effect 384
Gibbs energy 501, 519, 520
– difference 517
– function 504
Gibbs formation energy 519
Gibbs free energy 492, 509, 515, 519, 546, 562
– difference, temperature-dependence 517
– of surface structure (G_{sur}) 562
Gibbs free formation energy 519
Gibbs–Helmholtz equation 492
gigapascal 581
glass transitions 362
glycoenzymes 174
graphene 441
– electronic structures 440
– ELI-D/ELF and QTAIM space partitioning 453
– valence bonds in the -system 100
graphical unitary group method 170

graphite
– binary derivatives 443
Grassmann variables 126
Green's functions 19, 127, 128, 129, 134, 135, 136, 137, 139, 141, 216
– formalism 652
– impurity 133, 137
– noninteracting 136
– single-electron 128
– two-particle 127
Green-function operator 621
Grimme's approach 214
Grimm–Sommerfeld rule 460
Grimm–Sommerfeld valence rule 459, 460
ground-state energies 166
ground-state MO wavefunction of H_2 449
Grüneisen parameter 508
guanidine based on neutron diffraction data
– ORTEP plots 516
Gutzwiller approximation 104
Gutzwiller projection operator 102
GW approximation 135, 136
GW calculation 146
GW + DMFT equations 144
GW schemes
– flavours of 142
GW self-energy 145, 151
GW + DMFT
– approach, ad hoc assumption 136
– momentum-resolved t_{2g} spectral function 143
– spectral function 145
GW + DMFT calculations 151
GW + DMFT method 144
GW + DMFT scheme 134, 143, 153
GW + DMFT self-energy
– nonlocal dual part of 133
GW + DMFT spectra 145

h

half-Heusler alloys AA′X
– schematic orbital energy diagrams 462
half-Heusler compounds 464
Hamiltonian analysis 286
Hamiltonian dualism 100
Hamiltonian matrix 33, 44, 440, 443
– elements 50, 437
Hamiltonian operator 43, 421, 428, 431, 435
Hamiltonian transformation
– mean-field 109
Hamilton operator 2, 209
harmonic phonon free energy 502
harmonic theory 503

Hartee potential 129
Hartree–Fock approach 4, 5, 15
Hartree–Fock approximation 103, 171, 181
Hartree–Fock calculations 411, 578
Hartree–Fock equations 3, 652
Hartree–Fock exchange 296
Hartree–Fock expressions 21
Hartree–Fock like expression 25
Hartree–Fock (HF) method 203, 230, 430, 431, 648
Hartree–Fock scheme 165
Hartree–Fock type of wavefunction 451
Hartree part 134
heat capacities 520
heat of hydration 656
Heisenberg's uncertainty principle 495, 513
Heisenberg Hamiltonian 95, 376
Heisenberg model 346, 376, 377
– on triangular lattice 97
Heisenberg spin exchange 315
Heitler–London solution 410
Heitler–London variant 409
Hellmann–Feynman theorem 356, 693
Helmholtz double layer 217
Helmholtz free energy 502, 503
– schematic 504
– volume-dependence 492
Hermitian conjugates 98
heterocubic (HC) site 464
heterofullerenes, electronic spectra 183
Heusler alloy $Ni_{45}Co_5Mn_{34}In_{16}$
– exchange coupling constants 389
Heusler alloys
– magnetocaloric effects 381
Heusler alloys X_2YZ 374
Heusler compound Ni_2MnGa 369, 371, 375
Heusler crystal structure 381
Heusler-type structures 349
Heusler materials 364, 376
Heusler phases 376, 463
hexagonal Laves phase $MgZn_2$ 348
H-bonding interactions 176
H-bonds 78
highest-occupied molecular orbital (HOMO) 32, 582
high-level correlation methods 60
high-spin 371
– d^5 systems 332
high-spin Fe^{2+}
– spin-down electron configuration 323
high-temperature high-pressure (HTHP) 586
high-temperature superconductor 518
Hilbert space 123

Hirshfeld approach 13
Hohenberg–Kohn theorem 6
holy grail 583
HOMO 321, 458
homogeneous electron gas 407
HOMO–LUMO energy differences 337
HOMO–LUMO gap 24, 412
HOMO–LUMO interactions 322, 333
honeycomb (BN) 443
Hooke's law 497
H_3 system
– MO energy diagrams 470
Hubbard band 146
– forms 145
Hubbard correction 8
Hubbard-like term 218
Hubbard interactions 131, 150
Hubbard model 7, 124, 152, 154, 409
– to H_2 410
Hubbard–Anderson Hamiltonian scale 96
Hubbard–Anderson model 95
Hubbard–Stratonovich transformation 131
Hubbard on-site repulsion parameter 432
Hubbard parameter 143
Hubbard satellites 148
Hume-Rothery phases 345, 368
Hume-Rothery rules 346, 364
Hume-Rothery electron-counting rule 469
Hund's rule 121
Hund–Mulliken model 409
Hund–Mulliken molecular orbital approach 409
Hückel approximation 442
Hückel chain 126
Hückel-type solution 442
Hückel method 160, 367
Hückel models 94
– of benzene 94
Hückel Hamiltonian 367
Hückel setting 94
Hückel solution 91
– of cyclic polyene 92
Hückel theory 160
Hückel wave function 92, 93
hybrid functionals 529
– method 296
hybridization 472
– function 139
hybrid methods 573
hybrid quantum mechanical/molecular mechanical approach 649
– choice of method 652

– electronic structure, defect containing system 652
– fragmentation methods 653
– key questions for 649
– many-electron problem/electron groups/ localization/structural elements 651
– one-electron problem 650
hydrocarbon isomers 169
hydrogen atoms
– one-dimensional chain 498
hydrogen bonding 167
– energies 185
hydrogen bonds 407
hydrogen/lithium phase diagram 588
hydrogen–silane mixtures 577
hydroxide ion association 695
Hylleraas functional 64, 69
hyperpolarizabilities 19
hypervalent compounds
– heats of formation 166

i

ibuprofen adsorbed on a silica slab 78
icosahedral atom (B_p) 561
icosahedral clusters 537
imaginary phonons 575
– disadvantage 575
Imma
– phonon dispersion relations 511
– temperature-dependent Gibbs free energies 511
impurity correlation function 132
impurity Green's function 131, 132
impurity reference system 130
impurity self-energy 137
impurity susceptibility 132
inconsistent pseudopotentials
– factor affecting 47
independent gauge for localized orbitals (IGLO) 610
independent valence states
– quantum mechanical superposition 89
induced spin density 281
induced spin polarization 267
inelastic neutron scattering 493
infinite polyene resonance structures 98
infrared (IR) spectroscopy 505
initial convergence tests 179
inorganic solid materials 228
insulators 652
interaction energies 80, 81
interatomic-potential-based technique 649
interatomic potential parameters 573

intermediate neglect of differential overlap (INDO) 162
intermetallics 262
intermolecular hydrogen bonding interaction 176
intermolecular interactions 176
intermolecular reactions 176
intra-adsorbate correlation 79
intramolecular hydrogen bonds 584
intraslab correlation 79
inverse photoemission 144
ion-by-ion association 696
ionic bonding 416
ionic compounds 414
ionic solid NaCl, band diagram 416
Ir and Ni, interactions between 336
Ir—O—Ir linkages 333
iron aluminides 353
iron pnictide $BaFe_2As_2$
– dynamical screening 148
iron pnictide superconductors
– 122-family of 148
IrO_6 octahedron 327
irreducible Brillouin zone (IBZ) 245
IR spectroscopy 572
isolated $CoMnO_6$ chain
– schematic view 323
isolated oxygen vacancies in alumina 180
isolated spin dimer 286
– eigenstates use 297
– use of broken-symmetry states 301
isolated Sr_2IrO_4 layer 334
isomeric structures 547

j

Jacob's ladder 232
jadeite
– glass 521
– heat capacity 521, 522
Jahn–Teller (JT) 318
Janak's theorem 15
J-coupling tensors 611
jj-coupling scheme 337

k

Kawska–Zahn method 695
Kekulé-like structures 99
Kekulé's oscillation 88
Kekulé structures 94, 97
k-index 62
k-symmetry 66
k-vectors 227
kinetic contribution 269

kinetic core electrons 34
kinetic energy 10, 231, 278
– component 268
– density 232
Knight shift 615
Koelling–Harmon radial equation 625
Kohn–Sham orbitals 17
Kohn–Sham approach 5
Kohn–Sham band structure 152
Kohn–Sham density functional theory 29
Kohn–Sham equations 209, 219, 616, 617
Kohn–Sham Green function 138
Kohn–Sham Hamiltonian 264
Kohn–Sham-like equations 208
Kohn–Sham model system 24
Kohn–Sham–Dirac equation 209
Kohn–Sham operator 12
Kohn–Sham orbitals 24, 209, 211, 616, 618, 619
Kohn–Sham potential 129, 152
Kohn–Sham scheme 205
Kohn–Sham spin-polarized density functional theory 262, 269
Kohn–Sham spin orbitals 263
Kohn–Sham system 21
Kohn–Sham variant 59
Kolbe, H. 89
Koopmans' theorem 15
Korringa–Kohn–Rostoker coherent potential approximation (KKR-CPA) 377
Kramers degeneracy theorem 311
Kronecker deltas 497
Kröger–Vink notation 663
KS equations 231
K to Br ground-state structural features 473

l

Lagrange multipliers 106
Lam–Platzman correction 24
lanthanide complexes 168
Laplace-transformed MP2 method 61
LAPW-type methods 32, 33
large unit cell (LUC) 178
late transition metal monoxides 576
lattice fermions 131
lattice vectors 576
Laves phase 347, 348
– CN12-CN16, 347
– intermetalllics 347
– MgCu$_2$ 347
– MgZn$_2$ 348
– NiAlX 352
LDA/GGA density functional electronic structure 48
LDA + DMFT scheme 138
Legendre transform 128
Lennard-Jones clusters 533
– algorithms for, performance of 534
Lennard-Jones potentials 213
ligand-controlled spin orientation 331
ligand orbitals 261
light element systems 571
Li–Be system 481
linear combination of atomic orbital (LCAO) 87, 92, 238
– ansatz 410, 430, 436
– DFT band structure and 209
– MO theory and 448, 472
– to represent crystal orbitals (LCAO-CO) 366
linear equations 65
linearization 239
– energy 241
linearized augmented plane wave (LAPW) 240, 618
– and WIEN2k 238
linearized muffin tin orbital (LMTO) 351
liquids
– simulation 184
Li$_3$RuO$_4$ crystal lattice 573
lithium 587
lithium hydroxide (LiOH) 499, 581
lithium-rich region 588
lithium–beryllium phases 590
LMO. *See* localized molecular orbital (LMO)
LMP2 calculations 65
LMP2 energy 69, 73
– decay 71
LMP2 equations 72, 81
LMP2 interaction energies 79
LMP2 method 61
LMTO method 364
local fit-domains 75
local-density approximation (LDA) 205, 232, 495, 616
– band structure 149
– energies 146
– function of unit cell 249
– Hamiltonian 150
– Kohn–Sham state 144
localized molecular orbital (LMO) 176
local orbitals (LO) 241, 618
local spin density (LSD) 264

local spin density approximation (LSDA) 244
local spin hardness function 265
London dispersion 171
long-range electron correlation effects 171
Lorentz factor 236
Löwdin orbitals 161
lowest-unoccupied molecular orbitals 582
low-dimensional materials 527
– chemical bonding in 564
– chemical composition of 528
– prediction methods
– – cluster structure 532
– – evolutionary structure 528
– – polymer structure 533
– – surface reconstructions 535
– prediction of 564
low-dimensional nanomaterials 528
low-energy barriers 531
low-energy basin 575
low-energy hamiltonians 122
low-energy minima 575
low-energy structures 531
– bilayer 554
– monolayer 554
low-spin (LS) 371
low-spin Ir^{4+}
– spin-down states 327
low-spin–high-spin transition 121
LSD. See local spin density (LSD)
LS-coupling scheme 338
$LuAl_3$ 467
LUC. See large unit cell (LUC)
LUC–INDO method 179, 181
LUMO level 321, 458, 462

m

macroscopic atom picture 358
Madelung constant 419
Madelung energy 418, 420, 426
madelung potentials 333
MagAixTic program package 171
magic clusters 537
magic-angle spinning (MAS) 609
magnetic and orbital fluctuations
– schematic representation of 121
magnetic anisotropy 287, 309, 311, 320
– of S 326
magnetic coupling 274
– constants 346
magnetic dipole–dipole interactions 311
magnetic energy spectrum 302
magnetic exchange interactions 261

magnetic field-induced strain (MFIS) 379
magnetic field–temperature phase diagram 386
magnetic Heusler alloys 347, 352, 371, 382, 390
– martensitic transformation 371
magnetic Heusler materials 370
magnetic insulator 302, 303
magnetic ions 285, 320
– crystal-field split d-states 288
– preferred spin orientations 322
magnetic moments 251, 261, 263
magnetic Ni–Mn-based Heusler alloys 364
magnetic orbitals 288, 289, 332
magnetic phase transition 381
magnetic properties 285
– of 5d ion oxides 332
magnetic quantum numbers 290
magnetic shape memory alloy Ni–Mn–Ga
– intriguing martensitic instability 351
magnetic shape memory effect (MSME) 352, 377
magnetic susceptibility 286
magnetic system 301
– close-packed energy states 286
magnetic systems 269
magnetism, schematic illustration of types 262
magnetization curves
– cooling and heating across 384
magnetocrystalline anisotropy 311, 380
magnetostructural transformation 360
magnetovolume instabilities 368
many-body density matrix 47
many-body perturbation theory 129
martensitic transformations 358
Mataga–Nishimoto approximation 166
Matsubara frequencies
– fermionic and bosonic 130
Matsubara space 133
Maxwell relation 385
MB_2C_2-layered structures 629
M−L antibonding 292
M−L bonds 292
M−L π-antibonding 292
McWeeny's group 170
MD-GIPAW approach 634
mean absolute deviation (MAD)
– analysis 52
mechanical embedding 654
melting behavior 592
metadynamics 573, 577
– samples configuration 694
– technique 579

metaheuristics 573
metal clusters, aromaticity 538
metal-insulator transition 142
metallic bonding 262
metallic compounds 415
metallic hydrogen-rich phases 589
metallic superconductor 519
metallurgical process 577
metal–carbon bonding 367
metal–ligand
– interaction 168
– σ-antibonding 430
metal–organic framework (MOF) 179
metamagnetism 353
metastable prenucleation cluster 687
methanol
– configurations of 667
methoxylation energies 666
methylguanidinium 175
Meyer–de Giambiagi–Wiberg bond index 450
MgAgAs-type structure 464
Mg-rich complex metallic alloys 469
Mg–Al equilibrium system 345
MH. See minima hopping (MH)
micro-ball nanodiamond anvils 571
microscopic spin–phonon coupling 382
mictomagnetism 353
Miedema's semiempirical model 358
Miedema model 358
Millam–Scuseria conjugate gradient density matrix 172
MINDO 163
minima hopping (MH) 533, 573, 577, 578
– algorithm 533
missing xenon paradox 589
ML-SQC approach 173
ML$_6$ octahedron
– orbital character of d-states 294
M—L π-antibonding 292, 293
Mn—O bonds 325
MNDOC model 167
Mn-doped ZnO 180
Mn-excess alloys 384
modeling crystal nucleation 681
– molecular simulation techniques 681
– nonclassical nucleation processes 681
modeling point defects
– defect chemistry 661
– diffuse states 664
model spin density 270
modern acid–base theory 161
modified Becke–Johnson (mBJ) 244

modified embedded atom method (MEAM) 353
modified neglect of differential overlap (MNDO) 166
– advantage 166
modified SINDO 165
MOF. See metal–organic framework (MOF)
MO-type approach 424
MO LCAOs of benzene, pictorial representation 91
molecular adsorption on surfaces 60
molecular dynamics (MD) 181, 212, 220, 634
– simulations 685, 693, 697
molecular Gaussian computer code 631
molecular magnetism 274
molecular mechanics (MM) 204, 649
molecular orbital (MO) 159, 271
– MO diagram 432, 466
– MO energy 411
– MO picture 411
– MO theory 159, 410, 456, 466
molecular orbitals 91, 299
– tracking 170
molecular quantum chemistry 648
molecular simulation methods 692
molecular solids 573
molecular systems 31
molecules
– atomization energy 48
– solvation 185
molybdates AMoO$_4$
– ^{95}Mo isotropic nuclear shieldings 627
molybdenum cluster compounds
– ^{95}Mo nuclear shielding 626
momentum–energy dispersion relations 249
monomers, randomly selected 535
monopole–monopole interactions 165
Monte Carlo algorithm 695
Monte Carlo methods 29
– of Metropolis 578
Monte Carlo procedure 119, 578
Monte Carlo sampling 694
Monte Carlo simulations 384
Monte Carlo techniques 124
Møller–Plesset perturbation theory 81
Mott insulating state 103
Mott–Hubbard insulators 261, 271
Mott–Hubbard transition 478
Mössbauer-active species 507
MP2 energy 69
MP2 Hylleraas functional 63
MP2 method 61

MP2 program 82
MSINDO 164
MSINDO-BOMD 183
MSINDO-CCM-sCIS 182
MSINDO-CCM implementation 179
MSINDO-sCIS method 165, 170
MSINDO–CCM–CIS method 180
muffin-tin 237
muffin-tin region 33
– pseudo-wave function 44
muffin-tin spheres 33
muffin-tin wave function 36
muffin-tin approximation (MTA) 238
Mulliken approach 211
Mulliken approximation 161, 163
Mulliken population analysis 439
Mulliken transition charges 218
multicentered bonding schemes 584
multiplicative Coulomb kernel 74
multipolar approximation 72, 81
multipole–multipole interactions 166
multiquantum magic-angle spinning (MQMAS) 609
multireference configuration interaction approach 170
multireference configuration interaction with single and double excitations (MRCISD) 170
Mutations 581
mutually coupled equations 68

n

NaCl aggregation, in water 695
NaCl-type alkali metal halides 423
NaCl-type CaO 476
NaCl-type MX(s) solids
– chemical bonding characteristics 418
Na$\cdots$Na orbital interactions 417
Na$_2$IrO$_3$, edge-sharing 334
NaIrO$_3$ honeycomb layer
– projection view 335
Nambu notations 110
Nambu spinors 109
nanoclusters
– cerium oxide (Ce$_m$O$_n$) 538
– – chemical bonding in 538
– – stability/reactivity of 538
– cerium oxide (Ce$_m$O$_n$) stability/reactivity of
– – electronic structure analysis 538
– silicon (Si$_{10}$H$_{2m}$) 543
– silicon oxide (Si$_{10}$O$_{4n}$)
– – atomic structures of 548
– – ground-state atomic structures of 548

– silicon oxide (Si$_{10}$O$_{4n}$) 548
– titanium oxide (Ti$_m$O$_n$) 540
– – applications 540
nanoparticles 182, 537
nanotechnology 528
^{23}Na SSNMR spectrum of In$_2$Na$_2$Mo$_5$O$_{16}$ 608, 609
natural bond orbital (NBO) 539
Nb–O antibonding states 475
Nb–O bonding 475
Nb$_6$Cl$_{12}$Cl$_2$(H$_2$O)$_4$ 631
Nb–Nb cluster bonding 468
NBO. *See* natural bond orbital (NBO)
NDDO. *See* neglect of diatomic differential overlap (NDDO)
NDO methods 161
Néel state
– of antiferromagnet 96
nearest-neighbor COHP curves 445
nearest-neighbor superconducting order parameter 108
near-stoichiometric polycrystalline 383
neglect of diatomic differential overlap (NDDO) 162
– NDDO-G method 169
– NDDO methods 166
– NDDO model 184
– NDDO one-electron matrix 172
neglect of differential overlap (NDO) 431
N electrons 29
neurodegenerative disorders 177
neutral ion pairs 185
neutron diffraction 326, 571
– measurements 329
NiAl
– B2 CsCl-type crystal structure 350
– elastic modulus 349
– ternary additions 351
– Young's modulus 348
Ni$_{64}$Al$_{36}$ 355
– alloys 354
Ni$_{45}$Co$_5$Mn$_{36.7}$In$_{13.3}$
– isofield magnetization curves 387
Ni compositions 351
nido structures 466
Ni-based martensitic shape memory alloys 360
Ni-rich NiAl 351
Ni-rich side 358
Ni$_2$MnGa
– element-resolved magnetic exchange coupling 378
– Heusler structure 350, 370

– second-order magnetocrystalline anisotropy energy 379
Ni$_{46}$Mn$_{24}$Ga$_{22}$Co$_4$Cu$_4$ martensite 352
Ni$_2$MnGa, Ni–Mn–Al 352
Ni$_2$Mn$_{1+x}$Z$_{1-x}$
– phase diagram 372
Ni–Al-based alloys 353
Ni–Al-based intermetallics 345
Ni–Al–Mo superalloys 352
Ni–Al–Mo system 352
Ni–Mn COHP diagrams 375
Ni–Mn-based Heusler alloys 364
Ni–Mn–Ga
– structural phase diagram 373
Ni–Mn–In alloys 371
Ni–Mn–Z martensitic Heusler alloys 382
NiO$_6$ polyhedra 336
NiO theoretical spectrum 141
NiO$_6$ trigonal prisms 326
noble gases 537
– atoms 164
nodal relaxation 119
no free lunch theorems 574
no ghost states 47
NO-MNDO method 173
nonclassical nucleation 682
noncollinear magnetism 263
nonconducting systems 73
nonconstant solute concentration
– free energy diagram 691
noncrystal/classical nucleation, thermodynamics and kinetics of 682
– crystal growth at nonconstant solute concentration 690
– multistep nucleation/crystal polymorphism 685
– nucleation kinetics vs. diffusion-controlled crystal formation 689
– prenucleation clusters/nucleation from building blocks 685
noncrystal nucleation, thermodynamics and kinetics 682
nondegenerate perturbation 325
nonequilibrium Green's function (NEGF) formalisms 176
nonequilibrium Green's function theory (NEGF) 215
non-Fermi liquid 150
non-spin-polarized calculation 38
noninteracting harmonic oscillators 502
noninteracting system 120
nonmagnetic solid-state compounds
– relative magnitudes 608

nonmagnetic state 265
nonnegligible integrals 75
nonorthogonal basis, of PAOs 66
nonstoichiometric compositions constitutional disorder set 351
nontransition metal atoms 262
normalization factor 562
norm-conserving pseudopotentials
– mathematical form 42
Nosé–Hoover thermostat 356
no uniaxial magnetism 324
novel combinations 587
Nowotny–Juza phases 464
Nth order muffin-tin orbitals (NMTO) 138
nuclear explosions 571
nuclear inelastic scattering (NIS) 507
nuclear magnetic resonance (NMR) spectroscopy 607
– amorphous compounds 634
– Brillouin zone sampling 624
– chemical shift interaction 612
– cluster approach 610
– coloring problem in rare-earth metal boron carbides 628
– computation of parameters 615
– crystallography and structural characterization 611
– density functional theory 615
– – augmented plane wave (APW) 618
– – general basis 615
– – modeling of solids 616
– – pseudopotential plane wave approach 617
– DFT Kohn–Sham equations 611
– dipolar/metallic shift interactions 614
– electric field gradient (EFG) tensor 619
– exchange-correlation functional 625
– geometry 627
– GIPAW method 611
– ^{93}Nb solid-state parameters, characterization of 631
– physical basis 611
– quadrupolar coupling interaction 613
– tensor 620
– – augmented plane wave approach 622
– – converse approach 622
– – gauge-including projector augmented wave method (GIPAW) 620
– – J coupling 623
– – metallic shift 623
– treatment of relativistic effects 625
– user-friendly interfaces for quantum chemical softwares 624
nuclear quantum effects (NQE) 593

nuclear Schrödinger equation 611
nuclear shielding tensor 613
nucleation 687
– barrier 683
nucleobases
– π-stacks of of 175
nucleus–nucleus interactions 2, 166
nucleus transition 688
numerical atomic orbital (NAO) 32

o

OAO. See orthogonalized atomic orbital (OAO)
O–H···O interactions 176
occupation number (ON) 542
off-stoichiometric composition 352
Ohno–Klopman approximation 168
ω-technique 159
OM2 method 168
OM3 method 168
OMx 168
ON. See occupation number (ON)
one-dimensional chain of lithium hydride
– phonon dispersion relation 500
one-dimensional diatomic chain
– acoustical vibrational mode 496
– optical vibrational mode 496
one-dimensional monoatomic chain
– vibrational mode 499
one-electron states
– with electron correlation 294
– without electron correlation 292
ONIOM hybrid models 170
optical absorption spectra 170
orbital angular momentum 290
orbital density 263, 277
orbital energies 262
orbital energy diagrams 423, 426
orbital momentum 308
orbital rehybridization energy 274
ordered spin arrangement 302
order–fitness correlation 531
organic chemistry, aromaticity 537
organic molecules, electronic spectra 165
orthogonalized atomic orbital (OAO) 160
orthogonal local virtual orbitals 65
orthogonal magnetic orbital 272
orthogonal spin states 316
ortho-anisidine/beta-cyclodextrin 177
orthonormal magnetic orbital 271
orthonormal zeroth-order orbitals 266
oscillator strengths 18
Ostwald ripening 688

OSV–DF–LMP2 interaction energy, dependence of 79
OSV–LMP2 calculations of ibuprofen adsorbed on a silica slab 81
OSV truncation 71
oxalic acid, photoreaction 182
oxygen, chemical potential of dependence of surface energies on 563

p

pair approximation 69
pair natural orbital (PNO) 67
pairwise repulsive energy 207
PAO coefficients 66
PAO domain 68
PAO overlap matrix 68
parallelepipeds 532
parallelization of OSV code 80
parameterization 42
parametric method 3 (PM3), 167
parametrization strategies 173
partial wave expansion 45
particle-in-a-box model 408
particle swarm optimization (PSO) 577, 580
– technique 573
passive orbitals 266
path-integral quantum Monte Carlo schemes 121
path integral (PI) 174
Pauli exclusion forces 407
Pauli exclusion principle 230, 261, 434
Pauli matrices 109
Pauling's crystal chemistry rules 582
Pauling's benzene 93
Pauling model 94
Pauli paramagnetic 413, 478
Pauli principle 90
Pauli repulsion 413
PAW-type schemes 30
PAW method 553
PAW scheme 47
PBC. See periodic boundary conditions (PBC)
PBE. See Perdew, Burke and Ernzerhof (PBE)
Peierls distortion 476, 481, 593
Perdew, Burke and Ernzerhof (PBE) 348, 559
– using exchange–correlation functional 559
periodic boundary conditions (PBC) 178
periodic Coulomb potential 648
periodic local Møller–Plesset perturbation theory of second order for solids 59
– demonstrative application
– – ibuprofen on hydroxylated silica surface 78

– direct space
– – local density fitting 75
– – orbitals, choice of 65
– local approximations 68
– – domain approximation 68
– – pair approximation 71
– – truncated OSVs approximation 70
– multipolar approximation 76
– nomenclature 61
– periodic density fitting 73
– periodic LMP 2 energy 63
– periodic LMP 2 wave function 62
periodic quantum chemical treatment 61
periodic table 572
perovskite motif 509
perovskite surfaces
– oxygen defects, effect of 181
perturbation 267
perturbational molecular orbital 263
perturbational potential 280
perturbation theory 289, 310, 321
perturbed crystal techniques 650
phagraphene 552
– planar atomic density of 552
– structure of 554
phase transformation 482
phase X of H_2O 585
phonon band structure 498
phonon Brillouin zone 495
phonon dispersion 498, 499, 510
– curves 512
phonon frequencies 498
phononic wave vector 496
phonon mode eigenvectors 506
phonons 496
phosphine 592
photoemission intensity 120
photoemission spectra 149
p-valence orbitals 162
physical annealing procedure 578
planar unsaturated systems 161
plane wave (PW) 237, 245
– codes 30
– cutoff 51
– functions 279
– kinetic energy 434
– programs 46, 48
– scheme 239
PM6 direction 167
PMF. *See* potential of mean force (PMF)
PM3-MAIS model 185
PM3 method 167
Pmmm structure

– geometric structures of Pmmn and, similarities between 555
Pmmn–boron structure 555
– electronic structure of 556
Pmmn structure 554
PMx methods 167
Pnma
– – Imma transition 511
– – temperature-dependent Gibbs free energies 511
point defect formation
– calculated reaction energies 664
Poisson equation 75, 234
poisson-type orbital (PTO) 75
polar bonds 460
polarizabilities 17, 19
– tensors 171
polarized embedding 655
polar nucleophilic proton transfer 177
polyacetylene 25, 477
polyacetylene, π-band of
– Peierls distortion 476
polycrystalline $Ni_{45}Co_5Mn_{37}In_{13}$ 385
polycrystalline $Ni_{50}Mn_{25}ga_{25}$ 383
polycrystalline Ni–Co–Mn–In sample 386
polycrystalline sample, magnetization directions 380
polyene
– Hückel, Néel, and RVB states 98
polymers 548
polymorphic transformation 360
polynuclear transition metal complexes
– magnetic interactions 171
post-Hartree–Fock methods 173
post-HF schemes 230
potential energy component 268
potential energy surface (PES) 71, 573, 574
– local minima 574
– schematic diagram of one-dimensional 574
potential of mean force (PMF) 174
Potts model 346
powder X-ray diffraction 506
PPW method
– 4p monoelectronic wave function 618
preferred spin-orientation
– selection rules 321
preferred spin orientation
– perturbation theory analysis 337
prenucleation clusters 684, 687
– crystal nucleation 687
pressure-dependent zero-temperature enthalpy 494

pressure-induced band broadening 582
pressure-induced bond symmetrization 584
pressure-stabilized hydrides 585
principal quantum number 236
program packages 174
projected COHP (pCOHP) 440
projected crystal orbital Hamiltonian population 366
projected density of states (PDOS) 329
projector
– Fourier transform 48
– operator technique 74
projector augmented wave (PAW) 44, 138, 359, 610
propylene carbonate–dimethoxyethane mixtures 185
Protein Data Bank files 175
proton transfer energy profiles 185
pseudoatom-embedded cluster models 181
pseudocanonical orbitals 62, 67
pseudocanonical space 62
pseudocanonical virtuals to OSVs 68
pseudo-orbitals 34
– states 332
pseudo-wave function 39, 46
pseudopotential-based codes 49
pseudopotential plane wave (PPW) 617
pseudopotentials 29, 235, 237
– accuracy of 49
– basic concepts 34
– calculation 39
– construction methods 45
– general advantages 34
– methods 33
– numerical performance 47
– scheme 235
– vs. all-electron potential 47
pseudosymmetries 530
pseudowave functions 237
PSO algorithm 533
pyrolysis reaction, CO formation 183

q
QMC scheme
– fixed-node 122
QM/MM methodology 214
QTAIM atom 449, 453
– basins 454
QTAIM basins 448
– electron population 449
QTAIM framework 448
qualitative σ-only MO scheme 457
quantum chemical
– investigations 160
– models 61
– techniques 169
quantum mechanical
– calculations 227
– calculations/analyses 428
quantum mechanical (QM) cluster 649
quantum mechanical methods 416
quantum mechanical/molecular mechanical approaches
– applications of 657
– – active sites in zeolites 665
– – ionization potentials 657
– – modeling point defects 661
quantum mechanical/molecular mechanical calculation
– practicalities of 656
quantum mechanical/molecular mechanical calculations 659
quantum mechanical/molecular mechanical coupling
– classification 654
quantum mechanical/molecular mechanical model
– sketch of 658
quantum mechanics 59, 159, 290
– for treating electronic structure 230
quantum molecular dynamics 165
quantum Monte Carlo 529
quantum Monte Carlo calculations 52
quantum Monte Carlo scheme (QMC) 129
quantum Monte Carlo simulations
– finite-temperature 141
Quantum Theory of Atoms in Molecule (QTAIM) 447
quasicluster embedding scheme 178
quasi-harmonic approximation 504, 510, 522
quasi-Newton update 166
quasi-particle self-consistent GW (qsGW) scheme 151
quasilinear systems 25
quasimolecular LUC model 178
quasiparticle 18
– energies 19
quaternary Ti–Ni-based shape memory alloys 360

r
radial Schrödinger equation 33, 36, 42, 45
radial spin densities
– vs. total charge density 40
Raman spectra 589
Raman spectroscopy 572

randomly generated atomic coordinates 530
random phase approximation (RPA) 60, 232
random search 577
random (sensible) structure 576
random symmetric structures 530, 534
rare earth transition metal (RETM) 364
reactive force fields (ReaXXFF) 214
real-space codes 47
real-space grid points 49
real-space projectors 33
Re–O COHP curve 475
Re–O π-antibonding 443
recife model 1 (RM 1), 169
reference energies 41
rehybridization 268
relative average coordination number 164
relativistic effects 236
relaxed charge density 38
renormalization effect 146
ReO_3 443
– electronic structures 442
reparametrization 81
repulsive energy 208
resonance integral 166
resonance parameter 159
resonance valence bond (RVB) 90, 97
– mean field 108
– model 113
– order parameter 109
– phases 113
– wave function
– – Monte Carlo study 98
– – trial variational 100
resonating bonds 88
reversed approximation molecular orbital (raMO) 446
Ripening reactions
– zinc and hydroxide ions 696
RKKY. See Rudermann–Kittel–Kasuya–Yosida (RKKY)
RM1. See recife model 1 (RM 1)
rock-salt
– pressure-dependent phonon dispersion 514
ROHF method 183
root mean square deviation (RMSD) 52
Rudermann–Kittel–Kasuya–Yosida (RKKY) 376
– exchange interactions 278
– oscillation, of free electron metal 281
rutile TiO_2
– internal energy of 562
– structures of 565
– surface

– – atomic configurations of 562
– – reconstructions of 562
– – stable reconstructions of 563
– surface phase diagram of 564
– unreconstructed 563
$R_2V_2O_7$
– axially- compressed VO_6 octahedron 330

s
Sb_2Te_3
– partial densities of phonon states vs. nuclear inelastic scattering 506
– temperature-dependent bulk modulus 523
– topological insulator 506
scalar relativistic scheme 237
scanning tunneling microscopy (STM) 228
scattering boundary conditions 35
scattering properties 36, 46
SCF. See self consistent field (SCF)
schematic, of 1D PES 574
Schottky defects 181
Schrödinger equation 3, 29, 34, 46, 159, 230, 238, 416, 428, 429, 430, 433, 445, 468, 615, 616, 647
– for H_2 407
screened exchange dynamical mean field theory 150
– perspectives 150
second-order Møller–Plesset perturbation 179
self-consistent all-electron charge densities 37
self-consistent charge (SCC) 211
self-consistent charge (SCC) –DFTB implementations 213
self-consistent density functional tight-binding (SCC-DFTB) 175
self-consistent diagonalization approach 44
self-consistent field (SCF) 160, 167, 230
self-consistent polarized embedding 655
self-consistent radial all-electron charge density 38
self-diffusion coefficient 184
self-energy 216
self-interaction correction (SIC) 220, 271
self-interaction error 304
semiconductors 652
– surfaces of 562
semiempirical level of theory 175
semiempirical methods 163, 169
– parametrization 174
semiempirical molecular orbital methods
– biochemistry and pharmacology application 174

semiempirical quantum mechanics (SQM) 171, 174
semi-ab initio model 1 (SAM1), 168
semi-core states 235
semilocal potentials
– absolute value 53
SG15 PBE pseudopotentials
– nonsmooth local potential 52
shaking procedure 576
shape memory alloys 352
shape memory materials 349, 356
short-wavelength plane waves 41
σ-antibonding 292, 475
$Si_{10}H_{2m}$ clusters
– ground-state structures of 545
$Si_{10}H_{2n}$ clusters
– phase diagram of 547
Si-O clusters 548
silicon, electronic properties 182
silicon-oxygen clusters 548
silicon nanowires 548
simple metals surprising effect of pressure 583
simple spin lattices 286
simulated annealing 165, 577
– algorithm 577
Si–O–Si angles in silica 633
SINDO 163
– binding energies 164
single heat-capacity measurement 521
single-crystal neutron-diffraction data 514
single-determinant wavefunction 450
single-spin site
– SOC effect of 309
single-walled carbon nanotubes 183
singlet RVB (sRVB) 107
$Si_{10}O_{4n}$ clusters
– ground-state structures of 548
– phase diagram of 548, 549
size-induced solid–solid transformations 681
Slater determinant 91, 92, 102, 120, 230, 430
– of monoelectronic wave functions 616
Slater-type orbital (STO) 238
Slater insulator 333
Slater–Condon factors 166
small-displacement method 501
small periodic cluster 178
SOC-induced HOMO–LUMO interactions 290
SOC-induced interaction 313
SOC-induced Mott insulating state 333
sodium 594
soft pseudopotential 47
Soler–Williams APW method 248

solid-state chemistry 2, 228
solid-state embedding scheme 655
solid-state magnetic materials 288
solid-state modeler 647
solid-state NMR (SSNMR) 607
solid-state physics 29
solid-state properties 177
solid-state quantum-chemical simulations 492
solid-state software packages
– implementation 173
solid-state theory 177
solid–liquid equilibria 592
solid–liquid interphase 171
solids
– pressure-dependent phase diagrams 572
– structures of 572
solute diffusion 690
solvent effects 171
Sparkle model 169
spatially coherent slabs
– clustering 531
– heredity 531
– random selection and mutations 531
– soft mutation 531
specific reaction parameter (SRP) 174
spectral function 120
spectroscopy 175
spherical Bessel functions 41
spherical harmonics 289
spin canting 315
spin degree of freedom 287
spin density 262, 263, 265
spin dimer 301
spin exchange
– effect of SOC 312
spin exchange constants 286, 298
spin exchange interaction 297, 304
– energy 296
spin exchange parameters 286, 302
spin exchanges
– numerically extracted, general features 304
– orbital interactions, effect of 304
spin Hamiltonian 285, 288
– indirect SOC effect incorporation 308
spin hardness function 264
spin-dependent contributions 446
spin-down electron configurations 330
spin-half misconception 311, 326, 331
– effect of SOC 309
spin-half syndrome 287
spin-orbit contribution 625
spin-orbit coupling 623

spin-polarized relativistic
  Korringa–Kohn–Rostoker (SPR KKR) 377
spin-unpolarized state 269
spin-orbital entanglement 338
spin-orbit coupling 43, 237, 377
spin-orbit Mott insulating states 333, 337
spin orbital energy correction 267
spin orbitals 270
spin orbital shape and energy changes,
  approximate expressions of 265
spin-orbit coupling (SOC)
– coupling scheme, LS vs. jj 337
spin polarization 261, 562
– in isolated atoms and ions 269
spin polarization and exchange splitting
  270
spin polarization energy 274
spin polarization perturbational orbital theory
  (SPPOT) 262, 263, 266
– applications to simple systems 269
split d-states
– simulation 295
SPPOT. See spin polarization perturbational
  orbital theory (SPPOT)
SQM. See semiempirical quantum mechanics
  (SQM)
SQM/MM methods 185
squares fitting analysis 303
SRP. See specific reaction parameter (SRP)
$SrVO_3$
– electronic spectral function 140
SSAdNDP analysis 558
stabilization energy 337
stoichiometric NiAl 348
stoichiometric Ni_2MnGa 352
stoichiometric TiNi
– chemical bonding properties 364
stoichiometries 572
Stoner condition for spontaneous spin
  polarization 275
strain glass phases 357
strain glass transition 362
structeditor 243
struct files 243
structural polymorphism 509
Stuttgart pseudopotentials 42, 46
subtractive approach 654
superbasin 574
supercell 229, 243, 591
superconducting transition temperature
  590
superconductors 571
surface atoms 228, 560

surface energies 562
surface energy 562
surface-sensitive properties 229
surface ionization potentials 660
surface reconstructions
– prediction 535
surface slab 229
surface structures
– reactivity of solids
– – role of 559
symmetric algorithm 535
Synchrotron infrared absorption
  spectroscopy 588
systematic validation tests 49

t
Taylor series 102
TB-LMTO Hamiltonian 358
T–F–T bond angles 475
T–L σ-bonding 462
TD-DFT formalism 18
temperature-dependent heat capacity 493
Teo's topological electron counting rules
  467
ternary Fe–Mn–Al Heusler alloys 353
ternary Heusler alloys 346, 374
ternary magnetic Heusler intermetallics
  345
ternary system
– chemical frustration 369
ternary transition metal oxide $SrVO_3$ 144
tetrahedral cluster
– made up of four VO_6 octahedra 330
tetrahedral semiconductors
– band diagrams 421
– chemical bonding characteristics 423
– valence electrons per atom 459
theoretical best estimate (TBE-2), 170
theoretically generated energy versus volume
  (E–V) 494
thermal expansion coefficient 354
thermal ionization energies 665
thermodynamic quantity 581
thermodynamic rationale 686
three-dimensional periodicity
– advantages 648
three-index Coulomb integral 75
three-index one-electron integrals 75
three-index overlap integral decay 75
through-space (TS) 307
tight-binding density functional theory
  (DFTB) 203
– approximation 204

- computational implementations/ extensions 219
- copper, band structure of 211
- energy of Li_2 208
- Kohn–Sham (KS) realization 203
- molecular system, schematic diagram of 216
- periodic boundary conditions (PBC) 215
- practical realization 219
- weak interactions – QM/MM approach 213

tight-binding (TB-LMTO) electronic structure 358
tight-binding energy band diagrams 437
tight-binding (TB) methods 203
time-dependent density functional theory (TD-DFT) 217
- linear response DFT 218
time-dependent density functional theory (TDDFT) 119
time-dependent density-functional theory (TD-DFT) 15
time-dependent DFTB method (TDDFTB) 220
time-reversal symmetry 264
tin
- phase transition 521
Ti–Ni
- Ni-resolved electronic density 365
- phase diagram 357
Ti–Ni alloys 345, 356, 362
- ab initio calculations 356
Ti–Ni-based intermetallics 356
Ti–Ni-based shape memory alloys 357
Ti–Ni–Cu alloys 361
- Cu content dependence of martensitic transformation temperatures 362
Ti–Ni–Cu–Co thin film 361
Ti–Ni–Fe strain glass 362
Ti–Ni–X alloys 360
Ti–Ni–X intermetallics
- bonding and chemical frustration 363
Ti–Ni samples
- martensitic start temperature 359
Ti–Ni strain glass alloy
- shape memory effect and superelasticity 362
Ti–Ni systems 347
Ti–Ti
- ICOHP values 366
$Ti_{50}Ni_{50-x}Fe_x$
- phase diagram 363
TiNi wire
- thermomechanical response 361
tin pest 515

Ti_8O_{12} cluster
- chemical bonding elements of 544
- electronic structure of 541
- partial charge density of 543
Ti_8O_n clusters
- ground states of 542
- lowest-energy structures of 540
titania 660
T_3Ni_4 phases 357
total charge distribution 39
total electronic energy changes 264
total energy 11
total nucleonic charge 42
total pseudopotential 44
toy Hamiltonian 288
transferability 41
transformation magnetic fields 388
transformation matrix 68, 165
transition metal atom 426
transition metal compounds 261
- spectra of 166
transition metal impurity 360
transition metal magnetic ions 285, 287
transition metals 413, 427
- chemical bonding characteristics 425
transition metal systems 120
translational invariance 64
translational periodicity 165
translational symmetry 68, 530
triplet exciton self-trapping 178
triplet state 298
Troulier Martins pseudopotential 45
two adjacent spin sites
- effect of SOC 313
two-body interactions 302
two-dimensional materials
- applications 551
two-dimensional (2D) materials
- boron 552
- carbon 552
two-electron interactions 165
two-photon absorption spectra 176
T_6X_{12} cluster 468

u

ubiquitous characterization method 505
ultrasoft pseudopotentials 44, 49
unfrozen strain glass 362
uniaxial magnetism 318, 322, 323, 324
uniform RVB phase (uRVB) 107
unique electronic structures 582
unit cell deformations 576
unit cell, of the insulating structure 582

universal force field (UFF) 213
unpaired electron 297
unquenched orbital momentum 319
unrestricted natural orbital complete active space configuration interaction (UNO-CAS) 170
urea
– thermal ellipsoids 515
USPEX algorithm 528, 535, 538, 540, 548, 553, 559, 562, 563

v
vacuum energy level 648
valence band 556
valence band maximum (VBM) 249
valence bandwidths
– for alkali halides range 417
valence bond approach
– Heitler–London variant 409
valence charge density, of Cs-IV colored 582
valence electrons 34, 36, 235
valence-mixed (VM) 251
valence states 235
vanadium pentoxide 168, 181
van Arkel–Ketelaar diagram 414, 415, 420
van der Waals attractions 522
van der Waals correlation 71
van der Waals density functionals 23
van der Waals dispersion 60, 71
van der Waals forces 23, 407
van der Waals interactions 23, 213, 214, 495, 507, 522
variational Monte Carlo (VMC) 100
variational parameter (VP) 100
VASP code 358, 359, 553
VCA (virtual crystal approximation) calculations 359
vector 282
velocity autocorrelation function 182
vertical ionization potential 659
vibrational Brillouin zones 495
Virial Theorem 408
virtual functions 69
virtual orbitals 69, 73
virtual space 69
virtual WFs 66
volume depedence
– temperature-dependence 522
VSe$_2$
– Fermi surface of 250
VSEPR 458

w
Wade–Mingos rules 457, 465, 467, 537
– for closo-deltahedra 466
Wannier function (WF) 65, 67, 69, 70, 81, 123, 445, 586, 650
– decomposition 43
– Fourier coefficients 48
Wannier orbitals 125, 621
Wannier type construction 139
wavefunction expansion 178
wave functions 18
– boundary conditions 32
wave-function based approaches 21, 615
wave methods
– ultrasoft and projector augmented 44
wave vector
– 3N vibrational modes 496
weak magnetic anisotropy 325
Weiss dynamical mean field 137
wet chemical precipitation 518
WF–WF and PAO–PAO blocks 68
wide-gap semiconductors 662
WIEN2k community 242
Wigner–Seitz cell boundary electron density 358
Wigner–Seitz radii drop 427
Wigner–Seitz supercell (WSSC) 178
Wolfram's red salt [Pt(NH$_3$)$_4$Br]$^{2+}$ 477
Wolfsberg–Helmholz relation 411
WSSC interaction domains 179
WSSC. See Wigner–Seitz supercell (WSSC) 178
Wurtzite → rocksalt transition route 698
Wyckoff position 234

x
xenon oxides 590
xenon suboxides 581
X-ray absorption spectra (XAS) 245
X-ray crystallography 504
X-ray diffraction 228, 593
X-ray power diffraction data 252
x^2-y^2 magnetic orbital 306
x^2-y^2 σ-antibonding orbital 306

y
YBa$_2$Cu$_3$O$_{6.5}$
– Gibbs energy of formation 519
YBaFe$_2$O$_5$
– charge-ordered phase 252
– Verwey transition in 251

YB$_2$C
– crystalline powder 630
– structural arrangements 629

Y$_2$Nb$_2$O$_7$
– phonon dispersions of 253
– relaxed structure of 254
– structure relaxation of 252

z

ZDO. *See* differential overlap; zero-282
Zeeman energy 383
Zeeman interaction 319, 612
Zener's theory 376
zeolite-catalyzed conversion
– methanol to hydrocarbon 666
zeolites 655
– coordination of metal centers 668
– deprotonation energies, as measure of site reactivity 665
– deprotonation energies for 665
– REDOX potentials 667
zero-differential overlap (ZDO) 160
zero-field-cooled (ZFC) 362
zero-field spin Hamiltonian 309
zero point energy (ZPE) 593
zeroth-order regular approximation (ZORA) 625
Z-matrix 533
zinc oxide precipitation 692
ZINDO 165
Zintl concept 586
Zintl–Klemm concept 461
Zintl–Klemm formalism 460, 461, 464
Zintl phases 415
Zn$_4$O(acetate)$_6$ clusters 686, 689
ZnO crystal nucleation
– case studies 695
ZnO single crystal 697
ZnO/Zn(OH)$_2$ core 696
ZnO/Zn(OH)$_2$ core–shell nanocrystals 696
ZnO/Zn(OH)$_2$ core–shell nanoparticle
– scale-up model 697
ZORA GIPAW method 625